# QUANTUM PHYSICS

## of Atoms, Molecules, Solids, Nuclei, and Particles

**Robert Eisberg**

*University of California, Santa Barbara*

and

**Robert Resnick**

*Rensselaer Polytechnic Institute*

**JOHN WILEY & SONS, New York • Chichester • Brisbane • Toronto**

***Library of Congress Cataloging in Publication Data:***

Eisberg, Robert Martin.
Quantum physics of atoms, molecules, solids, nuclei, and particles.

Bibliography: p. L-1
1. Quantum theory. I. Resnick, Robert, 1923- joint author. II. Title.

QC174.12.E34 530.1′2 74-1195
ISBN 0-471-23464-8

Printed in the United States of America

20 19 18 17 16 15 14 13 12 11

# Preface

The basic purpose of this book is to present clear and valid treatments of the properties of almost all of the important quantum systems from the point of view of elementary quantum mechanics. Only as much quantum mechanics is developed as is required to accomplish the purpose. Thus we have chosen to emphasize the applications of the theory more than the theory itself. In so doing we hope that the book will be well adapted to the attitudes of contemporary students in a terminal course on the phenomena of quantum physics. As students obtain an insight into the tremendous explanatory power of quantum mechanics, they should be motivated to learn more about the theory. Hence we hope that the book will be equally well adapted to a course that is to be followed by a more advanced course in formal quantum mechanics.

The book is intended primarily to be used in a one year course for students who have been through substantial treatments of elementary differential and integral calculus and of calculus level elementary classical physics. But it can also be used in shorter courses. Chapters 1 through 4 introduce the various phenomena of early quantum physics and develop the essential ideas of the old quantum theory. These chapters can be gone through fairly rapidly, particularly for students who have had some prior exposure to quantum physics. The basic core of quantum mechanics, and its application to one- and two-electron atoms, is contained in Chapters 5 through 8 and the first four sections of Chapter 9. This core can be covered well in appreciably less than half a year. Thus the instructor can construct a variety of shorter courses by adding to the core material from the chapters covering the essentially independent topics: multielectron atoms and molecules, quantum statistics and solids, nuclei and particles.

Instructors who require a similar but more extensive and higher level treatment of quantum mechanics, and who can accept a much more restricted coverage of the applications of the theory, may want to use *Fundamentals of Modern Physics* by Robert Eisberg (John Wiley & Sons, 1961), instead of this book. For instructors requiring a more comprehensive treatment of special relativity than is given in Appendix A, but similar in level and pedagogic style to this book, we recommend using in addition *Introduction to Special Relativity* by Robert Resnick (John Wiley & Sons, 1968).

Successive preliminary editions of this book were developed by us through a procedure involving intensive classroom testing in our home institutions and four other schools. Robert Eisberg then completed the writing by significantly revising and extending the last preliminary edition. He is consequently the senior author of this book. Robert Resnick has taken the lead in developing and revising the last preliminary edition so as to prepare the manuscript for a modern physics counterpart at a somewhat lower level. He will consequently be that book's senior author.

The pedagogic features of the book, some of which are not usually found in books at this level, were proven in the classroom testing to be very successful. These features are: detailed outlines at the beginning of each chapter, numerous worked out examples in each chapter, optional sections in the chapters and optional appendices, summary sections and tables, sets of questions at the end of each chapter, and long and varied sets of thoroughly tested problems at the end of each chapter, with subsets of answers at the end of the book. The writing is careful and expansive. Hence we believe that the book is well suited to self-learning and to self-paced courses.

We have employed the MKS (or SI) system of units, but not slavishly so. Where general practice in a particular field involves the use of alternative units, they are used here.

It is a pleasure to express our appreciation to Drs. Harriet Forster, Russell Hobbie, Stuart Meyer, Gerhard Salinger, and Paul Yergin for constructive reviews, to Dr. David Swedlow for assistance with the evaluation and solutions of the problems, to Dr. Benjamin Chi for assistance with the figures, to Mr. Donald Deneck for editorial and other assistance, and to Mrs. Cassie Young and Mrs. Carolyn Clemente for typing and other secretarial services.

Santa Barbara, California — *Robert Eisberg*
Troy, New York — *Robert Resnick*

# Contents

## 13

## Solids—Conductors and Semiconductors 481



## 14

## Solids—Superconductors and Magnetic Properties 523



## 15

## Nuclear Models 551



## 16

## Nuclear Decay and Nuclear Reactions 601

# QUANTUM PHYSICS

# 1

# Thermal Radiation and Planck's Postulate

# Thermal Radiation and Planck's Postulate

# 1

## 1-1 Introduction

At a meeting of the German Physical Society on Dec. 14, 1900, Max Planck read his paper, "On the Theory of the Energy Distribution Law of the Normal Spectrum." This paper, which first attracted little attention, was the start of a revolution in physics. The date of its presentation is considered to be the birthday of quantum physics, although it was not until a quarter of a century later that modern quantum mechanics, the basis of our present understanding, was developed by Schroedinger and others. Many paths converged on this understanding, each showing another aspect of the breakdown of classical physics. In this and the following three chapters we shall examine the major milestones, of what is now called the *old quantum theory*, that led to modern quantum mechanics. The experimental phenomena which we shall discuss in connection with the old quantum theory span all the disciplines of classical physics: mechanics, thermodynamics, statistical mechanics, and electromagnetism. Their repeated contradiction of classical laws, and the resolution of these conflicts on the basis of quantum ideas, will show us the need for quantum mechanics. And our study of the old quantum theory will allow us to more easily obtain a deeper understanding of quantum mechanics when we begin to consider it in the fifth chapter.

As is true of relativity (which is treated very briefly in Appendix A), quantum physics represents a generalization of classical physics that includes the classical laws as special cases. Just as relativity extends the range of application of physical laws to the region of high velocities, so quantum physics extends that range to the region of small dimensions; and, just as a universal constant of fundamental significance, the velocity of light $c$, characterizes relativity, so a universal constant of fundamental significance, now called Planck's constant $h$, characterizes quantum physics. It was while trying to explain the observed properties of thermal radiation that Planck introduced this constant in his 1900 paper. Let us now begin to examine thermal radiation ourselves. We shall be led thereby to Planck's constant and the extremely significant related quantum concept of the discreteness of energy. We shall also find that thermal radiation has considerable importance and contemporary relevance in its own right. For instance, the phenomenon has recently helped astrophysicists decide among competing theories of the origin of the universe.

## 1-2 Thermal Radiation

The radiation emitted by a body as a result of its temperature is called *thermal radiation*. All bodies emit such radiation to their surroundings and absorb such radiation from them. If a body is at first hotter than its surroundings, it will cool off because its rate of emitting energy exceeds its rate of absorbing energy. When thermal equilibrium is reached the rates of emission and absorption are equal.

Matter in a condensed state (i.e., solid or liquid) emits a continuous spectrum of radiation. The details of the spectrum are almost independent of the particular material of which a body is composed, but they depend strongly on the temperature. At ordinary temperatures most bodies are visible to us not by their emitted light but by the light they reflect. If no light shines on them we cannot see them. At very high temperatures, however, bodies are self-luminous. We can see them glow in a darkened room; but even at temperatures as high as several thousand degrees Kelvin well over 90% of the emitted thermal radiation is invisible to us, being in the infrared part of the electromagnetic spectrum. Therefore, self-luminous bodies are quite hot.

Consider, for example, heating an iron poker to higher and higher temperatures in a fire, periodically withdrawing the poker from the fire long enough to observe its properties. When the poker is still at a relatively low temperature it radiates heat, but it is not visibly hot. With increasing temperature the amount of radiation that the poker emits increases very rapidly and visible effects are noted. The poker assumes a dull red color, then a bright red color, and, at very high temperatures, an intense blue-white color. That is, with increasing temperature the body emits more thermal radiation and the frequency of the most intense radiation becomes higher.

The relation between the temperature of a body and the frequency spectrum of the emitted radiation is used in a device called an optical pyrometer. This is essentially a rudimentary spectrometer that allows the operator to estimate the temperature of a hot body, such as a star, by observing the color, or frequency composition, of the thermal radiation that it emits. There is a continuous spectrum of radiation emitted, the eye seeing chiefly the color corresponding to the most intense emission in the visible region. Familiar examples of objects which emit visible radiation include hot coals, lamp filaments, and the sun.

Generally speaking, the detailed form of the spectrum of the thermal radiation emitted by a hot body depends somewhat upon the composition of the body. However, experiment shows that there is one class of hot bodies that emits thermal spectra of a universal character. These are called *blackbodies*, that is, bodies that have surfaces

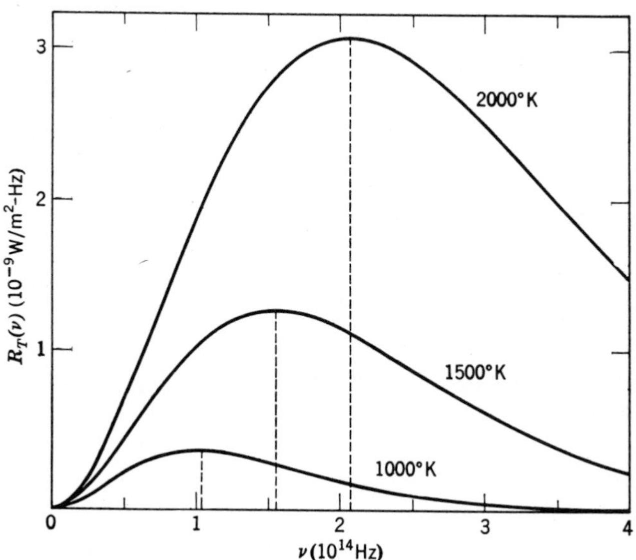


**FIGURE 1–1**

The spectral radiancy of a blackbody radiator as a function of the frequency of radiation, shown for temperatures of the radiator of 1000°K, 1500°K, and 2000°K. Note that the frequency at which the maximum radiancy occurs (dashed line) increases linearly with increasing temperature, and that the total power emitted per square meter of the radiator (area under curve) increases very rapidly with temperature.

which absorb all the thermal radiation incident upon them. The name is appropriate because such bodies do not reflect light and appear black. One example of a (nearly) blackbody would be *any* object coated with a diffuse layer of black pigment, such as lamp black or bismuth black. Another, quite different, example will be described shortly. Independent of the details of their composition, it is found that *all* blackbodies at the same temperature emit thermal radiation with the same spectrum. This general fact can be understood on the basis of classical arguments involving thermodynamic equilibrium. The specific form of the spectrum, however, cannot be obtained from thermodynamic arguments alone. The universal properties of the radiation emitted by blackbodies make them of particular theoretical interest and physicists sought to explain the specific features of their spectrum.

The spectral distribution of blackbody radiation is specified by the quantity $R_T(\nu)$, called the *spectral radiancy*, which is defined so that $R_T(\nu)\,d\nu$ is equal to the energy emitted per unit time in radiation of frequency in the interval $\nu$ to $\nu + d\nu$ from a unit area of the surface at absolute temperature $T$. The earliest accurate measurements of this quantity were made by Lummer and Pringsheim in 1899. They used an instrument essentially similar to the prism spectrometers used in measuring optical spectra, except that special materials were required for the lenses, prisms, etc., so that they would be transparent to the relatively low frequency thermal radiation. The experimentally observed dependence of $R_T(\nu)$ on $\nu$ and $T$ is shown in Figure 1-1.

*Distribution functions*, of which spectral radiancy is an example, are very common in physics. For example, the Maxwellian speed distribution function (which looks rather like one of the curves in Figure 1-1) tells us how the molecules in a gas at a fixed pressure and temperature are distributed according to their speed. Another distribution function that the student has probably already seen is the one (which has the form of a decreasing exponential) specifying the times of decay of radioactive nuclei in a sample containing nuclei of a given species, and he has certainly seen a distribution function for the grades received on a physics exam.

The spectral radiancy distribution function of Figure 1-1 for a blackbody of a given area and a particular temperature, say 1000°K, shows us that: (1) there is very little power radiated in a frequency interval of fixed size $d\nu$ if that interval is at a frequency $\nu$ which is very small compared to $10^{14}$ Hz. The power is zero for $\nu$ equal to zero. (2) The power radiated in the interval $d\nu$ increases rapidly as $\nu$ increases from very small values. (3) It maximizes for a value of $\nu \simeq 1.1 \times 10^{14}$ Hz. That is, the radiated power is most intense at that frequency. (4) Above $\simeq 1.1 \times 10^{14}$ Hz the radiated power drops slowly but continuously as $\nu$ increases. It is zero again when $\nu$ approaches infinitely large values.

The two distribution functions for the higher values of temperature, 1500°K and 2000°K, displayed in the figure show us that (5) the frequency at which the radiated power is most intense increases with increasing temperature. Inspection will verify that this frequency increases linearly with temperature. (6) The total power radiated in all frequencies increases with increasing temperature, and it does so more rapidly than linearly. The total power radiated at a particular temperature is given simply by the area under the curve for that temperature, $\int_0^\infty R_T(\nu)\,d\nu$, since $R_T(\nu)\,d\nu$ is the power radiated in the frequency interval from $\nu$ to $\nu + d\nu$.

The integral of the spectral radiancy $R_T(\nu)$ over all $\nu$ is the total energy emitted per unit time per unit area from a blackbody at temperature $T$. It is called the *radiancy* $R_T$. That is

$$R_T = \int_0^\infty R_T(\nu)\,d\nu \tag{1-1}$$

As we have seen in the preceding discussion of Figure 1-1, $R_T$ increases rapidly with increasing temperature. In fact, this result is called *Stefan's law*, and it was first stated in 1879 in the form of an empirical equation

$$R_T = \sigma T^4 \tag{1-2}$$

where

$$\sigma = 5.67 \times 10^{-8}\ \text{W/m}^2\text{-}^\circ\text{K}^4$$

is called the *Stefan-Boltzmann constant*. Figure 1-1 also shows us that the spectrum shifts toward higher frequencies as $T$ increases. This result is called *Wien's displacement law*

$$\nu_{\text{max}} \propto T \tag{1-3a}$$

where $\nu_{\text{max}}$ is the frequency $\nu$ at which $R_T(\nu)$ has its maximum value for a particular $T$. As $T$ increases, $\nu_{\text{max}}$ is displaced toward higher frequencies. All these results are in agreement with the familiar experiences discussed earlier, namely that the amount of thermal radiation emitted increases rapidly (the poker radiates much more heat energy at higher temperatures), and the principal frequency of the radiation becomes higher (the poker changes color from dull red to blue-white), with increasing temperature.

Another example of a blackbody, which we shall see to be particularly important, can be found by considering an object containing a cavity which is connected to the outside by a small hole, as in Figure 1-2. Radiation incident upon the hole from the outside enters the cavity and is reflected back and forth by the walls of the cavity, eventually being absorbed on these walls. If the area of the hole is very small compared to the area of the inner surface of the cavity, a negligible amount of the incident radiation will be reflected back through the hole. Essentially all the radiation incident upon the hole is absorbed; therefore, the *hole* must have the properties of the surface of a blackbody. Most blackbodies used in laboratory experiments are constructed along these lines.

Now assume that the walls of the cavity are uniformly heated to a temperature $T$. Then the walls will emit thermal radiation which will fill the cavity. The small fraction of this radiation incident from the inside upon the hole will pass through the hole. Thus the hole will act as an emitter of thermal radiation. Since the hole must have the properties of the surface of a blackbody, the radiation emitted by the hole must have a blackbody spectrum; but since the hole is merely sampling the thermal radiation present inside the cavity, it is clear that the radiation in the cavity must also have a blackbody spectrum. In fact, it will have a blackbody spectrum characteristic of the temperature $T$ on the walls, since this is the only temperature defined for the system. The spectrum emitted by the hole in the cavity is specified in terms of the energy flux

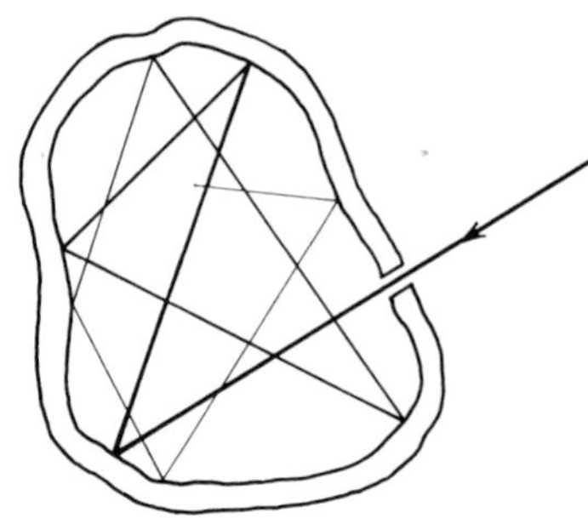

**FIGURE 1-2**
A cavity in a body connected by a small hole to the outside. Radiation incident on the hole is completely absorbed after successive reflections on the inner surface of the cavity. The hole absorbs like a blackbody. In the reverse process, in which radiation leaving the hole is built up of contributions emitted from the inner surface, the hole emits like a blackbody.

$R_T(\nu)$. It is more useful, however, to specify the spectrum of radiation inside the cavity, called *cavity radiation*, in terms of an *energy density*, $\rho_T(\nu)$, which is defined as the energy contained in a unit volume of the cavity at temperature $T$ in the frequency interval $\nu$ to $\nu + d\nu$. It is evident that these quantities are proportional to one another; that is

$$\rho_T(\nu) \propto R_T(\nu) \tag{1-4}$$

Hence, the radiation inside a cavity whose walls are at temperature $T$ has the same character as the radiation emitted by the surface of a blackbody at temperature $T$. It is convenient experimentally to produce a blackbody spectrum by means of a cavity in a heated body with a hole to the outside, and it is convenient in theoretical work to study blackbody radiation by analyzing the cavity radiation because it is possible to apply very general arguments to predict the properties of cavity radiation.

**Example 1-1.** (a) Since $\lambda\nu = c$, the constant velocity of light, Wien's displacement law (1-3a) can also be put in the form

$$\lambda_{max} T = \text{const} \tag{1-3b}$$

where $\lambda_{max}$ is the wavelength at which the spectral radiancy has its maximum value for a particular temperature $T$. The experimentally determined value of Wien's constant is $2.898 \times 10^{-3}$ m–°K. If we assume that stellar surfaces behave like blackbodies we can get a good estimate of their temperature by measuring $\lambda_{max}$. For the sun $\lambda_{max} = 5100$ Å, whereas for the North Star $\lambda_{max} = 3500$ Å. Find the surface temperature of these stars. (One *angstrom* $= 1$ Å $= 10^{-10}$ m.)

For the sun, $T = 2.898 \times 10^{-3}$ m–°K$/5100 \times 10^{-10}$ m $= 5700$°K. For the North Star, $T = 2.898 \times 10^{-3}$ m–°K$/3500 \times 10^{-10}$ m $= 8300$°K.

At 5700°K the sun's surface is near the temperature at which the greatest part of its radiation lies within the visible region of the spectrum. This suggests that over the ages of human evolution our eyes have adapted to the sun to become most sensitive to those wavelengths which it radiates most intensely.

(b) Using Stefan's law, (1-2), and the temperatures just obtained, determine the power radiated from 1 cm² of stellar surface.

For the sun

$$\begin{aligned} R_T = \sigma T^4 &= 5.67 \times 10^{-8}\ \text{W/m}^2\text{–°K}^4 \times (5700\text{°K})^4 \\ &= 5.90 \times 10^7\ \text{W/m}^2 \simeq 6000\ \text{W/cm}^2 \end{aligned}$$

For the North Star

$$\begin{aligned} R_T = \sigma T^4 &= 5.67 \times 10^{-8}\ \text{W/m}^2\text{–°K}^4 \times (8300\text{°K})^4 \\ &= 2.71 \times 10^8\ \text{W/m}^2 \simeq 27{,}000\ \text{W/cm}^2 \end{aligned}$$

◀

**Example 1-2.** Assume we have two small opaque bodies a large distance from one another supported by fine threads in a large evacuated enclosure whose walls are opaque and kept at a constant temperature. In such a case the bodies and walls can exchange energy only by means of radiation. Let $e$ represent the rate of emission of radiant energy by a body and let $a$ represent the rate of absorption of radiant energy by a body. Show that at equilibrium

$$\frac{e_1}{a_1} = \frac{e_2}{a_2} = 1 \tag{1-5}$$

This relation, (1-5), is known as *Kirchhoff's law for radiation.*

The equilibrium state is one of constant temperature throughout the enclosed system, and in that state the emission rate necessarily equals the absorption rate for each body. Hence

$$e_1 = a_1 \qquad \text{and} \qquad e_2 = a_2$$

Therefore

$$\frac{e_1}{a_1} = 1 = \frac{e_2}{a_2}$$

If one body, say body 2, is a blackbody, then $a_2 > a_1$ because a blackbody is a better absorber than a non-blackbody. Hence, it follows from (1-5) that $e_2 > e_1$. The observed fact that good absorbers are also good emitters is thus predicted by Kirchhoff's law. ◀

## 1-3 Classical Theory of Cavity Radiation

Shortly after the turn of the present century, Rayleigh, and also Jeans, made a calculation of the energy density of cavity (or blackbody) radiation that points up a serious conflict between classical physics and experimental results. This calculation is similar to calculations that arise in considering many other phenomena (e.g., specific heats of solids) to be treated later. We present the details here, but as an aid in guiding us through the calculations we first outline their general procedure.

Consider a cavity with metallic walls heated uniformly to temperature $T$. The walls emit electromagnetic radiation in the thermal range of frequencies. We know that this happens, basically, because of the accelerated motions of the electrons in the metallic walls that arise from thermal agitation (see Appendix B). However, it is not necessary to study the behavior of the electrons in the walls of the cavity in detail. Instead, attention is focused on the behavior of the electromagnetic waves in the interior of the cavity. Rayleigh and Jeans proceeded as follows. First, classical electromagnetic theory is used to show that the radiation inside the cavity must exist in the form of standing waves with nodes at the metallic surfaces. By using geometrical arguments, a count is made of the number of such standing waves in the frequency interval $\nu$ to $\nu + d\nu$, in order to determine how the number depends on $\nu$. Then a result of classical kinetic theory is used to calculate the average total energy of these waves when the system is in thermal equilibrium. The average total energy depends, in the classical theory, only on the temperature $T$. The number of standing waves in the frequency interval times the average energy of the waves, divided by the volume of the cavity, gives the average energy content per unit volume in the frequency interval $\nu$ to $\nu + d\nu$. This is the required quantity, the energy density $\rho_T(\nu)$. Let us now do all this.

We assume for simplicity that the metallic-walled cavity filled with electromagnetic radiation is in the form of a cube of edge length $a$, as shown in Figure 1-3. Then the

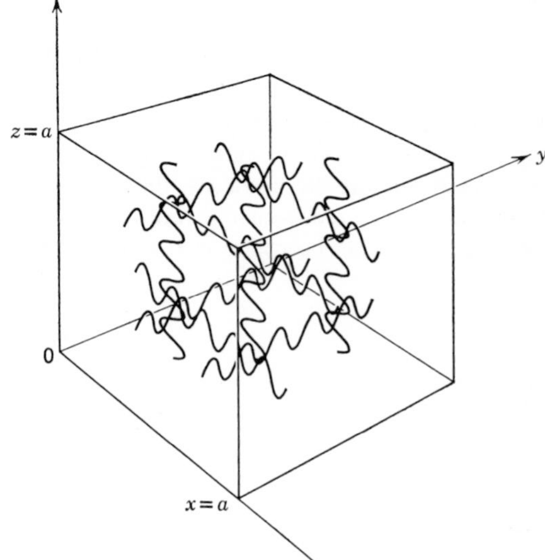


**FIGURE 1-3**

A metallic walled cubical cavity filled with electromagnetic radiation, showing three noninterfering components of that radiation bouncing back and forth between the walls and forming standing waves with nodes at each wall.

radiation reflecting back and forth between the walls can be analyzed into three components along the three mutually perpendicular directions defined by the edges of the cavity. Since the opposing walls are parallel to each other, the three components of the radiation do not mix, and we may treat them separately. Consider first the $x$ component and the metallic wall at $x = 0$. All the radiation of this component which is incident upon the wall is reflected by it, and the incident and reflected waves combine to form a standing wave. Now, since electromagnetic radiation is a transverse vibration with the electric field vector **E** perpendicular to the propagation direction, and since the propagation direction for this component is perpendicular to the wall in question, its electric field vector **E** is parallel to the wall. A metallic wall cannot, however, support an electric field parallel to the surface, since charges can always flow in such a way as to neutralize the electric field. Therefore, **E** for this component must always be zero at the wall. That is, the standing wave associated with the $x$-component of the radiation must have a node (zero amplitude) at $x = 0$. The standing wave must also have a node at $x = a$ because there can be no parallel electric field in the corresponding wall. Furthermore, similar conditions apply to the other two components; the standing wave associated with the $y$ component must have nodes at $y = 0$ and $y = a$, and the standing wave associated with the $z$ component must have nodes at $z = 0$ and $z = a$. These conditions put a limitation on the possible wavelengths, and therefore on the possible frequencies, of the electromagnetic radiation in the cavity.

Now we shall consider the question of counting the number of standing waves with nodes on the surfaces of the cavity, whose wavelengths lie in the interval $\lambda$ to $\lambda + d\lambda$ corresponding to the frequency interval $\nu$ to $\nu + d\nu$. To focus attention on the ideas involved in the calculation, we shall first treat the $x$ component alone; that is, we shall consider the simplified, but artificial, case of a "one-dimensional cavity" of length $a$. After we have worked through this case, we shall see that the procedure for generalizing to a real three-dimensional cavity is obvious.

The electric field for one-dimensional electromagnetic standing waves can be described mathematically by the function

$$E(x, t) = E_0 \sin (2\pi x/\lambda) \sin (2\pi\nu t) \tag{1-6}$$

where $\lambda$ is the wavelength of the wave, $\nu$ is its frequency, and $E_0$ is its maximum amplitude. The first two quantities are related by the equation

$$\nu = c/\lambda \tag{1-7}$$

where $c$ is the propagation velocity of electromagnetic waves. Equation (1-6) represents a wave whose amplitude has the sinusoidal space variation $\sin (2\pi x/\lambda)$ and which is oscillating in time sinusoidally with frequency $\nu$ like a simple harmonic oscillator. Since the amplitude is obviously zero, at all times $t$, for positions satisfying the relation

$$2x/\lambda = 0, 1, 2, 3, \ldots \tag{1-8}$$

the wave has fixed nodes; that is, it is a standing wave. In order to satisfy the requirement that the waves have nodes at both ends of the one-dimensional cavity, we choose the origin of the $x$ axis to be at one end of the cavity ($x = 0$) and then require that at the other end ($x = a$)

$$2x/\lambda = n \qquad \text{for } x = a \tag{1-9}$$

where

$$n = 1, 2, 3, 4, \ldots$$

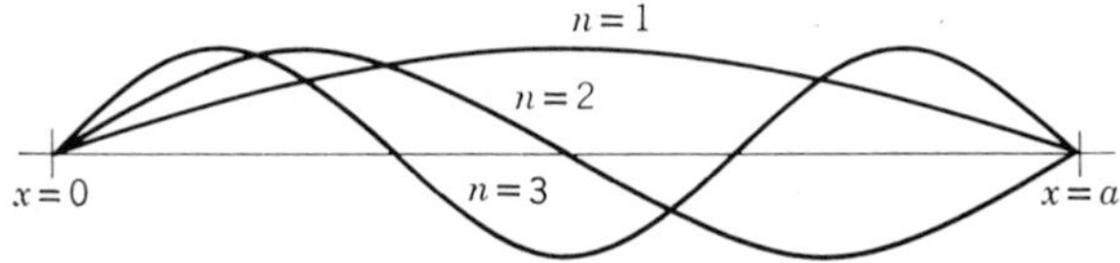


**FIGURE 1–4**

The amplitude patterns of standing waves in a one-dimensional cavity with walls at $x = 0$ and $x = a$, for the first three values of the index $n$.

This condition determines a set of allowed values of the wavelength $\lambda$. For these allowed values, the amplitude patterns of the standing waves have the appearance shown in Figure 1-4. These patterns may be recognized as the standing wave patterns for vibrations of a string fixed at both ends, a real physical system which also satisfies (1-6). In our case the patterns represent electromagnetic standing waves.

It is convenient to continue the discussion in terms of the allowed frequencies instead of the allowed wavelengths. These frequencies are $\nu = c/\lambda$, where $2a/\lambda = n$. That is

$$\nu = cn/2a \qquad n = 1, 2, 3, 4, \ldots \tag{1-10}$$

We can represent these allowed values of frequency in terms of a diagram consisting of an axis on which we plot a point at every integral value of $n$. On such a diagram, the value of the allowed frequency $\nu$ corresponding to a particular value of $n$ is, by (1-10), equal to $c/2a$ times the distance $d$ from the origin to the appropriate point, or the distance $d$ is $2a/c$ times the frequency $\nu$. These relations are shown in Figure 1-5. Such a diagram is useful in calculating the number of allowed values in frequency range $\nu$ to $\nu + d\nu$, which we call $N(\nu)\,d\nu$. To evaluate this quantity we simply count the number of points on the $n$ axis which fall between two limits which are constructed so as to correspond to the frequencies $\nu$ and $\nu + d\nu$, respectively. Since the points are distributed uniformly along the $n$ axis, it is apparent that the number of points falling between the two limits will be proportional to $d\nu$ but will not depend on $\nu$. In fact, it is easy to see that $N(\nu)\,d\nu = (2a/c)\,d\nu$. However, we must multiply this by an additional factor of 2 since, for each of the allowed frequencies, there are actually two independent waves corresponding to the two possible states of polarization of electromagnetic waves. Thus we have

$$N(\nu)\,d\nu = \frac{4a}{c}\,d\nu \tag{1-11}$$

This completes the calculation of the number of allowed standing waves for the artificial case of a one-dimensional cavity.

The above calculation makes apparent the procedures for extending the calculation to the real case of a three-dimensional cavity. This extension is indicated in Figure 1-6. Here the set of points uniformly distributed at integral values along a single $n$ axis is replaced by a uniform three-dimensional array of points whose three coordinates occur

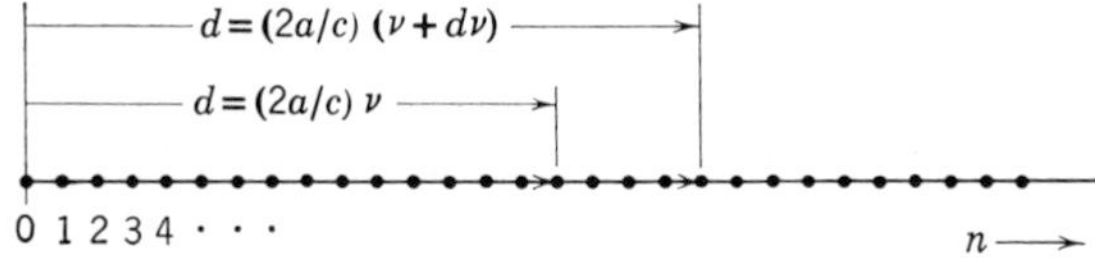


**FIGURE 1–5**

The allowed values of the index $n$, which determines the allowed values of the frequency, in a one-dimensional cavity of length $a$.

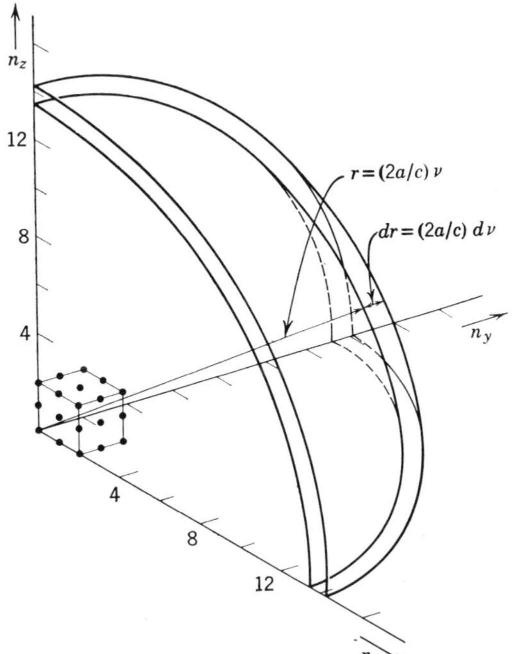


**FIGURE 1–6**
The allowed frequencies in a three-dimensional cavity in the form of a cube of edge length $a$ are determined by three indices $n_x$, $n_y$, $n_z$, which can each assume only integral values. For clarity, only a few of the very many points corresponding to sets of these indices are shown.

at integral values along each of three mutually perpendicular $n$ axes. *Each point* of the array corresponds to a particular allowed three-dimensional standing wave. The integral values of $n_x$, $n_y$, and $n_z$ specified by each point give the number of nodes of the $x$, $y$, and $z$ components, respectively, of the three-dimensional wave. The procedure is equivalent to analyzing a three-dimensional wave (i.e., one propagated in an arbitrary direction) into three one-dimensional component waves. Here the number of allowed frequencies in the frequency interval $\nu$ to $\nu + d\nu$ is equal to the number of points contained between shells of radii corresponding to frequencies $\nu$ and $\nu + d\nu$, respectively. This will be proportional to the volume contained between these two shells, since the points are uniformly distributed. Thus it is apparent that $N(\nu)\,d\nu$ will be proportional to $\nu^2\,d\nu$, the first factor, $\nu^2$, being proportional to the area of the shells and the second factor, $d\nu$, being the distance between them. In the following example we shall work out the details and find

$$N(\nu)\,d\nu = \frac{8\pi V}{c^3}\,\nu^2\,d\nu \tag{1-12}$$

where $V = a^3$, the volume of the cavity.

**Example 1-3.** Derive (1-12), which gives the number of allowed electromagnetic standing waves in each frequency interval for the case of a three-dimensional cavity in the form of a metallic-walled cube of edge length $a$.

Consider radiation of wavelength $\lambda$ and frequency $\nu = c/\lambda$, propagating in the direction defined by the three angles $\alpha$, $\beta$, $\gamma$, as shown in Figure 1-7. The radiation must be a standing wave since all three of its components are standing waves. We have indicated the locations of some of the fixed nodes of this standing wave by a set of planes perpendicular to the propagation direction $\alpha$, $\beta$, $\gamma$. The distance between these nodal planes of the radiation is just $\lambda/2$, where $\lambda$ is its wavelength. We have also indicated the locations at the three axes of the nodes of

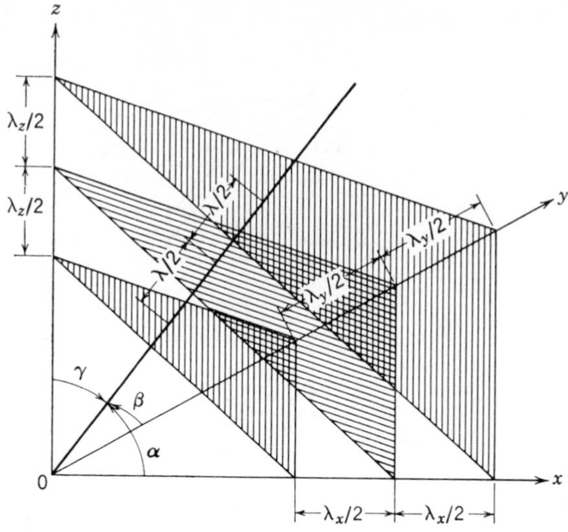


**FIGURE 1–7**

The nodal planes of a standing wave propagating in a certain direction in a cubical cavity.

the three components. The distances between these nodes are

$$\lambda_x/2 = \lambda/2\cos\alpha$$
$$\lambda_y/2 = \lambda/2\cos\beta \qquad (1\text{-}13)$$
$$\lambda_z/2 = \lambda/2\cos\gamma$$

Let us write expressions for the magnitudes at the three axes of the electric fields of the three components. They are

$$E(x, t) = E_{0_x} \sin(2\pi x/\lambda_x) \sin(2\pi\nu t)$$
$$E(y, t) = E_{0_y} \sin(2\pi y/\lambda_y) \sin(2\pi\nu t)$$
$$E(z, t) = E_{0_z} \sin(2\pi z/\lambda_z) \sin(2\pi\nu t)$$

The expression for the $x$ component represents a wave with a maximum amplitude $E_{0_x}$, with a space variation $\sin(2\pi x/\lambda_x)$, and which is oscillating with frequency $\nu$. As $\sin(2\pi x/\lambda_x)$ is zero for $2x/\lambda_x = 0, 1, 2, 3, \ldots$, the wave is a standing wave of wavelength $\lambda_x$ because it has fixed nodes separated by the distance $\Delta x = \lambda_x/2$. The expressions for the $y$ and $z$ components represent standing waves of maximum amplitudes $E_{0_y}$ and $E_{0_z}$ and wavelengths $\lambda_y$ and $\lambda_z$, but all three component standing waves oscillate with the frequency $\nu$ of the radiation. Note that these expressions automatically satisfy the requirement that the $x$ component have a node at $x = 0$, the $y$ component have a node at $y = 0$, and the $z$ component have a node at $z = 0$. To make them also satisfy the requirement that the $x$ component have a node at $x = a$, the $y$ component have a node at $y = a$, and the $z$ component have a node at $z = a$, set

$$2x/\lambda_x = n_x \qquad \text{for } x = a$$
$$2y/\lambda_y = n_y \qquad \text{for } y = a$$
$$2z/\lambda_z = n_z \qquad \text{for } z = a$$

where $n_x = 1, 2, 3, \ldots; n_y = 1, 2, 3, \ldots; n_z = 1, 2, 3, \ldots$. Using (1-13), these conditions become

$$(2a/\lambda)\cos\alpha = n_x \qquad (2a/\lambda)\cos\beta = n_y \qquad (2a/\lambda)\cos\gamma = n_z$$

Squaring both sides of these equations and adding, we obtain

$$(2a/\lambda)^2(\cos^2\alpha + \cos^2\beta + \cos^2\gamma) = n_x^2 + n_y^2 + n_z^2$$

but the angles $\alpha$, $\beta$, $\gamma$ have the property

$$\cos^2 \alpha + \cos^2 \beta + \cos^2 \gamma = 1$$

Thus

$$2a/\lambda = \sqrt{n_x^2 + n_y^2 + n_z^2}$$

where $n_x$, $n_y$, $n_z$ take on all possible integral values. This equation describes the limitation on the possible wavelengths of the electromagnetic radiation contained in the cavity.

We again continue the discussion in terms of the allowed frequencies instead of the allowed wavelengths. They are

$$\nu = \frac{c}{\lambda} = \frac{c}{2a}\sqrt{n_x^2 + n_y^2 + n_z^2} \tag{1-14a}$$

Now we shall count the number of allowed frequencies in a given frequency interval by constructing a uniform cubic lattice in one octant of a rectangular coordinate system in such a way that the three coordinates of each point of the lattice are equal to a possible set of the three integers $n_x$, $n_y$, $n_z$ (see Figure 1-6). By construction, each lattice point corresponds to an allowed frequency. Furthermore, $N(\nu)\,d\nu$, the number of allowed frequencies between $\nu$ and $\nu + d\nu$, is equal to $N(r)\,dr$, the number of points contained between concentric shells of radii $r$ and $r + dr$, where

$$r = \sqrt{n_x^2 + n_y^2 + n_z^2}$$

From (1–14a), this is

$$r = \frac{2a}{c}\,\nu \tag{1-14b}$$

Since $N(r)\,dr$ is equal to the volume enclosed by the shells times the density of lattice points, and since, by construction, the density is one, $N(r)\,dr$ is simply

$$N(r)\,dr = \frac{1}{8}\,4\pi r^2\,dr = \frac{\pi r^2\,dr}{2} \tag{1-15}$$

Setting this equal to $N(\nu)\,d\nu$, and evaluating $r^2\,dr$ from (1-14b), we have

$$N(\nu)\,d\nu = \frac{\pi}{2}\left(\frac{2a}{c}\right)^3 \nu^2\,d\nu$$

This completes the calculation except that we must multiply these results by a factor of 2 because, for each of the allowed frequencies we have enumerated, there are actually two independent waves corresponding to the two possible states of polarization of electromagnetic radiation. Thus we have derived (1-12). It can be shown that $N(\nu)$ is independent of the assumed shape of the cavity and depends only on its volume. ◀

Note that there is a very significant difference between the results obtained for the case of a real three-dimensional cavity and the results we obtained earlier for the artificial case of a one-dimensional cavity. The factor of $\nu^2$ found in (1-12), but not in (1-11), will be seen to play a fundamental role in the arguments that follow. This factor arises, basically, because we live in a three-dimensional world—the power of $\nu$ being one less than the dimensionality. Although Planck, in ultimately resolving the serious discrepancies between classical theory and experiment, had to question certain points which had been considered to be obviously true, neither he nor others working on the problem questioned (1-12). It was, and remains, generally agreed that (1-12) is valid.

We now have a count of the number of standing waves. The next step in the Rayleigh-Jeans classical theory of blackbody radiation is the evaluation of the average total energy contained in each standing wave of frequency $\nu$. According to classical physics, the energy of some particular wave can have any value from zero to infinity, the actual value being proportional to the square of the magnitude of its amplitude constant $E_0$. However, for a system containing a large number of physical entities of the same kind which are in thermal equilibrium with each other at temperature $T$, classical physics makes a very definite prediction about the *average* values of the energies of the entities. This applies to our case since the multitude of standing waves, which constitute the thermal radiation inside the cavity, are entities of the same kind which are in thermal equilibrium with each other at the temperature $T$ of the walls of the cavity. Thermal equilibrium is ensured by the fact that the walls of any real cavity will always absorb and reradiate, in different frequencies and directions, a small amount of the radiation incident upon them and, therefore, the different standing waves can gradually exchange energy as required to maintain equilibrium.

The prediction comes from classical kinetic theory, and it is called the *law of equipartition of energy*. This law states that for a system of gas molecules in thermal equilibrium at temperature $T$, the average kinetic energy of a molecule per degree of freedom is $kT/2$, where $k = 1.38 \times 10^{-23}$ joule/°K is called *Boltzmann's constant*. The law actually applies to any classical system containing, in equilibrium, a large number of entities of the same kind. For the case at hand the entities are standing waves which have one degree of freedom, their electric field amplitudes. Therefore, on the average their *kinetic* energies all have the same value, $kT/2$. However, each sinusoidally oscillating standing wave has a *total* energy which is twice its average kinetic energy. This is a common property of physical systems which have a single degree of freedom that execute simple harmonic oscillations in time; familiar cases are a pendulum or a coil spring. Thus each standing wave in the cavity has, according to the classical equipartition law, an average total energy

$$\overline{\mathscr{E}} = kT \tag{1-16}$$

The most important point to note is that the average total energy $\overline{\mathscr{E}}$ is predicted to have the same value for all standing waves in the cavity, independent of their frequencies.

The energy per unit volume in the frequency interval $\nu$ to $\nu + d\nu$ of the blackbody spectrum of a cavity at temperature $T$ is just the product of the average energy per standing wave times the number of standing waves in the frequency interval, divided by the volume of the cavity. From (1-15) and (1-16) we therefore finally obtain the result

$$\rho_T(\nu)\, d\nu = \frac{8\pi\nu^2 kT}{c^3}\, d\nu \tag{1-17}$$

This is the *Rayleigh-Jeans formula for blackbody radiation.*

In Figure 1-8 we compare the predictions of this equation with experimental data. The discrepancy is apparent. In the limit of low frequencies, the classical spectrum approaches the experimental results, but, as the frequency becomes large, the theoretical prediction goes to infinity! Experiment shows that the energy density always remains finite, as it obviously must, and, in fact, that the energy density goes to zero at very high frequencies. The grossly unrealistic behavior of the prediction of classical theory at high frequencies is known in physics as the "ultraviolet catastrophe." This term is suggestive of the importance of the failure of the theory.

## 1-4 Planck's Theory of Cavity Radiation

In trying to resolve the discrepancy between theory and experiment, Planck was led to consider the possibility of a violation of the law of equipartition of energy on

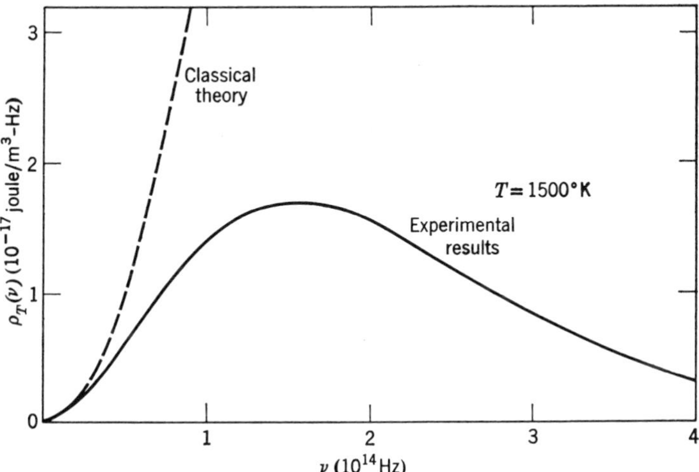


**FIGURE 1-8**

The Rayleigh-Jeans prediction (dashed line) compared with the experimental results (solid line) for the energy density of a blackbody cavity, showing the serious discrepancy called the ultraviolet catastrophe.

which the theory was based. From Figure 1-8 it is clear that the law gives satisfactory results for small frequencies. Thus we can assume

$$\bar{\mathscr{E}} \xrightarrow[\nu \to 0]{} kT \tag{1-18}$$

that is, the average total energy approaches $kT$ as the frequency approaches zero. The discrepancy at high frequencies could be eliminated if there is, for some reason, a cutoff, so that

$$\bar{\mathscr{E}} \xrightarrow[\nu \to \infty]{} 0 \tag{1-19}$$

that is, if the average total energy approaches zero as the frequency approaches infinity. In other words, Planck realized that, in the circumstances that prevail for the case of blackbody radiation, the average energy of the standing waves is a function of frequency $\bar{\mathscr{E}}(\nu)$ having the properties indicated by (1-18) and (1-19). This is in contrast to the law of equipartition of energy which assigns to the average energy $\bar{\mathscr{E}}$ a value independent of frequency.

Let us look at the origin of the equipartition law. It arises, basically, from a more comprehensive result of classical kinetic theory called the Boltzmann distribution. (Arguments leading to the Boltzmann distribution are given in Appendix C for students not already familiar with it.) Here we shall use a *special form of the Boltzmann distribution*

$$P(\mathscr{E}) = \frac{e^{-\mathscr{E}/kT}}{kT} \tag{1-20}$$

in which $P(\mathscr{E})\, d\mathscr{E}$ is the probability of finding a given entity of a system with energy in the interval between $\mathscr{E}$ and $\mathscr{E} + d\mathscr{E}$, when the number of energy states for the entity in that interval is independent of $\mathscr{E}$. The system is supposed to contain a large number of entities of the same kind in thermal equilibrium at temperature $T$, and $k$ represents Boltzmann's constant. The energies of the entities in the system we are considering, a set of simple harmonic oscillating standing waves in thermal equilibrium in a blackbody cavity, are governed by (1-20).

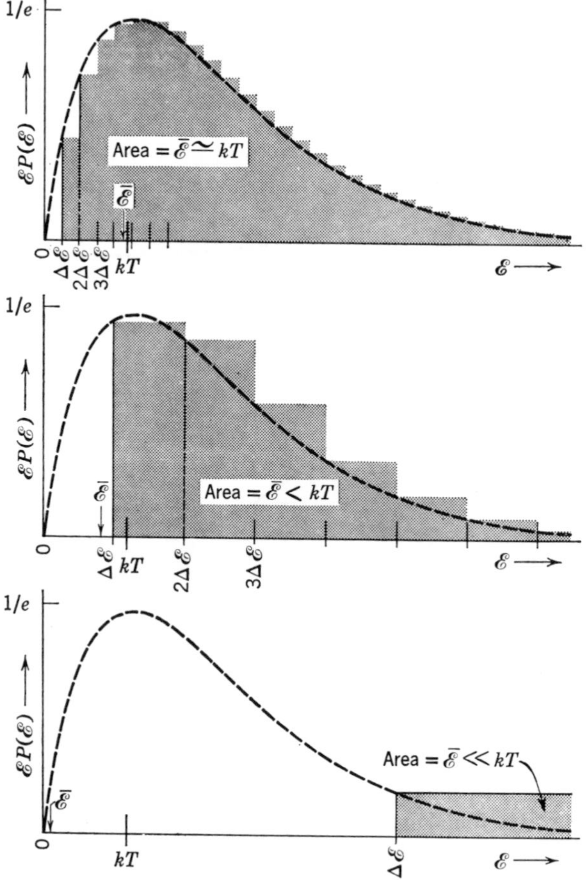


**FIGURE 1-10**

*Top:* If the energy $\mathscr{E}$ is not a continuous variable but is instead restricted to discrete values $0, \Delta\mathscr{E}, 2\Delta\mathscr{E}, 3\Delta\mathscr{E}, \ldots$, as indicated by the ticks on the $\mathscr{E}$ axis of the figure, the integral used to calculate the average value $\bar{\mathscr{E}}$ must be replaced by a summation. The average value is thus a sum of areas of rectangles, each of width $\Delta\mathscr{E}$, and with heights given by the allowed values of $\mathscr{E}$ times $P(\mathscr{E})$ at the beginning of each interval. In this figure $\Delta\mathscr{E} \ll kT$, and the allowed energies being closely spaced the area of all the rectangles differs but little from the area under the smooth curve. Thus the average value $\bar{\mathscr{E}}$ is nearly equal to $kT$, the value found in Figure 1-9. *Middle:* $\Delta\mathscr{E} \simeq kT$, and $\bar{\mathscr{E}}$ has a smaller value than it has in the case of the top figure. *Bottom:* $\Delta\mathscr{E} \gg kT$, and $\bar{\mathscr{E}}$ is further reduced. In all three figures the rectangles show the contribution to the total area of $\mathscr{E}P(\mathscr{E})$ for each allowed energy. The rectangle for $\mathscr{E} = 0$ of course is always of zero height. This will make a large effect on the total area if the widths of the rectangles are large.

Written as an equation instead of a proportionality, this is

$$\Delta\mathscr{E} = h\nu \tag{1-25}$$

where $h$ is the proportionality constant.

Further numerical work allowed Planck to determine the value of the constant $h$ by finding the value which produced the best fit of his theory with the experimental data. The value he obtained was very close to the currently accepted value

$$h = 6.63 \times 10^{-34} \text{ joule-sec}$$

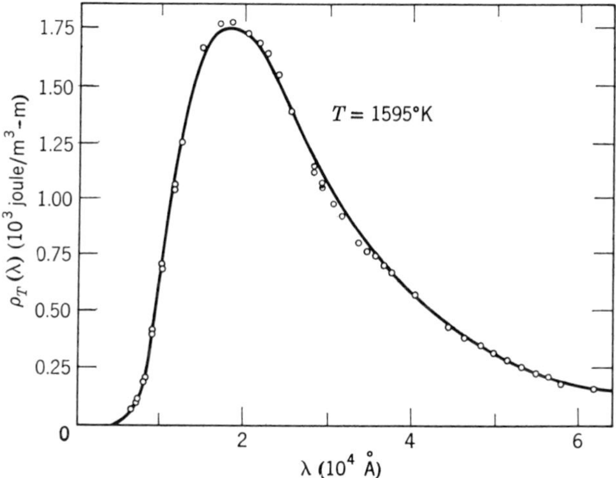


**FIGURE 1-11**

Planck's energy density prediction (solid line) compared to the experimental results (circles) for the energy density of a blackbody. The data were reported by Coblentz in 1916 and apply to a temperature of 1595°K. The author remarked in his paper that after drawing the spectral energy curves resulting from his measurements, "owing to eye fatigue it was impossible for months thereafter to give attention to the reduction of the data." The data, when finally reduced, led to a value for Planck's constant of 6.57 × $10^{-34}$ joule-sec.

This very famous constant is now called *Planck's constant.*

The formula Planck obtained for $\overline{\mathscr{E}}$ by evaluating the summation analogous to the integral in (1-21), and that we shall obtain in Example 1-4, is

$$\overline{\mathscr{E}}(\nu) = \frac{h\nu}{e^{h\nu/kT} - 1} \tag{1-26}$$

Since $e^{h\nu/kT} \to 1 + h\nu/kT$ for $h\nu/kT \to 0$, we see that $\overline{\mathscr{E}}(\nu) \to kT$ in this limit as predicted by (1-18). In the limit $h\nu/kT \to \infty$, $e^{h\nu/kT} \to \infty$, and $\overline{\mathscr{E}}(\nu) \to 0$, in agreement with the prediction of (1-19).

The formula which he then immediately obtained for the energy density in the blackbody spectrum, using his result for $\overline{\mathscr{E}}(\nu)$ rather than the classical value $\overline{\mathscr{E}} = kT$, is

$$\rho_T(\nu)\, d\nu = \frac{8\pi\nu^2}{c^3} \frac{h\nu}{e^{h\nu/kT} - 1}\, d\nu \tag{1-27}$$

This is *Planck's blackbody spectrum.* Figure 1-11 shows a comparison of this result of Planck's theory (expressed in terms of wavelength) with experimental results for a temperature $T = 1595°$K. The experimental results are in complete agreement with Planck's formula at all temperatures.

We should remember that Planck did not alter the Boltzmann distribution. "All" he did was to treat the energy of the electromagnetic standing waves, oscillating sinusoidally in time, as a discrete instead of a continuous quantity.

**Example 1-4.** Derive Planck's expression for the average energy $\overline{\mathscr{E}}$ and also his blackbody spectrum.

The quantity $\overline{\mathscr{E}}$ is evaluated from the ratio of sums

$$\overline{\mathscr{E}} = \frac{\sum_{n=0}^{\infty} \mathscr{E}P(\mathscr{E})}{\sum_{n=0}^{\infty} P(\mathscr{E})}$$

analogous to the ratio of integrals in (1-21). Sums must be used because with Planck's postulate the energy $\mathscr{E}$ becomes a discrete variable that takes on only the values $\mathscr{E} = 0$, $h\nu$, $2h\nu$, $3h\nu$, . . . . That is, $\mathscr{E} = nh\nu$ where $n = 0, 1, 2, 3, \ldots$. Evaluating the Boltzmann distribution $P(\mathscr{E}) = e^{-\mathscr{E}/kT}/kT$, we have

$$\overline{\mathscr{E}} = \frac{\sum_{n=0}^{\infty} \frac{nh\nu}{kT} e^{-nh\nu/kT}}{\sum_{n=0}^{\infty} \frac{1}{kT} e^{-nh\nu/kT}} = kT \frac{\sum_{n=0}^{\infty} n\alpha e^{-n\alpha}}{\sum_{n=0}^{\infty} e^{-n\alpha}} \qquad \text{where } \alpha = \frac{h\nu}{kT}$$

This, in turn, can be evaluated most easily by noting that

$$-\alpha \frac{d}{d\alpha} \ln \sum_{n=0}^{\infty} e^{-n\alpha} = \frac{-\alpha \frac{d}{d\alpha} \sum_{n=0}^{\infty} e^{-n\alpha}}{\sum_{n=0}^{\infty} e^{-n\alpha}} = \frac{-\sum_{n=0}^{\infty} \alpha \frac{d}{d\alpha} e^{-n\alpha}}{\sum_{n=0}^{\infty} e^{-n\alpha}} = \frac{\sum_{n=0}^{\infty} n\alpha e^{-n\alpha}}{\sum_{n=0}^{\infty} e^{-n\alpha}}$$

so that

$$\overline{\mathscr{E}} = kT\left(-\alpha \frac{d}{d\alpha} \ln \sum_{n=0}^{\infty} e^{-n\alpha}\right) = -h\nu \frac{d}{d\alpha} \ln \sum_{n=0}^{\infty} e^{-n\alpha}$$

Now

$$\sum_{n=0}^{\infty} e^{-n\alpha} = 1 + e^{-\alpha} + e^{-2\alpha} + e^{-3\alpha} + \cdots$$
$$= 1 + X + X^2 + X^3 + \cdots \qquad \text{where } X = e^{-\alpha}$$

but

$$(1 - X)^{-1} = 1 + X + X^2 + X^3 + \cdots$$

so

$$\overline{\mathscr{E}} = -h\nu \frac{d}{d\alpha} \ln (1 - e^{-\alpha})^{-1}$$
$$= \frac{-h\nu}{(1 - e^{-\alpha})^{-1}} (-1)(1 - e^{-\alpha})^{-2} e^{-\alpha}$$
$$= \frac{h\nu e^{-\alpha}}{1 - e^{-\alpha}} = \frac{h\nu}{e^{\alpha} - 1} = \frac{h\nu}{e^{h\nu/kT} - 1}$$

We have derived (1-26) for the average energy of an electromagnetic standing wave of frequency $\nu$. Multiplying this by (1-12), the number $N(\nu)\,d\nu$ of waves having this frequency derived in Example 1-3, we immediately obtain the Planck blackbody spectrum, (1-27). ◀

**Example 1-5.** It is convenient in analyzing experimental results, as in Figure 1-11, to express the Planck blackbody spectrum in terms of wavelength $\lambda$ rather than frequency $\nu$. Obtain $\rho_T(\lambda)$, the wavelength form of Planck's spectrum, from $\rho_T(\nu)$, the frequency form of the spectrum. The quantity $\rho_T(\lambda)$ is defined from the equality $\rho_T(\lambda)\,d\lambda = -\rho_T(\nu)\,d\nu$. The minus sign indicates that, though $\rho_T(\lambda)$ and $\rho_T(\nu)$ are both positive, $d\lambda$ and $d\nu$ have opposite signs. (An increase in frequency gives rise to a corresponding decrease in wavelength.)

From the relation $\nu = c/\lambda$ we have $d\nu = -(c/\lambda^2)\,d\lambda$, or $d\nu/d\lambda = -(c/\lambda^2)$, so that

$$\rho_T(\lambda) = -\rho_T(\nu)\frac{d\nu}{d\lambda} = \rho_T(\nu)\frac{c}{\lambda^2}$$

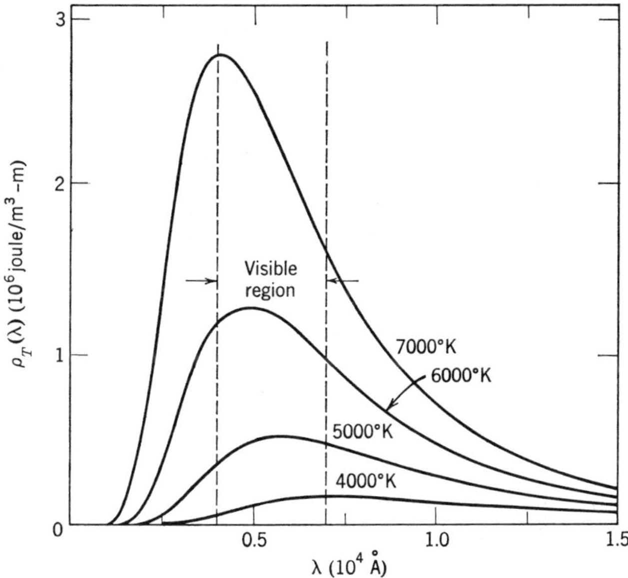


**FIGURE 1-12**

Planck's energy density of blackbody radiation at various temperatures as a function of wavelength. Note that the wavelength at which the curve is a maximum decreases as the temperature increases.

If now we set $\nu = c/\lambda$ in (1-27) for $\rho_T(\nu)$ we obtain

$$\rho_T(\lambda)\, d\lambda = \frac{8\pi hc}{\lambda^5} \frac{d\lambda}{e^{hc/\lambda kT} - 1} \tag{1-28}$$

In Figure 1-12 we show $\rho_T(\lambda)$ versus $\lambda$ for several different temperatures. The trend from "red heat" to "white heat" to "blue heat" radiation with rising temperatures becomes clear as the distribution of radiant energy with wavelength is studied for increasing temperatures. ◀

Stefan's law, (1-2), and Wien's displacement law, (1-3), can be derived from the Planck formula. By fitting them to the experimental results we can determine values of the constants $h$ and $k$. Stefan's law is obtained by integrating Planck's law over the entire spectrum of wavelengths. The radiancy is found to be proportional to the fourth power of the temperature, the proportionality constant $2\pi^5k^4/15c^2h^3$ being identified with $\sigma$, Stefan's constant, which has the experimentally determined value $5.67 \times 10^{-8}$ W/m$^2$–°K$^4$. Wien's displacement law is obtained by setting $d\rho(\lambda)/d\lambda = 0$. We find $\lambda_{\text{max}}T = 0.2014hc/k$ and identify the right-hand side of the equation with Wien's experimentally determined constant $2.898 \times 10^{-3}$ m–°K. Using these two measured values and assuming a value for the speed of light $c$, we can calculate the values of $h$ and $k$. Indeed, this was done by Planck, his values agreeing very well with those obtained subsequently by other methods.

## 1-5 The Use of Planck's Radiation Law in Thermometry

The radiation emitted from a hot body can be used to measure its temperature. If total radiation is used, then, from the Stefan-Boltzmann law, we know that the energies emitted by two sources are in the ratio of the fourth power of the temperature. However, it is difficult to measure total radiation from most sources so that we measure instead the radiancy over a

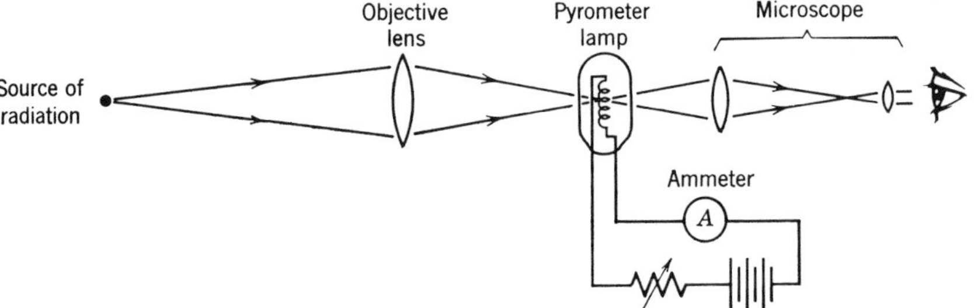


**FIGURE 1-13**
Schematic diagram of an optical pyrometer.

finite wavelength band. Here we use the Planck radiation law which gives the radiancy as a function of temperature and wavelength. For monochromatic radiation of wavelength $\lambda$ the ratio of the spectral intensities emitted by sources at $T_2$ °K and $T_1$ °K is given from Planck's law as

$$\frac{e^{hc/\lambda kT_1} - 1}{e^{hc/\lambda kT_2} - 1}$$

If $T_1$ is taken as a standard reference temperature, then $T_2$ can be determined relative to the standard from this expression by measuring the ratio experimentally. This procedure is used in the International Practical Temperature Scale, where the normal melting point of gold is taken as the standard fixed point, 1068°C. That is, the primary standard *optical pyrometer* is arranged to compare the spectral radiancy from a blackbody at an unknown temperature $T > 1068$°C with a blackbody at the gold point. Procedures must be adopted, and the theory developed, to allow for the practical circumstances that most sources are not blackbodies and that a finite spectral band is used instead of monochromatic radiation.

Most optical pyrometers use the eye as a detector and call for a large spectral bandwidth so that there will be enough energy for the eye to see. The simplest and most accurate type of instrument used above the gold point is the disappearing filament optical pyrometer (see Figure 1-13). The source whose temperature is to be measured is imaged on the filament of the pyrometer lamp, and the current in the lamp is varied until the filament seems to disappear into the background of the source image. Careful calibration and precision potentiometers insure accurate measurement of temperature.

A particularly interesting example in the general category of thermometry using blackbody radiation was discovered by Dicke, Penzias, and Wilson in the 1950s. Using a radio telescope operating in the several millimeter to several centimeter wavelength range, they found that a blackbody spectrum of electromagnetic radiation, with a characteristic temperature of about 3°K, is impinging on the earth with equal intensity from all directions. The uniformity in direction indicates that the radiation fills the universe uniformly. Astrophysicists consider these measurements as strong evidence in favor of the so-called *big-bang theory*, in which the universe was in the form of a very dense, and hot, fireball of particles and radiation around $10^{10}$ years ago. Due to subsequent expansion and the resulting Doppler shift, the temperature of the radiation would be expected to drop by now to something like the observed value of 3°K.

## 1-6 Planck's Postulate and Its Implications

Planck's contribution can be stated as a postulate, as follows:

*Any physical entity with one degree of freedom whose "coordinate" is a sinusoidal function of time (i.e., executes simple harmonic oscillations) can possess only total energies $\mathscr{E}$ which satisfy the relation*

$$\mathscr{E} = nh\nu \qquad n = 0, 1, 2, 3, \ldots$$

*where $\nu$ is the frequency of the oscillation, and $h$ is a universal constant.*

**FIGURE 1-14**

*Left:* The allowed energies in a classical system, oscillating sinusoidally with frequency $\nu$, are continuously distributed. *Right:* The allowed energies according to Planck's postulate are discretely distributed since they can only assume the values $nh\nu$. We say that the energy is quantized, $n$ being the quantum number of an allowed quantum state.

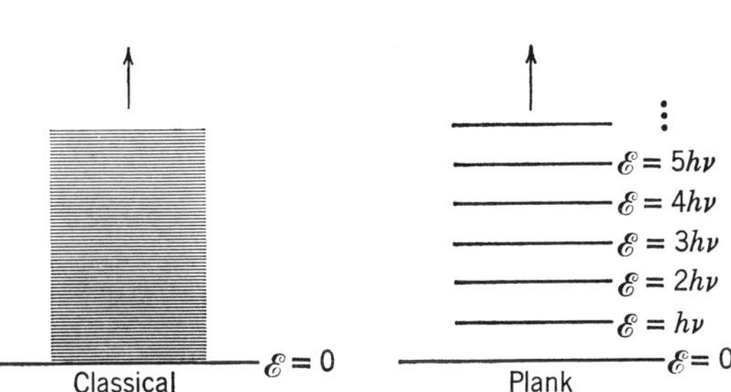


The word coordinate is used in its general sense to mean any quantity which describes the instantaneous condition of the entity. Examples are the length of a coil spring, the angular position of a pendulum bob, and the amplitude of a wave. All these examples happen also to be sinusoidal functions of time.

An energy-level diagram, as shown in Figure 1-14, provides a convenient way of illustrating the behavior of an entity governed by this postulate, and it is also useful in contrasting this behavior with what would be expected on the basis of classical physics. In such a diagram we indicate each of the possible energy states of the entity with a horizontal line. The distance from the line to the zero energy line is proportional to the total energy to which it corresponds. Since the entity may have any energy from zero to infinity according to classical physics, the classical energy-level diagram consists of a continuum of lines extending from zero up. However, the entity executing simple harmonic oscillations can have only one of the discrete total energies $\mathscr{E} = 0$, $h\nu$, $2h\nu$, $3h\nu$ . . . if it obeys Planck's postulate. This is indicated by the discrete set of lines in its energy-level diagram. The energy of the entity obeying Planck's constant is said to be *quantized*, the allowed energy states are called *quantum states*, and the integer $n$ is called the *quantum number*.

It may have occurred to the student that there are physical systems whose behavior seems to be obviously in disagreement with Planck's postulate. For instance, an ordinary pendulum executes simple harmonic oscillations, and yet this system certainly appears to be capable of possessing a continuous range of energies. Before we accept this argument, however, we should make some simple numerical calculations concerning such a system.

**Example 1-6.** A pendulum consisting of a 0.01 kg mass is suspended from a string 0.1 m in length. Let the amplitude of its oscillation be such that the string in its extreme positions makes an angle of 0.1 rad with the vertical. The energy of the pendulum decreases due, for instance, to frictional effects. Is the energy decrease observed to be continuous or discontinuous?

The oscillation frequency of the pendulum is

$$\nu = \frac{1}{2\pi}\sqrt{\frac{g}{l}} = \frac{1}{2\pi}\sqrt{\frac{9.8 \text{ m/sec}^2}{0.1 \text{ m}}} = 1.6/\text{sec}$$

The energy of the pendulum is its maximum potential energy

$$\begin{aligned} mgh &= mgl(1 - \cos\theta) = 0.01 \text{ kg} \times 9.8 \text{ m/sec}^2 \times 0.1 \text{ m} \times (1 - \cos 0.1) \\ &= 5 \times 10^{-5} \text{ joule} \end{aligned}$$

The energy of the pendulum is quantized so that changes in energy take place in discontinuous jumps of magnitude $\Delta E = h\nu$, but

$$\Delta E = h\nu = 6.63 \times 10^{-34}\,\text{joule-sec} \times 1.6/\text{sec} = 10^{-33}\,\text{joule}$$

whereas $E = 5 \times 10^{-5}$ joule. Therefore, $\Delta E/E = 2 \times 10^{-29}$. Hence, to measure the discreteness in the energy decrease we need to measure the energy to better than two parts in $10^{29}$. It is apparent that even the most sensitive experimental equipment is totally incapable of this energy resolution. ◀

We conclude that experiments involving an ordinary pendulum cannot determine whether Planck's postulate is valid or not. The same is true of experiments on all other macroscopic mechanical systems. The smallness of $h$ makes the graininess in the energy too fine to be distinguished from an energy continuum. Indeed, $h$ might as well be zero for classical systems and, in fact, one way to reduce quantum formulas to their classical limits would be to let $h \to 0$ in these formulas. Only where we consider systems in which $\nu$ is so large and/or $\mathcal{E}$ is so small that $\Delta\mathcal{E} = h\nu$ is of the order of $\mathcal{E}$ are we in a position to test Planck's postulate. One example is, of course, the high-frequency standing waves in blackbody radiation. Many other examples will be considered in following chapters.

## 1-7 A Bit of Quantum History

In its original form, Planck's postulate was not so far reaching as it is in the form we have given. Planck's initial work was done by treating, in detail, the behavior of the electrons in the walls of the blackbody and their coupling to the electromagnetic radiation within the cavity. This coupling leads to the same factor $\nu^2$ we obtained in (1-12) from the more general arguments due to Rayleigh and Jeans. Through this coupling, Planck related the energy in a particular frequency component of the blackbody radiation to the energy of an electron in the wall oscillating sinusoidally at the same frequency, and he postulated only that the energy of the oscillating particle is quantized. It was not until later that Planck accepted the idea that the oscillating electromagnetic waves were themselves quantized, and the postulate was broadened to include any entity whose single coordinate oscillates sinusoidally.

At first Planck was unsure whether his introduction of the constant $h$ was only a mathematical device or a matter of deep physical significance. In a letter to R. W. Wood, Planck called his limited postulate "an act of desperation." "I knew," he wrote, "that the problem (of the equilibrium of matter and radiation) is of fundamental significance for physics; I knew the formula that reproduces the energy distribution in the normal spectrum; a theoretical interpretation *had* to be found at any cost, no matter how high." For more than a decade Planck tried to fit the quantum idea into classical theory. With each attempt he appeared to retreat from his original boldness, but always he generated new ideas and techniques that quantum theory later adopted. What appears to have finally convinced him of the correctness and deep significance of his quantum hypothesis was its support of the definiteness of the statistical concept of entropy and the third law of thermodynamics.

It was during this period of doubt that Planck was editor of the German research journal *Annalen der Physik*. In 1905 he received Einstein's first relativity paper and stoutly defended Einstein's work. Thereafter he became one of young Einstein's patrons in scientific circles, but he resisted for some time the very ideas on the quantum theory of radiation advanced by Einstein that subsequently confirmed and extended Planck's own work. Einstein, whose deep insight into electromagnetism and statistical mechanics was perhaps unequalled by anyone at the time, saw as a result of Planck's work the need for a sweeping change in classical statistics and electromagnetism. He advanced predictions and interpretations of many physical phenomena which were later strikingly confirmed by experiment. In the next chapter we turn to one of these phenomena and follow another road on the way to quantum mechanics.

## QUESTIONS

1. Does a blackbody always appear black? Explain the term blackbody.

2. Pockets formed by coals in a coal fire seem brighter than the coals themselves. Is the temperature in such pockets appreciably higher than the surface temperature of an exposed glowing coal?

3. If we look into a cavity whose walls are kept at a constant temperature no details of the interior are visible. Explain.

4. The relation $R_T = \sigma T^4$ is exact for blackbodies and holds for all temperatures. Why is this relation not used as the basis of a definition of temperature at, for instance, 100°C?

5. A piece of metal glows with a bright red color at 1100°K. At this same temperature, however, a piece of quartz does not glow at all. Explain. (Hint: Quartz is transparent to visible light.)

6. Make a list of distribution functions commonly used in the social sciences (e.g., distribution of families with respect to income). In each case, state whether the variable whose distribution is described is discrete or continuous.

7. In (1-4) relating spectral radiancy and energy density, what dimensions would a proportionality constant need to have?

8. What is the origin of the ultraviolet catastrophe?

9. The law of equipartition of energy requires that the specific heat of gases be independent of the temperature, in disagreement with experiment. Here we have seen that it leads to the Rayleigh-Jeans radiation law, also in disagreement with experiment. How can you relate these two failures of the equipartition law?

10. Compare the definitions and dimensions of spectral radiancy $R_T(\nu)$, radiancy $R_T$, and energy density $\rho_T(\nu)$.

11. Why is optical pyrometry commonly used above the gold point and not below it? What objects typically have their temperatures measured in this way?

12. Are there quantized quantities in classical physics? Is energy quantized in classical physics?

13. Does it make sense to speak of charge quantization in physics? How is this different from energy quantization?

14. Elementary particles seem to have a discrete set of rest masses. Can this be regarded as quantization of mass?

15. In many classical systems the allowed frequencies are quantized. Name some of the systems. Is energy quantized there too?

16. Show that Planck's constant has the dimensions of angular momentum. Does this necessarily suggest that angular momentum is a quantized quantity?

17. For quantum effects to be everyday phenomena in our lives, what would be the minimum order of magnitude of $h$?

18. What, if anything, does the 3°K universal blackbody radiation tell us about the temperature of outer space?

19. Does Planck's theory suggest quantized atomic energy states?

20. Discuss the remarkable fact that discreteness in energy was first found in analyzing a continuous spectrum emitted by interacting atoms in a solid, rather than in analyzing a discrete spectrum such as is emitted by an isolated atom in a gas.

## PROBLEMS

1. At what wavelength does a cavity radiator at 6000°K radiate most per unit wavelength?

2. Show that the proportionality constant in (1-4) is $4/c$. That is, show that the relation between spectral radiancy $R_T(\nu)$ and energy density $\rho_T(\nu)$ is $R_T(\nu)\,d\nu = (c/4)\rho_T(\nu)\,d\nu$.

3. Consider two cavities of arbitrary shape and material, each at the same temperature $T$, connected by a narrow tube in which can be placed color filters (assumed ideal) which will allow only radiation of a specified frequency $\nu$ to pass through. (a) Suppose at a certain frequency $\nu'$, $\rho_T(\nu')\,d\nu$ for cavity 1 was greater than $\rho_T(\nu')\,d\nu$ for cavity 2. A color filter which passes only the frequency $\nu'$ is placed in the connecting tube. Discuss what will happen in terms of energy flow. (b) What will happen to their respective temperatures? (c) Show that this would violate the second law of thermodynamics; hence prove that all blackbodies at the same temperature must emit thermal radiation with the same spectrum independent of the details of their composition.

4. A cavity radiator at 6000°K has a hole 10.0 mm in diameter drilled in its wall. Find the power radiated through the hole in the range 5500–5510 Å. (Hint: See Problem 2.)

5. (a) Assuming the surface temperature of the sun to be 5700°K, use Stefan's law, (1-2), to determine the rest mass lost per second to radiation by the sun. Take the sun's diameter to be $1.4 \times 10^9$ m. (b) What fraction of the sun's rest mass is lost each year from electromagnetic radiation? Take the sun's rest mass to be $2.0 \times 10^{30}$ kg.

6. In a thermonuclear explosion the temperature in the fireball is momentarily $10^7$ °K. Find the wavelength at which the radiation emitted is a maximum.

7. At a given temperature $\lambda_{max} = 6500$ Å for a blackbody cavity. What will $\lambda_{max}$ be if the temperature of the cavity walls is increased so that the rate of emission of spectral radiation is doubled?

8. At what wavelength does the human body emit its maximum temperature radiation? List assumptions you make in arriving at an answer.

9. Assuming that $\lambda_{max}$ is in the near infrared for red heat and in the near ultraviolet for blue heat, approximately what temperature in Wien's displacement law corresponds to red heat? To blue heat?

10. The *average* rate of solar radiation incident per unit area on the earth is 0.485 cal/cm²-min (or 355 W/m²). (a) Explain the consistency of this number with the solar constant (the solar energy falling per unit time at normal incidence on unit area of the earth's surface) whose value is 1.94 cal/cm²-min (or 1340 W/m²). (b) Consider the earth to be a blackbody radiating energy into space at this same rate. What surface temperature would the earth have under these circumstances?

11. Show that the Rayleigh-Jeans radiation law, (1-17), is not consistent with the Wien displacement law $\nu_{max} \propto T$, (1-3a), or $\lambda_{max} T = \text{const}$, (1-3b).

12. We obtain $\nu_{max}$ in the blackbody spectrum by setting $d\rho_T(\nu)/d\nu = 0$ and $\lambda_{max}$ by setting $d\rho_T(\lambda)/d\lambda = 0$. Why is it not possible to get from $\lambda_{max} T = \text{const}$ to $\nu_{max} = \text{const} \times T$ simply by using $\lambda_{max} = c/\nu_{max}$? That is, why is it wrong to assume that $\nu_{max}\lambda_{max} = c$, where $c$ is the speed of light?

**13.** Consider the following numbers: 2, 3, 3, 4, 1, 2, 2, 1, 0 representing the number of hits garnered by each member of the Baltimore Orioles in a recent outing. (a) Calculate directly the average number of hits per man. (b) Let $x$ be a variable signifying the number of hits obtained by a man, and let $f(x)$ be the number of times the number $x$ appears. Show that the average number of hits per man can be written as

$$\bar{x} = \frac{\sum_0^4 xf(x)}{\sum_0^4 f(x)}$$

(c) Let $p(x)$ be the probability of the number $x$ being attained. Show that $\bar{x}$ is given by

$$\bar{x} = \sum_0^4 xp(x)$$

**14.** Consider the function

$$f(x) = \frac{1}{10}(10 - x)^2 \qquad 0 \leq x \leq 10$$

$$f(x) = 0 \qquad \text{all other } x$$

(a) From

$$\bar{x} = \frac{\int_{-\infty}^{\infty} xf(x)\,dx}{\int_{-\infty}^{\infty} f(x)\,dx}$$

find the average value of $x$. (b) Suppose the variable $x$ were discrete rather than continuous. Assume $\Delta x = 1$ so that $x$ takes on only integral values 0, 1, 2, . . . , 10. Compute $\bar{x}$ and compare to the result of part (a) (Hint: It may be easier to compute the appropriate sum directly rather than working with general summation formulas.) (c) Compute $\bar{x}$ for $\Delta x = 5$, i.e., $x = 0, 5, 10$. Compare to the result of part (a). (d) Draw analogies between the results obtained in this problem and the discussion of Section 1-4. Be sure you understand the roles played by $\bar{\mathscr{E}}$, $\Delta\mathscr{E}$, and $P(\mathscr{E})$.

**15.** Using the relations $P(\mathscr{E}) = e^{-\mathscr{E}/kT}/kT$ and $\int_0^\infty P(\mathscr{E})\,d\mathscr{E} = 1$, evaluate the integral of (1-21) to deduce (1-22), $\bar{\mathscr{E}} = kT$.

**16.** Use the relation $R_T(\nu)\,d\nu = (c/4)\rho_T(\nu)\,d\nu$ between spectral radiancy and energy density, together with Planck's radiation law, to derive Stefan's law. That is, show that

$$R_T = \int_0^\infty \frac{2\pi h}{c^2} \frac{\nu^3\,d\nu}{e^{h\nu/kT} - 1} = \sigma T^4$$

where $\sigma = 2\pi^5 k^4/15c^2h^3$.

$$\left(\text{Hint: } \int_0^\infty \frac{q^3\,dq}{e^q - 1} = \frac{\pi^4}{15}.\right)$$

**17.** Derive the Wien displacement law, $\lambda_{\text{max}} T = 0.2014\, hc/k$, by solving the equation $d\rho(\lambda)/d\lambda = 0$. (Hint: Set $hc/\lambda kT = x$ and show that the equation quoted leads to $e^{-x} + x/5 = 1$. Then show that $x = 4.965$ is the solution.)

**18.** To verify experimentally that the recently discovered 3°K universal radiation accurately fits a blackbody spectrum, it is decided to measure $R_T(\lambda)$ from a $\lambda$ below $\lambda_{\text{max}}$ where its value is $0.2R_T(\lambda_{\text{max}})$ to a $\lambda$ above $\lambda_{\text{max}}$ where its value is again $0.2R_T(\lambda_{\text{max}})$. Over what range of $\lambda$ must the measurements be made?

# 2

# Photons—Particlelike Properties of Radiation

# Photons—Particlelike Properties of Radiation

# 2

## 2-1 Introduction

In this chapter we shall examine processes in which radiation interacts with matter. Three processes (the photoelectric effect, the Compton effect, and pair production) involve the scattering or absorption of radiation in matter. Two processes (bremsstrahlung and pair annihilation) involve the production of radiation. In each case we shall obtain experimental evidence that radiation is particlelike in its interaction with matter, as distinguished from the wavelike nature of radiation when it propagates. In the following chapter we shall study a generalization of this result, due to de Broglie, which leads directly into quantum mechanics. Some of the material of these two chapters may be a review of topics the student has already come across in studying elementary physics.

## 2-2 The Photoelectric Effect

It was in 1886 and 1887 that Heinrich Hertz performed the experiments that first confirmed the existence of electromagnetic waves and Maxwell's electromagnetic theory of light propagation. It is one of those fascinating and paradoxical facts in the history of science that in the course of his experiments Hertz noted the effect that Einstein later used to contradict other aspects of the classical electromagnetic theory. Hertz discovered that an electric discharge between two electrodes occurs more readily when ultraviolet light falls on one of the electrodes. Lenard, following up some experiments of Hallwachs, showed soon after that the ultraviolet light facilitates the discharge by causing electrons to be emitted from the cathode surface. The ejection of electrons from a surface by the action of light is called the *photoelectric effect*.

Figure 2-1 shows an apparatus used to study the photoelectric effect. A glass envelope encloses the apparatus in an evacuated space. Monochromatic light, incident through a quartz window, falls on the metal plate $A$ and liberates electrons, called *photoelectrons*. The electrons can be detected as a current if they are attracted to the metal cup $B$ by means of a potential difference $V$ applied between $A$ and $B$. The sensitive ammeter $G$ serves to measure this photoelectric current.

Curve $a$ of Figure 2-2 is a plot of the photoelectric current, in an apparatus like that of Figure 2-1, as a function of the potential difference $V$. If $V$ is made large enough, the photoelectric current reaches a certain limiting (saturation) value at which all photoelectrons ejected from $A$ are collected by cup $B$.

If $V$ is reversed in sign, the photoelectric current does not immediately drop to zero, which suggests that the electrons are emitted from $A$ with kinetic energy. Some will reach cup $B$ in spite of the fact that the electric field opposes their motion. However, if this reversed potential difference is made large enough, a value $V_0$ called the *stopping potential* is reached at which the photoelectric current does drop to zero. This

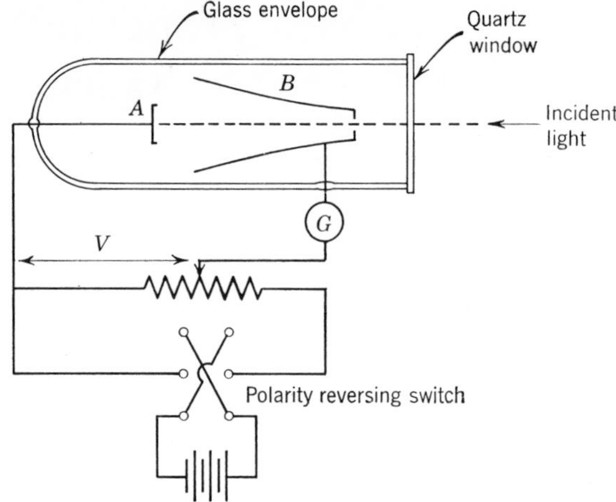


**FIGURE 2-1**

An apparatus used to study the photoelectric effect. The voltage $V$ can be varied continuously in magnitude, and also reversed in sign by the switching arrangement.

potential difference $V_0$, multiplied by electron charge, measures the kinetic energy $K_{max}$ of the *fastest* ejected photoelectron. That is

$$K_{max} = eV_0 \tag{2-1}$$

The quantity $K_{max}$ turns out experimentally to be independent of the intensity of the light, as is shown by curve $b$ in Figure 2-2 in which the light intensity has been reduced to one-half the value used in obtaining curve $a$.

Figure 2-3 shows the stopping potential $V_0$ as a function of the frequency of the light incident on sodium. Note that there is a definite *cutoff* frequency $\nu_0$, below which no photoelectric effect occurs. These data were taken in 1914 by Millikan whose painstaking work on the photoelectric effect won him the Nobel prize in 1923. Because the photoelectric effect for visible or near-visible light is largely a surface phenomenon, it is necessary in the experiments to avoid oxide films, grease, or other surface contaminants.

**FIGURE 2-2**

Graphs of current $i$ as a function of voltage $V$ from data taken with the apparatus of Figure 2-1. The applied potential difference $V$ is called positive when the cup $B$ in Figure 2-1 is positive with respect to the photoelectric surface $A$. In curve $b$ the incident light intensity has been reduced to one-half that of curve $a$. The stopping potential $V_0$ is independent of light intensity, but the saturation currents $I_a$ and $I_b$ are directly proportional to it.

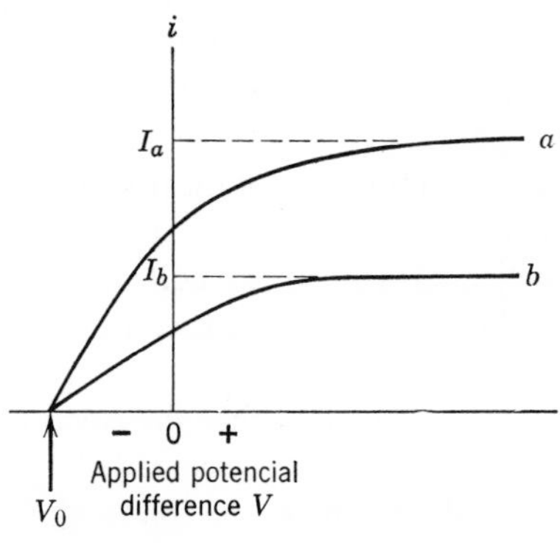

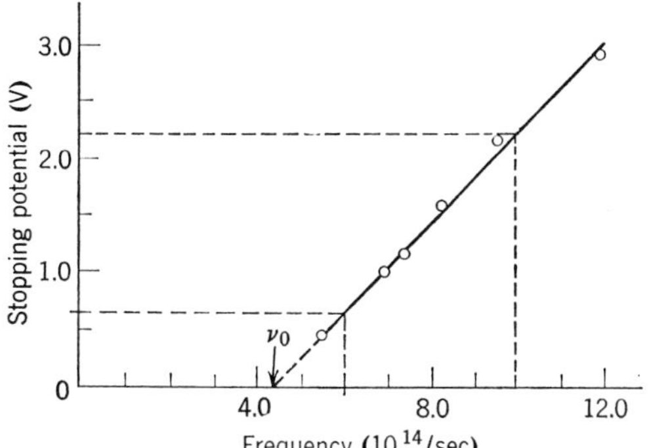


**FIGURE 2-3**

A plot of Millikan's measurements of the stopping potential at various frequencies for sodium. The cutoff frequency $\nu_0$ is $4.39 \times 10^{14}$ Hz.

There are three major features of the photoelectric effect that cannot be explained in terms of the classical wave theory of light:

1. Wave theory requires that the oscillating electric vector **E** of the light wave increase in amplitude as the intensity of the light beam is increased. Since the force applied to the electron is $e\mathbf{E}$, this suggests that the *kinetic energy* of the photoelectrons should also increase as the light beam is made more intense. However, Figure 2-2 shows that $K_{max}$, which equals $eV_0$, *is independent of the light intensity*. This has been tested over a range of intensities of $10^7$.

2. According to the wave theory the photoelectric effect should occur for any frequency of the light, provided only that the light is intense enough to give the energy needed to eject the photoelectrons. However, Figure 2-3 shows that there exists, for each surface, *a characteristic cutoff frequency* $\nu_0$. *For frequencies less than* $\nu_0$, *the photoelectric effect does not occur, no matter how intense the illumination.*

3. If the energy acquired by a photoelectron is absorbed from the wave incident on the metal plate, the "effective target area" for an electron in the metal is limited, and probably not much more than that of a circle having about an atomic diameter. In the classical theory the light energy is uniformly distributed over the wave front. Thus, if the light is feeble enough, there should be a measurable time lag, which we shall estimate in Example 2-1, between the time when light starts to impinge on the surface and the ejection of the photoelectron. During this interval the electron should be absorbing energy from the beam until it has accumulated enough to escape. *However, no detectable time lag has ever been measured.* This disagreement is particularly striking when the photoelectric substance is a gas; under these circumstances collective absorption mechanisms can be ruled out and the energy of the emitted photoelectron must certainly be soaked out of the light beam by a single atom or molecule.

**Example 2-1.** A potassium plate is placed 1 m from a feeble light source whose power is 1 W = 1 joule/sec. Assume that an ejected photoelectron may collect its energy from a circular area of the plate whose radius $r$ is, say, one atomic radius: $r \simeq 1 \times 10^{-10}$ m. The energy required to remove an electron through the potassium surface is about 2.1 eV = $3.4 \times 10^{-19}$ joule. (One electron volt = 1 eV = $1.60 \times 10^{-19}$ joule is the energy gained by an electron, of charge $1.60 \times 10^{-19}$ coul, in falling through a potential drop of 1 V.) How long would it take for such a target to absorb this much energy from the light source? Assume the light energy to be spread uniformly over the wave front.

The target area is $\pi r^2 = \pi \times 10^{-20}$ m$^2$. The area of a 1 m sphere centered on the source is $4\pi(1 \text{ m})^2 = 4\pi$ m$^2$. Thus if the source radiates uniformly in all directions (i.e., if the energy is uniformly distributed over spherical wave fronts spreading out from the source, in agreement with classical theory) the rate $R$ at which energy falls on the target is given by

$$R = 1 \text{ joule/sec} \times \frac{\pi \times 10^{-20} \text{ m}^2}{4\pi \text{ m}^2} = 2.5 \times 10^{-21} \text{ joule/sec}$$

Assuming that all this power is absorbed, we may calculate the time required for the electron to acquire enough energy to escape; we find

$$t = \frac{3.4 \times 10^{-19} \text{ joule}}{2.5 \times 10^{-21} \text{ joule/sec}} = 1.4 \times 10^{2} \text{ sec} \simeq 2 \text{ min}$$

Of course, we could modify the preceding picture to reduce the calculated time by assuming a larger effective target area. The most favorable assumption, that energy is transferred by a resonance process from light wave to electron, leads to a target area of $\lambda^2$, where $\lambda$ is the wavelength of the light, but we would still obtain a finite time lag which is well within our ability to measure experimentally. (For ultraviolet light of $\lambda = 100$ Å, for example, $t \simeq 10^{-2}$ sec.) However, no time lag has been detected under any circumstances, the early experiments setting an upper limit of $10^{-9}$ sec on any such possible delay! ◀

## 2-3 Einstein's Quantum Theory of the Photoelectric Effect

In 1905 Einstein called into question the classical theory of light, proposed a new theory, and cited the photoelectric effect as one application that could test which theory was correct. This was many years before Millikan's work, but Einstein was influenced by Lenard's experiment. As we have mentioned, Planck originally restricted his concept of energy quantization to the radiating electron in the walls of a blackbody cavity. Planck believed that electromagnetic energy, once radiated, spreads through space like water waves spread through water. Einstein proposed instead that radiant energy is quantized into concentrated bundles which later came to be called *photons*.

Einstein argued that the well-known optical experiments on interference and diffraction of electromagnetic radiation had been performed only in situations involving *very* large numbers of photons. These experiments yield results which are averages of the behaviors of the individual photons. The presence of the photons is not apparent in them any more than the presence of individual droplets of water is apparent in a fine spray from a garden hose, if the number of droplets is very high. Of course the interference and diffraction experiments definitely show that photons do not travel from where they are emitted to where they are absorbed in the simple ways that classical particles, like water droplets, do. They travel like classical waves, in the sense that calculations based on the way such waves propagate (and in particular the way two component waves reinforce or nullify each other depending on their relative phases) correctly explain measurements of the average way photons travel.

Einstein focused his attention not on the familiar wavelike way radiation propagates, but on what he first realized is the particlelike way it is emitted and absorbed. He reasoned that Planck's requirement that the energy content of the electromagnetic waves of frequency $\nu$ in a radiant source (e.g., an ultraviolet light source in a photoelectric experiment) can only be 0, or $h\nu$, or $2h\nu$, . . . , or $nh\nu$, . . . implies that in the process of going from energy state $nh\nu$ to energy state $(n - 1)h\nu$ the source would emit a discrete burst of electromagnetic energy of energy content $h\nu$.

*Einstein assumed that such a bundle of energy is initially localized in a small volume of space, and that it remains localized as it moves away from the source with velocity c. He assumed that the energy content E of the bundle, or photon, is related to its frequency ν by the equation*

$$E = h\nu \tag{2-2}$$

*He also assumed that in the photoelectric process one photon is completely absorbed by one electron in the photocathode.*

When the electron is emitted from the surface of the metal, its kinetic energy will be

$$K = h\nu - w \tag{2-3}$$

where $h\nu$ is the energy of the absorbed incident photon and $w$ is the work required to remove the electron from the metal. This work is needed to overcome the attractive fields of the atoms in the surface and losses of kinetic energy due to internal collisions of the electron. Some electrons are bound more tightly than others; some lose energy in collisions on the way out. In the case of loosest binding and no internal losses, the photoelectron will emerge with the maximum kinetic energy, $K_{max}$. Hence

$$K_{max} = h\nu - w_0 \tag{2-4}$$

where $w_0$, a characteristic energy of the metal called the *work function*, is the minimum energy needed by an electron to pass through the metal surface and escape the attractive forces that normally bind the electron to the metal.

Consider now how Einstein's photon hypothesis meets the three objections raised against the wave theory interpretation of the photoelectric effect. As for objection 1 (the lack of dependence of $K_{max}$ on the intensity of illumination), there is complete agreement of the photon theory with experiment. Doubling the light intensity merely doubles the number of photons and thus doubles the photoelectric current; it does *not* change the energy $h\nu$ of the individual photons or the nature of the individual photoelectric process described by (2-3).

Objection 2 (the existence of a cutoff frequency) is removed at once by (2-4). If $K_{max}$ equals zero we have

$$h\nu_0 = w_0 \tag{2-5}$$

which asserts that a photon of frequency $\nu_0$ has just enough energy to eject the photoelectrons and none extra to appear as kinetic energy. If the frequency is reduced below $\nu_0$, the individual photons, no matter how many of them there are (that is, no matter how intense the illumination), will not have enough energy individually to eject photoelectrons.

Objection 3 (the absence of a time lag) is eliminated in the photon theory because the required energy is supplied in concentrated bundles. It is *not* spread uniformly over a large area, as we assumed in Example 2-1, which is based on the assumption that the classical wave theory is true. If there is any illumination at all incident on the cathode, then there will be at least one photon that hits it; this photon will be immediately absorbed, by *some* atom, leading to the immediate emission of a photoelectron.

Let us rewrite Einstein's photoelectric equation, (2-4), by substituting $eV_0$ for $K_{max}$ from (2-1). This yields

$$V_0 = \frac{h\nu}{e} - \frac{w_0}{e}$$

Thus Einstein's theory predicts a linear relationship between the stopping potential $V_0$ and the frequency $\nu$, in complete agreement with experimental results as shown in Figure 2-3. The slope of the experimental curve in the figure should be $h/e$, or

$$\frac{h}{e} = \frac{2.20\ \text{V} - 0.65\ \text{V}}{10.0 \times 10^{14}/\text{sec} - 6.0 \times 10^{14}/\text{sec}} = 3.9 \times 10^{-15}\ \text{V-sec}$$

We can find $h$ by multiplying this ratio by the electronic charge $e$. Thus $h = 3.9 \times 10^{-15}$ V-sec $\times\ 1.6 \times 10^{-19}$ coul $= 6.2 \times 10^{-34}$ joule-sec. From a much more careful analysis of these and other data, including data taken with lithium surfaces, Millikan found the value $h = 6.57 \times 10^{-34}$ joule-sec, with an accuracy of about 0.5%. This early measurement was in good agreement with the value of $h$ derived from Planck's radiation formula. The numerical agreement in two determinations of $h$, using completely different phenomena and theories, is striking. A modern value of $h$, deduced from diverse experiments, is

$$h = 6.6262 \times 10^{-34} \text{ joule-sec}$$

To quote Millikan: "The photoelectric effect . . . furnishes a proof which is quite independent of the facts of blackbody radiation of the correctness of the fundamental assumption of the quantum theory, namely, the assumption of a discontinuous or explosive emission of the energy absorbed by the electronic constituents of atoms from . . . waves. It materializes, so to speak, the quantity $h$ discovered by Planck through the study of blackbody radiation and gives us a confidence inspired by no other type of phenomenon that the primary physical conception underlying Planck's work corresponds to reality."

**Example 2-2.** Deduce the work function for sodium from Figure 2-3.

The intersection of the straight line in Figure 2-3 with the horizontal axis is the cutoff frequency, $\nu_0 = 4.39 \times 10^{14}$/sec. Substituting this into (2-5) gives us

$$w_0 = h\nu_0 = 6.63 \times 10^{-34} \text{ joule-sec} \times 4.39 \times 10^{14}/\text{sec}$$

$$= 2.92 \times 10^{-19} \text{ joule} \times \frac{1 \text{ eV}}{1.60 \times 10^{-19} \text{ joule}}$$

$$= 1.82 \text{ eV}$$

The same value is obtained from Figure 2-3 as the magnitude of the intercept of the extended line with the vertical axis. However, modern experiments give the higher value 2.27 eV.

For most conducting metals the value of the work function is of the order of a few electron volts. It is the same as the work function for thermionic emission from these metals. ◀

**Example 2-3.** At what rate per unit area do photons strike the metal plate in Example 2-1? Assume that the light is monochromatic, of wavelength 5890 Å (yellow light).

The rate per unit area at which energy falls on a metal plate 1 m from a 1-W light source (see Example 2-1) is

$$R = \frac{1 \text{ joule/sec}}{4\pi(1 \text{ m})^2} = 8.0 \times 10^{-2} \text{ joule/m}^2\text{-sec}$$

$$= 5.0 \times 10^{17} \text{ eV/m}^2\text{-sec}$$

Each photon has an energy of

$$E = h\nu = \frac{hc}{\lambda} = \frac{6.63 \times 10^{-34} \text{ joule-sec} \times 3.00 \times 10^8 \text{ m/sec}}{5.89 \times 10^{-7} \text{ m}}$$

$$= 3.4 \times 10^{-19} \text{ joule}$$

$$= 2.1 \text{ eV}$$

Thus the rate $R$ at which photons strike a unit area of the plate is

$$R = 5.0 \times 10^{17} \text{ eV/m}^2\text{-sec} \times \frac{1 \text{ photon}}{2.1 \text{ eV}} = 2.4 \times 10^{17} \frac{\text{photon}}{\text{m}^2\text{-sec}}$$

The photoelectric effect is just able to occur because the photon energy just equals the 2.1 eV work function for the potassium surface (see Example 2-1). Note that if the wavelength is slightly increased (that is, if $\nu$ is slightly decreased) the photoelectric effect will not occur, no matter how large the rate $R$ might be.

This example suggests that the intensity of light $I$ can be regarded as the product of $N$, the number of photons per unit area per unit time, and $h\nu$, the energy of a single photon. We see that even at the relatively low intensity here ($\simeq 10^{-1}$ W/m$^2$) the number $N$ is extremely large ($\simeq 10^{17}$ photons/m$^2$-sec) so that the energy of any one photon is very small. This accounts for the extreme fineness of the granularity of radiation and suggests why ordinarily it is difficult to detect at all. It is analogous to detecting the atomic structure of bulk matter which for most purposes can be regarded as continuous, the discreteness being revealed only under special circumstances. ◀

In 1921 Einstein received the Nobel Prize for predicting theoretically the law of the photoelectric effect. Before Millikan's complete experimental validation of this law in 1914, Einstein was recommended to membership in the Prussian Academy of Science by Planck and others. Their early negative attitude toward the photon hypothesis is revealed in their signed affidavit, praising Einstein, in which they wrote: "Summing up, we may say that there is hardly one among the great problems, in which modern physics is so rich, to which Einstein has not made an important contribution. That he may have sometimes missed the target in his speculations, as, for example, in his hypothesis of light quanta (photons), cannot really be held too much against him, for it is not possible to introduce fundamentally new ideas, even in the most exact sciences, without occasionally taking a risk."

Today the photon hypothesis is used throughout the electromagnetic spectrum, not only in the light region (see Figure 2-4). A microwave cavity, for example, can be said to contain photons. At $\lambda = 10$ cm, a typical microwave wavelength, the photon energy can be computed as above to be $1.20 \times 10^{-5}$ eV. This energy is much too low to eject photoelectrons from metal surfaces. For x rays, or for energetic $\gamma$ rays such as are emitted from radioactive nuclei, the photon energy may be $10^6$ eV or higher. Such photons can eject electrons bound deep in heavy atoms by energies of the order of $10^5$ eV. The photons in the visible region of the electromagnetic spectrum are not energetic enough to do this, the photoelectrons which they eject being the so-called *conduction* electrons which are bound to the metal by energies of only a few electron volts.

Notice that the photons are absorbed in the photoelectric process. This requires the electrons to be bound to atoms, or solids, for a truly free electron cannot absorb a photon and conserve both total relativistic energy and momentum in the process. We must have a bound electron, therefore, the binding forces serving to transmit momentum to the atom or solid. Due to the large mass of an atom, or solid, compared to the electron, the system can absorb a large amount of momentum without acquiring a significant amount of energy. Our photoelectric energy equation remains valid, the effect being possible only because there is a heavy recoiling particle in addition to an ejected electron. The photoelectric effect is one important way in which photons, of energy up to and including x-ray energies, are absorbed by matter. At higher energies other photon absorption processes, soon to be discussed, become more important.

Finally, it should be emphasized here that in the Einstein picture a photon of frequency $\nu$ has exactly the energy $h\nu$; it does *not* have energies that are integral multiples of $h\nu$. Of course, there can be $n$ photons of frequency $\nu$ so that the energy at that frequency can be $nh\nu$. In treating blackbody cavity radiation in the Einstein picture, we deal with a "photon gas," because the radiant energy is localized in space in bundles rather than extended through space in standing waves. Years after the Planck deduction of the cavity radiation formula, Bose and Einstein derived the same formula on the basis of a photon gas.

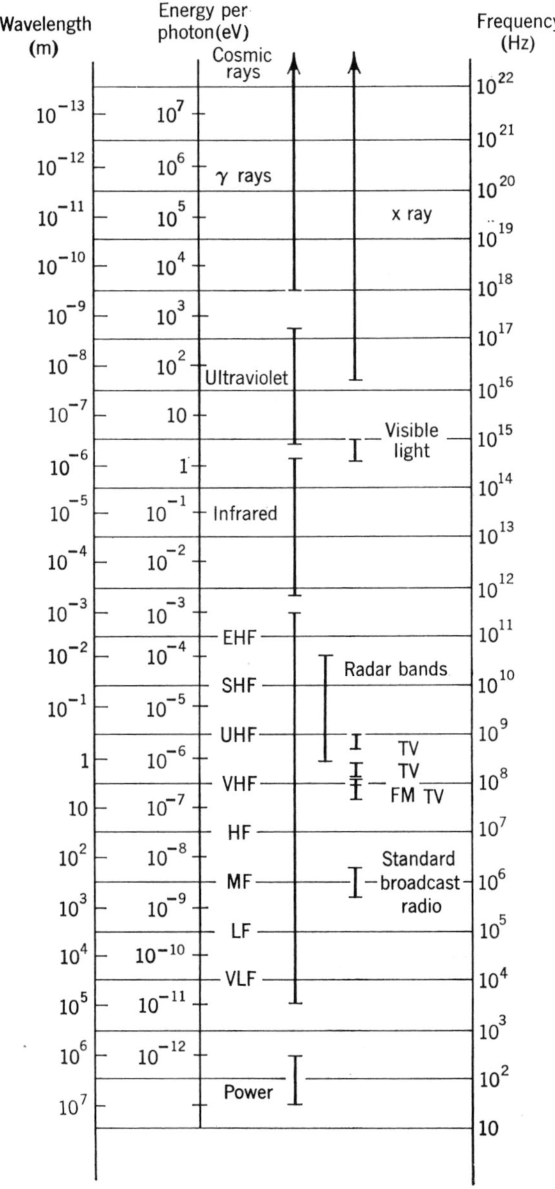


**FIGURE 2-4**
The electromagnetic spectrum, showing wavelength, frequency, and energy per photon on a logarithmic scale.

## 2-4 The Compton Effect

The corpuscular (particlelike) nature of radiation received dramatic confirmation in 1923 from the experiments of Compton. He allowed a beam of x rays of sharply defined wavelength $\lambda$ to fall on a graphite target, as shown in Figure 2-5. For various angles of scattering, he measured the intensity of the scattered x rays as a function of their wavelength. Figure 2-6 shows his experimental results. We see that, although the incident beam consists essentially of a single wavelength $\lambda$, the scattered x rays have

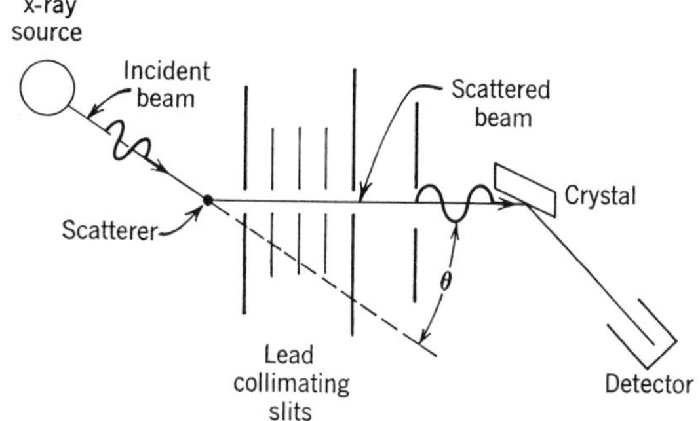


**FIGURE 2-5**

Compton's experimental arrangement. Monochromatic x rays of wavelength $\lambda$ fall on a graphite scatterer. The distribution of intensity with wavelength is measured for x rays scattered at any scattering angle $\theta$. The scattered wavelengths are measured by observing Bragg reflections from a crystal (see Figure 3-3). Their intensities are measured by a detector such as an ionization chamber.

intensity peaks at *two* wavelengths; one of them is the same as the incident wavelength, the other, $\lambda'$, being larger by an amount $\Delta\lambda$. This so-called *Compton shift* $\Delta\lambda = \lambda' - \lambda$ varies with the angle at which the scattered x rays are observed.

The presence of scattered wavelength $\lambda'$ cannot be understood if the incident x radiation is regarded as a classical electromagnetic wave. In the classical model the oscillating electric field vector in the incident wave of frequency $\nu$ acts on the free electrons in the scattering target and sets them oscillating at that same frequency. These oscillating electrons, like charges surging back and forth in a small radio transmitting antenna, radiate electromagnetic waves that again have this same frequency $\nu$. Hence,

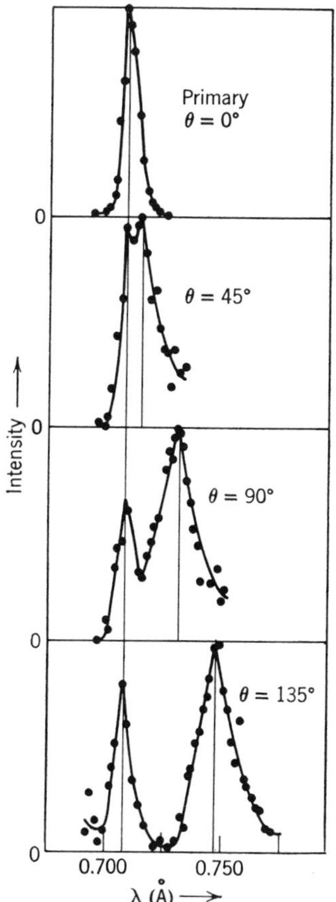


**FIGURE 2-6**

Compton's experimental results. The solid vertical line on the left corresponds to the wavelength $\lambda$, that on the right to $\lambda'$. Results are shown for four different angles of scattering $\theta$. Note that the Compton shift, $\Delta\lambda = \lambda' - \lambda$, for $\theta = 90°$, agrees well with the theoretical prediction $h/m_0c = 0.0243$ Å.

in the classical picture the scattered wave should have the same frequency $\nu$ and the same wavelength $\lambda$ as the incident wave.

Compton (and independently Debye) interpreted his experimental results by postulating that the incoming x-ray beam was not a wave of frequency $\nu$ but a collection of photons, each of energy $E = h\nu$, and that these photons collided with free electrons in the scattering target as in a collision between billiard balls. In this view, the "recoil" photons emerging from the target make up the scattered radiation. Since the incident photon transfers some of its energy to the electron with which it collides, the scattered photon must have a lower energy $E'$; it must therefore have a lower frequency $\nu' = E'/h$, which implies a longer wavelength $\lambda' = c/\nu'$. This point of view accounts qualitatively for the wavelength shift, $\Delta\lambda = \lambda' - \lambda$. Notice that in the interaction the x rays are regarded as particles, not as waves, and that, as distinguished from their behavior in the photoelectric process, the x-ray photons are scattered rather than absorbed. Let us now analyze a single photon-electron collision quantitatively.

For x radiation of frequency $\nu$, the energy of a photon in the incident beam is

$$E = h\nu$$

Taking the idea of a photon as a localized bundle of energy quite literally, we shall consider it to be a particle of energy $E$ and momentum $p$. Such a particle must, however, have certain quite specialized properties. Consider the equation (see Appendix A) giving the total relativistic energy of a particle in terms of its rest mass $m_0$ and its velocity $v$

$$E = m_0c^2/\sqrt{1 - v^2/c^2}$$

Since the velocity of a photon equals $c$, and since its energy content $E = h\nu$ is finite, it is apparent that the rest mass of a photon must be zero. Thus a photon can be considered to be a particle of zero rest mass, and of total relativistic energy $E$ which is entirely kinetic. The momentum of a photon can be evaluated from the general relation between the total relativistic energy $E$, momentum $p$, and rest mass $m_0$. This is

$$E^2 = c^2p^2 + (m_0c^2)^2 \tag{2-6}$$

For a photon the second term on the right is zero, and we have

$$p = E/c = h\nu/c \tag{2-7}$$

or

$$p = h/\lambda \tag{2-8}$$

where $\lambda = c/\nu$ is the wavelength of the electromagnetic radiation that the photon comprises. It is quite interesting to note that Maxwell's classical wave theory of electromagnetic radiation also leads to an equation $p = E/c$, with $p$ representing the momentum content per unit volume of radiation and $E$ representing its energy content per unit volume.

Now the frequency $\nu$ of the scattered radiation was observed to be independent of the material in the foil. This implies that the scattering does not involve entire atoms. Compton assumed that the scattering was due to collisions between the photon and an individual electron in the target. He also assumed that the electrons participating in this scattering process are free and initially stationary. Some *a priori* justification of these assumptions can be found from considering the fact that the energy of an x-ray photon is several orders of magnitude greater than the energy of an ultraviolet photon, and from our discussion of the photoelectric effect it is apparent that the energy of an ultraviolet photon is comparable to the minimum energy with which an electron is bound in a metal.

Consider, then, a collision between a photon and a free stationary electron, as in Figure 2-7. In the diagram on the left, a photon of total relativistic energy $E_0$ and momentum $p_0$ is incident on a stationary electron of rest mass energy $m_0c^2$. In the diagram on the right, the photon is scattered at an angle $\theta$ and moves off with total relativistic energy $E_1$ and momentum $p_1$, while the electron recoils at an angle $\varphi$ with kinetic energy $K$ and momentum $p$. Compton applied the conservation of momentum and total relativistic energy to this collision problem. Relativistic equations were used since the photon always moves at relativistic velocities, and the recoiling electron does too under most circumstances.

Momentum conservation requires

$$p_0 = p_1 \cos\theta + p \cos\varphi$$

and

$$p_1 \sin\theta = p \sin\varphi$$

Squaring these equations, we obtain

$$(p_0 - p_1 \cos\theta)^2 = p^2 \cos^2\varphi$$

and

$$p_1^2 \sin^2\theta = p^2 \sin^2\varphi$$

Adding, we find

$$p_0^2 + p_1^2 - 2p_0p_1 \cos\theta = p^2 \tag{2-9}$$

Conservation of total relativistic energy requires

$$E_0 + m_0c^2 = E_1 + K + m_0c^2$$

Thus

$$E_0 - E_1 = K$$

According to (2-7), this is

$$c(p_0 - p_1) = K \tag{2-10}$$

Writing $K + m_0c^2$ for $E$ in (2-6), that equation becomes

$$(K + m_0c^2)^2 = c^2p^2 + (m_0c^2)^2$$

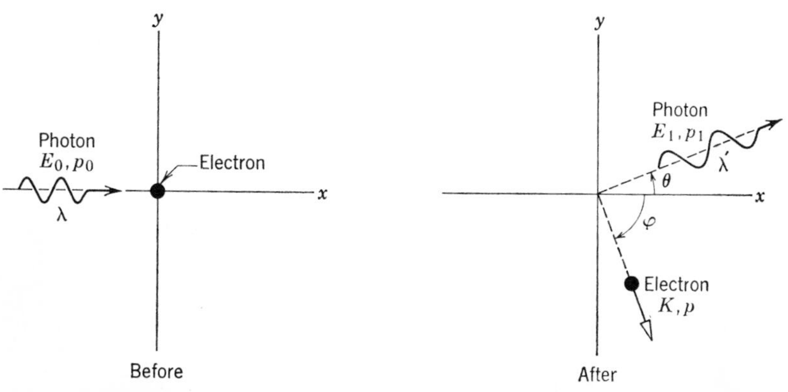


**FIGURE 2-7**

Compton's interpretation. A photon of wavelength $\lambda$ is incident on a free electron at rest. On collision, the photon is scattered at an angle $\theta$ with increased wavelength $\lambda'$, while the electron moves off at angle $\varphi$.

which simplifies to

$$K^2 + 2Km_0c^2 = c^2p^2$$

or

$$K^2/c^2 + 2Km_0 = p^2$$

Evaluating $p^2$ from (2-9) and $K$ from (2-10), we have

$$(p_0 - p_1)^2 + 2m_0c(p_0 - p_1) = p_0^2 + p_1^2 - 2p_0p_1 \cos\theta$$

which reduces to

$$m_0c(p_0 - p_1) = p_0p_1(1 - \cos\theta)$$

or

$$\frac{1}{p_1} - \frac{1}{p_0} = \frac{1}{m_0c}(1 - \cos\theta)$$

Multiplying through by $h$, and applying (2-8), we obtain the *Compton equation*

$$\Delta\lambda = \lambda_1 - \lambda_0 = \lambda_C(1 - \cos\theta) \tag{2-11}$$

where

$$\lambda_C \equiv h/m_0c = 2.43 \times 10^{-12}\ \text{m} = 0.0243\ \text{Å} \tag{2-12}$$

is the so-called *Compton wavelength.*

Notice that $\Delta\lambda$, the *Compton shift*, depends only on the scattering angle $\theta$, and *not* on the initial wavelength $\lambda$. Equation (2-11) predicts the experimentally observed Compton shifts of Figure 2-6 to within the experimental limits of accuracy. In (2-11) we see that $\Delta\lambda$ varies from zero (for $\theta = 0$, corresponding to a "grazing" collision with the incident photon being scarcely deflected) to $2h/m_0c = 0.049$ Å (for $\theta = 180°$, corresponding to a "head-on" collision, the incident photon being reversed in direction). Figure 2-8 is a plot of $\Delta\lambda$ versus $\theta$.

Subsequent experiments (by Compton, Simon, Wilson, Bothe, Geiger, and Blass) detected the recoil electron in the process, showed that it appeared simultaneously with the scattered x ray, and confirmed quantitatively the predicted electron energy and direction of scattering.

The presence of the peak in Figure 2-6 for which the photon wavelength does *not* change on scattering must still be explained. We have assumed heretofore that the

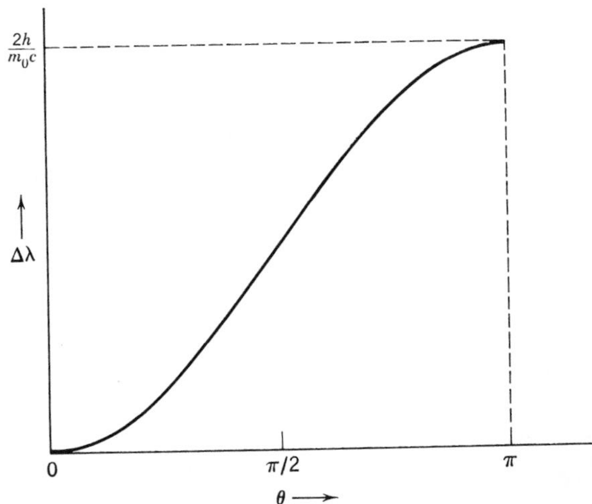


**FIGURE 2-8**

Compton's result $\Delta\lambda = (h/m_0c)(1 - \cos\theta)$.

electron with which the photon collides is free. Even though the electron is initially bound, this assumption is justifiable if the kinetic energy acquired by the electron in the collision is much larger than its binding energy. If the electron is particularly strongly bound to an atom in the target, however, or if the incident photon energy is very small, there is some chance that the electron will not be ejected from the atom. In this case, the collision can be regarded as taking place between the photon and the whole atom. The ionic core, to which the electron is bound in the scattering target, recoils as a whole during the collision. Then the mass $M$ of the atom is the characteristic mass for the process, and it must be substituted in the Compton shift equations for the electron mass $m_0$. Since $M \gg m_0$ ($M \simeq 22{,}000 m_0$ for carbon, for instance), the Compton shift for collisions with tightly bound electrons is seen, from (2-11) and (2-12), to be immeasurably small (one millionth of an angstrom for carbon), so that the scattered photon is essentially unmodified in wavelength. To summarize, some photons are scattered from electrons which are freed by the collision; these photons are modified in wavelength. Other photons are scattered from electrons which remain bound during the collision; these photons are not modified in wavelength.

The process that scatters photons without changing their wavelength is called *Thomson scattering*, after a physicist who developed a classical theory of x-ray scattering by atoms around the year 1900. Thomson considered the x rays to be a beam of electromagnetic waves whose oscillating electric field interacts with the charges of the atomic electrons in the target. This interaction produces forces on the electrons which cause oscillating accelerations. As a result of the accelerations, the electrons will radiate electromagnetic waves of the same frequency, and in phase with, the incident waves. Thus the atomic electrons absorb energy from the incident beam of x rays and scatter it in all directions, without modifying the wavelength. Although this classical explanation of Thomson scattering is different from the quantum explanation presented in the preceding paragraph, both explain the same feature observed in the measurements. Thus Thomson scattering is a case where classical and quantum results merge.

It is interesting to ask in what region of the electromagnetic spectrum Thomson scattering will be the dominant process, and in what region Compton scattering will dominate. If the incident radiation is in the visible, microwave, or radio part of the spectrum, then $\lambda$ is extremely large compared to the Compton shift $\Delta\lambda$, independent of whether an electron or an atomic mass is used in evaluating the Compton wavelength of (2-12). Thus the scattered radiation in this region of the spectrum will in all circumstances have a wavelength which is the same as the wavelength of the incident radiation within experimental accuracy. So, as $\lambda \to \infty$ the quantum results merge with the classical results, and Thomson scattering dominates. Moving into the x-ray region of the spectrum, Compton scattering starts to become important, particularly for scattering targets of low atomic number where the atomic electrons are not very tightly bound, and the wavelength shift in scattering from an electron which is freed in the process becomes easily measurable. In the $\gamma$-ray region where $\lambda \to 0$, the photon energy becomes so large, that an electron is always freed in a collision, and Compton scattering dominates.

It is in the short wavelength region that the classical results fail to explain the scattering of radiation, just as in the ultraviolet catastrophe of classical physics where predictions concerning the radiation in a cavity diverged radically from experimental results at short wavelengths. These circumstances are due to the size of Planck's constant $h$. At long wavelengths the frequency $\nu$ is small, and since $h$ is also small the granularity in electromagnetic energy, $h\nu$, is so small as to be virtually indistinguishable from the continuum of classical physics. But at sufficiently short wavelengths, where

$\nu$ is large enough, $h\nu$ is no longer too small to be negligible and quantum effects abound.

**Example 2-4.** Consider an x-ray beam, with $\lambda = 1.00$ Å and also a $\gamma$-ray beam from a $Cs^{137}$ sample, with $\lambda = 1.88 \times 10^{-2}$ Å. If the radiation scattered from free electrons is viewed at 90° to the incident beam: (a) What is the Compton wavelength shift in each case? (b) What kinetic energy is given to a recoiling electron in each case? (c) What percentage of the incident photon energy is lost in the collision in each case?

(a) The Compton shift, with $\theta = 90°$, is

$$\Delta\lambda = \frac{h}{m_0c}(1 - \cos\theta) = \frac{6.63 \times 10^{-34}\ \text{joule-sec}}{9.11 \times 10^{-31}\ \text{kg} \times 3.00 \times 10^{8}\ \text{m/sec}} \times (1 - \cos 90°)$$

$$= 2.43 \times 10^{-12}\ \text{m} = 0.0243\ \text{Å}$$

This result is independent of the incident wavelength, the same for the $\gamma$ rays as the x rays.

(b) Equation (2-10) can be written as

$$hc/\lambda = hc/\lambda' + K$$

Then, since $\lambda' = \lambda + \Delta\lambda$, we have

$$hc/\lambda = hc/(\lambda + \Delta\lambda) + K$$

so that $K = hc\ \Delta\lambda/\lambda(\lambda + \Delta\lambda)$.

For the x-ray beam, with $\lambda = 1.00$ Å, we have

$$K = \frac{6.63 \times 10^{-34}\ \text{joule-sec} \times 3.00 \times 10^{8}\ \text{m/sec} \times 2.43 \times 10^{-12}\ \text{m}}{1.00 \times 10^{-10}\ \text{m} \times (1.00 + 0.024) \times 10^{-10}\ \text{m}} = 4.73 \times 10^{-17}\ \text{joule}$$

$$= 295\ \text{eV} = 0.295\ \text{keV}$$

For the $\gamma$-ray beam, with $\lambda = 1.88 \times 10^{-2}$ Å, we have

$$K = \frac{6.63 \times 10^{-34}\ \text{joule-sec} \times 3.00 \times 10^{8}\ \text{m/sec} \times 2.43 \times 10^{-12}\ \text{m}}{1.88 \times 10^{-12}\ \text{m} \times (0.0188 + 0.0243) \times 10^{-10}\ \text{m}} = 5.98 \times 10^{-14}\ \text{joule}$$

$$= 378\ \text{keV}.$$

(c) The incident x-ray photon energy is

$$E = h\nu = \frac{hc}{\lambda} = \frac{6.63 \times 10^{-34}\ \text{joule-sec} \times 3.00 \times 10^{8}\ \text{m/sec}}{1.00 \times 10^{-10}\ \text{m}} = 1.99 \times 10^{-15}\ \text{joule}$$

$$= 12.4\ \text{keV}$$

The energy lost by the photon equals that gained by the electron, or 0.295 keV, so the percentage loss in energy is

$$\frac{0.295\ \text{keV}}{12.4\ \text{keV}} \times 100\% = 2.4\%$$

The incident $\gamma$-ray photon energy is

$$E = h\nu = \frac{hc}{\lambda} = \frac{6.63 \times 10^{-34}\ \text{joule-sec} \times 3.00 \times 10^{8}\ \text{m/sec}}{1.88 \times 10^{-12}\ \text{m}} = 1.06 \times 10^{-13}\ \text{joule}$$

$$= 660\ \text{keV}$$

The energy lost by the photon equals that gained by the electron, or 378 keV, so that the percentage loss in energy is

$$\frac{378\ \text{keV}}{660\ \text{keV}} \times 100\% = 57\%$$

Hence, the more energetic photons (which have small wavelengths) experience a larger *percent* loss in energy in Compton scattering. This corresponds to the fact that the photons of smaller wavelengths experience a larger *percent* increase in wavelength on being scattered. This becomes clear from the expression for fractional loss in energy, given simply by

$$\frac{K}{E} = \frac{hc\Delta\lambda/\lambda(\lambda + \Delta\lambda)}{hc/\lambda} = \frac{\Delta\lambda}{\lambda + \Delta\lambda}$$

From this it can be shown that at $\lambda = 5500$ Å, corresponding to visible protons, the percentage loss (for $\theta = 90°$) is less than one-thousandth of 1%, whereas at $\lambda = 1.25 \times 10^{-2}$ Å, corresponding to 1 MeV $\gamma$-ray photons, the percentage loss (for $\theta = 90°$) is 67%. ◀

## 2-5 The Dual Nature of Electromagnetic Radiation

In his paper, "A Quantum Theory of the Scattering of X-rays by Light Elements," Compton wrote: "The present theory depends essentially upon the assumption that each electron which is effective in the scattering scatters a complete quantum (photon). It involves also the hypothesis that the quanta of radiation are received from definite directions and are scattered in definite directions. The experimental support of the theory indicates very convincingly that a radiation quantum carries with it directed momentum as well as energy."

The need for a photon, or localized particle, interpretation of processes dealing with the interaction between radiation and matter is clear, but at the same time we need a wave theory of radiation to understand interference and diffraction phenomena. The idea that radiation is neither purely a wave phenomenon nor merely a stream of particles must therefore be taken seriously. Whatever radiation is, it behaves wavelike under some circumstances and particlelike under other circumstances. Indeed, the situation is revealed most forcefully in Compton's experimental work where (a) a crystal spectrometer is used to measure x-ray wavelengths, the measurement being interpreted by a wave theory of diffraction and (b) the scattering affects the wavelength in a way that can be understood only by treating the x rays as particles. It is in the very expressions $E = h\nu$ and $p = h/\lambda$ that the wave attributes ($\nu$ and $\lambda$) and the particle attributes ($E$ and $p$) are combined.

Although many physicists felt at first very uncomfortable when contemplating the "split personality" of electromagnetic radiation, the broader point of view provided by the development of quantum mechanics has caused the contemporary attitude to be quite different. The duality evident in the wave-particle nature of radiation is no longer considered at all unusual because it is now known to be a general characteristic of all physical entities. We shall see that electrons and protons, for example, have exactly the same dual nature as photons. We shall also see that it is possible to reconcile the existence of the wave aspects with the existence of the particle aspects, for any of these entities, with the aid of quantum mechanics.

## 2-6 Photons and X-Ray Production

X rays, so named by their discoverer Roentgen because their nature was then unknown, are radiations in the electromagnetic spectrum of wavelength less than about 1.0 Å. They show the typical transverse wave behavior of polarization, interference, and diffraction that is found in light and all other electromagnetic radiation. X rays are produced in the target of an x-ray tube, illustrated in Figure 2-9, when a beam of energetic electrons, accelerated through a potential difference of thousands of volts, is stopped upon striking the target. According to classical physics (see Appendix B), the deceleration of the electrons, brought to rest in the target material, results in the emission of a continuous spectrum of electromagnetic radiation.

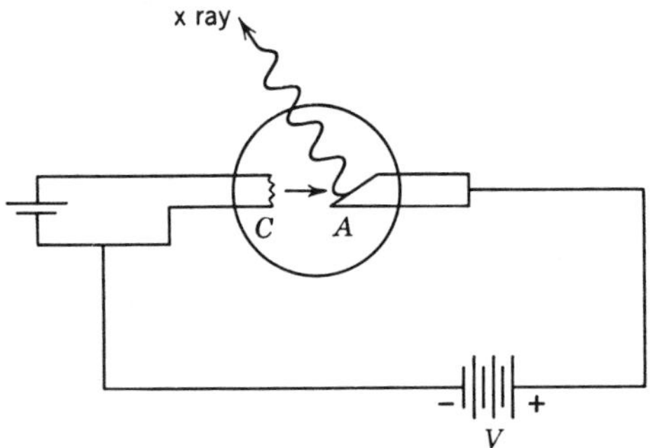


**FIGURE 2-9**
An x-ray tube. Electrons are emitted thermally from the heated cathode $C$ and are accelerated toward the anode target $A$ by the applied potential $V$. X rays are emitted from the target when electrons are stopped by striking it.

Figure 2-10 shows, for four different values of the incident electron energy, how the x rays emerging from a tungsten target are distributed in wavelength. (In addition to the continuous x-ray spectrum, x-ray lines characteristic of the target material are emitted. We shall discuss the lines in Chapter 9.) The most notable feature of these curves is that, for a given electron energy, there exists a well-defined minimum wavelength $\lambda_{\text{min}}$; for 40-keV electrons, for instance, $\lambda_{\text{min}}$ is 0.311 Å. Although the shape of the continuous x-ray distribution spectrum depends slightly on the choice of target material as well as on the electron accelerating potential $V$, the value of $\lambda_{\text{min}}$ depends only on $V$, being the same for all target materials. Classical electromagnetic theory cannot account for this fact, there being no reason why waves whose wavelength is less than a certain critical value should not emerge from the target.

A ready explanation appears, however, if we regard the x rays as photons. Figure 2-11 shows the elementary process that, on the photon view, is responsible for the continuous x-ray spectrum of Figure 2-10. An electron of initial kinetic energy $K$ is decelerated during an encounter with a heavy target nucleus, the energy it loses appearing in the form of radiation as an x-ray photon. The electron interacts with the charged nucleus via the Coulomb field, transferring momentum to the nucleus. The accompanying deceleration of the electron leads to photon emission. The target nucleus is so massive that the energy it acquires during the collision can safely be neglected. If $K'$ is the kinetic energy of the electron after the encounter, then the energy of the photon is

$$h\nu = K - K'$$

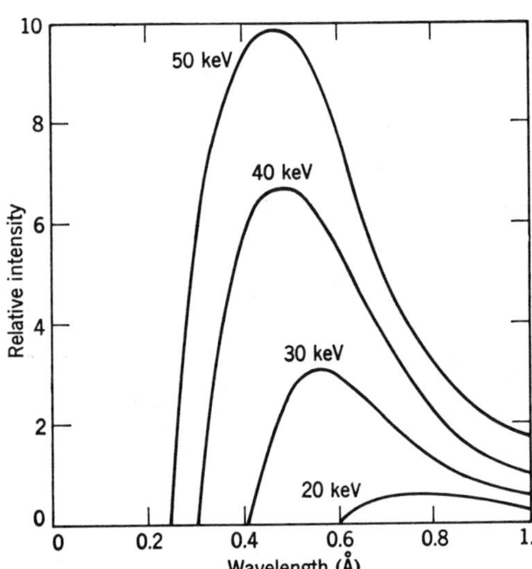


**FIGURE 2-10**
The continuous x-ray spectrum emitted from a tungsten target for four different values of $eV$, the incident electron energy.

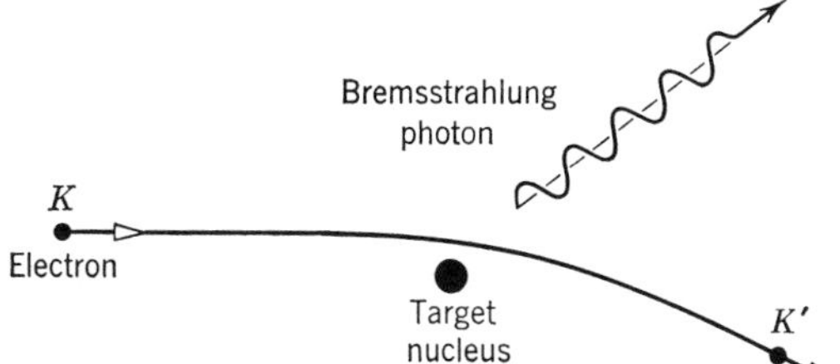


**FIGURE 2-11**
The bremsstrahlung process responsible for the production of x rays in the continuous spectrum.

and the photon wavelength follows from

$$hc/\lambda = K - K' \tag{2-13}$$

Electrons in the incident beam can lose different amounts of energy in such encounters and typically a single electron will be brought to rest only after many encounters. The x rays thus produced by many electrons make up the continuous spectrum of Figure 2-10 and are very many discrete photons whose wavelengths vary from $\lambda_{\min}$ to $\lambda \to \infty$, corresponding to the different energy losses in the individual encounters. The shortest wavelength photon would be emitted when an electron loses *all* its kinetic energy in one deceleration process; here $K' = 0$ so that $K = hc/\lambda_{\min}$. Since $K$ equals $eV$, the energy acquired by the electron in being accelerated through the potential difference $V$ applied to the x-ray tube, we have

$$eV = hc/\lambda_{\min}$$

or

$$\lambda_{\min} = hc/eV \tag{2-14}$$

Thus the minimum wavelength cutoff represents the complete conversion of the electron's kinetic energy to x radiation. Equation (2-14) shows clearly that if $h \to 0$ then $\lambda_{\min} \to 0$, which is the prediction of classical theory. This shows that the very existence of a minimum wavelength is a quantum phenomenon.

The continuous x radiation of Figure 2-10 is often called *bremsstrahlung*, from the German *brems* (= braking, i.e., decelerating) + *strahlung* (= radiation). The bremsstrahlung process occurs not only in x-ray tubes but wherever fast electrons collide with matter, as in cosmic rays, in the van Allen radiation belts which surround the earth, and in the stopping of electrons emerging from accelerators or radioactive nuclei. The bremsstrahlung process can be considered as an inverse photoelectric effect: in the photoelectric effect, a photon is absorbed, its energy and momentum going to an electron and a recoiling nucleus; in the bremsstrahlung process, a photon is created, its energy and momentum coming from a colliding electron and nucleus. We deal with the *creation* of photons in the bremsstrahlung process, rather than with their absorption or scattering by matter.

**Example 2-5.** Determine Planck's constant $h$ from the fact that the minimum x-ray wavelength produced by 40.0 keV electrons is $3.11 \times 10^{-11}$ m.

From (2-14), we have

$$\begin{aligned} h &= \frac{eV\lambda_{\min}}{c} \\ &= \frac{1.60 \times 10^{-19}\,\text{coul} \times 4.00 \times 10^{4}\,\text{V} \times 3.11 \times 10^{-11}\,\text{m}}{3.00 \times 10^{8}\,\text{m/sec}} \\ &= 6.64 \times 10^{-34}\,\text{joule-sec} \end{aligned}$$

This agrees well with the value of $h$ deduced from the photoelectric effect and the Compton effect.

Measurement of $V$, $\lambda_{\min}$, and $c$ provides one of the most accurate methods for evaluating the ratio $h/e$. Bearden, Johnson, and Watts at the Johns Hopkins University found in 1951, using this procedure, $h/e = 1.37028 \times 10^{-15}$ joule-sec/coul. This ratio is combined with many other measured combinations of physical constants, the assembly of data being analyzed by elaborate statistical methods to find the "best" value for the various physical constants. The best values change (but usually only within the *a priori* estimates of accuracy) and become increasingly precise as new experimental data and higher precision methods are used. ◀

## 2-7 Pair Production and Pair Annihilation

In addition to the photoelectric and Compton effects there is another process whereby photons lose their energy in interactions with matter, namely the process of *pair production*. Pair production is also an excellent example of the conversion of radiant energy into rest mass energy as well as into kinetic energy. In this process, illustrated schematically in Figure 2-12, a high energy photon loses all of its energy $h\nu$ in an encounter with a nucleus, creating an electron and a positron (the *pair*) and endowing them with kinetic energies. A *positron* is a particle which is identical in all of its properties with an electron, except that the sign of its charge (and of its magnetic moment) is opposite to that of an electron; a positron is a positively charged electron. In pair production the energy taken by the recoil of the nucleus is negligible because it is so massive, and thus the balance of total relativistic energy in the process is simply

$$h\nu = E_- + E_+ = (m_0c^2 + K_-) + (m_0c^2 + K_+) = K_- + K_+ + 2m_0c^2 \quad (2\text{-}15)$$

In this expression $E_-$ and $E_+$ are the total relativistic energies, and $K_-$ and $K_+$ are the kinetic energies of the electron and positron, respectively. Both particles have the same rest mass energy $m_0c^2$. The positron is produced with a slightly larger kinetic energy than the electron because the Coulomb interaction of the pair with the positively charged nucleus leads to an acceleration of the positron and a deceleration of the electron.

In analyzing this process here we ignore the details of the interaction itself, considering only the situation before and after the interaction. Our guiding principles are the conservation of total relativistic energy, conservation of momentum, and conservation of charge. From these conservation laws, it is not difficult to show that a photon cannot simply disappear in empty space, creating a pair as it vanishes. The presence of the massive nucleus (which can absorb momentum without appreciably affecting the energy balance) is necessary to allow both energy and momentum to be conserved in the process. Charge is automatically conserved, the photon having no charge and the created pair of particles having no net charge. From (2-15) we see that the minimum, or threshold, energy needed by a photon to create a pair is $2m_0c^2$ or 1.02 MeV (1 MeV = $10^6$ eV), which is a wavelength of 0.012 Å. If the wavelength is shorter than this, corresponding to an energy greater than the threshold value, the photon endows

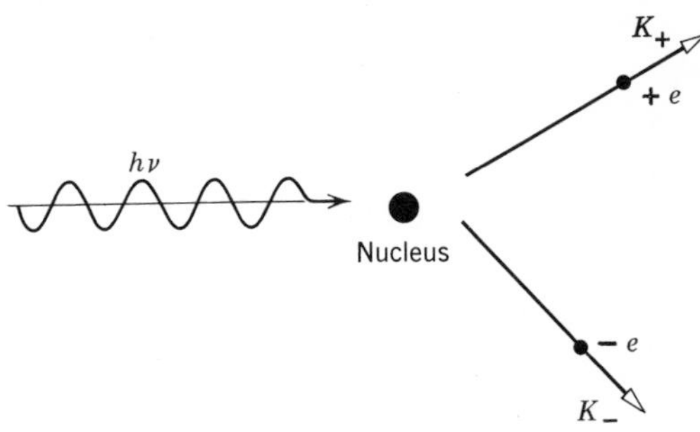


**FIGURE 2-12**
The pair production process.

positron tracks have opposite curvatures in the uniform magnetic field $B$ of 0.20 weber/m$^2$, their radii $r$ each being $2.5 \times 10^{-2}$ m. What was the energy and the wavelength of the pair producing photon?

The momentum $p$ of the electron is given by

$$\begin{aligned} p &= eBr = 1.6 \times 10^{-19}\text{ coul} \times 2.0 \times 10^{-1}\text{ weber/m}^2 \times 2.5 \times 10^{-2}\text{ m} \\ &= 8.0 \times 10^{-22}\text{ kg-m/sec} \end{aligned}$$

Its total relativistic energy $E_-$ is given by

$$E_-^2 = c^2p^2 + (m_0c^2)^2$$

Since $m_0c^2 = 0.51$ MeV, and $pc = 8.0 \times 10^{-22}$ kg-m/sec $\times 3.0 \times 10^8$ m/sec $= 2.4 \times 10^{-13}$ joule $= 1.5$ MeV, we have $E_-^2 = (1.5\text{ MeV})^2 + (0.51\text{ MeV})^2$ and $E_- = 1.6$ MeV.

The positron total relativistic energy had the same value since its track had the same radius, so the energy of the photon was

$$h\nu = E_- + E_+ = 3.2\text{ MeV}$$

The photon's wavelength follows from

$$E = h\nu = hc/\lambda$$

or

$$\lambda = \frac{hc}{E} = \frac{6.6 \times 10^{-34}\text{ joule-sec} \times 3.0 \times 10^8\text{ m/sec}}{3.2 \times 10^6\text{ eV} \times 1.6 \times 10^{-19}\text{ joule/eV}} = 3.9 \times 10^{-13}\text{ m} = 0.0039\text{ Å}$$ ◀

Closely related to pair production is the inverse process called *pair annihilation.* An electron and a positron, which are essentially at rest near one another, unite and are annihilated. Matter disappears and in its place we get radiant energy. Since the initial momentum of the system is zero and momentum must be conserved in the process, we cannot have only one photon created because a single photon cannot have zero momentum. The most probable process is the creation of two photons moving with equal momenta in opposite directions. Less probable, but possible, is the creation of three photons.

In the two-photon process illustrated by Figure 2-14, momentum conservation gives $0 = \mathbf{p}_1 + \mathbf{p}_2$ or $\mathbf{p}_1 = -\mathbf{p}_2$ so that the photon momenta are oppositely directed but equal in magnitude. Hence, $p_1 = p_2$ or $h\nu_1/c = h\nu_2/c$ and $\nu_1 = \nu_2 = \nu$. Total relativistic energy conservation then requires that $m_0c^2 + m_0c^2 = h\nu + h\nu$, the positron and electron having no initial kinetic energy and the photon energies being the same. Hence, $h\nu = m_0c^2 = 0.51$ MeV, corresponding to a photon wavelength of 0.024 Å. If the initial pair had some kinetic energy then the photon energy would exceed 0.51 MeV and its wavelength could be less than 0.024 Å.

Positrons are created in the pair production process. On passing through matter a positron loses energy in successive collisions until it combines with an electron to form a bound system called *positronium.* The positronium "atom" is short lived, decaying into photons within about $10^{-10}$ sec of its formation. The electron and positron presumably move about their common center of mass in a kind of death dance before mutual annihilation.

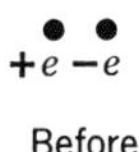


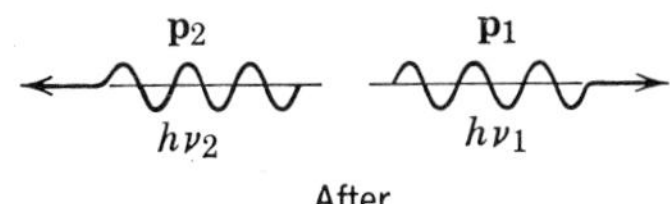


**FIGURE 2-14**
Pair annihilation producing two photons.

the pair with kinetic energy as well as rest energy. The pair production phenor is a high-energy one, the photons being in the very short x-ray or $\gamma$-ray regions electromagnetic spectrum (see Figure 2-4), where their energies $h\nu$ are equal greater than $2m_0c^2$. As we shall see in the next section, experimental results de strate that the absorption of photons in interaction with matter occurs principal the photoelectric process at low energies, by the Compton effect at medium enei and by pair production at high energies.

Electron-positron pairs are produced in nature by cosmic-ray photons and ir laboratory by bremsstrahlung photons from particle accelerators. Other particle p such as proton and antiproton, can be produced as well if the initiating photon sufficient energy. Because the electron and positron have the smallest rest mas known particles, the threshold energy of their production is the smallest. Experim verifies the quantum picture of the pair production process. There is no satisfact explanation whatever of this phenomenon in classical theory.

**Example 2-6.** Analysis of a bubble chamber photograph (as in Figure 2-13) reveals creation of an electron-positron pair as photons pass through matter. The electron a

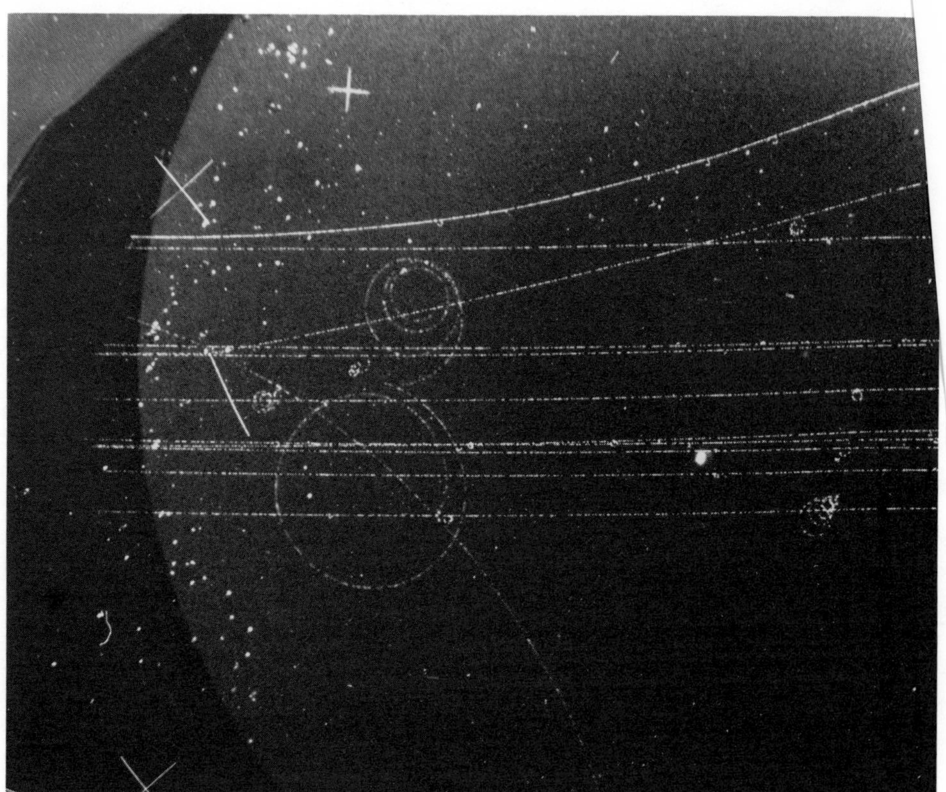

**FIGURE 2-13**

Electron pair production, as seen in a bubble chamber. The electron and positron tracks are the two spirals meeting at the point where the production took place in the liquid filling of the chamber. The student can determine which of the two spirals belongs to the positron by knowing that the long tracks are primarily positively charged deuterons which are incident from the left. (Courtesy of C. R. Sun, State University of New York at Albany)

**Example 2-7.** (a) Assume that Figure 2-14 represents the annihilation process in a reference frame $S$, the electron-positron pair being at rest there and the two annihilation photons moving along the $x$ axis. Find the wavelength $\lambda$ of these photons in terms of $m_0$, the rest mass of an electron or positron.

We saw that $p_1 = p_2$ and $h\nu_1 = h\nu_2$. Each photon has the same energy, the same frequency, and the same wavelength. We can drop the subscripts then and from the relation $h\nu = m_0c^2$ and $p = E/c$ we obtain

$$p = E/c = h\nu/c = m_0c^2/c = m_0c$$

But we also have the relation

$$p = h/\lambda$$

so that

$$\lambda = h/p = h/m_0c$$

Hence, in the rest frame of the positronium atom each photon has the same wavelength, $\lambda = h/m_0c$.

(b) Now consider the same annihilation event to be observed in frame $S'$, moving relative to $S$ with a velocity $\mathbf{v}$ to the left. What wavelength does this (moving) observer record for the annihilation photons?

Here, the pair has initial total relativistic energy $2mc^2$, where $m$ is relativistic mass, rather than merely the rest mass energy $2m_0c^2$, so that conservation of energy in the annihilation process gives us

$$2mc^2 = p_1'c + p_2'c$$

Also, the pair now moves with velocity $\mathbf{v}$ along the positive $x'$ axis so that its initial momentum is $2mv$, rather than zero as before. Conservation of momentum now gives us

$$2mv = p_1' - p_2'$$

the photons moving in opposite directions along the $x'$ axis. Let us combine these two expressions. We multiply the second by $c$ and add it to the first, obtaining, since $m = m_0/\sqrt{1 - v^2/c^2}$

$$p_1' = m(c + v) = \frac{m_0(c + v)}{\sqrt{1 - v^2/c^2}} = m_0c\sqrt{\frac{c + v}{c - v}}$$

But $p_1' = h/\lambda_1'$, so that

$$\lambda_1' = \frac{h}{p_1'} = \frac{h}{m_0c}\sqrt{\frac{c - v}{c + v}} = \lambda\sqrt{\frac{c - v}{c + v}} \tag{2-16a}$$

In a similar manner, by subtracting the second equation from the first, we obtain

$$\lambda_2' = \frac{h}{p_2'} = \frac{h}{m_0c}\sqrt{\frac{c + v}{c - v}} = \lambda\sqrt{\frac{c + v}{c - v}} \tag{2-16b}$$

The photons do *not* have the same wavelength, but they are *Doppler shifted* from the wavelength $\lambda$ they had in the rest frame of the source (the positronium atom). If an observer is situated on the $x'$ axis so that the source moves *toward* him, he will receive photon 1, having a frequency *higher* than the "rest" frequency. If an observer is situated on the $x'$ axis so that the source moves *away* from him, he will receive photon 2, having a frequency *lower* than the rest frequency. This Example is actually a derivation of the *longitudinal Doppler shift formula* of relativity theory. ◀

The first experimental evidence for the pair production process, and the existence of positrons, was obtained in 1933 by Anderson during an investigation of the cosmic radiation. This radiation consists of a flux of very high energy photons and charged particles incident upon the earth from extra-terrestrial sources. Anderson was using a cloud chamber containing a thin lead plate, with the entire apparatus in a magnetic field. Upon exposing this apparatus to the cosmic radiation, it was found that very infrequently a pair of charged particles were ejected from some point in the lead plate. These events were assumed to be the result of the interaction of a photon in the lead because no charged particle was seen to strike the point of ejection, whereas a photon being uncharged, could strike the point of ejection without being seen. The two charged particles ejected in these events were bent in opposite directions by the magnetic field. Therefore their charges were of the opposite sign. From other considerations it could be shown that the magnitudes of these charges were equal to one electronic charge and that the masses of the particles were approximately equal to one electronic mass.

The discovery of the pair production process explained the origin of a discrepancy between the then current theory of x-ray attenuation and the measured attenuation coefficients of several materials for 2.6 MeV x rays ($\gamma$ rays obtained from a radioactive source). As the theory originally did not include pair production, the predicted attenuation was too small; with the inclusion of the pair production process, good agreement is now obtained between experiment and theory. However, the real importance of Anderson's discovery was in the beautiful confirmation which it provided for Dirac's relativistic quantum mechanical theory of the electron.

The Dirac theory leads to the prediction that the allowed values of total relativistic energy $E$ for a free electron are

$$E = \pm\sqrt{c^2p^2 + (m_0c^2)^2} \tag{2-17}$$

where $m_0$ is the electron rest mass. These are simply the solutions for $E$ of (2-6), but the solution with the minus sign corresponds to a negative total relativistic energy—a concept as foreign to relativistic mechanics as a negative total energy is to classical mechanics. Instead of just throwing away the negative part on the grounds that it is not physically realistic, Dirac pursued the consequences of the entire equation. In doing this he was led to some very interesting conclusions. Consider Figure 2-15, which is an energy-level diagram representing (2-17). If the indicated continuum of negative energy levels exists, all free electrons of positive energy should be able to make transitions into these levels, accompanied by the emission of photons of the appropriate energies. This obviously disagrees with experiment because free electrons are not generally observed to emit spontaneously photons of energy $h\nu \geq 2m_0c^2$. However, Dirac pointed out that this difficulty can be removed by assuming that all the negative energy levels are normally filled at all points in space. According to this assumption, *a vacuum consists of a sea of electrons in negative energy levels.* This does not disagree with experiment. For instance, the negative charge could not be detected, as it is assumed to be uniformly distributed and therefore exerts no force on a charged body. Similar considerations will demonstrate that all the "usual" properties of a sea of negative energy electrons are such that its presence would not be apparent in any of the usual experiments. However, Dirac's

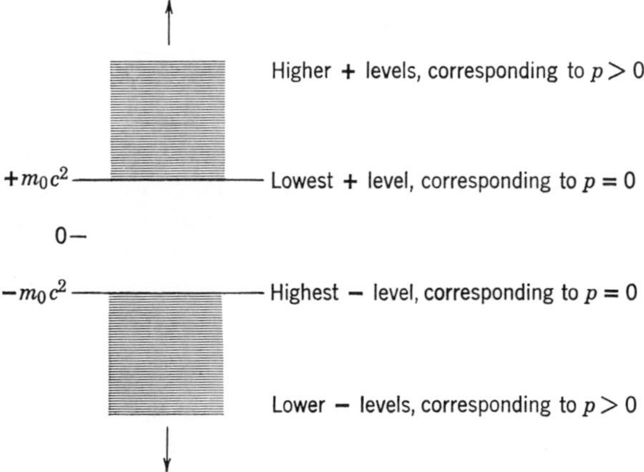


**FIGURE 2-15**
The energy levels of a free electron according to Dirac.

theory of the vacuum is not completely vacuous because it predicts certain new properties which can be tested by experiment.

The energy-level diagram for a free electron suggests the possibility of exciting an electron in a negative energy level by the absorption of a photon. Since all the negative energy levels are assumed to be fully occupied, the electron must be excited to one of the unoccupied positive energy levels. The minimum photon energy required for this process is obviously $h\nu = 2m_0c^2$, and the process results in the production of an electron in a positive energy level plus a hole in a negative energy level. We can demonstrate that a hole in a negative electron energy level has all the mechanical and electrical properties of a positron of positive energy. For instance, there is a positive charge $+e$ associated with the absence of an electron of negative charge $-e$. Consequently, this is the pair production process observed experimentally by Anderson three years after its theoretical prediction by Dirac.

## 2-8 Cross Sections for Photon Absorption and Scattering

Consider a parallel beam of photons passing through a slab of matter, as in Figure 2-16. The photons can interact with the atoms in the slab by four different processes: photoelectric, pair production, Thomson, and Compton. The first two absorb photons completely, while the last two only scatter them, but all the processes remove photons from the parallel beam. The question of what the chances of these processes happening are, in a given set of circumstances, is one of considerable theoretical and practical significance. For instance, it is very important to a medical physicist designing the shielding for an x-ray machine, or a nuclear engineer designing the shielding for a reactor. The answer to the question is expressed in terms of quantities called cross sections. We first meet cross sections here in connection with photons, but we shall encounter them again in other connections elsewhere in this book.

The probability that a photon of a given energy will be, for example, absorbed by the photoelectric process in passing an atom of the slab is specified by the value of the photoelectric *cross section* $\sigma_{PE}$. This measure of the likelihood of the photoelectric process occurring is defined so that the number $N_{PE}$ of photoelectric absorptions occurring is

$$N_{PE} = \sigma_{PE} I n \tag{2-18}$$

when a beam containing $I$ photons is incident on a slab containing $n$ atoms per unit area. It is assumed here that the slab is thin enough that the probability of a given photon being absorbed in passing through the slab is much smaller than one.

The definition of (2-18), which is a prototype of the definitions of all cross sections, is sufficiently important to warrant careful physical interpretation. First note that the

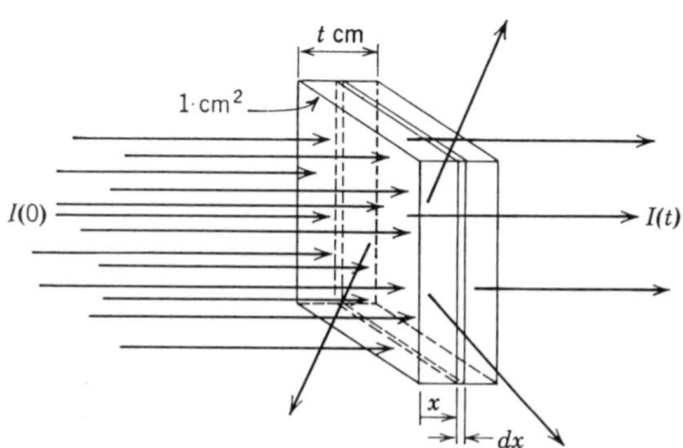


**FIGURE 2-16**

A beam of photons passing through a slab.

number $N_{PE}$ of absorptions should certainly increase in proportion to the number $I$ of photons incident on the slab. Furthermore, if the slab is thin in the sense specified previously the atoms in the slab will not appreciably "shadow" each other, as far as the incident photons are concerned. Then the number $N_{PE}$ of absorptions should also increase in proportion to the number $n$ of target atoms per unit area of the slab. Thus we should have

$$N_{PE} \propto In$$

If we write this proportionality as an equality, calling the proportionality constant $\sigma_{PE}$, we obtain the defining equation for that cross section. Thus we see that the cross section, which has a value depending on both the energy of the photon and the type of atom, measures how effective such atoms are in absorbing those photons by the photoelectric effect. Since the quantities $N_{PE}$ and $I$ in (2-18) are dimensionless, while $n$ has the dimensions of $(\text{area})^{-1}$, it is clear that $\sigma_{PE}$ must have the dimensions of (area). Thus it is reasonable to use the name cross section for $\sigma_{PE}$. It is often given a geometrical interpretation by imagining that a circle of area $\sigma_{PE}$ is centered on each atom in the slab in the plane of the slab, with the property that any photon entering the circular area is absorbed by the atom through the photoelectric effect. This geometrical interpretation is convenient for visualization and even for calculation, but it definitely should not be taken to be literally true. A cross section is really just a way of expressing numerically the probability that a certain type of atom will cause a photon of a given energy to undergo a particular process. The definitions and interpretations of the cross sections for the other absorption or scattering processes are completely analogous to those for the example we have considered.

Figure 2-17 shows the measured scattering ($\sigma_S$), photoelectric ($\sigma_{PE}$), pair production ($\sigma_{PR}$), and total ($\sigma$) cross sections for a lead atom as a function of the photon

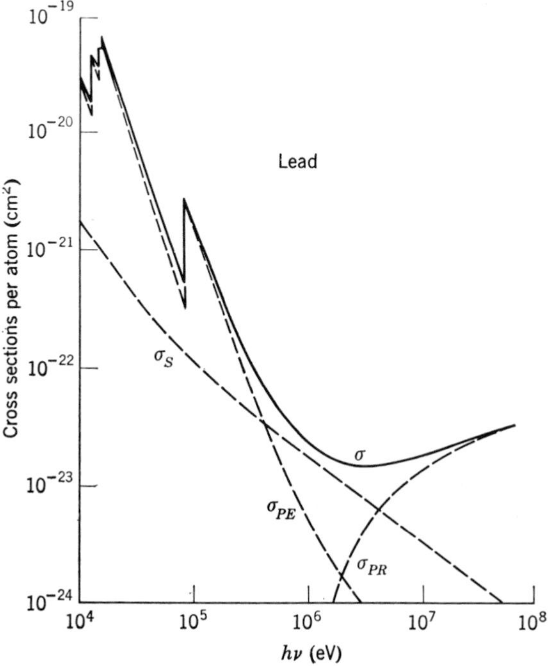


**FIGURE 2-17**

The scattering, photoelectric, pair production, and total cross sections for a lead atom.

energy $h\nu$. The scattering cross section specifies the probability of scattering occurring by either the Thomson or the Compton process. For lead, which has a high atomic number and thus tightly bound atomic electrons, Thomson scattering dominates Compton scattering when the photon energy is below about $h\nu = 10^5$ eV. The sharp breaks in the photoelectric cross section occur at the binding energies of the different electrons in the lead atom; when $h\nu$ drops below the binding energy of a particular electron a photoelectric process involving it is no longer energetically possible. The pair production cross section rises very rapidly from zero when $h\nu$ exceeds the threshold energy $2m_0c^2 \simeq 10^6$ eV required to materialize a pair. The total cross section $\sigma$ in Figure 2-17 is the sum of the scattering, photoelectric, and pair production cross sections. This quantity specifies the probability that a photon will make any kind of interaction with the atom. We see from the figure that the energy ranges in which each of the three processes makes the most important contribution to $\sigma$ are approximately, for lead:

Photoelectric effect: $h\nu < 5 \times 10^5$ eV
Scattering: $5 \times 10^5 \text{ eV} < h\nu < 5 \times 10^6$ eV
Pair production: $5 \times 10^6 \text{ eV} < h\nu$

Because these processes have probabilities with different dependences on atomic number, the energy ranges in which they dominate are quite different for atoms of low atomic number. The energy ranges are approximately, for aluminum:

Photoelectric effect: $h\nu < 5 \times 10^4$ eV
Scattering: $5 \times 10^4 \text{ eV} < h\nu < 1 \times 10^7$ eV
Pair production: $1 \times 10^7 \text{ eV} < h\nu$

**Example 2-8.** Evaluate, in terms of the total cross section $\sigma$, the attenuation of a parallel beam of x rays in passing through a *thick* slab of matter.

Referring to Figure 2-16, $I(0)$ photons are in the beam as it is incident on the front face of the slab of thickness $t$, which contains $\rho$ atoms per cm³. Assume, for simplicity, that the area of the slab is 1 cm². Because of scattering and absorption processes, the parallel beam contains a smaller number $I(x)$ of photons after penetrating $x$ cm into the slab. Consider a thin lamina of the slab, of width $dx$ located at $x$. The number of atoms per cm² in the lamina is $\rho$ times its volume $dx$, or $\rho\, dx$. The number of beam photons that will be scattered or absorbed in the lamina is specified by the total cross section $\sigma$, in a definition analogous to (2-18). It is $\sigma\, I(x) \rho\, dx$. Thus the number of beam photons emerging from the lamina, $I(x + dx)$, which equals the number incident minus the number removed, is

$$I(x + dx) = I(x) - \sigma\, I(x) \rho\, dx$$

or

$$dI(x) \equiv I(x + dx) - I(x) = -\sigma\, I(x) \rho\, dx$$

We find $I(t)$, the number of beam photons emerging from the rear face of the slab, by solving for $dI(x)/I(x)$ and then integrating over $x$

$$\frac{dI(x)}{I(x)} = -\sigma \rho\, dx$$

$$\int_0^t \frac{dI(x)}{I(x)} = -\sigma \rho \int_0^t dx$$

$$\ln I(x)]_0^t = -\sigma \rho\, t$$

$$\ln \frac{I(t)}{I(0)} = -\sigma \rho\, t$$

$$\frac{I(t)}{I(0)} = e^{-\sigma \rho t}$$

$$I(t) = I(0) e^{-\sigma \rho t} \tag{2-19}$$

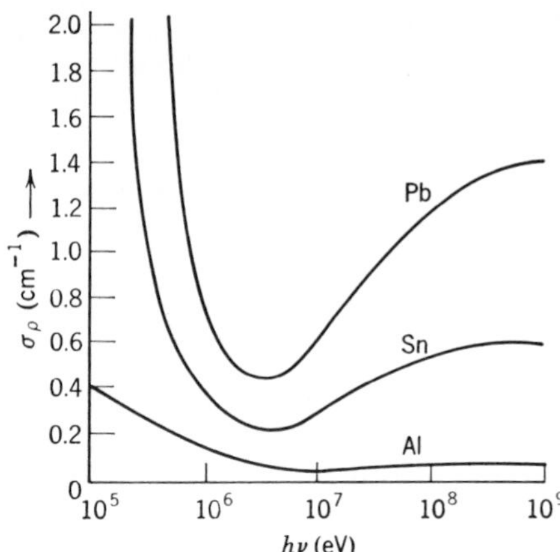


**FIGURE 2-18**

The attenuation coefficients for several atoms and a range of photon energies.

The intensity of the beam, as measured by the number $I$ of photons it contains, decreases exponentially as the thickness $t$ of the slab increases. The quantity $\sigma\rho$, which is called the *attenuation coefficient*, has the dimensions (cm$^{-1}$) and is the reciprocal of the thickness of slab required to attenuate the beam intensity by a factor of $e$. This thickness is called the *attenuation length* $\Lambda$. That is

$$\Lambda = 1/\sigma\rho \tag{2-20}$$

Of course, the attenuation coefficient has the same dependence on photon energy as the total cross section. Figure 2-18 shows measured attenuation coefficients of lead, tin, and aluminum for photons of relatively high energy. ◀

This section summarizes many of the practical aspects of the electromagnetic radiation emission and absorption phenomena we have studied in the present chapter. But the fundamental aspects of these phenomena are better summarized by saying that they show electromagnetic radiation to be quantized into particles of energy called photons. It should also be emphasized that the phenomena of interference and diffraction show photons do not travel through a system from where they are emitted to where they are absorbed in the simple way that classical particles do. Instead, photons act as if they were guided by classical waves because photons travel through a system such as a diffraction apparatus in a way that is best described by the way that classical waves would propagate through the apparatus.

## QUESTIONS

1. In the photoelectric experiments, the current (number of electrons emitted per unit time) is proportional to the intensity of light. Can this result alone be used to distinguish between the classical and quantum theories?

2. In Figure 2-2 why does the photoelectric current not rise vertically to its maximum (saturation) value when the applied potential difference is slightly more positive than $-V_0$?

3. Why is it that even for incident radiation that is monochromatic photoelectrons are emitted with a spread of velocities?

4. The existence of a cutoff frequency in the photoelectric effect is often regarded as the most potent objection to a wave theory. Explain.

5. Why are photoelectric measurements very sensitive to the nature of the photoelectric surface?

6. Do the results of photoelectric experiments invalidate Young's interference experiment?

7. Can you use the device of letting $h \to 0$ to obtain classical results from quantum results in the case of the photoelectric effect? Explain.

8. Assume that the emission of photons from a source of radiation is random in direction. Would you expect the intensity (or energy density) to vary inversely as the square of the distance from the source on the photon theory as it does in the wave theory?

9. Does a photon of energy $E$ have mass? If so, evaluate it.

10. Why, in Compton scattering, would you expect $\Delta\lambda$ to be independent of the materials of which the scatterer is composed?

11. Would you expect to observe the Compton effect more readily with scattering targets composed of atoms with high atomic number or those composed of atoms with low atomic number? Explain.

12. Do you observe a Compton effect with visible light? Why?

13. Would you expect a definite minimum wavelength in the emitted radiation for a given value of the energy of an electron incident on the target of an x-ray tube from the classical electromagnetic theory of the process?

14. Does a television tube emit x rays? Explain.

15. What effect(s) does decreasing the voltage across an x-ray tube have on the resulting x-ray spectrum?

16. Discuss the bremsstrahlung process as the inverse of the Compton process. Of the photoelectric process.

17. Describe several methods that can be used to determine experimentally the value of Planck's constant $h$.

18. From what factors would you expect to judge whether a photon will lose its energy in interactions with matter by the photoelectric process, the Compton process, or the pair production process?

19. Can you think of experimental evidence contradicting the idea that vacuum is a sea of electrons in negative energy states?

20. Can electron-positron annihilation occur with the creation of *one* photon if a nearby nucleus is available for recoil momentum?

21. Explain how pair annihilation with the creation of *three* photons is possible. Is it possible in principle to create even more than three photons in a single annihilation process?

22. What would be the inverse of the process in which two photons are created in electron-positron annihilation? Can it occur? Is it likely to occur?

23. What is wrong with taking the geometrical interpretation of a cross section as literally true?

## PROBLEMS

1. (a) The energy required to remove an electron from sodium is 2.3 eV. Does sodium show a photoelectric effect for yellow light, with $\lambda = 5890$ Å? (b) What is the cutoff wavelength for photoelectric emission from sodium?

2. Light of a wavelength 2000 Å falls on an aluminum surface. In aluminum 4.2 eV are required to remove an electron. What is the kinetic energy of (a) the fastest and (b) the slowest emitted photoelectrons? (c) What is the stopping potential? (d) What is the cutoff wavelength for aluminum? (e) If the intensity of the incident light is 2.0 W/m$^2$, what is the average number of photons per unit time per unit area that strike the surface?

3. The work function for a clean lithium surface is 2.3 eV. Make a rough plot of the stopping potential $V_0$ versus the frequency of the incident light for such a surface, indicating its important features.

4. The stopping potential for photoelectrons emitted from a surface illuminated by light of wavelength $\lambda = 4910$ Å is 0.71 V. When the incident wavelength is changed the stopping potential is found to be 1.43 V. What is the new wavelength?

5. In a photoelectric experiment in which monochromatic light and a sodium photocathode are used, we find a stopping potential of 1.85 V for $\lambda = 3000$ Å and of 0.82 V for $\lambda =$ 4000 Å. From these data determine (a) a value for Planck's constant, (b) the work function of sodium in electron volts, and (c) the threshold wavelength for sodium.

6. Consider light shining on a photographic plate. The light will be recorded if it dissociates an AgBr molecule in the plate. The minimum energy to dissociate this molecule is of the order of $10^{-19}$ joule. Evaluate the cutoff wavelength greater than which light will not be recorded.

7. The relativistic expression for kinetic energy should be used for the electron in the photoelectric effect when $v/c > 0.1$, if errors greater than about 1% are to be avoided. For photoelectrons ejected from an aluminum surface ($w_0 = 4.2$ eV) what is the smallest wavelength of an incident photon for which the classical expression may be used?

8. X rays with $\lambda = 0.71$ Å eject photoelectrons from a gold foil. The electrons form circular paths of radius $r$ in a region of magnetic induction $B$. Experiment shows that $rB =$ $1.88 \times 10^{-4}$ tesla-m. Find (a) the maximum kinetic energy of the photoelectrons and (b) the work done in removing the electron from the gold foil.

9. (a) Show that a free electron cannot absorb a photon and conserve both energy and momentum in the process. Hence, the photoelectric process requires a bound electron. (b) In the Compton effect, however, the electron *can* be free. Explain.

10. Under ideal conditions the normal human eye will record a visual sensation at 5500 Å if as few as 100 photons are absorbed per second. What power level does this correspond to?

11. Solar radiation falls on the earth at a rate of 1.94 cal/cm$^2$-min on a surface normal to the incoming rays. Assuming an average wavelength of 5500 Å, how many photons per cm$^2$-min is this?

12. What are the frequency, wavelength, and momentum of a photon whose energy equals the rest mass energy of an electron?

13. In the photon picture of radiation, show that if beams of radiation of two different wavelengths are to have the same intensity (or energy density) then the numbers of the photons per unit cross-sectional area per sec in the beams are in the same ratio as the wavelengths.

14. Derive the relation

$$\cot\frac{\theta}{2} = \left(1 + \frac{h\nu}{m_0c^2}\right)\tan\varphi$$

between the direction of motion of the scattered photon and the recoil electron in the Compton effect.

15. Derive a relation between the kinetic energy $K$ of the recoil electron and the energy $E$ of the incident photon in the Compton effect. One form of the relation is

$$\frac{K}{E} = \frac{\left(\frac{2h\nu}{m_0c^2}\right)\sin^2\frac{\theta}{2}}{1 + \left(\frac{2h\nu}{m_0c^2}\right)\sin^2\frac{\theta}{2}}$$

(Hint: See Example 2-4.)

16. Photons of wavelength 0.024 Å are incident on free electrons. (a) Find the wavelength of a photon which is scattered 30° from the incident direction and the kinetic energy imparted to the recoil electron. (b) Do the same if the scattering angle is 120°. (Hint: See Example 2-4.)

17. An x-ray photon of initial energy $1.0 \times 10^5$ eV traveling in the $+x$ direction is incident on a free electron at rest. The photon is scattered at right angles into the $+y$ direction. Find the components of momentum of the recoiling electron.

18. (a) Show that $\Delta E/E$, the fractional change in photon energy in the Compton effect, equals $(h\nu'/m_0c^2)(1 - \cos\theta)$. (b) Plot $\Delta E/E$ versus $\theta$ and interpret the curve physically.

19. What is the maximum possible kinetic energy of a recoiling Compton electron in terms of the incident photon energy $h\nu$ and the electron's rest energy $m_0c^2$?

20. Determine the maximum wavelength shift in the Compton scattering of photons from *protons*.

21. (a) Show that the short wavelength cutoff in the x-ray continuous spectrum is given by $\lambda_{\text{min}} = 12.4$ Å$/V$, where $V$ is applied voltage in kilovolts. (b) If the voltage across an x-ray tube is 186 kV what is $\lambda_{\text{min}}$?

22. (a) What is the minimum voltage across an x-ray tube that will produce an x ray having the Compton wavelength? A wavelength of 1 Å? (b) What is the minimum voltage needed across an x-ray tube if the subsequent bremsstrahlung radiation is to be capable of pair production?

23. A $\gamma$ ray creates an electron-positron pair. Show directly that, without the presence of a third body to take up some of the momentum, energy and momentum cannot both be conserved. (Hint: Set the energies equal and show that this leads to unequal momenta before and after the interaction.)

24. A $\gamma$ ray can produce an electron-positron pair in the neighborhood of an electron at rest as well as a nucleus. Show that in this case the threshold energy is $4m_0c^2$. (Hint: Do not ignore the recoil of the original electron, but assume that all three particles move off together. Also, use the center of mass frame of reference.)

25. A particular pair is produced such that the positron is at rest and the electron has a kinetic energy of 1.0 MeV moving in the direction of flight of the pair producing photon. (a) Neglecting the energy transferred to the nucleus of the nearby atom, find the energy of the incident photon. (b) What percentage of the photon's momentum is transferred to the nucleus?

26. Assume that an electron-positron pair is formed by a photon having the threshold energy for the process. (a) Calculate the momentum transferred to the nucleus in the process. (b) Assume the nucleus to be that of a lead atom and compute the kinetic energy of the recoil nucleus. Are we justified in neglecting this energy compared to the threshold energy assumed above?

**27.** Show that the results of Example 2-8, expressed in terms of $\rho$ and $t$, are valid independent of the assumed area of the slab.

**28.** Show that the attenuation length $\Lambda$ is just equal to the average distance a photon will travel before being scattered or absorbed.

**29.** Use the data of Figure 2-17 to calculate the thickness of a lead slab which will attenuate a beam of 10 keV x rays by a factor of 100.

# 3

# de Broglie's Postulate—Wavelike Properties of Particles

# de Broglie's Postulate—Wavelike Properties of Particles

# 3

## 3-1 Matter Waves

Maurice de Broglie was a French experimental physicist who, from the outset, had supported Compton's view of the particle nature of radiation. His experiments and discussions impressed his brother Louis so much with the philosophic problems of physics at the time that Louis changed his career from history to physics. In his doctoral thesis, presented in 1924 to the Faculty of Science at the University of Paris, Louis de Broglie proposed the existence of matter waves. The thoroughness and originality of his thesis was recognized at once, but because of the apparent lack of experimental evidence, de Broglie's ideas were not considered to have any physical reality. It was Albert Einstein who recognized their importance and validity and in turn called them to the attention of other physicists. Five years later de Broglie won the Nobel Prize in physics, his ideas having been dramatically confirmed by experiment.

The hypothesis of de Broglie was that the dual, that is wave-particle, behavior of radiation applies equally well to matter. Just as a photon has a light wave associated with it that governs its motion, so a material particle (e.g., an electron) has an associated matter wave that governs its motion. Since the universe is composed entirely of matter and radiation, de Broglie's suggestion is essentially a statement about a grand symmetry of nature. Indeed, he proposed that the wave aspects of matter are related to its particle aspects in exactly the same quantitative way that is the case for radiation. According to de Broglie, for matter *and* for radiation alike the total energy $E$ of an entity is related to the frequency $\nu$ of the wave associated with its motion by the equation

$$E = h\nu \tag{3-1a}$$

and the momentum $p$ of the entity is related to the wavelength $\lambda$ of the associated wave by the equation

$$p = h/\lambda \tag{3-1b}$$

Here the particle concepts, energy $E$ and momentum $p$, are connected through Planck's constant $h$ to the wave concepts, frequency $\nu$ and wavelength $\lambda$. Equation (3-1b), in the following form, is called the *de Broglie relation*

$$\lambda = h/p \tag{3-2}$$

It predicts the *de Broglie wavelength* $\lambda$ of a *matter wave* associated with the motion of a material particle having a momentum $p$.

**Example 3-1.** (a) What is the de Broglie wavelength of a baseball moving at a speed $v = 10$ m/sec?

Assume $m = 1.0$ kg. From (3-2)

$$\lambda = \frac{h}{p} = \frac{h}{mv} = \frac{6.6 \times 10^{-34}\text{ joule-sec}}{1.0\text{ kg} \times 10\text{ m/sec}} = 6.6 \times 10^{-35}\text{ m} = 6.6 \times 10^{-25}\text{ Å}$$

(b) What is the de Broglie wavelength of an electron whose kinetic energy is 100 eV? Here

$$\lambda = \frac{h}{p} = \frac{h}{\sqrt{2mK}} = \frac{6.6 \times 10^{-34}\text{ joule-sec}}{(2 \times 9.1 \times 10^{-31}\text{ kg} \times 100\text{ eV} \times 1.6 \times 10^{-19}\text{ joule/eV})^{1/2}}$$

$$= \frac{6.6 \times 10^{-34}\text{ joule-sec}}{5.4 \times 10^{-24}\text{ kg-m/sec}} = 1.2 \times 10^{-10}\text{ m} = 1.2\text{ Å}$$ ◀

The wave nature of light propagation is not revealed by experiments in geometrical optics, for the important dimensions of the apparatus used there are very large compared to the wavelength of light. If $a$ represents a characteristic dimension of an optical apparatus (e.g., the width of a lens, mirror, or slit) and $\lambda$ is the wavelength of the light passing through the apparatus, we are in the domain of geometrical optics when $\lambda/a \to 0$. The reason is that the diffraction effects in any apparatus are always confined to angles of about $\theta = \lambda/a$, so diffraction effects are completely negligible when $\lambda/a \to 0$. Note that geometrical optics involves ray propagation, which is similar to the trajectory motion of classical particles.

However, when the characteristic dimension $a$ of an optical apparatus becomes comparable to, or smaller than, the wavelength $\lambda$ of the light going through it, we are in the domain of physical optics. In this case, where $\lambda/a \gtrsim 1$, the diffraction angle $\theta = \lambda/a$ is large enough that diffraction effects are easily observed and the wave nature of light propagation becomes apparent. To observe wavelike aspects in the motion of matter, therefore, we need systems with apertures or obstacles of suitable small dimensions. The finest scale systems of apertures available to experimentalists at the time of de Broglie made use of the spacing between adjacent planes of atoms in a solid, where $a \simeq 1$ Å. (Now systems are available involving nuclear dimensions of $\simeq 10^{-4}$ Å.) Considering the de Broglie wavelengths evaluated in Example 3-1, we see that we cannot expect to detect *any* evidence of wavelike motion for a baseball, where $\lambda/a \simeq 10^{-25}$ for $a \simeq 1$ Å; but for a material particle of very much smaller mass than a baseball, the momentum $p$ is reduced, and the de Broglie wavelength $\lambda = h/p$ is increased sufficiently for diffraction effects to be observable. Using apparatus with characteristic dimensions $a = 1$ Å, wavelike aspects in the motion of the $\lambda = 1.2$ Å electron of Example 3-1 should be very apparent.

It was Elsasser who pointed out, in 1926, that the wave nature of matter might be tested in the same way that the wave nature of x rays was first tested, namely by allowing a beam of electrons of appropriate energy to fall on a crystalline solid. The atoms of the crystal serve as a three-dimensional array of diffracting centers for the electron wave, and so they should strongly scatter electrons in certain characteristic directions, just as for x-ray diffraction. This idea was confirmed in experiments by Davisson and Germer in the United States and by Thomson in Scotland.

Figure 3-1 shows schematically the apparatus of Davisson and Germer. Electrons from a heated filament are accelerated through a potential difference $V$ and emerge from the "electron gun" $G$ with kinetic energy $eV$. This electron beam falls at normal incidence on a single crystal of nickel at $C$. The detector $D$ is set at a particular angle $\theta$ and readings of the intensity of the scattered beam are taken at various values of the accelerating potential $V$. Figure 3-2, for example, shows that a strong scattered electron beam is detected at $\theta = 50°$ for $V = 54$ V. The existence of this *peak* in the electron scattering pattern demonstrates qualitatively the validity of de Broglie's postulate because *it can only be explained as a constructive interference of waves scattered by the periodic arrangement of the atoms into planes of the crystal.* The phenomenon is precisely analogous to the well-known "Bragg reflections" which occur in the scattering of x rays from the atomic planes of a crystal. It cannot be

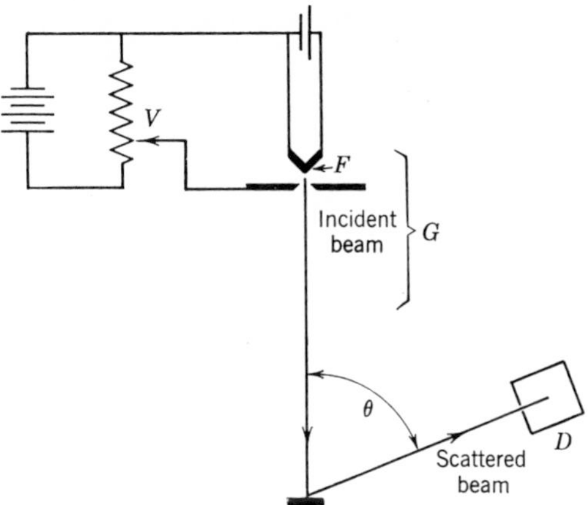


**FIGURE 3-1**

The apparatus of Davisson and Germer. Electrons from filament $F$ are accelerated by a variable potential difference $V$. After scattering from crystal $C$ they are collected by detector $D$.

understood on the basis of classical particle motion, but only on the basis of wave motion. Classical particles cannot exhibit interference, but waves can! The interference involved here is not between waves associated with one electron and waves associated with another. Instead, it is an interference between different parts of the wave associated with a single electron that have been scattered from various regions of the crystal. This can be demonstrated by using an electron beam of such low intensity that the electrons go through the apparatus one at a time, and by showing that the pattern of the scattered electrons remains the same.

Figure 3-3 shows the origin of a Bragg reflection, obeying the *Bragg relation* derived in the caption to that figure

$$n\lambda = 2\,d \sin \varphi \tag{3-3}$$

For the conditions of Figure 3-3 the effective interplanar spacing $d$ can be shown by x-ray scattering from the same crystal to be 0.91 Å. Since $\theta = 50°$, it follows that $\varphi = 90° - 50°/2 = 65°$. The wavelength calculated from (3-3), assuming $n = 1$, is

$$\lambda = 2\,d \sin \varphi = 2 \times 0.91\ \text{Å} \times \sin 65° = 1.65\ \text{Å}$$

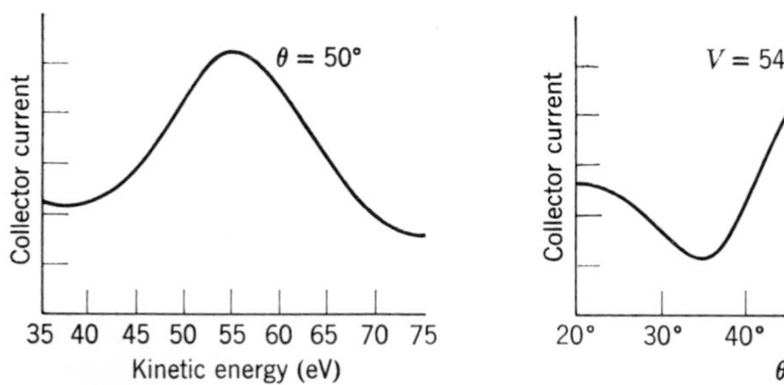


**FIGURE 3-2**

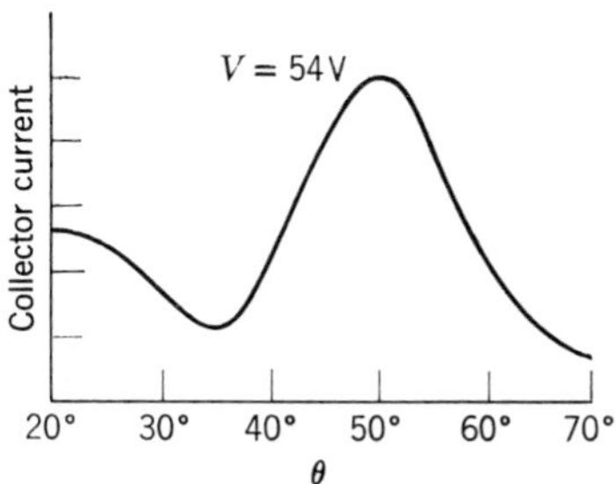

*Left:* The collector current in detector $D$ of Figure 3-1 as a function of the kinetic energy of the incident electrons, showing a diffraction maximum. The angle $\theta$ in Figure 3-1 is adjusted to 50°. If an appreciably smaller or larger value is used, the diffraction maximum disappears. *Right:* The current as a function of detector angle for the fixed value of electron kinetic energy 54 eV.

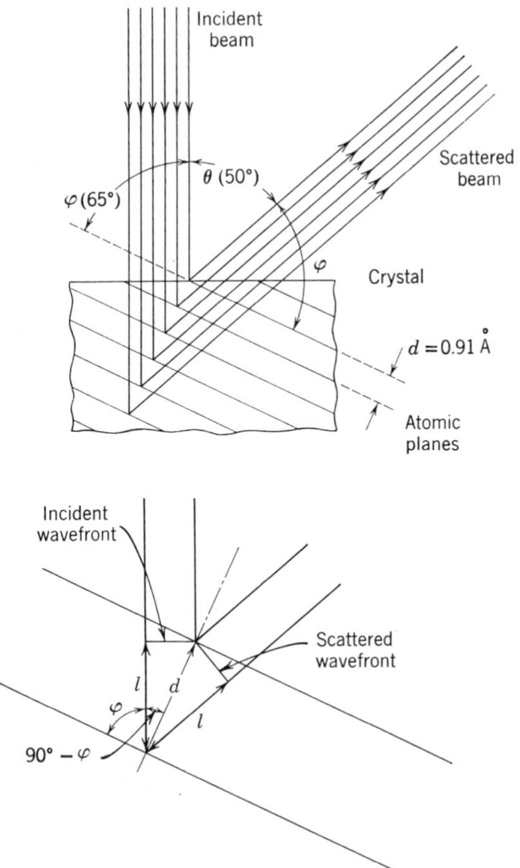


**FIGURE 3-3**
*Top:* The strong diffracted beam at $\theta = 50°$ and $V = 54$ V arises from wavelike scattering from the family of atomic planes shown, which have a separation distance $d = 0.91$ Å. The Bragg angle is $\varphi = 65°$. For simplicity, refraction of the scattered wave as it leaves the crystal surface is not indicated. *Bottom:* Derivation of the Bragg relation, showing only two atomic planes and two rays of the incident and scattered beams. If an integral number of wavelengths $n\lambda$ just fit into the distance $2l$ from incident to scattered wave fronts measured along the lower ray, then the contributions along the two rays to the scattered wave front will be in phase and a diffraction maximum will be obtained at the angle $\varphi$. Since $l/d = \cos(90° - \varphi) = \sin\varphi$, we have $2l = 2d \sin\varphi$, and so we obtain the Bragg relation $n\lambda = 2d \sin\varphi$. The "first order" diffraction maximum ($n = 1$) is usually most intense.

The de Broglie wavelength for 54 eV electrons, calculated from (3-2), is

$$\lambda = h/p = 6.6 \times 10^{-34}\,\text{joule-sec}/4.0 \times 10^{-24}\,\text{kg-m/sec} = 1.65\ \text{Å}$$

This impressive agreement gives quantitative confirmation of de Broglie's relation between $\lambda$, $p$, and $h$.

The breadth of the observed peak in Figure 3-2 is easily understood, also, for low-energy electrons cannot penetrate deeply into the crystal, so that only a small number of atomic planes contribute to the diffracted wave. Hence, the diffraction maximum is not sharp. Indeed, all the experimental results were in excellent qualitative and quantitative agreement with the de Broglie prediction, and they provided convincing evidence that material particles move according to the laws of wave motion.

In 1927, G. P. Thomson showed the diffraction of electron beams passing through thin films and independently confirmed the de Broglie relation $\lambda = h/p$ in detail. Whereas the Davisson-Germer experiment is like Laue's in x-ray diffraction (reflection from the regular array of atomic planes in a large single crystal), Thomson's experiment is similar to the Debye-Hull-Scherrer method of powder diffraction of x rays (transmission through an aggregate of very small crystals oriented at random). Thomson used higher-energy electrons, which are much more penetrating, so that many hundred atomic planes contribute to the diffracted wave. The resulting diffraction pattern has a sharp structure. In Figure 3-4 we show, for comparison, an x-ray

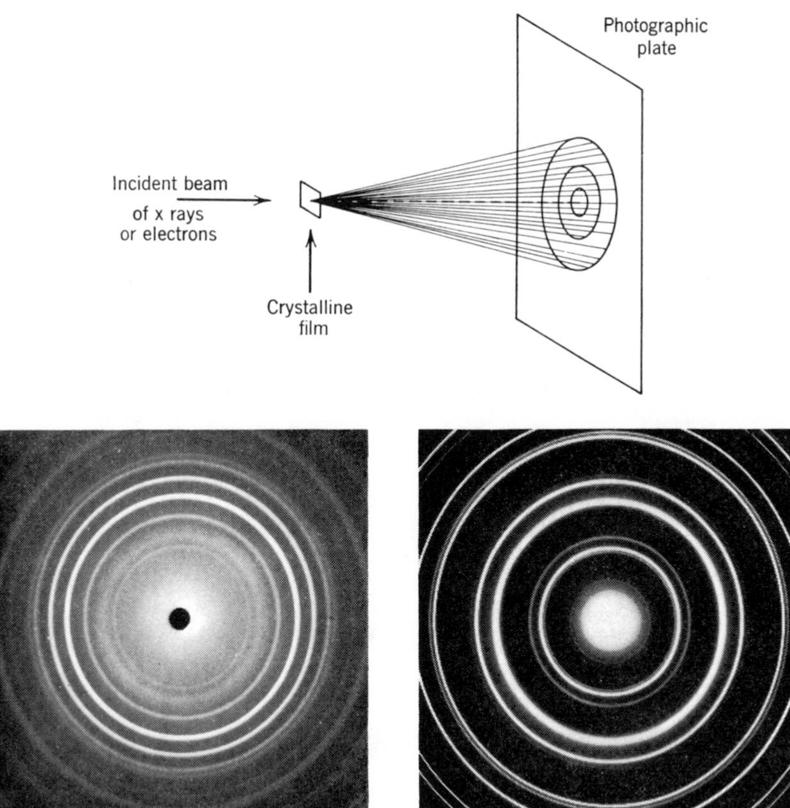


**FIGURE 3-4**

*Top:* The experimental arrangement for Debye-Scherrer diffraction of x rays or electrons by a polycrystalline material. *Bottom left:* Debye-Scherrer pattern of x-ray diffraction by zirconium oxide crystals. *Bottom right:* Debye-Scherrer pattern of electron diffraction by gold crystals.

diffraction pattern and an electron diffraction pattern from polycrystalline substances (substances in which a large number of microscopic crystals are oriented at random).

It is of interest that J. J. Thomson, who in 1897 discovered the electron (which he characterized as a particle with a definite charge-to-mass ratio) and was awarded the Nobel Prize in 1906, was the father of G. P. Thomson, who in 1927 experimentally discovered electron diffraction and was awarded the Nobel Prize (with Davisson) in 1937. Max Jammer writes of this, "One may feel inclined to say that Thomson, the father, was awarded the Nobel Prize for having shown that the electron is a particle, and Thomson, the son, for having shown that the electron is a wave."

Not only electrons but *all material objects*, charged or uncharged, show wavelike characteristics in their motion under the conditions of physical optics. For example, Estermann, Stern, and Frisch performed quantitative experiments on the diffraction of molecular beams of hydrogen and atomic beams of helium from a lithium fluoride crystal; and Fermi, Marshall, and Zinn showed interference and diffraction phenomena for slow neutrons. In Figure 3-5 we show a neutron diffraction pattern for a sodium chloride crystal. Even an interferometer operating with electron beams has been constructed. The existence of matter waves is well established.

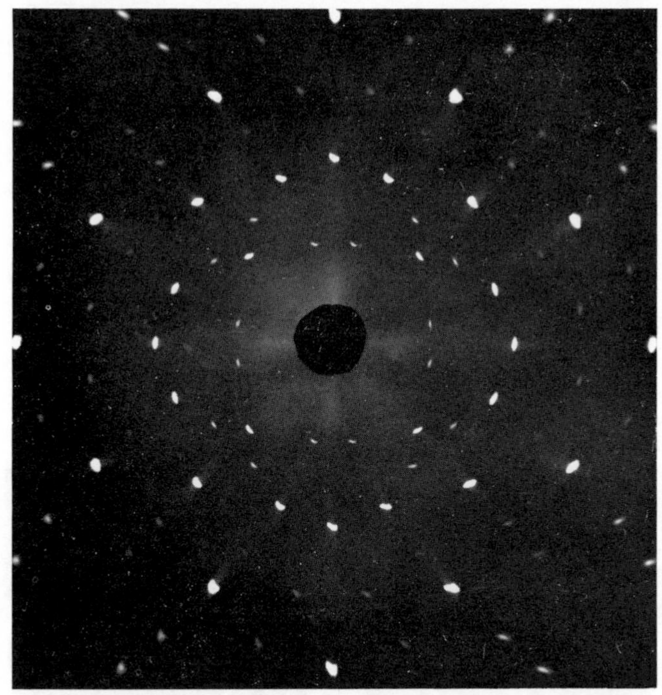

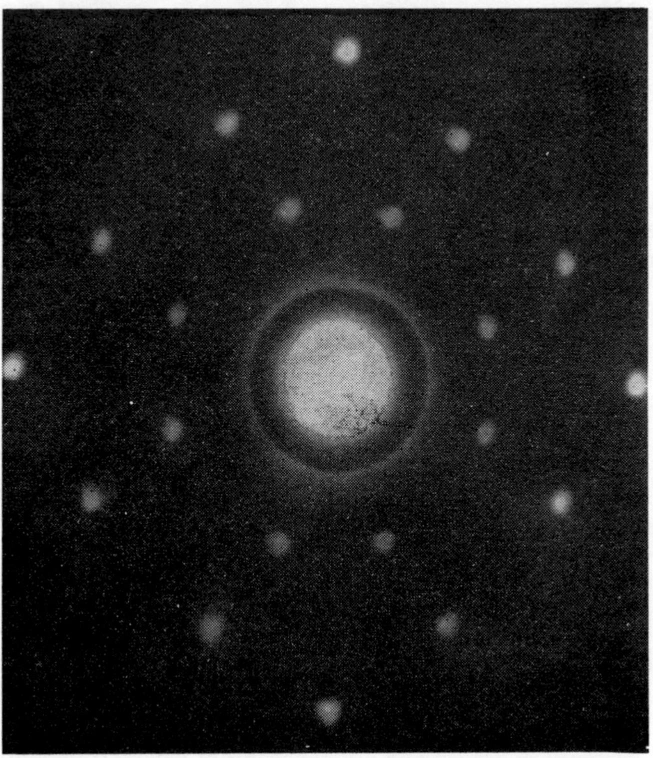

**FIGURE 3-5**

*Top:* Laue pattern of x-ray diffraction by a single sodium chloride crystal. *Bottom:* Laue pattern of diffraction of neutrons from a nuclear reactor by a single sodium chloride crystal.

It is instructive to note that we had to go to relatively long de Broglie wavelengths to find experimental evidence for the wave nature of matter. For both large and small wavelengths, both matter and radiation have both particle and wave aspects. The particle aspects are emphasized when their emission or absorption is studied, and the wave aspects are emphasized when their behavior in moving through a system is studied. But the wave aspects of their motion become more difficult to observe as their wavelengths become shorter. Once again we see the central role played by Planck's constant $h$. If $h$ were zero then in $\lambda = h/p$ we would obtain $\lambda = 0$ in all circumstances. All material particles would then always have a wavelength smaller than any characteristic dimension, and diffraction effects could never be observed. Although the value of $h$ is definitely not zero, it is small. It is the smallness of $h$ that obscures the existence of matter waves in the macroscopic world, for we must have very small momenta to obtain measurable wavelengths. For ordinary macroscopic particles the mass is so large that the momentum is always sufficiently large to make the de Broglie wavelength small enough to be beyond the range of experimental detection, and classical mechanics reigns supreme. In the microscopic world the masses of material particles are so small that their momenta are small even when their velocities are quite high. Thus their de Broglie wavelengths are large enough to be comparable to characteristic dimensions of systems of interest, such as atoms, and the wavelike properties are experimentally observable in their motion. But we should not forget that in their interaction, for instance when they are detected, their particlelike properties dominate even when their wavelengths are large.

**Example 3-2.** In the experiments with helium atoms referred to earlier, a beam of atoms of nearly uniform speed of $1.635 \times 10^5$ cm/sec was obtained by allowing helium gas to escape through a small hole in its enclosing vessel into an evacuated chamber and then through narrow slits in parallel rotating circular disks of small separation (a mechanical velocity selector). A strongly diffracted beam of helium atoms was observed to emerge from the lithium fluoride crystal surface upon which the atoms were incident. The diffracted beam was detected with a highly sensitive pressure gage. The usual crystal diffraction analysis of the experimental results indicated a wavelength of $0.600 \times 10^{-8}$ cm. How does this agree with the calculated de Broglie wavelength?

The mass of a helium atom is

$$m = \frac{M}{N_0} = \frac{4.00 \text{ g/mole}}{6.02 \times 10^{23} \text{ atom/mole}} = 6.65 \times 10^{-27} \text{ kg}$$

According to the de Broglie equation the wavelength then is

$$\lambda = \frac{h}{p} = \frac{h}{mv} = \frac{6.63 \times 10^{-34} \text{ joule-sec}}{6.65 \times 10^{-27} \text{ kg} \times 1.635 \times 10^{3} \text{ m/sec}} = 0.609 \times 10^{-10} \text{ m}$$

$$= 0.609 \times 10^{-8} \text{ cm}$$

This result, 1.5% greater than the value measured by crystal diffraction, is well within the limits of error of the experiment. ◀

Experiments like the one considered in Example 3-2 are very difficult since the intensities obtainable in atomic beams are quite low. Neutron diffraction experiments, using crystals of known lattice spacing, give confirmation of the existence of matter waves and precise confirmation of de Broglie's equation. The precision is due to the fact that the supply of neutrons from nuclear reactors is copious. Indeed, neutron diffraction is now an important method of studying crystal structure. Certain crystals, such as hydrogenous organic ones, are particularly well suited to neutron diffraction analysis, since neutrons are strongly scattered by hydrogen atoms whereas x rays are very weakly scattered by them. X rays interact chiefly with electrons in the atom, and electrons interact with the nuclear charge of the atom as well as the atomic electrons by electromagnetic forces, so that their interaction with hydrogen atoms is weak because the charge is small. Neutrons interact principally with the nucleus of the atom by nuclear forces, however, and the interaction is strong.

## 3-2 The Wave-Particle Duality

In classical physics energy is transported either by waves or by particles. Classical physicists observed water waves carrying energy over the water surface or bullets transferring energy from gun to target. From such experiences they built a wave model for certain macroscopic phenomena and a particle model for other macroscopic phenomena, and they quite naturally extended these models into visually less accessible regions. Thus they explained sound propagation in terms of a wave model and pressures of gases in terms of a particle model (kinetic theory). Their successes conditioned them to expect that all entities are either particles or waves. Indeed, these successes extended into the early twentieth century with applications of Maxwell's wave theory to radiation and the discovery of elementary particles of matter, such as the neutron and positron.

Hence, classical physicists were quite unprepared to find that to understand radiation they needed to invoke a particle model in some situations, as in the Compton effect, and a wave model in other situations, as in the diffraction of x rays. Perhaps more striking is the fact that this same wave-particle duality applies to matter as well as to radiation. The charge-to-mass ratio of the electron and its ionization trail in matter (a sequence of localized collisions) suggest a particle model, but electron diffraction suggests a wave model. Physicists now know that they are compelled to use both models for the same entity. It is very important to note, however, that in any given measurement only one model applies—both models are not used under the same circumstances. When the entity is detected by some kind of interaction, it acts like a particle in the sense that it is localized; when it is moving it acts like a wave in the sense that interference phenomena are observed, and, of course, a wave is extended, not localized.

Neils Bohr summarized the situation in his *principle of complementarity*. The wave and particle models are complementary; if a measurement proves the wave character of radiation or matter, then it is impossible to prove the particle character in the same measurement, and conversely. Which model we use is determined by the nature of the measurement. Furthermore, our understanding of radiation, or of matter, is incomplete unless we take into account measurements which reveal the wave aspects and also those that reveal the particle aspects. Hence, radiation and matter are not simply waves nor simply particles. A more general and, to the classical mind, a more complicated model is needed to describe their behavior, even though in extreme situations a simple wave model or a simple particle model may apply.

The link between wave model and particle model is provided by a probability interpretation of the wave-particle duality. In the case of radiation it was Einstein who united the wave and particle theories; subsequently Max Born applied a similar argument to unite wave and particle theories of matter.

In the wave picture the intensity of radiation, $I$, is proportional to $\overline{\mathscr{E}^2}$, where $\overline{\mathscr{E}^2}$ is the average value over one cycle of the square of the electric field strength of the wave.

($I$ is the average value of the so-called Poynting vector and we use the symbol $\mathscr{E}$ instead of $E$ for electric field to avoid confusion with the total energy $E$.) In the photon, or particle, picture the intensity of radiation is written as $I = Nh\nu$ where $N$ is the average number of photons per unit time crossing unit area perpendicular to the direction of propagation. It was Einstein who suggested that $\overline{\mathscr{E}^2}$, which in electromagnetic theory is proportional to the radiant energy in a unit volume, could be interpreted as a measure of the average number of photons per unit volume.

Recall that Einstein introduced a granularity to radiation, abandoning the continuum interpretation of Maxwell. This leads to a statistical view of intensity. In this view, a point source of radiation emits photons randomly in all directions. The average number of photons crossing a unit area will decrease with increasing distance from source to area. This is due to the fact that the photons spread over a sphere of larger area the farther they are from the source. Since the area of a sphere is proportional to the square of its radius, we obtain, on the average, an inverse square law of intensity just as in the wave picture. In the wave picture we imagine that spherical waves spread out from the source, the intensity dropping inversely as the square of the distance from the source. Here, these waves, whose strength can be measured by $\overline{\mathscr{E}^2}$, can be regarded as guiding waves for the photons; the waves themselves have no energy—there are only photons—but they are a construct whose intensity measures the average number of photons per unit volume.

We use the word "average" because the emission processes are statistical in nature. We do not specify exactly how many photons cross unit area in unit time, only their average number; the exact number can fluctuate in time and space, just as in kinetic theory of gases there are fluctuations about an average value from many quantities. We can say quite definitely, however, that the probability of having a photon cross unit area 3 m from the source is exactly one-ninth the probability that a photon will cross unit area 1 m from the source. In the formula $I = Nh\nu$, therefore, $N$ is an average value and is a measure of the probability of finding a photon crossing unit area in unit time. If we equate the wave expression to the particle expression we have

$$I = (1/\mu_0 c)\overline{\mathscr{E}^2} = h\nu N$$

so that $\overline{\mathscr{E}^2}$ is proportional to $N$. Einstein's interpretation of $\overline{\mathscr{E}^2}$ as a probability measure of photon density then becomes clear. We expect that, as in kinetic theory, fluctuations about an average will become more noticeable at low intensities than at high intensities, so that the granular quantum phenomena contradict the continuum classical view more dramatically there.

In analogy to Einstein's view of radiation, Max Born proposed a similar uniting of the wave-particle duality for matter. This came several years after Schroedinger developed his generalization of de Broglie's postulate, called quantum mechanics. We shall examine Schroedinger's theory quantitatively in later chapters. Here we wish merely to use Born's idea in a qualitative way to set the stage conceptually for the subsequent detailed analysis.

Let us associate more than just a wavelength and frequency with matter waves. We do this by introducing a function representing the de Broglie wave, called the *wave function* $\Psi$. For particles moving in the $x$ direction with a precise value of linear momentum and energy, for example, the wave function can be written as a simple sinusoidal function of amplitude $A$, such as

$$\Psi(x,t) = A \sin 2\pi\left(\frac{x}{\lambda} - \nu t\right) \tag{3-4a}$$

This is analogous to

$$\mathscr{E}(x,t) = A \sin 2\pi\left(\frac{x}{\lambda} - \nu t\right) \tag{3-4b}$$

for the electric field of a sinusoidal electromagnetic wave of wavelength $\lambda$, and frequency $\nu$, moving in the positive $x$ direction. The quantity $\overline{\Psi^2}$ will play a role for matter waves analogous to that played by $\overline{\mathscr{E}^2}$ for waves of radiation. That quantity, the average of the square of the wave function of matter waves, is a measure of the probability of finding a particle in unit volume at a given place and time. Just as $\mathscr{E}$ is a function of space and time, so is $\Psi$; and, as we shall see later, just as $\mathscr{E}$ satisfies a wave equation, so does $\Psi$ (Schroedinger's equation). The quantity $\mathscr{E}$ is a (radiation) wave associated with a photon, and $\Psi$ is a (matter) wave associated with a material particle.

As Born says: "According to this view, the whole course of events is determined by the laws of probability; to a state in space there corresponds a definite probability, which is given by the de Broglie wave associated with the state. A mechanical process is therefore accompanied by a wave process, the guiding wave, described by Schroedinger's equation, the significance of which is that it gives the probability of a definite course of the mechanical process. If, for example, the amplitude of the guiding wave is zero at a certain point in space, this means that the probability of finding the electron at this point is vanishingly small."

Just as in the Einstein view of radiation we do not specify the exact location of a photon at a given time, but specify instead by $\overline{\mathscr{E}^2}$ the probability of finding a photon at a certain location at a given time, so here in Born's view we do not specify the exact location of a particle at a given time, but specify instead by $\overline{\Psi^2}$ the probability of finding a particle at a certain location at a given time. Just as we are accustomed to adding wave functions ($\mathscr{E}_1 + \mathscr{E}_2 = \mathscr{E}$) for two superposed electromagnetic waves whose resultant intensity is given by $\mathscr{E}^2$, so we shall add wave functions for two superposed matter waves ($\Psi_1 + \Psi_2 = \Psi$) whose resultant intensity is given by $\Psi^2$. That is, a *principle of superposition* applies to matter as well as to radiation. This is in accordance with the striking experimental fact that matter exhibits interference and diffraction properties, a fact that simply cannot be understood on the basis of ideas in classical mechanics. Because waves can be superposed either constructively (in phase) or destructively (out of phase), two waves can combine either to yield a resultant wave of large intensity or to cancel, but two classical particles of matter cannot combine in such a way as to cancel.

The student might accept the logic of this fusion of wave and particle concepts but nevertheless ask whether a probabilistic or statistical interpretation is necessary. It was Heisenberg and Bohr who, in 1927, first showed how essential the concept of probability is to the union of wave and particle descriptions of matter and radiation. We investigate these matters in succeeding sections.

## 3-3 The Uncertainty Principle

The use of probability considerations is not foreign to classical physics. Classical statistical mechanics makes use of probability theory, for example. However, in classical physics the basic laws (such as Newton's laws) are deterministic, and statistical analysis is simply a practical device for treating very complicated systems. According to Heisenberg and Bohr, however, the probabilistic view is the fundamental one in quantum physics and determinism must be discarded. Let us see how this conclusion is reached.

In classical mechanics the equations of motion of a system with given forces can be solved to give us the position and momentum of a particle at all values of the time. All we need to know are the precise position and momentum of the particle at some value of the time $t = 0$ (the initial conditions) and the future motion is determined exactly. This mechanics has been used with great success in the macroscopic world, for example in astronomy, to predict the subsequent motions of objects in terms of their initial motions. Note, however, that in the process of making observations the observer interacts with the system. An example from contemporary astronomy is the precise measurement of the position of the moon by bouncing radar from it. The motion of the moon is disturbed by the measurement, but due to the very large mass of the moon the disturbance can be ignored. On a somewhat smaller scale, as in a very well-designed macroscopic experiment on earth, such disturbances are also usually small, or at least controllable, and they can be taken into account accurately ahead of time by suitable calculations. Hence, it was naturally assumed by classical physicists that in the realm of microscopic systems the position and momentum of an object, such as a electron, could be determined precisely by observations in a similar way. Heisenberg and Bohr questioned this assumption.

The situation is somewhat similar to that existing at the birth of relativity theory. Physicists spoke of length intervals and time intervals, i.e., space and time, without asking critically how one actually measures them. For example, they spoke of the simultaneity of two separated events without even asking how one would physically go about establishing simultaneity. In fact, Einstein showed that simultaneity was not an absolute concept at all, as had been assumed previously, but that two separated events that are simultaneous to one observer occur at different times to another observer moving with respect to the first. Simultaneity is a relative concept. Similarly then, we must ask ourselves how we actually measure position and momentum.

Can we determine by actual experiment at the same instant both the position and momentum of matter or radiation? The answer given by quantum theory, is: not more accurately than is allowed by the Heisenberg *uncertainty principle*. There are two parts to this principle, also called the indeterminacy principle. The first has to do with the simultaneous measurement of position and momentum. It states that experiment cannot simultaneously determine the exact value of a component of momentum, $p_x$ say, of a particle and also the exact value of its corresponding coordinate, $x$. Instead, our precision of measurement is inherently limited by the measurement process itself such that

$$\Delta p_x \Delta x \geq \hbar/2 \tag{3-5}$$

where the momentum $p_x$ is known to within an uncertainty of $\Delta p_x$ and the position $x$ at the same time to within an uncertainty $\Delta x$. Here $\hbar$ (read $h$-bar) is a shorthand symbol for $h/2\pi$, where $h$ is Planck's constant. That is

$$\hbar \equiv h/2\pi$$

There are corresponding relations for other components of momentum, namely $\Delta p_y \Delta y \geq \hbar/2$ and $\Delta p_z \Delta z \geq \hbar/2$, and for angular momentum as well. It is important to realize that this principle has nothing to do with improvements in instrumentation leading to better simultaneous determinations of $p_x$ and $x$. Rather the principle says that even with ideal instruments we can never in principle do better than $\Delta p_x \Delta x \geq \hbar/2$. Note also that the *product* of uncertainties is involved, so that, for example, the more we modify an experiment to improve our measure of $p_x$, the more we give up ability to determine $x$ accurately. If $p_x$ is known exactly we know nothing at all about $x$ (i.e., if $\Delta p_x = 0$, $\Delta x = \infty$). Hence, *the restriction is not on the accuracy to which $x$ or $p_x$ can be measured, but on the product $\Delta p_x \Delta x$ in a simultaneous measurement of both.*

The second part of the uncertainty principle has to do with the measurement of the energy $E$ and the time $t$ required for the measurements, as for example, the time interval $\Delta t$ during which a photon of energy spread $\Delta E$ is emitted from an atom. In this case

$$\Delta E \Delta t \geq \hbar/2 \tag{3-6}$$

where $\Delta E$ is the uncertainty in our knowledge of the energy $E$ of a system and $\Delta t$ the time interval characteristic of the rate of change in the system.

Heisenberg's relations will be shown later to follow from the de Broglie postulate plus simple properties common to all waves. Because the de Broglie postulate is verified by the experiments we have already discussed, it is fair to say that the uncertainty principle is grounded in experiment. We shall also consider soon the consistency of the principle with other experiments. Notice first, however, that it is Planck's constant $h$ that again distinguishes the quantum results from the classical ones. If $h$, or $\hbar$, in (3-5) and (3-6) were zero, there would be no basic limitation on our measurement at all, which is the classical view. Again it is the smallness of $h$ that takes the principle out of the range of our ordinary experiences. This is analogous to the smallness of the ratio $v/c$ in macroscopic situations taking relativity out of the range of ordinary experience. In principle, therefore, classical physics is of limited validity and in the microscopic domain it will lead to contradictions with experimental results. For if we cannot determine $x$ and $p$ simultaneously, then we cannot specify the initial conditions of motion exactly; therefore, we cannot precisely determine the future behavior of a system. Instead of making deterministic predictions, we can only state the possible results of an observation, giving the relative probabilities of their occurrence. Indeed, since the act of observing a system disturbs it in a manner that is not completely predictable, the observation changes the previous motion of the system to a new state of motion which cannot be completely known.

Let us now illustrate the physical origin of the uncertainty principle. With the insight thereby gained we shall better appreciate a more formal proof given in the following section. First, we use a thought experiment due to Bohr to verify (3-5). Let us say that we wish to measure as accurately as possible the position of a "point" particle, like an electron. For greatest precision we use a microscope to view the electron, as in Figure 3-6. To see the electron we must illuminate it, for it is actually the light photon scattered by the electron that the observer sees. At this stage, even before any calculations are made, we can see the uncertainty principle emerge. The very act of observing the electron disturbs it. The moment we illuminate the electron, it recoils because of the Compton effect, in a way that we shall soon find cannot be completely determined. If we don't illuminate the electron, however, we don't see (detect) it. Hence the uncertainty principle refers to the measuring process itself, and it expresses the fact that there is always an undetermined interaction between observer and observed; there is nothing we can do to avoid the interaction or to allow for it ahead of time. In the case at hand we can try to reduce the disturbance to the electron as much as possible by using a very weak source of light. The very weakest we can get is to assume that we can see the electron if only *one* scattered photon enters the objective lens of the microscope. The magnitude of the momentum of the photon is $p = h/\lambda$. But the photon may have been scattered *anywhere* within the angular range $2\theta'$ subtended by the objective lens at the electron. This is why the interaction cannot be allowed for. Hence, we find that the $x$ component of the momentum of the photon

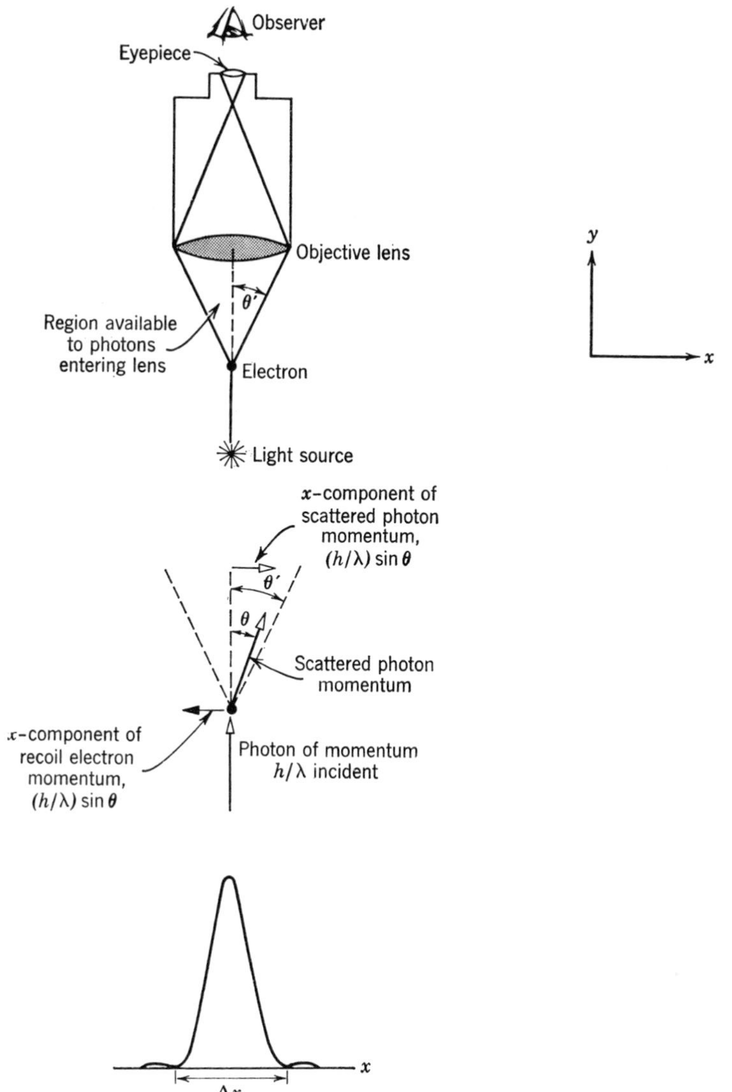


**FIGURE 3-6**

Bohr's microscope thought experiment. *Top:* The apparatus. *Middle:* The scattering of an illuminating photon by the electron. *Bottom:* The diffraction pattern image of the electron seen by the observer.

can vary from $+p \sin \theta'$ to $-p \sin \theta'$ and is uncertain after the scattering by an amount

$$\Delta p_x = 2p \sin \theta' = (2h/\lambda) \sin \theta'$$

Conservation of momentum then requires that the electron receive a recoil momentum in the $x$ direction that is equal in magnitude to the $x$ momentum change in the photon and, therefore, the $x$ momentum of the electron is uncertain by this same amount. Notice that to reduce $\Delta p_x$ we can use light of longer wavelength, or use a microscope with an objective lens subtending a smaller angle.

What about the location along $x$ of the electron? Recall that a microscope's image of a point object is not a point, but a diffraction pattern; the image of the electron is "fuzzy." The resolving power of a microscope determines the ultimate accuracy to which the electron can be located. If we take the width of the central diffraction maximum as a measure of the uncertainty in $x$, a well-known expression for the resolving power of a microscope gives

$$\Delta x = \lambda / \sin \theta'$$

(Note that, since $\sin \theta \simeq \theta$, this is an example of the general relation $a \simeq \lambda/\theta$ between the characteristic dimension in a diffraction apparatus, the wavelength of the diffracted waves, and the diffraction angle.) The one scattered photon at our disposal must have originated then *somewhere* within this range from the axis of the microscope, so the uncertainty in the electron's location is $\Delta x$. (We cannot be sure exactly where any one photon originates even though in a large number of repetitions of the experiment the photons forming the total image will produce the diffraction pattern shown in the figure.) Notice that to reduce $\Delta x$ we can use light of shorter wavelength, or a microscope with an objective lens subtending a larger angle.

If now we take the product of the uncertainties we find

$$\Delta p_x \Delta x = \left(\frac{2h}{\lambda} \sin \theta'\right)\left(\frac{\lambda}{\sin \theta'}\right) = 2h \tag{3-7}$$

in reasonable agreement with the ultimate limit $\hbar/2$ set by the uncertainty principle. We cannot *simultaneously* make $\Delta p_x$ and $\Delta x$ as small as we wish, for the procedure that makes one small makes the other large. For instance, if we use light of short wavelength (e.g., $\gamma$ rays) to reduce $\Delta x$ by obtaining better resolution, we increase the Compton recoil and increase $\Delta p_x$, and conversely. Indeed, the wavelength $\lambda$ and the angle $\theta'$ subtended by the objective lens do not even appear in the result. In practice an experiment might do much worse than (3-7) suggests, for that result represents the very ideal possible. We arrive at it, however, from genuinely measurable physical phenomena, namely the Compton effect and the resolving power of a lens.

There really should be no mystery in the student's mind about our result. It is a direct result of quantization of radiation. We had to have at least one photon illuminating the electron, or else no illumination at all; and even one photon carries a momentum of magnitude $p = h/\lambda$. It is this single scattered photon that provides the necessary interaction between the microscope and the electron. This interaction disturbs the particle in a way that cannot be exactly predicted or controlled. As a result, the coordinates and momentum of the particle cannot be completely known after the measurement. If classical physics were valid, then since radiation is regarded there as continuous rather than granular, we could reduce the illumination to arbitrarily small levels and deliver arbitrarily small momentum while using arbitrarily small wavelengths for "perfect" resolution. In principle there would be no simultaneous lower limit to resolution or momentum recoil and there would be no uncertainty principle. But we cannot do this; the single photon is indivisible. Again we see, from $\Delta p_x \Delta x \geq \hbar/2$, that Planck's constant is a measure of the minimum uncontrollable disturbance that distinguishes quantum physics from classical physics.

Now let us consider (3-6) relating energy and time uncertainties. For the case of a free particle we can obtain (3-6) from (3-5), which relates position and momentum, as follows. Consider an electron moving along the $x$ axis whose energy we can write as $E = p_x^2/2m$. If $p_x$ is uncertain by $\Delta p_x$, then the uncertainty in $E$ is given by $\Delta E = (p_x/m)\Delta p_x = v_x \Delta p_x$. Here $v_x$ can be interpreted as the recoil velocity along $x$ of the electron which is illuminated with light in a position measurement. If the time interval required for the measurement is $\Delta t$, then the uncertainty in its $x$ position is $\Delta x = v_x \Delta t$. Combining $\Delta t = \Delta x / v_x$ and $\Delta E = v_x \Delta p_x$, we obtain $\Delta E \Delta t = \Delta p_x \Delta x$. But

$\Delta p_x \Delta x \geq \hbar/2$. Hence

$$\Delta E \Delta t \geq \hbar/2$$

**Example 3-3.** The speed of a bullet ($m = 50$ g) and the speed of an electron ($m = 9.1 \times 10^{-28}$ g) are measured to be the same, namely 300 m/sec, with an uncertainty of 0.01%. With what fundamental accuracy could we have located the position of each, if the position is measured simultaneously with the speed in the same experiment?

For the electron

$$p = mv = 9.1 \times 10^{-31}\,\text{kg} \times 300\,\text{m/sec} = 2.7 \times 10^{-28}\,\text{kg-m/sec}$$

and

$$\Delta p = m\Delta v = 0.0001 \times 2.7 \times 10^{-28}\,\text{kg-m/sec} = 2.7 \times 10^{-32}\,\text{kg-m/sec}$$

so that

$$\Delta x \geq \frac{h}{4\pi \Delta p} = \frac{6.6 \times 10^{-34}\,\text{joule-sec}}{4\pi \times 2.7 \times 10^{-32}\,\text{kg-m/sec}} = 2 \times 10^{-3}\,\text{m} = 0.2\,\text{cm}$$

For the bullet

$$p = mv = 0.05\,\text{kg} \times 300\,\text{m/sec} = 15\,\text{kg-m/sec}$$

and

$$\Delta p = 0.0001 \times 15\,\text{kg-m/sec} = 1.5 \times 10^{-3}\,\text{kg-m/sec}$$

so that

$$\Delta x \geq \frac{h}{4\pi \Delta p} = \frac{6.6 \times 10^{-34}\,\text{joule-sec}}{4\pi \times 1.5 \times 10^{-3}\,\text{kg-m/sec}} = 3 \times 10^{-32}\,\text{m}$$

Hence, for macroscopic objects such as bullets the uncertainty principle sets no practical limit to our measuring procedure, $\Delta x$ in this example being about $10^{-17}$ times the diameter of a nucleus; but, for microscopic objects such as electrons, there are practical limits, $\Delta x$ in this example being about $10^7$ times the diameter of an atom. ◀

## 3-4 Properties of Matter Waves

In this section we shall derive the uncertainty principle relations by combining the de Broglie-Einstein relations, $p = h/\lambda$ and $E = h\nu$, with simple mathematical properties that are universal to all waves. We begin a development of these properties by calling attention to an apparent paradox.

The velocity of propagation $w$ of a wave with wavelength and frequency $\lambda$ and $\nu$ is given by the familiar relation, which we shall verify later

$$w = \lambda\nu \tag{3-8}$$

Let us evaluate $w$ for a de Broglie wave associated with a particle of momentum $p$ and total energy $E$. We obtain

$$w = \lambda\nu = \frac{h}{p}\frac{E}{h} = \frac{E}{p}$$

Now assume the particle is moving at nonrelativistic velocity $v$ in a region of zero potential energy. (The validity of our conclusions will not be limited by these assumptions.) Evaluating $p$ and $E$ in terms of $v$ and the mass $m$ of the particle, we find

$$w = \frac{E}{p} = \frac{mv^2/2}{mv} = \frac{v}{2} \tag{3-9}$$

This result seems disturbing because it appears that the matter wave would not be able to keep up with the particle whose motion it controls. However, there is really no difficulty, as the following argument shows.

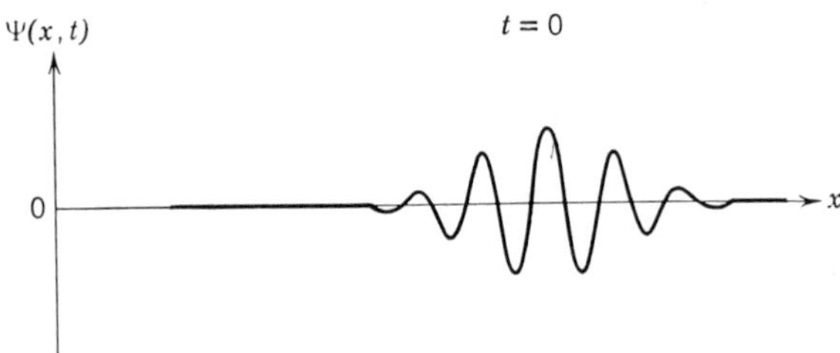


**FIGURE 3-7**
A de Broglie wave for a particle.

Imagine that a particle is moving along the $x$ axis under the influence of no force because its potential energy has the constant value zero. Moving along that axis is also its associated matter wave. Assume, for the sake of this thought experiment, that we have distributed along the axis a set of (hypothetical) instruments which are capable of measuring the amplitude of the matter wave. At some time, say $t = 0$, we record the readings of these instruments. The results of the experiment can be presented as a plot of the instantaneous values of the wave, which we designate by the symbol $\Psi(x,t)$, as a function of $x$ at a fixed time $t = 0$. It is not necessary to know much about matter waves at present to realize that the plot must look *qualitatively* like the one shown in Figure 3-7. The amplitude of the matter wave must be modulated in such a way that its value is nonzero only over some finite region of space in the vicinity of the particle. This is necessary because the matter wave must somehow be associated in space with the particle whose motion it controls. The matter wave is in the form of a *group* of waves and, as time passes, the group surely must move along the $x$ axis with the same velocity as the particle.

The student may recall, from his study of classical wave motion, that for such a moving group of waves it is necessary to distinguish between the velocity $g$ of the group and the quite different velocity $w$ of the individual oscillations of the waves. This is encouraging, but of course we must prove that $g$ is equal to the velocity of the particle. To do this, we develop a relation between $g$ and the quantities $\nu$ and $\lambda$ comparable to the relation of (3-8) between $w$ and these two quantities.

We start by considering the simplest type of wave motion, a sinusoidal wave of frequency $\nu$ and wavelength $\lambda$, which is of constant unit amplitude from $-\infty$ to $+\infty$, but which is moving with uniform velocity in the direction of increasing $x$. Such a wave can be represented mathematically by the function

$$\Psi(x,t) = \sin 2\pi\left(\frac{x}{\lambda} - \nu t\right) \tag{3-10a}$$

or, in a more convenient form

$$\Psi(x,t) = \sin 2\pi(\kappa x - \nu t) \qquad \text{where } \kappa \equiv 1/\lambda \tag{3-10b}$$

That this does represent the wave just described can be seen from the following considerations:

1. Holding $x$ fixed at any value, we see that the function oscillates in time sinusoidally with frequency $\nu$ and amplitude one.

2. Holding $t$ fixed, we see that the function has a sinusoidal dependence on $x$, with wavelength $\lambda$ or *reciprocal wavelength* $\kappa$.

3. The zeros of the function, which correspond to the nodes of the wave it represents, are found at positions $x_n$ for which

$$2\pi(\kappa x_n - \nu t) = \pi n \qquad n = 0, \pm 1, \pm 2, \ldots$$

or

$$x_n = \frac{n}{2\kappa} + \frac{\nu}{\kappa} t$$

Thus these nodes, and in fact all points on the wave, are moving in the direction of increasing $x$ with velocity

$$w = dx_n/dt$$

which is equal to

$$w = \nu/\kappa$$

Note that this is identical with (3-8) since $\kappa = 1/\lambda$.

Next we discuss the case in which the amplitude of the waves is modulated to form a group. We can obtain mathematically one group of waves moving in the direction of increasing $x$, similar to the group of matter waves pictured in Figure 3-7, by adding together an infinitely large number of waves of the form of (3-10b), each with infinitesimally differing frequencies $\nu$ and reciprocal wavelengths $\kappa$. (We shall soon explain how this happens.) The mathematical techniques become a little involved, however, and, for our purposes it will suffice to consider what happens when we add together only two such waves. Thus we take

$$\Psi(x,t) = \Psi_1(x,t) + \Psi_2(x,t) \tag{3-11}$$

where

$$\Psi_1(x,t) = \sin 2\pi[\kappa x - \nu t]$$

and

$$\Psi_2(x,t) = \sin 2\pi[(\kappa + d\kappa)x - (\nu + d\nu)t]$$

Now

$$\sin A + \sin B = 2\cos[(A - B)/2]\sin[(A + B)/2]$$

Applying this to the case at hand, we have

$$\Psi(x,t) = 2\cos 2\pi\left[\frac{d\kappa}{2}x - \frac{d\nu}{2}t\right]\sin 2\pi\left[\frac{(2\kappa + d\kappa)}{2}x - \frac{(2\nu + d\nu)}{2}t\right]$$

Since $d\nu \ll 2\nu$ and $d\kappa \ll 2\kappa$, this is

$$\Psi(x,t) = 2\cos 2\pi\left(\frac{d\kappa}{2}x - \frac{d\nu}{2}t\right)\sin 2\pi(\kappa x - \nu t) \tag{3-12}$$

A plot of $\Psi(x,t)$ as a function of $x$ for a fixed value of $t = 0$ is shown in Figure 3-8. The second term of $\Psi(x,t)$ is a wave of the same form as (3-10b), but this wave is

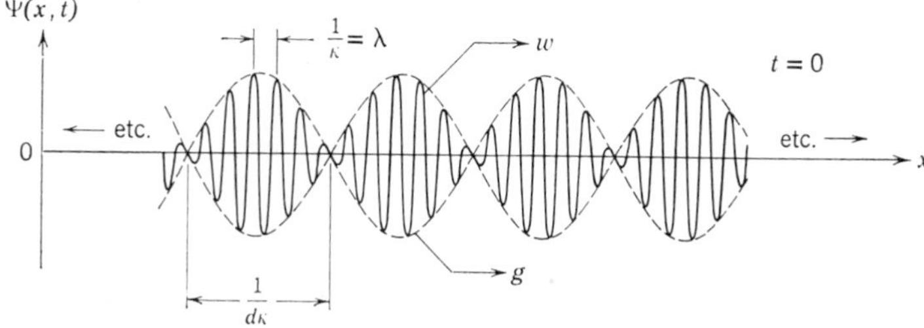


**FIGURE 3-8**

The sum of two sinusoidal waves of slightly different frequencies and reciprocal wavelengths $\kappa$.

modulated by the first term so that the oscillations of $\Psi(x,t)$ fall within an envelope of periodically varying amplitude. Two waves of slightly different frequency and reciprocal wavelength alternately interfere and reinforce in such a way as to produce a succession of groups. These groups, and the individual waves which they contain, are both moving in the direction of increasing $x$. The velocity $w$ of the individual waves can be evaluated by considering the second term of $\Psi(x,t)$, and the velocity $g$ of the groups can be evaluated from the first term. Proceeding as in consideration 3, we find again

$$w = \frac{\nu}{\kappa} \tag{3-13a}$$

and also the new result

$$g = \frac{d\nu/2}{d\kappa/2} = \frac{d\nu}{d\kappa} \tag{3-13b}$$

It can be shown that for an infinitely large number of waves that combine to form one moving group, the dependence of the *wave velocity* $w$, and the *group velocity* $g$, on $\nu$, $\kappa$, and $d\nu/d\kappa$ is exactly the same as for the simple case we have considered. Equations (3-13a) and (3-13b) have general validity.

Finally we are in a position to calculate the group velocity $g$ of the group of matter waves associated with the moving particle. From the Einstein and de Broglie relations, we have

$$\nu = E/h \qquad \text{and} \qquad \kappa \equiv 1/\lambda = p/h$$

so

$$d\nu = dE/h \qquad \text{and} \qquad d\kappa = dp/h$$

Thus the group velocity is

$$g = d\nu/d\kappa = dE/dp$$

Setting

$$E = \frac{mv^2}{2} \qquad \text{and} \qquad p = mv$$

we obtain

$$\frac{dE}{dp} = \frac{mv\,dv}{m\,dv} = v$$

which gives us the satisfying result that

$$g = v$$

*The velocity of the group of matter waves is just equal to the velocity of the particle whose motion they govern*, and de Broglie's postulate is internally consistent. The same conclusion is obtained when relativistic expressions for $E$ and $p$ are used in evaluating $dE/dp$.

Now we shall derive the uncertainty relations by combining the de Broglie-Einstein relations, $p = h/\lambda$ and $E = h\nu$, with properties of groups of waves. First consider a simple limiting case. Let $\lambda$ be the wavelength of a de Broglie wave associated with a particle. We can picture a definite (monochromatic) wavelength in terms of a single sinusoidal wave extending over all values of $x$, i.e., an infinitely long unmodulated wave like

$$\Psi = A \sin 2\pi(\kappa x - \nu t)$$

or

$$\Psi = A \cos 2\pi(\kappa x - \nu t)$$

If the wavelength has the definite value $\lambda$ there is no uncertainty $\Delta\lambda$ and the associated particle momentum $p = h/\lambda$ is also definite so $\Delta p_x = 0$. In such a wave the amplitude has the constant value $A$ everywhere; it is the same over the entire infinite range of $x$. Therefore, the probability of finding the particle, which Born tells us is to be related to the amplitude of the wave, is not concentrated in a particular range of $x$. In other words, the location of the particle is completely unknown. The particle can be anywhere, so that $\Delta x = \infty$. Analogous statements are that since $E = h\nu$, and since the frequency is definite, then $\Delta E = 0$. But to be sure that the amplitude of the wave is perfectly constant in time we must observe the wave for an infinite time, so that $\Delta t = \infty$. For this simple case we satisfy $\Delta p_x \Delta x \geq \hbar/2$, and $\Delta E \Delta t \geq \hbar/2$, in the limits $\Delta p_x = 0$, $\Delta x = \infty$, and $\Delta E = 0$, $\Delta t = \infty$.

In order to have a wave whose amplitude varies with $x$ or $t$, we must superpose several monochromatic waves of different wavelengths or frequencies. For two such waves superposed we obtain the familiar phenomenon of beats, as we have seen earlier in this section, with the amplitude being modulated in a regular way throughout space or time. If we wish to construct a wave having a finite extent in space (a single group with a definite beginning and end), then we must superpose sinusoidal waves having a continuous spectrum of wavelengths with a range $\Delta\lambda$. The amplitude of such a group will be zero everywhere outside a region of extent $\Delta x$.

To help visualize this, consider first a case in which we superpose a finite number of sinusoidal waves of slightly different wavelengths $\lambda$, or reciprocal wavelengths $\kappa$. Figure 3-9 shows seven component sinusoidal waves $\Psi_\kappa = A_\kappa \cos 2\pi(\kappa x - \nu t)$, at time $t = 0$. Their reciprocal wavelengths $\kappa = 1/\lambda$ take on integral values from $\kappa = 9$ to $\kappa = 15$. The amplitude of each is given by $A_\kappa$, with $A_{12} = 1$, $A_{13} = A_{11} = 1/2$, $A_{14} = A_{10} = 1/3$, and $A_{15} = A_9 = 1/4$, as shown in the figure. All the waves are in phase at $x = 0$ where they are centered (this is why cosines are used), but they get out of phase with one another proceeding in either direction from that point. As a result, their sum $\Psi = \Psi_9 + \cdots + \Psi_{15}$ oscillates with maximum amplitude at $x = 0$, but its oscillations die out with increasing or decreasing $x$ as the phase relations of the component waves get scrambled. The superposition thus contains a group whose extent in space $\Delta x$ has a value that can be read from the figure to be slightly larger than $1/12$, if we adopt the usual convention and measure from maximum amplitude to half-maximum amplitude. With an analogous convention, the range of reciprocal wavelengths used to compose the group, $\Delta\kappa$, has a value of 1. Note that the approximate value of the product $\Delta x \Delta\kappa$ equals $1/12$. Indicated on the right edge of the figure is the presence of an auxiliary group, of the same shape as the central group. Auxiliary groups are formed at uniformly spaced intervals along the positive and negative $x$ axis. They occur because, with only a finite number of component waves, there are points on the axis separated from $x = 0$ by a distance which is exactly some different integral number of wavelengths for each component. At these points the components are in phase again, and so the group is repeated. If the number of component waves spanning a fixed range $\Delta\kappa$ of reciprocal wavelengths is doubled, the width of the central group will be essentially unchanged but the distances separating it from the auxiliary groups will be doubled.

If we combine an infinitely large number of sinusoidal component waves, each with infinitesimally different reciprocal wavelength drawn from the same range $\kappa = 9$ to 15, we obtain a central group quite similar to the one shown in Figure 3-9, but the auxiliary groups will not be present. The reason is that in such a case there is no length of the $x$ axis into which an exactly integral number of wavelengths fits for every one of the infinite number of components. The components are all in phase at and near $x = 0$, and so they combine constructively to form the group. Proceeding away from this point, in either direction, the component waves begin to get out of phase with each other because their wavelengths or reciprocal wavelengths differ. Beyond certain

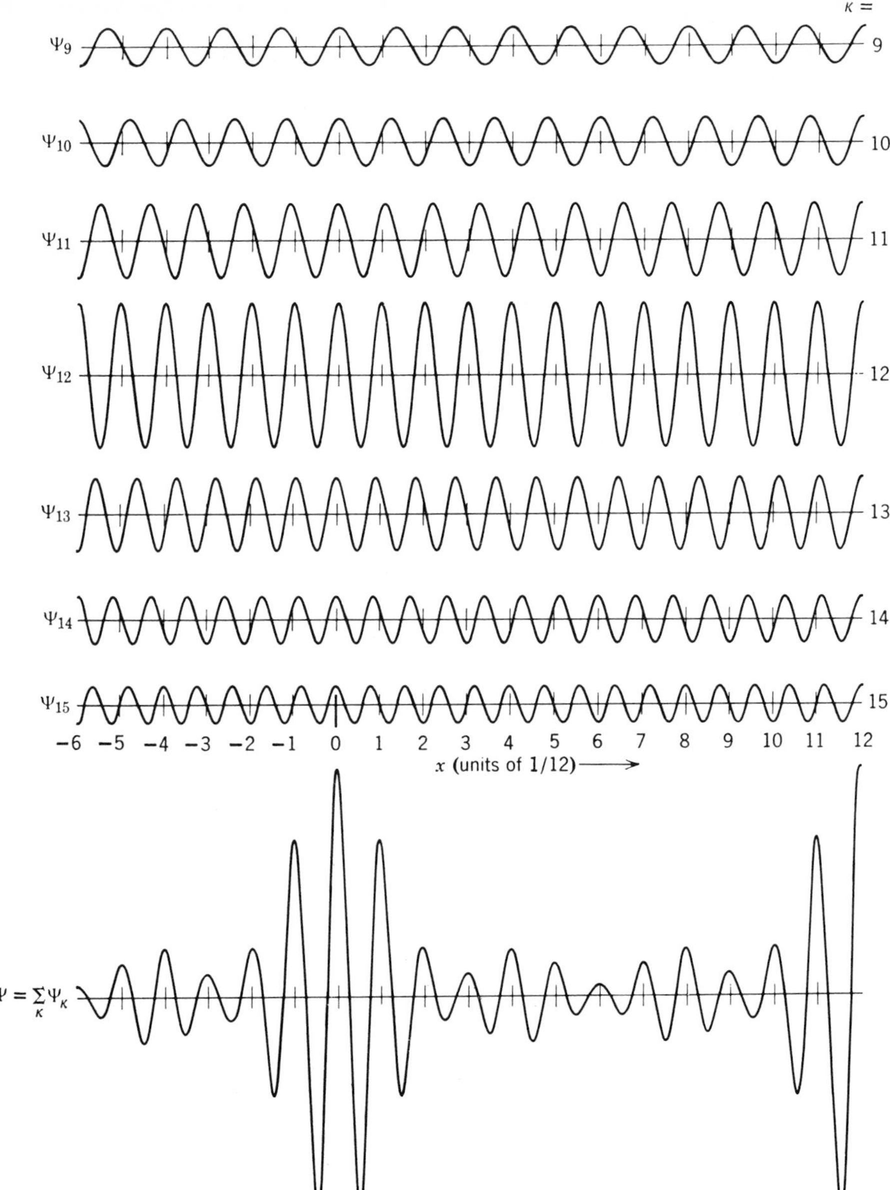


**FIGURE 3-9**

Showing, at $t = 0$, the superposition of seven cosine waves $\Psi_\kappa = A_\kappa \cos 2\pi(\kappa x - \nu t)$ with uniformly spaced reciprocal wavelengths drawn from the range $\kappa = 9$ to $\kappa = 15$. Their amplitudes $A_\kappa$ maximize at the value $A_{12} = 1$ for the wave whose $\kappa$ lies in the center of the range, and they decrease symmetrically through the values 1/2, 1/3, and 1/4 for the other waves as their $\kappa$ approach the ends of the range. The sum $\Psi = \sum_\kappa \Psi_\kappa$ of these waves consists of a group centered on $x = 0$, plus repeating groups of the same shape periodically spaced along the $x$ axis in both directions from $x = 0$. With $\Delta x$ defined as the maximum amplitude to half-maximum amplitude width of $\Psi$, and $\Delta\kappa$ defined as the range of reciprocal wavelengths of the components of $\Psi$ from maximum amplitude to half-maximum amplitude, we have $\Delta x \simeq 1/12$, $\Delta\kappa \simeq 1$, and $\Delta x\, \Delta\kappa \simeq 1/12$.

points the phases of the infinite number of components become completely random, and so the component waves sum up to zero. Furthermore, they never again get back into phase. Thus the components form one group of restricted length $\Delta x$. It is clear that the larger the range of reciprocal wavelengths $\Delta\kappa$ from which the components are drawn, the smaller the length $\Delta x$ of the group; the reason is simply that if the wavelengths cover a bigger span the phases will become random in a shorter distance. In fact, $\Delta x$ is just inversely proportional to $\Delta\kappa$. The exact value of the proportionality constant depends on the relative amplitudes of the component waves, as does the exact shape of the group that they form.

The mathematics used in carrying out the procedure just described involves the so-called Fourier integral and is a bit complicated, but it leads to numerical results that are very similar to the results we obtained from the construction in Figure 3-9. That is

$$\Delta x \Delta\kappa \geq 1/4\pi \tag{3-14}$$

The *optimum* job that can be done in composing a group of (half-width at half-maximum amplitude) length $\Delta x$ from components with reciprocal wavelengths covering a (half-width at half-maximum amplitude) range of $\Delta\kappa$ yields $\Delta x = 1/4\pi\Delta\kappa$, or $\Delta x\Delta\kappa = 1/4\pi$. Generally a somewhat larger value of this product is obtained.

A group of waves traveling through space of limited extent passes any given point of observation in a limited time. If $\Delta t$ is the duration of the group, or pulse, of waves then it necessarily must be composed from component sinusoidals whose frequencies span a range $\Delta\nu$, where

$$\Delta t \Delta\nu \geq 1/4\pi \tag{3-15}$$

Thus the frequency of the group is spread over the range $\Delta\nu$ if its duration covers the range $\Delta t$, just as its reciprocal wavelength is uncertain to within $\Delta\kappa$ if its width is $\Delta x$. Equation (3-15) is also obtained from a Fourier integral. It and (3-14) are different expressions of the same property; but the frequency-time relation, or at least some of its implications, may be more familiar to the student, as the following example shows.

**Example 3-4.** The signal from a television station contains pulses of full-width $\Delta t \sim 10^{-6}$ sec. Explain why it is not feasible to transmit television in the AM broadcasting band.

The full-width range of frequencies in the signal is, from (3-15), $\Delta\nu \sim 1/10^{-6}\text{ sec} = 10^6\text{ sec}^{-1} = 10^6$ Hz. Thus the entire broadcast band ($\nu \simeq 0.5 \times 10^6$ Hz to $\nu \simeq 1.5 \times 10^6$ Hz) would be able to accommodate only a single television "channel." There would also be serious difficulties in building transmitters and receivers with such a very large fractional bandpass. At the frequencies used in television transmission ($\nu \simeq 10^8$ Hz) many channels fit into a reasonable portion of the spectrum, and the bandpass requirements are nominal. ◀

Equations (3-14) and (3-15) are *universal properties of all waves.* If we apply them to matter waves by combining them with the de Broglie-Einstein relations, we immediately obtain the Heisenberg uncertainty relations. That is, if in

$$\Delta x\Delta\kappa = \Delta x\Delta(1/\lambda) \geq 1/4\pi$$

we set $p = h/\lambda$ or $1/\lambda = p/h$, we obtain

$$\Delta x\Delta(p/h) = (1/h)\,\Delta x\Delta p \geq 1/4\pi$$

or

$$\Delta p\Delta x \geq \hbar/2 \tag{3-16}$$

And if in

$$\Delta t\Delta\nu \geq 1/4\pi$$

we set $E = h\nu$ or $\nu = E/h$, we obtain

$$\Delta t \Delta (E/h) = (1/h)\, \Delta t \Delta E \geq 1/4\pi$$

or

$$\Delta E \Delta t \geq \hbar/2 \tag{3-17}$$

These results agree with our original statements of the relations in (3-5) and (3-6).

To summarize, we have seen that physical measurement necessarily involves interaction between the observer and the system being observed. Matter and radiation are the entities available to us for such measurements. The relations $p = h/\lambda$ and $E = h\nu$ apply to matter and to radiation, being the expression of the wave-particle duality. When we combine these relations with the properties universal to all waves we obtain the uncertainty relations. Hence, the uncertainty principle is a necessary consequence of this duality, that is, of the de Broglie-Einstein relations, and the uncertainty principle itself is the basis for the Heisenberg-Bohr contention that probability is fundamental to quantum physics.

**Example 3-5.** An atom can radiate at any time after it is excited. It is found that in a typical case the average excited atom has a life-time of about $10^{-8}$ sec. That is, during this period it emits a photon and is deexcited.

(a) What is the minimum uncertainty $\Delta\nu$ in the frequency of the photon?

From (3-15) we have

$$\Delta\nu \Delta t \geq 1/4\pi$$

or

$$\Delta\nu \geq 1/4\pi\Delta t$$

With $\Delta t = 10^{-8}$ sec we obtain $\Delta\nu \geq 8 \times 10^6\ \text{sec}^{-1}$.

(b) Most photons from sodium atoms are in two spectral lines at about $\lambda = 5890$ Å. What is the fractional width of either line, $\Delta\nu/\nu$?

For $\lambda = 5890$ Å, we obtain $\nu = c/\lambda = 3 \times 10^{10}\ \text{cm-sec}^{-1}/5890 \times 10^{-8}\ \text{cm} = 5.1 \times 10^{14}\ \text{sec}^{-1}$. Hence $\Delta\nu/\nu = 8 \times 10^6\ \text{sec}^{-1}/5.1 \times 10^{14}\ \text{sec}^{-1} = 1.6 \times 10^{-8}$ or about two parts in 100 million.

This is the so-called *natural width* of the spectral line. The line is much broader in practice because of the Doppler broadening and pressure broadening due to the motions and collisions of atoms in the source.

(c) Calculate the uncertainty $\Delta E$ in the energy of the excited state of the atom.

The energy of the excited state is not precisely measurable because only a finite time is available to make the measurement. That is, the atom does not stay in an excited state for an indefinite time but decays to its lowest energy state, emitting a photon in the process. The spread in energy of the photon equals the spread in energy of the excited state of the atom in accordance with the energy conservation principle. From (3-17), with $\Delta t$ equal to the mean life-time of the excited state, we have

$$\Delta E \geq \frac{h/4\pi}{\Delta t} = \frac{h}{4\pi\Delta t} = \frac{6.63 \times 10^{-34}\ \text{joule-sec}}{4\pi \times 10^{-8}\ \text{sec}}$$

$$= \frac{4.14 \times 10^{-15}\ \text{eV-sec}}{4\pi \times 10^{-8}\ \text{sec}} \simeq 3.3 \times 10^{-8}\ \text{eV}$$

This agrees, of course, with the value obtained from part (a) by multiplying the uncertainty in photon frequency $\Delta\nu$ by $h$ to obtain $\Delta E = h\Delta\nu$.

The energy spread of an excited state is usually called the *width* of the state.

(d) From the previous results determine, to within an accuracy $\Delta E$, the energy $E$ of the excited state of a sodium atom, relative to its lowest energy state, that emits a photon whose wavelength is centered at 5890 Å.

We have $\Delta\nu/\nu = h\Delta\nu/h\nu = \Delta E/E$. Hence, $E = \Delta E/(\Delta\nu/\nu) = 3.3 \times 10^{-8}\ \text{eV}/1.6 \times 10^{-8} = 2.1$ eV, in which we have used the results of the calculations in parts (b) and (c). ◀

**Example 3-6.** A measurement is made on the $y$ coordinate of an electron, which is a member of a broad parallel beam moving in the $x$ direction, by introducing into the beam a slit of narrow width $\Delta y$. Show that as a result an uncertainty $\Delta p_y$ is introduced in the $y$ component of momentum of the electron, such that $\Delta p_y \Delta y \geq \hbar/2$, as required by the uncertainty principle. Do this by considering the diffraction of the wave associated with the electron.

In propagating through the apparatus shown in Figure 3-10, the wave will be diffracted by the slit. The angle $\theta$ to the first minimum of the "single-slit" diffraction pattern sketched in the figure is given by $\sin \theta = \lambda/\Delta y$. (This is another example of the general relation $\theta \simeq \lambda/a$ between diffraction angle, wavelength, and characteristic dimension of a diffraction apparatus.) Since the propagation of the wave governs the motion of the associated particle, the diffraction pattern also gives the relative probabilities for the electron to arrive at different locations on the photographic plate. Thus the electron passing through the slit will be deflected through an angle which lies anywhere within a range from about $-\theta$ to $+\theta$. Even though its $y$ momentum was known with great precision to be zero before passing through the slit (because very little was then known about its $y$ position), after passing the slit where the measurement of its $y$ position was made its $y$ momentum can be anywhere within a range from about $-p_y$ to $+p_y$, where $\sin \theta = p_y/p$. So the $y$ momentum of the electron is made uncertain by the $y$ position measurement due to diffraction of the electron wave. The uncertainty is

$$\Delta p_y \simeq p_y = p \sin \theta = p\lambda/\Delta y$$

Using the de Broglie relation $p = h/\lambda$ to connect the momentum of the particle with the wavelength of the wave, we obtain

$$\Delta p_y = h/\Delta y$$

or

$$\Delta p_y \Delta y = h$$

Our result agrees with the limit set by the uncertainty principle. Diffraction, which refers to waves, and the uncertainty principle, which refers to particles, provide alternative but equivalent ways of treating this and all similar problems. ◀

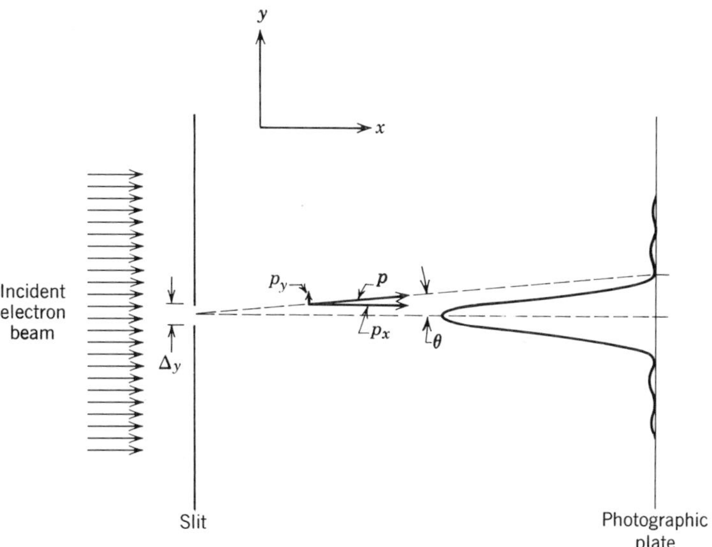


**FIGURE 3-10**

Measurement of the $y$ coordinate of an electron in a broad parallel beam, by requiring it to pass through a slit. The intensity pattern of the diffracted electron wave is indicated by using the line representing the photographic plate as an axis for a plot of the pattern.

Note that in Example 3-6 the wave associated with a *single* electron is regarded as being diffracted. The probability that the electron hits some point on the photographic plate is determined by the intensity of the electron wave. If only one electron goes through the apparatus it can hit anywhere except at the zero intensity locations of the diffraction pattern, and it will most likely hit somewhere near the principal maximum. If many electrons go through the apparatus each of their waves is diffracted independently in the same way and their points of arrival on the photographic plate are distributed according to the same pattern. The fact that diffraction phenomena involve interference between different parts of a wave belonging to a single particle, and not interference between waves belonging to different particles, was first shown experimentally by G. I. Taylor for the case of photons and light waves. Using light of such low intensity that the photons were known to be going through a diffraction apparatus one at a time, he obtained, after a very long exposure, a diffraction pattern. Then turning the intensity up to normal levels where many photons were in the apparatus at any time, he obtained the same diffraction pattern. Essentially the same experiment has subsequently been performed for electrons and other material particles.

## 3-5 Some Consequences of the Uncertainty Principle

The uncertainty principle allows us to understand why it is possible for radiation, and matter, to have a dual (wave-particle) nature. If we try experimentally to determine whether radiation is a wave or a particle, for example, we find that an experiment which forces radiation to reveal its wave character strongly suppresses its particle character. If we modify the experiment to bring out the particle character, its wave character is suppressed. We can never bring the wave and the particle view face to face in the same experimental situation. Radiation, and also matter, are like coins that can be made to display either face at will but not both simultaneously. This, of course, is the essence of Bohr's principle of complementarity; the ideas of wave and of particle complement rather than contradict one another.

Consider Young's two-slit interference experiment with light. On the wave picture the original wave front is split into two coherent wave fronts by the slits, and these overlapping wave fronts produce the interference fringes on the screen that are so characteristic of wave phenomena. Suppose now that we replace the screen by a photoelectric surface. Measurements of where the photoelectrons are ejected from the surface yield a pattern corresponding to the double-slit intensity pattern, so the wavelike aspects of the radiation seem to be present. But if the energy and time distributions of the ejected photoelectrons are measured, we obtain evidence which shows that the radiation consists of photons, so the particlelike aspects will seem to be present. If we then think of radiation as photons whose motion is governed by the wave propagation properties of certain associated (de Broglie) waves, we are faced with another apparent paradox. Each photon must pass through either one slit or the other; if this is the case, how can its motion beyond the slits be influenced by the interaction of its associated waves with a slit through which it did not pass?

The fallacy in the paradox lies in the statement that each photon must pass through either one slit or the other. How can we actually determine experimentally whether a photon detected at the screen has gone through the upper or the lower of the two slits? To do this we would have to set up a detector at each slit, but the detector that interacts with the photon at a slit throws it out of the path that it would otherwise follow. We can show from the uncertainty principle that a detector with enough space resolution to determine through which slit the photon passes disturbs its momentum so much that the double-slit interference pattern is destroyed. In other words, if we do prove that each photon actually passes through one slit or the other, we shall no longer obtain the interference pattern. If we wish to observe the interference pattern, we must refrain from disturbing the photons and not try to observe them as particles along their paths to the screen. We can observe either the wave or the particle behavior of radiation; but the uncertainty principle prevents us from observing both together, and so this dual behavior is not really self-contradictory. The same is true of the wave-particle behavior of matter.

The uncertainty principle also makes it clear that the mechanics of quantum systems must necessarily be expressed in terms of probabilities. In classical mechanics, if at any instant we know exactly the position and momentum of each particle in an isolated system, then we can predict the exact behavior of the particles of the system for all future time. In quantum mechanics, however, the uncertainty principle shows us that it is impossible to do this for systems involving small distances and momenta because it is impossible to know, with the required accuracy, the instantaneous positions and momenta of the particles. As a result, we shall be able to make predictions only of the *probable* behavior of these particles.

**Example 3-7.** Consider a microscopic particle moving freely along the $x$ axis. Assume that at the instant $t = 0$ the position of the particle is measured and is uncertain by the amount $\Delta x_0$. Calculate the uncertainty in the measured position of the particle at some later time $t$.

The uncertainty in the momentum of the particle at $t = 0$ is at least

$$\Delta p_x = \hbar/2\Delta x_0$$

Therefore, the velocity of the particle at that instant is uncertain by at least

$$\Delta v_x = \Delta p_x/m = \hbar/2m\Delta x_0$$

and the distance $x$ travelled by the particle in the time $t$ cannot be known more accurately than within

$$\Delta x = t\Delta v_x = \hbar t/2m\Delta x_0$$

If by a measurement at $t = 0$ we have localized the particle within the range $\Delta x_0$, then in a measurement of its position at time $t$ the particle could be found anywhere within the range at least as large as $\Delta x$.

Note that $\Delta x$ is inversely proportional to $\Delta x_0$, so that the *more* carefully we localize the particle at the initial instant, the *less* we shall know about its final position. Also, the uncertainty $\Delta x$ increases linearly with time $t$. This corresponds to a spreading out, as time goes on, of the group of waves associated with the motion of the particle. ◀

## 3-6 The Philosophy of Quantum Theory

Although there is agreement by all physicists that quantum theory works in the sense that it predicts results that are in excellent agreement with experiment, there is a growing controversy over its philosophic foundation. Neils Bohr has been the principal architect of the present interpretation, known as the *Copenhagen interpretation*, of quantum mechanics. His approach is supported by the vast majority of theoretical physicists today. Nevertheless, a sizable body of physicists, not all in agreement with one another, questions the Copenhagen interpretation. The principal critic of this interpretation was Albert Einstein. The Einstein-Bohr debates are a fascinating part of the history of physics. Bohr felt that he had met every challenge that Einstein invented by way of thought experiments intended to refute the uncertainty principle,

Einstein finally conceded the logical consistency of the theory and its agreement with the experimental facts, but he remained unconvinced to the end that it represented the ultimate physical reality. "God does not play dice with the universe," he said, referring to the abandonment of strict causality and individual events by quantum theory in favor of a fundamentally statistical interpretation.

Heisenberg has stated the commonly accepted view succinctly: "We have not assumed that the quantum theory, as opposed to the classical theory, is essentially a statistical theory, in the sense that only statistical conclusions can be drawn from exact data . . . . In the formulation of the causal law, namely, 'If we know the present exactly, we can predict the future,' it is not the conclusion, but rather the premise which is false. We *cannot* know, as a matter of principle, the present in all its details."

Among the critics of the Bohr-Heisenberg view of a fundamental indeterminacy in physics is Louis de Broglie. In a forword to a book by David Bohm, a young colleague of Einstein's whose attempts at a new theory revived interest in reexamining the philosophic basis of quantum theory, de Broglie writes: "We can reasonably accept that the attitude adopted for nearly 30 years by theoretical quantum physicists is, at least in appearance, the exact counterpart of information which experiment has given us of the atomic world. At the level now reached by research in microphysics it is certain that the methods of measurement do not allow us to determine simultaneously all the magnitudes which would be necessary to obtain a picture of the classical type of corpuscles (this can be deduced from Heisenberg's uncertainty principle), and that the perturbations introduced by the measurement, which are impossible to eliminate, prevent us in general from predicting precisely the result which it will produce and allow only statistical predictions. The construction of purely probabilistic formulae that all theoreticians use today was thus completely justified. However, the majority of them, often under the influence of preconceived ideas derived from positivist doctrine, have thought that they could go further and assert that the uncertain and incomplete character of the knowledge that experiment at its present stage gives us about what really happens in microphysics is the result of a real indeterminacy of the physical states and of their evolution. Such an extrapolation does not appear in any way to be justified. It is possible that looking into the future to a deeper level of physical reality we will be able to interpret the laws of probability and quantum physics as being the statistical results of the development of completely determined values of variables which are at present hidden from us. It may be that the powerful means we are beginning to use to break up the structure of the nucleus and to make new particles appear will give us one day a direct knowledge which we do not now have at this deeper level. To try to stop all attempts to pass beyond the present viewpoint of quantum physics could be very dangerous for the progress of science and would furthermore be contrary to the lessons we may learn from the history of science. This teaches us, in effect, that the actual state of our knowledge is always provisional and that there must be, beyond what is actually known, immense new regions to discover." (From *Causality and Chance in Modern Physics* by David Bohm, © 1957 D. Bohm; reprinted by permission of D. Van Nostrand Co.)

The student should notice here the acceptance of the correctness of quantum mechanics at the atomic and nuclear level. The search for a deeper level, where quantum mechanics might be superseded, is motivated much more by objection to its philosophic indeterminism than by other considerations. According to Einstein, "The belief in an external world independent of the perceiving subject is the basis of all natural science." Quantum mechanics, however, regards the interactions of object and observer as the ultimate reality. It uses the language of physical relations and processes rather than that of physical qualities and properties. It rejects as meaningless and useless the notion that behind the universe of our perception there lies a hidden objective world ruled by causality; instead it confines itself to the description of the relations among perceptions. Nevertheless, there is a reluctance by many to give up attributing objective properties to elementary particles, say, and dealing instead with our subjective knowledge of them, and this motivates their search for a new theory. According to de Broglie, such a search is in the interest of science. Whether it will lead to a new theory that in some currently unexplored realm contradicts quantum theory and also alters its philosophic foundations, no one knows.

## QUESTIONS

1. Why is the wave nature of matter not more apparent to us in our daily observations?
2. Does the de Broglie wavelength apply only to "elementary particles" such as an electron or neutron, or does it apply as well to compound systems of matter having internal structure? Give examples.
3. If, in the de Broglie formula, we let $m \to \infty$, do we get the classical result for macroscopic particles?
4. Can the de Broglie wavelength of a particle be smaller than a linear dimension of the particle? Larger? Is there necessarily any relation between such quantities?
5. Is the frequency of a de Broglie wave given by $E/h$? Is the velocity given by $\lambda\nu$? Is the velocity equal to $c$? Explain.
6. Can we measure the frequency $\nu$ for de Broglie waves? If so, how?
7. How can electron diffraction be used to study properties of the surface of a solid?
8. How do we account for regularly reflected beams in diffraction experiments with electrons and atoms?
9. Does the Bragg formula have to be modified for electrons to account for the refraction of electron waves at the crystal surface?
10. Do electron diffraction experiments give different information about crystals than can be obtained from x-ray diffraction experiments? From neutron diffraction experiments? Discuss.
11. Could crystallographic studies be carried out with protons? With neutrons?
12. Discuss the analogy: physical optics is to geometrical optics as wave mechanics is to classical mechanics.
13. Is an electron a particle? Is it a wave? Explain.
14. Does the de Broglie wavelength associated with a particle depend on the motion of the reference frame of the observer? What effect does this have on the wave-particle duality?
15. Give examples of how the process of measurement disturbs the system being measured.
16. Show the relation between the uncontrollable nature of the Compton recoil in Bohr's $\gamma$-ray microscope experiment and the fact that there are four unknowns and only three conservation equations in the Compton effect.
17. The uncertainty principle is sometimes stated in terms of angular quantities as $\Delta L_\varphi \Delta\varphi \geq \hbar/2$ where $\Delta L_\varphi$ is the uncertainty in a component of *angular* momentum and $\Delta\varphi$ is the uncertainty in the corresponding *angular* position. In some quantum mechanical systems the angular momentum is measured to have a definite (quantized) magnitude. Does this contradict this statement of the uncertainty principle?
18. Argue from the Heisenberg uncertainty principle that the lowest energy of an oscillator cannot be zero.
19. Discuss similarities and differences between a matter wave and an electromagnetic wave.
20. Explain qualitatively the results of Example 3–7 that the uncertainty in position of a particle increases the more accurately we localize the particle initially and that the uncertainty increases with time.

**21.** Does the fact that interference occurs between various parts of the wave associated with a *single* particle (as in the G. I. Taylor experiments) simplify or complicate quantum physics?

**22.** Games of chance contain events which are ruled by statistics. Do such games violate the strict determination of individual events? Do they violate cause and effect?

**23.** According to operational philosophy, if we cannot prescribe a feasible operation for determining a physical quantity, the quantity should be given up as having no physical reality. What are the merits and drawbacks of this point of view in your opinion?

**24.** Bohm and de Broglie suggest that there may be hidden variables at a level deeper than quantum theory which are strictly determined. Draw an analogy to the relation between statistical mechanics and Newton's law of motion.

**25.** In your opinion is there an objective physical reality independent of our subjective sense impressions? How is this question answered by defenders of the Copenhagen interpretation? By critics of the Copenhagen interpretation?

**26.** Are our concepts limited in principle by our everyday experiences or is this only our conceptual starting point? How is this question related to a resolution of the wave-particle duality?

## PROBLEMS

**1.** A bullet of mass 40 g travels at 1000 m/sec. (a) What wavelength can we associate with it? (b) Why does the wave nature of the bullet not reveal itself through diffraction effects?

**2.** The wavelength of the yellow spectral emission of sodium is 5890 Å. At what kinetic energy would an electron have the same de Broglie wavelength?

**3.** An electron and a photon each have a wavelength of 2.0 Å. What are their (a) momenta and (b) total energies? (c) Compare the kinetic energies of the electron and the photon.

**4.** A thermal neutron has a kinetic energy $(3/2)kT$ where $T$ is room temperature, 300°K. Such neutrons are in thermal equilibrium with normal surroundings. (a) What is the energy in electron volts of a thermal neutron? (b) What is its de Broglie wavelength?

**5.** (a) Show that the de Broglie wavelength of a particle, of charge $e$, rest mass $m_0$, moving at relativistic speeds is given as a function of the accelerating potential $V$ as

$$\lambda = \frac{h}{\sqrt{2m_0eV}}\left(1 + \frac{eV}{2m_0c^2}\right)^{-1/2}$$

(b) Show how this agrees with $\lambda = h/p$ in the nonrelativistic limit.

**6.** Determine at what energy, in electron volts, the nonrelativistic expression for the de Broglie wavelength will be in error by 1% for (a) an electron and (b) a neutron. (Hint: See Problem 5.)

**7.** The 50-GeV (i.e., $50 \times 10^9$ eV) electron accelerator at Stanford University provides an electron beam of very short wavelength, suitable for probing the fine details of nuclear structure by scattering experiments. What is this wavelength and how does it compare to the size of an average nucleus? (Hint: At these energies it is simpler to use the extreme relativistic relationship between momentum and energy, namely $p = E/c$. This is the same relationship used for photons, and it is justified whenever the kinetic energy of a particle is very much greater than its rest energy $m_0c^2$, as in this case.)

**8.** Make a plot of de Broglie wavelength against kinetic energy for (a) electrons and (b) protons. Restrict the range of energy values to those in which classical mechanics applies reasonably well. A convenient criterion is that the maximum kinetic energy on each plot be only about, say, 5% of the rest energy $m_0c^2$ for the particular particle.

**9.** In the experiment of Davisson and Germer, (a) show that the second- and third-order diffracted beams, corresponding to the strong first maximum of Figure 3-2 cannot occur and (b) find the angle at which the first-order diffracted beam would occur if the accelerating potential were changed from 54 to 60 V? (c) What accelerating potential is needed to produce a second-order diffracted beam at 50°?

**10.** What is the wavelength of a hydrogen atom moving with a velocity corresponding to the mean kinetic energy for thermal equilibrium at 20°C?

**11.** The principal planar spacing in a potassium chloride crystal is 3.14 Å. Compare the angle for first-order Bragg reflection from these planes of electrons of kinetic energy 40 keV to that of 40 keV photons.

**12.** Electrons incident on a crystal suffer refraction due to an attractive potential of about 15 V that crystals present to electrons (due to the ions in the crystal lattice). If the angle of incidence of an electron beam is 45° and the electrons have an incident energy of 100 eV, what is the angle of refraction?

**13.** What accelerating voltage would be required for electrons in an electron microscope to obtain the same ultimate resolving power as that which could be obtained from a "$\gamma$-ray microscope" using 0.2 MeV $\gamma$ rays?

**14.** The highest achievable resolving power of a microscope is limited only by the wavelength used; that is, the smallest detail that can be separated is about equal to the wavelength. Suppose we wish to "see" inside an atom. Assuming the atom to have a diameter of 1.0 Å, this means that we wish to resolve detail of separation about 0.1 Å. (a) If an electron microscope is used, what minimum energy of electrons is needed? (b) If a photon microscope is used, what energy of photons is needed? In what region of the electromagnetic spectrum are these photons? (c) Which microscope seems more practical for this purpose? Explain.

**15.** Show that for a free particle the uncertainty relation can also be written as

$$\Delta\lambda\Delta x \geq \lambda^2/4\pi$$

where $\Delta x$ is the uncertainty in location of the wave and $\Delta\lambda$ the simultaneous uncertainty in wavelength.

**16.** If $\Delta\lambda/\lambda = 10^{-7}$ for a photon, what is the simultaneous value of $\Delta x$ for (a) $\lambda = 5.00 \times 10^{-4}$ Å ($\gamma$ ray)? (b) $\lambda = 5.00$ Å (x ray)? (c) $\lambda = 5000$ Å (light)?

**17.** In a repetition of Thomson's experiment for measuring $e/m$ for the electron, a beam of $10^4$ eV electrons is collimated by passage through a slit of width 0.50 mm. Why is the beamlike character of the emergent electrons not destroyed by diffraction of the electron wave at this slit?

**18.** A 1 MeV electron leaves a track in a cloud chamber. The track is a series of water droplets each about $10^{-5}$ m in diameter. Show, from the ratio of the uncertainty in transverse momentum to the momentum of the electron, that the electron path should not noticeably differ from a straight line.

**19.** Show that if the uncertainty in the location of a particle is about equal to its de Broglie wavelength, then the uncertainty in its velocity is about equal to its velocity.

**20.** (a) Show that the smallest possible uncertainty in the position of an electron whose speed is given by $\beta = v/c$ is

$$\Delta x_{\text{min}} = \frac{h}{4\pi m_0 c}(1 - \beta^2)^{1/2} = \frac{\lambda_C}{4\pi}\sqrt{1 - \beta^2}$$

where $\lambda_C$ is the Compton wavelength $h/m_0c$. (b) What is the meaning of this equation for $\beta = 0$? For $\beta = 1$?

**21.** A microscope using photons is employed to locate an electron in an atom to within a distance of 0.2 Å. What is the uncertainty in the velocity of the electron located in this way?

**22.** (a) Consider an electron whose position is somewhere in an atom of diameter 1 Å. What is the uncertainty in the electron's momentum? Is this consistent with the binding energy of electrons in atoms? (b) Imagine an electron to be somewhere in a nucleus of diameter $10^{-12}$ cm. What is the uncertainty in the electron's momentum? Is this consistent with the binding energy of nuclear constituents? (c) Consider now a neutron, or a proton, to be in such a nucleus. What is the uncertainty in the neutron's, or proton's, momentum? Is this consistent with the binding energy of nuclear constituents?

**23.** The lifetime of an excited state of a nucleus is usually about $10^{-12}$ sec. What is the uncertainty in energy of the $\gamma$-ray photon emitted?

**24.** Use relativistic expressions for total energy and momentum to verify that the group velocity $g$ of a matter wave equals the velocity $v$ of the associated particle.

**25.** The energy of a linear harmonic oscillator is $E = p_x^2/2m + Cx^2/2$. (a) Show, using the uncertainty relation, that this can be written as

$$E = \frac{h^2}{32\pi^2 m x^2} + \frac{Cx^2}{2}$$

(b) Then show that the minimum energy of the oscillator is $h\nu/2$ where

$$\nu = \frac{1}{2\pi}\sqrt{\frac{C}{m}}$$

is the oscillatory frequency. (Hint: This result depends on the $\Delta x \Delta p_x$ product achieving its limiting value $\hbar/2$. Find $E$ in terms of $\Delta x$ or $\Delta p_x$ as in part (a), then minimize $E$ with respect to $\Delta x$ or $\Delta p_x$ in part (b). Note that classically the minimum energy would be zero.)

**26.** A TV tube manufacturer is attempting to improve the picture resolution, while keeping costs down, by designing an electron gun that produces an electron beam which will make the smallest possible spot on the face of the tube, using only an electron emitting cathode followed by a system of two well-spaced apertures. (a) Show that there is an optimum diameter for the second aperture. (b) Using reasonable TV tube parameters, estimate the minimum possible spot size.

**27.** A boy on top of a ladder of height $H$ is dropping marbles of mass $m$ to the floor and trying to hit a crack in the floor. To aim, he is using equipment of the highest possible precision. (a) Show that the marbles will miss the crack by an average distance of the order of $(\hbar/m)^{1/2}(H/g)^{1/4}$, where $g$ is the acceleration due to gravity. (b) Using reasonable values of $H$ and $m$, evaluate this distance.

**28.** Show that in order to be able to determine through which slit of a double-slit system each photon passes without destroying the double-slit diffraction pattern, the condition $\Delta y \Delta p_y \ll \hbar/2$ must be satisfied. Since this condition violates the uncertainty principle, it cannot be met.

# 4

# Bohr's Model of the Atom

# Bohr's Model of the Atom

# 4

## 4-1 Thomson's Model

Much experimental evidence existed by 1910 that atoms contain electrons (e.g., scattering of x rays by atoms, photoelectric effect, etc.). Such experiments provided an estimate of $Z$, the number of electrons in an atom, as roughly equal to $A/2$, where $A$ is the chemical atomic weight of the atom in question. Since atoms are normally neutral, they must also contain positive charge equal in magnitude to the negative charge carried by their normal complement of electrons. Thus a neutral atom has a negative charge $-Ze$, where $-e$ is the electron charge, and also a positive charge of the same magnitude. That the mass of an electron is very small compared to the mass of even the lightest atom implies that most of the mass of the atom must be associated with the positive charge.

These considerations naturally led to the question of the distribution of the positive and negative charges within the atom. J. J. Thomson proposed a tentative description, or *model*, of an atom according to which the negatively charged electrons were located within a continuous distribution of positive charge. The positive charge distribution was assumed to be spherical in shape with a radius of the known order of magnitude of the radius of an atom, $10^{-10}$ m. (This value can be obtained from the density of a typical solid, its atomic weight, and Avogadro's number.) Owing to their mutual repulsion, the electrons would be uniformly distributed through the sphere of positive charge. Figure 4-1 illustrates this "plum pudding" model of the atom. In an atom in its lowest possible energy state, the electrons would be fixed at their equilibrium positions. In excited atoms (e.g., atoms in a material at high temperature), the electrons would vibrate about their equilibrium positions. Since classical electromagnetic theory predicts that an accelerated charged body, such as a vibrating electron, emits electromagnetic radiation, it was possible to understand qualitatively the emission of such radiation by excited atoms on the basis of Thomson's model. Quantitative agreement with experimentally observed spectra was lacking, however.

**Example 4-1.** (a) Assume that there is one electron of charge $-e$ inside a spherical region of uniform positive charge density $\rho$ (a Thomson hydrogen atom). Show that its motion, if it has kinetic energy, can be simple harmonic oscillation about the center of the sphere.

Let the electron be displaced to a distance $a$ from the center, with $a$ less than the radius of the sphere. From Gauss's law, we know that we can calculate the force on it by using Coulomb's law

$$F = -\frac{1}{4\pi\epsilon_0}\left(\frac{4}{3}\pi a^3 \rho\right)\frac{e}{a^2} = -\frac{\rho e a}{3\epsilon_0}$$

where $(4/3)\pi a^3 \rho$ is the net positive charge in a sphere of radius $a$. Hence, we can write $F = -ka$, where the constant $k = \rho e/3\epsilon_0$. If the electron at $a$ is freed with no initial velocity, this force will produce simple harmonic motion along a diameter of the sphere since it is always directed towards the center and has a strength which is proportional to the displacement from the center.

(b) Let the total positive charge have the magnitude of one electron charge (so that the atom has no net charge), and let it be distributed over a sphere of radius $r' = 1.0 \times 10^{-10}$ m.

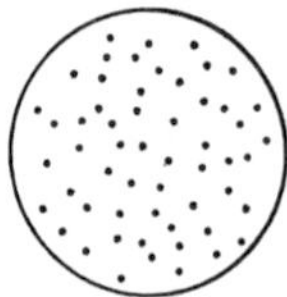

**FIGURE 4-1**
Thomson's model of the atom—a sphere of positive charge embedded with electrons.

Find the force constant $k$ and the frequency of the motion of the electron.

We have

$$\rho = \frac{e}{\frac{4}{3}\pi r'^3}$$

so that

$$k = \frac{\rho e}{3\epsilon_0} = \frac{e}{\frac{4}{3}\pi r'^3}\frac{e}{3\epsilon_0} = \frac{e^2}{4\pi\epsilon_0 r'^3}$$

$$= \frac{9.0 \times 10^9 \text{ nt-m}^2/\text{coul}^2 \times (1.6 \times 10^{-19} \text{ coul})^2}{(1.0 \times 10^{-10} \text{ m})^3} = 2.3 \times 10^2 \text{ nt/m}$$

The frequency of the simple harmonic motion is then

$$\nu = \frac{1}{2\pi}\sqrt{\frac{k}{m}} = \frac{1}{2\pi}\sqrt{\frac{2.3 \times 10^2 \text{ nt/m}}{9.11 \times 10^{-31} \text{ kg}}} = 2.5 \times 10^{15} \text{ sec}^{-1}$$

Since (in analogy to radiation emitted by electrons oscillating in an antenna) the radiation emitted by the atom will have this same frequency, it will correspond to a wavelength

$$\lambda = \frac{c}{\nu} = \frac{3.0 \times 10^8 \text{ m/sec}}{2.5 \times 10^{15}/\text{sec}} = 1.2 \times 10^{-7} \text{ m} = 1200 \text{ Å}$$

in the far ultraviolet portion of the electromagnetic spectrum. It is easy to show that an electron moving in a stable circular orbit of any radius inside the Thomson atom revolves at this same frequency, and so it would radiate at this frequency also.

Of course, a different assumed radius of the sphere of positive charge would give a different frequency. But the fact that a Thomson hydrogen atom has only one characteristic emission frequency conflicts with the very large number of different frequencies observed in the spectrum of hydrogen. ◀

Conclusive proof of the inadequacy of Thomson's model was obtained in 1911 by Ernest Rutherford, a former student of Thomson's, from the analysis of experiments on the scattering of $\alpha$ particles by atoms. Rutherford's analysis showed that, instead of being spread throughout the atom, the positive charge is concentrated in a very small region, or *nucleus*, at the center of the atom. This was one of the most important developments in atomic physics and was the foundation of the subject of nuclear physics.

Rutherford had already been awarded the Nobel Prize in 1908 for his "investigations in regard to the decay of elements and . . . the chemistry of radioactive substances." He was a talented, hard-working physicist with enormous drive and self-confidence. In a letter written later in life, the then Lord Rutherford wrote, "I've just been reading some of my early papers and, you know, when I'd finished, I said to myself, 'Rutherford, my boy, you used to be a damned clever fellow.' " Though pleased at winning a Nobel Prize he was not happy that it was a chemistry prize, rather than one in physics. (Any research in the elements was then

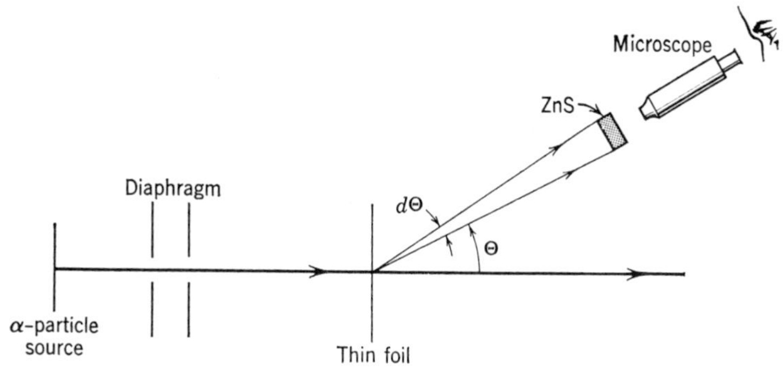


**FIGURE 4-2**

Arrangement of an α-particle scattering experiment. The region traversed by the α particles is evacuated.

considered chemistry.) In his speech accepting the prize he noted that he had observed many transformations in his work with radioactivity but never had seen one as rapid as his own, from physicist to chemist.

Rutherford already knew α particles to be doubly ionized helium atoms (i.e., He atoms with two electrons removed), emitted spontaneously from several radioactive materials at high speed. In Figure 4-2 we show a typical arrangement that he and his colleagues used to study the scattering of α particles on passing through thin foils of various substances. The radioactive source emits α particles which are collimated into a narrow parallel beam by a pair of diaphragms. The parallel beam is incident upon a foil of some substance, usually a metal. The foil is so thin that the particles pass completely through with only a small decrease in speed. In traversing the foil, however, each α particle experiences many small deflections due to the Coulomb force acting between its charge and the positive and negative charges of the atoms of the foil. Since the deflection of an α particle in passing through a single atom depends on the details of its trajectory through the atom, the net deflection in passing through the entire foil will be different for different α particles in the beam. As a result, the beam emerges from the foil not as a parallel beam but as a divergent beam. A quantitative measure of its divergence is found by measuring the number of α particles scattered into each angular range $\Theta$ to $\Theta + d\Theta$. The α particle detector consisted of a layer of the crystalline compound ZnS and a microscope. The crystal ZnS has the useful property of producing a small flash of light when struck by an α particle. If observed with a microscope, the flash due to the incidence of a *single* α particle can be distinguished. In the experiment an observer counts the number of light flashes produced per unit time as a function of the angular position of the detector.

Let $\mathcal{N}$ represent the number of atoms that deflect an α particle in its passage through the foil. If $\theta$ represents the angle of deflection in passing through one atom, as in Figure 4-3, and $\Theta$ is the net deflection in passing through all the atoms in its trajectory through the foil, then statistical theory shows that

$$(\overline{\Theta^2})^{1/2} = \sqrt{\mathcal{N}}\,(\overline{\theta^2})^{1/2} \tag{4-1}$$

Here $(\overline{\Theta^2})^{1/2}$ is the root mean square net deflection, or scattering, angle and $(\overline{\theta^2})^{1/2}$ is the root mean square scattering angle in a deflection from a single atom. The factor $\sqrt{\mathcal{N}}$ comes from the randomness of the deflection; if all deflections were in the same

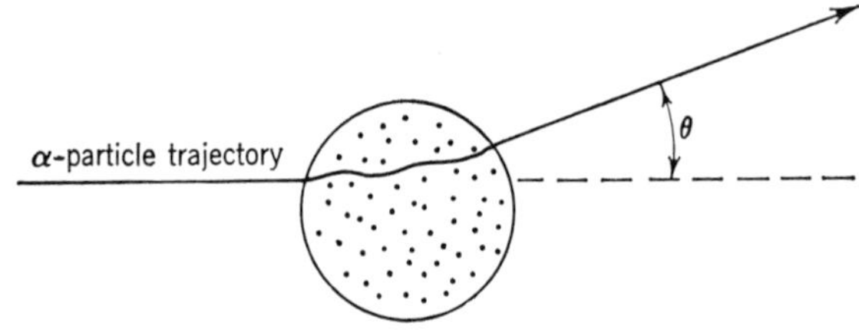


**FIGURE 4-3**

An $\alpha$ particle passing through a Thomson model atom. The angle $\theta$ specifies the deflection of the $\alpha$ particle.

direction, clearly we would obtain $\mathcal{N}$ instead of $\sqrt{\mathcal{N}}$. More generally, statistical theory gives the following angular distribution of the scattered $\alpha$ particles

$$N(\Theta)\, d\Theta = \frac{2I\Theta}{\overline{\Theta^2}} e^{-\Theta^2/\overline{\Theta^2}}\, d\Theta \tag{4-2}$$

where $N(\Theta)\, d\Theta$ is the number of $\alpha$'s scattered within the angular range $\Theta$ to $\Theta + d\Theta$, and $I$ is the number of $\alpha$'s passing through the foil.

Because electrons have a very small mass compared to the $\alpha$ particle, they can in any case produce only small $\alpha$-particle deflections; and because the positive charge is distributed over all the volume of the $r' \simeq 10^{-10}$ m radius Thomson atom it cannot provide a Coulomb repulsion intense enough to produce a large deflection of the $\alpha$ particle. Indeed, using Thomson's model we find that the deflection caused by one atom is $\theta \lessapprox 10^{-4}$ rad. This result and (4-1) and (4-2) comprise the $\alpha$-particle scattering predictions of the Thomson model of the atom. Rutherford and his group tested these predictions.

**Example 4-2.** (a) In a typical experiment (Geiger and Marsden, 1909), $\alpha$ particles were scattered by a gold foil $10^{-6}$ m thick. The average scattering angle was found to be $(\overline{\Theta^2})^{1/2} \simeq 1° \simeq 2 \times 10^{-2}$ rad. Calculate $(\overline{\theta^2})^{1/2}$.

The number of atoms traversed by the $\alpha$ particle is approximately equal to the thickness of the foil divided by the diameter of the atom. Hence

$$\mathcal{N} \simeq 10^{-6}\text{ m}/10^{-10}\text{ m} = 10^4$$

The average deflection angle in traversing a single atom then, from (4-1), is

$$(\overline{\theta^2})^{1/2} = \frac{(\overline{\Theta^2})^{1/2}}{\sqrt{\mathcal{N}}} \simeq \frac{2 \times 10^{-2}}{10^2} \simeq 2 \times 10^{-4}\text{ rad}$$

not in disagreement with the Thomson atom estimate $\theta \lessapprox 10^{-4}$ rad.

(b) More than 99% of the $\alpha$ particles were scattered at angles less than 3°. The measurements, using 1° for $(\overline{\Theta^2})^{1/2}$, were in agreement with (4-2) for $N(\Theta)\, d\Theta$ for angles $\Theta$ in this range; but the angular distribution of the small number of particles scattered at larger angles was in marked disagreement with (4-2). It was found, for example, that the fraction of $\alpha$'s scattered at angles greater than 90°, $N(\Theta > 90°)/I$, was about $10^{-4}$. What does (4-2) predict?

We have

$$\frac{N(\Theta > 90°)}{I} = \frac{\int_{90°}^{180°} N(\Theta)\, d\Theta}{I} = e^{-(90)^2} = 10^{-3500}$$

a strikingly different result than the experiment value of $10^{-4}$.

In general the number of scattered $\alpha$ particles was observed to be *very* much larger than the predicted number for all scattering angles greater than a few degrees. ◀

The existence of a small, but nonzero probability for scattering at large angles could not be explained at all in terms of Thomson's model of the atom, which basically involves *small* angle scattering from *many* atoms. To scientists accustomed to thinking in terms of this model it came as a great surprise that some $\alpha$ particles were deflected through very large angles, up to 180°. In Rutherford's words: "It was quite the most incredible event that ever happened to me in my life. It was as incredible as if you fired a 15-inch shell at a piece of tissue paper and it came back and hit you."

Experiments using foils of various thicknesses showed that the number of large angle scatterings was proportional to $\mathcal{N}$, the number of atoms traversed by the $\alpha$ particle. This is just the dependence on $\mathcal{N}$ that would arise if there were a small probability that an $\alpha$ particle could be scattered through a *large* angle in traversing a *single* atom. That cannot happen in Thomson's model of the atom, and this led Rutherford in 1911 to propose a new model.

## 4-2 Rutherford's Model

In Rutherford's model of the structure of the atom, all the positive charge of the atom, and consequently essentially all its mass, are assumed to be concentrated in a small region in the center called the *nucleus*. If the dimensions of the nucleus are small enough, an $\alpha$ particle passing very near it can be scattered by a strong Coulomb repulsion through a large angle in the traversal of a *single* atom. If, instead of using $r' = 10^{-10}$ m for the radius of the positive charge distribution of the Thomson atom, which leads to a maximum deflection angle $\theta \simeq 10^{-4}$ rad, we ask what the radius $r'$ of a nucleus should be to obtain $\theta \simeq 1$ rad, say, we find $r' = 10^{-14}$ m. This, as we shall see, turns out to be a good estimate of the radius of the atomic nucleus.

Rutherford made a detailed calculation of the angular distribution to be expected for the scattering of $\alpha$ particles from atoms of the type proposed in his model. The calculation was concerned only with scattering at angles greater than several degrees. Hence, scattering due to atomic electrons can be ignored. The scattering is then due to the repulsive Coulomb force acting between the positively charged $\alpha$ particle and the positively charged nucleus. Furthermore, the calculation considered only the scattering from heavy atoms, to permit the assumption that the mass of the nucleus is so large compared to that of the $\alpha$ particle that the nucleus does not recoil appreciably (remains fixed in space) during the scattering process. It was also assumed that the $\alpha$ particle does not actually penetrate the nuclear region, so that the particle and the nucleus (both assumed to be spherical) act like point charges as far as the Coulomb force is concerned. We shall see later that all these assumptions are quite valid except for the scattering of $\alpha$ particles from the lighter nuclei, and we can correct for the finite nuclear mass in such cases. The calculation, finally, uses nonrelativistic mechanics, since $v/c \simeq 1/20$.

Figure 4-4 illustrates the scattering of an $\alpha$ particle, of charge $+ze$ and mass $M$, in passing near a nucleus of charge $+Ze$. The nucleus is fixed at the origin of the coordinate system. When the particle is very far from the nucleus, the Coulomb force on it is negligible so that the particle approaches the nucleus along a straight line with constant speed $v$. After the scattering, the particle will move off finally along a straight line again with constant speed $v'$. The position of the particle relative to the nucleus is specified by the radial coordinate $r$ and the polar angle $\varphi$, with the latter measured from an axis drawn parallel to the initial trajectory line. The perpendicular distance

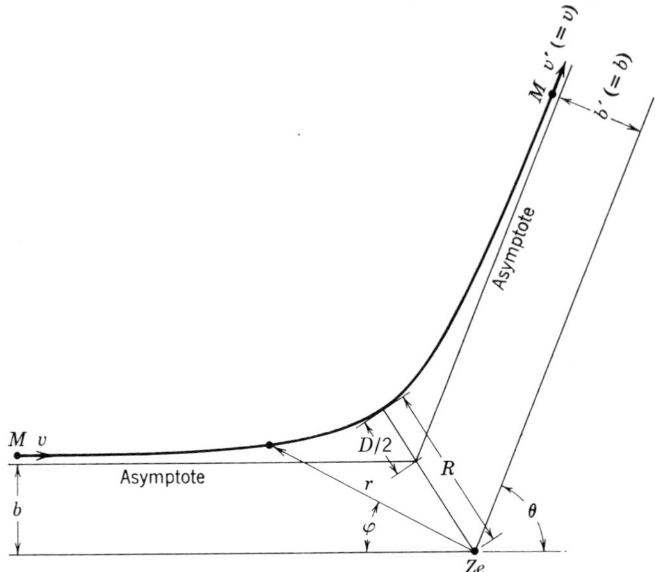


**FIGURE 4-4**
The hyperbolic Rutherford trajectory, showing the polar coordinates $r$, $\varphi$ and the parameters $b$, $D$. These two parameters completely determine the trajectory, in particular the scattering angle $\theta$ and the distance of closest approach $R$. The nuclear point charge $Ze$ lies at a focus of the branch of the hyperbola.

from that axis to the line of initial motion is called the *impact parameter*, specified by $b$. The scattering angle $\theta$ is just the angle between the axis and a line drawn through the origin parallel to the line of final motion; the perpendicular distance between these two lines is $b'$.

**Example 4-3.** Show that $v = v'$ and $b = b'$.

The force acting on the particle, being a Coulomb force, is always in the radial direction. Hence, the angular momentum of the particle about the origin has a constant value, $L$. Specifically then, the initial angular momentum is equal to the final angular momentum, or

$$Mvb = Mv'b' = L$$

Of course, the kinetic energy of the particle does not remain constant during the scattering, but the initial kinetic energy must be equal to the final kinetic energy since the nucleus is assumed to remain stationary. Thus

$$\frac{1}{2}Mv^2 = \frac{1}{2}Mv'^2$$

Therefore, $v = v'$ and so from the previous equation $b = b'$, as drawn in Figure 4-4. ◀

By a straightforward calculation of classical mechanics, using the repulsive Coulomb force $(1/4\pi\epsilon_0)(zZe^2/r^2)$, we can obtain the following equation for the trajectory of the $\alpha$ particle (see Appendix D for a derivation)

$$\frac{1}{r} = \frac{1}{b}\sin\varphi + \frac{D}{2b^2}(\cos\varphi - 1) \tag{4-3}$$

the equation of a hyperbola in polar coordinates. Here $D$ is a constant, defined by

$$D \equiv \frac{1}{4\pi\epsilon_0}\frac{zZe^2}{Mv^2/2} \tag{4-4}$$

It is a convenient parameter equal to the *distance of closest approach* to the nucleus in a *head-on* collision ($b = 0$), since $D$ is the distance at which the potential energy $(1/4\pi\epsilon_0)(zZe^2/D)$ is equal to the initial kinetic energy $Mv^2/2$ (simply equate the two and solve for $D$). At this point the particle would come to a stop and then reverse its direction of motion. The scattering angle $\theta$ follows from (4-3) by finding the value of $\varphi$ as $r \to \infty$ and setting $\theta = \pi - \varphi$. In this way we find

$$\cot\frac{\theta}{2} = \frac{2b}{D} \tag{4-5}$$

**Example 4-4.** Evaluate $R$, the distance of closest approach of the particle to the center of the nucleus (the origin in Figure 4-4).

The radial coordinate $r$ will equal $R$ when the polar angle is $\varphi = (\pi - \theta)/2$. Evaluating (4-3) for this angle, we get

$$\frac{1}{R} = \frac{1}{b}\sin\left(\frac{\pi - \theta}{2}\right) + \frac{D}{2b^2}\left[\cos\left(\frac{\pi - \theta}{2}\right) - 1\right]$$

Now, from (4-5) we can put

$$b = \frac{D}{2}\cot\frac{\theta}{2} = \frac{D}{2}\tan\left(\frac{\pi - \theta}{2}\right)$$

and, after some manipulation, obtain

$$R = \frac{D}{2}\left[1 + \frac{1}{\cos\left(\frac{\pi - \theta}{2}\right)}\right]$$

or

$$R = \frac{D}{2}\left[1 + \frac{1}{\sin(\theta/2)}\right] \tag{4-6}$$

This result can be checked physically. Note that as $\theta \to \pi$, corresponding to $b = 0$ or a head-on collision, $R \to D$, the distance of closest approach. Also, as $\theta \to 0$, corresponding to no deflection at all, both $b$ and $R$ go to infinity, as would be expected. ◀

From (4-5) we see that, in the scattering of an $\alpha$ particle by a single nucleus, if the impact parameter is in the range $b$ to $b + db$ then the scattering angle is in the range $\theta$ to $\theta + d\theta$, where the relation between $b$ and $\theta$ is given by the equation. This is illustrated in Figure 4-5. The problem of calculating the number $N(\Theta)\,d\Theta$ of $\alpha$ particles scattered into the angular range $\Theta$ to $\Theta + d\Theta$ in traversing the entire foil is therefore equivalent to the problem of calculating the number which are incident, with impact parameter from $b$ to $b + db$, upon the nuclei in the foil. As we show in the following example, the result is

$$N(\Theta)\,d\Theta = \left(\frac{1}{4\pi\epsilon_0}\right)^2\left(\frac{zZe^2}{2Mv^2}\right)^2\frac{I\rho t 2\pi\sin\Theta\,d\Theta}{\sin^4(\Theta/2)} \tag{4-7}$$

where $I$ is the number of $\alpha$ particles incident on a foil of thickness $t$ cm containing $\rho$ nuclei per cubic centimeter.

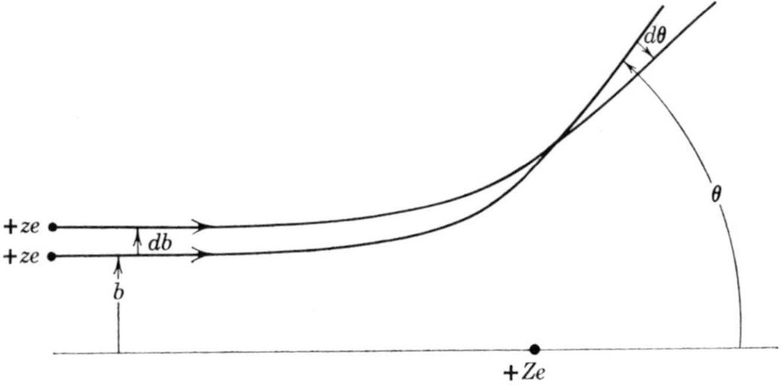


**FIGURE 4-5**

The relation between the impact parameter $b$ and the scattering angle $\theta$. As $b$ increases (less close nuclear approach) the angle $\theta$ decreases (smaller scattering angle). The $\alpha$ particles with impact parameters between $b$ and $b + db$ are scattered into the angular range between $\theta$ and $\theta + d\theta$.

**Example 4-5.** Verify (4-7).

Consider a segment of the foil with a cross-sectional area of 1 cm$^2$, as shown in Figure 4-6. A ring, of inner radius $b$ and outer radius $b + db$, is drawn around an incident axis passing through each nucleus, the area of each ring being $2\pi b\, db$. The number of such rings in this segment of the foil is $\rho t$. The probability that an $\alpha$ particle will pass through one of these rings, $P(b)\, db$, is equal to the total area obscured by the rings, as seen by the incident $\alpha$ particles, divided by the total area of the segment. We assume the foil to be thin enough that we can ignore overlapping of rings from different nuclei. The process involves *single scattering* and the probability for appreciable scattering by more than one nucleus is very low. Hence

$$P(b)\, db = \rho t 2\pi b\, db$$

but $b = (D/2) \cot(\theta/2)$ so that

$$db = -\frac{D}{2}\frac{d\theta/2}{\sin^2(\theta/2)}$$

and

$$b\, db = -\frac{D^2}{8}\frac{\cos(\theta/2)\, d\theta}{\sin^3(\theta/2)} = -\frac{D^2}{16}\frac{\sin\theta\, d\theta}{\sin^4(\theta/2)}$$

Thus

$$P(b)\, db = -\frac{\pi}{8}\rho t D^2 \sin\theta \frac{d\theta}{\sin^4(\theta/2)}$$

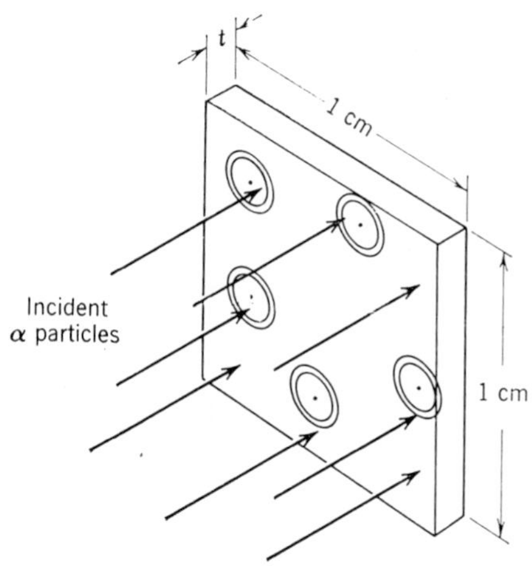


**FIGURE 4.6**

A beam of $\alpha$ particles incident on a foil of 1 cm$^2$ area and thickness $t$ cm. The rings, which are purely geometrical constructs and not anything physical, are centered on nuclei. Actually there are enormously many more rings than shown and the rings are very much smaller than shown.

But $-P(b)\,db$ is equal to the probability that the incident particles will be scattered into the angular range $\theta$ to $\theta + d\theta$. The minus sign arises from the fact that a decrease in $b$, i.e., $-db$, corresponds to an increase in $\theta$, i.e., $+d\theta$. Using our earlier notation $\Theta$ for the scattering angle in passing through the entire foil, this is

$$\frac{N(\Theta)\,d\Theta}{I} = -P(b)\,db = \frac{\pi}{8}\,\rho t D^2 \frac{\sin\Theta\,d\Theta}{\sin^4(\Theta/2)}$$

Finally, with $D = (1/4\pi\epsilon_0)zZe^2/(Mv^2/2)$, we obtain (4-7). ◀

If we compare the Rutherford atom result, (4-7), to the Thomson atom result, (4-2), we see that although the angular factor decreases rapidly with increasing angle in both, the decrease is very much less rapid for Rutherford's prediction. Large angle scattering is very much more probable in single scattering from a nuclear atom than in multiple small angle scattering from a plum pudding atom. Detailed experimental tests of (4-7) were performed within a few months of its derivation by Geiger and Marsden, with the following results:

1. The angular dependence was tested, using foils of Ag and Au, over the angular range 5° to 150°. Although $N(\Theta)\,d\Theta$ varies by a factor of about $10^5$ over this range, the experimental data remained proportional to the theoretical angular distribution to within a few percent.

2. The quantity $N(\Theta)\,d\Theta$ was found indeed to be proportional to the thickness $t$ of the foil for a range of about 10 in thickness for all the elements investigated.

3. Equation (4-7) predicts that the number of scattered $\alpha$'s will be inversely proportional to the square of their kinetic energy, $Mv^2/2$. This was tested by using $\alpha$ particles from several different radioactive sources and the predicted energy dependence was confirmed experimentally over an available energy variation of about a factor of 3.

4. Finally, the equation predicts $N(\Theta)\,d\Theta$ to be proportional to $(Ze)^2$, the square of the nuclear charge. At the time $Z$ was not known for the various atoms. Assuming (4-7) to be valid, the experiment was used to determine $Z$ and it was found that $Z$ was equal to the chemical *atomic number* of the target atoms. This implied that the first atom, H, in the periodic table contains one electron, the second atom, He, contains two electrons, the third atom, Li, contains three, etc., since $Z$ is also the number of electrons in the neutral atom. This result was soon independently confirmed by x-ray techniques that will be discussed in Chapter 9.

Rutherford, his model now confirmed, was able to put limits on the size of the nucleus. The distance of closest approach, $D$, is the smallest value that $R$ takes on, which is $R$ at $\Theta = 180°$. Hence

$$R_{180^\circ} = D = \frac{1}{4\pi\epsilon_0}\frac{zZe^2}{Mv^2/2}$$

The nucleus radius must be no larger than $D$ because the results are based on the assumption that the force acting on the $\alpha$ particle is always strictly a Coulomb force between two point charges. This assumption would not be true if the particle penetrated the nuclear region at its distance of closest approach. The previous equation shows that $R_{180^\circ}$ decreases as $Z$ decreases. The question arises: How much can $R_{180^\circ}$ decrease before $R_{180^\circ}$ is less than the nuclear radius? Departures from the predicted Rutherford scattering were actually observed from the very light (low $Z$) nuclei. Part of this was due to a violation, for the very light nuclei, of the assumption that the nuclear mass is large compared to the alpha particle mass; however, deviations remained even after the finite nuclear mass was taken into account in the theory. This suggests that penetration of the nucleus occurs in these cases thereby altering the

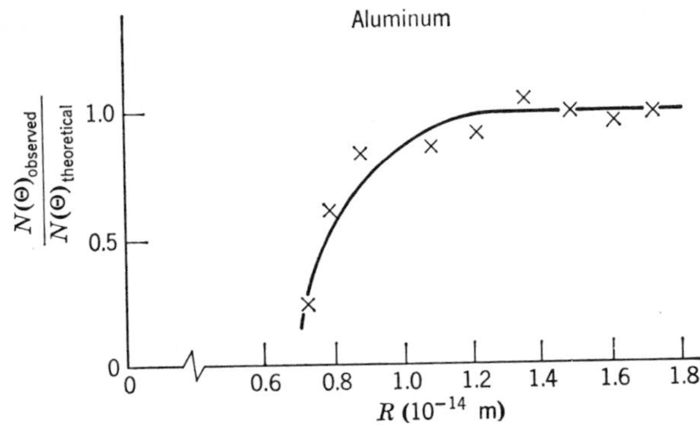


**FIGURE 4-7**
Some data obtained in the scattering of $\alpha$ particles from a radioactive source by aluminum. The abscissa is the distance of closest approach to the nuclear center.

predicted scattering. Hence, the nuclear radius can be defined as the value of $R$ at the limiting scattering angle, or limiting incident energy, at which deviations from Rutherford scattering set in. In Figure 4-7, for example, we show data from Rutherford's group for the scattering of $\alpha$ particles, of various energies, at a fixed large angle from an Al foil. The ordinate is the ratio of the observed number of scattered particles to the number predicted by the Rutherford theory (corrected for the finite nuclear mass). The abscissa is the distance of closest approach calculated from (4-6). These data imply that the radius of the Al nucleus is about $10^{-14}$ m $= 10$ F. (The unit of distance used in nuclear physics is the *fermi*, which equals $10^{-15}$ m. Note that 1 F $=$ $10^{-5}$ Å, where Å, the angstrom, is the unit used in atomic physics.)

The Rutherford scattering formula, (4-7), is usually expressed in terms of a *differential cross section* $d\sigma/d\Omega$. This quantity is defined so that the number $dN$ of $\alpha$ particles scattered into a solid angle $d\Omega$ at scattering angle $\Theta$ is

$$dN = \frac{d\sigma}{d\Omega} In \, d\Omega \tag{4-8}$$

if $I$ $\alpha$ particles are incident on a target foil containing $n$ nuclei per square centimeter. The definition is analogous to the definition of a cross section $\sigma$ in (2-18)

$$N = \sigma In$$

It is illustrated in Figure 4-8. The *solid angle* $d\Omega$, which is essentially a two-dimensional angular range, is measured numerically by the area which the angular range includes on a sphere of unit radius centered where the scatterings occur. For Rutherford scattering, which is symmetric about the axis of the incident beam, we are interested in the solid angle $d\Omega$ corresponding to all events in which the scattering angle lies in the range $d\Theta$ at $\Theta$. As is shown in the figure

$$d\Omega = 2\pi \sin \Theta \, d\Theta$$

Using this in (4-7), writing $N(\Theta) \, d\Theta$ in that equation as $dN$, and also writing the term $\rho t$ appearing there as $n$, we immediately obtain

$$dN = \left(\frac{1}{4\pi\epsilon_0}\right)^2 \left(\frac{zZe^2}{2Mv^2}\right)^2 \frac{1}{\sin^4(\Theta/2)} In \, d\Omega$$

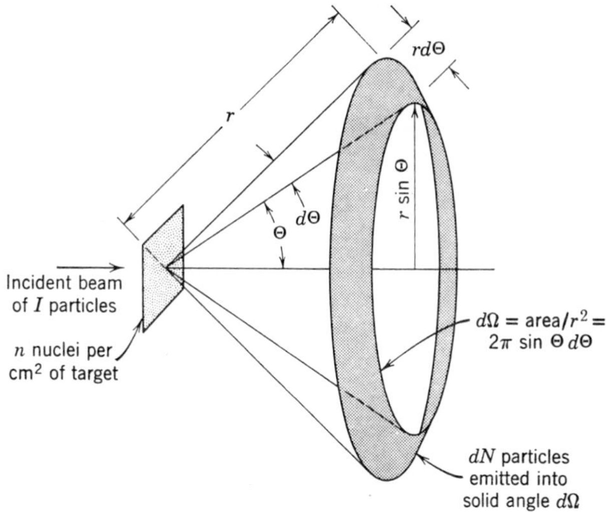


**FIGURE 4-8**
Illustrating the definition of the differential cross section $d\sigma/d\Omega$. If the target is thin enough for an incident particle to have negligible chance of interacting with more than one nucleus while passing through the target, then $dN = (d\sigma/d\Omega)In\,d\Omega$.

Comparison with the definition of (4-8) then shows that the *Rutherford scattering differential cross section* is

$$\frac{d\sigma}{d\Omega} = \left(\frac{1}{4\pi\epsilon_0}\right)^2\left(\frac{zZe^2}{2Mv^2}\right)^2\frac{1}{\sin^4(\Theta/2)} \tag{4-9}$$

## 4-3 The Stability of the Nuclear Atom

The detailed experimental verification of the predictions of Rutherford's nuclear model of the atom left little room for doubt concerning the validity of the model. At the center of the atom is a nucleus whose mass is approximately that of the entire atom and whose charge is equal to the atomic number $Z$ times $e$; around this nucleus there exist $Z$ electrons, neutralizing the atom as a whole. But serious questions emerge about the *stability* of such an atom. If we assume, for example, that the electrons in the atom are stationary, there exists no stable arrangement of the electrons which would prevent the electrons from falling into the nucleus under the influence of its Coulomb attraction. We cannot allow the atom to collapse (back to a nuclear-sized plum pudding) because then its radius would be of the order of a nuclear radius, which is four orders of magnitude smaller than diverse experiments show the radius of the atom to be.

At first glance it seems that we can simply allow the electrons to circulate about the nucleus in orbits similar to the orbits of the planets circulating about the sun. Such a system can be stable mechanically, as is the solar system. A serious difficulty arises, however, in trying to carry over this idea from the planetary system to the atomic system. The problem is that the charged electrons would be constantly accelerating in their motion around the nucleus and, according to classical electromagnetic theory, all accelerating charged bodies radiate energy in the form of electromagnetic radiation (see Appendix B). The energy would be emitted at the expense of the mechanical energy of the electron, and the electron would spiral into the nucleus. Again we have an atom which would rapidly collapse to nuclear dimensions. (For an atom of diameter $10^{-10}$ m the time of collapse can be computed to be $\simeq 10^{-12}$ sec!) Furthermore, the continuous spectrum of the radiation that would be emitted in this process is not in agreement with the discrete spectrum which is known to be emitted by atoms.

This difficult problem of the stability of atoms actually led to a simple model of atomic structure. A key feature of this very successful model, proposed by Niels Bohr in 1913, was the prediction of the spectrum of radiation emitted by certain atoms. Hence, it is appropriate at this point to describe some of the principal features of such spectra.

## 4-4 Atomic Spectra

A typical apparatus used in the measurement of atomic spectra is indicated in Figure 4-9. The source consists of an electric discharge passing through a region containing a monatomic gas. Owing to collisions with electrons, and with each other, some of the atoms in the discharge are put into a state in which their total energy is greater than it is in a normal atom. In returning to their normal energy state, the atoms give up their excess energy by emitting electromagnetic radiation. The radiation is collimated by the slit and then it passes through a prism (or diffraction grating for better resolution) where it is broken up into its wavelength spectrum which is recorded on the photographic plate.

The nature of the observed spectra is indicated on the photographic plate. In contrast to the continuous spectrum of electromagnetic radiation emitted, for instance, from the surface of solids at high temperature, *the electromagnetic radiation emitted by free atoms is concentrated at a number of discrete wavelengths*. Each of these wavelength components is called a *line* because of the line (image of the slit) which it produces on the photographic plate. Investigation of the spectra emitted from different kinds of atoms shows that each kind of atom has its own characteristic spectrum, i.e., a characteristic set of wavelengths at which the lines of the spectrum are found. This feature is of greatest practical importance because it makes *spectroscopy* a very useful addition to the usual techniques of chemical analysis. Chiefly for this

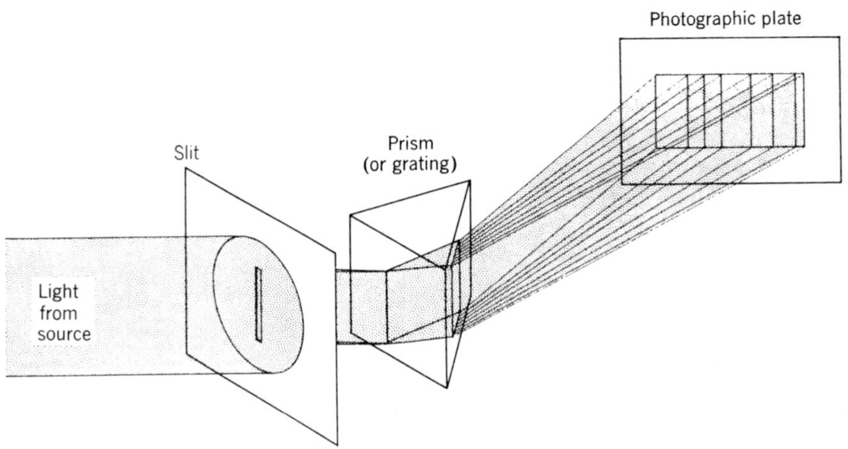


**FIGURE 4-9**
Schematic of an apparatus used to measure atomic spectra.

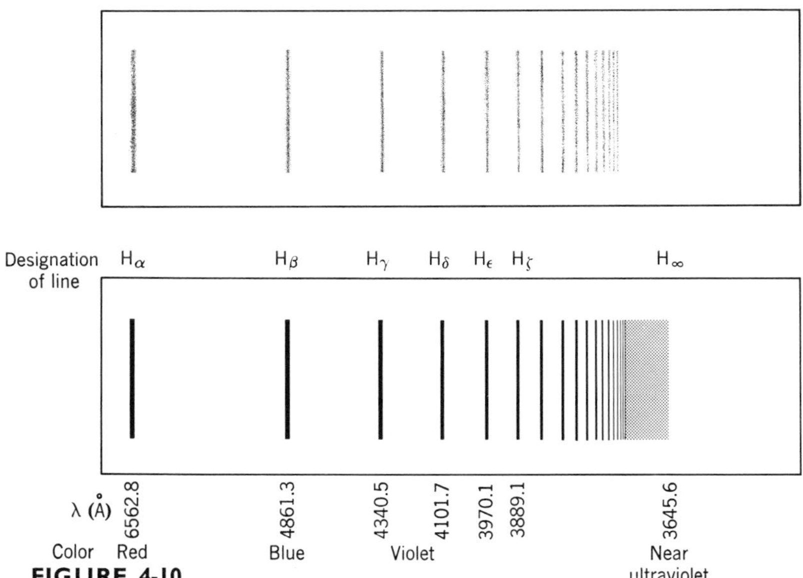


**FIGURE 4-10**

*Top:* A photograph of the visible part of the hydrogen spectrum. *Bottom:* A schematic of same, with the lines labeled.

reason much effort was devoted to the accurate measurement of atomic spectra, and, in fact, much effort was needed because the spectra consist of many hundreds of lines and in general are very complicated.

However, the spectrum of hydrogen is relatively simple. This is perhaps not surprising since hydrogen, which contains just one electron, is itself the simplest atom. Most of the universe consists of isolated hydrogen atoms so that the hydrogen spectrum is of considerable practical interest. There are historical and theoretical reasons as well for studying it, as will become apparent later. Figure 4-10 represents that part of the atomic hydrogen spectrum which falls approximately within the wavelength range of visible light. We see that the spacing, in wavelengths, between adjacent lines of the spectrum continuously decreases with decreasing wavelength of the lines, so that the *series of lines* converges to the so-called *series limit* at 3645.6 Å. The short wavelength lines, including the series limit, are hard to observe experimentally because of their close spacing and because they are in the ultraviolet.

The obvious regularity of the H spectrum tempted several people to look for an empirical formula which would represent the wavelength of the lines. Such a formula was discovered in 1885 by Balmer. He found that the simple equation

$$\lambda = 3646 \frac{n^2}{n^2 - 4} \qquad \text{(in Å units)}$$

where $n = 3$ for $H_\alpha$, $n = 4$ for $H_\beta$, $n = 5$ for $H_\gamma$, etc., was able to predict the wavelength of the first nine lines of the series, which were all that were known at the time, to better than one part in 1000. This discovery initiated a search for similar empirical formulas that would apply to series of lines which can sometimes be identified in the complicated distribution of lines that constitute the spectra of other elements. Most of this work was done around 1890 by Rydberg, who found it convenient to deal with the reciprocal of the wavelength of the lines, instead of their wavelength. In terms of reciprocal wavelength $\kappa$ the *Balmer formula* can be written

$$\kappa = 1/\lambda = R_H(1/2^2 - 1/n^2) \qquad n = 3, 4, 5, \ldots \quad (4\text{-}10)$$

where $R_H$ is the so-called *Rydberg constant* for hydrogen. From recent spectroscopic data, its value is known to be

$$R_H = 10967757.6 \pm 1.2 \text{ m}^{-1}$$

This indicates the accuracy possible in spectroscopic measurements.

Formulas of this type were found for a number of series. For instance, we now know of the existence of five series of lines in the hydrogen spectrum, as shown in Table 4-1.

For alkali element atoms (Li, Na, K, . . .) the series formulas are of the same general structure. That is

$$\kappa = \frac{1}{\lambda} = R\left[\frac{1}{(m - a)^2} - \frac{1}{(n - b)^2}\right] \quad (4\text{-}11)$$

where $R$ is the Rydberg constant for the particular element, $a$ and $b$ are constants for the particular series, $m$ is an integer which is fixed for the particular series, and $n$ is a variable integer. To within about 0.05% the Rydberg constant has the same value for all elements, although it does show a very slight systematic increase with increasing atomic weight.

We have been discussing the *emission spectrum* of an atom. A closely related property is the *absorption spectrum*. This may be measured with apparatus similar to that shown in Figure 4-9 except that a source emitting a continuous spectrum is used and a glass-walled cell, containing the monatomic gas to be investigated, is inserted somewhere between the source and the prism. After exposure and development, the photographic plate is found to be darkened everywhere except for a number of unexposed lines. These lines represent a set of discrete wavelength components which were missing from the otherwise continuous spectrum incident upon the prism, and which must have been absorbed by the atoms in the gas cell. It is observed that for every line in the absorption spectrum of an element there is a corresponding (same wavelength) line in its emission spectrum; however, the reverse is not true. Only certain emission lines show up in the absorption spectrum. For hydrogen gas, normally

**TABLE 4-1.** The Hydrogen Series

| Names | Wavelength Ranges | Formulas | |
|---|---|---|---|
| Lyman | Ultraviolet | $\kappa = R_H\left(\frac{1}{1^2} - \frac{1}{n^2}\right)$ | $n = 2, 3, 4, \ldots$ |
| Balmer | Near ultraviolet and visible | $\kappa = R_H\left(\frac{1}{2^2} - \frac{1}{n^2}\right)$ | $n = 3, 4, 5, \ldots$ |
| Paschen | Infrared | $\kappa = R_H\left(\frac{1}{3^2} - \frac{1}{n^2}\right)$ | $n = 4, 5, 6, \ldots$ |
| Brackett | Infrared | $\kappa = R_H\left(\frac{1}{4^2} - \frac{1}{n^2}\right)$ | $n = 5, 6, 7, \ldots$ |
| Pfund | Infrared | $\kappa = R_H\left(\frac{1}{5^2} - \frac{1}{n^2}\right)$ | $n = 6, 7, 8, \ldots$ |

only lines corresponding to the Lyman series appear in the absorption spectrum; but, when the gas is at very high temperatures, e.g., at the surface of a star, lines corresponding to the Balmer series are found.

## 4-5 Bohr's Postulates

All these features of atomic spectra, and many more which we have not discussed, must be explained by any successful model of atomic structure. Furthermore, the very great precision of spectroscopic measurements imposes severe requirements on the accuracy with which such a model must be able to predict the quantitative features of the spectra.

Nevertheless, in 1913 Niels Bohr developed a model which was in accurate quantitative agreement with certain of the spectroscopic data (e.g., the hydrogen spectrum). It had the additional attraction that the mathematics involved was very easy to understand. Although the student has probably seen something of Bohr's model in studying elementary physics, or chemistry, we shall consider it in detail here in order to obtain various results that we shall want to make comparisons with elsewhere in this book, and also in order to take a careful look at the rather confusing postulates on which the model is based. These postulates are:

1. *An electron in an atom moves in a circular orbit about the nucleus under the influence of the Coulomb attraction between the electron and the nucleus, obeying the laws of classical mechanics.*

2. *Instead of the infinity of orbits which would be possible in classical mechanics, it is only possible for an electron to move in an orbit for which its orbital angular momentum $L$ is an integral multiple of $\hbar$, Planck's constant divided by $2\pi$.*

3. *Despite the fact that it is constantly accelerating, an electron moving in such an allowed orbit does not radiate electromagnetic energy. Thus, its total energy $E$ remains constant.*

4. *Electromagnetic radiation is emitted if an electron, initially moving in an orbit of total energy $E_i$, discontinuously changes its motion so that it moves in an orbit of total energy $E_f$. The frequency of the emitted radiation $\nu$ is equal to the quantity $(E_i - E_f)$ divided by Planck's constant $h$.*

The first postulate bases Bohr's model on the existence of the atomic nucleus. The second postulate introduces quantization. Note the difference, however, between Bohr's quantization of the *orbital angular momentum* of an atomic electron moving under the influence of an *inverse square (Coulomb) force*

$$L = n\hbar \qquad n = 1, 2, 3, \ldots \tag{4-12}$$

and Planck's quantization of the *energy* of a particle, such as an electron, executing simple harmonic motion under the influence of a *harmonic restoring force:* $E = nh\nu$, $n = 0, 1, 2, \ldots$. We shall see in the next section that the quantization of the orbital angular momentum of the atomic electron does lead to the quantization of its total energy, but with an energy quantization equation which is different from Planck's equation. The third postulate removes the problem of the stability of an electron moving in a circular orbit, due to the emission of the electromagnetic radiation required of the electron by classical theory, by simply postulating that this particular feature of the classical theory is not valid for the case of an atomic electron. The postulate was based on the fact that *atoms are observed by experiment to be stable*—even though this is not predicted by the classical theory. The fourth postulate

$$\nu = \frac{E_i - E_f}{h} \tag{4-13}$$

is really just Einstein's postulate that the frequency of a photon of electromagnetic radiation is equal to the energy carried by the photon divided by Planck's constant.

These postulates do a thorough job of mixing classical and nonclassical physics. The electron moving in a circular orbit is assumed to obey classical mechanics, and yet the nonclassical idea of quantization of orbital angular momentum is included. The electron is assumed to obey one feature of classical electromagnetic theory (Coulomb's law), and yet not to obey another feature (emission of radiation by an accelerated charged body). However, we should not be surprised if the laws of classical physics, which are based on our experience with macroscopic systems, are not completely valid when dealing with microscopic systems such as the atom.

## 4-6 Bohr's Model

The justification of Bohr's postulates, or of any set of postulates, can be found only by comparing the predictions that can be derived from the postulates with the results of experiment. In this section we derive some of these predictions and compare them with the data of Section 4-4.

Consider an atom consisting of a nucleus of charge $+Ze$ and mass $M$, and a single electron of charge $-e$ and mass $m$. For a neutral hydrogen atom $Z = 1$, for a singly ionized helium atom $Z = 2$, for a doubly ionized lithium atom $Z = 3$, etc. We assume that the electron revolves in a circular orbit about the nucleus. Initially we suppose the mass of the electron to be completely negligible compared to the mass of the nucleus, and consequently assume that the nucleus remains fixed in space. The condition of mechanical stability of the electron is

$$\frac{1}{4\pi\epsilon_0}\frac{Ze^2}{r^2} = m\frac{v^2}{r} \tag{4-14}$$

where $v$ is the speed of the electron in its orbit, and $r$ is the radius of the orbit. The left side of this equation is the Coulomb force acting on the electron, and the right side is $ma$, where $a$ is the centripetal acceleration keeping the electron in its circular orbit. Now, the orbital angular momentum of the electron, $L = mvr$, must be a constant, because the force acting on the electron is entirely in the radial direction. Applying the quantization condition, (4-12), to $L$, we have

$$mvr = n\hbar \qquad n = 1, 2, 3, \ldots \tag{4-15}$$

Solving for $v$ and substituting into (4-14), we obtain

$$Ze^2 = 4\pi\epsilon_0 mv^2 r = 4\pi\epsilon_0 mr\left(\frac{n\hbar}{mr}\right)^2 = 4\pi\epsilon_0\frac{n^2\hbar^2}{mr}$$

so

$$r = 4\pi\epsilon_0\frac{n^2\hbar^2}{mZe^2} \qquad n = 1, 2, 3, \ldots \tag{4-16}$$

and

$$v = \frac{n\hbar}{mr} = \frac{1}{4\pi\epsilon_0}\frac{Ze^2}{n\hbar} \qquad n = 1, 2, 3 \ldots \tag{4-17}$$

The application of the angular momentum quantization condition has restricted the possible circular orbits to those of radii given by (4-16). Note that these radii are proportional to the square of the *quantum number n*. If we evaluate the radius of the smallest orbit ($n = 1$) for a hydrogen atom ($Z = 1$) by inserting the known values of $h$, $m$, and $e$, we obtain $r = 5.3 \times 10^{-11}$ m $\simeq 0.5$ Å. We shall show later that the electron has its minimum total energy when in the orbit corresponding to $n = 1$. Consequently we may interpret the radius of this orbit as a measure of the radius of a hydrogen atom in its normal state. It is in good agreement with the estimate, mentioned previously, that the order of magnitude of an atomic radius is 1 Å. Hence, Bohr's postulates predict a reasonable size for the atom. Evaluating the orbital velocity of an electron in the smallest orbit of a hydrogen atom from (4-17), we find $v = 2.2 \times 10^6$ m/sec. It is apparent from the equation that this is the largest velocity possible for a hydrogen atom electron. The fact that this velocity is less than 1% of the velocity of light is the justification for using classical mechanics instead of relativistic mechanics in the Bohr model. On the other hand, (4-17) shows that for large values of $Z$ the electron velocity becomes relativistic; the model could not be applied in such cases. That equation also makes it apparent why Bohr could not allow the quantum number $n$ ever to assume the value $n = 0$, as it may in Planck's quantization equation.

Next we calculate the total energy of an atomic electron moving in one of the allowed orbits. Let us define the potential energy to be zero when the electron is infinitely distant from the nucleus. Then the potential energy $V$ at any finite distance $r$ can be obtained by integrating the work that would be done by the Coulomb force acting from $r$ to $\infty$. Thus

$$V = -\int_r^\infty \frac{Ze^2}{4\pi\epsilon_0 r^2}\,dr = -\frac{Ze^2}{4\pi\epsilon_0 r}$$

The potential energy is negative because the Coulomb force is attractive; it takes work to move the electron from $r$ to infinity against this force. The kinetic energy of the electron, $K$, can be evaluated, with the aid of (4-14), to be

$$K = \frac{1}{2}mv^2 = \frac{Ze^2}{4\pi\epsilon_0 2r}$$

The total energy of the electron, $E$, is then

$$E = K + V = -\frac{Ze^2}{4\pi\epsilon_0 2r} = -K$$

Using (4-16) for $r$ in the preceding equation, we have

$$E = -\frac{mZ^2e^4}{(4\pi\epsilon_0)^2 2\hbar^2}\frac{1}{n^2} \qquad n = 1, 2, 3, \ldots \qquad (4\text{-}18)$$

We see that *the quantization of the orbital angular momentum of the electron leads to a quantization of its total energy.*

The information contained in (4-18) is presented as an energy-level diagram in Figure 4-11. The energy of each level, as evaluated from (4-18), is shown on the left, in terms of joules and electron volts, and the quantum number of the level is shown on the right. The diagram is so constructed that the distance from any level to the level of zero energy is proportional to the energy of that level. Note that the lowest (most negative) allowed value of total energy occurs for the smallest quantum number $n = 1$. As $n$ increases, the total energy of the quantum state becomes less negative, with $E$ approaching zero as $n$ approaches infinity. Since the state of lowest total energy is, of course, the most stable state for the electron, we see that the normal state of the electron in a one-electron atom is the state for which $n = 1$.

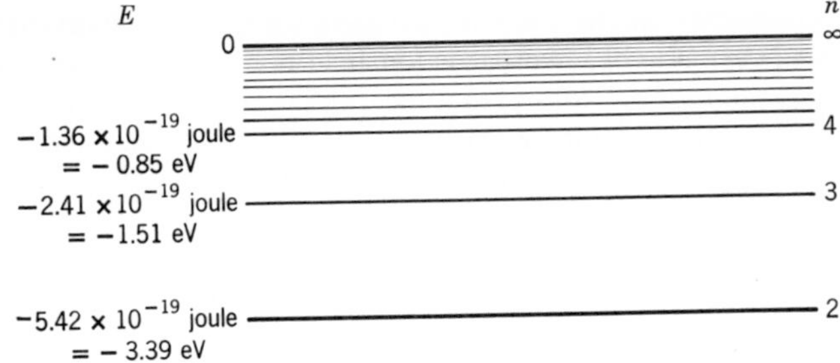


**FIGURE 4-11**
An energy-level diagram for the hydrogen atom.

**Example 4-6.** Calculate the binding energy of the hydrogen atom (the energy binding the electron to the nucleus) from (4-18).

The binding energy is numerically equal to the energy of the lowest state in Figure 4-11, corresponding to $n = 1$ in (4-18). This yields, with $Z = 1$

$$E = -\left(\frac{1}{4\pi\epsilon_0}\right)^2 \frac{me^4}{2\hbar^2}$$

$$= -\frac{(9.0 \times 10^9 \text{ nt-m}^2/\text{coul}^2)^2 \times 9.11 \times 10^{-31} \text{ kg} \times (1.60 \times 10^{-19} \text{ coul})^4}{2 \times (1.05 \times 10^{-34} \text{ joule-sec})^2}$$

$$= -2.17 \times 10^{-18} \text{ joule} = -13.6 \text{ eV}$$

which agrees very well with the experimentally observed binding energy for hydrogen. ◀

Next we calculate the frequency $\nu$ of the electromagnetic radiation emitted when the electron makes a transition from the quantum state $n_i$ to the quantum state $n_f$, that is, when an electron initially moving in an orbit characterized by the quantum number $n_i$ discontinuously changes its motion so that it moves in an orbit characterized by quantum number $n_f$. Using Bohr's fourth postulate (4-13), and (4-18), we have

$$\nu = \frac{E_i - E_f}{h} = +\left(\frac{1}{4\pi\epsilon_0}\right)^2 \frac{mZ^2e^4}{4\pi\hbar^3}\left(\frac{1}{n_f^2} - \frac{1}{n_i^2}\right)$$

In terms of the reciprocal wavelength $\kappa = 1/\lambda = \nu/c$, this is

$$\kappa = \left(\frac{1}{4\pi\epsilon_0}\right)^2 \frac{me^4}{4\pi\hbar^3 c} Z^2\left(\frac{1}{n_f^2} - \frac{1}{n_i^2}\right)$$

or

$$\kappa = R_\infty Z^2\left(\frac{1}{n_f^2} - \frac{1}{n_i^2}\right) \qquad \text{where } R_\infty \equiv \left(\frac{1}{4\pi\epsilon_0}\right)^2 \frac{me^4}{4\pi\hbar^3 c} \tag{4-19}$$

and where $n_i$ and $n_f$ are integers.

The essential predictions of the Bohr model are contained in (4-18) and (4-19). Let us first discuss the emission of electromagnetic radiation by a one-electron Bohr atom in terms of these equations:

1. The normal state of the atom will be the state in which the electron has the lowest energy, i.e., the state $n = 1$. This is called the *ground state*. (Ground state means fundamental state, the term originating from the German word *grund*, meaning fundamental.)

2. In an electric discharge, or in some other process, the atom receives energy due to collisions, etc. This means that the electron must make a transition to a state of higher energy, or *excited state*, in which $n > 1$.

3. Obeying the common tendency of all physical systems, the atom will emit its excess energy and return to the ground state. This is accomplished by a series of transitions in which the electron drops to excited states of successively lower energy, finally reaching the ground state. In each transition electromagnetic radiation is emitted with a wavelength which depends on the energy lost by the electron, i.e., on the initial and final quantum numbers. In a typical case, the electron might be excited into state $n = 7$ and drop successively through the states $n = 4$ and $n = 2$ to the ground state $n = 1$. Three lines of the atomic spectrum are emitted with reciprocal wavelengths given by (4-19) for $n_i = 7$ and $n_f = 4$, $n_i = 4$ and $n_f = 2$, and $n_i = 2$ and $n_f = 1$.

4. In the very large number of excitation and deexcitation processes which take place during a measurement of an atomic spectrum, all possible transitions occur and the complete spectrum is emitted. The reciprocal wavelengths, or wavelengths, of the set of lines which constitute the spectrum are given by (4-19), where we allow $n_i$ and $n_f$ to take on all possible integral values subject only to the restriction that $n_i > n_f$.

For hydrogen ($Z = 1$) let us consider the subset of spectral lines which arises from transitions in which $n_f = 2$. According to (4-19) the reciprocal wavelengths of these lines are given by

$$\kappa = R_\infty(1/n_f^2 - 1/n_i^2) \qquad n_f = 2 \text{ and } n_i > n_f$$

or

$$\kappa = R_\infty(1/2^2 - 1/n^2) \qquad n = 3, 4, 5, 6, \ldots$$

This is identical with the series formula for the Balmer series of the hydrogen spectrum (4-10), if $R_\infty$ is equal to $R_{\rm H}$. According to the Bohr model

$$R_\infty = \left(\frac{1}{4\pi\epsilon_0}\right)^2 \frac{me^4}{4\pi\hbar^3 c}$$

Although the numerical values of some of the quantities entering into this equation were not very accurately known at the time, Bohr evaluated $R_\infty$ in terms of these quantities and found that the resulting value was in quite good agreement with the experimental value of $R_{\rm H}$. In the next section we shall make a detailed comparison, using recent data, between the experimental value of $R_{\rm H}$ and Bohr's prediction, and we shall show that the two agree almost perfectly.

According to the Bohr model, each of the five known series of the hydrogen spectrum arises from a subset of transitions in which the electron goes to a certain final quantum state $n_f$. For the Lyman series $n_f = 1$; for the Balmer $n_f = 2$; for the Paschen $n_f = 3$; for the Brackett $n_f = 4$; and for the Pfund $n_f = 5$. The first three of these series are conveniently illustrated in terms of the energy-level diagram of Figure 4-12. The transition giving rise to a particular line of a series is indicated in this diagram by an arrow going from the initial quantum state $n_i$ to the final quantum state $n_f$. Only the arrows corresponding to the first few lines of each series and to the series limit are

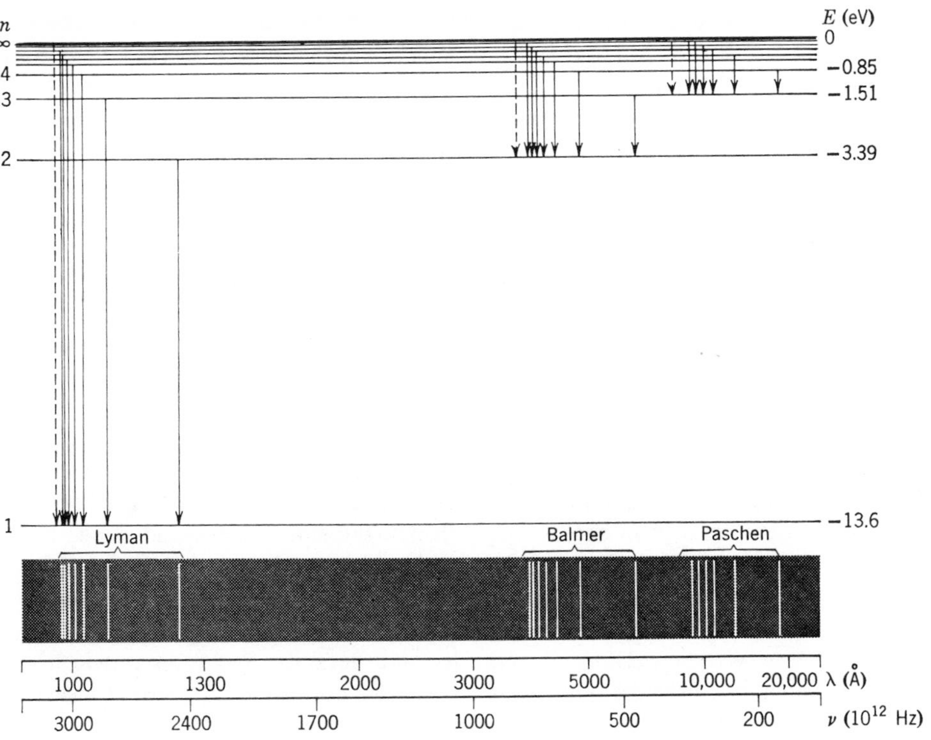


**FIGURE 4-12**

*Top:* The energy-level diagram for hydrogen with the quantum number $n$ for each level and some of the transitions that appear in the spectrum. An infinite number of levels is crowded in between the levels marked $n = 4$ and $n = \infty$. *Bottom:* The corresponding spectral lines for the three series indicated. Within each series the spectral lines follow a regular pattern, approaching the series limit at the shortwave end of the series. As drawn here, neither the wavelength nor frequency scales is linear, being chosen as they are merely for clarity of illustration. A linear wavelength scale would more nearly represent the actual appearance of the photographic plate obtained from a spectroscope. The Brackett and Pfund series, which are not shown, lie in the far infared part of the spectrum.

shown. Since the distance between any two energy levels in such a diagram is proportional to the difference between the energy of the two levels, and since (4-13) states that the frequency $\nu$ (or reciprocal wavelength) is proportional to the energy difference, the length of any arrow is proportional to the frequency (or reciprocal wavelength) for the corresponding spectral line.

The wavelengths of the lines of all these series are fitted very accurately by (4-19) by using the appropriate value of $n_f$. This was a great triumph for Bohr's model. The success of the model was particularly impressive because the Lyman, Brackett, and Pfund series had not been discovered at the time the model was developed by Bohr. The existence of these series was predicted, and the series were soon found experimentally by the persons after whom they are named.

The model worked equally well when applied to the case of one-electron atoms with $Z = 2$, i.e., singly ionized helium atoms $He^+$. Such atoms can be produced by passing a particularly violent electric discharge (a spark) through normal helium gas. They make their presence apparent by emitting a simpler spectrum than that emitted by normal helium atoms. In fact, the atomic spectrum of $He^+$ is exactly the same as the hydrogen spectrum except that the reciprocal wavelengths of all the lines are almost exactly four times as great. This is explained very easily, in terms of the Bohr model, by setting $Z^2 = 4$ in (4-19).

The properties of the absorption spectrum of one-electron atoms are also easy to understand in terms of the Bohr model. Since the atomic electron must have a total energy exactly equal to the energy of one of the allowed energy states, the atom can only absorb discrete amounts of energy from the incident electromagnetic radiation. This fact leads to the idea that we consider the incident radiation to be a beam of photons, and that only those photons can be absorbed whose frequency is given by $E = h\nu$, where $E$ is one of the discrete amounts of energy which can be absorbed by the atom. The process of absorbing electromagnetic radiation is then just the inverse of the normal emission process, and the lines of the absorption spectrum will have exactly the same wavelengths as the lines of the emission spectrum. Normally the atom is always initially in the ground state $n = 1$, so that only absorption processes from $n = 1$ to $n > 1$ can occur. Thus, only the absorption lines which correspond (for hydrogen) to the Lyman series will normally be observed. However, if the gas containing the absorbing atoms is at a very high temperature, then, owing to collisions, some of the atoms will initially be in the first excited state $n = 2$, and absorption lines corresponding to the Balmer series will be observed.

**Example 4-7.** Estimate the temperature of a gas containing hydrogen atoms at which the Balmer series lines will be observed in the absorption spectrum.

The Boltzmann probability distribution (see Appendix C) shows that the ratio of the number $n_2$ of atoms in the first excited state to the number $n_1$ of atoms in the ground state, in a large sample in equilibrium at temperature $T$, is

$$\frac{n_2}{n_1} = \frac{e^{-E_2/kT}}{e^{-E_1/kT}}$$

where $k$ is the Boltzmann's constant, $k = 1.38 \times 10^{-23}$ joule/°K $= 8.62 \times 10^{-5}$ eV/°K. For hydrogen atoms the energies of these two states are given in the energy-level diagram of Figure 4-11: $E_1 = -13.6$ eV, $E_2 = -3.39$ eV. Hence

$$\frac{n_2}{n_1} = e^{-(-3.39+13.6)\text{ eV}/(8.62\times10^{-5}\text{ eV/°K})T} = e^{-1.18\times10^{5}\text{ °K}/T}$$

Therefore, a significant fraction of the hydrogen atoms will initially be in the first excited state only when $T$ is of the order of, or greater than, $10^5$ °K; and only when they absorb from that state can they produce absorption lines of the Balmer series.

Balmer absorption lines are actually observed in the hydrogen gas of some stellar atmospheres. This gives us a way of estimating the temperature of the surface of a star. ◀

## 4-7 Correction for Finite Nuclear Mass

In the previous section we assumed the mass of the atomic nucleus to be infinitely large compared to the mass of the atomic electron, so that the nucleus remains fixed in space. This is a good approximation even for hydrogen, which contains the lightest nucleus, since the mass of that nucleus is about 2000 times larger than the electron mass. However, the spectroscopic data are so very accurate that before we make a detailed numerical comparison of these data with the Bohr model we must take into account the fact that the nuclear mass is actually finite. In such a case the electron and the nucleus move about their common center of mass. However, it is not difficult to show that in such a planetarylike system the electron moves relative to the nucleus as though the nucleus were fixed and the mass $m$ of the electron were slightly reduced to the value $\mu$, the *reduced mass* of the system. The equations of motion of the system are the same as those we have considered if we simply substitute $\mu$ for $m$, where

$$\mu = \frac{mM}{m + M} \tag{4-20}$$

is less than $m$ by a factor $1/(1 + m/M)$. Here $M$ is the mass of the nucleus.

To handle this situation Bohr modified his second postulate to require that *the total orbital angular momentum of the atom, L, is an integral multiple of Planck's constant divided by* $2\pi$. This is achieved by generalizing (4-15) to

$$\mu v r = n\hbar \qquad n = 1, 2, 3, \ldots \tag{4-21}$$

Using $\mu$ instead of $m$ in this equation takes into account the angular momentum of the nucleus as well as that of the electron. Making similar modifications to the rest of Bohr's derivation for the case of finite nuclear mass, we find that all the equations are identical with those derived before, except that the electron mass $m$ is replaced by the reduced mass $\mu$. In particular, the formula for the reciprocal wavelengths of the spectral lines becomes

$$\kappa = R_M Z^2\left(\frac{1}{n_f^2} - \frac{1}{n_i^2}\right) \qquad \text{where } R_M \equiv \frac{M}{m + M} R_\infty \equiv \frac{\mu}{m} R_\infty \tag{4-22}$$

The quantity $R_M$ is the *Rydberg constant for a nucleus of mass M*. As $M/m \to \infty$, it is apparent that $R_M \to R_\infty$, the Rydberg constant for an infinitely heavy nucleus which appears in (4-19). In general, the Rydberg constant $R_M$ is less than $R_\infty$ by the factor $1/(1 + m/M)$. For the most extreme case of hydrogen, $M/m = 1836$ and $R_M$ is less than $R_\infty$ by about one part in 2000.

If we evaluate $R_H$ from (4-22), using the currently accepted values of the quantities $m$, $M$, $e$, $c$, and $h$, we find $R_H = 10968100\ \text{m}^{-1}$. Comparing this with the experimental value of $R_H$ given in Section 4-4, we see that the Bohr model, corrected for finite nuclear mass, agrees with the spectroscopic data to within three parts in 100,000!

**Example 4-8.** In Chapter 2 we spoke of the *positronium* "atom," consisting of a positron and an electron revolving about their common center of mass, which lies halfway between them.

(a) If such a system were a normal atom, how would its emission spectrum compare to that of the hydrogen atom?

In this case the "nuclear" mass $M$ is that of the positron, which equals $m$, the mass of the electron. Hence, the reduced mass (4-20) is

$$\mu = \frac{mM}{m + M} = \frac{m^2}{2m} = \frac{m}{2}$$

The corresponding Rydberg constant $R_M$ is, according to (4-22)

$$R_M = \frac{m}{m + m} R_\infty = \frac{R_\infty}{2}$$

The energy states of the positronium atom then would be given by

$$E_{\text{positronium}} = -\frac{R_M hcZ^2}{n^2} = -\frac{R_\infty hcZ^2}{2n^2}$$

and the reciprocal wavelengths of the emitted spectral lines by

$$\kappa = \frac{1}{\lambda} = \frac{\nu}{c} = \frac{R_\infty}{2} Z^2\left(\frac{1}{n_f^2} - \frac{1}{n_i^2}\right)$$

The frequencies of the emitted lines would then be half, and the wavelengths double, that of a hydrogen atom (with infinitely heavy nucleus), $Z$ being equal to one for positronium and for hydrogen.

(b) What would be the radius of the ground state orbit of positronium?

In (4-16) we merely replace $m$ by $\mu = m/2$ and we find

$$r_{\text{positronium}} = \frac{4\pi\epsilon_0 n^2\hbar^2}{\mu Z e^2} = 2\,\frac{4\pi\epsilon_0 n^2\hbar^2}{mZe^2} = 2r_{\text{hydrogen}}$$

Hence, for any quantum state $n$ the radius of the electron relative to the "nucleus" is twice as great in the positronium atom as in the hydrogen atom (with infinitely heavy nucleus). ◀

**Example 4-9.** A *muonic* atom contains a nucleus of charge $Ze$ and a negative *muon*, $\mu^-$, moving about it. The $\mu^-$ is an elementary particle with charge $-e$ and a mass that is 207 times as large as an electron mass. Such an atom is formed when a proton, or some other nucleus, captures a $\mu^-$.

(a) Calculate the radius of the first Bohr orbit of a muonic atom with $Z = 1$.

The reduced mass of the system, with $m_{\mu^-} = 207m_e$ and $M = 1836m_e$, is, from (4-20)

$$\mu = \frac{207m_e \times 1836m_e}{207m_e + 1836m_e} = 186m_e$$

Then, from (4-16), with $n = 1, Z = 1$, and $m = 186m_e$, we obtain

$$r_1 = \frac{4\pi\epsilon_0\hbar^2}{186m_e e^2} = \frac{1}{186} \times 5.3 \times 10^{-11}\text{ m} = 2.8 \times 10^{-13}\text{ m} = 2.8 \times 10^{-3}\text{ Å}$$

Therefore the $\mu^-$ is much closer to the nuclear (proton) surface than is the electron in a hydrogen atom. It is this feature which makes such muonic atoms interesting, information about nuclear properties being revealed from their study.

(b) Calculate the binding energy of a muonic atom with $Z = 1$.

From (4-18), with $Z = 1$, $n = 1$, and $m = \mu = 186m_e$, we have

$$E = -186\frac{m_e e^4}{(4\pi\epsilon_0)^2 2\hbar^2} = -186 \times 13.6\text{ eV} = -2530\text{ eV}$$

as the ground state energy. Hence, the binding energy is 2530 eV.

(c) What is the wavelength of the first line in the Lyman series for such an atom?

From (4-22), with $Z = 1$, we have

$$\kappa = R_M\left(\frac{1}{n_f^2} - \frac{1}{n_i^2}\right)$$

For the first Lyman line, $n_i = 2$ and $n_f = 1$. In this case, $R_M = (\mu/m_e)R_\infty = 186R_\infty$. Hence

$$\kappa = \frac{1}{\lambda} = 186R_\infty\left(1 - \frac{1}{4}\right) = 139.5R_\infty$$

With $R_\infty = 109737\text{ cm}^{-1}$ we obtain

$$\lambda \simeq 6.5\text{ Å}$$

so that the Lyman lines lie in the x-ray part of the spectrum. X-ray techniques are necessary, therefore, to study the spectrum of muonic atoms. ◀

**Example 4-10.** Ordinary hydrogen contains about one part in 6000 of *deuterium*, or heavy hydrogen. This is a hydrogen atom whose nucleus contains a proton *and* a neutron. How does the doubled nuclear mass affect the atomic spectrum?

The spectrum would be identical if it were not for the correction for finite nuclear mass. For a normal hydrogen atom

$$R_{\rm H} = R_\infty \frac{\mu}{m} = \frac{R_\infty}{\left(1 + \dfrac{m}{M}\right)} = \frac{109737 \text{ cm}^{-1}}{\left(1 + \dfrac{1}{1836}\right)} = 109678 \text{ cm}^{-1}$$

For an atom of heavy hydrogen, or deuterium

$$R_{\rm D} = R_\infty \frac{\mu}{m} = \frac{R_\infty}{\left(1 + \dfrac{m}{M}\right)} = \frac{109737 \text{ cm}^{-1}}{\left(1 + \dfrac{1}{2 \times 1836}\right)} = 109707 \text{ cm}^{-1}$$

Hence, $R_{\rm D}$ is a bit larger than $R_{\rm H}$, so that the spectral lines of the deuterium atom are shifted to slightly shorter wavelengths compared to hydrogen.

Indeed, deuterium was discovered in 1932 by H. C. Urey following the observation of these shifted spectral lines. By increasing the concentration of the heavy isotope above its normal value in a hydrogen discharge tube, we now can enhance the intensity of the deuterium lines which, ordinarily, are difficult to detect. We then readily observe pairs of hydrogen lines; the shorter wavelength members of the pair correspond exactly to those predicted from $R_{\rm D}$ earlier. The resolution needed is easily obtained, the $H_\alpha$-line pair being separated by about 1.8 Å, for example, several thousand times greater than the minimum resolvable separation. ◀

## 4-8 Atomic Energy States

The Bohr model predicts that the total energy of an atomic electron is quantized. For example, (4-18) gives the allowed energy values for the electron in a one-electron atom. Although we have not attempted to derive similar expressions for the electrons in a multielectron atom, it is clear that according to the model the total energy of each of the electrons will also be quantized and, consequently, that the same must be true of the atom's total energy content. The Planck theory of blackbody radiation had also predicted that in the process of emission and absorption of radiation, the atoms in the cavity wall behaved as though they had quantized energy states. Hence, according to the old quantum theory every atom can have only certain discretely separated energy states.

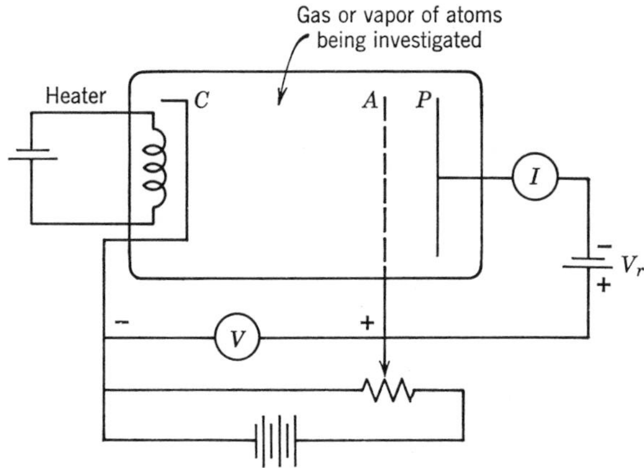


**FIGURE 4-13**

Schematic of the apparatus used by Franck and Hertz to prove that atomic energy states are quantized.

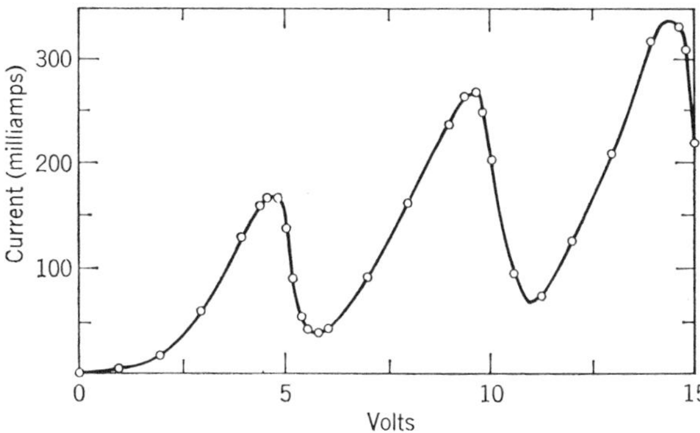


**FIGURE 4-14**
The voltage dependence of the current measured in the Franck-Hertz experiment.

Direct confirmation that the internal energy states of an atom are quantized came from a simple experiment performed by Franck and Hertz in 1914. The type of apparatus used by these investigators is indicated in Figure 4-13. Electrons are emitted thermally at low energy from the heated cathode $C$. They are accelerated to the anode $A$ by a potential $V$ applied between the two electrodes. Some of the electrons pass through holes in $A$ and travel to plate $P$, providing their kinetic energy upon leaving $A$ is enough to overcome a small retarding potential $V_r$ applied between $P$ and $A$. The entire tube is filled at a low pressure with a gas or vapor of the atoms to be investigated. The experiment involves measuring the electron current reaching $P$ (indicated by the current $I$ flowing through the meter) as a function of the accelerating voltage $V$.

The first experiment was performed with the tube containing Hg vapor. The nature of the results are indicated in Figure 4-14. At low accelerating voltage, the current $I$ is observed to increase with increasing voltage $V$. When $V$ reaches 4.9 V, the current abruptly drops. This was interpreted as indicating that some interaction between the electrons and the Hg atoms suddenly begins when the electrons attain a kinetic energy of 4.9 eV. Apparently a significant fraction of the electrons of this energy excite the Hg atoms and in so doing entirely lose their kinetic energy. If $V$ is only slightly more than 4.9 V, the excitation process must occur just in front of the anode $A$, and after the process the electrons cannot gain enough kinetic energy in falling toward $A$ to overcome the retarding potential $V_r$, and reach plate $P$. At somewhat larger $V$, the electrons can gain enough kinetic energy after the excitation process to overcome $V_r$, and reach $P$. The sharpness of the break in the curve indicates that electrons of energy less than 4.9 eV are not able to transfer their energy to an Hg atom. This interpretation is consistent with the existence of discrete energy states for the Hg atom. Assuming the first excited state of Hg to be 4.9 eV higher in energy than the ground state, an Hg atom would simply not be able to accept energy from the bombarding electrons unless these electrons had at least 4.9 eV.

Now, if the separation between the ground state and the first excited state is actually 4.9 eV, there should be a line in the Hg emission spectrum corresponding to the atom's loss of 4.9 eV in undergoing a transition from the first excited state to the ground state. Franck and Hertz found that when the energy of the bombarding electrons is less than 4.9 eV no spectral lines at all are emitted from the Hg vapor in the tube, and when the energy is not more than a few electron volts greater than this value only a single line is seen in the spectrum. This line is of wavelength 2536 Å, which corresponds exactly to a photon energy of 4.9 eV.

The Franck-Hertz experiment provided striking evidence for the quantization of the energy of atoms. It also provided a method for the direct measurement of the energy differences between the quantum states of an atom—the answers appear on the dial of a voltmeter! When the curve of $I$ versus $V$ is extended to higher voltages, additional breaks are found. Some are due to electrons exciting the first excited state of the atoms on several separate occasions in their trip from $C$ to $A$; but some are due to excitation of the higher excited states and, from the position of these breaks, the energy differences between the higher excited states and the ground state can be directly measured.

Another experimental method of determining the separations between the energy states of an atom is to measure its atomic spectrum and then empirically to construct a set of energy states which would lead to such a spectrum. In practice this is often quite difficult to do since the set of lines constituting the spectrum, as well as the set of energy states, is often very complicated; however, in common with all spectroscopic techniques, it is a very accurate method. In all cases in which determinations of the separations between the energy states of a certain atom have been made, using both this technique and the Franck-Hertz technique, the results have been found to be in excellent agreement.

In order to illustrate the preceding discussion, we show in Figure 4-15 a considerably simplified representation of the energy states of Hg in terms of an energy-level diagram. The separations between the ground state and the first and second excited states are known, from the Franck-Hertz experiment, to be 4.9 eV and 6.7 eV. These numbers can be confirmed, and in fact determined with much higher accuracy, by measuring the wavelengths of the two spectral lines corresponding to transitions of an electron in the Hg atom from these two states to the ground state. The energy $\mathscr{E} = -10.4$ eV, of the ground state relative to a state of zero total energy, is not determined by the Franck-Hertz experiment. However, it can be found by measuring the wavelength of the line corresponding to a transition of an atomic electron from a state of zero total energy to the ground state. This is the series limit of the series terminating on the ground state. The energy $\mathscr{E}$ can also be measured by measuring the energy which must be supplied to an Hg atom in order to send one of its electrons from the ground state to a state of zero total energy. Since an electron of zero total energy is no longer bound to the atom, $\mathscr{E}$ is the energy required to ionize the atom and is therefore called the *ionization energy*.

Lying above the highest discrete state at $E = 0$ are the energy states of the system consisting of an unbound electron plus an ionized Hg atom. *The total energy of an unbound electron (a free electron with $E > 0$) is not quantized.* Thus any energy $E > 0$ is possible for the electron, and the energy states form a continuum. The electron can be excited from its ground state to a continuum state if the Hg atom receives an energy

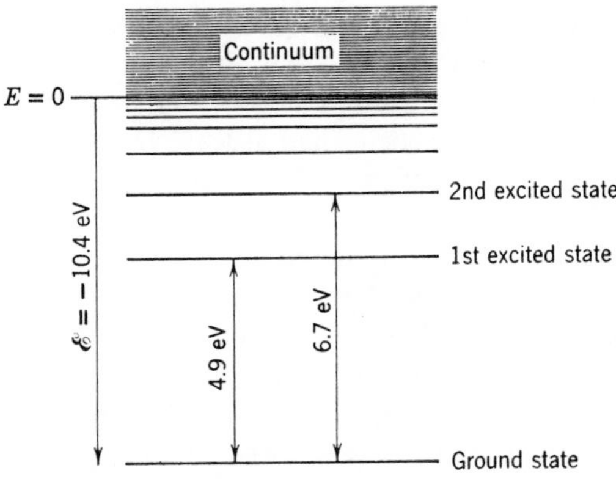


**FIGURE 4-15**

A considerably simplified energy-level diagram for mercury. Lying above the highest discrete energy level at $E = 0$ is a continuum of levels.

greater than 10.4 eV. Conversely, it is possible for an ionized Hg atom to capture a free electron into one of the quantized energy states of the neutral atom. In this process, radiation of frequency greater than the series limit corresponding to that state will be emitted. The exact value of the frequency depends on the initial energy $E$ of the free electron. Since $E$ can have any value, the spectrum of Hg should have a continuum extending beyond every series limit in the direction of increasing frequency. This can actually be seen experimentally, although with some difficulty. These comments concerning the continuum of energy states for $E > 0$, and its consequences, have been made in reference to the Hg atom, but they are equally true for all atoms.

## 4-9 Interpretation of the Quantization Rules

The success of the Bohr model, as measured by its agreement with experiment, was certainly very striking; but it only accentuated the mysterious nature of the postulates on which the model was based. One of the biggest mysteries was the question of the relation between Bohr's quantization of the angular momentum of an electron moving in a circular orbit and Planck's quantization of the total energy of an entity, such as an electron, executing simple harmonic motion. In 1916 some light was shed upon this by Wilson and Sommerfeld, who enunciated a set of rules for the quantization of any physical system for which the coordinates are periodic functions of time. These rules included both the Planck and the Bohr quantization as special cases. They were also of considerable use in broadening the range of applicability of the quantum theory. These rules can be stated as follows:

*For any physical system in which the coordinates are periodic functions of time, there exists a quantum condition for each coordinate. These quantum conditions are*

$$\oint p_q \, dq = n_q h \tag{4-23}$$

*where $q$ is one of the coordinates, $p_q$ is the momentum associated with that coordinate, $n_q$ is a quantum number which takes on integral values, and $\oint$ means that the integration is taken over one period of the coordinate $q$.*

The meaning of these rules can best be illustrated in terms of some specific examples. Consider a one-dimensional simple harmonic oscillator. Its total energy can be written, in terms of position and momentum, as

$$E = K + V = \frac{p_x^2}{2m} + \frac{kx^2}{2}$$

or

$$\frac{p_x^2}{2mE} + \frac{x^2}{2E/k} = 1$$

The quantization integral $\oint p_x \, dx$ is most easily evaluated, for the relation between $p_x$ and $x$ that is imposed by this equation, if we consider a geometric interpretation. The relation between $p_x$ and $x$ is the equation of an ellipse. Any instantaneous state of motion of the oscillator is represented by some point in a plot of this equation on a two-dimensional space having coordinates $p_x$ and $x$. We call such a space (the *p-q* plane) *phase space*, and the plot is a *phase diagram* of the linear oscillator, shown in Figure 4-16. During one cycle of oscillation the point representing the position and momentum of the particle travels once around the ellipse. The semiaxes $a$ and $b$ of the ellipse $p_x^2/b^2 + x^2/a^2 = 1$ are seen, by comparison with our equation, to be

$$b = \sqrt{2mE} \qquad \text{and} \qquad a = \sqrt{2E/k}$$

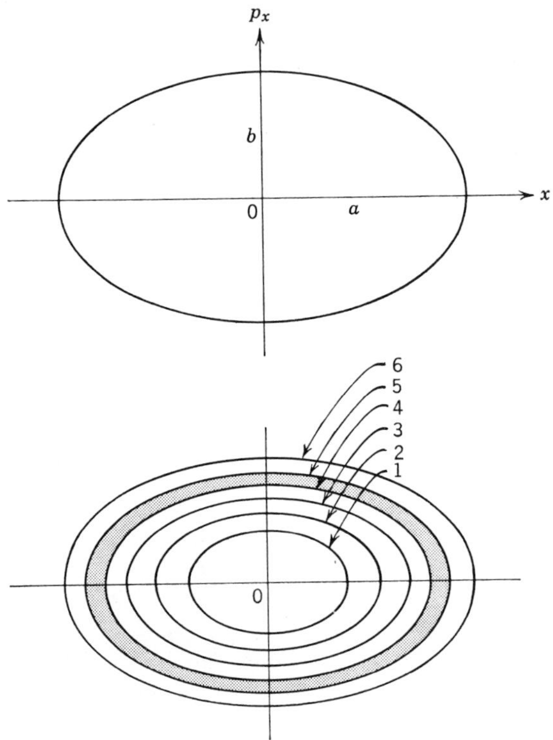


**FIGURE 4-16**

*Top:* A phase space diagram of the motion of the representative point for a linear simple harmonic oscillator. *Bottom:* The allowed energy states of the oscillator are represented by ellipses whose areas in phase space are given by *nh*. The space between adjacent ellipses (for example the shaded area) has an area *h*.

Now the area of an ellipse is $\pi ab$. Furthermore, the value of the integral $\oint p_x\,dx$ is just equal to that area. (To see this note that the integral over a complete oscillation equals an integral in which the representative point travels from $x = -a$ to $x = +a$ over the upper half of the ellipse plus an integral in which the point travels back to $x = -a$ over the lower half. In the first integral both $p_x$ and $dx$ are positive and its value equals the area enclosed between the upper half and the $x$ axis; in the second both $p_x$ and $dx$ are negative so the value of the integral is positive and equals the area enclosed between the lower half of the ellipse and the $x$ axis.) Thus we obtain

$$\oint p_x\,dx = \pi ab$$

In our case

$$\oint p_x\,dx = \frac{2\pi E}{\sqrt{k/m}}$$

but

$$\sqrt{k/m} = 2\pi\nu$$

where $\nu$ is the frequency of the oscillation, so that

$$\oint p_x\,dx = E/\nu$$

If we now use (4-23), the Wilson-Sommerfeld quantization rule, we have

$$\oint p_x \, dx = E/\nu = n_x h \equiv nh$$

or

$$E = nh\nu$$

which is identical with Planck's quantization law.

Note that the allowed states of oscillation are represented by a series of ellipses in phase space, the area enclosed between successive ellipses always being $h$ (see Figure 4-16). Again we find that the classical situation corresponds to $h \to 0$, all values of $E$ and hence all ellipses being allowed *if* that were true. The quantity $\oint p_x \, dx$ is sometimes called a *phase integral*; in classical physics it is the integral of the dynamical quantity called the *action* over one oscillation of the motion. Hence, the Planck energy quantization is equivalent to the *quantization of action.*

We can also deduce the Bohr quantization of angular momentum from the Wilson-Sommerfeld rule, (4-23). An electron moving in a circular orbit of radius $r$ has an angular momentum, $mvr = L$, which is constant. The angular coordinate is $\theta$, which is a periodic function of the time. That is, $\theta$ versus $t$ is a saw-tooth function, increasing linearly from zero to $2\pi$ rad in one period and repeating this pattern in each succeeding period. The quantization rule

$$\oint p_q \, dq = n_q h$$

becomes, in this case

$$\oint L \, d\theta = nh$$

and

$$\oint L \, d\theta = L \int_0^{2\pi} d\theta = 2\pi L$$

so that

$$2\pi L = nh$$

or

$$L = nh/2\pi \equiv n\hbar$$

which is identical with Bohr's quantization law.

A more physical interpretation of the Bohr quantization rule was given in 1924 by de Broglie. The Bohr quantization of angular momentum can be written as in (4-15) as

$$mvr = pr = nh/2\pi \qquad n = 1, 2, 3, \ldots$$

where $p$ is the linear momentum of an electron in an allowed orbit of radius $r$. If we substitute into this equation the expression for $p$ in terms of the corresponding de Broglie wavelength

$$p = h/\lambda$$

the Bohr equation becomes

$$hr/\lambda = nh/2\pi$$

or

$$2\pi r = n\lambda \qquad n = 1, 2, 3, \ldots \tag{4-24}$$

Thus *the allowed orbits are those in which the circumference of the orbit can contain exactly an integral number of de Broglie wavelengths.*

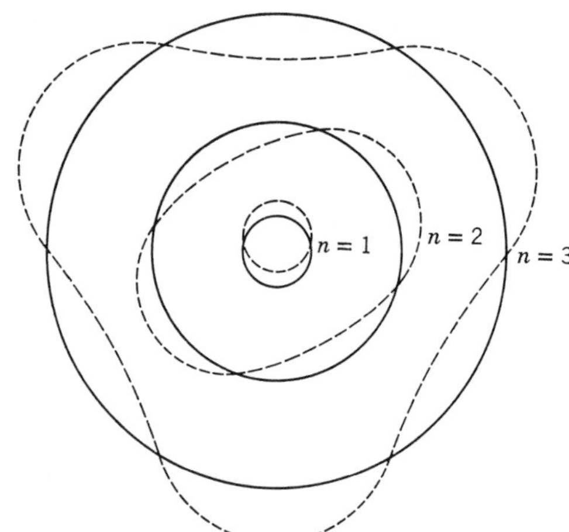


**FIGURE 4-17**

Illustrating standing de Broglie waves set up in the first three Bohr orbits. The locations of the nodes can, of course, be found anywhere on each orbit providing that their spacings are as shown.

Imagine the electron to be moving in a circular orbit at constant speed, with the associated wave following the electron. The wave, of wavelength $\lambda$, is then wrapped repeatedly around the circular orbit. The resultant wave that is produced will have zero intensity at any point unless the wave at each traversal is exactly in phase at that point with the wave in other traversals. If the waves in each traversal are exactly in phase, they join on perfectly in orbits that accommodate integral numbers of de Broglie wavelengths, as illustrated in Figure 4-17. But the condition that this happens is just the condition that (4-24) be satisfied. If this equation were violated, then in a large number of traversals the waves would interfere with each other in such a way that their average intensity would be zero. Since the average intensity of the waves, $\overline{\Psi^2}$, is supposed to be a measure of where the particle is located, we interpret this as meaning that an electron cannot be found in such an orbit.

This wave picture gives no suggestion of progressive motion. Rather, it suggests standing waves, as in a stretched string of a given length. In a stretched string only certain wavelengths, or frequencies of vibration, are permitted. Once such modes are excited, the vibration goes on indefinitely if there is no damping. To get standing waves, however, we need oppositely directed traveling waves of equal amplitude. For the atom this requirement is presumably satisfied by the fact that the electron can traverse an orbit in either direction and still have the magnitude of angular momentum required by Bohr. The de Broglie standing wave interpretation, illustrated in Figure 4-17, therefore provides a satisfying basis for Bohr's quantization rule and, for this case, of the more general Wilson-Sommerfeld rule.

There is another example of a system in which the origin of the Wilson-Sommerfeld quantization rule can be understood in terms of the requirement that *the de Broglie waves associated with a particle undergoing periodic motion form a set of standing waves.* Consider a particle which moves freely along the $x$ axis from $x = -a/2$ to $x = +a/2$, but which does not penetrate into the regions outside these limits. This system can be thought of as representing approximately the motion of a conduction electron in a one-dimensional piece of metal that extends from $-a/2$ to $+a/2$. The particle bounces back and forth between the ends of the region with momentum $p_x$ that changes sign at each bounce, but maintains a constant magnitude $p$. So the Wilson-Sommerfeld equation reads

$$\oint p_x \, dx = p \oint dx = p2a = nh$$

or

$$n\frac{h}{p} = 2a \tag{4-25}$$

But $h/p$ is just the de Broglie wavelength $\lambda$ of the particle, so we have

$$n\lambda = 2a$$

Thus an integral number of de Broglie wavelengths just fits into the distance covered by the particle in one traversal of the region, and this allows the waves associated with successive traversals to be in phase and so set up a standing wave.

We shall see in the following chapters that the properties of standing waves are equally important in the quantization conditions of Schroedinger's quantum mechanics. And the time-independent features of the standing wave associated with an electron in the ground state of an atom will make it possible to understand in a simple way the fundamental question of why the electron does not emit electromagnetic radiation and spiral into the nucleus.

## 4-10 Sommerfeld's Model

One of the important applications of the Wilson-Sommerfeld quantization rules is to the case of a hydrogen atom in which it was assumed that the electron could move in *elliptical* orbits. This was done by Sommerfeld in an attempt to explain the *fine structure* of the hydrogen spectrum. The fine structure is a splitting of the spectral lines, into several distinct components, which is found in all atomic spectra. It can be observed only by using equipment of very high resolution since the separation, in terms of reciprocal wavelength, between adjacent components of a single spectral line is of the order of $10^{-4}$ times the separation between adjacent lines. According to the Bohr model, this must mean that what we had thought was a single energy state of the hydrogen atom actually consists of several states which are very close together in energy.

Sommerfeld first evaluated the size and shape of the allowed elliptical orbits, as well as the total energy of an electron moving in such an orbit, using the formulas of classical mechanics. Describing the motion in terms of the polar coordinates $r$ and $\theta$, he applied the two quantum conditions

$$\oint L\, d\theta = n_\theta h$$

$$\oint p_r\, dr = n_r h$$

The first condition yields the same restriction on the orbital angular momentum

$$L = n_\theta \hbar \qquad n_\theta = 1, 2, 3, \ldots$$

that it does for the circular orbit theory. The second condition (which was not applicable in the limiting case of purely circular orbits) leads to the following relation between $L$ and $a/b$, the ratio of the semimajor axis to the semiminor axis of the ellipse

$$L(a/b - 1) = n_r \hbar \qquad n_r = 0, 1, 2, 3, \ldots$$

By applying the condition of mechanical stability analogous to (4-14), a third equation is obtained. From these equations Sommerfeld evaluated the semimajor and semiminor axes $a$ and $b$, which give the size and shape of the elliptical orbits, and also the total energy $E$ of an electron in such an orbit. The results are

$$a = \frac{4\pi\epsilon_0 n^2\hbar^2}{\mu Z e^2} \tag{4-26a}$$

$$b = a\,\frac{n_\theta}{n} \tag{4-26b}$$

$$E = -\left(\frac{1}{4\pi\epsilon_0}\right)^2 \frac{\mu Z^2 e^4}{2n^2\hbar^2} \tag{4-26c}$$

where $\mu$ is the reduced mass of the electron, and where the quantum number $n$ is defined by

$$n \equiv n_\theta + n_r$$

Since $n_\theta = 1, 2, 3, \ldots$ and $n_r = 0, 1, 2, 3, \ldots, n$ can take on the values

$$n = 1, 2, 3, 4, \ldots$$

For a given value $n$, $n_\theta$ can assume only the values

$$n_\theta = 1, 2, 3, \ldots, n$$

The integer $n$ is called the *principal quantum number*, and $n_\theta$ is called the *azimuthal quantum number*.

Equation (4-26b) shows that the shape of the orbit (the ratio of the semimajor to the semiminor axes) is determined by the ratio of $n_\theta$ to $n$. For $n_\theta = n$ the orbits are circles of radius $a$. Note that the equation giving $a$ in terms of $n$ is identical with (4-16), the equation giving the radius of the circular Bohr orbits. (Remember that (4-16) will have $m$ replaced by $\mu$ if proper account is taken of the finite nuclear mass.) Figure 4-18 shows, to scale, the possible orbits corresponding to the first three values of the principal quantum number. Corresponding to each value of the principal quantum number $n$ there are $n$ different allowed orbits. One of these, the circular orbit, is just the orbit described by the original Bohr model. The others are elliptical. But despite the very different paths followed by an electron moving in the different possible orbits for a given $n$, (4-26c) tells us that the total energy of the electron is the same. The total energy of the electron depends only on $n$. The several orbits characterized by a common value of $n$ are said to be *degenerate*. The energies of different states of motion "degenerate" to the same total energy.

This degeneracy in the total energy of an electron, following the orbits of very different shape but common $n$, is the result of a very delicate balance between potential and kinetic energy, which is characteristic of treating the inverse square Coulomb

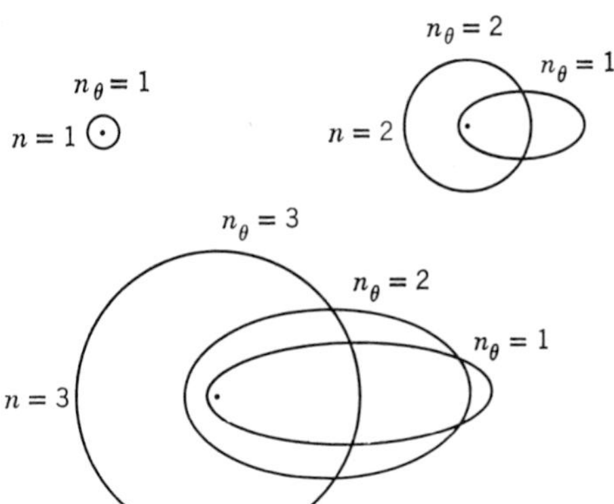


**FIGURE 4-18**
Some elliptical Bohr-Sommerfeld orbits. The nucleus is located at the common focus of the ellipses, indicated by the dot.

force by the methods of classical mechanics. Exactly the same phenomenon is found in planetary or satellite motion, which is governed by the inverse square gravitational force. For instance, a satellite may be launched into any one of a whole family of elliptical orbits, all of which correspond to the same total energy and have the same semimajor axis. Of course there is effectively no quantization of the orbit parameters in these macroscopic cases, but as far as degeneracy is concerned they are completely analogous to the case of a hydrogen atom.

Sommerfeld "removed the degeneracy" in the hydrogen atom by next treating the problem *relativistically*. In the discussion following (4-17) we showed that, for an electron in a hydrogen atom, $v/c \simeq 10^{-2}$ or less. Thus we would expect the relativistic corrections to the total energy, due to the relativistic variation of the electron mass which will be of the order of $(v/c)^2$, to be only of the order of $10^{-4}$; however, this is just the order of magnitude of the splitting in the energy states of hydrogen that would be needed to explain the fine structure of the hydrogen spectrum. The actual size of the correction depends on the average velocity of the electron which, in turn, depends on the ellipticity of the orbit. After a calculation which is much too tedious to reproduce here, Sommerfeld showed that the total energy of an electron in an orbit characterized by the quantum numbers $n$ and $n_\theta$ is equal to

$$E = -\frac{\mu Z^2 e^4}{(4\pi\epsilon_0)^2 2n^2\hbar^2}\left[1 + \frac{\alpha^2 Z^2}{n}\left(\frac{1}{n_\theta} - \frac{3}{4n}\right)\right] \tag{4-27a}$$

The quantity $\alpha$ is a pure number called the *fine structure constant*. Its value is

$$\alpha \equiv \frac{1}{4\pi\epsilon_0}\frac{e^2}{\hbar c} = 7.297 \times 10^{-3} \simeq \frac{1}{137} \tag{4-27b}$$

In Figure 4-19 we represent the first few energy states of the hydrogen atom in terms of an energy-level diagram. The separation between the several levels with a common value of $n$ has been greatly exaggerated for the sake of clarity. Arrows indicate transitions between the various energy states which produce the lines of the atomic spectrum. Lines corresponding to the transitions represented by the solid arrows are observed in the hydrogen spectrum. The wavelengths of these lines are in very good agreement with the predictions derived from (4-27a).

However, the lines corresponding to the transitions represented by dashed arrows in Figure 4-19 are not found in the spectrum. The transitions concerned do not take

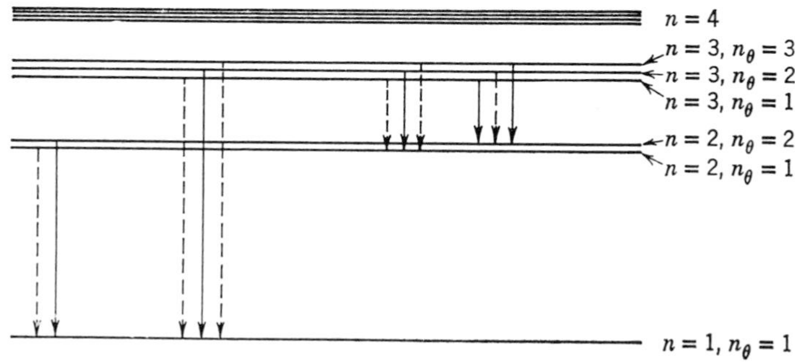


**FIGURE 4-19**

The fine-structure splitting of some energy levels of the hydrogen atom. The splitting is greatly exaggerated. Transitions which produce observed lines of the hydrogen spectrum are indicated by solid arrows.

place. Inspection of the figure will demonstrate that transitions only occur if

$$n_{\theta_i} - n_{\theta_f} = \pm 1 \tag{4-28}$$

This is called a *selection rule*. It selects from all the transitions those that actually occur.

## 4-11 The Correspondence Principle

A justification of selection rules could sometimes be found with the aid of an auxiliary postulate known as the *correspondence principle*. This principle, enunciated by Bohr in 1923, consists of two parts:

1. *The predictions of the quantum theory for the behavior of any physical system must correspond to the prediction of classical physics in the limit in which the quantum numbers specifying the state of the system become very large.*

2. *A selection rule holds true over the entire range of the quantum number concerned. Thus any selection rules which are necessary to obtain the required correspondence in the classical limit (large n) also apply in the quantum limit (small n).*

Concerning the first part, it is obvious that the quantum theory must correspond to the classical theory in the limit in which the system behaves classically. The only question is: Where is the classical limit? Bohr's assumption is that the classical limit is always to be found in the limit of large quantum numbers. In making this assumption he was guided by certain evidence available at the time. For instance, the classical Rayleigh-Jeans theory of the blackbody spectrum agrees with experiment in the limit of small $\nu$. Since Planck's quantum theory agrees with experiment everywhere, we see that correspondence between the quantum and classical theories is found, in this case, in the limit of small $\nu$. But it is easy to see that as $\nu$ becomes small the average value $\bar{n}$, of the quantum number specifying the energy state of blackbody electromagnetic waves of frequency $\nu$, will become large. (Since $\mathscr{E} = nh\nu$, we have $\overline{\mathscr{E}} = \bar{n}h\nu$. But as $\nu \to 0$, $\overline{\mathscr{E}} \to kT$, so in this limit $\bar{n}h\nu = kT$, which is a constant. Thus $\bar{n} \to \infty$ as $\nu \to 0$ in the classical limit. Note also that if we fix $\nu$ in the relation $\bar{n}h\nu = kT =$ const, and take $h \to 0$ as we frequently have in considering the classical limit, we again find $\bar{n} \to \infty$ in that limit.) The second part of the correspondence principle was purely an assumption, but certainly a reasonable one.

Let us illustrate the correspondence principle by applying it to a simple harmonic oscillator, such as a pendulum oscillating at frequency $\nu$. One prediction of quantum theory for this system is that the allowed energy states are given by $E = nh\nu$. In the discussion in Chapter 1, we saw that, in the limit of large $n$, this prediction is not in disagreement with what we actually know about the energy states of a classical pendulum. In this case of a simple harmonic oscillator, the quantum and classical theories do correspond for $n \to \infty$ insofar as the energy states are concerned. Next assume that the pendulum bob carries an electric charge, so that we can compare the predictions of the two theories concerning the emission and absorption of electromagnetic radiation by such a system. Classically the system would emit radiation due to the accelerated motion of the charge, and the frequency of the emitted radiation would be exactly $\nu$. According to the quantum physics, radiation is emitted as a result of the system making a transition from quantum state $n_i$ to quantum state $n_f$. The energy emitted in such a transition is equal to $E_i - E_f = (n_i - n_f)h\nu$. This energy is carried away by a photon of frequency $(E_i - E_f)/h = (n_i - n_f)\nu$. Thus, in order to obtain correspondence between the classical and quantum predictions of the frequency of the emitted radiation, we must require that the selection rule $n_i - n_f = 1$ be valid in the classical limit of large $n$. A similar argument concerning the absorption of radiation by the charged pendulum shows that in the classical limit there is also the possibility of a transition in which $n_i - n_f = -1$. The validity of these selection rules in the quantum limit of small $n$ can be tested by investigating the spectrum of radiation emitted by a vibrating diatomic molecule. The vibrational energy states for such a system are just those of a simple harmonic oscillator, since the force which leads to the equilibrium separation of the two atoms has the same form as a harmonic restoring force. From the vibrational spectrum it can be determined that the selection rule $n_i - n_f = \pm 1$ actually is in operation in the limit of small quantum numbers, in agreement with the second part of the correspondence principle.

A number of other selection rules were discovered empirically in the analysis of atomic and molecular spectra. Sometimes, but not always, it was possible to understand these selection rules in terms of a correspondence principle argument.

**Example 4-11.** Apply the correspondence principle to hydrogen atom radiation in the classical limit.

The frequency of revolution $\nu_0$ of an electron in a Bohr orbit follows from (4-16) and (4-17) and is given by

$$\nu_0 = \frac{v}{2\pi r} = \left(\frac{1}{4\pi\epsilon_0}\right)^2 \frac{me^4}{4\pi\hbar^3}\frac{2}{n^3}$$

According to classical physics the frequency of the light emitted in such a case is equal to $\nu_0$, the frequency of revolution.

Quantum physics predicts that the frequency $\nu$ of the emitted light is, from (4-19)

$$\nu = \frac{c}{\lambda} = c\kappa = \left(\frac{1}{4\pi\epsilon_0}\right)^2 \frac{me^4}{4\pi\hbar^3}\left[\frac{1}{n_f^2} - \frac{1}{n_i^2}\right]$$

But, if this is to agree with $\nu_0$ we must have $n_i - n_f = 1$, as a selection rule for large quantum numbers. To see this, take $n_i - n_f = 1$ and obtain

$$\nu = \left(\frac{1}{4\pi\epsilon_0}\right)^2 \frac{me^4}{4\pi\hbar^3}\left[\frac{1}{(n-1)^2} - \frac{1}{n^2}\right] = \left(\frac{1}{4\pi\epsilon_0}\right)^2 \frac{me^4}{4\pi\hbar^3}\left[\frac{2n-1}{(n-1)^2 n^2}\right]$$

where $n_i = n$ and $n_f = n - 1$. Then as $n \to \infty$ the expression in the square brackets above approaches $2/n^3$ so that $\nu \to \nu_0$ as $n \to \infty$.

In Table 4-2 we illustrate the correspondence for large $n$. ◀

It is instructive to note that although both parts of the correspondence principle lead to agreement with experiment for the simple harmonic oscillator, only the first part agrees with experiment in the hydrogen atom considered in the preceding example. For experiment shows that the selection rule $n_i - n_f = 1$, which was necessary to satisfy the first part of the principle for large $n$, does *not* apply to the hydrogen atom for small $n$. Transitions are observed to occur between states of low $n$, in which the quantum numbers differ in value by more than one. This illustrates the fact that the old quantum theory cannot always be made to agree with experiment, however it is patched up.

**TABLE 4-2.** The Correspondence Principle for Hydrogen

| $n$ | $\nu_0$ | $\nu$ | % Difference |
|---|---|---|---|
| 5 | $5.26 \times 10^{13}$ | $7.38 \times 10^{13}$ | 29 |
| 10 | $6.57 \times 10^{12}$ | $7.72 \times 10^{12}$ | 14 |
| 100 | $6.578 \times 10^{9}$ | $6.677 \times 10^{9}$ | 1.5 |
| 1,000 | $6.5779 \times 10^{6}$ | $6.5878 \times 10^{6}$ | 0.15 |
| 10,000 | $6.5779 \times 10^{3}$ | $6.5789 \times 10^{3}$ | 0.015 |

## 4-12 A Critique of the Old Quantum Theory

In the past four chapters we have discussed some of the developments which led to modern quantum mechanics. These developments are now referred to as the *old quantum theory*. In many respects this theory was very successful, even more so than may be apparent to the student because we have not mentioned a number of successful applications of the old quantum theory to phenomena, such as the heat capacity of solids at low temperature, which were inexplicable in terms of the classical theories. However, the old quantum theory certainly was not free of criticism. To complete our discussion of this theory we must indicate some of its undesirable aspects:

1. The theory only tells us how to treat systems which are periodic, by using the Wilson-Sommerfeld quantization rules, but there are many systems of physical interest which are not periodic. And the number of periodic systems for which a physical basis of these rules can be found in the de Broglie relation is very small.

2. Although the theory does tell us how to calculate the energies of the allowed states of certain systems, and the frequency of the photons emitted or absorbed when a system makes a transition between allowed states, it does not tell us how to calculate the *rate* at which such transitions take place. For example, it does not tell us how to calculate the intensities of spectral lines. And we have seen that the theory cannot always tell us even which transitions actually are observed to occur and which are not.

3. When applied to atoms, the theory is really only successful for one-electron atoms. The alkali elements (Li, Na, K, Rb, Cs) can be treated approximately, but only because they are in many respects similar to a one-electron atom. The theory fails badly even when applied to the neutral He atom, which contains only two electrons.

4. Finally we might mention the subjective criticism that the entire theory seems somehow to lack coherence—to be intellectually unsatisfying.

That some of these objections are really of a very fundamental nature was realized by everyone concerned, and much effort was expended in attempts to develop a quantum theory which would be free of these and other objections. The effort was well rewarded. In 1925 Erwin Schroedinger developed his theory of *quantum mechanics*. Although it is a generalization of the de Broglie postulate, the Schroedinger theory is in some respects very different from the old quantum theory. For instance, the picture of atomic structure provided by quantum mechanics is the antithesis of the picture, used in the old quantum theory, of electrons moving in well-defined orbits. Nevertheless, the old quantum theory is still frequently employed as a first approximation to the more accurate description of quantum phenomena provided by quantum mechanics. The reasons are that the old quantum theory is often capable of giving numerically correct results with mathematical procedures which are considerably less complicated than those used in quantum mechanics, and that the old quantum theory is often helpful in visualizing processes which are difficult to visualize in terms of the rather abstract language of quantum mechanics.

## QUESTIONS

1. In a collision between an $\alpha$ particle and an electron, what general considerations limit the momentum transfer? Does the fact that the force is Coulombic play any role in this respect?

2. How does the Thomson atom differ from a random distribution of protons and electrons in a spherical region?

3. List objections to the Thomson model of the atom.

4. Why do we specify that the foil be thin in experiments intended to check the Rutherford scattering formula?

5. The scattering of $\alpha$ particles at very small angles disagrees with the Rutherford formula for such angles. Explain.

6. How does the deduction of (4-3), which gives the trajectory of a particle moving under the influence of a repulsive inverse square Coulomb force, differ from the deduction of the trajectory of a planet moving under the influence of the gravitational field of the sun?

7. Could a differential scattering cross section, defined as in (4-8), be used to describe very small angle $\alpha$-particle scattering?

8. Did Bohr postulate the quantization of energy? What did he postulate?

9. For the Bohr hydrogen atom orbits, the potential energy is negative and greater in magnitude than the kinetic energy. What does this imply?

10. If only lines in the absorption spectrum of hydrogen need to be calculated, how would you modify (4-19) to obtain them?

11. On emitting a photon, the hydrogen atom recoils to conserve momentum. Explain the fact that the energy of the emitted photon is less than the energy difference between the energy levels involved in the emission process.

12. Can a hydrogen atom absorb a photon whose energy exceeds its binding energy, 13.6 eV?

13. Is it possible to get a continuous emission spectrum from hydrogen?

14. What minimum energy must a photon have to initiate the photoelectric effect in hydrogen gas? (Careful!)

15. Would you expect to observe all the lines of atomic hydrogen if such a gas were excited by electrons of energy 13.6 eV? Explain.

16. Assume that electron-positron annihilation takes place from the ground state of positronium. How, if at all, does this alter the $\gamma$-ray energies of the two-photon decay calculated in Chapter 2 by ignoring the bound system?

17. Is the ionization energy of deuterium different from that of hydrogen? Explain.

18. Why is the structure of the Franck-Hertz current versus voltage curve, Figure 4-14, not sharp?

19. Is the peak in Figure 4-14 just below 10 eV due to two consecutive excitations of the first excited state of mercury or to one excitation of the second excited state?

20. What examples of degeneracy in classical physics, other than planetary motion, can you think of?

21. The fine-structure constant $\alpha$ is dimensionless and relates $e$, $h$, and $c$, three of the fundamental constants of physics. Is any other combination of these constants dimensionless (other than powers of the same combination, of course)?

22. How can the correspondence principle be applied to the phase diagram of a linear oscillator, Figure 4-16?

23. According to classical mechanics, an electron moving in an atom should be able to do so with any angular momentum whatever. According to Bohr's theory of the hydrogen atom, however, the angular momentum is quantized to $L = nh/2\pi$. Can the correspondence principle reconcile these two statements?

## PROBLEMS

1. Show, for a Thomson atom, that an electron moving in a stable circular orbit rotates with the same frequency at which it would oscillate in an oscillation through the center along a diameter.

2. What radius must the Thomson model of a one-electron atom have if it is to radiate a spectral line of wavelength $\lambda = 6000$ Å? Comment on your results.

3. (a) An $\alpha$ particle of initial velocity $v$ collides with a free electron at rest. Show that, assuming the mass of the $\alpha$ particle to be about 7400 electronic masses, the maximum deflection of the $\alpha$ particle is about $10^{-4}$ rad. (b) Show that the maximum deflection of an $\alpha$ particle that interacts with the positive charge of a Thomson atom of radius 1.0 Å is also about $10^{-4}$ rad. Hence, argue that $\theta \lesssim 10^{-4}$ rad for the scattering of an $\alpha$ particle by a Thomson atom.

4. Derive (4-5) relating the distance of closest approach and the impact parameter to the scattering angle.

5. A 5.30 MeV $\alpha$ particle is scattered through 60° in passing through a thin gold foil. Calculate (a) the distance of closest approach, $D$, for a head-on collision and (b) the impact parameter, $b$, corresponding to the 60° scattering.

6. What is the distance of closest approach of a 5.30 MeV $\alpha$ particle to a copper nucleus in a head-on collision?

7. Show that the number of $\alpha$ particles scattered by an angle $\Theta$ or greater in Rutherford scattering is

$$\left(\frac{1}{4\pi\epsilon_0}\right)^2 \pi I \rho t \left(\frac{zZe^2}{Mv^2}\right)^2 \cot^2 (\Theta/2)$$

8. The fraction of 6.0 MeV *protons* scattered by a thin gold foil, of density 19.3 g/cm$^3$, from the incident beam into a region where scattering angles exceed 60° is equal to $2.0 \times 10^{-5}$. Calculate the thickness of the gold foil, using results of the previous problem.

9. A beam of $\alpha$ particles, of kinetic energy 5.30 MeV and intensity $10^4$ particle/sec, is incident normally on a gold foil of density 19.3 g/cm$^3$, atomic weight 197, and thickness $1.0 \times 10^{-5}$ cm. An $\alpha$ particle counter of area 1.0 cm$^2$ is placed at a distance 10 cm from the foil. If $\Theta$ is the angle between the incident beam and a line from the center of the foil to the center of the counter, use the Rutherford scattering differential cross section, (4-9), to find the number of counts per hour for $\Theta = 10°$ and for $\Theta = 45°$. The atomic number of gold is 79.

10. In the previous problem, a copper foil of density 8.9 g/cm$^3$, atomic weight 63.6 and thickness $1.0 \times 10^{-5}$ cm is used instead of gold. When $\Theta = 10°$ we get 820 counts per hour. Find the atomic number of copper.

11. Prove that Planck's constant has the dimensions of angular momentum.

12. Compare the gravitational attraction of an electron and proton in the ground state of a hydrogen atom to the Coulomb attraction. Are we justified in ignoring the gravitational force?

13. Show that the frequency of revolution of the electron in the Bohr model hydrogen atom is given by $\nu = 2\,|E|/hn$ where $E$ is the total energy of the electron.

14. Show that for all Bohr orbits the ratio of the magnetic dipole moment of the electronic orbit to its orbital angular momentum has the same value.

15. (a) Show that in the ground state of the hydrogen atom the speed of the electron can be written as $v = \alpha c$ where $\alpha$ is the fine-structure constant. (b) From the value of $\alpha$ what can you conclude about the neglect of relativistic effects in the Bohr calculations?

16. What is the energy, momentum, and wavelength of a photon that is emitted by a hydrogen atom making a direct transition from an excited state with $n = 10$ to the ground state? Find the recoil speed of the hydrogen atom in this process.

17. (a) Using Bohr's formula, calculate the three longest wavelengths in the Balmer series. (b) Between what wavelength limits does the Balmer series lie?

18. Calculate the shortest wavelength of the Lyman series lines in hydrogen. Of the Paschen series. Of the Pfund series. In what region of the electromagnetic spectrum does each lie?

19. In the ground state of the hydrogen atom, according to Bohr's model, what are (a) the quantum number, (b) the orbit radius, (c) the angular momentum, (d) the linear momentum, (e) the angular velocity, (f) the linear speed, (g) the force on the electron, (h) the acceleration of the electron, (i) the kinetic energy, (j) the potential energy, and (k) the total energy? How do the quantities (b) and (k) vary with the quantum number?

20. How much energy is required to remove an electron from a hydrogen atom in a state with $n = 8$?

21. A hydrogen atom is excited from a state with $n = 1$ to one with $n = 4$. (a) Calculate the energy that must be absorbed by the atom. (b) Calculate and display on an energy-level diagram the different photon energies that may be emitted if the atom returns to its $n = 1$ state. (c) Calculate the recoil speed of the hydrogen atom, assumed initially at rest, if it makes the transition from $n = 4$ to $n = 1$ in a single quantum jump.

22. A hydrogen atom in a state having a binding energy (this is the energy required to remove an electron) of 0.85 eV makes a transition to a state with an excitation energy (this is the difference in energy between the state and the ground state) of 10.2 eV. (a) Find the energy of the emitted photon. (b) Show this transition on an energy-level diagram for hydrogen, labeling the appropriate quantum numbers.

23. Show on an energy-level diagram for hydrogen the quantum numbers corresponding to a transition in which the wavelength of the emitted photon is 1216 Å.

24. (a) Show that when the recoil kinetic energy of the atom, $p^2/2M$, is taken into account the frequency of a photon emitted in a transition between two atomic levels of energy difference $\Delta E$ is reduced by a factor which is approximately $(1 - \Delta E/2Mc^2)$. (Hint: The recoil momentum is $p = h\nu/c$.) (b) Compare the wavelength of the light emitted from a hydrogen atom in the $3 \rightarrow 1$ transition when the recoil is taken into account to the wavelength without accounting for recoil.

25. What is the wavelength of the most energetic photon that can be emitted from a muonic atom with $Z = 1$?

26. Apply Bohr's model to singly ionized helium, that is, to a helium atom with one electron removed. What relationships exist between this spectrum and the hydrogen spectrum?

27. Using Bohr's model, calculate the energy required to remove the electron from singly ionized helium.

28. In a Franck-Hertz type of experiment atomic hydrogen is bombarded with electrons and excitation potentials are found at 10.21 V and 12.10 V. (a) Explain the observation that three different lines of spectral emission accompany these excitations. (Hint: Draw an energy-level diagram.) (b) Now assume that the energy differences can be expressed as $h\nu$ and find the three allowed values of $\nu$. (c) Assume that $\nu$ is the frequency of the emitted radiation and determine the wavelengths of the observed spectral lines.

**29.** Assume, in the Franck-Hertz experiment, that the electromagnetic energy emitted by an Hg atom, in giving up the energy absorbed from 4.9 eV electrons, equals $h\nu$, where $\nu$ is the frequency corresponding to the 2536 Å mercury resonance line. Calculate the value of $h$ according to the Franck-Hertz experiment and compare with Planck's value.

**30.** Radiation from a helium ion $He^+$ is nearly equal in wavelength to the $H_\alpha$ line (the first line of the Balmer series). (a) Between what states (values of $n$) does the transition in the helium ion occur? (b) Is the wavelength greater or smaller than that of the $H_\alpha$ line? (c) Compute the wavelength difference.

**31.** In stars the Pickering series is found in the $He^+$ spectrum. It is emitted when the electron in $He^+$ jumps from higher levels into the level with $n = 4$. (a) State the exact formula for the wavelength of lines belonging to this series. (b) In what region of the spectrum is the series? (c) Find the wavelength of the series limit. (d) Find the ionization potential, if $He^+$ is in the ground state, in electron volts.

**32.** Assuming that an amount of hydrogen of mass number three (tritium) sufficient for spectroscopic examination can be put into a tube containing ordinary hydrogen, determine the separation from the normal hydrogen line of the first line of the Balmer series that should be observed. Express the result as a difference in wavelength.

**33.** A gas discharge tube contains $H^1$, $H^2$, $He^3$, $He^4$, $Li^6$, and $Li^7$ ions and atoms (the superscript is the atomic mass), with the last four ionized so as to have only one electron. (a) As the potential across the tube is raised from zero, which spectral line should appear first? (b) Give, in order of increasing frequency, the origin of the lines corresponding to the first line of the Lyman series of $H^1$.

**34.** Consider a body rotating freely about a fixed axis. Apply the Wilson-Sommerfeld quantization rules, and show that the possible values of the total energy are predicted to be

$$E = \hbar^2 n^2/2I \qquad n = 0, 1, 2, 3, \ldots$$

where $I$ is its rotational inertia, or moment of inertia, about the axis of rotation.

**35.** Assume the angular momentum of the earth of mass $6.0 \times 10^{24}$ kg due to its motion around the sun at radius $1.5 \times 10^{11}$ m to be quantized according to Bohr's relation $L = nh/2\pi$. What is the value of the quantum number $n$? Could such quantization be detected?

# 5

# Schroedinger's Theory of Quantum Mechanics

# Schroedinger's Theory of Quantum Mechanics 5

## 5-1 Introduction

We have presented experimental evidence which shows conclusively that the particles of microscopic systems move according to the laws of some form of wave motion, and not according to the Newtonian laws of motion obeyed by the particles of macroscopic systems. Thus a microscopic particle acts as if certain aspects of its behavior are governed by the behavior of an associated de Broglie wave, or wave function. The experiments considered dealt only with simple cases (such as free particles, or simple harmonic oscillators, etc.) that can be analyzed with simple procedures (involving direct applications of the de Broglie postulate, Planck's postulate, etc.). But we certainly want to be able to treat the more complicated cases that occur in nature because they are interesting and important. To be able to do this we must have a more general procedure that can be used to treat the behavior of the particles of any microscopic system. *Schroedinger's theory of quantum mechanics* provides us with such a procedure.

The theory specifies the laws of wave motion which the particles of any microscopic system obey. This is done by specifying, for each system, the equation which controls the behavior of the wave function, and also by specifying the connection between the behavior of the wave function and the behavior of the particle. The theory is an extension of the de Broglie postulate. Furthermore, there is a close relation between it and Newton's theory of the motion of particles in macroscopic systems. Schroedinger's theory is a generalization that includes Newton's theory as a special case (in the macroscopic limit), much as Einstein's theory of relativity is a generalization that includes Newton's theory as a special case (in the low velocity limit).

We shall develop the essential points of the Schroedinger theory and use them to treat a number of important microscopic systems. For instance, we shall use the theory to obtain a detailed understanding of the properties of atoms. These properties form the basis of much of chemistry and solid state physics, and they are closely related to the properties of nuclei.

After we have applied Schroedinger's theory to a number of cases, the student should find that he is beginning to develop an intuition concerning the behavior of quantum mechanical systems, just as he has developed an intuitive feeling for classical systems from his study of Newton's theory and its applications to a number of cases. Actually, a better comparison can be made between the Schroedinger theory and Maxwell's theory of electromagnetism. The reason for this is that electromagnetic waves behave in a manner which is very analogous to the behavior of the wave functions of the Schroedinger theory. We shall use this analogy, when appropriate, to show how quantum mechanical results are related to results that are familiar from the study of electromagnetism, or of other forms of classical wave motion. We shall also discuss many experiments which directly confirm the quantum mechanical results that we obtain, just as we have discussed many experiments which set the stage for the theory. But the student will have to exercise a little patience because there is much to be done in developing the theory, and in working out its consequences, before we can make many comparisons between these consequences and experiment.

Now, we have seen that de Broglie's postulate provides a fundamental step in the development of Schroedinger's general theory of the behavior of microscopic particles. However, it is only a step. The postulate says the motion of a microscopic particle is governed by the propagation of an associated wave, but the postulate does not tell us how the wave propagates. The postulate does predict successfully the wavelength of the wave inferred from measurements of the diffraction pattern observed in the motion of the particle, but only in cases in which the wavelength is essentially constant. Furthermore, we must have a quantitative relation between the properties of the particle and the properties of the wave function that describes the wave. That is, we must know exactly how the wave governs the particle.

In this chapter we shall first study the equation, developed by Erwin Schroedinger in 1925, which tells us the behavior of any wave function of interest. Then we shall study the relation, developed by Max Born in the following year, which connects the behavior of the wave function to the behavior of the associated particle. Detailed solutions of the Schroedinger equation are deferred to the following chapters, but in this chapter we shall look at its solutions in a general way, and we shall see how they lead very naturally to the quantization of energy and other important phenomena.

We can appreciate some of the problems concerning the applicability of the de Broglie postulate, and also get some clues about what will have to be done to remove the problems, by considering again the case of a free particle. In this case we have been successful in doing much with the postulate. When, in Chapter 3, it was necessary to have a mathematical expression for a wave function, we used a simple sinusoidal traveling wave, such as

$$\Psi(x,t) = \sin 2\pi\left(\frac{x}{\lambda} - \nu t\right) \tag{5-1}$$

or else a wave function formed by adding several simple sinusoidals. The form in (5-1) was obtained essentially by guessing, with the guess being based on the fact that a free particle has a linear momentum $p$ of constant magnitude, since it is not acted on by a force, and therefore it has an associated de Broglie wavelength $\lambda = h/p$ of constant magnitude. Equation (5-1) is just the familiar form for a sinusoidal traveling wave of constant wavelength $\lambda$. It also has a constant frequency $\nu$, which we evaluated from the Einstein relation $\nu = E/h$, where $E$ is the total energy of the associated particle.

In Chapter 4 we were able to extend the use of a wave function like (5-1) to the case of a particle moving in a circular Bohr orbit by imagining such a sinusoidal wrapped around the orbit. But this was possible only because in a circular orbit the magnitude $p$ of the linear momentum remains constant so that $\lambda = h/p$, the de Broglie wavelength, is also constant, even though the particle is acted on by a force.

We shall not be able to make such simple extensions to treat cases where the linear momentum of the particle is of changing magnitude, and, of course, these cases are typical of what happens when a particle is acted on by a force. The point is that the de Broglie postulate, $\lambda = h/p$, says the wavelength $\lambda$ will change if $p$ changes; but a wavelength is not even well defined if it changes very rapidly. We illustrate this with the nonsinusoidal wave shown in Figure 5-1. For this wave it is difficult to define even a variable wavelength since the separation between adjacent maxima is not equal to the separation between adjacent minima. To put the point another way, if the linear momentum of a particle is not of constant magnitude because the particle is acted on by a force, functions which are more complicated than the sinusoidal of (5-1) are required to describe the associated wave. We shall need help to find these more complicated wave functions.

The *Schroedinger equation* will provide the required assistance. This is the equation

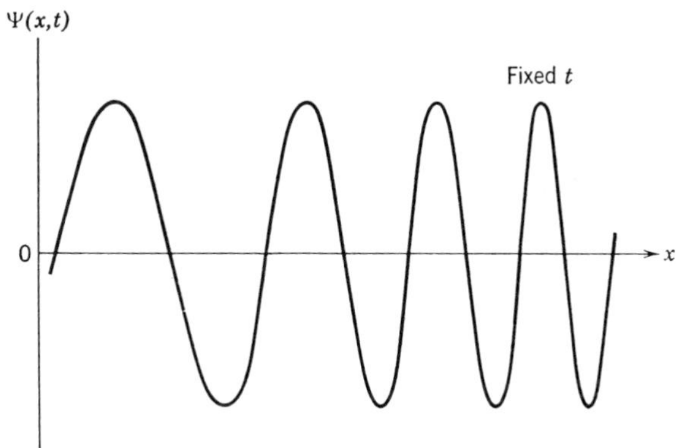


**FIGURE 5-1**

A non-sinusoidal wave. Inspection will show that the separation between an adjacent pair of maxima differs from that between the closest adjacent pair of minima. Therefore it is difficult to define a wavelength even for a single oscillation.

which tells us the form of the wave function $\Psi(x,t)$, if we tell it about the force acting on the associated particle by specifying the potential energy corresponding to the force. In other words, the wave function is a solution to the Schroedinger equation for that potential energy. The most common type of equation which has a function for a solution is a *differential equation.* In fact, the Schroedinger equation is a differential equation. That is, the equation is a relation between its solution $\Psi(x,t)$ and certain derivatives of $\Psi(x,t)$ with respect to the independent space and time variables $x$ and $t$. As there is more than one independent variable, these must be *partial derivatives*, such as

$$\frac{\partial\Psi(x,t)}{\partial x} \quad \text{or} \quad \frac{\partial\Psi(x,t)}{\partial t} \quad \text{or} \quad \frac{\partial^2\Psi(x,t)}{\partial x^2} \quad \text{or} \quad \frac{\partial^2\Psi(x,t)}{\partial t^2} \tag{5-2}$$

**Example 5-1.** Evaluate the partial derivatives listed above of the sinusoidal function, (5-1).

A partial derivative is a derivative of a function of several independent variables, which is evaluated by allowing one of the variables to vary, while holding all the others temporarily fixed. This is indicated by using a symbol such as $\partial\Psi(x, t)/\partial x$ instead of the usual symbol for the ordinary derivative $d\Psi(x, t)/dx$. The symbol means, for instance

$$\frac{\partial\Psi(x, t)}{\partial x} \equiv \left[\frac{d\Psi(x,t)}{dx}\right]_{\text{evaluated by treating } t \text{ as a constant}} \tag{5-3}$$

or

$$\frac{\partial\Psi(x, t)}{\partial t} \equiv \left[\frac{d\Psi(x,t)}{dt}\right]_{\text{evaluated by treating } x \text{ as a constant}} \tag{5-4}$$

Before applying this procedure on the sinusoidal function of (5-1), it is convenient to rewrite it in terms of the quantities $k = 2\pi/\lambda$ and $\omega = 2\pi\nu$. We obtain

$$\Psi(x, t) = \sin 2\pi\left(\frac{x}{\lambda} - \nu t\right) = \sin(kx - \omega t)$$

The partial differentiations then yield

$$\begin{aligned}
\frac{\partial\Psi(x,t)}{\partial x} &= \frac{\partial \sin(kx - \omega t)}{\partial x} = k\cos(kx - \omega t) \\
\frac{\partial^2\Psi(x,t)}{\partial x^2} &= k\,\frac{\partial \cos(kx - \omega t)}{\partial x} = -k^2 \sin(kx - \omega t) \\
\frac{\partial\Psi(x,t)}{\partial t} &= \frac{\partial \sin(kx - \omega t)}{\partial t} = -\omega\cos(kx - \omega t) \\
\frac{\partial^2\Psi(x,t)}{\partial t^2} &= -\omega\,\frac{\partial \cos(kx - \omega t)}{\partial t} = -\omega^2 \sin(kx - \omega t)
\end{aligned} \tag{5-5}$$

since $t$ can be treated as a constant in the first two differentiations, whereas $x$ can be treated as a constant in the last two. These results will prove to be useful shortly. ◀

The Schroedinger equation is a partial differential equation. We shall, in due course, study solutions of this equation, and we shall see that it is generally quite easy to decompose it into a set of ordinary differential equations (i.e., differential equations involving only ordinary derivatives). These ordinary differential equations will then be handled by the application of straightforward techniques. In all this work we shall assume no previous knowledge about differential equations of any type on the part of the student. We shall assume only that he knows how to differentiate and integrate. Of course, the student very probably has had some experience with ordinary differential equations in connection with his study of classical mechanics. He has probably even had a little experience with partial differential equations because the Schroedinger equation is a member of the class of partial differential equations called wave equations, which arise in many fields of classical as well as quantum physics. Examples from the former field are the wave equation for vibrations in a stretched string and the wave equation for electromagnetic radiation. We shall see that the quantum mechanical wave equation has many properties in common with the classical wave equation, and also that it has some very interesting differences.

## 5-2 Plausibility Argument Leading to Schroedinger's Equation

Now the first problem at hand is not how to solve a certain differential equation; instead, the problem is how to *find* the equation. That is, we are in the position of Newton when he was looking for the differential equation

$$F = \frac{dp}{dt} = m\,\frac{d^2x}{dt^2} \tag{5-6}$$

which is the basic equation of classical mechanics, or of Maxwell, when he was looking for the differential equations such as

$$\frac{\partial E_x}{\partial x} + \frac{\partial E_y}{\partial y} + \frac{\partial E_z}{\partial z} = \frac{\rho}{\epsilon_0} \tag{5-7}$$

that form the basis of classical electromagnetism.

The wave equation for a stretched string can be derived from Newton's law, and the electromagnetic wave equation can be derived from Maxwell's equations; but we cannot expect to be able to derive the *quantum mechanical* wave equation from any of the equations of *classical* physics. However, we can expect to receive some help from the de Broglie-Einstein postulates

$$\lambda = h/p \qquad \text{and} \qquad \nu = E/h \tag{5-8}$$

which connect the wavelength $\lambda$ of the wave function with the linear momentum $p$ of the associated particle, and also connect the frequency $\nu$ of the wave function with the total energy $E$ of the particle, for the case of a particle with essentially constant $p$ and $E$. That is, the quantum mechanical wave equation we seek must be consistent with these postulates, and we shall use this required consistency in our search. Equations (5-8), plus others that we shall have reason to accept, will be woven into an argument that is designed to make the quantum mechanical wave equation seem very plausible, but it must be emphasized that this *plausibility argument* will not constitute a derivation. In the final analysis, the quantum mechanical wave equation will be obtained by a *postulate*, whose justification is not that it has been deduced entirely from information already known experimentally, but that it correctly predicts results which can be verified experimentally.

We begin our plausibility argument by listing four reasonable assumptions concerning the properties of the desired quantum mechanical wave equation:

1. It must be consistent with the de Broglie-Einstein postulates, (5-8)

$$\lambda = h/p \qquad \text{and} \qquad \nu = E/h$$

2. It must be consistent with the equation

$$E = p^2/2m + V \tag{5-9}$$

relating the total energy $E$ of a particle of mass $m$ to its kinetic energy $p^2/2m$ and its potential energy $V$.

3. It must be *linear* in $\Psi(x,t)$. That is, if $\Psi_1(x,t)$ and $\Psi_2(x,t)$ are two different solutions to the equation for a given potential energy $V$ (we shall see that partial differential equations have many solutions), then any arbitrary linear combination of these solutions, $\Psi(x,t) = c_1\Psi_1(x,t) + c_2\Psi_2(x,t)$, is also a solution. This combination is said to be linear since it involves the first (linear) power of $\Psi_1(x,t)$ and $\Psi_2(x,t)$; it is said to be arbitrary since the constants $c_1$ and $c_2$ can have any (arbitrary) values. This *linearity* requirement ensures that we shall be able to *add together wave functions* to produce the constructive and destructive interferences that are so characteristic of waves. Interference phenomena are commonplace for electromagnetic waves; all the diffraction patterns of physical optics are understood in terms of the addition of electromagnetic waves. But the Davisson-Germer experiment, and others, show that diffraction patterns are also found in the motion of electrons, and other particles. Therefore, their wave functions also exhibit interferences, and so they should be capable of being added.

4. The potential energy $V$ is generally a function of $x$, and possibly even $t$. However, there is an important special case where

$$V(x,t) = V_0 \tag{5-10}$$

This is just the case of the free particle since the force acting on the particle is given by

$$F = -\partial V(x,t)/\partial x$$

which yields $F = 0$ if $V_0$ is a constant. In this case Newton's law of motion tells us that the linear momentum $p$ of the particle will be constant, and we also know that its total energy $E$ will be constant. We have here the situation of a free particle with constant values of $\lambda = h/p$ and $\nu = E/h$, discussed in Chapter 3. We therefore assume that, in this case, the desired differential equation will have sinusoidal traveling wave solutions of constant wavelength and frequency, similar to the sinusoidal wave function, (5-1), considered in that chapter.

Using the de Broglie-Einstein relations of assumption 1 to write the energy equation of assumption 2 in terms of $\lambda$ and $\nu$, we obtain

$$h^2/2m\lambda^2 + V(x,t) = h\nu$$

Before proceeding, it is convenient to introduce the quantities

$$k = 2\pi/\lambda \qquad \text{and} \qquad \omega = 2\pi\nu \tag{5-11}$$

As in Example 5-1, they are useful because they keep variables out of denominators and because they "absorb" a factor of $2\pi$ that would otherwise appear every time we write a sinusoidal wave function. The quantity $k$ is called the *wave number*; the quantity $\omega$ is called the *angular frequency*. Introducing them, we obtain

$$\hbar^2k^2/2m + V(x,t) = \hbar\omega \tag{5-12}$$

where

$$\hbar \equiv h/2\pi$$

is Planck's constant divided by $2\pi$. To satisfy assumptions 1 and 2, the wave equation we seek must be consistent with (5-12).

In order to satisfy the linearity assumption 3, it is necessary that every term in the differential equation be linear in $\Psi(x,t)$, i.e., be proportional to the first power of $\Psi(x,t)$. Note that any derivative of $\Psi(x,t)$ has this property. For instance, if we consider the change in the magnitude of $\partial^2\Psi(x,t)/\partial x^2$ that results if we change the magnitude of $\Psi(x,t)$, say by a factor of $c$, we see that the derivative increases by the same factor and thus is proportional to the first power of the function. This is true since

$$\frac{\partial^2[c\Psi(x,t)]}{\partial x^2} = c\frac{\partial^2\Psi(x,t)}{\partial x^2}$$

where $c$ is any constant. In order that the differential equation itself be linear in $\Psi(x,t)$, it cannot contain any term which is independent of $\Psi(x,t)$, i.e., which is proportional to $[\Psi(x,t)]^0$, or which is proportional to $[\Psi(x,t)]^2$ or any higher power. After obtaining the equation, we shall demonstrate explicitly that it is linear in $\Psi(x,t)$, and in the process the validity of these statements will become apparent.

Now let us use the assumption 4, which concerns the form of the free particle solution. As suggested by that assumption, we shall first try to write an equation containing the sinusoidal wave function, (5-1), and/or derivatives of that wave function. We have already evaluated some of the derivatives in Examples 5-1. Inspecting these, we see that the effect of taking the second space derivative is to introduce a factor of $-k^2$, and the effect of taking the first time derivative is to introduce a factor of $-\omega$. Since the differential equation we seek must be consistent with (5-12), which contains a factor of $k^2$ in one term and a factor of $\omega$ in another, these facts suggest that the differential equation should contain a second space derivative of $\Psi(x,t)$ and a first time derivative of $\Psi(x,t)$. But there must also be a term containing a factor of $V(x,t)$ because it is present in (5-12). In order to ensure linearity, this term must contain a factor of $\Psi(x,t)$. Putting all these ideas together, we try the following form for the differential equation

$$\alpha\frac{\partial^2\Psi(x,t)}{\partial x^2} + V(x,t)\Psi(x,t) = \beta\frac{\partial\Psi(x,t)}{\partial t} \tag{5-13}$$

The constants $\alpha$ and $\beta$ have values which remain to be determined. They are used to provide flexibility which, we might guess, will be needed in fitting (5-13) to the various requirements it must satisfy.

The form of (5-13) seems reasonable in general, but will it work in detail? To find out we consider the case of a constant potential, $V(x,t) = V_0$, and evaluate $\Psi(x,t)$ and its derivatives from (5-1) and (5-5). We obtain immediately

$$-\alpha \sin (kx - \omega t)k^2 + \sin (kx - \omega t)V_0 = -\beta \cos (kx - \omega t)\omega \tag{5-14}$$

Even though the constants $\alpha$ and $\beta$ are at our disposal, we cannot make this agree with (5-12), and thus satisfy assumptions 1 and 2, except for special combinations of the independent variables $x$ and $t$ for which $\sin (kx - \omega t) = \cos (kx - \omega t)$. It is true that we could obtain agreement if $\alpha$ and $\beta$ were not constants, but we reject this possibility in favor of the very much simpler one presented next.

The difficulty at hand arises because differentiation changes cosines into sines, and vice versa. This fact suggests that we try using for the free particle wave function not the single sinusoidal of (5-1), but instead the combination

$$\Psi(x,t) = \cos (kx - \omega t) + \gamma \sin (kx - \omega t) \tag{5-15}$$

where $\gamma$ is a constant, of as yet undetermined value, which is introduced for the purpose of providing additional flexibility. We hope to find the proper mixture of a cosine and a sine that will remove the difficulty. Evaluating the required derivatives, we find

$$\begin{aligned} \frac{\partial \Psi(x,t)}{\partial x} &= -k \sin (kx - \omega t) + k\gamma \cos (kx - \omega t) \\ \frac{\partial^2 \Psi(x,t)}{\partial x^2} &= -k^2 \cos (kx - \omega t) - k^2\gamma \sin (kx - \omega t) \\ \frac{\partial \Psi(x,t)}{\partial t} &= \omega \sin (kx - \omega t) - \omega\gamma \cos (kx - \omega t) \end{aligned} \tag{5-16}$$

Then we try again; substituting (5-15) and (5-16) into the same assumed form, (5-13), for the differential equation, and setting $V(x,t) = V_0$, we obtain

$$\begin{aligned} -\alpha k^2 \cos (kx - \omega t) - \alpha k^2\gamma \sin (kx - \omega t) + V_0 \cos (kx - \omega t) \\ + V_0\gamma \sin (kx - \omega t) = \beta\omega \sin (kx - \omega t) - \beta\omega\gamma \cos (kx - \omega t) \end{aligned}$$

or

$$[-\alpha k^2 + V_0 + \beta\omega\gamma] \cos (kx - \omega t) + [-\alpha k^2\gamma + V_0\gamma - \beta\omega] \sin (kx - \omega t) = 0$$

In order that the last equality hold for all possible combinations of the independent variables $x$ and $t$, it is necessary that the coefficients of both the cosine and the sine be zero. Thus we obtain

$$-\alpha k^2 + V_0 = -\beta\gamma\omega \tag{5-17}$$

and

$$-\alpha k^2 + V_0 = \beta\omega/\gamma \tag{5-18}$$

Now we have a problem that is easily handled; there are three algebraic equations that we must satisfy, (5-12), (5-17), and (5-18), but we have three free constants $\alpha$, $\beta$, and $\gamma$, at our disposal.

Subtracting (5-18) from (5-17), we find

$$0 = -\beta\gamma\omega - \beta\omega/\gamma$$

or

$$\gamma = -1/\gamma$$

so that

$$\gamma^2 = -1$$

or

$$\gamma = \pm\sqrt{-1} \equiv \pm i \tag{5-19}$$

where $i$ is the *imaginary number* (see Appendix E). Substituting this result into (5-17) we find

$$-\alpha k^2 + V_0 = \mp i\beta\omega$$

This can be compared directly with (5-12)

$$\hbar^2 k^2/2m + V_0 = \hbar\omega$$

to yield

$$\alpha = -\hbar^2/2m \tag{5-20}$$

and

$$\mp i\beta = \hbar$$

or

$$\beta = \pm i\hbar \tag{5-21}$$

There are two possible choices of the sign in (5-19). It turns out to be of no significant consequence which choice is made, and therefore we follow conventional usage and choose the plus sign. Then (5-21) yields $\beta = +i\hbar$ and, with (5-20), we finally can evaluate all the constants in the assumed form of the differential equation. Thus (5-13) becomes

$$-\frac{\hbar^2}{2m}\frac{\partial^2\Psi(x,t)}{\partial x^2} + V(x,t)\Psi(x,t) = i\hbar\frac{\partial\Psi(x,t)}{\partial t} \tag{5-22}$$

*This differential equation satisfies all four of our assumptions concerning the quantum mechanical wave equation.*

It should be emphasized that we have been led to (5-22) by treating a special case: the case of a free particle where $V(x,t) = V_0$, a constant. At this point it seems plausible to argue that the quantum mechanical wave equation might be expected to have the same form as (5-22) in the general case where the potential energy $V(x,t)$ does actually vary as a function of $x$ and/or $t$ (i.e., where the force is not zero); but we cannot prove this to be true. We can, however, *postulate* it to be true. We do this, and therefore take (5-22) as the quantum mechanical wave equation whose solutions $\Psi(x,t)$ give us the wave function which is to be associated with the motion of a particle of mass $m$ under the influence of forces which are described by the potential energy function $V(x,t)$. The validity of the postulate must be judged by comparing its implications with experiment, and we shall make many such comparisons later. Equation (5-22) was first obtained in 1926 by Erwin Schroedinger, and it is therefore called the *Schroedinger equation*.

Schroedinger was led to his equation by an argument different from ours (and more esoteric). We shall see the essential ideas of his argument in Section 5-4. However, he was as strongly influenced by the de Broglie postulate in his work as we have been in ours. This can be seen in the following quotation, in which the physicist Debye describes the circumstances surrounding Schroedinger's development of his equation.

> "Then de Broglie published his paper. At that time Schroedinger was my successor at the University in Zurich, and I was at the Technical University, which is a Federal Institute, and we had a colloquium together. We were talking about de Broglie's theory and agreed that we did not understand it, and that we should really think about his formulations and what they mean. So I called Schroedinger to give us a colloquium. And the preparation of that really got him started. There were only a few months between his talk and his publications."

It should be pointed out that we cannot expect the Schroedinger equation to be valid when applied to particles moving at relativistic velocities. This is the case because the equation has been designed to be consistent with (5-9), the classical energy equation, which is incorrect for velocities comparable to the velocity of light. In 1928 Dirac developed a relativistic theory of quantum mechanics utilizing essentially the same postulates as the Schroedinger theory, except that (5-9) was replaced by its relativistic analogue

$$E = \sqrt{c^2p^2 + (m_0c^2)^2} + V$$

The Dirac theory reduces to the Schroedinger theory, of course, in the low-velocity limit. Because of the serious complications introduced by the square root in the relativistic energy equation, a quantitative treatment of the Dirac theory would not be appropriate in this book. However, some of the more interesting features of the Dirac theory will be described qualitatively in the following chapters on occasions when relativistic quantum phenomena must be discussed; and one feature, pair production, has already been described. Fortunately, most of the interesting quantum phenomena can be studied in cases which are nonrelativistic.

**Example 5-2.** Verify that the Schroedinger equation is linear in the wave function $\Psi(x,t)$; i.e., that it is consistent with the linearity assumption 3.

We must show that, if $\Psi_1(x,t)$ and $\Psi_2(x,t)$ are two solutions to (5-22) for a particular $V(x,t)$, then

$$\Psi(x,t) = c_1\Psi_1(x,t) + c_2\Psi_2(x,t)$$

is also a solution to that equation, where $c_1$ and $c_2$ are constants of arbitrary value. Transposing (5-22), we have for the Schroedinger equation

$$-\frac{\hbar^2}{2m}\frac{\partial^2\Psi}{\partial x^2} + V\Psi - i\hbar\frac{\partial\Psi}{\partial t} = 0$$

Now we check the validity of the linear combination by substituting it into this equation it is supposed to satisfy. We obtain

$$-\frac{\hbar^2}{2m}\left(c_1\frac{\partial^2\Psi_1}{\partial x^2} + c_2\frac{\partial^2\Psi_2}{\partial x^2}\right) + V(c_1\Psi_1 + c_2\Psi_2) - i\hbar\left(c_1\frac{\partial\Psi_1}{\partial t} + c_2\frac{\partial\Psi_2}{\partial t}\right) = 0$$

which can be rewritten as

$$c_1\left[-\frac{\hbar^2}{2m}\frac{\partial^2\Psi_1}{\partial x^2} + V\Psi_1 - i\hbar\frac{\partial\Psi_1}{\partial t}\right] + c_2\left[-\frac{\hbar^2}{2m}\frac{\partial^2\Psi_2}{\partial x^2} + V\Psi_2 - i\hbar\frac{\partial\Psi_2}{\partial t}\right] = 0$$

If the linear combination actually is a solution to the Schroedinger equation then the last equality should be satisfied. It is, for all values of $c_1$ and $c_2$, because the Schroedinger equation says each bracket equals zero since $\Psi_1$ and $\Psi_2$ are solutions to that equation for the same $V$.

A little thought should convince the student that this essential result would not be obtained if the Schroedinger equation contained any terms which are not proportional to the first power of $\Psi(x,t)$. ◀

In following chapters we shall solve in a methodical way Schroedinger's equation for a number of important systems, and we shall obtain thereby the wave functions that describe the systems. But in this chapter we must use some of these wave functions in order to illustrate various properties of the Schroedinger theory. These wave functions will be "pulled out of the hat," as required. However, we shall give the student confidence in their validity by verifying that each is a solution to the Schroedinger equation, for the system it is supposed to describe, by the simple procedure of substituting it into that equation. In Example 5-3 we do this for a wave function which is particularly useful for illustrative purposes.

**Example 5-3.** The wave function $\Psi(x,t)$ for the lowest energy state of a simple harmonic oscillator, consisting of a particle of mass $m$ acted on by a linear restoring force of force constant $C$, can be expressed as

$$\Psi(x,t) = Ae^{-(\sqrt{Cm}/2\hbar)x^2}e^{-(i/2)\sqrt{C/m}\,t}$$

where the real constant $A$ can have any value. Verify that this expression is a solution to the Schroedinger equation for the appropriate potential. (The time-dependent term is a complex exponential; see Appendix E.)

The expression applies to the case in which the equilibrium point of the oscillator (the point at which the classical particle would rest if it were not oscillating) is at the origin of the $x$ axis ($x = 0$). In this case the time-independent potential energy is

$$V(x,t) = V(x) = Cx^2/2$$

as can be verified by noting that the corresponding force, $F = -dV(x)/dx = -Cx$, is a linear restoring force of force constant $C$. The Schroedinger equation for this potential is

$$-\frac{\hbar^2}{2m}\frac{\partial^2\Psi}{\partial x^2} + \frac{C}{2}x^2\Psi = i\hbar\frac{\partial\Psi}{\partial t}$$

To check the validity of the solution quoted, we evaluate its derivatives. We find

$$\frac{\partial\Psi}{\partial t} = -\frac{i}{2}\sqrt{\frac{C}{m}}\Psi$$

and

$$\frac{\partial\Psi}{\partial x} = -\frac{\sqrt{Cm}}{2\hbar}2x\Psi = -\frac{\sqrt{Cm}}{\hbar}x\Psi$$

$$\frac{\partial^2\Psi}{\partial x^2} = -\frac{\sqrt{Cm}}{\hbar}\Psi - \frac{\sqrt{Cm}}{\hbar}x\left(-\frac{\sqrt{Cm}}{\hbar}x\Psi\right)$$

$$= -\frac{\sqrt{Cm}}{\hbar}\Psi + \frac{Cm}{\hbar^2}x^2\Psi$$

Substituting into the Schroedinger equation yields

$$\frac{\hbar^2\sqrt{Cm}}{2m\hbar}\Psi - \frac{\hbar^2 Cm}{2m\hbar^2}x^2\Psi + \frac{C}{2}x^2\Psi = i\hbar\left(-\frac{i}{2}\right)\sqrt{\frac{C}{m}}\Psi$$

or

$$\frac{\hbar}{2}\sqrt{\frac{C}{m}}\Psi - \frac{C}{2}x^2\Psi + \frac{C}{2}x^2\Psi = \frac{\hbar}{2}\sqrt{\frac{C}{m}}\Psi$$

Since the last equality is obviously satisfied, the solution must be valid.

The general solution to the simple harmonic oscillator Schroedinger equation is treated in the following chapter. ◀

## 5-3 Born's Interpretation of Wave Functions

A very interesting and important property of wave functions can be seen by evaluating $\gamma = i$ in (5-15), which specifies the form of the free particle wave function. We obtain

$$\Psi(x,t) = \cos(kx - \omega t) + i\sin(kx - \omega t) \tag{5-23}$$

*The wave function is complex.* That is, it contains the imaginary number $i$. Recall that this behavior was forced upon us. We first tried to find a way of satisfying our four assumptions concerning the Schroedinger equation by using a purely real free particle wave function, (5-1), and we found that there was no reasonable way of doing this.

Only when we allowed the free particle wave function to have an imaginary part, by using the free particle wave function of (5-15) in which $\gamma$ turned out to be equal to $i$, did we succeed. In this process, we also ended up with an $i$ in the Schroedinger equation, (5-22). If the student looks carefully at our plausibility argument, it will become apparent that the equation contains an $i$ because it relates a *first time* derivative to a *second space* derivative. This is due, in turn, to the fact that the Schroedinger equation is based on the energy equation which relates the first power of total energy to the second power of momentum. The presence of an $i$ in the Schroedinger equation implies that in the general case (for any potential energy function) the wave functions which are its solutions will be complex. We shall shortly see that this is true.

Since a wave function of quantum mechanics is complex, it specifies simultaneously two real functions, its *real* part and its *imaginary* part (see Appendix E). This is in contrast to a "wave function" of classical mechanics. For instance, a wave in a string can be specified by one real function which gives the displacement of various elements of the string at various times. This classical wave function is not complex because the classical wave equation does not contain an $i$ since it relates a second time derivative to a second space derivative.

The fact that wave functions are complex functions should not be considered a weak point of the quantum mechanical theory. Actually, it is a desirable feature because it makes it immediately apparent that we should not attempt to give to wave functions a physical existence in the same sense that water waves have a physical existence. The reason is that a complex quantity cannot be measured by any actual physical instrument. The "real" world (using the term in its nonmathematical sense) is the world of "real" quantities (using the term in its mathematical sense).

Therefore, we should not try to answer, or even pose the question: Exactly what is waving, and what is it waving in? The student will remember that consideration of just such questions concerning the nature of electromagnetic waves led the nineteenth century physicists to the fallacious concept of the ether. As the wave functions are complex, there is no temptation to make the same mistake again. Instead, it is apparent from the outset that *the wave functions are computational devices* which have a significance only in the context of the Schroedinger theory of which they are a part. These comments should not be taken to imply that the wave functions have no physical interest. We shall see in this and the next sections that a wave function actually contains all the information which the uncertainty principle allows us to know about the associated particle.

The basic connection between the properties of the wave function $\Psi(x,t)$ and the behavior of the associated particle is expressed in terms of the *probability density* $P(x,t)$. This quantity specifies the probability, per unit length of the $x$ axis, of finding the particle near the coordinate $x$ at time $t$. According to a *postulate*, first stated in 1926 by Max Born, the relation between the probability density and the wave function is

$$P(x,t) = \Psi^*(x,t)\Psi(x,t) \tag{5-24}$$

where the symbol $\Psi^*(x,t)$ represents the *complex conjugate* of $\Psi(x,t)$ (see Appendix E). For emphasis, and clarification, we shall restate Born's postulate as follows:

*If, at the instant $t$, a measurement is made to locate the particle associated with the wave function $\Psi(x,t)$, then the probability $P(x,t)\,dx$ that the particle will be found at a coordinate between $x$ and $x + dx$ is equal to $\Psi^*(x,t)\Psi(x,t)\,dx$.*

Justification of the postulate can be found in the following considerations. Since the motion of a particle is connected with the propagation of an associated wave function (the de Broglie connection), these two entities must be associated in space.

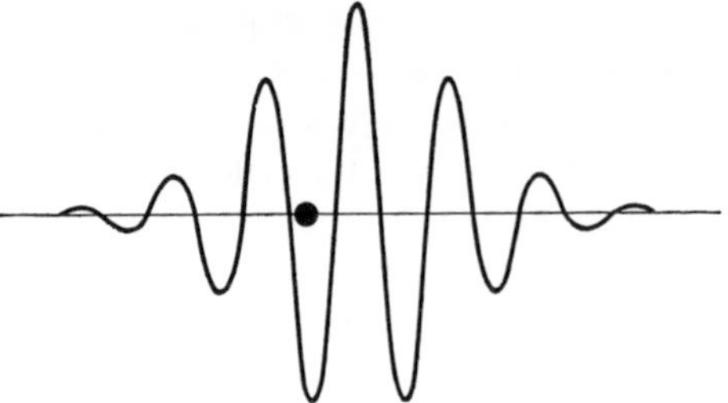

**FIGURE 5-2**
A very schematic picture of a wave function and its associated particle. The particle must be at some location where the wave function has an appreciable amplitude.

That is, the particle must be at some location where the waves have an appreciable amplitude. Therefore $P(x,t)$ must have an appreciable value where $\Psi(x,t)$ has an appreciable value. We attempt to illustrate schematically the situation in Figure 5-2. If the situation were otherwise, there would be serious difficulties with the theory. For instance, if the particle were separated in space from the wave, relativistic problems would arise because of the time required to transmit information between the two entities that are required to follow each other. Since the measurable quantity probability density $P(x,t)$ is real and non-negative, whereas the wave function $\Psi(x,t)$ is complex, it is obviously not possible to equate $P(x,t)$ to $\Psi(x,t)$. However, since $\Psi^*(x,t)\Psi(x,t)$ is always real and non-negative, Born was not inconsistent in equating it to $P(x,t)$.

**Example 5-4.** Prove that $\Psi^*(x,t)\Psi(x,t)$ is necessarily real, and either positive or zero.

Any complex function, such as $\Psi(x,t)$, can always be written

$$\Psi(x,t) = R(x,t) + iI(x,t) \tag{5-25a}$$

where $R(x,t)$ and $I(x,t)$ are both real functions that are called, respectively, its *real* and *imaginary* parts. The *complex conjugate* of $\Psi(x,t)$ is defined as

$$\Psi^*(x,t) \equiv R(x,t) - iI(x,t) \tag{5-25b}$$

Multiplying the two together, we obtain

$$\Psi^*\Psi = (R - iI)(R + iI)$$

or, since $i^2 = -1$

$$\Psi^*\Psi = R^2 - i^2I^2 = R^2 + I^2$$

Thus

$$\Psi^*(x,t)\Psi(x,t) = [R(x,t)]^2 + [I(x,t)]^2 \tag{5-26}$$

That is, it equals the sum of the squares of two real functions. Thus $\Psi^*(x,t)\Psi(x,t)$ must be real, and either positive or zero. ◀

Of course, there are other possible functions that can be generated from $\Psi(x,t)$ that are real. An example is the absolute value, or modulus, $|\Psi(x,t)|$. However, all these other possibilities can be ruled out by arguments, too lengthy to reproduce here, which show that they would lead to an unphysical behavior for $P(x,t)$.

It is worthwhile for us to consider again an analogy between electromagnetism and quantum mechanics, discussed in Section 3-2. The connection between the density of photons in a field of electromagnetic radiation and the square of the electric field vector is analogous to the connection between the probability density and the wave function multiplied by its complex conjugate. Consider, for instance, that the electric field vector is a solution to the electromagnetic wave equation, while the wave function is a solution to the quantum mechanical wave equation. Both quantities specify the amplitudes of waves, although the electric vector is real whereas the wave function is complex. Therefore, the square of the amplitude of the waves, $\mathscr{E}^2$, gives the intensity of the waves in the electromagnetic case, while it is necessary to take the amplitude times its complex conjugate, $\Psi^*\Psi$, to obtain a real intensity in the quantum mechanical case. In the electromagnetic case the intensity of the waves is proportional to their energy density. Since each photon in the electromagnetic field carries energy $h\nu$, the energy density is, in turn, proportional to the density of photons. For one dimension, this is the probability per unit length of finding a photon. In the quantum mechanical case the intensity of the waves gives directly the probability density which is, in one dimension, the probability per unit length of finding a particle.

**Example 5-5.** Evaluate the probability density for the simple harmonic oscillator lowest energy state wave function quoted in Example 5-3.

The wave function is

$$\Psi(x,t) = Ae^{-(\sqrt{Cm}/2\hbar)x^2}e^{-(i/2)\sqrt{C/m}\,t}$$

The probability density is therefore (see Appendix E for the evaluation of $\Psi^*$)

$$P = \Psi^*\Psi = Ae^{-(\sqrt{Cm}/2\hbar)x^2}e^{+(i/2)\sqrt{C/m}\,t}Ae^{-(\sqrt{Cm}/2\hbar)x^2}e^{-(i/2)\sqrt{C/m}\,t}$$

or

$$P = A^2e^{-(\sqrt{Cm}/\hbar)x^2}$$

Note that the probability density is independent of time, even though the wave function depends on time. We shall see later that this is true in any case in which the particle associated with the wave function is in a single energy state. The probability density $P$ predicted by quantum mechanics is plotted as a function of $x$ by the solid curve in the upper part of Figure 5-3. The probability that a measurement of the location of the oscillating particle will find it in an element of the $x$ axis between $x$ and $x + dx$ is equal to $P\,dx$.

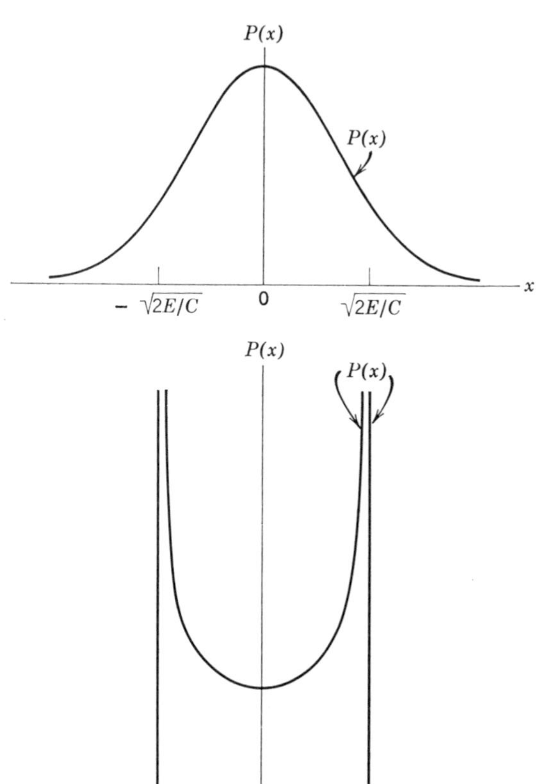


**FIGURE 5-3**

Quantum mechanical (*top*) and classical (*bottom*) probability densities for a particle in the lowest energy state of a simple harmonic oscillator. The quantum mechanical probability density peaks near the equilibrium point and extends beyond the sharp limits of motion predicted by classical physics. The classical probability density is inversely proportional to the classical velocity and is greatest at the endpoints of the motion, where the velocity vanishes.

Since $P$ has a maximum at $x = 0$, the equilibrium point of the oscillator, quantum mechanics predicts that the particle is most likely found in an element $dx$ located at the equilibrium point. Proceeding in either direction from that location, the chances of finding it in an element of the same length $dx$ decrease rather rapidly, but there are no well-defined limits beyond which the probability of finding the particle in an element of the $x$ axis is precisely zero. In the following example we shall find that these predictions are very different from what would be expected for the oscillating particle according to classical mechanics. ◀

**Example 5-6.** Evaluate the predictions of classical mechanics for the probability density of the simple harmonic oscillator of Example 5-5, and compare them with the quantum mechanical predictions found in that example.

In classical mechanics the oscillating particle has a definite momentum $p$, and therefore a definite velocity $v$, at every value of its displacement $x$ from the equilibrium point. The probability of finding it in an element of the $x$ axis of fixed length is proportional to the amount of time it spends in the element, and this is inversely proportional to its velocity when it passes through the element. That is

$$P = \frac{B^2}{v}$$

where $B^2$ is some constant. We obtain an expression for $v$ in terms of $x$ most simply by considering the energy equation

$$E = K + V = \frac{mv^2}{2} + \frac{Cx^2}{2}$$

where $E$, $K$, and $V$ are total, kinetic, and potential energies, and where the latter has been evaluated in terms of $x$ and the oscillator force constant $C$ from an equation justified in Example 5-3. We have then

$$\frac{mv^2}{2} = E - \frac{Cx^2}{2}$$

or

$$v = \sqrt{\frac{2}{m}}\sqrt{E - \frac{Cx^2}{2}}$$

So

$$P = \frac{B^2}{\sqrt{\frac{2}{m}}\sqrt{E - \frac{Cx^2}{2}}}$$

This expression for the classical probability density $P$ is plotted as the curve in the lower part of Figure 5-3. It has a minimum value at the equilibrium point $x = 0$, and it rises rapidly near the limits of the oscillation. The limits occur at values of $x$ where the particle has no kinetic energy so the potential energy equals its total energy

$$E = \frac{Cx^2}{2}$$

or

$$x = \pm\sqrt{\frac{2E}{C}}$$

Of course, the classical probability density drops abruptly to zero outside these limits of the particle's motion, as indicated by the straight lines in the figure. Simply put, the probability of finding the oscillating classical particle in an element of the $x$ axis of a given length is smallest near the equilibrium point, where it spends the least time, and it rises rapidly near the limits of its motion, where it lingers.

The value of the constant $B^2$ in the expression for the classical probability density can be determined by imposing the requirement that the total probability of finding the particle *somewhere* must equal one. The total probability is just the integral over all $x$ of $P$ so the expression

$$\int_{-\infty}^{\infty} P\,dx = \frac{B^2}{\sqrt{2/m}} \int_{-\sqrt{2E/C}}^{+\sqrt{2E/C}} \frac{dx}{\sqrt{E - Cx^2/2}} = 1$$

can be used to evaluate $B^2$. We shall not bother to carry out this so-called *normalization* procedure for the classical probability density, although it is not difficult to do after expressing $E$ in terms of $C$; but we shall carry out such a procedure in Example 5-7 to determine the value of the corresponding constant $A^2$ that occurs in the quantum mechanical probability density.

Figure 5-3 shows that the classical prediction for the probability density is very different from the quantum mechanical prediction. According to classical mechanics, measurements of the location of the particle in the simple harmonic oscillator will always find it within two well-defined limits, and they will usually find it near one or the other of these limits. According to quantum mechanics, when the simple harmonic oscillator is in the lowest energy state measurements will usually find the particle to be near the equilibrium point, but there are no well-defined limits beyond which the particle will never be found.

When the oscillator is in its lowest energy state we are very far from the range of validity of classical physics. Thus we expect that, of the two predictions, the one made by quantum mechanics is correct. As we shall see in Chapter 12, this can be confirmed by measuring properties of diatomic molecules that depend on the interatomic spacing, since in low-energy states the two atoms in such a molecule feel the linear restoring force characteristic of simple harmonic motion. Of course, the trouble with the classical calculation is that it neglects the uncertainty principle in associating a definite value of the velocity, or momentum, of the particle with a definite value of its position. In Example 5-12 we shall make a comparison between the classical and quantum mechanical predictions of the probability density function for a particle in a high-energy state of a simple harmonic oscillator, where the range of validity of classical physics is approached because the uncertainty principle is of no consequence. There we shall find the predictions of the two theories to be very similar, as would be expected from the correspondence principle. ◀

In Example 5-5 we saw one of the predictions of quantum mechanics concerning the behavior of a particle in a simple harmonic oscillator. The prediction is typical of the type of information that the theory can provide. It cannot tell us that a particle in a given energy state will be found in a precise location at a certain time, but only the relative probabilities that the particle will be found in various locations at that time. The predictions of quantum mechanics are *statistical.*

The uncertainty principle provides the fundamental reason why quantum mechanics expresses itself in probabilities, and not in certainties. For instance, consider investigating a harmonic oscillator in some typical energy state. In order to really know that the system is in a particular state, we must make a measurement of its energy. The measurement necessarily disturbs the system in a way that cannot be completely determined, so it is not surprising that we cannot predict with certainty where the particle will be found when we make a position measurement. In classical mechanics, even though the energy of the system is microscopic, we can make the energy measurement, plus any other measurements, without disturbing the system. So classical mechanics says we can predict precisely where the particle will be found in a subsequent measurement, if we wish. But, when applied to a microscopic system, classical mechanics is wrong. Not only is it impossible to predict from classical mechanics precisely where a particle in a microscopic system will be in a subsequent measurement, it is, as we found in Example 5-6, impossible even to predict accurately from that theory the relative probabilities of finding the particle in various locations.

Quantum mechanics does allow us to make accurate predictions about these relative probabilities because it takes into account quantitatively the fundamental fact of life of the microscopic world—the uncertainty principle.

Born has expressed the situation as follows:

"We describe the instantaneous state of the system by a quantity $\Psi$, which satisfies a differential equation, and therefore changes with time in a way which is completely determined by its form at a time $t = 0$, so that its behavior is rigorously causal. Since, however, physical significance is confined to the quantity $\Psi^*\Psi$, and to other similarly constructed quadratic expressions, which only partially define $\Psi$, it follows that, even when the physically determinable quantities are completely known at time $t = 0$, the initial value of the $\Psi$-function is necessarily not completely definable. This view of the matter is equivalent to the assertion that events happen indeed in a strictly causal way, but that we do not know the initial state exactly. In this sense the law of causation is therefore empty; physics is in the nature of the case indeterminate, and therefore the affair of statistics."

The first point that Born makes, about the space dependence of $\Psi$ at some initial time being sufficient to completely determine its space dependence at any subsequent time, is a consequence of the fact that $\Psi$ satisfies the Schroedinger equation which contains only a first time derivative.

His second point, about not being able to completely define the space dependence of the wave function at the initial time, can be seen by inspecting (5-25a) and (5-26). These show that if we know a probability density from an initial set of measurements on a system, we still cannot determine uniquely an initial wave function to associate with the system. All we can determine is the sum of the squares of the real and imaginary parts of the wave function.

We can summarize the ideas of the last few paragraphs by saying that the behavior of a given wave function of a system is predictable in the sense that the Schroedinger equation for the corresponding potential energy will determine exactly its form at some later time in terms of its form at some initial time; *but* its initial form cannot be specified completely by an initial set of measurements and its final form predicts only the relative probabilities of the results of the final set of measurements. Again quoting Born: "The motion of particles conforms to the laws of probability, but the probability itself is propagated in accordance with the law of causality."

**Example 5-7.** *Normalize* the wave function of Example 5-3, by determining the value of the arbitrary constant $A$ in that wave function for which the total probability of finding the associated particle somewhere on the $x$ axis equals one.

The total probability of finding the particle *somewhere* on the entire range of the $x$ axis is necessarily equal to one if the particle exists. This total probability can be obtained mathematically by integrating the probability density function $P$ over all $x$. Doing this, and setting the result equal to one, we have

$$\int_{-\infty}^{\infty} P\,dx = \int_{-\infty}^{\infty} \Psi^*\Psi\,dx = A^2 \int_{-\infty}^{\infty} e^{-(\sqrt{Cm}/\hbar)x^2}\,dx = 1$$

Since the integrand $e^{-(\sqrt{Cm}/\hbar)\,x^2}$ depends on $x^2$, it is an *even function* of $x$. That is, its value for a certain $x$ equals its value for $-x$, as can be seen in Figure 5-4. Thus the contribution to the total value of the integral obtained in the range $-\infty$ to 0 equals the contribution obtained in the range 0 to $+\infty$, and we have

$$A^2 \int_{-\infty}^{\infty} e^{-(\sqrt{Cm}/\hbar)x^2}\,dx = 2A^2 \int_{0}^{\infty} e^{-(\sqrt{Cm}/\hbar)x^2}\,dx = 1$$

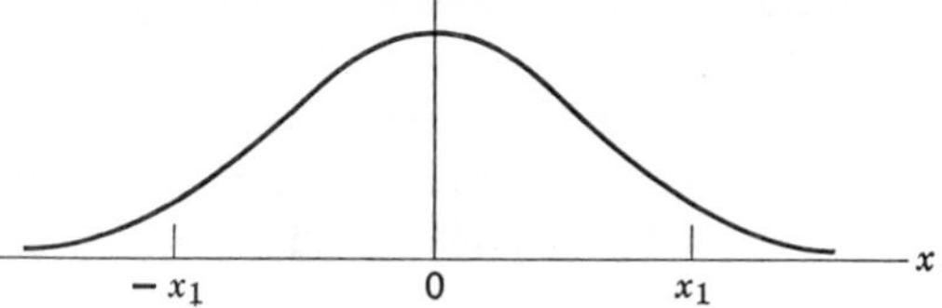


**FIGURE 5-4**
A plot of the *even* function $e^{-(\sqrt{Cm}/\hbar)x^2}$. Since the function depends on $x^2$, its value for any particular $x_1$ equals its value for $-x_1$.

The definite integral can be evaluated by consulting appropriate tables, and yields

$$\int_0^\infty e^{-(\sqrt{Cm}/\hbar)x^2}\, dx = \frac{(\pi\hbar)^{1/2}}{2(Cm)^{1/4}}$$

Then we find immediately that the required value of $A$ is

$$A = \frac{(Cm)^{1/8}}{(\pi\hbar)^{1/4}}$$

With this value of $A$, the wave function becomes

$$\Psi(x,t) = \frac{(Cm)^{1/8}}{(\pi\hbar)^{1/4}}\, e^{-(\sqrt{Cm}/2\hbar)x^2}\, e^{-(i/2)\sqrt{C/m}\, t}$$

◀

The procedure gone through in Example 5-7 is called *normalization* of a wave function, and the wave function quoted at the end of the example is said to be *normalized.* Before the procedure is carried out, the amplitude of a wave function is arbitrary because the linearity of the Schroedinger equation allows a wave function to be multiplied by a constant of arbitrary magnitude and still remain a solution to the equation. Normalizing has the effect of fixing the amplitude by fixing the value of the multiplicative constant, such as $A$ in Example 5-7. It is not always necessary to really carry through the calculation that leads to the value of the amplitude constant because useful results can often be obtained in terms of relative probabilities that are independent of the actual values of the amplitudes. But it should always be remembered that

$$\int_{-\infty}^{\infty} P\, dx = \int_{-\infty}^{\infty} \Psi^*\Psi\, dx = 1 \qquad (5\text{-}27)$$

since these integrals give the total probability of finding somewhere the particle described by the wave function, and the probability must equal one if there is a particle.

## 5-4 Expectation Values

In the previous section we saw that the wave function contains information about the behavior of the associated particle in that it specifies the probability density for the particle. In this section we shall see how to extract from the wave function a wide variety of additional information concerning the particle. That is, we shall learn how to obtain from the wave function detailed numerical information not only about the position of the particle but also about its momentum, energy, and all other quantities that characterize its behavior. For instance, we shall find out how to give quantitative evaluations of the terms $\Delta x$ and $\Delta p$ in the uncertainty principle. Wave functions are useful because they contain so much information about the behavior of the associated particle.

Consider a particle and its associated wave function $\Psi(x,t)$. In a measurement of the position of the particle in the system described by the wave function, there would be a finite probability of finding it at any $x$ coordinate in the interval $x$ to $x + dx$, as long as the wave function is nonzero in that interval. In general, the wave function is nonzero over an extended range of the $x$ axis. Thus we are generally not able to state that the $x$ coordinate of the particle has a certain definite value. However, it is possible to specify some sort of *average* position of the particle in the following way. Let us imagine making a measurement of the position of the particle at the instant $t$. The probability of finding it between $x$ and $x + dx$ is, according to Born's postulate, (5-24)

$$P(x,t)\,dx = \Psi^*(x,t)\Psi(x,t)\,dx$$

Imagine performing this measurement a number of times on identical systems described by the same wave function $\Psi(x,t)$, always at the same value of $t$, and recording the observed values of $x$ at which we find the particle. An example would be a set of measurements of the $x$ coordinates of particles in the lowest energy states of identical simple harmonic oscillators. In three dimensions, an example would be a set of measurements of the positions of electrons in hydrogen atoms, with all the atoms in their lowest energy states. We can use the average of the observed values to characterize the position at time $t$ of a particle associated with the wave function $\Psi(x,t)$. This average value we call the *expectation value* of the $x$ coordinate of the particle at the instant $t$. It is easy to see that the expectation value of $x$, which is written $\bar{x}$, will be given by

$$\bar{x} = \int_{-\infty}^{\infty} xP(x,t)\,dx$$

The reason is that the integrand in this expression is just the value of the $x$ coordinate weighted by the probability of observing that value. Therefore, we obtain upon integrating the average of the observed values. Using Born's postulate to evaluate the probability density in terms of the wave function, we obtain

$$\bar{x} = \int_{-\infty}^{\infty} \Psi^*(x,t)x\Psi(x,t)\,dx \tag{5-28}$$

The terms of the integrand are written in the order shown to preserve symmetry with a notation which will be developed later.

Some students may find these equations more familiar if they are written in the form

$$\bar{x} = \frac{\int_{-\infty}^{\infty} xP(x,t)\,dx}{\int_{-\infty}^{\infty} P(x,t)\,dx} = \frac{\int_{-\infty}^{\infty} \Psi^*(x,t)x\Psi(x,t)\,dx}{\int_{-\infty}^{\infty} \Psi^*(x,t)\Psi(x,t)\,dx}$$

but these are actually equivalent to the forms we use since (5-27) shows that the denominators equal one.

**Example 5-8.** Determine $\bar{x}$ for a particle in the lowest energy state of a simple harmonic oscillator, using the wave function and probability density considered in the preceding examples.

We can see immediately from Figures 5-3 and 5-4 that $\bar{x} = 0$. The reason is that $\bar{x}$ is the average value of $x$, with the average computed using a weighting factor $\Psi^*\Psi$ which is

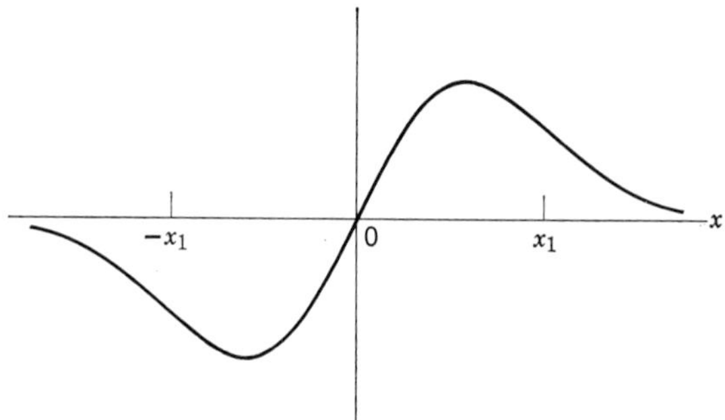


**FIGURE 5-5**
A plot of the *odd* function $xe^{-(\sqrt{Cm}/\hbar)x^2}$. The value of the function for any particular $x_1$ equals the negative of its value for $-x_1$.

symmetrical about $x = 0$; for every chance of observing a certain positive value of $x$ there is an exactly compensating chance of observing a negative value of $x$ of the same magnitude. The behavior of the particle in the oscillator is symmetrical about its equilibrium point at $x = 0$, so $\bar{x} = 0$.

More formally, we have

$$\bar{x} = \int_{-\infty}^{\infty} \Psi^* x \Psi \, dx$$

where the factor $\Psi^*\Psi$ in the integrand is plotted in Figures 5-3 and 5-4. Now this factor is an even function of $x$, and the remaining factor in the integrand is $x$ itself, which is an odd function of $x$. So the entire integrand is an *odd function* of $x$. That is, its value at a particular $x$ is exactly equal to the negative of its value at $-x$, as illustrated in Figure 5-5. From this it follows that the integral yields zero since for every contribution to its total value obtained from an element of the $x$ axis at some $x$ there is a compensating contribution of the opposite sign from the corresponding element at $-x$.

From arguments using a coordinate system in which the origin of the $x$ axis is chosen at the equilibrium point of the oscillator, we have concluded that $\bar{x}$ lies at the equilibrium point, as indicated in Figure 5-6*a*; but this conclusion is true, independent of the choice of the origin. That is, if the equilibrium point of the oscillator is located to the right of the origin, $\Psi^*\Psi$ is still centered on the equilibrium point so $\bar{x}$ is still located at that point, as indicated in Figure 5-6*b*. The reason is that the behavior of the oscillator is still symmetrical about its equilibrium point. If the oscillator is distorted by making the restoring force stronger in one direction than in the other, this symmetry is destroyed. (It will no longer be a simple harmonic oscillator.) Then $\Psi^*\Psi$ will lose its symmetry, and $\bar{x}$ will be displaced from the equilibrium point. Examples are shown in Figures 5-6*c* and 5-6*d*. ◀

It is apparent that an expression of the same form as (5-28) would be appropriate for the evaluation of the expectation value of any function of $x$. That is

$$\overline{x^2} = \int_{-\infty}^{\infty} \Psi^*(x,t) x^2 \Psi(x,t) \, dx$$

and

$$\overline{f(x)} = \int_{-\infty}^{\infty} \Psi^*(x,t) f(x) \Psi(x,t) \, dx$$

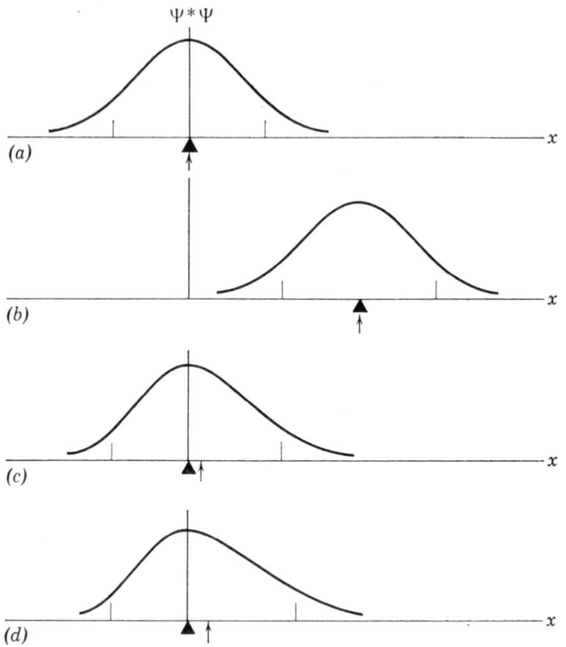


**FIGURE 5-6**

(*a*) The probability density for the ground state of a harmonic oscillator whose equilibrium point (marked with a triangle) lies at the origin. The expectation value $\bar{x}$ (marked with an arrow) also lies at the origin. (*b*) The oscillator is displaced along the $x$ axis, but the expectation value $\bar{x}$ remains coincident with the equilibrium point. (*c*) The restoring force is made weaker for positive displacements than for negative displacements, destroying the symmetry of the oscillator. The particle now would more likely be found to the right of the equilibrium point than to the left, so the expectation value $\bar{x}$ now lies to the right of that point. But the equilibrium point is still the location where the particle would most likely be found because it is still where the probability density maximizes. (*d*) As the restoring force is made even more asymmetric, $\bar{x}$ is further displaced to the right. In all figures the short vertical marks on the $x$ axis indicate the limits of the classical oscillation for the appropriate potential, or restoring force, and total energy.

where $f(x)$ is any function of $x$. Even for a function which may explicitly depend on the time, such as a potential energy $V(x,t)$, we may still write

$$\overline{V(x,t)} = \int_{-\infty}^{\infty} \Psi^*(x,t)V(x,t)\Psi(x,t)\,dx \tag{5-29}$$

because all measurements made to evaluate $V(x,t)$ are made at the same value of $t$, and so the preceding arguments would still hold.

The coordinate $x$ and the potential energy $V(x,t)$ are two examples of the *dynamical quantities* which can be used to characterize the behavior of the particle. Examples of other dynamical quantities are the momentum $p$ and the total energy $E$. The expectation value of these quantities is always given by the same type of expression. For example, the expectation value of the momentum is given by

$$\bar{p} = \int_{-\infty}^{\infty} \Psi^*(x,t) p \Psi(x,t)\, dx \tag{5-30}$$

However, in order to evaluate the integral in (5-30), the integrand $\Psi^*(x,t)p\Psi(x,t)$ must be expressed as a function of the variables $x$ and $t$. In classical mechanics, $p$ can always be written as a function of the variables $x$ and/or $t$. For instance, for a particle moving in a time-independent potential, $p$ can be written as a function of $x$ alone since its momentum is precisely known at every point on its path (after the problem has been solved). A moment's consideration of the behavior of a classical simple harmonic oscillator will verify this. But in quantum mechanics the uncertainty principle tells us that it is *not* possible to write $p$ as a function of $x$, because $p$ and $x$ cannot be simultaneously known with complete precision. Nor is it possible to write $p$ as a function of $t$. We must find some other way of expressing the integrand of (5-30) in terms of $x$ and $t$.

A clue can be found by considering the free particle wave function, (5-23), which is

$$\Psi(x,t) = \cos(kx - \omega t) + i \sin(kx - \omega t)$$

Differentiating with respect to $x$, we have

$$\frac{\partial \Psi(x,t)}{\partial x} = -k \sin(kx - \omega t) + ik \cos(kx - \omega t)$$

$$= ik[\cos(kx - \omega t) + i \sin(kx - \omega t)]$$

Since $k = p/\hbar$, this is

$$\frac{\partial \Psi(x,t)}{\partial x} = i\frac{p}{\hbar}\Psi(x,t)$$

which can be written

$$p[\Psi(x,t)] = -i\hbar \frac{\partial}{\partial x}[\Psi(x,t)]$$

This indicates that there is an association between the dynamical quantity $p$ and the *differential operator* $-i\hbar(\partial/\partial x)$. That is, the effect of multiplying the function $\Psi(x,t)$ by $p$ is the same as the effect of operating on it with the differential operator $-i\hbar(\partial/\partial x)$ (that is, of taking $-i\hbar$ times the partial derivative of the function with respect to $x$).

A similar association can be found between the dynamical quantity $E$ and the differential operator $i\hbar(\partial/\partial t)$ by differentiating the free particle wave function $\Psi(x,t)$ with respect to $t$. We obtain

$$\frac{\partial \Psi(x,t)}{\partial t} = +\omega \sin(kx - \omega t) - i\omega \cos(kx - \omega t)$$

$$= -i\omega[\cos(kx - \omega t) + i \sin(kx - \omega t)]$$

Since $\omega = E/\hbar$, this can be written

$$E[\Psi(x,t)] = i\hbar \frac{\partial}{\partial t}[\Psi(x,t)]$$

Are these relations restricted to the case of free particle wave functions? No! Consider (5-9), which relates the total energy $E$ to the momentum $p$ and the potential energy $V(x,t)$

$$\frac{p^2}{2m} + V(x,t) = E$$

Let us replace the dynamical quantities $p$ and $E$ by their associated differential operators. Then we have

$$\frac{1}{2m}\left(-i\hbar\frac{\partial}{\partial x}\right)^2 + V(x,t) = i\hbar\frac{\partial}{\partial t}$$

Since $(-i\hbar)^2 = -\hbar^2$, and $(\partial/\partial x)^2 = (\partial/\partial x)(\partial/\partial x) = \partial^2/\partial x^2$, we obtain

$$-\frac{\hbar^2}{2m}\frac{\partial^2}{\partial x^2} + V(x,t) = i\hbar\frac{\partial}{\partial t} \tag{5-31}$$

This is an *operator equation*. It has significance when applied to any wave function $\Psi(x,t)$, in the sense that identical results are obtained after performing on the wave function the operations indicated on either side of the equal sign. That is, (5-31) implies

$$-\frac{\hbar^2}{2m}\frac{\partial^2\Psi(x,t)}{\partial x^2} + V(x,t)\Psi(x,t) = i\hbar\frac{\partial\Psi(x,t)}{\partial t}$$

where $\Psi(x,t)$ is any wave function. Of course, this is just the Schroedinger equation. Therefore, we conclude that postulating the associations

$$p \leftrightarrow -i\hbar\frac{\partial}{\partial x} \qquad \text{and} \qquad E \leftrightarrow i\hbar\frac{\partial}{\partial t} \tag{5-32}$$

is equivalent to postulating the Schroedinger equation. The validity of these associations is unrestricted.

The procedure used in the last paragraph is essentially the one originally followed by Schroedinger in obtaining his equation. It provides us with a powerful method for obtaining the quantum mechanical wave equation for more complicated cases than the one-particle, one-dimensional case we treat in this chapter. We shall use it later to treat the systems we ultimately must deal with.

Now let us use the first of the operator associations to obtain an integrable expression for the expectation value of the momentum. We take (5-30), which is

$$\bar{p} = \int_{-\infty}^{\infty} \Psi^*(x,t)p\Psi(x,t)\,dx$$

and replace the $p$ in the integrand by $-i\hbar(\partial/\partial x)$. We obtain

$$\bar{p} = \int_{-\infty}^{\infty} \Psi^*(x,t)\left(-i\hbar\frac{\partial}{\partial x}\right)\Psi(x,t)\,dx$$

or

$$\bar{p} = -i\hbar\int_{-\infty}^{\infty} \Psi^*(x,t)\frac{\partial\Psi(x,t)}{\partial x}\,dx \tag{5-33}$$

We thus obtain an expression which can be integrated immediately if we know $\Psi(x,t)$.

At this point we can see the reason for the ordering of the terms in the integrands of (5-30) and (5-33). It would not be possible to have

$$\bar{p} = -i\hbar\int_{-\infty}^{\infty} \Psi^*(x,t)\Psi(x,t)\frac{\partial}{\partial x}\,dx$$

since this is meaningless. Nor would it be possible to have

$$\bar{p} = -i\hbar \int_{-\infty}^{\infty} \frac{\partial}{\partial x} [\Psi^*(x,t)\Psi(x,t)]\, dx$$
$$= -i\hbar[\Psi^*(x,t)\Psi(x,t)]_{-\infty}^{\infty}$$

because the right-hand side of the last equation always equals zero. This is true because, in any realistic situation, the particle would never be found at either $x = +\infty$ or $x = -\infty$, and therefore the probability density vanishes at both these limits. It should also be mentioned that using the expression

$$\bar{p} = -i\hbar \int_{-\infty}^{\infty} \Psi(x,t) \frac{\partial \Psi^*(x,t)}{\partial x}\, dx$$

is equivalent to using the minus sign in (5-19), and it adds nothing new to the theory.

The ordering of terms is of no consequence in integrands that occur in expressions for the expectation values of quantities that are functions of position and/or time, such as (5-28) and (5-29), because no derivatives are involved. Nevertheless, it is conventional to use the same ordering as is required in the expressions for the expectation value of the momentum.

Using the second of the operator associations of (5-32), we can evaluate the expectation value of the total energy $E$ of a particle in a state described by the wave function $\Psi(x,t)$, as follows

$$\bar{E} = \int_{-\infty}^{\infty} \Psi^*(x,t) E \Psi(x,t)\, dx$$
$$= \int_{-\infty}^{\infty} \Psi^*(x,t)\left(i\hbar \frac{\partial}{\partial t}\right)\Psi(x,t)\, dx$$
$$= i\hbar \int_{-\infty}^{\infty} \Psi^*(x,t) \frac{\partial \Psi(x,t)}{\partial t}\, dx$$

But note that we can also use the energy equation, (5-9), to write $E$ in terms of $p$ and $V(x,t)$, and then employ the first of the operator associations of (5-32) to convert $p$ into an operator, obtaining

$$\bar{E} = \int_{-\infty}^{\infty} \Psi^*(x,t)\left(-\frac{\hbar^2}{2m}\frac{\partial^2}{\partial x^2} + V(x,t)\right)\Psi(x,t)\, dx$$

In fact, the expectation value of any dynamical quantity can be evaluated by using only the first of the operator associations of (5-32). That is, *if $f(x,p,t)$ is any dynamical quantity which is a function of $x$, $p$, and possibly $t$, useful in describing the state of motion of the particle associated with the wave function $\Psi(x,t)$, then its expectation value $\overline{f(x,p,t)}$ is given by*

$$\overline{f(x,p,t)} = \int_{-\infty}^{\infty} \Psi^*(x,t) f_{op}\left(x, -i\hbar\frac{\partial}{\partial x}, t\right)\Psi(x,t)\, dx \qquad (5\text{-}34)$$

*where the operator $f_{op}(x, -i\hbar\, \partial/\partial x, t)$ is obtained from the function $f(x,p,t)$ by everywhere replacing $p$* by $-i\hbar\, \partial/\partial x$.

We have found that the wave function $\Psi(x,t)$ contains more information than just the probability density $P(x,t) = \Psi^*(x,t)\Psi(x,t)$. The wave function also contains, through (5-34), the expectation value of the coordinate $x$, the potential energy $V$, the momentum $p$, the total energy $E$, and, in general, the expectation value of *any* dynamical quantity $f(x,p,t)$. In fact, the wave function contains *all* the information that the uncertainty principle will allow us to learn about the associated particle.

**Example 5-9.** Consider a particle of mass $m$ which can move freely along the $x$ axis anywhere from $x = -a/2$ to $x = +a/2$, but which is *strictly* prohibited from being found outside this region. The particle bounces back and forth between the walls at $x = \pm a/2$ of a (one-dimensional) box. The walls are assumed to be completely impenetrable, no matter how energetic is the particle. Of course, this assumption is an idealization, but it is a very useful one. We shall study this problem in the following chapter, and we shall find that the wave function for the lowest energy state of the particle is

$$\Psi(x,t) = \begin{cases} A\cos\dfrac{\pi x}{a}e^{-iEt/\hbar} & -a/2 < x < +a/2 \\ 0 & x \leq -a/2 \text{ or } x \geq +a/2 \end{cases}$$

where $A$ is an arbitrary real constant, and $E$ is the total energy of the particle. This wave function is another one which is convenient for us to use in this chapter for illustrative purposes. Justify its use here by verifying that it is a solution to the Schroedinger equation in the region $-a/2 < x < +a/2$, and determine the value of $E$ for this lowest energy state.

If there are no forces acting on the particle in the region in question, the potential energy function must be constant in the region. As potential energies are always undefined to within an additive constant, we can take the value of the potential energy to be zero in the region. Then the Schroedinger equation in the region reads

$$-\frac{\hbar^2}{2m}\frac{\partial^2\Psi}{\partial x^2} = i\hbar\frac{\partial\Psi}{\partial t} \qquad -a/2 < x < +a/2$$

We verify the wave function by substituting its derivatives into the equation. With

$$\Psi = A\cos\frac{\pi x}{a}e^{-iEt/\hbar}$$

we obtain

$$\frac{\partial\Psi}{\partial x} = -\left(\frac{\pi}{a}\right)A\sin\frac{\pi x}{a}e^{-iEt/\hbar}$$

$$\frac{\partial^2\Psi}{\partial x^2} = -\left(\frac{\pi}{a}\right)^2 A\cos\frac{\pi x}{a}e^{-iEt/\hbar} = -\left(\frac{\pi}{a}\right)^2\Psi$$

and

$$\frac{\partial\Psi}{\partial t} = -\frac{iE}{\hbar}A\cos\frac{\pi x}{a}e^{-iEt/\hbar} = -\frac{iE}{\hbar}\Psi$$

Substitution yields

$$+\frac{\hbar^2}{2m}\frac{\pi^2}{a^2}\Psi = -i\hbar\frac{iE}{\hbar}\Psi$$

or

$$\frac{\hbar^2\pi^2}{2ma^2}\Psi = E\Psi$$

This is satisfied identically, providing $E$ has the value

$$E = \frac{\pi^2\hbar^2}{2ma^2}$$

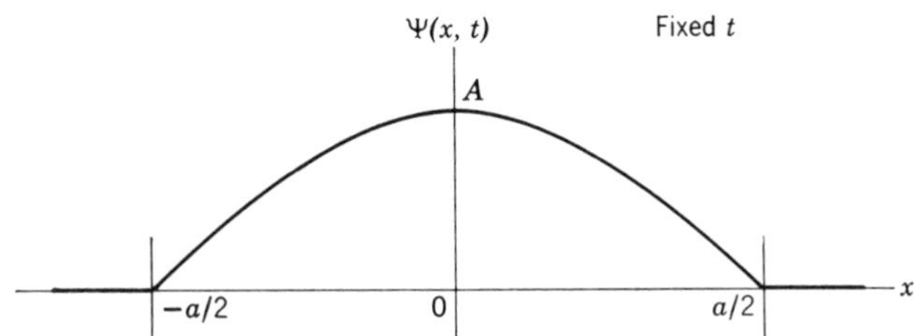


**FIGURE 5-7**

The $x$ dependence of a wave function for the lowest energy state of a particle strictly confined to a region of length $a$, but moving freely therein. Everywhere outside the region the value of the wave function is zero.

Thus we have determined the required value of $E$ corresponding to the wave function we are dealing with, and have also verified that the wave function is a solution of the Schroedinger equation.

Figure 5-7 illustrates the wave function by a plot of its space dependence. Note that the interior (inside the box) values of $\Psi(x,t)$ join onto the exterior (outside the box) values of zero at the boundaries of the region at $x = -a/2$ and $x = +a/2$ (walls of the box) because the cosine function goes to zero when $x$ approaches $\pm a/2$. The exterior values of $\Psi(x,t)$ are zero, of course, because the wave function describes a particle which is strictly prohibited from being found outside the region. ◀

**Example 5-10.** Use the "particle-in-a-box" wave function treated in Example 5-9 to evaluate the expectation values of $x$, $p$, $x^2$, and $p^2$ for the particle associated with the wave function.

To evaluate $\bar{x}$, we must evaluate

$$\bar{x} = \int_{-\infty}^{\infty} \Psi^* x \Psi \, dx$$

Using the wave function of Example 5-9, this is

$$\bar{x} = \int_{-a/2}^{+a/2} A \cos \frac{\pi x}{a} e^{+iEt/\hbar} x A \cos \frac{\pi x}{a} e^{-iEt/\hbar} \, dx$$

$$= A^2 \int_{-a/2}^{+a/2} x \cos^2 \frac{\pi x}{a} \, dx$$

where the integration has been restricted to the region from $-a/2$ to $+a/2$ since $\Psi(x,t)$ is zero outside this region. Now note that the integrand is a product of $\cos^2 (\pi x/a)$, which is an even function of $x$, times $x$ itself, which is an odd function of $x$. The integrand is therefore an odd function of $x$. From this conclusion it follows that

$$\int_{-a/2}^{+a/2} x \cos^2 \frac{\pi x}{a} \, dx = 0$$

because the integral of an integrand which is an odd function of the variable of integration is zero if the integration is taken over a range which is centered about its origin (see Example 5-8). Thus we obtain

$$\bar{x} = 0$$

A moment's thought should make it clear why measurements of the location of the particle which moves freely between $-a/2$ and $+a/2$ would be expected to average out to zero.

To evaluate $\bar{p}$, we evaluate

$$\bar{p} = \int_{-\infty}^{\infty} \Psi^*(-i\hbar)\frac{\partial \Psi}{\partial x}\,dx$$

Using the given $\Psi(x,t)$, and its $x$ derivative which has been calculated in Example 5-9, we obtain

$$\bar{p} = -i\hbar \int_{-a/2}^{+a/2} A\cos\frac{\pi x}{a}e^{+iEt/\hbar}\left(-\frac{\pi}{a}\right)A\sin\frac{\pi x}{a}e^{-iEt/\hbar}\,dx$$

or

$$\bar{p} = i\hbar\frac{\pi}{a}A^2\int_{-a/2}^{+a/2}\cos\frac{\pi x}{a}\sin\frac{\pi x}{a}\,dx$$

Again, the integrand is, in total, an odd function of the variable of integration since it is the product of an even function cos $(\pi x/a)$ times an odd function sin $(\pi x/a)$. Thus we obtain

$$\bar{p} = 0$$

because the integral is taken over a range centered on the origin, and consequently it yields zero. Physically, the expectation value of the momentum of the particle is zero because, if the particle is confined to the region from $-a/2$ to $+a/2$ and moving with total energy $E$, it must be bouncing back and forth between the ends of the region and constantly reversing the sign (i.e., the direction) of its momentum. That is, the magnitude of its momentum must be such that $p^2/2m = E$ but, since it is equally probable that the sign of the momentum will be either positive or negative, measurements of this quantity will average out to zero.

In evaluating $\overline{x^2}$, we must evaluate the integral

$$\overline{x^2} = \int_{-\infty}^{\infty}\Psi^* x^2\Psi\,dx = \int_{-a/2}^{+a/2} A\cos\frac{\pi x}{a}e^{+iEt/\hbar}x^2 A\cos\frac{\pi x}{a}e^{-iEt/\hbar}\,dx = A^2\int_{-a/2}^{+a/2}x^2\cos^2\frac{\pi x}{a}\,dx$$

This will not yield zero because the integrand is an even function of $x$. For the same reason we may, as in Example 5-7, immediately simplify the integral to obtain

$$\overline{x^2} = 2A^2\int_0^{+a/2}x^2\cos^2\frac{\pi x}{a}\,dx$$

If we multiply and divide by $(a/\pi)^3$, this can be written

$$\overline{x^2} = 2A^2\left(\frac{a}{\pi}\right)^3\int_0^{+\pi/2}\left(\frac{\pi x}{a}\right)^2\cos^2\frac{\pi x}{a}\,d\left(\frac{\pi x}{a}\right)$$

The integral can now be evaluated by consulting appropriate tables. We find

$$\overline{x^2} = A^2\frac{a^3}{4\pi^2}\left(\frac{\pi^2}{6} - 1\right)$$

In order to fully determine $\overline{x^2}$, we must also know the value of the constant $A$ that determines the amplitude of the wave function. As in Example 5-7, we can find the proper value by demanding that the wave function be *normalized*. That is, we adjust $A$ so that the total probability of finding the particle somewhere is equal to one. The condition gives

$$\int_{-\infty}^{\infty}\Psi^*\Psi\,dx = A^2\int_{-a/2}^{+a/2}\cos^2\frac{\pi x}{a}\,dx = 2A^2\frac{a}{\pi}\int_0^{+\pi/2}\cos^2\frac{\pi x}{a}\,d\left(\frac{\pi x}{a}\right) = 1$$

Integrating, we obtain

$$2A^2 \frac{a}{\pi}\frac{\pi}{4} = 1$$

or

$$A = \sqrt{\frac{2}{a}}$$

Thus we have

$$\overline{x^2} = \frac{2}{a}\frac{a^3}{4\pi^2}\left(\frac{\pi^2}{6} - 1\right) = \frac{a^2}{2\pi^2}\left(\frac{\pi^2}{6} - 1\right) = 0.033a^2$$

The quantity $\overline{x^2}$ is not zero, even though $\bar{x} = 0$, because any measurement of $x^2$ must necessarily yield a positive result. This quantity, or its square root $\sqrt{\overline{x^2}}$ (the root-mean-square position of statistical theory), can be taken as a measure of the fluctuations about the average, $\bar{x} = 0$, that would be observed in determinations of the position of the particle. The latter quantity has the value

$$\sqrt{\overline{x^2}} = 0.18a$$

The fluctuations arise because the particle is not always found at the same location, but instead at various locations, since the particle can be found wherever $\Psi^*\Psi$ has an appreciable value. (In this case where $\bar{x} = 0$, the quantity $\sqrt{\overline{x^2}}$ is a measure of the fluctuations. In a case where $\bar{x} \neq 0$, the quantity $\sqrt{\overline{x^2} - \bar{x}^2}$ is a measure of the fluctuations. Analogous comments apply to the momentum $p$.)

Finally, let us evaluate $\overline{p^2}$ from the expression

$$\overline{p^2} = \int_{-\infty}^{\infty} \Psi^*(-i\hbar)^2 \frac{\partial^2\Psi}{\partial x^2}\,dx = -\hbar^2 \int_{-\infty}^{\infty} \Psi^* \frac{\partial^2\Psi}{\partial x^2}\,dx$$

Using the value of $\partial^2\Psi/\partial x^2$ calculated in Example 5-9, we have

$$\overline{p^2} = \hbar^2 \frac{\pi^2}{a^2} \int_{-\infty}^{\infty} \Psi^*\Psi\,dx$$

Of course the integral equals one since it is just the probability of finding the particle somewhere. If we were interested only in evaluating $\overline{p^2}$, we would not find it necessary to actually carry through the normalization procedure to evaluate $A$ since we can make this statement and immediately conclude that

$$\overline{p^2} = \left(\frac{\hbar\pi}{a}\right)^2$$

The square root of this quantity (the root-mean-square momentum)

$$\sqrt{\overline{p^2}} = \frac{\hbar\pi}{a}$$

is a measure of the fluctuations about the average, $\bar{p} = 0$, that would be observed in determinations of the momentum of the particle. The fluctuations arise, as discussed above, because the particle can sometimes be found with momentum $p = +\sqrt{2mE}$ and sometimes with momentum $p = -\sqrt{2mE}$. If we evaluate

$$p = \sqrt{2mE} = \sqrt{\frac{2m\pi^2\hbar^2}{2ma^2}} = \frac{\pi\hbar}{a}$$

from Example 5-9, we note that $\sqrt{\overline{p^2}}$ is just equal to the magnitude of $p$.

If we define $\sqrt{\overline{x^2}}$ and $\sqrt{\overline{p^2}}$ as the *uncertainties* $\Delta x$ and $\Delta p$ in the position and momentum of the particle in the energy state we have been dealing with, we obtain

$$\Delta x \, \Delta p = \sqrt{\overline{x^2}}\sqrt{\overline{p^2}} = 0.18a\frac{\pi\hbar}{a} = 0.57\hbar$$

This is certainly consistent with the lower limit $\hbar/2$ set by the uncertainty principle. Note that this is the first time we have been able to become really quantitative when referring to the uncertainty principle. Expectation values calculated from wave functions make it possible to give quantitative definitions to the uncertainties. ◀

## 5-5 The Time-Independent Schroedinger Equation

The usefulness of wave functions more than justifies the work that is required to obtain them. This is done by solving Schroedinger's equation, (5-22)

$$-\frac{\hbar^2}{2m}\frac{\partial^2\Psi(x,t)}{\partial x^2} + V(x,t)\Psi(x,t) = i\hbar\frac{\partial\Psi(x,t)}{\partial t}$$

using the potential energy function $V(x,t)$ that properly describes the forces acting on the particle of interest. We shall now take the first step in solving this partial differential equation. As we promised, we shall carefully develop the required mathematical procedures, assuming no previous knowledge of differential equations on the part of the student.

The standard technique for solving partial differential equations consists of searching for solutions in the form of products of functions, each of which contains only a single one of the independent variables that are involved in the equation. The technique, called the *separation of variables*, is used because it immediately reduces the partial differential equation to a set of ordinary differential equations. As we shall see, this is a significant simplification. Here we are dealing with a partial differential equation involving a single space variable $x$ plus the time variable $t$. Thus the technique consists in searching for solutions in which the wave function $\Psi(x,t)$ can be written as the product

$$\Psi(x,t) = \psi(x)\varphi(t) \tag{5-35}$$

where the first term on the right side is a function of $x$ alone and the second term is a function of $t$ alone. We shall assume the existence of solutions of this form, substitute these solutions into the Schroedinger equation that they are supposed to satisfy, and see what happens. If our assumed form is invalid we shall, of course, soon find out. However, we shall actually find that solutions of the assumed form do exist, *providing that the potential energy does not depend explicitly on the time t* so that the function can be written as $V(x)$. Since in quantum mechanics, as in classical mechanics, almost all systems have potential energies of this form, the condition is not a very serious restriction.

Separation of variables will lead to the conclusion that the function $\psi(x)$, which specifies the space dependence of the wave function $\Psi(x,t) = \psi(x)\varphi(t)$, is a solution to the differential equation

$$-\frac{\hbar^2}{2m}\frac{d^2\psi(x)}{dx^2} + V(x)\psi(x) = E\psi(x)$$

called the *time-independent Schroedinger equation*. Note that this equation is simpler than the Schroedinger equation for the same potential energy because it involves only one independent variable, $x$, and it is therefore an ordinary differential equation instead of a partial differential equation. The technique will give us even more information about the function $\varphi(t)$ specifying the time dependence of the wave function. In fact, it will show that $\varphi(t)$ satisfies a simple ordinary differential equation that can be solved immediately to yield the simple expression

$$\varphi(t) = e^{-iEt/\hbar}$$

where $E$ is the total energy of the particle in the system. Separation of variables is such a useful technique that we shall employ it on a number of occasions in the remainder of this book. Let us now carry through the details of its application to the Schroedinger equation.

Substituting the assumed form of the solution, $\Psi(x,t) = \psi(x)\varphi(t)$, into the Schroedinger equation, and also restricting ourselves to time-independent potential energies that can be written as $V(x)$, we obtain

$$-\frac{\hbar^2}{2m}\frac{\partial^2\psi(x)\varphi(t)}{\partial x^2} + V(x)\psi(x)\varphi(t) = i\hbar\frac{\partial\psi(x)\varphi(t)}{\partial t}$$

Now

$$\frac{\partial^2\psi(x)\varphi(t)}{\partial x^2} = \varphi(t)\frac{\partial^2\psi(x)}{\partial x^2} = \varphi(t)\frac{d^2\psi(x)}{dx^2}$$

the notation $\partial^2\psi(x)/\partial x^2$ being redundant with $d^2\psi(x)/dx^2$ since $\psi(x)$ is a function of $x$ alone. Similarly

$$\frac{\partial\psi(x)\varphi(t)}{\partial t} = \psi(x)\frac{\partial\varphi(t)}{\partial t} = \psi(x)\frac{d\varphi(t)}{dt}$$

Therefore, we have

$$-\frac{\hbar^2}{2m}\varphi(t)\frac{d^2\psi(x)}{dx^2} + V(x)\psi(x)\varphi(t) = i\hbar\psi(x)\frac{d\varphi(t)}{dt}$$

Dividing both sides of this equation by $\psi(x)\varphi(t)$, we obtain

$$\frac{1}{\psi(x)}\left[-\frac{\hbar^2}{2m}\frac{d^2\psi(x)}{dx^2} + V(x)\psi(x)\right] = i\hbar\frac{1}{\varphi(t)}\frac{d\varphi(t)}{dt} \tag{5-36}$$

Note that the right side of (5-36) does not depend on $x$, while the left side does not depend on $t$. Consequently, their common value cannot depend on either $x$ or $t$. In other words, the common value must be a constant, which we shall call $G$. The result of this consideration is that (5-36) leads to two separate equations. One equation is obtained by setting the left side equal to the common value

$$\frac{1}{\psi(x)}\left[-\frac{\hbar^2}{2m}\frac{d^2\psi(x)}{dx^2} + V(x)\psi(x)\right] = G \tag{5-37}$$

The other equation is obtained by setting the right side equal to the common value

$$i\hbar\frac{1}{\varphi(t)}\frac{d\varphi(t)}{dt} = G \tag{5-38}$$

The constant $G$ is called the *separation constant*, for the same reason that this technique for solving partial differential equations is called the separation of variables.

In retrospect, we see that the effect of employing the technique has been to convert the single partial differential equation, involving two independent variables $x$ and $t$, into a pair of ordinary differential equations, one involving $x$ alone and the other involving $t$ alone. These equations are coupled in the sense that they both contain the same separation constant $G$, but this type of coupling does not lead to any difficulty in obtaining solutions to the equations. We shall find that the time equation, (5-38), has a very simple solution. Furthermore, when we demand that this solution agree with the de Broglie-Einstein postulate, we shall see that the value of the separation constant $G$ becomes determined. Substituting this value of $G$ into the space equation, (5-37), we then have an ordinary differential equation, whose solutions can be obtained by employing one of the several standard techniques that have been developed for solving such equations. What we have done, in effect, is to reduce the problem from that of solving the partial differential space-time Schroedinger equation, (5-22), to that of solving the ordinary differential space equation. The product of the solution of that equation and the solution of the time equation is the desired solution of the Schroedinger equation.

We can see that the product form $\Psi(x,t) = \psi(x)\varphi(t)$, which we assumed for the wave function, is justified because we shall be able to carry out the procedure just outlined. We can also see that we cannot carry through the separation of (5-36), into the pair of equations that follow from it, if the potential energy function depends on both $x$ and $t$, as stated earlier. The reason is that we cannot then separate terms so that one side of the equation does not depend on $x$ while the other side does not depend on $t$.

The time equation, (5-38), is a simple first-order ordinary differential equation for $\varphi$ as a function of $t$. There are several general techniques available for finding the solutions to such equations. All these techniques have a common feature; they involve assuming a general form for the solution, substituting this form into the differential equation and, from the resulting equation, determining the specific form required for the solution. After studying these techniques, it is often possible to develop enough intuition to be able to guess the specific form of the solution in the first instance, at least for fairly simple differential equations. This is a time saving and perfectly legitimate procedure, providing the guess is verified by substituting it into the differential equation and showing that the equation is satisfied, and this is the procedure that will usually be employed in this book. Consider (5-38) which, upon transposition, can be written as

$$\frac{d\varphi(t)}{dt} = -\frac{iG}{\hbar}\varphi(t) \tag{5-39}$$

This differential equation tells us that the function $\varphi(t)$, which is its solution, has the property that its first derivative is proportional to the function itself. Anyone with much experience in differentiating would not have difficulty in guessing that $\varphi(t)$ must be an exponential function. Therefore, let us assume that the solution to the differential equation is of the form

$$\varphi(t) = e^{\alpha t}$$

where $\alpha$ is a constant that will be determined shortly. We verify this assumed solution by differentiating it, to obtain

$$\frac{d\varphi(t)}{dt} = \alpha e^{\alpha t} = \alpha\varphi(t)$$

which we then substitute into (5-39). This yields

$$\alpha\varphi(t) = -\frac{iG}{\hbar}\varphi(t)$$

If we set

$$\alpha = -\frac{iG}{\hbar}$$

the assumed solution obviously satisfies the equation. Therefore

$$\varphi(t) = e^{-iGt/\hbar} \tag{5-40}$$

is a solution to (5-38) or (5-39).

The solution $\varphi(t)$ is written in (5-40) as a complex exponential, but it can be written as

$$\varphi(t) = e^{-iGt/\hbar} = \cos\frac{Gt}{\hbar} - i\sin\frac{Gt}{\hbar} \tag{5-41a}$$

or

$$\varphi(t) = \cos 2\pi\frac{G}{h}t - i\sin 2\pi\frac{G}{h}t \tag{5-41b}$$

We see that $\varphi(t)$ is an oscillatory function of time of frequency $\nu = G/h$. But, according to the de Broglie-Einstein postulates of (5-8), the frequency must also be given by $\nu = E/h$, where $E$ is the total energy of the particle associated with the wave function corresponding to $\varphi(t)$. The reason is, of course, that $\varphi(t)$ is the function that specifies the time dependence of the wave function. Comparing these expressions, we see that the separation constant must be equal to the total energy of the particle. That is

$$G = E \tag{5-42}$$

Using this value of $G$ in the space equation, (5-37), that we obtained from the separation of variables, we have

$$-\frac{\hbar^2}{2m}\frac{d^2\psi(x)}{dx^2} + V(x)\psi(x) = E\psi(x) \tag{5-43}$$

Using this value of $G$ in the solution (5-40) to the time equation, so that we complete the specification of $\varphi(t)$, the product form of the wave function becomes

$$\Psi(x,t) = \psi(x)e^{-iEt/\hbar} \tag{5-44}$$

where $E$ is the total energy of the particle.

Equation (5-43) is called the time-independent Schroedinger equation, because the time variable $t$ does not enter the equation. Its time-independent solutions $\psi(x)$ determine, through (5-44), the space dependence of the solutions $\Psi(x,t)$ to the Schroedinger equation. For the one-dimensional cases that we have been treating in this chapter, the time-independent Schroedinger equation can involve only one independent variable $x$, and it must, therefore, be an ordinary differential equation. However, if there are more space dimensions, the time-independent Schroedinger equation will involve more independent variables and will therefore be a partial differential equation. (It can usually be reduced to a set of ordinary differential equations, in such cases, by applying the technique of separation of variables.)

In all cases the time-independent Schroedinger equation does not contain the imaginary number $i$, and its solutions $\psi(x)$ are therefore not necessarily complex functions. (That is, $\psi(x)$ need not be complex, but it can be if convenience dictates.) This equation, and its solutions, are essentially identical to the time-independent differential equation for classical wave motion, and its solutions.

The functions $\psi(x)$ are called *eigenfunctions*. The first part, *eigen*, is the German word for characteristic. We shall subsequently get a better idea of why characteristic is appropriate terminology. Here it will suffice to say that its use is conventional. It is also conventional not to translate it into English, perhaps in honor of the dominant role played by German speaking physicists in the development of quantum mechanics.

The student is cautioned to keep clearly in mind the difference between the eigenfunctions $\psi(x)$ and the wave functions $\Psi(x,t)$, and also the difference between the time-independent Schroedinger equation and the Schroedinger equation itself. Wave functions will always be represented by a capital letter $\Psi$; eigenfunctions will always be represented by a lower case letter $\psi$.

**Example 5-11.** Develop a plausibility argument, similar to the one given in Section 5-2, which leads directly to the time-independent Schroedinger equation.

We assume the equation must be consistent with the classical energy equation

$$\frac{p^2}{2m} + V = E$$

and also with the de Broglie postulate

$$p = \frac{h}{\lambda} = \hbar k$$

These two relations combine to yield

$$\frac{\hbar^2 k^2}{2m} + V = E$$

or

$$k^2 = \frac{2m}{\hbar^2}(E - V)$$

Then we assume that the space dependence of the wave function for a free particle is given by the sinusoidal

$$\psi(x) = \sin\frac{2\pi x}{\lambda} = \sin kx$$

The wave number $k$ is constant since the potential energy $V$ is constant for the case of a free particle, and since the total energy is constant also. Differentiating $\psi(x)$ twice with respect to its only independent variable, we obtain

$$\frac{d\psi(x)}{dx} = k\cos kx$$

$$\frac{d^2\psi(x)}{dx^2} = -k^2 \sin kx = -k^2\psi(x)$$

since $k$ is a constant. Now we substitute for $k^2$ the value found above, and obtain

$$\frac{d^2\psi(x)}{dx^2} = -\frac{2m}{\hbar^2}(E - V)\psi(x)$$

or

$$-\frac{\hbar^2}{2m}\frac{d^2\psi(x)}{dx^2} + V\psi(x) = E\psi(x)$$

This is the time-independent Schroedinger equation, but we have obtained it from an argument specific to the case of a free particle where $V$ is a constant. If, as in Section 5-2, we postulate that the equation is valid even in the general case where $V = V(x)$, we obtain the time-independent Schroedinger equation for a particle acted on by a force.

We have followed a much longer route in the text to obtain the same equation, but we have, of course, learned much along the way that is not contained in the time-independent Schroedinger equation. For instance, we know about the time dependence of the wave function $\Psi(x,t) = \psi(x)e^{-iEt/\hbar}$, which is responsible for its necessarily complex character and the many consequences resulting therefrom. ◀

## 5-6 Required Properties of Eigenfunctions

In the following section we shall consider, in a very general way, the problem of finding solutions to the time-independent Schroedinger equation. These considerations will show that energy quantization appears quite naturally in the Schroedinger theory. We shall see that this extremely significant property results from the fact that *acceptable solutions* to the time-independent Schroedinger equation can be found only for certain values of the total energy $E$.

To be an acceptable solution, an eigenfunction $\psi(x)$ and its derivative $d\psi(x)/dx$ are required to have the following properties:

| | |
|---|---|
| $\psi(x)$ must be *finite*. | $d\psi(x)/dx$ must be *finite*. |
| $\psi(x)$ must be *single valued*. | $d\psi(x)/dx$ must be *single valued*. |
| $\psi(x)$ must be *continuous*. | $d\psi(x)/dx$ must be *continuous*. |

These requirements are imposed in order to ensure that the eigenfunction be a mathematically "well-behaved" function so that measurable quantities which can be evaluated from the eigenfunction will also be well-behaved. Figure 5-8 illustrates the

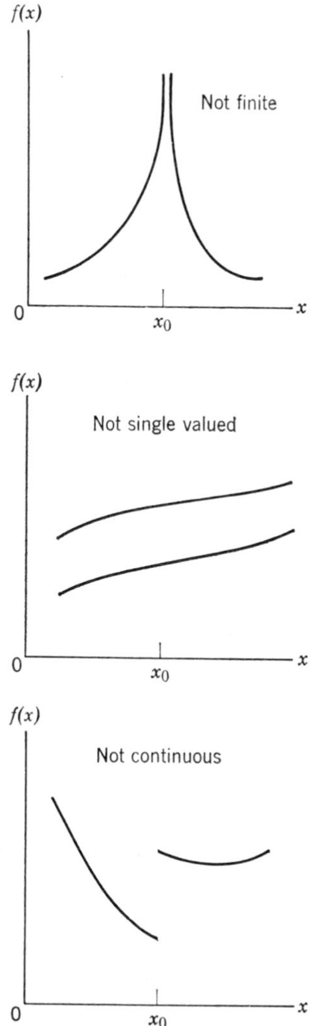


**FIGURE 5-8**
Illustrating functions which are not finite, not single valued, or not continuous, at a point $x_0$.

meaning of these properties by plotting functions which are not finite, not single valued, or not continuous, at the point $x_0$.

If $\psi(x)$ or $d\psi(x)/dx$ were not finite, or not single valued, then the same would be true for $\Psi(x,t) = e^{-iEt/\hbar}\psi(x)$ or $\partial\Psi(x,t)/\partial x = e^{-iEt/\hbar}\, d\psi(x)/dx$. Since the general formula for calculating expectation values of position or momentum, etc., (5-34), contains $\Psi(x,t)$ and $\partial\Psi(x,t)/\partial x$, we see that in any of these cases *we might not obtain finite and definite values when we evaluate measurable quantities*. This would be completely unacceptable because measurable quantities, like the expectation value of position $\bar{x}$, or of momentum $\bar{p}$, do not behave in unreasonable ways. (In very rare circumstances, which we shall not encounter, $\psi(x)$ may actually go to infinity at a point, providing it does so slowly enough to keep finite the integral of $\psi^*(x)\psi(x)$ over a region containing that point.)

In order that $d\psi(x)/dx$ be finite, it is necessary that $\psi(x)$ be continuous. The reason is that any function always has an infinite first derivative wherever it has a discontinuity. The necessity for $d\psi(x)/dx$ to be continuous can be demonstrated by considering the time-independent Schroedinger equation, which we write as

$$\frac{d^2\psi(x)}{dx^2} = \frac{2m}{\hbar^2}\,[V(x) - E]\psi(x)$$

For finite $V(x)$, $E$, and $\psi(x)$, we see that $d^2\psi(x)/dx^2$ must be finite. This in turn, demands that we require $d\psi(x)/dx$ to be continuous because any function that has a discontinuity in the first derivative will have an infinite second derivative at the same point. (Note that there are discontinuities in the first derivative of the eigenfunction for the particle in a box, considered in Example 5-9. They occur at the walls of the box, and they arise from the fact that the system is an idealization in which the walls are assumed to be completely impenetrable, no matter how high the energy of the particle. That is, the potential energy is assumed to become infinite at the walls. This is discussed at length in the next chapter.)

The importance of these requirements on the properties of acceptable solutions to the time-independent Schroedinger equation cannot be overemphasized. Differential equations have a wide variety of possible solutions. It is only when we select from all the possible solutions those that conform to these requirements that we obtain energy quantization, or other equally significant properties of the Schroedinger theory that will be treated in the following chapter. The requirements of finiteness and continuity will be used immediately; single valuedness will not be used until later, but it is of equal importance.

## 5-7 Energy Quantization in the Schroedinger Theory

It is educational to study the problem of obtaining acceptable solutions to the time-independent Schroedinger equation with qualitative arguments that concern the curvatures and slopes of curves obtained by plotting the solution. As we shall see, these arguments are both very general and very simple. They can teach us about many important properties of the time-independent Schroedinger equation, while avoiding any involved mathematics. In fact, the point of view that we shall use in this section is very useful for making a preliminary investigation of the properties of almost any differential equation, and it also provides an intuitive understanding of the behavior of such equations.

We shall obtain only qualitative conclusions from these arguments, but they will be quite valuable. A number of quantitative solutions to the time-independent Schroedinger equation for various potentials will be found in the following chapters. We shall obtain those solutions from standard analytical techniques for solving differential equations. A quantitative solution to the time-independent Schroedinger equation will also be found in Appendix F. That solution is obtained by using a numerical technique that is based on the same ideas used in the qualitative arguments of this section, and so the student may wish to read that appendix after reading this section.

We begin our arguments by writing the time-independent Schroedinger equation as

$$\frac{d^2\psi}{dx^2} = \frac{2m}{\hbar^2}[V(x) - E]\psi \tag{5-45}$$

The properties of this differential equation depend, among other things, upon the form of the potential energy function $V(x)$. This is as it should be since $V(x)$ determines the force acting on the particle whose behavior is supposed to be described by the solutions to the differential equation. We consequently cannot say much about the properties of the differential equation until we say something about $V(x)$, so we shall do this first.

In Figure 5-9 we specify the form of $V(x)$ that we shall use in our arguments by plotting $V$ versus its independent variable $x$. The form has been chosen so that it contains features which will allow us to illustrate several interesting points, but the form also has physical significance. It represents the potential energy for an atom that can be bound to a similar atom and form a diatomic molecule. In this case the $x$ coordinate represents the separation between the centers of the two atoms. The minimum in $V(x)$ occurs at the equilibrium separation, and at the minimum the force acting on the atom is $F = -dV(x)/dx = 0$. As the separation decreases from the equilibrium value a repulsive force develops in the direction of increasing separation, and it becomes larger as the atoms get closer. As the separation increases from the equilibrium value an attractive force develops in the direction of decreasing separation. But if the separation exceeds the disassociation separation indicated in Figure 5-9, the force drops to zero since the molecule is broken and the atoms no longer interact.

With our choice of $V(x)$ the time-independent Schroedinger equation, (5-45), begins to assume a specific form. Since this differential equation contains the total energy $E$ in a crucial location, however, we must also choose its value in order that the equation

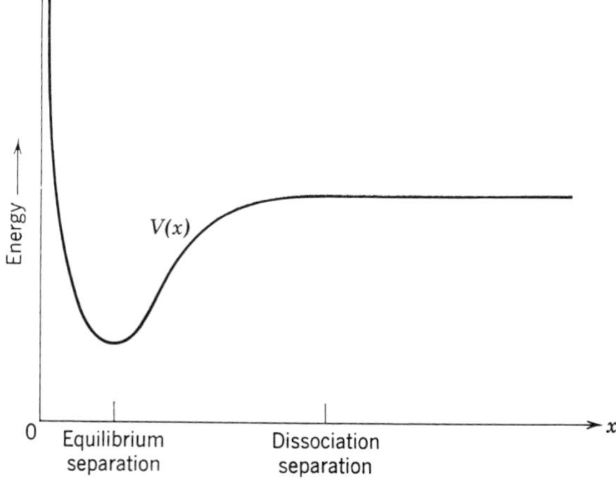


**FIGURE 5-9**

The potential energy $V(x)$ for an atom that can be bound to a similar atom to form a diatomic molecule, plotted as a function of the separation between the centers of the two atoms.

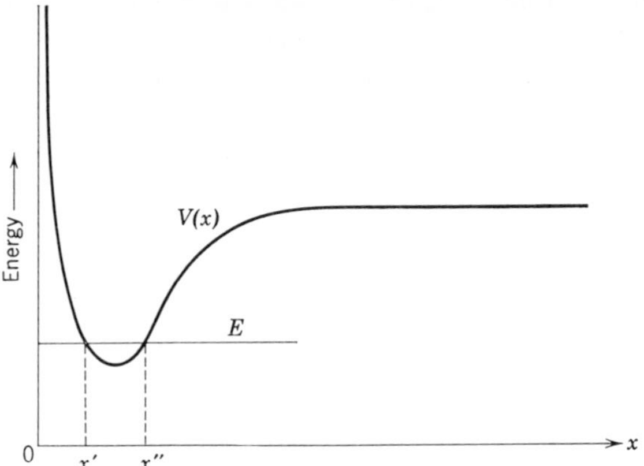


**FIGURE 5-10**

The potential energy $V(x)$ used in qualitative arguments concerning the solutions to the time-independent Schroedinger equation, and the total energy $E$ chosen for these arguments.

has properties which are specific enough to make them easy to discuss. The value that we choose is indicated in Figure 5-10 by the horizontal line: energy $= E =$ const. This figure also replots the curve: energy $= V(x)$. We choose the total energy $E$ in such a way that the molecule is bound (classically the separation distance $x$ between the atoms must be between the values $x'$ and $x''$ shown in the figure), but the exact value of $E$ that we choose is, at this stage, arbitrary. We shall not have to say anything about the combination of parameters $2m/\hbar^2$, appearing in the differential equation, other than that it has a positive value.

Our argument will consider the differential equation, (5-45), as a prescription which determines the value of the second derivative $d^2\psi/dx^2$ of the solution, at a certain $x$, in terms of the values of $(2m/\hbar^2)[V(x) - E]$ and of the solution $\psi$ itself, at that $x$. This will allow us to study important properties of the equation in terms of the general shape of the curve traced by a plot of $\psi$ versus $x$. Thus we shall obtain a geometrical interpretation of the differential equation.

We shall be particularly concerned with the sign of $d^2\psi/dx^2$ because it is a property of second derivatives that a curve, of the dependent variable plotted versus the independent variable, is concave upwards wherever the second derivative is positive and concave downwards wherever the second derivative is negative. Students not already familiar with this property should inspect Figure 5-11, which shows a case in which the slope of the curve of $\psi$ versus $x$ is negative for small $x$, becomes less negative with

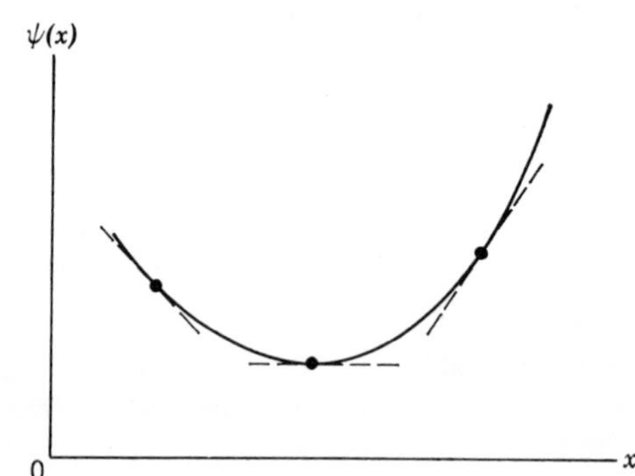


**FIGURE 5-11**

A curve which is *concave upwards*. The value of the first derivative of the function plotted by the curve increases with increasing $x$, so the second derivative is positive.

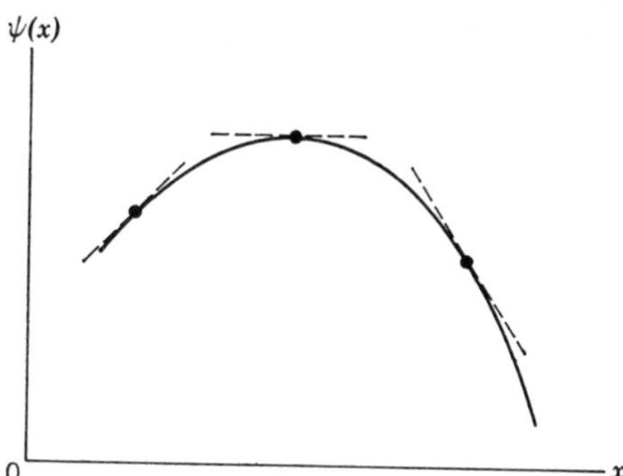


A curve which is *concave downwards*. The value of the first derivative of the function decreases with increasing $x$, so the second derivative is negative.

increasing $x$, goes through zero, and then becomes positive as $x$ continues to increase. The slope, which is equal to $d\psi/dx$, always increases in numerical value with increasing $x$. Therefore the rate of change of slope, which is equal to $d^2\psi/dx^2$, is always positive. The curve in this figure is said to be *concave upwards*. Figure 5-12 shows a case in which the curve is said to be *concave downwards*. Similar considerations prove that in this case $d^2\psi/dx^2$ is always negative.

Now note that in Figure 5-10 there are two intersections of the line energy $= E$ and the curve energy $= V(x)$. These intersections occur at $x = x'$ and $x = x''$, which divide the $x$ axis into three regions: $x < x'$, $x' < x < x''$, and $x > x''$. In the first and third regions the quantity $[V(x) - E]$ is positive since the value of $V(x)$ is everywhere greater than the value of $E$ in these regions. In the second region $[V(x) - E]$ is negative. Inspection of (5-45) then shows that the sign of $d^2\psi/dx^2$ is the same as the sign of $\psi$ in the first and third regions, and it is opposite to the sign of $\psi$ in the second region, since the sign of $2m/\hbar^2$ is positive. This means that in the first and third regions the curve of $\psi$ versus $x$ will be concave upwards if the value of $\psi$ itself is positive, and it will be concave downwards if the value of $\psi$ is negative. In the second region the curve will be concave downwards if $\psi$ is positive, and it will be concave upwards if $\psi$ is negative. The various possibilities are shown in Figure 5-13. We have now laid the groundwork for our geometrical interpretation of the time-independent Schroedinger equation.

For a given form of the potential energy $V(x)$, the differential equation enforces a relation between $d^2\psi/dx^2$ and $\psi$ that determines the *general* behavior of $\psi$. If we also specify the value of $\psi$ and its first derivative $d\psi/dx$ at some value of the independent variable $x$, then the *particular* behavior of the dependent variable $\psi$ is determined for

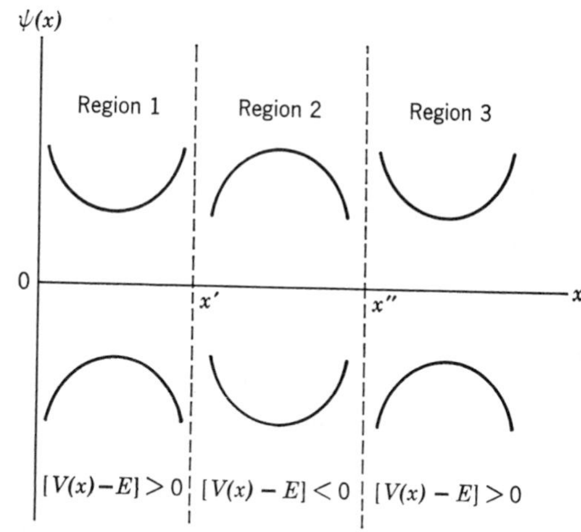


**FIGURE 5-13**

Illustrating the relation between the sign of $\psi$ and the sign of $d^2\psi/dx^2$ in regions defined by the sign of $[V(x) - E]$. The relation can be summarized by stating that $\psi$ is *concave away* from the $x$ axis wherever $[V(x) - E] > 0$, and *concave toward* the $x$ axis wherever $[V(x) - E] < 0$.

all values of $x$. The situation is completely analogous to situations found in classical mechanics. Consider the differential equation for a classical simple harmonic oscillator

$$\frac{d^2x}{dt^2} = -\frac{Cx}{m}$$

This is just Newton's law of motion, $a = F/m$, with a linear restoring force of force constant $C$. In this case $x$ is the dependent variable, and the independent variable is $t$, but otherwise the analogy is complete. The differential equation enforces a relation between $x$ and its second derivative, which determines the general behavior of $x$ as a function of $t$. And if we also specify the value of $x$ and its first derivative $dx/dt$ at some value of $t$ (the initial conditions of the motion), then the particular behavior of $x$ is determined for all values of $t$.

Thus it should be possible to use the time-independent Schroedinger equation, for the $V(x)$ and $E$ we have chosen, to determine the behavior of $\psi$ for all $x$ in terms of assumed values of $\psi$ and $d\psi/dx$ for some particular $x$. Quantitative calculations that do this are found in the next chapters and, particularly, in Appendix F. Here we shall obtain qualitative results from arguments based upon the features of the differential equation just developed. The arguments will be presented as "thought calculations," in the same spirit as the thought experiments of Einstein or Bohr.

On curve 1 of Figure 5-14 we indicate qualitatively the results of a thought calculation, which started with assumed values of $\psi$ and $d\psi/dx$ at a convenient point $x_0$ in the second region, and then traced out the behavior $\psi$ in the direction of increasing $x$. Since we took the initial value of $\psi$ to be positive in the region $x' < x < x''$, we found the curve describing $\psi$ initially to be concave downwards. It remained concave downwards until it passed into the third region, $x > x''$, where $[V(x) - E]$ changes sign. Although the slope of the curve was negative at $x = x''$, it soon became zero, and then positive. Then $\psi$ started to increase in value, and matters rapidly went from bad to worse. The reason is that the differential equation shows that the rate of change of slope, i.e., $d^2\psi/dx^2$, is proportional to the distance from the curve to the axis, i.e., $\psi$. This first calculation produced a $\psi$ that goes to infinity as $x$ becomes large. We found (part of) a solution to the differential equation, but it was not an acceptable solution because an acceptable eigenfunction remains finite.

Curve 2 of Figure 5-14 indicates the results of another attempt made to find an acceptable solution. There was no point in changing the assumed initial value of $\psi$ as

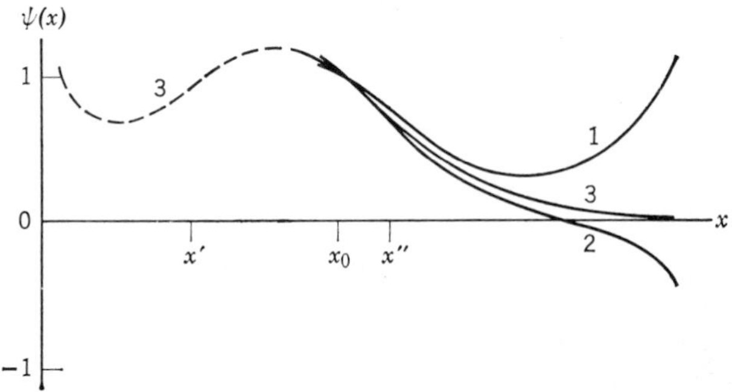


**FIGURE 5-14**

Three attempts at finding an acceptable solution to a time-independent Schroedinger equation for an assumed value of the total energy $E$. The first two (1, 2) failed because the solution became infinite at large $x$. The third (3) gave a solution with acceptable behavior at large $x$, but failed because the solution became infinite at small $x$ (dashed curve).

this would only expand or contract the vertical scale of the curve because of the linearity of the differential equation. What was done was to change the assumed initial value of $d\psi/dx$. The attempt was not successful because $\psi$ became negative in the region where $[V(x) - E]$ is positive. The curve became concave downwards and went to negative infinity.

The difficulty in obtaining an acceptable eigenfunction should now be apparent. It should also be apparent that, by making *exactly* the right choice for the initial value of $d\psi/dx$, it is possible to find a $\psi$ whose acceptable behavior with increasing $x$ is as indicated by curve 3 of Figure 5-14. For this $\psi$ the curve is concave upwards in the third region because it remains above the $x$ axis. Nevertheless, the curve does not turn up because it gets closer and closer to the axis with increasing $x$, and the closer it gets the less concave upwards it becomes. That is, $d^2\psi/dx^2$ approaches zero as $\psi$ approaches zero because the differential equation says these two quantities are proportional.

In Figure 5-14 we also indicate with a dashed curve the results of extending the $\psi$ of curve 3 in the direction of decreasing $x$. From the preceding discussion we must expect that, in general, $\psi$ will go to either positive or negative infinity when extended to decreasing $x$. This cannot be prevented by adjusting the initial choice of $d\psi/dx$, as that would disturb the acceptable behavior for large $x$. Nor can the infinite value of $\psi$ at small $x$ be prevented by joining two different $\psi$ functions with different slopes at $x = x_0$. This is ruled out by the requirement that for an acceptable eigenfunction $d\psi/dx$ is everywhere continuous. For a similar reason we cannot try a discontinuity in $\psi$ itself. We are forced to conclude that, for the particular value of the total energy $E$ that was initially chosen, there is *no* acceptable solution to the time-independent Schroedinger equation. The relation between $\psi$ and its second derivative $d^2\psi/dx^2$, imposed by the differential equation for the given $V(x)$ and that $E$, is such that $\psi$ will approach $\pm\infty$ at either large $x$ or small $x$ (or both). The solution to the equation is unstable, in the sense that it has a pronounced tendency to go to infinity in regions where $E < V$.

By repeating this procedure for many different choices of the energy $E$, however, it will eventually be possible to find a value $E_1$ for which the time-independent Schroedinger equation has an acceptable solution $\psi_1$. In fact, there will, in general, be a number of *allowed values of total energy*, $E_1, E_2, E_3, \ldots$ for which the time-independent Schroedinger equation has *acceptable solutions* $\psi_1, \psi_2, \psi_3, \ldots$. In Figure 5-15 we indicate the form of the first three acceptable solutions. The behavior of $\psi_1$ for both small and large $x$ is the same as the behavior of the function shown in curve 3 of Figure 5-14 for large $x$. For $x < x_0$, the behavior of $\psi_2$ is at first similar to the behavior of $\psi_1$, but, since its second derivative is relatively larger in magnitude, $\psi_2$ crosses the axis at some value of $x$ less than $x_0$ but greater than $x'$. When this happens, the sign of the second derivative reverses and the function becomes concave upwards. At $x = x'$ the second derivative reverses again and, for $x < x'$, the function gradually approaches the $x$ axis.

From Figure 5-15 we can see that the allowed energy $E_2$ is larger than the allowed energy $E_1$. Consider the point $x_0$ where both $\psi_1$ and $\psi_2$ have the same value. It is apparent from the figure that at this point the rate of change of the slope for the latter exceeds the same quantity for the former, i.e.

$$\left|\frac{d^2\psi_2}{dx^2}\right| > \left|\frac{d^2\psi_1}{dx^2}\right|.$$

Using this in the time-independent Schroedinger equation, (5-45), we find that

$$|V(x) - E_2| > |V(x) - E_1|$$

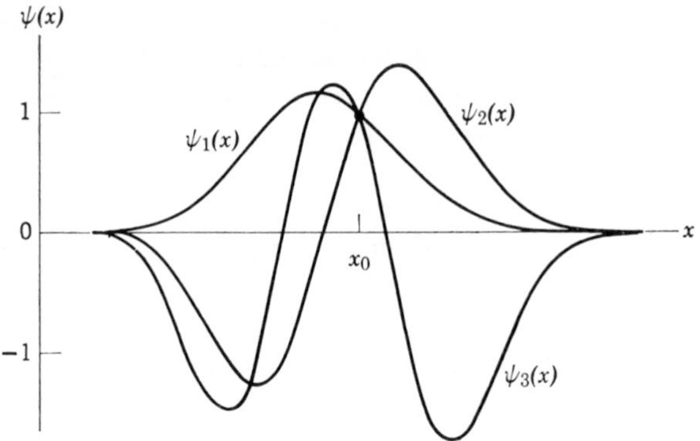


**FIGURE 5-15**

The form of the acceptable eigenfunctions corresponding to the three lowest allowed energy states for a potential with a minimum. At $x = x_0$ all three eigenfunctions have the same value, but $\psi_3$ has the largest curvature because it corresponds to the highest energy of the three. The solutions are for the potential in Figure 5-10, and they are not accurately left-right symmetric because the potential is not symmetric about its minimum.

Consulting Figure 5-10, it is clear that if this is true at $x_0$ then

$$E_2 > E_1$$

since $E > V(x)$ at $x_0$. From a similar argument we can show that $E_3 > E_2$. It is also apparent that the energy differences $E_2 - E_1$, $E_3 - E_2$, etc., are not infinitesimals since, for example, the difference in the first inequality above is not an infinitesimal. Thus the allowed values of energy are *well separated* and form a *discrete set* of energies. For a particle moving under the influence of a time-independent potential $V(x)$, acceptable solutions to the time-independent Schroedinger equation exist only if the energy of the particle is *quantized*, that is, restricted to a discrete set of energies $E_1, E_2, E_3, \ldots$.

This statement is true as long as the relation between the potential energy $V(x)$ and the total energy $E$ is similar to that shown in Figure 5-10, in the sense that there are two values of the coordinate, $x'$ and $x''$, with $[V(x) - E]$ positive for all $x < x'$ and also positive for all $x > x''$. But for a potential of the type illustrated in Figure 5-9, that is, a potential which has a finite limiting value $V_l$ as $x$ becomes very large, there is generally room only for a finite number of discrete allowed energy values which

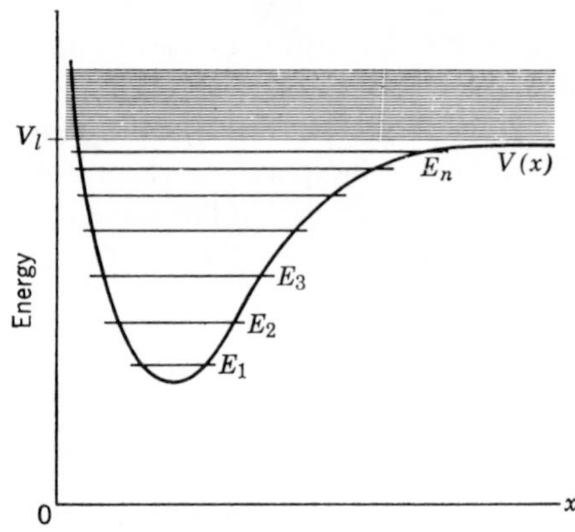


**FIGURE 5-16**

Illustrating discretely separated allowed energies $E_n$ lying below the limiting value $V_l$ of a potential $V(x)$, and the continuum of $E_n$ lying above. Since $E_{n+1} - E_n$ decreases as $V(x)$ approaches $V_l$, if the approach is gradual enough there can be an infinite number of $E_n < V_l$. But generally there are only a finite number.

satisfy the relation $E < V_l$. This is illustrated in Figure 5-16. For $E > V_l$, the situation changes. Now the molecule is unbound (classically the separation distance $x$ between the atoms could be any value larger than $x'$). As far as the time-independent Schroedinger equation is concerned, there are now only two regions of the $x$ axis: $x < x'$ and $x > x'$. In the second region $[V(x) - E]$ will be negative for all values of $x$, no matter how large. But, when $[V(x) - E]$ is negative, $\psi$ is concave downwards if its value is positive, and concave upwards if its value is negative. It always tends to return to the axis and is, therefore, an oscillatory function. Consequently, there will be no problem of $\psi(x)$ going to infinity for large values of $x$. Since we can always make $\psi(x)$ gradually approach the axis for small values of $x$ by a proper initial choice of $d\psi/dx$, we shall be able to find an acceptable eigenfunction for *any* value of $E > V_l$. Thus the allowed energy values for $E_l$ are *continuously distributed*, and are said to form a *continuum*. It is evident that if the potential $V(x)$ is restricted in value for small values of $x$, or for both large and small values of $x$, then the allowed energies will form a continuum for all energies greater than the lowest $V_l$.

The conclusion of our arguments can be stated concisely as follows:

*When the relation between the total energy of a particle and its potential energy is such that classically the particle would be bound to a limited region of space because the potential energy would exceed the total energy outside the region, then Schroedinger theory predicts that the total energy is quantized. When that relation is such that the particle is not bound to a limited region, then the theory predicts the total energy can have any value.*

Since in classical mechanics a particle bound to a limited region would move periodically between the limits of the region, the Wilson-Sommerfeld rules of the old quantum theory would also predict a quantization of the particle's energy in such circumstances; but these quantization rules were a postulate of the old quantum theory, which had a justification in the de Broglie relation only for certain special cases. In his first paper on quantum mechanics, Schroedinger wrote:

> "The essential point is the fact that the mysterious 'requirement of integralness' no longer enters into the quantization rules but has been traced, so to speak, a step further back having been shown to result from the finiteness and single-valuedness of a certain space function (an eigenfunction)."

**Example 5-12.** Use the arguments developed in this section to draw *qualitative* conclusions concerning the form of the eigenfunction for one of the higher energy states of a simple harmonic oscillator. Then compare the corresponding probability density function with what would be predicted for a classical simple harmonic oscillator of the same energy.

The potential $V(x)$ for a simple harmonic oscillator (see Example 5-3) is plotted by the curve in Figure 5-17. In the same figure one of the higher allowed values of the total energy $E$ is plotted by a horizontal line. According to the time-independent Schroedinger equation, (5-45)

$$\frac{d^2\Psi}{dx^2} = \frac{2m}{\hbar^2}[V(x) - E]\psi$$

the eigenfunction $\psi$ will be an oscillatory function throughout the region where $[V(x) - E]$ is negative since $d^2\psi/dx^2$ will be negative (concave downward) if $\psi$ is positive in that region, while $d^2\psi/dx^2$ will be positive (concave upwards) if $\psi$ is negative in that region. However, $\psi$ will oscillate less rapidly near the ends of the region than it does near the center since the magnitude of $d^2\psi/dx^2$, which determines the rapidity of oscillation of $\psi$, is proportional to the magnitude of $[V(x) - E]$, and the difference between $V(x)$ and $E$ becomes smaller as the ends of the region are approached. Therefore, the separation between the nodes of the oscillatory function increases near the ends of the region, in the manner indicated in Figure 5-18. The figure shows the amplitude of the oscillations in $\psi$ increasing as the ends of the region are

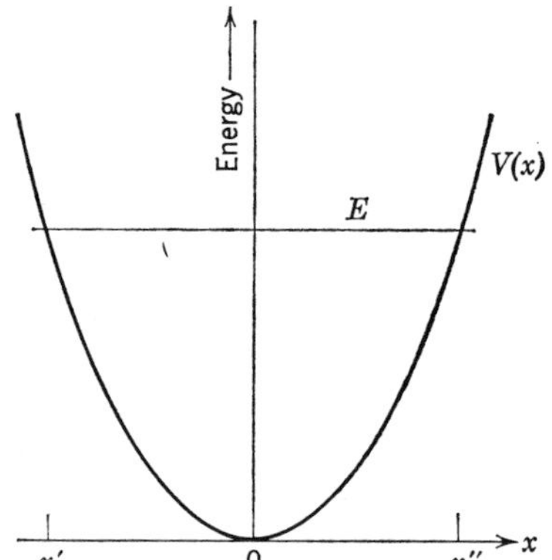


**FIGURE 5-17**

The potential energy $V(x)$ and one of the higher allowed values of the total energy $E$ for a simple harmonic oscillator.

approached. The reason is that $\psi$ must become larger in magnitude where it "bends over," if $[V(x) - E]$ becomes smaller in magnitude, in order that $d^2\psi/dx^2$, which is proportional to their product, continue to have a large enough magnitude to make it bend. Note that Figure 5-18 indicates $\psi$ gradually approaches the axis outside the region where $[V(x) - E]$ is negative, as is required for an acceptable bound state eigenfunction. Also note that as $\psi$ crosses the points where $[V(x) - E]$ changes sign, it has no curvature because both that quantity and $d^2\psi/dx^2$ are zero at these points.

The probability density function is essentially the square of $\psi$, and is indicated in Figure 5-19 by a solid curve. The dashed curve in the same figure indicates the probability density that would be expected in classical mechanics for a particle executing simple harmonic oscillations in the same potential with the same total energy. As we discussed at length in Example 5-6, the classical probability density becomes relatively large near the ends of the region where $[V(x) - E]$ is negative since the particle moves most slowly near the ends. The figure actually shows the classical and quantum mechanical probability densities for a state of only moderately large energy $E$ (actually $E_{13}$), but it makes quite apparent the nature of the correspondence between the probability densities found in the classical limit of very large values of $E$ ($E_n$ as $n \to \infty$). In this limit the quantum mechanical probability density fluctuates within such small distances that only its average behavior, which agrees with the classical

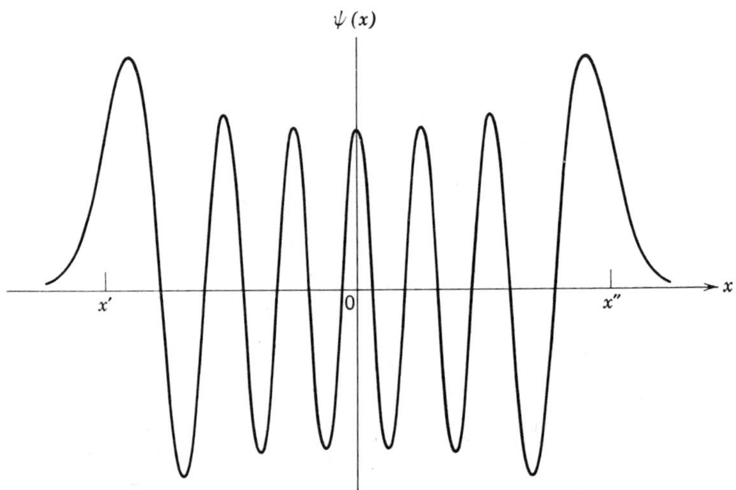


**FIGURE 5-18**

The eigenfunction for the thirteenth allowed energy of the simple harmonic oscillator. The classical limits of motion are indicated by $x'$ and $x''$.

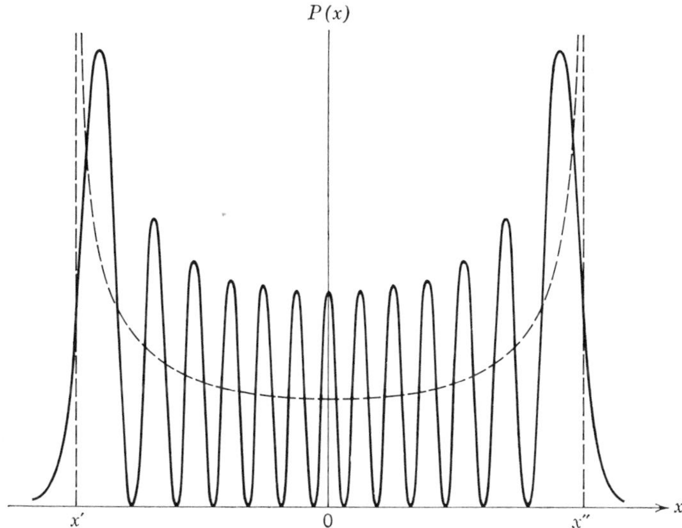


**FIGURE 5-19**

The solid curve is the probability density function for the thirteenth allowed energy of the simple harmonic oscillator. The dashed curve is the classical probability density function for simple harmonic motion with the same energy, and it follows closely the average value of the fluctuating quantum mechanical function. Compare with these functions for the first allowed energy shown in Figure 5-3.

prediction, can be detected experimentally. Also, in the classical limit the quantum mechanical probability density does not penetrate a measurable distance outside the region where $[V(x) - E]$ is negative because the penetration distance is comparable to the distance in which it fluctuates. This agrees with the sharp cutoff predicted by the classical probability density. For an idealized simple harmonic oscillator, $V(x)$ remains proportional to $x^2$ even for very large values of $x^2$, and so all the allowed energies are discretely separated. ◀

## 5-8 Summary

A particular quantum mechanical system is described by a particular potential energy function. We have found that if the potential is time-independent, i.e., can be written $V(x)$, the Schroedinger equation for the potential leads immediately to the corresponding time-independent Schroedinger equation. We have also found that acceptable solutions to the time-independent Schroedinger equation exist only for certain values of the energy, which we list in order of increasing energy as

$$E_1, E_2, E_3, \ldots, E_n, \ldots$$

These energies are called the *eigenvalues* of the potential $V(x)$; a particular potential has a particular set of eigenvalues. The eigenvalues early in the list may be discretely separated in energy. However, unless the potential increases without limit for both very large and very small values of $x$, the eigenvalues become continuously distributed in energy beyond a certain energy.

Corresponding to each eigenvalue is an *eigenfunction*

$$\psi_1(x), \psi_2(x), \psi_3(x), \ldots, \psi_n(x), \ldots$$

which is a solution to the time-independent Schroedinger equation for the potential $V(x)$.

For each eigenvalue there is also a corresponding *wave function*

$$\Psi_1(x,t), \Psi_2(x,t), \Psi_3(x,t), \ldots, \Psi_n(x,t), \ldots$$

From (5-44) we know that these wave functions are

$$\psi_1(x)e^{-iE_1t/\hbar}, \psi_2(x)e^{-iE_2t/\hbar}, \psi_3(x)e^{-iE_3t/\hbar}, \ldots, \psi_n(x)e^{-iE_nt/\hbar}, \ldots$$

Each wave function is a solution to the Schroedinger equation for the potential $V(x)$.

The index $n$, which takes on successive integral values, and which is employed to designate a particular eigenvalue and its corresponding eigenfunction and wave function, is called the *quantum number*. If the system is described by the wave function $\Psi_n(x,t)$, it is said to be in the *quantum state n*.

Each of the wave functions $\Psi_n(x,t)$ is a particular solution to the Schroedinger equation for the potential $V(x)$. Since that equation is linear in the wave function, we expect that any linear combination of these functions will also be a solution. This was verified in Example 5-2 for the case of a linear combination of two wave functions, but the proof can clearly be extended to show that an arbitrary linear combination of *all* wave functions which are solutions to the Schroedinger equation for a particular potential $V(x)$, i.e.

$$\Psi(x,t) = c_1\Psi_1(x,t) + c_2\Psi_2(x,t) + \cdots + c_n\Psi_n(x,t) + \cdots \tag{5-46}$$

is also a solution to that Schroedinger equation. In fact, this expression gives the most general form of the solution to the Schroedinger equation for a potential $V(x)$. Its generality can be appreciated by noting that it is a function which is composed of a very large number of different functions combined in proportions governed by the adjustable constants $c_n$.

It should be noted that the time-independent Schroedinger equation is also a linear equation but, in contrast to the Schroedinger equation, it contains explicitly the total energy $E$. Therefore, an arbitrary linear combination of different solutions will satisfy the equation *only* if they all correspond to the same value of $E$. We shall see in the next chapter that there are two different solutions to the time-independent Schroedinger equation that do correspond to the same value of $E$ because the equation involves a second derivative. We shall also see that both solutions are not always acceptable, even for an allowed value of $E$.

**Example 5-13.** When a particle is in a state such that a measurement of its total energy can lead only to a single result, the eigenvalue $E$, it is described by the wave function

$$\Psi = \psi(x)e^{-iEt/\hbar}$$

An example (whose three-dimensionality makes no difference here) would be an electron in the ground state of a hydrogen atom. In this case, the probability density function

$$\Psi^*\Psi = \psi^*(x)e^{+iEt/\hbar}\psi(x)e^{-iEt/\hbar} = \psi^*(x)\psi(x)$$

does not depend on time, as we have seen before. Consider a particle in a state such that a measurement of its total energy could lead to either of two results, the eigenvalue $E_1$ or the eigenvalue $E_2$. Then the wave function describing the particle is

$$\Psi = c_1\psi_1(x)e^{-iE_1t/\hbar} + c_2\psi_2(x)e^{-iE_2t/\hbar}$$

An example would be an electron that is in the process of making a transition from an excited state to the ground state of the atom. Show that in this case the probability density function is an oscillatory function of time, and calculate the oscillation frequency.

We have for the probability density

$$\Psi^*\Psi = [c_1^*\psi_1^*(x)e^{+iE_1t/\hbar} + c_2^*\psi_2^*(x)e^{+iE_2t/\hbar}][c_1\psi_1(x)e^{-iE_1t/\hbar} + c_2\psi_2(x)e^{-iE_2t/\hbar}]$$

Multiplying the two terms in brackets, we obtain four terms

$$\begin{aligned}\Psi^*\Psi = {} & c_1^* c_1 \psi_1^*(x)\psi_1(x) + c_2^* c_2 \psi_2^*(x)\psi_2(x) \\ & + c_2^* c_1 \psi_2^*(x)\psi_1(x) e^{i(E_2 - E_1)t/\hbar} \\ & + c_1^* c_2 \psi_1^*(x)\psi_2(x) e^{-i(E_2 - E_1)t/\hbar}\end{aligned} \tag{5-47}$$

The time dependences cancel in the first two, but not in the last two. These two terms contain complex exponentials that oscillate in time at frequency $\nu$. By rewriting the complex exponentials as in (5-41a) and (5-41b), we see immediately that

$$\nu = \frac{E_2 - E_1}{2\pi\hbar} = \frac{E_2 - E_1}{h} \tag{5-48}$$

◀

Some very interesting comments can be made about the results of Example 5-13. Consider an electron in the ground state of a hydrogen atom. Since the electron could be found at any location where the probability density has an appreciable value, the charge it carries would not be confined to a particular location. Thus, when speaking of average properties of the electron in the atom, it is appropriate to speak of its charge distribution, which is proportional to its probability density. Since the probability density is independent of time in the ground state, the charge distribution is also. But even in classical electromagnetism a static distribution of charge does not emit radiation. We see that quantum mechanics provides a way of resolving the paradox of old quantum theory concerning the stability, against the emission of radiation, of atoms in their ground states.

Atoms that are excited do emit radiation, and they eventually return to their ground states. Consider an electron in the process of making a transition from an excited state to the ground state of a hydrogen atom. Its probability density, and therefore the associated charge distribution, are oscillating in time at the frequency given by (5-48)

$$\nu = \frac{E_2 - E_1}{h}$$

where $E_2$ is the energy of the excited state and $E_1$ is the energy of the ground state. According to classical electromagnetism, this charge distribution would be expected to emit radiation at the same frequency; but this is also precisely the frequency of the photon that Bohr and Einstein say should be emitted, since the energy carried by the photon is $E_2 - E_1$. Of course this cannot happen for an electron in the ground state of the atom because there is *no* state of lower energy for the ground state to mix with and produce an oscillatory probability density or charge distribution.

In addition to predicting correctly the frequencies of the photons emitted in atomic transitions, quantum mechanics also predicts correctly the probabilities per second that the transitions will take place. We shall obtain these predictions in Chapter 8 by a simple extension of the calculation of Example 5-13. It will be seen there that the perplexing selection rules of old quantum theory follow as an immediate consequence of these predictions.

Schroedinger stressed the fact that his theory provides a physical picture of the process of emission of radiation by excited atoms that is very much more appealing than that provided by the Bohr theory. In discussing the advantages of his theory, he wrote: "It is hardly necessary to point out how much more gratifying it would be to conceive of a quantum transition as an energy change from one vibrational mode to another than to regard it as a jumping of electrons."

## QUESTIONS

1. Why are there difficulties in applying the de Broglie postulate, $\lambda = h/p$, to a particle whose linear momentum is of changing magnitude?

2. How does the de Broglie postulate enter into the Schroedinger theory?

3. Is the experimental evidence that the de Broglie-Einstein relation, $\nu = E/h$, applies to wave functions for material particles as firm as the evidence that it applies to electromagnetic waves and photons? Is the evidence that it applies to wave functions as firm as the evidence that $\lambda = h/p$ applies to wave functions?

4. What would be the effect on the Schroedinger theory of changing the definition of total energy in the relation $\nu = E/h$ by adding the constant rest mass energy of the particle?

5. Why is the Schroedinger equation not valid for relativistic particles?

6. Did Newton derive his laws of motion, or did he obtain them from plausibility arguments?

7. Give a reason why the Schroedinger equation is written in terms of the potential energy, and not in terms of the force.

8. Why is it so important for the Schroedinger equation to be linear in the wave function?

9. The mass $m$ of a particle appears explicitly in Schroedinger's equation, but its charge $e$ does not, even though both may effect its motion. Why?

10. The wave equations of classical physics contain a second space derivative and a second time derivative. The Schroedinger equation contains a second space derivative and a first time derivative. Use these facts to explain why the solutions to the classical wave equations can be real functions, while the solutions to the Schroedinger equation must be complex functions.

11. Why does the Schroedinger equation contain a first time derivative?

12. Explain why it is not possible to measure the value of a complex quantity.

13. In electromagnetism we compute the intensity of a wave by taking the square of its amplitude. Why do we not do exactly the same thing with quantum mechanical waves?

14. Consider a water wave traveling across the surface of the ocean. If no one were observing the wave, or even thinking about it, would you say that the wave exists? Would you automatically give the same answer for a quantum mechanical wave? If not, why not?

15. What is the basic connection between the properties of a wave function and the behavior of the associated particle?

16. Why does the probability density function have to be everywhere real, non-negative, and of finite and definite value?

17. Explain in words what is meant by normalization of a wave function.

18. If the normalization condition is not applied, why can a wave function be multiplied by any constant factor and still remain a solution to the Schroedinger equation?

19. Why does Schroedinger quantum mechanics provide only statistical information? In your opinion, does this reflect a failing of the theory, or a property of nature?

20. Since the wave function describing the behavior of a particle satisfies a differential equation, its evolution in time is perfectly predictable. How does this fact fit in with the uncertainty principle?

21. State in words the meaning of the expectation value of $x$.

22. Why is it necessary to use a differential operator in calculating the expectation value of $p$?

23. Are there other examples in science, engineering, or mathematics in which differential operators are related to physical quantities?

24. Do you think it is legitimate to say that we have *solved* a differential equation by guessing the form of the solution and then verifying the guess by substitution?

25. Explain briefly the meaning of a well-behaved eigenfunction in the context of Schroedinger quantum mechanics.

26. Why must an eigenfunction be well behaved in order to be acceptable in the Schroedinger theory?

27. Explain in two or three sentences how the quantization of energy is related to the well-behaved character of acceptable eigenfunctions.

28. Why is $\psi$ necessarily an oscillatory function if $V(x) < E$?

29. Why does $\psi$ tend to go to infinity if $V(x) > E$?

30. Is it ever possible for an allowed value of the total energy $E$ of a system to be less than the minimum value of its potential energy $V(x)$? Give a qualitative argument, along the lines of the arguments in Section 5-7, to justify your answer.

31. We have seen several examples of the general result that the lowest allowed value of the total energy $E$, for a particle bound in a potential $V(x)$, lies above the minimum value of $V(x)$. Use the uncertainty principle in a qualitative argument to explain why this must be so.

32. If a particle is not bound in a potential, its total energy is not quantized. Does this mean the potential has *no* effect on the behavior of the particle? What effect would you expect it to have?

## PROBLEMS

1. If the wave functions $\Psi_1(x,t)$, $\Psi_2(x,t)$, and $\Psi_3(x,t)$ are three solutions to the Schroedinger equation for a particular potential $V(x,t)$, show that the arbitrary linear combination $\Psi(x,t) = c_1\Psi_1(x,t) + c_2\Psi_2(x,t) + c_3\Psi_3(x,t)$ is also a solution to that equation.

2. At a certain instant of time, the dependence of a wave function on position is as shown in Figure 5-20. (a) If a measurement that could locate the associated particle in an element $dx$ of the $x$ axis were made at that instant, where would it most likely be found? (b) Where would it least likely be found? (c) Are the chances better that it would be found at *any* positive value of $x$, or are they better that it would be found at *any* negative value of $x$?

3. (a) Determine the frequency $\nu$ of the time-dependent part of the wave function, quoted in Example 5-3, for the lowest energy state of a simple harmonic oscillator. (b) Use this value of $\nu$, and the de Broglie-Einstein relation $E = h\nu$, to evaluate the total energy $E$ of the oscillator. (c) Use this value of $E$ to show that the limits of the classical motion of the oscillator, found in Example 5-6, can be written as $x = \pm\hbar^{1/2}/(Cm)^{1/4}$.

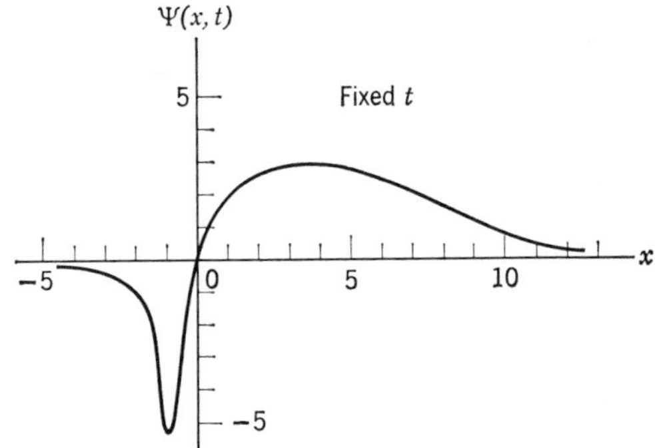


**FIGURE 5-20**
The space dependence of a wave function considered in Problem 2, evaluated at a certain instant of time.

4. By evaluating the classical normalization integral in Example 5-6, determine the value of the constant $B^2$ which satisfies the requirement that the total probability of finding the particle in the classical oscillator somewhere between its limits of motion must equal one.

5. Use the results of Examples 5-5, 5-6, and 5-7 to evaluate the probability of finding a particle, in the lowest energy state of a quantum mechanical simple harmonic oscillator, within the limits of the classical motion. (Hint: (i) The classical limits of motion are expressed in a convenient form in the statement of Problem 3c. (ii) The definite integral that will be obtained can be expressed as a normal probability integral, or an error function. It can then be evaluated immediately by consulting mathematical handbooks which tabulate these quantities. Alternatively, the definite integral can be evaluated by plotting the integrand on graph paper, and counting squares to find the area enclosed between the integrand, the axis, and the limits.)

6. At sufficiently low temperature, an atom of a vibrating diatomic molecule is a simple harmonic oscillator in its lowest energy state because it is bound to the other atom by a linear restoring force. (The restoring force is linear, at least approximately, because the molecular vibrations are very small.) The force constant $C$ for a typical molecule has a value of about $C \sim 10^3$ nt/m. The mass of the atom is about $m \sim 10^{-26}$ kg. (a) Use these numbers to evaluate the limits of the classical motion from the formula quoted in Problem 3c. (b) Compare the distance between these limits to the dimensions of a typical diatomic molecule, and comment on what this comparison implies concerning the behavior of such a molecule at very low temperatures.

7. Use the particle in a box wave function verified in Example 5-9, with the value of $A$ determined in Example 5-10, to calculate the probability that the particle associated with the wave function would be found in a measurement within a distance of $a/3$ from the right-hand end of the box of length $a$. The particle is in its lowest energy state. (b) Compare with the probability that would be predicted classically from a very simple calculation related to the one in Example 5-6.

8. Use the results of Example 5-9 to estimate the total energy of a neutron of mass about $10^{-27}$ kg which is assumed to move freely through a nucleus of linear dimensions of about $10^{-14}$ m, but which is strictly confined to the nucleus. Express the estimate in MeV. It will be close to the actual energy of a neutron in the lowest energy state of a typical nucleus.

9. (a) Following the procedure of Example 5-9, verify that the wave function

$$\Psi(x,t) = \begin{cases} A \sin \dfrac{2\pi x}{a} e^{-iEt/\hbar} & -a/2 < x < +a/2 \\ 0 & x < -a/2 \text{ or } x > +a/2 \end{cases}$$

is a solution to the Schroedinger equation in the region $-a/2 < x < +a/2$ for a particle which moves freely through the region but which is strictly confined to it. (b) Also determine the value of the total energy $E$ of the particle in this first excited state of the system, and compare with the total energy of the ground state found in Example 5-9. (c) Plot the space dependence of this wave function. Compare with the ground state wave function of Figure 5-7, and give a qualitative argument relating the difference in the two wave functions to the difference in the total energies of the two states.

**10.** (a) Normalize the wave function of Problem 9, by adjusting the value of the multiplicative constant $A$ so that the total probability of finding the associated particle somewhere in the region of length $a$ equals one. (b) Compare with the value of $A$ obtained in Example 5-7 by normalizing the ground state wave function. Discuss the comparison.

**11.** Calculate the expectation value of $x$, and the expectation value of $x^2$, for the particle associated with the wave function of Problem 10.

**12.** Calculate the expectation value of $p$, and the expectation value of $p^2$, for the particle associated with the wave function of Problem 10.

**13.** (a) Use quantities calculated in the preceding two problems to calculate the product of the uncertainties in position and momentum of the particle in the first excited state of the system being considered. (b) Compare with the uncertainty product when the particle is in the lowest energy state of the system, obtained in Example 5-10. Explain why the uncertainty products differ.

**14.** (a) Calculate the expectation values of the kinetic energy and the potential energy for a particle in the lowest energy state of a simple harmonic oscillator, using the wave function of Example 5-7. (b) Compare with the time-averaged kinetic and potential energies for a classical simple harmonic oscillator of the same total energy.

**15.** In calculating the expectation value of the product of position times momentum, an ambiguity arises because it is not apparent which of the two expressions

$$\overline{xp} = \int_{-\infty}^{\infty} \Psi^* x\left(-i\hbar\frac{\partial}{\partial x}\right)\Psi\, dx$$

$$\overline{px} = \int_{-\infty}^{\infty} \Psi^* \left(-i\hbar\frac{\partial}{\partial x}\right)x\Psi\, dx$$

should be used. (In the first expression $\partial/\partial x$ operates on $\Psi$; in the second it operates on $x\Psi$.) (a) Show that neither is acceptable because both violate the obvious requirement that $\overline{xp}$ should be real since it is measurable. (b) Then show that the expression

$$\overline{xp} = \int_{-\infty}^{\infty} \Psi^* \left[\frac{x\left(-i\hbar\frac{\partial}{\partial x}\right) + \left(-i\hbar\frac{\partial}{\partial x}\right)x}{2}\right]\Psi\, dx$$

is acceptable because it does satisfy this requirement. (Hint: (i) A quantity is real if it equals its own complex conjugate. (ii) Try integrating by parts. (iii) In any realistic case the wave function will always vanish at $x = \pm\infty$.)

**16.** Show by direct substitution into the Schroedinger equation that the wave function

$$\Psi(x,t) = \psi(x)e^{-iEt/\hbar}$$

satisfies that equation if the eigenfunction $\psi(x)$ satisfies the time-independent Schroedinger equation for a potential $V(x)$.

**17.** (a) Write the classical wave equation for a string of density per unit length which varies with $x$. (b) Then separate it into two ordinary differential equations, and show that the equation in $x$ is very analogous to the time-independent Schroedinger equation.

**18.** By using an extension of the procedure leading to (5-31), obtain the Schroedinger equation for a particle of mass $m$ moving in three dimensions (described by rectangular coordinates $x$, $y$, $z$).

**19.** (a) Separate the Schroedinger equation of Problem 18 for a time-independent potential, into a time-independent Schroedinger equation and an equation for the time dependence of the wave function. (b) Compare to the corresponding one-dimensional equations, (5-37) and (5-38), and explain the similarities and the differences.

**20.** (a) Separate the time-independent Schroedinger equation of Problem 19 into three time-independent Schroedinger equations, one in each of the coordinates. (b) Compare them with (5-37). (c) Explain clearly what must be assumed about the form of the potential energy in order to make the separation possible, and what the physical significance of this assumption is. (d) Give an example of a system that would have such a potential.

**21.** Consider a particle moving under the influence of the potential $V(x) = C|x|$, where $C$ is a constant, which is illustrated in Figure 5-21. (a) Use qualitative arguments, very similar to

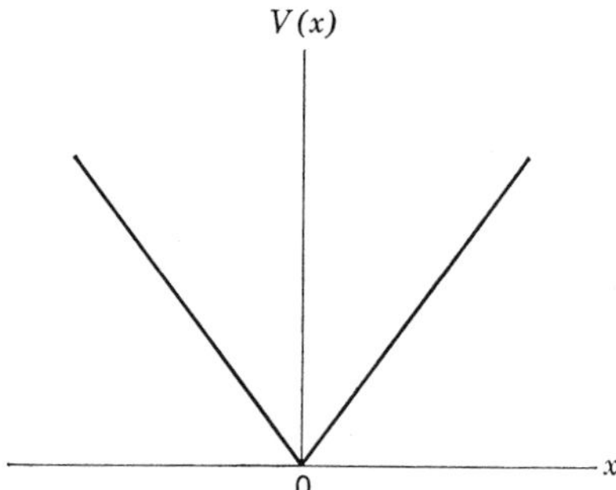


**FIGURE 5-21**
A potential function considered in Problem 21.

those of Example 5-12, to make a sketch of the first eigenfunction and of the tenth eigenfunction for the system. (b) Sketch both of the corresponding probability density functions. (c) Then use the classical mechanics to calculate, in the manner of Example 5-6, the probability density functions predicted by that theory. (d) Plot the classical probability density functions with the quantum mechanical probability density functions, and discuss briefly their comparison.

**22.** Consider a particle moving in the potential $V(x)$ plotted in Figure 5-22. For the following ranges of the total energy $E$, state whether there are any allowed values of $E$ and if so, whether they are discretely separated or continuously distributed. (a) $E < V_0$, (b) $V_0 < E < V_1$, (c) $V_1 < E < V_2$, (d) $V_2 < E < V_3$, (e) $V_3 < E$.

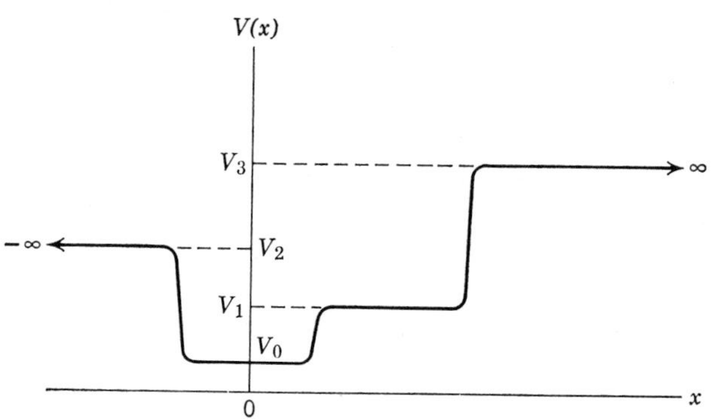


**FIGURE 5-22**
A potential function considered in Problem 22.

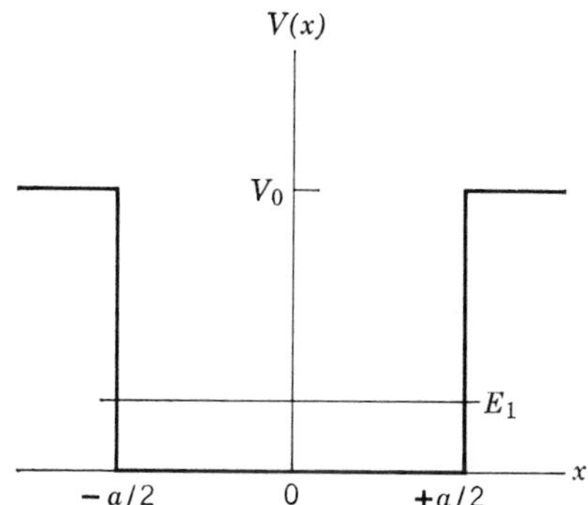


**FIGURE 5-23**

A potential function considered in Problem 23.

23. Consider a particle moving in the potential $V(x)$ illustrated in Figure 5-23, that has a rectangular region of depth $V_0$, and width $a$, in which the particle can be bound. These parameters are related to the mass $m$ of the particle in such a way that the lowest allowed energy $E_1$ is found at an energy about $V_0/4$ above the "bottom." Use qualitative arguments to sketch the approximate shape of the corresponding eigenfunction $\psi_1(x)$.

24. Suppose the bottom of the potential function of Problem 23 is changed by adding a bump in the center of height about $V_0/10$ and width $a/4$. That is, suppose the potential now looks like the illustration of Figure 5-24. Consider qualitatively what will happen to the curvature of the eigenfunction in the region of the bump, and how this will, in turn, effect the problem of obtaining an acceptable behavior of the eigenfunction in the region outside the binding region. From these considerations predict, qualitatively, what the bump will do to the value of the lowest allowed energy $E_1$.

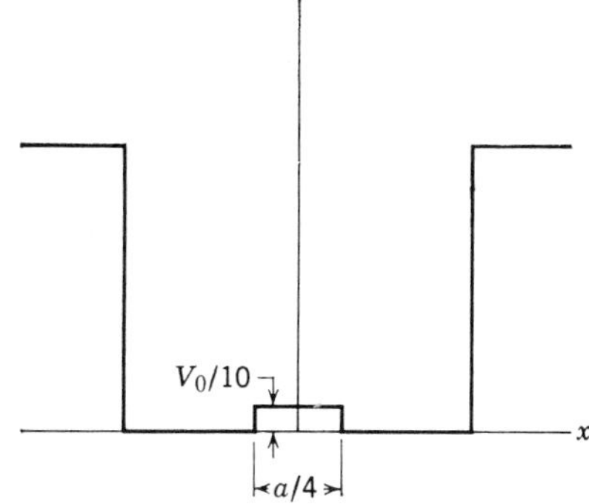


**FIGURE 5-24**

A rectangular bump added to the bottom of the potential of Figure 5-23.

25. Because the bump in Problem 24 is small, a good approximation to the lowest allowed energy of the particle in the presence of the bump can be obtained by taking it as the sum

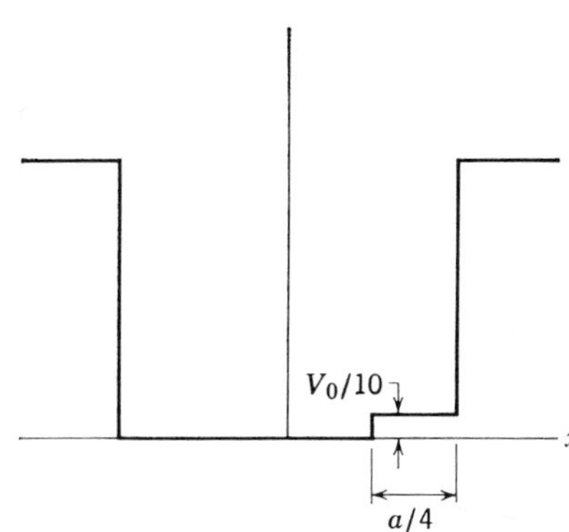


**FIGURE 5-25**

The same rectangular bump as in Figure 5-24, but moved to the edge of the potential.

of the energy in the absence of the bump plus the expectation value of the extra potential energy represented by the bump, taking the $\Psi$ corresponding to no bump to calculate the expectation value. Using this point of view, predict whether a bump of the same "size," but located at the edge of the bottom as in Figure 5-25, would have a larger, smaller, or equal effect on the lowest allowed energy of the particle, compared to the effect of a centered bump. (Hint: Make a rough sketch of the product of $\Psi^*\Psi$ and the potential energy function that describes the centered bump. Then consider qualitatively the effect of moving the bump to the edge on the integral of this product.)

**26.** By substitution into the time-independent Schroedinger equation for the potential illustrated in Figure 5-23, show that in the region to the right of the binding region the eigenfunction has the mathematical form

$$\psi(x) = Ae^{-[\sqrt{2m(V_0-E)}/\hbar]\,x} \qquad x > +a/2$$

**27.** Using the probability density corresponding to the eigenfunction of Problem 26, write an expression to estimate the distance $D$ outside the binding region of the potential within which there would be an appreciable probability of finding the particle. (Hint: Take $D$ to extend to the point at which $\Psi^*\Psi$ is smaller than its value at the edge of the binding region by a factor of $e^{-1}$. This $e^{-1}$ criteria is similar to one often used in the study of electrical circuits.)

**28.** The potential illustrated in Figure 5-23 gives a good description of the forces acting on an electron moving through a block of metal. The energy difference $V_0 - E$, for the highest energy electron, is the work function for the metal. Typically, $V_0 - E \simeq 5$ eV. (a) Use this value to estimate the distance $D$ of Problem 27. (b) Comment on the results of the estimate.

**29.** Consider the eigenfunction illustrated in the top part of Figure 5-26. (a) Which of the three potentials illustrated in the bottom part of the figure could lead to such an eigenfunction? Give qualitative arguments to justify your answer. (b) The eigenfunction shown is not the one corresponding to the lowest allowed energy for the potential. Sketch the form of the eigenfunction which does correspond to the lowest allowed energy. (c) Indicate on another sketch the range of energies where you would expect discretely separated allowed energy states, and the range of energies where you would expect the allowed energies to be continuously distributed.

**30.** (a) Use the numerical integration procedure, developed in Appendix F, to find the lowest allowed energy value $E_1$, and the form of the corresponding eigenfunction $\psi_1(x)$, for a particle of mass $m$ moving in the potential

$$V(x) = \begin{cases} \infty & x < -a/2 \text{ or } x > +a/2 \\ 0 & -a/2 < x < +a/2 \end{cases}$$

As will be proven in Chapter 6, since $V(x)$ increases without limit when $x$ is outside the region of length $a$ the particle is strictly prohibited from being found outside that region. Therefore $\psi_1(x)$ goes to zero at $x = \pm a/2$. Symmetry arguments show that for the lowest eigenfunction $d\psi_1(x)/dx$ is zero at $x = 0$. (Hint: The parameter $\alpha$ cannot be defined in this problem, but the parameter $C$ can still be defined directly in terms of $E_1$.) (b) Compare the value of $E_1$ you obtain with the exact solution to this problem obtained analytically in Example 5-9.

**31.** Make the same calculation indicated in Problem 30, except for a potential containing a rectangular bump of height $v_0$ and width $a/2$, centered at the bottom of the binding region. That is

$$V(x) = \begin{cases} \infty & x < -a/2 \text{ or } x > +a/2 \\ 0 & -a/2 < x < -a/4 \text{ or } +a/4 < x < +a/2 \\ v_0 & -a/4 < x < +a/4 \end{cases}$$

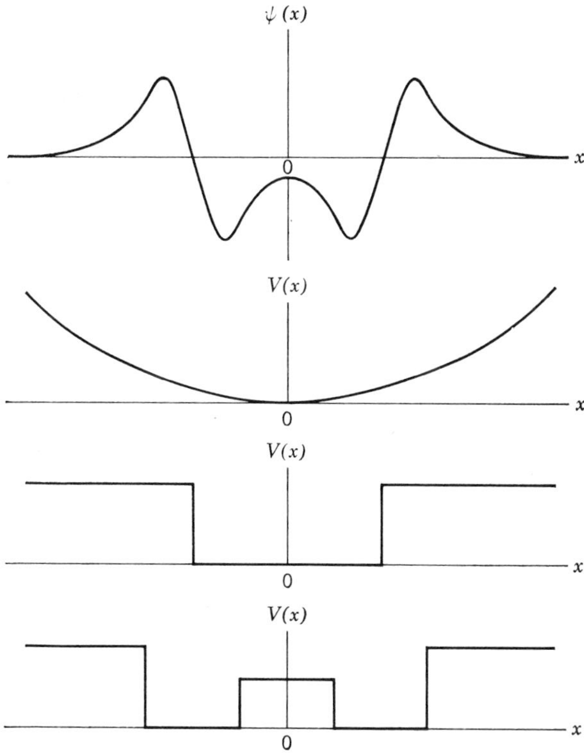


**FIGURE 5-26**
An eigenfunction (*top* curve) and three possible forms (*bottom* curves) of the potential energy function considered in Problem 29.

Take $v_0$ to have the value

$$v_0 = \frac{\pi^2\hbar^2}{8ma^2}$$

Problem 17 of the next chapter asks for an analytical solution to the time-independent Schroedinger equation for this potential. (Hint: A guess concerning an appropriate initial choice of $E_1$ can be obtained from the qualitative considerations of Problem 24.)

**32.** Repeat the numerical integration of Appendix F for assumed values of $E$ of higher energy, and find the first excited state of the potential treated there. (Hint: (i) For this state, $\psi = 0$ at $u = 0$. (ii) Take $d\psi/du = +1$ at that point, since linearity allows it to have any value. (iii) The eigenfunction looks something like a full sine wave fitted into the region of the well.)

**33.** Using the first two normalized wave functions $\Psi_1(x,t)$ and $\Psi_2(x,t)$ for a particle moving freely in a region of length $a$, but strictly confined to that region, construct the linear combination

$$\Psi(x,t) = c_1\Psi_1(x,t) + c_2\Psi_2(x,t)$$

Then derive a relation involving the adjustable constants $c_1$ and $c_2$ which, when satisfied, will ensure that $\Psi(x,t)$ is also normalized. The normalized $\Psi_1(x,t)$ and $\Psi_2(x,t)$ are obtained in Example 5-10 and Problem 10.

**34.** (a) Using the normalized "mixed" wave function of Problem 33, calculate the expectation value of the total energy $E$ of the particle in terms of the energies $E_1$ and $E_2$ of the two states and of the values $c_1$ and $c_2$ of the mixing parameters. (b) Interpret carefully the meaning of your result.

**35.** If the particle described by the wave function of Problem 33 is a proton moving in a nucleus, it will give rise to a charge distribution which oscillates in time at the same frequency as the oscillations of its probability density. (a) Evaluate this frequency for values of $E_1$ and $E_2$ corresponding to a proton mass of $10^{-27}$ kg and a nuclear dimension of $10^{-14}$ m. (b) Also evaluate the frequency and energy of the photon that would be emitted by this oscillating charge distribution as the proton drops from the excited state to the ground state. (c) In what region of the electromagnetic spectrum is such a photon?

# 6

# Solutions of Time-Independent Schroedinger Equations

# Solutions of Time-Independent Schroedinger Equations 6

## 6-1 Introduction

In this chapter we shall obtain many interesting predictions concerning quantum mechanical phenomena. We shall also discuss some of the experiments confirming the predictions, and some of the important practical applications of the phenomena. The predictions will be obtained by solving the time-independent Schroedinger equation for different forms of the potential energy function $V(x)$, to find the eigenfunctions, eigenvalues, and wave functions, and then using the procedures developed in the previous chapter to interpret the physical significance of these quantities.

Our approach will be very systematic. We shall start by treating the simplest possible form of the potential, namely $V(x) = 0$. Then we shall gradually add complexity to the potential. With each new potential treated, the student will obtain new insight into quantum mechanics and into the behavior of microscopic systems. In this process the student should begin to develop an intuition for quantum mechanics, just as he has developed an intuition for classical mechanics by repeated use of that theory.

The potentials considered in the first sections of this chapter are not able to bind a particle because there is no region in which they have a depression. Although discrete quantization of energy will not be found for these potentials, other fundamental phenomena will be found. In addition to the fact that they naturally fit in at the beginning of our systematic approach, another reason for treating nonbinding potentials first is that it emphasizes their importance. Probably half of the work *currently* being done in quantum mechanics concerns unbound particles.

It is true, however, that most of the applications of quantum mechanics that were made *initially* concerned bound particles. Most aspects of the structure of atoms, molecules, and solids are examples of bound particle problems, as are many aspects of nuclear structure. Since these are the topics we shall concentrate on in the following chapters of this book, some students (or instructors) may prefer to go directly to Section 6-7, which is the first to treat binding potentials, or to Section 6-8, which treats an important special case. Those sections are sufficiently self-contained to make such short cuts feasible without too much difficulty.

Throughout this chapter we deal only with time-independent potentials, since only for such potentials does the time-independent Schroedinger equation have significance. We further restrict ourselves to a single dimension because this simplifies the mathematics while still allowing us to demonstrate most of the interesting quantum phenomena. Obvious exceptions are phenomena involving angular momentum, since this quantity has no meaning in one dimension. Because angular momentum plays a dominant role in atomic structure, the following chapter begins by extending our development of quantum mechanics to three dimensions.

## 6-2 The Zero Potential

The simplest time-independent Schroedinger equation is the one for the case: $V(x) =$ const. A particle moving under the influence of such a potential is a *free particle* since the force acting on it is $F = -dV(x)/dx = 0$. As this is true regardless of the value of the constant, we do not lose generality by choosing the arbitrary additive constant, that always arises in the definition of a potential energy, in such a way as to obtain

$$V(x) = 0 \tag{6-1}$$

We know that in *classical mechanics* a free particle may be either at rest or moving with constant momentum $p$. In either case its total energy $E$ is a constant.

To find the behavior predicted by *quantum mechanics* for a free particle, we solve the time-independent Schroedinger equation, (5-43), setting $V(x) = 0$. With this form for the potential, the equation is

$$-\frac{\hbar^2}{2m}\frac{d^2\psi(x)}{dx^2} = E\psi(x) \tag{6-2}$$

The solutions are the eigenfunctions $\psi(x)$, and the wave functions $\Psi(x,t)$ according to (5-44) are

$$\Psi(x,t) = \psi(x)e^{-iEt/\hbar} \tag{6-3}$$

The eigenvalues $E$ are equal to the total energy of the particle. From the qualitative discussion of Section 5-7, we know that an acceptable solution of the time-independent Schroedinger equation for this nonbinding potential should exist for *any* value of $E \geq 0$.

Of course, we already know a form of the free particle wave function from our plausibility argument leading to the Schroedinger equation. That wave function, (5-23), is

$$\Psi(x,t) = \cos(kx - \omega t) + i\sin(kx - \omega t)$$

Rewriting it as a complex exponential, we have

$$\Psi(x,t) = e^{i(kx-\omega t)} \tag{6-4a}$$

The wave number $k$ and angular frequency $\omega$ are

$$k = \frac{p}{\hbar} = \frac{\sqrt{2mE}}{\hbar} \quad \text{and} \quad \omega = \frac{E}{\hbar} \tag{6-4b}$$

We break the exponential into the product of two factors

$$\Psi(x,t) = e^{ikx}e^{-i\omega t} = e^{ikx}e^{-iEt/\hbar}$$

Then we compare with the general form of the wave function quoted in (6-3)

$$\Psi(x,t) = \psi(x)e^{-iEt/\hbar}$$

This comparison makes it apparent that

$$\psi(x) = e^{ikx} \qquad \text{where } k = \frac{\sqrt{2mE}}{\hbar} \tag{6-5}$$

That is, the complex exponential of (6-5) gives the form of a free particle eigenfunction corresponding to the eigenvalue $E$.

More specifically, it is a traveling wave free particle eigenfunction because the corresponding wave function, $\Psi(x,t) = e^{i(kx-\omega t)}$, represents a traveling wave. This can be seen, for example, from the fact that the nodes of the real part of the oscillatory wave function are located at positions where $kx - \omega t = (n + 1/2)\pi$, with $n = 0, \pm 1, \pm 2, \ldots$. The reason is that the real part of $\Psi(x,t)$, which is $\cos(kx - \omega t)$, has the value zero wherever $kx - \omega t = (n + 1/2)\pi$. Thus the nodes occur wherever

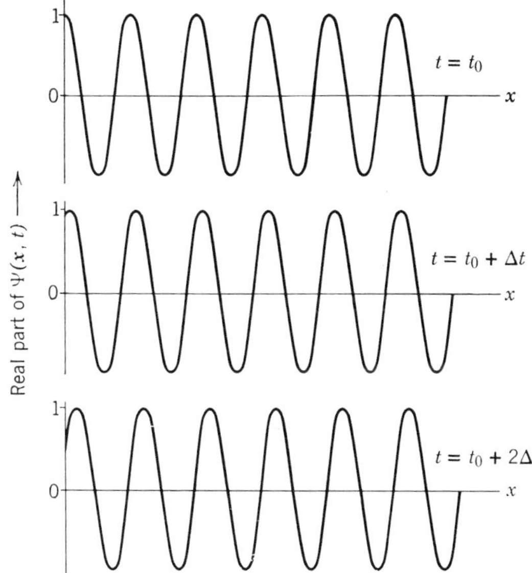


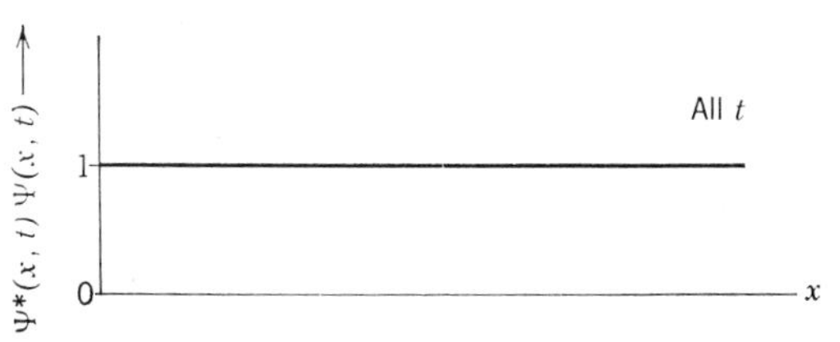


**FIGURE 6-1**

*Top:* The real part, $\cos(kx - \omega t)$, of a complex exponential traveling wave function, $\Psi = e^{i(kx-\omega t)}$, for a free particle. With increasing time the nodes move in the direction of increasing $x$. *Bottom:* For this wave function a sense of motion is not conveyed by plotting the probability density $\Psi^*\Psi = e^{-i(kx-\omega t)}e^{i(kx-\omega t)} = 1$ since it is constant for all $t$ (and all $x$). Of course, we cannot plot $\Psi$ itself, as it is complex.

$x = (n + 1/2)\pi/k + \omega t/k$ and, since these values of $x$ increase with increasing $t$, the nodes travel in the direction of increasing $x$. The conclusion is illustrated in the top part of Figure 6-1 which shows plots of the real part of $\Psi(x,t)$ at successively later times. For this wave function, the probability density $\Psi^*(x,t)\Psi(x,t)$, illustrated in the bottom of Figure 6-1, conveys no sense of motion.

Intuition suggests that, for the same value of $E$, there should also be a wave function representing a wave traveling in the direction of decreasing $x$. The preceding argument indicates that this wave function would be written with the sign of $kx$ reversed, that is

$$\Psi(x,t) = e^{i(-kx-\omega t)} \tag{6-6}$$

The corresponding eigenfunction would be

$$\psi(x) = e^{-ikx} \qquad \text{where } k = \frac{\sqrt{2mE}}{\hbar} \tag{6-7}$$

It is easy to see that this eigenfunction is also a solution to the time-independent Schroedinger equation for $V(x) = 0$. In fact, any arbitrary linear combination of the two eigenfunctions of (6-5) and (6-7), for the *same* value of the total energy $E$, is also a solution to the equation. To prove these statements, we take the linear combination

$$\psi(x) = Ae^{ikx} + Be^{-ikx} \qquad \text{where } k = \frac{\sqrt{2mE}}{\hbar} \tag{6-8}$$

in which $A$ and $B$ are arbitrary constants, and substitute it into the time-independent Schroedinger equation, (6-2). Since

$$\frac{d^2\psi(x)}{dx^2} = i^2k^2Ae^{ikx} + i^2k^2Be^{-ikx} = -k^2\psi(x) = -\frac{2mE}{\hbar^2}\psi(x)$$

substitution into the equation yields

$$-\frac{\hbar^2}{2m}\left(-\frac{2mE}{\hbar^2}\right)\psi(x) = E\psi(x)$$

Since this is obviously satisfied, the linear combination is a valid solution to the time-independent Schroedinger equation.

The most general form of the solution to an ordinary (i.e., not partial) differential equation involving a second derivative contains two arbitrary constants. The reason is that obtaining the solution from such an equation basically amounts to performing two successive integrations to remove the second derivative, and each step yields a constant of integration. Examples familiar to the student are found in general solutions of Newton's equation of motion, which involve two arbitrary constants such as initial position and velocity. Since the linear combination of (6-8) is a solution containing two arbitrary constants to (6-2), it is its *general* solution. The general solution is useful because it allows us to describe *any* possible eigenfunction associated with the eigenvalue $E$. For instance, if we set $B = 0$, we obtain an eigenfunction for a wave traveling in the direction of increasing $x$. If we set $A = 0$, the wave is traveling in the direction of decreasing $x$. If we set $|A| = |B|$, there are two oppositely directed traveling waves that combine to form a standing wave. Standing wave eigenfunctions will be used in Section 6-3.

Let us consider now the question of giving physical interpretation to the free particle eigenfunctions and wave functions. Take first the case of a wave traveling in the direction of increasing $x$. The eigenfunction and wave function for this case are

$$\psi(x) = Ae^{ikx} \qquad \text{and} \qquad \Psi(x,t) = Ae^{i(kx-\omega t)} \tag{6-9}$$

An obvious guess is that the particle whose motion is described by these functions is also traveling in the direction of increasing $x$. To verify this, let us calculate the expectation value of the momentum, $\bar{p}$, for the particle. According to the general expectation value formula, (5-34)

$$\bar{p} = \int_{-\infty}^{\infty} \Psi^* p_{op} \Psi \, dx$$

where the operator for momentum is

$$p_{op} = -i\hbar\frac{\partial}{\partial x}$$

Now, for the wave function in question, we have

$$p_{op}\Psi = -i\hbar\frac{\partial}{\partial x}Ae^{i(kx-\omega t)} = -i\hbar(ik)Ae^{i(kx-\omega t)} = +\hbar k\Psi = +\sqrt{2mE}\,\Psi$$

so

$$\bar{p} = +\int_{-\infty}^{\infty} \Psi^*\sqrt{2mE}\,\Psi\, dx = +\sqrt{2mE}\int_{-\infty}^{\infty} \Psi^*\Psi\, dx$$

The integral on the right is the probability density integrated over the entire range of the $x$ axis. This is just the probability that the particle will be found *somewhere*, which must equal one. Therefore, we obtain

$$\bar{p} = +\sqrt{2mE}$$

This is exactly the momentum that we would expect for a particle moving in the direction of increasing $x$ with total energy $E$ in a region of zero potential energy.

For the case of a wave traveling in the direction of decreasing $x$, the eigenfunction and wave function are

$$\psi(x) = Be^{-ikx} \quad \text{and} \quad \Psi(x,t) = Be^{i(-kx-\omega t)} \tag{6-10}$$

When we operate on $\Psi$ with $p_{\text{op}}$, the sign reversal of the $kx$ term in the former leads to a sign reversal in the result. This, in turn, leads to a momentum expectation value of

$$\bar{p} = -\sqrt{2mE}$$

Therefore, we interpret the eigenfunction, and wave function, as describing the motion of a particle which is moving in the direction of decreasing $x$ with negative momentum of the magnitude that would be expected in consideration of its energy.

The eigenfunctions and wave functions just considered represent the *idealized* situations of a particle moving, in one direction or the other, in a beam of infinite length. Its $x$ coordinate is completely unknown because the amplitudes of the waves are the same in all regions of the $x$ axis. That is, the probability densities, for instance

$$\Psi^*\Psi = A^*e^{-i(kx-\omega t)}Ae^{i(kx-\omega t)} = A^*A$$

are constants independent of $x$. Thus the particle is equally likely to be found anywhere, and the uncertainty in its position is $\Delta x = \infty$. The uncertainty principle states that in these situations we may know the value of the momentum $p$ of the particle with complete precision, since

$$\Delta p \, \Delta x \geq \hbar/2$$

can be satisfied for an uncertainty in its momentum of $\Delta p = 0$, if $\Delta x = \infty$. Perfectly precise values of $p$ are also indicated by the de Broglie relation, $p = \hbar k$, because these wave functions contain only a single value of the wave number $k$. Since there is an infinite amount of time available to measure the energy of a particle traveling through a beam of infinite length, the energy-time uncertainty principle $\Delta E \, \Delta t \geq \hbar/2$ allows its energy to be known with complete precision. This agrees with the presence of a single value of the angular frequency $\omega$ in these wave functions, because the de Broglie-Einstein relation $E = \hbar\omega$ shows this means a single value of the energy $E$.

A physical example approximating the idealized situation represented by these wave functions would be a proton moving in a highly monoenergetic beam emerging from a cyclotron. Such beams are used to study the scattering of protons by targets of nuclei inserted in the beam. From the point of view of the target nucleus, and in terms of distances of the order of its nuclear radius $r'$, the $x$ position of a proton in the beam may be for all practical purposes completely unknown. That is $\Delta x \gg r'$. Thus the free particle wave functions of (6-9) and (6-10) can give a good approximation to the description of the beam proton in the region of interest near the nucleus where the scattering takes place. In other words, near a nucleus the wave function of (6-9)

$$\Psi = Ae^{i(kx-\omega t)}$$

can be used to describe a proton in a cyclotron beam directed towards increasing $x$, providing the beam is extremely long compared to the dimensions of the nucleus—a condition which is always satisfied in practice since nuclei are extremely small. The wave function describes a particle moving with momentum precisely $p = \hbar k$ and total energy precisely $E = \hbar\omega$, where these quantities are related by the equation $p = \sqrt{2mE}$ appropriate to a particle of mass $m$ moving in a region of zero potential energy.

There is a difficulty concerning the normalization of the wave functions of (6-9) and (6-10). In order to have, for instance

$$\int_{-\infty}^{\infty} \Psi^*\Psi \, dx = \int_{-\infty}^{\infty} A^*A \, dx = A^*A \int_{-\infty}^{\infty} dx = 1$$

the amplitude $A$ must be zero as $\int_{-\infty}^{\infty} dx$ has an infinite value. The difficulty arises from the unrealistic statement made by the wave function that the particle can be found with equal probability anywhere in a beam of infinite length. This is never really true since real beams are always of finite length. The proton beam is limited on one end by the cyclotron and on the other end by a laboratory wall. Although the uncertainty $\Delta x$ in location of a proton is very much larger than a nuclear radius $r'$, it is not larger than the distance $L$ from the cyclotron to the wall. That is, even though $\Delta x \gg r'$, it is also true that $\Delta x < L$. This suggests that normalization can be obtained by setting $\Psi = 0$ outside of the range $-L/2 < x < +L/2$, or else by restricting $x$ to be within that range. In either way we obtain a more realistic description of the actual physical situation, and we can also normalize the wave function with a nonvanishing amplitude $A$. The procedure is called *box normalization*. Despite the fact that the value of $A$ obtained depends on the length $L$ of the box, it always turns out that the final result of calculation of a measurable quantity is independent of the actual value of $L$ used. Furthermore, we shall see that it is usually not necessary to carry through box normalization in detail because quantities of physical interest can be expressed as ratios in which the value of $A$ cancels.

The situation is quite analogous to ones commonly encountered in classical physics. For instance, in solving a problem of electrostatics, a straight charged wire of infinite length is often used to approximate one of finite length in a system where "end effects" are not important. This idealization very much simplifies the geometry of the problem, but it leads to the difficulty that an infinite amount of energy is required to charge the infinitely long wire, unless its charge density is zero. It is usually possible, however, to get around this difficulty simply by expressing the quantities that arise in the problem in terms of ratios.

It is possible to obtain a much more realistic sense of motion than is seen in either part of Figure 6-1 by using a large number of wave functions of the form of (6-9) to generate a *group* of traveling waves. Figure 6-2 shows the probability density $\Psi^*\Psi$ for a particularly simple group, its motion in the direction of increasing $x$, and the ever increasing width of the group. At any instant the location of the group can be well characterized by the expectation value $\bar{x}$, calculated from the probability density. The constant velocity of the group, $d\bar{x}/dt$, equals the constant velocity of the free particle, $v = p/m = \sqrt{2mE}/m = \sqrt{2E/m}$, in agreement with the conclusions of Chapter 3. The spreading of the group is a characteristic property of waves that is intimately related to the uncertainty principle, as discussed in that chapter. Of course the behavior of the group wave function is easier to interpret than the behavior of a purely sinusoidal wave function, such as that of (6-9), because the corresponding probability density is closer to the description of particle motion we are familiar with from classical mechanics. However the mathematics required to describe the group, and treat its behavior analytically, is much more complicated. The reason is that a group must necessarily involve a distribution of wave numbers $k$, and therefore a distribution of energies $E = \hbar^2k^2/2m$. In order to compose even as simple a group as the one shown in the figure, a very large number of sinusoidal waves, with very small differences in wave numbers or energies, must be summed in the manner described in Chapter 3. These mathematical complications far outweigh any advantages involved

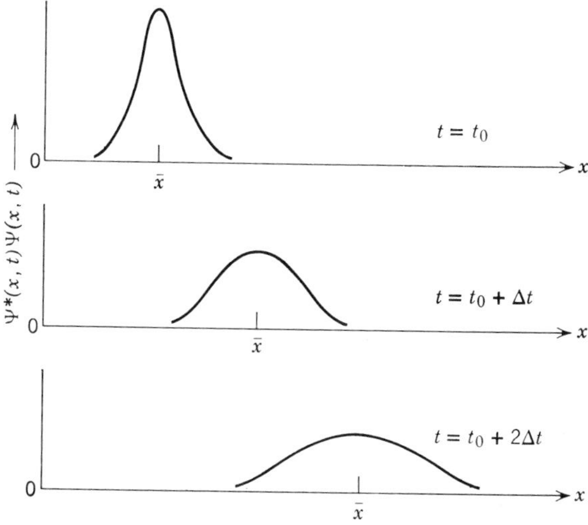


**FIGURE 6-2**

The probability density $\Psi^*\Psi$ for a group traveling wave function of a free particle. With increasing time the group moves in the direction of increasing $x$, and also spreads.

in the ease of interpretation. Consequently, groups are rarely used in practical quantum mechanical calculations, and most such calculations are performed with wave functions involving a single wave number and energy.

Our consideration of the motion of the group in Figure 6-2 leads us to discuss briefly a related case of great interest. If, instead of having the constant value zero, the potential function $V(x)$ changes so slowly that its value is almost constant over a distance of the order of the de Broglie wavelength of the particle, the group wave function will still propagate in a manner similar to that illustrated in the figure, but the velocity of the group will now also change slowly. Calculations, starting from the Schroedinger equation, lead to an expression relating the change in the velocity, $d\bar{x}/dt$, of the group to the change in the potential, $V(x)$. The expression is

$$\frac{d}{dt}\left(\frac{d\bar{x}}{dt}\right) = \overline{\frac{d}{dx}\left(-\frac{V(x)}{m}\right)}$$

or

$$\frac{d^2\bar{x}}{dt^2} = \frac{-\overline{\dfrac{dV(x)}{dx}}}{m} = \frac{\overline{F(x)}}{m}$$

where the bars denote expectation values and $F(x)$ is the force corresponding to the potential $V(x)$. It is unfortunate that the calculations are too complicated to reproduce here. They are very significant because they show that the acceleration of the average location of the particle associated with the group wave function equals the average force acting on the particle, divided by its mass. That is, Schroedinger's equation leads to the result that Newton's law of motion is obeyed, on the average, by a particle of a microscopic system. The fluctuations from its average behavior reflect the uncertainty principle, and they are very important in the microscopic limit. But these fluctuations become negligible in the macroscopic limit where the uncertainty principle is of no consequence, and it is no longer necessary to speak of averages in talking about locations in that limit. Also, in the macroscopic limit any realistic potential changes by only a small amount in a distance as short as a de Broglie wavelength. So it is also not necessary, in that limit, to speak of averages when discussing potentials. Thus, in the macroscopic limit we can ignore the bars representing expectation values, or averages, in the equations just displayed. We then conclude that *Newton's law of motion can be derived from the Schroedinger equation, in the classical limit of macroscopic systems. Newton's law of motion is a special case of Schroedinger's equation.*

## 6-3 The Step Potential (Energy Less Than Step Height)

In the next sections we shall study solutions to the time-independent Schroedinger equation for a particle whose potential energy can be represented by a function $V(x)$ which has a different constant value in each of several adjacent ranges of the $x$ axis. These potentials change in value abruptly in going from one range to the adjacent range. Of course potentials which change abruptly (i.e., are discontinuous functions of $x$) do not really exist in nature. Nevertheless, these idealized potentials are used frequently in quantum mechanics to approximate real situations because, being constant in each range, they are easy to treat mathematically. The results we obtain for these potentials will allow us to illustrate a number of characteristic quantum mechanical phenomena.

An analogy, that is surely familiar to the student, is found in the procedure used in studying electromagnetism. This involves treating many idealized systems like the infinite wire, the capacitor without edges, etc. These systems are studied because they are relatively easy to handle, because they are excellent approximations to real ones, and because real systems are usually complicated to treat mathematically since they have complicated geometries. The idealized potentials we treat in this chapter are used in the same way and with the same justification.

The simplest case is the *step potential*, illustrated in Figure 6-3. If we choose the origin of the $x$ axis to be at the step, and the arbitrary additive constant that always occurs in the definition of a potential energy so that the potential energy of the particle is zero when it is to the left of the step, $V(x)$ can be written

$$V(x) = \begin{matrix} V_0 & \quad x > 0 \\ 0 & \quad x < 0 \end{matrix} \tag{6-11}$$

where $V_0$ is a constant. We may think of $V(x)$ as an approximate representation of the potential energy function for a charged particle moving along the axis of a system of two electrodes, separated by a very narrow gap, which are held at different voltages. The upper half of Figure 6-4 illustrates this system, and the lower half illustrates the corresponding potential energy function. As the gap decreases, the potential function approaches the idealization illustrated in Figure 6-3. In Example 6-2 we shall see that the potential energy for an electron moving near the surface of a metal is very much like a step potential since it rapidly increases at the surface from an essentially constant interior value to a higher constant exterior value.

Assume that a particle of mass $m$ and total energy $E$ is in the region $x < 0$, and that it is moving toward the point $x = 0$ at which the step potential $V(x)$ abruptly changes its value. According to classical mechanics, the particle will move freely in that region until it reaches $x = 0$, where it is subjected to an impulsive force $F = -dV(x)/dx$ acting in the direction of decreasing $x$. The idealized potential, (6-11), yields an impulsive force of infinite magnitude acting only at the point $x = 0$. However, as it acts on the particle only for an infinitesimal time, the quantity $\int F\,dt$ (the

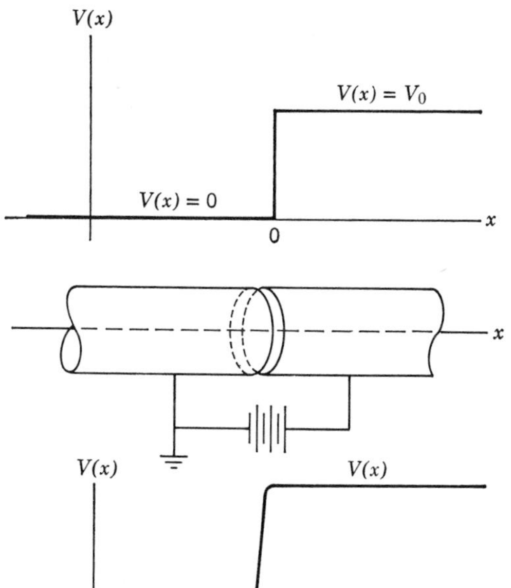

**FIGURE 6-3**
A step potential.

**FIGURE 6-4**
Illustrating a physical system with a potential energy function that can be approximated by a step potential. A charged particle moves along the axis of two cylindrical electrodes held at different voltages. Its potential energy is constant when it is inside either electrode, but it changes very rapidly when passing from one to the other.

impulse), which determines the change in its momentum, is finite. In fact, the momentum change is not affected by the idealization.

The motion of the particle subsequent to experiencing the force at $x = 0$ depends, in classical mechanics, on the relation between $E$ and $V_0$. This is also true in quantum mechanics. In the present section we treat the case where $E < V_0$, i.e., where the total energy is less than the height of the potential step as illustrated in Figure 6-5. (The case where $E > V_0$ is treated in the following section.) Since the total energy $E$ is a constant, *classical mechanics* says that the particle cannot enter the region $x > 0$. The reason is that in that region

$$E = \frac{p^2}{2m} + V(x) < V(x)$$

or

$$\frac{p^2}{2m} < 0$$

Thus the kinetic energy $p^2/2m$ would be negative in the region $x > 0$, which would lead to an imaginary value for the linear momentum $p$ in that region. Neither is allowed, or even makes physical sense, in classical mechanics. According to classical mechanics, the impulsive force will change the momentum of the particle in such a way that it will exactly reverse its motion, traveling off in the direction of decreasing $x$ with momentum in the direction opposite to its initial momentum. The magnitude of the momentum $p$ will be the same before and after the reversal since the total energy $E = p^2/2m$ is constant.

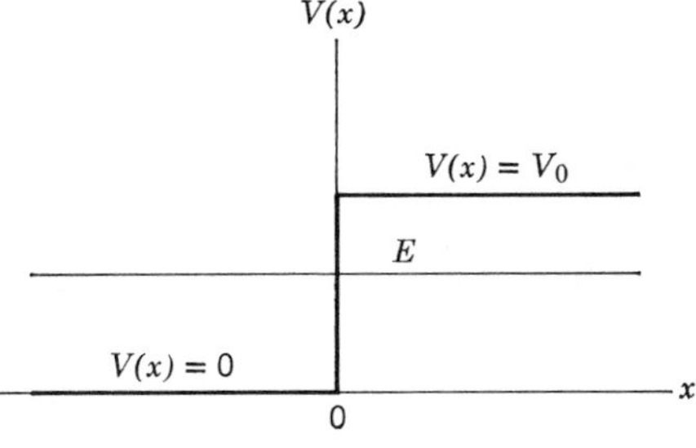


**FIGURE 6-5**
The relation between total and potential energies for a particle incident upon a potential step with total energy less than the height of the step.

To determine the motion of the particle according to *quantum mechanics*, we must find the wave function which is a solution, for the total energy $E < V_0$, to the Schroedinger equation for the step potential of (6-11). Since this potential is independent of time, the actual problem is to solve the time-independent Schroedinger equation. From our qualitative discussion of the previous chapter, we know that an acceptable solution should exist for *any* value of $E \geq 0$, since the potential cannot bind the particle to a limited range of the $x$ axis.

For the step potential, the $x$ axis breaks up into two regions. In the region where $x < 0$ (left of the step), we have $V(x) = 0$, so the eigenfunction that will tell us about the behavior of the particle is a solution to the simple time-independent Schroedinger equation

$$-\frac{\hbar^2}{2m}\frac{d^2\psi(x)}{dx^2} = E\psi(x) \qquad x < 0 \quad (6\text{-}12)$$

In the region where $x > 0$ (right of the step), we have $V(x) = V_0$, and the eigenfunction is a solution to a time-independent Schroedinger equation which is almost as simple

$$-\frac{\hbar^2}{2m}\frac{d^2\psi(x)}{dx^2} + V_0\psi(x) = E\psi(x) \qquad x > 0 \quad (6\text{-}13)$$

The two equations are solved separately. Then an eigenfunction valid for the entire range of $x$ is constructed by joining the two solutions together at $x = 0$ in such a way as to satisfy the requirements, of Section 5-6, that the eigenfunction and its first derivative are everywhere finite, single valued, and continuous.

Consider the differential equation valid for the region in which $V(x) = 0$, (6-12). Since this is precisely the time-independent Schroedinger equation for a free particle, we take for its *general* solution the traveling wave eigenfunction of (6-8). We write that eigenfunction as

$$\psi(x) = Ae^{ik_1x} + Be^{-ik_1x} \qquad \text{where } k_1 = \frac{\sqrt{2mE}}{\hbar} \qquad x < 0 \quad (6\text{-}14)$$

Next consider the differential equation valid for the region in which $V(x) = V_0$, (6-13). From the qualitative considerations of Section 5-7, we do not expect an oscillatory function, such as in (6-14), to be a solution since the total energy $E$ is less than the potential energy $V_0$ in the region of interest. In fact, those considerations tell us that the solution will be a function which "gradually approaches the $x$ axis." The simplest function with this property is the decreasing *real* exponential, which can be written

$$\psi(x) = e^{-k_2x} \qquad x > 0 \quad (6\text{-}15)$$

Let us find out if this is a solution and, if so, also find the required value of $k_2$, by substituting it into (6-13), which it is supposed to satisfy. We first evaluate

$$\frac{d^2\psi(x)}{dx^2} = (-k_2)^2e^{-k_2x} = k_2^2\psi(x)$$

Then the substitution yields

$$-\frac{\hbar^2}{2m}k_2^2\psi(x) + V_0\psi(x) = E\psi(x)$$

This satisfies the equation, and therefore verifies the solution, providing

$$k_2 = \frac{\sqrt{2m(V_0 - E)}}{\hbar} \qquad E < V_0 \quad (6\text{-}16)$$

The solution we have just verified is not a *general* solution to the time-independent Schroedinger equation, (6-13). The reason is that the equation contains a second derivative, so the general solution must contain two arbitrary constants. However, if we can find a solution to the equation for the same value of $E$, which is different in form from the one we have just found, we can make an arbitrary linear combination of these two so-called *particular* solutions. The linear combination will also be a solution and, since it will contain two arbitrary constants, it will be a general solution.

A clue to the form of another particular solution is found by noting that $k_2$ enters as a square in the equation preceding (6-16). Therefore, its sign is immaterial, and the increasing exponential

$$\psi(x) = e^{+k_2 x} \qquad \text{where } k_2 = \frac{\sqrt{2m(V_0 - E)}}{\hbar} \qquad x > 0 \quad (6\text{-}17)$$

should also be a solution to the time independent Schroedinger equation that we are dealing with. It is equally easy to verify this, by substitution into the equation. But let us instead verify that the arbitrary linear combination of the two particular solutions

$$\psi(x) = Ce^{k_2 x} + De^{-k_2 x} \qquad \text{where } k_2 = \frac{\sqrt{2m(V_0 - E)}}{\hbar} \qquad x > 0 \quad (6\text{-}18)$$

and where $C$ and $D$ are arbitrary constants, is a solution to (6-13). We calculate

$$\frac{d^2\psi(x)}{dx^2} = Ck_2^2 e^{k_2 x} + D(-k_2)^2 e^{-k_2 x} = k_2^2 \psi(x) = \frac{2m(V_0 - E)}{\hbar^2}\psi(x)$$

and substitute the result into the equation. We obtain

$$-\frac{\hbar^2}{2m}\frac{2m}{\hbar^2}(V_0 - E)\psi(x) + V_0\psi(x) = E\psi(x)$$

Since this is obviously satisfied, we have verified that (6-18) is a solution. Since it contains two arbitrary constants, it is the *general* solution to the time-independent Schroedinger equation for the region of the step potential where $V(x) = V_0$, with $E < V_0$. Although the increasing exponential part will not actually be used in the present section, it will be used in a subsequent section.

The arbitrary constants $A$, $B$, $C$, and $D$ of (6-14) and (6-18) must be so chosen that the total eigenfunction satisfies the requirements concerning finiteness, single valuedness, and continuity, of $\psi(x)$ and $d\psi(x)/dx$. Consider first the behavior of $\psi(x)$ as $x \rightarrow +\infty$. In this region of the $x$ axis the general form of $\psi(x)$ is given by (6-18). Inspection shows that it will generally increase without limit as $x \rightarrow +\infty$, because of the presence of the first term, $Ce^{k_2 x}$. In order to prevent this, and keep $\psi(x)$ finite, we must set the arbitrary coefficient $C$ of the first term equal to zero. Thus we find

$$C = 0 \qquad (6\text{-}19)$$

Single valuedness is satisfied automatically by these functions. To study their continuity, we consider the point $x = 0$. At this point the two forms of $\psi(x)$, given by (6-14) and (6-18), must join in such a way that $\psi(x)$ and $d\psi(x)/dx$ are continuous. Continuity of $\psi(x)$ is obtained by satisfying the relation

$$D(e^{-k_2 x})_{x=0} = A(e^{ik_1 x})_{x=0} + B(e^{-ik_1 x})_{x=0}$$

which comes from equating the two forms at $x = 0$. This relation yields

$$D = A + B \qquad (6\text{-}20)$$

Continuity of the derivative of the two forms

$$\frac{d\psi(x)}{dx} = -k_2 D e^{-k_2 x} \qquad x > 0$$

and

$$\frac{d\psi(x)}{dx} = ik_1 A e^{ik_1 x} - ik_1 B e^{-ik_1 x} \qquad x < 0$$

is obtained by equating these derivatives at $x = 0$. Thus we set

$$-k_2 D(e^{-k_2 x})_{x=0} = ik_1 A(e^{ik_1 x})_{x=0} - ik_1 B(e^{-ik_1 x})_{x=0}$$

This yields

$$\frac{ik_2}{k_1} D = A - B \tag{6-21}$$

Adding (6-20) and (6-21) gives

$$A = \frac{D}{2}\left(1 + \frac{ik_2}{k_1}\right) \tag{6-22}$$

Subtracting gives

$$B = \frac{D}{2}\left(1 - \frac{ik_2}{k_1}\right) \tag{6-23}$$

We have now determined $A$, $B$, and $C$ in terms of $D$. Thus the eigenfunction for the step potential, and for the energy $E < V_0$, is

$$\psi(x) = \begin{cases} \frac{D}{2}(1 + ik_2/k_1)e^{ik_1 x} + \frac{D}{2}(1 - ik_2/k_1)e^{-ik_1 x} & x \leq 0 \\ De^{-k_2 x} & x \geq 0 \end{cases} \tag{6-24}$$

The one remaining arbitrary constant, $D$, determines the amplitude of the eigenfunction, but it is not involved in any of its more important characteristics. The presence of this constant reflects the fact that the time-independent Schroedinger equation is linear in $\psi(x)$, and so solutions of any amplitude are allowed by the equation. We shall see that useful results can be usually obtained without bothering to carry through the normalization procedure that would specify $D$. The reason is that the measurable quantities that we shall obtain as predictions of the theory contain $D$ in both the numerator and the denominator of a ratio, and so it cancels out.

The wave function corresponding to the eigenfunction is

$$\Psi(x,t) = \begin{cases} Ae^{ik_1 x}e^{-iEt/\hbar} + Be^{-ik_1 x}e^{-iEt/\hbar} = Ae^{i(k_1 x - Et/\hbar)} + Be^{i(-k_1 x - Et/\hbar)} & x \leq 0 \\ De^{-k_2 x}e^{-iEt/\hbar} & x \geq 0 \end{cases} \tag{6-25}$$

Consider the region $x < 0$. The first term in the wave function for this region is a traveling wave propagating in the direction of increasing $x$. This term describes a particle moving in the direction of increasing $x$. The second term in the wave function for $x < 0$ is a traveling wave propagating in the direction of decreasing $x$, and it describes a particle moving in that direction. This information, plus the classical predictions described earlier, suggests that we should associate the first term with the incidence of the particle on the potential step and the second term with the reflection of the particle from the step. Let us use this association to calculate the probability that the incident particle is reflected, which we call the *reflection coefficient R*. Obviously, $R$ depends on the ratio $B/A$, which specifies the amplitude of the reflected part of the wave function relative to the amplitude of the incident part. But in quantum mechanics probabilities depend on intensities, such as $B^*B$ and $A^*A$, not on amplitudes. Thus, we must evaluate $R$ from the formula

$$R = \frac{B^*B}{A^*A} \tag{6-26}$$

That is, the reflection coefficient is equal to the ratio of the intensity of the part of the wave that describes the reflected particle to the intensity of the part that describes the incident particle. We obtain

$$R = \frac{B^*B}{A^*A} = \frac{(1 - ik_2/k_1)^*(1 - ik_2/k_1)}{(1 + ik_2/k_1)^*(1 + ik_2/k_1)}$$

or

$$R = \frac{(1 + ik_2/k_1)(1 - ik_2/k_1)}{(1 - ik_2/k_1)(1 + ik_2/k_1)} = 1 \qquad E < V_0 \tag{6-27}$$

The fact that this ratio equals one means that a particle incident upon the potential step, with total energy less than the height of the step, has probability one of being reflected—it is always reflected. This is in agreement with the predictions of classical mechanics.

Consider now the eigenfunction of (6-24). Using the relation

$$e^{ik_1x} = \cos k_1x + i \sin k_1x \tag{6-28}$$

it is easy to show that the eigenfunction can be expressed as

$$\psi(x) = \begin{cases} D \cos k_1x - D\dfrac{k_2}{k_1} \sin k_1x & x \leq 0 \\ De^{-k_2x} & x \geq 0 \end{cases} \tag{6-29}$$

If we generate the wave function by multiplying $\psi(x)$ by $e^{-iEt/\hbar}$, we see immediately that we actually have a standing wave because the locations of the nodes do not change in time. In this problem the incident and reflected traveling waves for $x < 0$ combine to form a standing wave because they are of equal intensity. Figure 6-6 illustrates this schematically.

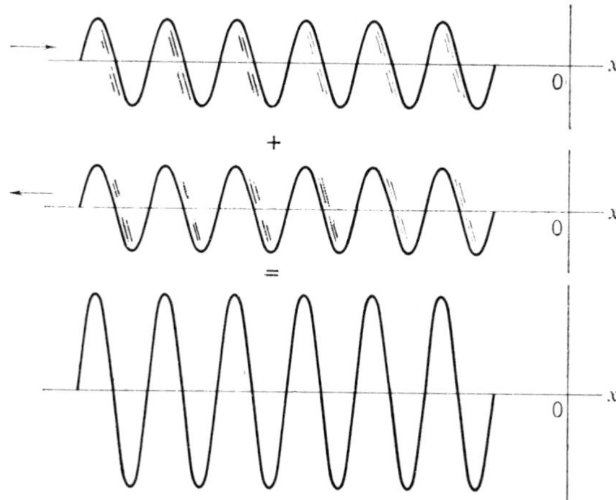


**FIGURE 6-6**

Illustrating schematically the combination of an incident and a reflected wave of equal intensities to form a standing wave. The wave function is reflected from a potential step at $x = 0$. Note that the nodes of the traveling waves move to the right or left, but those of the standing wave are stationary.

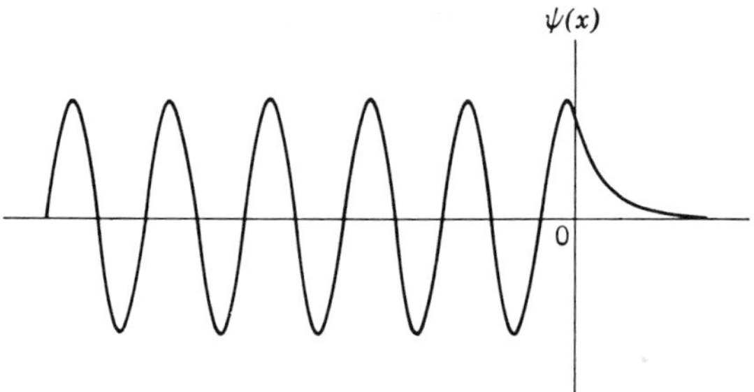


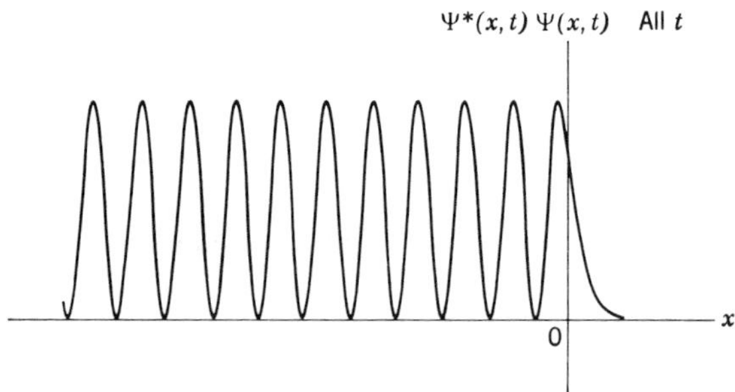


**FIGURE 6-7**

*Top:* The eigenfunction $\psi(x)$ for a particle incident upon a potential step at $x = 0$, with total energy less than the height of the step. Note the penetration of the eigenfunction into the classically excluded region $x > 0$. *Bottom:* The probability density $\Psi^*\Psi = \psi^*\psi = \psi^2$ corresponding to this eigenfunction. The spacing between the peaks of $\psi^2$ is twice as close as the spacing between the peaks of $\psi$.

In the top part of Figure 6-7 we illustrate the wave function by plotting the eigenfunction, (6-29), which is a real function of $x$ if we take $D$ real. The wave function can be thought of as oscillating in time according to $e^{-iEt/\hbar}$, with an amplitude whose space dependence is given by $\psi(x)$. Here we find a feature which is in sharp contrast to the classical predictions. Although in the region $x > 0$ the probability density

$$\Psi^*\Psi = D^*e^{-k_2x}e^{+iEt/\hbar}De^{-k_2x}e^{-iEt/\hbar} = D^*De^{-2k_2x} \tag{6-30}$$

illustrated in the bottom of Figure 6-7, decreases rapidly with increasing $x$, there *is* a finite probability of finding the particle in the region $x > 0$. In classical mechanics it would be absolutely impossible to find the particle in the region $x > 0$ because there the total energy is less than the potential energy, so the kinetic energy $p^2/2m$ is negative and the momentum $p$ is imaginary. This phenomena, called *penetration of the classically excluded region*, is one of the more striking predictions of quantum mechanics.

We shall discuss later certain experiments which confirm this prediction, but here we should like to make several points about it. One is that penetration does *not* mean that the particle is stored in the classically excluded region. Indeed, we have seen that the incident particle is definitely reflected from the step.

Another point is that penetration of the excluded region, which obeys (6-30), is *not* in conflict with the experiments of classical mechanics. It is apparent from the equation that the probability of finding the particle with a coordinate $x > 0$ is only appreciable in a region starting at $x = 0$ and extending in a *penetration distance* $\Delta x$, which equals $1/k_2$. The reason is that $e^{-2k_2x}$ goes very rapidly to zero when $x$ is very much larger than $1/k_2$. Since $k_2 = \sqrt{2m(V_0 - E)}/\hbar$, we have

$$\Delta x = \frac{\hbar}{\sqrt{2m(V_0 - E)}}$$

In the classical limit, the product of $m$ and $(V_0 - E)$ is so large, compared to $\hbar^2$, that $\Delta x$ is immeasurably small.

**Example 6-1.** Estimate the penetration distance $\Delta x$ for a very small dust particle, of radius $r = 10^{-6}$ m and density $\rho = 10^4$ kg/m$^3$, moving at the very low velocity $v = 10^{-2}$ m/sec, if the particle impinges on a potential step of height equal to twice its kinetic energy in the region to the left of the step.

The mass of the particle is

$$m = \frac{4}{3}\pi r^3 \rho \simeq 4 \times 10^{-18}\ \text{m}^3 \times 10^4\ \text{kg/m}^3 = 4 \times 10^{-14}\ \text{kg}$$

Its kinetic energy before hitting the step is

$$\frac{1}{2}mv^2 \simeq \frac{1}{2} \times 4 \times 10^{-14}\ \text{kg} \times 10^{-4}\ \text{m}^2/\text{sec}^2 = 2 \times 10^{-18}\ \text{joule}$$

and this is also the value of $(V_0 - E)$. The penetration distance is

$$\Delta x = \frac{\hbar}{\sqrt{2m(V_0 - E)}} \simeq \frac{10^{-34}\ \text{joule-sec}}{\sqrt{2 \times 4 \times 10^{-14}\ \text{kg} \times 2 \times 10^{-18}\ \text{joule}}}$$
$$\simeq 2 \times 10^{-19}\ \text{m}$$

Of course, this is many orders of magnitude smaller than could be detected in any possible measurement. For the more massive particles and higher energies typically considered in classical mechanics, $\Delta x$ is even smaller. ◀

Furthermore, we should like to point out that the uncertainty principle shows the wavelike properties exhibited by an entity in penetrating the classically excluded region are really *not* in conflict with its particlelike properties. Consider an experiment capable of proving that the particle is located somewhere in the region $x > 0$. Since the probability density for $x > 0$ is appreciable only in a range of length $\Delta x$, the experiment amounts to localizing the particle within that range. In doing this, the experiment necessarily leads to an uncertainty $\Delta p$ in the momentum, which must be at least as large as

$$\Delta p \simeq \frac{\hbar}{\Delta x} \simeq \sqrt{2m(V_0 - E)}$$

Consequently, the energy of the particle is uncertain by an amount

$$\Delta E \simeq \frac{(\Delta p)^2}{2m} \simeq V_0 - E$$

and it is no longer possible to say that the total energy $E$ of the particle is definitely less than the potential energy $V_0$. This removes the conflict alluded to.

Penetration of the classically excluded region *can* lead to measurable consequences. We shall see this later for a potential that steps up to a height $V_0 > E$, but remains up only for a distance not much larger than the penetration distance $\Delta x$, and then steps down. In fact, the phenomenon has significant practical consequences. One example, which we shall refer to soon, is the tunnel diode used in modern electronics.

**Example 6-2.** A conduction electron moves through a block of Cu at total energy $E$ under the influence of a potential which, to a good approximation, has a constant value of zero in the interior of the block and abruptly steps up to the constant value $V_0 > E$ outside the block. The interior value of the potential is essentially constant, at a value that can be taken as zero, since a conduction electron inside the metal feels little net Coulomb force exerted by the approximately uniform charge distributions that surround it. The potential increases very rapidly at the surface of the metal, to its exterior value $V_0$, because there the electron feels a strong force exerted by the nonuniform charge distributions present in that region. This force tends to attract the electron back into the metal and is, of course, what causes the conduction electron to be bound to the metal. Because the electron is bound, $V_0$ must be greater than its total energy $E$. The exterior value of the potential is constant, if the metal has no total charge, since outside the metal the electron would feel no force at all. The mass of the electron is $m = 9 \times 10^{-31}$ kg. Measurements of the energy required to permanently remove it from the block, i.e., measurements of the work function, show that $V_0 - E = 4$ eV. From these data estimate the distance $\Delta x$ that the electron can penetrate into the classically excluded region outside the block.

In the mks system

$$V_0 - E = 4\text{ eV} \times \frac{1.6 \times 10^{-19}\text{ joule}}{1\text{ eV}} \simeq 6 \times 10^{-19}\text{ joule}$$

So

$$\Delta x = \frac{\hbar}{\sqrt{2m(V_0 - E)}}$$

$$\simeq \frac{10^{-34}\text{ joule-sec}}{\sqrt{2 \times 9 \times 10^{-31}\text{ kg} \times 6 \times 10^{-19}\text{ joule}}} \simeq 10^{-10}\text{ m}$$

The penetration distance is of the order of atomic dimensions. Therefore, the effect can be of consequence in atomic systems. We shall find soon that, in certain circumstances, the effect is very important indeed. ◀

Let us finally make the point that penetration of the classically excluded region is nonclassical in the sense that an entity that does it is not behaving like a classical *particle*. But it is behaving like a classical *wave* since, as we shall see later, the phenomenon has been known to occur with light waves since the time of Newton. Penetration of the classically excluded region by material particles is just another manifestation of the wavelike nature of material particles.

Figure 6-8 shows the probability density for a wave function in the form of a *group*, for the problem of a particle incident in the direction of increasing $x$ upon a potential step with an average value of the total energy less than the step height. The wave function can be obtained by summing, over the total energy $E$, a very large number of wave functions of the form we have obtained in (6-25). It can also be obtained by a direct numerical solution of the Schroedinger equation. Either way involves a large amount of work on a high-speed computer, as can be guessed from the complications indicated in the figure. The results of the calculations certainly convey a realistic sense of the particle motion; but note that these results show, again, that the particle associated with the wave function is reflected from the step with probability one, and that there is some penetration of the classically excluded region. The fact that we have been able to learn these basic results from simple calculations, involving only the wave function of (6-25) which contains a single value of $E$, is an example of the fact that it is generally not necessary in quantum mechanics to use wave functions in the form of groups. Of course, we must be willing to learn how to interpret the simple wave functions.

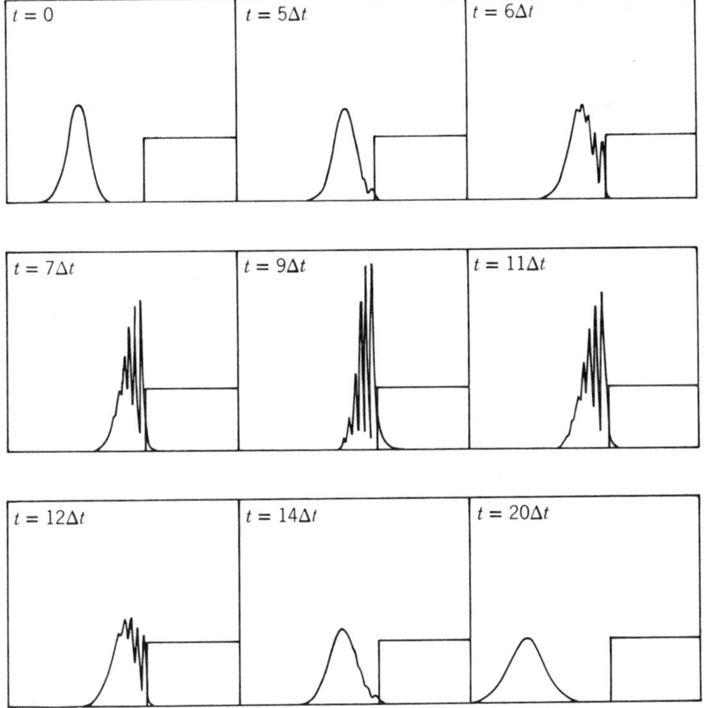


**FIGURE 6-8**

A potential step, and the probability density $\Psi^*\Psi$ for a group wave function describing a particle incident on the step with total energy less than the step height. As time evolves, the group moves up to the step, penetrates slightly into the classically excluded region, and then is completely reflected from the step. The complications of the mathematical treatment using a group are indicated by the complications of its structure during reflection.

## 6-4 The Step Potential (Energy Greater Than Step Height)

In this section we consider the motion of a particle under the influence of a step potential, (6-11), when its total energy $E$ is greater than the height $V_0$ of the step. That is, we take $E > V_0$, as illustrated in Figure 6-9.

In *classical mechanics*, a particle of total energy $E$ traveling in the region $x < 0$, in the direction of increasing $x$, will suffer an impulsive retarding force $F = -dV(x)/dx$ at the point $x = 0$. But the impulse will only slow the particle, and it will enter the region $x > 0$, continuing its motion in the direction of increasing $x$. Its total energy $E$

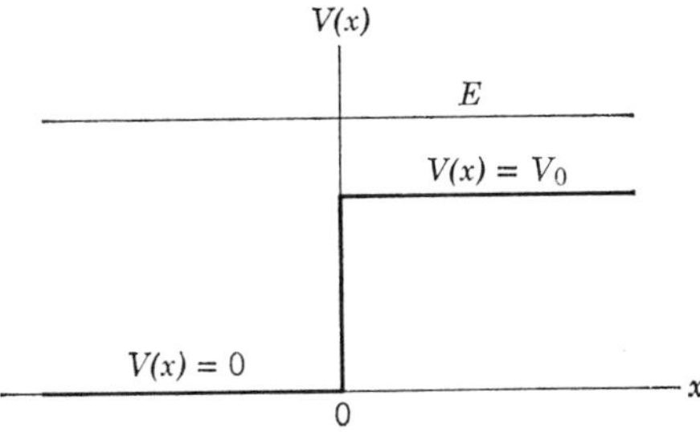


**FIGURE 6-9**

The relation between total and potential energies for a particle incident upon a potential step with total energy greater than the height of the step.

remains constant; its momentum in the region $x < 0$ is $p_1$, where $p_1^2/2m = E$; its momentum in the region $x > 0$ is $p_2$, where $p_2^2/2m = E - V_0$.

We shall see that the predictions of *quantum mechanics* are not so simple. If $E$ is not too much larger than $V_0$, the theory predicts that the particle has an appreciable chance of being reflected at the step back into the region $x < 0$, even though it has enough energy to pass over the step into the region $x > 0$.

One example of this is found in the case of an electron in the cathode of a photoelectric cell, which has received energy from absorbing a photon, and which is trying to escape the surface of the metallic cathode. If its energy is not much higher than the height of the step in the potential that it feels at the surface of the metal, it may be reflected back and not succeed in escaping. This leads to a significant reduction in the efficiency of photocells for light of frequencies not far above the cutoff frequency.

A more important example of reflection occurring when a particle tries to pass over a potential step is found in the motion of a neutron in a nucleus. To a good approximation, the potential acting on the neutron near the nuclear surface is a step potential. The potential rises very rapidly at the nuclear surface because a nucleus tends to bind a neutron. If the neutron has received energy, in one way or another, and is trying to escape the nucleus, it will probably be reflected back into the nucleus at the surface if its energy is only a little greater than the step height. This has the effect of inhibiting the emission of lower energy neutrons from nuclei, and thereby considerably increases the stability of nuclei in low-lying excited states. The effect is a manifestation of the wavelike properties of neutrons that is very significant in the processes taking place in nuclear reactions, as we shall see near the end of this book.

In quantum mechanics, the motion of the particle under the influence of the step potential is described by the wave function

$$\Psi(x,t) = \psi(x)e^{-iEt/\hbar}$$

where the eigenfunction $\psi(x)$ satisfies the time-independent Schroedinger equation for the potential. This equation has different forms in the regions to the left and right of the potential step, namely

$$-\frac{\hbar^2}{2m}\frac{d^2\psi(x)}{dx^2} = E\psi(x) \qquad x < 0 \quad (6\text{-}31)$$

and

$$-\frac{\hbar^2}{2m}\frac{d^2\psi(x)}{dx^2} = (E - V_0)\psi(x) \qquad x > 0 \quad (6\text{-}32)$$

The eigenfunction $\psi(x)$ also satisfies the conditions requiring finiteness, single valuedness, and continuity, for it and its derivative, particularly at the joining point $x = 0$.

Equation (6-31) describes the motion of a free particle of momentum $p_1$. Its general solution is

$$\psi(x) = Ae^{ik_1x} + Be^{-ik_1x} \qquad x < 0 \quad (6\text{-}33)$$

where

$$k_1 = \frac{\sqrt{2mE}}{\hbar} = \frac{p_1}{\hbar}$$

Equation (6-32) describes the motion of a free particle of momentum $p_2$. Its general solution is

$$\psi(x) = Ce^{ik_2x} + De^{-ik_2x} \qquad x > 0 \quad (6\text{-}34)$$

where

$$k_2 = \frac{\sqrt{2m(E - V_0)}}{\hbar} = \frac{p_2}{\hbar} \qquad E > V_0$$

The wave function specified by these two forms consists of traveling waves of de Broglie wavelength $\lambda_1 = h/p_1 = 2\pi/k_1$ in the region $x < 0$, and of longer de Broglie wavelength $\lambda_2 = h/p_2 = 2\pi/k_2$ in the region $x > 0$. Note that the functions we deal with here already satisfy the requirements of finiteness and single valuedness; but we must explicitly consider their continuity, and we shall do so shortly.

A particle initially in the region $x < 0$, and moving towards $x = 0$ would, in classical mechanics, have probability one of passing the point $x = 0$ and entering the region $x > 0$. This is not true in quantum mechanics. Because of the wavelike properties of the particle, there is a certain probability that the particle will be reflected at the point $x = 0$, where there is a discontinuous change in the de Broglie wavelength. Thus we need to take both terms of the general solution of (6-33) to describe the incident and reflected traveling waves in the region $x < 0$. We do not, however, need to take the second term of the general solution of (6-34). This term describes a wave traveling in the direction of decreasing $x$ in the region $x > 0$. Since the particle is incident in the direction of increasing $x$, such a wave could arise only from a reflection at some point with a large positive $x$ coordinate (well beyond the discontinuity at $x = 0$). As there is nothing out there to cause a reflection, we know that there is only a transmitted traveling wave in the region $x > 0$, and so we set the arbitrary constant

$$D = 0 \tag{6-35}$$

The arbitrary constants $A$, $B$, and $C$ must be chosen to make $\psi(x)$ and $d\psi(x)/dx$ continuous at $x = 0$. The first requirement, that the values of $\psi(x)$ expressed by (6-33) and (6-34) be the same at $x = 0$, is satisfied if

$$A(e^{ik_1x})_{x=0} + B(e^{-ik_1x})_{x=0} = C(e^{ik_2x})_{x=0}$$

or

$$A + B = C \tag{6-36}$$

The second requirement, that the values of the derivatives of the two expressions for $\psi(x)$ be the same at $x = 0$, is satisfied if

$$ik_1A(e^{ik_1x})_{x=0} - ik_1B(e^{-ik_1x})_{x=0} = ik_2C(e^{ik_2x})_{x=0}$$

or

$$k_1(A - B) = k_2C \tag{6-37}$$

From the last two numbered equations, we find

$$B = \frac{k_1 - k_2}{k_1 + k_2}A \qquad \text{and} \qquad C = \frac{2k_1}{k_1 + k_2}A \tag{6-38}$$

Thus the eigenfunction is

$$\psi(x) = \begin{cases} Ae^{ik_1x} + A\dfrac{k_1 - k_2}{k_1 + k_2}e^{-ik_1x} & x \leq 0 \\ A\dfrac{2k_1}{k_1 + k_2}e^{ik_2x} & x \geq 0 \end{cases} \tag{6-39}$$

As before, it will not be necessary to evaluate the arbitrary constant $A$ that determines the amplitude of the eigenfunction.

It is clear that an eigenfunction satisfying the two continuity conditions could not have been found if we had initially set the coefficient $B$ of the reflected wave equal to zero. We would then

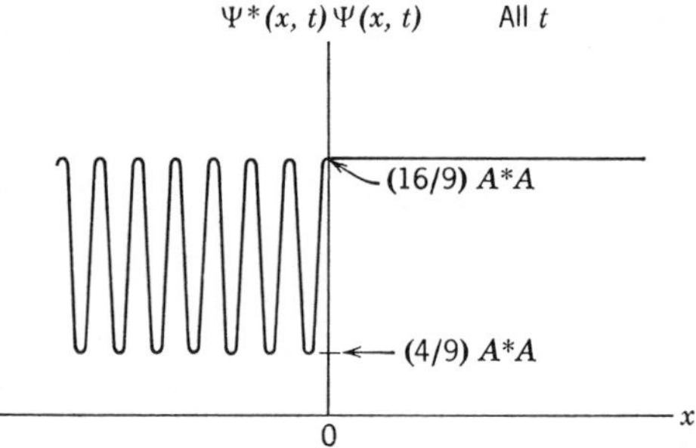


**FIGURE 6-10**
The probability density $\Psi^*\Psi$ for the eigenfunction of (6-39), when $k_1 = 2k_2$.

have had only two arbitrary constants to satisfy the two continuity conditions, and we would not have had one left over to play the role, demanded by the linearity of the time-independent Schroedinger equation, of an arbitrary constant that determines the amplitude of the eigenfunction.

By analogy with our interpretation of the eigenfunction of (6-24), we recognize that the first term in the expression of (6-39) valid for $x < 0$ (left of the discontinuity) represents the incident traveling wave; the second term in the expression valid for $x < 0$ represents the reflected traveling wave; and the expression valid for $x > 0$ (right of the discontinuity) represents the transmitted traveling wave.

Figure 6-10 illustrates the probability density $\Psi^*(x,t)\Psi(x,t) = \psi^*(x)\psi(x)$ for the wave function $\Psi(x,t)$ corresponding to the eigenfunction $\psi(x)$ of (6-39) (in the representative case $k_1 = 2k_2$). We do not plot either the eigenfunction or wave function as both are complex. In the region $x > 0$ the wave function is a pure traveling wave (of amplitude $4A/3$ in this case) traveling to the right, and so the probability density is constant as in the bottom part of Figure 6-1. In the region $x < 0$ the wave function is a combination of the incident traveling wave (of amplitude $A$) moving to the right, and a reflected traveling wave (of amplitude $A/3$) moving to the left. As the amplitude of the reflected wave is necessarily smaller than that of the incident wave, the two cannot combine to yield a pure standing wave. Their sum $\Psi(x,t)$ in that region is, instead, something between a standing wave and a traveling wave. This is seen in the behavior of $\Psi^*(x,t)\Psi(x,t)$ for $x < 0$, which looks like something between the pure standing wave probability density of Figure 6-7 and the pure traveling wave probability density of Figure 6-1 in that it oscillates but has minimum values greater than zero.

The ratio of the intensity of the reflected wave to the intensity of the incident wave gives the probability that the particle will be reflected by the potential step back into the region $x < 0$. This probability is the *reflection coefficient* $R$. That is

$$R = \frac{B^*B}{A^*A} = \left(\frac{k_1 - k_2}{k_1 + k_2}\right)^* \left(\frac{k_1 - k_2}{k_1 + k_2}\right) = \left(\frac{k_1 - k_2}{k_1 + k_2}\right)^2 \qquad E > V_0 \qquad (6\text{-}40)$$

We see from this result that $R < 1$ when $E > V_0$, i.e., when the total energy of the particle is greater than the height of the potential step. This is in contrast to the value $R = 1$ when $E < V_0$, that we obtained from the result of Section 6-3. Of course, the thing that is surprising about the present result is not that $R < 1$, but that $R > 0$. It is surprising because a classical particle would definitely not be reflected if it had enough energy to pass the potential discontinuity. On the other hand, at a corresponding discontinuity a classical wave would be reflected, as we shall discuss shortly.

Also of interest is the *transmission coefficient* $T$, which specifies the probability that the particle will be transmitted past the potential step from the region $x < 0$ into the region $x > 0$. The evaluation of $T$ is slightly more complicated than the evaluation of $R$ because the velocity of the particle is different in the two regions. According to accepted convention, transmission and reflection coefficients are actually defined in terms of the ratios of probability fluxes. A *probability flux* is the probability per second that a particle will be found crossing some reference point traveling in a particular direction. The incident probability flux is the probability per second of finding a particle crossing a point at $x < 0$ in the direction of increasing $x$; the reflected probability flux is the probability per second of finding a particle crossing a point at $x < 0$ in the direction of decreasing $x$; and the transmitted probability flux is the probability per second of finding a particle crossing a point at $x > 0$ in the direction of increasing $x$. Since the probability per second that a particle will cross a given point is proportional to the distance it travels per second, the probability flux is proportional not only to the intensity of the appropriate wave but also to the appropriate velocity of the particle. Thus, according to the strict definition, the reflection coefficient $R$ is

$$R = \frac{v_1 B^* B}{v_1 A^* A} = \frac{B^* B}{A^* A} \tag{6-41}$$

where $v_1$ is the velocity of the particle in the region $x < 0$. Since the velocities cancel, what remains is identical to the formula we have used previously for $R$. For $T$, the velocities do *not* cancel, and we have

$$T = \frac{v_2 C^* C}{v_1 A^* A} = \frac{v_2}{v_1}\left(\frac{2k_1}{k_1 + k_2}\right)^2$$

where $v_2$ is the velocity of the particle in the region $x > 0$. Now

$$v_1 = \frac{p_1}{m} = \frac{\hbar k_1}{m} \qquad \text{and} \qquad v_2 = \frac{p_2}{m} = \frac{\hbar k_2}{m}$$

So the above expression gives

$$T = \frac{k_2}{k_1}\frac{(2k_1)^2}{(k_1 + k_2)^2} = \frac{4k_1 k_2}{(k_1 + k_2)^2} \qquad E > V_0 \tag{6-42}$$

It is easy to show by evaluating $R$ and $T$ from (6-40) and (6-42) that

$$R + T = 1 \tag{6-43}$$

This useful relation is the motivation for defining the reflection and transmission coefficients in terms of probability fluxes.

The probability flux incident upon the potential step is split into a transmitted flux and a reflected flux. But (6-43) says their sum equals the incident flux; i.e., the probability that the particle is either transmitted *or* reflected is one. The particle does not vanish at the step; nor does the particle itself split at the step. In any particular trial the particle will go one way or the other. For a large number of trials, the average probability of going in the direction of decreasing $x$ is measured by $R$, and the average probability of going in the direction of increasing $x$ is measured by $T$.

Note that $R$ and $T$ are both unchanged in value if $k_1$ and $k_2$ are exchanged in (6-40) and (6-42). A moment's consideration should convince the student that this means the same values of $R$ and $T$ would be obtained if the particle were incident upon the potential step in the direction of decreasing $x$ from the region $x > 0$. The wave function describing the motion of the particle, and consequently the probability flux, is partially reflected simply because there is a discontinuous change in $V(x)$, and not because $V(x)$ becomes larger in the direction of the incidence of the particle. The behavior of $R$ and $T$ when $k_1$ and $k_2$ are exchanged involves a characteristic property

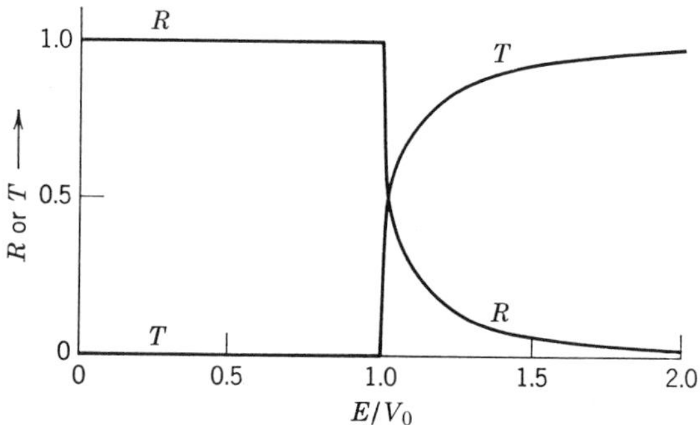


**FIGURE 6-11**

The reflection and transmission coefficients $R$ and $T$ for a particle incident upon a potential step. The abscissa $E/V_0$ is the ratio of the total energy of the particle to the increase in its potential energy at the step. The case $k_1 = 2k_2$, illustrated in Figure 6-10, corresponds to $E/V_0 = 1.33$.

of all waves that, in optics, is sometimes called the *reciprocity* property. When light passes perpendicularly through a sharp interface between media with different indices of refraction, a fraction of the light is reflected because of the abrupt change in its wavelength, and the same fraction is reflected independent of whether it is incident from one side of the interface or from the other. Exactly the same thing happens when a microscopic particle experiences an abrupt change in its de Broglie wavelength. In fact, the equations governing the two phenomena are identical in form. We see, once again, that a microscopic particle moves in a wavelike manner.

In Figure 6-11 the reflection and transmission coefficients are plotted as a function of the convenient ratio $E/V_0$. By evaluating $k_1$ and $k_2$ in (6-40) and (6-42), we find that these expressions for the reflection and transmission coefficients can be written in terms of the ratio as

$$R = 1 - T = \left(\frac{1 - \sqrt{1 - V_0/E}}{1 + \sqrt{1 - V_0/E}}\right)^2 \qquad \frac{E}{V_0} > 1 \qquad (6\text{-}44)$$

The figure also plots the results

$$R = 1 - T = 1 \qquad \frac{E}{V_0} < 1$$

obtained in (6-27) of the preceding section for a step potential when $E/V_0 < 1$.

As an example, for $E/V_0 = 1.33$ the transmission coefficient has the value $T = 0.88$. This $E/V_0$ ratio corresponds to the case $k_2 = k_1/2$ whose probability density pattern is illustrated in Figure 6-10. Note from that figure that the probability of finding the particle in a given length of the $x$ axis, which is long enough to average over the quantum mechanical fluctuations in the probability density, is nearly twice as large to the right of the potential step as it is to the left of the step. From a classical point of view, which is appropriate to discussing an average over quantum mechanical fluctuations, it can be said that the reasons for this are: (a) the probability that the particle will pass the step and proceed into the region to its right is almost equal to one, and (b) the particle's velocity is halved when it enters the region to the right of the step since $k = p/\hbar = mv/\hbar$ and $k_2 = k_1/2$, so it spends twice as much time in any given length of the axis in that region.

From Figure 6-11 we see that the energy of the particle must be appreciably higher than the height of the potential step before the probability of reflection becomes negligible. However, the case in which $E$ becomes very large is not necessarily the case of the classical limit for which we know there will be no reflection at all. The point is that (6-44) says $R$ depends only on the ratio $E/V_0$, so that it will keep the same value if $V_0$ increases as rapidly as $E$. This seems paradoxical until we realize that, in the limit of large energies, our basic assumption that the change in the value of the step potential $V(x)$ is perfectly sharp can no longer be even an approximation to a real physical situation. If the potential function changes only very gradually with $x$, then the de Broglie wavelength will change only very gradually. In this case the reflection will be negligible because the change in wavelength is gradual, and reflection arises from an abrupt change in the wavelength. Specifically, if the fractional change in $V(x)$ is very small when $x$ changes by one de Broglie wavelength, then the reflection coefficient will be very small. This gives rise to the classical limit since in that limit the de Broglie wavelength is so short that any physically realistic potential $V(x)$ changes only by a negligible fraction in one wavelength.

For particles in atomic or nuclear systems, the de Broglie wavelength can be long relative to the distance in which the potential experienced by the particle changes value significantly. Then the step potential is a very good approximation. For these microscopic particles, the probability of reflection can be large.

**Example 6-3.** When a neutron enters a nucleus, it experiences a potential energy which drops at the nuclear surface very rapidly from a constant external value $V = 0$ to a constant internal value of about $V = -50$ MeV. The decrease in the potential is what makes it possible for a neutron to be bound in a nucleus. Consider a neutron incident upon a nucleus with an external kinetic energy $K = 5$ MeV, which is typical for a neutron that has just been emitted from a nuclear fission. Estimate the probability that the neutron will be reflected at the nuclear surface, thereby failing to enter and have its chance at inducing another nuclear fission.

For an estimate, we may take the neutron-nucleus potential to be a one-dimensional step potential, as illustrated in Figure 6-12. Because of the reciprocity property of the reflection coefficient, we may evaluate it from (6-44), using $V_0 = 50$ MeV and $E = 55$ MeV for reasons that can be seen by inspection of the figure. We have

$$R = \left(\frac{1 - \sqrt{1 - 50/55}}{1 + \sqrt{1 - 50/55}}\right)^2 \simeq 0.29$$

This estimate gives a correct impression of the great importance of the reflection phenomenon when low energy neutrons collide with nuclei. But the numerical value we have obtained for the reflection coefficient is not very accurate since the actual neutron-nucleus potential does not drop quite as rapidly at the nuclear surface, in comparison to the de Broglie wavelength, as a step potential. ◀

## 6-5 The Barrier Potential

In this section we consider a *barrier potential*, illustrated in Figure 6-13. The potential can be written as follows

$$V(x) = \begin{cases} V_0 & 0 < x < a \\ 0 & x < 0 \text{ or } x > a \end{cases} \tag{6-45}$$

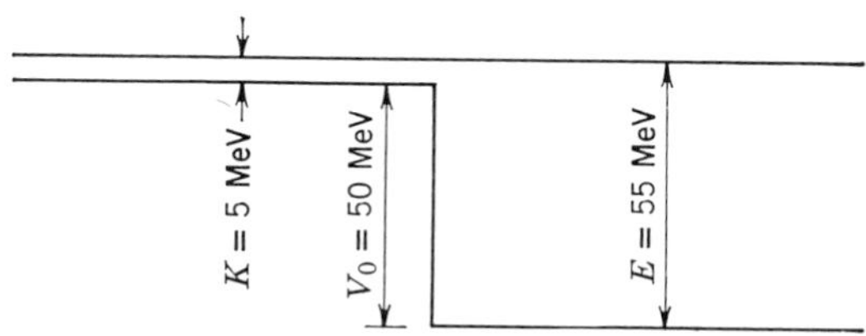


**FIGURE 6-12**

A neutron of external kinetic energy $K$ incident upon a decreasing potential step of depth $V_0$, which approximates the potential it feels upon entering a nucleus. Its total energy, measured from the bottom of the step potential, is $E$.

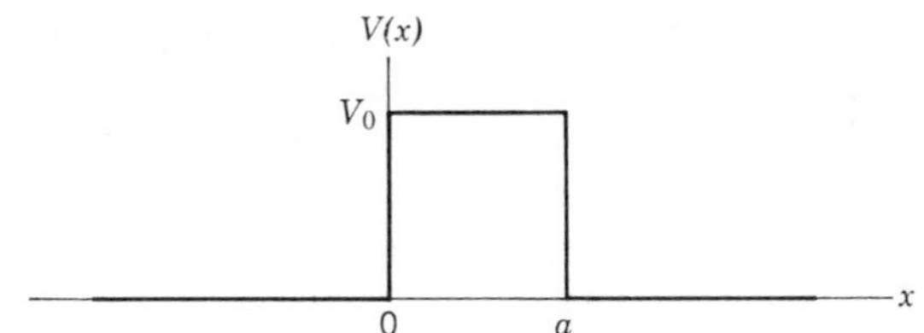


**FIGURE 6-13**
A barrier potential.

According to *classical mechanics*, a particle of total energy $E$ in the region $x < 0$, which is incident upon the barrier in the direction of increasing $x$, will have probability one of being reflected if $E < V_0$, and probability one of being transmitted into the region $x > a$ if $E > V_0$.

Neither of these statements describes accurately the *quantum mechanical* results. If $E$ is not much larger than $V_0$, the theory predicts that there will be some reflection, except for certain values of $E$. If $E$ is not much smaller than $V_0$, quantum mechanics predicts that there is a certain probability that the particle will be transmitted through the barrier into the region $x > a$.

In "tunneling" through a barrier whose height exceeds its total energy, a material particle is behaving purely like a wave. But in the region beyond the barrier it can be detected as a localized particle, without introducing a significant uncertainty in the knowledge of its energy. Thus penetration of a classically excluded region of limited width by a particle *can* be observed, in the sense that the particle can be observed to be a particle, of total energy less than the potential energy in the excluded region, both before and after it penetrates the region. We shall discuss some consequences of this fascinating effect in the present section, as well as some consequences of the reflection of particles attempting to pass over a barrier. The following section is devoted completely to examples of tunneling through barriers, and considers three of particular importance: (1) the emission of $\alpha$ particles from radioactive nuclei through the potential barrier they experience in the vicinity of the nuclei, (2) the inversion of the ammonia molecule which provides a frequency standard for atomic clocks, and (3) the tunnel diode used as a switching unit in fast electronic circuits.

For the barrier potential of (6-45), we know from the qualitative arguments of the last chapter that acceptable solutions to the time-independent Schroedinger equation should exist for *all* values of the total energy $E \geq 0$. We also know that the equation breaks up into three separate equations for the three regions: $x < 0$ (left of the barrier), $0 < x < a$ (within the barrier), and $x > a$ (right of the barrier). In the regions to the left and to the right of the barrier the equations are those for a free particle of total energy $E$. Their general solutions are

$$\begin{aligned} \psi(x) &= Ae^{ik_{\mathrm{I}}x} + Be^{-ik_{\mathrm{I}}x} && x < 0 \\ \psi(x) &= Ce^{ik_{\mathrm{I}}x} + De^{-ik_{\mathrm{I}}x} && x > a \end{aligned} \tag{6-46}$$

where

$$k_{\mathrm{I}} = \frac{\sqrt{2mE}}{\hbar}$$

In the region within the barrier, the form of the equation, and of its general solution, depends on whether $E < V_0$ or $E > V_0$. Both of these cases have been treated in the previous sections. In the first case, $E < V_0$, the general solution is

$$\psi(x) = Fe^{-k_{\mathrm{II}}x} + Ge^{k_{\mathrm{II}}x} \qquad 0 < x < a \tag{6-47}$$

where

$$k_{\mathrm{II}} = \frac{\sqrt{2m(V_0 - E)}}{\hbar} \qquad E < V_0$$

In the second case, $E > V_0$, it is

$$\psi(x) = Fe^{ik_{\text{III}}x} + Ge^{-ik_{\text{III}}x} \qquad 0 < x < a \quad \text{(6-48)}$$

where

$$k_{\text{III}} = \frac{\sqrt{2m(E - V_0)}}{\hbar} \qquad E > V$$

Note that (6-47) involves *real* exponentials, whereas (6-46) and (6-48) involve *complex* exponentials.

Since we are considering the case of a particle incident on the barrier from the left, in the region to the right of the barrier there can be only a transmitted wave as there is nothing in that region to produce a reflection. Thus we can set

$$D = 0$$

In the present situation, however, we cannot set $G = 0$ in (6-47) since the value of $x$ is limited in the barrier region, $0 < x < a$, so $\psi(x)$ for $E < V_0$ cannot become infinitely large even if the increasing exponential is present. Nor can we set $G = 0$ in (6-48) since $\psi(x)$ for $E > V_0$ will have a reflected component in the barrier region that arises from the potential discontinuity at $x = a$.

We consider first the case in which the energy of the particle is less than the height of the barrier, i.e., the case:

$E < V_0$

In matching $\psi(x)$ and $d\psi(x)/dx$ at the points $x = 0$ and $x = a$, four equations in the arbitrary constants $A$, $B$, $C$, $F$, and $G$ will be obtained. These equations can be used to evaluate $B$, $C$, $F$, and $G$ in terms of $A$. The value of $A$ determines the amplitude of the eigenfunction, and it can be left arbitrary. The form of the probability density corresponding to the eigenfunction obtained is indicated in Figure 6-14 for a typical situation. In the region $x > a$ the wave function is a pure traveling wave and so the probability density is constant, as for $x > 0$ in Figure 6-10. In the region $x < 0$ the wave function is principally a standing wave but has a small traveling wave component because the reflected traveling wave has an amplitude less than that of the incident wave. So the probability density in that region oscillates but has minimum values somewhat greater than zero, as for $x < 0$ in Figure 6-10. In the region $0 < x < a$ the wave function has components of both types, but it is principally a standing wave of exponentially decreasing amplitude, and this behavior can be seen in the behavior of the probability density in the region.

The most interesting result of the calculation is the ratio $T$, of the probability flux transmitted through the barrier into the region $x > a$, to the probability flux incident

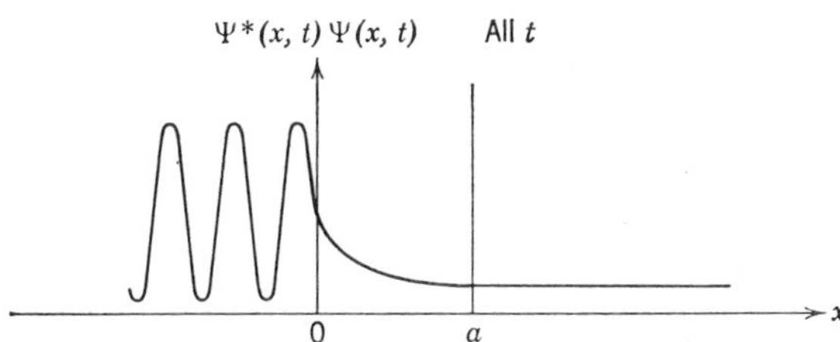


**FIGURE 6-14**

The probability density function $\Psi^*\Psi$ for a typical barrier penetration situation.

upon the barrier. This transmission coefficient is found to be

$$T = \frac{v_1 C^* C}{v_1 A^* A} = \left[1 + \frac{(e^{k_{II}a} - e^{-k_{II}a})^2}{16\frac{E}{V_0}\left(1 - \frac{E}{V_0}\right)}\right]^{-1} = \left[1 + \frac{\sinh^2 k_{II}a}{4\frac{E}{V_0}\left(1 - \frac{E}{V_0}\right)}\right]^{-1} \tag{6-49}$$

where

$$k_{II}a = \sqrt{\frac{2mV_0a^2}{\hbar^2}\left(1 - \frac{E}{V_0}\right)} \qquad E < V_0$$

If the exponents are very large, this formula reduces to

$$T \simeq 16\frac{E}{V_0}\left(1 - \frac{E}{V_0}\right)e^{-2k_{II}a} \qquad k_{II}a \gg 1 \tag{6-50}$$

as can be verified with ease. When (6-50) is a good approximation, $T$ is extremely small.

These equations make a prediction which is, from the point of view of classical mechanics, very remarkable. They say that a particle of mass $m$ and total energy $E$, incident on a potential barrier of height $V_0 > E$ and finite thickness $a$, actually has a certain probability $T$ of penetrating the barrier and appearing on the other side. This phenomenon is called *barrier penetration*, and the particle is said to *tunnel* through the barrier. Of course, $T$ is vanishingly small in the classical limit because in that limit the quantity $2mV_0a^2/\hbar^2$, which is a measure of the opacity of the barrier, is extremely large.

We shall discuss barrier penetration in detail shortly, but let us first finish describing the calculations by considering the case in which the energy of the particle is greater than the height of the barrier, i.e., the case:

$E > V_0$

In this case the eigenfunction is oscillatory in all three regions, but of longer wavelength in the barrier region, $0 < x < a$. Evaluation of the constants $B$, $C$, $F$, and $G$ by application of the continuity conditions at $x = 0$ and $x = a$, leads to the following formula for the transmission coefficient

$$T = \frac{v_1 C^* C}{v_1 A^* A} = \left[1 - \frac{(e^{ik_{III}a} - e^{-ik_{III}a})^2}{16\frac{E}{V_0}\left(\frac{E}{V_0} - 1\right)}\right]^{-1} = \left[1 + \frac{\sin^2 k_{III}a}{4\frac{E}{V}\left(\frac{E}{V_0} - 1\right)}\right]^{-1} \tag{6-51}$$

where

$$k_{III}a = \sqrt{\frac{2mV_0a^2}{\hbar^2}\left(\frac{E}{V_0} - 1\right)} \qquad E > V_0$$

**Example 6-4.** An electron is incident upon a rectangular barrier of height $V_0 = 10$ eV and thickness $a = 1.8 \times 10^{-10}$ m. This rectangular barrier is an idealization of the barrier encountered by an electron that is scattering from a negatively ionized gas atom in the "plasma" of a gas discharge tube. The actual barrier is not rectangular, of course, but it is about the height and thickness quoted. Evaluate the transmission coefficient $T$ and the reflection coefficient $R$, as a function of the total energy $E$ of the electron.

From Example 6-2 we can see that if $E$ is a reasonable fraction of $V_0$ the penetration length $\Delta x$ will be comparable to the barrier thickness $a$. Thus we can expect appreciable transmission through the barrier. To determine exactly how much, we use the numbers given to evaluate

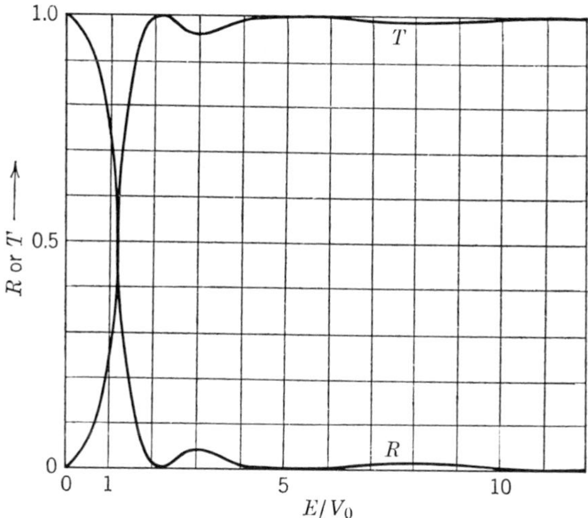


**FIGURE 6-15**

The reflection and transmission coefficients $R$ and $T$ for a particle incident upon a potential barrier of height $V_0$ and thickness $a$, such that $2mV_0a^2/\hbar^2 = 9$. The abscissa $E/V_0$ is the ratio of the total energy of the particle to the height of the potential barrier.

the combination of parameters

$$\frac{2mV_0a^2}{\hbar^2} \simeq \frac{2 \times 9 \times 10^{-31}\,\text{kg} \times 10\,\text{eV} \times 1.6 \times 10^{-19}\,\text{joule/eV} \times (1.8)^2 \times 10^{-20}\,\text{m}^2}{10^{-68}\,\text{joule}^2\text{-sec}^2} \simeq 9$$

which enters (6-49). From this we can plot $T$, and also $R = 1 - T$, versus $E/V_0$, in the range $0 \leq E/V_0 \leq 1$. The plot is shown in Figure 6-15. We see that $T$ is very small when $E/V_0 \ll 1$. But, when $E/V_0$ is only somewhat smaller than one, so that $E$ is nearly as large as $V_0$, $T$ is not at all negligible. For instance, when $E$ is half as large as $V_0$ so that $E/V_0 = 0.5$, the transmission coefficient has the appreciable value $T \simeq 0.05$. It is apparent that electrons can penetrate this barrier with relative ease.

For $E/V_0 > 1$, we evaluate $T$, and $R = 1 - T$, from (6-51), using the same combination of parameters as before. The results are also shown in Figure 6-15. For $E/V_0 > 1$, the transmission coefficient $T$ is in general somewhat less than one, owing to reflection at the discontinuities in the potential. However, from (6-51) it can be seen that $T = 1$ whenever $k_{III}a = \pi, 2\pi, 3\pi, \ldots$. This is simply the condition that the length of the barrier region, $a$, is equal to an integral or half-integral number of de Broglie wavelengths $\lambda_{III} = 2\pi/k_{III}$ in that region. For this particular barrier, electrons of energy $E \simeq 21$ eV, 53 eV, etc., satisfy the condition $k_{III}a = \pi, 2\pi$, etc., and so pass into the region $x > a$ without any reflection. The effect is a result of constructive interference between reflections at $x = 0$ and $x = a$. It is closely related to the *Ramsauer effect* observed in the scattering of low-energy electrons by noble gas atoms, in which electrons of certain energies in the range of a few electron volts pass through these atoms as if they were not there, and so have transmission coefficients equal to one. Essentially the same effect is seen in scattering of neutrons, with energies of a few MeV, from all nuclei. The nuclear effect, called *size resonance*, will be discussed later in the book. ◀

We can bring together the results of the last three sections by comparing the plot of the energy dependence of the reflection coefficient $R$ for a barrier potential, in Figure 6-15, with the plot of the same thing for a step potential, in Figure 6-11. The comparison shows that for both potentials $R \to 1$ as $E/V_0 \to 0$, and $R \to 0$ as $E/V_0 \to \infty$, with the decrease in $R$ occurring around $E/V_0 = 1$. But for the barrier potential the reflection coefficient approaches one gradually, at small energies, since the finite thickness of the classically excluded region allows some transmission. Also, the barrier potential reflection coefficient oscillates, at large energies, because of interferences in the reflections from its two discontinuities. As the step potential can be considered to be a limiting case of a barrier of very great width, we can see from our comparison the behavior of the barrier potential reflection coefficient in this limit.

Now we shall discuss in some detail the origins of these results. They all involve phenomena which arise from the wavelike behavior of the motion of microscopic particles, and each phenomenon is also observed in other types of wave motion. As we remarked in Chapter 5, the time-independent differential equation governing classical wave motion is of the same form as the time-independent Schroedinger equation. For instance, electromagnetic radiation of frequency $\nu$ propagating through a medium with index of refraction $\mu$ obeys the equation

$$\frac{d^2\psi(x)}{dx^2} + \left(\frac{2\pi\nu}{c}\mu\right)^2 \psi(x) = 0 \tag{6-52}$$

where the function $\psi(x)$ specifies the magnitude of the electric or magnetic field. When we compare this with the time-independent Schroedinger equation, written in the form

$$\frac{d^2\psi(x)}{dx^2} + \frac{2m}{\hbar^2}[E - V(x)]\psi(x) = 0$$

we see that they are identical if the index of refraction in the former is connected with the potential energy function in the latter by the relation

$$\mu(x) = \frac{c}{2\pi\nu}\sqrt{\frac{2m}{\hbar^2}[E - V(x)]} \tag{6-53}$$

Thus the behavior of an optical system with index of refraction $\mu(x)$ should be identical to the behavior of a mechanical system with potential energy $V(x)$, providing the two functions are related as in (6-53). Indeed, there are optical phenomena which are exactly analogous to each of the quantum mechanical phenomena that arise in considering the motion of an unbound particle. An optical phenomenon, completely analogous to the total transmission of particles over barriers of length equal to an integral or half-integral number of wavelengths, is used in the coating of lenses to obtain very high light transmissions and in thin film optical filters.

An optical analogue to the penetration of barriers by particles is found in the imaginary indices of refraction that arise in total internal reflection. Consider a ray of light incident upon a glass-to-air interface at an angle greater than the critical angle $\theta_c$. The resulting behavior of the light ray is called *total internal reflection*, and it is illustrated in the top of Figure 6-16. A detailed treatment of the process in terms of electromagnetic theory shows that the index of refraction, measured along the line $ABC$, is real in the region $AB$ but imaginary in the region $BC$. Note that an imaginary $\mu(x)$ is suggested by (6-53) for a region analogous to one in which $E < V(x)$. Furthermore, electromagnetic theory shows that there are electromagnetic vibrations in the region $BC$ of exactly the same form as the decreasing exponential standing wave of (6-29) for the region where $E < V(x)$. The flux of energy (the Poynting vector) is zero in this electromagnetic standing wave, just as the flux of probability is zero in the quantum mechanical standing wave, so the light ray is totally reflected. However, if a

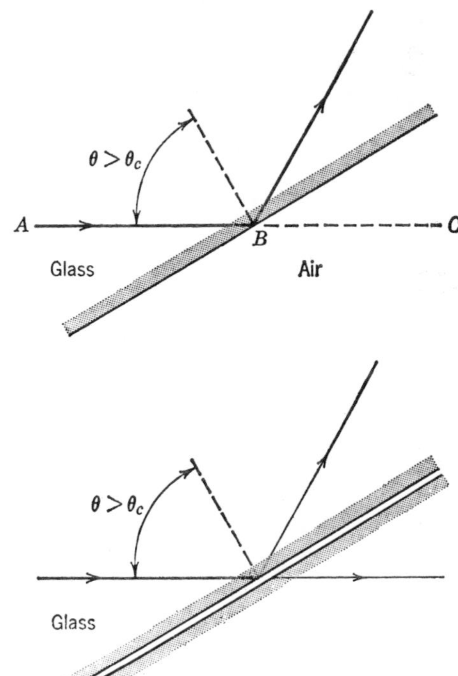


**FIGURE 6-16**

*Top:* Illustrating total internal reflection of a light ray. The angle of incidence is greater than the critical angle. *Bottom:* Illustrating frustrated total internal reflection. Some of the light ray is transmitted through the air gap if the gap is sufficiently narrow.

second block of glass is placed near enough to the first block to be in the region in which the electromagnetic vibrations are still appreciable, these vibrations are picked up and propagate through the second block. Furthermore, the electromagnetic vibrations in the air gap now carry a flux of energy through to the second block. This phenomenon, called *frustrated total internal reflection*, is illustrated in the bottom of Figure 6-16. Essentially the same thing happens in the quantum mechanical case when the region in which $E < V(x)$ is reduced from infinite thickness (step potential) to finite thickness (barrier potential). The transmission of light through an air gap, at an angle of incidence greater than the critical angle, was first observed by Newton around 1700. The equation relating the intensity of the transmitted beam to the thickness of the air gap, and other parameters, is identical in form to (6-49), and it has been verified experimentally.

It is particularly easy to observe frustrated total internal reflection of electromagnetic waves, using the microwave region of the spectrum and two blocks of paraffin separated by an air gap. Furthermore, careful inspection of the "ripple tank" photographs in Figures 6-17 and 6-18 will show that the phenomenon can even be observed with water waves. Frustrated total internal reflection, or its quantum mechanical equivalent barrier penetration, arises from properties common to all forms of classical or quantum mechanical wave motion.

## 6-6 Examples of Barrier Penetration by Particles

**There are a number of interesting, and important, examples of barrier penetration by microscopic particles. A widespread, but not widely recognized, example occurs in aluminum household wiring. The usual way for an electrician to join two wires is to twist them together. Often there is a layer of aluminum oxide between the two wires, and this material is quite an effective insulator. Fortunately, the layer is extremely thin so the electrons flowing through the wire are able to tunnel through the layer by barrier penetration.**

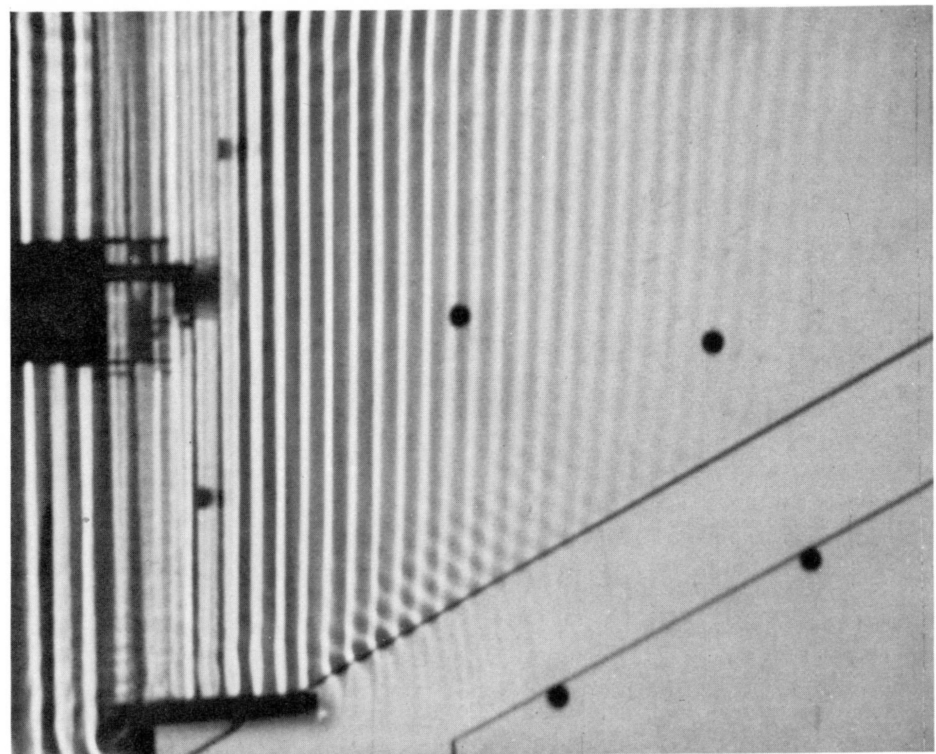

**FIGURE 6-17**

The total internal reflection of water waves. A long vibrating plunger on the left produces a set of waves in a region of shallow water, the waves being illuminated so as to make their crests easily visible. The waves are totally internally reflected at the diagonal boundary of a region where the layer of water abruptly becomes deeper, this reflection occurring because the velocity of water waves depends on the depth of the water. Note that the intensity of the waves decreases rapidly when they try to penetrate into the region of deeper water, but there is some penetration of that region. (Courtesy Film Studio, Education Development Center)

Historically, the first application of the quantum mechanical theory of barrier penetration by particles was to explain a long standing paradox concerning the emission of $\alpha$ particles in the decay of radioactive nuclei. As a typical example, consider the $U^{238}$ nucleus. The potential energy $V(r)$ of an $\alpha$ particle at a distance $r$ from the center of the nucleus had been investigated around 1910 by Rutherford, and others, who performed scattering experiments. Using as a probe the 8.8 MeV $\alpha$ particles emitted from the radioactive nuclei of $Po^{212}$, it was observed that their probability of scattering at various angles from $U^{238}$ nuclei agreed with the predictions of Rutherford's scattering formula (see Chapter 4). The student will recall that that formula was based on the assumption that the interaction between the $\alpha$ particle and the nucleus strictly followed the Coulomb law repulsion that would be expected to operate between the two positively charged spherical objects. Thus Rutherford was able to conclude that, for the $U^{238}$ nucleus, the potential function $V(r)$ felt by a neighboring $\alpha$ particle followed Coulomb's law, $V(r) = 2Ze^2/4\pi\epsilon_0 r$, where $2e$ is the $\alpha$-particle charge and $Ze$ is the nuclear charge—at least for distances greater than $r'' = 3 \times 10^{-14}$ m where $V(r'') = 8.8$ MeV, the probe $\alpha$-particle energy. It was also known by scattering $\alpha$ particles from nuclei of light atoms that $V(r)$ eventually departs from a $1/r$ law when $r < r'$, the nuclear radius, although the exact value of $r'$ was not known for the nuclei of heavy atoms at that time. Furthermore, since $\alpha$

**FIGURE 6-18**

Frustrated total internal reflection of water waves. When the region of deeper water becomes a sufficiently narrow gap, the waves that have penetrated into the deeper water are picked up and transmitted into a second region of shallow water. (Courtesy Film Studio, Education Development Center)

particles are occasionally emitted by $U^{238}$ nuclei, it was assumed that they exist inside such nuclei, to which they are normally bound by the potential $V(r)$. From these arguments it was concluded that the form of $V(r)$ in the region $r < r''$ must be qualitatively as depicted in Figure 6-19. This conclusion has been verified by modern experiments involving the scattering of $\alpha$ particles produced by cyclotrons at energies high enough to allow the investigation of the potential over the entire range of $r$.

The paradox was connected with the fact that it was also known that the kinetic energy of $\alpha$ particles emitted in radioactive decay by $U^{238}$ was 4.2 MeV. The kinetic energy was, of course, measured at a very large distance from the nucleus where $V(r) = 0$ and the kinetic energy equals the total energy $E$. This value of the constant total energy of the decay $\alpha$ particles emitted by $U^{238}$ is also shown in Figure 6-19. From the point of view of classical mechanics, the situation was certainly paradoxical. An $\alpha$ particle of total energy $E$ is initially in the region $r < r'$. This region is separated from the rest of space by a potential barrier of height which was known to be at least twice $E$. Yet it was observed that on occasion the $\alpha$ particle penetrates the barrier and moves off to large values of $r$.

To put it another way, according to classical mechanics an $\alpha$ particle emitted from a region where the potential energy function has the form shown in Figure 6-19 must, necessarily, have a much higher kinetic energy than was actually observed when it is far from the region. The reason is simply that in classical mechanics the total energy must be greater than the maximum value of the potential energy, if the particle is to escape the barrier. Consider the following analogy. You are walking beneath the span of a tall bridge, not looking up. Suddenly a brick

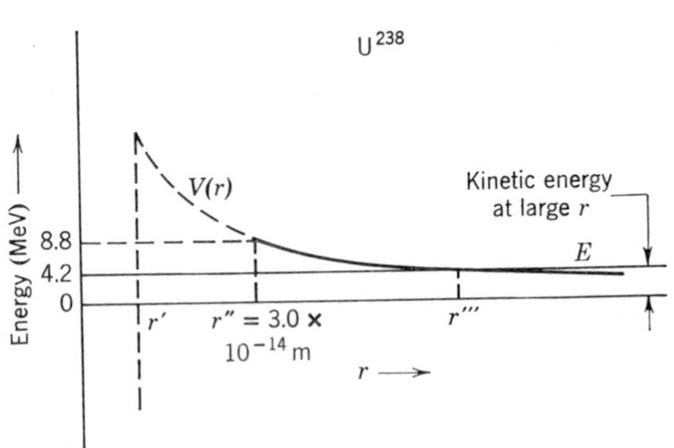


**FIGURE 6-19**

The potential energy $V$ acting on an $\alpha$ particle at a distance $r$ from the center of a $U^{238}$ nucleus, and the total energy $E$ of an $\alpha$ particle emitted from that radioactive nucleus. The solid part of the potential curve was known from scattering measurements to follow Coulomb's law into the distance of closest approach $r''$ of an 8.8 MeV $\alpha$ particle. The dashed part of the curve shows that the potential was assumed to continue to follow Coulomb's law into the nuclear radius $r'$, where it must drop very rapidly to form a binding region. A 4.2 MeV $\alpha$ particle emitted from the radioactive nucleus must penetrate the potential barrier from the nuclear radius $r'$ to the point at distance $r'''$ from the center where its potential energy $V$ becomes less than its total energy $E$.

hits you on the head, but gently, with a light tap. There is no place for the brick to come from, other than the bridge, but a brick falling from such a height would have developed enough kinetic energy to kill you!

In 1928 Gamow, Condon, and Gurney treated $\alpha$ particle emission as a quantum mechanical barrier penetration problem. They assumed that $V(r) = 2Ze^2/4\pi\epsilon_0 r$ for $r > r'$, where $2e$ is the $\alpha$-particle charge and $Ze$ is the charge of the nucleus remaining after the $\alpha$ particle is emitted. They also assumed that $V(r) < E$ for $r < r'$, as shown in Figure 6-19. Equation (6-50) was used to evaluate the transmission coefficient $T$ since the exponent $k_{II}a$, which determines $T$, has a value large compared to one. In fact, the exponent is so large that the exponential completely dominates the behavior of $T$, and it was sufficient to take

$$T \simeq e^{-2k_{II}a} = e^{-2\sqrt{(2m/\hbar^2)(V_0-E)}\,a} \tag{6-54}$$

This expression was derived for a rectangular barrier of height $V_0$ and width $a$, but when the expression is valid it can be applied to the barrier $V(r)$ by considering it to be a set of adjacent rectangular barriers of height $V(r_i)$ and very small width $\Delta r_i$. This reasoning leads, in the limit, to the expression

$$T \simeq e^{-2\int_{r'}^{r'''}\sqrt{(2m/\hbar^2)[V(r)-E]}\,dr} \tag{6-55}$$

where the integration is taken from the nuclear radius $r'$, where $V(r)$ rises above $E$, to the radius $r'''$, where $V(r)$ drops below $E$. The use of (6-54), which was derived for a one-dimensional case, in (6-55) that concerns a three-dimensional problem, was justified because the $\alpha$ particles are almost always emitted with zero angular momentum. That is, they move out along essentially linear paths emanating from the nuclear center, obeying equations which are essentially one dimensional.

The quantity $T$ gives the probability that in one trial an $\alpha$ particle will penetrate the barrier. The number of trials per second could be estimated to be

$$N \simeq \frac{v}{2r'} \tag{6-56}$$

if it were assumed that an $\alpha$ particle is bouncing back and forth with velocity $v$ inside the nucleus of diameter $2r'$. Then the probability per second that the nucleus will decay by emitting an $\alpha$ particle, called the decay rate $R$, would be

$$R \simeq \frac{v}{2r'} e^{-2\int_{r'}^{r'''} \sqrt{(2m/\hbar^2)(2Ze^2/4\pi\epsilon_0 r - E)}\, dr} \tag{6-57}$$

Today we know that (6-56) is not a very accurate estimate, but this function, or its more correct form, varies so slowly compared to the rapid variation in the exponential that the result expressed by (6-57) is an accurate estimate.

In applying (6-57) to a particular radioactive nucleus, Gamow, Condon, and Gurney took all the quantities in the expression as known, except $v$ and $r'$ ($r'''$ can be evaluated from $Z$ and $E$). Assuming $v$ to be comparable to the velocity of the $\alpha$ particle after emission (i.e., $mv^2/2 = E$), the decay rate $R$ is then a function only of the nuclear radius $r'$. Using $r' = 9 \times 10^{-15}$ m, which was certainly in line with the values obtained from Rutherford's analysis of $\alpha$-particle scattering from light nuclei, they obtained values of $R$ which were in good agreement with those measured experimentally, although the decay rate varies over a *tremendously* large range. As an example, for $U^{238}$, the decay rate is $R = 5 \times 10^{-18}\ \text{sec}^{-1}$. An example at the other extreme is $Po^{212}$, for which $R = 2 \times 10^{6}\ \text{sec}^{-1}$. This variation in $R$ is due primarily to the variation, from one radioactive nucleus to the next, of the energy $E$ of the emitted $\alpha$ particles. The height of the barrier and the nuclear radius do not change significantly for nuclei in the limited range of the periodic table in which $\alpha$-emitting nuclei are found. A comparison between experiment and theory is shown in Figure 6-20. The successful application of Schroedinger quantum mechanics to the $\alpha$-particle emission paradox provided one of its earliest, and most convincing, verifications.

Barrier penetration of atoms takes place in the periodic *inversion* of the ammonia molecule, $NH_3$. Figure 6-21 illustrates schematically the structure of this molecule. It consists of three H atoms arranged in a plane, and equidistant from the N atom. There are two completely equivalent equilibrium positions for the N atom, one on either side of the plane containing the H atoms. Figure 6-22 indicates the potential energy acting on the N atom, as a function of its distance $x$ from that plane. The potential function $V(x)$ has two minima, corresponding to the two equilibrium positions, which are symmetrically disposed about a low maximum located at $x = 0$. This maximum, which constitutes a barrier separating the two binding regions, arises from the repulsive Coulomb forces that act on the N atom if it penetrates the plane of the H atoms. The forces are strong enough that in classical mechanics the N atom is not able to cross the barrier, if the molecule is in one of its low-lying energy states; that is, the

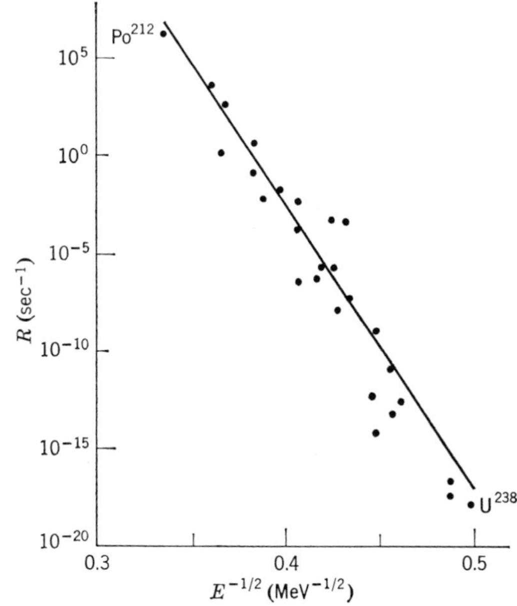

**FIGURE 6-20**

The probability per second $R$ that a radioactive nucleus will emit an $\alpha$ particle of energy $E$. The points are experimental measurements and the solid curve is the prediction of (6-57), a result of barrier penetration theory.

**FIGURE 6-21**

A schematic illustration of the $NH_3$ molecule. The light spheres represent the three H atoms arranged in a plane. The dark spheres represent two equivalent equilibrium positions of the single N atom.

lower allowed energies of this binding potential are below the top of the barrier, as indicated in the figure. But penetration of the classically excluded region allows the N atom to tunnel through the barrier. If it is initially on one side, it will tunnel through and eventually appear on the other side. Then it will do it again in the opposite direction. The N atom actually oscillates slowly back and forth across the plane of the H atoms. The oscillation frequency is $\nu = 2.3786 \times 10^{10}$ Hz, when the molecule is in its ground state. This frequency is much lower than those found in molecular vibrations not involving barrier penetration, or in other atomic or molecular phenomena. Due to the resulting technical simplifications, the frequency was used as a standard in the first atomic clocks which measure time with maximum precision.

A recent, and very useful, example of barrier penetration of electrons is found in the *tunnel diode*. This is a semiconductor device, like a transistor, which is used in fast electronic circuits since its high frequency response is much better than that of any transistor. The operation of a tunnel diode will be explained in Chapter 13, in the context of a discussion of semiconductors. So here we shall say only that the device employs controllable barrier penetration to switch currents on or off so rapidly that it can be used to make an oscillator that can operate at frequencies above $10^{11}$ Hz.

## 6-7 The Square Well Potential

In the preceding sections we have treated the motion of particles in potentials which are not capable of binding them to limited regions of space. Although a number of

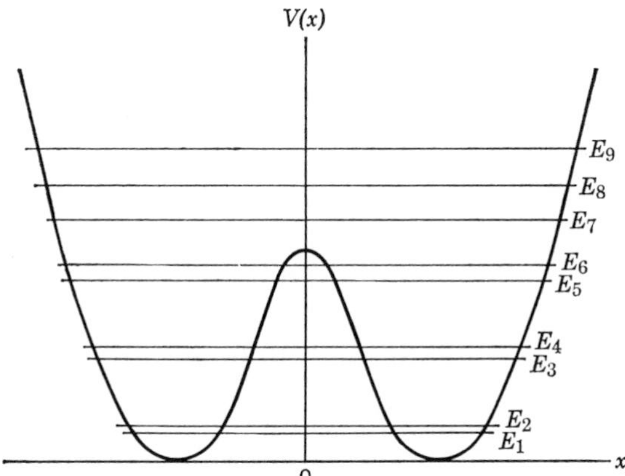


**FIGURE 6-22**

The potential energy of the N atom in the $NH_3$ molecule, as a function of its distance from the plane containing the three H atoms, which lies at $x = 0$. In its lower energy states, the total energy of the molecule lies below the top of the barrier separating the two minima, as indicated by the eigenvalues of the potential shown in the figure.

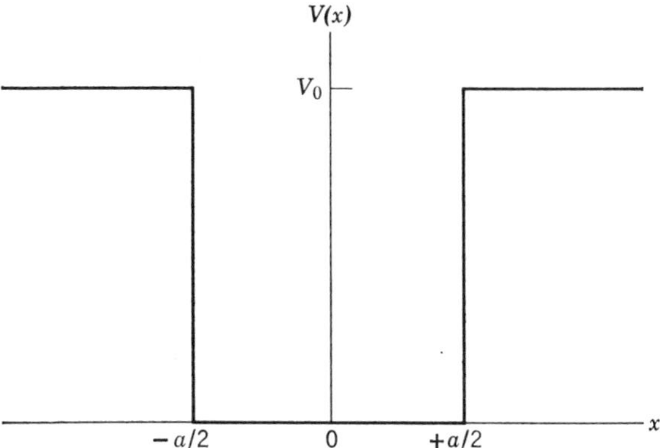


**FIGURE 6-23**
A square well potential.

interesting quantum phenomena showed up, energy quantization did not. Of course we know, from the qualitative discussion of the last chapter, that energy quantization can be expected only for potentials which are capable of binding a particle. In this section we shall discuss one of the simplest potentials having this property, the *square well potential.*

The potential can be written

$$V(x) = \begin{matrix} V_0 & x < -a/2 \text{ or } x > +a/2 \\ 0 & -a/2 < x < +a/2 \end{matrix} \qquad (6\text{-}58)$$

The illustration in Figure 6-23 indicates the origin of its name. If the particle has total energy $E < V_0$, then in *classical mechanics* it can be only in the region $-a/2 < x < +a/2$ (within the well). The particle is bound to that region and bounces back and forth between the ends of the region with momentum of constant magnitude but alternating direction. Furthermore, *any* value $E \geq 0$ of the total energy is possible. But in *quantum mechanics* only *certain* discretely separated values of the total energy are possible.

The square well potential is often used in quantum mechanics to represent a situation in which a particle moves in a restricted region of space under the influence of forces which hold it in that region. Although this simplified potential loses some details of the motion, it retains the essential feature of binding the particle by forces of a certain strength to a region of a certain size. From the discussion in Example 6-2 it is apparent that it is a good approximation to represent the potential acting on a conduction electron in a block of metal by a square well. The depth of the square well is around 10 eV, and its width equals the width of the block. Figure 6-24 indicates, from a point of view different from that used in Example 6-2, how something like a square well can be obtained by superimposing the potentials produced by the closely spaced positive ions in the metal. In Example 6-3, we indicated that the motion of a neutron in a nucleus can be approximated by assuming that the particle is in a square well potential with a depth of about 50 MeV. The linear dimensions of the potential equal the nuclear diameter, which is about $10^{-14}$ m.

We begin our treatment by considering, qualitatively, the form of the eigenfunctions which are solutions to the time-independent Schroedinger equation for the square well potential of (6-58). As in the preceding sections, the problem decomposes itself into three regions: $x < -a/2$ (left of the well), $-a/2 < x < +a/2$ (within the well), and $x > +a/2$ (right of the well). The so-called *general solution* to the equation for the

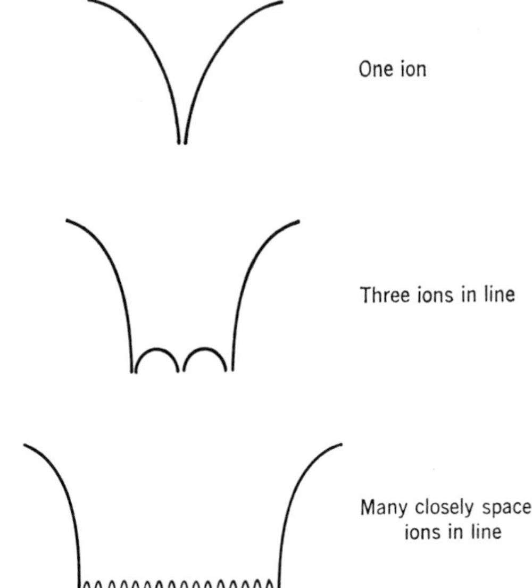


**FIGURE 6-24**
A qualitative indication of how an approximation to a square well potential results from superimposing the potentials acting on a conduction electron in a metal. The potentials are due to the closely spaced positive ions in the metal.

region within the well is

$$\psi(x) = Ae^{ik_I x} + Be^{-ik_I x} \qquad \text{where } k_I = \frac{\sqrt{2mE}}{\hbar} \qquad -a/2 < x < +a/2 \quad (6\text{-}59)$$

The first term describes waves traveling in the direction of increasing $x$, and the second term describes waves traveling in the direction of decreasing $x$. (This solution was derived in Section 6-2. If the student has not studied that section, he can easily show that it is a solution to the time-independent Schroedinger equation, for any values of the arbitrary constants $A$ and $B$, by substituting it into (6-2).)

Now, the classical description of the particle bouncing back and forth within the well suggests that the eigenfunction in that region should correspond to an equal mixture of waves traveling in both directions. The two oppositely directed traveling waves of equal amplitude will combine to form a standing wave. We can obtain such behavior by setting the arbitrary constants equal to one another, so that $A = B$. This yields

$$\psi(x) = B(e^{ik_I x} + e^{-ik_I x})$$

which we write as

$$\psi(x) = B' \frac{e^{ik_I x} + e^{-ik_I x}}{2}$$

where $B'$ is a new arbitrary constant defined by the relation $B' = 2B$. But this combination of complex exponentials gives us simply

$$\psi(x) = B' \cos k_I x \qquad \text{where } k_I = \frac{\sqrt{2mE}}{\hbar} \quad (6\text{-}60)$$

This eigenfunction describes a standing wave since inspection of the associated wave function $\Psi(x,t) = \psi(x)e^{-iEt/\hbar}$ shows that it has nodes in the fixed locations where $\cos k_I x = 0$.

We can also obtain a standing wave by setting $-A = B$. This gives

$$\psi(x) = A(e^{ik_I x} - e^{-ik_I x})$$

which we write as

$$\psi(x) = A' \frac{e^{ik_I x} - e^{-ik_I x}}{2i}$$

where $A'$ is a new arbitrary constant defined by $A' = 2iA$. But this is just

$$\psi(x) = A' \sin k_I x \qquad \text{where } k_I = \frac{\sqrt{2mE}}{\hbar} \qquad (6\text{-}61)$$

Since both (6-60) and (6-61) specify solutions to the time-independent Schroedinger equation for the same value of $E$, and since that differential equation is linear in $\psi(x)$, their sum

$$\psi(x) = A' \sin k_I x + B' \cos k_I x \qquad \text{where } k_I = \frac{\sqrt{2mE}}{\hbar} \qquad -a/2 < x < +a/2 \quad (6\text{-}62)$$

is also a solution, as can be verified by direct substitution. In fact, this is a *general solution* to the differential equation for the region *within* the well because it contains two arbitrary constants—it is just as general as the solution of (6-59). Mathematically, the two are completely equivalent. However, (6-62) is more convenient to use in problems involving the motion of bound particles. Physically, (6-62) can be thought of as describing a situation in which a particle is moving in such a way that the magnitude of its momentum is known to be precisely $p = \hbar k_I = \sqrt{2mE}$, but the direction of the momentum could either be in the direction of increasing or decreasing $x$.

Now consider the solutions to the time-independent Schroedinger equation in the two regions *outside* the potential well: $x < -a/2$ and $x > +a/2$. In these regions the *general solutions* have the forms

$$\psi(x) = Ce^{k_{II}x} + De^{-k_{II}x} \qquad \text{where } k_{II} = \frac{\sqrt{2m(V_0 - E)}}{\hbar} \qquad x < -a/2 \quad (6\text{-}63)$$

and

$$\psi(x) = Fe^{k_{II}x} + Ge^{-k_{II}x} \qquad \text{where } k_{II} = \frac{\sqrt{2m(V_0 - E)}}{\hbar} \qquad x > +a/2 \quad (6\text{-}64)$$

The two forms of $\psi(x)$ describe standing waves in the region outside the well, since in the associated wave function $\Psi(x,t) = \psi(x)e^{-iEt/\hbar}$ the $x$ and $t$ dependences occur as separate factors. These standing waves have no nodes, but they will be joined onto the standing waves inside the well which do have nodes. (The general solutions were derived in Section 6-3. Their validity, for any values of the arbitrary constants $C$, $D$, $F$, and $G$, can easily be verified by students who skipped that section by substitution in (6-13).)

Eigenfunctions valid for all $x$ can be constructed by joining the forms assumed, in each of the three regions of $x$, by the general solutions to the time independent Schroedinger equation. These three forms involve six arbitrary constants: $A'$, $B'$, $C$, $D$, $F$, and $G$. Now since an acceptable eigenfunction must everywhere remain finite, we can immediately see that we must set $D = 0$ and $F = 0$. If this were not done the second exponential in (6-63) would make $\psi(x) \to \infty$ as $x \to -\infty$, and the first exponential in (6-64) would make $\psi(x) \to \infty$ as $x \to +\infty$. Four more equations involving the remaining arbitrary constants can be obtained by demanding that $\psi(x)$ and $d\psi(x)/dx$ be continuous at the two boundaries between the regions, $x = -a/2$ and $x = +a/2$, as is required for acceptable eigenfunctions. (They are already single valued.) But we cannot allow all four of the remaining arbitrary constants to be specified by these four equations. One of them must remain unspecified so that the

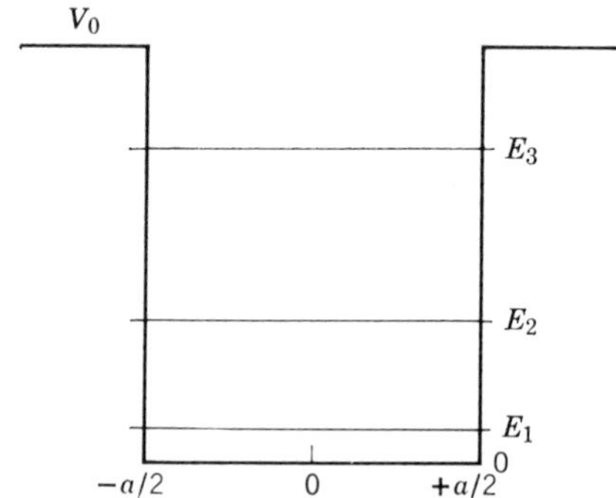


**FIGURE 6-25**
A square well potential and its three bound eigenvalues. Not shown is a continuum of eigenvalues of energy $E > V_0$.

amplitude of the eigenfunction can be arbitrary. Arbitrary amplitude is required because the differential equation is linear in the eigenfunction $\psi(x)$. Thus there seems to be a discrepancy between the number of equations to be satisfied and the number of constants that can be adjusted, but it is resolved by treating the total energy $E$ as an additional constant that can be adjusted, as needed. We shall find that this procedure works, but only for certain values of $E$. That is, there will emerge a certain set of possible values of the total energy $E$, and so the energy will be *quantized* to a set of eigenvalues. Only for these values of the total energy does the Schroedinger equation have acceptable solutions.

It is not difficult to carry through this procedure, as we shall see shortly in treating a special case. But the general case leads to a solution involving a complicated transcendental equation (an equation in which the unknown is contained in the argument of a function such as a sinusoidal), which precludes expressing the solution mathematically in a concise way. Therefore, we relegate the details of the general solution to Appendix G, and here continue for a while with our qualitative discussion.

Figures 6-25 and 6-26 show, respectively, the eigenvalues and eigenfunctions for the three bound states of a particle in a particular square well potential. Not shown are a continuum of eigenvalues which extend from the top of the well on up, since any value of total energy $E$ that is greater than the height of the potential walls $V_0$ is allowed. Also not shown are the continuum eigenfunctions. Focusing attention first on the region of $x$ within the well, we note that the curvature of the sinusoidal part of the eigenfunction increases as the energy of the corresponding eigenvalue increases. As a consequence, the higher the energy of the eigenvalue the more numerous are the

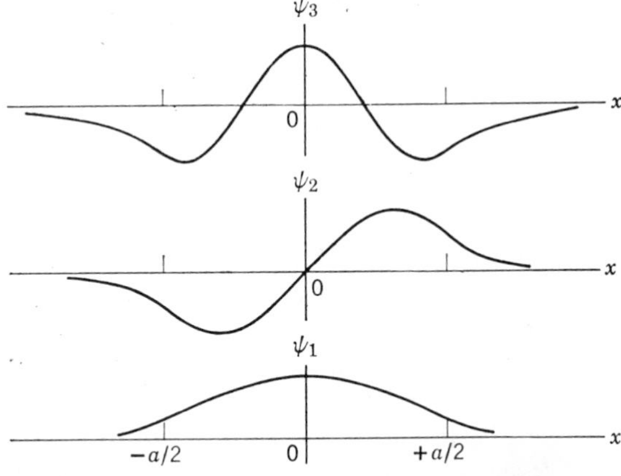


**FIGURE 6-26**
The three bound eigenfunctions for the square well of Figure 6-25.

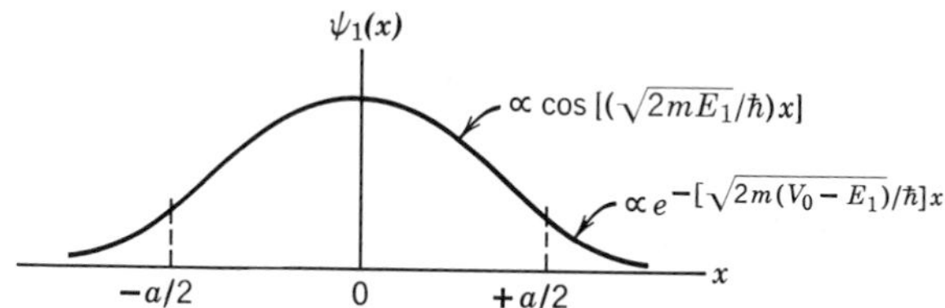


**FIGURE 6-27**
The first eigenfunction for a square well with walls of moderate height.

oscillations of the corresponding eigenfunction and the higher is its wave number. These results reflect the fact that the wave number $k_{\text{I}}$, in the solution of (6-62) for the region inside the well, is proportional to $E^{1/2}$. The square well potential depicted in the figure does not have a fourth *bound* eigenvalue because the associated value of $k_{\text{I}}$, and therefore of $E^{1/2}$, would be too large to satisfy the binding condition $E < V_0$.

Now consider the parts of the eigenfunctions that extend into the regions outside the well. In classical mechanics a particle could never be found in these regions since its kinetic energy is $p^2/2m = E - V(x)$, which is negative where $E < V(x)$. Note that the eigenfunctions go to zero in these *classically excluded regions* more rapidly the lower the energy of the corresponding eigenvalue. This agrees with the fact that the exponential parameter $k_{\text{II}}$, in the solutions of (6-63) and (6-64) for the region outside the well, is proportional to $(V_0 - E)^{1/2}$. It also agrees with the idea that the more serious the violation of the classical restriction, that the total energy $E$ must be at least as large as the potential energy $V(x)$, the more reluctant the eigenfunctions are to penetrate the classically excluded regions.

It is instructive to consider the effect on the eigenfunctions of letting the walls of the square well become very high, i.e., letting $V_0 \rightarrow \infty$. Shown in Figure 6-27 is the first eigenfunction for a square well potential. As $V_0 \rightarrow \infty$, $E_1$ will increase, but it will do so very slowly compared to the increase in $V_0$. This is true because $E_1$ is determined essentially by the requirement that approximately half an oscillation of the eigenfunction must fit into the length of the well. Therefore, the exponential parameter $k_{\text{II}} = \sqrt{2m(V_0 - E)}/\hbar$, which determines the behavior of the eigenfunction in the regions outside of the well, will become very large as $V_0$ becomes very large, and the eigenfunction will go to zero very rapidly outside the well. In the limit, $\psi_1(x)$ must be zero for all $x < -a/2$ and for all $x > +a/2$. For a square well with infinitely high walls, $\psi_1(x)$ has the form shown in Figure 6-28. It is apparent that this argument holds for all the eigenfunctions of such a potential. That is, for all values of $n$, in an infinite square well potential

$$\psi_n(x) = 0 \qquad x \leq -a/2 \text{ or } x \geq +a/2 \quad (6\text{-}65)$$

This condition for infinite square well eigenfunctions can only be satisfied by violating at $x = \pm a/2$ the requirement of Section 5-6 that the derivative $d\psi_n(x)/dx$ of an eigenfunction be continuous everywhere; but if the student will review the argument which

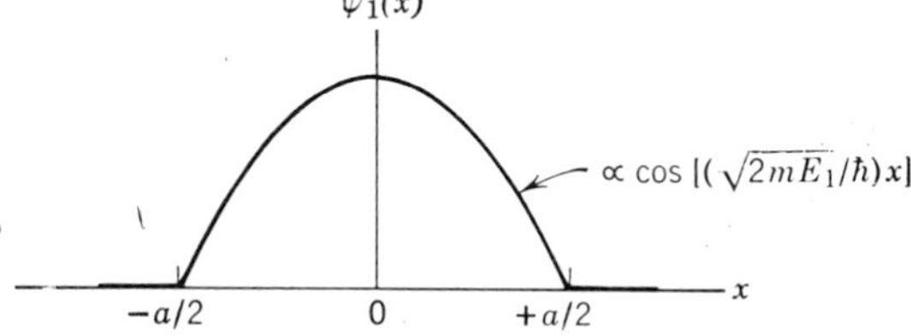


**FIGURE 6-28**
The first eigenfunction of a square well with walls of infinite height.

was presented to justify the requirement, he will find that the derivative must be continuous only when the potential is finite.

## 6-8 The Infinite Square Well Potential

The infinite square well potential is written as

$$V(x) = \begin{matrix} \infty & x < -a/2 \text{ or } x > +a/2 \\ 0 & -a/2 < x < +a/2 \end{matrix} \tag{6-66}$$

and is illustrated in Figure 6-29. It has the feature that it will bind a particle with any finite total energy $E \geq 0$. In *classical mechanics*, *any* of these energies are possible, but in *quantum mechanics* only *certain* discrete eigenvalues $E_n$ are allowed.

We shall see that it is very easy to find simple and concise expressions for the eigenvalues and eigenfunctions of this potential because the transcendental equation that arises in the solution of its time-independent Schroedinger equation happens to have simple solutions. For values of the quantum number $n$ which are not too large, these eigenvalues and eigenfunctions can often be used to approximate the corresponding (same $n$) eigenvalues and eigenfunctions of a square well potential with large but finite $V_0$. For instance, we mentioned before that it is a very good approximation to take the potential for a conduction electron in a block of metal to be a finite square well. In Example 6-2 we showed that for the typical metal Cu the eigenfunctions penetrate into the classically excluded regions exterior to the well by a distance of about $10^{-10}$ m. This distance is so small compared to the width of the square well, which is the width of the Cu block, that for many purposes it is an equally good approximation to use the corresponding eigenfunctions and eigenvalues for an infinite square well, and we shall do so later. We shall also use infinite square well potentials to discuss the quantum mechanical properties of a system of gas molecules, and other particles, which are strictly confined within a box of certain dimensions. A particle moving under the influence of an infinite square well potential is often called a *particle in a box*.

In the region within the well the *general solution* to the time-independent Schroedinger equation for the infinite square well potential can be written as the standing wave of (6-62), which we simplify, by dropping the primes, into the form

$$\psi(x) = A \sin kx + B \cos kx \qquad \text{where } k = \frac{\sqrt{2mE}}{\hbar} \qquad -a/2 \leq x \leq +a/2 \tag{6-67}$$

(Students who have skipped the preceding sections can see that this $\psi(x)$ represents a standing wave by noting that the associated wave function $\Psi(x,t) = \psi(x)e^{-iEt/\hbar}$ has fixed nodes. They can verify that the $\psi(x)$ is actually a solution to the applicable

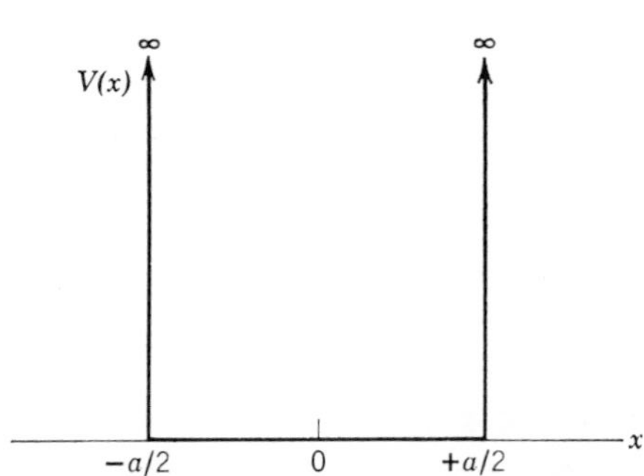


**FIGURE 6-29**
An infinite square well potential.

time-independent Schroedinger equation by substituting it into (6-2).) According to the condition of (6-65), $\psi(x)$ has the value zero in the regions outside the well. Of course, this must be true so that the probability density will be zero in these regions, since the particle is strictly confined within the well by its infinitely high potential walls. In particular, at the boundaries of the well

$$\psi(x) = 0 \qquad\qquad x = \pm a/2 \quad (6\text{-}68)$$

That is, the standing wave has nodes at the walls of the box.

Now we develop relations which are satisfied by the arbitrary constants $A$ and $B$, and by the parameter $k$. Applying the boundary conditions of (6-68) at $x = +a/2$, we obtain

$$A \sin \frac{ka}{2} + B \cos \frac{ka}{2} = 0 \qquad (6\text{-}69)$$

At $x = -a/2$, (6-68) yields

$$A \sin\left(-\frac{ka}{2}\right) + B \cos\left(-\frac{ka}{2}\right) = 0$$

or

$$-A \sin \frac{ka}{2} + B \cos \frac{ka}{2} = 0 \qquad (6\text{-}70)$$

Addition of the last two numbered equations gives

$$2B \cos \frac{ka}{2} = 0 \qquad (6\text{-}71)$$

Subtraction gives

$$2A \sin \frac{ka}{2} = 0 \qquad (6\text{-}72)$$

Both (6-71) and (6-72) must be satisfied. When this is done, $\psi(x)$ and $d\psi(x)/dx$ will be everywhere finite and single valued, and $\psi(x)$ will be everywhere continuous. As discussed at the end of the preceding section, $d\psi(x)/dx$ will be discontinuous at $x = \pm a/2$.

There is no value of the parameter $k$ for which both $\cos(ka/2)$ and $\sin(ka/2)$ are simultaneously zero. And we certainly do not want to satisfy the two equations by setting both $A$ and $B$ equal to zero, for then $\psi(x) = 0$ everywhere and the eigenfunction would be of no interest because the associated particle would not be in the box! However, we can satisfy these equations *either* by choosing $k$ so that $\cos(ka/2)$ is zero and also setting $A$ equal to zero, *or* by choosing $k$ so that $\sin(ka/2)$ is zero and also setting $B$ equal to zero. That is, we take either

$$A = 0 \qquad \text{and} \qquad \cos \frac{ka}{2} = 0 \qquad (6\text{-}73)$$

or

$$B = 0 \qquad \text{and} \qquad \sin \frac{ka}{2} = 0 \qquad (6\text{-}74)$$

Thus there are two classes of solutions.

For the *first class*

$$\psi(x) = B \cos kx \qquad \text{where} \quad \cos \frac{ka}{2} = 0 \quad (6\text{-}75)$$

For the *second class*

$$\psi(x) = A \sin kx \qquad \text{where} \quad \sin \frac{ka}{2} = 0 \quad (6\text{-}76)$$

The conditions on the wave number $k$, expressed in (6-75) and (6-76), are in the form of transcendental equations since the unknown, $k$, occurs in the arguments of the sinusoidals; but these transcendental equations happen to be so simple that their solutions can be written in concise form immediately. The allowed values of $k$ for the first class, (6-75), are

$$\frac{ka}{2} = \frac{\pi}{2}, \frac{3\pi}{2}, \frac{5\pi}{2}, \ldots$$

since $\cos(\pi/2) = \cos(3\pi/2) = \cos(5\pi/2) = \cdots = 0$. It is convenient to express this as

$$k_n = \frac{n\pi}{a} \qquad n = 1, 3, 5, \ldots \tag{6-77}$$

The allowed values of $k$ for the second class, (6-76), are

$$\frac{ka}{2} = \pi, 2\pi, 3\pi, \ldots$$

since $\sin \pi = \sin 2\pi = \sin 3\pi = \cdots = 0$. This can also be expressed as

$$k_n = \frac{n\pi}{a} \qquad n = 2, 4, 6, \ldots \tag{6-78}$$

Knowing the allowed values of $k$, we can then obtain the solutions to the time-independent Schroedinger equation for the infinite square well from (6-75) and (6-76). We find

$$\psi_n(x) = B_n \cos k_n x \qquad \text{where } k_n = \frac{n\pi}{a} \qquad n = 1, 3, 5, \ldots \tag{6-79}$$

and

$$\psi_n(x) = A_n \sin k_n x \qquad \text{where } k_n = \frac{n\pi}{a} \qquad n = 2, 4, 6, \ldots \tag{6-80}$$

The solution corresponding to $n = 0$ is $\psi_0(x) = A \sin 0 = 0$; it is ruled out because it does not describe a particle in a box. The quantum number $n$ has been used to label the different solutions of the transcendental equations, and the corresponding eigenfunctions. If it is necessary to apply the normalization condition, the constants $A_n$ and $B_n$, which specify the amplitudes of the eigenfunctions, will thereby be determined (see Example 5-10); but it is not usually necessary to do this.

The quantum number $n$ is also used to label the corresponding eigenvalues. Using the relation $k = \sqrt{2mE}/\hbar$ of (6-67), and the expression $k_n = n\pi/a$ in (6-79) and (6-80) for the allowed values of $k$, we find

$$E_n = \frac{\hbar^2 k_n^2}{2m} = \frac{\pi^2 \hbar^2 n^2}{2ma^2} \qquad n = 1, 2, 3, 4, 5, \ldots \tag{6-81}$$

Thus we conclude that only certain values of the total energy $E$ are allowed. The total energy of the particle in the box is *quantized*.

The quantitative treatment of the finite square well, discussed in the preceding section and carried out in Appendix G, is essentially the same as what we have just gone through. But the penetration of the eigenfunction into the regions outside the well, which varies with the energy of the associated eigenvalue, leads to more complicated transcendental equations for $k$ that must be solved graphically or numerically.

Figure 6-30 illustrates the infinite square well potential and its first few eigenvalues specified by (6-81). Of course, all the eigenvalues are discretely separated for an

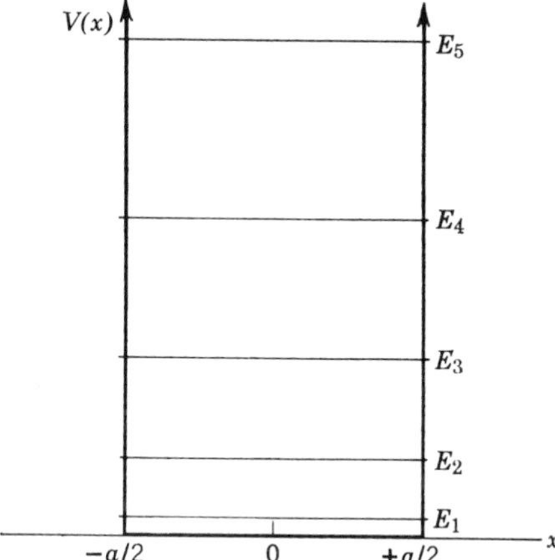


**FIGURE 6-30**
The first few eigenvalues of an infinite square well potential.

infinite square well potential since the particle is bound for any finite eigenvalue. Note that the pattern formed by the first three eigenvalues of the infinite square well is quite similar to that formed by the three bound eigenvalues of the finite square well shown in Figure 6-25. In this regard, the infinite square well results provide an approximation to the finite square well results. However, in detail each potential energy function $V(x)$ has its own characteristic set of bound eigenvalues $E_n$.

Of particular interest is the energy of the first eigenvalue. For the infinite square well it is

$$E_1 = \frac{\pi^2\hbar^2}{2ma^2} \tag{6-82}$$

This is called the *zero-point energy*. It is the lowest possible total energy the particle can have if it is bound by the infinite square well potential to the region $-a/2 \leq x \leq +a/2$. *The particle cannot have zero total energy*. The phenomenon is basically a result of the uncertainty principle. To see this, consider the fact that if the particle is bound by the potential, then we know its $x$ coordinate to within an uncertainty of about $\Delta x \simeq a$. Consequently, the uncertainty in its $x$ momentum must be at least $\Delta p \simeq \hbar/2\Delta x \simeq \hbar/2a$. The uncertainty principle cannot allow the particle to be bound by the potential with zero total energy since that would mean the uncertainty in the momentum would be zero. For the particular case of eigenvalue $E_1$, the magnitude of the momentum is $p_1 \simeq \sqrt{2mE_1} = \pi\hbar/a$. Since the particle is in a state of motion described by a standing wave eigenfunction, it can be moving in either direction and the actual value of the momentum is uncertain by an amount which is about $\Delta p \simeq 2p_1 \simeq 2\pi\hbar/a$. The uncertainty product $\Delta x\, \Delta p \simeq a2\pi\hbar/a \simeq 2\pi\hbar$ is roughly in agreement with the lower limit $\hbar/2$ set by the uncertainty principle. (Compare with the accurate calculation of Example 5-10.)

We conclude that there must be a zero-point energy because there must be a *zero-point motion*. This is in sharp contrast to the idea, of classical physics, that all motion ceases when a system has its minimum energy content at the temperature of absolute zero. The zero-point energy is responsible for several interesting quantum phenomena that are seen in the behavior of matter at very low temperatures. A striking example is the fact that helium will not solidify even at the lowest attainable temperature ($\sim$0.001°K), unless a very high pressure is applied.

The first few eigenfunctions of the infinite square well potential are shown in Figure 6-31. Note that the number of half wavelengths of each eigenfunction is equal to its

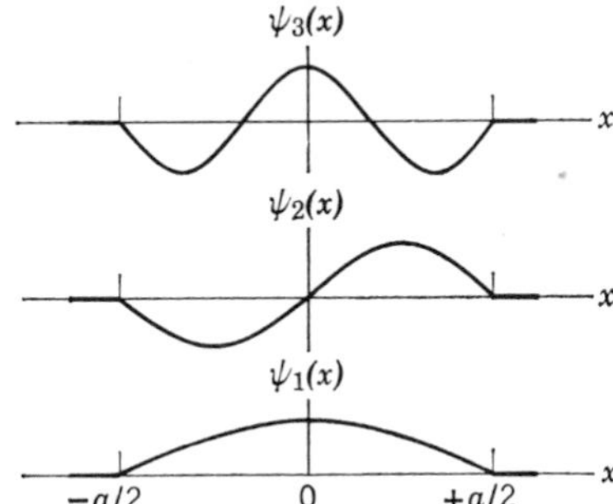


**FIGURE 6-31**
The first few eigenfunctions of an infinite square well potential.

quantum number $n$, and that therefore the number of nodes is $n + 1$. By comparing these eigenfunctions with the corresponding eigenfunctions of the finite square well shown in Figure 6-26, the student can see again how the results obtained for the simple potential can be used to approximate those of the more complicated potential (most accurately for eigenfunctions of lowest $n$ value).

Students familiar with stringed musical instruments may notice that the eigenfunctions for a particle strictly confined between two points at the ends of the box look like the functions describing the possible shapes assumed by a vibrating string fixed at two points at the ends of the string. The reason is that both systems obey time-independent differential equations of analogous form, and they satisfy analogous conditions at the two points. Here is yet another example of the relation between quantum mechanics and classical wave motion. Musically inclined students may also notice that the frequencies, $\nu_n = E_n/h$, of the time-dependent factor in the wave functions for the confined particle satisfy the relation $\nu_n \propto n^2$ (since $E_n = \pi^2\hbar^2n^2/2ma^2$), whereas the frequencies of the vibrating string satisfy the "harmonic progression" $\nu_n \propto n$. This difference arises because the two systems obey time-dependent differential equations which are not at all analogous.

**Example 6-5.** Derive the infinite square well energy quantization law, (6-81), directly from the de Broglie relation $p = h/\lambda$, by fitting an integral number of half de Broglie wavelengths $\lambda/2$ into the width $a$ of the well.

It is clear from Figure 6-31 that the infinite square well eigenfunctions satisfy the following relation between the de Broglie wavelengths and the length of the well

$$n\frac{\lambda}{2} = a \qquad n = 1, 2, 3, \ldots$$

That is, an integral number of half-wavelengths fits into the well. This means

$$\lambda = \frac{2a}{n} \qquad n = 1, 2, 3, \ldots$$

So according to de Broglie, the corresponding values of the momentum of the particle are

$$p = \frac{h}{\lambda} = \frac{hn}{2a} \qquad n = 1, 2, 3, \ldots$$

As the potential energy of the particle is zero within the well, its total energy equals its kinetic energy. Thus

$$E = \frac{p^2}{2m} = \frac{h^2n^2}{2m4a^2} = \frac{\pi^2\hbar^2n^2}{2ma^2} \qquad n = 1, 2, 3, \ldots$$

in agreement with (6-81). This trivial calculation can be used only for the simplest case of a bound particle—the case of an infinite square well potential. It cannot be applied to find the eigenvalues or eigenfunctions of a more complicated potential such as a finite square well. (See also the discussion, in connection with (4-25), of the application of the Wilson-Sommerfeld quantization rule to the infinite square well.) ◀

**Example 6-6.** Before the discovery of the neutron, it was thought that a nucleus of atomic number $Z$ and atomic weight $A$ was composed of $A$ protons and $(A - Z)$ electrons, but there was a serious problem concerning the magnitude of the zero-point energy for a particle as light as an electron confined to a region as small as a nucleus. Estimate the zero-point energy $E$.

Setting the electron mass $m$ equal to $10^{-30}$ kg and the width of the well equal to $10^{-14}$ m (a typical nuclear dimension), from (6-82) we obtain

$$E = \frac{\pi^2\hbar^2}{2ma^2} \simeq \frac{10 \times 10^{-68}\ \text{joule}^2\text{-sec}^2}{2 \times 10^{-30}\ \text{kg} \times 10^{-28}\ \text{m}^2} \simeq \frac{10^{-9}}{2}\ \text{joule}$$

$$\simeq \frac{10^{-9}\ \text{joule}}{2} \times \frac{1\ \text{eV}}{1.6 \times 10^{-19}\ \text{joule}} \sim 10^9\ \text{eV}$$

$$= 10^3\ \text{MeV}$$

For estimating the zero-point energy, we are certainly justified in treating the electron as if it were confined to an infinite square well. We are also justified in ignoring the three-dimensional character of the actual system. But we would not be justified in quoting the value of $E$ just obtained because it is extremely large compared to the electron rest mass energy $m_0c^2 \simeq 0.5$ MeV. A relativistically valid analogue of (6-82) must be used in this particular problem.

The required formula can be obtained from the technique used in Example 6-5. Both of the equations $\lambda = 2a/n$ and $p = h/\lambda$ retain their validity in the extreme relativistic range. So, if we replace $E = p^2/2m$ by $E = cp$ (the energy-momentum relation $E^2 = c^2p^2 + m_0^2c^4$ in the limit $E \gg m_0c^2$), we immediately obtain for $n = 1$

$$E = cp = \frac{ch}{\lambda} = \frac{chn}{2a} = \frac{\pi c\hbar}{a} \simeq \frac{3 \times 3 \times 10^8\ \text{m/sec} \times 10^{-34}\ \text{joule-sec}}{10^{-14}\ \text{m}}$$

$$\times \frac{1\ \text{eV}}{1.6 \times 10^{-19}\ \text{joule}} \sim 10^8\ \text{eV} = 10^2\ \text{MeV}$$

An electron *could* be bound in a nucleus with this zero-point energy, if the magnitude of the depth of the binding potential were greater than the magnitude of the zero-point energy. There is a binding potential acting on the electron due to the Coulomb attraction of the positive charge of the nucleus, but the magnitude of the potential is not great enough. We may estimate this magnitude by setting $r = 10^{-14}$ m, and $Q_1 = Ae$, $Q_2 = -e$, where $e$ is the magnitude of the electron charge, in the Coulomb potential formula. We obtain, for a typical value of $A = 100$

$$\frac{Q_1Q_2}{4\pi\epsilon_0 r} = -\frac{Ae^2}{4\pi\epsilon_0 r} \simeq -\frac{10^2 \times (1.6 \times 10^{-19}\ \text{coul})^2}{10^{-10}\ \text{coul}^2/\text{nt-m}^2 \times 10^{-14}\ \text{m}} \times \frac{1\ \text{eV}}{1.6 \times 10^{-19}\ \text{joule}}$$

$$\sim -10^7\ \text{eV} = -10\ \text{MeV}$$

This is ten times smaller than the required binding energy. So an electron could not be bound in a nucleus because of the zero-point energy required by the uncertainty principle.

In 1932 Chadwick, motivated by a suggestion of Rutherford, discovered the neutron. We now know that a nucleus is composed of $Z$ protons and $(A - Z)$ neutrons. Because neutrons are heavy particles, like protons, their zero-point energy in a nucleus is relatively low so they can be bound without difficulty. Indeed, we shall see in Chapter 15 that some of the most important properties of nuclei can be explained in terms of the quantum states of neutrons, and protons, moving in a (finite) square well potential. ◀

Figure 6-31 makes quite apparent the essential difference between the two classes of standing wave eigenfunctions specified by (6-79) and (6-80). The eigenfunctions of the first class, $\psi_1(x)$, $\psi_3(x)$, $\psi_5(x)$, . . . , are *even functions* of $x$; that is

$$\psi(-x) = +\psi(x) \tag{6-83}$$

In quantum mechanics, these functions are said to be an *even parity*. The eigenfunctions of the second class, $\psi_2(x)$, $\psi_4(x)$, $\psi_6(x)$, . . . , are *odd functions* of $x$; that is

$$\psi(-x) = -\psi(x) \tag{6-84}$$

and are said to be of *odd parity*.

The eigenfunctions have a *definite parity*, either even or odd, because we have chosen the origin of the $x$ axis so that the *symmetrical* square well potential $V(x)$ is an *even* function of $x$. Note that if we redefine the origin of the $x$ axis in Figure 6-31 to be at, say, the point $x = -a/2$, the eigenfunctions will no longer have a definite parity.

These results are obtained for the square well potential, and for any other symmetrical potential, since measurable quantities describing the motion of a particle in *bound* states of such potentials must also be symmetrical about the point of symmetry of the potential. If the origin of the $x$ axis is chosen to be at that symmetry point, then the function describing the measurable quantity must be an even function. As an example, this is true for the probability density function $P(x,t)$, for both even and odd parity eigenfunctions, since

$$P(-x,t) = \psi^*(-x)\psi(-x) = [\pm\psi^*(x)][\pm\psi(x)] = \psi^*(x)\psi(x) = P(x,t) \tag{6-85}$$

This is not true for the wave function itself in the case of an odd parity eigenfunction; such a wave function is an odd function of $x$, but this is not a contradiction because the wave function itself is not measurable. Eigenfunctions for *unbound* states of potentials that are even functions of $x$ do not necessarily have definite parities since they do not necessarily describe symmetrical motions of the particle.

In one dimension, the fact that standing wave eigenfunctions have definite parities, if $V(-x) = V(x)$, is of importance largely because it simplifies certain calculations. In three dimensions, the property has a deeper significance that will be seen first in Chapter 8 in connection with the emission of radiation by an atom making a transition from an excited state to its ground state.

The probability density functions, corresponding to the first few eigenfunctions of the infinite square well, are plotted in Figure 6-32. Also illustrated in the figure is the probability density that would be predicted by classical mechanics for a bound particle bouncing back and forth between $-a/2$ and $+a/2$. Since the classical particle would spend an equal amount of time in any element of the $x$ axis in that region, it would be equally likely found in any such element. The quantum mechanical probability density oscillates more and more as $n$ increases. In the limit that $n$ approaches infinity, that is for eigenvalues of very high energy, the oscillations are so compressed that no experiment could possibly have the resolution to observe anything other than the average behavior of the probability density predicted by quantum mechanics.

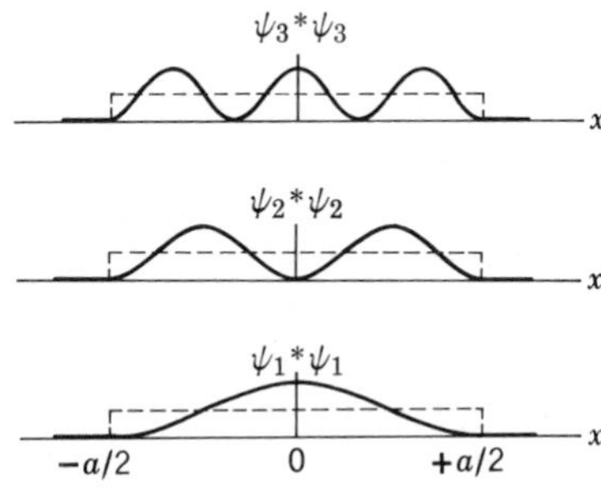


**FIGURE 6-32**

The first few probability density functions for an infinite square well potential. The dashed curves are the predictions of classical mechanics.

Furthermore, the *fractional* separation of the eigenvalues approaches zero as $n$ approaches infinity, so in that limit their discreteness cannot be resolved. Thus we see that the quantum mechanical predictions approach the predictions of classical mechanics in the large quantum number, or high-energy, limit. This is what would be expected from the correspondence principle of the old quantum theory.

## 6-9 The Simple Harmonic Oscillator Potential

We have discussed several potentials which are discontinuous functions of position with constant values in adjacent regions. Now we turn to the more realistic cases of potentials which are continuous functions of position. It turns out that there are only a limited number of such potentials for which it is possible to obtain solutions to the Schroedinger equation by analytical techniques. But, fortunately, these potentials include some of the most important cases, such as the Coulomb potential, $V(r) \propto r^{-1}$, discussed in the following chapter, and the simple harmonic oscillator potential, $V(x) \propto x^2$, discussed in this section. (In this connection, we should remind the student that solutions to the Schroedinger equations for potentials of *any* form can always be obtained by the numerical techniques developed in Appendix F.)

The simple harmonic oscillator is of tremendous importance in physics, and all fields based on physics, because it is the prototype for any system involving oscillations. For instance, it is used in the study of: the vibration of atoms in diatomic molecules, the acoustic and thermal properties of solids which arise from atomic vibrations, magnetic properties of solids that involve vibrations in the orientation of nuclei, and the electrodynamics of quantum systems in which electromagnetic waves are vibrating. Generally speaking, the simple harmonic oscillator can be used to describe almost any system in which an entity is executing *small vibrations* about a point of *stable equilibrium*.

At a position of stable equilibrium, the potential function $V(x)$ must have a minimum. Since any realistic potential function is continuous, the function in the region near its minimum can almost always be well approximated by a parabola, as illustrated in Figure 6-33. But for *small* vibrations the only thing that counts is what $V(x)$ does near its minimum. If we choose the origins of the $x$ axis and the energy axis to be at the minimum, we can write the equation for this parabolic potential function as

$$V(x) = \frac{C}{2} x^2 \tag{6-86}$$

where $C$ is a constant. Such a potential is illustrated in Figure 6-34. A particle moving under its influence experiences a linear (or Hooke's law) restoring force $F(x) = -dV(x)/dx = -Cx$, with $C$ being the force constant.

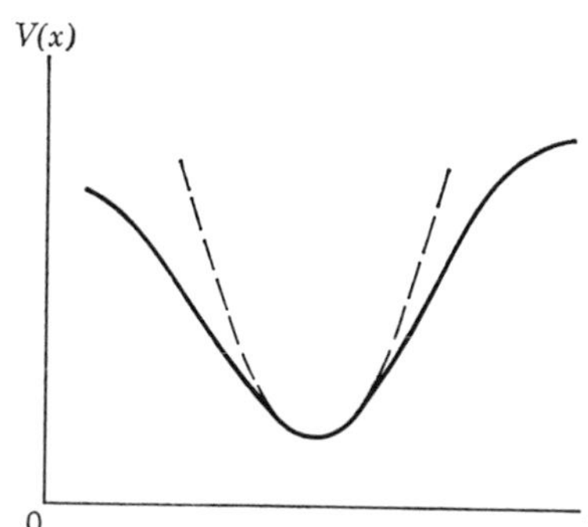


**FIGURE 6-33**

Illustrating the fact that any continuous potential with a minimum (solid curve) can be approximated near the minimum very well by a parabolic potential (dashed curve).

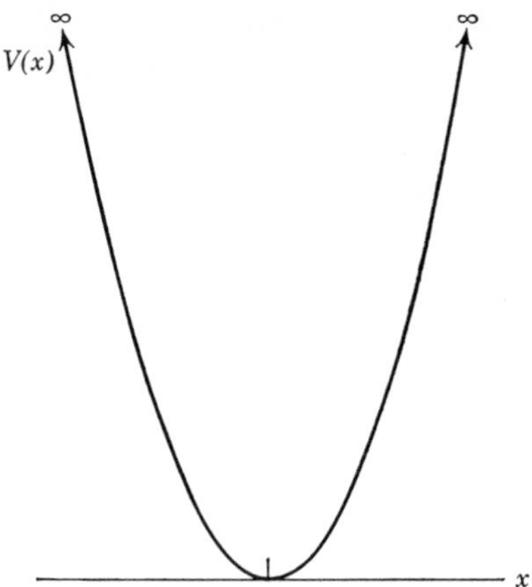


**FIGURE 6-34**
The simple harmonic oscillator potential.

*Classical mechanics* predicts that a particle under the influence of the linear restoring force exerted by the potential of (6-86), which is displaced by an amount $x_0$ from the equilibrium position and then released, will oscillate in simple harmonic motion about the equilibrium position with frequency

$$\nu = \frac{1}{2\pi}\sqrt{\frac{C}{m}} \tag{6-87}$$

where $m$ is its mass. According to that theory, the total energy $E$ of the particle is proportional to $x_0^2$, and can have *any* value since $x_0$ is arbitrary.

*Quantum mechanics* predicts that the total energy $E$ can assume *only* a discrete set of values because the particle is bound by the potential to a region of finite extent. Even in the old quantum theory this was known. The student will recall that Planck's postulate predicts the energy of a particle executing simple harmonic oscillations can assume only one of the values

$$E_n = nh\nu \qquad n = 0, 1, 2, 3, \ldots \tag{6-88}$$

What are the allowed energy values predicted by Schroedinger quantum mechanics for this very important potential? To find out, the time-independent Schroedinger equation for the simple harmonic oscillator potential must be solved.

The mathematics used in the analytical solution to the equation is not difficult to follow, and it is quite interesting; but since the solution is very lengthy it has been placed in Appendix H. Other than verifying by substitution a typical eigenfunction and eigenvalue obtained from the solution, here we shall concentrate on describing the results of the solution and discussing their physical significance.

It is found that the eigenvalues for the simple harmonic oscillator potential are given by the formula

$$E_n = (n + 1/2)h\nu \qquad n = 0, 1, 2, 3, \ldots \tag{6-89}$$

where $\nu$ is the classical oscillation frequency of the particle in the potential. All the eigenvalues are discrete since the particle is bound for any of them. The potential, and the eigenvalues, are shown in Figure 6-35.

If we compare the Schroedinger results with the Planck postulate, we see that in quantum mechanics all the eigenvalues are shifted up by an amount $h\nu/2$. As a consequence, the minimum possible total energy for a particle bound to the potential is $E_0 = h\nu/2$. This is the zero-point energy for the potential, the existence of which is required by the uncertainty principle. Therefore, Planck's postulated energy

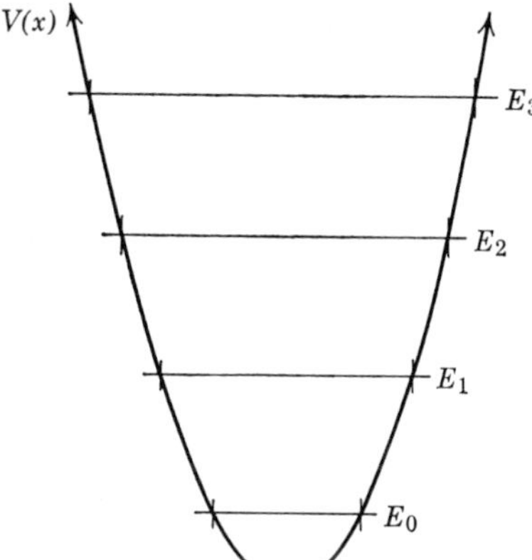


**FIGURE 6-35**

The first few eigenvalues of the simple harmonic oscillator potential. Note that the classically allowed regions (between the intersections of $V(x)$ and $E_n$) expand with increasing values of $E_n$.

quantization of the simple harmonic oscillator was actually in error by the additive constant $h\nu/2$. This constant cancels out in most applications of Planck's postulate because they involve only differences between two energy values. As an example, consider the electromagnetic radiation emitted by the vibrating charge distribution of a diatomic molecule whose interatomic spacing is executing simple harmonic oscillations. Since the frequencies of the emitted photons depend only on the differences in the allowed energies of the molecule, the additive constant has no effect on the *frequencies* of the photons.

But there are observable quantities that show Planck's postulate is in error because it does not contain the zero-point energy. The most important example is also connected with the emission of radiation by a vibrating molecule, or atom. When we study this subject in a subsequent chapter, we shall see that the *rate of emission* of the photons would not agree with experiment unless simple harmonic oscillators have zero-point energies. In fact, we shall find the only reason why the molecule emits *any* radiation is that its vibrations have been stimulated by a surrounding electromagnetic field whose field strengths are executing simple harmonic oscillations because of the zero-point energy of the field.

In addition to providing completely correct eigenvalues, quantum mechanics also provides the eigenfunctions for the simple harmonic oscillator. The eigenfunctions $\psi_n$, corresponding to the first few eigenvalues $E_n$, are listed in Table 6-1 and plotted in Figure 6-36. The eigenfunctions are expressed in terms of the dimensionless variable $u = [(Cm)^{1/4}/\hbar^{1/2}]x$, which differs from $x$ only by a proportionality constant that depends on the properties of the oscillator. For all values of $x$, the eigenfunction is

**TABLE 6-1.** Some Eigenfunctions $\psi(u)$ for the Simple Harmonic Oscillator Potential, where $u$ is Related to the Coordinate $x$ by the Equation $u = [(Cm)^{1/4}/\hbar^{1/2}]x$

| Quantum Number | Eigenfunctions |
|---|---|
| 0 | $\psi_0 = A_0 e^{-u^2/2}$ |
| 1 | $\psi_1 = A_1 u e^{-u^2/2}$ |
| 2 | $\psi_2 = A_2(1 - 2u^2)e^{-u^2/2}$ |
| 3 | $\psi_3 = A_3(3u - 2u^3)e^{-u^2/2}$ |
| 4 | $\psi_4 = A_4(3 - 12u^2 + 4u^4)e^{-u^2/2}$ |
| 5 | $\psi_5 = A_5(15u - 20u^3 + 4u^5)e^{-u^2/2}$ |

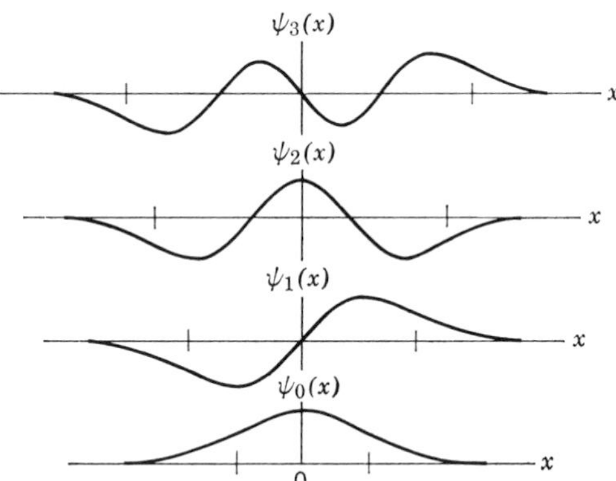


**FIGURE 6-36**
The first few eigenfunctions of the simple harmonic oscillator potential. The vertical ticks on the $x$ axes indicate the limits of classical motion shown in Figure 6-35.

given by the product of an exponential, whose exponent is proportional to $-x^2$, times a simple polynomial of order $x^n$. The polynomial is responsible for the oscillatory behavior of $\psi_n$ in the classically allowed region where $E_n < V(x)$. The number of oscillations increases with increasing $n$ because there are $n$ values of $x$ for which a polynomial of the order $x^n$ has the value zero. These values of $x$ are the locations of the nodes of $\psi_n$. The classically allowed regions lie within the vertical marks shown in Figure 6-36. These regions become wider with increasing $n$ because of the shape of the simple harmonic oscillator potential $V(x)$, as can be seen by inspecting Figure 6-35 which also indicates the classically allowed regions for each $E_n$. Outside these regions, the eigenfunctions decrease very rapidly because their behavior is dominated by the decreasing exponential. Since the relation $V(-x) = V(x)$ is satisfied by the potential, we expect that its eigenfunctions should have definite parities. Inspection of Table 6-1 shows this is true, and that the parity is even for even $n$ and odd for odd $n$. Thus the eigenfunction for the lowest allowed energy is of even parity, as in the case of a square well potential. The multiplicative constants $A_n$ determine the amplitudes of the eigenfunctions. If necessary, the normalization procedure can be used to fix their values, as in Example 5-7; but this is usually not necessary.

The simple harmonic oscillator eigenfunctions contain a wealth of information about the behavior of the system. Some of this information was extracted in Chapter 5. For instance, Figures 5-3 and 5-18 gave accurate representations of the probability density functions for the $n = 1$ and $n = 13$ quantum states of the oscillator. In Chapter 8 we shall show how the eigenfunctions can be used to calculate the rate of emission of radiation by a charged simple harmonic oscillator, and derive the $n_i - n_f = \pm 1$ selection rule that had to be introduced in the old quantum theory by arguments based on the rather unreliable correspondence principle.

**Example 6-7.** Because the simple harmonic oscillator eigenfunctions for small $n$ have fairly simple mathematical forms, it is not too difficult to verify by direct substitution that they satisfy the time-independent Schroedinger equation, for the potential of (6-86), and for the eigenvalues of (6-89). Make such a verification for $n = 1$. (For $n = 0$ the wave function was verified by direct substitution in the Schroedinger equation in Example 5-3.)

The time-independent Schroedinger equation is

$$-\frac{\hbar^2}{2m}\frac{d^2\psi}{dx^2} + \frac{C}{2}x^2\psi = E\psi$$

To verify that the eigenvalue

$$E_1 = \frac{3}{2}h\nu = \frac{3}{2}\frac{h}{2\pi}\left(\frac{C}{m}\right)^{1/2} = \frac{3}{2}\hbar\left(\frac{C}{m}\right)^{1/2}$$

**TABLE 6-2.** A Summary of the Systems Studied in Chapter 6

| Name of System | Physical Example | Potential and Total Energies | Probability Density | Significant Feature |
|---|---|---|---|---|
| Zero potential | Proton in beam from cyclotron | 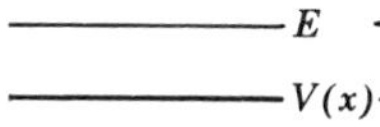 $E$, $V(x)$ | 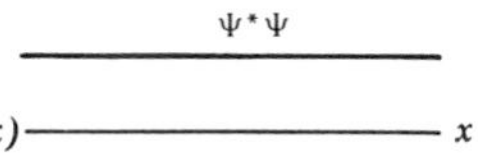 $\Psi^*\Psi$, $x$ | Results used for other systems |
| Step potential (energy below top) | Conduction electron near surface of metal | 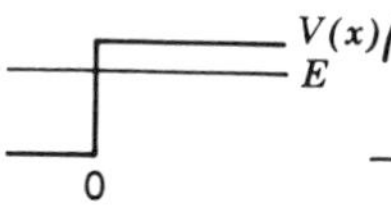 $V(x)$, $E$, 0 | 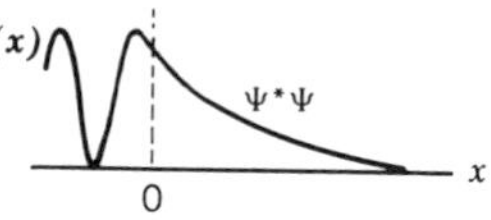 $\Psi^*\Psi$, 0, $x$ | Penetration of excluded region |
| Step potential (energy above top) | Neutron trying to escape nucleus | 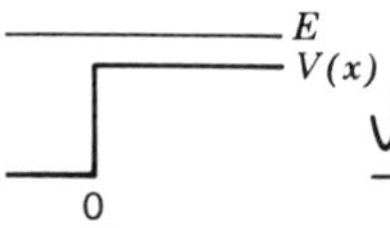 $E$, $V(x)$, 0 | 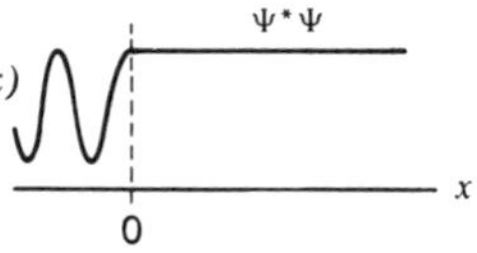 $\Psi^*\Psi$, 0, $x$ | Partial reflection at potential discontinuity |
| Barrier potential (energy below top) | α particle trying to escape Coloumb barrier | 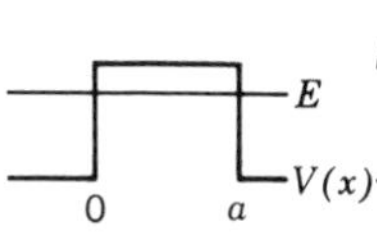 $E$, $V(x)$, 0, $a$ | 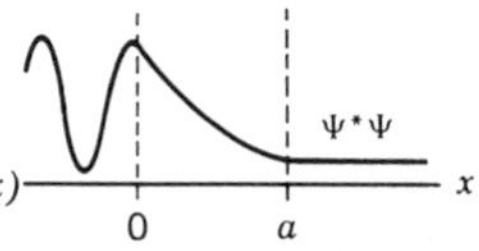 $\Psi^*\Psi$, 0, $a$, $x$ | Tunneling |
| Barrier potential (energy above top) | Electron scattering from negatively ionized atom | 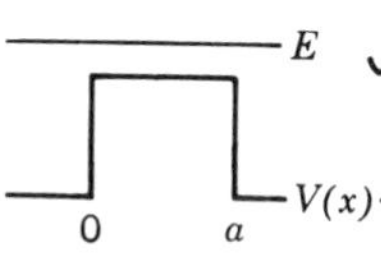 $E$, $V(x)$, 0, $a$ | 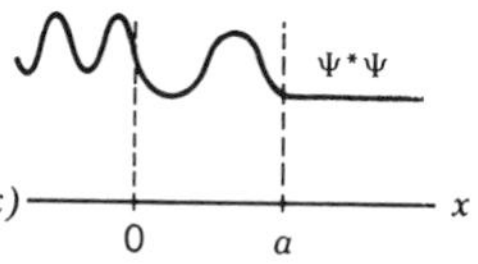 $\Psi^*\Psi$, 0, $a$, $x$ | No reflection at certain energies |
| Finite square well potential | Neutron bound in nucleus | 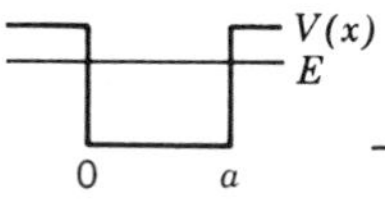 $V(x)$, $E$, 0, $a$ | 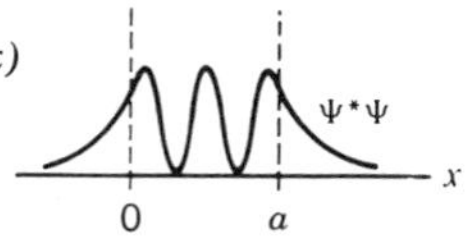 $\Psi^*\Psi$, 0, $a$, $x$ | Energy quantization |
| Infinite square well potential | Molecule strictly confined to box | 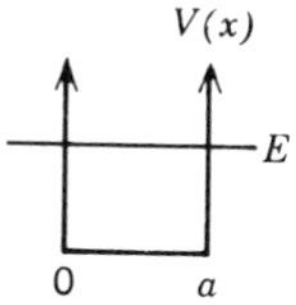 $V(x)$, $E$, 0, $a$ | 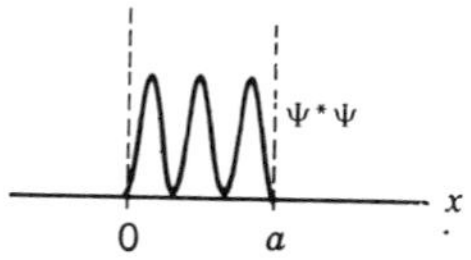 $\Psi^*\Psi$, 0, $a$, $x$ | Approximation to finite square well |
| Simple harmonic oscillator potential | Atom of vibrating diatomic molecule | 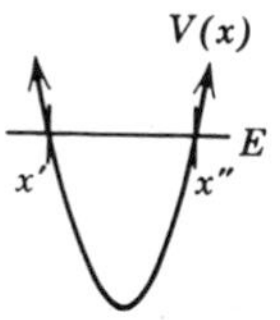 $V(x)$, $E$, $x'$, $x''$ | 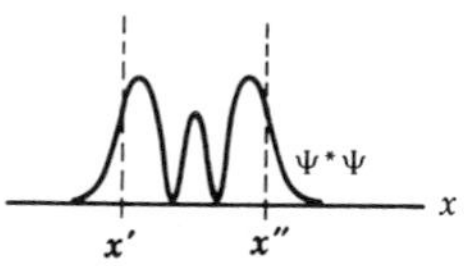 $\Psi^*\Psi$, $x'$, $x''$, $x$ | Zero-point energy |

and the eigenfunction

$$\psi_1 = A_1 u e^{-u^2/2} \qquad \text{where } u = \frac{(Cm)^{1/4}}{\hbar^{1/2}} x$$

satisfy the equation, we evaluate the derivatives

$$\frac{d\psi_1}{dx} = \frac{du}{dx}\frac{d\psi}{du} = \frac{(Cm)^{1/4}}{\hbar^{1/2}} [A_1 e^{-u^2/2} + A_1 u(-u) e^{-u^2/2}]$$

$$= \frac{(Cm)^{1/4}}{\hbar^{1/2}} A_1 e^{-u^2/2} [1 - u^2]$$

and

$$\frac{d^2\psi_1}{dx^2} = \frac{du}{dx}\frac{d}{du}\frac{d\psi_1}{dx} = \frac{(Cm)^{1/4}}{\hbar^{1/2}} \left\{ \frac{d}{du} \frac{(Cm)^{1/4}}{\hbar^{1/2}} A_1 e^{-u^2/2} [1 - u^2] \right\}$$

$$= \frac{(Cm)^{1/2}}{\hbar} A_1 \{-u e^{-u^2/2} [1 - u^2] + e^{-u^2/2} [-2u]\}$$

$$= \frac{(Cm)^{1/2}}{\hbar} A_1 u e^{-u^2/2} \{u^2 - 3\}$$

$$= \frac{(Cm)^{1/2}}{\hbar} \{u^2 - 3\} \psi_1 = \frac{(Cm)^{1/2}}{\hbar} \left\{ \frac{(Cm)^{1/2}}{\hbar} x^2 - 3 \right\} \psi_1$$

Substitution of $d^2\psi_1/dx^2$ and $E_1$ into the equation they are supposed to satisfy yields

$$-\frac{\hbar^2}{2m} \frac{(Cm)^{1/2}}{\hbar} \left\{ \frac{(Cm)^{1/2}}{\hbar} x^2 - 3 \right\} \psi_1 + \frac{C}{2} x^2 \psi_1 = \frac{3\hbar}{2} \left( \frac{C}{m} \right)^{1/2} \psi_1$$

Since inspection shows this is satisfied, the verification is completed. ◀

## 6-10 Summary

In Table 6-2 we summarize some of the properties of the systems studied in this chapter. The table gives an abbreviated name for each idealized system, and an example of a physical system whose potential and total energies are approximated by the idealization. It also gives sketches of the forms of the potential and total energies, and corresponding probability density functions, for each system. If the particle is not bound, it is incident from the left. We have chosen one significant feature of each system to list in the table, but there are many other significant features that we have discussed, which are not listed. In fact, in this chapter we have obtained most of the important predictions of quantum mechanics for systems involving one particle moving in a one-dimensional potential. In the following chapters we shall obtain predictions from the theory for systems involving three dimensions and several particles.

## QUESTIONS

1. Can there be solutions with $E < 0$ to the time-independent Schroedinger equation for the zero potential?
2. Why is it never possible in classical mechanics to have $E < V(x)$? Why is it possible in quantum mechanics, providing there is some region in which $E > V(x)$?

3. Explain why the general solution to a one-dimensional time-independent Schroedinger equation contains two different functions, while the general solution to the corresponding Schroedinger equation contains many different functions.

4. Consider a particle in a long beam of very accurately known momentum. Does a wave function in the form of a group provide a more or a less realistic description of the particle than a single complex exponential wavefunction like (6-9)?

5. Under what circumstances is a discontinuous potential function a reasonable approximation to an actual system?

6. If a potential function has a discontinuity at a certain point, do its eigenfunctions have discontinuities at that point? If not, why not?

7. By combining oppositely directed traveling waves of equal amplitudes, we obtain a standing wave. What kind of a wave do we get if the amplitudes are not equal?

8. Just what is a probability flux, and why is it useful?

9. How can it be that a probability flux is split at a potential discontinuity, although the associated particle is not split?

10. Is there an analogy between the splitting of a probability flux that characterizes the behavior of an unbound particle in a one-dimensional system, and the alternative paths that can be followed by an unbound particle moving in two dimensions through a diffraction apparatus? Why?

11. Exactly what is meant by the statement that the reflection coefficient is one for a particle incident on a potential step with total energy less than the step height? What is meant by the statement that the reflection coefficient is less than one if the total energy is greater than the step height? Can the reflection coefficient ever be greater than one?

12. Since a real exponential is a nonoscillatory function, why is a complex exponential an oscillatory function?

13. What do you think causes the rapid oscillations in the group wave function of Figure 6-8 as it reflects from the potential step?

14. What is the fallacy in the following statement? "Since a particle cannot be detected while tunneling through a barrier, it is senseless to say that the process actually happens."

15. A particle is incident on a potential barrier, with total energy less than the barrier height, and it is reflected. Does the reflection involve only the potential discontinuity facing its direction of incidence? If the other discontinuity were removed, so that the barrier were changed into a step, is the reflection coefficient changed?

16. In the sun, two nuclei of low mass in violent thermal motion can collide by penetrating the Coulomb barrier which separates them. The mass of the single nucleus formed is less than the sum of the masses of the two nuclei, so energy is liberated. This *fusion* process is responsible for the heat output of the sun. What would be the consequences to life on earth if it could not happen because barriers were impenetrable?

17. Are there any *measurable* consequences of the penetration of a classically excluded region which is of infinite length? Consider a bound particle in a finite square well potential.

18. Show from a qualitative argument that a one-dimensional finite square well potential always has one bound eigenvalue, no matter how shallow the binding region. What would the eigenfunction look like if the binding region were very shallow?

19. Why do finite square wells have only a finite number of bound eigenvalues? What are the characteristics of the unbound eigenvalues?

20. What would a standing wave eigenfunction for an unbound eigenvalue of a finite square well look like?

21. Why do the lowest eigenvalues and eigenfunctions of an infinite square well provide the best approximation to the corresponding eigenvalues and eigenfunctions of a finite square well?

22. In the $n = 3$ state, the probability density function for a particle in a box is zero at two positions between the walls of the box. How then can the particle ever move across these positions?

23. Explain in simplest terms the relation between the zero-point energy and the uncertainty principle.

24. Would you expect the zero-point energy to have much effect on the heat capacity of matter at very low temperatures? Justify your answer.

25. If the eigenfunctions of a potential have definite parities, the one of lowest energy always has even parity. Explain why.

26. Are there analogies in classical physics to the quantum mechanical concept of parity?

27. Are there unbound states for a simple harmonic oscillator potential? How many bound states are there? How realistic is the potential?

28. Explain all aspects of the behavior of all the probability densities of Table 6-2; in particular explain the probability density for the barrier potential with energy above the top.

29. What are the other significant features of the systems of Table 6-2?

30. Considering separately each system treated in this chapter, state which of its properties agree, and disagree, with classical mechanics in the microscopic limit. Which agree and disagree, with classical wave motion in that limit. Make the same classifications for the properties of the systems in the macroscopic limit.

## PROBLEMS

1. Show that the step potential eigenfunction, for $E < V_0$, can be converted in form from the sum of two traveling waves, as in (6-24), to a standing wave, as in (6-29).

2. Repeat the step potential calculation of Section 6-4, but with the particle initially in the region $x > 0$ where $V(x) = V_0$, and traveling in the direction of decreasing $x$ towards the point $x = 0$ where the potential steps down to its value $V(x) = 0$ in the region $x < 0$. Show that the transmission and reflection coefficients are the same as those obtained in Section 6-4.

3. Prove (6-43) stating that the sum of the reflection and transmission coefficients equals one, for the case of a step potential with $E > V_0$.

4. Prove (6-44) which expresses the reflection and transmission coefficients in terms of the ratio $E/V_0$.

5. Consider a particle tunneling through a rectangular potential barrier. Write the general solutions presented in Section 6-5, which give the form of $\psi$ in the different regions of the potential. (a) Then find four relations between the five arbitrary constants by matching $\psi$ and $d\psi/dx$ at the boundaries between these regions. (b) Use these relations to evaluate the transmission coefficient $T$, thereby verifying (6-49). (Hint: First eliminate $F$ and $G$, leaving relations between $A$, $B$, and $C$. Then eliminate $B$.)

6. Show that the expression of (6-49), for the transmission coefficient in tunneling through a rectangular potential barrier, reduces to the form quoted in (6-50) if the exponents are very large.

7. Consider a particle passing over a rectangular potential barrier. Write the general solutions, presented in Section 6-5, which give the form of $\psi$ in the different regions of the potential. (a) Then find four relations between the five arbitrary constants by matching $\psi$ and $d\psi/dx$ at the boundaries between these regions. (b) Use these relations to evaluate the transmission coefficient $T$, thereby verifying (6-51). (Hint: Note that the four relations become exactly the same as those found in the first part of Problem 5, if $k_{II}$ is replaced by $ik_{III}$. Make this substitution in (6-49) to obtain directly (6-51).)

8. (a) Evaluate the transmission coefficient for an electron of total energy 2 eV incident upon a rectangular potential barrier of height 4 eV and thickness $10^{-10}$ m, using (6-49) and then using (6-50). (b) Repeat the evaluation for a barrier thickness of $10^{-9}$ m.

9. A proton and a deuteron (a particle with the same charge as a proton, but twice the mass) attempt to penetrate a rectangular potential barrier of height 10 MeV and thickness $10^{-14}$ m. Both particles have total energies of 3 MeV. (a) Use qualitative arguments to predict which particle has the highest probability of succeeding. (b) Evaluate quantitatively the probability of success for both particles.

10. A fusion reaction important in solar energy production (see Question 16) involves capture of a proton by a carbon nucleus, which has six times the charge of a proton and a radius of $r' \simeq 2 \times 10^{-15}$ m. (a) Estimate the Coulomb potential $V$ experienced by the proton if it is at the nuclear surface. (b) The proton is incident upon the nucleus because of its thermal motion. Its total energy cannot realistically be assumed to be much higher than $10\,kT$, where $k$ is Boltzmann's constant (see Chapter 1) and where $T$ is the internal temperature of the sun of about $10^7$ °K. Estimate this total energy, and compare it with the height of the Coulomb barrier. (c) Calculate the probability that the proton can penetrate a rectangular barrier potential of height $V$ extending from $r'$ to $2r'$, the point at which the Coulomb barrier potential drops to $V/2$. (d) Is the penetration through the actual Coulomb barrier potential greater or less than through the rectangular barrier potential of part (c)?

11. Verify by substitution that the standing wave general solution, (6-62), satisfies the time-independent Schroedinger equation, (6-2), for the finite square well potential in the region inside the well.

12. Verify by substitution that the exponential general solutions, (6-63) and (6-64), satisfy the time-independent Schroedinger equation (6-13) for the finite square well potential in the regions outside the well.

13. (a) From qualitative arguments, make a sketch of the form of a typical unbound *standing wave* eigenfunction for a finite square well potential. (b) Is the amplitude of the oscillation the same in all regions? (c) What does the behavior of the amplitude predict about the probabilities of finding the particle in a unit length of the $x$ axis in various regions? (d) Does the prediction agree with what would be expected from classical mechanics?

14. Use the qualitative arguments of Problem 13 to develop a condition on the total energy of the particle, in an unbound state of a finite square well potential, which makes the probability of finding it in a unit length of the $x$ axis the same inside the well as outside the well. (Hint: What counts is the relation between the de Broglie wavelength inside the well and the width of the well.)

15. (a) Make a quantitative calculation of the transmission coefficient for an unbound particle moving over a finite square well potential. (Hint: Use a trick similar to the one indicated in Problem 7.) (b) Find a condition on the total energy of the particle which makes the transmission coefficient equal to one. (c) Compare with the condition found in Problem 14, and explain why they are the same. (d) Give an example of an optical analogue to this system.

16. An atom of the noble gas krypton exerts an attractive potential on an unbound electron, which has a very abrupt onset. Because of this it is a reasonable approximation to

describe the potential as an attractive square well, of radius equal to the $4 \times 10^{-10}$ m radius of the atom. Experiments show that an electron of kinetic energy 0.7 eV, in regions outside the atom, can travel through the atom with essentially no reflection. The phenomenon is called the *Ramsauer effect*. Use this information in the conditions of Problem 14 or 15 to determine the depth of the square well potential. (Hint: One de Broglie wavelength just fits into the width of the well. Why not one-half a de Broglie wavelength?)

**17.** Obtain an analytical solution, as in Appendix G, to find the first eigenvalue of the potential

$$V(x) = \begin{cases} \infty & x < -a/2 \text{ or } x > +a/2 \\ 0 & -a/2 < x < -a/4 \text{ or } +a/4 < x < +a/2 \\ v_0 & -a/4 < x < +a/4 \end{cases}$$

where

$$v_0 = \frac{\pi^2 \hbar^2}{8ma^2}$$

Compare with the numerical integration of Problem 31 of Chapter 5. (Hint: (i) Because of the symmetry of $V(x)$, the first eigenfunction $\psi$ must be of even parity. This means there can be no sine term in the form assumed by $\psi$ in the region $-a/4 < x < +a/4$ surrounding $x = 0$. (ii) Because of this symmetry, it is necessary only to match $\psi$ and $d\psi/dx$ at $x = +a/4$, and to make $\psi = 0$ at $x = +a/2$.)

**18.** Verify by substitution that the standing wave general solution, (6-67), satisfies the time-independent Schroedinger equation, (6-2), for the infinite square well potential in the region inside the well.

**19.** Two possible eigenfunctions for a particle moving freely in a region of length $a$, but strictly confined to that region, are shown in Figure 6-37. When the particle is in the state

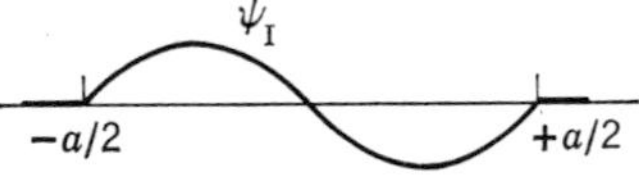


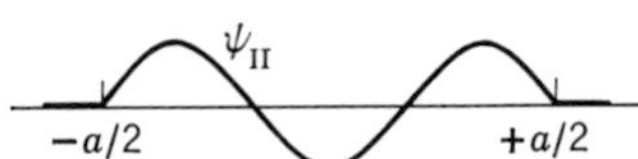


**FIGURE 6-37**
Two eigenfunctions considered in Problem 19.

corresponding to the eigenfunction $\psi_{\text{I}}$, its total energy is 4 eV. (a) What is its total energy in the state corresponding to $\psi_{\text{II}}$? (b) What is the lowest possible total energy for the particle in this system?

**20.** (a) Estimate the zero-point energy for a neutron in a nucleus, by treating it as if it were in an infinite square well of width equal to a nuclear diameter of $10^{-14}$ m. (b) Compare your answer with the electron zero-point energy of Example 6-6.

**21.** (a) Solve the classical wave equation governing the vibrations of a stretched string, for a string fixed at both its ends. Thereby show that the functions describing the possible shapes assumed by the string are essentially the same as the eigenfunctions for an infinite square well potential. (b) Also show that the possible frequencies of vibration of the string are essentially different from the frequencies of the wave functions for the potential.

**22.** (a) For a particle in a box, show that the fractional difference in the energy between adjacent eigenvalues is

$$\frac{\Delta E_n}{E_n} = \frac{2n + 1}{n^2}$$

(b) Use this formula to discuss the classical limit of the system.

23. Apply the normalization condition to show that the value of the multiplicative constant for the $n = 3$ eigenfunction of the infinite square well potential, (6-79), is $B_3 = \sqrt{2/a}$.

24. Use the eigenfunction of Problem 23 to calculate the following expectation values, and comment on each result: (a) $\bar{x}$, (b) $\bar{p}$, (c) $\overline{x^2}$, (d) $\overline{p^2}$.

25. (a) Use the results of Problem 24 to evaluate the product of the uncertainty in position times the uncertainty in momentum, for a particle in the $n = 3$ state of an infinite square well potential. (b) Compare with the results of Example 5-10 and Problem 13 of Chapter 5, and comment on the relative size of the uncertainty products for the $n = 1$, $n = 2$, and $n = 3$ states.

26. Form the product of the eigenfunction for the $n = 1$ state of an infinite square well potential times the eigenfunction for the $n = 3$ state of that potential. Then integrate it over all $x$, and show that the result is equal to zero. In other words, prove that

$$\int_{-\infty}^{\infty} \psi_1(x)\psi_3(x)\, dx = 0$$

(Hint: Use the relation: $\cos u \cos v = [\cos(u + v) + \cos(u - v)]/2$.) Students who have worked Problem 33 of Chapter 5 have already proved that the integral over all $x$ of the $n = 1$ eigenfunction times the $n = 2$ eigenfunction also equals zero. It can be proved that the integral over all $x$ of any two different eigenfunctions of the potential equals zero. Furthermore, this is true for any two different eigenfunctions of any other potential. (If the eigenfunctions are complex, the complex conjugate of one is taken in the integrand.) This property is called *orthogonality*.

27. Apply the results of Problem 20 of Chapter 5 to the case of a particle in a three-dimensional box. That is, solve the time-independent Schroedinger equation for a particle moving in a three-dimensional potential that is zero inside a cubical region of edge length $a$, and becomes infinitely large outside that region. Determine the eigenvalues and eigenfunctions for the system.

28. Airline passengers frequently observe the wingtips of their planes oscillating up and down with periods of the order of 1 sec and amplitudes of about 0.1 m. (a) Prove that this is definitely not due to the zero-point motion of the wings by comparing the zero-point energy with the energy obtained from the quoted values plus an estimated mass for the wings. (b) Calculate the order of magnitude of the quantum number $n$ of the observed oscillation.

29. The restoring force constant $C$ for the vibrations of the interatomic spacing of a typical diatomic molecule is about $10^3$ joules/m$^2$. Use this value to estimate the zero-point energy of the molecular vibrations.

30. (a) Estimate the difference in energy between the ground state and first excited state of the vibrating molecule considered in Problem 29. (b) From this estimate determine the energy of the photon emitted by the vibrations in the charge distribution when the system makes a transition between the first excited state and the ground state. (c) Determine also the frequency of the photon, and compare it with the classical oscillation frequency of the system. (d) In what range of the electromagnetic spectrum is it?

31. A pendulum, consisting of a weight of 1 kg at the end of a light 1 m rod, is oscillating with an amplitude of 0.1 m. Evaluate the following quantities: (a) frequency of oscillation, (b) energy of oscillation, (c) approximate value of quantum number for oscillation, (d) separation in energy between adjacent allowed energies, (e) separation in distance between adjacent bumps in the probability density function near the equilibrium point.

32. Devise a simple argument verifying that the exponent in the decreasing exponential, which governs the behavior of simple harmonic oscillator eigenfunctions in the classically

excluded region, is proportional to $x^2$. (Hint: Take the finite square well eigenfunctions of (6-63) and (6-64), and treat the quantity $(V_0 - E)$ as if it increased with increasing $x$ in proportion to $x^2$.)

**33.** Verify the eigenfunction and eigenvalue for the $n = 2$ state of a simple harmonic oscillator by direct substitution into the time-independent Schroedinger equation, as in Example 6-7.

**34.** Determine the forms of the first five simple harmonic oscillator eigenfunctions by evaluating the coefficients of the polynomials from the recursion relation developed in Appendix H.

**35.** Carry through, as far as possible, an attempt to make a direct series solution of (H-7) of Appendix H. Explain clearly why the attempt fails.

# 7

# One-Electron Atoms

# One-Electron Atoms 7

## 7-1 Introduction

In this chapter we begin our quantum mechanical study of atoms by treating the simplest case, the *one-electron atom*. This is also the most important case. For instance, the one-electron atom hydrogen is of historical importance because it was the first system which Schroedinger treated with his theory of quantum mechanics. We shall see that the eigenvalues which the theory predicts for the hydrogen atom agree with those predicted by the Bohr model and observed by experiment. This provided the first verification of the Schroedinger theory.

There is much more to the Schroedinger theory of the one-electron atom than its prediction of the eigenvalues, because it also predicts the eigenfunctions. Using the eigenfunctions, we shall learn about the following properties of the atom: (1) the probability density functions, which give us detailed pictures of the structure of the atom that do not violate the uncertainty principle as do the precise orbits of the Bohr model, (2) the orbital angular momenta of the atom, which were incorrectly predicted by the Bohr model, (3) the electron spin and other effects of relativity on the atom, which were also incorrectly predicted by the Bohr model, and (4) the rates at which the atom makes transitions from its excited states to its ground state—measurable quantities that were not predictable at all by the Bohr model.

Above and beyond its historical and intrinsic importance, the Schroedinger theory of the one-electron atom is of great practical importance because it forms the foundation of the quantum mechanical treatment of all multielectron atoms, as well as of molecules and nuclei. In later chapters this will become very apparent.

The one-electron atom is the simplest bound system that occurs in nature. But it is more complicated than the systems we have dealt with in the preceding chapters because it contains *two particles*, and because it is *three dimensional*. The system consists of a positively charged nucleus and a negatively charged electron, moving under the influence of their mutual Coulomb attraction and bound together by that attraction. The three-dimensional character of the system allows it to have angular momentum. We shall see that interesting new quantum mechanical phenomena arise as a consequence. Quantum mechanical phenomena involving angular momentum could not arise in our earlier considerations, which dealt with only one-dimensional systems.

The three-dimensional character of the atom causes difficulty because it complicates the mathematical procedures that must be used in its treatment. However, the procedures are straightforward extensions of the simpler ones we have used on one-dimensional systems, so no conceptual problems should arise. We shall avoid practical problems by avoiding the actual solution of the more difficult equations. And certain other details, of interest to some but not all students, will be relegated to appendices. We shall present in this chapter enough of the mathematics to make it apparent how it is related to that used in the preceding chapters. But here we shall emphasize the physical considerations underlying the mathematics, the results which it yields, and the interpretation of the results.

The fact that the one-electron atom contains two particles causes no difficulty at all, if use is made of the reduced mass technique. This technique, discussed in Section

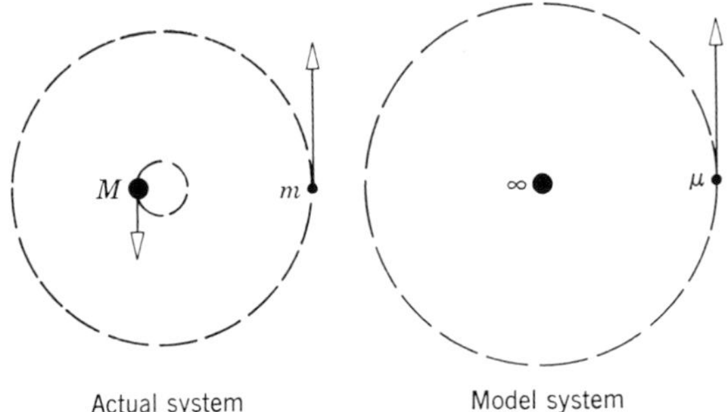


**FIGURE 7-1**

*Left:* In an actual one-electron atom, an electron of mass $m$ and nucleus of mass $M$ move about their fixed center of mass. *Right:* In the equivalent model atom, a particle of reduced mass $\mu$ moves about a stationary nucleus of infinite mass.

4-7, models the actual atom by an atom in which the nucleus is infinitely massive and the electron has the *reduced mass* $\mu$ given by

$$\mu = \left(\frac{M}{m + M}\right) m \tag{7-1}$$

where $m$ is the true mass of the electron and $M$ is the true mass of the nucleus. The reduced mass electron moves about the infinitely massive nucleus with the same electron-nucleus separation as in the actual atom. Since the infinitely massive nucleus must be completely stationary, it is necessary to treat *only* the motion of the reduced mass electron in the model atom, and the problem is therefore simplified from one involving a pair of moving particles to one involving only a single moving particle.

In classical mechanics, the motion of the reduced mass electron about the stationary nucleus in the model atom exactly duplicates the motion of the electron relative to the nucleus in the actual atom. Furthermore, the total energy of the model atom, which is just the total energy of its reduced mass electron, equals the total energy of the actual atom in a frame of reference in which its center of mass is at rest. The student may have seen a proof of these results of classical mechanics in connection with the motion of a planet about the sun, or some other system involving the motion of two particles. It is not difficult to prove that the same results are obtained in quantum mechanics, but we shall not bother to do so here. Figure 7-1 indicates the behavior of the electron and the nucleus in the actual atom and in the model atom. In both cases the center of mass of the atom is at rest.

## 7-2 Development of the Schroedinger Equation

We consider, therefore, an electron of reduced mass $\mu$ which is moving under the influence of the *Coulomb potential*

$$V = V(x,y,z) = \frac{-Ze^2}{4\pi\epsilon_0\sqrt{x^2 + y^2 + z^2}} \tag{7-2}$$

where $x$, $y$, $z$ are the rectangular coordinates of the electron of charge $-e$ relative to the nucleus, which is fixed at the origin. The square root in the denominator is just the electron-nucleus separation distance $r$. The nuclear charge is $+Ze$ ($Z = 1$ for neutral hydrogen, $Z = 2$ for singly ionized helium, etc.).

As a first step, we must develop the Schroedinger equation for this three-dimensional system. We do this by using the procedure indicated in Section 5-4. We first write the classical expression for the total energy $E$ of the system

$$\frac{1}{2\mu}(p_x^2 + p_y^2 + p_z^2) + V(x,y,z) = E \tag{7-3}$$

The quantities $p_x$, $p_y$, $p_z$ are the $x$, $y$, $z$ components of the linear momentum of the electron. Thus the first term on the left is the kinetic energy of the system, while the second term is its potential energy. Now we replace the *dynamical quantities* $p_x$, $p_y$, $p_z$, and $E$, by their associated *differential operators*, using an obvious three-dimensional extension of the scheme in (5-32). This gives us the *operator equation*

$$-\frac{\hbar^2}{2\mu}\left(\frac{\partial^2}{\partial x^2} + \frac{\partial^2}{\partial y^2} + \frac{\partial^2}{\partial z^2}\right) + V(x,y,z) = i\hbar\frac{\partial}{\partial t} \tag{7-4}$$

Operating with each term on the *wave function*

$$\Psi = \Psi(x,y,z,t) \tag{7-5}$$

we obtain the *Schroedinger equation* for the system

$$-\frac{\hbar^2}{2\mu}\left[\frac{\partial^2\Psi(x,y,z,t)}{\partial x^2} + \frac{\partial^2\Psi(x,y,z,t)}{\partial y^2} + \frac{\partial^2\Psi(x,y,z,t)}{\partial z^2}\right] + V(x,y,z)\Psi(x,y,z,t) = i\hbar\frac{\partial\Psi(x,y,z,t)}{\partial t} \tag{7-6}$$

It is often convenient to write this as

$$-\frac{\hbar^2}{2\mu}\nabla^2\Psi + V\Psi = i\hbar\frac{\partial\Psi}{\partial t} \tag{7-7}$$

where we use the symbol

$$\nabla^2 = \frac{\partial^2}{\partial x^2} + \frac{\partial^2}{\partial y^2} + \frac{\partial^2}{\partial z^2} \tag{7-8}$$

which is called the *Laplacian operator*, or "del squared," in rectangular coordinates.

Many of the properties of the three-dimensional Schroedinger equation, and of the wave functions which are its solutions, can be obtained by obvious extensions of the properties developed in the preceding chapters. For instance, it is easy to show by the technique of separation of variables, used in Section 5-5, that since the potential function $V(x,y,z)$ does not depend on time there are solutions to the Schroedinger equation which have the form

$$\Psi(x,y,z,t) = \psi(x,y,z)e^{-iEt/\hbar} \tag{7-9}$$

where the *eigenfunction* $\psi(x,y,z)$ is a solution to the *time-independent Schroedinger equation*

$$-\frac{\hbar^2}{2\mu}\nabla^2\psi(x,y,z) + V(x,y,z)\psi(x,y,z) = E\psi(x,y,z) \tag{7-10}$$

Note that in three dimensions this equation is a *partial* differential equation because it contains three independent variables, the space coordinates $x$, $y$, $z$.

## 7-3 Separation of the Time-Independent Equation

The time-independent Schroedinger equation for the Coulomb potential can be solved by making repeated applications of the technique of separation of variables to split

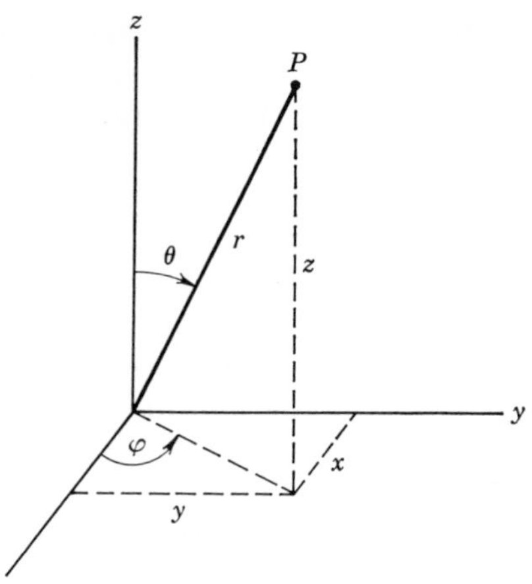


**FIGURE 7-2**
The spherical coordinates $r$, $\theta$, $\varphi$ of a point $P$, and its rectangular coordinates $x$, $y$, $z$.

the partial differential equation into a set of three ordinary differential equations, each involving only one coordinate, and then using standard procedures to solve these equations. However, separation of variables *cannot* be carried out when *rectangular coordinates* are employed because the Coulomb potential energy is a function $V(x,y,z) = -Ze^2/4\pi\epsilon_0\sqrt{x^2 + y^2 + z^2}$ of all three of these coordinates. Separation of variables will not work in rectangular coordinates because the potential itself cannot be split into terms, each of which involves only one such coordinate.

The difficulty is removed by changing to *spherical polar coordinates*. These are the coordinates $r$, $\theta$, $\varphi$, illustrated in Figure 7-2. The length of the straight line connecting the electron with the origin (the nucleus) is $r$, and $\theta$ and $\varphi$ are the polar and azimuthal angles specifying the orientation of that line. Now the distance between the electron and the nucleus is just $r$. So in spherical polar coordinates the Coulomb potential can be expressed as a function of a single coordinate $r = \sqrt{x^2 + y^2 + z^2}$, as follows

$$V = V(r) = \frac{-Ze^2}{4\pi\epsilon_0 r} \tag{7-11}$$

Because of this great simplification in the form of the potential, it then becomes possible to carry out the separation of variables on the time-independent Schroedinger equation, as we shall soon see.

The space derivatives in the time-independent Schroedinger equation also change form when the coordinates are changed from rectangular to spherical. A straightforward, but tedious, application of the rules of differential calculus shows that the time-independent Schroedinger equation can be written as

$$-\frac{\hbar^2}{2\mu}\nabla^2\psi(r,\theta,\varphi) + V(r)\psi(r,\theta,\varphi) = E\psi(r,\theta,\varphi) \tag{7-12}$$

where

$$\nabla^2 = \frac{1}{r^2}\frac{\partial}{\partial r}\left(r^2\frac{\partial}{\partial r}\right) + \frac{1}{r^2\sin^2\theta}\frac{\partial^2}{\partial\varphi^2} + \frac{1}{r^2\sin\theta}\frac{\partial}{\partial\theta}\left(\sin\theta\frac{\partial}{\partial\theta}\right) \tag{7-13}$$

is the Laplacian operator in the spherical polar coordinates $r$, $\theta$, $\varphi$. For the details of the coordinate transformation leading to (7-12) and (7-13), the student should consult Appendix I. A comparison of the forms of the Laplacian operator in rectangular and spherical polar coordinates, (7-8) and (7-13), shows that we have simplified the expression of the potential energy function at the expense of considerably complicating the expression of the Laplacian operator in the time-independent Schroedinger equation that must be solved.

Nevertheless, the change of coordinates is worthwhile because it will allow us to find solutions to the time-independent Schroedinger equation of the form

$$\psi(r,\theta,\varphi) = R(r)\Theta(\theta)\Phi(\varphi) \tag{7-14}$$

That is, we shall show that there are solutions $\psi(r,\theta,\varphi)$ to (7-12) that split into products of three functions, $R(r)$, $\Theta(\theta)$, and $\Phi(\varphi)$, each of which depends on only one of the coordinates. The advantage lies in the fact that these three functions can be found by solving *ordinary* differential equations. We show this by substituting the product form, $\psi(r,\theta,\varphi) = R(r)\Theta(\theta)\Phi(\varphi)$, into the time-independent Schroedinger equation obtained by evaluating the Laplacian operator in (7-12) from (7-13). This yields

$$-\frac{\hbar^2}{2\mu}\left[\frac{1}{r^2}\frac{\partial}{\partial r}\left(r^2\frac{\partial R\Theta\Phi}{\partial r}\right) + \frac{1}{r^2\sin^2\theta}\frac{\partial^2 R\Theta\Phi}{\partial\varphi^2} + \frac{1}{r^2\sin\theta}\frac{\partial}{\partial\theta}\left(\sin\theta\frac{\partial R\Theta\Phi}{\partial\theta}\right)\right] + V(r)R\Theta\Phi = ER\Theta\Phi$$

Carrying out the partial differentiations, we have

$$-\frac{\hbar^2}{2\mu}\left[\frac{\Theta\Phi}{r^2}\frac{d}{dr}\left(r^2\frac{dR}{dr}\right) + \frac{R\Theta}{r^2\sin^2\theta}\frac{d^2\Phi}{d\varphi^2} + \frac{R\Phi}{r^2\sin\theta}\frac{d}{d\theta}\left(\sin\theta\frac{d\Theta}{d\theta}\right)\right] + V(r)R\Theta\Phi = ER\Theta\Phi$$

In this equation we have written the partial derivative $\partial R/\partial r$ as the total derivative $dR/dr$ since the two are equivalent because $R$ is a function of $r$ alone. The same comment applies to the other derivatives. If we now multiply through by $-2\mu r^2\sin^2\theta/R\Theta\Phi\hbar^2$, and transpose, we obtain

$$\frac{1}{\Phi}\frac{d^2\Phi}{d\varphi^2} = -\frac{\sin^2\theta}{R}\frac{d}{dr}\left(r^2\frac{dR}{dr}\right) - \frac{\sin\theta}{\Theta}\frac{d}{d\theta}\left(\sin\theta\frac{d\Theta}{d\theta}\right) - \frac{2\mu}{\hbar^2}r^2\sin^2\theta[E - V(r)]$$

As the left side of this equation does not depend on $r$ or $\theta$, whereas the right side does not depend on $\varphi$, their common value cannot depend on any of these variables. The common value must therefore be a constant, which we shall find it convenient to designate as $-m_l^2$. Thus we obtain two equations by setting each side equal to this constant

$$\frac{d^2\Phi}{d\varphi^2} = -m_l^2\Phi \tag{7-15}$$

and

$$-\frac{1}{R}\frac{d}{dr}\left(r^2\frac{dR}{dr}\right) - \frac{1}{\Theta\sin\theta}\frac{d}{d\theta}\left(\sin\theta\frac{d\Theta}{d\theta}\right) - \frac{2\mu}{\hbar^2}r^2[E - V(r)] = -\frac{m_l^2}{\sin^2\theta}$$

By transposing, we can rewrite the second equation as

$$\frac{1}{R}\frac{d}{dr}\left(r^2\frac{dR}{dr}\right) + \frac{2\mu r^2}{\hbar^2}[E - V(r)] = \frac{m_l^2}{\sin^2\theta} - \frac{1}{\Theta\sin\theta}\frac{d}{d\theta}\left(\sin\theta\frac{d\Theta}{d\theta}\right)$$

Since we have here an equation whose left side does not depend on one of the variables and whose right side does not depend on the other, we conclude again that both sides must equal a constant. It is convenient to designate this constant as $l(l+1)$. Thus we obtain, by setting each side equal to $l(l+1)$, two more equations

$$-\frac{1}{\sin\theta}\frac{d}{d\theta}\left(\sin\theta\frac{d\Theta}{d\theta}\right) + \frac{m_l^2\Theta}{\sin^2\theta} = l(l+1)\Theta \tag{7-16}$$

and

$$\frac{1}{r^2}\frac{d}{dr}\left(r^2\frac{dR}{dr}\right) + \frac{2\mu}{\hbar^2}[E - V(r)]R = l(l+1)\frac{R}{r^2} \tag{7-17}$$

We see that the assumed product form of the solution, $\psi(r,\theta,\varphi) = R(r)\Theta(\theta)\Phi(\varphi)$, is valid because it works! We also see that the problem has been reduced to that of solving the ordinary differential equations, (7-15), (7-16), and (7-17), for $\Phi(\varphi)$, $\Theta(\theta)$, and $R(r)$.

In solving these equations, we shall find that the equation for $\Phi(\varphi)$ has *acceptable* solutions only for *certain values of* $m_l$. Using these values of $m_l$ in the equation for $\Theta(\theta)$, it turns out that this equation has *acceptable* solutions only for *certain values of* $l$. With these values of $l$ in the equation for $R(r)$, this equation is found to have *acceptable* solutions only for *certain values of the total energy* $E$; that is, the energy of the atom is quantized.

## 7-4 Solution of the Equations

Consider (7-15) for $\Phi(\varphi)$. By differentiation and substitution, the student may easily verify that it has a particular solution

$$\Phi(\varphi) = e^{im_l\varphi}$$

(The discussion following Example 7-5 explains why this particular solution is used.) Here we must, for the first time, explicitly consider the requirement of Section 5-6 that the eigenfunctions be *single valued.* This demands that the function $\Phi(\varphi)$ be single valued, and the demand must be considered explicitly because the azimuthal angles $\varphi = 0$ and $\varphi = 2\pi$ are actually the same angle. Thus, we must require that $\Phi(\varphi)$ has the same value at $\varphi = 0$ as it does at $\varphi = 2\pi$, that is

$$\Phi(0) = \Phi(2\pi)$$

Evaluating the exponential in the particular solution $\Phi(\varphi)$, we obtain

$$e^{im_l 0} = e^{im_l 2\pi}$$

or

$$1 = \cos m_l 2\pi + i \sin m_l 2\pi$$

The requirement is satisfied only if the absolute value of $m_l$ has one of the values

$$|m_l| = 0, 1, 2, 3, \ldots \tag{7-18}$$

In other words, $m_l$ can be only a positive or negative *integer*. Thus the set of functions which are *acceptable* solutions to (7-15) are

$$\Phi_{m_l}(\varphi) = e^{im_l\varphi} \tag{7-19}$$

where $m_l$ has one of the integral values specified by (7-18). The *quantum number* $m_l$ is used as a subscript to identify the specific form of an acceptable solution.

In solving (7-16) for the functions $\Theta(\theta)$, the procedure is very nearly the same as that used to obtain analytical solutions of the time-independent Schroedinger equation for the simple harmonic oscillator potential. Interested students are referred to Appendix H, which explains this quite lengthy procedure. Here we shall only quote the results. It is found that solutions to (7-16) which are *acceptable* (remain finite) are obtained only if the constant $l$ is equal to one of the *integers*

$$l = |m_l|, |m_l| + 1, |m_l| + 2, |m_l| + 3, \ldots \tag{7-20}$$

The acceptable solutions can be written

$$\Theta_{lm_l}(\theta) = \sin^{|m_l|}\theta\, F_{l|m_l|}(\cos\theta) \tag{7-21}$$

The $F_{l|m_l|}(\cos\theta)$ are polynomials in $\cos\theta$, which have forms that depend on the value of the *quantum number* $l$ and on the absolute value of the quantum number $m_l$. Thus it is necessary to use both of these quantum numbers to identify the functions $\Theta_{lm_l}(\theta)$ that are acceptable solutions to the equation. Examples of these functions will be presented in Section 7-6.

The procedure used in the solution of (7-17) for the functions $R(r)$ is also very similar to that used for the simple harmonic oscillator potential. It is found that there are bound-state solutions which are *acceptable* (remain finite) only if the constant $E$ (the total energy) has one of the values $E_n$, where

$$E_n = -\frac{\mu Z^2 e^4}{(4\pi\epsilon_0)^2 2\hbar^2 n^2} \tag{7-22}$$

In this expression the *quantum number n* is one of the *integers*

$$n = l+1, l+2, l+3, \ldots \tag{7-23}$$

The acceptable solutions are most conveniently written as

$$R_{nl}(r) = e^{-Zr/na_0}\left(\frac{Zr}{a_0}\right)^l G_{nl}\left(\frac{Zr}{a_0}\right) \tag{7-24}$$

where the parameter $a_0$ is

$$a_0 = \frac{4\pi\epsilon_0\hbar^2}{\mu e^2} \tag{7-25}$$

The $G_{nl}(Zr/a_0)$ are polynomials in $Zr/a_0$, with different forms for different values of $n$ and $l$. Thus both of these quantum numbers are required to identify the different functions $R_{nl}(r)$ that are acceptable solutions to the equation. But the allowed values $E_n$ of the total energy carry only the quantum number $n$ as a label since they depend only on the value of that quantum number. Examples of the functions $R_{nl}(r)$ will be presented in Section 7-6.

## 7-5 Eigenvalues, Quantum Numbers, and Degeneracy

One of the important results of the Schroedinger theory of the one-electron atom is the prediction of (7-22) for the allowed values of total energy of the bound states of the atom. Comparing this prediction for the eigenvalues

$$E_n = -\frac{\mu Z^2 e^4}{(4\pi\epsilon_0)^2 2\hbar^2 n^2} = -\frac{13.6 \text{ eV}}{n^2}$$

with the predictions of the Bohr model (see (4-18)), we find that *identical* allowed energies are predicted by these treatments. Both predictions are in excellent agreement with experiment. Schroedinger's derivation of (7-22) provided the first convincing verification of his theory of quantum mechanics. Figure 7-3 illustrates the Coulomb potential $V(r)$ for the one-electron atom, and its *eigenvalues* $E_n$.

What is the relation between the Coulomb potential and its eigenvalues, and the potentials studied in Chapter 6 and their eigenvalues? One obvious difference is that the quantum mechanical calculations leading to the eigenvalues of the Coulomb potential are appreciably more complicated. But the Coulomb potential is an exact description of a real three-dimensional system. The potentials previously treated are approximate descriptions of idealized one-dimensional systems, which are designed to simplify the calculations. Part of the complication for the Coulomb potential is also due to its spherical symmetry, which forces the use of spherical polar coordinates instead of rectangular coordinates.

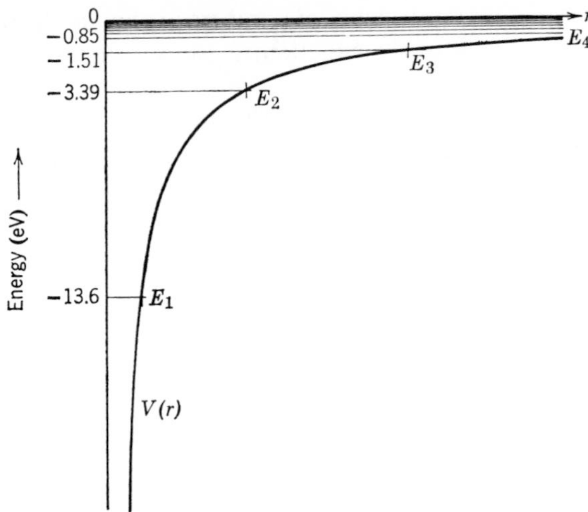


**FIGURE 7-3**

The Coulomb potential $V(r)$ and its eigenvalues $E_n$. For large values of $n$ the eigenvalues become very closely spaced in energy since $E_n$ approaches zero as $n$ approaches infinity. Note that the intersection of $V(r)$ and $E_n$, which defines the location of one end of the classically allowed region, moves out as $n$ increases. Not shown in this figure is the continuum of eigenvalues at positive energies corresponding to unbound states.

The similarities are much more fundamental than the differences. For the Coulomb potential, as for any other binding potential, the allowed total energies of a particle bound to the potential are *discretely quantized.* Figure 7-4 makes a comparison between the allowed energies for a Coulomb potential and for several one-dimensional binding potentials. In this figure the Coulomb potential represents a crosscut along a diameter through the one-electron atom. Note that all the binding potentials have a *zero-point energy*. That is, in all cases the lowest allowed value of total energy lies above the minimum value of the potential energy. Associated with its zero-point energy, the one-electron atom has a zero-point motion like other systems described by binding potentials. In the following section we shall see that this phenomena can give us a basic explanation of the stability of the ground state of the atom.

Although the eigenvalues of the one-electron atom depend on only the quantum number $n$, the eigenfunctions depend on all three quantum numbers $n$, $l$, $m_l$ since they are products of the three functions $R_{nl}(r)$, $\Theta_{lm_l}(\theta)$, and $\Phi_{m_l}(\varphi)$. *The fact that three*

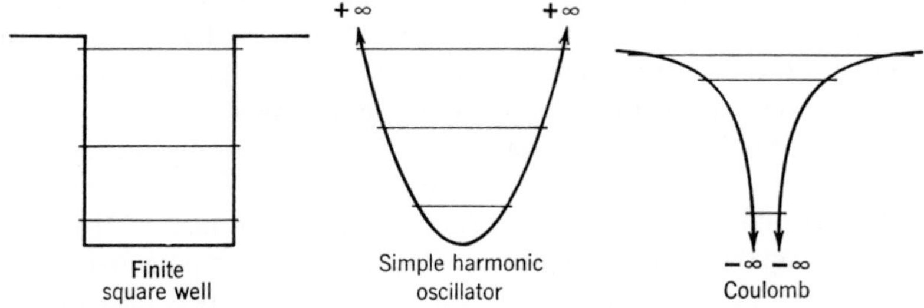


**FIGURE 7-4**

A comparison between the allowed energies of several binding potentials. The three-dimensional Coulomb potential is shown in a cross-sectional view along a diameter; the other potentials are one-dimensional.

*quantum numbers arise is a consequence of the fact that the time-independent Schroedinger equation contains three independent variables, one for each space coordinate.* Gathering together the conditions which the quantum numbers satisfy, we have

$$\begin{aligned} |m_l| &= 0, 1, 2, 3, \ldots \\ l &= |m_l|, |m_l| + 1, |m_l| + 2, |m_l| + 3, \ldots \\ n &= l + 1, l + 2, l + 3, \ldots \end{aligned} \tag{7-26}$$

These conditions are more conveniently expressed as

$$\begin{aligned} n &= 1, 2, 3, \ldots \\ l &= 0, 1, 2, \ldots, n - 1 \\ m_l &= -l, -l + 1, \ldots, 0, \ldots, +l - 1, l \end{aligned} \tag{7-27}$$

**Example 7-1.** Show that the conditions of (7-27) are equivalent to those of (7-26).

According to (7-26) the minimum value of $l$ is equal to $|m_l|$, and the minimum value of $|m_l|$ is 0. Thus the minimum value of $l$ is 0 and the minimum value of $n$, which is equal to $l + 1$, is $0 + 1 = 1$. Since $n$ increases by integers without limit, the possible values of $n$ are $n = 1, 2, 3, \ldots$. For a given $n$, the maximum value of $l$ is the one satisfying the relation $n = l + 1$, that is, $l = n - 1$. Consequently the possible values of $l$ are $l = 0, 1, 2, \ldots, n - 1$. Finally, for a given $l$, the largest value which $|m_l|$ can assume is $|m_l| = l$. Thus the maximum value of $m_l$ is $+l$ and the minimum value is $-l$, and it can assume only the values $m_l = -l, -l + 1, \ldots, 0, \ldots, +l - 1, +l$. ◀

Because of its role in specifying the total energy of the atom, $n$ is sometimes called the *principal* quantum number. Because the azimuthal, or orbital, angular momentum of the atom depends on $l$, as we shall soon see, $l$ is sometimes called the *azimuthal* quantum number. We shall also see that if the atom is in an external magnetic field there is a dependence of its energy on $m_l$. Consequently, $m_l$ is sometimes called the *magnetic* quantum number.

The conditions of (7-27) make it apparent that for a given value of $n$ there are generally several different possible values of $l$ and $m_l$. Since the form of the eigenfunctions depends on all three quantum numbers, it is apparent that there will be situations in which two or more completely different eigenfunctions correspond to exactly the same eigenvalue $E_n$. As the eigenfunctions described the behavior of the atom, we see that it has states with *completely different behavior* that nevertheless have the *same total energy*. In physics the word used to characterize this phenomenon is *degeneracy*, and eigenfunctions corresponding to the same eigenvalue are said to be *degenerate*. There is little relation to the common usage of the word; degenerate eigenfunctions are not at all reprehensible!

Degeneracy also occurs in classical mechanics and in the related old quantum theory. In the discussion of elliptical orbits of the Bohr-Sommerfeld atom in Section 4-10, we indicated that the total energy of the atom is independent of the semiminor axis of the ellipse. Thus the atom has states with very different behavior, that is, with the electron traveling in very different orbits, which nevertheless have the same total energy. Exactly the same phenomenon occurs in planetary motion. This classical degeneracy is comparable to the $l$ degeneracy that arises in the quantum mechanical one-electron atom. The energy of a Bohr-Sommerfeld atom, or of a planetary system, is also independent of the orientation in space of the plane of the orbit. This is comparable to the $m_l$ degeneracy of the quantum mechanical atom.

In either classical or quantum mechanics, degeneracy is a result of certain properties of the potential energy function that describes the system. In the quantum

**TABLE 7-1.** Possible Values of $l$ and $m_l$ for $n = 1, 2, 3$

<table>
<tr><td>$n$</td><td>1</td><td colspan="2">2</td><td colspan="3">3</td></tr>
<tr><td>$l$</td><td>0</td><td>0</td><td>1</td><td>0</td><td>1</td><td>2</td></tr>
<tr><td>$m_l$</td><td>0</td><td>0</td><td>$-1, 0, +1$</td><td>0</td><td>$-1, 0, +1$</td><td>$-2, -1, 0, +1, +2$</td></tr>
<tr><td>Number of degenerate eigenfunctions for each $l$</td><td>1</td><td>1</td><td>3</td><td>1</td><td>3</td><td>5</td></tr>
<tr><td>Number of degenerate eigenfunctions for each $n$</td><td>1</td><td colspan="2">4</td><td colspan="3">9</td></tr>
</table>

mechanical one-electron atom, the degeneracy with respect to $m_l$ arises because the potential depends only on the coordinate $r$, so the potential is spherically symmetrical and the total energy of the atom is independent of its orientation in space. The $l$ degeneracy is a consequence of the particular form of the $r$ dependence of the Coulomb potential.

If an external magnetic field is applied to the atom, then its total energy *will* depend on its orientation in space because of an interaction between currents in the atom and the applied field. We shall study this later, and we shall find that the orientation in space is determined by the quantum number $m_l$. Thus in an external magnetic field the degeneracy with respect to $m_l$ is removed and the atom has different energy levels for different $m_l$ values. If the external magnetic field is gradually reduced in intensity, the dependence of the total energy of the atom on $m_l$ is reduced in proportion. When the field is reduced to zero the energy levels that correspond to different values of $m_l$ degenerate into a single energy level, and the corresponding eigenfunctions become degenerate.

Many properties of alkali atoms can be discussed in terms of the motion of a single "valence" electron in a potential which is spherically symmetrical, but which does not have the $1/r$ behavior of the Coulomb potential. The energy of this electron *does* depend on $l$. Thus the degeneracy with respect to $l$ is removed if the form of the $r$ dependence of the potential is changed. We shall study this phenomenon on a number of occasions later in this book, and in the process more insight into the origin of the $l$ degeneracy of the Coulomb potential will be obtained.

From (7-27) it is easy to see how many degenerate eigenfunctions there are, for an isolated one-electron atom, which correspond to a particular eigenvalue $E_n$. The possible values of the quantum numbers for $n = 1$, 2, and 3 are shown in Table 7-1. Inspection of this table makes it apparent that:

1. For each value of $n$, there are $n$ possible values of $l$.
2. For each value of $l$, there are $(2l + 1)$ possible values of $m_l$.
3. For each value of $n$, there are a total of $n^2$ degenerate eigenfunctions.

## 7-6 Eigenfunctions

The mathematical techniques used in quantum mechanics to obtain (7-22) for the eigenvalues of the one-electron atom are, admittedly, quite complicated compared to

**TABLE 7-2.** Some Eigenfunctions for the One-Electron Atom

| Quantum Numbers $n$ | $l$ | $m_l$ | Eigenfunctions |
|---|---|---|---|
| 1 | 0 | 0 | $\psi_{100} = \frac{1}{\sqrt{\pi}}\left(\frac{Z}{a_0}\right)^{3/2} e^{-Zr/a_0}$ |
| 2 | 0 | 0 | $\psi_{200} = \frac{1}{4\sqrt{2\pi}}\left(\frac{Z}{a_0}\right)^{3/2}\left(2 - \frac{Zr}{a_0}\right) e^{-Zr/2a_0}$ |
| 2 | 1 | 0 | $\psi_{210} = \frac{1}{4\sqrt{2\pi}}\left(\frac{Z}{a_0}\right)^{3/2} \frac{Zr}{a_0} e^{-Zr/2a_0} \cos\theta$ |
| 2 | 1 | $\pm1$ | $\psi_{21\pm1} = \frac{1}{8\sqrt{\pi}}\left(\frac{Z}{a_0}\right)^{3/2} \frac{Zr}{a_0} e^{-Zr/2a_0} \sin\theta e^{\pm i\varphi}$ |
| 3 | 0 | 0 | $\psi_{300} = \frac{1}{81\sqrt{3\pi}}\left(\frac{Z}{a_0}\right)^{3/2}\left(27 - 18\frac{Zr}{a_0} + 2\frac{Z^2r^2}{a_0^2}\right) e^{-Zr/3a_0}$ |
| 3 | 1 | 0 | $\psi_{310} = \frac{\sqrt{2}}{81\sqrt{\pi}}\left(\frac{Z}{a_0}\right)^{3/2}\left(6 - \frac{Zr}{a_0}\right)\frac{Zr}{a_0} e^{-Zr/3a_0} \cos\theta$ |
| 3 | 1 | $\pm1$ | $\psi_{31\pm1} = \frac{1}{81\sqrt{\pi}}\left(\frac{Z}{a_0}\right)^{3/2}\left(6 - \frac{Zr}{a_0}\right)\frac{Zr}{a_0} e^{-Zr/3a_0} \sin\theta e^{\pm i\varphi}$ |
| 3 | 2 | 0 | $\psi_{320} = \frac{1}{81\sqrt{6\pi}}\left(\frac{Z}{a_0}\right)^{3/2} \frac{Z^2r^2}{a_0^2} e^{-Zr/3a_0}(3\cos^2\theta - 1)$ |
| 3 | 2 | $\pm1$ | $\psi_{32\pm1} = \frac{1}{81\sqrt{\pi}}\left(\frac{Z}{a_0}\right)^{3/2} \frac{Z^2r^2}{a_0^2} e^{-Zr/3a_0} \sin\theta\cos\theta e^{\pm i\varphi}$ |
| 3 | 2 | $\pm2$ | $\psi_{32\pm2} = \frac{1}{162\sqrt{\pi}}\left(\frac{Z}{a_0}\right)^{3/2} \frac{Z^2r^2}{a_0^2} e^{-Zr/3a_0} \sin^2\theta e^{\pm 2i\varphi}$ |

those used in the Bohr model to obtain the same equation. Putting aside questions concerning the logical consistency of the postulates of the Bohr model, it is still reasonable to question whether all the extra work involved in the quantum mechanical treatment of the one-electron atom is justified by the results obtained. The answer is, overwhelmingly, yes! We can now find out much more about the one-electron atom than we possibly can from the Bohr model, because we have the *eigenfunctions* as well as the eigenvalues. The eigenfunctions contain a wealth of additional information about the properties of the atom. The remainder of this chapter, and the following chapter, will be devoted largely to studying the eigenfunctions and extracting this information from them.

We know that the eigenfunctions are formed by taking the product

$$\psi_{nlm_l}(r,\theta,\varphi) = R_{nl}(r)\Theta_{lm_l}(\theta)\Phi_{m_l}(\varphi)$$

We also know, from (7-19), (7-21), and (7-23) that for any bound state

$$\Phi_{m_l}(\varphi) = e^{im_l\varphi}$$

$$\Theta_{lm_l}(\theta) = \sin^{|m_l|}\theta\,(\text{polynomial in } \cos\theta)$$

and

$$R_{nl}(r) = e^{-(\text{constant})r/n}\, r^l\,(\text{polynomial in } r)$$

All the eigenfunctions have basically the same mathematical structure, except that with increasing values of $n$ and $l$ the polynomials in $r$ and $\cos\theta$ become increasingly more complicated. Table 7-2 lists the one-electron atom eigenfunctions for the first three values of $n$. They are expressed in terms of the parameter

$$a_0 = \frac{4\pi\epsilon_0\hbar^2}{\mu e^2} = 0.529 \times 10^{-10}\ \text{m} = 0.529\ \text{Å}$$

which the student may recognize as equal to the radius of the smallest orbit of a Bohr hydrogen atom. The multiplicative constant in front of each eigenfunction has been adjusted so that it is normalized. In other words, the integral over all space of the corresponding probability density function equals one, so that in each quantum state there is probability one of finding the atomic electron somewhere.

**Example 7-2.** Verify that the eigenfunction $\psi_{211}$, and the associated eigenvalue $E_2$, satisfy the time-independent Schroedinger equation, (7-12), for the one-electron atom with $Z = 1$.

Since the differential equation is linear in $\psi$, for the purposes of this verification we can ignore completely the multiplicative constant $1/8\pi^{1/2}a_0^{5/2}$, and write the eigenfunction as

$$\psi = re^{-r/2a_0}\sin\theta e^{i\varphi}$$

This is the simplest case with a nontrivial dependence on all three coordinates. Nevertheless, the verification of this case should give the student some confidence in the validity of all the eigenfunctions quoted in Table 7-2.

Before beginning, let us introduce the convenient notation

$$\psi = f(r,\varphi)\sin\theta = f\sin\theta$$

and

$$\psi = g(\theta,\varphi)\, re^{-r/2a_0} = gre^{-r/2a_0}$$

This notation will be useful in evaluating the derivatives that enter in (7-12), which is

$$-\frac{\hbar^2}{2\mu}\left[\frac{1}{r^2}\frac{\partial}{\partial r}\left(r^2\frac{\partial\psi}{\partial r}\right) + \frac{1}{r^2\sin^2\theta}\frac{\partial^2\psi}{\partial\varphi^2} + \frac{1}{r^2\sin\theta}\frac{\partial}{\partial\theta}\left(\sin\theta\frac{\partial\psi}{\partial\theta}\right)\right] + V\psi = E\psi$$

First we calculate

$$\frac{\partial\psi}{\partial\theta} = \frac{\partial}{\partial\theta}(f\sin\theta) = f\cos\theta$$

$$\sin\theta\frac{\partial\psi}{\partial\theta} = f\sin\theta\cos\theta$$

$$\frac{\partial}{\partial\theta}\left(\sin\theta\frac{\partial\psi}{\partial\theta}\right) = f(\cos^2\theta - \sin^2\theta)$$

$$\frac{1}{r^2\sin\theta}\frac{\partial}{\partial\theta}\left(\sin\theta\frac{\partial\psi}{\partial\theta}\right) = \frac{f}{r^2}\left(\frac{\cos^2\theta - \sin^2\theta}{\sin\theta}\right)$$

Next we calculate

$$\frac{\partial^2\psi}{\partial\varphi^2} = (i)^2\psi = -\psi = -f\sin\theta$$

$$\frac{1}{r^2\sin^2\theta}\frac{\partial^2\psi}{\partial\varphi^2} = -\frac{f}{r^2\sin\theta}$$

Adding these two results, we obtain

$$\frac{1}{r^2\sin^2\theta}\frac{\partial^2\psi}{\partial\varphi^2} + \frac{1}{r^2\sin\theta}\frac{\partial}{\partial\theta}\left(\sin\theta\frac{\partial\psi}{\partial\theta}\right)$$

$$= \frac{f}{r^2\sin\theta}(\cos^2\theta - \sin^2\theta - 1) = -\frac{2f\sin^2\theta}{r^2\sin\theta} = -\frac{2f\sin\theta}{r^2} = -\frac{2\psi}{r^2}$$

Then we calculate

$$\frac{\partial \psi}{\partial r} = g\left(e^{-r/2a_0} - \frac{r}{2a_0}e^{-r/2a_0}\right)$$

$$r^2\frac{\partial \psi}{\partial r} = g\left(r^2e^{-r/2a_0} - \frac{r^3}{2a_0}e^{-r/2a_0}\right)$$

$$\frac{\partial}{\partial r}\left(r^2\frac{\partial \psi}{\partial r}\right) = g\left(2re^{-r/2a_0} - \frac{r^2}{2a_0}e^{-r/2a_0} - \frac{3r^2}{2a_0}e^{-r/2a_0} + \frac{r^3}{4a_0^2}e^{-r/2a_0}\right)$$

$$= 2gre^{-r/2a_0}\left(1 - \frac{r}{a_0} + \frac{r^2}{8a_0^2}\right) = 2\left(1 - \frac{r}{a_0} + \frac{r^2}{8a_0^2}\right)\psi$$

$$\frac{1}{r^2}\frac{\partial}{\partial r}\left(r^2\frac{\partial \psi}{\partial r}\right) = 2\left(\frac{1}{r^2} - \frac{1}{ra_0} + \frac{1}{8a_0^2}\right)\psi$$

Substituting this term, and the term coming from the $\theta$ and $\varphi$ derivatives, into the differential equation that is supposed to be satisfied, we obtain

$$-\frac{\hbar^2}{2\mu}\left[2\left(\frac{1}{r^2} - \frac{1}{ra_0} + \frac{1}{8a_0^2}\right) - \frac{2}{r^2}\right]\psi + V\psi = E\psi$$

or

$$\frac{\hbar^2}{\mu a_0}\left(\frac{1}{r} - \frac{1}{8a_0}\right) + V = E$$

Now

$$E = E_2 = -\frac{\mu e^4}{8(4\pi\epsilon_0)^2\hbar^2}$$

Also

$$V = -\frac{e^2}{4\pi\epsilon_0 r}$$

and

$$a_0 = \frac{4\pi\epsilon_0\hbar^2}{\mu e^2}$$

So we have

$$\frac{\hbar^2}{\mu}\frac{\mu e^2}{4\pi\epsilon_0\hbar^2}\left(\frac{1}{r} - \frac{\mu e^2}{8(4\pi\epsilon_0)\hbar^2}\right) - \frac{e^2}{4\pi\epsilon_0 r} = -\frac{\mu e^4}{8(4\pi\epsilon_0)^2\hbar^2}$$

Since inspection demonstrates that this equation is satisfied identically, we have completed the verification. ◀

## 7-7 Probability Densities

We begin to extract information from the one-electron atom eigenfunctions by studying the forms of the corresponding *probability density functions*

$$\Psi^*\Psi = \psi^*_{nlm_l}e^{iE_nt/\hbar}\psi_{nlm_l}e^{-iE_nt/\hbar} = \psi^*_{nlm_l}\psi_{nlm_l} = R^*_{nl}\Theta^*_{lm_l}\Phi^*_{m_l}R_{nl}\Theta_{lm_l}\Phi_{m_l}$$

As these are functions of three coordinates, we cannot directly plot them in two dimensions. Nevertheless, we can study their three-dimensional behavior by considering separately their dependence on each coordinate. We treat first the $r$ dependence in terms of the *radial probability density* $P(r)$, defined so that $P(r)\,dr$ is the probability of finding the electron at any location with radial coordinate between $r$ and $r + dr$. By integrating the probability density $\Psi^*\Psi$, which is a probability per unit volume, over the volume enclosed between spheres of radii $r$ and $r + dr$, it is easy to show that

$$P_{nl}(r)\,dr = R^*_{nl}(r)R_{nl}(r)r^2\,dr \tag{7-28}$$

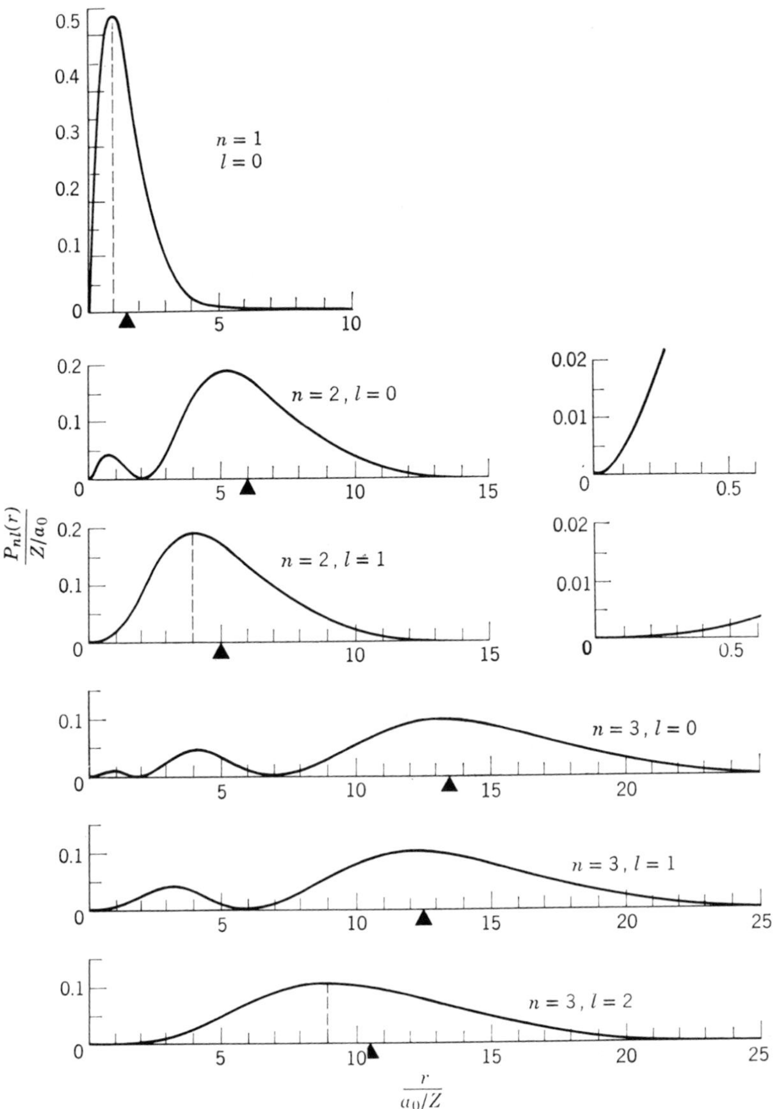


**FIGURE 7-5**
The radial probability density for the electron in a one-electron atom for $n = 1, 2, 3$ and the values of $l$ shown. The triangle on each abscissa indicates the value of $\overline{r_{nl}}$ as given by (7-29). For $n = 2$ the plots are redrawn with abscissa and ordinate scales expanded by a factor of 10 to show the behavior of $P_{nl}(r)$ near the origin. Note that in the three cases for which $l = l_{\text{max}} = n - 1$ the maximum of $P_{nl}(r)$ occurs at $r_{\text{Bohr}} = n^2a_0/Z$, which is indicated by the location of the dashed line.

The factor of $r^2$ is present on the right side because the volume enclosed between the spheres is proportional to that factor. The use of the quantum numbers $n$ and $l$ as labels to specify the form of a particular radial probability density function is obviously appropriate, but the form of these functions does not depend on the quantum number $m_l$. Figure 7-5 plots several $P_{nl}(r)$, using dimensionless quantities for each axis.

Inspection of the figure shows that the radial probability densities, for each set of the pertinent quantum numbers, have appreciable values only in reasonably restricted ranges of the radial coordinate. Thus, when the atom is in one of its quantum states, specified by a particular set of its quantum numbers, there is a high probability that the radial coordinate of the electron will be found within a reasonably restricted range. The electron would quite probably be found within a certain so-called *shell* contained within two concentric spheres centered on the nucleus. A study of the figure will demonstrate that the characteristic radii of these shells is determined primarily by the quantum number $n$, although there is a small $l$ dependence.

This property can be seen in a more quantitative way by using the expectation value of the radial coordinate of the electron to characterize the radius of the shell. An obvious extension of the arguments of Section 5-4 to three dimensions shows that the expectation value is given by the expression

$$\overline{r_{nl}} = \int_0^\infty r P_{nl}(r)\, dr$$

If the integral is evaluated, this yields

$$\overline{r_{nl}} = \frac{n^2 a_0}{Z}\left\{1 + \frac{1}{2}\left[1 - \frac{l(l+1)}{n^2}\right]\right\} \tag{7-29}$$

The values of $\overline{r_{nl}}$ are indicated in Figure 7-5 with small triangles. It is apparent that $\overline{r_{nl}}$ depends primarily on $n$, since the $l$ dependence is suppressed by the factor of 1/2 and the factor of $1/n^2$ in (7-29).

An interesting comparison can be made between (7-29) and (4-16)

$$r_{\text{Bohr}} = \frac{n^2 a_0}{Z}$$

which gives the radii of the circular orbits of a Bohr atom. Quantum mechanics shows that the radii of the shells are of approximately the same size as the radii of the circular Bohr orbits. These radii increase rapidly with increasing $n$. The basic reason is that the total energy $E_n$ of the atom becomes more positive with increasing $n$, so the region of the coordinate $r$ for which $E_n$ is greater than $V(r)$ expands with increasing $n$, as can be seen in Figure 7-3. That is, the shells expand with increasing $n$ because the classically allowed regions expand.

**Example 7-3.** (a) Calculate the location at which the radial probability density is a maximum for the ground state of the hydrogen atom. (b) Next calculate the expectation value for the radial coordinate in this state. (c) Then interpret these results in terms of the results of measurements of the location of the electron in the atom.

(a) The radial probability density for the $n = 1$, $l = 0$ ground state is

$$P_{10}(r) = R_{10}^*(r) R_{10}(r) r^2$$

We take $R_{10}(r)$ from the $r$-dependent factor of the first eigenfunction listed in Table 7-2, with $Z = 1$, and obtain

$$P_{10}(r) = e^{-r/a_0}\, e^{-r/a_0}\, r^2 = e^{-2r/a_0}\, r^2$$

We have ignored normalization (i.e., for simplicity taken the multiplicative constant equal to one) since it has no effect on what we are about to do. This is to find the maximum in $P_{10}(r)$ by evaluating its derivative with respect to $r$ and setting the result equal to zero. That is

$$\frac{dP_{10}(r)}{dr} = -\frac{2}{a_0} e^{-2r/a_0}\, r^2 + e^{-2r/a_0}\, 2r$$

$$= \left(1 - \frac{r}{a_0}\right) e^{-2r/a_0}\, 2r = 0$$

The solution to the equation we have obtained is

$$1 - \frac{r}{a_0} = 0$$

or

$$r = a_0$$

This is the location of the maximum in the radial probability density.

(b) To calculate the expectation value of the radial coordinate $r$, we evaluate (7-29), with $n = 1, l = 0$, and $Z = 1$. We obtain

$$\overline{r_{10}} = a_0\{1 + (1/2)[1]\} = 1.5a_0$$

(c) We have found that the expectation value of $r$ is somewhat larger than the value of $r$ at which the radial probability density is a maximum. The reason is that the radial probability density is asymmetrical about its maximum in such a way that there is a small, but not negligible, probability of finding fairly large values of $r$ in measurements of the location of the electron in the atom. So, although the most likely location of the electron is at $r = a_0$ (i.e., at the ground state Bohr orbit radius), the average value obtained in measurements of the location is $\bar{r} = 1.5a_0$. All these features can be seen by inspecting the top curve of Figure 7-5. ◀

**Example 7-4.** In its ground state, the size of the hydrogen atom can be taken to be the radius of the $n = 1$ shell for $Z = 1$, which is essentially $a_0 = 4\pi\epsilon_0\hbar^2/\mu e^2 \simeq 0.5$ Å. Show that this fundamental atomic dimension can be obtained directly from consideration of the uncertainty principle.

The form of the potential function

$$V(r) = \frac{-e^2}{4\pi\epsilon_0 r}$$

tends to cause the atom to collapse since the smaller the distance from the electron to the nucleus the more negative is the potential energy. This tendency is opposed by the effect of the uncertainty principle, as follows.

If the electron is located within a region of size $R$, then any component of its linear momentum must have an uncertainty of approximately

$$\Delta p = \frac{\hbar}{R}$$

This uncertainty reflects the fact that the linear momentum of magnitude $p$ can be in any direction, so the components can have values ranging from $-p$ to $+p$. Thus the uncertainty in any component of the linear momentum also satisfies approximately the relation

$$\Delta p = p$$

Therefore, the electron must have a kinetic energy approximately equal to

$$K = \frac{p^2}{2\mu} = \frac{(\Delta p)^2}{2\mu} = \frac{\hbar^2}{2\mu R^2}$$

We see that the kinetic energy becomes more positive with decreasing $R$, which opposes the effect of the potential energy to cause collapse.

If the size of the atom is $R$, its potential energy is approximately

$$V = \frac{-e^2}{4\pi\epsilon_0 R}$$

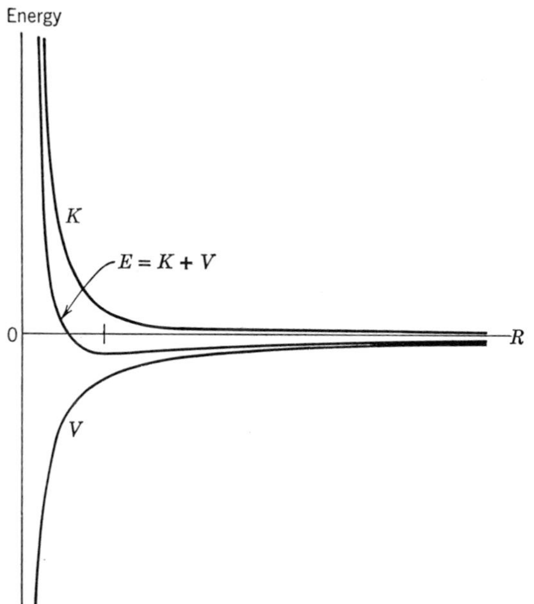


**FIGURE 7-6**
The qualitative behavior of the kinetic energy $K$, potential energy $V$ and total energy $E$ of a hydrogen atom, as functions of the size $R$ of the atom. For small $R$, $K$ increases more rapidly than $V$ decreases because $K \propto 1/R^2$ while $V \propto -1/R$. For large $R$, $K$ becomes negligible compared to $V$. As a result, $E$ has a minimum at a certain value of $R$ (indicated by the mark on the $R$ axis), and at this size the atom is most stable.

Then the total energy of the atom is approximately

$$E = K + V = \frac{\hbar^2}{2\mu R^2} - \frac{e^2}{4\pi\epsilon_0 R}$$

Obeying the common tendency of all physical systems to be as stable as possible, the atom will adjust its size so as to minimize its total energy. The existence of an optimum size can be seen qualitatively by inspecting Figure 7-6, which plots $K$, $V$, and $E$ as functions of $R$. (Note that $R$ is *not* the radial coordinate; it is the size of the atom which we are treating as a variable in order to determine its optimum value.) We can find the most energetically favorable size quantitatively by differentiating $E$ with respect to $R$, and setting the derivative equal to zero. That is

$$\frac{dE}{dR} = -\frac{2\hbar^2}{2\mu R^3} + \frac{e^2}{4\pi\epsilon_0 R^2} = 0$$

Solving this equation for $R$, we find

$$R = \frac{4\pi\epsilon_0\hbar^2}{\mu e^2} = a_0$$

the size which gives minimum total energy, and therefore the most stable atom.

The uncertainty principle governs the minimum size of the atom because its governs its minimum energy. This is the zero-point energy of the ground state, which has a size that arises from its zero-point motion. These simple ideas provide a very satisfactory answer to the question of the stability of the ground state of the atom. And this is particularly so if we also consider the discussion following Example 5-13, which shows that in its ground state the atom does not radiate. ◀

Figure 7-5 shows that the *details* of the structure of the radial probability density functions do depend on the value of the quantum number $l$. For a given $n$, the function has a single strong maximum when $l$ takes on its largest possible value; but additional weaker maxima develop inside the strong one when $l$ takes on smaller values. Generally, these weaker maxima are not so important. However, there is a related property that can be very important. Inspection of the figure, particularly the expanded plots for $n = 2, l = 0$, and $n = 2, l = 1$, will demonstrate that the radial probability density functions have appreciable values near the origin at $r = 0$ only for $l = 0$. This means that only for $l = 0$ will there be an appreciable probability of finding the electron near the nucleus.

Another way of seeing this property is to consider the probability density, $\Psi^*\Psi = \psi^*\psi$, itself. Inspection of the eigenfunctions listed in Table 7-2 will show that for values of $r$ which are small compared to $a_0/Z$, where the exponential term is slowly varying, the radial dependence of all the eigenfunctions has the behavior

$$\psi \propto r^l \qquad r \to 0 \quad (7\text{-}30)$$

This behavior can easily be verified by direct substitution into (7-17), the equation that determines the radial dependence of the $\psi$. As a consequence, the radial dependence of the probability densities for small $r$ is

$$\psi^*\psi \propto r^{2l} \qquad r \to 0 \quad (7\text{-}31)$$

From this it follows that the value of $\psi^*\psi$ in a small volume near $r = 0$ is relatively large only for $l = 0$, and decreases very rapidly with increasing $l$. The reason is that $r^0 \gg r^2 \gg r^4 \gg \ldots$, for $r \to 0$.

We see that there is some probability that the electron will be near the nucleus if $l = 0$, but very much less probability that this will happen if $l = 1$, and even less if $l = 2$, etc. This can have important effects in certain circumstances because the potential energy of the atom becomes very large in magnitude if the electron is near the nucleus. We shall see later that this is particularly true for the case of multielectron atoms, which have essentially the same property. In fact the $r^l$ behavior of the eigenfunctions for small $r$ is of predominant importance in the structure of multielectron atoms. We shall also see later that the $r^l$ behavior is due physically to the angular momentum of the atom, which depends on $l$.

Now let us proceed to the study of the *angular* dependence of the probability density functions

$$\psi^*_{nlm_l}\psi_{nlm_l} = R^*_{nl}R_{nl}\Theta^*_{lm_l}\Theta_{lm_l}\Phi^*_{m_l}\Phi_{m_l}$$

From (7-19) we have

$$\Phi^*_{m_l}(\varphi)\Phi_{m_l}(\varphi) = e^{-im_l\varphi}e^{im_l\varphi} = 1$$

Thus the probability density does not depend on the coordinate $\varphi$. The three-dimensional behavior of $\psi^*_{nlm_l}\psi_{nlm_l}$ is therefore completely specified by the product of the quantity $R^*_{nl}(r)R_{nl}(r) = P_{nl}(r)/r^2$ and the quantity $\Theta^*_{lm_l}(\theta)\Theta_{lm_l}(\theta)$, which plays the role of a directionally dependent modulation factor.

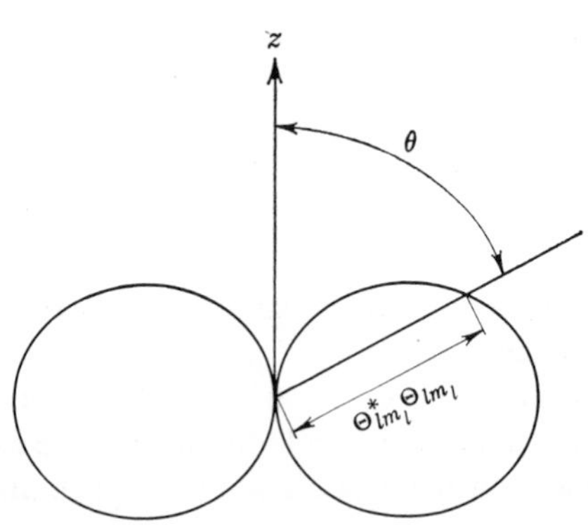


**FIGURE 7-7**

A polar diagram of the factor which determines the directional dependence of the one-electron atom probability density.

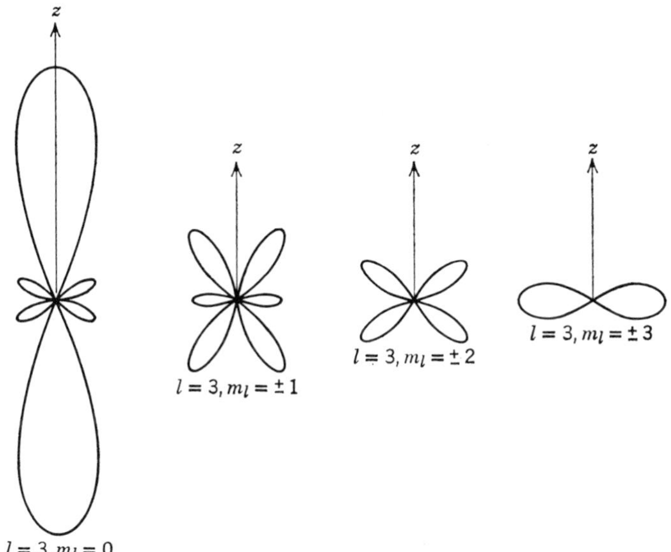


**FIGURE 7-8**

Polar diagrams of the directional dependence of the one-electron atom probability densities for $l = 3$; $m_l = 0, \pm 1, \pm 2, \pm 3$.

The form of the factor $\Theta^*_{lm_l}(\theta)\Theta_{lm_l}(\theta)$ is conveniently presented in terms of polar diagrams, of which one is shown in Figure 7-7. The origin of the diagram is at the point $r = 0$ (the nucleus), and the $z$ axis is taken along the direction from which the angle $\theta$ is measured. The distance from the origin to the curve, measured at the angle $\theta$, is equal to the value of $\Theta^*_{lm_l}(\theta)\Theta_{lm_l}(\theta)$ for that angle. Such a diagram can also be thought of as representing the complete directional dependence of $\psi^*_{nlm_l}\psi_{nlm_l}$ by visualizing the three-dimensional surface obtained by rotating the diagram about the $z$ axis through the 360° range of the angle $\varphi$. The distance, measured in the direction specified by the angles $\theta$ and $\varphi$, from the origin to a point on the surface, is equal to $\Theta^*_{lm_l}(\theta)\Theta_{lm_l}(\theta)\Phi^*_{m_l}(\varphi)\Phi_{m_l}(\varphi)$ for those values of $\theta$ and $\varphi$.

In Figure 7-8 we illustrate an example of the dependence of the form of $\Theta^*_{lm_l}(\theta)\Theta_{lm_l}(\theta)$ on the quantum number $m_l$, by a set of polar diagrams for $l = 3$, and the seven possible values of $m_l$ for this value of $l$, i.e., for $m_l = -3, -2, -1, 0, 1, 2, 3$. Note the way in which the region of concentration of $\Theta^*_{lm_l}(\theta)\Theta_{lm_l}(\theta)$, and therefore $\psi^*_{nlm_l}\psi_{nlm_l}$, shifts from the $z$ axis to the plane perpendicular to the $z$ axis as the absolute value of $m_l$ increases. Some features of the dependence of $\Theta^*_{lm_l}(\theta)\Theta_{lm_l}(\theta)$ on the qua[illegible] tum number $l$ are indicated in Figure 7-9 in terms of a set of polar diagram[illegible] $m_l = \pm l$ and $l = 0, 1, 2, 3, 4$. In the case $n = 1$, $l = m_l = 0$, which is the state of the atom, $\psi^*_{nlm_l}\psi_{nlm_l}$ depends on neither $\theta$ nor $\varphi$ and the probabil[illegible] is spherically symmetrical. For the other states, the concentration o[illegible] density in the plane perpendicular to the $z$ axis, when $m_l = \pm l$, be[illegible] more pronounced with increasing $l$. Figure 7-10 is an attempt to o[illegible] tions of the two-dimensional printed page using shading to [illegible] dimensional appearance of the probability density functions [illegible] one-electron atom.

The probability density functions displayed in these [illegible] spherical and conical surfaces, defined by certain v[illegible] equal zero. These nodal surfaces are analogous [illegible]

[illegible]ree- [illegible]s of the [illegible]ly have a set of [illegible]d $\theta$, on which they [illegible]al points at which the

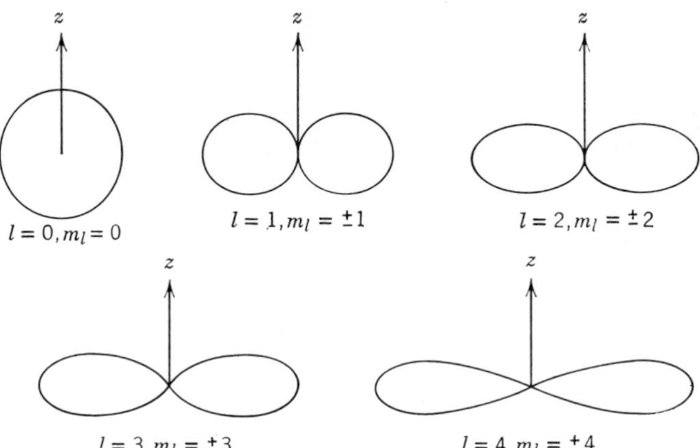


**FIGURE 7-9**

Polar diagrams of the directional dependence of the one-electron probability densities for $l = 0, 1, 2, 3, 4$; $m_l = \pm l$.

probability density for a particle bound in a one-dimensional potential equals zero (see, for example, Figure 6-32). They are a consequence of the fact that the wave functions for a bound particle must be standing waves with fixed nodes.

However, if a collection of hydrogen atoms has been completely isolated from its environment, it is not possible to then make measurements on the locations of the electron in each atom, knowing that they are all in a quantum state with a particular set of quantum numbers $n$, $l$, $m_l$, and thereby locate the nodal surfaces for that state. If it could be done it would certainly be remarkable, because it would allow the determination of the direction of the $z$ axis. And this would amount to finding for each atom a preferred direction in a space which should be spherically symmetrical, because the Coulomb potential of the atom $V = -Ze^2/4\pi\epsilon_0 r$ is spherically symmetrical. In fact, it cannot be done because it is generally not possible to observe any of the probability density patterns of Figure 7-10 in actual measurements on *free* atoms (i.e., atoms in the complete absence of external magnetic or electric fields). The only exception is the spherically symmetrical state for $n = 1$, $l = m_l = 0$. The reason is that, with the exception of the state just mentioned, every state is degenerate with several er states of the same $n$ value. Because the energies of atoms in degenerate states are cal, it is not possible experimentally to separate them from each other with es that leave the probability density unchanged. Thus, all that can be measured age probability density of the atoms for the entire set of states which are dege ith each other. It turns out that the probability density functions, when We h er in this manner, always yield a spherically symmetrical function.

uate the average of the probability density functions for the set of ponding to the energy $E_2$.

$$\frac{1}{4}[\psi^*_{200}\psi_{200} \ldots \psi^*_{210}\psi_{210} + \psi^*_{211}\psi_{211}]$$

$$= \frac{1}{128\pi}\left(\frac{Z}{a_0}\right)^3 e^{-Zr/a_0}\left[\ldots \frac{Zr}{a_0}\right)^2 + \left(\frac{Zr}{a_0}\right)^2\left(\frac{1}{2}\sin^2\theta + \frac{1}{2}\sin^2\theta + \cos^2\theta\right)\right]$$

$$\ldots\Big)^2 + \left(\frac{Zr}{a_0}\right)^2\Big] \qquad (7\text{-}32)$$

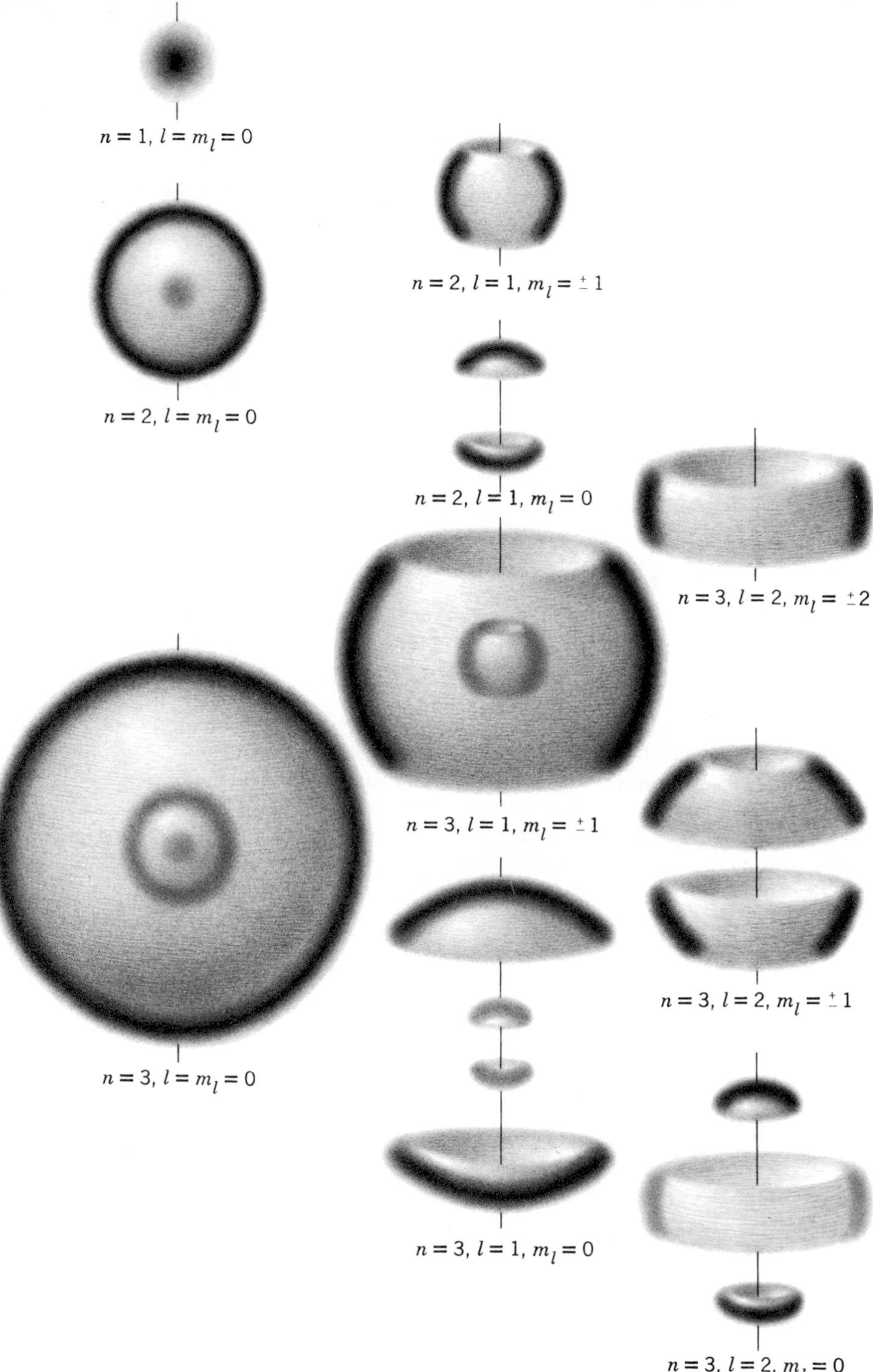


**FIGURE 7-10**

An artist's conception of the three-dimensional appearance of several one-electron atom probability density functions. For each of the drawings a line represents the $z$ axis. If all the probability densities for a given $n$ and $l$ are combined, the result is spherically symmetrical.

This spherically symmetrical distribution would be the result of a sequence of measurements on the locations of the electrons in one-electron atoms of total energy $E_2$. Of course, it cannot be used to determine the direction of the $z$ axis, and so there is no contradiction with the fact that this direction was initially chosen in a completely arbitrary way.

Note that even for each subset of states including all possible values of $m_l$ for a given $n$ and $l$ (a "subshell") the sum of the probability densities is spherically symmetrical. That is, $\psi^*_{200}\psi_{200}$ is spherically symmetrical, and also $\psi^*_{21-1}\psi_{21-1} + \psi^*_{210}\psi_{210} + \psi^*_{211}\psi_{211}$ is spherically symmetrical. This important property is illustrated in Figure 7-10. It will be used later in arguments concerning multielectron atoms, and nuclei. ◀

On the other hand, consider a situation in which the orientation of the $z$ axis is not arbitrary because there *is* a preferred direction defined, for instance, by an external magnetic or electric field applied in that direction to the collection of hydrogen atoms. In such a field the quantum states are not degenerate, as we shall see later, and measurements of the probability density of atoms in a particular state can be performed. In fact, such measurements can be used to determine the direction of the external field.

To help the student understand the ideas just discussed, let us restate them as follows:

1. If the behavior of an atom is governed by a potential which has spherical symmetry, like the Coulomb potential which depends only on the *distance* from the electron to the nucleus, none of the properties of the atom should single out any particular *direction* in space because all directions are equivalent.

2. If the atom is placed in an external electric or magnetic field, the spherical symmetry is destroyed and the direction defined by the external field becomes unique.

3. When one direction is unique, we choose one axis of our coordinate system to be in that preferred direction because it simplifies the description of the physical situation. We can choose other directions, but this unnecessarily complicates the mathematical description. (In electromagnetism, as an example, when treating a cylindrical wire it is very advantageous to take one axis of the coordinate system along the axis of the cylinder.)

4. By convention, we call the preferred axis the $z$ axis. (The convention probably comes from cylindrical coordinates, in which the axis about which the angular coordinate varies is called the $z$ axis.) But we could have called the preferred axis the $x$ or $y$ axis, just as well.

5. Even if there is no preferred direction, because no external field is applied to the atom, we still must choose *some* arbitrary direction in space for the $z$ axis of our coordinate system. But in this case the $z$ axis is not unique physically; it is merely a mathematical construct. Therefore, its choice should have no measurable consequences.

We should also point out that a *uniform* applied field can serve to define for the atom only a *single* preferred direction. As we have indicated, such a field will generally remove part of the degeneracy of the eigenfunctions, and probability densities that depend on the angle $\theta$ can be measured. But the probability densities remain independent of the angle $\varphi$, since $\psi^*\psi \propto \Phi^*_{m_l}(\varphi)\Phi_{m_l}(\varphi) = e^{-im_l\varphi}e^{im_l\varphi} = 1$ for every eigenfunction. That is, the probability densities retain their axial rotation symmetry about the direction of the applied field, as certainly must be the case.

A *nonuniform* applied field can serve to define *additional* preferred directions. It is not surprising that such fields can destroy the axial rotation symmetry of the probability density of an atom under their influence. Although we have not allowed for this possibility in our development, because we shall not need to, it is easy to do if necessary by taking particular solutions to (7-15) in the form $\Phi_{m_l}(\varphi) = \cos m_l\varphi$ or $\Phi_{m_l}(\varphi) = \sin m_l\varphi$, instead of in the form we have taken. With no applied field, or with uniform applied field, the eigenfunction associated with $\cos m_l\varphi$ is degenerate with the eigenfunction associated with $\sin m_l\varphi$, so measurement of the probability density will always yield a $\varphi$-independent combination $\propto \cos^2 m_l\varphi + \sin^2 m_l\varphi = 1$, just as with the eigenfunctions that we use. In the nonuniform applied field the degeneracy can be removed, however, and probability densities that do not have axial rotation symmetry can be observed. The solutions $\Phi_{m_l}(\varphi) = \cos m_l\varphi$ and $\Phi_{m_l}(\varphi) = \sin m_l\varphi$ are frequently used in chemistry since one atom in a molecule is acted on by a highly nonuniform field produced by the other atoms.

In the next section we shall show that the quantum numbers $l$ and $m_l$ are related to the magnitude $L$ of the orbital angular momentum of the electron, and to its $z$ component $L_z$, by the relations

$$L = \sqrt{l(l+1)}\,\hbar$$

$$L_z = m_l\hbar$$

We mention this now because it is an important clue to the interpretation of the dependence of $\psi^*_{nlm_l}\psi_{nlm_l}$ on $l$ and $m_l$. Consider the case $m_l = l$. Then $L_z = l\hbar$, which is almost equal to $L = \sqrt{l(l+1)}\,\hbar$. In this case the angular momentum vector must point nearly in the direction of the $z$ axis. For a Bohr atom this would mean that the orbit of the electron would lie nearly in the plane perpendicular to the $z$ axis, as illustrated in Figure 7-11. With increasing values of $l$, the value of $l\hbar$ approaches the value of $\sqrt{l(l+1)}\,\hbar$, so that $L_z$ approaches $L$. This means the angle between the angular momentum vector and the $z$ axis decreases. In terms of the Bohr picture, this demands that the orbit lie more nearly in the plane perpendicular to the $z$ axis. An inspection of the polar diagrams of Figure 7-9 will demonstrate the correspondence between these features of $\psi^*_{nlm_l}\psi_{nlm_l}$ and the picture of a Bohr orbit. For $m_l = 0$ we have $L_z = 0$, and the angular momentum vector must be perpendicular to the $z$ axis. In a Bohr atom this would mean that the plane of the orbit contained the $z$ axis. Some indication of this behavior can be seen in the polar diagram for $l = 3$, $m_l = 0$ of Figure 7-8.

Although there are many points at which the quantum mechanical theory of the one-electron atom corresponds quite closely to the Bohr model, there are certain striking differences. In both treatments the ground state corresponds to the quantum number $n = 1$, and it has the same value of total energy. But in the Bohr model the orbital angular momentum for this state is $L = n\hbar = \hbar$, whereas in quantum mechanics it is $L = \sqrt{l(l+1)}\,\hbar = 0$, since $l = 0$ when $n = 1$. There is an overwhelming amount of evidence, from measurements of atomic spectra and elsewhere, that shows the quantum mechanical prediction for zero orbital angular momentum in the ground state to be the correct one. This prediction is also in agreement with one obtained by using the techniques we developed earlier to calculate the expectation values of the total kinetic energy of the electron in the ground state and of the kinetic energy associated only with radial motion. The two values are found to be equal, implying

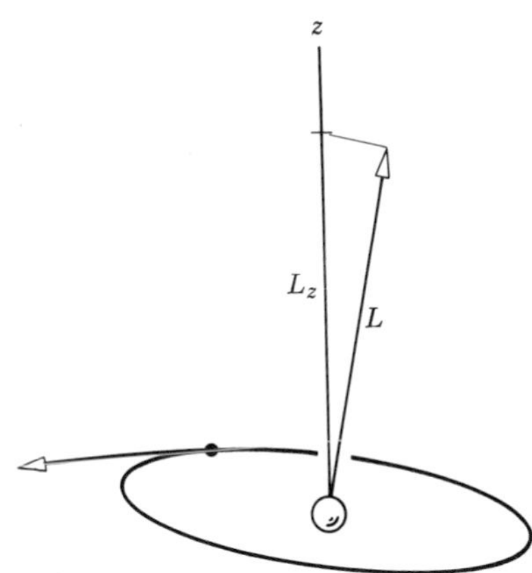


**FIGURE 7-11**

A Bohr orbit lying in a plane nearly perpendicular to the $z$ axis.

that the motion is entirely radial in that state. If the Bohr model were modified in a way that would allow for zero angular momentum states, the orbit for such a state would be a radial oscillation in which the electron passes directly through the nucleus, and the oscillation could take place along any direction in space. This would correspond, in a sense, to a spherical symmetrical probability density or charge distribution, similar to that which is predicted by quantum mechanics and is observed experimentally. Nevertheless, it is difficult to visualize the motion of an electron in the ground state of the quantum mechanical atom. That is, it is difficult to make an analogy to a classical picture, such as the Bohr picture. But this situation is not unique; it is equally difficult to visualize the motion of an electron traveling through a two-slit diffraction apparatus.

## 7-8 Orbital Angular Momentum

We shall now proceed to justify the relations

$$L_z = m_l \hbar \tag{7-33}$$

$$L = \sqrt{l(l+1)}\,\hbar \tag{7-34}$$

between the quantum numbers $m_l$ and $l$, and the $z$ component $L_z$ and magnitude $L$ of the angular momentum of an electron in its "orbital" motion about the center of an atom. The justification will take a little effort, but it will be well worth it. We have just seen that these relations are very useful in interpreting the angular dependence of the probability density functions for a one-electron atom. As we continue our study of quantum physics, we shall see that the angular momentum relations are extremely important in the study of all atoms (and nuclei). The basic reason is that in most circumstances the $z$ component and magnitude of the angular momenta of the particles in microscopic systems remain constant. From a classical point of view, this happens because in most systems the particles move in spherically symmetrical potentials that cannot exert torques on them. We shall find that, of all the quantities that can be used to describe atoms (and nuclei), angular momentum and total energy are about the only ones that do remain constant. A consequence is that most experiments on such systems involve measuring angular momentum and total energy. Therefore, quantum mechanics must be able to make predictions about angular momentum, as well as total energy. Another parallel between these two is that both are quantized. In other words, the relations of (7-33) and (7-34), stating that $L_z$ and $L$ have the *precise* values $m_l\hbar$ and $\sqrt{l(l+1)}\,\hbar$, are quantization relations just like the energy quantization relation stating that the total energy $E$ of a one-electron atom has the precise values $-\mu Z^2 e^4/(4\pi\epsilon_0)^2 2\hbar^2 n^2$. Angular momentum quantization is certainly as important as energy quantization. The only reason that it has not appeared before in our treatment of Schroedinger quantum mechanics is that the treatment was restricted to one-dimensional systems. Of course, angular momentum is the dynamical quantity that sets real three-dimensional systems apart from one-dimensional idealizations in which it has no meaning.

The *angular momentum* of a particle, relative to the origin of a certain coordinate system, is the vector quantity **L** defined by the equation

$$\mathbf{L} = \mathbf{r} \times \mathbf{p} \tag{7-35a}$$

where **r** is the position vector of the particle relative to the origin, and **p** is the linear momentum vector for the particle. By evaluating the components in rectangular coordinates of the vector, or cross, product, it is easy to show that the three rectangular components of **L** are

$$L_x = yp_z - zp_y$$
$$L_y = zp_x - xp_z \qquad (7\text{-}35b)$$
$$L_z = xp_y - yp_x$$

where $x$, $y$, $z$ are the components of $\mathbf{r}$, and $p_x$, $p_y$, $p_z$ are the components of $\mathbf{p}$.

In order to study the *dynamical quantity* angular momentum in quantum mechanics, we construct the associated *operators*. This is done by replacing $p_x$, $p_y$, $p_z$ by their quantum mechanical equivalents $-i\hbar\,\partial/\partial x$, $-i\hbar\,\partial/\partial y$, $-i\hbar\,\partial/\partial z$, according to an obvious three-dimensional extension of (5-32). Thus the operators for the three components of angular momentum are

$$L_{x_{\text{op}}} = -i\hbar\left(y\frac{\partial}{\partial z} - z\frac{\partial}{\partial y}\right)$$
$$L_{y_{\text{op}}} = -i\hbar\left(z\frac{\partial}{\partial x} - x\frac{\partial}{\partial z}\right) \qquad (7\text{-}36)$$
$$L_{z_{\text{op}}} = -i\hbar\left(x\frac{\partial}{\partial y} - y\frac{\partial}{\partial x}\right)$$

Because we must use spherical polar coordinates, these expressions must be transformed into these coordinates. Appendix I shows how this can be done. The results are

$$L_{x_{\text{op}}} = i\hbar\left(\sin\varphi\frac{\partial}{\partial\theta} + \cot\theta\cos\varphi\frac{\partial}{\partial\varphi}\right)$$
$$L_{y_{\text{op}}} = i\hbar\left(-\cos\varphi\frac{\partial}{\partial\theta} + \cot\theta\sin\varphi\frac{\partial}{\partial\varphi}\right) \qquad (7\text{-}37)$$
$$L_{z_{\text{op}}} = -i\hbar\frac{\partial}{\partial\varphi}$$

We shall also be interested in the square of the magnitude of the angular momentum vector **L**, which is

$$L^2 = L_x^2 + L_y^2 + L_z^2$$

As is indicated in Appendix I, in spherical polar coordinates the associated operator is

$$L_{\text{op}}^2 = -\hbar^2\left[\frac{1}{\sin\theta}\frac{\partial}{\partial\theta}\left(\sin\theta\frac{\partial}{\partial\theta}\right) + \frac{1}{\sin^2\theta}\frac{\partial^2}{\partial\varphi^2}\right] \qquad (7\text{-}38)$$

The first step in deriving the angular momentum quantization equations involves using the operators to calculate the expectation values of the $z$ component of **L**, and of the square of its magnitude, for an electron in the $n$, $l$, $m_l$ quantum state of a one-electron atom. According to the three-dimensional extension of the prescription of (5-34), the expectation value $L_z$ is

$$\overline{L_z} = \int_0^\infty\int_0^\pi\int_0^{2\pi}\Psi^* L_{z_{\text{op}}}\Psi r^2 \sin\theta\, dr\, d\theta\, d\varphi$$

The quantity $r^2 \sin\theta\, dr\, d\theta\, d\varphi$ is the element of volume in spherical polar coordinates, and the integrations are taken over the complete ranges of all three coordinates. Because it will simplify the notation, without causing confusion, we shall write this expression as

$$\overline{L_z} = \int \Psi^* L_{z_{op}} \Psi \, d\tau$$

Here $d\tau$ stands for the three-dimensional volume element $r^2 \sin\theta \, dr \, d\theta \, d\varphi$, and $\int$ stands for the three definite integrals $\int_0^\infty \int_0^\pi \int_0^{2\pi}$. The same shorthand notation will be used in the remainder of this chapter, and in the following chapters. Continuing our calculation of $\overline{L_z}$, by expressing the wave function as a product of the eigenfunction and the exponential time factor we obtain

$$\overline{L_z} = \int e^{iE_n t/\hbar} \psi^*_{nlm_l} L_{z_{op}} e^{-iE_n t/\hbar} \psi_{nlm_l} \, d\tau$$

or

$$\overline{L_z} = \int \psi^*_{nlm_l} L_{z_{op}} \psi_{nlm_l} \, d\tau \tag{7-39}$$

Similarly, the expectation value of $L^2$ is

$$\overline{L^2} = \int \psi^*_{nlm_l} L^2_{op} \psi_{nlm_l} \, d\tau \tag{7-40}$$

To evaluate the integrals in the two numbered equations above, we must first evaluate $L_{z_{op}} \psi_{nlm_l}$ and $L^2_{op} \psi_{nlm_l}$.

**Example 7-6.** Evaluate $L_{z_{op}} \psi_{nlm_l}$, where $L_{z_{op}} = -i\hbar \partial/\partial\varphi$, and where $\psi_{nlm_l}$ is a one-electron atom eigenfunction.

We have

$$L_{z_{op}} \psi_{nlm_l} = -i\hbar \frac{\partial \psi_{nlm_l}}{\partial \varphi}$$

Since

$$\psi_{nlm_l} = R_{nl}(r) \Theta_{lm_l}(\theta) \Phi_{m_l}(\varphi)$$

we obtain

$$-i\hbar \frac{\partial \psi_{nlm_l}}{\partial \varphi} = R_{nl}(r) \Theta_{lm_l}(\theta) \left[ -i\hbar \frac{d\Phi_{m_l}(\varphi)}{d\varphi} \right]$$

According to (7-19)

$$\Phi_{m_l}(\varphi) = e^{im_l\varphi}$$

so

$$\frac{d\Phi_{m_l}(\varphi)}{d\varphi} = im_l e^{im_l\varphi} = im_l \Phi_{m_l}(\varphi)$$

Thus

$$-i\hbar \frac{\partial \psi_{nlm_l}}{\partial \varphi} = R_{nl}(r) \Theta_{lm_l}(\theta) [-i\hbar i m_l \Phi_{m_l}(\varphi)]$$
$$= m_l \hbar R_{nl}(r) \Theta_{lm_l}(\theta) \Phi_{m_l}(\varphi)$$

and we obtain the answer

$$L_{z_{op}} \psi_{nlm_l} = m_l \hbar \psi_{nlm_l} \tag{7-41}$$

◀

Although we do not have a concise expression for the functions $\Theta_{lm_l}(\theta)$, which must be differentiated to evaluate $L^2_{op} \psi_{nlm_l}$, we know that these functions satisfy the differential equation (7-16). Using this fact, it is not difficult to show that

$$L^2_{op} \psi_{nlm_l} = l(l+1)\hbar^2 \psi_{nlm} \tag{7-42}$$

Using (7-41) from Example 7-6 in (7-39), which is

$$\overline{L_z} = \int \psi^*_{nlm_l} L_{z_{\text{op}}} \psi_{nlm_l}\, d\tau$$

it is trivial to evaluate $\overline{L_z}$. We have

$$\overline{L_z} = m_l\hbar \int \psi^*_{nlm_l} \psi_{nlm_l}\, d\tau$$

But we know that this integral has the value one because it is equal to the probability density integrated over all space, i.e., the probability of finding the electron somewhere. Thus we obtain

$$\overline{L_z} = m_l\hbar \tag{7-43}$$

In a similar fashion we use (7-42) in (7-40), which is

$$\overline{L^2} = \int \psi^*_{nlm_l} L^2_{\text{op}} \psi_{n\ m_l}\, d\tau$$

to obtain

$$\overline{L^2} = l(l+1)\hbar^2 \int \psi^*_{nlm_l} \psi_{nlm_l}\, d\tau$$

$$\overline{L^2} = l(l+1)\hbar^2 \tag{7-44}$$

Let us compare the results of our expectation value calculations, (7-43) and (7-44), with the quantization relations we are trying to verify, that can be written

$$L_z = m_l\hbar \tag{7-45}$$
$$L^2 = l(l+1)\hbar^2 \tag{7-46}$$

The former are certainly consistent with the latter, but they are not proofs of the latter. The quantization relations make stronger statements about the values of $L_z$ and $L^2$. These relations say that *any* measurement of the angular momentum of an electron in the $n$, $l$, $m_l$ state of the atom will *always* yield $L_z = m_l\hbar$ and $L^2 = l(l+1)\hbar^2$ since, in that state, these quantities have *precisely* the values quoted. But the expectation value relations say only that the values quoted will be obtained on the average, that is, when the results of a large number of measurements of $L_z$ and $L^2$ are averaged.

To complete the proof of the quantization relations is a matter of continuing along the line we have been following. For example, by calculating the expectation value of some power of $L_z$, say the square $L_z^2$, it is found that $\overline{L_z^2} = (m_l\hbar)^2$. This immediately leads to the conclusion that not only must $L_z$ equal $m\hbar$ on the average, i.e., $\overline{L_z} = m_l\hbar$, but that $L_z$ must equal $m\hbar$ always, i.e., $L_z = m_l\hbar$. The point is that if $L_z$ fluctuated about its average $m_l\hbar$ it would not be possible to obtain $\overline{L_z^2} = (m_l\hbar)^2$ because when averaging a power of $L_z$ higher than the first more weight is given to fluctuations above the average than to fluctuations below the average. In order to proceed with our interpretation of the angular momentum of one-electron atoms, we defer the details of this proof to the following section. There we shall also obtain the interesting conclusion that $L_x$ and $L_y$, the $x$ and $y$ components of the orbital angular momentum, do *not* obey quantization relations.

The fact that $\psi_{nlm_l}$ does not describe a state with a definite $x$ and $y$ component of orbital angular momentum, because these quantities are not quantized, is mysterious from the point of view of classical mechanics. According to the angular momentum conservation law of classical mechanics, the orbital angular momentum vector of an electron moving under the influence of a spherically symmetrical potential $V(r)$ of a one-electron atom in free space would be completely fixed in direction and magnitude, and all three components of the vector would have definite values. The reason is that there would be no torques acting on the electron. The fact that this result is not obtained in the quantum mechanical theory is a consequence of the fact that there is an uncertainty principle relation which states that no two components of an angular momentum can be known simultaneously with complete precision. Because the $z$ component of orbital angular momentum has the precise value $m_l\hbar$, the relation requires that the values of the $x$ and $y$ components be indefinite. Upon evaluating $\overline{L_x}$ and $\overline{L_y}$, the average values of these components, it is found that both equal zero. Thus the orientation of the orbital angular momentum vector of an electron moving in a spherically symmetrical potential can be thought of as always changing so that its $x$ and $y$ components fluctuate about an average value of zero, while its $z$ component and magnitude remain fixed. This result might be called the *quantum mechanical orbital angular momentum conservation law.*

Many of the properties of the orbital angular momentum can be conveniently represented by such a *vector model.* Consider the set of states having a common value of the quantum number $l$. For each of these states the length of the orbital angular momentum vector, in units of $\hbar$, is $L/\hbar = \sqrt{l(l+1)}$. In the same units, the $z$ component of this vector is $L_z/\hbar = m_l$. The $z$ component can assume any integral value from $L_z/\hbar = -l$ to $L_z/\hbar = +l$, depending on the value of $m_l$. The case of $l = 2$ is illustrated in Figure 7-12. The figure depicts the angular momentum vectors for each of the five states corresponding to the five possible values of $m_l$ for this value of $l$. If in any one of these states $L_x$ and $L_y$ fluctuate about their average values of zero, the vectors describing the states precess randomly in the conical surface surrounding the $z$ axis, satisfying the quantum mechanical angular momentum conservation law. The actual orientation in space of the angular momentum vector is known with the greatest precision for the states with $m_l = \pm l$. But even for these states there is some uncertainty since the vector can be anywhere on a cone of half-angle $\cos^{-1}[l/\sqrt{l(l+1)}]$. In the classical limit $l \to \infty$, and this angle becomes vanishingly small. Thus, in the classical limit the angular momentum vector for the states $m_l = \pm l$ is constrained to lie almost along the $z$ axis and is therefore essentially fixed in space. This agrees with the behavior predicted by the classical theory, i.e., with the classical orbital angular momentum conservation law.

The quantum number $m_l$ determines the space orientation of the orbital angular momentum vector of the one-electron atom. Therefore, in a sense it determines the orientation in space of the atom itself. As the spherically symmetrical Coulomb potential implies that there is no preferred direction in the space in which the atom is

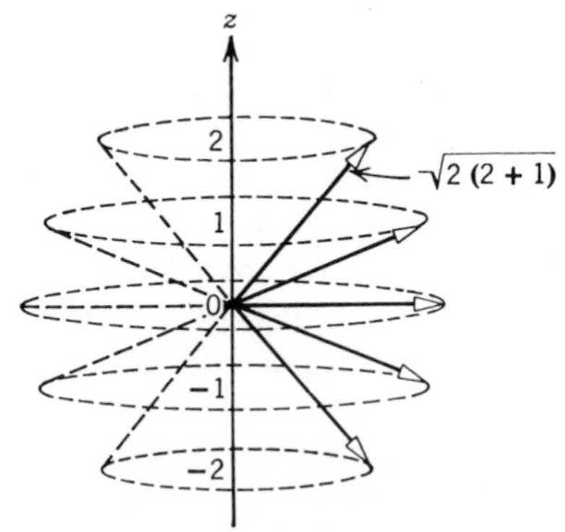


**FIGURE 7-12**

Representing the angular momentum vectors (measured in units of $\hbar$) for the possible states with $l = 2$. In each state the vector precesses randomly about the $z$ axis, maintaining a constant magnitude and a constant $z$ component.

situated, we can understand why the theory predicts that the total energy of the atom does not depend on $m_l$, which determines this orientation. Thus we can understand why the eigenfunctions are degenerate with respect to the quantum number $m_l$. The energy of the atom simply does not depend on its orientation in empty space.

## 7-9 Eigenvalue Equations

Here we shall complete the derivation, started in the previous section, of the orbital angular momentum quantization conditions. Then we shall generalize the results of the derivation to point out an interesting feature of Schroedinger's theory of quantum mechanics.

To study the quantization of the orbital angular momentum, we focus attention first on its $z$ component, $L_z$. Now, if the $z$ component quantization condition of (7-45) is valid, then any measurement of $L_z$ will always yield the same precise value specified by that quantization condition

$$L_z = m_l\hbar \tag{7-47}$$

Furthermore, measurements of some higher power of $L_z$, say the square $L_z^2$, will always yield the same value $L_z^2 = (m_l\hbar)^2$. As a consequence, the expectation value of the square of $L_z$ will be just $\overline{L_z^2} = (m_l\hbar)^2$. Note that, since we also have $\overline{L_z} = m_l\hbar$, this means

$$\overline{L_z^2} = \overline{L_z}^2 \tag{7-48}$$

That is, the expectation value of the square of $L_z$ equals the square of the expectation value of $L_z$, if the quantization condition of (7-47) is valid.

On the other hand, if (7-47) is not valid then measurements of $L_z$ can lead to various values, subject, however, to the constraint that the values average out to yield $m_l\hbar$ because we have proven in (7-43) that $\overline{L_z} = m_l\hbar$ in any case. If the measured values of $L_z$ fluctuate about the average value $m_l\hbar$, then the expectation value of the square of $L_z$ will no longer equal the square of $m_l\hbar$. The reason is that when averaging a higher power of $L_z$, like its square $L_z^2$, we give much more weight to the cases in which $L_z$ is larger than $\overline{L_z}$, and much less weight to the equally numerous cases in which $L_z$ is smaller than $\overline{L_z}$. In this situation $\overline{L_z^2} \neq (m_l\hbar)^2$, so $\overline{L_z^2} \neq \overline{L_z}^2$.

An example is shown in Table 7-3, which applies the ideas just discussed to calculating the square of the average, and the average of the squares, of the ages of a group of children whose individual ages are 1, 2, and 3 years. Inspection of the table shows that when the ages are first squared, and then averaged, a larger result is obtained than when the ages are first averaged, and then squared. This will be true in any case in which a power of the ages higher than the first is averaged, and in which the ages fluctuate. But if all the children in the group have ages

**TABLE 7-3.** The Square of the Average, and the Average of the Squares, of a Set of Fluctuating Numbers

$$A = 1, 2, 3$$

$$\overline{A} = \frac{1 + 2 + 3}{3} = \frac{6}{3} = 2$$

$$\overline{A}^2 = 4$$

$$A^2 = 1, 4, 9$$

$$\overline{A^2} = \frac{1 + 4 + 9}{3} = \frac{14}{3} = 4.67$$

$$\Delta A \equiv \sqrt{\overline{A^2} - \overline{A}^2} = \sqrt{4.67 - 4} = \sqrt{0.67} = 0.82$$

**TABLE 7-4.** The Square of the Average, and the Average of the Squares, of a Set of Nonfluctuating Numbers

$$A = 2, 2, 2$$

$$\bar{A} = \frac{2+2+2}{3} = \frac{6}{3} = 2$$

$$\bar{A}^2 = 4$$

$$A^2 = 4, 4, 4$$

$$\overline{A^2} = \frac{4+4+4}{3} = \frac{12}{3} = 4$$

$$\Delta A \equiv \sqrt{\overline{A^2} - \bar{A}^2} = \sqrt{4-4} = 0$$

precisely equal to each other, and therefore to the average age, then it makes no difference in which order the operations are carried out and the average of the squares equals the square of the averages. An example of that situation is shown in Table 7-4.

For another illustration of these ideas, consider the quantity $\Delta x = \sqrt{\overline{x^2} - \bar{x}^2}$. As mentioned in Example 5-10, this quantity is used as a measure of the fluctuations that would be observed in measurements of the $x$ coordinate of a particle. If there were no fluctuations, then $\overline{x^2} = \bar{x}^2$. But the uncertainty principle demands that there be fluctuations in $x$ (which are larger the smaller the fluctuations in the linear momentum $p$). As a result $\overline{x^2} > \bar{x}^2$, and the difference between $\overline{x^2}$ and $\bar{x}^2$ increases as the fluctuations in $x$ increase so $\sqrt{\overline{x^2} - \bar{x}^2}$ is a measure of these fluctuations.

Now, it is easy to prove the validity of the relation expressed by (7-48), $\overline{L_z^2} = \overline{L_z}^2$, and therefore also the validity of the quantization condition $L_z = m_l\hbar$ of (7-47). To do this we twice use (7-41), $L_{z_{op}}\psi_{nlm_l} = m_l\hbar\psi_{nlm_l}$, to calculate $\overline{L_z^2}$. According to the three-dimensional extension of the prescription for calculating expectation values, we have

$$\overline{L_z^2} = \int \Psi^* L_{z_{op}}^2 \Psi \, d\tau$$

This immediately gives

$$\overline{L_z^2} = \int \psi_{nlm_l}^* L_{z_{op}}^2 \psi_{nlm_l} \, d\tau$$

The dynamical quantity $L_z^2$ is the product of two factors of the form $L_z$

$$L_z^2 = L_z \cdot L_z$$

According to the expectation value prescription, the operator $L_{z_{op}}^2$ obtained from that dynamical quantity is thus the product of two operators of the form $L_{z_{op}}$. Therefore

$$L_{z_{op}}^2 \psi_{nlm_l} = L_{z_{op}} \cdot L_{z_{op}} \psi_{nlm_l}$$

In other words, $L_{z_{op}}^2 \psi_{nlm_l}$ means that $L_{z_{op}}$ operates twice on $\psi_{nlm_l}$. But according to (7-41)

$$L_{z_{op}} \psi_{nlm_l} = m_l\hbar\, \psi_{nlm_l}$$

Thus each operation of $L_{z_{op}}$ on $\psi_{nlm_l}$ yields the same function $\psi_{nlm_l}$, multiplied by a constant factor $m_l\hbar$. Therefore, the result of two operations is simply to multiply $\psi_{nlm_l}$ by two factors of $m_l\hbar$. That is

$$L_{z_{op}}^2 \psi_{nlm_l} = (m_l\hbar)^2 \psi_{nlm_l}$$

Knowing this, we immediately obtain

$$\begin{aligned}\overline{L_z^2} &= \int \psi_{nlm_l}^* (m_l\hbar)^2 \psi_{nlm_l}\, d\tau \\ &= (m_l\hbar)^2 \int \psi_{nlm_l}^* \psi_{nlm_l}\, d\tau \\ &= (m_l\hbar)^2 \\ &= \overline{L_z}^2\end{aligned}$$

where we have made use of the fact that the integral over all space of $\psi_{nlm_l}^*\psi_{nlm_l}$ equals one because of the normalization condition. Since we have verified (7-48), we have completed our verification of the quantization condition $L_z = m_l\hbar$. The proof of the validity of the quantization condition $L^2 = l(l+1)\hbar^2$ is carried through in a completely parallel manner.

Note that these proofs depend on (7-41) and (7-42), $L_{z_{op}}\psi_{nlm_l} = m_l\hbar\psi_{nlm_l}$ and $L_{op}^2\psi_{nlm_l} = l(l+1)\hbar^2\psi_{nlm_l}$. The equations state the surprising facts that the result of operating on the one-electron atom eigenfunction $\psi_{nlm_l}$ with the differential operator $L_{z_{op}}$ is simply to multiply that eigenfunction by the constant $m_l\hbar$, while the result of operating on it with the differential operator $L_{op}^2$ is simply to multiply it by the constant $l(l+1)\hbar^2$. These results are certainly not typical of what happens when a differential operator operates on a function. For instance, if we operate on a function, say $f(x) = x^2$, with the differential operator $d/dx$, we obtain a very different function $f'(x) = 2x$. As another example, it is not difficult to show that the results of operating on $\psi_{nlm_l}$ with the operators $L_{x_{op}}$ or $L_{y_{op}}$ is to produce new functions of $r$, $\theta$, $\varphi$ in which these variables enter quite differently from the way they enter in the function $\psi_{nlm_l}$. That is

$$L_{x_{op}}\psi_{nlm_l} \neq (\text{const})\psi_{nlm_l} \tag{7-49}$$

$$L_{y_{op}}\psi_{nlm_l} \neq (\text{const})\psi_{nlm_l} \tag{7-50}$$

The ideas that we have developed, in the process of verifying the angular momentum quantization conditions, can be extended to provide a deeper insight into the theory of Schroedinger quantum mechanics. They can also be used to lead into the more sophisticated theories, such as Heisenberg's *matrix mechanics.* We must leave these matters for more advanced books. Here we shall say only that the properties associated with (7-41) and (7-42) are perfectly general. That is, *whenever the dynamical quantity $f$ has the precise value $F$ in the quantum state described by the function $\psi$, then that function satisfies the relation*

$$f_{op}\psi = F\psi \tag{7-51}$$

*where $f_{op}$ is the operator corresponding to $f$.*

We shall also show that the time-independent Schroedinger equation can be written in the form of (7-51). To do this, consider the time-independent Schroedinger equation in rectangular coordinates

$$-\frac{\hbar^2}{2\mu}\left(\frac{\partial^2\psi}{\partial x^2} + \frac{\partial^2\psi}{\partial y^2} + \frac{\partial^2\psi}{\partial z^2}\right) + V\psi = E\psi$$

Rewrite it as

$$\left[-\frac{\hbar^2}{2\mu}\left(\frac{\partial^2}{\partial x^2} + \frac{\partial^2}{\partial y^2} + \frac{\partial^2}{\partial z^2}\right) + V\right]\psi = E\psi$$

By comparing (7-3) with (7-4), we see that the square bracket is just the operator $e_{op}$ for the total energy. Thus we have

$$e_{op}\psi = E\psi$$

Here $E$ is one of the precise allowed values of the total energy of the system described by the potential $V$. The system is also described by the total energy operator $e_{op}$.

The general relation of (7-51) is called an *eigenvalue equation*, $\psi$ is said to be an *eigenfunction of the operator* $f_{op}$, and $F$ is said to be the corresponding *eigenvalue*. This is the same terminology as is used in the particular case of the eigenvalue equation for the total energy operator—that is, in the case of the time-independent Schroedinger equation. The total energy operator $e_{op}$ is sometimes called the *Hamiltonian.*

These considerations lead to the important conclusion that, since (7-49) and (7-50) show $\psi_{nlm_l}$ is not an eigenfunction of the operators $L_{x_{op}}$ or $L_{y_{op}}$, the corresponding dynamical quantities $L_x$ and $L_y$ do not have precise values in the one-electron atom. That is, $L_x$ and $L_y$ do not obey quantization conditions.

## QUESTIONS

1. If a hydrogen atom were not at rest, but moving freely through space, how would the quantum mechanical description of the atom be modified?

2. Since it is well known that the Coulomb potential has a much simpler form in spherical polar coordinates, why did we begin our treatment of the one-electron atom in rectangular coordinates?

3. In what important equations of classical physics does the Laplacian operator enter?

4. Would the results of the calculations be affected if we took different forms for the separation constants that arise in the splitting of the time-independent Schroedinger equation, for the one-electron atom, into three ordinary differential equations?

5. Why must $\Phi(\varphi)$ be single valued? How does this lead to the restriction that $m_l$ must be an integer?

6. What would happen if we took $e^{-im_l\varphi}$ as the particular solution to the $\Phi(\varphi)$ equation? What about $\cos m_l\varphi$ or $\sin m_l\varphi$?

7. Why do three quantum numbers arise in the treatment of the (spinless) one-electron atom?

8. Can you say what the functions $\Theta(\theta)$ and $\Phi(\varphi)$ would be like if $V$ were a function of $r$, but not proportional to $-1/r$? (This is the case for the valence electron of an alkali atom.)

9. Just what is degeneracy?

10. What is the relation between the size of a Bohr atom and the size of a Schroedinger atom?

11. What is the fundamental reason why the size of the hydrogen atom in its ground state has the value it does?

12. For a one-electron atom in free space, what would be the mathematical consequences of changing the choice of direction of the $z$ axis? The physical consequences? What if the atom is in an external electric or magnetic field?

13. Why does a uniform electric or magnetic field define only one unique direction is space?

14. How do the predictions of the Bohr and Schroedinger treatments of the hydrogen atom (ignoring spin and other relativistic effects) compare with regard to the location of the electron, its total energy, and its orbital angular momentum?

15. Devise an explanation for the obvious relation between the last two terms of the Laplacian operator, in spherical polar coordinates, and the operator for the square of the magnitude of the orbital angular momentum.

16. Using the connection between $L$ and $l$, explain physically why $\psi^*\psi$ is very small near $r = 0$, unless $l = 0$.

17. Exactly why do we say that for a hydrogen atom in free space the orbital angular momentum vector precesses randomly about the $z$ axis (ignoring spin)?

18. What is the quantum mechanical orbital angular momentum conservation law?

19. Is every eigenfunction of angular momentum magnitude necessarily also an eigenfunction of total energy? Is the reciprocal statement true?

20. Are examples of eigenvalue equations found in classical physics? If so, what are they?

## PROBLEMS

1. Using the technique of separation of variables, show that there are solutions to the three-dimensional Schroedinger equation for a time-independent potential, which can be written

$$\Psi(x,y,z,t) = \psi(x,y,z)e^{-iEt/\hbar}$$

where $\psi(x,y,z)$ is a solution to the time-independent Schroedinger equation.

2. Verify that $\Phi(\varphi) = e^{im_l\varphi}$ is the solution to the equation for $\Phi(\varphi)$, (7-15).

3. Hydrogen, deuterium, and singly ionized helium are all examples of one-electron atoms. The deuterium nucleus has the same charge as the hydrogen nucleus, and almost exactly twice the mass. The helium nucleus has twice the charge of the hydrogen nucleus, and almost exactly four times the mass. Make an accurate prediction of the ratios of the ground state energies of these atoms. (Hint: Remember the variation in the reduced mass.)

4. (a) Evaluate, in electron volts, the energies of the three levels of the hydrogen atom in the states for $n = 1, 2, 3$. (b) Then calculate the frequencies in hertz, and the wavelengths in angstroms, of all the photons that can be emitted by the atom in transitions between these levels. (c) In what range of the electromagnetic spectrum are these photons?

5. Verify by substitution that the ground state eigenfunction $\psi_{100}$, and the ground state eigenvalue $E_1$, satisfy the time-independent Schroedinger equation for the hydrogen atom.

6. (a) Extend Example 7-4 to obtain from the uncertainty principle a prediction of the total energy of the ground state of the hydrogen atom. (b) Compare with the energy predicted by (7-22).

7. (a) Calculate the location at which the radial probability density is a maximum for the $n = 2, l = 1$ state of the hydrogen atom. (b) Then calculate the expectation value of the radial coordinate in this state. (c) Explain the physical significance of the difference in the answers to (a) and (b). (Hint: See Figure 7-5.)

8. (a) Calculate the expectation value $\bar{V}$ for the potential energy in the ground state of the hydrogen atom. (b) Show that in the ground state $E = \bar{V}/2$, where $E$ is the total energy. (c) Use the relation $E = K + V$ to calculate the expectation value $\bar{K}$ of the kinetic energy in the ground state, and show that $\bar{K} = -\bar{V}/2$. These relations are obtained for any state of motion of any quantum mechanical (or classical) system with a potential in the form $V(r) \propto -1/r$. They are sometimes called the *virial theorem*.

9. (a) Calculate the expectation value $\bar{V}$ of the potential energy in the $n = 2, l = 1$ state of the hydrogen atom. (b) Do the same for the $n = 2, l = 0$ state. (c) Discuss the results of (a) and (b), in connection with the virial theorem of Problem 8, and explain how they bear on the origin of the $l$ degeneracy.

10. By substituting into the equation for $R(r)$, (7-17), the form $R(r) \propto r^l$, show that it is a solution for $r \rightarrow 0$. (Hint: Ignore terms that become negligible relative to others as $r \rightarrow 0$.)

11. Show that the sum of hydrogen atom probability densities for the $n = 3$ quantum states, analogous to the sum in Example 7-5, is spherically symmetrical.

12. Show that $\Phi(\varphi) = \cos m_l\varphi$, and $\Phi(\varphi) = \sin m_l\varphi$, are particular solutions to the equation for $\Phi(\varphi)$, (7-15).

13. By using the techniques of Appendix I show that $L_{x_{op}}$ has the form stated in (7-37).

14. Prove that $L^2_{op}\psi_{nlm_l} = l(l+1)\hbar^2\psi_{nlm_l}$. (Hint: Use the differential equation satisfied by $\Theta_{lm_l}(\theta)$, (7-16).)

15. We know that $\psi = e^{ikx}$ is an eigenfunction of the total energy operator $e_{op}$ for the one-dimensional problem of the zero potential. (a) Show that it is also an eigenfunction of the linear momentum operator $p_{op}$, and determine the associated momentum eigenvalue. (b) Repeat for $\psi = e^{-ikx}$. (c) Interpret what the results of (a) and (b) mean concerning measurements of the linear momentum. (d) We also know that $\psi = \cos kx$ and $\psi = \sin kx$ are eigenfunctions of the zero potential $e_{op}$. Are they eigenfunctions of $p_{op}$? (e) Interpret the results of (d).

16. All four of the functions $e^{im_l\varphi}$, $e^{-im_l\varphi}$, $\cos m_l\varphi$, and $\sin m_l\varphi$ are particular solutions to the equation for $\Phi(\varphi)$, (7-15) (see Problem 12). (a) Find which are also eigenfunctions of the operator for the $z$ component of angular momentum $L_{z_{op}}$. (b) Interpret your results.

17. A particle of mass $\mu$ is fixed at one end of a rigid rod of negligible mass and length $R$. The other end of the rod rotates in the $x$-$y$ plane about a bearing located at the origin, whose axis is in the $z$ direction. This two-dimensional "rigid rotator" is illustrated in Figure 7-13. (a) Write an expression for the total energy of the system in terms of its

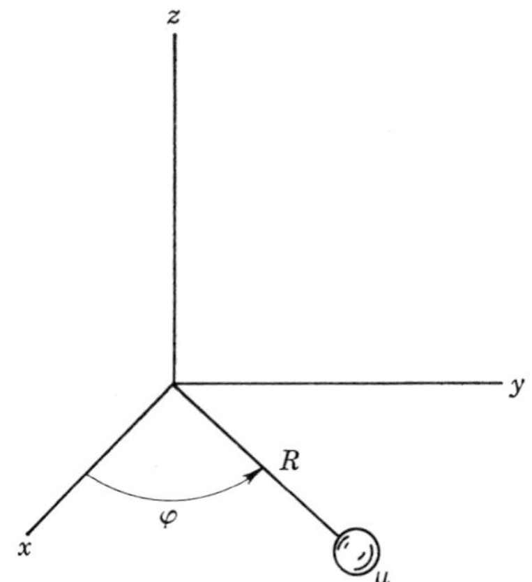


**FIGURE 7-13**
The rigid rotator moving in the $x$-$y$ plane considered in Problem 17.

angular momentum $L$. (Hint: Set the constant potential energy equal to zero, and then express the kinetic energy in terms of $L$.) (b) By introducing the appropriate operators into the energy equation, convert it into the Schroedinger equation

$$-\frac{\hbar^2}{2I}\frac{\partial^2\Psi(\varphi,t)}{\partial\varphi^2} = i\hbar\frac{\partial\Psi(\varphi,t)}{\partial t}$$

where $I = \mu R^2$ is the rotational inertia, or moment of inertia, and $\Psi(\varphi, t)$ is the wave function written in terms of the angular coordinate $\varphi$ and the time $t$. (Hint: Since the angular momentum is entirely in the $z$ direction, $L = L_z$ and the corresponding operator is $L_{z_{op}} = -i\hbar\partial/\partial\varphi$.)

18. By applying the technique of separation of variables, split the rigid rotator Schroedinger

equation of Problem 17 to obtain: (a) the time-independent Schroedinger equation

$$-\frac{\hbar^2}{2I}\frac{d^2\Phi(\varphi)}{d\varphi^2} = E\Phi(\varphi)$$

and (b) the equation for the time dependence of the wave function

$$\frac{dT(t)}{dt} = -\frac{iE}{\hbar}T(t)$$

In these equations $E =$ the separation constant, and $\Phi(\varphi)T(t) = \Psi(\varphi, t)$, the wave function.

**19.** (a) Solve the equation for the time dependence of the wave function obtained in Problem 18. (b) Then show that the separation constant $E$ is the total energy.

**20.** Show that a particular solution to the time-independent Schroedinger equation for the rigid rotator of Problem 18 is $\Phi(\varphi) = e^{im\varphi}$ where $m = \sqrt{2IE}/\hbar$.

**21.** (a) Apply the condition of single valuedness to the particular solution of Problem 20. (b) Then show that the allowed values of the total energy $E$ for the two-dimensional quantum mechanical rigid rotator are

$$E = \frac{\hbar^2 m^2}{2I} \qquad |m| = 0, 1, 2, 3, \ldots$$

(c) Compare the results of quantum mechanics with those of the old quantum theory obtained in Problem 34 of Chapter 4. (d) Explain why the two-dimensional quantum mechanical rigid rotator has no zero-point energy. Also explain why it is not a completely realistic model for a microscopic system.

**22.** Normalize the functions $\Phi(\varphi) = e^{im\varphi}$ found in Problem 21.

**23.** (a) Calculate the expectation value of the angular momentum, $\overline{L}$, for a two-dimensional rigid rotator in a typical quantum state, using the eigenfunctions found in Problem 22. (b) Then calculate $\overline{L^2}$ and $\overline{L}^2$, and interpret what your results have to say about the values of $L$ that would be obtained in a series of measurements on the system.

# 8

# Magnetic Dipole Moments, Spin, and Transition Rates

# Magnetic Dipole Moments, Spin, and Transition Rates 8

## 8-1 Introduction

In this chapter we continue our study of the one-electron atom. First we shall discuss experiments which measure the orbital angular momentum **L** of an atomic electron. These experiments do not actually measure **L** directly. Instead they measure a related quantity $\boldsymbol{\mu}_l$, the orbital magnetic dipole moment, by measuring its interaction with a magnetic field applied to the atom. We shall develop the relation between $\boldsymbol{\mu}_l$ and **L** that forms the basis of the measurements. We shall also remind the student of some of the properties of the interaction between a magnetic dipole and a magnetic field used in the measurements, and in others frequently carried out in atomic, solid state, and nuclear physics.

When considering the results of measurements of atomic magnetic dipole moments, we shall discover the very important fact that electrons have an intrinsic angular momentum called spin, and an associated spin magnetic dipole moment. The effect that electron spin has on the energy levels of a one-electron atom will then be explored. Finally, we shall develop a procedure for calculating the rate at which excited one-electron atoms make transitions to lower-lying states by emitting the photons that form their line spectrum.

Our treatments in this chapter will employ a combination of simple electromagnetic theory, partly classical physics such as the Bohr model, and quantum mechanics. Completely quantum mechanical treatments will not be presented because they require a more advanced knowledge of electromagnetic theory than has been assumed in this book. This procedure is justified by the fact that the results agree with those of completely quantum mechanical treatments. Of course, the justification is available to us only because someone has taken the trouble to work out the completely quantum mechanical treatments.

## 8-2 Orbital Magnetic Dipole Moments

Consider an electron of mass $m$ and charge $-e$ moving with velocity of magnitude $v$ in a circular Bohr orbit of radius $r$, as illustrated in Figure 8-1. (Since it is conventional to use $\mu$ for magnetic dipole moment, here we do not use it for the reduced electron mass. No confusion will arise because the inherent accuracy of the experiments, and calculations, generally does not warrant making a distinction between the reduced electron mass and the electron mass $m$.) The charge circulating in a loop constitutes a current of magnitude

$$i = \frac{e}{T} = \frac{ev}{2\pi r} \tag{8-1}$$

where $T$ is the orbital period of the electron whose charge has magnitude $e$. In elementary electromagnetic theory, it is shown that such a current loop produces a magnetic field which is the same at large distances from the loop as that of a magnetic dipole located at the center of the loop and oriented perpendicular to its plane. For a

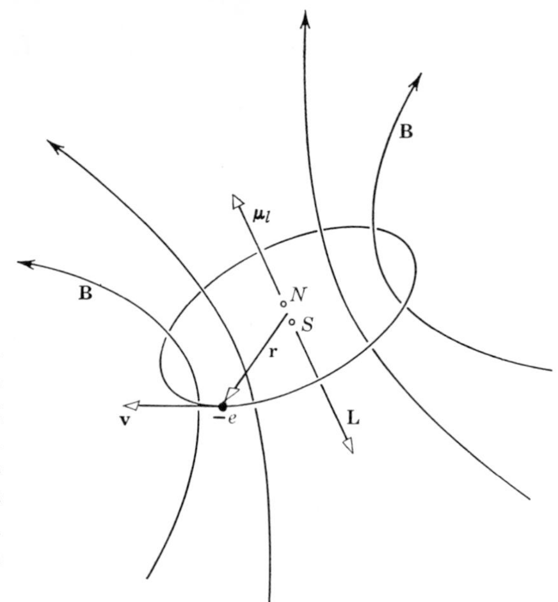


**FIGURE 8-1**
The orbital angular momentum $\mathbf{L}$ and the orbital magnetic dipole moment $\boldsymbol{\mu}_l$ of an electron $-e$ moving in a Bohr orbit. The magnetic field $\mathbf{B}$ produced by the circulating charge is indicated by the curved lines. The fictitious magnetic dipole that would produce an identical field far from the loop is indicated by its poles $N$, $S$.

current $i$ in a loop of area $A$, the magnitude of the *orbital magnetic dipole moment* $\boldsymbol{\mu}_l$ of the equivalent dipole is

$$\mu_l = iA \tag{8-2}$$

and the direction of the magnetic dipole moment is perpendicular to the plane of the orbit, in the sense indicated in Figure 8-1. The figure shows the magnetic field produced by the current loop. It also indicates the two fictitious poles of a dipole that would produce a magnetic field which becomes identical to the actual field far from the loop. The quantity $\mu_l$ specifies the strength of this magnetic dipole; it equals the product of the poles strength times their separation. Because the electron has a negative charge, its magnetic dipole moment $\boldsymbol{\mu}_l$ is antiparallel to its orbital angular momentum $\mathbf{L}$, whose magnitude is given by

$$L = mvr \tag{8-3}$$

and whose direction is illustrated in Figure 8-1.

Evaluating $i$ from (8-1), and $A$ for a circular Bohr orbit, (8-2) yields

$$\mu_l = iA = \frac{ev}{2\pi r}\pi r^2 = \frac{evr}{2} \tag{8-4}$$

Dividing by (8-3), we obtain

$$\frac{\mu_l}{L} = \frac{evr}{2mvr} = \frac{e}{2m} \tag{8-5}$$

We see that the ratio of the magnitude $\mu_l$ of the orbital magnetic dipole moment to the magnitude $L$ of the orbital angular momentum for the electron is a combination of universal constants. It is usual to write this ratio as

$$\frac{\mu_l}{L} = \frac{g_l \mu_b}{\hbar} \tag{8-6}$$

where

$$\mu_b = \frac{e\hbar}{2m} = 0.927 \times 10^{-23} \text{ amp-m}^2 \tag{8-7}$$

and

$$g_l = 1 \tag{8-8}$$

The quantity $\mu_b$ forms a natural unit for the measurement of atomic magnetic dipole moments, and is called the *Bohr magneton*. The quantity $g_l$ is called the *orbital g factor*. It is introduced, even though it appears here to be redundant, to preserve symmetry with equations we shall develop later in treating cases involving $g$ factors which are not equal to one. In terms of these quantities, we may rewrite (8-5) as a vector equation specifying both the magnitude of $\boldsymbol{\mu}_l$ and its orientation relative to $\mathbf{L}$. That is

$$\boldsymbol{\mu}_l = -\frac{g_l\mu_b}{\hbar}\mathbf{L} \tag{8-9}$$

The ratio of $\mu_l$ to $L$ does not depend on the size of the orbit or on the orbital frequency. By making a calculation similar to the one above for an elliptical orbit, it can be shown that $\mu_l/L$ is independent of the shape of the orbit. That this ratio is completely independent of the details of the orbit suggests its value might not depend on the details of the mechanical theory used to evaluate it, and this is actually the case. Upon evaluation of $\mu_l$ quantum mechanically (which cannot be done here because the electromagnetic theory required is too sophisticated), and dividing by the quantum mechanical expression $L = \sqrt{l(l+1)}\,\hbar$, the ratio of $\mu_l$ to $L$ is found to have the same value that we have obtained. Granting this, the student will accept that the correct quantum mechanical expressions for the magnitude and $z$ component of the orbital magnetic dipole moment are

$$\mu_l = \frac{g_l\mu_b}{\hbar}L = \frac{g_l\mu_b}{\hbar}\sqrt{l(l+1)}\,\hbar = g_l\mu_b\sqrt{l(l+1)} \tag{8-10}$$

and

$$\mu_{l_z} = -\frac{g_l\mu_b}{\hbar}L_z = -\frac{g_l\mu_b}{\hbar}m_l\hbar = -g_l\mu_b m_l \tag{8-11}$$

The minus sign in the last equation reflects the fact that the vector $\boldsymbol{\mu}_l$ is antiparallel to the vector $\mathbf{L}$.

Now we shall remind the student of the behavior of a magnetic dipole of moment $\boldsymbol{\mu}_l$ when it is placed in an applied magnetic field $\mathbf{B}$. In elementary electromagnetic theory it is shown that the dipole will experience a torque

$$\boldsymbol{\tau} = \boldsymbol{\mu}_l \times \mathbf{B} \tag{8-12}$$

tending to align the dipole with the field, and that, associated with this torque, there is a potential energy of orientation

$$\Delta E = -\boldsymbol{\mu}_l \cdot \mathbf{B} \tag{8-13}$$

**Example 8-1.** Assume that a magnetic dipole, whose moment has magnitude $\mu_l$, is aligned parallel to an external magnetic field, whose strength has magnitude $B$. Take $\mu_l = 1$ Bohr magneton (typical of the magnetic dipole moment of an atom), and $B = 1$ tesla (typical of the field produced by a fairly powerful electromagnet). Calculate the energy required to turn the magnetic dipole so that it is aligned antiparallel to the field.

According to (8-13), the orientational potential energy when the dipole is parallel to the field is $-\mu_l B$, and it is $+\mu_l B$ when the dipole is antiparallel to the field. So the energy that must be supplied to turn the dipole is

$$\begin{aligned}2\mu_l B &= 2 \times 0.927 \times 10^{-23}\,\text{amp-m}^2 \times 1\,\text{joule/amp-m}^2 \\ &= 1.85 \times 10^{-23}\,\text{joule} = 1.16 \times 10^{-4}\,\text{eV}\end{aligned}$$

Although this energy is very small, even by atomic standards, the dipole cannot turn unless it is supplied the energy. Conversely, if the dipole is originally aligned antiparallel to the field, it cannot turn to align itself parallel to the field unless it can get rid of the same amount of energy. ◀

If there is no way for a system, consisting of a magnetic dipole moment $\boldsymbol{\mu}_l$ in a magnetic field $\mathbf{B}$, to dissipate energy, the orientational potential energy $\Delta E$ of the system must remain constant. In these circumstances, $\boldsymbol{\mu}_l$ cannot align itself with $\mathbf{B}$. Instead $\boldsymbol{\mu}_l$ will precess around $\mathbf{B}$ in such a way that the angle between these two vectors remains constant, and that the magnitudes of both vectors remain constant. The precessional motion is a consequence of the fact that, according to (8-9) and (8-12), the torque acting on the dipole is always perpendicular to its angular momentum, in complete analogy to the case of a spinning top. The precession, and its explanation, are illustrated in Figure 8-2. It is easy to show (see the figure caption) that the magnitude of the angular frequency of precession of $\boldsymbol{\mu}_l$ about $\mathbf{B}$ is given by

$$\boldsymbol{\omega} = \frac{g_l \mu_b}{\hbar} \mathbf{B} \tag{8-14}$$

This equation also indicates that the sense of the precession is in the direction of $\mathbf{B}$. The phenomenon is known as the *Larmor precession*, and $\boldsymbol{\omega}$ is called the *Larmor frequency*.

Equation (8-14) is obtained from a classical treatment. But a quantum mechanical treatment leads to the same result, in the sense that the *expectation values* of the components perpendicular to the magnetic field of a quantum mechanical magnetic dipole moment change cyclically in time in the same way as do the actual components perpendicular to the magnetic field of a classical magnetic dipole moment. To simplify the discussion in subsequent sections,

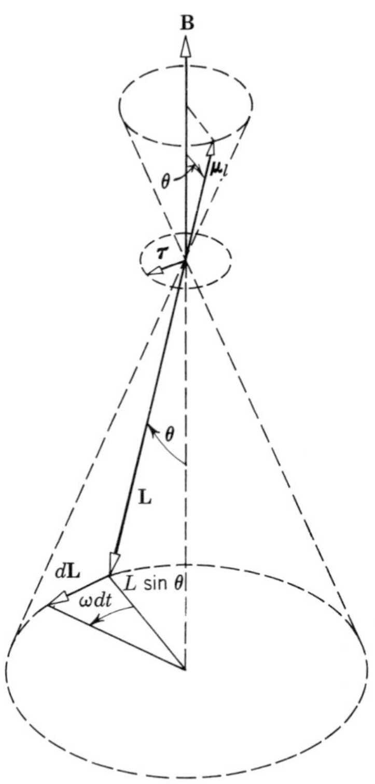


**FIGURE 8-2**

A torque $\boldsymbol{\tau} = \boldsymbol{\mu}_l \times \mathbf{B} = -(g_l\mu_b/\hbar)\mathbf{L} \times \mathbf{B}$ arises as the atom's magnetic dipole moment $\boldsymbol{\mu}_l$ interacts with the applied field $\mathbf{B}$. This torque gives rise to a change $d\mathbf{L}$ in the angular momentum during time $dt$, according to a form of Newton's law, $d\mathbf{L}/dt = \boldsymbol{\tau}$. The change $d\mathbf{L}$ causes $\mathbf{L}$ to precess through an angle $\omega\, dt$, where $\omega$ is the precessional angular velocity. From the diagram, we see that $dL = L \sin\theta\omega\, dt$, or $L\omega \sin\theta = dL/dt = \tau = (g_l\mu_b/\hbar)LB \sin\theta$. So $\omega = g_l\mu_b B/\hbar$, as in (8-14).

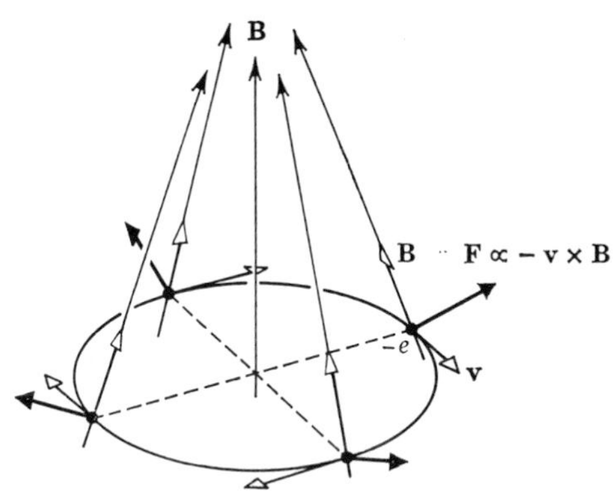


**FIGURE 8-3**

In a region where an applied field **B** is converging, an electron moves in a Bohr orbit with velocity **v**, the field exerting force **F** on the electron. Because the electron charge is negative, $\mathbf{F} \propto -\mathbf{v} \times \mathbf{B}$. Regardless of the position of the electron in the orbit, this force has a component that is radially outward and a component in the direction towards which **B** becomes more intense. Averaged over the orbit, the radial component cancels, and the average force is in that direction (upward).

we shall frequently speak of the precession of a quantum mechanical magnetic dipole moment in a magnetic field, although to be strictly correct we should speak of the cyclic change in the expectation values of its perpendicular components.

If the applied magnetic field is uniform in space, there will be no net *translational* force acting on the magnetic dipole (although there is certainly a torque). But if the field is nonuniform, there will be such a translational force (in addition to the torque). What really happens is illustrated in Figure 8-3. This figure shows that an electron moving with velocity **v** through a circular orbit, in a region in which the **B** field is converging, feels a force proportional to $-\mathbf{v} \times \mathbf{B}$ that always has a component in the direction in which the field becomes more intense. The effect can also be seen via the analogy between a fictitious magnetic dipole in a nonuniform magnetic field, and an electric dipole in a nonuniform electric field, as illustrated in Figure 8-4. Using this analogy, it is easy to show that the average force acting on the magnetic dipole is

$$\overline{F_z} = \frac{\partial B_z}{\partial z} \mu_{l_z} \tag{8-15}$$

where $z$ is the coordinate axis in the direction of increase of the field strength, and $\partial B_z/\partial z$ is the rate at which it increases. We conclude that a magnetic dipole in a nonuniform magnetic field experiences a torque, which will cause precession, and a force, which will cause displacement.

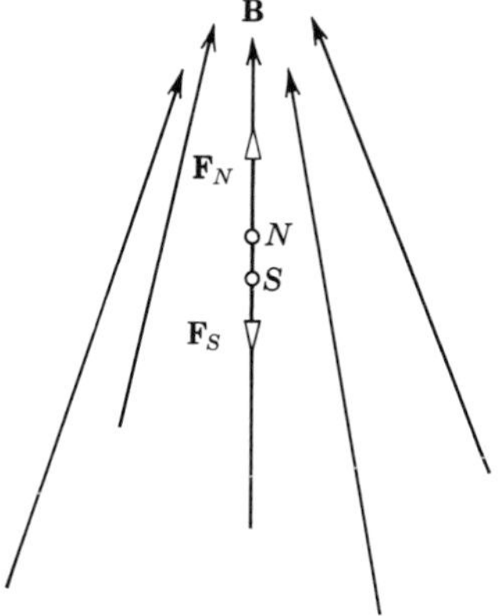


**FIGURE 8-4**

Illustrating the forces $\mathbf{F}_N$ and $\mathbf{F}_S$ acting on the poles of a fictitious magnetic dipole, equivalent to the circulating electron of Figure 8-3, located in a region where the applied field **B** is converging. Since $\mathbf{F}_N$ is greater in magnitude than $\mathbf{F}_S$, the net force on the dipole is in the direction in which **B** becomes more intense. This situation may be familiar to the student in the case in which the fields and dipole moment are electric instead of magnetic.

## 8-3 The Stern-Gerlach Experiment and Electron Spin

In 1922 Stern and Gerlach measured the possible values of the magnetic dipole moment for silver atoms by sending a beam of these atoms through a nonuniform magnetic field. A drawing of their apparatus is shown in Figure 8-5. A beam of neutral atoms is formed by evaporating silver from an oven. The beam is collimated by a diaphragm, and it enters a magnet. The cross-sectional view of the magnet shows that it produces a field that increases in intensity in the $z$ direction defined in the figure, which is also the direction of the magnetic field itself in the region of the beam. As the atoms are neutral overall, the only net force acting on them is the force $\overline{F_z}$ of (8-15), which is proportional to $\mu_{l_z}$. Since the force acting on each atom of the beam is proportional to its value of $\mu_{l_z}$, each atom is deflected in passing through the magnetic field by an amount which is proportional to $\mu_{l_z}$. Thus the beam is *analyzed* into components according to the various values of $\mu_{l_z}$. The deflected atoms strike a metallic plate, upon which they condense and leave a visible trace.

If the orbital magnetic moment vector of the atom has a magnitude $\mu_l$, then in classical physics the $z$ component $\mu_{l_z}$ of this quantity can have *any* value from $-\mu_l$ to $+\mu_l$. The reason is that classically the atom can have any orientation relative to the $z$ axis, and so this will also be true of its orbital angular momentum and its magnetic dipole moment. The predictions of quantum mechanics, as summarized by (8-11), are that $\mu_{l_z}$ can have only the *discretely quantized* values

$$\mu_{l_z} = -g_l \mu_b m_l \tag{8-16a}$$

where $m_l$ is one of the integers

$$m_l = -l, -l + 1, \ldots, 0, \ldots, +l - 1, +l \tag{8-16b}$$

Thus the classical prediction is that the deflected beam would be spread into a continuous band, corresponding to a continuous distribution of values of $\mu_{l_z}$ from one atom to the next. The quantum mechanical prediction is that the deflected beam would be split into several discrete components. Furthermore, quantum mechanics predicts that this should happen for all orientations of the analyzing magnet. That is, the magnet is essentially acting as a measuring device which investigates the quantization of the component of the magnetic dipole moment along a $z$ axis, which it defines by

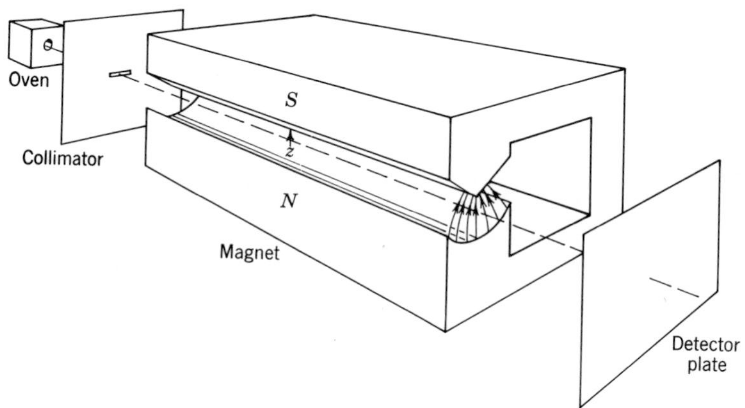


**FIGURE 8-5**

The Stern-Gerlach apparatus. The field between the two magnet pole pieces is indicated by the field lines drawn at the near end of the magnet. The field intensity increases in the positive $z$ direction (upward).

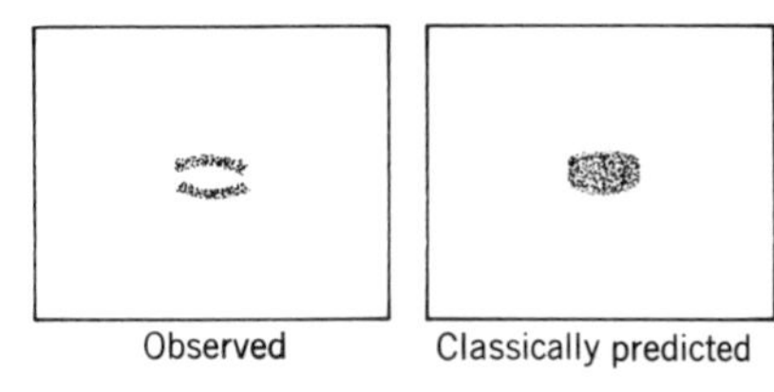


**FIGURE 8-6**

The deflection pattern recorded on the detecting plate in a Stern-Gerlach measurement of the $z$ component of the magnetic dipole moment of silver atoms. Maximum deflection occurs at the center of the beam because the atoms there pass through the region of maximum field gradient, $\partial B_z/\partial z$. The observed pattern consists of two discrete components due to space quantization. According to the classical prediction a continuous band would be expected.

the direction in which its field increases in intensity. Since, according to quantum mechanics, $\mu_{l_z}$ should be quantized for any choice of the $z$ direction because $L_z$ is quantized for any choice of that direction, the same results should be obtained for all positions of the analyzing magnet.

Stern and Gerlach found that the beam of silver atoms is split into two discrete components, one component being bent in the positive $z$ direction and the other bent in the negative $z$ direction. Figure 8-6 shows the type of pattern observed on the detecting plate. They also found that these results were obtained independent of the choice of the $z$ direction. The experiment was repeated using several other species of atoms, and in each case investigated it was found that the deflected beam is split into two, or more, discrete components. The results are, *qualitatively*, very direct experimental proof of the quantization of the $z$ component of the magnetic dipole moments of atoms and, therefore of their angular momenta. In other words, the experiments showed that the orientation in space of atoms is quantized. The phenomenon is called *space quantization.*

But the results of the Stern-Gerlach experiment are not *quantitatively* in agreement with (8-16a) and (8-16b), the equations summarizing the predictions of the theory we have developed. According to these equations, the number of possible values of $\mu_{l_z}$ is equal to the number of possible values of $m_l$, which is $2l + 1$. Since $l$ is an integer, this is always an odd number. Also for any value of $l$ one of the possible values of $m_l$ is zero. Thus the fact that the beam of silver atoms is split into only two components, both of which are deflected, indicates either that something is wrong with the Schroedinger theory of the atom, or that the theory is incomplete.

The theory is not wrong (we shall see later that atoms do have orbital angular momenta and magnetic dipole moments with the predicted properties); but, as it stands, the Schroedinger theory of the atom is *incomplete*. This is shown most clearly by an experiment performed in 1927 by Phipps and Taylor, who used the Stern-Gerlach technique on a beam of hydrogen atoms. The experiment is particularly significant because the atoms contain a single electron, so the theory we have developed makes unambiguous predictions. Since the atoms in the beam are in their ground state because of the relativity low temperature of the oven, the theory predicts that the quantum number $l$ has the value $l = 0$. Then there is only one possible value of $m_l$, namely $m_l = 0$, and we expect that the beam will be unaffected by the magnetic field since $\mu_{l_z}$ will be equal to zero. However, Phipps and Taylor found that the beam is split into two symmetrically deflected components. Thus there is certainly some magnetic dipole moment in the atom which we have not hitherto considered.

One possibility is a magnetic dipole moment associated with motion of charges in the nucleus. The magnitude of such a magnetic dipole moment would be of the order of $e\hbar/2M$, where $M$ is the mass of a proton. But the magnetic dipole moment measured experimentally from the size of the splitting is of the order of $\mu_b = e\hbar/2m$, where $m$ is the mass of an electron, which is about 2000 times larger. Therefore, the nucleus cannot be responsible for the observed magnetic dipole moment. Its source must be the electron.

This leads us to some reasonable assumptions, which are also supported by other evidence to be discussed shortly. We assume that an electron has an intrinsic (built-in) magnetic dipole moment $\boldsymbol{\mu}_s$, due to the fact that it has an intrinsic angular momentum $\mathbf{S}$ called its *spin*. From a classical point of view, we can think, at least crudely, of the electron producing the external magnetic field of a magnetic dipole because of the current loops associated with its spinning charge. We also assume that the magnitude $S$ and $z$ component $S_z$ of the *spin angular momentum* are related to two quantum numbers, $s$ and $m_s$, by quantization relations which are identical to those for orbital angular momentum. That is

$$S = \sqrt{s(s+1)}\,\hbar \tag{8-17}$$

$$S_z = m_s\hbar \tag{8-18}$$

(Note that $S_x$ and $S_y$ are not quantized, as is also the case for $L_x$ and $L_y$.) We further assume that the relation between the *spin magnetic dipole moment* and the spin angular momentum is of the same form as the relation for the orbital case. That is

$$\boldsymbol{\mu}_s = -\frac{g_s\mu_b}{\hbar}\mathbf{S} \tag{8-19}$$

$$\mu_{s_z} = -g_s\mu_b m_s \tag{8-20}$$

The quantity $g_s$ is called the *spin g factor*.

From the experimental observation that the beam of hydrogen atoms is split into two symmetrically deflected components, it is apparent that $\mu_{s_z}$ can assume just two values, which are equal in magnitude but opposite in sign. If we make the final assumption that the possible values of $m_s$ differ by one and range from $-s$ to $+s$, as is true of the quantum numbers $m_l$ and $l$ for orbital angular momentum, then we can conclude that the two possible values of $m_s$ are

$$m_s = -1/2,\ +1/2 \tag{8-21}$$

and that $s$ has the single value

$$s = 1/2 \tag{8-22}$$

By measuring the splitting of the beam of hydrogen atoms, it is possible to evaluate the net force $\overline{F_z}$ they feel while traversing the magnetic field. From analogy to (8-15), and from (8-20), this is $\overline{F_z} = -(\partial B_z/\partial z)\mu_b g_s m_s$. Since $\mu_b$ is known and $\partial B_z/\partial z$ can be measured, the experiments determine the value of the quantity $g_s m_s$. Within their accuracy, it was found that $g_s m_s = \pm 1$. Since we have concluded that $m_s = \pm 1/2$, this implies

$$g_s = 2 \tag{8-23}$$

These conclusions are confirmed by many different experiments. For instance, in the *Zeeman effect* a uniform external magnetic field is applied to a collection of atoms, and measurements are made of the potential energies of orientation in the field of the magnetic dipole moments of the atoms. As we shall discuss in detail in Chapter 10, this is done by measuring the splitting of the spectral line emitted when the atoms decay from some higher energy level to their ground state energy level. The splitting of the line occurs because the levels themselves are split according to the different values assumed by the orientational potential energy of the atoms. A simple example is the Zeeman effect for hydrogen atoms. In their ground state these atoms have no orbital angular momentum, and therefore no orbital magnetic dipole moments. But the measurements show that their ground state energy level is split by the applied magnetic field into two components, symmetrically disposed about the energy of the ground state in the absence of a field. This splitting reflects the two possible values of the orientational potential energy

$$\begin{aligned}\Delta E &= -\boldsymbol{\mu}_s \cdot \mathbf{B} = -\mu_{s_z} B \\ &= g_s \mu_b m_s B \\ &= \pm g_s \mu_b B/2\end{aligned}$$

where the $z$ axis is taken in the direction of the applied field. The fact that the level is symmetrically split into two components confirms the conclusion that $m_s = \pm 1/2$, and the measured magnitude of the splitting confirms the conclusion that $g_s = 2$.

Recent spectroscopic measurements of Lamb, using a technique of extreme accuracy, actually have shown that $g_s = 2.00232$. However in almost all situations it is quite adequate to say simply that the spin $g$ factor for an electron is twice as large as its orbital $g$ factor; i.e., that the spin magnetic dipole moment is *twice as large*, compared to the spin angular momentum, as the orbital magnetic dipole moment is compared to the orbital angular momentum. On the other hand, $\boldsymbol{\mu}_s$ and $\mathbf{S}$ are antiparallel, just like $\boldsymbol{\mu}_l$ and $\mathbf{L}$, because the relative orientation of either pair of vectors depends only on the fact that the electron has a negative charge.

**Example 8-2.** A beam of hydrogen atoms, emitted from an oven running at a temperature $T = 400°\text{K}$, is sent through a Stern-Gerlach magnet of length $X = 1$ m. The atoms experience a magnetic field with a gradient of 10 tesla/m. Calculate the transverse deflection of a typical atom in each component of the beam, due to the force exerted on its spin magnetic dipole moment, at the point where the beam leaves the magnet.

At this temperature, the atoms are in their ground state and have no orbital angular momentum or orbital magnetic dipole moment. They typically have kinetic energy $(3/2)kT$, where $k$ is Boltzmann's constant. From (8-15) and (8-20), they experience a transverse force

$$F_z = -\frac{\partial B_z}{\partial z}\, \mu_b g_s m_s$$

Since $g_s m_s = \pm 1$, this is

$$F_z = \pm \frac{\partial B_z}{\partial z}\, \mu_b$$

The typical longitudinal velocity $v_x$ of an atom of mass $M$ in traveling through the magnet can be evaluated by setting

$$\frac{1}{2} M v_x^2 = \frac{3}{2} kT$$

So

$$v_x = \sqrt{\frac{3kT}{M}}$$

Thus the time $t$ the atom experiences the transverse force in traveling through the magnet of length $X$ is

$$t = \frac{X}{v_x} = \frac{X}{\sqrt{\dfrac{3kT}{M}}} = X\sqrt{\frac{M}{3kT}}$$

Because of the force they have a transverse acceleration $a_z = F_z/M$, and so suffer a transverse deflection

$$Z = \frac{1}{2}a_z t^2 = \frac{1}{2}\frac{F_z}{M}\frac{X^2 M}{3kT}$$

$$= \pm \frac{\frac{\partial B_z}{\partial z}\mu_b X^2}{6kT}$$

$$= \pm \frac{10 \text{ tesla/m} \times 0.927 \times 10^{-23} \text{ amp-m}^2 \times 1 \text{ m}^2}{6 \times 1.38 \times 10^{-23} \text{ joule/}^\circ\text{K} \times 400^\circ\text{K}}$$

$$= \pm 2.8 \times 10^{-3} \text{ m}$$

The separation of the two components is more than half a centimeter, which is very easy to observe. ◀

The idea of electron spin was introduced some time before the work of Phipps and Taylor. In the final sentence of a research paper on the scattering of x rays by atoms, published in 1921, Compton had written, "May I then conclude that the electron itself, spinning like a tiny gyroscope, is probably the ultimate magnetic particle." This was really more of a speculation than a conclusion, and Compton apparently never followed it further.

Credit for the introduction of electron spin is generally given to Goudsmit and Uhlenbeck. In 1925, as graduate students, they were trying to understand why certain lines of the optical spectra of hydrogen and the alkali atoms are composed of a closely spaced pair of lines. This is the *fine structure*, which had been treated by Sommerfeld in terms of the Bohr model as due to a splitting of the atomic energy levels because of a small (about one part in $10^4$) contribution to the total energy resulting from the relativistic variation of electron mass with velocity (see Section 4-10). The results of Sommerfeld were in good numerical agreement with the observed fine structure of hydrogen. But the situation was not so satisfactory for the alkalis. In these atoms the electron responsible for the optical spectrum would be expected to move in a Bohr-like orbit of large radius at low velocity, so the relativistic variation of mass would be expected to be small. However, the fine structure splitting was observed to be very much larger than in hydrogen. Consequently, doubt arose concerning the validity of Sommerfeld's explanation of the origin of fine structure. In considering other possibilities, Goudsmit and Uhlenbeck proposed that an electron has an intrinsic angular momentum and magnetic dipole moment, whose $z$ components are specified by a fourth quantum number $m_s$, which can assume either of two values, $-1/2$ and $+1/2$. The splitting of the atomic energy levels could then be understood as due to a potential energy of orientation of the magnetic dipole moment of the electron in the magnetic field that is present in the atom because it contains moving charged particles. The energy of orientation would be either positive or negative depending on the sign of $m_s$, i.e., depending on whether the spin is "up" or "down" relative to the direction of the *internal* magnetic field of the atom. (This should not be confused with the previously mentioned Zeeman effect, which involves the splitting of energy levels of an atom due to the orientational potential energy of its magnetic dipole moment in an *external* magnetic field applied to the atom.) Uhlenbeck has described the circumstances as follows:

"Goudsmit and myself hit upon this idea by studying a paper of Pauli, in which the famous exclusion principle (to be treated in Chapter 9) was formulated and in which, for the first time, *four* quantum numbers were ascribed to the electron. This was done rather formally; no concrete picture was connected with it. To us this was a mystery. We were so conversant with the proposition that every quantum number corresponds to a degree of freedom (an independent coordinate), and on the other hand with the idea of a point electron, which obviously had three degrees of freedom only, that we could not place the fourth quantum number. We could understand it only if the electron was assumed to be a small sphere that could rotate....

Somewhat later we found in a paper of Abraham, to which Ehrenfest drew our attention, that for a rotating sphere with surface charge the necessary factor two in the magnetic moment ($g_s = 2$) could be understood classically. This encouraged us, but our enthusiasm was considerably reduced when we saw that the rotational velocity at the surface of the electron had to be many times the velocity of light! I remember that most of these thoughts came to us on an afternoon at the end of September 1925. We were excited, but we had not the slightest intention of publishing anything. It seemed so speculative and bold, that something ought to be wrong with it, especially since Bohr, Heisenberg, and Pauli, our great authorities, had never proposed anything of the kind. But of course we told Ehrenfest. He was impressed at once, mainly, I feel, because of the visual character of our hypothesis, which was very much in his line. He called our attention to several points, e.g., to the fact that in 1921 A. H. Compton already had suggested the idea of a spinning electron as a possible explanation of the natural unit of magnetism, and finally said that it was either highly important or nonsense, and that we should write a short note for *Naturwissenschaften* (a physics research journal) and give it to him. He ended with the words 'and then we will ask Lorentz.' This was done. Lorentz received us with his well known great kindness, and he was very much interested, although, I feel, somewhat skeptical too. He promised to think it over. And in fact, already next week he gave us a manuscript, written in his beautiful handwriting, containing long calculations on the electromagnetic properties of rotating electrons. We could not fully understand it, but it was quite clear that the picture of the rotating electron, if taken seriously, would give rise to serious difficulties. For one thing, the magnetic energy would be so large that by the equivalence of mass and energy the electron would have a larger mass than the proton, or, if one sticks to the known mass, the electron would be bigger than the whole atom! In any case, it seemed to be nonsense. Goudsmit and myself both felt that it might be better for the present not to publish anything; but when we said this to Ehrenfest, he answered: 'I have already sent your letter in long ago; you are both young enough to allow yourselves some foolishness!' " (from *The Conceptual Development of Quantum Mechanics* by Max Jammer, McGraw-Hill, 1966)

The most recent experimental evidence indicates that the electron is a point particle, and certainly not "bigger than the whole atom." One set of experiments studies the scattering of electrons by electrons at very high kinetic energies. If these objects had appreciable extent in space, in collisions which were so close that they overlap the force acting between them would be modified—just as in the close collision of an $\alpha$ particle and a nucleus. It was found that the electrons always act like two point objects, with charge $-e$ and magnetic dipole moment $\boldsymbol{\mu}_s$, even in the closest collisions investigated. Thus electrons have an extent less than this collision distance, which is about $10^{-16}$ m. In comparison to the dimensions of an atom ($10^{-10}$ m), or even the dimensions of a nucleus ($10^{-14}$ m), electrons have negligible dimensions.

Although the electron seems to be a point particle, four quantum numbers are required to specify its quantum states. The first three arise because three independent coordinates are required to describe its location in three-dimensional space. The fourth arises because it is also necessary to describe the orientation in space of its spin, which can be either "up" or "down" relative to some $z$ axis. For a classical point particle, there is room only for the first three quantum numbers. But the electron is not a classical particle.

Schroedinger quantum mechanics is completely compatible with the existence of electron spin; but it does not predict it, so spin must be introduced as a separate postulate. The reason for this is that the theory is an approximation which ignores relativistic effects. The student will recall that the theory is based on the nonrelativistic energy equation, $E = p^2/2m + V$. The student may also recall reading in Chapter 5 brief mention of the fact that Dirac developed a relativistic theory of quantum mechanics in 1929. Using the same postulates as the Schroedinger theory, but replacing the energy equation by its relativistic form, $E = (c^2p^2 + m_0^2c^4)^{1/2} + V$, Dirac showed that an electron *must* have an intrinsic $s = 1/2$ angular momentum, an intrinsic magnetic dipole moment with a $g$ factor of 2, and all the other properties we have stated previously. This was a great triumph for relativity theory; it put electron spin on a firm theoretical foundation and showed that *electron spin is intimately connected with relativity*. A quantitative treatment of the Dirac theory would, unfortunately, be beyond the level of this book, but we shall from time to time describe qualitatively its results.

Another aspect of the nonclassical character of spin can be seen by noting that the quantum number $s$, which specifies the magnitude of the spin angular momentum $S$, has the fixed value 1/2. Therefore, we cannot take $S$ to the classical limit by letting $s \rightarrow \infty$, as we did in Section 7-8 for the magnitude of the orbital angular momentum $L$ by letting its quantum number $l \rightarrow \infty$. An equivalent statement is that in the classical limit the magnitude of $S$ is completely negligible because $\hbar$ is so small, so spin is essentially nonclassical. This being the case, it is sometimes more harmful than helpful to think of spin in terms of a classical model like a small spinning sphere; but it must be admitted that it is difficult to avoid thinking in such terms.

## 8-4 The Spin-Orbit Interaction

Although spin itself is subtle, there is nothing subtle about many of the effects it produces. Perhaps the most important is that it doubles the number of electrons which the "exclusion principle" allows to populate the quantum states of multielectron atoms. When we study this effect in Chapter 10, we shall see that the ground states of atoms would be very much altered if electrons did not have spin. This would have profound consequences on the periodic properties of atoms, and therefore on all of chemistry and solid state physics.

In the present section we shall study the interaction between an electron's spin magnetic dipole moment and the *internal* magnetic field of a one-electron atom. Since the internal magnetic field is related to the electron's orbital angular momentum, this is called the *spin-orbit interaction*. It is a relatively weak interaction which is responsible, in part, for the *fine structure* of the excited states of one-electron atoms.

The spin-orbit interaction also occurs in multielectron atoms, but in such atoms it is reasonably strong because the internal magnetic fields are very strong. Furthermore, an effect completely analogous to the spin-orbit interaction occurs in nuclei. The nuclear spin-orbit interaction is so strong that it governs the periodic properties of nuclei.

The origin of the internal magnetic field experienced by an electron moving in a one-electron atom is easy to understand, if we consider the motion of the nucleus from the point of view of the electron. In a frame of reference fixed on the electron, the charged nucleus moves around the electron and the electron is, in effect, located inside a current loop which produces the magnetic field. The argument is illustrated qualitatively in Figure 8-7. To make the argument quantitative, we note that the charged nucleus moving with velocity $-\mathbf{v}$ constitutes a current element $\mathbf{j}$, where

$$\mathbf{j} = -Ze\mathbf{v}$$

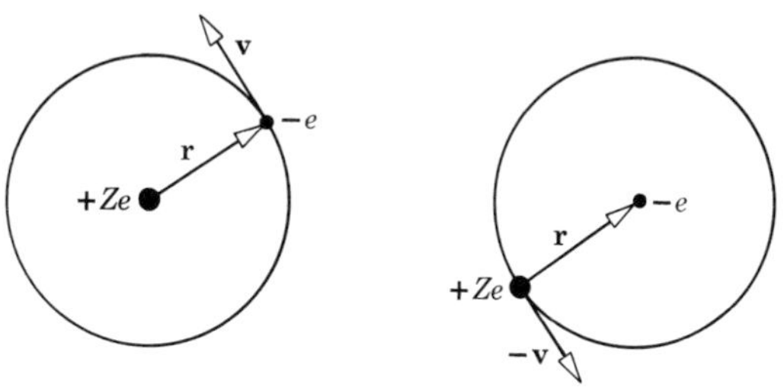


**FIGURE 8-7**

*Left:* An electron moves in a circular Bohr orbit, the motion as seen by the nucleus. *Right:* The same motion, but as seen by the electron. From the point of view of the electron, the nucleus moves around it. The magnetic field **B** experienced by the electron is in the direction out of the page at the electron's location.

According to Ampere's law, this produces a magnetic field **B** which, at the position of the electron, is

$$\mathbf{B} = \frac{\mu_0}{4\pi}\frac{\mathbf{j} \times \mathbf{r}}{r^3} = -\frac{Ze\mu_0}{4\pi}\frac{\mathbf{v} \times \mathbf{r}}{r^3}$$

It is convenient to express this in terms of the electric field **E** acting on the electron. According to Coulomb's law

$$\mathbf{E} = \frac{Ze}{4\pi\epsilon_0}\frac{\mathbf{r}}{r^3}$$

From the last two equations, we have

$$\mathbf{B} = -\epsilon_0\mu_0\mathbf{v} \times \mathbf{E}$$

or

$$\mathbf{B} = -\frac{1}{c^2}\mathbf{v} \times \mathbf{E} \tag{8-24}$$

since $c = 1/\sqrt{\epsilon_0\mu_0}$. The quantity **B** is the magnetic field strength experienced by the electron when it is moving with velocity **v** relative to the nucleus, and therefore through the electric field of strength **E** which the nucleus exerts on it. Equation (8-24) is actually of very general validity, and it can be derived from relativistic considerations.

The electron and its spin magnetic dipole moment can assume different orientations in the internal magnetic field of the atom, and its potential energy is different for each of these orientations. If we evaluate the orientational potential energy of the magnetic dipole moment in this magnetic field, from an equation analogous to (8-13), we have

$$\Delta E = -\boldsymbol{\mu}_s \cdot \mathbf{B}$$

Using (8-19), this can be written in terms of the electron's spin angular momentum **S** as

$$\Delta E = \frac{g_s\mu_b}{\hbar}\mathbf{S} \cdot \mathbf{B}$$

But this energy has been evaluated in a frame of reference in which the electron is at rest, whereas we are interested in the energy as measured in the normal frame of reference in which the nucleus is at rest. Because of an effect of the relativistic transformation of velocities, called the *Thomas precession*, the transformation back to the nuclear rest frame results in a reduction of the orientational potential energy by a factor of 2. Thus, the spin-orbit interaction energy is

$$\Delta E = \frac{1}{2}\frac{g_s\mu_b}{\hbar}\mathbf{S} \cdot \mathbf{B} \tag{8-25}$$

The transformation leading to the factor of 2 is interesting, but rather complicated, so we shall not carry it out here. (It is carried out in Appendix J.)

We shall find it convenient to express (8-25) in terms of $\mathbf{S} \cdot \mathbf{L}$, the scalar, or dot, product of the spin and orbital angular momentum vectors. To this end, we use, in (8-24), the relation

$$-e\mathbf{E} = \mathbf{F}$$

between the electric field $\mathbf{E}$ and the force $\mathbf{F}$ acting on the electron of charge $-e$. We also use the relation

$$\mathbf{F} = -\frac{dV(r)}{dr}\frac{\mathbf{r}}{r}$$

between the force and the potential. (The term $\mathbf{r}/r$ is a unit vector in the radial direction which gives $\mathbf{F}$ its proper direction.) With these relations, (8-24) becomes

$$\mathbf{B} = -\frac{1}{ec^2}\frac{1}{r}\frac{dV(r)}{dr}\mathbf{v} \times \mathbf{r}$$

Multiplying and dividing by the electron mass $m$ allows us to write this in terms of the orbital angular momentum, $\mathbf{L} = \mathbf{r} \times m\mathbf{v} = -m\mathbf{v} \times \mathbf{r}$, as

$$\mathbf{B} = \frac{1}{emc^2}\frac{1}{r}\frac{dV(r)}{dr}\mathbf{L} \tag{8-26}$$

Note that the strength of the magnetic field $\mathbf{B}$, experienced by the electron because it is moving about the nucleus with orbital angular momentum $\mathbf{L}$, is proportional to the magnitude of $\mathbf{L}$, and also that the magnetic field vector is in the same direction as the angular momentum vector. With this result, we can express the spin-orbit interaction energy, (8-25), as

$$\Delta E = \frac{g_s\mu_b}{2emc^2\hbar}\frac{1}{r}\frac{dV(r)}{dr}\mathbf{S} \cdot \mathbf{L}$$

Evaluating $g_s$ and $\mu_b$, we obtain

$$\Delta E = \frac{1}{2m^2c^2}\frac{1}{r}\frac{dV(r)}{dr}\mathbf{S} \cdot \mathbf{L} \tag{8-27}$$

This equation was first derived in 1926 by Thomas, using as we have a combination of the Bohr model, Schroedinger quantum mechanics, and relativistic kinematics. However, it is in complete agreement with the results of the relativistic quantum mechanics of Dirac. It is important in the theory of multielectron atoms as well as of one-electron atoms. Furthermore, a similar equation is central to the understanding of the theory of the structure of nuclei, as we shall see later in the book.

**Example 8-3.** Estimate the magnitude of the orientational potential energy $\Delta E$ for the $n = 2, l = 1$ state of the hydrogen atom, to check whether it is of the same order of magnitude as the observed fine-structure splitting of the corresponding energy level. (There is no spin-orbit energy in the $n = 1$ state, since for $n = 1$ the only possible value for $l$ is $l = 0$, which means $L = 0$.)

The potential is

$$V(r) = -\frac{e^2}{4\pi\epsilon_0}r^{-1}$$

So

$$\frac{dV(r)}{dr} = \frac{e^2}{4\pi\epsilon_0}r^{-2}$$

and

$$\Delta E = \frac{e^2}{4\pi\epsilon_0 2m^2c^2}\frac{1}{r^3}\mathbf{S}\cdot\mathbf{L}$$

The magnitude of $\mathbf{S}\cdot\mathbf{L}$ is approximately $\hbar^2$ since each of these angular momentum vectors has a magnitude of approximately $\hbar$. The expectation value of $1/r^3$ for the $n = 2$ state is approximately $1/(3a_0)^3$. Thus

$$|\Delta E| \sim \frac{e^2}{4\pi\epsilon_0 2m^2c^2}\frac{1}{3^3}\frac{m^3e^6}{(4\pi\epsilon_0)^3\hbar^6}\hbar^2 = \frac{me^8}{54 \times (4\pi\epsilon_0)^4c^2\hbar^4}$$

$$= \frac{(9 \times 10^9 \text{nt-m}^2/\text{coul}^2)^4 \times 9 \times 10^{-31}\text{ kg} \times (1.6 \times 10^{-19}\text{ coul})^8}{54 \times (3 \times 10^8\text{ m/sec})^2 \times (1.1 \times 10^{-34}\text{ joule-sec})^4}$$

$$\sim 10^{-23}\text{ joule} \sim 10^{-4}\text{ eV}$$

Since $\mathbf{S}\cdot\mathbf{L}$ can be either positive or negative, depending on the relative orientation of the two vectors, the energy level is split by roughly $2 \times 10^{-4}$ eV.

Comparing this with the energy of the $n = 2$, $l = 1$ level of hydrogen, $E_2 = -3.4$ eV, we see that the ratio of the predicted energy splitting to the energy itself, $|\Delta E/E|$, is about one part in $10^4$. This is in reasonable agreement with the splitting required to explain the fine structure of the lines of the hydrogen spectrum associated with this level, as discussed in Section 4-10, and therefore it provides some confirmation of the theory we have developed. A more detailed comparison of the theory with experiment will be made shortly. ◀

**Example 8-4.** Estimate the magnitude of the magnetic field **B** acting on the spin magnetic dipole moment of the electron in Example 8-3.

From an equation analogous to (8-13), we have $\Delta E = -\boldsymbol{\mu}_s\cdot\mathbf{B}$. So

$$|\Delta E| \sim \mu_s B$$

where

$$\mu_s \sim \mu_b \sim 10^{-23}\text{ amp-m}^2$$

Therefore

$$B \sim \frac{10^{-23}\text{ joule}}{10^{-23}\text{ amp-m}^2} \sim 1\text{ tesla}$$

This is about equal to the field produced by an electromagnet operating at the limit at which its iron core saturates. We see that the electron's spin magnetic dipole moment feels a strong magnetic field because it is moving at a high velocity through the strong electric field surrounding the nucleus. ◀

## 8-5 Total Angular Momentum

If there were no spin-orbit interaction, the orbital and spin angular momenta **L** and **S** of an atomic electron would be independent of each other, and so they would independently obey the quantum mechanical angular momentum conservation law discussed in Section 7-8. That is, when an atom without spin-orbit interaction is in free space there would be no torques acting on either **L** or **S**, so both these vectors would precess randomly about the $z$ axis in such a way that their magnitude and $z$ components, $L$, $L_z$, $S$, $S_z$, would have fixed values. These fixed values are the ones specified by the quantum numbers $l$, $m_l$, $s$, $m_s$.

However, there *is* a spin-orbit interaction. That is, a strong internal magnetic field is acting on the atomic electron, the orientation of which is determined by **L**, and produces a torque on its spin magnetic dipole moment, the orientation of which is determined by **S**. As in the case of the Larmor precession of Section 8-2, the torque will not change the magnitude of **S**. Nor will the reaction torque acting on **L** change its

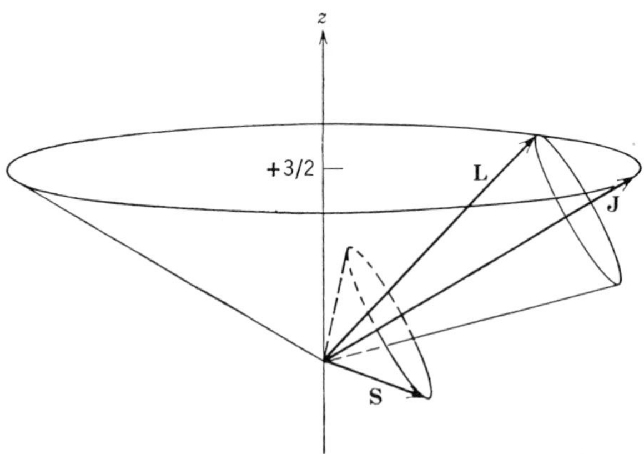


**FIGURE 8-8**

The angular momentum vectors **L**, **S**, and **J** for a typical case of a state with $l = 2, j = 5/2, m_j = 3/2$. The vectors **L** and **S** precess uniformly about their sum **J**, as **J** precesses randomly about the $z$ axis.

magnitude. But the torque does enforce a coupling between **L** and **S** which makes the orientation of each dependent on the orientation of the other. As a result, these angular momentum vectors undergo a different motion than they would if there were no spin-orbit interaction. They precess around their sum, instead of both precessing around the $z$ axis. Since these vectors do not precess around the $z$ axis, their $z$ components, $L_z$ and $S_z$, do not have fixed values when there is a spin-orbit interaction.

The situation is illustrated in Figure 8-8, which shows **L** and **S** precessing due to the spin-orbit interaction coupling. Their motion is involved, but not as involved as it might be because they must move in such a way that their sum, the *total angular momentum* **J**, satisfies the quantum mechanical angular momentum conservation law. That is, if the atom is in free space so that no external torques act on it, its total angular momentum

$$\mathbf{J} = \mathbf{L} + \mathbf{S} \tag{8-28}$$

maintains a fixed magnitude $J$ and a fixed $z$ component $J_z$, while its $x$ and $y$ components $J_x$ and $J_y$ fluctuate about zero. The vectors **L** and **S** precess around their sum **J**, and their components in the direction of **J** remain fixed so that its magnitude $J$ is fixed. Simultaneously, **J** precesses around the $z$ axis, maintaining a fixed component $J_z$. As we continue our studies of atoms, we shall find the total angular momentum to be quite useful because of the simple behavior of its magnitude and $z$ component. This is particularly so in the case of multielectron atoms, where the many orbital and spin angular momenta, that compose the total angular momentum, have very complicated behaviors.

By using techniques closely related to those we used in Section 7-8 to study the properties of the orbital angular momentum, it can be shown that the magnitude and $z$ component of the total angular momentum **J** are specified by two quantum numbers $j$ and $m_j$, according to the usual quantization conditions

$$J = \sqrt{j(j+1)}\,\hbar \tag{8-29}$$

and

$$J_z = m_j\hbar \tag{8-30}$$

The possible values of the quantum number $m_j$ are, as would be expected

$$m_j = -j, -j+1, \ldots, +j-1, +j \tag{8-31}$$

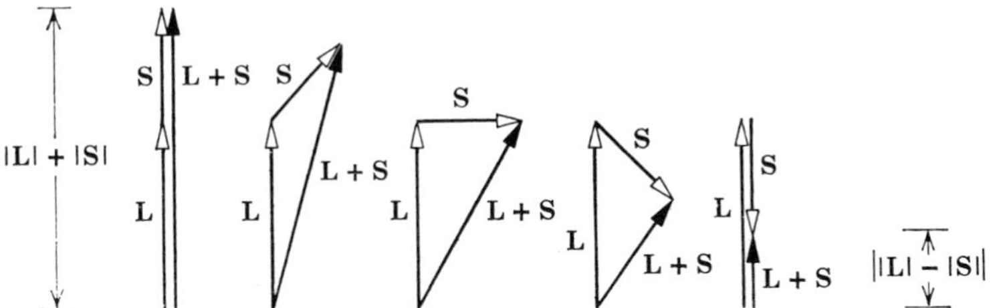


**FIGURE 8-9**
Vector diagrams which show that for any two vectors **L** and **S** the magnitude $|\mathbf{L} + \mathbf{S}|$ of their sum is always at least as large as the magnitude of the difference in their magnitudes, $||\mathbf{L}| - |\mathbf{S}||$. The case for which $|\mathbf{L}| > |\mathbf{S}|$ is shown; the student can show in his own diagram that the conclusion is unaltered if $|\mathbf{L}| < |\mathbf{S}|$.

We may determine the possible values of the quantum number $j$ by taking the $z$ component of (8-28), which defines **J**. This gives

$$J_z = L_z + S_z$$

which is

$$m_j\hbar = m_l\hbar + m_s\hbar$$

or

$$m_j = m_l + m_s \tag{8-32}$$

Since the maximum possible value of $m_l$ is $l$, and the maximum possible value of $m_s$ is $s = 1/2$, the maximum possible value of $m_j$ is

$$(m_j)_{\text{max}} = l + 1/2$$

According to (8-31), this is also the maximum possible value of $j$. In common with the other angular momentum quantum numbers, the possible values of $j$ differ by integers. Therefore these values must be members of the decreasing series

$$j = l + 1/2, l - 1/2, l - 3/2, l - 5/2, \ldots$$

To determine where the series terminates, we may use the vector inequality

$$|\mathbf{L} + \mathbf{S}| \geq \big||\mathbf{L}| - |\mathbf{S}|\big|$$

whose validity the student may easily demonstrate by inspecting Figure 8-9. Writing $\mathbf{L} + \mathbf{S}$ as $\mathbf{J}$, we have from the inequality

$$|\mathbf{J}| \geq \big||\mathbf{L}| - |\mathbf{S}|\big|$$

or

$$\sqrt{j(j+1)}\,\hbar \geq \left|\sqrt{l(l+1)}\,\hbar - \sqrt{s(s+1)}\,\hbar\right|$$

From this it can be shown with no difficulty that since $s = 1/2$ there are generally two members of the series which satisfy the inequality. These are

$$j = l + 1/2, l - 1/2 \tag{8-33a}$$

It is even more apparent that if $l = 0$ there is only one possible value of $j$, namely

$$j = 1/2 \qquad \text{if } l = 0 \tag{8-33b}$$

The content of the equations stating the possible values of the quantum numbers $m_j$ and $j$ can be represented in terms of the rules of vector addition, by constructing a set of vectors whose lengths are proportional to the values of the quantum numbers $l$, $s$, and $j$. This is illustrated in the following example.

**Example 8-5.** Enumerate the possible values of the quantum numbers $j$ and $m_j$, for states in which $l = 2$ and, of course, $s = 1/2$.

According to (8-33a), the two possible values of $j$ are 5/2 and 3/2. According to (8-31), for $j = 5/2$ the possible values of $m_j$ are $-5/2$, $-3/2$, $-1/2$, 1/2, 3/2, 5/2. The same equation states that for $j = 3/2$ the possible values of $m_j$ are $-3/2$, $-1/2$, 1/2, 3/2. Vector diagrams for this case are shown in Figure 8-10. Inspection should make their interpretation obvious. ◀

Vector diagrams of the type shown in Figure 8-10 represent only the rules for adding the *quantum numbers* $l$ and $s$ to obtain the possible values of the quantum numbers $j$ and $m_j$. *If* the relation between the magnitude of an angular momentum vector, such as $L$, and its associated quantum number were $L = l\hbar$, instead of $L = \sqrt{l(l+1)}\,\hbar$, these diagrams would also represent the addition of the *angular momenta* **L** and **S** to obtain the angular momentum **J** and its $z$ component $J_z$. Since

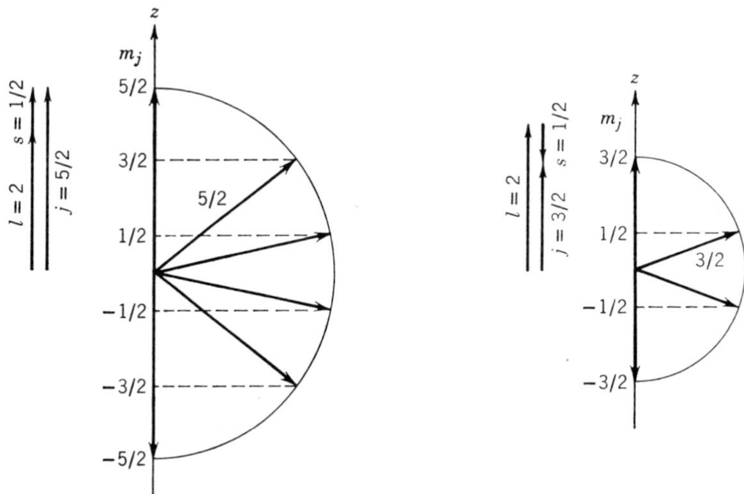


**FIGURE 8-10**

Vector diagrams representing the rules for adding the quantum numbers $l = 2$ and $s = 1/2$ to obtain the possible values for the quantum numbers $j$ and $m_j$. *Left:* The maximum possible value of $j$ is obtained when a vector of magnitude $l$ is added to a parallel vector of magnitude $s$, yielding $j = l + s = 2 + 1/2 = 5/2$. The maximum possible $z$ component of this vector gives the maximum possible value of the quantum number $m_j$, and the minimum possible $z$ component gives the minimum possible value of $m_j$. The intermediate values of $m_j$ differ by integers. Thus the possible values are $m_j = -5/2, -3/2, -1/2, 1/2, 3/2, 5/2$. *Right:* A vector of magnitude $l = 2$ is added to an antiparallel vector of magnitude $s = 1/2$ to yield a vector of magnitude $j = l - s = 2 - 1/2 = 3/2$, which represents the minimum possible value of the quantum number $j$. The possible $z$ components of the vector of magnitude $j = 3/2$, which differ in value by integers, correspond to the possible values $m_j = -3/2, -1/2, 1/2, 3/2$. There are no values of $j$ intermediate between 5/2 and 3/2 since its possible values also may differ only by integers. Note that these diagrams do *not* accurately represent the addition of the angular momenta associated with the quantum numbers.

this relation is approximately valid, such diagrams are sometimes used in discussions of atomic structure as a simplified description of the addition of the angular momentum vectors themselves. The description is another form of the *vector model.* The description is useful, but it must be remembered that it is only approximate. An accurate description of the behavior of the angular momenta would have an appearance similar to that previously shown in Figure 8-8, which illustrates the angular momentum vectors for the case $l = 2, j = 5/2, m_j = 3/2$.

## 8-6 Spin-Orbit Interaction Energy and the Hydrogen Energy Levels

In the first part of this section we shall obtain an expression for the spin-orbit interaction energy in terms of the potential function $V(r)$ and the quantum numbers $l$, $s$, and $j$. In the second part we shall explain how the expression is used to predict the detailed structure of the energy levels of the hydrogen atoms. The expression for the spin-orbit interaction energy will also enter, on several occasions, into our subsequent discussion of multielectron atoms, and it will enter into our discussion of nuclei, since they have very strong spin-orbit interactions.

According to (8-27), the spin-orbit interaction energy is

$$\Delta E = \frac{1}{2m^2c^2}\frac{1}{r}\frac{dV(r)}{dr}\mathbf{S}\cdot\mathbf{L}$$

To express this in terms of $l$, $s$, and $j$, we first write

$$\mathbf{J} = \mathbf{L} + \mathbf{S}$$

Taking the dot product of this equality times itself, and employing the fact that $\mathbf{L}\cdot\mathbf{S} = \mathbf{S}\cdot\mathbf{L}$, we have

$$\mathbf{J}\cdot\mathbf{J} = \mathbf{L}\cdot\mathbf{L} + \mathbf{S}\cdot\mathbf{S} + 2\mathbf{S}\cdot\mathbf{L}$$

So

$$\mathbf{S}\cdot\mathbf{L} = (\mathbf{J}\cdot\mathbf{J} - \mathbf{L}\cdot\mathbf{L} - \mathbf{S}\cdot\mathbf{S})/2$$

or

$$\mathbf{S}\cdot\mathbf{L} = (J^2 - L^2 - S^2)/2 \tag{8-34}$$

In a quantum state associated with the quantum numbers $l$, $s$, and $j$, each term on the right has a fixed value, and $\mathbf{S}\cdot\mathbf{L}$ has the fixed value

$$\mathbf{S}\cdot\mathbf{L} = \frac{\hbar^2}{2}[j(j+1) - l(l+1) - s(s+1)]$$

Thus

$$\Delta E = \frac{\hbar^2}{4m^2c^2}[j(j+1) - l(l+1) - s(s+1)]\frac{1}{r}\frac{dV(r)}{dr}$$

The spin-orbit energy for the state is just the expectation value of this quantity. That is, the energy arising from the spin-orbit interaction is

$$\overline{\Delta E} = \frac{\hbar^2}{4m^2c^2}[j(j+1) - l(l+1) - s(s+1)]\overline{\frac{1}{r}\frac{dV(r)}{dr}} \tag{8-35}$$

where the expectation value $\overline{(1/r)dV(r)/dr}$ is calculated using the potential function $V(r)$ for the system and the probability density (actually the radial probability density $r^2R_{nl}^*R_{nl}$) for the state of interest. As was indicated earlier, (8-35) gives a convenient expression of an important result.

Now we consider the energy levels of the hydrogen atom. In Section 7-5 we obtained the predictions of quantum mechanics for the energy levels of a hydrogen atom in which the spin-orbit interaction is not considered, and found that they are simply the predictions of the Bohr model. In Example 8-3 we estimated the change in the energy of a typical one of these levels due to the presence of the spin-orbit interaction. We found that the energy is shifted up by about one part in $10^4$ if **L** is approximately parallel to **S** (if $j = l + 1/2$), and that it is shifted down about that amount if **L** is approximately antiparallel to **S** (if $j = l - 1/2$). We also saw that there is obviously no spin-orbit energy shift if $\mathbf{L} = 0$ (if $j = 1/2$).

To obtain quantitative predictions of the hydrogen atom spin-orbit interaction energy-level shifts from the general expression of (8-35), the potential function is equated to the Coulomb potential $V(r) = -e^2/4\pi\epsilon_0 r$, and then the expectation value $\overline{(1/r)dV(r)/dr}$ is calculated using the hydrogen atom eigenfunctions. However, before these predictions can be compared with experiments other effects, of comparable importance in the hydrogen atom, must be taken into account. In discussing Sommerfeld's relativistic modification of the Bohr model in Section 4-10, we estimated that the shift in a typical hydrogen atom energy level, due to the relativistic dependence of mass on velocity, is about one part in $10^4$. So this relativistic effect produces energy shifts in the hydrogen atom comparable to those produced by the spin-orbit interaction, which is really also a relativistic effect but a different one. A complete treatment of all the effects of relativity on the energy levels of the hydrogen atom can be given only in terms of the Dirac theory. But results which are almost (i.e., except for $l = 0$ states) complete can be obtained from the Schroedinger theory by adding to the simple hydrogen energy-level formula both the expectation value of the correction to the energy due to the spin-orbit interaction and the expectation value of the correction to the energy due to the dependence of mass on velocity. We shall not do this for two reasons: (1) it would get us into some fairly lengthy calculations, and (2) relativistic effects, other than the spin-orbit interaction, are significant only for hydrogen and a few more atoms of very small atomic number $Z$. For typical atoms of medium and large values of $Z$, and the levels involved in their optical spectra, the energy associated with these relativistic effects remains of the order of $10^{-4}$ times the energy of a level. But we shall see later that the spin-orbit interaction energy increases very rapidly with increasing $Z$. The spin-orbit interaction is the only effect we have considered that is generally important in a typical atom, and we have already said enough about it here. Therefore, we do no more than present the results of Dirac's completely relativistic treatment of the hydrogen atom energy levels, which predicts that the energies are

$$E = -\frac{\mu e^4}{(4\pi\epsilon_0)^2 2\hbar^2 n^2}\left[1 + \frac{\alpha^2}{n}\left(\frac{1}{j + 1/2} - \frac{3}{4n}\right)\right] \tag{8-36}$$

In this equation $\mu$ stands for the reduced electron mass, $\mu = mM/(m + M)$, and $\alpha$ is the *fine-structure constant*, $\alpha = e^2/4\pi\epsilon_0\hbar c \simeq 1/137$.

If the student will compare these results of the Dirac theory with the results of the Sommerfeld model expressed in (4-27a) and (4-27b), he will see that they are essentially the same. (Both $j + 1/2$ and $n_\theta$ are integers ranging from 1 to $n$.) Since the Sommerfeld model is based on the Bohr model, it is only a very rough approximation to physical reality. In contrast, the Dirac theory represents an extremely refined expression of our understanding of physical reality. That these two theories lead to essentially the same results for the hydrogen atom is a coincidence that caused much confusion in the 1920s, when the modern quantum theories were being developed. The coincidence occurs because the errors made by the Sommerfeld model, in ignoring the spin-orbit

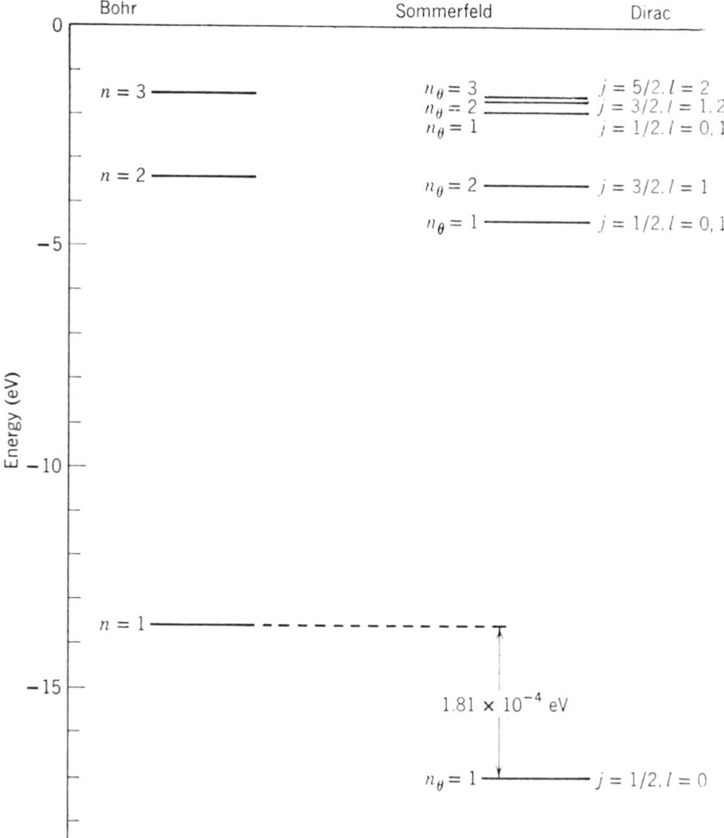


**FIGURE 8-11**

The energy levels of the hydrogen atom for $n = 1, 2, 3$ according to Bohr, Sommerfeld, and Dirac. The displacements of the Sommerfeld and Dirac levels from those given by Bohr have been exaggerated by a factor of $(1/\alpha)^2 \simeq (137)^2 \simeq 1.88 \times 10^4$.

interaction and in using classical mechanics to evaluate the average energy shift due to the relativistic dependence of mass on velocity, happen to cancel for the case of the hydrogen atom.

The energy levels of the hydrogen atom, as predicted by Bohr, Sommerfeld, and Dirac are shown in Figure 8-11. In order to make visible the energy-level splittings, called the *fine structure*, the shifts of the Sommerfeld and Dirac energy levels from those given by Bohr have been exaggerated by a factor of $(137)^2 = 1.88 \times 10^4$. Thus the diagrams would be completely to scale if the value of the fine-structure constant $\alpha$ were 1 instead of $\simeq 1/137$. Not shown on the Dirac energy-level diagram are the values of the quantum number $m_j$, which specify the orientation in space of the atom, since its energy is independent of the orientation if there are no external fields. There is a similar space orientation quantum number in the Sommerfeld model, whose values are not shown on the Sommerfeld energy levels, since the quantum number is of no consequence unless an external field is applied to the atom. Also not shown are the energy levels of hydrogen measured by optical spectroscopy. They are in very good agreement with the levels of both Sommerfeld and Dirac.

The only difference between the results of these two treatments is that Dirac, but not Sommerfeld, predicts that for most levels there is a degeneracy (in addition to

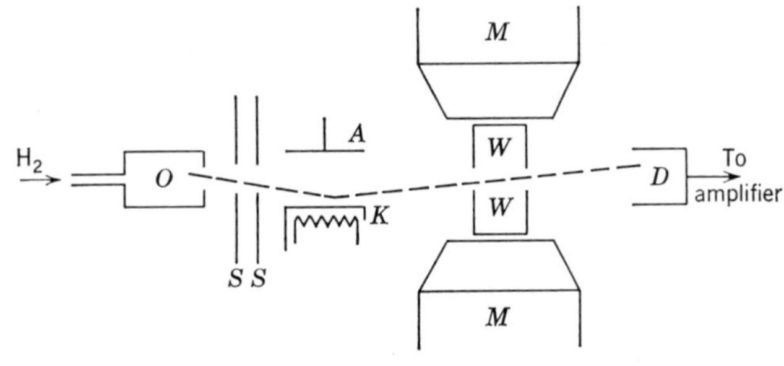


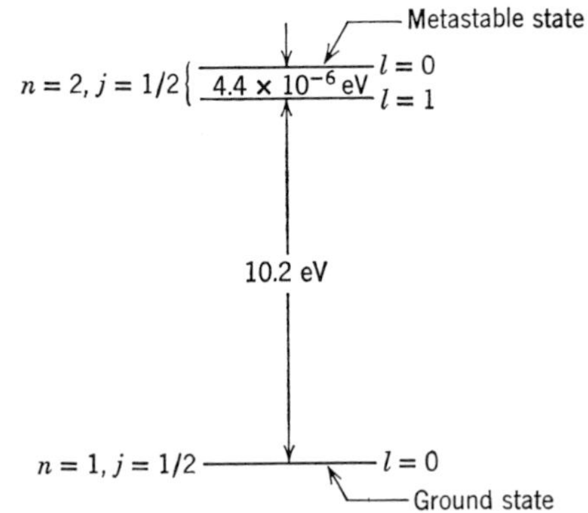


**FIGURE 8-12**

The apparatus of Lamb and Retherford. Molecular hydrogen ($H_2$) entering oven $O$ is largely dissociated into atomic hydrogen which leaves the oven, passing through slits $S$, $S$. The arrangement $K$, $A$ is essentially a vacuum diode, electrons being emitted from heated cathode $K$ and accelerated toward anode $A$. As the hydrogen passes through this region, some atoms collide with the electrons and are excited into the $n = 2$, $l = 0$ state described in the text. This state is called a metastable state because decay from it to the ground state ($n = 1$, $l = 0$) is highly inhibited by the $\Delta l$ selection rule and because all other states lie above it except the $n = 2$, $l = 1$, $j = 1/2$ state, which, according to the Dirac theory, has exactly the same energy as the metastable state. The experiment showed, however, that the $l = 1$ state was in fact about 4.4 $\mu$eV below the metastable state. These levels are shown below the apparatus.

The metastable atoms pass out of the collision region $K$, $A$ and are detected by detector $D$. Any mechanism which causes these atoms to undergo a transition to the $l = 1$ state (transitions to the ground state are forbidden) will result in a decreased signal from $D$, which is sensitive only to metastable atoms. Such transitions can be induced by passing the atoms through a region where there is an alternating electric field whose frequency $\nu$ is such that $h\nu \simeq 4.4$ $\mu$eV, or $\nu \simeq 1060$ MHz. Such an alternating field is provided by a waveguide $WW$, through whose walls the beam is passed.

To measure exactly the energy difference (Lamb shift) between the metastable ($l = 0$) and $l = 1$ states (both $n = 2, j = 1/2$), we could in principle merely vary the frequency $\nu$, searching for a value that maximized transitions from the former to the latter state, thereby minimizing the signal from $D$. In practice, the frequency is not easily adjusted and the levels themselves are adjusted instead by a known amount by means of a magnet $MM$, this shifting being due to the Zeeman effect.

the trivial degeneracy with respect to space orientation just mentioned) because the energy depends on the quantum numbers $n$ and $j$ but not on the quantum number $l$. Since there are generally two values of $l$ corresponding to the same value of $j$, the Dirac theory predicts that most levels are really double. This prediction was verified experimentally in 1947 by Lamb, who showed that for $n = 2$ and $j = 1/2$ there are two levels, which actually do not quite coincide. The $l = 0$ level lies above the $l = 1$ level by about one-tenth the separation between that level and the $n = 2$, $j = 3/2$ $l = 1$, level. The experiments involved measuring the frequency of photons absorbed in transitions between the two levels, using the apparatus shown in Figure 8-12. The energy separation between these levels is so small that the frequency is in the microwave radio range. Since measurements of radio frequencies can be made very accurately, it is possible to obtain the energy separation to five significant figures. These very accurate measurements of the so-called *Lamb shift* can be explained with precision in terms of the theory of *quantum electrodynamics*, as can the slight departure of the spin $g$ factor from 2 mentioned in Section 8-3. We cannot develop this quite sophisticated theory here, but we shall discuss it in the following section in connection with radiation by excited atoms, and in Chapter 17 in connection with the properties of the elementary particles.

Even with its exaggerated scale, Figure 8-11 cannot show the *hyperfine splitting* of the energy levels, which in hydrogen is due to an interaction between the internal magnetic field produced by the motion of the electron and a spin magnetic dipole moment of the *nucleus*. As nuclear magnetic dipole moments are smaller than electronic magnetic dipole moments by $\sim 10^{-3}$, the hyperfine splitting is smaller than the spin-orbit splitting by the same factor. Nevertheless, we shall see later that this effect can be understood quantitatively in terms of Schroedinger quantum mechanics, and that it can be used to measure nuclear spins and magnetic moments. In fact, every aspect of the behavior of a hydrogen atom can be explained in detail by the theories of quantum physics!

## 8-7 Transition Rates and Selection Rules

If hydrogen atoms are excited to their higher energy levels, e.g., in collisions with energetic electrons in a gas discharge tube, the atoms will in due course spontaneously make transitions to successively lower energy levels. In each transition between a pair of levels, a photon is emitted of frequency equal to the difference in their energies divided by Planck's constant. The discrete frequencies emitted in all the transitions that take place constitute the "lines" of the spectrum, but measurements show that not all conceivable transitions do take place. Photons are observed only with frequencies corresponding to transitions between energy levels whose quantum numbers satisfy the *selection rules:*

$$\Delta l = \pm 1 \tag{8-37}$$

$$\Delta j = 0, \pm 1 \tag{8-38}$$

That is, transitions take place only between levels whose $l$ quantum numbers differ by one and whose $j$ quantum numbers differ by zero or one. Measurements of the spectra of other one-electron atoms show that these selection rules apply to transitions in all such atoms.

As discussed in Section 4-11, some of the selection rules could be given some justification in the old quantum theory by using the *correspondence principle* to invoke certain restrictions that apply in the classical limit; but the predictions of this technique were not reliable. Furthermore, the old quantum theory had nothing at all to say about atomic *transition rates.* A transition rate is the probability per second that an atom in a certain energy level will make a transition to some other energy level. It is easy to measure a transition rate by measuring the probability per second of detecting a photon of the corresponding frequency, since this is proportional to the intensity of the corresponding spectral line. So it should certainly be possible to calculate a transition rate from atomic theory. An impressive feature of the Schroedinger quantum mechanics is that this can be done with no difficulty, using the atomic eigenfunctions. Of course all the selection rules can be obtained from transition rate calculations, since *a selection rule just specifies which transitions have rates so small that they are not normally observed.*

We have already used elementary quantum mechanics, in Example 5-13 and the discussion following, to develop much of the physical picture that the theory provides for the emission of photons by excited atoms. According to that example, if the wave function describing an atom is the wave function associated with a single quantum state, then the probability density function for the atom will be constant in time. But if the wave function is a mixture of the wave functions associated with two quantum states, corresponding to the two energy levels $E_2$ and $E_1$, then the probability density contains terms which oscillate in time at frequency $\nu = (E_2 - E_1)/h$. Since the atomic electron can be found at any location where the probability density has an appreciable value, the charge it carries is not confined to a particular location. In effect, the atom has a charge distribution which is proportional to its probability density. Thus when the atom is in a mixture of two quantum states its charge distribution oscillates at precisely the frequency of the photon emitted in the transition between the states. This is true since the photon carries away the excess energy $E_2 - E_1$, and so has frequency $\nu = (E_2 - E_1)/h$.

The simplest aspect of the atom's charge distribution that can be oscillating is the *electric dipole moment.* This is the product of the electron charge and the expectation value of its displacement vector from the essentially fixed massive nucleus. The electric dipole moment is a measure of the separation of the center of the electron charge distribution from the nuclear center of the atom. Even in classical physics, a charge distribution that is constant in time will not emit electromagnetic radiation, while a charge distribution with an oscillating electric dipole moment emits radiation of frequency equal to the oscillation frequency. In fact, an oscillating electric dipole is the most efficient radiator.

We can actually use the classical formula for the rate of emission of energy by an oscillating electric dipole to obtain the important factors in the formula for atomic transition rates. In Appendix B it is shown that the dipole radiates electromagnetic energy at the average rate $\bar{R}$, where

$$\bar{R} = \frac{4\pi^3\nu^4}{3\epsilon_0 c^3} p^2 \tag{8-39}$$

with $p$ the amplitude of its oscillating electric dipole moment and $\nu$ the frequency of oscillation. Since the energy is carried off by photons whose energies are of magnitude $h\nu$, the rate of emission of photons, $R$, is

$$R = \frac{\bar{R}}{h\nu} = \frac{4\pi^3\nu^3}{3\epsilon_0 hc^3} p^2 \tag{8-40}$$

This probability per second that a photon is emitted is just equal to the probability per second that the atom has undergone the transition. Thus $R$ is also the atomic *transition rate.*

Relative to an origin at the essentially fixed nucleus, the electric dipole moment **p** of the one-electron atom is defined as

$$\mathbf{p} = -e\mathbf{r} \tag{8-41}$$

where $-e$ is the charge of the electron and **r** is its position vector from the nucleus at the origin. To obtain an expression for the amplitude of the oscillating electric dipole moment of the atom when it is in a mixture of two states, we calculate the expectation value of **p**, using the mixed state probability density obtained in Example 5-13

$$\begin{aligned}\Psi^*\Psi &= c_1^*c_1\psi_1^*\psi_1 + c_2^*c_2\psi_2^*\psi_2 \\ &\quad + c_2^*c_1\psi_2^*\psi_1 e^{i(E_2-E_1)t/\hbar} + c_1^*c_2\psi_1^*\psi_2 e^{-i(E_2-E_1)t/\hbar}\end{aligned}$$

There is no way, from the present argument, for us to determine precisely what values of the adjustable constants $c_1$ and $c_2$ should be used to specify how much of the two quantum states are mixed together. But the results we seek are independent of their values, as will be seen shortly, so for simplicity we set them both equal to 1. Then we have

$$\Psi^*\Psi = \psi_f^*\psi_f + \psi_i^*\psi_i + \psi_i^*\psi_f e^{i(E_i-E_f)t/\hbar} + \psi_f^*\psi_i e^{-i(E_i-E_f)t/\hbar}$$

where we have replaced the labels 2 and 1 by $i$ and $f$, for initial and final. As this probability density is not normalized, when we use it to evaluate the expectation value of **p** we obtain only a proportionality, but this will suffice. That is, we have

$$\bar{\mathbf{p}} \propto \int \Psi^*(-e\mathbf{r})\Psi \, d\tau \propto \int \Psi^* e\mathbf{r}\Psi \, d\tau$$

or

$$\begin{aligned}\bar{\mathbf{p}} \propto & \int \psi_f^* e\mathbf{r}\psi_f \, d\tau + \int \psi_i^* e\mathbf{r}\psi_i \, d\tau \\ &+ e^{i(E_i-E_f)t/\hbar}\int \psi_i^* e\mathbf{r}\psi_f \, d\tau \\ &+ e^{-i(E_i-E_f)t/\hbar}\int \psi_f^* e\mathbf{r}\psi_i \, d\tau\end{aligned}$$

where we have sandwiched the term $e\mathbf{r}$ between the other terms of the integrands to conform with accepted notation, and where the integrals are three-dimensional. Now the first two integrals on the right are not associated with an oscillating $\bar{\mathbf{p}}$; in fact both integrals yield zero. The last two integrals are each multiplied by complex exponentials with a time dependence that oscillates at the frequency $\nu = (1/2\pi)(E_i - E_f)/\hbar = (E_i - E_f)/h$. These two terms describe oscillations in the electric dipole moment expectation value, of amplitude which is measured by the magnitude of the integral in either term. Thus we find that the amplitude of the oscillating electric dipole moment is proportional to the quantity $p_{fi}$, where

$$p_{fi} \equiv \left| \int \psi_f^* e\mathbf{r}\psi_i \, d\tau \right| \tag{8-42}$$

This quantity is called the *matrix element of the electric dipole moment taken between the initial and final states.* Note that its value depends on the behavior of the atom in both the initial state, through $\psi_i$, and in the final state, through $\psi_f^*$. This is reasonable because the radiating atom is in a mixture of the two states. Setting the $p$ in (8-40) proportional to $p_{fi}$, we obtain

$$R \propto \frac{\nu^3 p_{fi}^2}{\epsilon_0 h c^3}$$

where $R$ is the transition rate.

We have obtained the factors $\nu^3$ and $p_{fi}^2$, as well as the constants $\epsilon_0 hc^3$, in the expression for the transition rate by a partly classical argument. A much more sophisticated argument which uses only Schroedinger quantum mechanics leads to the same result, except that the numerical proportionality constant is determined. This result is

$$R = \frac{16\pi^3\nu^3 p_{fi}^2}{3\epsilon_0 hc^3} \tag{8-43}$$

The same equation can be derived in an even more rigorous manner from the theory of *quantum electrodynamics*, which provides an exact treatment of the quantization properties of electromagnetic fields. Although the results are not different, quantum electrodynamics gives a more complete picture of the emission of photons by excited atoms. In particular, it explains how the radiating atom gets into the mixed state. This happens through a kind of resonance interaction between vibrations of the appropriate frequency, in a surrounding field of electromagnetic radiation, and an atom in the initial state. The interaction induces the charge oscillations of that frequency, which are characteristic of the mixed state, and then the atom emits electromagnetic radiation of the same frequency. The process is indicated schematically in Figure 8-13.

The emission of photons by atoms, under the influence of the photons that comprise an electromagnetic field *applied* to the atom, is a phenomenon called *stimulated emission.* Atoms also emit photons when an electromagnetic field is *not applied*, in a phenomenon called *spontaneous emission.* Quantum electrodynamics shows that spontaneous emission takes place because there is always some electromagnetic field present in the vicinity of an atom, even if a field is not applied! The reason is that the electromagnetic field has an energy content which is *discretely quantized* because the energy, at any particular frequency, is given by the *number* of photons of that frequency. Like any other system with discretely quantized energy, the electromagnetic field has a *zero-point energy*. Thus quantum electrodynamics shows that there will always be some electromagnetic field vibrations present, of whatever frequency is required to induce the charge oscillations that cause the atom to radiate "spontaneously." We can see that spontaneous and stimulated emission are qualitatively similar. In spontaneous emission, the electromagnetic field surrounding the atom is in its zero-point energy state. In stimulated emission an additional field is applied so that the electromagnetic field surrounding the atom is in a higher energy state. Then more intense field vibrations of the required frequency are present, and there is more chance that the atom will be stimulated to radiate.

From this argument, it is apparent that the transition rate for stimulated emission is proportional to the intensity of the applied electromagnetic field. For intense fields it becomes very large and the atom radiates very efficiently. This has important practical consequences in the *laser*, a device to produce extremely bright beams of coherent light that will be discussed in Chapter 11. In that chapter we shall go more deeply into the relation between stimulated and spontaneous emission, but here we shall consider only spontaneous emission.

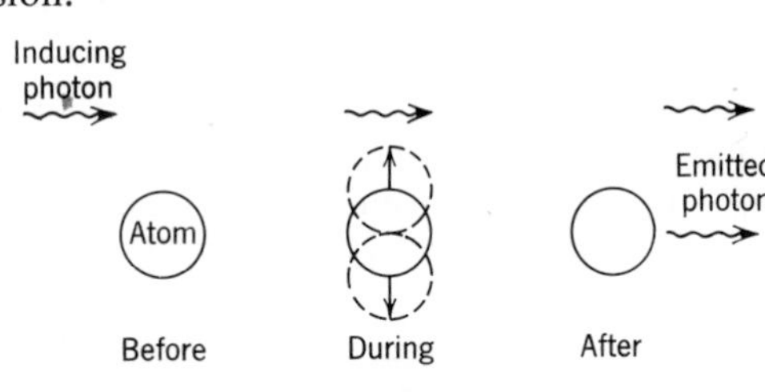


**FIGURE 8-13**

A schematic illustration of the emission of a photon by an atom. Electromagnetic radiation impinging on the atom induces dipole charge oscillations in the atom. Then the atom emits electromagnetic radiation.

The transition rate for spontaneous emission, evaluated in (8-43), is independent of whether or not an external field is applied. It depends only on the properties of the atomic eigenfunctions. Since the eigenfunctions are known, the electric dipole moment matrix elements between various pairs of levels can be obtained by calculating the value of the associated integral (8-42). Then the rates for transitions between these levels can be calculated from (8-43).

It is found that the agreement between the predictions and the measurements is quite good, even though the transition rates vary appreciably from one case to the next. For the transition of the hydrogen atom from its first excited state to its ground state, the transition rate has the value $R \simeq 10^8$ sec$^{-1}$. This means that in about $10^{-8}$ sec the probability that the transition has occurred is about equal to one. It is said that the first excited state has a *lifetime* $t = 1/R \simeq 10^{-8}$ sec. The value quoted is typical of the orders of magnitude encountered in atomic transition rates—except that the transition rates between certain pairs of levels are essentially zero. These are the transitions for which the spectral lines are observed to be absent, or extremely weak. The transition rates are predicted to be zero in these cases because the integral in the electric dipole matrix element yields zero. Thus *the selection rules are a set of conditions on the quantum numbers of the eigenfunctions of the initial and final energy levels, such that the electric dipole matrix elements are zero when calculated with a pair of eigenfunctions whose quantum numbers violate these conditions.*

**Example 8-6.** When a hydrogen atom is placed in a very strong external magnetic field, the spin-orbit interaction coupling of its orbital angular momentum $\mathbf{L}$ to its spin angular momentum $\mathbf{S}$ is overwhelmed, and both vectors precess independently about the direction of the external field with constant $z$ components $L_z = m_l\hbar$ and $S_z = m_s\hbar$. That is, $m_l$ and $m_s$ are good quantum numbers under these circumstances. Spectrum measurements made on such atoms show the existence of a selection rule $\Delta m_l = 0, \pm 1$. Obtain this selection rule by evaluating the appropriate electric dipole matrix element.

Written in full, the matrix element is

$$p_{fi} = \left| \int_0^\infty \int_0^\pi \int_0^{2\pi} \psi_f^*(r,\theta,\varphi) e\mathbf{r} \psi_i(r,\theta,\varphi) r^2 \sin\theta \, dr \, d\theta \, d\varphi \right|$$

The triple integral factors into the product of three single integrals. The one that is interesting, because it leads to the selection rule, is

$$\mathbf{I} = \int_0^{2\pi} \Phi_f^*(\varphi) \mathbf{r} \Phi_i(\varphi) \, d\varphi$$

This is a vector quantity, which has components

$$I_x = \int_0^{2\pi} \Phi_f^*(\varphi) x \Phi_i(\varphi) \, d\varphi$$

$$I_y = \int_0^{2\pi} \Phi_f^*(\varphi) y \Phi_i(\varphi) \, d\varphi$$

$$I_z = \int_0^{2\pi} \Phi_f^*(\varphi) z \Phi_i(\varphi) \, d\varphi$$

If we use the relations

$$x = r \sin\theta \cos\varphi$$
$$y = r \sin\theta \sin\varphi$$
$$z = r \cos\theta$$

which can be verified by inspecting Figure 7-2, and also evaluate $\Phi_i(\varphi)$ and $\Phi_f^*(\varphi)$ from (7-19), we obtain

$$I_x = r \sin\theta \int_0^{2\pi} \cos\varphi e^{i(m_{l_i}-m_{l_f})\varphi}\, d\varphi$$

$$I_y = r \sin\theta \int_0^{2\pi} \sin\varphi e^{i(m_{l_i}-m_{l_f})\varphi}\, d\varphi$$

$$I_z = r \cos\theta \int_0^{2\pi} e^{i(m_{l_i}-m_{l_f})\varphi}\, d\varphi$$

Any table of definite integrals will show that the integral in $I_z$ equals zero, unless

$$m_{l_i} - m_{l_f} = 0 \qquad \text{or} \qquad \Delta m_l = 0$$

The integral in $I_x$ can be rewritten, to yield

$$I_x = \frac{1}{2} r \sin\theta \int_0^{2\pi} [e^{i(m_{l_i}-m_{l_f}-1)\varphi} + e^{i(m_{l_i}-m_{l_f}+1)\varphi}]\, d\varphi$$

This definite integral equals zero, unless

$$m_{l_i} - m_{l_f} = \pm 1 \qquad \text{or} \qquad \Delta m_l = \pm 1$$

The same result is obtained from the integral in $I_y$. Therefore, unless $\Delta m_l = 0$, or $\pm 1$, there will be no components of **I** that are not zero. Since this will also be true of the electric dipole matrix element, we have obtained the selection rule. ◀

Physically, the selection rules arise because of symmetry properties of the oscillating charge distribution of the atom. The atom cannot radiate like an electric dipole unless the electric dipole moment of its electron charge distribution is oscillating. A classical analogy is found in a very short antenna, which is center-fed from high frequency sources of alternating current, as illustrated in Figure 8-14. If the leads to the antenna are fed out of phase, so that charge flows into one end at the same time it flows out of the other, the antenna will radiate relatively efficiently. But if the leads are fed in phase, so that charge flows into or out of both ends in unison, the antenna will hardly radiate at all.

Mathematically, it is the symmetry properties of the eigenfunctions in the matrix element that are responsible for the selection rules. Some idea of this can be obtained in an easy way by considering the *parities* of the eigenfunctions. In Section 6-8 we defined the parity of a one-dimensional eigenfunction as the quantity which describes the behavior of the eigenfunction when the sign of the coordinate is changed. The definition can be extended immediately to three dimensions. That is, eigenfunctions satisfying the relation

$$\psi(-x,-y,-z) = +\psi(x,y,z) \tag{8-44}$$

are said to be of *even parity*, and eigenfunctions satisfying the relation

$$\psi(-x,-y,-z) = -\psi(x,y,z) \tag{8-45}$$

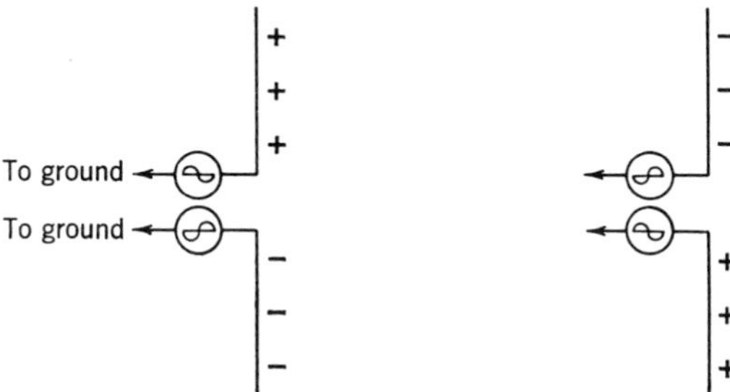


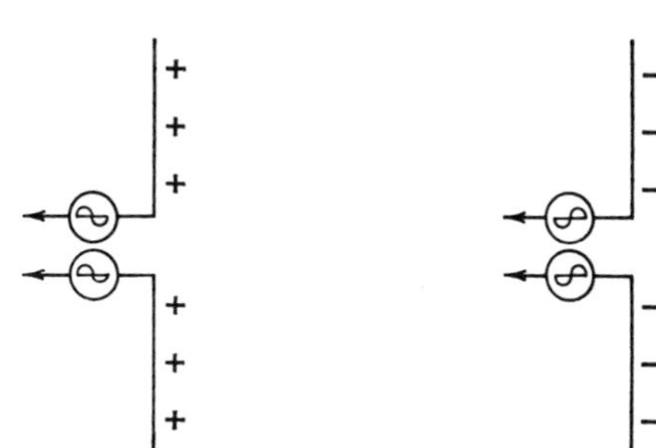

**FIGURE 8-14**

*Upper* diagrams: Center-fed antennas driven out of phase. *Lower* diagrams: Driven in phase. *Left* diagrams: The charge distributions are shown at some initial time. *Right* diagrams: At half a period later. The antenna driven in phase will emit very little radiation if its length is short compared to a wavelength, and if the distance to the ground plane is long compared to a wavelength.

are said to be of *odd parity*. All eigenfunctions that are bound-state solutions to time-independent Schroedinger equations for a potential that can be written as $V(r)$, like the Coulomb potential, have definite parities, either even or odd. The reason is that the probability densities $\psi^*\psi$ will then have the same value at the point $(-x,-y,-z)$ that they have at the point $(x,y,z)$, which is a requirement of the fact that the potential has the same value at these points.

An example is found in the one-electron atom eigenfunctions of Table 7-2. To see this, inspect Figure 8-15, which shows that when the signs of the rectangular coordinates are changed the behavior of the spherical polar coordinates is

$$r \rightarrow r, \qquad \theta \rightarrow \pi - \theta, \qquad \varphi \rightarrow \pi + \varphi \tag{8-46}$$

By carrying out these changes on several of the eigenfunctions, it is easy to demonstrate that

$$\psi_{nlm_l}(r,\pi - \theta,\pi + \varphi) = (-1)^l \psi_{nlm_l}(r,\theta,\varphi) \tag{8-47}$$

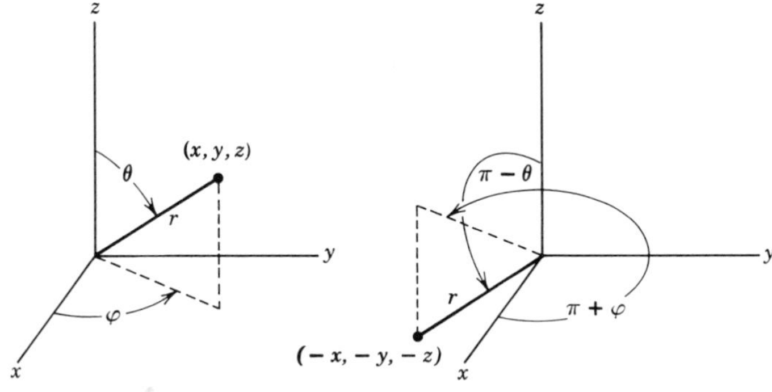


**FIGURE 8-15**

Illustrating the parity operation.

*The parity is determined by* $(-1)^l$; *it is even if the orbital angular momentum quantum-number* $l$ *is even, and odd if* $l$ *is odd.* This is true for *all* eigenfunctions, bound or unbound, of any spherically symmetrical potential $V(r)$, since the only significant assumption that is used to obtain (8-47) is that $V$ can be written as $V(r)$.

Now consider the matrix element of the electric dipole moment

$$p_{fi} = \left| \int \psi_f^* e\mathbf{r}\psi_i \, d\tau \right|$$

The parity of $e\mathbf{r}$ is odd since the vector $\mathbf{r}$ changes into its negative when the signs of the rectangular coordinates are changed. Therefore, if the initial and final eigenfunctions $\psi_i$ and $\psi_f$ are of the same parity, both even or both odd, the entire integrand will be of odd parity. If this is the case the integral will yield zero because the contribution from any volume element will be cancelled by the contribution from the diametrically opposite volume element. Then the transition rate will also be zero. Therefore, the parity of the final eigenfunction must differ from the parity of the initial eigenfunction in an electric dipole transition. Since the parities are determined by $(-1)^l$, we can understand why transitions for $\Delta l = 0$, or $\pm 2$, are not allowed, in agreement with the $\Delta l = \pm 1$ selection rule of (8-37). The reason is that in such transitions the parities of the initial and final eigenfunctions would be the same.

Quantum electrodynamics shows, and experiments verify, that a photon carries angular momentum as well as linear momentum. In particular, the theory shows that the angular momentum carried by a photon emitted in an electric dipole transition is, in units of $\hbar$, equal to 1. From this point of view, the total angular momentum quantum number selection rule $\Delta j = 0, \pm 1$ of (8-38) represents the requirements of angular momentum conservation, which is fundamentally a symmetry property, by restricting electric dipole transitions to pairs of states where the change in the total angular momentum of the atom can be compensated for by the angular momentum carried by the photon it emits. (When $\Delta j = 0$ angular momentum conservation is satisfied by a change in the orientation in space of the total angular momentum vector of the atom at the time the photon is emitted.) This point of view also makes it apparent that $\Delta l = \pm 3$ electric dipole transitions cannot occur because they would lead to too large a change in the total angular momentum, even though they would be all right as far as parity is concerned.

It should be mentioned that selection rules do not absolutely prohibit transitions that violate them, but only make such transitions very unlikely. If a transition cannot take place by the normal means of emission of radiation from an oscillating electric dipole moment, there is a very small probability (typically reduced by a factor of about $10^{-4}$) that it will take place by emission of radiation from an oscillating magnetic dipole moment. This may occur through oscillations in orientation of electron spin angular momentum and magnetic dipole moment. Transitions can also take place with very small probabilities (typically reduced by approximately a factor of $10^{-6}$) by emission of radiation from an oscillating electric quadrupole moment. This involves oscillations in the electron charge distribution of the atom between an elongated ellipsoid and a flattened ellipsoid.

If an atom is excited to a state from which it can return to its ground state only by one of these highly inhibited transitions, it may remain in the excited state for an appreciable fraction of a second, instead of the lifetime of $10^{-8}$ sec corresponding to the typical transition rate of $10^8$ sec$^{-1}$. The excited state is said to be *metastable*, and the delayed emission of a photon is a form of *phosphorescence*. In practice, phosphorescence of atoms is rarely observed because the metastable state is deexcited, without the emission of a photon, when the atom collides with the wall of its container and gives up its excess energy directly to the atoms of the wall. A process completely analogous to phosphorescence is commonly observed in nuclei, however.

## 8-8 A Comparison of the Modern and Old Quantum Theories

We shall very briefly summarize the last chapters by making a comparison between the modern quantum theories (Schroedinger, Dirac, and quantum electrodynamics) and the old quantum theories (Bohr and Sommerfeld).

One of the most striking aspects of the modern quantum theories is the way they lead progressively to more and more accurate treatments of the hydrogen atom. The Schroedinger theory without electron spin accounts for the energy levels of the atom that are observed in spectroscopic measurements of moderate resolution. Measurements of high resolution reveal the fine-structure splitting of the energy levels. They can be explained almost completely by adding to the Schroedinger theory corrections for the electron spin-orbit interaction and for the relativistic dependence of mass on velocity. They can be explained completely by the Dirac theory. Spectroscopic measurements of very high resolution show the Lamb shift, which can be understood in terms of quantum electrodynamics. Extremely high-resolution measurements show the hyperfine splittings, which can be accounted for in the Schroedinger theory by an interaction involving the nuclear spin. Another great success of the modern quantum theories is their ability to give very satisfactory treatments of the transition rates and selection rules observed in the measurements of the spectra emitted by hydrogen atoms, and all other one-electron and multielectron atoms.

The record of the old quantum theory is spotty. The Bohr model leads to correct values for the energies of the unsplit hydrogen atom levels. Sommerfeld's relativistic modification of the model agrees with the fine-structure splittings in hydrogen, but the agreement is accidental. The relativistic modification cannot account for the Lamb shift, nor for hyperfine splittings. Furthermore, it disagrees by orders of magnitude with the fine-structure splittings seen in typical multielectron atoms. In fact, the Bohr model itself fails completely to explain many of the most obvious features of the energy levels of multielectron atoms; it is already in serious trouble with the helium atom that contains only two electrons. The old quantum theory is unreliable in explaining selection rules, and incapable of explaining transition rates.

A particularly helpful feature of the Schroedinger theory is that almost all of the work done in applying it to one-electron atoms carries over directly when it is applied to multielectron atoms. And the theory is certainly accurate enough to explain every important feature of multielectron atoms. Furthermore, it is not very much more complicated to apply Schroedinger quantum mechanics to such atoms than it is to apply it to one-electron atoms. As we shall see in the next two chapters, part of the reason that this is true is that most of the electrons in a multielectron atom group together with other electrons to form symmetrical and inert shells in which they do not have to be treated individually. Only the few electrons in the atom which are not in such shells require detailed treatment.

## QUESTIONS

**1.** Why, in discussing Figures 8-1 and 8-4, do we speak of fictitious magnetic poles?

**2.** Why does the torque acting on a magnetic dipole in a magnetic field cause the dipole to precess about the field, instead of lining up with the field?

3. It is not possible to do a Stern-Gerlach experiment on a free electron to measure its spin magnetic dipole moment; it is only possible if the electron is in a neutral atom. Explain why. (Hint: There is a superficial answer, which has a superficial rebuttal. A complete answer involves the uncertainty principle.)

4. Exactly why do we conclude that the spin quantum numbers are half-integral?

5. Is it fair to criticize Schroedinger quantum mechanics for not predicting electron spin?

6. Are there conceptual difficulties with the idea of a point electron?

7. Is the electron the "ultimate magnetic particle"?

8. Explain in simple terms why an electron in a hydrogen atom experiences a magnetic field. Does it experience a field in all quantum states?

9. Just what is the spin-orbit interaction? How does it lead to the observed fine-structure splitting of the spectral lines of the hydrogen atom?

10. When the spin-orbit interaction is taken into account, it is sometimes said that $m_l$ and $m_s$ are no longer "good quantum numbers." Explain why this terminology is appropriate. What are the good quantum numbers for the one-electron atom when the spin-orbit interaction is taken into account?

11. What are good quantum numbers for a one-electron atom in an external magnetic field which, compared to the internal field, is very weak? Extremely strong?

12. Why is the spin-orbit interaction particularly sensitive to the form of the potential $V(r)$ for small $r$? How can this be used to study experimentally the potentials of multielectron atoms?

13. What is the justification of performing vector additions, as in Figure 8-10, with vectors whose lengths are proportional to the quantum numbers specifying the angular momenta, instead of with the angular momenta vectors themselves?

14. Describe briefly all the features of the hydrogen atom energy-level diagram in Figure 8-11, and explain the origin of these features. What features are not shown?

15. Can there be electromagnetic radiation emitted from an oscillating electric monopole (i.e., emitted from a charge of oscillating magnitude at a fixed location)?

16. There are similarities between the emission of electromagnetic radiation by a system of oscillating charges, and the emission of gravitational radiation, by a system of oscillating masses, but dipole gravitational radiation cannot be emitted. Why?

17. What experimental evidence do you know of that is in contradiction to the presence of zero-point energy vibrations of the electromagnetic field? In support of its presence?

18. What is the relation between spontaneous and stimulated emission?

19. Explain in physical terms the origin of the selection rules.

20. Do all atoms take the same time to make a transition between a certain pair of levels?

## PROBLEMS

1. Evaluate the magnetic field produced by a circular current loop at a point on the axis of symmetry far from the loop. Then evaluate the magnetic field produced at the same point by a dipole formed from two separated magnetic monopoles located at the center of the loop and lying along the axis of symmetry. Show that the fields are the same if the current

in the loop and its area are related to the magnetic moment of the dipole by (8-2). Can you see how to extend the argument to show that the fields will be the same at all points far from the loop or dipole, and independent of the shape of the loop?

2. (a) Evaluate the ratio of the orbital magnetic dipole moment to the orbital angular momentum, $\mu_l/L$, for an electron moving in an elliptical orbit of the Bohr-Sommerfeld atom discussed in Section 4-10. (Hint: The area swept out by the radius vector of length $r$, when the angular coordinate increases by the increment $d\theta$, is $dA = r^2\,d\theta/2$. Use $L = mr^2\,d\theta/dt$ to evaluate $d\theta$ in terms of the time increment $dt$, and then make the trivial integration.) (b) Compare the results with those of (8-5) for a circular orbit.

3. A beam of hydrogen atoms in their ground state is sent through a Stern-Gerlach magnet, which splits it into two components according to the two spin orientations. One component is stopped by a diaphragm at the end of the magnet, and the other continues into a second Stern-Gerlach magnet which is coaxial with the beam leaving the first magnet, but is rotated relative to the first magnet about their approximately common axes through an angle $\alpha$. There is a second diaphragm fixed on the end of the second magnet which also allows only one component to pass. Describe qualitatively how the intensity of the beam passing the second diaphragm depends on $\alpha$.

4. Determine the field gradient of a 50 cm long Stern-Gerlach magnet that would produce a 1 mm separation at the end of the magnet between the two components of a beam of silver atoms emitted with typical kinetic energy from a 960°C oven. The magnetic dipole moment of silver is due to a single $l = 0$ electron, just as for hydrogen.

5. If a hydrogen atom is placed in a magnetic field which is very strong compared to its internal field, its orbital and spin magnetic dipole moments precess independently about the external field, and its energy depends on the quantum numbers $m_l$ and $m_s$ which specify their components along the external field direction. (a) Evaluate the splitting of the energy levels according to the values of $m_l$ and $m_s$. (b) Draw the pattern of split levels originating from the $n = 2$ level, enumerating the quantum numbers of each component of the pattern. (c) Calculate the strength of the external magnetic field that would produce an energy difference between the most widely separated $n = 2$ levels which equals the difference between the energies of the $n = 1$ and $n = 2$ levels in the absence of the field.

6. The Thomas precession can also be described in terms of a time dilation between the reference frame in which the nucleus is at rest and the reference frames in which the electron is instantaneously at rest, which leads to a disagreement between an observer at the nucleus and the observers at the electron concerning the time required for each to make a complete revolution about the other. Work out the details of this description, and compare with the results of Appendix J.

7. Use the procedure of Example 8-3 to estimate the spin-orbit interaction energy in the $n = 2$, $l = 1$ state of a muonic atom, defined in Example 4-9.

8. Prove that the only possible values of the quantum number $j$ from the series $j = l + 1/2$, $l - 1/2, l - 3/2, \ldots$, that satisfy the inequality $\sqrt{j(j+1)} \geq |\sqrt{l(l+1)} - \sqrt{s(s+1)}|$ with $s = 1/2$, are $j = l + 1/2, l - 1/2$, if $l \neq 0$, or $j = 1/2$, if $l = 0$.

9. (a) Enumerate the possible values of $j$ and $m_j$, for the states in which $l = 1$, and, of course, $s = 1/2$. (b) Draw the corresponding "vector model" figures. (c) Draw a figure illustrating the angular momentum vectors for a typical state. (d) Show also the spin and orbital magnetic dipole moment vectors, and their sum the total magnetic dipole moment vector. (e) Is the total magnetic dipole moment vector antiparallel to the total angular momentum vector?

10. Enumerate the possible values of $j$ and $m_j$ for states in which $l = 3$ and $s = 1/2$.

11. The relativistic shift in the energy levels of a hydrogen atom due to the relativistic dependence of mass on velocity can be determined by using the atomic eigenfunctions to

calculate the expectation value $\overline{\Delta E_{\rm rel}}$ of the quantity $\Delta E_{\rm rel} = E_{\rm rel} - E_{\rm clas}$, the difference between the relativistic and classical expressions for the total energy $E$. Show that for $p$ not too large

$$\Delta E_{\rm rel} \simeq -\frac{p^4}{8m^3c^2} = -\frac{E^2 + V^2 - 2EV}{2mc^2}$$

so that

$$\overline{\Delta E_{\rm rel}} = -\frac{E_n^2}{2mc^2} - \frac{e^4}{(4\pi\epsilon_0)^2 2mc^2}\int \psi^*_{nljm_j}\frac{1}{r^2}\psi_{nljm_j}\,d\tau - \frac{E_n e^2}{4\pi\epsilon_0 mc^2}\int \psi^*_{nljm_j}\frac{1}{r}\psi_{nljm_j}\,d\tau$$

**12.** (a) Draw the hydrogen energy-level diagram for all states through $n = 2$ as in the right-hand part of Figure 8-11, but with the splitting according to $l$ also shown. (b) With arrows connecting pairs of levels, show all the transitions that are allowed by the selection rules.

**13.** Verify that the parities of the one-electron atom eigenfunctions $\psi_{300}$, $\psi_{310}$, $\psi_{320}$, and $\psi_{322}$ are determined by $(-1)^l$.

**14.** (a) Use parity considerations to prove that the first two integrals of the display equation preceding (8-42) both yield zero. (b) Interpret what this means about the existence of atomic electric dipole moments which are static in time.

**15.** By a straightforward evaluation of the electric dipole matrix elements for the eigenfunctions of Table 7-2, show that the selection rule $\Delta l = \pm 1$ of (8-37) is valid for the $n = 2 \rightarrow n = 1$ transitions of the hydrogen atom.

**16.** Consider the electric dipole moment matrix elements for a charged one-dimensional simple harmonic oscillator making the transitions $n_i = 3$, $n_f = 0$; $n_i = 2$, $n_f = 0$; $n_i = 1$, $n_f = 0$. Use the eigenfunctions of Table 6-1 to show that the matrix elements which are not zero agree with the selection rule $\Delta n = \pm 1$, discussed in Section 4-11. (Hint: Use parity considerations whenever you can.)

**17.** (a) Calculate the rate for spontaneous transitions between the $n = 1$ and $n = 0$ states of a simple harmonic oscillator, carrying charge $e$. Take the mass of the oscillator to be equal to the mass of an atom of some typical ionic molecule, and the restoring force constant $C$ to be $10^3$ joules/m$^2$, which is typical for such a molecule. (Hint: Normalized eigenfunctions must be used.) (b) From the transition rate, estimate the average time required to complete the transition. This is the lifetime of the $n = 1$ vibrational state of the molecule.

**18.** Consider enough of the electric dipole moment matrix elements for a charged particle in an infinite square well potential, using the eigenfunctions of Section 6-8, to see if there is a selection rule for this system and, if so, to determine what it is.

# 9

# Multielectron Atoms—Ground States and X-Ray Excitations

# Multielectron Atoms—Ground States and X-Ray Excitations

# 9

## 9-1 Introduction

In this chapter we shall use Schroedinger's quantum mechanics to study multielectron atoms from helium to uranium. First we shall discuss in a general way the interesting properties of quantum mechanical systems containing several identical particles, such as electrons. This will lead us to the so-called exclusion principle, which is of dominant importance in determining the structure of multielectron atoms. Then we shall consider multielectron atoms in their ground states, and the systematic description of these atoms provided by the periodic table of the elements. We shall see that quantum mechanics gives a complete explanation of the periodic table, which is the basis of inorganic chemistry and much of organic chemistry and solid state physics. Finally, we shall consider the high-energy excited states of multielectron atoms that are involved in the emission of x rays by these atoms.

A multielectron atom of atomic number $Z$ contains a nucleus of charge $+Ze$ surrounded by $Z$ electrons each of charge $-e$. Every electron moves under the influence of an attractive Coulomb interaction exerted by the nucleus and the repulsive Coulomb interactions exerted by all the other $Z - 1$ electrons, as well as certain weaker interactions involving the angular momenta. The quantum mechanical treatment of this complicated system is easier than might be supposed. One reason is that the various interactions experienced by an atomic electron are of different strengths, so it is possible to deal with them one or two at a time in order of decreasing strength. In the first step, which we consider in this chapter, an approximate description which takes into account only the strongest interactions is developed. In subsequent steps, which we consider in the next chapter, the description is made more and more exact by successively taking into account the weaker interactions. We shall find that with this procedure it is not difficult to obtain a qualitative understanding of the behavior of multielectron atoms.

Quantitative information about multielectron atoms can be obtained from this approximation procedure, but the required calculations must be carried out on large computers. Of course, we shall not be able to reproduce such calculations. However, in this chapter and the next we shall describe the calculations and their results. We shall also compare the results with the properties of multielectron atoms observed by experiment. Our description will be based, in major part, on the theory of the one-electron atom developed in the preceding chapters.

## 9-2 Identical Particles

Before studying multielectron atoms, we must discuss an important topic of quantum mechanics that does not enter into the theory of one-electron atoms. This concerns the question of how to give an accurate quantum mechanical description of a system containing two or more *identical* particles, such as electrons. Discussing this question will lead us to quantum mechanical phenomena that have absolutely no classical analogues. In fact, the discussion will bring out some of the most striking differences between classical and quantum mechanics.

The nature of the question can best be illustrated by a specific example. Consider a box containing two electrons. These two identical particles move around in the box, bouncing from the walls and occasionally scattering from each other. In a classical description of this system, the electrons travel in sharply defined trajectories so that constant observation of the system allows us to distinguish between the two electrons, even though they are identical particles. For instance, in classical physics we can follow the development of the system, without disturbing it, by taking motion pictures of the system. If on a certain frame of the film we label the image of one of the electrons 1, and label the image of the other electron 2, we can follow the motion of the electrons through subsequent frames and always be able to say which electron is 1 and which electron is 2. The procedure is indicated in Figure 9-1. Of course, we cannot label the electrons themselves any more than we can paint one red and the other green. Electrons are identical particles—any electron is exactly the same as any other electron. Nevertheless, *in classical physics identical particles can be distinguished from each other by procedures which do not otherwise affect their behavior, and so it is possible to assign labels to the particles.*

In quantum mechanics this cannot be done because the uncertainty principle does not allow us to observe constantly the motion of the electrons without changing their behavior. As we have seen in Section 3-3, the photons which we must use to illuminate the scene for the motion picture camera interact with the electrons in a significant and unpredictable manner. The behavior of the electrons is seriously affected by any attempt to distinguish them.

An equivalent, but more formal, statement is that in quantum mechanics the finite extent of the wave functions associated with each electron may lead to an overlapping

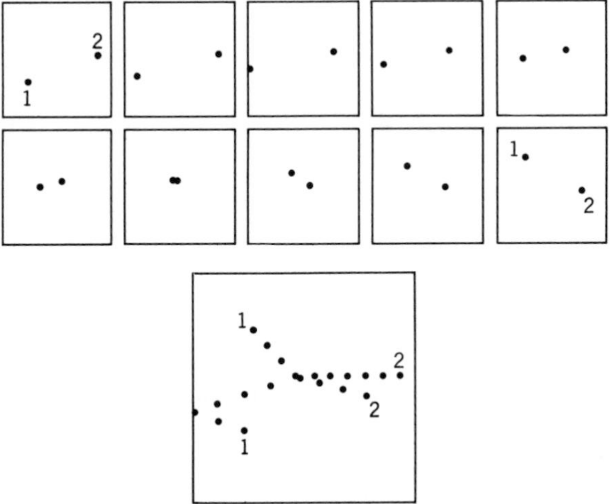


**FIGURE 9-1**

*Top:* A sequence of ten frames from a motion picture of two electrons moving in a box, according to classical physics. If labels were assigned to their images in the first frame, there would be no ambiguity in assigning the same labels to their images in any subsequent frame, although it may be necessary to use high magnification and "slow motion." *Bottom:* An enlarged superposition of all ten frames, showing the trajectories of the electrons.

of these wave functions that makes it difficult to tell which wave function was associated with which electron. A good example is provided by the helium atom. The wave functions of the two electrons overlap highly in all quantum states, and so the electrons cannot be distinguished. There is also an overlap of the wave functions associated with the electron and the proton of a hydrogen atom. But this does not lead to any problems in distinguishing one particle from the other because an electron and a proton are not identical—they can be distinguished by the differences in their mass, charge, etc.

We see that there is a fundamental distinction between the classical and quantum mechanical description of a system containing identical particles. An accurate quantum mechanical treatment of these systems must be formulated in such a way that the *indistinguishability* of identical particles is explicitly taken into account. That is, *measurable results obtained from accurate quantum mechanical calculations should not depend on the assignment of labels to identical particles*. This property leads to important effects which have no classical analogies because indistinguishability itself is purely quantum mechanical.

Since it is the eigenfunctions that carry the burden of describing quantum mechanical systems, we must look for a way of writing them so that they contain a mathematical expression of the qualitative ideas developed above. We continue considering two identical particles (e.g., two electrons, or two protons, or two $\alpha$ particles, or two helium atoms) in a box. To simplify the argument, we assume that we can neglect the interactions between the particles. Then they will bounce between the walls of the box, but they will not scatter from each other. Despite this simplification, the results of the following discussion are of quite general validity.

The time-independent Schroedinger equation for our system of two noninteracting particles in three dimensions can be written

$$-\frac{\hbar^2}{2m}\left(\frac{\partial^2\psi_T}{\partial x_1^2}+\frac{\partial^2\psi_T}{\partial y_1^2}+\frac{\partial^2\psi_T}{\partial z_1^2}\right)-\frac{\hbar^2}{2m}\left(\frac{\partial^2\psi_T}{\partial x_2^2}+\frac{\partial^2\psi_T}{\partial y_2^2}+\frac{\partial^2\psi_T}{\partial z_2^2}\right)+V_T\psi_T=E_T\psi_T \tag{9-1}$$

where

$m$ = the mass of either particle

$x_1, y_1, z_1$ = the coordinates of particle 1

$x_2, y_2, z_2$ = the coordinates of particle 2

This equation can be obtained immediately by writing the classical expression for the total energy of the system, replacing the dynamical quantities by their associated quantum mechanical operators to obtain the Schroedinger equation, and then separating out the time dependence. Since the procedure is a simple extension of that used to obtain the time-independent Schroedinger equation for one particle in three dimensions, (7-10), and since the validity of (9-1) is quite obvious anyway, we shall not include the details here. It is more important to point out that (9-1) does use labels, which specify the identity of the two particles as 1 and 2. The language of mathematics forces us to use such labels because there would otherwise be hopeless confusion between the symbols; we challenge the student to devise a way to write an unambiguous equation, analogous to (9-1), without employing particle labels. In using (9-1), we clearly stand a chance of violating the quantum mechanical requirements of indistinguishability. We shall see later that this does happen, but that it is possible to arrange things in such a way as to remove the difficulty. We shall do this by finding certain linear combinations of labeled eigenfunctions which lead to measurable predictions that are independent of the assignment of the labels.

In the time-independent Schroedinger equation, (9-1)

$$\psi_T(x_1, \ldots, z_2) = \text{the eigenfunction for the total system}$$
$$V_T(x_1, \ldots, z_2) = \text{the potential energy for the total system}$$
$$E_T = \text{the total energy for the total system}$$

Since we have assumed that there is no interaction between the two particles, the particles move *independently*. The potential energy of the total system is then simply the sum of the potential energies of each particle in its interaction with the walls of the box. Each potential energy will depend only on the coordinates of one particle and, since the particles are identical, the two potential energy functions are the same. Thus

$$V_T(x_1, \ldots, z_2) = V(x_1, y_1, z_1) + V(x_2, y_2, z_2) \tag{9-2}$$

It is easy to show, by applying the technique of separation of variables, that for the potential of (9-2), there are solutions to (9-1) of the form

$$\psi_T(x_1, \ldots, z_2) = \psi(x_1, y_1, z_1)\psi(x_2, y_2, z_2) \tag{9-3}$$

where $\psi(x_1, y_1, z_1)$ and $\psi(x_2, y_2, z_2)$ satisfy identical one-particle time-independent Schroedinger equations. Note that the total eigenfunction is written as a *product* of the two eigenfunctions describing the *independently* moving particles.

Each of the eigenfunctions describing one of the particles requires three quantum numbers to specify the mathematical form of its dependence on its three space coordinates. In addition, each requires one more quantum number to specify the orientation of the spin of the particle. We shall shorten the notation by using a single symbol, such as $\alpha$, or $\beta$, or $\gamma$, etc., to designate a particular set of the four quantum numbers required to specify the space and spin quantum state of one of the particles. Thus $\alpha$, for example, stands for a certain set of values of the four quantum numbers. Then a particular eigenfunction for particle 1 would be written

$$\psi_\alpha(x_1, y_1, z_1)$$

We further shorten the notation by writing this as

$$\psi_\alpha(1)$$

This eigenfunction contains the information that particle 1 is in the space and spin quantum state described by $\alpha$. Numerically, it is the function of the form specified by $\psi_\alpha$, evaluated at the coordinates of particle 1. An eigenfunction indicating that particle 2 is in the space and spin quantum state $\beta$ would be written

$$\psi_\beta(2)$$

The total eigenfunction $\psi_T(x_1, \ldots, z_2)$ for the case in which particle 1 is in the state $\alpha$, and particle 2 is in the state $\beta$, is

$$\psi_T(x_1, \ldots, z_2) = \psi_\alpha(1)\psi_\beta(2) \tag{9-4}$$

An eigenfunction indicating that particle 1 is in the state $\beta$, and particle 2 is in the state $\alpha$, has the quantum number symbols interchanged

$$\psi_T(x_1, \ldots, z_2) = \psi_\beta(1)\psi_\alpha(2) \tag{9-5}$$

Now let us see whether measurable quantities, evaluated from these total eigenfunctions, depend on the assignment of the particle labels. The simplest measurable is the probability density function. For the eigenfunction of (9-4), it is

$$\psi_T^*\psi_T = \psi_\alpha^*(1)\psi_\beta^*(2)\psi_\alpha(1)\psi_\beta(2) \tag{9-6}$$

and for the eigenfunction of (9-5), it is

$$\psi_T^*\psi_T = \psi_\beta^*(1)\psi_\alpha^*(2)\psi_\beta(1)\psi_\alpha(2) \tag{9-7}$$

Since the two identical particles are indistinguishable, we should be able to exchange their labels without changing a measurable quantity such as the probability density. As an example, we carry out this operation on (9-6), obtaining

$$\psi_\alpha^*(1)\psi_\beta^*(2)\psi_\alpha(1)\psi_\beta(2) \underset{\substack{1\to 2\\ 2\to 1}}{\longrightarrow} \psi_\alpha^*(2)\psi_\beta^*(1)\psi_\alpha(2)\psi_\beta(1)$$

where the arrows mean that the expression on the left changes into the expression on the right when 1 changes into 2 and 2 changes into 1. But it is apparent that the relabeled probability density function is not equal to the original probability density function. For instance, the first term in the relabeled function (expression on the right) is $\psi_\alpha^*$ evaluated at the coordinates $x_2$, $y_2$, $z_2$, while the first term in the original function (expression on the left) is $\psi_\alpha^*$ evaluated at the coordinates $x_1$, $y_1$, $z_1$. Thus a relabeling of the particles actually does change the probability density function calculated from the eigenfunction of (9-4). The same is true for the eigenfunction of (9-5). Therefore, we must conclude that these are not acceptable eigenfunctions for the accurate description of a system containing two identical particles. The suspicion which we expressed after writing the time-independent Schroedinger equation, (9-1), has been justified.

It is, however, possible to construct an eigenfunction which satisfies the time-independent Schroedinger equation, and yet has the acceptable property that its probability density function is not changed by a relabeling of the particles. In fact, there are two ways of doing this. Consider the following two linear combinations of the eigenfunctions of (9-4) and (9-5)

$$\psi_S = \frac{1}{\sqrt{2}}\,[\psi_\alpha(1)\psi_\beta(2) + \psi_\beta(1)\psi_\alpha(2)] \tag{9-8}$$

and

$$\psi_A = \frac{1}{\sqrt{2}}\,[\psi_\alpha(1)\psi_\beta(2) - \psi_\beta(1)\psi_\alpha(2)] \tag{9-9}$$

The first is called the *symmetric* total eigenfunction, and the second the *antisymmetric* total eigenfunction (for reasons that will become apparent soon). Now the total energy of a system containing a particle in a quantum state $\alpha$ and another particle in a quantum state $\beta$ will not depend on which particle is in which state, if the particles are identical. Thus both $\psi_T = \psi_\alpha(1)\psi_\beta(2)$ and $\psi_T = \psi_\beta(1)\psi_\alpha(2)$ are solutions to the time-independent Schroedinger equation, (9-1), corresponding to the *same* value of the total energy $E_T$. Because that equation is linear in $\psi_T$, it follows immediately that the linear combinations $\psi_S$ and $\psi_A$, of the two forms of $\psi_T$, are also solutions. Since they correspond to the same value of $E_T$, they are degenerate solutions—that is $\psi_S$ and $\psi_A$ are *different eigenfunctions* corresponding to precisely the *same eigenvalue*. The phenomenon is called *exchange degeneracy* since the difference between the degenerate eigenfunctions has to do with exchange of the particle labels. The factor of $1/\sqrt{2}$ ensures that $\psi_S$ and $\psi_A$ will be normalized if $\psi_T = \psi_\alpha(1)\psi_\beta(2)$ and $\psi_T = \psi_\beta(1)\psi_\alpha(2)$ are normalized.

It is easy to evaluate the probability density functions for $\psi_S$ and $\psi_A$, and then show that in both cases their values are not changed by an exchange of the particle labels. We shall obtain this result by investigating the effect of an exchange of the particle labels on the eigenfunctions themselves. Carrying out the operation, we have

$$\psi_S = \frac{1}{\sqrt{2}}[\psi_\alpha(1)\psi_\beta(2) + \psi_\beta(1)\psi_\alpha(2)] \underset{\substack{1\to2\\2\to1}}{\longrightarrow} \frac{1}{\sqrt{2}}[\psi_\alpha(2)\psi_\beta(1) + \psi_\beta(2)\psi_\alpha(1)] = \psi_S$$

and (9-10)

$$\psi_A = \frac{1}{\sqrt{2}}[\psi_\alpha(1)\psi_\beta(2) - \psi_\beta(1)\psi_\alpha(2)] \underset{\substack{1\to2\\2\to1}}{\longrightarrow} \frac{1}{\sqrt{2}}[\psi_\alpha(2)\psi_\beta(1) - \psi_\beta(2)\psi_\alpha(1)] = -\psi_A \tag{9-11}$$

We see that the symmetric total eigenfunction $\psi_S$ is unchanged by an exchange of the particle labels, and that the antisymmetric total eigenfunction $\psi_A$ is multiplied by minus one by an exchange of the particle labels. (These properties give rise to their names.) We then have for the probability densities

$$\psi_S^*\psi_S \underset{\substack{1\to2\\2\to1}}{\longrightarrow} \psi_S^*\psi_S \tag{9-12}$$

and

$$\psi_A^*\psi_A \underset{\substack{1\to2\\2\to1}}{\longrightarrow} (-1)^2\psi_A^*\psi_A = \psi_A^*\psi_A \tag{9-13}$$

Hence, for both the symmetric and antisymmetric total eigenfunctions, the probability density functions are not changed by an exchange of the particle labels. The change in sign of the antisymmetric eigenfunction under an exchange of the particle labels is, of course, not objectionable since an eigenfunction itself is not measurable.

It can be shown that *any* measurable quantity that can be obtained from the symmetric, or antisymmetric, total eigenfunctions is not affected by an exchange of the particle labels. Thus these two eigenfunctions provide an accurate description of a system containing two identical particles. *Although the labels* 1 *and* 2 *do appear in the expressions for* $\psi_S$ *and* $\psi_A$, *this labeling does not violate the requirements of indistinguishability because the value of any measurable quantity obtained from the eigenfunctions is independent of the assignment of the labels.*

**Example 9-1.** Two identical particles move independently in a one-dimensional box of length $a$, one being in the ground state of the infinite square well potential describing the box and the other being in the first excited state of that potential. For simplicity, assume that the particles have no spin, so that the total eigenfunctions for the system are just space eigenfunctions. (a) Evaluate the symmetric and antisymmetric total eigenfunctions of (9-8) and (9-9), and verify that the factor $1/\sqrt{2}$ in these equations does properly normalize them.

Using the general forms of (6-79) and (6-80) for the eigenfunctions for one particle in an infinite square well potential, and also using the normalization constant evaluated in Example 5-10, we find that the normalized space eigenfunction of the particle in the ground state is $\sqrt{2/a}\cos(\pi x/a)$ and the normalized space eigenfunction for the particle in the first excited state is $\sqrt{2/a}\sin(2\pi x/a)$. Thus writing the symmetric and antisymmetric space eigenfunctions for the two particle system as $\psi_+$ and $\psi_-$, we have from (9-8) and (9-9)

$$\psi_+ = \frac{1}{\sqrt{2}}\frac{2}{a}\left[\cos\frac{\pi x_1}{a}\sin\frac{2\pi x_2}{a} + \sin\frac{2\pi x_1}{a}\cos\frac{\pi x_2}{a}\right]$$

$$\psi_- = \frac{1}{\sqrt{2}}\frac{2}{a}\left[\cos\frac{\pi x_1}{a}\sin\frac{2\pi x_2}{a} - \sin\frac{2\pi x_1}{a}\cos\frac{\pi x_2}{a}\right]$$

when both $x_1$ and $x_2$ lie within the range $-a/2$ to $a/2$. When either $x_1$ or $x_2$ lie outside that range both $\psi_+$ and $\psi_-$ are zero since the one particle eigenfunctions have zero value there.

The normalization integral of $\psi_+$ is

$$\int_{-\infty}^{\infty}\int_{-\infty}^{\infty} \psi_+^* \psi_+ \, dx_1 \, dx_2$$

$$= \int_{-a/2}^{a/2}\int_{-a/2}^{a/2} \frac{1}{2}\left(\frac{2}{a}\right)^2 \left[\cos^2\frac{\pi x_1}{a}\sin^2\frac{2\pi x_2}{a} + \sin^2\frac{2\pi x_1}{a}\cos^2\frac{\pi x_2}{a}\right.$$

$$+ \cos\frac{\pi x_1}{a}\sin\frac{2\pi x_2}{a}\sin\frac{2\pi x_1}{a}\cos\frac{\pi x_2}{a}$$

$$\left. + \sin\frac{2\pi x_1}{a}\cos\frac{\pi x_2}{a}\cos\frac{\pi x_1}{a}\sin\frac{2\pi x_2}{a}\right] dx_1\, dx_2$$

$$= \frac{1}{2}\left[\int_{-a/2}^{a/2} \frac{2}{a}\cos^2\frac{\pi x_1}{a}\, dx_1 \int_{-a/2}^{a/2} \frac{2}{a}\sin^2\frac{2\pi x_2}{a}\, dx_2 + \int_{-a/2}^{a/2} \frac{2}{a}\sin^2\frac{2\pi x_1}{a}\, dx_1 \int_{-a/2}^{a/2} \cos^2\frac{\pi x_2}{a}\, dx_2\right.$$

$$+ \int_{-a/2}^{a/2} \frac{2}{a}\cos\frac{\pi x_1}{a}\sin\frac{2\pi x_1}{a}\, dx_1 \int_{-a/2}^{a/2} \frac{2}{a}\sin\frac{2\pi x_2}{a}\cos\frac{\pi x_2}{a}\, dx_2$$

$$\left. + \int_{-a/2}^{a/2} \frac{2}{a}\sin\frac{2\pi x_1}{a}\cos\frac{\pi x_1}{a}\, dx_1 \int_{-a/2}^{a/2} \frac{2}{a}\cos\frac{\pi x_2}{a}\sin\frac{2\pi x_2}{a}\, dx_2\right]$$

Now each of the first two terms in the bracket yields one since in each both integrals are just the normalization integrals for the normalized one particle eigenfunctions $\sqrt{2/a}\cos(\pi x/a)$ and $\sqrt{2/a}\sin(2\pi x/a)$. Furthermore, each of the last two terms in the bracket yields zero since both are the product of two integrals of the form, and value

$$\int_{-a/2}^{a/2} \cos\frac{\pi x}{a}\sin\frac{2\pi x}{a}\, dx = 0$$

The value can be verified in any table of definite integrals. Thus the normalization integral for $\psi_+$ yields $(1/2)[1 + 1]$, where the $1/2$ came from squaring the factor $1/\sqrt{2}$ in (9-8). So we find that that factor does properly normalize $\psi_+$ by making its normalization integral equal one. We can also immediately show that the same conclusion is obtained for $\psi_-$.

Inspection of a table of definite integrals will further show that the integral from $-a/2$ to $a/2$ of any two different sinusoidal eigenfunctions for a particle in an infinite square well potential has the value zero. In fact, it can be proven from general considerations that the integral over all $x$ of any two *different* eigenfunctions of any particular potential has the value zero. This property is called *orthogonality*. Because of the orthogonality of one-particle eigenfunctions, only 2 of the $2^2$ terms in the normalization integral for any symmetric or antisymmetric two-particle eigenfunction have nonzero values; and because of the normalization of one-particle eigenfunctions, those two values are both equal to 1. Therefore, the factor $1/\sqrt{2}$ in (9-8) and (9-9) ensures that these total eigenfunctions are normalized in all cases.

(b) Write expressions for the expectation value of the separation distance $D$ between the particles for the case in which the space eigenfunction for the two particle system is symmetric, and for the case in which it is antisymmetric. Then show that in neither of these cases is this expectation value affected by an exchange of the particle labels.

The separation distance $D$ is the absolute value of the difference in their $x$ coordinates. That is $D = |x_2 - x_1| = |x_1 - x_2|$. The expectation value $\overline{D}$ is, for the case of $\psi_+$

$$\overline{D} = \int_{-\infty}^{\infty}\int_{-\infty}^{\infty} \psi_+^* D \psi_+ \, dx_1 \, dx_2 = \int_{-a/2}^{a/2}\int_{-a/2}^{a/2} D\psi_+^2 \, dx_1 \, dx_2$$

$$= \frac{2}{a^2}\int_{-a/2}^{a/2}\int_{-a/2}^{a/2} |x_2 - x_1| \left[\cos^2\frac{\pi x_1}{a}\sin^2\frac{2\pi x_2}{a} + \sin^2\frac{2\pi x_1}{a}\cos^2\frac{\pi x_2}{a}\right.$$

$$\left. + 2\cos\frac{\pi x_1}{a}\sin\frac{2\pi x_2}{a}\sin\frac{2\pi x_1}{a}\cos\frac{\pi x_2}{a}\right] dx_1 \, dx_2$$

Similarly, for the case of $\psi_-$

$$\overline{D} = \frac{2}{a^2}\int_{-a/2}^{a/2}\int_{-a/2}^{a/2} |x_2 - x_1| \left[\cos^2\frac{\pi x_1}{a}\sin^2\frac{2\pi x_2}{a} + \sin^2\frac{2\pi x_1}{a}\cos^2\frac{\pi x_2}{a}\right.$$

$$\left. - 2\cos\frac{\pi x_1}{a}\sin\frac{2\pi x_2}{a}\sin\frac{2\pi x_1}{a}\cos\frac{\pi x_2}{a}\right] dx_1 \, dx_2$$

Some work would be required to evaluate the integrals for these two cases. But we can see immediately that in both the values are not affected by exchanging the particle labels. The reason is that in both integrals neither the factor $|x_2 - x_1|$ nor the third term in the square brackets are changed and, although the first term in the square bracket changes into the second term, the second term changes into the first term.

We can also see that the value of $\overline{D}$ obtained with the symmetric space eigenfunction is different from the value obtained with the antisymmetric space eigenfunction, because of the difference of the sign of the third term in the square bracket. In other words, the average separation between the particles in a state in which the space eigenfunction is symmetric is different from what it is in a state in which the space eigenfunction is antisymmetric. In Section 9-4 we shall give further interpretation to these results, and we shall see that they have very interesting consequences. ◀

## 9-3 The Exclusion Principle

As a result of an analysis of data concerning the energy levels of atoms, which we shall study soon, in 1925 Pauli was led to his famous *exclusion principle* (weaker condition):

*In a multielectron atom there can never be more than one electron in the same quantum state.*

He then established from the analysis of other experimental data that the exclusion principle represents a property of electrons and not, specifically, of atoms. The exclusion principle operates in any system containing electrons.

Now consider the antisymmetric total eigenfunction of (9-9), for a case in which both particles are in the same space and spin quantum state $\alpha$. It is

$$\psi_A = \frac{1}{\sqrt{2}}[\psi_\alpha(1)\psi_\alpha(2) - \psi_\alpha(1)\psi_\alpha(2)] \equiv 0 \tag{9-14}$$

The eigenfunction is identically equal to zero. Hence, if two particles are described by the antisymmetric total eigenfunction, they cannot both be in a state with the same space and spin quantum numbers. The eigenfunctions we have been dealing with were obtained under the assumption that there are two identical particles, and that the interactions between them can be neglected. If there are more than two identical particles and/or if their interactions must be taken into account, the total eigenfunctions have different forms, as we shall see in Examples 9-2 and 9-3. But they can still be used to make linear combinations of definite symmetry, either symmetric or antisymmetric, and the antisymmetric linear combinations still have values identically equal to zero if *any two* particles are in the same quantum state. In other words, all antisymmetric total eigenfunctions have properties which conform to the requirements of the exclusion principle. So we conclude there is an alternative expression of the *exclusion principle* (stronger condition):

*A system containing several electrons must be described by an antisymmetric total eigenfunction.*

The condition specified by the second statement of the exclusion principle is stronger than the condition specified by the first statement, because it satisfies that condition, and it also satisfies the requirements of indistinguishability which demand total eigenfunctions of a definite symmetry. The stronger condition must be used in quantum mechanical calculations that aim for complete accuracy, but the weaker condition, which is much easier to apply, is often used in approximate calculations. In Section 9-5 we shall discuss the use of these conditions in the treatment of multielectron atoms, and we shall compare the results obtained from the stronger one with those obtained from the weaker.

In discovering the exclusion principle, Pauli found the answer to a long-standing problem concerning the structure of multielectron atoms. He has written:

"The question as to why all electrons for an atom in its ground state were not bound in the innermost shell had already been emphasized by Bohr as a fundamental problem in his earlier works . . . . However, no convincing explanation of this phenomenon could be given on the basis of classical mechanics. It made a strong impression on me that Bohr at that time and in later discussions was looking for a general explanation."

Pauli's explanation of the problem was certainly general. All the electrons cannot be bound in the same quantum state represented by the innermost shell of the atom because the system must be described by antisymmetric total eigenfunctions, which vanish if even two electrons are in the same quantum state. To emphasize just how fundamental the problem is, we jump a little ahead of our development to state that if all the electrons in an atom were in the innermost shell, then the atom would be essentially like a noble gas. The atom would be inert, and it would not combine with other atoms to form molecules. If electrons did not obey the exclusion principle this would be true of all atoms. Then the entire universe would be radically different. For instance, with no molecules there would be no life!

**Example 9-2.** Determine the form of the normalized antisymmetric total eigenfunction for a system of three particles, in which the interactions between the particles can be ignored.

This is easy to do if it is noted that the two-particle antisymmetric total eigenfunction

$$\psi_A = \frac{1}{\sqrt{2}}\left[\psi_\alpha(1)\psi_\beta(2) - \psi_\beta(1)\psi_\alpha(2)\right]$$

can also be written as a so-called *Slater determinant*

$$\psi_A = \frac{1}{\sqrt{2!}}\begin{vmatrix} \psi_\alpha(1) & \psi_\alpha(2) \\ \psi_\beta(1) & \psi_\beta(2) \end{vmatrix}$$

where $2! = 2 \times 1 = 2$. The identity of these two expressions can be verified by expanding the determinant. In determinantal form, the extension to three particles is obvious

$$\psi_A = \frac{1}{\sqrt{3!}}\begin{vmatrix} \psi_\alpha(1) & \psi_\alpha(2) & \psi_\alpha(3) \\ \psi_\beta(1) & \psi_\beta(2) & \psi_\beta(3) \\ \psi_\gamma(1) & \psi_\gamma(2) & \psi_\gamma(3) \end{vmatrix}$$

where $3! = 3 \times 2 \times 1 = 6$. Expansion of this determinant yields

$$\psi_A = \frac{1}{\sqrt{3!}} [\psi_\alpha(1)\psi_\beta(2)\psi_\gamma(3) + \psi_\beta(1)\psi_\gamma(2)\psi_\alpha(3) + \psi_\gamma(1)\psi_\alpha(2)\psi_\beta(3) - \psi_\gamma(1)\psi_\beta(2)\psi_\alpha(3) - \psi_\beta(1)\psi_\alpha(2)\psi_\gamma(3) - \psi_\alpha(1)\psi_\gamma(2)\psi_\beta(3)]$$

Each term of this linear combination is a solution, for the same total energy, to the time-independent Schroedinger equation for a potential energy function in which the variables can be grouped into a sum of terms that each depend on the coordinates of only one particle, as in (9-2). Therefore, the linear combination is also a solution. By exchanging the appropriate particle labels, as we did in (9-11) for a system of two particles, it is easy to verify that it is antisymmetric with respect to the exchange of any pair of labels. It also has the property of being identically equal to zero if any two particles are in the same space and spin quantum state. This can be seen most easily from the determinant itself, since it is a well known property of determinants that they vanish if any two rows are identical. It is not difficult to follow the procedures of Example 9-1 and to show that $\psi_A$ is normalized if $\psi_\alpha(1)\psi_\beta(2)\psi_\gamma(3)$, and similar terms, are normalized. ◀

As is the case for electrons, the symmetry character of other kinds of particles is a question settled by experiment. It is found that systems of protons, or of neutrons, or of certain other particles, must also be described by antisymmetric total eigenfunctions. On the other hand, it is found that systems of photons, helium atoms, and certain other particles, must be described by symmetric total eigenfunctions. There are important phenomena associated with the symmetry character of the symmetric particles. The most spectacular example is the "superfluid" behavior of liquid helium at temperatures near absolute zero. This, and other examples, will be discussed in Chapter 11, which treats the general properties of systems containing a large number of symmetric, or antisymmetric, particles.

Table 9-1 lists several kinds of particles, their symmetry character, and also the value of the quantum number $s$ that specifies the magnitude of their spin angular momentum. Also indicated are the two names, *fermion* and *boson*, sometimes used to distinguish the two classes of particles according to their symmetry character. It is very interesting to note that there must be some connection between the symmetry character of a particle and its spin. The point is that all the antisymmetric particles have half-integral spin just as the electron has, while all the symmetric particles have

**TABLE 9-1.** The Symmetry Character of Various Particles

| Particle | Symmetry | Generic Name | Spin ($s$) |
|---|---|---|---|
| Electron | Antisymmetric | Fermion | 1/2 |
| Positron | Antisymmetric | Fermion | 1/2 |
| Proton | Antisymmetric | Fermion | 1/2 |
| Neutron | Antisymmetric | Fermion | 1/2 |
| Muon | Antisymmetric | Fermion | 1/2 |
| $\alpha$ particle | Symmetric | Boson | 0 |
| He atom (ground state) | Symmetric | Boson | 0 |
| $\pi$ meson | Symmetric | Boson | 0 |
| Photon | Symmetric | Boson | 1 |
| Deuteron | Symmetric | Boson | 1 |

zero or integral spin. This connection has been studied by Pauli, and others, using very sophisticated forms of quantum mechanics. Some understanding of its origin has been obtained, but at the level of this book it is appropriate to say that the symmetry character of a particle should be considered as a basic property, like mass, charge, and spin, which is determined by experiment. An exception to this statement is that the symmetry of a well-bound composite particle, like a helium atom, can be predicted immediately from the symmetries of its constituents. (If the composite particle has an even number of antisymmetric constituents, it is symmetric.)

**Example 9-3.** Determine the form of the normalized symmetric total eigenfunction for a system of three particles, in which the interactions between the particles can be ignored.

In analogy to the relation between (9-8) and (9-9), the required eigenfunction can be obtained immediately by writing the linear combination found in Example 9-2 with all the signs positive. That is

$$\begin{aligned}\psi_S = \frac{1}{\sqrt{3!}} [&\psi_\alpha(1)\psi_\beta(2)\psi_\gamma(3) + \psi_\beta(1)\psi_\gamma(2)\psi_\alpha(3) \\ &+ \psi_\gamma(1)\psi_\alpha(2)\psi_\beta(3) + \psi_\gamma(1)\psi_\beta(2)\psi_\alpha(3) \\ &+ \psi_\beta(1)\psi_\alpha(2)\psi_\gamma(3) + \psi_\alpha(1)\psi_\gamma(2)\psi_\beta(3)]\end{aligned}$$

It is immediately apparent that this linear combination is symmetric with respect to the exchange of any two particle labels. The normalization can be verified by the procedure used in Example 9-1. ◀

## 9-4 Exchange Forces and the Helium Atom

We turn now to a property of indistinguishable particles which is, to say the least, very strange. Consider a pair of electrons in a system in which we can ignore any explicit interactions (like the Coulomb interaction) between the two particles. According to (9-9), the total eigenfunction for the system can be written

$$\psi_A = \frac{1}{\sqrt{2}} [\psi_\alpha(1)\psi_\beta(2) - \psi_\beta(1)\psi_\alpha(2)]$$

This antisymmetric total eigenfunction depends on both the space variables and the spin variables of the two electrons since the symbols $\alpha, \beta, \gamma, \ldots$ specify sets of three space quantum numbers plus one spin quantum number. For the present discussion we rewrite it in such a way that the space and spin variables occur in separate factors, i.e.

$$(\text{total eigenfunction}) = (\text{space eigenfunction}) \times (\text{spin eigenfunction})$$

We also make both factors have a definite symmetry with respect to exchange of the particle labels. Antisymmetry of the total eigenfunction can then be obtained by multiplying a symmetric space eigenfunction times an antisymmetric spin eigenfunction, or by multiplying an antisymmetric space eigenfunction times a symmetric spin eigenfunction.

The normalized symmetric and antisymmetric space eigenfunctions have the forms we used in Example 9-1

symmetric space eigenfunction: $$\frac{1}{\sqrt{2}} [\psi_a(1)\psi_b(2) + \psi_b(1)\psi_a(2)] \qquad (9\text{-}15)$$

antisymmetric space eigenfunction: $$\frac{1}{\sqrt{2}} [\psi_a(1)\psi_b(2) - \psi_b(1)\psi_a(2)] \qquad (9\text{-}16)$$

where $\psi_a(1)\psi_b(2)$ and $\psi_b(1)\psi_a(2)$ are normalized. Each symbol from the series $a$, $b$, $c, \ldots$ represents a particular set of the three space quantum numbers only (in contrast to the $\alpha, \beta, \gamma, \ldots$, which represent sets of three space and one spin quantum number). Of course these forms are very *general*, there being a wide variety of different $\psi_a$ and $\psi_b$ for different systems.

The forms of the symmetric and antisymmetric spin eigenfunctions are quite another matter. The reason is that the spin variable is not continuous like a space variable, but instead is discrete. For instance, the spin of a single electron can have only two discrete orientations relative to any $z$ axis since its $z$ component is either $+1/2$ or $-1/2$, in units of $\hbar$. Continuous functions, such as those displayed in the one-electron atom space eigenfunctions of Table 7-2, therefore cannot be used for spin eigenfunctions. For the case of two noninteracting electrons, each of which has two possible spin orientations, there are only four possible spin states for the system, and therefore only four possible spin eigenfunctions. Because there are so few we can display their *specific* forms. If these four spin eigenfunctions for the system are written so as to have definite symmetries, then one will be antisymmetric and the other three symmetric. Matrices are frequently employed to write mathematical expressions for the spin eigenfunctions, but here we shall write them in terms of combinations of the symbols $+1/2$ and $-1/2$ because their interpretations will be more obvious.

The only possible antisymmetric spin eigenfunction for two noninteracting electrons is

$$\begin{array}{r}\text{antisymmetric spin}\\ \text{eigenfunction:}\end{array} \quad \frac{1}{\sqrt{2}}[(+1/2,-1/2)-(-1/2,+1/2)] \quad \text{(singlet)} \tag{9-17}$$

This is a linear combination of a symbol $(+1/2,-1/2)$ that specifies a state where the $z$ components of the spins have values, in units of $\hbar$, of $+1/2$ for electron 1 and $-1/2$ for electron 2, minus a symbol $(-1/2,+1/2)$ that specifies a state where the $z$ components are $-1/2$ for electron 1 and $+1/2$ for electron 2. Due to the minus sign between the symbols, the linear combination is antisymmetric in an exchange of the labels of the two electrons since such an exchange would convert the first symbol to $(-1/2,+1/2)$ and the second symbol to $(+1/2,-1/2)$, thereby changing the overall sign of the linear combination. We shall not need to further manipulate these symbols and their linear combinations, and we shall only use them to describe spin states. So it will not be necessary for us to further specify their mathematical (i.e., matrix) properties.

There are three possible symmetric spin eigenfunctions

$$\begin{array}{r}\\ \text{symmetric spin}\\ \text{eigenfunctions:}\\ \\ \end{array} \quad \begin{array}{l}(+1/2,+1/2)\\ \frac{1}{\sqrt{2}}[(+1/2,-1/2)+(-1/2,+1/2)] \quad \text{(triplet)}\\ (-1/2,-1/2)\end{array} \tag{9-18}$$

Their symmetry is obvious since for each an exchange of labels results in no change in the eigenfunction. These three describe the so-called *triplet* states, and the antisymmetric eigenfunction describes the so-called *singlet* state. All four of these spin eigenfunctions are normalized.

A physical interpretation of the singlet and triplet states can be obtained by evaluating, for each state, the magnitude $S'$ and $z$ component $S'_z$ of the *total spin angular momentum* $\mathbf{S}'$. This vector is

$$\mathbf{S}' = \mathbf{S}_1 + \mathbf{S}_2 \tag{9-19}$$

the sum of the spin angular momenta of the two electrons. As is true for all angular momenta in quantum mechanics, $S'$ and $S'_z$ are quantized according to the relations

$$S' = \sqrt{s'(s'+1)}\hbar \qquad S'_z = m'_s\hbar \tag{9-20}$$

The quantum numbers satisfy the relations

$$m'_s = -s', \ldots, +s' \qquad s' = 0, 1 \tag{9-21}$$

The relations between the quantum numbers, obtained when $S'$ and $S'_z$ are evaluated, can be represented and explained by the rules of vector addition used in Section 8-5. Figure 9-2 shows two vectors of length $s = 1/2$ added to form a vector of length $s' = 0$ or $1$, which can have, in the latter case, $z$ components of $+1, 0, -1$. As we have warned the student before, these vector addition diagrams must be interpreted cautiously since the vectors are not really angular momenta. But they do convey correctly the impression that in the three *triplet states*, which correspond to $s' = 1$, $m'_s = +1$; $s' = 1$, $m'_s = 0$; $s' = 1$, $m'_s = -1$, the electron spins are *essentially parallel.* In the *singlet state*, $s' = 0$, $m'_s = 0$, the electron spins are *essentially anti-parallel.* Figure 9-3 attempts to show the angular momenta; but as it cannot truly represent the linear combinations in (9-17) and (9-18) it oversimplifies somewhat.

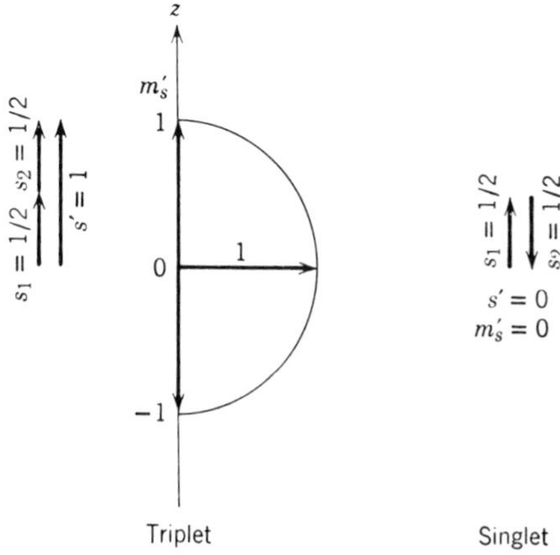


**FIGURE 9-2**

Vector diagrams representing the rules for adding the quantum numbers $s_1 = 1/2$ and $s_2 = 1/2$ to obtain the possible values for the quantum numbers $s'$ and $m'_s$. *Left:* The maximum possible value of $s'$ is obtained when a vector of magnitude $s_1$ is added to a parallel vector of magnitude $s_2$, yielding $s' = s_1 + s_2 = 1/2 + 1/2 = 1$. The maximum possible $z$ component of this vector gives the maximum possible value of the quantum number $m'_s$, and the minimum possible $z$ component gives the minimum possible value of $m'_s$. The intermediate values of $m'_s$ (only one in this case) differ by integers. Thus the possible values are $m'_s = +1, 0, -1$. *Right:* A vector of magnitude $s_1 = 1/2$ is added to an antiparallel vector of magnitude $s_2 = 1/2$ to yield a vector of magnitude $s' = s_1 - s_2 = 1/2 - 1/2 = 0$. A vector whose length is zero must have $z$ component zero as well, so the only possible value for $m'_s$ is zero. The term triplet refers to the state $s' = 1$ where three possible values of $m'_s$ arise; the term singlet refers to the state $s' = 0$ where only one possible value of $m'_s$ arises.

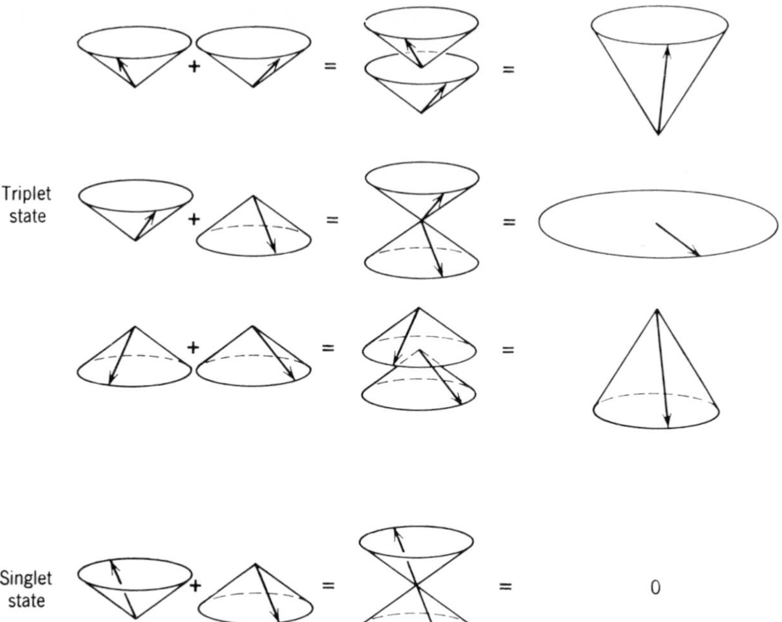


**FIGURE 9-3**

*Triplet state:* Two spin angular momentum vectors of magnitudes $S_1 = S_2 = \sqrt{(1/2)(1/2 + 1)}\,\hbar$ precess randomly, but in step so as to remain pointing in the same general direction, about the vertical $z$ axis. If their $z$ components are both positive, $S_{1_z} = S_{2_z} = +(1/2)\,\hbar$, or both negative, $S_{1_z} = S_{2_z} = -(1/2)\,\hbar$, their sum is a total spin vector of magnitude $S' = \sqrt{1(1+1)}\,\hbar$ and positive $z$ component, $S'_z = +1\hbar$, or negative $z$ component, $S'_z = -1\hbar$. If the spin vectors have $z$ components of opposite sign, but precess so as to point in the same general direction, the total spin vector has a zero $z$ component, $S'_z = 0$, but still has the magnitude $S' = \sqrt{1(1+1)}\,\hbar$, because it precesses in the plane perpendicular to the $z$ axis. These possibilities are the three which can occur in the triplet state. *Singlet state:* If the two spin vectors have $z$ components of opposite sign and precess so as to remain pointing in essentially opposite directions the total spin vector has zero $z$ component, $S'_z = 0$, because it has zero magnitude, $S' = 0$. This is the singlet state. In a certain sense, the two spin vectors are out of phase in this state. In the same sense, the two vectors are in phase in the $S'_z = 0$ triplet state. These phases are related to the minus and plus signs occurring between the terms in the linear combinations of the total spin eigenfunctions of (9-17) and (9-18).

Now we shall employ these ideas to explain a fundamental property of a system containing two electrons. If the spins of the two electrons are "parallel" and the spin eigenfunction is one of the symmetric triplets of (9-18), the space eigenfunction must be antisymmetric as in (9-16), in order to have the *total eigenfunction antisymmetric.* Let us consider such a situation for a case in which the space variables of the two electrons happen to have almost the same values. Then $\psi_a(1) \simeq \psi_a(2)$ since the left-hand side is evaluated at the coordinates of electron 1, which are almost equal to the coordinates of electron 2 where the right-hand side is evaluated. For the same reason, $\psi_b(1) \simeq \psi_b(2)$. As a consequence

$$\psi_a(1)\psi_b(2) \simeq \psi_b(1)\psi_a(2)$$

In this case the value of the antisymmetric space eigenfunction is

$$\frac{1}{\sqrt{2}}[\psi_a(1)\psi_b(2) - \psi_b(1)\psi_a(2)] \simeq \frac{1}{\sqrt{2}}[\psi_b(1)\psi_a(2) - \psi_b(1)\psi_a(2)] = 0$$

The result is that the probability density will be very small when the triplet state electrons have similar coordinates, i.e., when they are close together. Since there is little chance of finding them close together, the *triplet state electrons act as if they repel each other*. This has nothing to do with a Coulomb repulsion because we assumed at the very beginning of our treatment that there is no explicit interaction between the electrons. Instead, it has to do with the properties of antisymmetric space eigenfunctions.

Symmetric space eigenfunctions have inverse properties. If the space eigenfunction for the two electrons is symmetric, and they happen to have almost the same coordinates, then that eigenfunction is

$$\frac{1}{\sqrt{2}}[\psi_a(1)\psi_b(2) + \psi_b(1)\psi_a(2)] \simeq \frac{1}{\sqrt{2}}[\psi_b(1)\psi_a(2) + \psi_b(1)\psi_a(2)] = \sqrt{2}\,\psi_b(1)\psi_a(2)$$

since we shall again have $\psi_a(1) \simeq \psi_a(2)$ and $\psi_b(1) \simeq \psi_b(2)$. Thus the probability density will have the value $2\,\psi_b^*(1)\psi_a^*(2)\psi_b(1)\psi_a(2)$ when the two electrons with a symmetric space eigenfunction are close together. This is twice the average value over all space of the probability density for the symmetric space eigenfunction (because $\psi_b(1)\psi_a(2)$ is normalized so the integral of $\psi_b^*(1)\psi_a^*(2)\psi_b(1)\psi_a(2)$ over all space equals one, as does the integral over all space of the symmetric space eigenfunction probability density). So there is a particularly large chance of finding the two noninteracting electrons close together if their space eigenfunction is symmetric. Thus, if the spins of the two electrons are "antiparallel" and the spin eigenfunction is the antisymmetric singlet, as in (9-17), the space eigenfunction must be symmetric, as in (9-15), and the *singlet state electrons act as if they attract each other* since there is a large chance of finding them close together.

Figure 9-4 illustrates the symmetries of surfaces representing the $x_1$ and $x_2$ dependences of a typical antisymmetric, or symmetric, space eigenfunction for a one-dimensional system containing two identical noninteracting particles. The particular simple case shown is for one particle being in the ground state of an infinite square well potential of width *a*, for which the eigenfunction has the form of one-half of a cosine wave, and the other particle being in the first excited state of that potential, for which the eigenfunction has the form of one full sine wave. The top surface represents a situation in which the particle whose coordinate is written $x_1$ is in the ground state (note the half cosine in the $x_1$ direction), and the particle whose coordinate is $x_2$ is in the first excited state (note the full sine in the $x_2$ direction). Since identical particles are indistinguishable, it is equally possible that the system is in a situation in which the particle with coordinate $x_1$ is in the first excited state and the particle with coordinate $x_2$ is in the ground state. This situation is described by the second surface from the top. In quantum mechanics, both situations are allowed for by taking the *eigenfunction* for the system to be a linear combination of equal parts of the eigenfunctions describing either of them. This can be done either by adding or subtracting. In subtracting, we obtain the antisymmetric space eigenfunction for the system, which is illustrated by the third surface; in adding, we obtain the symmetric space eigenfunction for the system, illustrated by the bottom surface. The point of particular interest here is that the antisymmetric space eigenfunction is zero along the line $x_1 = x_2$ corresponding to the two particles being in the same location, while the symmetric space eigenfunction has its maximum magnitudes along the line. Thus the probability density $\psi^*\psi$ will be very small for the antisymmetric case, and very large for the symmetric case, when evaluated for coordinates of the two particles which are nearly the same.

In classical mechanics a roughly analogous situation could arise in a system containing two

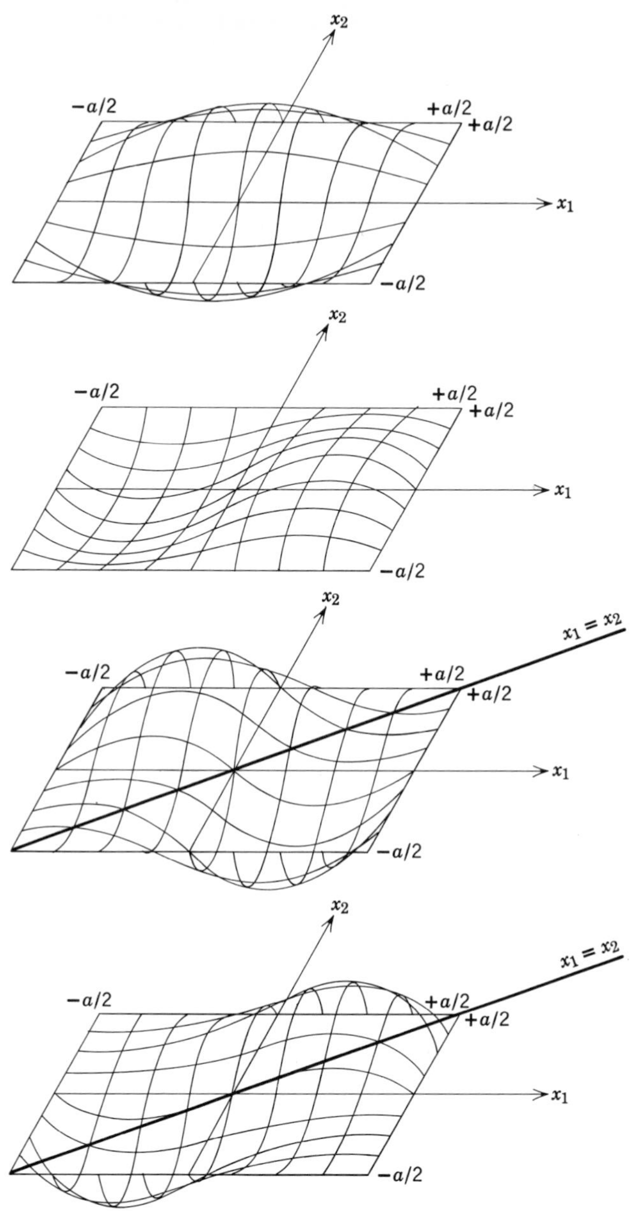


**FIGURE 9-4**

Depicting the antisymmetric space eigenfunction $\psi_-$, of Example 9-1, for a system of two noninteracting identical particles in a one-dimensional infinite square well potential of width $a$ when one particle is in the ground state with eigenfunction $\sqrt{2/a}\cos(\pi x/a)$ and the other is in the first excited state with eigenfunction $\sqrt{2/a}\sin(2\pi x/a)$. *Top:* The first term of $\psi_-$ is shown by constructing the surface whose distance above or below the $x_1, x_2$ plane is the positive or negative value of $(2/a)\cos(\pi x_1/a)\sin(2\pi x_2/a)$. *Upper middle:* The surface describing the second term of $\psi_-$, i.e., $(2/a)\sin(2\pi x_1/a)\cos(\pi x_2/a)$. *Lower middle:* $1/\sqrt{2}$ times the first term minus the second term, which shows the geometry of $\psi_-$ itself. It is apparent that the value of $\psi_-$ is zero along the line $x_1 = x_2$, and it is small everywhere near that line. Thus the probability density $\psi_-^*\psi_-$ is very small wherever $x_1 \simeq x_2$, and so the probability is very small that this condition will be achieved.

**FIGURE 9-5**

A schematic illustration of the tendency for electrons in a triplet spin state to be relatively far apart, and the tendency for electrons in a singlet spin state to be relatively close together.

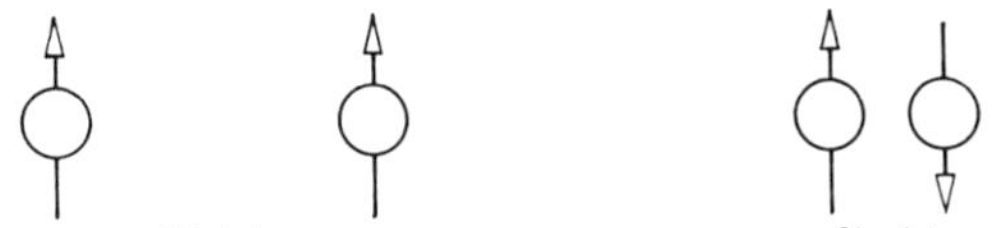

identical particles, if no effort were made to distinguish them by measurement, in that the *probability function* describing the system would be a linear combination of equal parts (one for particle 1 being in a lower energy state and particle 2 in a higher energy state and the other for particle 1 being in the higher state and particle 2 being in the lower state). But the single possible result for this situation has no analogy to the two distinctly different quantum results, because in quantum mechanics we deal with eigenfunctions that can exhibit interferences since they can be of either sign (or even complex), and then we calculate probabilities from them, whereas in classical mechanics we deal directly with probabilities which are necessarily positive and so cannot interfere.

If the student visualizes similar figures, he will be able to see why the same striking difference between the antisymmetric and symmetric space eigenfunctions is found when the particles are in any two different states of the infinite square well potential, or any other one-, two-, or three-dimensional potential. For a system containing more than two identical particles, these conclusions are also obtained for space eigenfunctions which are antisymmetric, or symmetric, with respect to the exchange of any two particle labels, since the geometry of the terms in the eigenfunctions that involve the two labels can be analyzed in the same way as for a system containing only two particles.

The triplet and singlet cases for a system of two electrons is illustrated schematically in Figure 9-5. The requirement that an accurate description of the system must use a *total* eigenfunction which is *antisymmetric* in an exchange of their labels, leads to a coupling between their spin and space variables. They act as if they move under the influence of a force whose sign depends on the relative orientation of their spins. This is called an *exchange force*. It is a purely quantum mechanical effect and has no classical analogy.

Exchange forces do not arise between two electrons which are always constrained to remain far apart. An example is the electrons in two hydrogen atoms which are well separated from each other. In fact, none of the requirements of indistinguishability need be taken into account for a pair of identical particles which are so widely separated that their wave functions do not overlap. The reason is simply that these particles can be distinguished from each other by appropriate measurements.

Exchange forces do arise between two electrons in the same atom, or two neutrons or protons in the same nucleus. We shall show this by considering the low-lying energy levels of the helium atom.

**Example 9-4.** The simplest, but least accurate, treatment of the helium atom involves ignoring the Coulomb interaction between its two electrons, and taking the total energy of the atom to be the sum of the one-electron atom energies of each electron moving about the $Z = 2$

---

*Bottom:* $1/\sqrt{2}$ times the sum of the term $(2/a)\cos(\pi x_1/a)\sin(2\pi x_2/a)$ and the term $(2/a)\sin(2\pi x_1/a)\cos(\pi x_2/a)$, showing the symmetric space eigenfunction $\psi_+$ for the system. This eigenfunction has its maximum magnitudes along the line $x_1 = x_2$. The probability density $\psi_+^*\psi_+$ therefore has its largest magnitudes if the two particles are in the same location in their one-dimensional well, and so we conclude that there is a large chance of finding them close together.

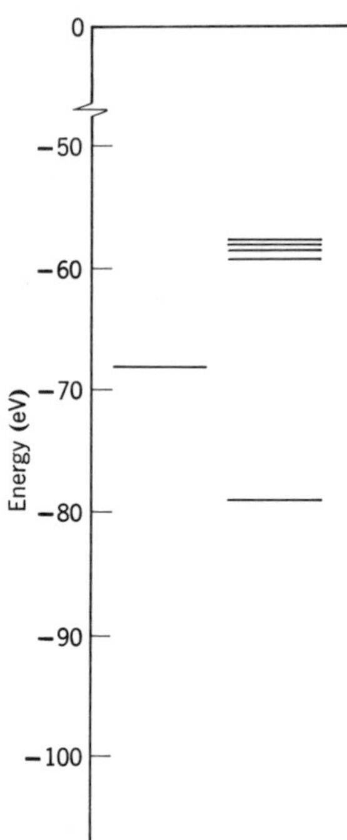


**FIGURE 9-6**
*Left:* Helium energy levels predicted by a treatment in which the electron -electron interaction is ignored. *Right:* The ground state and first four excited states of helium, as determined from the observed spectrum.

nucleus. Use this treatment to predict the energies of the ground and first excited states of the atom.

From (7-22) for the one-electron atom eigenvalues, we have

$$E = -\frac{\mu Z^2 e^4}{(4\pi\epsilon_0)^2 2\hbar^2 n_1^2} - \frac{\mu Z^2 e^4}{(4\pi\epsilon_0)^2 2\hbar^2 n_2^2}$$

$$= -\frac{4 \times 13.6 \text{ eV}}{n_1^2} - \frac{4 \times 13.6 \text{ eV}}{n_2^2}$$

where we have set $Z = 2$. In the ground state, the quantum numbers $n_1$ and $n_2$ are both equal to 1, and we obtain

$$E = -(4 + 4) \times 13.6 \text{ eV} = -109 \text{ eV}$$

In the first excited state, one of these quantum numbers equals 1, and the other equals 2. For this we obtain

$$E = -(4 + 1) \times 13.6 \text{ eV} = -68 \text{ eV}$$

The energies predicted are shown on the left side of the energy-level diagram of Figure 9-6. The right side of that figure shows the energies of the first few levels of helium obtained from measurements of the optical spectrum emitted by that atom. The predictions are quite inaccurate because the Coulomb interaction between the two electrons in the atom is really not negligible compared to the Coulomb interactions between each electron and the nucleus, as was assumed in this simple treatment, and also because the treatment ignores exchange forces. ◀

Figure 9-7 indicates the origin of the first few energy levels of the helium atom. The left side of the figure shows the energies of the levels that would be found, as in Example 9-4, if there were no Coulomb interaction between its electrons. If this were the case, the total energy would be just the sum of the one-electron atom energies of each electron moving about the $Z = 2$ nucleus in states described by the one-electron atom eigenfunctions with the quantum numbers indicated. The center of the figure shows, in part, the effect of the Coulomb interaction between the electrons. Since this interaction energy is positive because both electron charges have the same sign, the

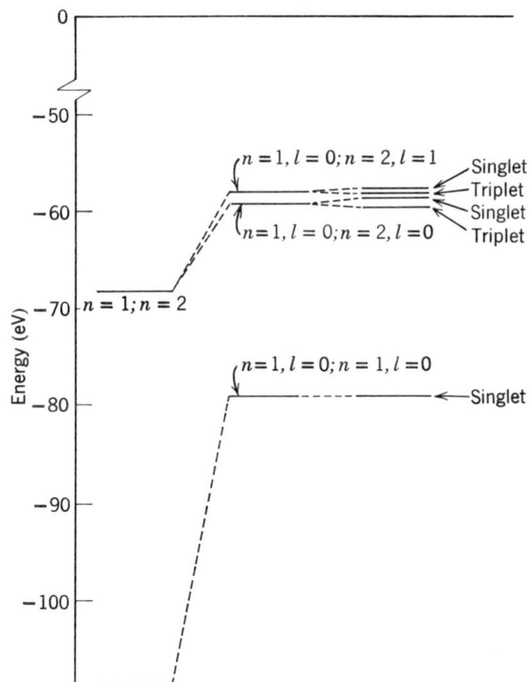


**FIGURE 9-7**

The low-lying energy levels of helium *Left:* The levels that would be found if there were no Coulomb interaction between its electrons. *Center:* The levels that would be found if there were a Coulomb interaction but no exchange force. *Right:* The levels that would be found if there were a Coulomb interaction and an exchange force. These levels are in excellent agreement with the experimentally observed levels shown on the right in Figure 9-6.

levels are raised. Furthermore, the upper level is split into two. The reason is that the two electrons are somewhat more widely separated on the average when one has $n = 1$, $l = 0$, and the other has $n = 2$, $l = 0$, than when one has $n = 1$, $l = 0$ and the other has $n = 2$, $l = 1$. This can be seen by inspecting the one-electron atom radial probability densities of Figure 7-5. As the energy associated with the Coulomb interaction between the electrons is inversely proportional to their separation, the energy of the atom is raised less for the first set of quantum numbers, and the degeneracy with respect to the $l$ quantum number (found in one-electron atoms) is removed by this interaction. The right side of Figure 9-7 shows the effect of the exchange force. In the triplet states the electrons tend to keep apart, and in the singlet state they tend to keep together. Therefore, the Coulomb interaction between them is relatively less effective in raising the energy of the atom in the triplet states, and relatively more effective in the singlet state. Part of the $m_s$ degeneracy (of one-electron atoms) is also removed by the Coulomb interaction between the electrons, and the levels are further split into singlet state and triplet state levels. These are the energy levels that are observed from measurements of the spectrum of the helium atom. Quantitative results in good agreement with the measurements can be obtained from quantum mechanics by adding to the energies obtained in Example 9-4 the expectation values of the energies due to the Coulomb repulsion between the two electrons. Antisymmetric total eigenfunctions, composed of one-electron atom eigenfunctions for $Z = 2$, are used to calculate the expectation values.

It is particularly interesting to note from Figure 9-7 that there is no triplet level corresponding to the singlet level in the ground state of helium. It is absent because the antisymmetric space eigenfunction, which must be used to multiply the symmetric triplet spin eigenfunction, has the form

$$\frac{1}{\sqrt{2}}[\psi_a(1)\psi_a(2) - \psi_a(1)\psi_a(2)] \equiv 0$$

The value is identically equal to zero in the ground state since the space quantum numbers for both electrons have the same values, $n = 1$, $l = 0$, $m_l = 0$. In agreement with the exclusion principle, only the singlet level is found in the ground state since the spin quantum numbers of the two electrons must be different, i.e., the two electrons must have "antiparallel" spins. Historically the argument was made in the opposite order. The experimental fact that the helium spectrum shows this triplet level to be absent provided the primary evidence that led Pauli to the discovery of the exclusion principle.

## 9-5 The Hartree Theory

We begin here the quantum mechanical study of multielectron atoms that will occupy us for the remainder of this chapter, and the next chapter. Compared to simplified one-dimensional systems, or even to the one-electron atom, multielectron atoms are quite complicated. But it is possible to treat them in a reasonable way by using a succession of approximations. Only the most important interactions experienced by the atomic electrons are treated in the first approximation, and then the treatment is made more exact in succeeding approximations that take into account the less important interactions. In this way the treatment is broken into a series of steps, none of which is too difficult. The results obtained will certainly justify the effort expended because we shall have a detailed understanding of the atoms that are the constituents of everything in the universe. Furthermore, the procedures used are worth studying for their own sake because they are typical of those used in solving the real problems of professional science and engineering, in contrast to the artificial problems of much textbook science and engineering.

In the first approximation used in treating a multielectron atom of *atomic number* $Z$, we must consider the Coulomb interaction between each of its $Z$ electrons of charge $-e$ and its nucleus of charge $+Ze$. Due to the magnitude of the nuclear charge, this is the strongest single interaction felt by each electron. But even in the first approximation *we must also consider the Coulomb interactions between each electron and all the other electrons in the atom.* These interactions are individually weaker than the interaction between each electron and the nucleus, but, as we saw for the case of the helium atom in Example 9-4, they are certainly not negligible. Furthermore, in a typical multielectron atom there are so many interactions between an electron and all the other electrons that their net effect is very strong except if the electron is quite near the nucleus. This is illustrated in Figure 9-8.

On the other hand, the first approximation must not be so complicated that the Schroedinger equation to which it leads is unsolvable. In practice, this requirement means that in the first approximation *the atomic electrons must be treated as moving independently* so that the motion of one electron does not depend on the motion of the others. Then the time-independent Schroedinger equation for the system can be separated into a set of equations, one for each electron, which can be solved without too much difficulty since each involves the coordinates of a single electron only. Note that this is how the solutions, (9-3), were obtained to the time-independent Schroedinger equation, (9-1), for two particles moving independently in a box.

The requirements of the last two paragraphs are in conflict—the Coulomb interactions between the electrons must be considered, but the electrons must be treated as moving independently. A compromise between the requirements is obtained by assuming each electron to move *independently* in a spherically symmetrical *net potential* $V(r)$, where $r$ is the radial coordinate of the electron with respect to the nucleus. The net potential is the sum of the spherically symmetrical attractive Coulomb

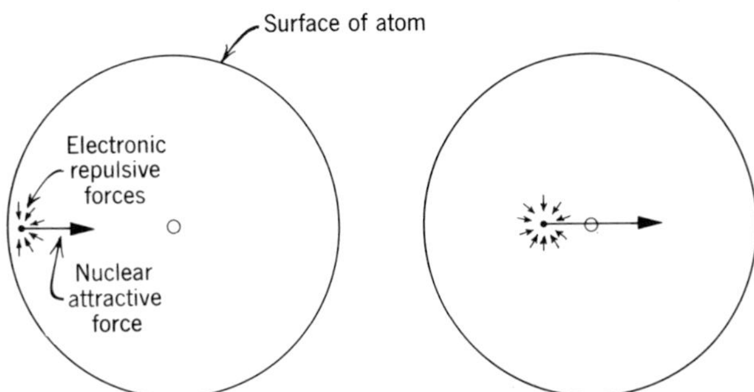


**FIGURE 9-8**

*Left:* The strong attractive force exerted by the nucleus on an electron near the surface of an atom, and the weak repulsive forces exerted by the other electrons. The net effect of the repulsive forces is important because they tend to reinforce each other. *Right:* The very strong attractive force exerted by the nucleus on an electron near the center of an atom, and the weak repulsive forces exerted by the other electrons. Here the repulsive forces tend to cancel each other.

potential due to the nucleus and a spherically symmetrical repulsive potential which represents the *average* effect of the repulsive Coulomb interactions between a typical electron and its $Z - 1$ colleagues. It can be seen from Figure 9-8 that very near the center of the atom the behavior of the net potential acting on an electron should be essentially like that of the Coulomb potential due to the nuclear charge $+Ze$. The reason is that in this region the interactions of the electron with the other electrons tend to cancel. It can also be seen from the figure that very far from the center the behavior of the net potential should be essentially like that of the Coulomb potential due to a net charge $+e$, which represents the nuclear charge $+Ze$ shielded by the charge $-(Z - 1)e$ of the other electrons.

The procedure of introducing a net potential is one that is encountered in the study of many fields of physics. For instance, in Chapter 15 we shall find that a net potential is the basis of the "shell model" which provides a relatively simple, but very useful, description of the behavior of neutrons and protons in a nucleus.

It might seem that there is no way to find the net potential of an atom at intermediate distances from its center. The problem is that it obviously depends on the details of the charge distribution of the atomic electrons, and this is not known until solutions have been obtained to the Schroedinger equation that contains the net potential. But it can be taken care of by demanding that the net potential be *self-consistent*. That is, if we calculate the electron charge distribution from the correct net potential, and then evaluate the net potential from the charge distribution, we demand that the potential with which we end up must be the same as the potential with which we started. As we shall see, this condition of self-consistency is enough to determine the correct net potential.

Most of the work in this field has been done by Douglas Hartree and collaborators, starting in 1928 and continuing to this day. It involves solving the time-independent Schroedinger equation for a system of $Z$ electrons moving *independently* in the atom. This equation is analogous to the equation for two electrons moving independently in a box, (9-1), in that the total potential of the atom can be written as the sum of a set of $Z$ identical net potentials $V(r)$, each depending on the radial coordinate $r$ of one electron only. Consequently, the equation can be separated into a set of $Z$ time-independent Schroedinger equations, all of which are of the same form, and each of which describes one electron moving independently in its net potential. A typical time-independent Schroedinger equation for one electron is

$$-\frac{\hbar^2}{2m}\nabla^2\psi(r,\theta,\varphi) + V(r)\psi(r,\theta,\varphi) = E\psi(r,\theta,\varphi) \tag{9-22}$$

Here $r$, $\theta$, $\varphi$ are the spherical polar coordinates of the typical electron; $\nabla^2$ is the Laplacian operator in these coordinates, of (7-13); $E$ is the total energy of the electron; $V(r)$ is its net potential; and $\psi(r,\theta,\varphi)$ is the eigenfunction of the electron. The total energy of the atom is the sum of $Z$ of these total energies. The total eigenfunction for the atom is composed of products of $Z$ of these eigenfunctions that describe the independently moving electrons.

Initially, the exact form of the net potential $V(r)$ experienced by the typical electron is not known, but it can be found by going through a self-consistent treatment comprised of the following steps:

1. A first guess at the form of $V(r)$ is obtained by taking

$$V(r) = \begin{cases} -\dfrac{Ze^2}{4\pi\epsilon_0 r} & r \to 0 \\[2ex] -\dfrac{e^2}{4\pi\epsilon_0 r} & r \to \infty \end{cases} \tag{9-23}$$

and by taking any reasonable interpolation for intermediate values of $r$. This guess is based on the idea, mentioned previously, that an electron very near the nucleus feels the full Coulomb attraction of its charge $+Ze$, while an electron very far from the nucleus feels a net charge of $+e$ because the nuclear charge is shielded by the charge $-(Z-1)e$ of the other electrons surrounding the nucleus.

2. The time-independent Schroedinger equation for a typical electron, (9-22), is solved for the net potential $V(r)$ obtained in the previous step. This is not easy to do because the radial part of the equation must be solved by numerical integration, as in Appendix F, since $V(r)$ is a complicated function. The eigenfunctions for a typical electron, found in this step, are: $\psi_\alpha(r,\theta,\varphi)$, $\psi_\beta(r,\theta,\varphi)$, $\psi_\gamma(r,\theta,\varphi)$, . . . . They are listed in order of increasing energy of the corresponding eigenvalues: $E_\alpha$, $E_\beta$, $E_\gamma$, . . . . Each of the symbols, $\alpha$, $\beta$, $\gamma$, . . . , stands for a complete set of three space and one spin quantum numbers for the electron.

3. To obtain the ground state of the atom, the quantum states of its electrons are filled in such a way as to minimize the total energy and yet satisfy the weaker condition of the exclusion principle. That is, the states are filled in order of increasing energy, with one electron in each state, as illustrated schematically in Figure 9-9. Then the eigenfunction for the first electron will be $\psi_\alpha(r_1,\theta_1,\varphi_1)$, the eigenfunction for the second will be $\psi_\beta(r_2,\theta_2,\varphi_2)$, and so forth through the $Z$ eigenfunctions corresponding to the $Z$ lowest eigenvalues, obtained in the previous step.

4. The electron charge distributions of the atom are then evaluated from the eigenfunctions specified in the previous step. This is done by taking the charge distribution for each electron as the product of its charge $-e$ times its probability density function $\psi^*\psi$. The justification is that $\psi^*\psi$ determines the probability that the charge would be found in various locations in the atom. The charge distributions of $Z-1$ representative electrons are added to the nuclear charge distribution, a point charge $+Ze$ at the origin, to determine the total charge distribution of the atom as seen by a typical electron.

5. Gauss's law of electrostatics is used to calculate the electric field produced by the total charge distribution obtained in the previous step. The integral of this electric

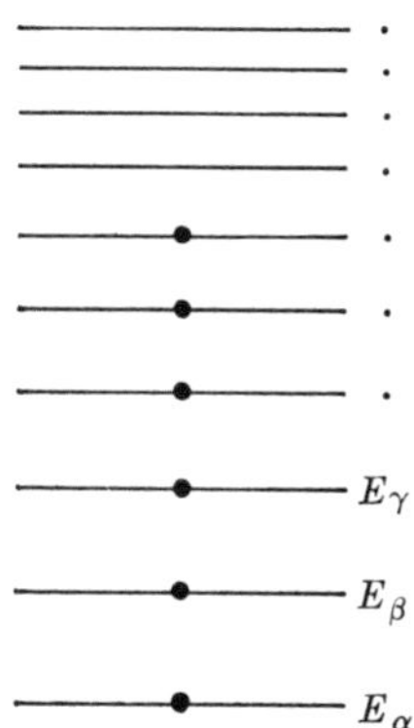


**FIGURE 9-9**

A schematic energy-level diagram illustrating the effect of the exclusion principle in limiting the population of each quantum state of an atom with six electrons. Note that the total energy of the atom would be much more negative if the exclusion principle did not operate. The diagram does not indicate that many quantum states are actually degenerate, nor are the spacings between the levels meant to be realistic.

field is then evaluated to obtain a more accurate estimate of the net potential $V(r)$ experienced by a typical electron. The new $V(r)$ that is found generally differs from the estimate made in step 1.

6. If it is appreciably different, the entire procedure is repeated, starting at step 2 and using the new $V(r)$. After several cycles $(2 \to 3 \to 4 \to 5 \to 2 \to 3 \to 4 \to 5 \to \cdots)$ the $V(r)$ obtained at the end of a cycle is essentially the same as that used in the beginning. Then this $V(r)$ is the self-consistent net potential, and the eigenfunctions calculated from this potential describe the electrons in the ground state of the multi-electron atom.

In the Hartree procedure, the weaker condition of the exclusion principle is satisfied by the requirement of step 3 that only one electron populates each quantum state. But the stronger condition is not satisfied since antisymmetric total eigenfunctions are not used. The reason is that an antisymmetric eigenfunction would involve a linear combination of $Z! = Z(Z - 1)(Z - 2) \cdots 1$ terms, which is an extremely large number for all atoms except those of very small $Z$. The procedure is difficult enough as is, and the use of antisymmetric eigenfunctions would make it very much more difficult. Anyway, the main effect of using antisymmetric total eigenfunctions would be to decrease the separation between certain pairs of electrons, and increase it between others. This leaves the average electron charge distribution of the atom essentially unchanged. Since the average electron charge distribution is the important quantity in the approximation treated by Hartree, the use of eigenfunctions which are not of a definite symmetry does not introduce a significant error. This has been verified by Fock. He made calculations using antisymmetric total eigenfunctions for a restricted selection of atoms, and he compared his results with those obtained by Hartree. When we discuss in the next chapter the excited states of atoms, however, it will be necessary for us to take into account the fact that antisymmetric total eigenfunctions must be used to give a completely accurate description of a system of electrons. Fock's calculations, and the ones we shall consider in the next chapter, are feasible because, for reasons we shall see, it is really only necessary to antisymmetrize the part of the total eigenfunction describing the behavior of a limited number of electrons in a "partially filled subshell."

It is an interesting bit of history to recall that one of the first large digital computers was employed to perform Hartree calculations. It used relays as switching elements, instead of the transistors of modern computers. But even with modern computers the calculations are so time consuming that results for a wide variety of atoms were obtained only in the 1960s by Herman and Skillman. These results provide a very satisfactory explanation of the essential features of all multielectron atoms in their ground states. As we shall find, the explanation is not unduly complicated.

## 9-6 Results of the Hartree Theory

The eigenfunctions that are found in the Hartree theory, for the electron in the spherically symmetrical net potential of a multielectron atom, are closely related to the eigenfunctions discussed in Chapter 7 for the electron in a one-electron atom. In fact, the Hartree eigenfunctions can be written

$$\psi_{nlm_lm_s}(r,\theta,\varphi) = R_{nl}(r)\Theta_{lm_l}(\theta)\Phi_{m_l}(\varphi)(m_s) \tag{9-24}$$

The eigenfunctions are labeled by the same set of quantum numbers $n$, $l$, $m_l$, $m_s$, as are used for the one-electron atom eigenfunctions, and these quantum numbers are related to each other just as before. The spin eigenfunction, which we indicate schematically as $(m_s)$, is exactly the same as for a one-electron atom. Furthermore, the functions describing the angular dependence, $\Theta_{lm_l}(\theta)$ and $\Phi_{m_l}(\varphi)$, are also exactly the same. The reason is that the time-independent Schroedinger equation for an electron in a spherically symmetrical net potential, (9-22), is of exactly the same form as the time-independent Schroedinger equation for an electron in the spherically symmetrical Coulomb potential, (7-12), as far as $\theta$ and $\varphi$ are concerned. Therefore, (9-22) leads directly to (7-15) and (7-16), whose solutions are $\Theta_{lm_l}(\theta)$ and $\Phi_{m_l}(\varphi)$. Consequently, *all the discussion of Chapter 7 concerning the $\theta$ and $\varphi$ dependence of the eigenfunctions for an electron in a one-electron atom applies directly to the $\theta$ and $\varphi$ dependence of the eigenfunctions for an electron in a multielectron atom.*

As an example, (7-32) shows that the sum of the probability densities for the one-electron atom eigenfunctions with $n = 2$, $l = 1$, and all possible values of $m_l$, is spherically symmetrical. This statement is certainly also true for $n = 2$, $l = 0$, and it can be shown to be true for any given $n$ and $l$. From the previous discussion, we conclude that the same statement applies to the eigenfunctions for a multielectron atom. Now, when a multielectron atom is in its ground state, the lowest energy quantum states of its electrons are completely filled. This means that for almost all values of $n$ and $l$ there are electrons in states with all possible values of $m_l$. Since the sum of the probability densities for these electrons is spherically symmetrical, their total charge distribution is also. At most, only a few electrons in the highest energy states, for which states with all possible values of $m_l$ might not be filled, can contribute to any asymmetries in the charge distribution. In step 4 of the Hartree procedure, the charge distribution used is taken to be completely spherically symmetrical; i.e., it is the best fit of a spherically symmetrical distribution to the distribution actually obtained.

*The $r$ dependence of the eigenfunctions for an electron in a multielectron atom is not the same as for an electron in a one-electron atom.* The reason is that the net potential $V(r)$, which enters the differential equation that determines the functions $R_{nl}(r)$, does not have the same $r$ dependence as the Coulomb potential. Typical examples of the radial behavior of the multielectron atom eigenfunctions are shown in Figure 9-10. In this figure we plot the results of a Hartree calculation for the argon atom, $Z = 18$, in terms of the quantities $2(2l + 1)r^2R_{nl}^2(r) = 2(2l + 1)P_{nl}(r)$. Here $P_{nl}(r)$ is the *radial probability density* of (7-28), which specifies the probability of finding an electron, with quantum numbers $n$ and $l$, in a location with a radial coordinate near $r$. Since there are $(2l + 1)$ possible values of $m_l$ for each $l$, and since for each of these there are 2 possible values of $m_s$, the quantity $2(2l + 1)P_{nl}(r)$ is the radial probability density for the quantum states with quantum numbers $n$ and $l$, times the total number of electrons which the exclusion principle allows to populate those states.

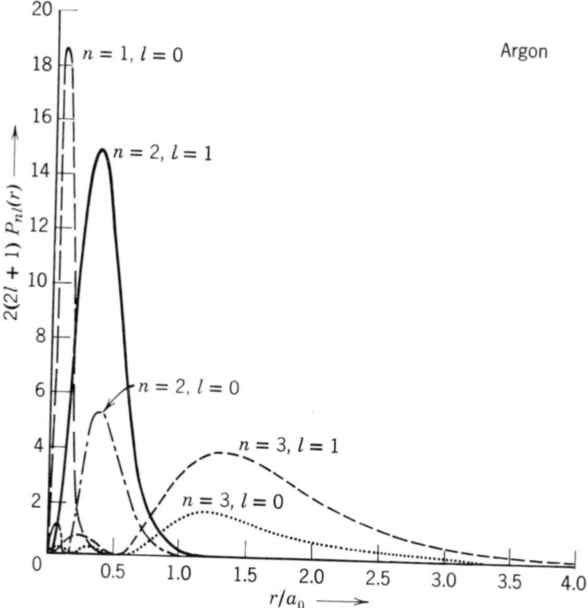


**FIGURE 9-10**

The Hartree theory radial probability densities for the filled quantum states of the argon atom, plotted as a function of $r/a_0$, the radial coordinate in units of the hydrogen atom first Bohr orbit radius $a_0$. For each $n$ the probability density is largely concentrated in a restricted range of $r/a_0$, called a shell. Note that the characteristic radius of the outermost shell ($n = 3$) has an $r/a_0$ value only a little larger than 1.0, while the characteristic radius of the innermost shell ($n = 1$) has an $r/a_0$ value much smaller than 1.0. That is, the outermost shell of argon is only a little larger in radius than $a_0$, which is the radius of the single shell in hydrogen. The innermost shell of argon is of much smaller radius than the hydrogen shell.

In the ground state of argon, two electrons populate the states for $n = 1$, $l = 0$; two for $n = 2$, $l = 0$; six for $n = 2$, $l = 1$; two for $n = 3$, $l = 0$; and six for $n = 3$, $l = 1$. These are the states which are filled in the ground state of the atom because, as we shall see later, they have the lowest energy.

Figure 9-11 shows the *total radial probability density* $P(r)$ for the argon atom. This is the sum, over the $n$ and $l$ values populated in the atom, of the radial probability density for each state times the number of electrons it contains. That is, $P(r)$ gives the probability of finding *some* electron with radial coordinate in the region of $r$.

Figure 9-11 also shows the radial dependence of the net potential $V(r)$ in which each electron of the argon atom is moving, as obtained from Hartree calculations for that atom. The net potential is not displayed directly, but indirectly in terms of a convenient quantity $Z(r)$. The relation between the two is given by the equation

$$V(r) = -\frac{Z(r)e^2}{4\pi\epsilon_0 r} \tag{9-25}$$

Note that the figure shows $Z(r) \to Z$ as $r \to 0$, and $Z(r) \to 1$ as $r \to \infty$, in agreement with the ideas discussed in connection with (9-23).

By inspecting the plots of $P_{nl}(r)$ in Figure 9-10, we see that, for all the electrons in states with common values of the quantum number $n$, the probability densities are

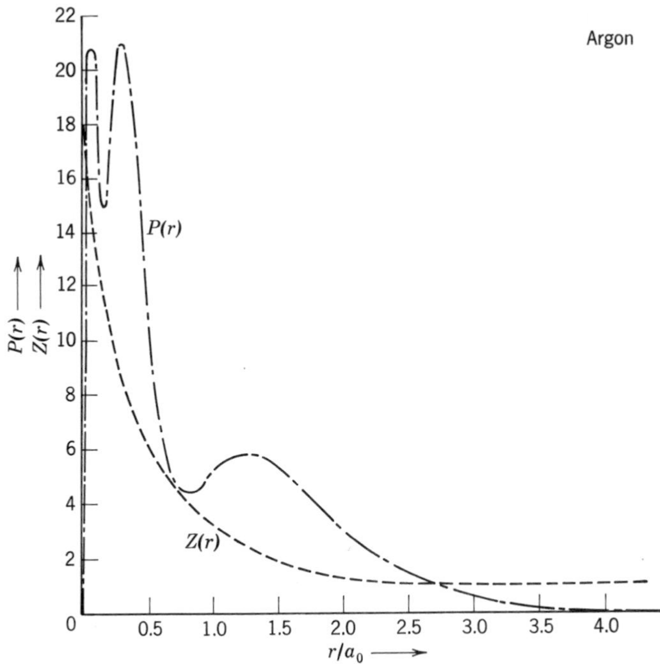


**FIGURE 9-11**

The total radial probability density $P(r)$ of the argon atom, and the quantity $Z(r)$ that specifies its net potential.

large only in essentially the same range of $r$. All these electrons are said to be in the same *shell*—terminology we have used before in connection with one-electron atoms. Furthermore, the range of $r$ in which the probability densities are large (the "thickness" of each shell) is restricted enough that $Z(r)$ has a reasonably well-defined value in that range.

These circumstances form the basis of a crude, but useful, approximate description of the results of the Hartree theory, in which all the electrons in the shell labeled by $n$ of a multielectron atom are considered to be moving in a Coulomb potential

$$V_n(r) = -\frac{Z_n e^2}{4\pi\epsilon_0 r} \tag{9-26}$$

where $Z_n$ is a constant equal to $Z(r)$ evaluated at the average value of $r$ for the shell (the "radius" of the shell). In the crude approximation of (9-26), the one-electron atom equations specifying the total energy, and other quantities of interest, can be used if we replace $Z$ by $Z_n$. The quantity $Z_n$ is sometimes called the *effective* $Z$ for the shell. This approximation is useful because it allows us to discuss many results of the Hartree theory in terms of some very simple equations with easily understandable properties, although the Hartree theory actually uses purely numerical procedures and so leads to results which must be expressed in cumbersome tables or graphs.

**Example 9-5.** Determine the values of $Z_n$ for the argon atom, and then use these values to estimate the total energy of the electrons in the three shells populated in the ground state of the atom.

Inspecting Figure 9-11 to estimate the average values of $r$ characteristic of the populated shells, obtaining the values of $Z(r)$ for these $r$ from the same figure, and setting the $Z_n$ equal to these values of $Z(r)$, we find that for the argon atom with $Z = 18$

$$Z_1 \simeq 16 \quad \text{and} \quad Z_2 \simeq 8 \quad \text{and} \quad Z_3 \simeq 3$$

As indicated earlier, we may use the one-electron atom energy formula, (7-22), with $Z = Z_n$

$$E \simeq -\frac{\mu Z_n^2 e^4}{(4\pi\epsilon_0)^2 2\hbar^2 n^2} = -\left(\frac{Z_n}{n}\right)^2 \times 13.6 \text{ eV}$$

to obtain an estimate to the electron energies yielded by the Hartree theory calculations. Doing this, we obtain

$$E_1 \simeq -\left(\frac{16}{1}\right)^2 \times 13.6 \text{ eV} = -3500 \text{ eV}$$

$$E_2 \simeq -\left(\frac{8}{2}\right)^2 \times 13.6 \text{ eV} = -220 \text{ eV}$$

$$E_3 \simeq -\left(\frac{3}{3}\right)^2 \times 13.6 \text{ eV} = -14 \text{ eV}$$

These energies agree within something like 20% with the Hartree results. ◀

In Example 9-5 we found that for the argon atom, with $Z = 18$, the effective $Z$ of the innermost shell ($n = 1$) is $Z_1 \simeq 16$. Hartree calculations show that *in all multi-electron atoms $Z_1$ has a value of about $Z_1 \simeq Z - 2$*. The reason is that for all atoms a sphere surrounding the nucleus, of radius equal to the average radial coordinate of an electron in the $n = 1$ shell, contains a negative charge of about $-2e$, due to the charge distributions of all the other electrons. According to Gauss's law of electrostatics, this spherically symmetrical distribution of negative charge *shields* the $n = 1$ electron from part of the nuclear charge $+Ze$, effectively reducing it to about $+Ze - 2e = +(Z - 2)e$. Thus the $n = 1$ electron experiences an effective $Z$ of about $Z_1 = Z - 2$.

We also found in Example 9-5 that for the outermost shell of the argon atom ($n = 3$ for that atom), the effective $Z$ has the small value $Z_n \simeq 3$. This is because an electron in the outermost shell is almost completely shielded from the nuclear charge by the intervening charge distributions of all the other electrons. The result is comparable to what is found in all Hartree calculations. But with increasing $Z$ the value of $Z_n$ obtained from the calculations for the outermost shell slowly increases; i.e., it increases about as slowly as the increase in $n$ itself. The reason it increases is that the shielding of the nuclear charge by the electrons in the intervening shells is not perfect. To an accuracy consistent with the crude approximation we are considering, we may describe these results by saying that *in all multielectron atoms $Z_n$ has a value of about $Z_n \simeq n$, if $n$ specifies the outermost shell populated in the atom.*

We shall now use the facts stated in the last two paragraphs to describe and explain a number of important results of the Hartree theory:

1. In multielectron atoms the inner shells of small $n$ are of very small radii because for these shells there is little shielding, and the electrons feel the full Coulomb attraction of the highly charged nucleus. In fact, the Hartree theory predicts that *the radius of the $n = 1$ shell is smaller than that of the $n = 1$ shell of hydrogen by approximately a factor of* $1/(Z - 2)$. (This prediction is not too accurate for atoms of very large $Z$ because of relativistic effects, not taken into account in the Hartree theory, which become important because inner shell electrons in large $Z$ atoms have energies comparable to their rest mass energies $mc^2 \simeq 5 \times 10^5$ eV.) The prediction can be understood in our crude description of the Hartree theory results by setting $Z = Z_1 \simeq Z - 2$ and $n = 1$ in the one-electron atom equation for the radial coordinate

expectation value, (7-29)

$$\bar{r} \simeq \frac{n^2 a_0}{Z}$$

yielding

$$\bar{r} \simeq \frac{\bar{r}_{\text{hydrogen}}}{Z_1} \simeq \frac{\bar{r}_{\text{hydrogen}}}{Z - 2}$$

2. The electrons in the inner shells are in a region of large negative potential energy, so their total energies are correspondingly large and negative. The results of the Hartree theory predict that *the magnitude of the total energy of an electron in the* $n = 1$ *shell is more negative than that of an electron in the* $n = 1$ *shell of hydrogen by approximately a factor of* $(Z - 2)^2$. (Relativistic effects limit the accuracy for high $Z$.) This can be understood by setting $Z = Z_1 \simeq Z - 2$ and $n = 1$ in the one-electron atom energy equation, (7-22)

$$E = -\frac{\mu Z^2 e^4}{(4\pi\epsilon_0)^2 2\hbar^2 n^2}$$

yielding

$$E \simeq Z_1^2 E_{\text{hydrogen}} \simeq (Z - 2)^2 E_{\text{hydrogen}}$$

3. Electrons in the outer shells of large $n$ are almost completely shielded from the nucleus, and so they feel an attraction to it not so different from that felt by an electron to the singly charged nucleus of a hydrogen atom. The radius of the outermost shell can be obtained from our crude description by setting $Z = Z_n \simeq n$ in the one-electron atom radial expectation value equation, yielding

$$\bar{r} \simeq \frac{n^2 a_0}{Z_n} \simeq \frac{n^2 a_0}{n} \simeq n a_0$$

If we check the predictions of this equation with the actual Hartree results for the argon atom shown in Figure 9-10, we see that the equation overestimates by a factor of 2. About the same factor of 2 overestimate is found in a similar comparison with Hartree results for elements of the highest atomic number. The effective $Z$ description of the Hartree results is crude, but still useful, because it correctly describes the fact that *the radius of the outermost populated shell increases only very slowly with increasing atomic number*. The Hartree results themselves show that *this radius is only about three times larger for elements of the highest atomic number than it is for hydrogen.*

Since the radius of the outermost populated shell is essentially the size of the atom, the previous statements apply directly to the sizes of various atoms. Nevertheless, it is a common misconception to think that atoms of high atomic number are very much larger than atoms of low atomic number. Measurements made on atoms, molecules, and solids show this is not true. The Hartree theory explains that it is not true, basically because as the nuclear charge $Z$ increases in going from one atom to the next, the inner atomic shells rapidly contract.

4. We can also see, from our crude description of the Hartree theory results, that the theory predicts that *the total energy of an electron in the outermost populated shell of any atom is comparable to that of an electron in the ground state of hydrogen.* If we set $Z = Z_n$ in the one-electron atom energy equation to obtain

$$E \simeq -\frac{\mu Z_n^2 e^4}{(4\pi\epsilon_0)^2 2\hbar^2 n^2} \tag{9-27}$$

and in this set $Z_n \simeq n$, we obtain a predicted energy which is approximately equal to the ground state hydrogen energy. The basic reason for this is the shielding of the outer shell electron from the full nuclear charge by the charges of the intervening inner shell electrons.

5. Finally, we can use (9-27) to describe crudely the dependence, for a *given* atom, of the total energy of an electron on its quantum number $n$. Due both to the $Z_n^2$ in the numerator and the $n^2$ in the denominator, $E$ becomes less negative with increasing $n$ in going through the shells of a given atom. *The total energy of an electron in a given multielectron atom becomes less negative very rapidly with increasing n for small n, but much less rapidly for large n.* The behavior for large $n$ reflects the fact that the energy cannot become positive since the electron is bound. This prediction of the Hartree theory, and all the others just mentioned, are verified by experiment.

We close our discussion of the results of the Hartree theory by describing its predictions for the total energies of the atomic electrons more accurately than can be done on the basis of the crude description we have been using. In a one-electron atom, all the quantum states corresponding to a certain shell have exactly the same total energy, if the very small energy associated with the spin-orbit interaction is ignored. That is, all states in a shell of a particular $n$ are *degenerate* since the total energy depends only on $n$. But in a multielectron atom this is not the case. As mentioned in Section 7-5, the fact that the total energy of a one-electron atom does not depend on $l$ is a consequence of the fact that its potential is Coulombic, i.e., exactly proportional to $-1/r$. In a multielectron atom the electrons are moving in a net potential $\bar{V}(r)$ which is definitely not proportional to $-1/r$, and so the total energy of these electrons depends on $l$ as well as on $n$. (Since we are here ignoring the spin-orbit and certain other weak interactions, the total energy of the electrons does not depend on the quantum number $m_s$ which determines the space orientation of the spin, nor on the quantum number $m_l$ which determines the space orientation of the "orbit".)

The results of the Hartree theory show that the total energy of an atomic electron is actually somewhat more negative than would be predicted from (9-27), the energy equation obtained from our crude description of the theory. The difference is largest for $l = 0$, and it diminishes progressively with increasing $l$. Thus in the Hartree approximation we write the energy of an electron in a multielectron atom as $E_{nl}$, to indicate that it depends on both $n$ and $l$.

The explanation for the $l$ dependence concerns the behavior of the electron probability density $\psi^*\psi$, in the region of small $r$ near the nucleus of the multielectron atom. According to (7-31)

$$\psi^*\psi \propto r^{2l} \qquad r \to 0$$

This was demonstrated for one-electron atom eigenfunctions, but it is equally true for multielectron atom eigenfunctions. The reason can be seen by inspecting (7-17), which is the differential equation for the function $R$ governing the radial behavior of the eigenfunctions. Note that as $r \to 0$ the term $[l(l + 1)/r^2]R$ completely dominates the other term $(2\mu/\hbar^2)[E - V(r)]R$ since the factor $1/r^2$ makes it increase so rapidly with decreasing $r$ for small $r$. Consequently, for small $r$ the exact form of $V(r)$ is unimportant as long as it increases in magnitude less rapidly than $1/r^2$. In *all* atoms the eigenfunctions have a radial dependence proportional to $r^l$ for small $r$, and therefore the probability density is proportional to $r^{2l}$ for small $r$. So if we consider, as an example, two electrons in the same shell $n$ of a multielectron atom, one with $l = 0$ and the other with $l = 1$, there is much more chance of finding the $l = 0$ electron in the region of small $r$ than of finding the $l = 1$ electron in that region. This is true since $r^0 \gg r^2$ for small $r$. Similarly, the chance of finding an $l = 1$ electron is much larger than the chance of finding an $l = 2$ electron of the same $n$ at small $r$ since there $r^2 \gg r^4$, etc. This property can be seen by carefully inspecting Figure 9-10.

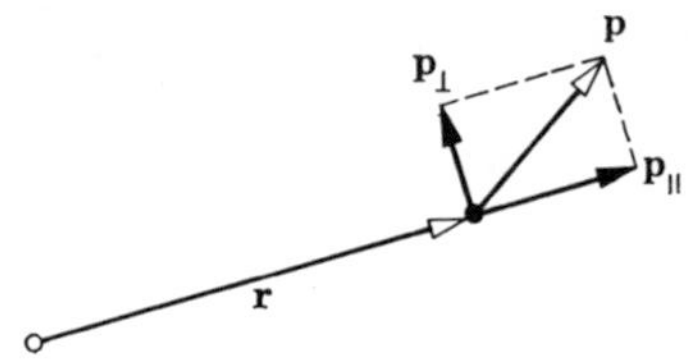


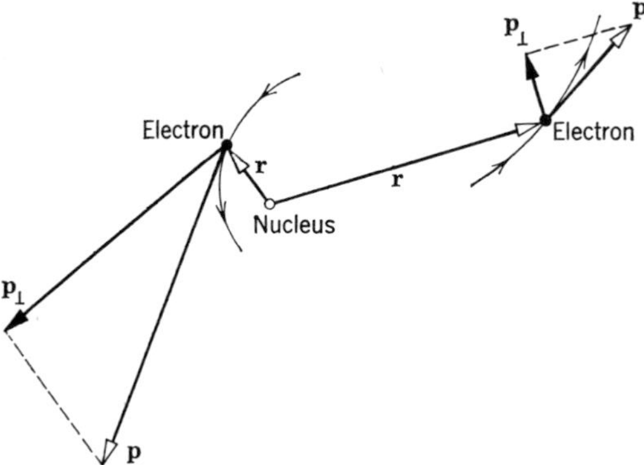


**FIGURE 9-12**

*Top:* The linear momentum **p** of an electron can be decomposed into a component $p_{\|}$ parallel to the radial vector from the nucleus **r**, and a component $p_{\perp}$ perpendicular to the radial vector. The product of $p_{\perp}$ and $r$ is equal to the constant magnitude of the angular momentum $L$. *Bottom:* An electron moving about a nucleus with constant $L$. When the electron is relatively near the nucleus (illustrated on the left), $r$ is small so $p_{\perp}$ must be large. When the electron is relatively far away (illustrated on the right), $p_{\perp}$ is smaller. Note that the magnitude of the total momentum $p$ will also be large when $p_{\perp}$ is large. Therefore the kinetic energy of the electron will be large when it is near the nucleus, in order to allow the angular momentum to be a constant of the motion.

Before using the property to explain the dependence of $E_{nl}$ on $l$, we indicate its physical origin by going through a semiclassical argument involving Figure 9-12. An electron with quantum number $l$ has an orbital angular momentum of fixed magnitude $L = \sqrt{l(l+1)}\,\hbar$. But $L = rp_{\perp}$, where $p_{\perp}$ is the magnitude of its component of linear momentum perpendicular to its radial coordinate vector whose length is $r$. If the electron moves into a region where $r$ becomes small, then $p_{\perp}$ must become large. Since the kinetic energy $K$ of the electron contains a term proportional to $p_{\perp}^2$, it becomes more positive with decreasing $r$ in proportion to $1/r^2$, for small $r$. But for small $r$ the net potential approaches the Coulomb potential of an unshielded nuclear charge, so the potential energy $V$ of the electron becomes more negative with decreasing $r$ in proportion to $1/r$. Since $K \propto +1/r^2$ and $V \propto -1/r$ for small $r$, its kinetic energy increases more rapidly than its potential energy decreases, as $r \rightarrow 0$. Thus the electron avoids that region because there it cannot maintain a constant value of its total energy $E = K + V$, as is required by energy conservation. However, the tendency to avoid the region of small $r$ is not present for $l = 0$ since then $L = 0$. So there is much more chance of finding an $l = 0$ electron at small $r$ than of finding an $l = 1$ electron in that region. Since the tendency to avoid small $r$ is more pronounced with increasing $l$, there is much more chance of finding an $l = 1$ electron than an $l = 2$ electron at small $r$, etc.

Now we can understand the $l$ dependence of $E_{nl}$. The crude description of the results of the Hartree theory underestimates how negative the total energy of an atomic electron is because it assumes essentially that the electron stays within its shell. In fact, there is a small probability that the electron will be found inside its shell in the region of small $r$ near the nucleus. When the electron is in this region it has penetrated the intervening charge distributions of the other electrons, and it feels nearly the full unshielded nuclear charge. Then it has a *very* much more negative potential energy than it has when it is in its shell. The electron will also occasionally be found outside its shell where its potential energy is less negative than in its shell, but the change is considerably smaller than the change in potential energy occurring when it is inside its shell. The overall effect of the excursions of an electron inside and outside its shell is to make the expectation value of its potential energy somewhat more negative, and therefore to make its total energy somewhat more negative than it would be if it stayed in its shell. Since we have learned that the probability of an electron with a given $n$ being inside the shell in the region near the nucleus is larger the smaller its value of $l$, we can see that for a given value of $n$, *the total energy $E_{nl}$ of an electron in a multielectron atom is more negative for $l = 0$ than for $l = 1$, more negative for $l = 1$ than for $l = 2$, etc.* For outer shells with large values of $n$, where the $n$ dependence is not very strong, the values of $E_{nl}$ can actually depend in a more sensitive way on $l$ than on $n$. But for a one-electron atom there is no $l$ dependence at all in the total energy because there is never any shielding so an electron always feels the full nuclear charge, and the expectation value of its potential energy is independent of $l$.

All the electrons in a particular shell have radial probability densities which are of approximately the same form in the region of the shell, but which are significantly different in the region of small $r$. We have seen that the second property causes the total energies of the electrons in the shell to depend on $l$. Consequently, it is convenient to speak of each shell as being composed of a number of *subshells*, one for each value of $l$. All the electrons in the same subshell have the same quantum numbers $n$ and $l$. Therefore, all have exactly the same total energy (in the Hartree approximation which neglects spin-orbit and other weak interactions). Also, all the electrons in the same subshell have exactly the same radial probability density $P_{nl}(r)$.

## 9-7 Ground States of Multielectron Atoms and the Periodic Table

Most of the properties of the chemical elements are periodic functions of the atomic number $Z$ that specifies the number of electrons in an atom of the element. It was first emphasized by Mendeleev in 1869 that these periodicities can be made most apparent by constructing a *periodic table* of the elements. A modern version of his table is presented in Figure 9-13. Each element is represented in the table by its chemical symbol, and also by its atomic number. Elements with similar chemical and physical properties are in the same column. For instance, all elements in the first column are alkalis and have a valence of plus one; all elements in the last column are noble gases and have a valence of zero. The discovery of the periodic table was a great breakthrough of chemistry. Its interpretation was an equally significant development of physics.

We assume that the student has some familarity with the periodic properties of the elements from his study of elementary chemistry. For this reason, we do not need to stress their importance to chemistry. Our task here is to interpret these properties in terms of the Hartree theory of multielectron atoms. That is, in this section we shall

| | $s^1$ | $s^2$ | | $d^1$ | $d^2$ | $d^3$ | $d^4$ | $d^5$ | $d^6$ | $d^7$ | $d^8$ | $d^9$ | $d^{10}$ | | $p^1$ | $p^2$ | $p^3$ | $p^4$ | $p^5$ | $p^6$ |
|---|---|---|---|---|---|---|---|---|---|---|---|---|---|---|---|---|---|---|---|---|
| 1s | 1 H | | | | | | | | | | | | | | | | | | | 2 He |
| 2s | 3 Li | 4 Be | | | | | | | | | | | | 2p | 5 B | 6 C | 7 N | 8 O | 9 F | 10 Ne |
| 3s | 11 Na | 12 Mg | | | | | | | | | | | | 3p | 13 Al | 14 Si | 15 P | 16 S | 17 Cl | 18 A |
| 4s | 19 K | 20 Ca | 3d | 21 Sc | 22 Ti | 23 V | 24 Cr $4s^1 3d^5$ | 25 Mn | 26 Fe | 27 Co | 28 Ni | 29 Cu $4s^1 3d^{10}$ | 30 Zn | 4p | 31 Ga | 32 Ge | 33 As | 34 Se | 35 Br | 36 Kr |
| 5s | 37 Rb | 38 Sr | 4d | 39 Y | 40 Zr | 41 Nb $5s^1 4d^4$ | 42 Mo | 43 Tc | 44 Ru $5s^1 4d^7$ | 45 Rh $5s^1 4d^8$ | 46 Pd $5s^0 4d^{10}$ | 47 Ag $5s^1 4d^{10}$ | 48 Cd | 5p | 49 In | 50 Sn | 51 Sb | 52 Te | 53 I | 54 Xe |
| 6s | 55 Cs | 56 Ba | 5d | 57 La Lantha-nides | 72 Hf | 73 Ta | 74 W | 75 Re | 76 Os | 77 Ir | 78 Pt $6s^1 5d^9$ | 79 Au $6s^1 5d^{10}$ | 80 Hg | 6p | 81 Ti | 82 Pb | 83 Bi | 84 Po | 85 At | 86 Rn |
| 7s | 87 Fr | 88 Ra | 6d | 89 Ac Actinides | | | | | | | | | | 7p | | | | | | |

| | $f^1$ | $f^2$ | $f^3$ | $f^4$ | $f^5$ | $f^6$ | $f^7$ | $f^8$ | $f^9$ | $f^{10}$ | $f^{11}$ | $f^{12}$ | $f^{13}$ | $f^{14}$ |
|---|---|---|---|---|---|---|---|---|---|---|---|---|---|---|
| 4f Lanthanides | 58 Ce $5d^0 4f^2$ | 59 Pr $5d^0 4f^3$ | 60 Nd $5d^0 4f^4$ | 61 Pm $5d^0 4f^5$ | 62 Sm $5d^0 4f^6$ | 63 Eu $5d^0 4f^7$ | 64 Gd $5d^0 4f^7$ | 65 Tb $5d^0 4f^9$ | 66 Dy $5d^0 4f^{10}$ | 67 Ho $5d^0 4f^{11}$ | 68 Er $5d^0 4f^{12}$ | 69 Tm $5d^0 4f^{13}$ | 70 Yb $5d^0 4f^{14}$ | 71 Lu $5d^1 4f^{14}$ |
| 5f Actinides | 90 Th $6d^2 5f^0$ | 91 Pa $6d^1 5f^2$ | 92 U $6d^1 5f^3$ | 93 Np $6d^1 5f^4$ | 94 Pu $6d^1 5f^5$ | 95 Am $6d^1 5f^6$ | 96 Cm $6d^1 5f^7$ | 97 Bk $6d^1 5f^8$ | 98 Cf $6d^0 5f^{10}$ | 99 Es $6d^0 5f^{11}$ | 100 Fm $6d^0 5f^{12}$ | 101 Md $6d^0 5f^{13}$ | 102 No $6d^0 5f^{14}$ | 103 Lw $6d^1 5f^{14}$ |

**FIGURE 9-13**

The periodic table of the elements, showing the electron configuration for each element.

**TABLE 9-2.** The Energy Ordering of the Outer Filled Subshells

| Quantum Numbers $n, l$ | Designation of Subshell | Capacity of Subshell $2(2l+1)$ | |
|---|---|---|---|
| — | — | — | |
| — | — | — | |
| 6, 2 | $6d$ | 10 | |
| 5, 3 | $5f$ | 14 | |
| 7, 0 | $7s$ | 2 | |
| 6, 1 | $6p$ | 6 | |
| 5, 2 | $5d$ | 10 | |
| 4, 3 | $4f$ | 14 | |
| 6, 0 | $6s$ | 2 | ↑ |
| 5, 1 | $5p$ | 6 | |
| 4, 2 | $4d$ | 10 | Increasing energy |
| 5, 0 | $5s$ | 2 | (less negative) |
| 4, 1 | $4p$ | 6 | |
| 3, 2 | $3d$ | 10 | |
| 4, 0 | $4s$ | 2 | |
| 3, 1 | $3p$ | 6 | |
| 3, 0 | $3s$ | 2 | |
| 2, 1 | $2p$ | 6 | |
| 2, 0 | $2s$ | 2 | |
| 1, 0 | $1s$ | 2 | ←Lowest energy (most negative) |

present the quantum mechanical interpretation of the basis of inorganic chemistry, plus that of much organic chemistry and solid state physics.

The interpretation of the periodic table is based on information about the ordering according to energy of the outer filled subshells of multielectron atoms. The required information can be obtained from the results of the Hartree calculations, described in the last section, which yield the ordering according to energy of the outer filled subshells as is shown in Table 9-2. The first column identifies the subshell by the quantum numbers $n$ and $l$.

The second column of Table 9-2 identifies the subshells by giving the *spectroscopic notation* for $n$ and $l$. This notation is commonly used in discussing the spectra and energy levels of atoms. The number gives the value of $n$, and the letter gives the value of $l$ according to the scheme shown in Table 9-3. In this scheme the $l = 0$ state is called an $s$ state; the $l = 1$ state is called a $p$ state; etc.

The third column of Table 9-2 is equal to $2(2l + 1)$. As mentioned in the last section, that quantity is the number of possible combinations of $m_l$ and $m_s$, for the value of $l$ characteristic of the subshell. Thus the third column gives the maximum

**TABLE 9-3.** The Spectroscopic Notation for $l$

| $l$ | 0 | 1 | 2 | 3 | 4 | 5 | 6 | ··· |
|---|---|---|---|---|---|---|---|---|
| Spectroscopic notation | $s$ | $p$ | $d$ | $f$ | $g$ | $h$ | $i$ | ··· |

number of electrons that can occupy different states in the same subshell without violating the exclusion principle.

In our discussion of the last section we found that the Hartree theory predicts that the energy of the subshell becomes more negative with decreasing values of $n$ and with decreasing values of $l$. We see this immediately in Table 9-2. The $1s$ subshell, which is the only subshell in the $n = 1$ shell, has the lowest energy. The two subshells of the $n = 2$ shell are both of higher energy and, of these, the $2s$ subshell is of lower energy than the $2p$ subshell. In the $n = 3$ shell the subshells $3s$, $3p$, $3d$ are also ordered in energy according to the predictions of the Hartree theory. However, the energy of the $4s$ subshell is actually lower than the energy of the $3d$ subshell because, for reasons described in the last section, the $l$ dependence of the energy $E_{nl}$ of the subshells can be more important than the $n$ dependence for outer subshells with large values of $n$. Continuing up the list, we see that the ordering of the outer subshells always satisfy the following *rule:*

*For a given n, the outer subshell with the lowest l has the lowest energy. For a given l, the outer subshell with the lowest n has the lowest energy.*

Near the top of the list, the $l$ dependence of $E_{nl}$ becomes so much stronger than the $n$ dependence that the energy of the $7s$ subshell is lower than the energy of the $5f$ subshell.

It should be noted that Table 9-2 does not necessarily give the energy ordering of *all* subshells in any particular atom, but only the energy ordering of the subshells which happen to be the outer subshells for that atom. For instance, the energy of the $4s$ subshell is lower than that of the $3d$ subshell for K atoms and the next few atoms of the periodic table. But for atoms further up in the periodic table the $3d$ subshell is of lower energy than the $4s$ subshell because for these atoms they are inner subshells and the $n$ dependence of $E_{nl}$ is so strong that it dominates the $l$ dependence. Additional information of this type is presented in Figure 9-14.

Now the characteristics of an atom depend on the behavior of its electrons. The behavior of an electron is specified by the set of four quantum numbers which specify its quantum state. However, in the approximation represented by the Hartree theory only the quantum numbers $n$ and $l$ are important. Therefore, in this approximation an atom can be characterized by specifying the $n$ and $l$ quantum numbers of all the electrons. This specification of the subshells occupied by the various electrons is called the *configuration* of the atom. The ordering according to energy of the outer filled subshells being known, it is trivial to determine the configuration of any atom in its ground state. In the ground state the electrons must fill all the subshells in such a way as to minimize the total energy of the atom, and yet not exceed the capacity $2(2l + 1)$ of any subshell. The subshells will fill in order of increasing energy, as listed in Table 9-2.

Consider first the H atom. The single electron occupies the $1s$ subshell, with its spin either "up" or "down". For the He atom both electrons are in the $1s$ subshell, one spin "up" and the other with spin "down". The configuration of H is written

$$^{1}\text{H}:\quad 1s^1$$

The configuration of He is written

$$^{2}\text{He}:\quad 1s^2$$

The superscript on the subshell designation specifies the number of electrons which it contains; the superscript on the chemical symbol specifies the $Z$ value for the atom. In the $^{3}$Li atom one of the electrons must be in the $2s$ subshell because the capacity of the $1s$ subshell is only 2. The configuration of this atom is

$$^{3}\text{Li}:\quad 1s^2 2s^1$$

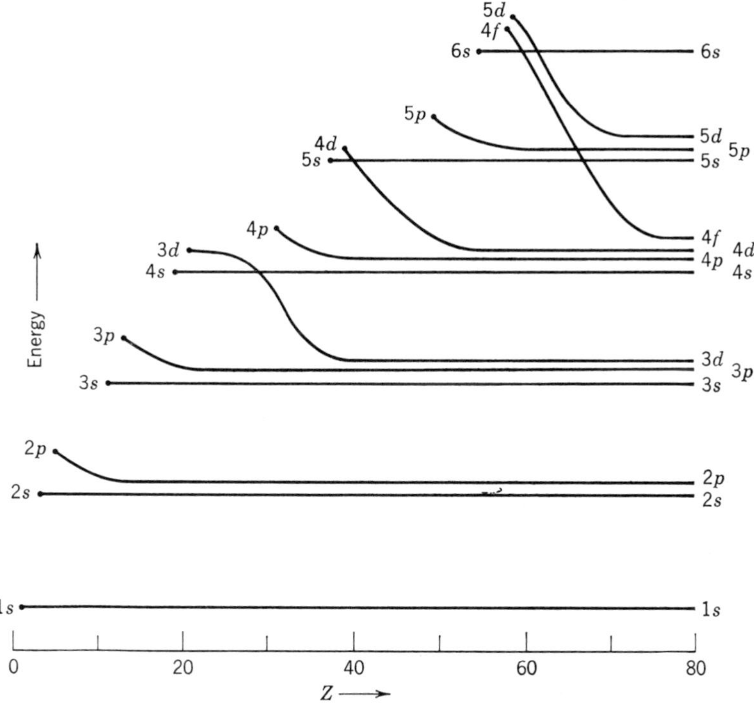


**FIGURE 9-14**

A schematic representation of the energy ordering of all the subshells in an atom, as a function of its atomic number $Z$. Each curve begins at the $Z$ for which the subshell begins to be occupied. Only subshells occupied in atoms through mercury are shown, so all curves stop at $Z = 80$. The ordering of the outer filled subshells in various atoms is found on the left side of the diagram. The ordering of all filled subshells in mercury is found on the right side of the diagram. The energy scale is non-linear and, furthermore, varies with $Z$.

The $^4$Be atom completes the $2s$ subshell and has the configuration

$$^4\text{Be}: \quad 1s^2 2s^2$$

In the six elements from $^5$B to $^{10}$Ne the additional electrons fill the $2p$ subshell. The configurations of $^5$B and $^{10}$Ne are

$$^5\text{B}: \quad 1s^2 2s^2 2p^1$$

$$^{10}\text{Ne}: \quad 1s^2 2s^2 2p^6$$

Note that the periodic table of the elements presented in Figure 9-13 is divided vertically into a series of blocks with each row labeled by the subshell which, according to Table 9-2, the elements of the row are filling. Knowing this, it is easy to write the configuration of any atom, with a procedure that will become more apparent in Example 9-6. But there are certain atoms for which the last few electrons are observed to be in different subshells than would be predicted by this scheme. The configurations for these atoms are indicated in the periodic table by the entries below their chemical symbol.

**Example 9-6.** Write the configurations for the ground states of $^{19}$K, $^{23}$V, $^{24}$Cr, $^{43}$Tc, $^{44}$Ru, $^{46}$Pd, $^{57}$La, $^{58}$Ce, and $^{59}$Pr.

From the absence of any entry below $^{19}$K in the periodic table of Figure 9-13, we conclude that there is nothing exceptional about its configuration. The configuration is then obtained by inspecting the periodic table and listing in order the lowest energy subshells, and their populations, for the 19 electrons of the atom. It is

$$^{19}\text{K}: \quad 1s^2 2s^2 2p^6 3s^2 3p^6 4s^1$$

The first 18 electrons completely fill the subshells of lowest energy, and the last electron partly fills the $4s$ subshell. Adding four more electrons to obtain $^{23}$V completes the filling of the $4s$ subshell and puts three electrons in the $3d$ subshell, which is the one of next highest energy. The configuration is

$$^{23}\text{V}: \quad 1s^2 2s^2 2p^6 3s^2 3p^6 4s^2 3d^3$$

The entry $4s^1 3d^5$ for $^{24}$Cr in Figure 9-13 means that the configuration of this atom does not end with the symbols $4s^2 3d^4$, as would be expected, but instead is

$$^{24}\text{Cr}: \quad 1s^2 2s^2 2p^6 3s^2 3p^6 4s^1 3d^5$$

The reason for this behavior will be explained later. Inspection shows that the configurations of the other atoms of interest are

$$^{43}\text{Tc}: \quad 1s^2 2s^2 2p^6 3s^2 3p^6 4s^2 3d^{10} 4p^6 5s^2 4d^5$$
$$^{44}\text{Ru}: \quad 1s^2 2s^2 2p^6 3s^2 3p^6 4s^2 3d^{10} 4p^6 5s^1 4d^7$$
$$^{46}\text{Pd}: \quad 1s^2 2s^2 2p^6 3s^2 3p^6 4s^2 3d^{10} 4p^6 4d^{10}$$
$$^{57}\text{La}: \quad 1s^2 2s^2 2p^6 3s^2 3p^6 4s^2 3d^{10} 4p^6 5s^2 4d^{10} 5p^6 6s^2 5d^1$$
$$^{58}\text{Ce}: \quad 1s^2 2s^2 2p^6 3s^2 3p^6 4s^2 3d^{10} 4p^6 5s^2 4d^{10} 5p^6 6s^2 4f^2$$
$$^{59}\text{Pr}: \quad 1s^2 2s^2 2p^6 3s^2 3p^6 4s^2 3d^{10} 4p^6 5s^2 4d^{10} 5p^6 6s^2 4f^3$$

◀

We see from Example 9-6 that in certain cases the actual configurations observed for the elements do not strictly adhere to the predictions of Table 9-2. For instance, this table says that the energy of the $3d$ subshell is greater than the energy of the $4s$ subshell when these subshells are filling. Yet in $^{24}$Cr, and also in $^{29}$Cu, one of the electrons that could be in the $4s$ subshell is actually in the $3d$ subshell. Similar situations are observed to occur for the $5s$ and $4d$ subshells. In $^{43}$Tc the $5s$ subshell is filled in the normal manner. But in $^{45}$Rh there is only one electron in the $5s$ subshell; in $^{46}$Pd both electrons have left the $5s$ subshell and moved to the $4d$ subshell. The $^{78}$Pt and $^{79}$Au configurations show that the same kind of thing can happen for the $6s$ and $5d$ subshells. From these circumstances we conclude that the energy separations between the $4s$ and $3d$, the $5s$ and $4d$, and the $6s$ and $5d$ subshells must be so small while they are being filled that, although generally the ordering of these subshells is as shown in Table 9-2, in certain cases the ordering can actually be reversed. This can be seen in Figure 9-14. Configurations which disagree with Table 9-2 are also observed in $^{57}$La and in the *lanthanides* ($Z = 58$ to 71), more commonly called the *rare earths*. Table 9-2 predicts that after the completion of the $6s$ subshell the $4f$ subshell should fill, but in two of the rare earths there is one $5d$ electron. A similar situation occurs in the group of elements following $^{89}$Ac, which are called the *actinides* ($Z = 90$ to 103). From the same argument we used previously, we interpret these observations to mean that the energy differences between the $5d$ and $4f$ subshells, and between the $6d$ and $5f$ subshells, are very small while these subshells are being filled.

On the other hand, certain predictions of Table 9-2 are always obeyed. Since none of the configurations is exceptional for elements in the first two and last six columns of the periodic table, we conclude that every $p$ subshell is always of higher energy than the preceding $s$ or $d$ subshell while these subshells are being filled, and that in these circumstances every $s$ subshell is always of higher energy than the preceding $p$ subshell.

Therefore there must be large energy differences between the subshells concerned while they are being filled. In fact, the energy differences between every $s$ subshell and the preceding $p$ subshell are particularly large as can be seen in Figure 9-14, and it is easy to understand why. Since for a given $n$ the energy of a subshell becomes higher with increasing $l$, an $s$ subshell is always the first subshell to be occupied in a new shell. Consequently, when an electron is added to a configuration with a completed $p$ subshell and goes into the subshell of next highest energy, which according to Table 9-2 is always an $s$ subshell, the electron will be the first one in a new shell. Compared to the electrons in the preceding subshell, its average radial coordinate will be considerably larger, its average potential energy will be considerably less negative, and its total energy will be considerably higher—much higher than for the usual increase in total energy in going from one subshell to the next.

The fact that there is a particularly large energy difference between every $s$ subshell and the preceding $p$ subshell has some important consequences. Consider atoms of the elements $^{10}$Ne, $^{18}$A, $^{36}$Kr, $^{54}$Xe, and $^{86}$Rn, in which a $p$ subshell is just completed. Because of the very large difference between the energy of an electron in the $p$ subshell and the energy it would have if it were in the $s$ subshell, the first excited state of these atoms is unusually far above the ground state. As a result, these atoms are particularly difficult to excite. In their ground state, Gauss's law shows they produce no electric field external to the atom since they consist of sets of completely filled subshells, and so they have spherically symmetrical charge distributions with zero net charge because they are neutral overall. Furthermore, these atoms produce no external magnetic fields in their ground state since, as we shall see later, the total angular momenta of electrons in completely filled subshells couple to zero, and this coupling yields zero total magnetic dipole moment. Because of the absence of external fields (at least on a time-averaged basis), it is very difficult for these atoms to interact with other atoms to produce chemical compounds. They also have very low boiling and freezing points because they have little tendency to condense into liquids or solid form. These are the *noble gas* elements.

The atom $^{2}$He is also a noble gas because for it the first unfilled subshell is an $s$ subshell (even though it does not contain a filled $p$ subshell) so it has an unusually high first excited state, and because in its ground state the atom consists of completely filled subshells and so produces no external fields. That $^{2}$He is a noble gas is indicated by its being listed in the last column of the periodic table instead of the second column. An element such as $^{20}$Ca is not a noble gas, even though it consists of completely filled subshells, because in its first excited state an electron goes to a $3d$ subshell. So the excited state is not far above the ground state and very little energy is required to make the atom produce an external field which will allow it to interact with other atoms.

Another aspect of the particular inertness of the noble gases can be obtained by plotting, for the various elements, the measured values of the magnitude of the total energy of an electron in the highest energy filled subshell. This is equal to the energy required to remove the electron from the atom, which is the *ionization energy* of the atom. Figure 9-15 shows such a plot. We see that the ionization energy oscillates about an average value which is essentially independent of $Z$, in agreement with our conclusion of the previous section that the total energy of electrons in the outer shells is roughly the same throughout the periodic table. The oscillations are quite pronounced, however, and it is apparent that the total energy of an electron in the highest energy filled subshell of a noble gas is considerably more negative than average. These electrons are very tightly bound, and the atoms are very difficult to ionize.

We also see that the ionization energy is particularly small for the elements $^{3}$Li, $^{11}$Na, $^{19}$K, $^{37}$Rb, $^{55}$Cs, and $^{87}$Fr. These are the *alkalis*. They contain a single weakly

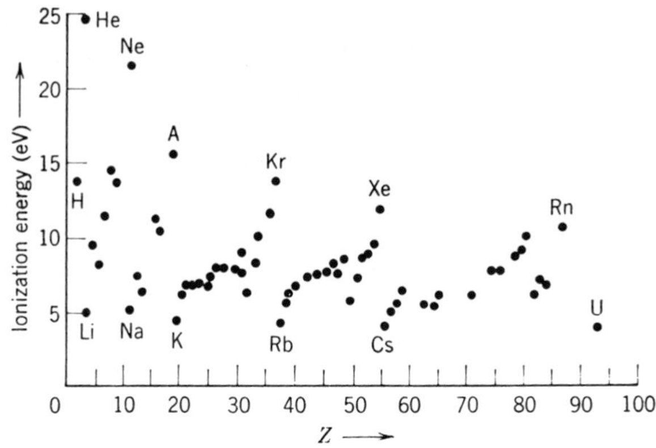


**FIGURE 9-15**

The measured ionization energies of the elements.

bound electron in an $s$ subshell. Alkali elements are very active chemically because it is energetically favorable for them to get rid of the weakly bound electron and revert to the more stable arrangement obtained with completely filled subshells. These elements are said to have one *valence electron*, and a *valence* of plus one.

At the other extreme are the *halogens*, $^{9}$F, $^{17}$Cl, $^{35}$Br, $^{53}$I, and $^{85}$At, which have one less electron than is required to fill their $p$ subshell. These elements have a high *electron affinity*; i.e., they are very prone to capture an electron. They have a valence of minus one. In 1962 it was discovered that in special circumstances noble gases could be made to combine with the halogen $^{9}$F to form stable molecules. Before that time it was believed that the noble gases were completely inert. These molecules can be formed only because $^{9}$F has such a high electron affinity that it can remove one of the very tightly bound electrons from the filled outer subshells of the noble gases.

For the first three rows of the periodic table, the properties of the elements, such as valence and ionization energy, change uniformly from the alkali element with which the row begins to the noble gas with which it ends. In the fourth row of the periodic table this situation is no longer always true. The elements $^{21}$Sc through $^{28}$Ni, which are called the first *transition group*, have quite similar chemical properties and almost the same ionization energies. These elements occur during the filling of the $3d$ subshell. The radius of this subshell is considerably less than that of the $4s$ subshell, which is completely filled for all the first transition group except $^{24}$Cr. The filled $4s$ subshell tends to shield the $3d$ electrons from external influences, and so the chemical properties of these elements are all quite similar, independent of exactly how many $3d$ electrons they contain. The point is that the chemical properties of the elements depend on the electrons in the *outer* subshells of their atoms, since these are the electrons responsible for producing the electric and magnetic fields that interact with electrons in other atoms. The chemical properties of $^{29}$Cu are somewhat different from those of the first transition group because it has only a single $4s$ electron in the outermost subshell. To a lesser extent this is also true for $^{24}$Cr. The element $^{30}$Zn consists of a set of completely filled subshells and so is somewhat more inert, as can be seen from its ionization energy. Similar transition groups occur in the filling of the $4d$ and $5d$ subshells.

An extreme example of the same situation is found in the rare earths $^{58}$Ce through $^{71}$Lu. These are the elements in which the $4f$ subshell is filling. This subshell lies deep within the $6s$ subshell, which is completely filled in all the rare earths. The $4f$ electrons are so well shielded from the external environment that the chemical properties of these elements are almost identical. The same thing happens in the actinides, $^{90}$Th through $^{103}$Lw. In this group the 5*f* subshell is filling inside the filled 7*s* subshell. Some of the most exciting work in contemporary chemistry is the study of the actinides of highest atomic number, which have only recently been discovered.

It is appropriate to close our discussion by emphasizing the importance of the *exclusion principle*. If it were not obeyed, all the electrons in a multielectron atom would be in the 1*s* subshell because this is the subshell of lowest energy. If this were the case, all atoms would have spherically symmetrical charge distributions of very small radii that would produce no external electric fields, and furthermore they would also have very high first excited states. Then *all* atoms would be much like noble gases, and therefore there would be no molecules. In fact, the entire universe would be completely different if electrons did not obey the exclusion principle!

**Example 9-7.** Make an order of magnitude estimate of the ionization energy of $^{92}$U, if the exclusion principle did not operate so that all of its electrons were in its $n = 1$ shell. For this purpose assume that the typical electron feels the nuclear charge shielded by the charge of half the other electrons in the shell. Compare the results of the estimate with the actual value of the ionization energy shown in Figure 9-15.

An estimate of the total energy of a typical electron can be obtained from the one-electron atom energy formula

$$E = -\frac{\mu Z^2 e^4}{(4\pi\epsilon_0)^2 2\hbar^2 n^2} = -\frac{Z^2}{n^2} \times 13.6 \text{ eV}$$

If we set $n = 1$ and use an effective $Z$ with the value $Z_1 = Z/2 = 92/2 = 46$, the absolute value of the result is the ionization energy. So we obtain

$$|E| = (46)^2 \times 13.6 \text{ eV} \simeq 3 \times 10^4 \text{ eV}$$

From Figure 9-15 we find that the actual ionization energy is

$$|E| = 4 \text{ eV}$$

Without the exclusion principle the ionization energy of $^{92}$U would be something like four orders of magnitude larger than it actually is. ◀

## 9-8 X-Ray Line Spectra

In an x-ray tube such as the one shown in Figure 2-9, electrons are emitted from a heated cathode, accelerated in a beam to kinetic energies of the order of $10^4$ eV by a voltage applied between the cathode and anode, and then strike the anode. While traveling through the atoms of the anode, a beam electron occasionally passes near an electron in an *inner* subshell. By means of the Coulomb interaction between the energetic beam electron and the atomic electron, the latter can be given enough energy to remove it from its very negative energy level and eject it from the atom. This leaves the atom in a highly excited state because one of its electrons that had a very negative energy is missing. The atom will eventually return to its ground state by emitting a set of high energy, and therefore high-frequency, photons which are members of its *x-ray line spectrum*. (The interaction between a beam electron and an outer subshell atomic electron leading to low-energy excited states, and the production of the optical spectrum, is discussed in the next chapter.) The total spectrum of x radiation emitted by an x-ray tube consists of the discrete line spectrum, superimposed on a continuum, as is illustrated for a typical case in Figure 9-16. The continuum is due to the bremsstrahlung processes occurring when the beam electrons

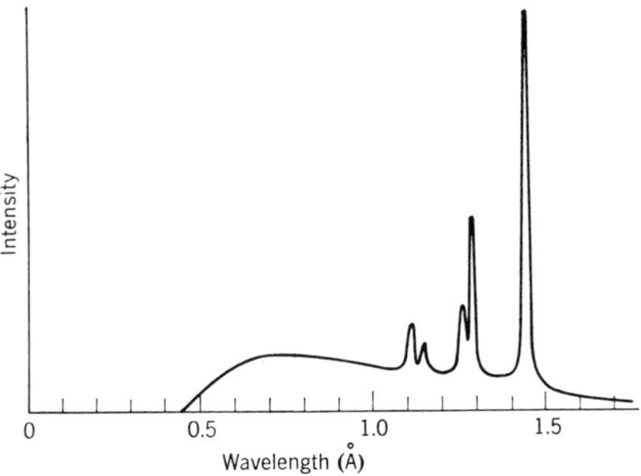


**FIGURE 9-16**

A typical x-ray spectrum. The lines are characteristic of the atoms of the x-ray tube anode (tungsten for the case illustrated). The continuum arises from bremsstrahlung by electrons accelerated in scattering from the nuclei of these atoms.

suffer accelerations in scattering from the nuclei of the atoms in the anode. As we saw in Section 2-6, the shape of the bremsstrahlung continuum depends mainly on the energy of the electron beam. But the shape of the x-ray line spectrum is characteristic of the particular atoms composing the anode.

X-ray line spectra are of practical interest because they are significant features of x rays, which have so many useful applications in technology and science. These spectra are of theoretical interest because they provide information about the energies of electrons in the inner subshells of atoms. We shall see that this information is in good agreement with the predictions of the Hartree theory.

As an example of the production of an x-ray line spectrum, assume that an electron is initially removed from the $1s$ subshell of an atom in the anode of the tube. In the first step of the deexcitation process an electron from one of the subshells of less negative energy drops into the hole in the $1s$ subshell; for instance, a $2p$ electron could drop into the hole. This would leave a hole in the $2p$ subshell, but the excitation energy of the atom would be considerably reduced. Energy is conserved by the emission of a photon of energy equal to the decrease in the excitation energy of the atom; that is, the difference between the energies associated with an electron missing from the $1s$ and $2p$ subshells. Typically there would be several subsequent steps in the deexcitation process. For instance, the hole in the $2p$ subshell could be filled by a $3d$ electron, leaving a hole in the $3d$ subshell which is then filled by a $4p$ electron, etc. The net effect of each step is that a *hole* jumps to a subshell of less negative energy. When the hole works its way to the subshell of the atom of least negative energy, which is usually the outermost shell, it is filled by the electron initially ejected from the $1s$ subshell or, more typically, by some other electron in the anode. The atom is then neutral again, and in its ground state.

The energy levels of an atom which are involved in the emission of its x-ray line spectrum are most conveniently represented in terms of an energy-level diagram that is rather different from the standard type with which we have become familiar. Figure 9-17 shows such a diagram for the $^{92}$U atom, including all its x-ray energy levels through $n = 4$. Because of the wide range of energies involved, it is conventional

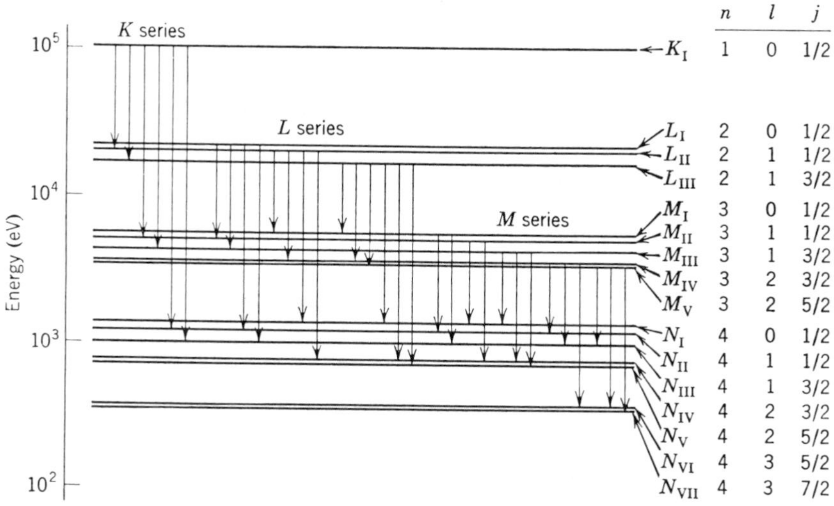


**FIGURE 9-17**

The higher energy x-ray levels for the uranium atom and the transitions between these levels allowed by the selection rules.

to use a logarithmic energy scale. Because it simplifies the discussion, it is also conventional to define the total energy of the atom to be zero when the atom is in its ground state. Since the energy scale is logarithmic, the zero energy level representing the ground state cannot be displayed on the diagram, but this does no harm. The most important difference between an x-ray energy-level diagram and a standard energy-level diagram is that the x-ray diagram gives the energy of the atom when one electron of the indicated quantum numbers $n$, $l$, $j$ is *missing*. That is, the diagram describes the energy levels of the *hole*, with quantum numbers $n$, $l$, $j$, that jumps from one subshell to the next when the atom emits its x-ray line spectrum. As a hole represents the absence of an electron of negative energy, the energy associated with a hole is positive. So the energies of all the levels of an x-ray diagram are positive.

The energy levels in Figure 9-17 are also identified by a notation commonly used in discussing x-ray spectra. In this notation the value of the quantum number $n$ is specified by capital letters, according to the scheme shown in Table 9-4. That is, an $n = 1$ level is called a $K$ level, an $n = 2$ level is called an $L$ level, etc. Similarly, the $n = 1$ shell is called the $K$ shell, etc. Roman numeral subscripts are used to label levels of the same $n$, according to decreasing energy. That is, in order of decreasing energy the three $L$ levels are called $L_I$, $L_{II}$, and $L_{III}$.

If the energy of an atom with an electron of quantum numbers $n$, $l$, $j$ is particularly negative, the energy of an atom with a hole of the same quantum numbers is particularly positive since more energy must be given to the atom to remove the electron. In other words, the lack of a large negative energy is equivalent to the presence of a large positive energy. Keeping this inversion in mind, we see from Figure 9-17, which

**TABLE 9-4.** The Spectroscopic Notation for $n$

| $n$ | 1 | 2 | 3 | 4 | 5 | $\cdots$ |
|---|---|---|---|---|---|---|
| Spectroscopic notation | $K$ | $L$ | $M$ | $N$ | $O$ | $\cdots$ |

was obtained from an analysis of the measured x-ray line spectrum of $^{92}$U, that the $n$, $l, j$ dependences of the x-ray energy levels are as would be expected from the Hartree theory. The energies of these levels increase with decreasing values of $n$ and of $l$, in agreement with an inversion of the rule describing the theoretical predictions that was stated in the preceding section. The x-ray energy level for $j = l + 1/2$ has lower energy, and the level for the other possibility, $j = l - 1/2$, has higher energy. This is the expected inversion of the splitting of the energy levels according to $j$, discussed in connection with one-electron atoms in Section 8-6. In the $L$ shell ($n = 2$) of $^{92}$U this splitting is more than 2000 eV, and it is larger than the dependence on $l$. So it is hardly appropriate to call the $j$ dependence of x-ray energy levels "fine-structure splitting." The strong $j$ dependence, which is characteristic of the inner shells of all atoms except those of very low $Z$, is partly due to the increase in the magnitude of the spin-orbit interaction because of the high value of the term $\overline{(1/r)dV(r)/dr}$ in (8-35). It also involves the other relativistic effects that become very large for the high velocity electrons that populate the inner shells of these atoms.

As we have indicated, it is convenient to think of the production of the x-ray line spectra in terms of the creation of a hole in one of its higher-energy levels, and the subsequent jumping of the hole through its lower-energy levels. With each jump, an x-ray photon is emitted that carries off the excess energy. The frequency $\nu$ of the photon bears the usual relation to the energy $E$ which it carries, $E = h\nu$. But not all transitions occur. There are the following set of *selection rules* for the change in quantum numbers of the hole:

$$\Delta l = \pm 1 \tag{9-28}$$

$$\Delta j = 0, \pm 1 \tag{9-29}$$

These are the same as the selection rules of (8-37) and (8-38), for an electron in a one-electron atom, and they have the same explanation as presented in Section 8-7. The x-ray energy-level diagram for $^{92}$U, of Figure 9-17, shows the transitions that obey these selection rules. The totality of x rays which are emitted in such transitions (plus a few which are observed to be emitted very infrequently in violation of the selection rules) constitute the x-ray line spectrum of the atom. All transitions from the $K$ shell produce lines of the so-called *K series*, with $K_\alpha$ corresponding to a transition to the $L$ shell, $K_\beta$ to the $M$ shell, etc. All transitions from the $L$ shell produce lines of the $L$ *series*, and so forth.

**Example 9-8.** Estimate the minimum accelerating voltage required for an x-ray tube with a $^{26}$Fe anode to emit a $K_\alpha$ line of its spectrum. Also estimate the wavelength of a $K_\alpha$ photon.

We can use the crude description of the results of the Hartree theory to estimate the excitation energy of a $^{26}$Fe atom with a hole in its $K$ shell. Equation (9-27) tells us that this energy is

$$E_K \simeq +\frac{\mu Z_n^2 e^4}{(4\pi\epsilon_0)^2 2\hbar^2 n^2} \simeq 13.6 \frac{Z_n^2}{n^2} \text{ eV}$$

$$\simeq 13.6(Z-2)^2 \text{ eV} = 13.6 \times (24)^2 \text{ eV}$$

$$\simeq +7.8 \times 10^3 \text{ eV}$$

where we have set $n = 1$ and $Z_n = Z_1 = Z - 2$. A beam electron bombarding an atom in the anode must have this much energy to produce the hole. The voltage $V$ required to accelerate the beam electron to this energy is just

$$V \simeq 7.8 \times 10^3 \text{ V}$$

After the atom emits a $K_\alpha$ photon, the hole is in its $L$ shell. Then its energy is

$$E_L \simeq +13.6 \frac{Z_n^2}{n^2} \text{ eV} \simeq 13.6 \frac{(26 - 10)^2}{4} \text{ eV} \simeq +8.7 \times 10^2 \text{ eV}$$

where we have set $n = 2$ and, following the results of Example 9-5, set $Z_n = Z_2 = Z - 10$. The photon carries away energy

$$h\nu = E_K - E_L$$

But since the value of $E_L$ is only about 10 % of the value of $E_K$, and since the crude approximation we have used to obtain $E_K$ is generally not accurate to 10 %, we might as well take

$$h\nu \simeq E_K$$

The wavelength $\lambda$ of the photon is related to its frequency $\nu$ and its velocity $c$ by the expression

$$\frac{1}{\lambda} = \frac{\nu}{c} = \frac{h\nu}{hc}$$

so

$$\frac{1}{\lambda} \simeq \frac{E_K}{hc} \simeq \frac{\mu e^4}{(4\pi\epsilon_0)^2 4\pi c\hbar^3}(Z - 2)^2$$

The term multiplying $(Z - 2)^2$ is Rydberg's constant, $R_M$, defined in (4-22). Therefore

$$\frac{1}{\lambda} \simeq R_M(Z - 2)^2 \simeq 1.1 \times 10^7 \times (24)^2 \text{ m}^{-1} = 6.3 \times 10^9 \text{ m}^{-1} \tag{9-30}$$

and

$$\lambda \simeq 1.6 \times 10^{-10} \text{ m} = 1.6 \text{ Å}$$

This wavelength is about the size of a typical molecule, or the spacing of atoms or molecules in a crystal. Thus the $K_\alpha$ x rays from $^{26}$Fe can be used in diffraction experiments to study the structure of molecules or crystals. ◀

A striking feature of x-ray line spectra is that the frequencies and wavelengths of the lines vary smoothly from element to element. There are none of the abrupt changes from one element to the next which occur in atomic spectra in the optical frequency range. The reason is that the characteristics of x-ray spectra depend on the binding energies of the electrons in the inner shells. With increasing atomic number $Z$, these binding energies simply increase uniformly, owing to the higher nuclear charge, and they are not affected by the periodic changes in the number of electrons in the outer shells of the atom that affect the optical spectra. The regularity of x-ray spectra was first observed by Moseley. In 1913 he made a survey of x-ray spectra and obtained data for a number of elements on the wavelengths of the $K_\alpha$ line. (There are really two closely spaced $K_\alpha$ lines, as can be seen from Figure 9-17, but it was difficult for Moseley to resolve this structure.) The measured wavelengths could be fitted within experimental accuracy by the empirical formula

$$\frac{1}{\lambda} \simeq C(Z - a)^2 \tag{9-31}$$

where $C$ is a constant with a value approximately equal to the Rydberg constant $R_M$, and $a$ is a constant with a value of about 1 or 2. This formula, and some of the data, are plotted in Figure 9-18.

Moseley interpreted the empirical formula on the basis of the Bohr model, which had been proposed just before he made his measurements. He performed a calculation essentially the same as our calculation in Example 9-8 to obtain (9-30), which agrees well enough with (9-31), but he took the basic energy equation, (9-27), from the Bohr

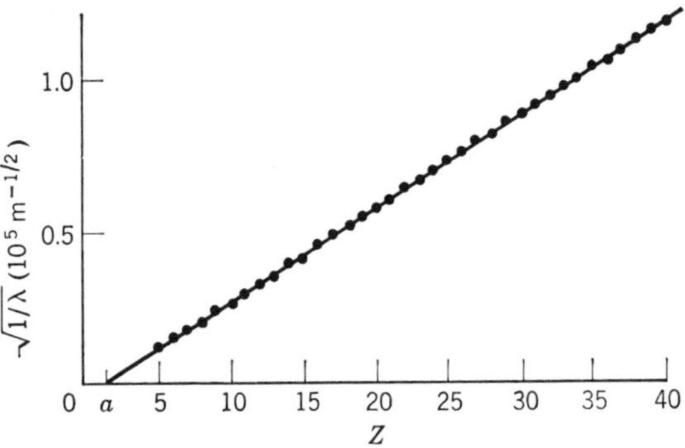


**FIGURE 9-18**

Points representing Moseley's data, and a curve representing his empirical formula. The curve is a straight line since the square root of the reciprocal of the wave lengths of the x-ray lines is plotted versus the atomic number of the atoms producing the lines.

model instead of the Hartree theory. That is, he adapted the Bohr energy equation into (9-27) by replacing $Z$ by $Z_n$, as a way of describing the shielding of the nuclear charge by electron charges in a multielectron atom. His arguments concerning shielding were similar to ours of Section 9-6, except that he thought the electrons travel in well-defined Bohr orbits and concluded that $Z_1 \simeq Z - 1$ instead of $Z_1 \simeq Z - 2$.

Moseley's work, carried out when he was a graduate student, was an important step in the development of quantum physics. His simple and successful application of the Bohr model to x-ray line spectra provided one of its earliest confirmations. By using the empirical formula to determine $Z$, he established unambiguously the correlation between the nuclear charge of an atom and its ordering in the periodic table of the elements. For instance, he found that the atomic number of $^{27}$Co is one less than that of $^{28}$Ni, even though its atomic weight is greater. He also showed that there were gaps in the periodic table, as it was then known, at $Z = 43$, 61, 72, and 75. Elements of these atomic numbers have subsequently been discovered. Moseley's contributions were brought to a halt by service in World War I, from which he did not return.

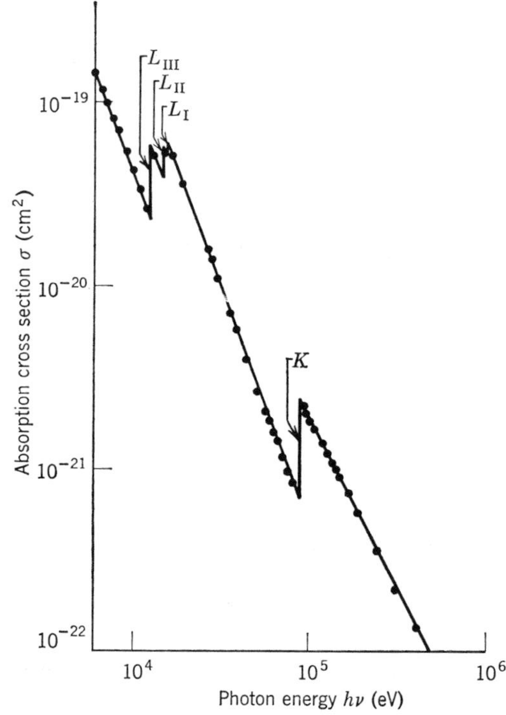

**FIGURE 9-19**

The probability that a lead atom will absorb an x-ray photon by the photoelectric effect, as a function of the energy of the photon. The probability is expressed in terms of the absorption cross section.

**Example 9-9.** Measured values of the probability that a $^{82}$Pb atom will absorb by the photoelectric effect an x-ray photon from an incident beam of photons, are displayed in Figure 9-19 by plotting the absorption cross section as a function of the energy $h\nu$ of the photon. The prominent discontinuity just below $10^5$ eV is called the *K absorption edge*. Show that it occurs at an energy for which the incident photon can just produce a hole in the $K$ shell of $^{82}$Pb. Then explain the origin of the discontinuities a little above $10^4$ eV.

According to (9-27), the energy required to produce a hole in the $K$ shell of $^{82}$Pb is approximately

$$E_K \simeq +13.6 \frac{Z_n^2}{n^2} \text{ eV} \simeq 13.6(Z - 2)^2 \text{ eV} = 13.6 \times (80)^2 \text{ eV} = 8.7 \times 10^4 \text{ eV}$$

This agrees within a few percent with the measured energy of the $K$ absorption edge. A photon whose energy is slightly above this edge can be absorbed by the photoelectric effect on any electron of the atom. But a photon of energy slightly below the $K$ absorption edge does not have enough energy to eject a $K$ shell electron, so for it the photoelectric effect cannot occur on a $K$ shell electron. Thus the photoelectric absorption cross section drops abruptly at the $K$ absorption edge.

At energies a little above $10^4$ eV there are three $L$ absorption edges. These occur at the energies required to produce holes in the $L$ shell of the atom. There are three because "fine structure", due to spin-orbit and other relativistic effects, splits the $L$ level into three levels, $L_{\text{I}}$, $L_{\text{II}}$, $L_{\text{III}}$, as can be seen in Figure 9-17. ◀

## QUESTIONS

1. Why is there difficulty in distinguishing the two electrons in a helium atom from each other, but not the two electrons in separated hydrogen atoms? What about a diatomic hydrogen molecule?

2. Explain, without reference to the time-independent Schroedinger equation, why the product form of the eigenfunction of (9-3) immediately implies that the two particles it describes move independently.

3. Can *you* write a time-independent Schroedinger equation for two identical particles, without using particle labels?

4. Are particle labels themselves objectionable, in working with quantum mechanical systems containing identical particles? If not, explain precisely what care must be exercised in using them.

5. Since the value of an antisymmetric total eigenfunction changes when its particle labels are exchanged, why can such eigenfunctions be used to give an accurate description of a system of electrons?

6. Does the exchange degeneracy increase the number of degenerate states in an atom containing two electrons? Explain.

7. Do you think the sign of the charge of an elementary particle, like an electron or proton, is a more, or less, fundamental property than the "sign" of its symmetry?

8. Would atoms be affected more by reversing the signs of the charges of all their constituent particles, or by reversing all their symmetries?

9. Exactly what is meant by the statement that the spin variable is not continuous?

10. Would it be possible to measure effects of the exchange force acting between two electrons if there were no Coulomb interaction between them to produce an interaction energy of magnitude dependent on the sign of the exchange force?

**11.** Why would it be much more difficult to solve the time-independent Schroedinger equation for a system of interacting particles than for a system of independently moving particles?

**12.** Describe the steps in a cycle of the self-consistent Hartree treatment of a multielectron atom. Why is the estimate of the net potential $V(r)$ obtained at the end of a cycle more accurate than the estimate used at the beginning?

**13.** Why is the angular dependence of multielectron atom eigenfunctions the same as for one-electron atom eigenfunctions? Why is the radial dependence different, except near the origin where it is the same?

**14.** Just what is the justification for using one-electron atom equations with an effective $Z$ to discuss multielectron atoms?

**15.** What are the consequences of the fact that the sizes of all atoms are about the same? What are the reasons for this fact?

**16.** Devise a purely mechanical system in which a classical particle would exhibit the tendency, illustrated in Figure 9-12, to avoid the point about which it rotates.

**17.** Explain all aspects of the $Z$ dependence of the subshell energies, plotted in Figure 9-14.

**18.** Why is it particularly difficult to separate mixtures of the rare earth elements by chemical techniques?

**19.** How can we be sure that if there were no molecules there would be no life?

**20.** What property of x rays makes them so useful in seeing otherwise invisible internal structures?

**21.** Give an example in the classical world where the concept of a hole might be used in a way comparable to the way it is used in discussing x-ray line spectra.

**22.** What argument might Moseley have used to conclude that the effective $Z$ for the $K$ shell is $Z_1 \simeq Z - 1$? Can Gauss's law of electrostatics be applied to evaluate the shielding produced by electrons moving in Bohr orbits?

**23.** What features of the periodic table of Figure 9-13 would Mendeleev fail to recognize?

**24.** Do the properties of the electrons in multielectron atoms provide any explanation of why the element of highest atomic number found in nature is ${}^{92}$U?

**25.** In your opinion, what is the most important consequence of the exclusion principle?

## PROBLEMS

**1.** By going through the procedure indicated in the text, develop the time-independent Schroedinger equation for two noninteracting identical particles in a box, (9-1).

**2.** By applying the technique of separation of variables, show that for a potential of the additive form of (9-2), there are solutions to the two-particle time-independent Schroedinger equation, (9-1), in the product form of (9-3).

**3.** Exchange the particle labels in the two probability density functions, obtained from the symmetric and antisymmetric eigenfunctions of (9-8) and (9-9), and show that neither is affected by the exchange.

**4.** Verify that the expanded form of the three particle eigenfunction of Example 9-2 is antisymmetric with respect to an exchange of the labels of two particles.

5. Verify that the expanded form of the three-particle eigenfunction of Example 9-2 is identically equal to zero if two particles are in the same space and spin quantum state.

6. Verify that the $1/\sqrt{3!}$ normalization factor quoted in Example 9-2 is correct.

7. Verify that the expanded form of the three-particle eigenfunction of Example 9-3 is symmetric with respect to an exchange of the labels of two particles.

8. An $\alpha$ particle contains two protons and two neutrons. Show that if each of its constituents is antisymmetric then it must be symmetric, as stated in Table 9-1. (Hint: Consider a pair of $\alpha$ particles, and the effect of exchanging the labels of all the constituents in one with those of all the constituents in the other.)

9. Write an expression for the expectation value of the energy associated with the Coulomb interaction between the two electrons of a helium atom in its ground state. Use a space eigenfunction for the system composed of products of one-electron atom eigenfunctions, each of which describes an electron moving independently about the $Z = 2$ nucleus. Do not bother to evaluate the expectation value integral, but instead comment on its relation to the energy levels shown in Figure 9-7.

10. Prove that any two different nondegenerate bound eigenfunctions $\psi_i(x)$ and $\psi_j(x)$ that are solutions to the time-independent Schroedinger equation for the same potential $V(x)$ obey the *orthogonality* relation

$$\int_{-\infty}^{\infty} \psi_j^*(x)\psi_i(x)\, dx = 0 \qquad i \neq j$$

(Hint: (i) Write the equations to which $\psi_i$ and $\psi_j$ are solutions, and then take the complex conjugate of the second one to obtain the equation satisfied by $\psi_j^*$. (ii) Multiply the equation in $\psi_i$ by $\psi_j^*$, the equation in $\psi_j^*$ by $\psi_i$, and then subtract. (iii) Integrate, using a relation such as $\psi_j^*\, d^2\psi_i/dx^2 - \psi_i\, d^2\psi_j^*/dx^2 = (d/dx)(\psi_j^*\, d\psi_i/dx - \psi_i\, d\psi_j^*/dx)$.) The proof can be extended to include degenerate eigenfunctions, and also unbound eigenfunctions that are properly normalized. Can you see how to do this?

11. (a) By going through the procedure indicated in Section 9-5, develop the time-independent Schroedinger equation for a system of $Z$ electrons of an atom moving independently in a set of identical net potentials $V(r)$. (b) Then separate it into a set of $Z$ identical time-independent Schroedinger equations, one for each electron. (c) Verify that the form of a typical one is as stated in (9-22). (d) Compare this form with the time-independent Schroedinger equation for a one-electron atom, (7-12).

12. (a) Show that there are $N!$ terms in the linear combination for an antisymmetric total eigenfunction describing a system of $N$ independent electrons. (Hint: Consider Example 9-2, and use the mathematical technique of induction.) (b) Evaluate the number of such terms for the case of the argon atom with $Z = 18$. (Hint: Use a mathematical table to evaluate $N!$, or use Stirling's formula, found in most mathematical references, to approximate it.) (c) State briefly the connection between the results of (b) and the procedure used by Hartree to treat the argon atom.

13. (a) Use information from Figure 9-11 to make a sketch, on semilog paper, of the net potential $V(r)$ for the argon atom. Be sure to determine several values for $r/a_0$ between 0 and 0.25, as this information will be used in Problem 17. (b) Also show the energy levels $E_1$ and $E_2$, using estimates from Example 9-5, and the energy level $E_3$, using measured data from Figure 9-15.

14. (a) Find the value of $Z_1$ for the helium atom which, when used in the energy equation, (9-27), leads to agreement with the ground state energy shown in Figure 9-6. (b) Compare $Z_1$ with $Z$. (c) Is $Z_1$ meaningful for an atom with as few electrons as helium? Explain briefly.

**15.** (a) Use the $Z_n$ for the argon atom obtained in Example 9-5 in the one-electron atom equation for the radial coordinate expectation value, to estimate the radii of the $n = 1, 2$, and 3 shells of the atom. (b) Compare the results with Figure 9-10.

**16.** Develop a mathematical argument for the tendency, illustrated in Figure 9-12, of an atomic electron with angular momentum $L$ to avoid the point about which it rotates. Treat the electron semiclassically by assuming that it moves around an orbit in a fixed plane passing through the nucleus. (a) Show that its total energy can be written

$$E = \frac{p_{\parallel}^2}{2m} + \left[V(r) + \frac{L^2}{2mr^2}\right] = \frac{p_{\parallel}^2}{2m} + V'(r)$$

where $p_{\parallel}$ is its component of linear momentum parallel to its radial coordinate vector of length $r$. (b) Explain why this indicates that its radial motion is like it would be in a one-dimensional system with potential $V'(r)$. (c) Then show that $V'(r)$ becomes repulsive for small $r$ because of the dominant behavior of the term $L^2/2mr^2$, sometimes called the *centrifugal potential.*

**17.** (a) Sketch the potentials $V'(r)$ for the argon atom with $l = 0$ and $l = 1$, defined in Problem 16, by adding the corresponding centrifugal potentials to the $V(r)$ obtained in Problem 13. (b) Also sketch the energy level $E_2$. (c) Show the classical limits of motion, within which $E_2 \geq V'(r)$. (d) Compare these limits with the radial probability densities of Figure 9-10, for $n = 2, l = 0$, and $n = 2, l = 1$.

**18.** Write the configurations for the ground states of $^{28}$Ni, $^{29}$Cu, $^{30}$Zn, $^{31}$Ga.

**19.** Write the configurations for the ground states of all the lanthanides, making as much use as possible of ditto marks.

**20.** Recent work in nuclear physics has led to the prediction that nuclei of atomic number $Z = 110$ might be sufficiently stable to allow some of the element $Z = 110$ to have survived from the time the elements were created. (a) Predict a likely configuration for this element. (b) Make a prediction of the chemical properties of the element. (c) Where would be a likely place to start searching for traces of it?

**21.** (a) From information contained in Figures 9-6 and 9-15, determine the energy required to remove the remaining electron from the ground state of a singly ionized helium atom. (b) Compare this energy with the energy predicted by the quantum mechanics of one-electron atoms.

**22.** (a) Draw a schematic representation of a standard energy-level diagram for the $^{22}$Ti atom, showing the states populated by electrons for a case in which one electron is missing from the $K$ shell. The diagram should be comparable to the one in Figure 9-9 in that it should not attempt to give the energies of the levels to an accurate scale, and no distinction should be made between $L_{\text{I}}$, $L_{\text{II}}$, and $L_{\text{III}}$ levels, etc. (b) Do the same for a case in which one electron is missing from the $L$ shell. (c) Draw a schematic representation of an x-ray energy-level diagram showing the energies of the atom when a hole is in the $K$ or $L$ shells. (d) Compare the utility of the standard and x-ray energy-level diagrams for cases in which a hole is in an inner shell. (e) Also make such a comparison for cases in which a hole is in an outer shell.

**23.** The wavelengths of the lines of the $K$ series of $^{74}$W are (ignoring fine structure): for $K_\alpha$, $\lambda = 0.210$ Å; for $K_\beta$, $\lambda = 0.184$ Å; for $K_\gamma$, $\lambda = 0.179$ Å. The wavelength corresponding to the $K$ absorption edge is $\lambda = 0.178$ Å. Use this information to construct an x-ray energy-level diagram for $^{74}$W.

**24.** (a) Make a rough estimate of the minimum accelerating voltage required for an x-ray tube with a $^{26}$Fe anode to emit a $L_\alpha$ line of its spectrum. (Hint: As in Example 9-5, $Z_2 \simeq Z - 10$.) (b) Also estimate the wavelength of the $L_\alpha$ photon.

**25.** (a) Use Moseley's data of Figure 9-18 to determine the values of the constants $C$ and $a$ in his empirical formula, (9-31). (b) Compare these values with those of (9-30), which was derived from the results of the Hartree theory.

**26.** It is suspected that the cobalt is very poorly mixed with the iron in a block of alloy. To see regions of high cobalt concentration, an x-ray is taken of the block. (a) Predict the energies of the $K$ absorption edges of its constituents. (b) Then determine an x-ray photon energy that would give good contrast. That is, determine an energy of the photon for which the probability of absorption by a cobalt atom would be very different from the probability of absorption by an iron atom.

# 10

# Multielectron Atoms—Optical Excitations

# Multielectron Atoms—Optical Excitations 10

## 10-1 Introduction

A description of the behavior of electrons in multielectron atoms involves a succession of increasingly accurate approximations. In the first step only the strongest interactions felt by the atomic electrons are considered. This is the Hartree approximation, discussed in the preceding chapter, in which each electron is treated as if it were moving independently in a spherically symmetrical net potential that describes the average of its Coulomb interactions with the nucleus and the other electrons. In the next steps the description is made more and more accurate by taking into account successively the weaker interactions which the electrons feel. In a typical multielectron atom these weaker interactions include two that involve departures of the actual Coulomb interactions experienced by an atomic electron from the average described by the net potential. One of these leads to couplings between the orbital angular momenta of the electrons, and the other leads to couplings between the spin angular momenta of the electrons through an interesting effect of the exchange force. A third weaker interaction involves the internal magnetic fields of the atom, and leads to couplings between the spin and orbital angular momenta. A fourth weaker interaction is present if the atom is placed in an external magnetic field, as in the so-called Zeeman effect. In this chapter we discuss qualitatively the steps in this succession of approximations, and we use the discussion to describe the behavior of the atomic electrons. That is, we shall consider the four weaker interactions experienced by these electrons, and we shall see that they provide a very satisfactory explanation of the important properties of the ground states and low-energy excited states of all atoms.

An atom is raised from its ground state to one of its low-energy excited states when an electron in one of its outer subshells is given a small amount of energy. As an example, this can happen when an atom collides with another atom in a gas discharge tube. The Coulomb field of the incident atom can act on an electron in an outer subshell of the struck atom and give it a few electron volts of excitation energy. In the deexcitation process, the atom that has received energy goes from the state initially excited to its ground state by emitting a set of low-energy photons whose frequencies constitute its *optical line spectrum*. The initial excitation is therefore called an *optical excitation*. Note the contrast between an optical excitation, which involves giving a small amount of energy to an electron in an outer subshell, and an x-ray excitation, which involves giving a large amount of energy to an electron in an inner subshell.

The low-energy excited states of atoms that enter into the production of optical line spectra are certainly worth studying. One reason is that a study of these excited states of atoms leads to an extremely complete description of their ground states. Another reason is that the general ideas behind the successive approximation procedure used in the study are similar to those behind the procedures used throughout science and engineering to break down a complicated problem into a sequence of not too complicated steps. The details of the procedure are of particular interest to students who will continue in physics beyond the level of this book because they are closely related to those used in the theory of molecules, nuclei, and elementary particles. Furthermore, optical line spectra are themselves of great practical interest because they are valuable experimental tools in many fields. Certainly the best example is astronomy. Much of what is known about the stars has come from measurements and analysis of optical line spectra. The pattern of lines observed in emission spectra is used to identify the composition of stars; the intensity of lines observed in absorption spectra is used to measure the temperatures of stellar surfaces; the Doppler shift of the spectral lines is used to measure the velocities of stars; and the Zeeman effect is used to measure the magnetic fields produced by stars.

## 10-2 Alkali Atoms

We begin our study of the optical excitations of multielectron atoms with the simplest case, *alkali atoms*. In their ground states, these atoms contain a set of completely filled subshells, the highest energy one being a $p$ subshell, plus a single additional electron in the next $s$ subshell. As discussed in Section 9-7, the energy of the electrons in a filled $p$ subshell is quite a bit more negative than the energy of an electron in the next $s$ subshell. Consequently, the $p$ subshell electrons are not excited in any of the low-energy processes which lead to the production of the optical spectra. In essence, an alkali atom consists of an inert noble gas core plus a single electron moving in an external subshell. The analysis of the optical line spectrum of an alkali atom in terms of its excited states is fairly simple since the excited states can be described completely by describing the single so-called *optically active* electron, and the core of filled subshells can be ignored. The total energy of the core does not change, so the total energy of the atom is a constant plus the total energy of the optically active electron. It is convenient in discussing the excited states of an alkali atom to define the zero of total energy in such a way that the total energy of the atom is equal to that of the optically active electron. Using this definition, we present in Figure 10-1 diagrams showing the energies of the ground state and the first few excited states of the alkali atoms $^{3}$Li and $^{11}$Na, obtained from an analysis of the optical line spectra of these elements, and also the energy levels of $^{1}$H for $n = 2, 3, 4, 5$, and 6. Each energy level is labeled by the quantum numbers $n$ and $l$ of the optically active electron, i.e., by its configuration. These diagrams do not show fine-structure splittings, which will be discussed shortly.

The Hartree theory works particularly well as a first step in calculating the energy levels of the optically active electron of an alkali element because the net potential $V(r)$, due to the nucleus plus the electrons of the core, actually is spherically symmetrical as assumed in the theory. The energies predicted by the theory are in excellent agreement with those shown in Figure 10-1. Furthermore, the theory makes it easy to understand the structure of these energy-level diagrams and their relation to the diagram for $^{1}$H. The dependence of the energy of the optically active electron on its quantum numbers $n$ and $l$ is just as we have described in the previous chapter. For a given $n$, the energy is most negative for the smallest value of $l$ because the electron spends more time near the center of the atom, where it feels the full nuclear charge. In the ground state of the $^{3}$Li atom, the optically active electron is in the $2s$ subshell and its energy is about 2 eV more negative than an $n = 2$ electron in a $^{1}$H atom. In the first excited state, the optically active electron is in the $2p$ subshell and its energy is only about 0.2 eV more negative than an $n = 2$ electron in $^{1}$H. For $^{11}$Na the $l$ dependence makes the $4s$ level more negative than the $3d$ level. However, for the large radii subshells with large values of $n$, the $l$ dependence becomes less important, and the energy levels of the optically active electron become very close to the energy levels of an electron in a $^{1}$H atom. The reason is that the shielding of the nuclear charge $+Ze$ by the charge $-(Z - 1)e$ of the electrons in the core of the alkali atom becomes

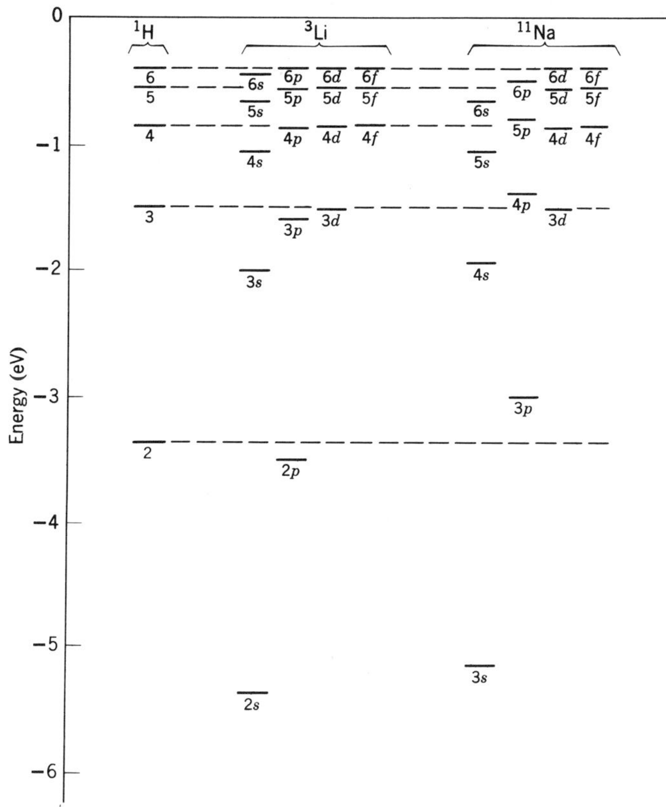


**FIGURE 10-1**
Some of the energy levels of hydrogen, lithium, and sodium atoms.

practically complete for an electron in a subshell of radius large compared to the radius of the core, so the electron experiences essentially the same Coulomb potential due to a single charge $+e$ as an electron in a $^1$H atom.

The lines of the optical spectra emitted by alkali elements show a fine-structure splitting which indicates that all energy levels are double, except those for $l = 0$. This is due to a spin-orbit interaction acting on the optically active electron, i.e., due to the coupling between the magnetic dipole moment of the electron and the internal magnetic field it feels because it moves through the electric field of the atom. Other relativistic effects, which are just as important as the spin-orbit interaction in the case of a one-electron atom, are generally quite negligible for the optically active electrons in *all* multielectron atoms. We can see this by using the Bohr model result of (4-17)

$$v = \frac{Ze^2}{4\pi\epsilon_0 n\hbar}$$

to estimate the average velocity $v$ of an optically active electron, providing we replace $Z$ by $Z_n$. As $Z_n/n$ is about equal to one for the optically active electrons of all atoms, the equation shows that the average value of $v/c$ is about equal to its value in the ground state of the $^1$H atom; that is $v/c \simeq 10^{-2}$. The associated relativistic effects for optically active electrons thus are of the same order of magnitude throughout the periodic table. In contrast, we shall see below that the spin-orbit interaction increases in magnitude rapidly in going from $^1$H to elements further up the periodic table, and so it dominates the other relativistic effects.

The splitting of the energy levels of an alkali element due to the spin-orbit interaction acting on the optically active electron can be understood by considering the interaction energy, (8-35)

$$\overline{\Delta E} = \frac{\hbar^2}{4m^2c^2}[j(j+1) - l(l+1) - s(s+1)]\,\overline{\frac{1}{r}\frac{dV(r)}{dr}}$$

The arguments leading to this equation apply as well to the optically active electron in an alkali atom as to the electron of a one-electron atom, providing that $V(r)$ is equated to the Hartree net potential and the expectation value of $(1/r)dV(r)/dr$ is calculated using the probability density obtained from the Hartree eigenfunctions. As is true for a one-electron atom, when the spin-orbit interaction is included the eigenfunctions describing the optically active electron of an alkali atom are labeled by the quantum numbers $n$, $l$, $j$, $m_j$. These quantum numbers obey the same rules as before. Specifically

$$s = 1/2 \tag{10-1}$$

$$j = \begin{cases} l - 1/2,\ l + 1/2 & l \neq 0 \\ 1/2 & l = 0 \end{cases} \tag{10-2}$$

$$m_j = -j, -j+1, \ldots, +j-1, +j \tag{10-3}$$

For $l = 0$, (8-35) shows that the spin-orbit interaction energy is $\overline{\Delta E} = 0$. For other values of $l$, it shows that $\overline{\Delta E}$ assumes two different values, one positive and the other negative, according to whether $j = l + 1/2$ or $j = l - 1/2$. Except for $l = 0$, each energy level is thus split into two components, one of slightly higher energy for the spin and orbital angular momenta "parallel," and one of slightly lower energy for these angular momenta "antiparallel." The energy difference is the work required to turn the electron magnetic dipole moment from one orientation to the other in the internal magnetic field of the atom. The magnitude of the energy splitting is proportional to the expectation value of $(1/r)dV(r)/dr$, which determines the strength of the magnetic field. Since both $1/r$ and the derivative of the net potential $V(r)$ become large for small $r$, the expectation value is dependent primarily on the behavior of $V(r)$ near $r = 0$.

According to (9-25) for the net potential $V(r)$ of the Hartree theory, the larger the value of $Z$ the more rapidly $V(r)$ becomes negative as $r$ becomes small. Thus the magnitude of $dV(r)/dr$ increases with increasing $Z$, near $r = 0$. Consequently $\overline{(1/r)dV(r)/dr}$, and also the spin-orbit splitting, should increase in magnitude with increasing $Z$. This behavior can be found in the experimental data of Table 10-1, which lists the observed splittings of the energy levels of an electron excited to the first $p$ subshell of various alkali atoms.

The spectral lines of an alkali atom are emitted in transitions between energy levels whose quantum numbers satisfy the *selection rules*:

$$\Delta l = \pm 1 \tag{10-4}$$

$$\Delta j = 0, \pm 1 \tag{10-5}$$

**TABLE 10-1.** Spin-orbit Splittings in a Number of Alkali Atoms

| Element | $^{3}$Li | $^{11}$Na | $^{19}$K | $^{37}$Rb | $^{55}$Cs |
|---|---|---|---|---|---|
| Subshell | $2p$ | $3p$ | $4p$ | $5p$ | $6p$ |
| Spin-orbit splitting (eV) | $0.42 \times 10^{-4}$ | $21 \times 10^{-4}$ | $72 \times 10^{-4}$ | $295 \times 10^{-4}$ | $687 \times 10^{-4}$ |

These selection rules for the transitions of the single optically active electron of an alkali atom are the same as those for the electron of a one-electron atom, and they have the same explanation. Of course, the frequencies of the spectral lines are the energy differences of the levels involved in the transitions, divided by Planck's constant.

If an alkali atom is not placed in an external magnetic field, only one of the weaker interactions, mentioned in Section 10-1, acts on the optically active electron. This is the spin-orbit interaction that arises from the presence of the internal magnetic field of the atom. There are no weaker interactions arising from departures of the actual Coulomb interactions experienced by the optically active electron from the average described by the spherically symmetrical net potential $V(r)$. The reason is that the potential experienced by the optically active electron really is spherically symmetrical since all the other electrons in the alkali atom are in the spherically symmetrical core. We shall soon see that this simplification does not hold for a typical atom.

**Example 10-1.** The yellow light of sodium vapor lamps frequently employed in highway illumination is a spectral line arising from the $3p$ to $3s$ transitions in $^{11}$Na. (a) Evaluate the wavelength of this line by using information contained in Figure 10-1. (b) The line is split by the spin-orbit interaction. Evaluate the separation in wavelength of its two components from information contained in Table 10-1. (c) Also comment on the application of the selection rules to the transitions involved in emission of the two components of the line.

(a) Careful inspection of Figure 10-1 shows that the energy difference between the $3p$ and $3s$ levels of $^{11}$Na is

$$E_{3p} - E_{3s} \simeq (-3.0 \text{ eV}) - (-5.1 \text{ eV}) = 2.1 \text{ eV}$$

The photons emitted in transitions between these levels carry away energy $h\nu = E_{3p} - E_{3s}$, and have frequency $\nu$ and wavelength $\lambda$, where

$$\lambda = \frac{c}{\nu} = \frac{hc}{h\nu} \simeq \frac{6.6 \times 10^{-34} \text{ joule-sec} \times 3.0 \times 10^{8} \text{ m/sec}}{2.1 \text{ eV} \times 1.6 \times 10^{-19} \text{ joule/eV}} = 5.9 \times 10^{-7} \text{ m} = 5900 \text{ Å}$$

The value obtained directly from accurate measurements is $\lambda = 5893$ Å.

(b) According to Table 10-1, the spin-orbit interaction splits the $3p$ level by an energy $dE = 2.1 \times 10^{-3}$ eV. Since

$$\lambda = c\nu^{-1}$$

it follows that

$$d\lambda = -c\nu^{-2}\, d\nu$$

and that the magnitude of the separation in wavelength of the two components of the spectral line is

$$d\lambda = \frac{c}{\nu^2} d\nu = \frac{hch\, d\nu}{(h\nu)^2} = \frac{hc\, dE}{(h\nu)^2}$$

$$\simeq \frac{6.6 \times 10^{-34} \text{ joule-sec} \times 3 \times 10^{8} \text{ m/sec} \times 2.1 \times 10^{-3} \text{ eV} \times 1.6 \times 10^{-19} \text{ joule/eV}}{(2.1 \text{ eV} \times 1.6 \times 10^{-19} \text{ joule/eV})^2}$$

$$= 5.7 \times 10^{-10} \text{ m} = 5.7 \text{ Å}$$

(c) The $3p$ level of higher energy corresponds to $j = l + 1/2 = 1 + 1/2 = 3/2$, and the $3p$ level of lower energy corresponds to $j = l - 1/2 = 1 - 1/2 = 1/2$. The $3s$ level is not split since $l = 0$, and $j = 1/2$ only. For transitions from the higher $3p$ level to the $3s$ level, $\Delta l = -1$ and $\Delta j = -1$; for transitions from the lower $3p$ level to the $3s$ level, $\Delta l = -1$ and $\Delta j = 0$. So both of these transitions are allowed by the selection rules of (10-4) and (10-5). ◀

## 10-3 Atoms with Several Optically Active Electrons

We turn now to the more typical case of an atom containing a core of completely filled subshells surrounding the nucleus, plus several electrons in a partially filled outer subshell. Since any of these electrons can participate in the excitations leading to the emission of the optical spectrum of the atom, all the electrons in the partially filled subshell are optically active. The excited states of such an atom are treated by first using the Hartree approximation, which accounts for the stronger interactions felt by its optically active electrons, and by then including the effects of other interactions which are weaker but still important.

It should be emphasized that we shall consider here, and in the remainder of the chapter, only atoms in which the outer subshell is *less than half filled*. If the subshell is more than half filled, the optical excitations of the atom are discussed in terms of the behavior of holes—not electrons—as in our discussion of x-ray line spectra. Since a hole is the absence of a negative charge, it is equivalent to the presence of a positive charge. Because of this sign reversal, certain effects that we shall deal with have a sign reversal in atoms with outer subshells that are more than half filled.

In the Hartree approximation, the energy of each independently moving optically active electron is determined by its quantum numbers $n$ and $l$. The dependence of its energy $E_{nl}$ on these two quantum numbers is similar to that of a single optically active electron in an alkali atom with the same core, since its net potential is not very different from the net potential due to the core alone. The total energy of the atom is the constant total energy of the core, plus the sum of the total energies of the optically active electrons. Consequently, the energy of the atom is determined completely in the Hartree approximation by the configuration of the optically active electrons, which specifies the $n$ and $l$ quantum numbers of each of these electrons. Since there are $2l + 1$ possible values of $m_l$ for every $l$, and since there are also 2 possible values of $m_s$, every configuration has a number of different quantum states of the same energy. Thus, in the Hartree approximation there are a number of degenerate energy levels associated with each configuration. Many of these degeneracies are removed when weaker interactions, ignored in the Hartree approximation, are finally taken into account. This is just what happens when the spin-orbit interaction is applied to alkali atoms, removing some of the degeneracies of its energy levels.

The weaker interactions experienced by optically active electrons must be included in a treatment of the low-energy excited states of typical atoms. They can be thought of as corrections for effects ignored in the Hartree approximation. The two most important corrections are for:

1. *The residual Coulomb interaction*, an electric interaction which compensates for the fact that the Hartree net potential $V(r)$ acting on each optically active electron describes only the *average* effect of the Coulomb interactions between that electron and all the other optically active electrons.

2. *The spin-orbit interaction*, a magnetic interaction which couples the spin angular momentum of each optically active electron with its own orbital angular momentum.

There are also relativistic corrections, corrections for interactions between the spin of one optically active electron and another because of magnetic interactions between the associated magnetic moments, etc.; but these are all very small and can usually be ignored.

We are by now quite familiar with the spin-orbit interaction since it is found in studying the optical excitations of one-electron atoms and alkali atoms. The residual Coulomb interaction is something new (except for our brief discussion of the $^2$He atom in Section 9-4) since it is found only in studying the optical excitations of atoms with two or more optically active electrons. In such atoms the Coulomb interactions felt by an optically active electron include those due to the presence of the other optically active electrons in the same subshell. Since the charge distribution of the other optically active electrons is not spherically symmetrical because the subshell is only partly filled, the effect of their Coulomb interactions is not spherically symmetrical. Therefore, the spherically symmetrical net Hartree potential $V(r)$ cannot accurately describe the actual Coulomb interactions felt by an optically active electron, but only the best spherically symmetrical average of these interactions. For accuracy, we must consider the departures from this average of the actual Coulomb interactions. We must also take into account the requirement that an eigenfunction describing accurately the optically active electrons be antisymmetric in an exchange of the labels of any two of them, since this requirement alters their charge distribution.

A quantitative treatment can be given by adding, to the energies obtained from the Hartree theory, the expectation values of the energies of the residual Coulomb and spin-orbit interactions. This is rather like the treatment of the $^1$H atom energy levels described in Section 8-6, but in the present case antisymmetric eigenfunctions must be used for the optically active electrons. Since there are, at most, only a few optically active electrons, these antisymmetric eigenfunctions are not too complicated to be handled by a large computer. Of course, we cannot present the quantitative treatment here; we present instead a qualitative discussion of the excited states of typical atoms.

We have laid the groundwork for a qualitative discussion of *one* aspect of the *residual Coulomb interaction* in Section 9-4. The student will recall that the requirement that the total eigenfunction describing two electrons be antisymmetric, in an exchange of their labels, introduces a connection between the relative orientation of the spins of the electrons and their relative space coordinates (the exchange force). The average distance between the two electrons is larger in the triplet states where the spins are "parallel" than it is in the singlet state where they are "antiparallel". Consequently, the positive Coulomb repulsion energy acting between the two electrons is smaller in the triplet states, for which the magnitude of the total spin has the constant value of $S' = \sqrt{1(1+1)}\,\hbar$ than it is in the singlet state, for which it has the constant value $S' = 0$. We have seen an example of this in our consideration of the low-energy excited states of the $^2$He atom at the end of Section 9-4. In that atom the spin angular momenta of the two optically active electrons couple together so as to yield a total spin angular momentum with either the constant magnitude $S' = \sqrt{1(1+1)}\,\hbar$ or the constant magnitude $S' = 0$, while maintaining constant magnitudes for their individual spin angular momenta. Due to the connection between the spin orientation and space coordinates, and also to what we now call the residual Coulomb interaction, the energy of the atom is lowest for the state in which $S'$ is largest and the electrons are furthest apart. It is found in analyses of the experimentally observed spectra, and it is also found in the quantitative theoretical treatment, that essentially the same effect is important in all atoms with two or more optically active electrons. That is, for such atoms *the residual Coulomb interaction produces a tendency for the spin angular momenta of the optically active electrons to couple in such a way that the magnitude of the total spin angular momentum $S'$ is constant, and the energy is usually lowest for the state in which $S'$ is largest.*

It is easy to see that *another* aspect of the *residual Coulomb interaction* is to produce a tendency for the orbital angular momenta of the optically active electrons to couple in such a way that the magnitude of the total orbital angular momentum $L'$ is constant. This happens simply because in most quantum states the charge distributions of the electrons are not spherically symmetrical, and so they exert torques on each other.

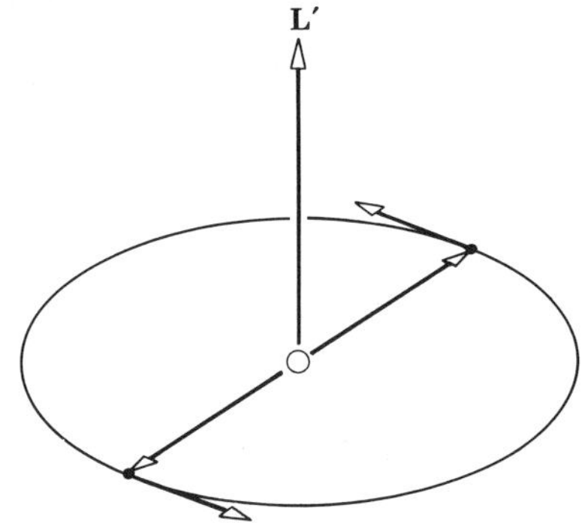


**FIGURE 10-2**
Two optically active electrons moving in the same Bohr orbit tend to remain at opposite ends of a diameter so as to minimize their Coulomb repulsion. As a result, their orbital angular momenta tend to couple in such a way as to yield a maximum total orbital angular momentum.

Since the space orientation of the charge distribution of an electron is related to the space orientation of its orbital angular momentum vector, there are torques acting between the angular momentum vectors. The torques do not tend to change the magnitude of the individual orbital angular momentum vectors, but only tend to make them precess about the total orbital angular momentum vector in such a way that its magnitude $L'$ remains constant.

The question then arises: Which of the possible values of $L'$ corresponds to the state of lowest energy? There are opposing tendencies, but the basis of the one which usually dominates can be understood even from classical physics by considering two electrons in a Bohr atom, as illustrated in Figure 10-2. Because of the Coulomb repulsion between the electrons, the most stable arrangement is obtained when the electrons stay at the opposite ends of a diameter. In this state of lowest energy, the electrons rotate together with individual orbital angular momentum vectors parallel, and therefore with the magnitude $L'$ of the total angular momentum vector a maximum. This conclusion is confirmed by an analysis of the spectra produced by atoms with several optically active electrons. That is, for such atoms the *residual Coulomb interaction produces a tendency for the orbital angular momenta of the optically active electrons to couple in such a way that the magnitude of the total orbital angular momentum $L'$ is constant, and the energy is usually lowest for the state in which $L'$ is largest.*

In contrast to the tendencies produced by the residual Coulomb interaction, *the spin-orbit interaction produces a tendency for the spin angular momentum of each optically active electron to couple with its own orbital angular momentum, in such a way as to leave the magnitudes of these vectors constant, while they precess about their resultant total angular momentum vector that is of constant magnitude $J$.* We are familiar with this tendency in one-electron atoms and in alkali atoms. We know that it is due to torques arising from the interaction of the magnetic dipole moment connected with the spin angular momentum and the magnetic field connected with the orbital angular momentum. We also know that *the energy is lowest for the state in which $J$ is smallest* (for a less than half-filled subshell).

The residual Coulomb and spin-orbit interactions tend to produce effects which are in opposition to each other. But for atoms of small and intermediate $Z$ the effects of the residual Coulomb interaction are much larger than the effects of the spin-orbit interaction. *Except for atoms of large $Z$, the residual Coulomb interaction is treated first*, since it is the most important, and the spin-orbit interaction is temporarily ignored. Then the individual spin angular momenta $\mathbf{S}_i$ of the optically active electrons are considered to couple to form a *total spin angular momentum* $\mathbf{S}'$, where

$$\mathbf{S}' = \mathbf{S}_1 + \mathbf{S}_2 + \cdots + \mathbf{S}_i + \cdots \qquad (10\text{-}6)$$

and where $\mathbf{S}'$ has a constant magnitude satisfying the quantization condition

$$S' = \sqrt{s'(s'+1)}\,\hbar \tag{10-7}$$

Also, the individual orbital angular momenta $\mathbf{L}_i$ of the optically active electrons are considered to couple to form a *total orbital angular momentum* $\mathbf{L}'$, where

$$\mathbf{L}' = \mathbf{L}_1 + \mathbf{L}_2 + \cdots + \mathbf{L}_i + \cdots \tag{10-8}$$

and where $\mathbf{L}'$ has a constant magnitude satisfying the quantization condition

$$L' = \sqrt{l'(l'+1)}\,\hbar \tag{10-9}$$

These vectors couple in such a way that all their magnitudes $S_i$ and $L_i$ also remain constant. Because of the residual Coulomb interaction, the energy of the atom depends on $S'$ and $L'$, so quantum states of the same configuration, but associated with different values of $S'$ and $L'$, no longer have the same energy. The state with the maximum possible values of $S'$ and $L'$ usually has the minimum energy.

Having taken the dominant residual Coulomb interaction into account, the weaker spin-orbit interaction is then included. This is done by considering a spin-orbit interaction between the angular momentum vectors $\mathbf{S}'$ and $\mathbf{L}'$. The interaction couples these two vectors in such a way that the magnitude $J'$ of the *total angular momentum*

$$\mathbf{J}' = \mathbf{L}' + \mathbf{S}' \tag{10-10}$$

is constant, and $S'$ and $L'$ remain constant. The magnitude of $J'$ is also quantized according to the usual condition

$$J' = \sqrt{j'(j'+1)}\,\hbar \tag{10-11}$$

As a result of the spin-orbit interaction, the energy of the atom depends also on $J'$. The state with the minimum possible value of $J'$ has the minimum energy. The procedure described in the last two paragraphs is commonly named *LS coupling*. But sometimes it is named *Russell-Saunders coupling* after the two astronomers who first used it in studying atomic spectra emitted by stars. The procedure is valid except for atoms of large $Z$.

The student should be warned that the common name frequently causes confusion because it seems to imply that the coupling between the $L$ and $S$ vectors is the most important. In fact, just the opposite is true. In $LS$ coupling the coupling of the individual $L$ vectors to form the total $L$ vector, and also the coupling of the individual $S$ vectors to form the total $S$ vector, are the most important because they have the largest effect on the energy. The coupling of the total $L$ vector to the total $S$ vector is less important because it has a smaller effect on the total energy.

If $Z$ is large, the spin-orbit interaction is too strong (see Table 10-1) to justify ignoring it even temporarily. This complicates the situation because both the residual Coulomb and the spin-orbit interactions must then be treated simultaneously. For atoms of the largest $Z$, the spin-orbit interaction begins to dominate the residual Coulomb interaction, and the treatment simplifies because a sequential procedure again becomes possible. This procedure, called *JJ coupling*, involves first treating the relatively strong coupling of the spin and orbital angular momenta of each optically active electron of the large $Z$ atom, to form its total angular momentum, and then treating the relatively weak coupling of these angular momenta to form the total angular momentum for all the electrons. Since most atoms are either good or fair examples of $LS$ coupling, it is the only procedure we shall consider in this chapter. In Chapter 15, we shall consider $JJ$ coupling in connection with the behavior of protons and neutrons in nuclei, since in all nuclei these particles move under the influence of a very strong spin-orbit interaction.

## 10-4 *LS* Coupling

Figure 10-3 illustrates the way the various angular momentum vectors combine in *LS* coupling in the state which is normally the one of *minimum energy* for two optically active electrons with quantum numbers $l_1 = 1$, $s_1 = 1/2$, and $l_2 = 2$, $s_2 = 1/2$. The spin angular momenta $\mathbf{S}_1$ and $\mathbf{S}_2$ precess about their sum $\mathbf{S}'$, and $\mathbf{S}'$ has its maximum possible magnitude (corresponding to $s' = 1$). The precession is rapid because their coupling is relatively strong. The orbital angular momenta $\mathbf{L}_1$ and $\mathbf{L}_2$ precess rapidly about their sum $\mathbf{L}'$ because their coupling is also relatively strong, and $\mathbf{L}'$ also has its maximum possible magnitude (corresponding to $l' = 3$). In addition, there is a slow precession of $\mathbf{S}'$ and $\mathbf{L}'$ about their sum $\mathbf{J}'$, with $\mathbf{J}'$ having its minimum possible magnitude (corresponding to $j' = 2$). This precession is slow because the coupling between $\mathbf{S}'$ and $\mathbf{L}'$ is relatively weak. Finally, $\mathbf{J}'$ precesses randomly about the $z$ axis, to satisfy the angular momentum uncertainty principle, with its component $J_z'$ along that axis a constant given by the quantization condition

$$J_z' = m_j'\hbar \tag{10-12}$$

where

$$m_j' = -j', -j' + 1, \ldots, +j' - 1, +j \tag{10-13}$$

Figure 10-3 is drawn for $m_j' = j'$. The quantization of the magnitude of the total angular momentum $J'$, and of its $z$ component $J_z'$, is a necessary requirement of the absence of external torques acting on the atom; i.e., it is a consequence of the quantum mechanical angular momentum conservation law.

Figure 10-3 shows only one of the quantum states that can be formed in *LS* coupling by two optically active electrons with quantum numbers $l_1 = 1$, $s_1 = 1/2$, and $l_2 = 2$, $s_2 = 1/2$. In fact, there are twelve different sets of states, with different quantum numbers $s'$, $l'$, $j'$, that can be formed by these two electrons; and each of these

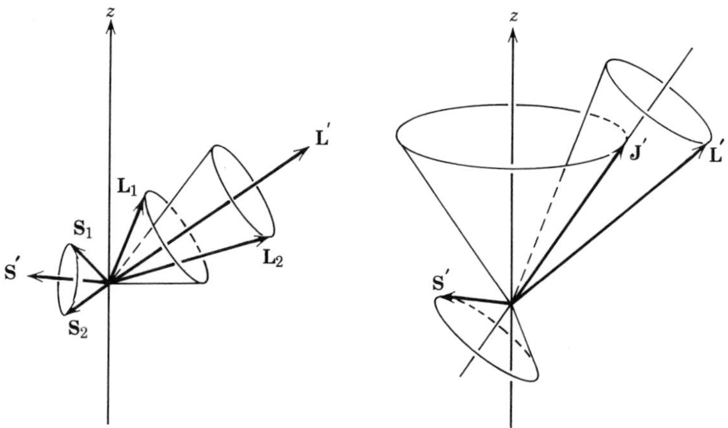


**FIGURE 10-3**

The coupling of various angular momentum vectors in a typical *LS* coupling state of minimum energy. *Left:* The orbital angular momenta $\mathbf{L}_1$ and $\mathbf{L}_2$ of the two electrons precess rapidly about their vector sum $\mathbf{L}'$. Similarly, their spins $\mathbf{S}_1$ and $\mathbf{S}_2$ precess rapidly about their sum $\mathbf{S}'$. *Right:* The total orbital angular momentum $\mathbf{L}'$ and the total spin angular momentum $\mathbf{S}'$ precess slowly about their sum $\mathbf{J}'$, the total angular momentum. Finally, $\mathbf{J}'$ precesses randomly about the $z$ axis.

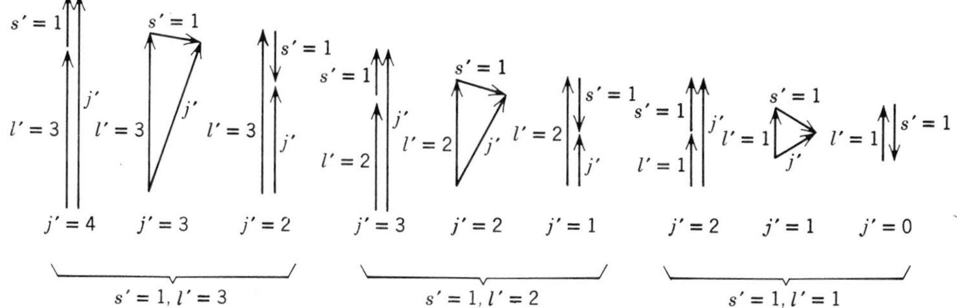


**FIGURE 10-4**

Vector addition diagrams for the quantum numbers $l_1 = 1$, $s_1 = 1/2$; $l_2 = 2$, $s_2 = 1/2$.

twelve sets contains states of $2j' + 1$ different possible values of $m_j'$. The rule specifying the possible values of $m_j'$ is expressed by (10-13). The rules specifying the possible values of $s', l', j'$ are conveniently expressed with reference to vector addition diagrams employing vectors whose lengths are proportional to the quantum numbers, just as we have done in Section 8-5. For the two electrons in question, these diagrams have the form indicated in Figure 10-4. The student may verify that the possible values of $s'$, $l'$, $j'$ shown in the vector diagrams agree with those obtained from the equations

$$\begin{aligned} s' &= |s_1 - s_2|, |s_1 - s_2| + 1, \ldots, s_1 + s_2 \\ l' &= |l_1 - l_2|, |l_1 - l_2| + 1, \ldots, l_1 + l_2 \\ j' &= |s' - l'|, |s' - l'| + 1, \ldots, s' + l' \end{aligned} \tag{10-14}$$

Since $s_1 = s_2 = 1/2$, the first equation gives

$$s' = 0, 1$$

This is the same as (9-21). The other two equations can be proved by the same type of vector inequality arguments we used to prove (8-33). Obvious generalizations of the vector diagrams can be used to find the possible quantum numbers for cases with more than two optically active electrons.

**Example 10-2.** Find the possible values of $s'$, $l'$, and $j'$ for a configuration with three optically active electrons of quantum numbers $l_1 = 1$, $l_2 = 2$, and $l_3 = 4$.

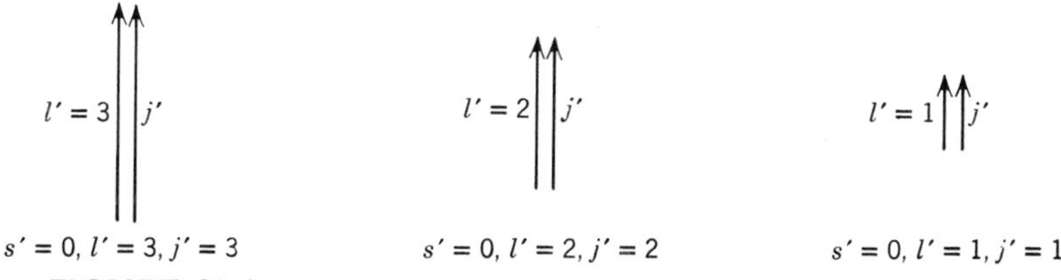

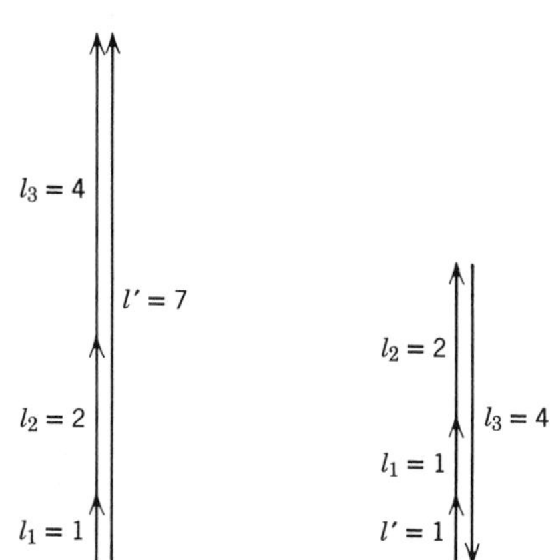


**FIGURE 10-5**
Vector addition diagrams for the maximum and minimum values of $s'$ and $l'$ in a configuration of three optically active electrons with $l_1 = 1$, $l_2 = 2$, $l_3 = 4$.

With the aid of the constructions shown in Figure 10-5, we conclude that the minimum value of $s'$ is 1/2 and that the maximum value of $s'$ is 3/2. Therefore, the possible values are $s' = 1/2, 3/2$. The constructions also show that the minimum value of $l'$ is 1, and that the maximum value of $l'$ is 7. So the possible values are $l' = 1, 2, 3, 4, 5, 6, 7$. The possible values of $j'$ are then $j' = 1/2, 3/2, 5/2, 7/2, 9/2, 11/2, 13/2, 15/2, 17/2$. Not indicated in Figure 10-5, or in Figure 10-4, are the $2j' + 1$ possible values of $m_j'$ for each value of $j'$. In the absence of external fields, the energy of the atom does not depend on $m_j'$. ◀

Figure 10-6 illustrates the splitting of the single degenerate level of a particular configuration of an atom with two optically active electrons, due to the residual Coulomb and spin-orbit interactions. The configuration is $3d^1 4p^1$, or in abbreviated form $3d4p$, which involves the same quantum numbers, $l_1 = 1, s_1 = 1/2; l_2 = 2, s_2 = 1/2$, considered in Figures 10-3 and 10-4. Also illustrated in the figure is the notation used by spectroscopists to label the quantum numbers of the levels. For instance, the lowest energy level is identified by the symbol $3d4p\ ^3F_2$. The first part of the symbol

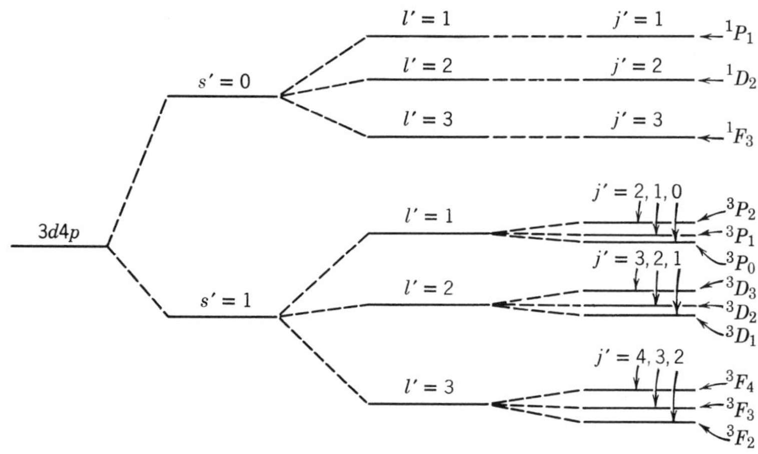


**FIGURE 10-6**
The splitting of the energy levels in a typical *LS* coupling configuration.

gives the configuration. The second part gives the values of $s'$, $l'$, $j'$. The letter specifies the value of $l'$ according to the scheme of Table 9-3 (except that it is conventional to use capitals); that is, $F$ means $l' = 3$. The subscript gives the value of $j'$; that is, $j' = 2$. The superscript is equal to $2s' + 1$ (and, if $s' \leq l'$, is also equal to the number of components into which the levels are split by the spin-orbit interaction); that is, $2s' + 1 = 3$ so $s' = 1$. The second part of the symbol is read "triplet $F$ 2."

We cannot present explicit equations from which the energies of all the levels in Figure 10-6 can be evaluated, but we can write an equation which gives the $j'$ dependence of the spin-orbit interaction energy. This dependence splits the levels for $s' = 1$, and a given $l'$, into triplets of levels. We consider again (8-35) for the spin-orbit interaction energy, writing it as

$$\overline{\Delta E} = K[j'(j' + 1) - l'(l' + 1) - s'(s' + 1)] \tag{10-15}$$

This equation predicts the expectation value of the interaction energy of the total spin and orbital angular momentum vectors $\mathbf{S}'$ and $\mathbf{L}'$, providing $LS$ coupling is valid so that these vectors are meaningful. The quantity $K$ is not simply proportional to a term like $\overline{(1/r)\, dV(r)/dr}$, as might be expected from earlier applications of (8-35), because the potential is more complicated in the present situation. However, $K$ does have the same value for all the energy levels of a so-called *multiplet*; i.e., for all the energy levels of a configuration with common values of $s'$ and $l'$. Therefore, we can calculate from (10-15) the separation in energy between the adjacent levels of a multiplet. If the quantum number associated with the level of lower energy is $j'$, the quantum number associated with the level of higher energy is $j' + 1$, and the separation $\mathscr{E}$ in the energy of the two levels is

$$\begin{aligned} \mathscr{E} &= K[(j' + 1)(j' + 2) - l'(l' + 1) - s'(s' + 1)] \\ &\quad - K[j'(j' + 1) - l'(l' + 1) - s'(s' + 1)] \\ &= K[(j' + 1)(j' + 2) - j'(j' + 1)] \end{aligned}$$

This yields the simple result

$$\mathscr{E} = 2K(j' + 1) \tag{10-16}$$

Thus we see that *the separation $\mathscr{E}$ in the energy of adjacent levels of a multiplet is proportional to the total angular momentum quantum number of the level of higher energy*. This prediction of (10-16) is called the *Landé interval rule*. It is widely used in atomic physics, as we shall see in Examples 10-3 and 10-4. Essentially the same rule is used in molecular and nuclear physics.

**Example 10-3.** In the $3d3d$ configuration of the $^{20}$Ca atom there is a multiplet (in this case a triplet) of levels: $^3P_0$, $^3P_1$, $^3P_2$. The lowest energy level is observed to be $^3P_0$, the next is $^3P_1$, and the highest is $^3P_2$. The measured separation $\mathscr{E}$ in energy between the $^3P_1$ and $^3P_0$ levels is $16.7 \times 10^{-4}$ eV, and $\mathscr{E}$ between the $^3P_2$ and $^3P_1$ levels is measured to be $33.3 \times 10^{-4}$ eV. Compare these values of $\mathscr{E}$ with the predictions of the Landé interval rule, (10-16).

The theory does not predict an accurate value for the $K$ in (10-16), but it does predict that $K$ has the same value for all the levels of a multiplet. So we can obtain an accurate prediction for the ratio of the two values of $\mathscr{E}$. For the lowest energy level $j' = 0$; for the next $j' = 1$; and for the highest $j' = 2$. Thus the Landé interval rule predicts

$$\frac{\mathscr{E}(^3P_2, {}^3P_1)}{\mathscr{E}(^3P_1, {}^3P_0)} = \frac{2K(j' + 1)_{j'=1}}{2K(j' + 1)_{j'=0}} = \frac{2}{1}$$

The ratio of the measured values of $\mathscr{E}$ is

$$\frac{\mathscr{E}(^3P_2, {}^3P_1)}{\mathscr{E}(^3P_1, {}^3P_0)} = \frac{33.3 \times 10^{-4}\ \text{eV}}{16.7 \times 10^{-4}\ \text{eV}} = 1.99$$

**TABLE 10-2.** Fine-Structure Splittings in the Calcium Atom

| Configuration | Levels | Separation | Levels | Separation | Ratio Exp. | Ratio Theo. |
|---|---|---|---|---|---|---|
| $3d3d$ | ${}^3P_1, {}^3P_0$ | $16.7 \times 10^{-4}$ eV | ${}^3P_2, {}^3P_1$ | $33.3 \times 10^{-4}$ eV | 1.99 | 2/1 |
| $4s4p$ | ${}^3P_1, {}^3P_0$ | $64.9 \times 10^{-4}$ eV | ${}^3P_2, {}^3P_1$ | $131.2 \times 10^{-4}$ eV | 2.02 | 2/1 |
| $4s3d$ | ${}^3D_2, {}^3D_1$ | $16.9 \times 10^{-4}$ eV | ${}^3D_3, {}^3D_2$ | $26.9 \times 10^{-4}$ eV | 1.59 | 3/2 |
| $3d4p$ | ${}^3D_2, {}^3D_1$ | $33.1 \times 10^{-4}$ eV | ${}^3D_3, {}^3D_2$ | $49.6 \times 10^{-4}$ eV | 1.50 | 3/2 |

This excellent agreement between the experimentally measured and theoretically predicted ratios of $\mathscr{E}$ provides evidence for $LS$ coupling in the ${}^{20}$Ca atom. In other words, the Landé interval rule can be used as a test for the presence of $LS$ coupling. ◀

The first row in Table 10-2 summarizes the successful Landé interval rule test for the presence of $LS$ coupling, carried out in Example 10-3, for a triplet in one of the configurations of the ${}^{20}$Ca atom. The other rows show the equally successful results of the same test applied to triplets in other configurations of that atom. All together, these tests provide convincing evidence for the presence of $LS$ coupling in the ${}^{20}$Ca atom. When the same tests are applied to multiplets in various configurations of other atoms with more than one optically active electron, they show that $LS$ coupling is present in all such atoms of small and intermediate $Z$.

**Example 10-4.** Measurements made on the line spectrum emitted by a certain atom of intermediate $Z$ show that the separations between adjacent energy levels of increasing energy, in a particular multiplet, are approximately in the ratio 3 to 5. Use the Landé interval rule to assign the quantum numbers $s', l', j'$ to these levels. This example gives some insight into the procedure used by the experimental spectroscopist in analyzing his measurements.

The experimental information is indicated in the energy-level diagram of Figure 10-7. If the separation between the lowest energy pair of levels is $\mathscr{E}$, then the separation between the higher energy pair is approximately $(5/3)\mathscr{E}$. Although the values of $j'$ for the levels are not initially known, it is known that the possible values differ by one, and that the lowest energy level is obtained for the lowest $j'$. So if that quantum number has the value $j'$ for the lowest level, it has the values $j' + 1$ and $j' + 2$ for the successively higher levels.

Now the Landé interval rule says that the separation between adjacent levels is proportional to the $j'$ value of the upper level. So the separation between the lower pair of levels should be

$$\mathscr{E} = 2K(j' + 1)$$

and the separation between the higher pair of levels should be

$$(5/3)\mathscr{E} = 2K(j' + 2)$$

Dividing the first equation by the second, to eliminate the unknown $K$, we obtain

$$\frac{3\mathscr{E}}{5\mathscr{E}} = \frac{2K(j' + 1)}{2K(j' + 2)}$$

**FIGURE 10-7**
Illustrating the assignment of quantum numbers in a multiplet from the observed level separations.

which gives

$$5j' + 5 = 3j' + 6$$

or

$$2j' = 1$$

and

$$j' = 1/2$$

Thus the $j'$ values of the levels are, in order of increasing energy, $j' = 1/2, 3/2, 5/2$.

To determine the values of $s'$ and $l'$ for the multiplet, we use the third of equations (10-14)

$$j' = |s' - l'|, |s' - l'| + 1, \ldots, s' + l'$$

Since the minimum value of $j'$ is 1/2 and the maximum is 5/2, we have

$$|s' - l'| = 1/2$$

and

$$s' + l' = 5/2$$

To handle the absolute value, we consider two cases. In the first case $s' \geq l'$, and these two equations are

$$s' - l' = 1/2$$

and

$$s' + l' = 5/2$$

Adding gives

$$2s' = 6/2 \quad \text{or} \quad s' = 3/2$$

Subtracting gives

$$2l' = 4/2 \quad \text{or} \quad l' = 1$$

In the second case $s' \leq l'$, and the equations we must solve are

$$-(s' - l') = 1/2$$

and

$$s' + l' = 5/2$$

Adding gives

$$2l' = 6/2 \quad \text{or} \quad l' = 3/2$$

But this is not possible since the total orbital angular momentum quantum number $l'$ cannot have a half-integral value. Therefore, the first case, $s' \geq l'$, is the correct one, and we conclude that $s' = 3/2$ and $l' = 1$.

The spectroscopist carries out this procedure on all the muliplets of a particular configuration, the levels being grouped into configurations by the similarity of their energies. Having thereby obtained the $l'$ values for the multiplets of the configuration, the $l$ quantum numbers of the configuration are identified by using the second of (10-14) (or by using an obvious extension of the equation if he knows that there are more than two optically active electrons because some of the $s'$ values are larger than 1). Identification of the $n$ quantum numbers associated with the various $l$ quantum numbers is not difficult, if the $n$ quantum numbers of the ground state configuration are known, by making use of the fact that the energy of the subshells with common values of $l$ increases monotonically with increasing $n$. The identification of the $n$ quantum numbers of the ground state configuration of the atoms is based on the same fact. ◀

## 10-5 Energy Levels of the Carbon Atom

As yet another example of *LS* coupling, we consider in this section the energy-level diagram of the $^6$C atom, shown in Figure 10-8. The ground state of this atom has the configuration $1s^2 2s^2 2p^2$, so that there are two $p$ electrons which are optically active. The zero of the energy scale in the diagram is defined such that the magnitude of the total energy of the atom in its ground state is equal to the energy required to singly

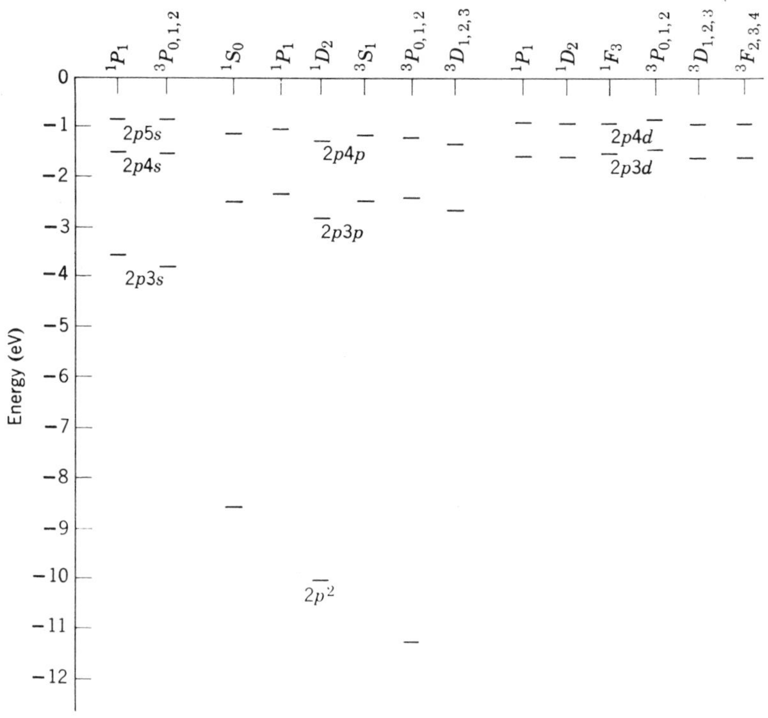


**FIGURE 10-8**

Some energy levels of the carbon atom.

ionize the atom. Consequently, the diagram is directly comparable with energy-level diagrams for alkali atoms and $^1$H, in which the zero of energy is defined in the same way. The energy levels are labeled by the configuration of the two optically active electrons, and by the spectroscopic symbol specifying $s'$, $l'$, $j'$.

Consider first the *average* energy of the levels of the various configurations. In the configuration of lowest energy, $2p^2$, both electrons remain in the same subshell that they occupy in the ground state of the atom. In other configurations, one electron remains in that subshell and one is in a subshell of higher energy. Note that the average energies of the configurations depend on the $n$ and $l$ quantum numbers of the electron in the higher energy subshell in essentially the same way as if this electron were the single optically active electron in an alkali atom.

In the $2p^2$ configuration, the one of lowest average energy, the $^3P_{0,1,2}$ states are of lower energy than the $^1S_0$ and $^1D_2$ states because they correspond to a larger value of $s'$, and the $^1D_2$ states are of lower energy than the $^1S_0$ state because they correspond to a larger value of $l'$. Note that the $s'$ dependence is stronger than the $l'$ dependence. It is almost always found that the energy associated with the residual Coulomb interaction coupling of the spin angular momenta is somewhat larger than the energy associated with the residual Coulomb interaction coupling of the orbital angular momenta. Of the three closely spaced energy levels for the $^3P_{0,1,2}$ states that would be resolved on a larger diagram, the one for the $^3P_0$ state is of lowest energy because it corresponds to the smallest value of $j'$. Thus the ground state of the atom is the state $2p^2\,{}^3P_0$. That is, in the ground state of carbon there are two electrons in the partially filled third subshell (the $2p$ subshell), which are coupled so that they have one unit of total spin angular momentum, one unit of total orbital angular momentum, and zero total angular momentum. The study of the low-energy excited states of atoms leads to an extremely complete description of their ground states!

In the $2p3s$ configuration of $^6$C the level corresponding to maximum $s'$ is lowest in energy, just as in the $2p^2$ configuration. Deviations from this rule, and from the rule that the maximum $l'$ gives the lowest energy, are seen in the configurations of higher average energy, but in $^6$C there are no deviations from the rule that the minimum $j'$ gives the minimum energy.

Not shown in Figure 10-8 are a few energy levels of the configuration $2s2p^3$, which are not usually excited. Also not shown is the spin-orbit splitting of the energy levels, since it is much too small to be seen on the scale of the diagram.

Although not present in $^6$C, in many atoms there is a *hyperfine splitting* of the energy levels. It is smaller than the spin-orbit splitting by about three orders of magnitude. Hyperfine splitting is due to either or both of the following: (1) the interaction between an intrinsic *magnetic dipole moment of the nucleus* and a magnetic field produced by the atomic electrons, and/or (2) the interaction between a *nonspherically symmetrical nuclear charge distribution* and a nonspherically symmetrical electric field produced by the atomic electrons. These effects are of interest principally because they can provide very useful information about the nucleus, and they will be discussed in Chapter 15.

Note the absence in the $^6$C energy-level diagram, of Figure 10-8, of levels for the $^1P_1$ and $^3S_1$ states in the $2p^2$ configuration. This is an effect of the *exclusion principle.* In all other configurations of the diagram the exclusion principle is automatically satisfied by the fact that the $n$ quantum numbers of the optically active electrons differ. But in the $2p^2$ configuration both the $n$ and $l$ quantum numbers are the same, so the exclusion principle puts restrictions on the possible values of the remaining quantum numbers. In the Hartree approximation these are sets of the quantum numbers $m_l$, $m_s$, one set for each of the *independent* optically active electrons having common values of the quantum numbers $n$ and $l$. In this approximation the restrictions of the exclusion principle are simply that no two electrons can have the same set of all four quantum numbers. In *LS* coupling, where the $m_l$ and $m_s$ are not useful and the quantum numbers $l'$, $s'$, $j'$, $m_j'$ are used instead to specify the way the optically active electrons are interacting, the restrictions of the exclusion principle are more complicated. For the general situation the arguments used to work out the *LS coupling exclusion principle* restrictions are very involved, and even in simpler special situations they are somewhat involved. (Interested students will find a sample of these arguments, and a complete statement of their conclusions, in Appendix K.) Here we shall only mention two of the conclusions obtained from the arguments. One is that the absence of the $^1P_1$ and $^3S_1$ states in a $2p^2$ configuration, and of other states in other configurations in which the electrons have the same $n$ and $l$ quantum numbers, can be understood on the basis of the exclusion principle. Another conclusion is that when there are as many electrons having the same $n$ and $l$ quantum numbers as is allowed by the exclusion principle, then the only state that occurs is $^1S_0$. This restriction can be expressed by saying that *when a subshell is completely filled, the only allowed state is one in which the total spin angular momentum, total orbital angular momentum, and total angular momentum, are all zero*. A consequence of the fact that there are no total angular momenta in a completely filled subshell is that it has no net magnetic dipole moment. Therefore, only the few electrons in an atom that are not in filled subshells are involved in its interaction with external magnetic fields—an important simplification.

This particular restriction of the exclusion principle applied to *LS* coupling is exactly what would be expected from the exclusion principle applied to the Hartree approximation. To see that this is so, assume that the electrons in a completely filled subshell are not interacting at all with each other. Then the behavior of each can be described by values of the quantum numbers $m_l$ and $m_s$. Since the subshell is filled, electrons would be found with all possible combinations of $m_l$ and $m_s$, but since all the electrons have the same $n$ and $l$, each combination of $m_l$ and $m_s$ would occur only once. The result is that for each electron having a certain positive $z$ component of orbital angular momentum (because it has a certain positive $m_l$), there would be an electron having the corresponding negative $z$ component (because it has the corresponding negative $m_l$). Thus the total orbital angular momentum of the electrons in the filled subshell would sum up to zero. The same would be true for their total spin angular momentum. Therefore, their total angular momentum would also have to be zero.

The optical line spectrum of the $^6$C atom, or of any other $LS$ coupling atom, can be constructed from its energy-level diagram by evaluating the energy and frequency of photons emitted in all possible transitions that do not violate the following $LS$ *coupling selection rules:*

1. Transitions can occur only between configurations which differ in the $n$ and $l$ quantum numbers of a *single* electron. This means that two or more electrons cannot simultaneously make transitions between subshells.

2. Transitions can occur only between configurations in which the change in the $l$ quantum number of that electron satisfies the same restriction that applies to one-electron atoms, (8-37)

$$\Delta l = \pm 1$$

3. Transitions can occur only between states in these configurations for which the changes in the $s'$, $l'$, $j'$ quantum numbers satisfy the restrictions

$$\begin{aligned} \Delta s' &= 0 \\ \Delta l' &= 0, \pm 1 \\ \Delta j' &= 0, \pm 1 \quad (\text{but not } j' = 0 \text{ to } j' = 0) \end{aligned} \tag{10-17}$$

The first of (10-17) prohibits transitions between singlet ($s' = 0$) and triplet ($s' = 1$) states, and vice versa. Nevertheless, transitions are observed between the $2p^2\ {}^1D_2$ states and the $2p^2\ {}^3P_{0,1,2}$ states of $^6$C. The reason is that all excitations of that atom to singlet states eventually lead to the population of its $2p^2\ {}^1D_2$ states, since Figure 10-8 shows them to be the lowest energy singlet states. When they are highly populated, the total number of transitions per second to the $2p^2\ {}^3P_{0,1,2}$ states becomes appreciable, even though the probability is very small that any single atom will make this transition since it violates the $\Delta s' = 0$ selection rule. Physically, this rule says that if the coupling of the electron spins change in an atomic transition, the atom cannot emit radiation of the type produced by oscillating *electric dipole* moments. If the spin coupling does change, radiation is emitted, but at a very low rate. The radiation is produced inefficiently by oscillating spin *magnetic dipole* moments, associated with the change in the spin coupling. The last two selection rules of (10-17) are similar to those of (8-37) and (8-38).

## 10-6 The Zeeman Effect

In 1896 it was observed by Zeeman that, when an atom is placed in an *external* magnetic field, and then excited, the spectral lines it emits in the deexcitation process are split into several components. Examples of the *Zeeman effect* are illustrated in Figure 10-9. For fields less than several tenths of 1 tesla, the splitting is proportional to the strength of the field. The Zeeman splitting in such fields is smaller than the fine-structure splitting, which is proportional to the strength of the more intense internal

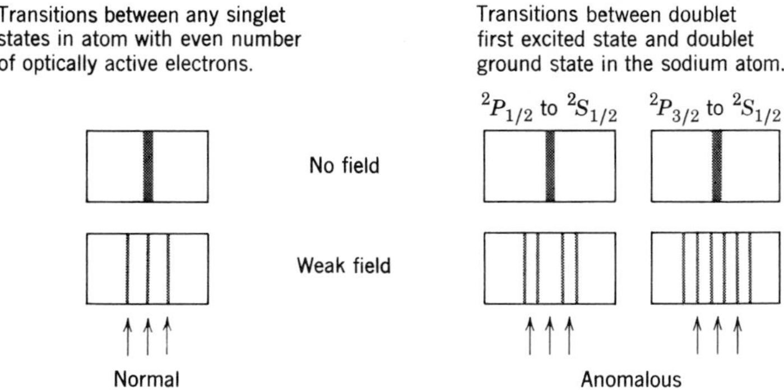


**FIGURE 10-9**

Representations of photographic plates showing the splitting of several spectral lines in the normal and anomalous Zeeman effect. The arrows show the splittings predicted by a classical theory of Lorentz.

fields of the atom. Clearly, the Zeeman effect indicates that the energy levels of the atom are split into several components in the presence of an external magnetic field. In certain special cases, which were called "*normal*," these energy-level splittings could be understood in terms of a classical theory developed by Lorentz. But in general cases, which were called "*anomalous*," even a qualitative explanation of the observed splittings could not be given until the development of quantum mechanics and the introduction of electron spin.

In terms of the modern theory, both the normal and the anomalous Zeeman splittings are easy to understand. Except when it is in an $^1S_0$ state, an atom will have a *total magnetic dipole moment*, $\boldsymbol{\mu}$, due to the orbital and spin magnetic dipole moments, $\boldsymbol{\mu}_l$ and $\boldsymbol{\mu}_s$, of its optically active electrons. (The other electrons are in completely filled subshells which have no net magnetic dipole moments.) When this magnetic dipole moment of the atom is in the external magnetic field $\mathbf{B}$ it will have the usual potential energy of orientation

$$\Delta E = -\boldsymbol{\mu} \cdot \mathbf{B} \tag{10-18}$$

Each of the atom's energy levels will be split into several discrete components corresponding to the various values of $\Delta E$ associated with the different *quantized* orientations of $\boldsymbol{\mu}$ relative to the direction of $\mathbf{B}$. In other words, because it has a magnetic dipole moment the energy of the atom depends upon which of the possible orientations it assumes in the external magnetic field.

To see qualitatively what is behind the distinction between normal and anomalous splittings, we evaluate $\boldsymbol{\mu}$ by using (8-9) and (8-19) to obtain $\boldsymbol{\mu}_l$ and $\boldsymbol{\mu}_s$ for each optically active electron in terms of its orbital and spin angular momenta, and then summing over all these electrons. That is, we take

$$\begin{aligned}\boldsymbol{\mu} &= -\frac{g_l\mu_b}{\hbar}\mathbf{L}_1 - \frac{g_l\mu_b}{\hbar}\mathbf{L}_2 - \cdots \\ &\quad -\frac{g_s\mu_b}{\hbar}\mathbf{S}_1 - \frac{g_s\mu_b}{\hbar}\mathbf{S}_2 - \cdots \\ &= -\frac{\mu_b}{\hbar}[(\mathbf{L}_1 + \mathbf{L}_2 + \cdots) + 2(\mathbf{S}_1 + \mathbf{S}_2 + \cdots)]\end{aligned}$$

We have inserted the values $g_l = 1$ and $g_s = 2$ for the orbital and spin $g$ factors that determine the ratios of the magnetic dipole moments to the angular momenta. Now, if the atom obeys $LS$ coupling, the individual orbital angular momenta couple to give the total orbital angular momentum $\mathbf{L}'$, and the individual spin angular momenta couple to give the total spin angular momentum $\mathbf{S}'$. Then the expression for the total magnetic dipole moment of the atom immediately simplifies to

$$\boldsymbol{\mu} = -\frac{\mu_b}{\hbar}[\mathbf{L}' + 2\mathbf{S}'] \tag{10-19}$$

We see that the total magnetic dipole moment of the atom is not antiparallel to its *total angular momentum*

$$\mathbf{J}' = \mathbf{L}' + \mathbf{S}' \tag{10-20}$$

The basic reason is that the orbital and spin $g$ factor have different values. The result is that the behavior of $\boldsymbol{\mu}$ is quite complicated because its orientation is not simply related to the orientation of $\mathbf{J}'$. But if $\mathbf{S}' = 0$, i.e., if the spin angular momenta of the optically active electrons couple to zero, then $\boldsymbol{\mu}$ is antiparallel to $\mathbf{J}'$, and the behavior of $\boldsymbol{\mu}$, and thus the term $\boldsymbol{\mu} \cdot \mathbf{B}$ that produces the energy level splittings, is simpler. In fact, in this case where the nonclassical phenomenon of spin is effectively not involved, the behavior of $\boldsymbol{\mu} \cdot \mathbf{B}$ can be explained satisfactorily by the old theory of Lorentz. This is the case of normal Zeeman splitting. In the general case, $\mathbf{S}' \neq 0$ and the theory of Lorentz fails. This is the case of anomalous Zeeman splitting. The terminology was introduced long before quantum theory provided a complete understanding of all aspects of the Zeeman splittings and, from the modern point of view, it is not very appropriate because there is really nothing anomalous about any of the splittings. It is interesting to note that the anomalous splittings could have been used at a very early date to show that spin exists and to show that the spin $g$ factor differs from the orbital $g$ factor.

Now we shall evaluate quantitatively the Zeeman splittings for typical energy levels of $LS$ coupling atoms by applying what we have learned about the behavior of the various angular momentum vectors in such atoms. From (10-20) we see that $\mathbf{L}'$, $\mathbf{S}'$, and $\mathbf{J}'$ always lie in a common plane. But that plane precesses about $\mathbf{J}'$ because of the Larmor precession of $\mathbf{S}'$ in the internal atomic magnetic field associated with $\mathbf{L}'$ (i.e., because of the spin-orbit interaction). Equation (8-14) shows that this precessional frequency is proportional to the strength of the internal magnetic field of the atom. From (10-19) we see that $\boldsymbol{\mu}$ also lies in the precessing plane, and is typically not antiparallel to $\mathbf{J}'$. So $\boldsymbol{\mu}$ must also precess about $\mathbf{J}'$ with a precessional frequency proportional to the internal magnetic field of the atom. If an external magnetic field $\mathbf{B}$ is applied to the atom, there will in addition be a tendency for $\boldsymbol{\mu}$ to precess about the direction of this field, with a precessional frequency proportional to its strength. If the external field is weak compared to the atomic field, the precession of $\boldsymbol{\mu}$ about $\mathbf{B}$ will be slow compared to its precession about $\mathbf{J}'$. Then the motion of $\boldsymbol{\mu}$ is something like that illustrated in Figure 10-10. Even in the case of a relatively weak external field the motion of $\boldsymbol{\mu}$ is complicated, but not too complicated to prevent the evaluation of the orientational potential energy $\Delta E$.

In Example 8-4 we saw that the strength of an internal magnetic field acting on an optically active electron is typically of the order of 1 tesla. So we assume that the external magnetic field $\mathbf{B}$ is weak compared to 1 tesla. To evaluate the potential energy $\Delta E$ of the orientation of $\boldsymbol{\mu}$ in the field $\mathbf{B}$, we must evaluate $-\boldsymbol{\mu} \cdot \mathbf{B} = -\mu_{\mathbf{B}} B$, where $\mu_{\mathbf{B}}$ is the component of $\boldsymbol{\mu}$ along the direction of $\mathbf{B}$. Since $\boldsymbol{\mu}$ precesses much more rapidly about $\mathbf{J}'$ than about $\mathbf{B}$, we may evaluate $\mu_{\mathbf{B}}$ by first finding $\mu_{J'}$, which is the

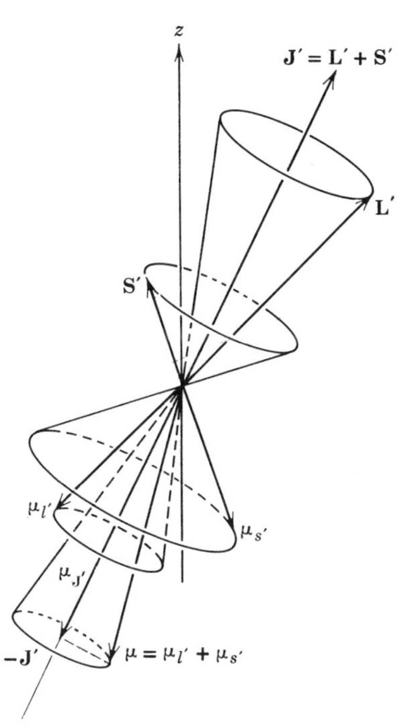


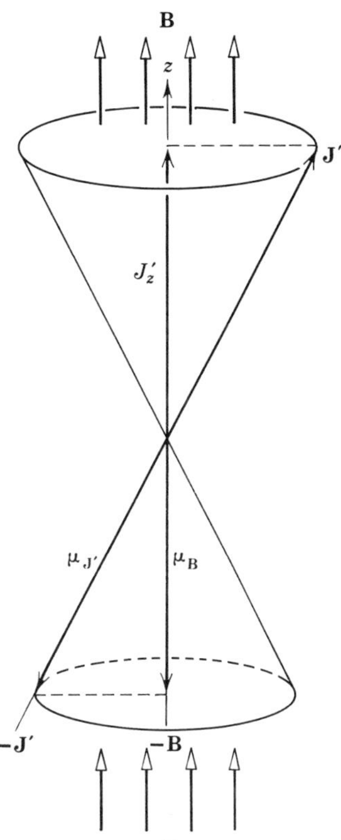


**FIGURE 10-10**

*Left:* The total orbital angular momentum $\mathbf{L}'$ and total spin $\mathbf{S}'$ couple together to form the total angular momentum $\mathbf{J}'$ of a typical atom. The total orbital magnetic dipole moment $\boldsymbol{\mu}_{l'}$ and total spin magnetic dipole moment $\boldsymbol{\mu}_{s'}$ similarly couple together to form the total magnetic dipole moment $\boldsymbol{\mu}$. Since the proportionality constant connecting $\mathbf{L}'$ and $\boldsymbol{\mu}_{l'}$ is only half the magnitude of the constant connecting $\mathbf{S}'$ and $\boldsymbol{\mu}_{s'}$, the total dipole moment will not be exactly antiparallel to $\mathbf{J}'$. And since $\mathbf{L}'$ and $\mathbf{S}'$ precess rapidly about $\mathbf{J}'$, $\boldsymbol{\mu}_{l'}$ and $\boldsymbol{\mu}_{s'}$ precess rapidly as well, causing $\boldsymbol{\mu}$ to precess about $-\mathbf{J}'$ at the same rate. Thus the component of $\boldsymbol{\mu}$ perpendicular to $-\mathbf{J}'$ averages to zero, and the component parallel to $-\mathbf{J}'$ remains a constant of magnitude $\mu_{\mathbf{J}'}$. *Right:* In a weak applied magnetic field $\mathbf{B}$, a torque is exerted which causes the direction $-\mathbf{J}'$, on which $\boldsymbol{\mu}$ has the constant average component $\mu_{\mathbf{J}'}$, to precess about the direction of $-\mathbf{B}$. So the average magnitude of this component on the direction of the field has the magnitude $\mu_{\mathbf{B}}$ indicated in the figure.

average component of $\boldsymbol{\mu}$ in the direction of $\mathbf{J}'$. We do this by multiplying $\mu$ by the cosine of the angle between $\boldsymbol{\mu}$ and $\mathbf{J}'$. Then we find $\mu_{\mathbf{B}}$ by multiplying $\mu_{\mathbf{J}'}$ by the cosine of the angle between $\mathbf{J}'$ and $\mathbf{B}$. That is

$$\mu_{\mathbf{J}'} = \mu \frac{\boldsymbol{\mu} \cdot \mathbf{J}'}{\mu J'} = -\frac{\mu_b}{\hbar} \frac{(\mathbf{L}' + 2\mathbf{S}') \cdot (\mathbf{L}' + \mathbf{S}')}{J'}$$

and

$$\mu_{\mathbf{B}} = \mu_{\mathbf{J}'} \frac{\mathbf{J}' \cdot \mathbf{B}}{J'B} = \mu_{\mathbf{J}'} \frac{J_z'}{J'} = -\frac{\mu_b}{\hbar} \frac{(\mathbf{L}' + 2\mathbf{S}') \cdot (\mathbf{L}' + \mathbf{S}')J_z'}{J'^2}$$

where we have chosen the $z$ axis to be in the direction of **B**. Evaluating the dot product gives

$$\mu_{\mathbf{B}} = -\frac{\mu_b}{\hbar}(L'^2 + 2S'^2 + 3\mathbf{L}'\cdot\mathbf{S}')\frac{J'_z}{J'^2}$$

Writing (8-34) with primes, we have

$$3\mathbf{L}'\cdot\mathbf{S}' = 3(J'^2 - L'^2 - S'^2)/2$$

So

$$\mu_{\mathbf{B}} = -\frac{\mu_b}{\hbar}[L'^2 + 2S'^2 + 3(J'^2 - L'^2 - S'^2)/2]\frac{J'_z}{J'^2}$$

$$= -\frac{\mu_b}{\hbar}\frac{(3J'^2 + S'^2 - L'^2)}{2J'^2}J'_z$$

Then, according to (10-18)

$$\Delta E = -\boldsymbol{\mu}\cdot\mathbf{B} = -\mu_{\mathbf{B}}B$$

the orientational potential energy is

$$\Delta E = \frac{\mu_b B}{\hbar}\frac{(3J'^2 + S'^2 - L'^2)}{2J'^2}J'_z \qquad (10\text{-}21)$$

In the state specified by the quantum numbers $s'$, $l'$, $j'$, $m'_j$ the dynamical quantities $S'^2$, $L'^2$, $J'^2$, $J'_z$ have the precise values $s'(s'+1)\hbar^2$, $l'(l'+1)\hbar^2$, $j'(j'+1)\hbar^2$, $m'_j\,\hbar$, respectively. Using these values in (10-21) we obtain an expression for the Zeeman effect energy splitting that is most conveniently written as

$$\Delta E = \mu_b B g m'_j \qquad (10\text{-}22)$$

where

$$g = 1 + \frac{j'(j'+1) + s'(s'+1) - l'(l'+1)}{2j'(j'+1)} \qquad (10\text{-}23)$$

The quantity $g$ is called the *Landé g factor*. Note that its value is $g = 1 = g_l$, when $s' = 0$ so $j' = l'$. Its value is $g = 2 = g_s$, when $l' = 0$ so $j' = s'$. These are just the values that would be expected since if $s' = 0$ the angular momentum is purely orbital, and if $l' = 0$ it is purely spin. Thus the Landé $g$ factor is a kind of variable $g$ factor that determines the ratio of the total magnetic dipole moment to the total angular momentum in states where that angular momentum is partly spin and partly orbital. From (10-22) we see that in an external field of strength $B$ each energy level will split into $2j' + 1$ components, one for each value of $m'_j$. We also see that the magnitude of the splitting will be different for levels with different Landé $g$ factors.

**Example 10-5.** Evaluate the Landé $g$ factor for the $^3P_1$ level in the $2p3s$ configuration of the $^6C$ atom, and use the result to predict the splitting of the level when the atom is in an external magnetic field of 0.1 tesla.

For the $^3P_1$ state $s' = l' = j' = 1$. So

$$g = 1 + \frac{1(1+1) + 1(1+1) - 1(1+1)}{2 \times 1(1+1)} = 1 + \frac{2}{2 \times 2} = \frac{3}{2}$$

For $j' = 1$ the possible values of $m'_j$ are $-1, 0, 1$, so the level is split into three components, one with the same energy and the others displaced in energy by

$$\begin{aligned}\Delta E = \mu_b B g m'_j &= \pm\mu_b B g = \pm 9.3 \times 10^{-24}\ \text{amp-m}^2 \times 10^{-1}\ \text{tesla} \times 1.5 \\ &= \pm 1.4 \times 10^{-24}\ \text{joule} \\ &= \pm 8.7 \times 10^{-6}\ \text{eV}\end{aligned}$$

◀

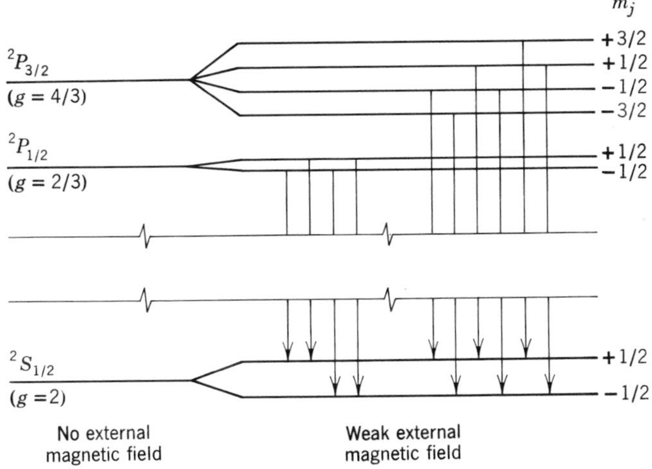


**FIGURE 10-11**

The Zeeman splittings of the $^2P_{1/2,3/2}$ first excited state levels of sodium, and of its $^2S_{1/2}$ ground state level. The transitions allowed by the selection rules are shown. Compare the resulting spectral lines with those shown in Figure 10-9.

Figure 10-11 shows, to scale, the splittings of the $^2S_{1/2}$ ground state energy level and the $^2P_{1/2}$ and $^2P_{3/2}$ lowest-excited-state energy levels of the $^{11}$Na atom, when it is placed in a weak external magnetic field. Note that *the external magnetic field removes the last vestige of degeneracy of the levels*, since the energy depends on $m_j'$. The figure also shows the transitions allowed by the *selection rule* for $m_j'$:

$$\Delta m_j' = 0, \pm 1 \quad (\text{but not } m_j' = 0 \text{ to } m_j' = 0 \text{ if } \Delta j' = 0) \tag{10-24}$$

This selection rule is very closely related to the one we derived in Example 8-6. Even with its restrictions on the allowed transitions, the Zeeman effect splits each spectral line emitted by the atom into a pattern that generally contains a number of components. The student should compare the allowed transitions, indicated by arrows in Figure 10-11, with the anomalous pattern of lines emitted by $^{11}$Na in these transitions, shown in Figure 10-9.

All spectral lines arising from transitions between singlet states are split into a simple pattern of two components symmetrically disposed about a third component that has the same frequency as the single zero-field line, as can be seen in the normal pattern of lines shown in Figure 10-9. The reason is that $s' = 0$ for singlet states, so all the $g$ factors have the same value $g = 1$. It is easy to show that this leads to spectral lines with only three components, by constructing a diagram similar to Figure 10-11.

**Example 10-6.** The most easily interpreted evidence for the splitting of atomic energy levels in an external magnetic field is *electron spin resonance*. If $^{11}$Na atoms in their ground state are placed in a region containing electromagnetic radiation of frequency $\nu$, and a magnetic field of strength $B$ is applied to the region, electromagnetic energy will be strongly absorbed when the photons have energy $h\nu$ which just equals the Zeeman splitting of the two components of the ground state energy level. The reason is that these photons are able to induce transitions between the components, indicated in Figure 10-12, in which they are absorbed. In a typical experiment $\nu = 1.0 \times 10^{10}$ Hz. Determine the value of $B$ at which the frequency defined by the Zeeman splitting is in resonance with this microwave frequency.

The ground state of $^{11}$Na is a $^2S_{1/2}$ state, for which $g = 2$ and $m_j' = \pm 1/2$. So (10-22) predicts that the displacement in energy of the components of the ground state level in an

**FIGURE 10-12**

Illustrating the transition observed in electron spin resonance involving the ground state energy levels of sodium, split by an external magnetic field.

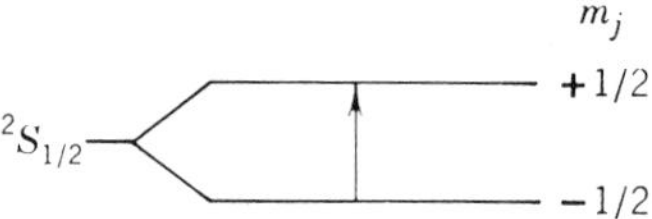


external field $B$ will be

$$\Delta E = \mu_b B g m_j' = \mu_b B 2(\pm 1/2) = \pm \mu_b B$$

Equating $h\nu$ to the separation in energy between these two components, we have

$$h\nu = 2\mu_b B$$

So

$$B = \frac{h\nu}{2\mu_b} = \frac{6.6 \times 10^{-34}\,\text{joule-sec} \times 1.0 \times 10^{10}/\text{sec}}{2 \times 9.3 \times 10^{-24}\,\text{amp-m}^2} = 0.35\ \text{tesla}$$

This effect is widely used by chemists to measure the magnetic fields experienced by an optically active electron in an atom that is part of a molecule. The electromagnetic radiation is supplied by a microwave oscillator, and the power drawn from the oscillator is monitored while its frequency is varied until the resonance condition is observed. ◀

The Zeeman effect is very useful in experimental spectroscopy. By analyzing the Zeeman splittings of the spectral lines of an atom, the spectroscopist determines the Zeeman splittings of the energy levels of the atom. These can conclusively confirm the assignment of the quantum number $j'$ of each level, because $2j' + 1$ is equal to the number of components into which the level is split. Furthermore, the magnitude of the splitting between any two components gives the value of $\mu_b B g$ and, $\mu_b$ and $B$ being known, this gives the value of $g$ for the energy level. Since the value of $g$ depends on $s'$, $l'$, $j'$ if the atom obeys $LS$ coupling, it can be used to confirm the assignment of $s'$ and $l'$. The initial assignment of values to these three quantum numbers usually comes from application of the Landé internal rule to measured separations of the levels of a multiplet, as in Example 10-4.

An external magnetic field **B**, which is weak compared to the internal atomic magnetic fields that couple **S**′ and **L**′ to form **J**′, cannot disturb this coupling and only causes a relatively slow precession of **J**′ about the direction of **B**. However, if **B** is stronger than the atomic magnetic field, it overpowers the field and destroys the coupling of **S**′ to **L**′. In this case **S**′ and **L**′ precess independently about the direction of **B**. This is the case of the *Paschen-Bach effect*, which is observed for external fields somewhat larger than 1 tesla. If the atom obeys $LS$ coupling, its total magnetic dipole moment is still given by (10-19)

$$\boldsymbol{\mu} = -\frac{\mu_b}{\hbar}[\mathbf{L}' + 2\mathbf{S}']$$

since neither the coupling of the individual spin angular momenta to form **S**′ nor the coupling of the individual orbital angular momenta to form **L**′ are destroyed by such an external field. But in this case $\mu_{\mathbf{B}}$ is simply

$$\mu_{\mathbf{B}} = -\frac{\mu_b}{\hbar}(L_z' + 2S_z')$$

where we have chosen the $z$ axis in the direction of **B**. Then

$$\Delta E = -\boldsymbol{\mu}\cdot\mathbf{B} = -\mu_{\mathbf{B}} B = \frac{\mu_b B}{\hbar}(L_z' + 2S_z')$$

and we obtain immediately

$$\Delta E = \mu_b B(m_l' + 2m_s') \tag{10-25}$$

The quantum numbers $m_l'$ and $m_s'$ are useful for an atom in an external magnetic field somewhat stronger than the internal magnetic field, because $L_z'$ and $S_z'$ have definite values in these circumstances. It is observed that the *selection rules* for the two quantum numbers are:

$$\Delta m_s' = 0 \tag{10-26}$$

$$\Delta m_l' = 0, \pm 1 \tag{10-27}$$

The first selection rule says that the total spin angular momentum, and magnetic dipole moment, do not change orientation in an atomic transition. Since such transitions involve the emission of electric dipole radiation, whereas a magnetic dipole moment of changing orientation would lead to the emission of magnetic dipole radiation, the origin of the selection rule is obvious. The second selection rule was derived in Example 8-6. All the spectral lines are split by the Paschen-Bach effect into three components, just as in the normal Zeeman effect.

## 10-7 Summary

This chapter is summarized in Table 10-3, which lists, in order of decreasing importance in determining the energy, all of the significant interactions experienced by the optically active electrons in a typical multielectron atom placed in a weak external magnetic field. By typical, we mean an atom with a less than half-filled outer subshell, whose atomic number $Z$ is low enough that it obeys $LS$ coupling. If $Z$ is very high, the atom obeys $JJ$ coupling and the most important weaker interaction is the spin-orbit interaction. If the external magnetic field is stronger than the internal magnetic field,

**TABLE 10-3.** Interactions in a Typical (*LS* Coupling; Less Than Half-Filled Subshell) Atom Placed in a Weak External Magnetic Field

| Importance in Determining Energy | Name | Nature of Interaction | Quantum Numbers Determining Energy | Energy Lowest For |
|---|---|---|---|---|
| Dominant interaction | Hartree | Electric; average potential | a set of $n$, $l$ | Minimum $n$<br>Minimum $l$ |
| Most important weaker interaction | Residual Coulomb; spin coupling | Electric; departures from average potential | $s'$ | Maximum $s'$ |
| Slightly less important | Residual Coulomb; orbital coupling | Electric; departures from average potential | $l'$ | Maximum $l'$ |
| Appreciably less important | Spin-orbit | Magnetic; internal field | $j'$ | Minimum $j'$ |
| Least important | Zeeman | Magnetic; external field | $m_j'$ | Most negative $m_j'$ |

the interaction it produces is called the Paschen-Bach interaction, and it is more important than the spin-orbit interaction in *LS* coupling. External electric fields have effects similar to, but more complicated than, external magnetic fields.

If the optically active electrons are in a more than half-filled subshell the sign of the spin-orbit interaction is reversed because the atom acts as if it had positively charged holes instead of negatively charged electrons, which reverses the relative orientation of the magnetic dipole moment and angular momentum vectors. This results in the energy level with maximum instead of minimum $j'$ lying lowest. But for such atoms maximum $s'$ and maximum $l'$ still give the lowest energy level because the sign of the residual Coulomb interaction is unchanged; it is repulsive between positive holes just as it is between negative electrons.

## QUESTIONS

**1.** Give an example of a system studied in science or engineering, other than a multielectron atom, which is best treated by a succession of increasingly accurate approximations.

**2.** Why are astronomers so dependent on information obtained from optical spectra?

**3.** Why is it not possible to give a small amount of energy to an electron in an inner subshell of an atom? What happens if a large amount of energy is given to an electron in an outer subshell?

**4.** Where in the Hartree approximation is the assumption made that the net potential is spherically symmetrical?

**5.** Explain, in simple terms, why the spin-orbit interaction becomes stronger with increasing $Z$.

**6.** Do atoms of high $Z$ generally have more optically active electrons than atoms of low $Z$?

**7.** Chemists usually speak of valence electrons. What is the corresponding term usually employed by physicists?

**8.** In studying the residual Coulomb interaction, eigenfunctions are used which are antisymmetric with respect to exchange of the labels of pairs of optically active electrons. What is the justification for not using eigenfunctions which are antisymmetric with respect to the exchange of labels for any pair of electrons in the atom?

**9.** Does the coupling of the spin angular momentum of one optically active electron in a typical atom to the spin angular momentum of another optically active electron involve a magnetic interaction between their spin magnetic dipole moments? If not, explain why not, and also explain in simple terms what the coupling is due to.

**10.** Explain the physical origin of the coupling between the orbital angular momenta of the optically active electrons in a typical atom.

**11.** Why is there a classical explanation for the coupling of orbital angular momenta of optically active electrons, but not for the coupling of their spin angular momenta?

**12.** In a multiplet with $s' > l'$, into how many components are the levels split by the spin-orbit interaction? Consider the multiplet discussed in Example 10-4.

**13.** What is the difference between *LS* coupling and *JJ* coupling?

**14.** What is the relation between the quantum states allowed by the *LS* coupling exclusion principle for a subshell with one hole (i.e., completely filled except for one electron) and the quantum states allowed for a subshell with one electron? Would there be a simple relation between the optical excitations of a halogen atom and the optical excitations of an alkali atom?

**15.** What would the exclusion principle be like for *JJ* coupling?

**16.** Is it possible for a Landé $g$ factor to have a value smaller than 1? Larger than 2?

**17.** What would be the effect of placing an atom in an external magnetic field of strength *very* much larger than the strength of the internal magnetic field?

**18.** Is it possible to completely remove the degeneracy of atomic energy levels without using an external magnetic field?

## PROBLEMS

**1.** (a) Construct an energy-level diagram for $^{11}$Na, similar to Figure 10-1, showing all levels lower in energy than the 5$s$ level. (b) Devise a way of indicating the spin-orbit splitting of the levels. (Hint: See Figure 10-8.) (c) Indicate which transitions these levels are allowed by the selection rules.

**2.** (a) Predict the values of $s'$, $l'$, $j'$ in the state of *maximum* energy of two optically active electrons with the quantum numbers $l_1 = 1$, $s_1 = 1/2$; $l_2 = 2$, $s_2 = 1/2$. (b) Make a sketch, similar to Figure 10-3, which shows the motion of the angular momentum vectors in this state.

**3.** Find the possible values of $s'$, $l'$, $j'$ for a configuration with two optically active electrons with quantum numbers $l_1 = 2, s_1 = 1/2; l_2 = 3, s_2 = 1/2$. Specify which $j'$ go with each $l'$ and $s'$ combination.

**4.** (a) Write down the quantum numbers for the states described in spectroscopic notation as $^2S_{3/2}$, $^3D_2$, and $^5P_3$. (b) Determine if any of these states are impossible, and if so explain why.

**5.** Make a sketch, similar to Figure 10-6, which illustrates the *LS* coupling splittings of the energy levels of a 4$s$3$d$ configuration. Use the Landé interval rule to predict the ratios of the fine-structure splittings of each multiplet, so that they can be drawn to scale. Label the levels with spectroscopic notation.

**6.** (a) Use the periodic table of Figure 9-13 to determine the ground state configurations for the atoms $^{12}$Mg, $^{13}$Al, and $^{14}$Si. (b) Then predict the *LS* coupling quantum numbers for the ground state of each atom. Express your result in spectroscopic notation.

**7.** Use the procedure of Example 10-3 to verify the theoretical prediction of Table 10-2 for the Landé interval rule test for the presence of *LS* coupling in the 4$s$3$d$ configuration of the $^{20}$Ca atom.

**8.** In an atom which obeys *LS* coupling, the separations between adjacent energy levels of increasing energy in the five levels of a particular multiplet are in the ratios 1 :2 :3 :4. Use the procedure of Example 10-4 to assign the quantum numbers $s'$, $l'$, $j'$ to these levels.

**9.** Consider a completely filled $d$ subshell, i.e., one containing the ten electrons allowed by the exclusion principle. Ignore the interactions between the electrons, so that the Hartree approximation quantum numbers $n$, $l$, $m_l$, $m_s$ can be used to describe each electron. (a) Show that there is only one possible quantum state for the system that satisfies the exclusion principle. (b) Show that in this state the $z$ components of the total spin angular momentum, the total orbital angular momentum, and the total angular momentum, are all zero. (c) Give an argument showing that these conclusions imply that the magnitudes of the total spin angular momentum, the total orbital angular momentum, and the total angular momentum, are also all zero. (Hint: If an angular momentum vector is not of zero magnitude, but has zero $z$ component in one quantum state, then there are other quantum states in which it has a nonzero $z$ component.) (d) Now consider the interactions between the electrons that are actually present. Can they change the conclusion about the total angular momentum of the subshell? What about the total spin angular momentum and total orbital angular momentum?

**10.** (a) Make a rough sketch of the $^6$C energy levels in the $2p^2$ and $2p3s$ configurations, using information from Figure 10-8. Indicate the fine-structure splittings of the levels by exaggerating their magnitude. (b) Show all the transitions allowed by the $LS$ coupling selection rules.

**11.** (a) Find a state with $s'$, $l'$, $j'$ quantum numbers for which the value of the Landé $g$ factor lies outside the range $g = 1$ to $g = 2$. (b) Make a sketch, similar to Figure 10-10, which illustrates the angular momentum and magnetic dipole moment vectors for this state.

**12.** Consider the $2p3s$ configuration of the $^6$C atom, in which the ordering of the energy levels according to $s'$, $l'$, $j'$, and the relative strengths of the dependences of the energy on these quantum numbers, are what is normal for $LS$ coupling. Draw a schematic energy-level diagram for this configuration, like Figure 10-6. Use the same (exaggerated) scale for the fine-structure splitting, given by the Landé interval rule, for all the levels within a given multiplet. (b) Label each level with the spectroscopic notation.

**13.** On the energy-level diagram of Problem 12, draw to the same (highly exaggerated) scale the Zeeman effect splitting, given by the Landé $g$ factor, for each level under the influence of a weak external magnetic field.

**14.** (a) Count the total number of components obtained in Problem 13, i.e., the total number of different quantum states in the configuration. (b) Show that this equals the degeneracy of the configuration in the Hartree approximation, i.e., the product of degeneracy factors $2(2l + 1)$ for each of the two optically active electrons in the configuration.

**15.** Derive an expression for the Zeeman effect splitting of the levels of a singlet. (Hint: Start at the beginning, and take $s' = 0$ so that a simple expression is obtained for the total magnetic dipole moment.)

**16.** (a) Construct a diagram, similar to Figure 10-11, which shows transitions allowed by the selection rules between the singlet states $2p3s\,^1P_1$ and $2p^2\,^1D_2$ of the $^6$C atom. (b) Verify that the normal Zeeman pattern of three spectral lines will be produced in these transitions. (c) Evaluate the differences in wavelength of these three spectral lines when the atom is in an external field of 0.1 tesla. (Hint: Use a formula for the difference in wavelength derived in Example 10-1.) (d) Evaluate the wavelength of the single line obtained when there is no external field, using information from Figure 10-8.

**17.** (a) Redraw the energy levels of Figure 10-11, for a case in which the strength of the external magnetic field is increased to the point where the splitting is described by the Paschen-Bach effect. (Hint: Here $j'$ is no longer a useful quantum number.) (b) Redraw the transitions allowed by the $m_s'$ and $m_l'$ selection rules, as in Figure 10-11, and show that they then produce spectral lines which are split into only three components.

**18.** (a) Use the information contained in Figure 10-8 to estimate the magnitude of the energy associated with the coupling of the two spin angular momenta to form the total spin angular momentum, and with the coupling of the two orbital angular momenta to form the total orbital angular momentum, in the $2p^2$ configuration of the $^6$C atom. (b) Then estimate the strength of an external field which will produce an energy of orientation with the magnetic dipole moment of each optically active electron larger than the energy estimated in (a). In such a field the couplings of the angular momenta of the optically active electrons are completely destroyed. (c) Is such a field available in the laboratory?

# 11

## Quantum Statistics

# Quantum Statistics 11

## 11-1 Introduction

As the number of constituents of a physical system increases, a detailed description of the behavior of the system becomes more complex. Thus as we proceed in our studies from one-electron atoms to multielectron atoms, and then to molecules, and finally to solids, we anticipate increasing complexity and difficulty in treating in detail these systems. For a familiar example, consider what would be involved in trying to describe the motion of one molecule of a gas in a system containing a liter of that gas under standard conditions (containing $\simeq 10^{22}$ molecules). Fortunately, it is generally unnecessary to have such detailed information to determine the most important properties of the system—that is, to determine the measurable properties, like the pressure and temperature of a gas. Furthermore, the very complexity of a system containing a large number of constituents is often responsible for many of the simple properties that we observe, as we now explain.

If we apply the general principles of mechanics (such as the conservation laws) to a system of many particles, we can ignore the detailed motion or interaction of each particle and deduce simple properties of the behavior of the system from *statistical* considerations alone. In fact, even an elementary statistical approach enables us to describe and explain a wide range of physical phenomena and gives us a good deal of insight into the behavior of real physical systems. The reason is that there is a relationship between the observed properties and the probable behavior of the system, if the system contains enough particles for statistical considerations to be valid. Consider, for instance, an isolated system containing a large number of classical particles in thermal equilibrium with each other at temperature $T$. To achieve, and maintain, this equilibrium, the particles must be able to exchange energy with each other. In the exchanges, the energy of any one of the particles will fluctuate, sometimes having a larger value and sometimes a smaller value than the average value of the energy of a particle in the system. However, the classical theory of statistical mechanics demands that the energies successively assumed by the particle, or the energies of the various particles of the system at some particular time, be distributed according to a definite probability distribution function, called the *Boltzmann distribution*, which has a form that depends on the temperature $T$. Knowing the probabilities that the particles of the system will occupy the various energy states, we can then determine a variety of important properties of the entire system by using these occupation probabilities to calculate averages over the system of the corresponding properties of the particles when they are in those states.

A more specific example that the student has likely encountered earlier in his studies of physics is the relation between the properties of a classical gas and the Maxwell distribution of speeds of the molecules of the gas. The Maxwell distribution is a special case of the Boltzmann distribution. It is described by a distribution function $N(v)$, where $N(v)\,dv$ gives the probability that a molecule has a speed in the interval between $v$ and $v + dv$. From it we can calculate quantities such as the average speed (which is related to the momentum carried by the molecules), the average squared speed (which is related to the energy they carry), etc., and from these *average* quantities we calculate observable properties such as the pressure (which is related to the momentum) and temperature (which is related to the energy), etc.

Statistical treatments are also applicable as an approximation in systems that contain only moderately large numbers of particles. For instance, we shall in Chapter 15 apply a statistical treatment to a nucleus (containing $\simeq 10^2$ nucleons) in the so-called Fermi gas model of nuclei. But that treatment will not use the Boltzmann distribution, since it is not valid for quantum particles like those found in a nucleus.

In this chapter we seek distribution functions that are valid for quantum particles. We shall find that there are two: the *Bose distribution*, which applies to particles that must be described by eigenfunctions which are symmetric with respect to an exchange of any two particle labels (like $\alpha$ particles or photons); and the *Fermi distribution*, which applies to particles that must be described by eigenfunctions which are antisymmetric in such an exchange of labels (like electrons, protons, and neutrons).

First we shall review the procedures of classical statistical mechanics, developed in Appendix C and used in Chapter 1, that lead to the Boltzmann distribution. Then we shall see how quantum considerations force significant changes in the classical procedures. Next we shall derive the quantum distribution functions in simple equilibrium arguments that start from the Boltzmann distribution. Then we shall obtain useful insights by comparing all the distribution functions with one another. Finally we shall give a variety of examples of the application of each of them, and compare their predictions with experiment. In this process we shall examine many important phenomena, such as superfluidity, electronic and lattice specific heats of solids, and light amplification by stimulated emission of radiation (the laser).

## 11-2 Indistinguishability and Quantum Statistics

The Boltzmann distribution is a fundamental result of classical physics, not quantum physics. It is, nevertheless, frequently used in discussing quantum physics, as we have seen before and shall see again. For these reasons, in this book we have included two quite different arguments that each lead to the Boltzmann distribution, but we have put these arguments in Appendix C. The student would be well advised to read, or reread, that appendix now.

Our first argument in Appendix C involves counting the number of distinguishable ways the identical entities of a system in thermal equilibrium can divide between them the fixed total energy of the system. The Boltzmann distribution follows from assuming that all possible divisions occur with the same probability. In this procedure, an energy division is counted as distinguishable from some other division if it differs from that division only by a rearrangement of identical entities between different energy states. That is, identical entities are treated as if they are distinguishable in such rearrangements. In the second argument leading to the Boltzmann distribution, we assume that the presence of one entity in some particular energy state in no way inhibits or enhances the chance that another identical entity will be in that state and, again, that all possible divisions of the system's energy occur with the same probability.

These assumptions are perfectly acceptable in classical physics. In quantum physics the assumption that all possible divisions occur with the same probability remains acceptable; but the other assumptions do not. As we saw in Section 9-2, if there is appreciable overlapping of the wave functions of two identical particles in a system, very important nonclassical effects arise from the *indistinguishability* of identical particles (i.e., identical entities). One is that measurable results cannot depend on the assignment of labels to identical particles. So the classical definition of distinguishable divisions of the energy of a system is in error because if there is no unambiguous way to label the identical particles of the system there is no way to distinguish between two divisions which differ only by rearranging them, even in rearrangements between different quantum states (i.e., energy states). Another effect of the indistinguishability of quantum particles is that the presence of one in a particular quantum state very definitely influences the chance that another identical particle will be in that state. We have seen that if two identical particles are described by an antisymmetric total eigenfunction, that is, if they are particles like electrons which obey the exclusion principle, then the presence of one in some quantum state totally inhibits the other from being in that state. We shall see soon that if the two identical particles are described by a symmetric total eigenfunction, that is, if they are like $\alpha$ particles in that they do not obey the exclusion principle, then the presence of one in some quantum state considerably enhances the chance that the other will be in the same state.

Of course, if a system contains identical quantum particles, but the circumstances are such that there is negligible overlap of the wave functions of any two, the particles actually can be distinguished experimentally. In these circumstances the effects of indistinguishability become negligible, as we mentioned before in Sections 9-2 and 9-4, and the assumptions underlying the Boltzmann distribution become valid. An example of such a system is, again, a gas. In the range of density normally encountered in the laboratory, the wave functions of the molecules, which are certainly identical quantum particles, do not overlap appreciably, and so the Boltzmann distribution can be accurately applied to predict the properties of the system.

In quantum statistics, particles which are described by antisymmetric eigenfunctions are called *fermions*, and particles which are described by symmetric eigenfunctions are called *bosons*. That is, the eigenfunction for a system of several identical fermions changes sign if the labels of any two of them are exchanged, while the eigenfunction for a system of several identical bosons does not change sign in such a label exchange. A partial list of fermions and bosons is found in Table 9-1. These names honor two physicists, Fermi and Bose, who were prominent in the development of quantum statistics.

The fact that one fermion prevents another identical fermion from joining it in the same quantum state, i.e., the exclusion principle, and certain of its extremely important consequences, is something we are familiar with from our study of multi-electron atoms. This can be described, somewhat formally, by saying that *if there are already n fermions in a quantum state the probability of one more joining them is smaller by an inhibition factor of* $(1 - n)$ *than it would be if there were no quantum mechanical indistinguishability requirements.* If $n = 0$, the factor has the value $(1 - 0) = 1$, and so there is no inhibition of the probability for the first fermion entering the state. But for $n = 1$, the factor has the value $(1 - 1) = 0$, and so a second fermion is strictly inhibited from entering the same state. Note that the factor automatically limits the number $n$ of fermions in any particular quantum state to the values $n = 0$ or $n = 1$, in agreement with the exclusion principle. The use of the plural in the preceding italicized statement may therefore seem somewhat inappropriate; it is used to make the statement analogous to one concerning bosons that will follow, and because otherwise the argument immediately below the statement would be circular.

We have not had occasion to show that the presence of one boson in a quantum state enhances the probability of a second identical boson being found in that state, because we have done little with bosons since developing the quantum mechanics of indistinguishable particles. Let us show this now.

Consider the symmetric eigenfunction for a system of two identical bosons, (9-8)

$$\psi_S = \frac{1}{\sqrt{2}}\left[\psi_\alpha(1)\psi_\beta(2) + \psi_\beta(1)\psi_\alpha(2)\right]$$

Recall that $\psi_\alpha(1)$ means the particle labeled 1 is in the quantum state $\alpha$, $\psi_\beta(2)$ means particle 2 is in state $\beta$, etc., and that although particle labels are actually used measurable quantities like the probability density $\psi_S^*\psi_S$ have values which are independent of the assignment of labels to particles. Recall also that $\psi_S$ is normalized, by the normalization factor $1/\sqrt{2}$, if we assume that $\psi_\alpha(1)\psi_\beta(2)$ and $\psi_\beta(1)\psi_\alpha(2)$ are normalized. Now we place both bosons in the same state, say the state $\beta$, by setting $\alpha = \beta$. Then the eigenfunction is

$$\psi_S = \frac{1}{\sqrt{2}}[\psi_\beta(1)\psi_\beta(2) + \psi_\beta(1)\psi_\beta(2)]$$

$$= \frac{2}{\sqrt{2}}\psi_\beta(1)\psi_\beta(2) = \sqrt{2}\,\psi_\beta(1)\psi_\beta(2)$$

and the probability density is

$$\psi_S^*\psi_S = 2\,\psi_\beta^*(1)\psi_\beta^*(2)\psi_\beta(1)\psi_\beta(2) \tag{11-1}$$

What would the eigenfunction and probability density for this two identical particle system be like if we had not taken into account the quantum mechanical requirements of indistinguishability of identical particles? The eigenfunction would be in the form given by (9-4) or (9-5), since we obtained those directly from the Schroedinger equation before applying indistinguishability requirements. Let us take (9-4)

$$\psi = \psi_\alpha(1)\psi_\beta(2)$$

This eigenfunction $\psi$ is normalized since we have assumed that $\psi_\alpha(1)\psi_\beta(2)$ is normalized. For the case at hand, where $\alpha = \beta$, we have

$$\psi = \psi_\beta(1)\psi_\beta(2)$$

and the normalized probability density is

$$\psi^*\psi = \psi_\beta^*(1)\psi_\beta^*(2)\psi_\beta(1)\psi_\beta(2) \tag{11-2}$$

It is fair to compare the probability densities of (11-1) and (11-2), since both are properly normalized. Doing so, we see that the probability $\psi_S^*\psi_S$ of having two bosons in the same quantum state has twice the value of the probability $\psi^*\psi$ of this situation occurring if the system is described by an eigenfunction that does not satisfy the quantum mechanical requirements of indistinguishability. We can express this by saying that the probability of having two bosons in the same state is twice what it would be for classical particles. Thus the presence of one boson in a particular quantum state doubles the chance that the second boson will be in that state, compared to the case of classical particles where there is no particular correlation between the quantum states occupied by the particles.

**Example 11-1.** Compare the probability for three bosons to be in a particular quantum state with the probability for three classical particles to be in the same state.

Inspection of the symmetric eigenfunction for a three boson system, found in Example 9-3, shows that it contains $3! = 3 \times 2 \times 1 = 6$ terms like $\psi_\alpha(1)\psi_\beta(2)\psi_\gamma(3)$, and that the normalization constant is $1/\sqrt{3!}$. After setting $\alpha = \beta = \gamma$ to put all the bosons in the same state, the probability density contains $(3!)^2$ equal terms, but it is multiplied by the square of the normalization constant, $(1/\sqrt{3!})^2$. So the probability is larger by a factor of $(3!)^2/3!$ than it would be if there were three identical classical particles in the state. The probability for the boson case consequently is larger by a factor of $3!$. ◀

The results of Example 11-1 can obviously be extended to the case of $n$ identical bosons in the same quantum state, and show that the probability of this occurring is larger by a factor of $n! = n(n-1)(n-2)\cdots 1$, compared to the probability that it would occur in the case of $n$ identical classical particles. These results can be looked at from a most useful point of view by answering the following question. If there are already $n$ bosons in a particular final quantum state of a system in which bosons are making transitions from various initial to various final states, what is the probability that one more boson will make a transition to that particular final state?

Let $P_1$ represent the probability that the first boson is added to the originally empty state of particular interest. If the enhancement effect we are discussing did not exist, the probability that there be $n$ bosons in that state would be just the $n$th power of $P_1$ since the probabilities of adding successive bosons would all be the same, and since the additions would take place independently and independent probabilities are multiplicative. That is

$$P_n = (P_1)^n$$

But the actual probability that there are $n$ bosons in the state is enhanced to the value

$$P_n^{\text{boson}} = n!\,P_n = n!\,(P_1)^n$$

The actual probability that there are $n+1$ bosons in the state is

$$P_{n+1}^{\text{boson}} = (n+1)!\,P_{n+1}$$

Since $(n+1)! = (n+1)n!$, and since $P_{n+1} = (P_1)^{n+1} = (P_1)^n P_1 = P_n P_1$, we have

$$P_{n+1}^{\text{boson}} = (n+1)n!\,P_n P_1$$

or

$$P_{n+1}^{\text{boson}} = (1+n)P_1 P_n^{\text{boson}} \tag{11-3}$$

Now $P_n^{\text{boson}}$ is the probability that there actually are $n$ bosons in the state. So the answer to the question posed, "*If* there are already $n$ bosons in a particular final quantum state. . . ?," is $(1+n)P_1$. But $P_1$ is the probability of adding any one of the bosons if there were no enhancement. So we conclude that, *if there are already n bosons in a quantum state the probability of one more joining them is larger by an enhancement factor of* $(1+n)$ *than it would be if there were no quantum mechanical indistinguishability requirements.*

## 11-3 The Quantum Distribution Functions

The most frequently used procedure for obtaining distribution functions that are consistent with the requirements of the indistinguishability of quantum particles involves modifying the first argument of Appendix C so as to satisfy these requirements, and then extending the calculations to the case of a large number of particles and energy states. Here we shall use a much simpler procedure that is in the spirit of the second argument of Appendix C.

As a preliminary, consider a system of identical *classical particles* in thermal equilibrium. The particles exchange energy, but they act independently in that one does not influence the specific behavior of another. Focus attention on two particular energy states of these particles $\mathcal{E}_1$ and $\mathcal{E}_2$, and let the average numbers of particles occupying them be $n_1$ and $n_2$. Also let the average rate at which a particle of the system that is in state 1 makes a transition to state 2 be $R_{1\to2}$, and the rate at which a particle that is in state 2 makes a transition to state 1 be $R_{2\to1}$. Both $R_{1\to2}$ and $R_{2\to1}$ are rates per particle, i.e., probabilities per second per particle. So the total rate at which particles of the system will be making $1 \rightarrow 2$ transitions is $n_1 R_{1\rightarrow 2}$, since $n_1$ is the number of particles that have an opportunity to do so and $R_{1\rightarrow 2}$ is the probability per second that each will take the opportunity. The total rate at which particles in the system will make $2 \rightarrow 1$ transitions is $n_2 R_{2\rightarrow 1}$.

If these total transition rates are equal, that is if

$$n_1 R_{1\rightarrow 2} = n_2 R_{2\rightarrow 1} \tag{11-4}$$

and if the same is true of "forward" and "backward" total transition rates between all pairs of particle energy states, then the average population of each of these states will obviously remain constant in time. But constant *average* state populations is the condition that characterizes thermal equilibrium. Equation (11-4) is a condition which ensures that the equilibrium we assume in all of our arguments is maintained. In principle, equilibrium can also be maintained by balancing interlocking sets of transition cycles, each involving several energy states, without balancing individual pairs of total transition rates as in (11-4); but there is no evidence that this situation arises in practice. To put the matter another way, (11-4) can be taken as a postulate, called *detailed balancing*, whose justification is found in the fact that it leads to results which agree with experiment.

Note that (11-4) implies

$$\frac{n_1}{n_2} = \frac{R_{2\rightarrow 1}}{R_{1\rightarrow 2}} \tag{11-5}$$

Now in thermal equilibrium the average, or probable, number $n_1$ of particles in our classical system that will be found in state 1 is given by the Boltzmann distribution, derived in Appendix C, evaluated at the state energy $\mathscr{E}_1$. So

$$n_1 = n(\mathscr{E}_1) = Ae^{-\mathscr{E}_1/kT} \tag{11-6}$$

and similarly for $n_2$. Thus the population ratio has the value

$$\frac{n_1}{n_2} = \frac{e^{-\mathscr{E}_1/kT}}{e^{-\mathscr{E}_2/kT}} \tag{11-7}$$

Hence, (11-5) and (11-7) show that the transition rates per particle must be in the ratio

$$\frac{R_{2\rightarrow 1}}{R_{1\rightarrow 2}} = \frac{e^{-\mathscr{E}_1/kT}}{e^{-\mathscr{E}_2/kT}} \tag{11-8}$$

for classical particles.

Now we shall apply the thermal equilibrium condition of (11-4) to a system of *bosons*. We write it as

$$n_1 R^{\text{boson}}_{1\rightarrow 2} = n_2 R^{\text{boson}}_{2\rightarrow 1} \tag{11-9}$$

where $n_1$ and $n_2$ are the average boson populations of two quantum states of interest, and $R^{\text{boson}}_{1\rightarrow 2}$ and $R^{\text{boson}}_{2\rightarrow 1}$ are the transition rates per boson between these states. These rates can be expressed in terms of the rates for the case of classical particles simply by multiplying the classical rates by the $(1 + n)$ enhancement factor derived at the end of Section 11-2. That is, since there are on the average $n_2$ bosons in quantum state 2 when the $1 \rightarrow 2$ transition takes place, the actual probability per second per particle, $R^{\text{boson}}_{1\rightarrow 2}$, is larger by a factor of $(1 + n_2)$ than the value $R_{1\rightarrow 2}$, the rate a classical particle that does not satisfy the indistinguishability requirements would have. As $n$ ranges from $\simeq 0$ (for a state which almost never contains a boson) to larger and larger values (for a state which contains more and more bosons), the enhancement factor ranges from $\simeq 1$ (almost no enhancement) to ever larger values (ever larger enhancement).

To summarize, we have

$$R_{1\to 2}^{\text{boson}} = (1 + n_2)R_{1\to 2} \tag{11-10}$$

and, similarly

$$R_{2\to 1}^{\text{boson}} = (1 + n_1)R_{2\to 1} \tag{11-11}$$

Combining (11-9), (11-10), and (11-11), we obtain

$$n_1(1 + n_2)R_{1\to 2} = n_2(1 + n_1)R_{2\to 1}$$

or

$$\frac{n_1(1 + n_2)}{n_2(1 + n_1)} = \frac{R_{2\to 1}}{R_{1\to 2}} = \frac{e^{-\mathscr{E}_1/kT}}{e^{-\mathscr{E}_2/kT}} \tag{11-12}$$

where we have used (11-8) to evaluate the ratio of the classical transition rates per particle in terms of the Boltzmann distribution. Equation (11-12) can be expressed as

$$\frac{n_1}{1 + n_1} e^{\mathscr{E}_1/kT} = \frac{n_2}{1 + n_2} e^{\mathscr{E}_2/kT} \tag{11-13}$$

The left side of this equality does not involve properties of state 2, and the right side does not involve properties of state 1. So the common value of both sides cannot involve properties particular to either state, but only a property common to both. It obviously does since the common equilibrium temperature $T$ is found on both sides. Thus we conclude that both sides of (11-13) are equal to some function of $T$, which is most conveniently written as $e^{-\alpha}$, where $\alpha = \alpha(T)$. Equating the left side to the common value, we have

$$\frac{n_1}{1 + n_1} e^{\mathscr{E}_1/kT} = e^{-\alpha}$$

or

$$\frac{n_1}{1 + n_1} = e^{-(\alpha + \mathscr{E}_1/kT)}$$

so

$$n_1 = n_1 e^{-(\alpha + \mathscr{E}_1/kT)} + e^{-(\alpha + \mathscr{E}_1/kT)}$$

or

$$n_1[1 - e^{-(\alpha + \mathscr{E}_1/kT)}] = e^{-(\alpha + \mathscr{E}_1/kT)}$$

Thus

$$n_1 = \frac{e^{-(\alpha + \mathscr{E}_1/kT)}}{1 - e^{-(\alpha + \mathscr{E}_1/kT)}} = \frac{1}{e^{\alpha} e^{\mathscr{E}_1/kT} - 1}$$

If we use the right side of (11-13), we obtain a completely similar result for the dependence of $n_2$ on $\mathscr{E}_2$. In fact, this result is obtained for the average, or probable, number of bosons occupying a quantum state of any energy $\mathscr{E}$. So we have

$$n(\mathscr{E}) = \frac{1}{e^{\alpha} e^{\mathscr{E}/kT} - 1} \tag{11-14}$$

This is the *Bose distribution* which specifies the probable number of bosons, of a system in equilibrium at temperature $T$, that will be in a quantum state of energy $\mathscr{E}$.

The same sort of argument can be applied to an equilibrium system of *fermions.* For these particles we write the thermal equilibrium condition, (11-4), as

$$n_1 R_{1\to 2}^{\text{fermion}} = n_2 R_{2\to 1}^{\text{fermion}} \tag{11-15}$$

Here $R_{1\to 2}^{\text{fermion}}$ is the rate per fermion for transitions between quantum states 1 and 2, $R_{2\to 1}^{\text{fermion}}$ is the same for $2 \to 1$ transitions, and $n_1$ and $n_2$ are the average fermion populations of these states. Because of the exclusion principle, the instantaneous populations of either state can be only zero or one. The populations fluctuate in time, due to the statistical nature of the processes that maintain thermal equilibrium, and they have average values given by $n_1$ and $n_2$. The fermion transition rates can be expressed in terms of the rates for classical particles simply by multiplying the classical rates by the $(1 - n)$ inhibition factor discussed in the middle of Section 11-2. With $n$ being interpreted as the average population of a quantum state, $(1 - n)$ is the average value of the inhibition factor, and this is just what is needed here. As $n$ ranges from $\simeq 0$ (for a state which almost never contains a fermion) to $\simeq 1$ (for a state which almost always contains a fermion), the inhibition factor ranges from $\simeq 1$ (almost no inhibition) to $\simeq 0$ (almost complete inhibition), in agreement with the exclusion principle. Thus we have

$$R_{1\to 2}^{\text{fermion}} = (1 - n_2)R_{1\to 2} \tag{11-16}$$

and

$$R_{2\to 1}^{\text{fermion}} = (1 - n_1)R_{2\to 1} \tag{11-17}$$

where $R_{1\to 2}$ and $R_{2\to 1}$ are the rates for a classical particle that does not satisfy the indistinguishability requirements leading to the exclusion principle for fermions.

Combining (11-15), (11-16), and (11-17), we obtain

$$n_1(1 - n_2)R_{1\to 2} = n_2(1 - n_1)R_{2\to 1}$$

or

$$\frac{n_1(1 - n_2)}{n_2(1 - n_1)} = \frac{R_{2\to 1}}{R_{1\to 2}} = \frac{e^{-\mathscr{E}_1/kT}}{e^{-\mathscr{E}_2/kT}} \tag{11-18}$$

where we have used (11-8) to evaluate the ratio of the classical transition rates per particle in terms of the Boltzmann probabilities. Equation (11-18) can be expressed as

$$\frac{n_1}{1 - n_1} e^{\mathscr{E}_1/kT} = \frac{n_2}{1 - n_2} e^{\mathscr{E}_2/kT} \tag{11-19}$$

By the same reasoning that we used previously, we see that both sides of this equation are equal to some function of $T$, which we again write as $e^{-\alpha}$, where $\alpha = \alpha(T)$. Equating the left side to the common value, we have

$$\frac{n_1}{1 - n_1} e^{\mathscr{E}_1/kT} = e^{-\alpha}$$

or

$$\frac{n_1}{1 - n_1} = e^{-(\alpha + \mathscr{E}_1/kT)}$$

so

$$n_1 = -n_1 e^{-(\alpha + \mathscr{E}_1/kT)} + e^{-(\alpha + \mathscr{E}_1/kT)}$$

or

$$n_1[1 + e^{-(\alpha + \mathscr{E}_1/kT)}] = e^{-(\alpha + \mathscr{E}_1/kT)}$$

Thus

$$n_1 = \frac{e^{-(\alpha + \mathscr{E}_1/kT)}}{1 + e^{-(\alpha + \mathscr{E}_1/kT)}} = \frac{1}{e^{\alpha} e^{\mathscr{E}_1/kT} + 1}$$

We write this as

$$n(\mathscr{E}) = \frac{1}{e^{\alpha} e^{\mathscr{E}/kT} + 1} \tag{11-20}$$

where we again drop the subscript 1 because the same results are obtained for all quantum states. This is the *Fermi distribution* which gives the average, or probable, number of fermions, of a system in equilibrium at temperature $T$, to be found in a quantum state of energy $\mathscr{E}$.

## 11-4 Comparison of the Distribution Functions

Consider first the Boltzmann distribution of (11-6)

$$n(\mathscr{E}) = Ae^{-\mathscr{E}/kT}$$

If we set the multiplicative constant $A$ equal to $e^{-\alpha}$, the *Boltzmann distribution* is

$$n_{\text{Boltz}}(\mathscr{E}) = \frac{1}{e^{\alpha}e^{\mathscr{E}/kT}} \tag{11-21}$$

From (11-14), we know that the *Bose distribution* is

$$n_{\text{Bose}}(\mathscr{E}) = \frac{1}{e^{\alpha}e^{\mathscr{E}/kT} - 1} \tag{11-22}$$

and (11-20) tells us that the *Fermi distribution* is

$$n_{\text{Fermi}}(\mathscr{E}) = \frac{1}{e^{\alpha}e^{\mathscr{E}/kT} + 1} \tag{11-23}$$

In these relations, $k$ is Boltzmann's constant and $T$ is the equilibrium temperature of the system. The parameter $\alpha$, for a given temperature and system, is specified by the total number of particles it contains. For instance, at the end of Appendix C we evaluated $A = e^{-\alpha}$ for a special form of the Boltzmann distribution that applies to a system of simple harmonic oscillators where we defined $n_{\text{Boltz}}(\mathscr{E})$ to be a measure of the probability of finding a *particular* one of them in a state at energy $\mathscr{E}$. The result was $A = 1/kT$. If there we defined $n_{\text{Boltz}}(\mathscr{E})$ in terms of the probability of finding *any* one of the oscillators in the state, or the probable number in the state, we would obviously have found $A = \mathscr{N}/kT$, where $\mathscr{N}$ is the total number of oscillators in the system. This is essentially the way we define $n_{\text{Boltz}}(\mathscr{E})$ here, since it gives the probable number of classical particles in the state of energy $\mathscr{E}$. In other words, $A$ is a normalization constant whose value for a given $T$ is specified by the total number of particles in the system described by the Boltzmann distribution. So the same is true for the parameter $\alpha$ appearing in that distribution. It is also true that the $\alpha$ appearing in the Bose distribution for a given $T$ is specified by the total number of Bosons in the system, and that the distribution gives the probable number of bosons in the state of energy $\mathscr{E}$. The corresponding statements apply as well for the Fermi distribution.

In Figure 11-1 we plot the Boltzmann distribution function versus energy for three different values of $T$ and $\alpha$. Note that this distribution is a pure exponential which falls by a factor of $1/e$ for each increase of $kT$ in the energy $\mathscr{E}$, as we discussed at some length in Chapter 1.

In Figure 11-2 we plot the Bose distribution function versus energy for three different values of $T$. We choose $\alpha = 0$ in each case, so that $e^{\alpha} = 1$, a case applicable to the photon gas to be discussed later. Notice that at energies small compared to $kT$ the number of particles per quantum state is greater for the Bose distribution than for the Boltzmann distribution. This is a result of the presence of the $-1$ term in the denominator of the Bose distribution law. At energies large compared to $kT$, however, the distribution approaches the exponential form characteristic of the

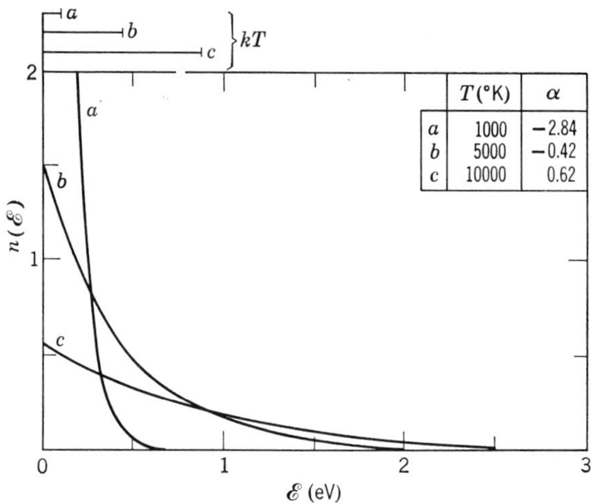


| | $T$(°K) | $\alpha$ |
|---|---|---|
| $a$ | 1000 | −2.84 |
| $b$ | 5000 | −0.42 |
| $c$ | 10000 | 0.62 |

**FIGURE 11-1**

The Boltzmann distribution function versus energy for three different values of $T$ and $\alpha$. This function is a pure exponential, falling by a factor of $1/e$ with each increase $kT$ in energy. The energy $kT$ is shown for each temperature at the top of the figure. The figure is drawn for a system of particles with the same density as that used in Figure 11-3. Choosing the density fixes $\alpha$ for any temperature $T$.

Boltzmann distribution, for in this range the exponential factor in (11-22) overwhelms the term $-1$. This is the region in which the average number of particles per quantum state is much less than one.

In Figure 11-3 we plot the Fermi distribution function versus energy for four different values of $T$ and $\alpha$. Because the exclusion principle applies here we cannot have more than one particle per quantum state. This accounts for the distinctly

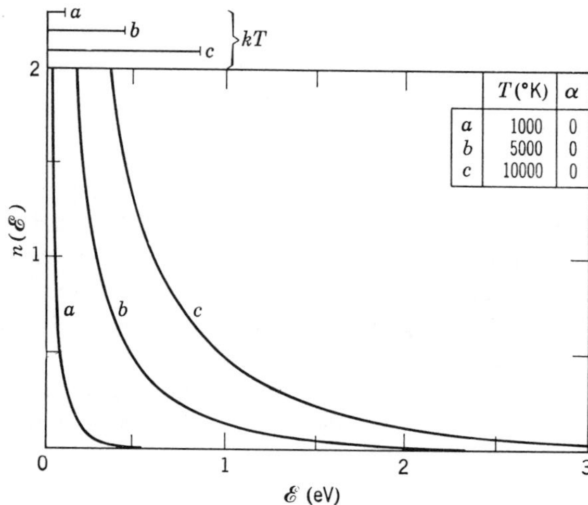


| | $T$(°K) | $\alpha$ |
|---|---|---|
| $a$ | 1000 | 0 |
| $b$ | 5000 | 0 |
| $c$ | 10000 | 0 |

**FIGURE 11-2**

The Bose distribution function versus energy for three different values of $T$, all with $\alpha = 0$. At energies large compared to $kT$ this function approaches the exponential form of the Boltzmann distribution, but at energies small compared to $kT$ it exceeds the Boltzmann values, tending to infinity as the energy goes to zero. The energy $kT$ for each temperature is shown at the top of the figure.

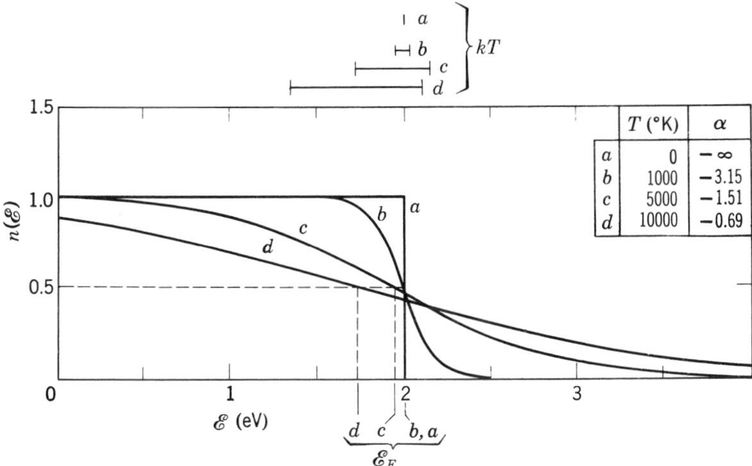


**FIGURE 11-3**

The Fermi distribution function versus energy for four different values of $T$ and $\alpha$. The exclusion principle sets the limit of one particle per quantum state. The Fermi energy $\mathscr{E}_F$ is shown for each curve at the bottom of the figure, and the energy $kT$ is shown at the top. The drop, occurring in a region of width about $kT$ centered on $\mathscr{E}_F$, becomes more gradual as the temperature increases. At high temperatures and energies, the function approaches the Boltzmann distribution function. The figure is drawn for a material with electron density similar to that of potassium, whose Fermi energy is about 2.1 eV. Choosing the density fixes the Fermi energy and, for any given $T$, fixes $\alpha$ as well.

different shape of the curves at low energies compared to the other two distributions in which there was no restriction against multiple occupancy of states. If we define the *Fermi energy* as $\mathscr{E}_F = -\alpha kT$, so that $\alpha = -\mathscr{E}_F/kT$, we can write (11-23) conveniently as

$$n_{\text{Fermi}}(\mathscr{E}) = \frac{1}{e^{(\mathscr{E}-\mathscr{E}_F)/kT} + 1} \tag{11-24}$$

This facilitates interpretation of the distribution function. For example, for states with $\mathscr{E} \ll \mathscr{E}_F$ the exponential term in the above equation is essentially zero at low temperatures and $n_{\text{Fermi}} = 1$. These states contain one fermion. For states with $\mathscr{E} \gg \mathscr{E}_F$, the exponential dominates the denominator at low temperatures and the Fermi distribution approaches the Boltzmann distribution. Note that in this region the average number of particles per quantum state is much less than one. At $\mathscr{E} = \mathscr{E}_F$, the average number of particles per quantum state is exactly one-half because of the way $\mathscr{E}_F$ is defined.

When $T = 0$, the Fermi distribution gives $n_{\text{Fermi}} = 1$ for all states with energies below $\mathscr{E}_F$ and $n_{\text{Fermi}} = 0$ for all states with energies above $\mathscr{E}_F$. Thus at $T = 0$ the lowest energy states are filled, starting from the bottom and putting one fermion in each successively higher state, until the last fermion in the system goes into the highest energy filled state at $\mathscr{E}_F$. This obviously minimizes the total energy content of the system, as would be expected at absolute zero temperature. Note from Figure 11-3 that for $T \ll \mathscr{E}_F/k$, $\mathscr{E}_F$ is at nearly the same energy as it is at for $T = 0$. At these relatively low temperatures, the thermal energy of the system has gone into promoting fermions from states of energy somewhat below the zero temperature $\mathscr{E}_F$ energy to states somewhat above that energy. The population changes are restricted to a region

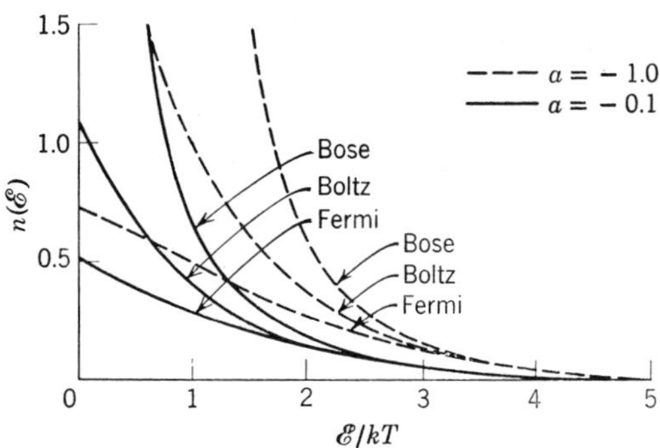


**FIGURE 11-4**
The Boltzmann, Bose, and Fermi distribution functions plotted versus $\mathscr{E}/kT$ for two different values of $\alpha$, $-0.1$ and $-1.0$. It should be noted that the dashed curves, if moved to the left $(-0.1) - (-1.0) = 0.9$ units, would coincide exactly with the solid curves. This observation may provide some further insight into the physical interpretation of $\alpha$.

of width about equal to $kT$, since $kT$ is a measure of the thermal energy content per particle of the system. The depopulation below the zero temperature $\mathscr{E}_F$ energy is quite symmetrical to the population above that energy for very small temperatures, and so $\mathscr{E}_F$, which is always the energy where $n_{\text{Fermi}} = 0.5$, hardly changes energy. For increasing temperatures, $\mathscr{E}_F$ begins to shift downward in energy as this symmetry begins to be lost.

Certain general features of the three distribution functions should be cited. At high energies ($\mathscr{E} \gg kT$) where the probable number of particles per quantum state for the

**TABLE 11-1.** Comparison of the Three Distribution Functions

| | Boltzmann | Bose | Fermi |
|---|---|---|---|
| Basic characteristic | Applies to distinguishable particles | Applies to indistinguishable particles not obeying the exclusion principle | Applies to indistinguishable particles obeying the exclusion principle |
| Example of system | Distinguishable particles, or approximation to quantum distributions at $\mathscr{E} \gg kT$ | Bosons—identical particles of zero or integral spin | Fermions—identical particles of odd half integral spin |
| Eigenfunctions of particles | No symmetry requirements | Symmetric under exchange of particle labels | Antisymmetric under exchange of particle labels |
| Distribution function | $Ae^{-\mathscr{E}/kT}$ | $\dfrac{1}{e^{\alpha}e^{\mathscr{E}/kT} - 1}$ | $\dfrac{1}{e^{(\mathscr{E}-\mathscr{E}_F)/kT} + 1}$ |
| Behavior of distribution function versus $\mathscr{E}/kT$ | Exponential | For $\mathscr{E} \gg kT$, exponential<br>For $\mathscr{E} \ll kT$, lies above Boltzmann | For $\mathscr{E} \gg kT$, exponential where $\mathscr{E} \gg \mathscr{E}_F$<br>If $\mathscr{E}_F \gg kT$, decreases abruptly near $\mathscr{E}_F$ |
| Specific problems applied to in this chapter | Gases at essentially any temperature; modes of vibration in an isothermal enclosure | Photon gas (cavity radiation); phonon gas (heat capacity); liquid Helium | Electron gas (electronic specific heat, contact potential, thermionic emission) |

classical distribution is much less than one, the quantum distributions merge with the classical distribution. That is, $n_{\text{Fermi}} \simeq n_{\text{Boltz}} \simeq n_{\text{Bose}}$, if $n_{\text{Boltz}} \ll 1$. At low energies ($\mathscr{E} \ll kT$) where this number is comparable to or larger than one, the quantum distributions fall on opposite sides of the classical distribution. That is, $n_{\text{Fermi}} < n_{\text{Boltz}} < n_{\text{Bose}}$, if $n_{\text{Boltz}} \gtrsim 1$. These features are most easily seen in Figure 11-4, which plots the three distribution functions against the energy ratio $\mathscr{E}/kT$ for the same value of $\alpha$. These features are just what would be expected from our considerations of Section 11-2. When $n_{\text{Boltz}} \ll 1$ the effects of the indistinguishability of two identical particles will have very little chance to manifest themselves because there is very little chance anyway that two particles will be in the same quantum state. So we expect the quantum distributions to join with the classical distribution for $n_{\text{Boltz}} \ll 1$. When the classical distribution predicts an appreciable probability of there being more than one particle per quantum state, i.e., when $n_{\text{Boltz}} \gtrsim 1$, then this probability will be inhibited for fermions and enhanced for bosons, and we expect the quantum distributions to diverge from the classical distribution in the manner indicated in Figure 11-4. Table 11-1 summarizes the most important attributes of the three distribution functions.

## 11-5 The Specific Heat of a Crystalline Solid

In this section we present the first of several examples of applications of the Boltzmann distribution to *quantum* systems. The specific heat of a solid was found in the early (room temperature) experiments of Dulong and Petit to be very similar for all materials, about 6 cal/mole-°K. That is, the amount of heat energy required per molecule to raise the temperature of a solid by a given amount seemed to be about the same regardless of the chemical element of which it is composed. At the time this result could be understood on the basis of the following classical statistical ideas. There are Avagadro's number, $N_0$, atoms in a mole. Each atom is regarded as executing simple harmonic oscillations about its lattice site in three dimensions, so one mole of the solid has $3N_0$ degrees of freedom. Each degree of freedom is assigned an average total energy $kT$, according to the classical law of equipartition of energy, so that

$$E = 3N_0kT = 3RT$$

where $R$ is the universal gas constant. Then, the heat capacity at constant volume is

$$c_v = \frac{dE}{dT} = 3R = 6 \text{ cal/mole-°K}$$

This is called the *law of Dulong and Petit.*

Later experiments showed conclusively, however, that as we lower the temperature the molar heat capacities vary. In fact, the specific heats of all solids tend to zero as the temperature decreases, and near absolute zero the specific heat varies as $T^3$. It was Einstein who saw that the $kT$ factor, from classical equipartition, had to be replaced by a factor that takes into account the energy quantization of a simple harmonic oscillator much as Planck had done in the cavity radiation problem. He represented a solid body as a collection of $3N_0$ simple harmonic oscillators of the same fundamental frequency and replaced $kT$ with the result $h\nu/(e^{h\nu/kT} - 1)$ of (1-26), which was obtained by combining Planck's energy quantization and the Boltzmann distribution. He thus found

$$E = \frac{3N_0h\nu}{e^{h\nu/kT} - 1} = 3RT\frac{h\nu/kT}{e^{h\nu/kT} - 1} \tag{11-25}$$

From this he calculated the specific heat as $c_v = dE/dT$ and found qualitative agreement with experiment at reasonably low temperatures. Although all substances do have curves of $c_v$ versus $T$ of the same form, we must choose a different characteristic frequency $\nu$ for each substance to match the experimental results. Furthermore, at very low temperatures the Einstein formula does not contain the $T^3$ temperature dependence required by experiment.

Peter Debye, in a general and simple way, found the theoretical approach that successfully yields the exact experimental results. Earlier treatments dealt with the individual atoms in a solid as if they vibrated independently of one another. Actually, of course, the atoms are strongly coupled together. Rather than $N_0$ atoms vibrating in three dimensions independently at the same frequency, we should deal with a system of $3N_0$ coupled vibrations. Such a dynamical problem would not only be difficult to handle directly, but, because the atoms do interact strongly, we could not use the statistics of noninteracting particles. Debye pointed out, however, that a superposition of elastic modes of longitudinal vibration of the solid as a whole—each mode independent of the others like the independent modes of two coupled pendulums —gives the same individual atom motions as the actual coupling. The temperature vibrations of the atoms of a solid are equivalent to a large combination of standing elastic waves of a great range of frequencies. The atomic vibrations of a crystal lattice appear as macroscopic elastic vibrations of the whole crystal. The problem remains to determine the frequency spectrum of the elastic modes of longitudinal vibration. Thereafter each mode can be treated as an independent harmonic oscillator, whose quantized eigenvalues we already know. Then by summing we can obtain the total energy of the system.

Before carrying out the calculation, we should point out that the Boltzmann distribution *is* applicable here. The individual atoms, in the original formulation of the problem, may be treated as *distinguishable* particles; the atoms are distinguished from one another by their *location in space* at the lattice sites of the crystal. However, the assumption of the earlier formulations that these particles do not interact clearly is wrong. In the Debye model, the atoms are replaced by elastic modes of vibration of the solid as a whole. These are independent, noninteracting elements—independent harmonic oscillators. These elements, furthermore, *are* distinguishable from one another, for each mode of vibration (standing wave) is characterized by a different set of numbers $(n_x, n_y, n_z)$ which correspond essentially to the different number of nodes of each mode of vibration. No two modes of vibration can have identical sets of these numbers.

In order to get the frequency spectrum of the modes of vibration, Debye assumed that the solid behaved like a continuous, elastic, three-dimensional body, the allowed modes corresponding to longitudinal vibrations with nodes at the boundaries. This is identical in principle to the calculation of the modes of vibration of electromagnetic waves in a cavity, considered in Section 1-3. Thus the number of modes with frequencies between $\nu$ and $\nu + d\nu$ is

$$N(\nu)\,d\nu = \frac{4\pi V}{v^3}\nu^2\,d\nu \tag{11-26}$$

where $v$ is the speed of elastic waves and $V$ is the volume of the solid. This is identical to (1-12), except that $v$ replaces $c$, and that a factor of 2 is removed because, with longitudinal rather than transverse waves, we do not have two states of polarization. Debye further assumed that the number of modes is limited to $3N_0$ per mole, the number of translational degrees of freedom of $N_0$ atoms, to account for the actual atomic nature of a crystalline solid. The allowed modes varied in frequency then from zero to some maximum $\nu_m$. To get $\nu_m$ Debye set

$$\int_0^{\nu m} N(\nu)\, d\nu = 3N_0$$

obtaining

$$\frac{4\pi V}{3v^3}\nu_m^3 = 3N_0 \tag{11-27a}$$

or

$$\nu_m = v\left(\frac{9N_0}{4\pi V}\right)^{1/3} \tag{11-27b}$$

If now each mode is treated as a one-dimensional oscillator of average energy $\bar{\mathscr{E}}$ given by Planck's quantization and the Boltzmann distribution

$$\bar{\mathscr{E}} = \frac{h\nu}{e^{h\nu/kT} - 1}$$

the total elastic energy in the solid is then

$$E = \int_0^{\nu m} \frac{h\nu}{e^{h\nu/kT} - 1}\, \frac{4\pi V}{v^3}\nu^2\, d\nu \tag{11-28}$$

This expression can be put in a more compact form if we change to a dimensionless variable of integration $x = h\nu/kT$ so that $x_m = h\nu_m/kT$. Then

$$E = \frac{4\pi V}{v^3}\int_0^{\nu m} \frac{h\nu^3\, d\nu}{e^{h\nu/kT} - 1} = \frac{4\pi V}{v^3}\left(\frac{kT}{h}\right)^4 h\int_0^{xm} \frac{x^3\, dx}{e^x - 1}$$

and, after substituting $4\pi V/v^3 = 9N_0/\nu_m^3$ and consolidating symbols, we obtain

$$E = 3RT\,\frac{3}{x_m^3}\int_0^{xm} \frac{x^3\, dx}{e^x - 1} \tag{11-29}$$

which is Debye's formula.

Because $x$ is a dimensionless quantity, $h\nu_m/k$ has the dimensions of a temperature. It is often called the *Debye characteristic temperature*, $\Theta$, of the substance involved. Hence, with $x_m = \Theta/T$, (11-29) becomes

$$E = 9R\,\frac{T^4}{\Theta^3}\int_0^{\Theta/T} \frac{x^3}{e^x - 1}\, dx \tag{11-30}$$

and *Debye's formula for the specific heat of a solid is*

$$c_v = \frac{dE}{dT} = 9R\left[4\left(\frac{T}{\Theta}\right)^3 \int_0^{\Theta/T} \frac{x^3}{e^x - 1}\, dx - \frac{\Theta}{T}\,\frac{1}{e^{\Theta/T} - 1}\right] \tag{11-31}$$

Debye's theory involves a parameter $\Theta$ which, because of its connection to the elastic properties of the solid, can be determined independently of specific heat measurements, as we shall see in Example 11-2. Using these independently determined values in the theory, we obtain the excellent agreement with experimental measurements of specific heat illustrated in Figure 11-5. In particular, the theory agrees with the observed $T^3$ law at very low temperatures.

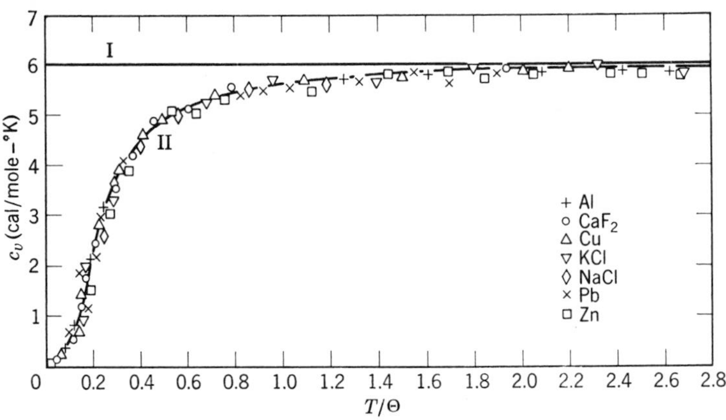


**FIGURE 11-5**
The measured specific heat at constant volume, as a function of temperature, for several materials. Horizontal line I represents the Dulong-Petit law, and curve II represents the predictions of the Debye theory.

**Example 11-2.** (a) Show how $\Theta$ can be obtained directly from the elastic properties of a solid.

Because $\Theta = h\nu_m/k$ we must find $\nu_m$ first. From (11-27b), $\nu_m = v\,(9N_0/4\pi V)^{1/3}$ so we have $\Theta = (hv/k)(9N_0/4\pi V)^{1/3}$. All quantities are measurable experimentally so that $\Theta$ can be found from measurements of $V$ (the molar volume) and $v$ (the speed of elastic waves).

Actually, since both longitudinal (compressional) and transverse (shear) waves can be transmitted by the solid, and since their speeds are different, we replace $v$ by a more general expression. In particular, if $v_l$ represents the speed of longitudinal waves and $v_t$ the speed of transverse waves in the solid, we require $3N_0 = (4\pi V/3v_l^3)\nu_m^3 + (4\pi V/3v_t^3)2\nu_m^3$ instead of (11-27a), where allowance is now made for the two polarization states of transverse waves, as well. Then we use in (11-27b)

$$\frac{1}{v^3} = \left(\frac{1}{v_l^3} + \frac{2}{v_t^3}\right)$$

From measurements of $v_l$ and $v_t$,$v$ is computed. Some calculated results for $\Theta$ and $\nu_m$ are:

| | | |
|---|---|---|
| Iron | $\Theta = 465°$K | $\nu_m = 9.7 \times 10^{12}\ \text{sec}^{-1}$ |
| Aluminum | $\Theta = 395°$K | $\nu_m = 8.3 \times 10^{12}\ \text{sec}^{-1}$ |
| Silver | $\Theta = 210°$K | $\nu_m = 4.4 \times 10^{12}\ \text{sec}^{-1}$ |

(b) Show that as $T \to 0$, $c_v \to \text{const} \times T^3$ in Debye's (11-31).

We have

$$c_v = 9R\left[4\left(\frac{T}{\Theta}\right)^3 \int_0^{\Theta/T} \frac{x^3}{e^x - 1}\,dx - \frac{\Theta}{T}\frac{1}{e^{\Theta/T} - 1}\right]$$

As $T$ decreases, $\Theta/T$ becomes very large. Indeed, as $T \to 0$, $\Theta/T \to \infty$, and the last term goes to zero. Hence

$$c_v \to 9R4\left(\frac{T}{\Theta}\right)^3 \int_0^{\infty} \frac{x^3}{e^x - 1}\,dx$$

which, because $\int_0^\infty x^3/(e^x - 1)\,dx = \pi^4/15$, yields

$$c_v = \frac{12\pi^4}{5}\frac{R}{\Theta^3}T^3$$

the required $T^3$ law for very low temperatures.

(c) Show how $\Theta$ can be obtained from specific heat measurements.
If $T = \Theta$, then from (11-31) we have

$$c_v = 9R\left[4\int_0^1 \frac{x^3}{e^x - 1}\,dx - \frac{1}{e-1}\right] = 2.856R = 5.67 \text{ cal/mole-°K}$$

so that the Debye temperature $\Theta$ can be defined as that temperature at which $c_v = 2.856R$. For comparison with part (a), the values so obtained are 455°K for iron, 420°K for aluminum, and 215°K for silver. ◀

It is remarkable that so simple a model as Debye's yields such excellent results. The true frequency spectrum of the modes of vibration should depend on the actual lattice structure of the crystalline solid and may differ from the results of Debye's continuum model. Such differences as have been found between experiment and Debye's predictions can indeed be accounted for by expected differences between the actual spectrum and Debye's so that the experimental facts of the specific heats of solids seem to be completely understood. Here we have considered the contributions to the specific heat of a solid from the lattice vibrations alone. In Section 11-11 we shall consider the contribution made by free electrons to the specific heat of a solid conductor.

## 11-6 The Boltzmann Distribution as an Approximation to Quantum Distributions

We have seen that where the average number of particles per quantum state is much less than one, the quantum distributions merge with the classical distribution. Particularly useful in this region is the *Boltzmann factor*

$$\frac{n_{\text{Boltz}}(\mathscr{E}_2)}{n_{\text{Boltz}}(\mathscr{E}_1)} = e^{-(\mathscr{E}_2 - \mathscr{E}_1)/kT} \tag{11-32}$$

giving the relative number of particles per quantum state at two different energies $\mathscr{E}_2$ and $\mathscr{E}_1$, for a system in equilibrium at temperature $T$. We have already made use of this approximation in Example 4-7. It can be applied in all quantum systems at energies more than several $kT$ above the ground state—the states are sparsely occupied so that $n_{\text{Boltz}}$ is very much less than one. For example, when we consider thermal collisions of atoms in a gas in equilibrium at temperature $T$ the excited states of the atoms are normally sparsely populated. Hence we can obtain the relative equilibrium populations of the various excited states as a function of temperature by using the Boltzmann factor. Since the intensities of the spectral lines depend on these populations, we can then predict the variation of spectral intensities with temperature. More often the procedure is reversed; that is, starting with the known relative intensities we can determine the temperature of the source, such as the star considered in Example 4-7. The same idea is applicable to molecular spectra, as we shall see in Chapter 12.

The Maxwell distribution of speeds of gas molecules moving freely inside a box is validly deduced from the Boltzmann distribution because $n_{\text{Boltz}}$ for all the free particle states is very small under the conditions usually existing in nature for ordinary gases.

**Example 11-3.** The technique of *nuclear magnetic resonance* is used to obtain information about internal magnetic fields in solids. It is more sensitive than chemical techniques, for example, in identifying magnetic impurities in a crystal. Principally, however, it enables us to use the nucleus as a probe to get information about solids, much as radioactive tracers are used in biological systems. For nuclei of nonzero spin the degeneracy of the energy levels with respect to the orientation of the nuclear spin is lifted by the magnetic field. (This is analogous to the Zeeman effect.) A resonance absorption of electromagnetic power occurs when photons bombarding the solid have the proper energy to excite transitions between these levels. The strength of the absorption depends upon the difference in population of the levels involved. To illustrate the sensitivity of the technique, use the Boltzmann factor to compute the difference between the populations $n_1$ and $n_2$ of two levels at room temperature, if the resonant absorption is detected at a frequency of 10 MHz.

The Boltzmann factor is

$$\frac{n_{\text{Boltz}}(\mathscr{E}_2)}{n_{\text{Boltz}}(\mathscr{E}_1)} = e^{-(\mathscr{E}_2-\mathscr{E}_1)/kT} = \frac{n_2}{n_1}$$

We have $T = 300°\text{K}$, $\mathscr{E}_2 - \mathscr{E}_1 = h\nu$, and $\nu = 10^7 \text{ sec}^{-1}$. Hence

$$\frac{n_2}{n_1} = e^{-h\nu/kT}$$

$$= e^{-6.6\times10^{-34}\text{joule-sec}\times10^7\text{sec}^{-1}/1.4\times10^{-23}\text{joule-}°\text{K}^{-1}\times300°\text{K}}$$

$$= e^{-1.6\times10^{-6}} \simeq 1 - 1.6 \times 10^{-6}$$

Therefore

$$1 - \frac{n_2}{n_1} = 1.6 \times 10^{-6}$$

or

$$\frac{n_1 - n_2}{n_1} = 1.6 \times 10^{-6}$$

So a difference in populations of less than two parts in a million is detectable, a result which reveals the high sensitivity of the NMR technique. The Boltzmann factor is applicable here since the population is spread over several close levels, so both $n_1$ and $n_2$ are small. ◀

## 11-7 The Laser

We saw in the previous section that the relative number of particles per quantum state at two different energies for a system in thermal equilibrium at temperature $T$ is given, in certain circumstances, by the Boltzmann factor, $e^{-(\mathscr{E}_2-\mathscr{E}_1)/kT}$. We use this result now to explain the behavior of a very important device called a *laser*, an acronym for "*l*ight *a*mplification by *s*timulated *e*mission of *r*adiation." A *maser* is the corresponding system operating in the *m*icrowave region of the electromagnetic spectrum.

Consider transitions between two energy states of an atom in the presence of an electromagnetic field. In Figure 11-6 we illustrate schematically the three transition processes, namely, *spontaneous emission*, *stimulated absorption*, and *stimulated emission*. In the spontaneous emission process, the atom is initially in the upper state of energy $\mathscr{E}_2$ and decays to the lower state of energy $\mathscr{E}_1$ by the emission of a photon of frequency $\nu = (\mathscr{E}_2 - \mathscr{E}_1)/h$. (The mean lifetime of an atom in most excited states is about $10^{-8}$ sec. But some decays may be much slower, the excited states then being called metastable; the mean lifetime in such cases may be as long as $10^{-3}$ sec.) In the stimulated absorption process, an incident photon of frequency $\nu$, from an electromagnetic field applied to the atom, stimulates the atom to make a transition from the lower to the higher energy state, the photon being absorbed by the atom. In the stimulated emission process, an incident photon of frequency $\nu$ stimulates the atom to make a transition from the higher to the lower energy state; the atom is left in this lower state as two photons of the same frequency, the incident one and the emitted one, emerge.

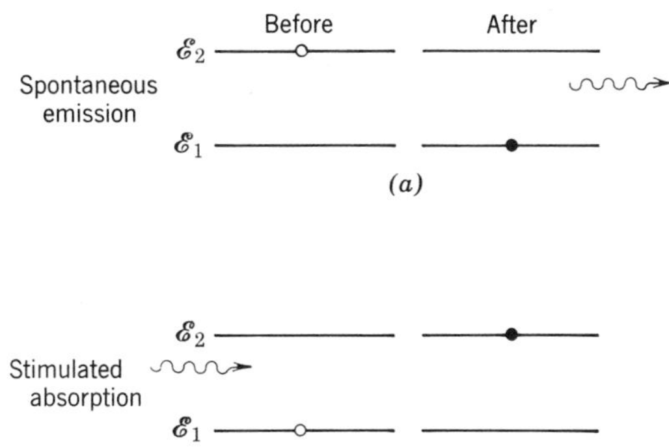


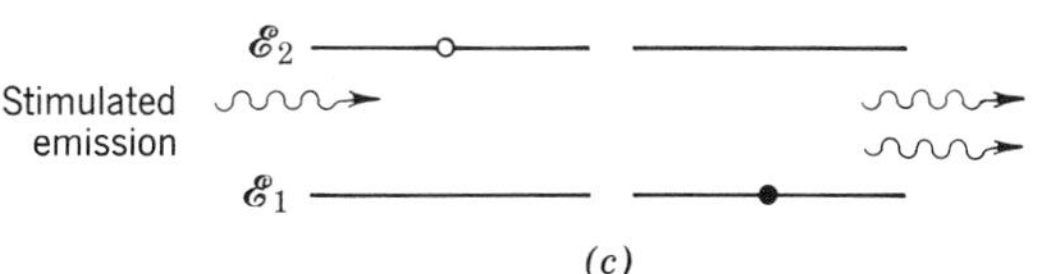


**FIGURE 11-6**

Illustrating (*a*) the spontaneous emission process, (*b*) the stimulated absorption process, and (*c*) the stimulated emission process for two energy states of an atom.

The processes of stimulated absorption and emission of electromagnetic energy in quantized systems can be regarded as analogous to the stimulated absorption or emission of mechanical energy in classical resonating systems upon which a periodic mechanical force of the same frequency as the natural frequency of the system is impressed. In such a mechanical system, energy can be put in or taken out depending on the relative phases of motion of the system and the impressed force. The spontaneous emission process, however, is a strictly quantum effect. As discussed in Section 8-7, quantum electrodynamics shows that there are fluctuations in the electromagnetic field. Because of the zero-point energy of the electromagnetic field, these fluctuations occur even when classically there is no field. It is these fluctuations that induce the so-called spontaneous emission of radiation from atoms in excited states. In all three processes, then, we deal with the interaction of radiation with the atom.

We wish to show now how these processes are related quantitatively. Let the spectral energy density of the electromagnetic radiation applied to the atoms be $\rho(\nu)$. Consider that there are $n_1$ atoms in energy state $\mathscr{E}_1$ and $n_2$ in state $\mathscr{E}_2$, where $\mathscr{E}_2 > \mathscr{E}_1$. The probability per atom per unit time, or transition rate per atom, that an atom in state 1 will undergo a transition to state 2 (stimulated absorption) clearly will be proportional to the energy density $\rho(\nu)$ of the applied radiation at frequency $\nu = (\mathscr{E}_2 - \mathscr{E}_1)/h$. In Section 8-7 we argued that the transition rate for stimulated emission is also proportional to $\rho(\nu)$. But as we explained in Section 8-7, the transition rate for spontaneous emission does not contain $\rho(\nu)$ because that process does not involve the applied electromagnetic field.

The transition rates also depend on the detailed properties of the atomic states 1 and 2 through the electric dipole moment matrix element of (8-42). Hence, the probability per unit time for a transition from state 1 to state 2 can be written as

$$R_{1\to 2} = B_{12}\rho(\nu) \tag{11-33}$$

in which $B_{12}$ is a coefficient that includes the dependence on properties of the states 1 and 2. The total probability per unit time that an atom in state 2 will undergo a transition to state 1 is the sum of two terms, the probability per unit time $A_{21}$ of spontaneous emission and the probability per unit time $B_{21}\rho(\nu)$ of stimulated emission.

Again, $A_{21}$ and $B_{21}$ are coefficients whose values depend on the properties of states 1 and 2, through the appropriate matrix elements. Hence

$$R_{2\to 1} = A_{21} + B_{21}\rho(\nu) \tag{11-34}$$

Note again that spontaneous emission occurs at a rate independent of $\rho(\nu)$, whereas stimulated emission occurs at a rate proportional to $\rho(\nu)$.

If now we consider that the $n_1$ atoms in state 1 and the $n_2$ atoms in state 2 of the system are in thermal equilibrium at temperature $T$ with the radiation field of energy density $\rho(\nu)$, then the total absorption rate for the system $n_1R_{1\to 2}$ and the total emission rate $n_2R_{2\to 1}$ must be equal, as in (11-4). That is

$$n_1R_{1\to 2} = n_2R_{2\to 1} \tag{11-35}$$

Thus we have

$$n_1B_{12}\rho(\nu) = n_2[A_{21} + B_{21}\rho(\nu)]$$

If we solve this equation for $\rho(\nu)$ we obtain

$$\rho(\nu) = \frac{\dfrac{A_{21}}{B_{21}}}{\dfrac{n_1}{n_2}\dfrac{B_{12}}{B_{21}} - 1} \tag{11-36}$$

We now assume we can use the Boltzmann factor, (11-32), with $h\nu = \mathscr{E}_2 - \mathscr{E}_1$, to obtain

$$\frac{n_1}{n_2} = e^{(\mathscr{E}_2-\mathscr{E}_1)/kT} = e^{h\nu/kT}$$

so that (11-36) becomes

$$\rho(\nu) = \frac{\dfrac{A_{21}}{B_{21}}}{\dfrac{B_{12}}{B_{21}}e^{h\nu/kT} - 1} \tag{11-37}$$

This equation, giving the spectral energy density of radiation of frequency $\nu$ that is in thermal equilibrium at temperature $T$ with atoms of energies $\mathscr{E}_1$ and $\mathscr{E}_2$, must be consistent with Planck's blackbody spectrum, (1-27)

$$\rho_T(\nu) = \frac{8\pi h\nu^3}{c^3}\left(\frac{1}{e^{h\nu/kT} - 1}\right)$$

Hence, we conclude that

$$\frac{B_{12}}{B_{21}} = 1 \tag{11-38}$$

and

$$\frac{A_{21}}{B_{21}} = \frac{8\pi h\nu^3}{c^3} \tag{11-39}$$

These results were first obtained by Einstein in 1917, and therefore the coefficients are called the *Einstein A and B coefficients*. Note that the argument does not give us values of the coefficients, but only their ratios. However, if we compute the spontaneous emission coefficient $A_{21}$ from quantum mechanics, using the techniques of Section 8-7, we then can obtain the other coefficients from these formulas.

There is much of physical interest here. For one thing, we find from (11-38) that the coefficients of stimulated emission and stimulated absorption are equal. For another, we see from (11-39) that the ratio of the spontaneous emission coefficient to the stimulated emission coefficient varies with frequency as $\nu^3$. This means, for example, that the bigger the energy difference between the two states, the much more likely is spontaneous emission compared to stimulated emission. Equation (8-43) shows that the $\nu^3$ is present in this ratio because $A_{21}$ itself is proportional to $\nu^3$. Still another result is that we can obtain the ratio of the probability $A_{21}$ of spontaneous emission to the probability $B_{21}\rho(\nu)$ of stimulated emission, namely

$$\frac{A_{21}}{B_{21}\rho(\nu)} = e^{h\nu/kT} - 1 \tag{11-40}$$

This shows that, for atoms in thermal equilibrium with the radiation, spontaneous emission is far more probable than stimulated emission if $h\nu \gg kT$. Since this condition applies to electronic transitions in both atoms and molecules, stimulated emission can be ignored in such transitions. Stimulated emission can become significant, however, if $h\nu \simeq kT$, and it may be dominant if $h\nu \ll kT$, a condition that applies at room temperature to atomic transitions in the microwave region of the spectrum where $\nu$ is relatively small.

We are now in a position to understand the concept behind lasers and masers. In general, the ratio of the emission rate to the absorption rate can be written as $n_2R_{2\to1}/n_1R_{1\to2}$ or

$$\frac{\text{rate of emission}}{\text{rate of absorption}} = \frac{n_2A_{21} + n_2B_{21}\rho(\nu)}{n_1B_{12}\rho(\nu)}$$

$$= \left[1 + \frac{A_{21}}{B_{21}\rho(\nu)}\right]\frac{n_2}{n_1} \tag{11-41}$$

If we have energy states such that $\mathscr{E}_2 - \mathscr{E}_1 \ll kT$, or $h\nu \ll kT$, then (11-40) shows that we can ignore the second term in the parenthesis as very much smaller than one, and obtain

$$\frac{\text{rate of emission}}{\text{rate of absorption}} \simeq \frac{n_2}{n_1} \tag{11-42}$$

This result is general in the sense that we have not assumed an equilibrium situation. In situations of thermal equilibrium, where the Boltzmann factor applies, we expect $n_2 < n_1$. But in nonequilibrium situations any ratio is possible in principle. If now we have a means of inverting the normal population of states so that $n_2 > n_1$, then the emission would exceed the absorption rate. This means that the applied radiation of frequency $\nu = (\mathscr{E}_2 - \mathscr{E}_1)/h$ will be *amplified* in intensity by the interaction process, more such radiation emerging than entering. Of course, such a process will reduce the population of the upper state until equilibrium is reestablished. In order to sustain the process, therefore, we must use some method to maintain the *population inversion* of the states. Devices that do this are called lasers or masers, depending upon the portion of the electromagnetic spectrum in which they operate. Energy must be injected into the system, most commonly by a method described later called *optical pumping*, and the output is an intense, coherent, monochromatic beam of radiation, as we now explain.

In the ordinary atomic light sources there is a random relationship between the phases of the photons emitted by different atoms so that the resulting radiation is incoherent. The reason is that there is no correlation in the times that the atoms make their transitions. In laser light sources, on the other hand, atoms radiate in phase with the inducing radiation because their charge oscillations are in phase with that radiation. Since in a laser the inducing radiation is a coherent parallel beam formed by reflection between the ends of a resonant cell, the emitted photons are all in phase and act *coherently*. The resulting intensity, which is the square of the constructively combined amplitudes, is correspondingly high. The states between which transitions are made are an upper metastable state, whose relatively long lifetime allows it to be highly populated, and the lower ground state of infinitely long lifetime. From the uncertainty relation $\Delta E \, \Delta t \simeq h$, with $\Delta t$ equal to the long lifetime of the upper state, we conclude that the energy uncertainty in the energy difference of the states is small and the emitted transition frequency is sharp, giving a highly *monochromatic* beam. In practical devices the beam is also *unidirectional*, the coherence property making it possible to obtain essentially perfect collimation, or focusing. This further enhances the concentration of energy density. Some indication of the concentration of energy in a laser beam is given by the fact that a laser with less power than a typical light bulb can burn a hole in a metal plate.

In the solid state laser that operates with a ruby crystal, some Al atoms in the $Al_2O_3$ molecules are replaced by Cr atoms. These "impurity" chromium atoms account for the laser action. In Figure 11-7 we show a simplified version of the appropriate energy-level scheme of chromium. (The uppermost level is really a multiplet.) The level of energy $\mathscr{E}_1$ is the ground state and the level of energy $\mathscr{E}_3$ is the unstable upper state with a short lifetime ($\simeq 10^{-8}$ sec), the energy difference $\mathscr{E}_3 - \mathscr{E}_1$ corresponding to a wavelength of about 5500 Å. Level $\mathscr{E}_2$ is an intermediate excited state which is metastable, its lifetime against spontaneous decay being about $3 \times 10^{-3}$ sec. If the chromium atoms are in thermal equilibrium, the population numbers of the states are such that $n_3 < n_2 < n_1$. By pumping in radiation of wavelength 5500 Å, however, we stimulate absorption of incoming photons by Cr atoms in the ground state, thereby raising the population number of energy state $\mathscr{E}_3$ and depleting energy state $\mathscr{E}_1$ of occupants. Spontaneous emission, bringing atoms from state 3 to state 2, then enhances the occupancy of state 2, which is relatively long-lived. The result of this optical pumping is to decrease $n_1$ and increase $n_2$, such that $n_2 > n_1$ and population inversion exists. Now, when an atom does make a transition from state 2 to state 1, the emitted photon of wavelength 6943 Å will stimulate further transitions. Stimulated emission will dominate stimulated absorption (because $n_2 > n_1$) and the output of photons of wavelength 6943 Å is much enhanced. We obtain an intensified coherent monochromatic beam.

In practice, the ruby laser is a cylindrical rod with parallel, optically flat reflecting ends, one of which is only partly reflecting as shown in Figure 11-7. The emitted photons that do not travel along the axis escape through the sides before they are able to cause much stimulated emission. But those photons that move exactly in the direction of the axis are reflected several times, and they are capable of stimulating emission repeatedly. Thus the number of photons is built up rapidly, those escaping from the partially reflecting end giving a unidirectional beam of great intensity and sharply defined wavelength.

Note that this is reminiscent of the conclusion of Section 11-2 that $n$ bosons already in a quantum state will enhance the probability of one more joining them by a factor of $(1 + n)$. The conclusion is applicable to the photons in the quantum states of the cylindrical rod, since photons are bosons. It is possible to develop the basic theory of the laser by applying the Bose distribution to the quantum states of the photons, instead of by applying the Boltzmann distribution to the quantum states of the atoms as we have done here. But the treatments are very closely related (as they should be since they lead to the same results) because the energy density $\rho(\nu)$ of (11-34) is proportional to the number $n$ of photons in a state at energy $h\nu$ so that equation is very

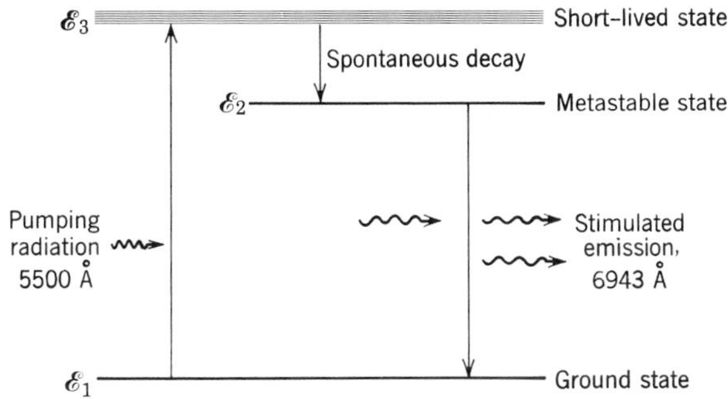


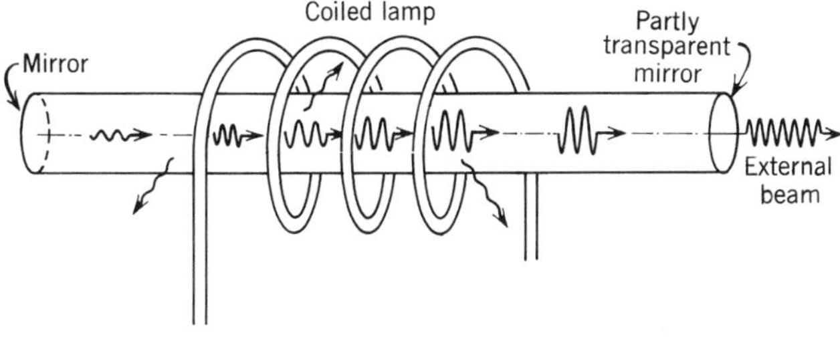


**FIGURE 11-7**

*Top:* The relevant energy levels of chromium atoms in a ruby laser. State 3 is very broad (large $\Delta E$) because it is short lived (small $\Delta t$). State 2 is very sharp (small $\Delta E$) because it is long lived (large $\Delta t$). Optical pumping raises the atom from ground state 1 to excited state 3, the latter's breadth facilitating the process. Then spontaneous decay occurs to state 2, the energy released usually going into mechanical energy in the ruby crystal rather than into photon radiation. Finally, state 2 decays to the ground state, either through spontaneous emission or through stimulated emission due to photons from other such transitions. Since state 2 is very sharply defined and the ground state is infinitely sharply defined, this radiation will be very monochromatic. *Bottom:* A schematic of the ruby laser, showing the optical pumping lamp, the escape of photons not moving axially, suggesting the buildup of repeatedly reflected axially moving photons which stimulate further emission, and indicating the escape of a fraction of the axial photons through the partially reflecting mirror at one end.

similar to the enhancement equations, (11-10) or (11-11), that we used in deriving the Bose distribution. Furthermore, (11-35) is identical to the thermal equilibrium condition of (11-4) that was also used in the Bose distribution derivation.

Generally speaking, a laser is a device in which a material is prepared so that the higher of two energy levels is more highly populated than the lower energy level, the material being enclosed in an appropriate resonator of sharp response. The system produces coherent radiation at those frequencies common to the resonator and the difference in energy of the levels. There is now a wide variety of lasers—gas lasers, liquid lasers, and solid state lasers—covering various regions of the electromagnetic spectrum. The intense coherent nature of the radiation they provide has led to increasing application of lasers in fields such as radio astronomy, microwave spectroscopy, photography, biophysics, and communications.

## 11-8 The Photon Gas

We begin in this section to study applications of the Bose distribution. The first will be a derivation of Planck's blackbody cavity radiation spectrum, in which the photons in thermal equilibrium at temperature $T$ with the walls of the cavity are treated as a gas that is governed by the Bose distribution. According to (11-22), that distribution is

$$n(\mathscr{E}) = \frac{1}{e^{\alpha}e^{\mathscr{E}/kT} - 1}$$

The discussion following (11-22) indicated that the value of the parameter $\alpha$ is specified by the total number of particles the system governed by the distribution contains. But for the case at hand the total number of particles in the system is not constant. A photon can be completely absorbed when it strikes a wall of the cavity, or the hot wall may at some other time emit a new photon. Thus for a system of photons the distribution cannot contain the term $e^{\alpha}$. That is, the *Bose distribution for photons* (or other bosons that can be created or destroyed within the system) must have the form

$$n(\mathscr{E}) = \frac{1}{e^{\mathscr{E}/kT} - 1} \tag{11-43}$$

The number of particles in the system has indeed specified the value of $\alpha$; because that number varies it is necessary that $\alpha = 0$ so that $e^{\alpha} = 1$. Confirmation of the validity of this argument will be obtained soon.

If $N(\mathscr{E})\, d\mathscr{E}$ is the number of quantum states for photons in the cavity within the energy interval $\mathscr{E}$ to $\mathscr{E} + d\mathscr{E}$, then since $n(\mathscr{E})$ is the probable number of photons per quantum state their product $n(\mathscr{E})N(\mathscr{E})\, d\mathscr{E}$ gives the number of photons in the energy interval. However, $N(\mathscr{E})\, d\mathscr{E}$ for radiation confined to a cavity has already been evaluated by geometrical arguments in Example 1-3, except that the language used there is different from that which we are currently using; there we spoke of the radiation as waves and here we speak of it as particles (photons). We found there that

$$N(\nu)\, d\nu = \frac{8\pi V}{c^3}\nu^2\, d\nu$$

where $V$ is the volume of the cavity and $\nu$ is the frequency of a wave contained in the cavity. Using the familiar relation $\mathscr{E} = h\nu$ to evaluate the energy of the associated photon, here we find, after multiplying and dividing the term $\nu^2\, d\nu$ by $h^3$, that

$$N(\mathscr{E})\, d\mathscr{E} = \frac{8\pi V}{c^3}\frac{\mathscr{E}^2\, d\mathscr{E}}{h^3} \tag{11-44}$$

Taking the product of this expression times $n(\mathscr{E})$, multiplying by the energy $\mathscr{E}$ carried by each photon, and then dividing by the volume $V$ of the cavity, we have

$$\rho_T(\mathscr{E})\, d\mathscr{E} = \frac{\mathscr{E}n(\mathscr{E})N(\mathscr{E})\, d\mathscr{E}}{V} = \frac{8\pi\mathscr{E}^3\, d\mathscr{E}}{c^3h^3(e^{\mathscr{E}/kT} - 1)}$$

where $\rho_T(\mathscr{E})\, d\mathscr{E}$ is the energy per unit volume in the energy interval $\mathscr{E}$ to $\mathscr{E} + d\mathscr{E}$. Planck's spectrum follows at once by using the relation $\mathscr{E} = h\nu$ to convert from $\mathscr{E}$ to $\nu$. Thus

$$\rho_T(\nu)\, d\nu = \frac{8\pi\nu^2}{c^3}\frac{h\nu}{e^{h\nu/kT} - 1}h\nu \tag{11-45}$$

Equation (11-45) is identical to (1-27), obtained in Chapter 1 and verified there by comparison with experiment. Note that this agreement confirms the validity of the Bose distribution for photons, (11-43). In the Planck derivation the radiation is a set of waves confined to the cavity. Each of these standing waves is a mode of vibration that is distinguishable from all the others, just as for the lattice vibration modes in the Debye model, so it is valid to apply the Boltzmann distribution to them. In the present derivation the cavity radiation is a set of indistinguishable particles—photons to which the Bose distribution must be applied.

## 11-9 The Phonon Gas

We were able to use the wave-particle duality for electromagnetic radiation to derive the thermally excited distribution of radiation in a cavity either on a wave picture or a particle picture. Similarly, the thermally excited distribution of elastic vibrations in a solid can be deduced by applying a wave-particle duality for acoustic radiation. Just as photons are the quanta of electromagnetic radiation, so *phonons* are the quanta of *acoustic radiation*. Just as photons are emitted and absorbed by vibrations of the atoms in a cavity wall, so phonons are emitted and absorbed by vibrating atoms at the lattice points in the solid. The sources of each type of radiation are quantized so that the energy gain or loss is discrete; the discrete energy transferred through the system has an energy $h\nu$, where $\nu$ is the frequency of the acoustic vibration for phonons and of the electromagnetic vibration for photons. Just as the number of photons is not fixed or conserved, so the number of phonons is not fixed or conserved. The Bose distribution with $\alpha = 0$, i.e., (11-43), applies to phonon and to photon. There are differences, of course between the photons and phonons. For example, the photon propagates through vacuum whereas the phonon propagates through a crystal lattice. This leads to different energy-momentum relations, a matter we return to in a subsequent chapter.

It should be clear that the Debye specific heat formula can be deduced on the phonon picture from the Bose distribution in a way analogous to the photon deduction of the Planck spectrum formula using the Bose distribution. That is, the wave-particle duality for acoustic radiation is used just as before we used the wave-particle duality for electromagnetic radiation. The phonon calculation will not be reproduced here because it is completely analogous to the photon calculation and leads to no new results. The solid contains a *gas of phonons* just as the cavity contained a gas of photons.

## 11-10 Bose Condensation and Liquid Helium

Here we sketch an application of the Bose distribution to an ideal gas in order to compare quantum and classical gas behavior. As a practical application we shall then consider the remarkable properties of liquid helium.

The general form of the Bose distribution is

$$n(\mathscr{E}) = \frac{1}{e^{\alpha}e^{\mathscr{E}/kT} - 1} \tag{11-46}$$

To apply this to bosons whose total number $\mathcal{N}$ in a system remains fixed, like helium atoms, we must first determine the parameter $\alpha$. This is done by setting

$$\mathcal{N} = \int_0^\infty n(\mathscr{E})N(\mathscr{E})\, d\mathscr{E}$$

where $N(\mathscr{E})\,d\mathscr{E}$ is the number of quantum states of the system in an energy interval $\mathscr{E}$ to $\mathscr{E} + d\mathscr{E}$, and $n(\mathscr{E})$ is the number of bosons per quantum state, so that the integral is just the total number $\mathscr{N}$. Using (11-46), we have

$$\mathscr{N} = \int_0^\infty \frac{N(\mathscr{E})\,d\mathscr{E}}{e^\alpha e^{\mathscr{E}/kT} - 1} \tag{11-47}$$

To proceed we must determine $N(\mathscr{E})\,d\mathscr{E}$, the number of states in the energy interval $\mathscr{E}$ to $\mathscr{E} + d\mathscr{E}$, for an ideal gas. Consider the gas particles to be in a cubical box of side $a$. The potential energy for a particle in such a three-dimensional box is that of a three-dimensional infinite square well. The Schroedinger equation for a one-dimensional infinite square well was solved in Section 6-8, giving allowed energies $\mathscr{E}_n = (h^2/8ma^2)n^2$. By a simple extension of the calculation we find the allowed energies $\mathscr{E}$ for a three-dimensional well to be

$$\mathscr{E} = \frac{h^2}{8ma^2}(n_x^2 + n_y^2 + n_z^2) \tag{11-48}$$

in which the quantum numbers $n_x$, $n_y$, $n_z$ are positive integers. The number of states in an energy interval can be obtained by plotting, in a space formed by axes $n_x$, $n_y$, $n_z$, the allowed states (which are points where $n_x$, $n_y$, $n_z$ take on positive integral values) and counting them. We have done this, in a different context, for the calculation of Example 1-3. There we defined $r = \sqrt{n_x^2 + n_y^2 + n_z^2}$, and we found in (1-15) that the number of states for $r$ lying between $r$ and $r + dr$ is

$$N(r)\,dr = \frac{\pi r^2\,dr}{2}$$

The same is true here. We convert this into the desired form, $N(\mathscr{E})\,d\mathscr{E}$, by using (11-48) to write

$$\mathscr{E} = \frac{h^2}{8ma^2}r^2$$

and then taking this equation, and its differential, to evaluate

$$\frac{\pi r^2\,dr}{2} = \frac{\pi}{4}\left(\frac{h^2}{8ma^2}\right)^{-3/2}\mathscr{E}^{1/2}\,d\mathscr{E} = \frac{4\pi a^3}{h^3}(2m^3)^{1/2}\mathscr{E}^{1/2}\,d\mathscr{E}$$

So the number of states for $\mathscr{E}$ lying between $\mathscr{E}$ and $\mathscr{E} + d\mathscr{E}$ is

$$N(\mathscr{E})\,d\mathscr{E} = \frac{4\pi V}{h^3}(2m^3)^{1/2}\mathscr{E}^{1/2}\,d\mathscr{E} \tag{11-49}$$

where $V = a^3$, the volume of the box.

If now we combine this result with (11-47) and carry out the integration we obtain

$$\mathscr{N} = \frac{(2\pi mkT)^{3/2}V}{h^3}e^{-\alpha}\left(1 + \frac{1}{2^{3/2}}e^{-\alpha} + \frac{1}{3^{3/2}}e^{-2\alpha} + \cdots\right)$$

To simplify the appearance of this equation, let $e^{-\alpha} = A$ so that we can write

$$\mathscr{N} = \frac{(2\pi mkT)^{3/2}V}{h^3}A\left(1 + \frac{1}{2^{3/2}}A + \frac{1}{3^{3/2}}A^2 + \cdots\right) \tag{11-50}$$

For large mass $m$ and high temperature $T$, $A$ must be very small since $\mathscr{N}$ is fixed. In these circumstances, terms beyond the first power in $A$ can be dropped. But large $m$

and high $T$ should be the classical region. Indeed, we find that the first term gives the classical Boltzmann result

$$\mathcal{N} = \frac{(2\pi mkT)^{3/2}V}{h^3} A$$

or

$$A = \frac{\mathcal{N}h^3}{(2\pi mkT)^{3/2}V} = e^{-\alpha} \tag{11-51}$$

Note that $A = e^{-\alpha}$ is proportional to $\mathcal{N}$, as in the Boltzmann result for a system of classical oscillators discussed after (11-23). Also note that here we conclude that since $\mathcal{N}$ is fixed $\alpha$ must be very large (as $A$ is very small), in contrast to our conclusion that $\alpha$ is zero for a system of bosons in which $\mathcal{N}$ varies.

If we now compute the total energy $E$ of the ideal gas from

$$E = \int_0^\infty \mathscr{E} n(\mathscr{E}) N(\mathscr{E})\, d\mathscr{E}$$

we obtain

$$E = \frac{(2\pi mkT)^{3/2}}{h^3} V\left(\frac{3}{2}kT\right) A\left(1 + \frac{1}{2^{5/2}}A + \frac{1}{3^{5/2}}A^2 + \cdots\right) \tag{11-52}$$

Once again the classical result follows for very small values of $A$. Neglecting terms beyond the first power in $A$, and using (11-51), we have $E = (3/2)\mathcal{N}kT$. This corresponds to an average energy per particle $E/\mathcal{N}$ equal to $(3/2)kT$, which is the classical equipartition of energy result for three-dimensional translational motion. The general Bose result for the average energy per particle, obtained by dividing (11-52) by (11-50), is, including terms up to $A^2$

$$\bar{E} = \frac{E}{\mathcal{N}} = \frac{3}{2}kT\left[1 - \frac{1}{2^{5/2}} \frac{\mathcal{N}h^3}{V(2\pi mkT)^{3/2}}\right] \tag{11-53}$$

The term beyond 1 in the bracketed expression of (11-53) represents the deviation of the Bose gas from the classical gas. This is sometimes called *the degeneracy effect*. (This degeneracy effect, or gas degeneration, is not related to the degeneracy that describes different quantum states having the same energy.) Equation (11-53), which neglects higher order terms, pertains to the case of weak degeneracy. Note that the degeneracy term is *negative* so that the average particle energy is less for a Bose gas than for a classical gas. This corresponds to previous results in which we found a greater probability of two particles to an energy state for the Bose distribution than for the Boltzmann distribution, the lower energy states being relatively fuller in the Bose gas than in the classical gas as a consequence. Physically, this manifests itself, for example, as a lower gas pressure (lower average momentum) at the same temperature for a Bose gas than for a classical gas.

**Example 11-4.** Whenever the mean interparticle distance is comparable to or smaller than the de Broglie wavelength assigned to particles on the basis of their temperature, we should expect to observe wave effects, that is quantum effects, in the system of particles. Show that this criterion leads to the requirement that the degeneracy term $\mathcal{N}h^3/V(2\pi mkT)^{3/2}$ not be negligible compared to 1 if deviations from classical behavior are to be detected.

The de Broglie wavelength of a particle is $\lambda = h/p$. In a gas in equilibrium at temperature $T$ the mean kinetic energy is $(3/2)kT$ so that $p = \sqrt{2mK} = \sqrt{3mkT}$. Hence

$$\lambda = \frac{h}{(3mkT)^{1/2}}$$

If the volume of gas is $V$ and there are $\mathcal{N}$ atoms of gas, the volume per particle $V/\mathcal{N}$ can be set equal to $d^3$, where $d$ is the mean interatomic separation. Hence

$$d = \left(\frac{V}{\mathcal{N}}\right)^{1/3}$$

Now, if $\lambda \geq d$ we expect wave effects to be important. This requires

$$\frac{h}{(3mkT)^{1/2}} \geq \left(\frac{V}{\mathcal{N}}\right)^{1/3}$$

or, cubing each side

$$\frac{h^3}{(3mkT)^{3/2}} \geq \frac{V}{\mathcal{N}}$$

which is the same as

$$\frac{\mathcal{N}h^3}{V(3mkT)^{3/2}} \geq 1$$

Hence, $\mathcal{N}h^3/V(2\pi mkT)^{3/2}$ should exceed about 1/3 and so the term beyond 1 in the bracketed part of (11-53) should exceed about 1/16 to meet our criterion. ◀

Under what circumstances might we detect the degeneracy effect experimentally? The degeneracy term is negligible in practice for most gases, having a value of about $10^{-5}$, so that the Boltzmann distribution applies almost universally to them. Note that the degeneracy term, $\mathcal{N}h^3/V(2\pi mkT)^{3/2}$, becomes more important the smaller the mass $m$, the lower the temperature $T$, and the higher the density $\mathcal{N}/V$. The smallest mass gases obeying the Bose distribution (zero or integral spin angular momentum) are $H_2$ and He. If we prepare such a gas to be at high density and low temperature we bring it near its condensation point. For this reason, and another to be mentioned shortly, the degeneracy effect is sometimes called the *Bose condensation*. For $H_2$ the degeneracy term at its normal condensation point is less than 1/100, whereas for He near its normal condensation point (4.2°K) the degeneracy term is about 1/7. Hence, we should get observable effects more easily for helium. The theory would be approximate in this case, for at such high densities the behavior is like a real gas of interacting particles rather than an ideal gas of noninteracting particles. Indeed, in the liquid, or condensed phase, we observe the most striking nonclassical effects in the behavior of helium. Let us now describe these effects.

Ordinary helium gas is composed almost wholly of neutral atoms of the isotope $He^4$. The spin angular momentum of such an atom is zero so that the Bose distribution must be used to treat the behavior of this gas. At normal atmospheric pressure helium gas condenses to a liquid at 4.18°K. It remains as a liquid, i.e., it does not freeze into a solid, down to the absolute zero of temperature if it is cooled at a pressure equal to its own vapor pressure. (To obtain solid helium it is necessary to pressurize the liquid, about 26 atm of pressure being needed near absolute zero.) If, by pumping off the vapor, the temperature of liquid helium is reduced to 2.18°K, a dramatic change in its properties is observed. The temperature 2.18°K is called the *λ point* because the shape of the graph of specific heat versus temperature resembles the letter λ with the anomaly at 2.18°K. Liquid helium is called He I when it is above this temperature and He II when below. He I is essentially a classical fluid, its behavior not being unusual, but He II contains a *superfluid* component which causes it to show spectacular large scale quantum effects, including the following:

1. As the temperature of liquid helium is lowered by evaporation and the vapor is pumped away, the liquid boils in the usual manner. But as the λ point is reached and

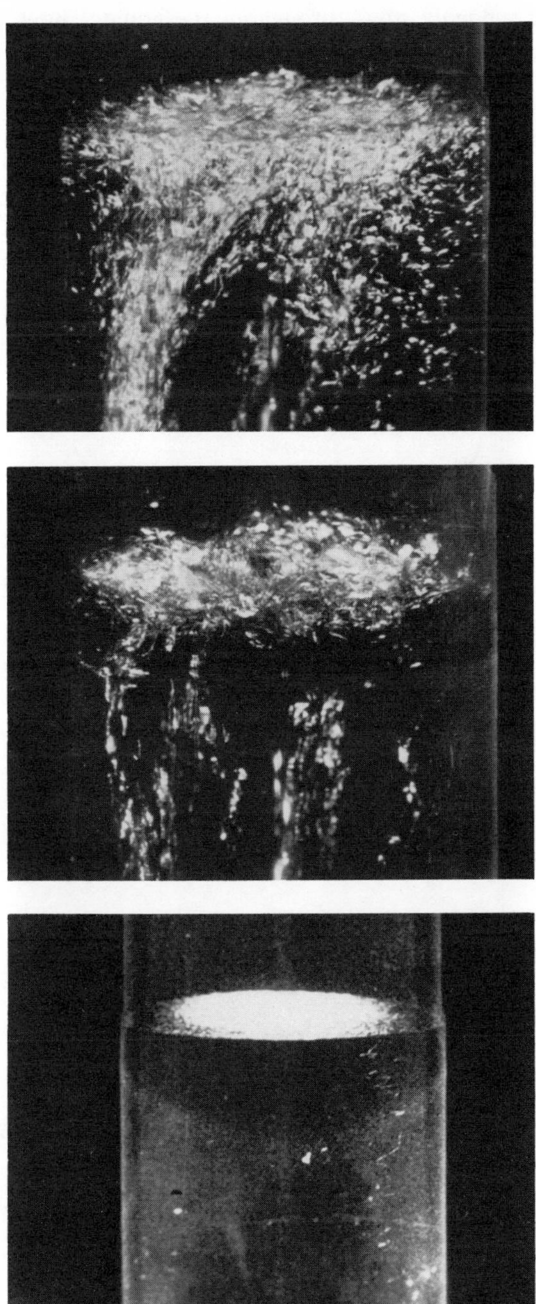

**FIGURE 11-8**

The $\lambda$ point transition in liquid helium. As liquid helium is cooled from its normal boiling point at 4.2°K by evaporation, with the use of a vacuum pump, it boils normally with small bubbles. As it undergoes the phase transition from He I to He II at the $\lambda$ point, 2.18°K, it suddenly and briefly boils up violently (see *top* and *middle* pictures), and equally suddenly stops boiling altogether (see *bottom* picture). Below this transition point liquid helium cannot boil, even when pumping, evaporation, and cooling continue. (Courtesy of A. Leitner, Rensselaer Polytechnic Institute)

passed the boiling suddenly stops throughout the liquid. Though evaporation continues, and the temperature and vapor pressure fall, the liquid is completely calm (see Figure 11-8). This is explained by the fact that heat can be conducted out of the liquid with practically no resistance, since the heat conductivity is measured to increase by a factor of about one million below the $\lambda$ point.

2. We can determine the viscosity of liquid helium by measuring its rate of flow through a fine capillary tube. At the $\lambda$ point, the measured value of the viscosity drops by a factor of about one million.

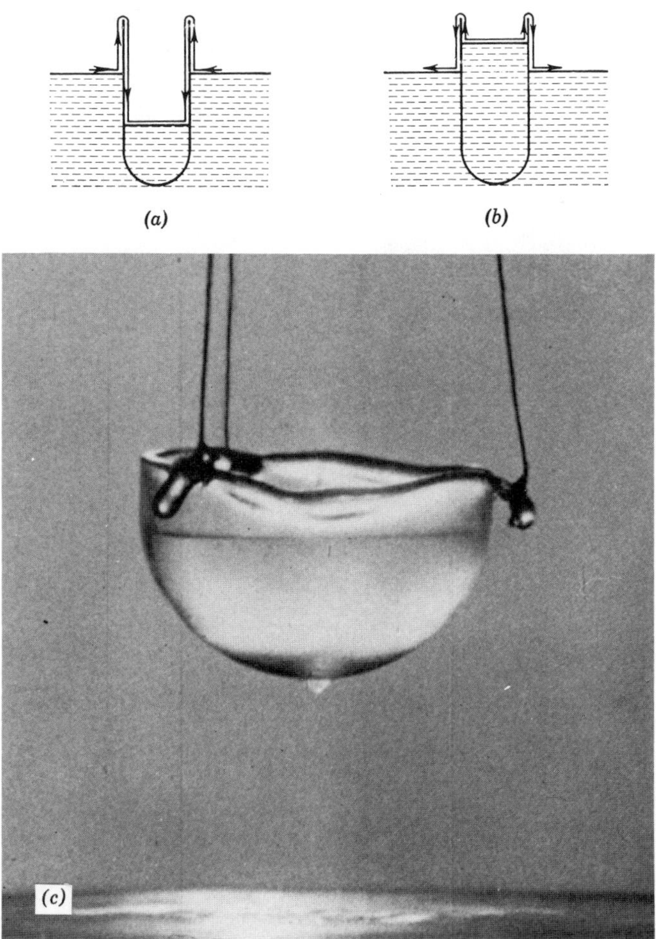


**FIGURE 11-9**

The creeping motion of a film of liquid $He^4$ below the transition temperature demonstrates the superfluidity of He II. The film behavior, suggestive of liquid flow through a siphon, is shown schematically for liquid levels in the container (*a*) below and (*b*) above the level of the liquid helium reservoir. In (*c*) is a photograph of a glass vessel partially filled with liquid He II and suspended by threads above the surface of the same liquid seen at the bottom of the picture. He II creeps up along the inside wall, over the rim, and down along the outside wall as a thin film, collecting as a drop on the bottom. When this drop falls another will form, and so on, until the vessel is empty. (Courtesy of A. Leitner, Rensselaer Polytechnic Institute)

3. Most unusual and spectacular is the ability of liquid helium, below the $\lambda$ point, to creep as a thin film along the walls of its container, as shown in Figure 11-9. The speed of this ordered mass motion may be 30 cm or more per second. The effect involves helium first adsorbing on the entire surface of the cold container to form a thin film. The film then acts like a siphon through which the liquid flows with almost no viscosity.

K. Mendelssohn has written of the film flow as follows:

"If the beaker is withdrawn from the bath, the level will drop until it has reached the level of the bath. If the beaker is pulled out completely, the level will still drop, and one can see little drops of helium forming at the bottom of the beaker and falling back into the bath. This is the sort of thing that makes one look twice and rub his eyes and wonder whether it is quite true. I remember well the night when we first observed this film transfer. It was well after dinner, and we looked around the building and finally found two nuclear physicists still at work. When they, too, saw the drops, we were happier."

All of the properties of He II indicate that it has a very high degree of order. For instance, the almost complete absence of viscosity means that, when flowing, He II does not develop the small scale turbulences that cause the frictional energy loss responsible for the viscosity of ordinary fluids. The order is imposed by the $(1 + n)$ enhancement factor that we often find when studying the low-energy behavior of a system of bosons. When the temperature becomes low enough to allow it, all the helium atoms in a system tend to condense into the same lowest energy quantum state. The superfluid component, whose concentration rapidly approaches 100% as the temperature decreases below the $\lambda$ point, is comprised of those atoms which are in that quantum state. To the extent that all the atoms do get into the same microscopic state, it becomes *the* state of the entire macroscopic system and the system can only behave in a completely ordered way in which the action of any atom is correlated with the action of all the others. This tendency is extremely pronounced because the factor $(1 + n)$ has an extremely large value if $n$ is anything like the total number of atoms in a beaker of liquid helium.

## 11-11 The Free Electron Gas

In this and the following section we apply the Fermi distribution to quantum systems.

In a manner analogous to that used for a boson gas, we could deduce the behavior of an ideal gas of fermions. To the same degree of approximation we would find, for example, that the average energy per particle is

$$\bar{E} = \frac{E}{\mathscr{N}} = \frac{3}{2}kT\left[1 + \frac{1}{2^{5/2}}\frac{\mathscr{N}h^3}{V(2\pi mkT)^{3/2}}\right] \tag{11-54}$$

which is the Fermi result corresponding to the Bose result of (11-53). The degeneracy term here (second term in brackets) is *positive* so that the average particle energy is greater for a Fermi gas than for a classical gas. This corresponds to a lower probability (strictly zero) of finding two particles in the same quantum state for the Fermi distribution than for the Boltzmann distribution, the lower energy states being relatively fuller in the classical gas than in the Fermi gas as a result. Physically, this manifests itself as a higher gas pressure (higher average momentum) at the same temperature for a *Fermi gas* than for a classical gas. Notice again how the Bose and Fermi results fall on opposite sides of the classical result.

It is natural to ask for an example of a Fermi gas whose degeneracy effect we can detect. In Chapter 15 we shall find an example in the neutrons, and the protons, confined to a nucleus. Helium gas containing only the isotope $He^3$ also obeys the Fermi distribution, as do all particles with odd half-integral spin angular momentum, and it remains a gas without condensing to a low enough temperature that the degeneracy term of (11-54) is detectable. This isotope is rare and more difficult to get in large quantities, but the behavior of $He^3$ atoms has been shown to be markedly different from that of $He^4$ atoms in ways predicted by the different distribution functions applicable to them. For example, the vapor pressure of liquid $He^3$ at a given temperature is much higher than that of liquid $He^4$. Indeed, this is the basis for a practical method of cooling to 0.02°K.

It would be quite easy to detect the effect of the degeneracy term for fermions, however, if we could obtain a gas of *electrons*. The degeneracy term can be written as $nh^3/(2\pi mkT)^{3/2}$, in which $n = \mathscr{N}/V$ is the number density of the particles. Notice that a small mass $m$ and a high density $n$ can increase the importance of this term, as well as a low temperature $T$. Because the electronic mass is several thousand times smaller than that of atoms, the degeneracy effect for electrons should actually be detectable even at high temperatures. For electrons in a metal the number density $n$ of conduction electrons is also very high, so that conduction electrons in a metal show quantum degeneracy effects. The question remains whether we can regard such electrons, even approximately, as a gas of free electrons, i.e., an ideal gas.

In a crystalline solid most of the atomic electrons are bound to the nuclei at the lattice points, but if it is a metallic conductor electrons from outer subshells of the atoms are relatively free to move through the solid. These are the *conduction electrons*. Because their mutual repulsion is cancelled, on the average, by the attractions of the atomic cores, we may regard the conduction electrons as approximately free particles and can treat them to good approximation as an ideal electron gas (see Figure 6-24). Indeed, we can regard the interior of the solid as a region of approximately constant potential for these electrons with the metal boundaries acting as high potential walls. The electron then behaves as a particle in a box whose quantum states we already know (see Section 6-8).

To get the number $N(\mathscr{E})\,d\mathscr{E}$ of states in an energy interval $\mathscr{E}$ to $\mathscr{E} + d\mathscr{E}$ we simply count the number of standing waves, each representing a definite state of the motion, in that energy interval. We have made this calculation before for an ideal gas in a box, with results described in (11-49). The results here are the same, after taking into account the two possible spin orientations for an electron having a given space eigenfunction. That is

$$N(\mathscr{E})\,d\mathscr{E} = \frac{8\pi V(2m^3)^{1/2}}{h^3}\,\mathscr{E}^{1/2}\,d\mathscr{E} \tag{11-55}$$

Multiplying by $n(\mathscr{E})$, the probable number of electrons per quantum state, we obtain

$$n(\mathscr{E})N(\mathscr{E})\,d\mathscr{E} = \frac{8\pi V(2m^3)^{1/2}}{h^3}\,\frac{\mathscr{E}^{1/2}\,d\mathscr{E}}{e^{(\mathscr{E}-\mathscr{E}_F)/kT}+1} \qquad \mathscr{E}_F = -\alpha kT \tag{11-56}$$

This is the *electron gas energy distribution of conduction electrons in a metal.*

If now we assume that the temperature is very low (strictly speaking, $T = 0$), we know that all the quantum states up to the Fermi energy $\mathscr{E}_F$ are occupied and that none of the higher states are occupied. In that case the total number of free electrons equals the total number of distinct states up to energy $\mathscr{E}_F$, and we have a way of calculating the Fermi energy. That is

$$\mathcal{N} = \int_0^{\mathscr{E}_F} N(\mathscr{E})\, d\mathscr{E} = \frac{8\pi V(2m^3)^{1/2}}{h^3} \int_0^{\mathscr{E}_F} \mathscr{E}^{1/2}\, d\mathscr{E} = \frac{16\pi V(2m^3)^{1/2}}{3h^3} \mathscr{E}_F^{3/2}$$

or

$$\mathscr{E}_F = \frac{h^2}{8m}\left(\frac{3\mathcal{N}}{\pi V}\right)^{2/3} \tag{11-57}$$

For temperatures such that $kT \ll \mathscr{E}_F$ this result is an excellent approximation. For ordinary metals we need temperatures of the order of several thousand degrees before the approximation breaks down.

**Example 11-5.** Consider silver in the metallic state, with one free (conduction) electron per atom.

(a) Calculate the Fermi energy from (11-57).

The density of silver is 10.5 g/cm³ and its atomic weight is 108. Hence,

$$n = \frac{\mathcal{N}}{V} = \frac{6.02 \times 10^{23}\ \text{atom/mole} \times 10.5\ \text{g/cm}^3}{108\ \text{g/mole}} \times 1\ \text{free electron/atom}$$

$$= 5.9 \times 10^{22}\ \text{free electron/cm}^3 = 5.9 \times 10^{28}/\text{m}^3$$

Therefore

$$\mathscr{E}_F = \frac{h^2}{8m}\left(\frac{3n}{\pi}\right)^{2/3} = \frac{(6.6 \times 10^{-34}\ \text{joule-sec})^2}{8 \times 9.1 \times 10^{-31}\ \text{kg}} \times \left(\frac{3 \times 5.9 \times 10^{-28}/\text{m}^3}{\pi}\right)^{2/3}$$

$$= 8.8 \times 10^{-19}\ \text{joule} = 5.5\ \text{eV}$$

(b) Calculate the degeneracy term for the conduction electrons in metallic silver at 300°K. We have

$$\frac{nh^3}{(2\pi mkT)^{3/2}} = \frac{5.9 \times 10^{28}/\text{m}^3 \times (6.6 \times 10^{-34}\ \text{joule-sec})^3}{(2\pi \times 9.1 \times 10^{-31}\ \text{kg} \times 1.38 \times 10^{-23}\ \text{joule/°K} \times 300\text{°K})^{3/2}}$$

$$\simeq 4700$$

so that the second term in the brackets of (11-54) has the value

$$\frac{1}{2^{5/2}} \frac{nh^3}{(2\pi mkT)^{3/2}} \simeq 820$$

Hence, the degeneracy term is extremely large and completely overwhelms the leading (classical) term of (11-54). The electron gas is said to be a completely degenerate Fermi gas; that is, it behaves as if $T \simeq 0$°K with the electrons in the configuration of lowest energy. Such a gas shows quantum behavior (i.e., is nonclassical) up to the highest attainable metallic temperature, the electron gas in silver remaining almost completely degenerate until the temperature is of the order of $10^5$ °K. At those temperatures and higher the degeneracy term becomes small compared to one. ◀

We can now understand a result that classical physics was unable to explain, namely the experimental observation that the conduction electrons do not contribute to the specific heat of metals at ordinary temperatures. According to the classical view the free electrons take part in the thermal motion in a metal, each free electron having a mean energy $(3/2)kT$. Therefore, the specific heat for a metal should be not simply $3R$, due to the vibrations of the *atoms* at the lattice sites, but it should be $(3 + 3/2)R$ instead, in which the $(3/2)R$ term is the contribution per mole of the *electron* gas. The origin of this term is seen by noting that if $E = (3/2)kTN_0 = (3/2)RT$, then $c_v = dE/dT = (3/2)R$, where $N_0$ is Avagadros number. According to the Fermi model of an electron gas, the electrons do *not* exhibit this classical behavior until the temperature reaches about $10^5$ °K. That is, there is no equipartition of energy between electrons and lattice contributions, the electron gas in this sense not being anywhere near thermal equilibrium with the atoms of the metal confining it. As the temperature is raised, the Fermi distribution of electrons among available energy levels is affected only slightly at the high-energy end (see Figure 11-3) so that the average electron energy is hardly changed at all. This means that at ordinary temperatures the electron gas does not contribute to the specific heat of the metal in an appreciable way. That is, $E \neq (3/2)kTN_0$, but instead it is approximately independent of temperature, so that $c_v = 0$. Hence, the Fermi distribution is in accord with experimental facts concerning electrons at ordinary temperatures.

At ordinary temperatures, and even at temperatures high enough to make the $c_v = 3R$ law of Dulong and Petit a good approximation to the specific heat contribution of the lattice vibrations of a solid, the electronic specific heat term is too small relative to the atomic specific heat term to be detected. At temperatures near absolute zero, where the atomic specific heat is very small, the electronic contribution will exceed the atomic contribution. It is in the region of a few degrees Kelvin that the electronic specific heat dependence is observed experimentally, again in agreement with the Fermi distribution predictions.

## 11-12 Contact Potential and Thermionic Emission

Up to now we have treated the electron in a metal as a particle in a box, that is we have implicitly assumed the electron does not escape the metal, the potential box having very high walls. We know, however, that electrons can escape from metals, as in the photoelectric effect, thermionic emission, etc., so that we should modify the potential function somewhat. Inside the metal the potential function is approximately constant, and near the metal boundary it increases rapidly to reach its higher constant value outside the metal. If we take the zero of potential energy to correspond to the electron being far outside the metal, then we can let $-V_0$ represent the depth of the resulting potential energy well illustrated in Figure 11-10.

We can determine $V_0$ from photoelectric experiments, specifically from the fact that there is a cutoff frequency $\nu_0$ below which photons cannot eject electrons from the metal (see Section 2-2). This suggests that the most energetic electrons in the metal are an energy interval $h\nu_0$ below the top of the potential well. The fact that the photoelectric current rises rapidly as the photon energy rises above the threshold value suggests an abrupt rise in the number of electrons with decreasing kinetic energy inside the metal. This corresponds to the features of the Fermi distribution, the most energetic electrons having kinetic energy $\mathscr{E}_F$ and many electrons having nearby smaller kinetic energies. Therefore, we can retain the distribution of quantum states with energy that we found for the particle in a box. (See Section 6-8 for a discussion of the similarity in energy levels of an infinite and a finite square well potential.) At $T = 0$ all states are filled up to an energy $\mathscr{E}_F$ above the bottom of the well, this highest state having a total energy $-h\nu_0$. That is, $-V_0 + \mathscr{E}_F = -h\nu_0$. Recall now that $h\nu_0 = w_0$, the work function of the metal, so that $-V_0 + \mathscr{E}_F = -w_0$ or

$$V_0 = \mathscr{E}_F + w_0 \tag{11-58}$$

For silver the work function is 4.7 eV and $\mathscr{E}_F$ is 5.5 eV, so that $V_0$ is 10.2 eV. For most metals $V_0$ lies between 5 and 15 eV, as can be seen in Table 11-2. Of course, at ordinary

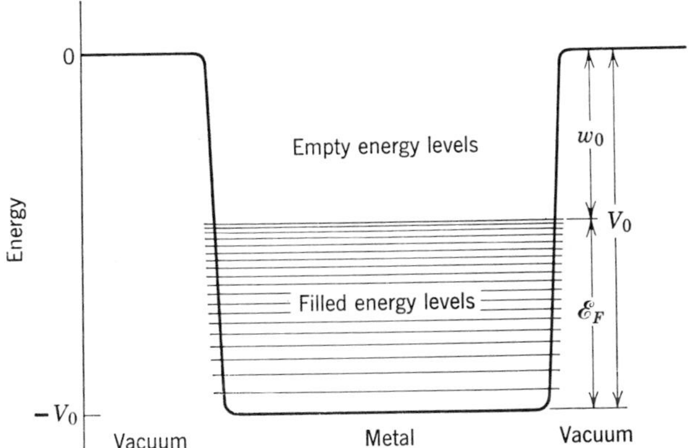


**FIGURE 11-10**

The average potential energy for a conduction electron in a metal. The potential is a well of depth $V_0$ that rises rapidly near the metal boundaries to zero. The energy levels increase in density in proportion to $\mathscr{E}^{1/2}$, and are filled up to the Fermi energy $\mathscr{E}_F$. The work function is $w_0$, and $V_0 = w_0 + \mathscr{E}_F$.

temperatures the Fermi distribution does not give a sharp cutoff at $\mathscr{E}_F$ but is spread out continuously over a narrow energy region near $\mathscr{E}_F$. In a region of the order of $kT$ on each side of the Fermi energy, i.e., in a transition region of width $2kT$, the number of particles per quantum state goes from a value near one to a value near zero. In the limit when $T \to 0$ this transition region becomes infinitisimally narrow.

With this model for the behavior of electrons in a metal we can explain the *contact potential difference* of two metals and understand the *thermionic emission* process. First, consider the thermionic emission process, which is of great practical importance because it is responsible for the emission of electrons from the heated filament of a vacuum tube. At high temperatures (i.e., for large values of $kT$) the distribution of electrons among available energy states in a metal extends to energies well above $\mathscr{E}_F$. At sufficiently high temperature some electrons may acquire a kinetic energy greater than $V_0$ (i.e., greater than $\mathscr{E}_F + w_0$) and thereby escape from the metal. We can calculate the thermoelectric current density emitted from a metal surface as a function of temperature from the Fermi distribution and from the Boltzmann distribution. The calculation involves determining how many electrons will arrive at the metal surface moving in the required direction and with enough kinetic energy to escape. The two distributions give a different temperature dependence for the current density, and experiment rules in favor of the Fermi distribution for electrons.

**TABLE 11-2.** Work Function and Fermi Level Energy for Some Metals

| Metal | $w_0$ (eV) | $\mathscr{E}_F$ (eV) |
|---|---|---|
| Ag | 4.7 | 5.5 |
| Au | 4.8 | 5.5 |
| Ca | 3.2 | 4.7 |
| Cu | 4.1 | 7.1 |
| K | 2.1 | 2.1 |
| Li | 2.3 | 4.7 |
| Na | 2.3 | 3.1 |

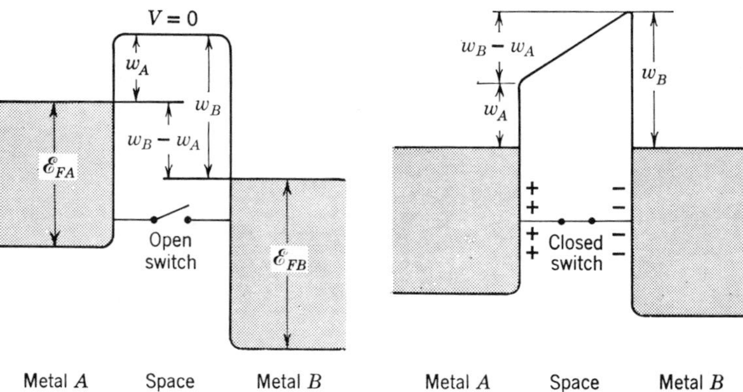


**FIGURE 11-11**

*Left:* Showing the potential energy for an electron in two separated metals $A$ and $B$ with different work functions. *Right:* The metals are now connected electrically by a wire, becoming oppositely charged and exhibiting a contact potential difference.

As for the contact potential difference between metals, consider two metals $A$ and $B$ which at first are not in contact, as is indicated schematically in the left part of Figure 11-11. Outside the metals the potential energy of an electron is zero. Inside the metals the Fermi level of metal $A$ is $w_A$ below zero and the Fermi level of metal $B$ is $w_B$ below zero. Let $w_B > w_A$ so that the Fermi level of metal $A$ is higher than that of $B$. Now let the metals be connected electrically, as illustrated in the right part of Figure 11-11. Then the most energetic electrons in metal $A$ will flow into metal $B$, filling the energy levels in $B$ just above its Fermi energy and depleting the upper levels in $A$. The process continues until equilibrium is reached; that is, until the highest filled levels in $A$ and $B$ are at the same energy, because the total energy of the system is minimized when this situation is achieved. The result is that metal $A$ becomes positively charged in the process and metal $B$ becomes negatively charged. Consequently there is a potential difference of $(w_B - w_A)/e$ between the metals when they are connected electrically, a result in essential agreement with experimental values.

## 11-13 Classical and Quantum Descriptions of the State of a System

We saw in Section 4-9 an example of how the instantaneous state of the motion of a classical particle can be represented by a point in *phase space*. For the one-dimensional motion considered there, the phase space was a two-dimensional space whose abscissa was the position $x$ and whose ordinate was the momentum $p_x$. For a three-dimensional motion phase space is a six-dimensional space of coordinates $x$, $y$, $z$, $p_x$, $p_y$, $p_z$. As the particle moves, the point representing it in phase space traces out a path, the path being an ellipse in our earlier example of a one-dimensional harmonic oscillator. If we had a large number of such oscillators we would have a large number of representative points in phase space corresponding to the instantaneous distribution of oscillators. For most systems of interest we can write the total energy of each member as $E = K + V = (p_x^2 + p_y^2 + p_z^2)/2m + V(x, y, z)$ so that the location of a point $(x,y,z,p_x,p_y,p_z)$ in phase space gives the total energy of that member of the system which the point represents. The distribution of points gives the distribution in energy of all members of the system.

Thus, in classical statistics we can characterize the energy distribution of a system by giving the number of points in each small volume of phase space, say $\Delta x \Delta y \Delta z \Delta p_x \Delta p_y \Delta p_z$. We call such a small volume element a *cell* in phase space, and points in that cell have total energy

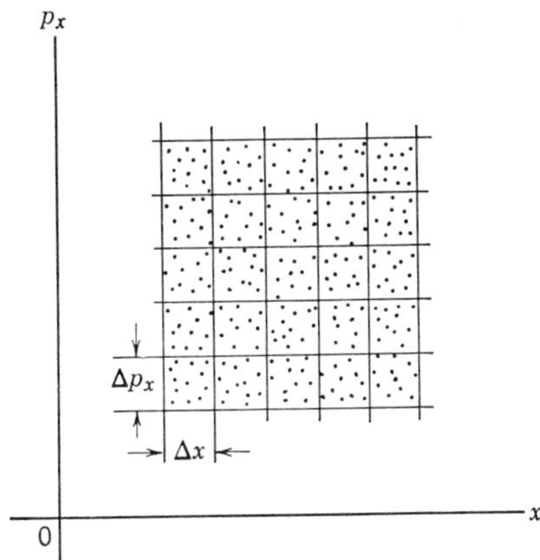


**FIGURE 11-12**
Phase space and representative points for a one-dimensional system.

between $E$ and $E + dE$, corresponding to momentum values between $p_x$ and $p_x + \Delta p_x$, etc. and position values between $x$ and $x + \Delta x$, etc. The cell is chosen to be small enough that the average total energy of its representative points differs little from the energy of any one of them; it is chosen large enough so that there are many points in a cell, thereby permitting the application of statistical ideas. Hence, the size of a cell is somewhat arbitrary and indefinite, but once it is chosen the cell is characterized by an average total energy and a population number. The cell then is the classical statistical analogue to the quantum state of quantum statistics. In Figure 11-12 we illustrate the situation for a one-dimensional system.

In quantum mechanics we must modify the preceding picture because of the uncertainty principle. For one thing we cannot describe the trajectory of a particle by giving the path of a representative point in two-dimensional phase space because we cannot simultaneously know the exact values of $x$ and $p_x$ for the particle. The best we can do is locate the representative point at any time between $x$ and $x + \Delta x$ and $p_x$ and $p_x + \Delta p_x$ where $\Delta x \Delta p_x \simeq h$, so that instead of a representative point tracing out a line we have a small area tracing out a ribbonlike path in two-dimensional phase space. More important, however, is the fact that there is a definite smallest size to a cell in the quantum description. For a cell in which $\Delta x \Delta p_x$ is less than $h$ is meaningless, such a specification being more precise than allowed by the uncertainty principle. For the general six-dimensional phase space, therefore, the smallest cell has a "volume" of $h^3$.

It is therefore possible in the quantum description to remove the arbitrariness and indefiniteness of the volume element in phase space. Because the size of the cell obviously affects the counting of distinguishable divisions of the total energy of the system, there is a certain indefiniteness in the results of classical statistics. For example, the entropy of a system can be written as $S = k \log P$ where $P$ is the number of distinguishable divisions of its energy content (i.e., $P$ is a measure of the probability that it has the particular energy). However, the classical entropy has an arbitrary constant in it basically because of the indefiniteness of the cell size. The quantum value is exact, because of the definiteness of the cell size, and it gives an absolute entropy constant in agreement with experiment and the laws of thermodynamics. Indeed, it was this result, and not the results concerning the cavity radiation, that convinced Max Planck of the correctness of his ideas concerning energy quantization and the constant $h$. And it is this smallest size of a cell in phase space in quantum statistics that is the origin of the factor $h^3$ displayed in many of the equations in this chapter.

From considerations discussed here we can also understand the applicability of the classical Boltzmann distribution to so many quantum problems. If there is no definite smallest size to a cell in phase space then we can always get a situation in which there is not more than one particle per state. But this is just the high temperature case wherein classical and quantum statistics agree. The classical distribution function is valid in this case, regardless of the indistinguishability of particles. Of course, the real quantum world does set a limit to the smallness of a cell so that the classical distribution will not apply when the number of particles per cell is more than one.

## QUESTIONS

**1.** Exactly what do the inhibition and enhancement factors describe? What are their origins?

**2.** Can you devise a cycle of transitions between three states which would maintain an equilibrium in the populations of these states, with transitions that violate detailed balancing? Does it seem reasonable to extend this to a system with many states?

**3.** What is the basic reason why the quantum distributions merge with the classical distribution at energies much larger than $kT$?

**4.** Explain why the behavior of the Boltzmann distribution is intermediate to that of the Bose and Fermi distributions.

**5.** Give examples of systems to which the Boltzmann distribution is applicable in principle. As a good approximation.

**6.** What factors determine the value of $\alpha$ for the three distributions?

**7.** Interpret physically the Fermi energy $\mathscr{E}_F$.

**8.** Thermal expansion is related to the anharmonic nature of the vibrations of atoms in a solid. Would the Debye model be appropriate to studying thermal expansion of solids?

**9.** In Debye's model of a solid, the maximum frequency $\nu_m$ corresponds to a minimum wavelength. Because of the discrete nature of a solid this minimum wavelength corresponds to a vibration in which adjacent atoms move 180° out of phase with one another; that is, the interatomic spacing is half a wavelength. Is this plausible? Explain.

**10.** Interpret the Debye characteristic temperature $\Theta$ physically.

**11.** In our analysis of emission and absorption processes of an atom in an electromagnetic field we neglected recoil effects. How does this affect our results? Are we justified in ignoring recoils?

**12.** What are the dimensions of the Einstein $A$ and $B$ coefficients?

**13.** It is said that a laser is not a source of energy but a converter of energy. Explain.

**14.** We have ignored the possible degeneracy of the states involved in laser action. How would you take this into account? What effect does it have?

**15.** Make a step-by-step comparison of the deduction of the Planck radiation law on the basis of the Maxwell distribution and the Bose distribution.

**16.** List similarities and differences between phonons and photons.

**17.** At low densities and high temperatures the Bose gas behaves like a classical ideal gas. Make this result plausible physically.

**18.** In writing about experiments on the scattering of $\alpha$ particles in helium Rutherford said, "On account of the impossibility of distinguishing between the scattered alpha particles and the projected He nuclei, the results are subject to a certain ambiguity." Explain how an awareness of quantum statistics could have removed the ambiguity. What determines whether a gas obeys Bose or Fermi distributions?

**19.** How can the ordered state of He II explain its lack of resistance to heat conduction?

**20.** What examples of a Fermi gas are there other than an electron gas and a gas of $He^3$ atoms?

**21.** In the ideal gas equations we use the rest mass of particles. Should we ever use the relativistic mass instead? Consider the effect of temperature and the nature of the particle.

**22.** Give a plausibility argument for the relation, (11-57), between the Fermi energy $\mathscr{E}_F$ and the density of free electrons in a metal.

**23.** In the Fermi distribution we obtain the result that at the Fermi energy $\mathscr{E}_F$ the average number of particles per quantum state is exactly one-half. This is definitely *not* the same as saying that 50% of the particles are at energies above the Fermi energy and 50% below. Explain.

**24.** Justify the assumption that conduction electrons behave approximately as a system of free noninteracting particles.

**25.** Is there a connection between $V_0$, the depth of the potential well for conduction electrons in a metal, and electron diffraction experiments of the Davisson-Germer type? Can we determine $V_0$ from such experiments?

**26.** Explain physically the effect of letting $h \to 0$ in the expression for density of states, such as (11-49). Explain physically the effect of letting $h \to 0$ in equations involving the quantum degeneracy term, such as (11-53).

## PROBLEMS

**1.** The equilibrium state is one of maximum entropy $S$ in thermodynamics and one of maximum probability $P$ in statistics. Assuming then that $S$ is a function of $P$, show that we should expect $S = k \ln P$, where $k$ is a universal constant. This relation is sometimes called the Boltzmann postulate. (Hint: Consider the effect on $S$ and $P$ of combining two systems.)

**2.** The Maxwell distribution can be developed by looking at elastic collisions between two particles. If initially these particles have energies $\mathscr{E}_1$ and $\mathscr{E}_2$, and finally $\mathscr{E}_3$ and $\mathscr{E}_4$, then

$$\mathscr{E}_3 + \mathscr{E}_4 = (\mathscr{E}_1 - \delta) + (\mathscr{E}_2 + \delta)$$

If all possible states are equally probable, the number of collisions per second $P$ is proportional to the number of particles in each initial state, i.e.

$$P_{1,2} = CP(\mathscr{E}_1)P(\mathscr{E}_2)$$

where $P(\mathscr{E}_i)$ is the probability of a state being occupied, and $C$ is a constant. Similarly $P_{3,4} = CP(\mathscr{E}_3)P(\mathscr{E}_4)$. In equilibrium, for each collision $(1,2) \to (3,4)$ there must be a collision $(3,4) \to (1,2)$. Thus $P_{1,2} = P_{3,4}$. (a) Show that $P(\mathscr{E}_i) = e^{-\mathscr{E}_i/kT}$ solves this equation. (b) Use similar reasoning to derive the Fermi distribution. Here, however, the initial states must be filled and the final states must be empty, and the number of collisions becomes

$$P_{1,2} = CP(\mathscr{E}_1)P(\mathscr{E}_2)[1 - P(\mathscr{E}_3)][1 - P(\mathscr{E}_4)]$$

Then show that the equation $P_{1,2} = P_{3,4}$ can be solved by

$$\left[\frac{1 - P(\mathscr{E}_i)}{P(\mathscr{E}_i)}\right] = Ce^{\mathscr{E}_i/kT}$$

which yields (11-23).

**3.** (a) From (11-25), show that the Einstein model of a solid gives the specific heat as

$$c_v = 3R\left[\frac{e^{h\nu/kT}}{(e^{h\nu/kT} - 1)^2}\left(\frac{h\nu}{kT}\right)^2\right]$$

(b) Show that $c_v \to 0$ as $T \to 0$ but that at low $T$, $c_v$ increases as $e^{-h\nu/kT}$ rather than as the required $T^3$ law.

**4.** Show that the Debye specific heat result, (11-31), reduces to the classical law of Dulong and Petit at high temperatures. (Hint: First expand both exponentials and retain only first order terms. Justify.)

**5.** Imagine a cavity at temperature $T$. Show that $c_v$, the specific heat of the enclosed radiation, is given by $(32\pi^5 kV/15)(kT/hc)^3$. Explain why $c_v$ does not have an upper limit in this case whereas it does for solids.

**6.** In some temperature region graphite can be considered a two-dimensional Debye solid, but there are still $3N_0$ modes per mole. (a) Show that $N(\nu)\,d\nu = (2\pi A/v^2)\nu\,d\nu$ where $A$ is the area of the sample. (b) Find an expression for $\nu_m$ and $\Theta$ for graphite. (c) Show that at low temperatures the heat capacity is proportional to $T^2$.

**7.** $\mathcal{N}$ distinguishable atoms are distributed over two energy levels $\mathscr{E}_1 = 0$ and $\mathscr{E}_2 = \mathscr{E}$. (a) Show that the energy of the system is given by

$$E = \frac{\mathcal{N}\mathscr{E}e^{-\mathscr{E}/kT}}{1 + e^{-\mathscr{E}/kT}}$$

(b) Show that $c_v$ is given by

$$c_v = \frac{\mathcal{N}k\left(\dfrac{\mathscr{E}}{kT}\right)^2 e^{-\mathscr{E}/kT}}{(1 + e^{-\mathscr{E}/kT})^2}$$

(This is the *Schottky specific heat* and is observed for paramagnetic solids at low temperatures. The energy levels correspond to the magnetic moments being aligned parallel or antiparallel to the magnetic field.) (c) Sketch the heat capacity as a function of temperature, being careful to have the correct temperature dependence at high and low temperatures.

**8.** The variation of density $\rho$ with altitude $y$ of the gaseous atmosphere of the earth can be written as $\rho = \rho_0 e^{-g(\rho_0/P_0)y}$, where $\rho_0$ and $P_0$ are sea level density and pressure. (a) From the ideal gas laws show that this can be put into the form $\rho = \rho_0 e^{-mgy/kT}$. (b) Show that this has the form of the Boltzmann distribution,

**9.** (a) By combining $n(\mathscr{E})$ of (11-21) and $N(\mathscr{E})$ of (11-49) for an ideal gas of classical particles, with

$$A = e^{-\alpha} = \frac{Nh^3}{(2\pi mkT)^{3/2}V}$$

Show that

$$n(\mathscr{E})N(\mathscr{E})\,d\mathscr{E} = \frac{2N}{(kT)^{3/2}\pi^{1/2}}\,\mathscr{E}^{1/2}e^{-\mathscr{E}/kT}\,d\mathscr{E}$$

is the energy distribution of particles in an ideal gas. (b) Show that Maxwell's speed distribution of molecules in a gas, which has the form $n(v)\,dv = Cv^2e^{-mv^2/2kT}\,dv$, where $C$ is a constant, follows directly from this.

**10.** Assume that the thermal neutrons emerging from a nuclear reactor have an energy distribution corresponding to a classical ideal gas at a temperature of 300°K. Calculate the density of neutrons in a beam of flux $10^{13}$/m²-sec. (Hint: Consider the average velocity, and justify its use.)

**11.** In a certain nucleus the magnetic moment is $1.4 \times 10^{-26}$ joule-m²/weber. Calculate the fractional difference in population of the nuclear Zeeman levels in a magnetic field of 1 weber/m², (a) at room temperature and (b) at 4°K.

12. Electron spin resonance is much like nuclear magnetic resonance except that electronic transitions are excited between atomic Zeeman levels. These experiments are done at microwave frequencies. If the electromagnetic wave has a frequency of 32 KMHz (K band) calculate the fractional difference in population between two atomic Zeeman levels (a) at room temperature and (b) at 4°K.

13. (a) Determine the order of magnitude of the fraction of hydrogen atoms in a state with principle quantum number $n = 2$ to those in state $n = 1$ in a gas at 300°K. (b) Take into account the degeneracy of the states corresponding to quantum numbers $n = 1$ and 2 of atomic hydrogen and determine at what temperature approximately one atom in a hundred is in a state with $n = 2$.

14. Consider the relation $n_1/n_2 = e^{(\mathscr{E}_2 - \mathscr{E}_1)/kT}$, the Boltzmann factor for nondegenerate states for systems in equilibrium, where $\mathscr{E}_2 > \mathscr{E}_1$. (a) Show that $n_2 = 0$ at $T = 0$. (b) Show that $n_1 = n_2$ at $T = \infty$ or $T = -\infty$. (c) Show that $n_2 > n_1$ at finite negative temperature $T$. (d) Show that $n_1 \to 0$ as $T \to -0$. (e) Hence, explain the statements, "Negative absolute temperatures are not colder than absolute zero but hotter than infinite temperature," and "One approaches negative temperatures through infinity, not through zero." (f) Can you suggest a change in temperature scale that would avoid temperatures that are negative in this sense?

15. Determine approximately the ratio of the probability of spontaneous emission to the probability of stimulated emission at room temperature in (a) the x-ray region of the electromagnetic spectrum, (b) the visible region, (c) the microwave region.

16. (a) Show that at $T = 0$, in the Fermi distribution $n(\mathscr{E}) = 1$ for all energy states in which $\mathscr{E} \leq \mathscr{E}_F$ and $n(\mathscr{E}) = 0$ for all energy states in which $\mathscr{E} > \mathscr{E}_F$. (b) Show that $n(\mathscr{E}) = 1/2$ for $\mathscr{E} = \mathscr{E}_F$.

17. Consider the Fermi distribution of (11-24), $n(\mathscr{E}) = 1/[e^{(\mathscr{E} - \mathscr{E}_F)/kT} + 1]$. (a) Show that $n(\mathscr{E}) = 1 - n(2\mathscr{E}_F - \mathscr{E})$; that is, with $\mathscr{E} - \mathscr{E}_F = \delta$, show that $n(\mathscr{E}_F + \delta) = 1 - n(\mathscr{E}_F - \delta)$. This proves that the distribution has a symmetry about $n(\mathscr{E}_F) = 1/2$. (b) Find $n(\mathscr{E})$ for $\delta = \mathscr{E} - \mathscr{E}_F = kT$, or $2kT$, or $4kT$, or $10kT$. Make a rough sketch of $n(\mathscr{E})$ versus $\mathscr{E}$ for any $T > 0$. (c) What percent error is made by approximating the Fermi distribution by the Boltzmann distribution when $\delta/kT = 1, 2, 4, 10$?

18. Combine (11-49) and (11-47) to obtain (11-50), as follows. Let $x = \mathscr{E}/kT$ and obtain

$$\mathscr{N} = \frac{2\pi V(2mkT)^{3/2}}{h^3} \int_0^\infty \frac{x^{1/2}\, dx}{e^{\alpha + x} - 1}$$

Then, with $\alpha$ positive, use the relation $(e^{\alpha+x} - 1)^{-1} = e^{-\alpha-x}(1 - e^{-\alpha-x})^{-1} = e^{-\alpha}(e^{-x} + e^{-\alpha-2x} + \cdots)$ to obtain (11-50).

19. Obtain (11-52) as follows. Let $x = \mathscr{E}/kT$ and show that

$$E = \frac{2\pi kTV(2mkT)^{3/2}}{h^3} \int_0^\infty \frac{x^{3/2}\, dx}{e^{\alpha+x} - 1} = \frac{3}{2} kT \frac{V(2\pi mkT)^{3/2}}{h^3} e^{-\alpha}\left(1 + \frac{1}{2^{5/2}} e^{-\alpha} + \cdots\right)$$

20. Show that the quantum degeneracy in a Fermi gas occurs if $kT \ll \mathscr{E}_F$. (Hint: See Example 11-4 and use (11-57).)

21. Show from the Fermi distribution that in a metal at $T = 0$°K the average energy of an electron is $3\mathscr{E}_F/5$.

22. Using 23 as the atomic weight and $9.7 \times 10^2$ kg/m$^3$ as the density of metallic sodium, compute the Fermi energy on the assumption that each sodium atom gives one electron to the conduction band. (Hint: See Example 11-5.)

**23.** Using 197 as the atomic weight and $19.3 \times 10^3$ kg/m$^3$ as the density of gold, compute the depth of the potential well for free electrons in gold. The work function is 4.8 eV and there is one free electron per atom.

**24.** In a one-dimensional system the number of energy states per unit energy is $(l/h)\sqrt{2m/\mathscr{E}}$, where $l$ is the length of the sample and $m$ is the mass of the electron. There are $\mathscr{N}$ electrons in the sample and each state can be occupied by two electrons. (a) Determine the Fermi energy at 0°K. (b) Find the average energy per electron at 0°K.

**25.** Show that about one conduction electron in a thousand in metallic silver has an energy greater than the Fermi energy at room temperature.

# 12

# Molecules

# Molecules 12

## 12-1 Introduction

The subject matter of the previous chapters is considered to be common to all of quantum physics. The concepts and techniques we have developed in these chapters for the purpose of studying atoms prove to be necessary, or at least useful, in studying most of the areas to which quantum physics is applied. But from atoms the applications of quantum physics branch into two well-defined, and fairly well-separated, channels. One of these leads to the systems larger than atoms; i.e., it goes from atoms to molecules and then to solids. The other channel leads from atoms to the smaller systems; i.e., to nuclei and then to their constituents, the elementary particles. In the next three chapters we shall follow the first channel, and in the last three chapters of this book we shall explore the second.

We know that two or more atoms can combine to form a stable molecule. Here we seek a description of the interatomic forces which bind atoms into molecules, and also an understanding of the nature of energy levels and spectra of molecules. Since a very large number of atoms may join together to make a solid, in much the same way as a few do to form a molecule, the phenomenon of molecular binding is very relevant to the properties of solids. The motivation for studying molecular spectra, in addition to its intrinsic interest, is found in practical considerations. For example, a new but rapidly expanding field of science is molecular astronomy, which involves the measurement of molecular spectra originating in interstellar, or intergalactic, matter, for the purpose of determining its composition and condition. And as we shall see, measurements of molecular spectra have for a long time provided the primary source of information about important properties of the nuclei contained in the molecule.

## 12-2 Ionic Bonds

From one point of view a molecule is a stable arrangement of a group of nuclei and electrons. The exact arrangement is determined by electromagnetic forces and the laws of quantum mechanics. This concept of a molecule is a natural extension of the concept of an atom. Another view regards a molecule as a stable structure formed by the association of two or more atoms. In this view the atoms retain their identity whereas in the first-mentioned view they do not. Of course, both views are useful and there are situations wherein each is directly applicable. In general, however, the structure and properties of molecules are best described by a combination of both views. When a molecule is formed from two atoms, the inner shell electrons of each atom remain tightly bound to the original nucleus and are barely disturbed at all. The outermost loosely bound electrons, however, are strongly disturbed and are influenced by all the particles (ions + electrons) of the system. Their wave functions are significantly modified when the atoms are brought together. Indeed, it is this very interaction that leads to binding, i.e., to a lower total energy, when the nuclei or ions are close together. This interaction, called the *interatomic force*, is of *electromagnetic* origin. Hence, we see that valence electrons play the central role in molecular binding.

There are two principal types of molecular binding, the *ionic* bond and the *covalent* bond. The NaCl molecule is an example of ionic binding and the $H_2$ molecule an

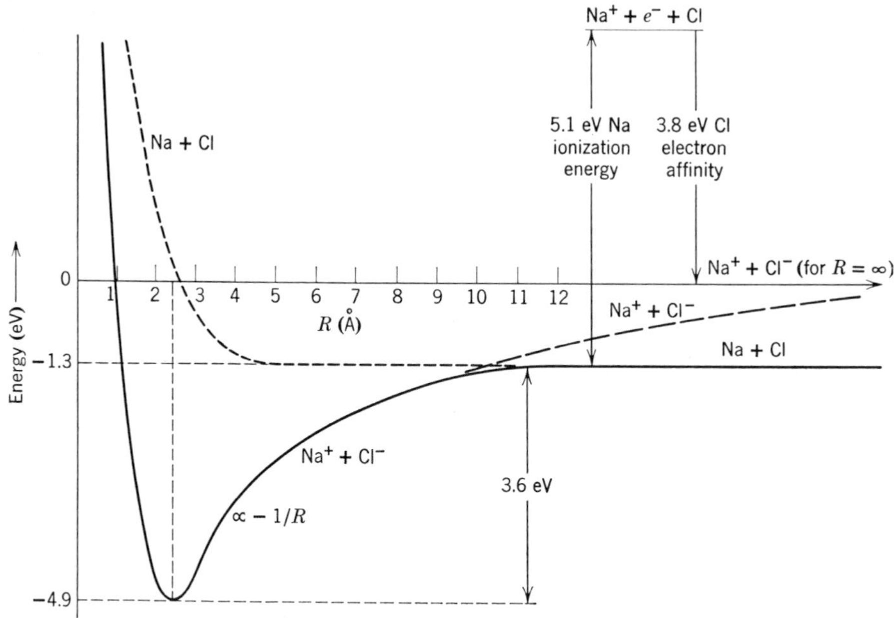


**FIGURE 12-1**

The energy for the neutral atoms Na and Cl, and for the ions $Na^+$ and $Cl^-$, as a function of the internuclear separation $R$. The ionic combination has lower energy at small separation, while the neutral atom combination has lower energy at large separation. Thus, as the two neutral atoms are brought together, they go over to ionic form when their separation becomes less than a certain value.

example of covalent binding. Consider the formation of a NaCl molecule from an atom of Na and an atom of Cl which are far apart initially. Figure 9-15 shows that to remove the outermost 3*s* electron from Na and form the $Na^+$ ion requires an ionization energy of 5.1 eV. The atomic binding in the alkali Na is relatively weak because its filled inner subshells are effective in shielding the valence electron electrically from the nucleus so that it moves in a weakened field at an outlying position. If now we attach this electron to the halogen Cl atom it will complete a previously unfilled 3*p* shell in Cl to form a $Cl^-$ ion. The halogen has a relatively high electron affinity; that is, the closed shell ion is more stable than the neutral atom, its energy being lower by 3.8 eV. Hence, at the cost of 1.3 eV of energy (5.1 eV − 3.8 eV), we have formed two distinct separate ions, $Na^+$ and $Cl^-$; but these ions exert attractive Coulomb forces on one another, and the energy of attraction is greater than 1.3 eV. Now, since the mutual Coulomb potential energy of the ions is negative, the potential energy of the combined system initially decreases as the separation of the ions is steadily reduced. As the ions are brought still closer together the electron charge distributions begin to overlap. This has two effects, each of which increases the potential energy: (1) the nuclei are not as well shielded from one another as before and they begin to repel one another and (2) at small internuclear separation we effectively have a single system to which the exclusion principle applies, and some electrons must be in higher energy states than before to avoid violating this principle. The potential energy curve therefore yields a repulsive force at small interatomic separations and an attractive force at large separations. There is a separation at which this energy is a minimum, the energy being 4.9 eV lower at this proximity than for distantly separated ions. Hence, compared to two neutral atoms, Na + Cl, the combined system NaCl is lower in energy by 3.6 eV (that is, $E = 1.3 \text{ eV} - 4.9 \text{ eV} = -3.6 \text{ eV}$) so that a bound state is energetically favored, as illustrated in Figure 12-1. The equilibrium nuclear separation in NaCl is 2.4 Å.

**Example 12-1.** Evaluate approximately the depth of the minimum in Figure 12-1 by assuming that at the 2.4 Å equilibrium nuclear separation $R$ of NaCl the $Na^+$ and $Cl^-$ ions have spherically symmetrical charge distributions that do not yet overlap.

With this assumption, Gauss's law of electrostatics allows us to evaluate the Coulomb binding energy of the unit charge ions from the simple expression

$$V = -\frac{1}{4\pi\epsilon_0}\frac{e^2}{R}$$

where $R = 2.4$ Å. We obtain

$$V = -\frac{9.0 \times 10^9 \text{ nt-m}^2/\text{coul}^2 \times (1.6 \times 10^{-19} \text{ coul})^2}{2.4 \times 10^{-10} \text{ m}}$$

$$= -9.7 \times 10^{-19} \text{ joule} \times \frac{1 \text{ eV}}{1.6 \times 10^{-19} \text{ joule}}$$

$$= -6.0 \text{ eV}$$

If the student extrapolates slightly the $1/R$ behavior in Figure 12-1 to $R = 2.4$ Å, he will see that the results of this evaluation are consistent with its assumptions. ◀

NaCl is a molecule held together by ionic binding. Because the region of positive charge ($Na^+$) and the region of negative charge ($Cl^-$) are separated, there is a permanent electric dipole moment. An ionic molecule is thus said to be a *polar* molecule. Ionic binding is also called *heteropolar* binding. Ionic bonds are not directional, for each ion has a closed shell configuration which is spherically symmetrical. Ionic bonds can be formed with more than one valence electron, as in the case of the $MgCl_2$ molecule, when the molecular state is energetically lower than the state of separated atoms. The number of ionic bonds that an atom can form depends on the shell structure of the atom, i.e., on the ionization potentials for successively removing electrons. It will be energetically favorable to form ionic bonds only for those (few) outer subshell electrons that have ionization potentials in certain ranges. Compounds of elements from the first column, and the second from last column, of the periodic table (the alkali halides, such as KCl, LiBr, etc.) are ionic, as are many of those from the second column and the third from last column (the alkaline-earth oxides, sulfides, etc.).

## 12-3 Covalent Bonds

Let us consider now the formation of the $H_2$ molecule. If in the case of $H_2$ we were to calculate the energy required to form positive and negative hydrogen ions by moving an electron from one hydrogen atom to the other, and then added to this the energy of the Coulomb interaction of the ions, we would find that there is no distance of separation at which the total energy is negative. That is, ionic bonding does *not* result in a bound $H_2$ molecule. The fact that $H_2$ *is* bound is explained quantum mechanically by the behavior of the electronic eigenfunction describing the charge distribution of the system, as two hydrogen atoms approach one another. As we shall see soon, the resulting charge distribution does lead to electrostatic attraction, but it is a charge distribution that can be interpreted as a sharing of electrons by both atoms. The binding is called *covalent*.

We can best understand the covalent bond by treating first the simpler case of $H_2^+$, the hydrogen molecular ion. In this case we have two nuclei each exerting a Coulomb repulsion on the other, and both exerting a Coulomb attraction on the single electron. Since the electron motion is very rapid compared to the nuclear motions, the procedure is to assume that the nuclei are at rest a distance $R$ apart, with the single electron moving in their Coulomb fields, and then determine the electron energy from the Schroedinger equation. We next treat $R$ as a variable and consider both the electron energy, and the internuclear Coulomb repulsion energy, as a function of the internuclear separation. The total energy of the system is the sum of these two energies, and the system will be bound if the total energy exhibits a minimum at some value of internuclear separation.

The top of Figure 12-2 indicates the potential energy in which the electron moves by plotting its value along an $x$ axis passing through the two nuclei, for an internuclear separation $R = 1.1$ Å. The potential energy is symmetrical with respect to a plane perpendicular to the line connecting the two nuclei and passing through its middle, since the potential is just the sum of a Coulomb potential centered on one end of that line and an equal Coulomb potential centered on the other end. Because the motion of the electron in a bound state of this potential will have the same symmetry, the electron's bound state probability densities $\psi^*\psi$ will have equal values at two points on either side of the plane and equidistant from it. But this requires each of its eigenfunctions $\psi$ to have either precisely the same value at the two points, or else to have at one point a value precisely the negative of its value at the other point. That is, the eigenfunctions must be either *even* or *odd* with respect to reflection in the plane. The situation is shown schematically in the bottom of Figure 12-2 by plotting the lowest energy even and odd normalized eigenfunctions along a line passing through the two nuclei. The important idea is that the odd eigenfunction must necessarily have zero value at the center of this line since it obeys the equation $\psi(-x) = -\psi(x)$, which would otherwise be internally inconsistent at the center where $x = 0$. But the

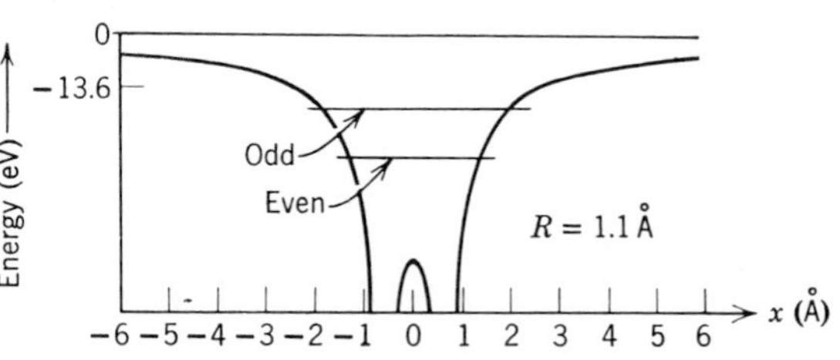


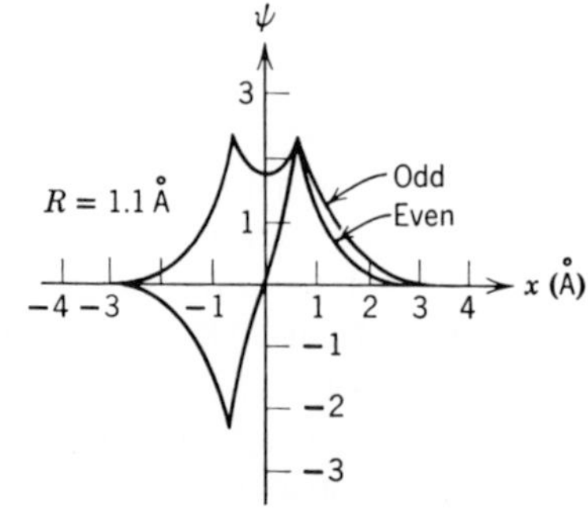


**FIGURE 12-2**

*Top:* The potential function, and the two lowest energy levels, for an electron in a $H_2^+$ molecule with internuclear separation $R = 1.1$ Å. The potential function is evaluated along the line passing through the two nuclei. *Bottom:* The even and odd eigenfunctions corresponding to the two energy levels, evaluated along the internuclear line. Near each nucleus, both eigenfunctions have magnitudes that are decreasing exponentials of the distance from the nucleus, as in the ground state of the hydrogen atom.

even eigenfunction is not so constrained, and thus it has an appreciable value at $x = 0$.

Because an electron with probability density $\psi^*\psi$ for the odd eigenfunction must avoid the center of the molecule, to a certain extent it avoids the central region. And since the integral over all space of $\psi^*\psi$ equals one, if that quantity is relatively small in the region between the nuclei, it must be relatively large in the regions outside the nuclei. These outside regions are where the potential is least binding, however, so such an electron is relatively loosely bound. The odd eigenfunction could be more tightly concentrated in the regions near the nuclei, while still being zero at the center, but only if its curvature were higher. Since higher curvature requires higher kinetic energy, this would not decrease the total energy of the electron. An electron whose behavior is described by the probability density for the even eigenfunction has a relatively high probability of being found in the region where the potential is most binding—that is, in the region from near one nucleus, through the center of the molecule, to near the other nucleus. Thus such an electron is relatively tightly bound. The two lowest energy levels for an electron in the potential are shown in Figure 12-2. We can now understand why the lowest of these is for the quantum state in which the eigenfunction is even.

Figure 12-3 shows the sum of the electron energy and the internuclear Coulomb repulsion energy for the two lowest energy states of the $H_2^+$ molecule, as a function of the internuclear separation distance $R$. For very large $R$, the electron will bind to one nucleus or the other in the lowest energy state of an H atom, and the repulsion energy will be negligible, so the energy of the system will have the familiar value $-13.6$ eV. For the quantum state with the even eigenfunction, the energy of the system at first decreases with decreasing $R$. The reason is that the binding energy exerted on the electron already near one nucleus becomes negative more rapidly, as the other nucleus moves into proximity, than the repulsion energy between the two nuclei becomes positive. (The electron in the even eigenfunction state at moderate internuclear separation tends to be between the nuclei, so its distance to either nucleus is smaller than the distance separating the nuclei.) As the internuclear separation continues to decrease, the energy of the system passes through a minimum and then begins to increase rapidly. This happens because the electron binding energy when the nuclei overlap can become no more negative than $-(2)^2 \times 13.6 \text{ eV} = -54.4 \text{ eV}$, the ground state energy of a singly ionized helium atom, whereas the internuclear repulsion energy increases without limit as the internuclear separation decreases. For the even eigenfunction case the molecule is stably bound by a rudimentary covalent bond. At equilibrium it has $R \simeq 1.1$ Å, which is where the energy as a function of $R$ has a minimum that is about 2.7 eV deep. The measured binding energy, i.e., the

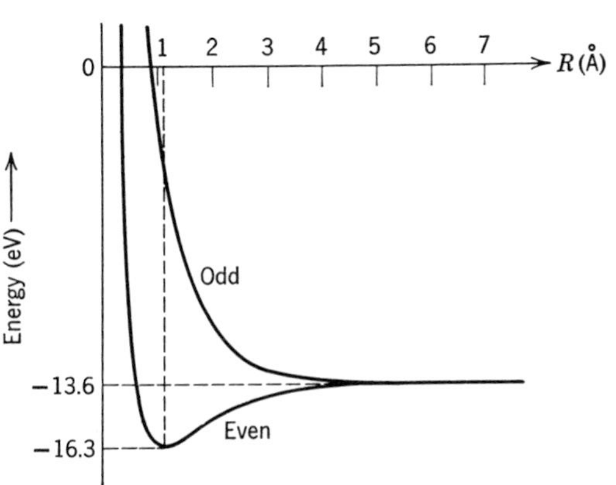


**FIGURE 12-3**
The total energy of the $H_2^+$ molecule for the two lowest electron energy levels, as a function of the internuclear separation. The molecule binds only in the state where the electron eigenfunction is even.

energy required to dissociate $H_2^+$ into H and $H^+$, is in good agreement with this value. Because of the significantly weaker binding of the electron in the odd eigenfunction state, the corresponding total molecular energy curve does not have a minimum at any value of $R$. Thus the molecule will not bind if the eigenfunction of the electron is odd since its energy always decreases as the nuclear separation increases.

If we now add a second electron to $H_2^+$ to form $H_2$, the energy of the system is decreased further, the two additional attractive forces acting between this electron and the nuclei more than counteracting the electron-electron repulsion. For $H_2$ the binding energy is about 4.7 eV, and the equilibrium internuclear separation is about 0.7 Å. So $H_2$ is more compact, and more tightly bound, than $H_2^+$. The second electron in $H_2$ goes into a quantum state whose eigenfunction has the same space properties as the eigenfunction for the first electron. That is, in the lowest energy state of $H_2$ both electrons are in a state with the same space eigenfunction, and that eigenfunction is even with respect to reflection in the plane halfway between the two nuclei. So for both the probability density shows some concentration in the region between the two nuclei. Of course the exclusion principle demands that the two electrons have different spin eigenfunctions; thus they have spins with opposite $z$ components. Using the more precise terms of Section 9-3, the eigenfunction describing the system of two indistinguishable electrons is a product of a symmetric space eigenfunction and the antisymmetric (i.e., singlet) spin eigenfunction. In that section we found that the two electrons may be relatively close together when the system is described by such an eigenfunction. Of course this is consistent with the idea that both have a reasonable chance of being located near the point halfway between the nuclei.

Because of the complete space overlap of the wave functions of the indistinguishable electrons in $H_2$, it is definitely not possible to associate a particular electron with a particular atom of the molecule. Instead, the two electrons, which are responsible for the bond that holds the atoms together as a molecule, are shared by the molecule, or shared by the bond itself. This is the idea of the *shared pair of electrons, with "anti-parallel" spins, that form a covalent bond.* Note that if the two electrons had essentially parallel spins they could not both be in the region between the two nuclei. Then they could not both be where they optimize the attraction exerted on them by both nuclei. If we imagined trying to form $H_2$ by bringing two separated H atoms together, it would make a decisive difference whether the electrons spins were "parallel" or "anti-parallel." In Figure 12-4 we show the prediction of quantum mechanics for the total energy of the system as a function of internuclear separation in the two possibilities; binding is obtained only for "antiparallel" spins. The calculations that produced the curves in Figure 12-4 take into account the electron-electron repulsion.

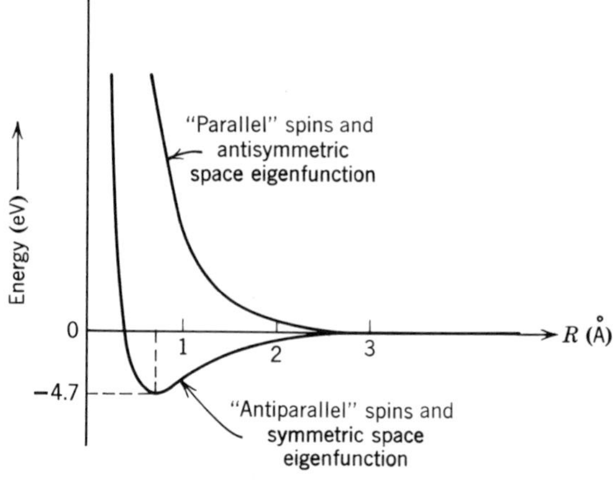


**FIGURE 12-4**

The total energy of the $H_2$ molecule for "parallel" and "antiparallel" electron spins, as a function of the internuclear separation. The molecule binds only in the state where the electron spins are "antiparallel".

This has a quantitative effect in reducing the binding, but it does not make a qualitative change in the description we have presented of the origin of the covalent bond.

No more than two electrons can form one covalent bond. We say an electron from one atom *pairs* up with an electron of "*antiparallel*" spin from another atom. If an atom has several electrons in an uncompleted outer subshell, i.e., if it has several valence electrons, each may try to form a covalent bond with a valence electron in a nearby atom. However, if there are two valence electrons with "antiparallel" spins in one atom, an additional valence electron from another atom will not succeed in forming a bond with either of them since they are already paired with each other. That is, if the spin of the additional electron is "antiparallel" to the spin of one of these electrons, it is "parallel" to the spin of the other. Since the exclusion principle acts in the molecule in such a way as to prevent two electrons with "parallel" spins from having the same space eigenfunction, the additional electron may not occupy the same energetically favorable molecular region as the electrons of the preexisting pair. Therefore the valence electrons of an atom that are effective in forming covalent bonds are those which the action of the exclusion principle in the atom has not already forced into pairs with "antiparallel" spins. For instance, in the Hartree theory all of the three $2p$ electrons in N can have "parallel" spins because there are three possible values of the quantum number $m_l$ for $l = 1$, so none of them are forced to pair in that atom. (In the residual Coulomb interaction theory the three electrons *do* have "parallel" spins in the ground state of the $LS$ coupling atom N.) The result is that the molecule $N_2$ has three covalent bonds. But O has a fourth electron in the $2p$ subshell, and the exclusion principle forces it to have its spin "antiparallel" to the spin of one of the other three. So there are only two unpaired valence electrons in O, and the molecule $O_2$ has only two covalent bonds. In general, the number of unpaired valence electrons equals the number of electrons in the subshell up to the point where it is half filled, and it equals the number of vacancies, or holes, in the subshell beyond that point.

As in ionic binding, the forces *saturate* in covalent binding. That is, a given atom strongly interacts with only a limited number of other atoms. Saturation is due to the limited number of electrons or vacancies in the outermost occupied subshell of the atom. As distinguished from the ionic bond, the covalent bond is *directional*. The directional property is not present in $H_2$ since the probability density of the valence electron in each separated H atom is spherically symmetrical, so the only defined direction in the $H_2$ molecule is the one connecting the two nuclei, and the covalent bond acts along that direction, whatever it may be. In a more typical case the probability density of a valence electron has its own directional dependence and certain preferred directions for forming covalent bonds. The directional properties of covalent bonds are manifested in the structural properties of covalently bonded molecules, and so form the basis of organic chemistry. The charge distribution of the paired electrons in a covalent bond has a symmetry about the center of the molecule, as we discussed in the case of $H_2$, so there is no permanent electric dipole moment associated with the covalent bond. The bond is therefore sometimes called *homopolar*. Because the binding in molecules other than those containing two identical nuclei may be partly ionic, even though principally covalent, only molecules like $O_2$ or $N_2$ are strictly homopolar.

## 12-4 Molecular Spectra

Molecules can remain bound in excited states as well as in the ground state. The emission and absorption spectra of molecules are due to transitions between allowed energy states. The energy-level scheme is relatively complicated and differs in many respects from the atomic case. For one thing, we can no longer classify states according to the electronic orbital angular momentum. Because the force on an electron is not a central force (in a diatomic molecule, e.g., there are two separated nuclear attracting centers), the magnitude of its orbital angular momentum $L$ is not conserved. In the words of Section 7-9, the energy eigenfunctions are not eigenfunctions of the operator $L^2_{op}$. However, in a diatomic molecule the total charge distribution is symmetrical about an axis connecting the nuclei, say the $z$ axis, so that the component of angular momentum about this axis, $L_z$, is conserved. We find then that the molecular energy eigenfunctions are eigenfunctions of $L_{z_{op}}$ and that $L_z$ has allowed values which are integral multiples of $\hbar$, in analogy to the values $m_l\hbar$ of atomic states.

Another difference between the molecular and atomic cases is that we could neglect the nuclear motion in an atom, or else we could take it into account easily by using the reduced electron mass. Of course, in a molecule, as well as in an atom, we do not need to consider the translational motion because that motion, being free particle motion, is not quantized. However, the nuclei in a molecule can move relative to one another. In a diatomic molecule, for example, the nuclei can vibrate about the equilibrium separation, and in addition the whole system can rotate about its center of mass. The energy in each of these motions, vibrational and rotational, is quantized so that we expect many more energy levels in a molecule than in an atom. Indeed, these motions interact or couple with one another and an exact analysis would have to take this into account.

Of course, the solution of the Schroedinger equation for any but the simplest molecules is very difficult. However, empirical results of molecular spectroscopy show that we can consider the energy of a molecule to be made up of three principal parts—electronic, vibrational, and rotational. The molecular energy levels fall into widely separated groups, each group being said to correspond to a different electronic state of the molecule. For a given electronic state the levels again fall into groups separated by nearly equal energy intervals; these are said to correspond to successive states of vibration of the nuclei. Within a vibrational state is a fine structure of levels ascribed to different states of rotation of the molecule. This level structure suggests that we can obtain an approximate solution to the Schroedinger equation by separating it into three equations, one describing the motion of the electrons, one the vibration of the nuclei, and one the rotation of the nuclei. In the next approximation we can take into account the coupling between the electronic and the nuclear motions, such as that between the electronic angular momentum and the rotation of the molecule, and the coupling between the nuclear vibrational and rotational motions.

The spectrum emitted by a molecule can be divided into three spectral ranges corresponding to the different types of transitions between molecular quantum states. In the far infrared we observe the rotation spectra, corresponding to radiation emitted in transitions between rotational states of a molecule having an electric dipole moment. In the near infrared we observe the vibration-rotation spectra, corresponding to radiation emitted in vibrational transitions of molecules having electric dipole moments, within which there are changes in rotational states as well. In the visible and ultraviolet part of the spectrum we observe electronic spectra, corresponding to radiation emitted in electronic transitions. The electronic vibrations undergo many cycles in the time required for the nuclear configuration to change (this being the physical reason that permits us to separate the eigenfunction into an electronic and nuclear factor to begin with), so that the electronic spectra have a fine structure determined by the rotational and vibrational state of the nuclei during electronic transitions.

In the succeeding sections we shall examine the motion and spectra of diatomic molecules and from this extract valuable information about their properties.

## 12-5 Rotational Spectra

The rotational motion of a diatomic molecule can be visualized as the rotation of a rigid body about its center of mass, illustrated in Figure 12-5. The center of mass lies on the axis connecting the nuclei, and the angular momentum associated with the rotation is a vector passing through the center of mass on the axis of rotation perpendicular to the internuclear axis. Rotation about the internuclear axis itself is negligible. The rotational inertia, or moment of inertia, about the axis of rotation due to the nuclei is $I = \mu R_0^2$, where $R_0$ is the (equilibrium) separation of the nuclei and $\mu$ is the reduced mass of the system. As in proven in the caption to Figure 12-5, the rotational energy is, classically, $E_r = L^2/2I$ where $L$ is the angular momentum of the system about the axis of rotation. Quantization of the magnitude of the angular momentum gives $L^2 = r(r + 1)\hbar^2$ with the *rotational quantum number* $r = 0$, 1,

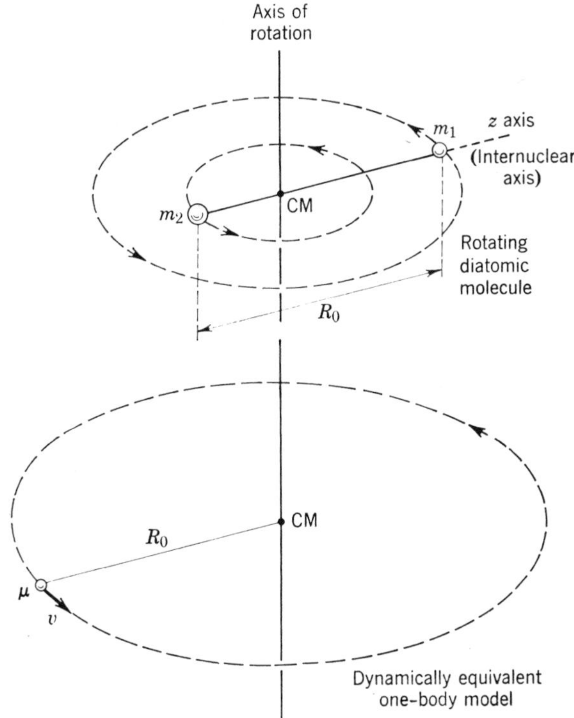


**FIGURE 12-5**

*Top:* A simplified picture of a diatomic molecule consisting of two masses $m_1$ and $m_2$ rotating about their common center of mass (CM) with separation $R_0$. *Bottom:* A dynamically equivalent model consisting of a reduced mass $\mu = m_1m_2/(m_1 + m_2)$ rotating at distance $R_0$ about a fixed point. If $v$ is the speed of the reduced mass $\mu$, then its kinetic energy of rotation is $E_r = \mu v^2/2$ and its angular momentum is $L = \mu v R_0$. So $E_r = \mu L^2/2\mu^2R_0^2 = L^2/2\mu R_0^2 = L^2/2I$, where $I \equiv \mu R_0^2$ is its rotational inertia, or moment of inertia.

2, . . . , so that

$$E_r = \frac{\hbar^2}{2I} r(r + 1) \tag{12-1}$$

Successive rotational levels will be separated in energy by

$$\Delta E_r = E_r - E_{r-1} = \frac{\hbar^2}{2I} [r(r + 1) - (r - 1)r] = \frac{\hbar^2}{I} r \tag{12-2}$$

The quantity $\hbar^2/I$ for the typical molecule has a value of about $10^{-4}$ eV to $10^{-3}$ eV, so little energy is needed to raise a molecule to an excited rotational state. At room temperature, for example, the translational thermal energy of molecules is $2.5 \times 10^{-2}$ eV, so that ordinary collisions can transfer the necessary energy of excitation. At

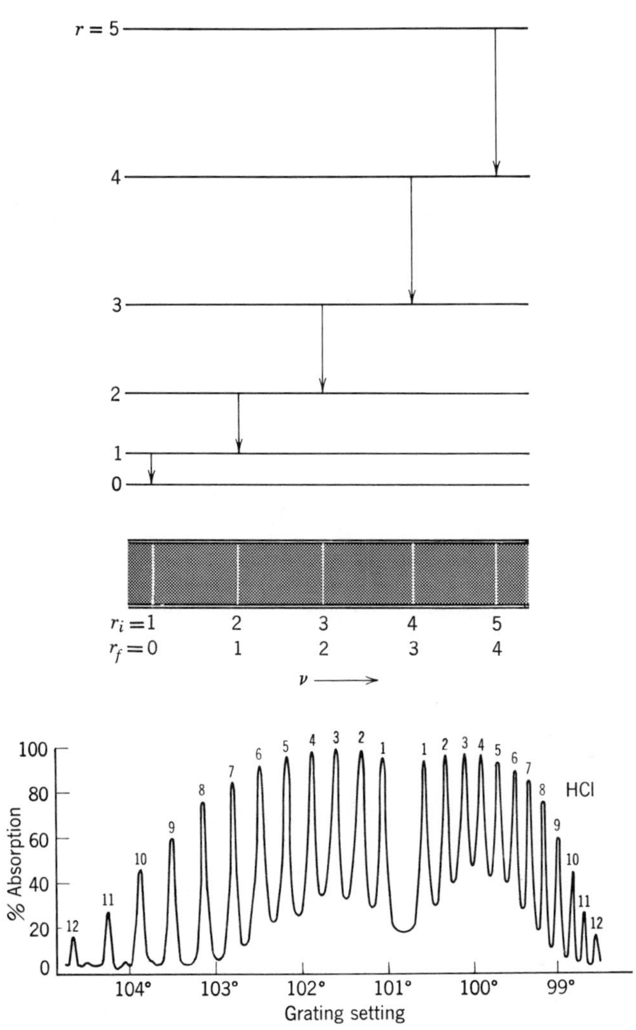


**FIGURE 12-6**

*Top:* Schematic energy-level diagram for the rotational energy states of a diatomic molecule, and the corresponding frequency emission spectrum for allowed transitions. *Bottom:* The rotational absorption spectrum for gaseous HCl, giving the percent absorption versus a measure of the reciprocal wavelength.

any given temperature the rotational state populations obey the Boltzmann distribution, since they are spread over many states so each population is small.

If the molecule has a permanent electric dipole moment, as do all diatomic molecules that do not have identical nuclei, rotational emission and absorption spectra may be observed. The emission of radiation is due to the rotation of the electric dipole and the absorption of radiation is due to the interaction of this dipole with the electric field of the incident radiation. For electric dipole radiation, the allowed transitions between states are given by the selection rule analogous to that for orbital angular momentum in atomic transitions, namely $\Delta r = \pm 1$. The spectral wavelengths $\lambda$ follow from (12-2), and

$$\Delta E_r = h\nu$$

That is,

$$\frac{\hbar^2}{I} r = \frac{hc}{\lambda}$$

or

$$\frac{1}{\lambda} = \frac{\hbar}{2\pi Ic} r \tag{12-3}$$

in which $r$ is the quantum number of the upper rotational state. With $\Delta r = \pm 1$, the separation between spectral lines (in terms of reciprocal wavelength) then is $\Delta(1/\lambda) = \hbar/2\pi Ic$, a constant. This is illustrated in Figure 12-6. Measurement of the separation gives the value of $I$, the rotational inertia of the molecule, and from this we can estimate the value of the equilibrium internuclear separation $R_0$. In the case of HCl, for example, we find $\hbar/2\pi Ic = 2079.4 \text{ m}^{-1}$, which gives $I = 2.66 \times 10^{-47}$ kg-m$^2$; from the known masses of H and Cl we then obtain $R_0 = 1.27 \times 10^{-10}$ m as a measure of the separation of the atoms in the molecule. Pure rotational spectra fall in the extreme infrared or the microwave regions, the corresponding wavelengths $\lambda$ being about 1 mm to 1 cm. An example is shown in Figure 12-6. Diatomic molecules with identical nuclei, like $O_2$, having no permanent electric dipole moment, do not exhibit pure rotational spectra.

**Example 12-2.** (a) Find the ratio of $n_r$, the number of molecules in rotational level $r$, to $n_0$, the number in the $r = 0$ level, in a sample in equilibrium at temperature $T$.

From the Boltzmann factor we have

$$\frac{n_r}{n_0} = \frac{\mathcal{N}_r}{\mathcal{N}_0} e^{-(E_r - E_0)/kT}$$

in which the $\mathcal{N}$'s are the degeneracy factors, or number of degenerate quantum states for each energy level. For energy $E_r$ there are $2r + 1$ states, corresponding to the number of possible values of the $z$ component quantum number $m_r$ associated with each value of $r$. Hence, $\mathcal{N}_r = 2r + 1$ and $\mathcal{N}_0 = 1$, so that

$$\frac{n_r}{n_0} = (2r + 1)e^{-(E_r - E_0)/kT}$$

(b) Show that the population of rotational energy levels first increases with $r$ and then decreases as $r$ continues to increase.

From (12-1) we have $E_r = (\hbar^2/2I)r(r + 1)$ and $E_0 = 0$, so that

$$n_r = n_0(2r + 1)e^{-(\hbar^2/2IkT)r(r+1)}$$

Now as $r$ increases the factor $2r + 1$ increases whereas the exponential factor decreases. For large $r$ the exponential term dominates so that at first $n_r$ increases with $r$, but soon the exponential suppresses the increase and $n_r$ decreases for larger $r$. For example, for HBr at room temperature $n_r$ is a maximum at $r = 3$ with $n_3/n_0 \simeq 4$, whereas by $r = 9$ we have $n_9/n_0 \simeq 1/2$.

(c) Relate these populations to the intensities of the rotational lines.

Consider the absorption spectrum. The probability that a particular frequency will be absorbed is proportional to the number of molecules in the initial rotational energy level. Hence the intensity variation of the absorption lines ($\Delta r = +1$) are proportional to the populations of the initial rotational energy levels (see Figure 12-6). The student should construct a similar argument for the emission spectrum. ◀

## 12-6 Vibration-Rotation Spectra

The nuclei do not maintain a fixed separation, of course, as we assumed previously, so that the molecule is not like a rotating rigid body except in approximation. Indeed, the rotational inertia $I$ changes from the value assumed previously when the molecule rotates because of the stretching of the internuclear distance. Also the nuclei vibrate about some equilibrium separation and this vibrational motion is quantized. Let us now consider the vibrational motion.

For a given electronic configuration, we have a potential energy curve whose minimum is at an equilibrium separation $R_0$. Near $R_0$ the curve is nearly a parabola so that small oscillations are simple harmonic. According to (6-89) the energy of such oscillations is quantized to satisfy

$$E_v = (v + 1/2)h\nu_0 \tag{12-4}$$

with the *vibrational quantum number* $v = 0, 1, 2, 3, \ldots$, and where the classical vibration frequency is $\nu_0 = (1/2\pi)\sqrt{C/\mu}$. Note that the energy levels here are equally spaced and that there is a zero-point energy $(1/2)h\nu_0$. The separation $h\nu_0$ equals 0.04 eV for NaCl and, because the dissociation energy is about 1 eV, there are approximately 20 vibrational levels in the potential well. Actually as the energy rises the potential energy curve becomes anharmonic so that the levels are not equally separated but get somewhat closer to one another. The rotational levels are spaced much closer still, as we saw earlier, there being about 40 rotational levels of NaCl, and about 50 of HCl, between each pair of vibrational levels.

**Example 12-3.** (a) Given that the equivalent force constant $C$ of a vibrating HCl molecule is about 470 nt/m, estimate the energy difference between the lowest and the first vibrational state of HCl.

We have for HCl

$$\mu = \frac{35}{36}m_{\rm H} \qquad \text{and} \qquad C = 470 \text{ nt/m}$$

and also

$$m_{\rm H} = \frac{1}{6.02 \times 10^{23}} \text{ g} = \frac{1}{6.02 \times 10^{26}} \text{ kg}$$

From (12-4) we have that $\Delta E = h\nu_0$, where $\nu_0 = (1/2\pi)\sqrt{C/\mu}$. Hence, using these data, we get the energy difference to be $h\nu_0 = (h/2\pi)\sqrt{C/\mu} = 0.59 \times 10^{-19}$ joule $= 0.37$ eV.

(b) Given that the rotational inertia of HCl has the value $I = 2.66 \times 10^{-47}$ kg-m$^2$, estimate the energy difference between the lowest and first excited rotational state of HCl.

Since $E_r = (\hbar^2/2I)r(r+1)$, the lowest rotational state has an energy $E_0 = 0$ and the first excited rotational state has an energy $E_1 = (\hbar^2/2I)2 = \hbar^2/I$. The required energy difference then is $\Delta E = \hbar^2/I$. Hence

$$\frac{\hbar^2}{I} = \frac{(6.63 \times 10^{-34}\text{ joule-sec})^2}{(2\pi)^2 \times 2.66 \times 10^{-47}\text{ kg-m}^2} = 4.2 \times 10^{-22}\text{ joule} = 2.6 \times 10^{-3}\text{ eV}$$

Thus the energy difference between the two lowest vibrational levels is greater by a factor 142 (i.e., $0.37/2.6 \times 10^{-3}$) than that between the two lowest rotational levels in HCl.

(c) At room temperature, collisions of HCl molecules in a gas can transfer sufficient kinetic energy to internal energy to excite many *rotational* states. At what temperature would the number of molecules in the first excited *vibrational* state be equal to $1/e$ (about 37%) of the number in the ground vibrational state?

We have

$$\frac{n_1}{n_0} = \frac{\mathcal{N}_1}{\mathcal{N}_0} e^{-(E_1 - E_0)/kT}$$

where the subscripts refer to $v = 1$ or $v = 0$. The vibrational states are not degenerate so that $\mathcal{N}_1 = 1 = \mathcal{N}_0$. Also $(E_1 - E_0) = h\nu_0$ so that

$$\frac{n_1}{n_0} = e^{-h\nu_0/kT}$$

and if $kT = h\nu_0$

$$n_1 = n_0 e^{-1}$$

Hence

$$T = \frac{h\nu_0}{k} = \frac{0.59 \times 10^{-19}\,\text{joule}}{1.38 \times 10^{-23}\,\text{joule}/^\circ\text{K}} \simeq 4300^\circ\text{K}$$

is the temperature at which the number of HCl molecules in the first excited vibrational state is about 37% of the number in the ground state. Clearly the number of HCl molecules in the $v = 1$ state at room temperature is negligible compared to the number in the ground state. ◀

If the molecule, like HCl or NaCl, has a permanent electric dipole moment at the equilibrium internuclear separation, it will exhibit vibrational emission and absorption spectra due to the oscillations in the electric dipole moment arising from oscillations in the nuclear separation. The selection rule for electric dipole transitions is $\Delta v = \pm 1$ so that $\Delta E_v \simeq h\nu_0$. The resulting spectral lines lie in the infrared, between 8000 Å and 50,000 Å for most molecules. Diatomic molecules with identical nuclei do not have vibrational spectra because they have no electric dipole moment at any nuclear separation. In a vibrational transition the molecule may also change its rotational state so that vibrational changes really result in a combined vibration-rotation spectrum. The vibrational transition determines the wavelength region of the spectrum and the rotational transitions determine the separation of the lines. The spectrum consists of a band of lines, as in Figure 12-7.

Among the interesting results that can be obtained from analysis of vibrational states and spectra are the relative abundance of nuclear isotopes. The frequency of vibration, $\nu_0 = (1/2\pi)\sqrt{C/\mu}$, depends on the masses of the atoms in the molecule through the reduced mass $\mu$. If in a sample of HCl molecules, for example, the isotopes $Cl^{35}$ and $Cl^{37}$ are each present, then the vibrational frequencies and resulting energy levels will be slightly different for the two types of molecule (see Figure 12-7). Their spectral lines, consequently, will be shifted with respect to one another, and from a measurement of spectral intensities we can obtain the relative abundance of the isotopes $Cl^{35}$ and $Cl^{37}$.

In a somewhat related way we obtain experimental evidence for the finite zero-point energy of an oscillator. Consider the molecules $H_2$, HD, and $D_2$ in which D stands for a deuterium atom. Because the electric forces are identical in all cases we obtain for all the same potential energy curve $V(R)$, illustrated in Figure 12-8. The energy required to dissociate the molecule is $E_d = V_0 - \delta$. If the ground state energy $\delta$ were zero, then the dissociation energies would be the same, $E_d = V_0$, for each type of

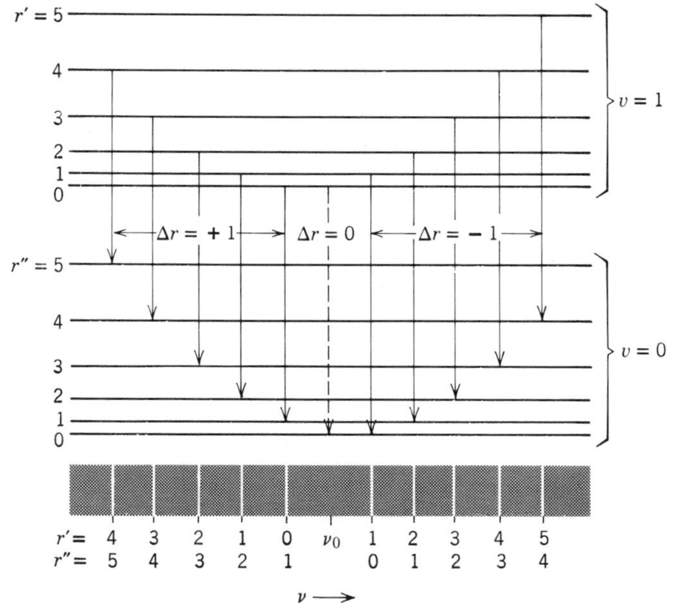


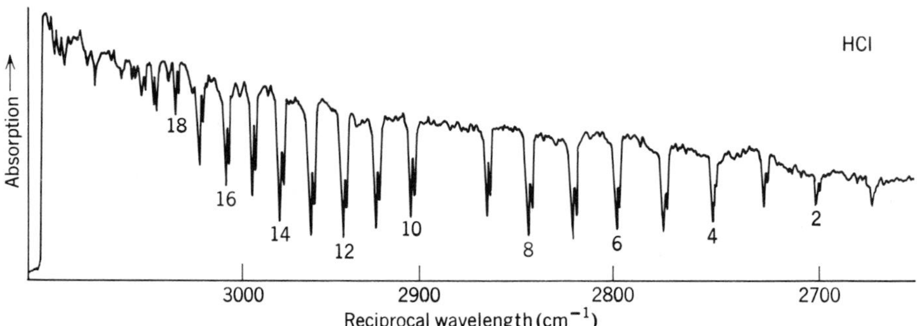


**FIGURE 12-7**

*Top:* Energy-level diagram for vibrational and rotational states of a diatomic molecule, showing allowed transitions and the formation of a band of equally spaced lines, as indicated in the spectrum below. Note that all $\Delta r = 0$ transitions would yield photons of the same frequency $\nu_0$, but being forbidden, that line is missing in the spectrum. *Bottom:* A recorder trace of the vibration-rotation absorption spectrum in HCl. Again note that the central transition is missing. The slightly different frequencies at each absorption line are due to the presence of two isotopes of chlorine.

molecule. Quantum theory gives a finite zero-point energy, namely $\delta = (1/2)h\nu_0$. However, because the reduced mass $\mu$ enters the formula for $\nu_0$, $\delta$ has a different value for each type of molecule so that their dissociation energies should differ. In fact, with

$$\mu_{\mathrm{D}} = 2\mu_{\mathrm{H}_2} \qquad \text{and} \qquad \mu_{\mathrm{HD}} = (4/3)\mu_{\mathrm{H}_2}$$

we can predict the difference, and we find that the observed dissociation energies differ exactly as predicted, thereby verifying the existence of a zero-point energy in agreement with the requirements of the uncertainty principle.

In Table 12-1 we list the rotational and vibrational constants of some diatomic molecules.

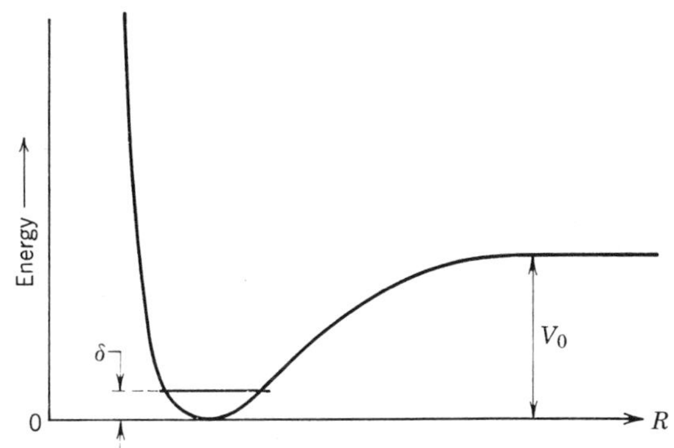


**FIGURE 12-8**
The energy for $H_2$, HD, and $D_2$ is the same function of the internuclear separation $R$. But the ground state vibrational energy $\delta$ differs for each molecule.

## 12-7 Electronic Spectra

The rotational and vibrational states in molecules are due to the motion of the nuclei. There can be also electronic excited states, of course. For each of the electronic states, corresponding to different electron configurations, there is a different dependence of the molecule's energy on its internuclear separation. Because the atoms are more loosely bound in the excited states, the curves representing the molecule's potential energy as a function of nuclear separation become shallower and broader, and the equilibrium separation $R_0$ increases, with increasing electronic excitation, as illustrated in Figure 12-9. The energy separation between different electronic states is from 1 to 10 eV, so that transitions between electronic states give radiation in the visible or ultraviolet portion of the electromagnetic spectrum.

To each electronic state $E_e$ there are many bound vibrational states of energy $E_v$, and to each vibrational state there are many bound rotational states of energy $E_r$. Neglecting interactions between these modes, we can write the total energy as $E = E_e + E_v + E_r$. The energies of all three modes may change in an electronic transition so that in general we can write

$$\Delta E = \Delta E_e + (E_v' - E_v'') + (E_r' - E_r'') \tag{12-5}$$

The initial (primed) and final (double-primed) vibrational and rotational states differ in their binding so that the equilibrium spacing, the rotational inertia, and the fundamental vibrational frequency change. A great many transitions are possible and they produce a complex spectrum of lines, which appear in a series of bands as illustrated in Figure 12-10. Hence the term *band spectra.*

The term $\Delta E_e$ is the energy difference of the minima of the two electronic states. The vibrational term is $E_v' - E_v'' = (v' + 1/2)h\nu_0' - (v'' + 1/2)h\nu_0''$ and the rotational

**TABLE 12-1.** Rotational and Vibrational Constants of Some Diatomic Molecules

| Molecule | $R_0$(Å) | $\nu_0$ (cm$^{-1}$) | $\frac{\hbar^2}{2I}$ (eV) | Molecule | $R_0$(Å) | $\nu_0$ (cm$^{-1}$) | $\frac{\hbar^2}{2I}$ (eV) |
|---|---|---|---|---|---|---|---|
| $H_2$ | 0.74 | 4395 | $7.56 \times 10^{-3}$ | LiH | 1.60 | 1406 | $9.27 \times 10^{-4}$ |
| HD | 0.74 | 3817 | $5.69 \times 10^{-3}$ | $HCl^{35}$ | 1.27 | 2990 | $1.32 \times 10^{-3}$ |
| $D_2$ | 0.74 | 3118 | $3.79 \times 10^{-3}$ | $NaCl^{35}$ | 2.51 | 380 | $2.36 \times 10^{-5}$ |
| $Li_2$ | 2.67 | 351 | $8.39 \times 10^{-5}$ | $KCl^{35}$ | 2.79 | 280 | $1.43 \times 10^{-5}$ |
| $N_2$ | 1.09 | 2360 | $2.48 \times 10^{-4}$ | $KBr^{79}$ | 2.94 | 231 | $9.1 \times 10^{-6}$ |
| $O_2$ | 1.21 | 1580 | $1.78 \times 10^{-4}$ | $HBr^{79}$ | 1.41 | 2650 | $1.06 \times 10^{-3}$ |

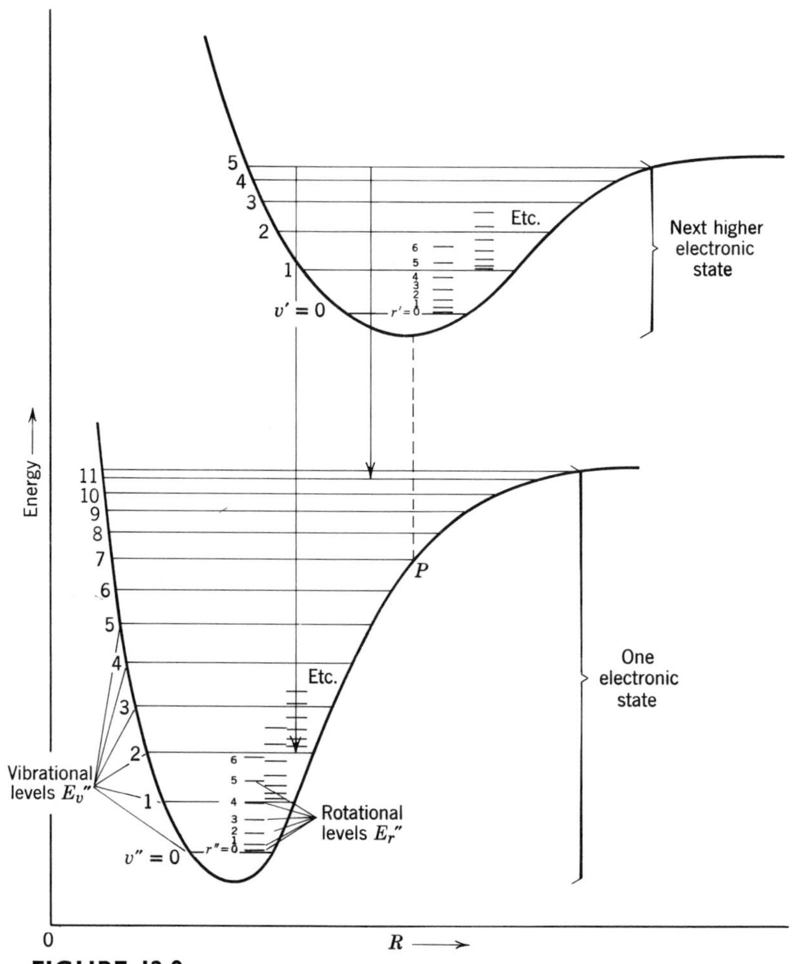


**FIGURE 12-9**

Illustrating the molecular energy versus internuclear separation curves for two electronic states. Each electronic state has its own set of vibrational levels, and each vibrational level has its own set of rotational levels.

term is $E_r' - E_r'' = (\hbar^2/2I')r'(r' + 1) - (\hbar^2/2I'')r''(r'' + 1)$. For a given electronic transition the spectrum consists of bands, where each band corresponds to given values of $v'$ and $v''$ and all possible values of $r'$ and $r''$. The selection rules determine the possible combination of values of $v'$, $v''$, and $r'$, $r''$. The rotational selection rule here is $\Delta r = 0, \pm 1$ for electric dipole radiation. This rule is broader than for pure rotation in that $\Delta r = 0$ is now allowed. The reason is that the change in the electronic configuration accompanying the rotational change eliminates the parity considerations which earlier excluded $\Delta r = 0$ (see Section 8-7). The vibrational selection rule for electric dipole radiation is $\Delta v = \pm 1$ for a simple harmonic oscillator. If, however, the potential deviates from the simple harmonic, i.e., if it is anharmonic, then $\Delta v = 2, 3, \ldots$, etc., are also allowed. These vibrational rules apply only if the electronic state does not change and they apply to pure vibration-rotation bands. If there is a change in electronic state then the selection rules are determined from the so-called *Franck-Condon principle*, which we explain next.

We have seen that there is little interaction between the electronic motion and the nuclear motion in a molecule. Furthermore, the characteristic time for an electronic

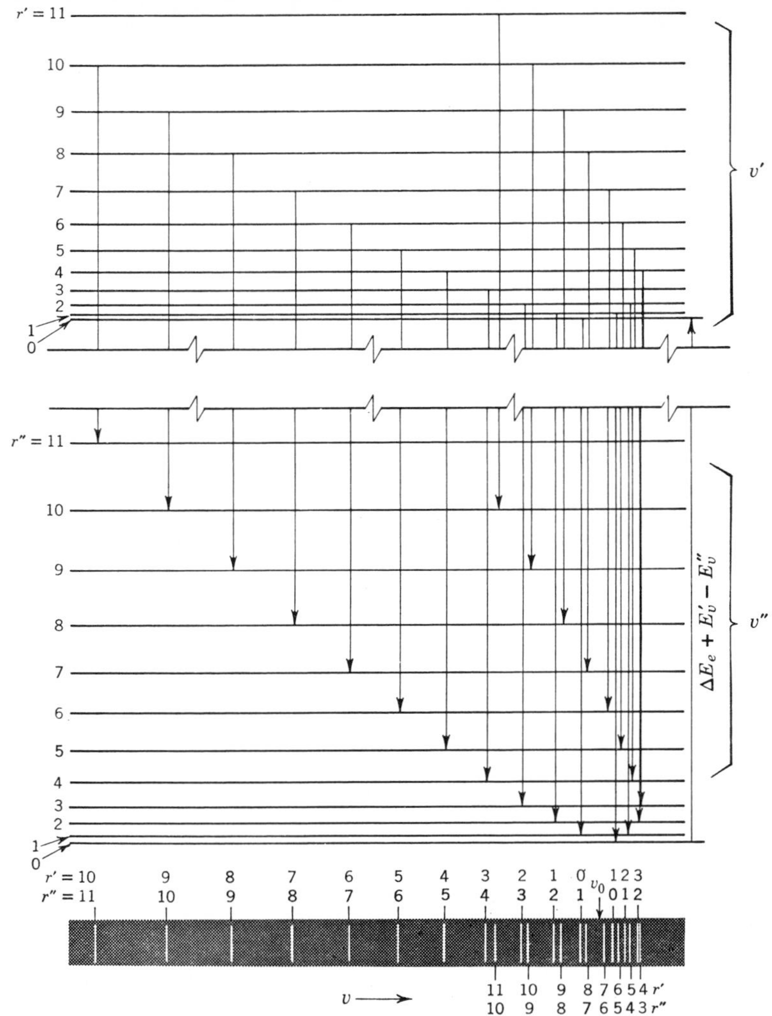


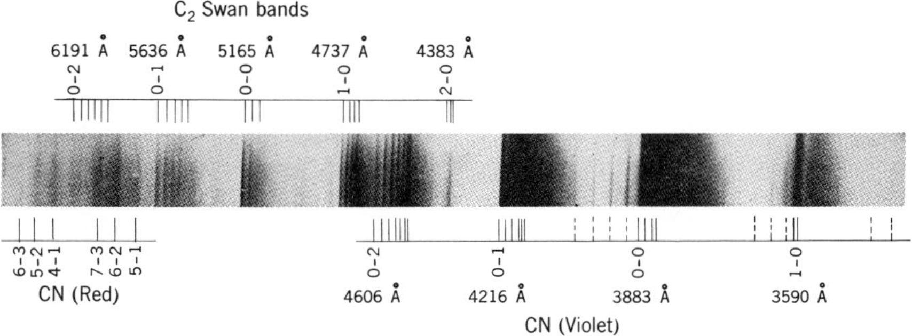


**FIGURE 12-10**

*Top:* Energy-level diagram and transitions leading to the formation of an electronic band. Unlike Figure 12-7, the band spectrum indicated folds back on itself, giving rise to a band head at the right end of the spectrum. Again note that the transition of frequency $\nu_0$ is missing. *Bottom:* Bands of the CN and $C_2$ molecules in a carbon arc in air. (From Herzberg, *Spectra of Diatomic Molecules*, 1950. D. Van Nostrand Co., Inc., New York)

transition is $\Delta t \simeq 10^{-16}$ sec, whereas for a nuclear vibration the time has the much longer value $\Delta t \simeq 10^{-13}$ sec. As a result the internuclear distance stays about the same during an electronic transition, and a vertical line (a line of constant $R$) in Figure 12-9 accurately represents such a transition. If the upper state corresponds to $v' = 0$, then the probability distribution function for the oscillator is large only near the equilibrium separation, and an electronic transition to the lower state leaves the molecule at about the point $P$ on the potential curve in that figure. This corresponds to $v'' = 7$ for the lower state. Notice that classically the nuclei have small kinetic energy in each case, because $v' = 0$ initially, and because $P$ corresponds to the end point of the vibrational motion for $v'' = 7$. This meets the requirement that the relative nuclear velocity be about the same in both states at the time of a transition in order that the nuclear motion be able to adjust quickly to the new electronic conditions. Transitions are most favorable under these conditions. Quantum mechanically we get the same result because in the ground state of an oscillator, as in $v' = 0$, the maximum amplitude of the eigenfunction occurs at the center of the motion, whereas for the upper states, such as in $v'' = 7$, the eigenfunction has maximum amplitude near the ends of the oscillation. Since the integral in the electric dipole matrix element, (8-42), that determines the relative intensities, or selection rules, involves a product of the eigenfunctions of the upper and lower states, the intensities will be large only where both these eigenfunctions have significant space overlap. In general, the most favored transitions are those that, from a classical point of view, can occur with the internuclear distance for both initial and final states the same and the nuclei at end points of their oscillations. Examples in Figure 12-9 are shown by vertical lines from $v' = 5$ to $v'' = 2$ or $v'' = 11$. These rules were deduced by Franck from classical considerations and put on a firm quantum mechanical basis by Condon.

If the excited electronic state is not bound, the molecule dissociates. Because such unbound states have a continuum of possible energies, the corresponding spectrum gives a continuous band. The appearance of a continuum in the absorption spectrum of a molecule is therefore experimental evidence for photochemical dissociation.

## 12-8 The Raman Effect

An interesting effect which gives much information about molecular quantum states was discovered experimentally in 1928 by Raman. This is the scattering of light by molecules with a frequency change. The student may be familiar with other light scattering processes. In ordinary Rayleigh scattering by molecules, the scattered frequency is the same as the incident frequency. In the fluorescence process, the frequency of the incident light coincides with an absorption frequency of the scattering gas molecules; this is a resonance phenomenon in which the molecule is raised to an excited state and, after a short lifetime there, reemits light at a different frequency. In the Raman effect, the scattered frequency is different from the incident frequency, and the incident frequency is *not* related to a characteristic frequency of the scattering molecule.

If the incident radiation is intense and monochromatic with a frequency $\nu$, it is found that the light scattered at right angles to the incident direction contains not only radiation of frequency $\nu$ (Rayleigh scattering), but also weaker radiation of frequency $\nu \pm \nu'$ (Raman scattering). The scattered spectrum therefore has weak Raman lines on each side of the Rayleigh line. If we change the incident frequency, we again find weak lines on each side of the Rayleigh line in the scattered spectrum with the same frequency difference as before. The frequency *difference* $\nu'$ between the incident and scattered light in the Raman effect is characteristic of transitions in the scattering molecule. During the scattering process the molecule may have its state changed from one allowed energy to another. To conserve energy in the process the scattered photon must then have an energy different from the incident photon by an amount equal but opposite to the molecular energy change.

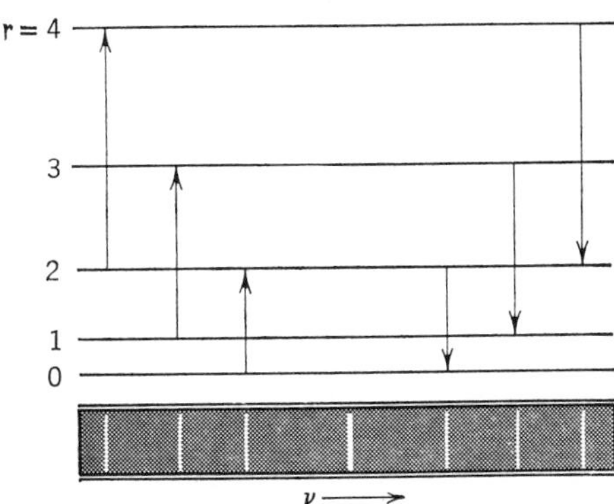


**FIGURE 12-11**
Schematic diagram showing the origin of rotational Raman lines on each side of the Rayleigh scattering line.

Consider a scattering molecule in a rotational state $r$. In the ordinary rotational spectrum, lines will be found corresponding to transitions with $\Delta r = \pm 1$. In the scattered Raman spectrum, however, we find frequency shifts from the incident frequency that correspond to rotational transitions in the scattering molecule with $\Delta r = \pm 2$. Hence, transitions that are not allowed in the ordinary emission or absorption spectrum are allowed in the Raman process. A quantum mechanical analysis of the Raman process leads to the conclusion that a Raman transition between states $\alpha$ and $\beta$ can occur only if there is a state $\gamma$ such that ordinary transitions are allowed between $\alpha$ and $\gamma$ and $\beta$ and $\gamma$. It is as though we get from $\alpha$ to $\beta$ by going through $\gamma$. In this case, if $\alpha$ has quantum number $r$ then $\gamma$ has $r \pm 1$. An ordinary transition from $\gamma$ to $\beta$, however, requires another change $\Delta r = \pm 1$, so that the total change in $r$ from $\alpha$ to $\beta$ is $\Delta r = 0, \pm 2$. The $\Delta r = 0$ selection rule gives Rayleigh scattering, and the $\Delta r = \pm 2$ selection rule gives Raman scattering. Hence in the scattered spectrum we have lines on each side of the incident line which are spaced about twice as far apart in frequency as the lines in the ordinary rotational spectrum. This is shown schematically in Figure 12-11.

There is a Raman effect with vibrational states as well. In the process of scattering a photon of frequency $\nu$ a molecule may change its vibrational state. Because $\Delta v = \pm 1$, the final vibrational level of the molecule may be one just above or just below the initial level. Therefore the Raman scattering frequency will be $\nu \pm \nu'$, where the frequency change $\nu'$ is a characteristic vibrational frequency of the molecule. At ordinary temperature, however, most molecules are in the ground vibrational state, $v = 0$, so that the molecule absorbs energy in changing to state $v = 1$. Hence, only the lower frequency line $\nu - \nu'$ appears in the Raman spectrum. However, the higher frequency line $\nu + \nu'$ may be observed if the $v = 1$ level is sufficiently populated so that enough transitions from $v = 1$ to $v = 0$ occur to give detectable intensities. This is more likely the lower the energy of the $v = 1$ state and the higher the temperature of the scattering gas.

As an example of the utility of Raman scattering, consider molecules with two identical nuclei, such as $O_2$ and $N_2$. We cannot directly observe rotational spectra or vibration-rotation spectra for such molecules because they have no electric dipole moment. We can, however, obtain a spectrum corresponding to vibration and rotation of such molecules in the Raman scattering. It is as though the incident radiation polarizes the molecule, thereby inducing an electric dipole moment; this permits absorption and emission of radiation corresponding to rotational and vibrational motions of the molecule. Of course, in an electronic transition in $O_2$ or $N_2$ the fine structure of the spectrum reveals the vibrational and rotational structure, but such a spectrum lies in the ultraviolet and the fine structure is very difficult to resolve. Historically, Rasetti used the Raman spectrum to make the first determination of the rotational inertia, or moment of inertia, of the $N_2$ molecule.

## 12-9 Determination of Nuclear Spin and Symmetry Character

We have ignored the weaker interactions that enter in the detailed structure of molecular spectra, such as the effect of nuclear spin on the energy states of a molecule. But we cannot ignore a very important effect that nuclear spin has on the spectrum of a molecule even when the spin interaction itself is negligible. For a diatomic molecule with *identical* nuclei, the states that can be occupied and the transitions that are allowed are restricted by symmetry requirements. If the nuclear spins are integral (0,1,2,. . .) then the complete eigenfunction of the molecule must be symmetric with respect to exchange of the labels of the two identical boson nuclei. If the nuclear spins are half-integral (1/2,3/2,. . .) then this eigenfunction must be antisymmetric in an exchange of the labels of the two nuclei because they are identical fermions.

If we neglect the small interactions between the modes associated with the electronic, vibrational, rotational, and nuclear spin behavior of the molecule, we can write the molecular eigenfunction as a product of four factors. Since it is usually the case, we henceforth assume the electronic factor is symmetric in an exchange of the labels of the two nuclei because it is even in a reflection in the plane half way between them (as in $H_2$). The vibrational factor is always symmetric since it can be written

$$\psi_v = \psi_v(|x_1 - x_2|)$$

where $x_1$ and $x_2$ are the coordinates of the nuclei labeled 1 and 2, measured along their center to center line. That is, the independent variable in the vibrational eigenfunction is the *magnitude* of the distance between the two identical nuclei. Since this does not change when the nuclear labels are exchanged, $\psi_v$ itself does not change and so is symmetric with respect to the exchange. Thus the symmetry of the molecular eigenfunction is governed by the symmetry of the product of its rotational factor and its nuclear spin factor.

The question of what happens to the sign of the rotational factor $\psi_r$ when we exchange the labels of the identical nuclei is intimately related to the question of what happens to the sign when we change the signs of all the coordinates, providing we are wise enough to choose the origin of coordinates at the center of the molecule (i.e., at its center of mass, halfway between the nuclei). With this choice, the parity questioning operation of (8-44) ($x \rightarrow -x, y \rightarrow -y, z \rightarrow -z$) obviously accomplishes the same thing as the symmetry questioning operation ($1 \rightarrow 2, 2 \rightarrow 1$), and the symmetry of $\psi_r$ becomes the same as its parity. Furthermore, we can immediately apply the interpretation of (8-47) to determine the parity of $\psi_r$, if we change from the orbital angular momentum quantum number $l$ used there to the rotational quantum number $r$ used here, and conclude that the parity of $\psi_r$ is even if $r$ is even and the parity of $\psi_r$ is odd if $r$ is odd. The justification is that if the rotational angular momentum of the molecule *is* quantized then there can be no external torques acting on it, so the potential energy function describing the external environment (if any) in which the molecular rotation takes place must be spherically symmetrical about our origin of coordinates; this is the only requirement for the validity of (8-47). Putting it all together, we see that the rotational eigenfunction $\psi_r$ is symmetric if $r$ is even, and antisymmetric if $r$ is odd.

Now let us consider a situation in which the *nuclear spin angular momentum quantum number* $i$ has one of the values $i = 1/2, 3/2, 5/2, \ldots$. Then the complete molecular eigenfunction must be antisymmetric in a nuclear label exchange. There are two ways this can come about: (1) either the nuclear spin eigenfunction is antisymmetric and the rotational eigenfunction is symmetric, or (2) the nuclear spin eigenfunction is symmetric and the rotational eigenfunction is antisymmetric. Both possibilities will occur, but not in the same molecule. The reasons are: (1) the symmetry of the nuclear spin eigenfunction factor is determined by the relative orientation of the two nuclear spins (e.g., for $i = 1/2$, the symmetric case corresponds to the two spins being essentially parallel while the antisymmetric case corresponds to them being essentially antiparallel, exactly as for two electrons with spin quantum number $s = 1/2$), and (2) the interaction between the nuclear spins is very small so that if the spins have a particular relative orientation, they will maintain it for a very long time (as long as years).

Practically, it is as though there are two distinctly different species of molecules. The species with *symmetric nuclear spin eigenfunctions* is called *ortho* and the species with *antisymmetric nuclear spin eigenfunctions* is called *para* as, for example, orthohydrogen and parahydrogen. The same terminology is used in the same way, whether $i$ is half-integral or integral. But if $i$ is half-integral, the ortho species has only antisymmetric rotational eigenfunctions and the para species only symmetric rotational eigenfunctions, as we have been considering; while if $i$ is integral, the symmetry of the complete molecular eigenfunction is reversed so the ortho species has only symmetric rotational eigenfunctions and the para species has only antisymmetric rotational eigenfunctions. These relations are summarized in the rotational energy-level diagrams of Figure 12-12. The pair on the left are for molecules whose nuclei have half-integral spin. For the ortho species of such molecules only odd-$r$ rotational states can be populated because the rotational eigenfunction must be antisymmetric, and it is only for odd $r$. In the para species only the symmetric rotational states can be populated, and these are the ones for even $r$. The relations are reversed for molecules with integral spin nuclei, as is indicated in the pair of energy-level diagrams on the right side of Figure 12-12. The dots in the figure show the energy levels that can be populated, and the arrows show the possible transitions between these levels.

Since molecules with two identical nuclei have no electric dipole moments, we cannot directly observe the rotational spectra emitted in such transitions; but we can indirectly observe transitions between rotational states in Raman scattering, or in band spectra, as explained in earlier sections.

Measurements of the number of transitions made by the para species of such molecules, relative to the number of transitions made by the ortho species, constitute

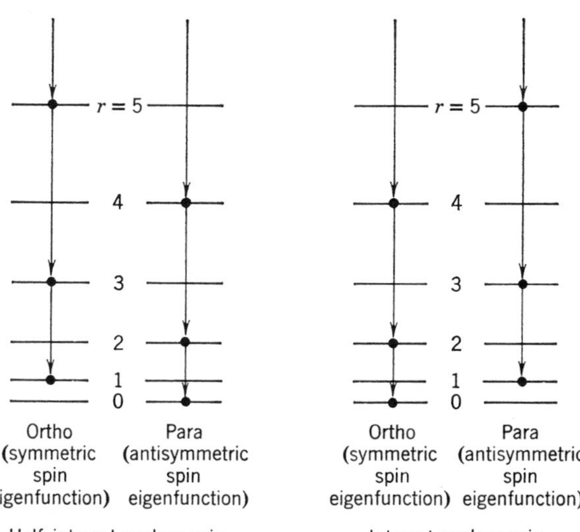


**FIGURE 12-12**

Illustrating the relation between the rotational and spin states that can be populated in molecules with identical half-integral, and integral, spin nuclei. The dots indicate the possible states and the arrows indicate transitions between these states.

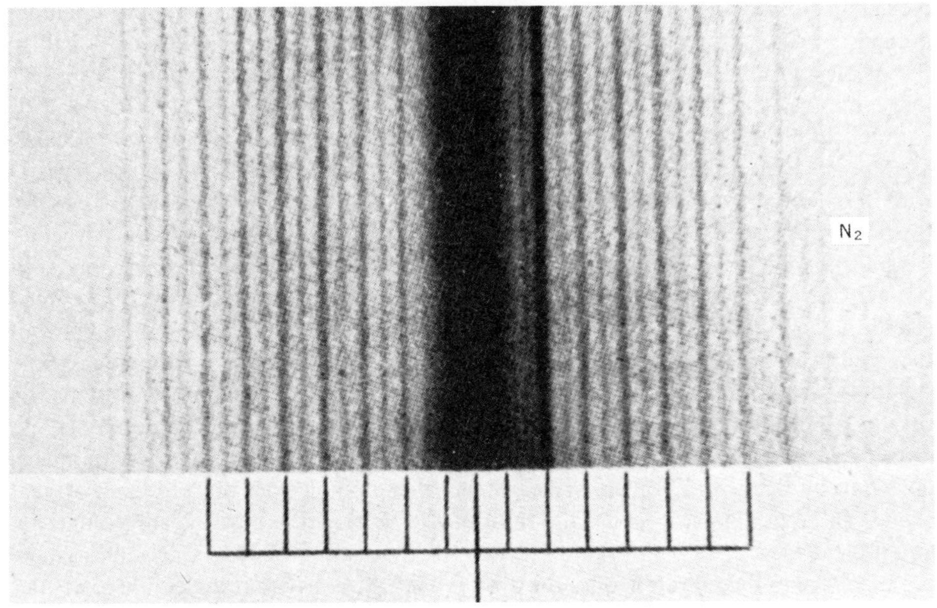


**FIGURE 12-13**

Alternating intensities in a rotational Raman spectrum of $N_2$, excited by the Hg line 2536.5 Å.

a quite frequently used procedure for determining the value of the spin quantum number $i$ of the nuclei forming the molecules. These numbers are in proportion to the relative amounts of the two species present in the sample and, at ordinary temperatures where many rotational states are excited, the relative amounts are in proportion to the numbers of nuclear spin states for the two species. We shall show in Example 12-6 that the ratio of the number of antisymmetric spin states, $\mathcal{N}_{\text{para}}$, to the number of symmetric spin states, $\mathcal{N}_{\text{ortho}}$, is

$$\frac{\mathcal{N}_{\text{para}}}{\mathcal{N}_{\text{ortho}}} = \frac{i}{i+1} \tag{12-6}$$

The number of transitions should be in this ratio, so that we get an *alternation of intensities* in the Raman spectra or band spectra, of diatomic molecules with identical nuclei. This can be seen in the photograph of the $N_2$ rotational Raman spectrum, shown in Figure 12-13, for which the intensities of alternate lines are measured to be quite accurately in the ratio 1/2. Even more dramatic is the spectrum of $C_2$, for which the ratio is 0/1 because alternate lines are completely missing! We do not show that spectrum because the drama is not apparent until a careful comparison between the measured and predicted frequencies of the lines demonstrates that half are absent.

**Example 12-4.** Determine the values of the nuclear spin quantum number $i$ for the nuclei in $N_2$ and $C_2$, by using the measured intensity ratios 1/2 and 0/1 in (12-6).

Since the possible values of $i$ are restricted to $i = 0, 1/2, 1, 3/2, 2, \ldots$, inspection immediately demonstrates that the solution to

$$\frac{1}{2} = \frac{i}{i+1}$$

is $i = 1$. This is the spin of the N nucleus (i.e., of its overwhelmingly abundant isotope $N^{14}$).

For

$$\frac{0}{1} = \frac{i}{i + 1}$$

the solution is obviously $i = 0$. This is the spin of the C nucleus (actually, of its most abundant isotope $C^{12}$, since the other isotopes, $C^{13}$ and $C^{14}$, are so rare that the abundant one completely dominates the spectrum). ◀

The reason for the complete absence of half of the transitions involving rotational levels of molecules with two identical $i = 0$ nuclei is simply that $i = 0$ means the nuclei are bosons that have no spin, so the molecular eigenfunction is necessarily symmetric and has no spin factor in it. Therefore its rotational factor must always be symmetric, which requires that the molecule only be in even-$r$ rotational levels. Proof that these symmetry considerations are very real indeed comes from that fact that if in $C_2$ the nuclei are not identical (e.g., if we have $C^{12} - C^{13}$), then half the transitions are *not* missing. This experimental fact actually led to the discovery of the isotope $C^{13}$.

As we have said, the procedure of Example 12-4 has been widely applied. It was used in the first determination of the spin $i = 1/2$ of the proton, from the measured intensity ratio of 1/3 in the spectrum of $H_2$. The measurements are difficult to make only when $i$ becomes very large.

The determination of the symmetry character of the identical nuclei in molecules like $N_2$ is a matter of keeping track of *which* lines of the spectrum are the more intense.

**Example 12-5.** In $N_2$ it is observed that transitions involving even-$r$ rotational states yield the most intense lines. Determine the symmetry character of the nuclei in that molecule.

Since (12-6) shows that the highest population is for nuclear spin states that are symmetric (ortho), and since even-$r$ rotational states are also symmetric, the symmetric nuclear spin states are associated with the symmetric rotational states. Therefore the $N^{14}$ nucleus must be a boson. ◀

Symmetry character determinations made in this manner on a number of nuclei provided some of the earliest evidence for the correlation, seen in Table 9-1, between symmetry character and spin. Furthermore, we shall see in Chapter 15 how the fact that the particular nucleus $N^{14}$ is an $i = 1$ boson was used at an early date to show that nuclei must contain protons and neutrons, instead of protons and electrons.

**Example 12–6.** Show that the ratio of the number of antisymmetric spin states to the number of symmetric spin states is $i/(i + 1)$, in agreement with (12-6).

The number of possible individual states of spin for a particle of a given spin quantum number $i$ is equal to the number of possible values of its $z$ component quantum number $m_i$. Since, as usual, the values of $m_i$ differ by integers and range from $-i$ to $+i$, this number is the familiar $(2i + 1)$. So the total number of possible independent combinations of spin states for two identical particles of spin $i$ is $(2i + 1)(2i + 1) = (2i + 1)^2$. In $(2i + 1)$ of these states both particles will have the same $m_i$, and so are in identical spin states. For these the spin eigenfunction of the two particle system is symmetric with respect to particle label exchange (like the top and bottom members of (9-18) in the case of $i = 1/2$). Of the $(2i + 1)^2 - (2i + 1) = 2i(2i + 1)$ remaining states, half will be symmetric and half will be antisymmetric in such an exchange, since half will involve the sums of products of individual spin eigenfunctions and the other half will involve the differences of the same products (like the center member of (9-18), and (9-17), in the case of $i = 1/2$). So the total number of symmetric eigenfunctions is

$$\mathcal{N}_{\text{symmetric}} = \mathcal{N}_{\text{ortho}} = (2i + 1) + (1/2)2i(2i + 1) = (i + 1)(2i + 1)$$

and the total number of antisymmetric eigenfunctions is

$$\mathcal{N}_{\text{antisymmetric}} = \mathcal{N}_{\text{para}} = (1/2)2i(2i + 1) = i(2i + 1)$$

The ratio of the number of eigenfunctions, or spin states, is

$$\frac{\mathcal{N}_{\text{para}}}{\mathcal{N}_{\text{ortho}}} = \frac{i}{i + 1}$$

in agreement with (12-6). ◀

## QUESTIONS

**1.** Discuss the statement that the interatomic force law must be attractive to permit condensed phases and must be repulsive to avoid zero volume.

**2.** Would you expect $H_3$ to exist in a bound state? $He_2$? Explain.

**3.** Of the so-called inert gases, which might most easily form molecules with other elements? Explain.

**4.** How would you explain the existence of bound states of $XeF_4$, in view of the absence of valence electrons in a Xe atom?

**5.** Do the even, or odd, $H_2^+$ eigenfunctions have even, or odd, parity?

**6.** Explain why only two electrons can form a covalent bond.

**7.** Would you predict ionic binding or covalent binding in $H_2O$? In $NH_3$? In $CH_4$? Does experiment decide the issue or can you rule out one or the other types of binding independently?

**8.** From the fact that $CO_2$ does not have a permanent electric dipole moment, what can you conclude about the binding and the arrangements of the atoms in the molecule?

**9.** Of the molecules $H_2$, $D_2$, and HD which has the greatest binding energy? The least?

**10.** What does it mean to say that a molecule is in an excited state?

**11.** Explain how the existence of a finite zero-point vibrational energy is related to the uncertainty principle.

**12.** The fundamental vibrational energy for HCl is about ten times that for NaCl. Considering the factors determining this quantity, make this plausible.

**13.** What effect, if any, does the increasing angular momentum of higher rotational states of a diatomic molecule have on the vibrational energy of the molecule?

**14.** What effect does the change in internuclear separation in a diatomic molecule due to its vibration (the binding energy curve is asymmetric) have on the rotational energy levels of the molecule?

**15.** The asymmetry in the binding energy curve accounts for thermal expansion of solids. How can information from molecular spectra be used to determine the shape of this curve?

**16.** Explain why the separation between vibrational levels is somewhat smaller in an excited electronic state than in the ground electronic state (see Figure 12-9). Explain the same effect for rotational states.

**17.** If Raman rotational lines arise from an induced electric dipole moment how can we explain that the selection rule is $\Delta r = \pm 2$ rather than $\Delta r = \pm 1$?

18. Since it is known to take a very long time for the para and ortho species of a molecule to convert themselves into each other, the interaction between the two nuclear spins in a molecule must be very small. Why would you expect this to be the case?

## PROBLEMS

1. From the following data, find the energy required to dissociate a KCl molecule into a K atom and a Cl atom. The first ionization potential of K is 4.34 eV; the electron affinity of Cl is 3.82 eV; the equilibrium separation of KCl is 2.79 Å. (Hint: Show that the mutual potential energy of $K^+$ and $Cl^-$ is $-(14.40/R)$ eV if $R$ is given in Angstroms).

2. The first ionization potential for K is 4.3 eV; the ion $Br^-$ is lower in energy by 3.5 eV than the neutral bromine atom. Compute the largest separation of $K^+$ and $Br^-$ ions that gives a bound KBr molecule.

3. For a system which executes simple harmonic motion about a position of stable equilibrium, the force, $F$, is given by

$$F = -\left(\frac{\partial^2 V}{\partial R^2}\right)_{R_0}(R - R_0)$$

where $V$ is the potential energy and $R - R_0$ is the deviation from equilibrium. Show that the zero-point vibration of a molecule is given by

$$\frac{1}{2}h\nu_0 = \frac{h}{4\pi\mu^{1/2}}\left[\left(\frac{\partial^2 V}{\partial R^2}\right)_{R_0}\right]^{1/2}$$

4. The potential energy $V$ of NaCl can be described empirically by

$$V = -\frac{e^2}{4\pi\epsilon_0 R} + Ae^{-R/\rho}$$

where $R$ is the internuclear separation. The equilibrium separation of the nuclei $R_0$ is 2.5 Å and the dissociation energy is 3.6 eV. (a) Calculate $A$ and $\rho/R_0$, neglecting zero-point vibrations. (b) Sketch $V$ and each of the terms in $V$ on one graph. (c) Give the physical significance of $A$ and $\rho$.

5. Taking the rotational inertia of $H_2$ from Table 12-1, find the temperature at which the average translational kinetic energy of an $H_2$ molecule equals the energy between the ground rotational state and first excited rotational state. What can you conclude about the occupation of rotational excited states in $H_2$ at room temperature?

6. Determine $\delta$, the zero-point vibrational energy, for a NaCl molecule, given that its fundamental vibrational frequency is $1.14 \times 10^{13}$ vib/sec.

7. (a) Show that, if $E_d$ is the dissociation energy of a molecule, the fraction of the molecules that dissociate at a temperature $T$ is $e^{-E_d/kT}$. (b) It is found (from electron diffraction studies) that as $T$ increases, the internuclear separation increases. Explain what effect this has on the potential energy curve and on the result of part (a).

8. For NaCl, the separation of two vibrational levels is about $4 \times 10^{-2}$ eV. Using Table 12-1, and noting that the rotational levels are not equally spaced, show that there are about 40 rotational levels between a pair of vibrational levels.

9. The potential energies of two diatomic molecules of the same reduced mass are shown in Figure 12-14. From the graph determine which molecule has the larger (a) internuclear distance, (b) rotational inertia (moment of inertia), (c) separation between rotational

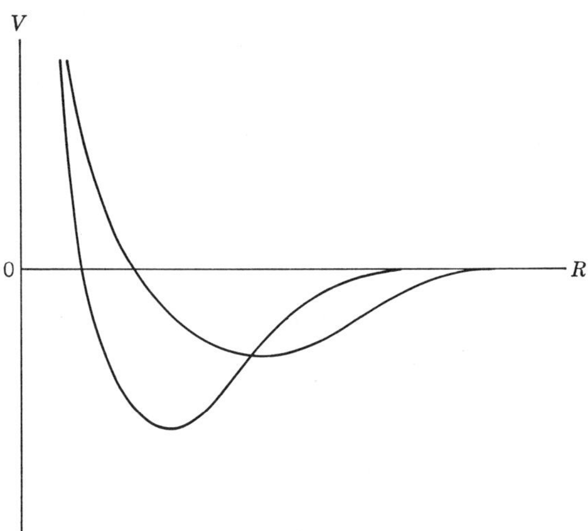


**FIGURE 12-14**
Potential energy curves considered in Problem 9.

energy levels of the same $r$ and $v$, (d) binding energy, (e) zero-point energy (Hint: See Problem 3), (f) separation between low-lying vibrational states.

**10.** (a) What fraction of HCl molecules at 1000°K will be found in the first excited vibrational state? (Hint: Use the Boltzmann factor.) (b) Find the ratio of HCl molecules in the first excited rotational state to those in the first excited vibrational state at 1000°K. (Hint: Remember the degeneracy factors.)

**11.** (a) Derive an expression giving the ratio of the energy of a transition from the lowest to the first excited vibrational level to the energy of a transition from the lowest to the first excited rotational level for a diatomic molecule. (b) What is this ratio for NaCl? For $H_2$? (Hint: See Example 12-3.)

**12.** (a) Show that the relative frequency shift of a spectral line in a rotational band arising from a mixture of two isotopic diatomic molecules is given by $\Delta\nu/\nu = -\Delta\mu/\mu$, where $\mu$ is the reduced mass of the molecule. (b) What is this ratio for a mixture of $HCl^{35}$ and $HCl^{37}$?

**13.** From the value 2940.8 cm$^{-1}$ for the reciprocal wavelength equivalent to the fundamental vibration of a molecule $Cl_2$, each of whose atoms has an atomic weight 35, determine the corresponding reciprocal wavelength for $Cl_2$ in which one atom has atomic weight 35 and the other 37. What is the separation of spectral lines, in reciprocal wavelengths, due to this isotope effect?

**14.** (a) Specify the resolution, $\Delta\lambda/\lambda$, of a spectrometer which can just resolve the rotational spectra of $Na^{23}Cl^{35}$ and $Na^{23}Cl^{37}$ assuming $R_0$ to be the same for both molecules. (b) Would this spectrometer also resolve the vibrational spectra of the two molecules, assuming the force constants are the same?

**15.** Calculate the difference in dissociation energies of $H_2$ and $D_2$ from the value 4395.2 cm$^{-1}$ for the reciprocal wavelength equivalent to the fundamental vibration of an $H_2$ molecule.

**16.** The zero-point vibrational energy for $H_2$ is 0.265 eV. Compare the vibrational energy levels of $H_2$, $D_2$, and HD numerically for the low-lying states.

**17.** From the fact that the lowest electronic excited state in $O_2$ and $N_2$ molecules is over 3 eV above the ground state, explain why air is transparent in the visible.

**18.** In the vibrational Raman spectrum of HF are adjacent Raman lines of wavelength 2670 Å and 3430 Å. (a) What is the fundamental vibrational frequency of the molecule? (b) What is the equivalent force constant for HF?

**19.** A ruby laser ($\lambda = 6943$ Å) is used to excite the Raman spectrum of $N_2$. (a) What are the wavelengths of the lines which result from the lowest energy allowed transitions in the pure rotational spectrum of $N_2$? (b) What is the ratio of the intensities of the lines of part (a) at room temperature? (c) What are the wavelengths of the lines which result from the allowed transitions to and from the ground state vibrational level? (d) What is the ratio of the intensities of the lines of part (c) at room temperature? (e) How do the answers to parts (a) and (c) change if the laser is used to excite the Raman spectrum of diatomic molecules with nonidentical nuclei having the same rotational inertia and force constant as $N_2$?

**20.** The energy-level diagram for the rotational levels in each of the two lowest vibrational states of the electronic ground state is given in Figure 12-15 for a diatomic molecule.

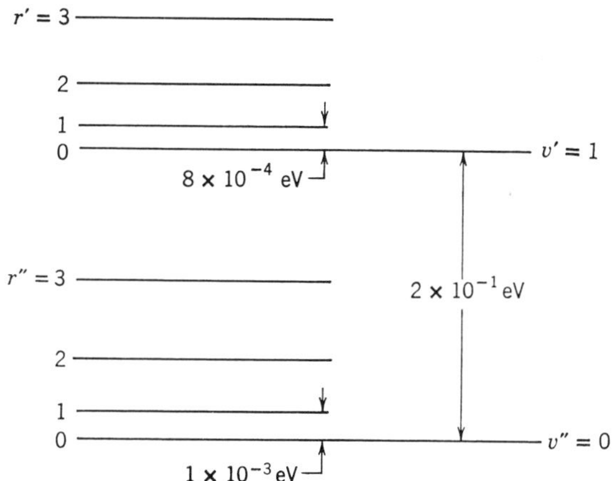


**FIGURE 12-15**
Energy levels considered in Problems 20, 21, and 22.

Find the energies of the transitions that give rise to the allowed spectral lines in the infrared *and* Raman spectra, (a) for molecules containing two identical $i = 0$ nuclei, (b) for molecules containing two identical $i = 1/2$ nuclei, and (c) for molecules containing two nonidentical nuclei.

**21.** Calculate the relative intensities at room temperature for the lines found in parts (a) and (b) of Problem 20.

**22.** Using the information in Figure 12-15, (a) calculate the rotational inertia, or moment of of inertia, of the molecule in each vibrational level, and (b) calculate the zero-point energy.

**23.** (a) How many rotational degrees of freedom do you expect in a polyatomic molecule? Translational degrees? If the molecule has $N$ atoms ($N > 2$) there should be $3N - 6$ vibrational degrees of freedom, i.e., independent modes of vibration. Explain. (b) How many vibrational degrees of freedom are there in an $H_2O$ molecule? A $CH_4$ molecule?

**24.** Consider the relative intensities of the spectra of $H_2$ and $D_2$ to determine which Raman rotation spectrum will yield lines alternating in intensity and having a relative intensity of 1/2.

**25.** Band spectrum measurements of diatomic molecules containing $Cl^{35}$ nuclei yield an alternating intensity ratio of 3/5. What is the spin of the $Cl^{35}$ nucleus?

# 13

# Solids—Conductors and Semiconductors

# Solids—Conductors and Semiconductors 13

## 13-1 Introduction

Solid state physics is a vast area of quantum physics in which we are concerned with understanding the mechanical, thermal, electrical, magnetic, and optical properties of solid matter. Some aspects have been discussed in earlier chapters, such as the lattice and electronic contributions to the specific heats of solids, radiation from a blackbody, thermionic emission, and contact potentials. Here we shall focus on the origin of the forces that hold atoms together in a solid and on the allowed energy levels of the electrons in the solid. This will lead us to the band theory of solids. That theory will then be applied to phenomena of much practical and theoretical interest, including semiconductors and semiconductor devices. Many electrical, thermal, and optical properties of solids will thereby become more clearly understood. In the next chapter we extend the theory to the phenomenon of superconductivity and consider magnetic properties of solids as well.

## 13-2 Types of Solids

In the gaseous state the average distance between molecules is large compared to the size of a molecule, so the molecules may be regarded as isolated from one another. Many substances, however, are in the solid state at ordinary temperatures and pressures. In that state molecules (or atoms) can no longer be regarded as isolated. Their separation is comparable to the molecular size, and the strength of the forces holding them together is of the same order of magnitude as the forces binding the atoms into a molecule. Hence, the properties of a molecule are altered by the presence of neighboring molecules. Characteristic of crystalline solids is the regular arrangement of atoms, a recurrent or periodic pattern called a *crystal lattice*. The solid can be regarded as a large molecule, the forces between atoms being due to interaction between atomic electrons, and the structure of the solid being determined as that arrangement of nuclei and electrons which yields a quantum mechanically stable system. Although the number of atoms involved is very large, they are arranged in a regular pattern. In noncrystalline solids, such as concrete and plastic, the perfectly regular pattern does not hold over long distances, but there is an orderly pattern in the neighborhood of any one atom. We shall discuss only crystalline solids in this book. Such solids are classified according to the predominant type of binding, the principal types being *molecular*, *ionic*, *covalent*, and *metallic*.

*Molecular solids* consist of molecules which are so stable that they retain much of their individuality when brought in close proximity. The electrons in the molecule are all paired so that atoms in different molecules cannot form covalent bonds with one another. The intermolecular binding force is the weak van der Waals attraction that is present between such molecules in the gaseous phase. The physical mechanism involved in the van der Waals attraction is an interaction between electric dipoles. Because of the fluctuating quantum mechanical behavior of the electrons in a molecule, all molecules have a fluctuating electric dipole moment, even though for many of them symmetry considerations require that it fluctuate about an average value of zero. At a time when a molecule has a certain instantaneous electric dipole moment, the external electric field that it produces will induce in the charge distribution of a nearby molecule a dipole moment. By drawing rudimentary sketches of the charges and field in various cases, the student can immediately convince himself that the force exerted between the inducing and the induced electric dipole is always attractive. The interaction energy is proportional to the mean square of the inducing electric dipole moment. The resulting attraction is weak, the binding energies being of the order of $10^{-2}$ eV and the force varying with the inverse seventh power of the intermolecular separation. In the solid, successive molecules have electric dipole moments which alternate in orientation so as to produce successive attractions. Many organic compounds, inert gases, and ordinary gases such as oxygen, nitrogen, and hydrogen form molecular solids in the solid state. Because the binding is weak, solidification takes place only at very low temperatures where the disruptive effects of thermal agitation are very small. (The melting point of solid hydrogen is 14°K, for example.) The weak binding makes molecular solids easy to deform and compress, and the absence of free electrons makes them very poor conductors of heat or electricity.

*Ionic solids*, such as sodium chloride, consist of a close regular three-dimensional array of alternating positive and negative ions having a lower energy than the separated ions. The structure is stable because the binding energy due to the net electrostatic attraction exceeds the energy spent in transferring electrons to create the isolated ions from neutral atoms, just as for ionic binding in molecules. Ionic binding in solids is not directional because spherically symmetrical closed shell ions are involved. Hence the ions are arranged like close-packed spheres. The actual crystal geometry depends on which arrangement minimizes the energy, and this in turn depends principally on the relative sizes of the ions involved. Because there are no free electrons to carry energy or charge from one part of the solid to another, such solids are poor conductors of heat or electricity. Because of the strong electrostatic forces between the ions, ionic solids are usually hard and have high melting points. Lattice vibrations can be excited by energies corresponding to radiation in the far infrared, so that ionic solids show strong optical absorption properties in that region. But optical absorption by excitation of electrons requires energies in the ultraviolet, so that ionic crystals are transparent to visible radiation.

*Covalent solids* contain atoms that are bound by shared valence electrons, as in covalent binding of molecules. The bonds are directional and determine the geometrical arrangement of atoms in the crystal structure. The rigidity of their electronic structure makes covalent solids hard and difficult to deform, and it accounts for their high melting points. Because there are no free electrons, covalent solids are not good heat or electrical conductors. Sometimes, as for silicon and germanium, they are semiconductors. At room temperature some covalent solids, such as diamond, are transparent; the energy required to excite their electronic states exceeds that of photons in the visible region of the spectrum so that such photons are not absorbed. But most covalent solids absorb in the visible and are therefore opaque.

*Metallic solids* exhibit a binding that can be thought of as a limiting case of covalent binding in which electrons are shared by all the ions in the crystal. When a crystal is formed of atoms having a few weakly bound electrons in the outermost subshells, electrons can be freed from the individual atoms by the energy released in binding. These electrons move in the combined potential of all the positive ions and are shared by all the atoms in the crystal. We speak of an electron gas interspersed between the positive ions and exerting attractive forces on each ion that exceed the repulsive forces of other ions, hence the binding. The atoms have vacancies in their outermost electron subshell, and there are not enough valence electrons per atom to form tight covalent bonds. The electrons are shared by all the atoms and are free to wander through the crystal from atom to atom, there being many unoccupied electronic states. In this sense they behave like a gas, an "electron gas." A metallic solid is a regular lattice of spherically symmetrical positive ions, arranged like close-packed spheres, through which the electrons move. Metallic solids are obviously excellent conductors of electricity, or heat, the electrons easily absorbing energy from incident radiation, or lattice vibrations, and moving under the influence of an applied electric field, or thermal gradient. Because radiation in the visible portion of the electromagnetic spectrum is easily absorbed, such solids are opaque. All the alkalies form metallic solids.

The type of binding that a particular solid has is determined experimentally by studies of x-ray diffraction, dielectric properties, optical emissions, and so forth. There are some solids whose binding must be interpreted as a mixture of the principal types we have described. In addition, not all solids have the ideal structure implied by the discussion so far. Indeed, the so-called lattice imperfections, or deviations from ideal crystal structure, lead to many properties of solids which have practical consequences.

## 13-3 Band Theory of Solids

To understand the effect of putting a great many atoms close together in a solid, consider first two atoms only that are initially far apart. All of the energy levels of this two-atom system have a twofold exchange degeneracy. That is, for the combined system the space part of the eigenfunction for the electrons can contain either a combination of the individual atom space eigenfunctions which is symmetric in an exchange of pairs of electron labels, or which is antisymmetric in such a label exchange. (The total eigenfunction of the system of electrons is, of course, antisymmetric, since the symmetric space eigenfunction is associated with an antisymmetric spin eigenfunction, and vice versa.) When the atoms are widely separated, the two different types of eigenfunctions lead to the same energy, and so each of the energy levels is said to have a twofold exchange degeneracy. But when the atoms are brought together, the exchange degeneracy is removed. Because the electron charge density in the important region between the atoms depends on whether the space eigenfunction is symmetric or antisymmetric, when the atoms are close enough together that the wave functions of the individual atoms overlap, the energy of the system depends on the symmetry of the space eigenfunction. Hence, a given energy level of the system is split into two distinct energy levels as overlap commences, and the splitting increases as the separation of the atoms decreases. Of course a famous example of this phenomenon is found in the ground state energy level of the system containing two hydrogen atoms, as we saw in Section 12-3. Figure 12-4 shows this splitting for the ground state level only, but each of the higher levels of the system splits in the same way, and for the same reason, as the atoms are brought together.

If we had started with three isolated atoms, we would have had a threefold exchange degeneracy of the energy levels. When the atoms are brought together in a uniform linear lattice, each of the levels splits into three distinct levels. Figure 13-1 illustrates this schematically for a typical energy level of a system of six atoms. The splitting commences when the center-to-center atomic separation $R$ becomes small enough for the atoms to begin overlapping. As $R$ decreases from this value there is a decrease in the energy of the levels for which the symmetry of the space eigenfunction leads

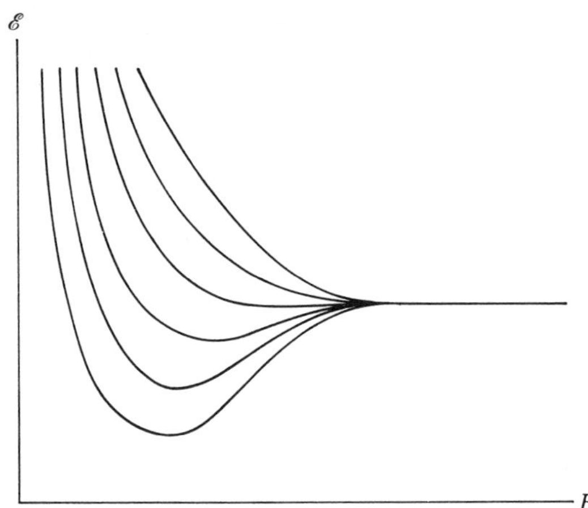


**FIGURE 13-1**

Schematic drawing of the splitting of an energy level in a system of six atoms, as a function of the separation distance $R$ between adjacent atoms.

to a favorable electron charge distribution (i.e., which puts electron charge where the ions exert the strongest binding), and an increase in the energy of the levels associated with space eigenfunctions whose symmetry leads to an unfavorable charge distribution. The more favorable, or unfavorable, the charge distribution is, the greater is the decrease, or increase, in the energy. So the levels are spread, by the quantum mechanical requirements of indistinguishability, about an average energy equal to the energy the system would have at a given $R$ if there were no such requirements. Note that this average energy begins to increase rapidly for sufficiently small $R$. This is due to the Coulomb repulsion that the ions exert on each other.

As we go to a system containing $N$ atoms of a given species, each level of one of these atoms leads to an $N$-fold degenerate level of the system when the atoms are well separated. With decreasing separation, each of these splits into a set of $N$ levels. The spread in energy between the lowest and highest level of a particular set depends on the separation distance $R$, since $R$ specifies the amount of overlap that causes the splitting. But it does not depend significantly on the number of atoms in the system if the same separation distance is maintained. Thus, as more and more atoms are added to the system each set of split levels contains more and more levels spread over about the same energy range at a particular $R$. At the values of $R$ found in a solid, a few angstroms, the energy spread is of the order of a few electron volts (see Figure 12-4). If we then consider that a solid contains something like $10^{23}$ atoms/mole, we see that the levels of each set in a solid are so extremely closely spaced in energy that they form a practically continuous energy *band*.

The process we have just described is indicated in Figure 13-2. We see from this figure that the lower-lying energy levels are spread less than those that lie higher. The reason is that the electrons in lower levels are electrons in inner subshells of the atoms, which are not significantly influenced by the presence of nearby atoms. These electrons are localized on particular atoms, even when $R$ is small, because the potential barriers between the atoms are for them relatively high and wide. The valence electrons, on the other hand, are not localized at all for small $R$, but they become part of the whole system. The overlapping of their wave functions results in a spreading of their energy levels. It should be pointed out that the $1s$ level of an individual atom becomes a band of $N$ levels, as does the $2s$ level, if we count in such a way that each of these can

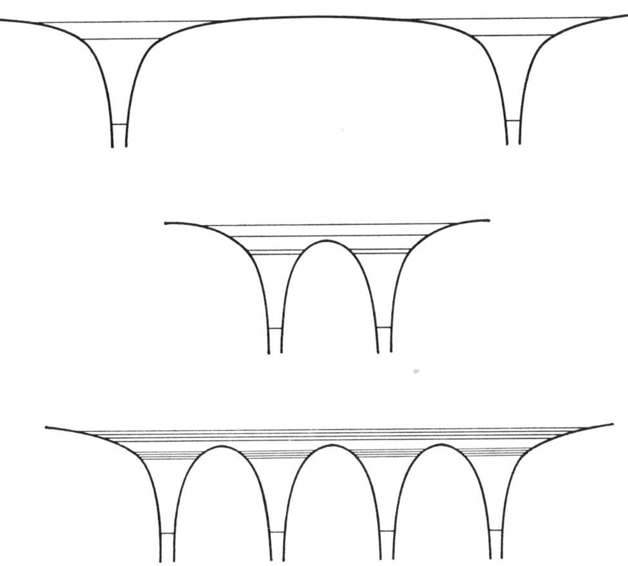

**FIGURE 13-2**

*Top:* Energy-level scheme for two isolated atoms. *Middle:* Energy-level scheme for the same two atoms in a diatomic molecule. *Bottom:* Energy-level scheme for four of the same atoms in a rudimentary one-dimensional crystal. Note that the lowest lying levels are not split appreciably because the atomic eigenfunctions for these levels do not overlap significantly.

accommodate two electrons of opposite spin. But the $2p$ level is triply degenerate in the space quantum number $m_l$ in the isolated atom, since $m_l$ can assume any of the values $-1$, 0, $+1$. Thus the $2p$ level in the atom leads to $3N$ levels in the solid. As we shall discuss soon, these can be thought of as forming three bands of $N$ levels, whose energy ranges may or may not coincide.

In Figure 13-3 we show the band formation for the higher levels of sodium, whose ground state atomic configuration is $1s^2 2s^2 2p^6 3s^1$. Several general features of *allowed bands* (the continuous bands of energy levels for electrons) and *forbidden bands* (the regions where there are no electron energy levels) are illustrated in this figure. Allowed bands corresponding to inner subshells, such as $2p$ in sodium, are extremely narrow until the interatomic spacing becomes smaller than the value actually found in the crystal. As we go through the outer occupied subshells and into the unoccupied subshells of the atom in its ground state, however, the bands become progressively wider at a given interatomic separation. The reason is, again, that the greater the energy of the electrons the larger the regions in which they can move and the more they are affected by nearby ions. As the energy increases, therefore, the successive allowed bands widen and overlap each other in energy.

Direct experimental verification of energy bands comes from observations of x-ray spectra in solids. For example, the $3s \rightarrow 2p$ transition in sodium gives the $L$ series x-ray lines. A very sharp line spectrum is observed for gaseous sodium in which the $3s$ and $2p$ levels are narrow. But the same x-ray lines from solid sodium are broadened because, although the low-lying $2p$ level remains narrow, the $3s$ level has now become an energy band. The observed shape of x-ray lines from solids agrees with the energy band picture.

Consider now the occupation of the energy levels. Those bands which originated in levels of closed subshell electrons of an isolated atom have all their levels occupied.

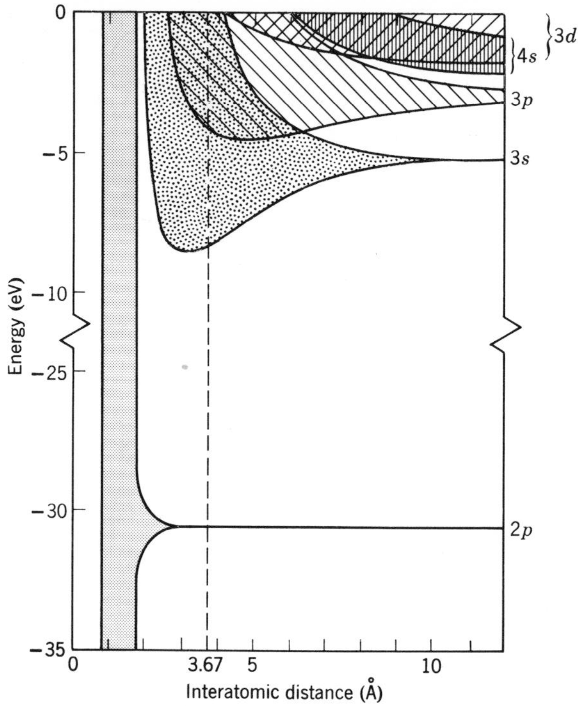


**FIGURE 13-3**

Showing the formation of energy bands from the energy levels of isolated sodium atoms as the interatomic separation decreases. The dashed line indicates the observed interatomic separation in solid sodium. The several overlapping bands that constitute each $p$ or $d$ band are not indicated.

The bands that originated from valence electrons may or may not be fully occupied. If an electric field is applied to the solid the electrons will acquire extra energy only if there are available empty levels within the range of energy that the strength of the applied field allows the electron to gain. If there are no nearby empty levels, then the electron will not be able to gain any energy at all and the solid behaves like an *insulator*. What counts in determining the emptiness, or fullness, of the bands containing valence electrons is the valence of the atoms forming the solid, and the geometry of the crystal lattice into which they solidify. An *isolated* band will be full if a unit cell of the crystal lattice contains two valence electrons, one for each of the two possible values of the spin quantum number $m_s$.

Crystal structure geometry is a complex subject which is very important in any detailed study of solid state physics. We shall be able to avoid it in this book by restricting ourselves to particularly simple (usually one-dimensional) crystal lattices. But we shall at least define a *unit cell* as the smallest geometrical arrangement of atoms that by periodic repetition along the coordinate axes can fully describe the geometrical arrangement of the atoms in the complete crystal. We shall also say that in a crystal lattice some or all of the degeneracy of the atomic valence electron levels with respect to the quantum number $m_l$ is removed because these electrons are not in the spherically symmetrical potential of an atom in free space, but in a potential whose more complicated symmetry depends on the crystal geometry. For this reason, the three degenerate levels from a $p$ subshell of a single atom lead to three bands of $N$ levels, each capable of holding two electrons of opposite spin, in a crystal containing $N$ of these atoms.

These bands may be completely nonoverlapping, partly overlapping, or completely overlapping in energy, depending on the crystal geometry. The term *isolated* band, used in expressing the condition for a full band, refers to a case in which these bands do not overlap each other or bands from other subshells. Then if there are two valence electrons per unit cell, each of the $N$ levels in the lowest lying band will have its full complement of two electrons. Note that the quantity determining occupation is the number of valence electrons per unit cell, and not per atom. In a uniform one-dimensional lattice of identical atoms, such as we considered in the argument from which we concluded that a band contains $N$ levels, if the crystal contains $N$ atoms, a unit cell contains one atom and there is no distinction to be made. When that argument is extended to three-dimensional crystals containing atoms of different species, it is found that the conclusion remains the same, providing $N$ is the number of unit cells in the crystal. Thus if there are two valence electrons per unit cell there will be two in each of the levels of the band, and the band will be fully occupied.

The problem in predicting whether or not a solid is an insulator is that the question of band overlap is all important, and this depends on details of the geometry of the crystal structure (and of the geometry of the atomic eigenfunctions). If what, as far as valence is concerned, might have been a completely filled band actually overlaps what might have been a completely empty band, then there will be two partly filled bands. The result is that a solid that might have been an insulator will actually be a conductor. But it is at least possible to say that *a solid can certainly not be an insulator unless one of its unit cells contains an even number of valence electrons*, because an odd valence electron can never be in a filled band. Most covalent solids like diamond, or ionic solids like sodium chloride, are insulators; they all have an even number of valence electrons per unit cell. In diamond each carbon atom has four valence electrons, and there are two atoms in each unit cell. The eight valence electrons per unit cell fully occupy the $4N$ levels of four bands, one originating from the $2s$ subshell of the atom and three originating from the three $2p$ subshells. These bands overlap each other, but they are well separated from empty higher energy bands. Sodium chloride contains one sodium ion and one chlorine ion per unit cell, and the valence band consists of a set of completely filled bands that overlap each other but do not overlap unfilled bands. Alkali-earth atoms like beryllium are divalent and form crystals with an even number of valence electrons per unit cell, but these solids are metals, not insulators, because overlapping bands make slightly higher unfilled levels energetically available to the electrons.

In solids formed from the monovalent alkali atoms like sodium, the band containing the valence electrons cannot be filled, and so the solid behaves like a *conductor*. Only half of the levels of the isolated $3s$ allowed band of sodium are filled because a sodium atom has a single electron in the $3s$ level, whereas the exclusion principle allows such a level to accommodate two electrons. Hence electrons in the solid can easily acquire a small amount of additional energy. Thus any applied electric field will be effective in giving electrons energy, and the solid will be a conductor. As we mentioned in the previous paragraph, conductors are also found in cases where bands containing valence electrons overlap.

It is worthwhile putting the distinction between conductors and insulators into momentum, instead of energy, language. Without an applied electric field there are as many electrons in the solid with momentum vectors in one direction as there are with momentum vectors in the opposite direction, since there is no net current. When an electric field is applied, this equilibrium can be upset causing a current to flow, if some of the electrons can go into quantum states with changed momentum vectors. This is quite possible for electrons in a partially filled band, but it cannot be done by electrons in a completely filled band.

At temperatures above absolute zero it is, of course, possible for some electrons to gain enough thermal energy to jump over the energy gap of a forbidden band of energy into a higher allowed band, thereby creating vacancies in the lower allowed band and making a new allowed band available. We speak of the nearly filled band as a *valence band* and the nearly empty band as a *conduction band*. The probability of this happening increases with temperature, and it depends strongly on the width of the forbidden band. Substances in which the width of the energy gap is small are called *semiconductors*. An example is silicon, a covalent solid with a diamondlike structure, but with a forbidden band only about 1 eV wide. It becomes reasonably conducting at room temperature though at low temperatures it is an insulator. On the other hand the gap between the filled and empty allowed bands in diamond is about 7 eV. Thus diamond is an insulator even at relatively high temperatures.

## 13-4 Electrical Conduction in Metals

Some useful results concerning conduction electrons in metals can be obtained from classical ideas. In the absence of an applied electric field, the directions in which these electrons move are random. The reason is that the electrons frequently collide with imperfections in the crystal lattice of the metal, which arise from thermal motion of the ions about their equilibrium positions in the lattice or from the presence of impurity ions in the lattice. In colliding with these imperfections, the electrons suffer changes in speed and direction, and this makes their motion random. As in the case of molecular collisions in a classical gas, we can describe the frequency of electron-lattice imperfection collisions by a *mean free path* $\lambda$, where $\lambda$ is the average distance that an electron travels between collisions. When an electric field is applied to a metal, the electrons modify their random motion in such a way that, on the average, they drift slowly in the direction opposite to that of the field, because their charge is negative, with a drift speed $v_d$. This drift speed is very much less than the effective instantaneous speed $\bar{v}$ of the random motion. In copper $v_d$ is of the order of $10^{-2}$ cm/sec, whereas $\bar{v}$ is of the order of $10^8$ cm/sec.

The drift speed can be calculated in terms of the applied electric field $E$ and of $\bar{v}$ and $\lambda$. When a field is applied to an electron in the metal, it will experience a force of magnitude $eE$ which will give it an acceleration of magnitude $a$ given by $a = eE/m$. Consider now an electron that has just collided with a lattice imperfection. In general, the collision will momentarily destroy the tendency to drift and the electron will move in a truly random direction after the collision. Just before its next collision the electron will have changed its velocity, on the average, by $a\lambda/\bar{v}$ where $\lambda/\bar{v}$ is the mean time between collisions. We call this the drift speed $v_d$, so that

$$v_d = \frac{a\lambda}{\bar{v}} = \frac{eE\lambda}{m\bar{v}}$$

If $n$ is the number of conduction electrons per unit volume and $j$ is the current density, we have $v_d = j/ne = eE\lambda/m\bar{v}$. Combining this with the definition of *resistivity*, $\rho = E/j$, gives us

$$\rho = \frac{m\bar{v}}{ne^2\lambda} \tag{13-1a}$$

Equation (13-1a) can be taken as a statement that metals obey Ohm's law, for the quantities $\bar{v}$ and $\lambda$ that determine the resistivity $\rho$ do not depend on the applied electric field, which is the criterion that the law is obeyed.

Often we deal with the *conductivity*

$$\sigma = \frac{1}{\rho} = \frac{ne^2\lambda}{m\bar{v}} \tag{13-1b}$$

This can be put in a more useful form by defining a measurable quantity, the *mobility* $\mu$, given by the ratio of the drift velocity to the applied electric field, i.e.

$$\mu = \frac{v_d}{E} = \frac{e\lambda}{m\bar{v}} \tag{13-1c}$$

Then since $\sigma = ne^2\lambda/m\bar{v}$, we have $\mu = \sigma/ne$ or

$$\sigma = ne\mu \tag{13-2}$$

If we have conduction by positive carriers as well as negative carriers, the conductivity is given by

$$\sigma = nq_n\mu_n + pq_p\mu_p$$

in which $\mu_n$ and $\mu_p$ are the mobilities of negative and positive carriers, $q_n$ and $q_p$ are their charges, and $n$ and $p$ are the numbers of these carriers per unit volume. If conduction is by negative charge carriers the charge $q$ of the carrier is negative, whereas $q$ is positive if conduction is by positive carriers. Since the sign of $\mu$ also depends on the sign of $q$, each term in the expression for $\sigma$ is always positive.

The sign of the charge carrier of electric current in a metal can be determined from measurements of the *Hall effect*. That is, when a current carrying conducting sheet is placed perpendicular to a magnetic field, an electric field is set up perpendicular both to the magnetic field and the flow of current. By measuring the potential difference between the two surfaces of the conductor, it is possible to deduce the sign and value of the quantity $1/nq$, called the *Hall coefficient*. Here $n$ is the number of charge carriers per unit volume and $q$ is the charge of the carrier. The electric field arises from an accumulation of charge carriers on one surface due to the $\mathbf{v}_d \times \mathbf{B}$ force exerted on them when they move with velocity $\mathbf{v}_d$ through the magnetic field $\mathbf{B}$.

In some metals, as zinc and beryllium for example, the Hall effect indicates net positive charge carriers. This is interpreted as being due to transitions of electrons from the filled valence band to the conduction band leaving *holes* (unoccupied energy levels) in the valence band. Such holes correspond to the absence of an electron and behave much like positive charges. As these vacancies are filled by electrons, moving under the influence of an electric field, the holes move in a direction opposite to the electrons just as though positive charge carriers were moving in the field direction. In

**TABLE 13-1.** Observed Hall Coefficient and Calculated Number of Free Electrons per Atom

| Metal | $1/nq$ ($10^{-10}$ m$^3$/coul) | No./atom | Metal | $1/nq$ ($10^{-10}$ m$^3$/coul) | No./atom |
|---|---|---|---|---|---|
| Na | −2.5 | 0.99 | Be | +2.4 | −2.2 |
| K | −4.2 | 1.1 | Zn | +0.33 | −2.9 |
| Cu | −0.55 | 1.3 | Cd | +0.60 | −2.5 |
| Ag | −0.84 | 1.3 | As | +40 | −0.04 |
| Al | −0.30 | 3.5 | Sb | −20 | 0.09 |
| Li | −1.70 | 1.0 | Bi | −5000 | 0.0005 |

the case of metals with an $s^2$ atomic configuration, such as zinc and beryllium, the mobility of the $s$-band holes is much greater than that of the $p$-band electrons. Since the sign of the Hall coefficient depends on which type of carrier has the higher mobility, the Hall coefficient is positive for these metals.

In Table 13-1 we list the Hall coefficients of some metals and also the number of free electrons per atom. The latter is computed from the value of the Hall coefficient, $1/nq$, and the density of the metal. For the alkalis and other monovalent metals, Hall measurements agree with one conduction electron per atom. Of course, the free-electron model on which the simple Hall effect analysis is based is not expected to be valid for all metals.

## 13-5 The Quantum Free-Electron Model

Let us now recall our application in Section 11-11 of quantum theory and the Fermi distribution to conduction electrons in a metal. There we saw that the potential in which the electron moves can be approximated by a rectangular potential well. This constant potential smooths out the actual periodic variation due to the ion cores and includes the average effect of all the remaining electrons. It is equivalent to treating the electrons as an ideal gas of fermions inside the solid. This approximation, which greatly simplifies quantum mechanical calculations, turns out to be surprisingly good in determining many of the observed properties of solids, as we saw in Section 11-12 when we used it in describing phenomena such as contact potential and electronic specific heats. In connection with our present discussion we can use the result, (11-56), for the distribution with energy of free conduction electrons in a metal, namely

$$n(\mathscr{E})N(\mathscr{E})\,d\mathscr{E} = \frac{8\pi V(2m^3)^{1/2}}{h^3}\,\frac{\mathscr{E}^{1/2}\,d\mathscr{E}}{e^{(\mathscr{E}-\mathscr{E}_F)/kT}+1} \tag{13-3}$$

where $n(\mathscr{E})N(\mathscr{E})\,d\mathscr{E}$ is the number of electrons with energy from $\mathscr{E}$ to $\mathscr{E} + d\mathscr{E}$ in a metal at temperature $T$. The justification is that the distribution of energy states in a band is nearly the same as that for free electrons if the Fermi energy $\mathscr{E}_F$ is not close to the top of the band. This condition applies to the alkali metals, for example, and accounts for the success that the free electron model has in describing their electrical properties.

On the left side of Figure 13-4 we show the prediction of (13-3) for the absolute zero temperature energy distribution of electrons in a partly filled band, with energy being measured from the lowest energy in the band. The maximum energy allowed in the band is $\mathscr{E}_{\max}$ and $\mathscr{E}_F < \mathscr{E}_{\max}$, as shown in that figure. At a temperature greater than zero, the uppermost electrons are excited to occupy nearby available higher states, and the distribution function takes the form shown on the right side of Figure 13-4. The number of quantum states in an energy interval $\mathscr{E}$ to $\mathscr{E} + d\mathscr{E}$ is the factor $N(\mathscr{E})\,d\mathscr{E}$ of (13-3), namely

$$N(\mathscr{E})\,d\mathscr{E} = \frac{8\pi V(2m^3)^{1/2}}{h^3}\,\mathscr{E}^{1/2}\,d\mathscr{E} \tag{13-4}$$

In Figure 13-4 $N(\mathscr{E})$ is shown by a dashed curve and, for unit volume, is the *density of states*. The dash-dot curve is $n(\mathscr{E})$, the Fermi distribution for the number of electrons per state. The solid curve gives the product $n(\mathscr{E})N(\mathscr{E})$, the energy distribution of electrons, or number of electrons per unit energy interval.

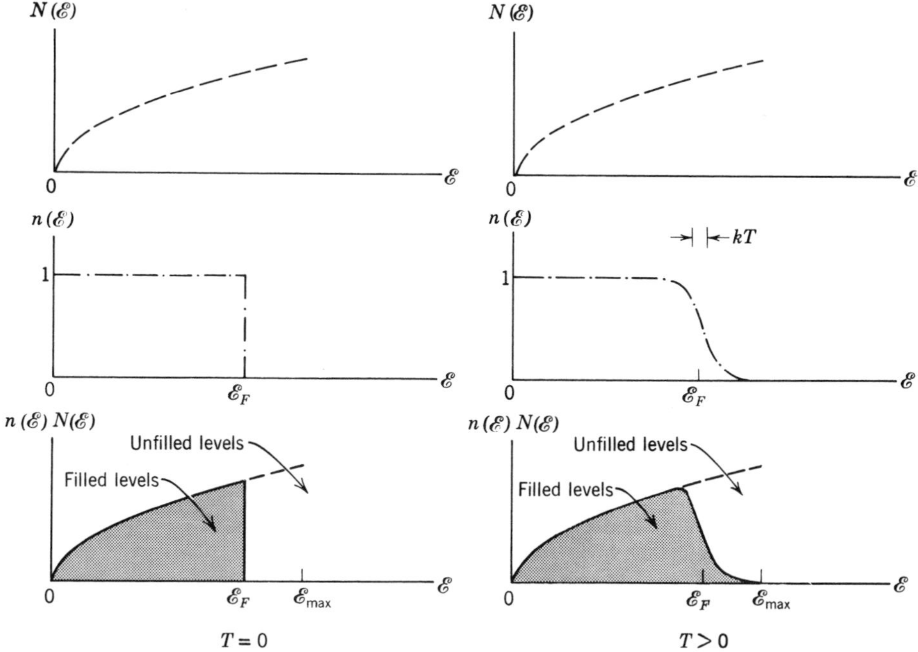


**FIGURE 13-4**

*Left:* The distribution with energy of conduction electrons in an unfilled band of width $\mathscr{E}_{\text{max}}$ in a solid at $T = 0$, according to the free electron model. *Right:* The same at a higher temperature.

**Example 13-1.** The Fermi energy, $\mathscr{E}_F$, for lithium is 4.72 eV at $T = 0$. Calculate the number of conduction electrons per unit volume in lithium.

From (11-57) we have

$$\mathscr{E}_F = \frac{h^2}{8m}\left(\frac{3\mathscr{N}}{\pi V}\right)^{2/3} \qquad \text{for } kT \ll \mathscr{E}_F \qquad (13\text{-}5)$$

so that the number of free electrons per unit volume is

$$n = \frac{\mathscr{N}}{V} = \left(\frac{8m}{h^2}\right)^{3/2} \mathscr{E}_F^{3/2}\,\frac{\pi}{3}$$

in which $m$ is the mass of the electron. Then, with $\mathscr{E}_F = 4.72$ eV, we have

$$n = \frac{\mathscr{N}}{V} = \left[\frac{8 \times 9.11 \times 10^{-31}\,\text{kg}}{(6.63 \times 10^{-34}\,\text{joule-sec})^2}\right]^{3/2} \frac{(4.72 \times 1.60 \times 10^{-19}\,\text{joule})^{3/2}}{3}\,\pi$$
$$= 4.64 \times 10^{28}/\text{m}^3 = 4.64 \times 10^{22}/\text{cm}^3$$

as the number of conduction electrons per unit volume in lithium.

This corresponds exactly to one free electron per lithium atom, since the number of lithium atoms per unit volume, in solid lithium of density 0.534 g/cm³, is

$$0.534\,\frac{\text{g}}{\text{cm}^3} \times \frac{1\text{ mole}}{6.94\text{ g}} \times 6.02 \times 10^{23}\,\frac{\text{atom}}{\text{mole}} = 4.64 \times 10^{22}\,\text{atom/cm}^3$$ ◀

**Example 13-2.** Make an estimate of the relative number of conduction electrons in a metal which are thermally excited to higher energy states.

Figure 13-4 shows that most of the excited electrons are in a range $\Delta\mathscr{E}$ above the Fermi energy $\mathscr{E}_F$, where $\Delta\mathscr{E} \simeq 2kT$. Assuming that $kT \ll \mathscr{E}_F$, the number $\Delta\mathscr{N}$ of excited electrons can be calculated from

$$\Delta\mathcal{N} \simeq N(\mathscr{E}_F)n(\mathscr{E}_F)\Delta\mathscr{E} \simeq N(\mathscr{E}_F)(1/2)2kT \simeq N(\mathscr{E}_F)kT$$

Equation (13-5) shows that, for $kT \ll \mathscr{E}_F$

$$\mathcal{N} = \frac{\pi V}{3}\left(\frac{8m}{h^2}\right)^{3/2}\mathscr{E}_F^{3/2}$$

and (13-4) shows that

$$N(\mathscr{E}_F) = \frac{\pi V}{2}\left(\frac{8m}{h^2}\right)^{3/2}\mathscr{E}_F^{1/2}$$

Hence

$$\begin{aligned}\frac{\Delta\mathcal{N}}{\mathcal{N}} &= \frac{N(\mathscr{E}_F)kT}{\mathcal{N}} \\ &= \frac{\dfrac{\pi V}{2}\left(\dfrac{8m}{h^2}\right)^{3/2}\mathscr{E}_F^{1/2}kT}{\dfrac{\pi V}{3}\left(\dfrac{8m}{h^2}\right)^{3/2}\mathscr{E}_F^{3/2}} \\ &= \frac{3}{2}\frac{kT}{\mathscr{E}_F} \\ &\simeq \frac{kT}{\mathscr{E}_F}\end{aligned}$$

The fraction of conduction electrons that is thermally excited is small. At room temperature $kT \simeq 0.025$ eV and typically $\mathscr{E}_F \simeq 4$ eV, so that $\Delta\mathcal{N}/\mathcal{N} \simeq 1/160$. The absolute number of excited conduction electrons is large, however, because $\mathcal{N}$ itself is so large. ◀

Now we shall use the free electron model to evaluate the width in energy of a band for the simple case of a one-dimensional metal. The eigenfunctions for an electron in the deep square well, representing the smoothed out attraction of the ion cores distributed uniformly along the $x$ axis plus the average repulsion of the remaining electrons, are essentially sinusoidal standing waves like

$$\psi \propto \cos\frac{2\pi x}{\lambda} = \cos kx \qquad \text{and} \qquad \psi \propto \sin\frac{2\pi x}{\lambda} = \sin kx \tag{13-6}$$

where $\lambda$ is the wavelength and $k = 2\pi/\lambda$ is the wave number. The eigenfunctions have nodes at each end of the well since their values go to zero outside the well. These boundary conditions lead immediately to the requirement that $n\lambda/2 = L$, where $L$ is the length of the well. Each value of the integer $n = 1, 2, 3, \ldots$, corresponds to a different eigenfunction, or energy level if we allow two electrons of opposite spin per level. Since for free electrons the energy is $\mathscr{E} = p^2/2m = h^2/2m\lambda^2 = h^2n^2/8mL^2$, the minimum value of $n$ corresponds to the level of essentially zero energy at the bottom of the band, and the maximum value of $n$ corresponds to the level of maximum energy at the top, the width of the band being approximately equal to that maximum energy. If there are $N$ ions each separated by distance $a$ in the one-dimensional metal of length $L$, then $N = L/a$. As we have explained before, the number of levels in the band is just equal to $N$, so the maximum value of $n$ will also be equal to $N$. Thus the maximum energy, or energy width of the band in our one-dimensional metal, is

$$\mathscr{E}_{\max} = \frac{h^2N^2}{8mL^2} = \frac{h^2L^2}{8mL^2a^2}$$

or

$$\mathscr{E}_{\text{max}} = \frac{\hbar^2\pi^2}{2ma^2} \tag{13-7}$$

This result, which depends on $a$ but is independent of $N$, confirms the statement made earlier that the width of a band depends on the separation of the ions and not on the number of ions in the lattice.

The free-electron model gives very good results for many metals. It is especially good for the alkali metals where the overlap of bands (as in Figure 13-3 for sodium) is so complete that the density of states $N(\mathscr{E})$ behaves like the curves of Figure 13-4. The $\mathscr{E}^{1/2}$ dependence of $N(\mathscr{E})$ on $\mathscr{E}$ is not correct, however, in the case of an isolated band. Although the actual shape of the curve of density of states depends on the position of the band and the structure of the lattice, its shape is roughly symmetric, as shown in the upper part of Figure 13-5, in that it decreases to zero at the top of the band.

To understand how this comes about, we consider a one-dimensional crystal which is so long that we first ignore the boundary conditions at its end. Then the most convenient eigenfunctions for a free electron are sinusoidal traveling waves like

$$\psi \propto e^{ikx} \qquad \text{and} \qquad \psi \propto e^{-ikx} \tag{13-8}$$

where the forms with positive, or negative, exponents describe an electron moving in the positive, or negative, direction of the $x$ axis. It is even more convenient to take only the form $\psi \propto e^{ikx}$, and let $k$ be either positive or negative. Now we write the energy $\mathscr{E}$ of a free electron in terms of its wave number $k = p/\hbar$, where $p$ is its momentum. That is

$$\mathscr{E} = \frac{p^2}{2m} = \frac{\hbar^2k^2}{2m} \tag{13-9}$$

This relation is plotted in Figure 13-6, over a range of $k$ including both positive and negative values. A positive value of $k$ corresponds to an electron moving in the positive $x$ direction, and a negative $k$ corresponds to motion in the opposite direction. The energy depends on $k^2$, so the curve is symmetrical about $k = 0$. It can be seen immediately by comparing (13-7) and (13-9) that

$$-\pi/a \leq k \leq +\pi/a \tag{13-10}$$

That is, the values of $k$ corresponding to the maximum value of $\mathscr{E}$ found in the band are $-\pi/a$ and $+\pi/a$, and the value of $k$ corresponding to the minimum value $\mathscr{E} = 0$

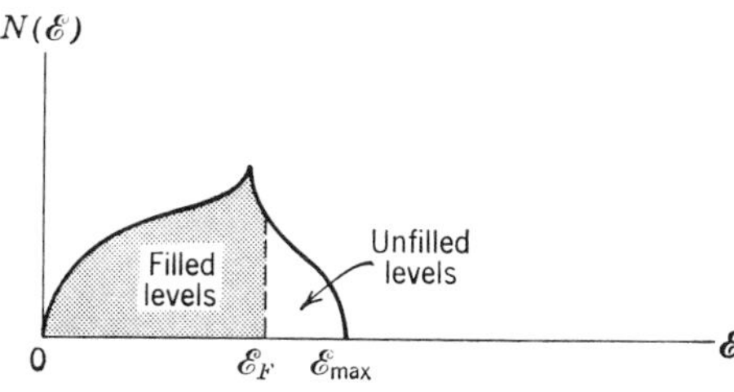


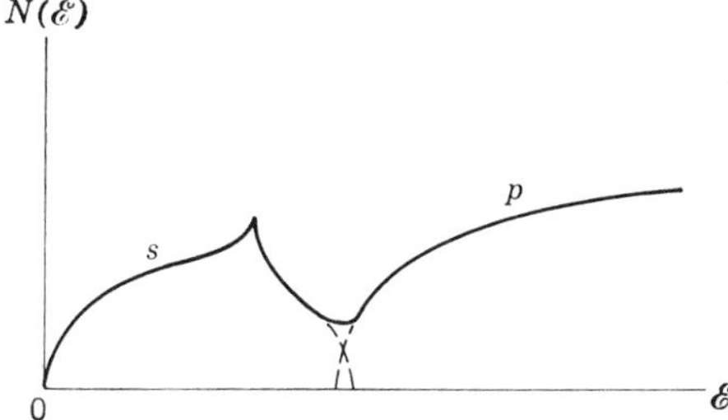


**FIGURE 13-5**

*Top:* A qualitative representation of the density of states as a function of energy in an unfilled isolated band. *Bottom:* The same for the case of two barely overlapping bands.

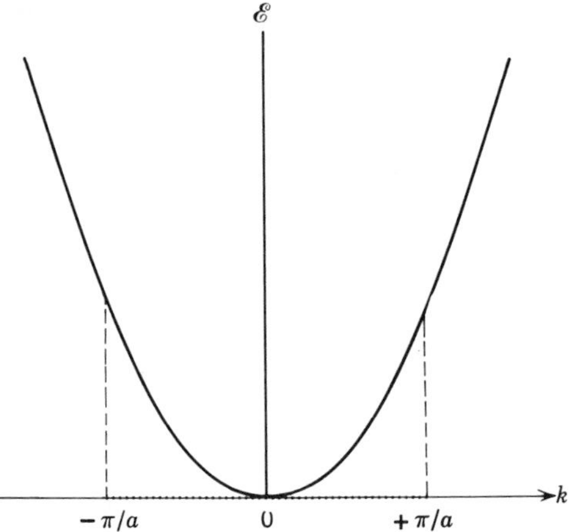


**FIGURE 13-6**
The energy of a free electron plotted as a function of its wave number $k$. The points indicate schematically the uniformly spaced allowed values of $k$. For the first band of the crystal they fall within the range $-\pi/a \leq k \leq +\pi/a$, where $a$ is the ion separation of the one-dimensional lattice in which the electron moves freely.

is the value $k = 0$ in the middle of this range. Since $k \propto 1/\lambda \propto n$ and $n = 1, 2, 3, \ldots$, the values of $k$ allowed by the boundary conditions are evenly spread throughout this range. Each of them is associated with a different quantum state for the electron.

Next consider a two-dimensional metal with ions spaced by the same distance $a$ in both the $x$ and $y$ directions. In a band the allowed values of both the $x$ and $y$ component wave numbers, $k_x$ and $k_y$, are uniformly distributed over ranges extending from $-\pi/a$ to $+\pi/a$, as shown in Figure 13-7. Each pair of $k_x$ and $k_y$ values defines a point that specifies a quantum state for a free electron of the metal; these points are uniformly distributed within the square. A circle surrounding the origin of radius $k$, where $k^2 = k_x^2 + k_y^2$, passes through all states having the same energy since in two dimensions (13-9) reads

$$\mathscr{E} = \frac{\hbar^2(k_x^2 + k_y^2)}{2m} = \frac{\hbar^2 k^2}{2m}$$

The number of states $dN$, for values of $k$ ranging from $k$ to $k + dk$ is equal to the number of points contained within the area limited by $k$ and $k + dk$. As the points are uniformly distributed, this number will be proportional to the area. The figure shows that as long as $k < \pi/a$, $dN$ increases with increasing $k$; specifically $dN = 2\pi k\, dk$. When $k$ begins to exceed $\pi/a$, further increase in $k$ causes $dN$ to decrease. Thus $dN/dk = N(k)$, the number of states per unit range of wave number, increases from zero for small $k$, reaches a maximum, and then decreases back to zero when $k$ reaches the largest allowed value for the band of our two-dimensional metal.

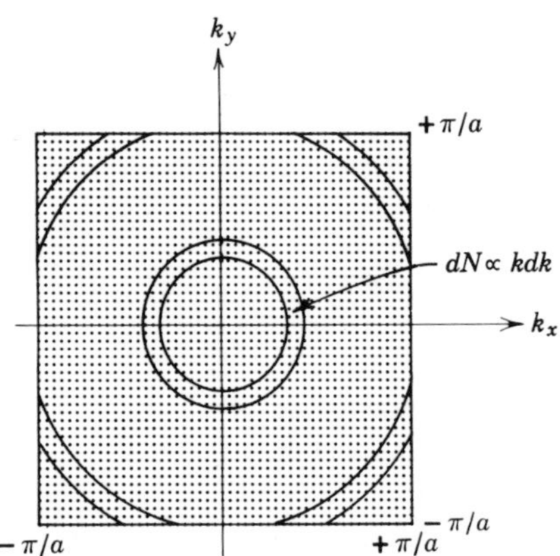


**FIGURE 13-7**
Illustrating the uniformly distributed allowed values of the $x$ and $y$ component wave numbers for a free electron in the first band of a two-dimensional square lattice with ion separation $a$.

The same general behavior is found when these results are converted from $N(k)$ to $N(\mathscr{E})$, the number of states per unit energy. In a real three-dimensional metal it is also true. That is, the density of states $N(\mathscr{E})$ increases from zero for small $\mathscr{E}$ (the bottom of the band), reaches a maximum, and then decreases back to zero at the largest allowed value $\mathscr{E}_{max}$ found in the band (the top of the band). The detailed behavior of $N(\mathscr{E})$ depends on the geometrical details of the arrangements of ions in the crystalline metal, as does the exact value of $\mathscr{E}_{max}$. But the general behavior is always about as we have indicated, and the approximate value of $\mathscr{E}_{max}$ is given by (13-7) if $a$ is interpreted as the characteristic ion spacing in the crystal.

## 13-6 The Motion of Electrons in a Periodic Lattice

The free-electron model that we have used ignores the effects of electrons interacting with the crystal lattice. Let us begin to consider this by making some general remarks about the effect of the periodic variation in the potential. For one thing, the lattice periodicity has the effect that the wave functions for an infinitely long lattice are no longer sinusoidal traveling waves of constant amplitude, but they exhibit the lattice periodicity in their amplitudes. In addition, electrons may be scattered by the lattice. Just as an electromagnetic wave suffers a Bragg "reflection" when the Bragg condition is satisfied, so also when the de Broglie wavelength of the electron corresponds to a periodicity in the spacing of the ions the electron interacts particularly strongly with the lattice. We shall see that these modifications result, among other things, in changing the resistance of the crystal to the conduction of electricity.

Our approach in finding the allowed energies of electrons in solids has been to consider the effect of forming a solid as the individual constituent atoms are brought together. If, instead, we had begun by modelling the periodic potential seen by an electron in the crystal lattice by a succession of rectangular wells and barriers, and had then solved the Schroedinger equation for such a potential, we would have found sinusoidal wave solutions in certain energy ranges (the allowed bands) and real decaying exponential wave solutions in the other energy ranges (the forbidden bands). This approach permits detailed quantitative calculations, but we present it here only qualitatively.

Although the electrons tend to smooth out the variations in the potential due to the ions, the potential is not constant but varies in a periodic way. The actual shape of the potential determines the exact solution to the Schroedinger equation for an electron in a crystal lattice, but the most important feature of the potential is its periodicity. The effect of periodicity is to change the free particle traveling wave eigenfunction in such a way that instead of constant amplitude it has a varying amplitude which changes with the period of the lattice. If the space periodicity of the lattice is $a$, then, according to Bloch, the eigenfunctions for a one-dimensional system do not have the free particle traveling wave form $\psi(x) = Ae^{ikx}$ of (13-8), but instead they have the form

$$\psi(x) = u_k(x)e^{ikx} \tag{13-11a}$$

where the periodicity of the lattice requires that

$$u_k(x) = u_k(x + a) = u_k(x + na) \tag{13-11b}$$

$n$ being an integer. Hence, the effect of the periodicity is to modulate periodically the free-electron solution amplitude. The wave function is

$$\Psi(x,t) = u_k(x)e^{i(kx-\omega t)} \tag{13-12}$$

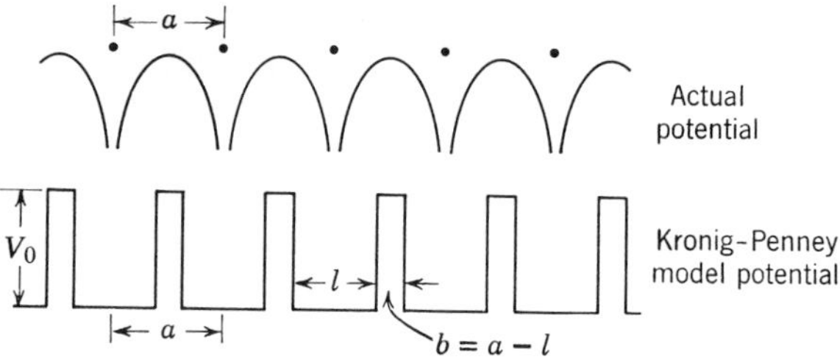


**FIGURE 13-8**

Illustrating how the potential for an electron moving in a periodic lattice can be approximated by the Kronig-Penney model of an array of rectangular potential wells and barriers.

where the second (exponential) factor describes a wave of wavelength $\lambda = 2\pi/k$ that travels toward $+x$ if $k > 0$ and toward $-x$ if $k < 0$, and the first factor $u_k(x)$ describes the modulation. The function $u_k(x)$ resembles the eigenfunction for an isolated atom. Its exact form depends on the particular potential assumed and the value of $k$. A very good approximation to $V(x)$ for a crystal is an array of rectangular potential wells and barriers having the lattice periodicity, as in Figure 13-8. Each well represents an approximation to the potential produced by one ion. This is the *Kronig-Penney model* which is, of course, easier to treat mathematically than the real case, but which retains all of its important features. Let us now examine the model in more detail.

For wells that are deep and widely spaced, the electron of not too high energy is practically bound within one of the wells, so that the lower energy eigenvalues are those of a single well. For wells that are closer together the eigenfunctions can penetrate the potential barriers more easily. This results in the spreading of a previously single energy level into a band of energy levels. As the separation of the wells is reduced the band becomes wider. Indeed, in the limit of zero barrier thickness we obtain an infinitely wide single well in which all energies are allowed, i.e., we obtain the free-electron model. In Figure 13-9 we compare the allowed energies of a single well with those of the Kronig-Penney model of an array of wells and barriers. Notice that each allowed band corresponds to a discrete level of the single well, and that forbidden bands appear even for energies $\mathscr{E}$ greater than the well depth $V_0$. The band widths can be made to approach the level width as $a$ increases (the width of the individual wells, $l$, remaining fixed) and to approach a continuum as $a$ decreases.

In solving the Schroedinger equation for the Kronig-Penney model, we must satisfy the conditions on the continuity of $\psi$ and $d\psi/dx$, just as we had to do for the single rectangular well. This restricts the validity of the Bloch solution, (13-11a) and (13-11b), to certain ranges of energy and gives the allowed bands. For energy values in the forbidden bands, the eigenfunctions are rapidly damped by a real decaying exponential factor. The expression $\mathscr{E}(k)$ for the allowed energies in terms of the wave number $k$ of the electron is more complicated than that for the free electron, but the gaps or discontinuities in energy occur at values of $k$ given simply by

$$k = \pm\frac{\pi}{a}, \pm\frac{2\pi}{a}, \pm\frac{3\pi}{a}, \ldots \tag{13-13}$$

in which $a$ is the space periodicity of the lattice. In Figure 13-10 we plot the function $\mathscr{E}(k)$. At values of $k$ equal to the values specified in (13-13) we get energy gaps, whereas for values of $k$ not near those values the energies are much like that of a free

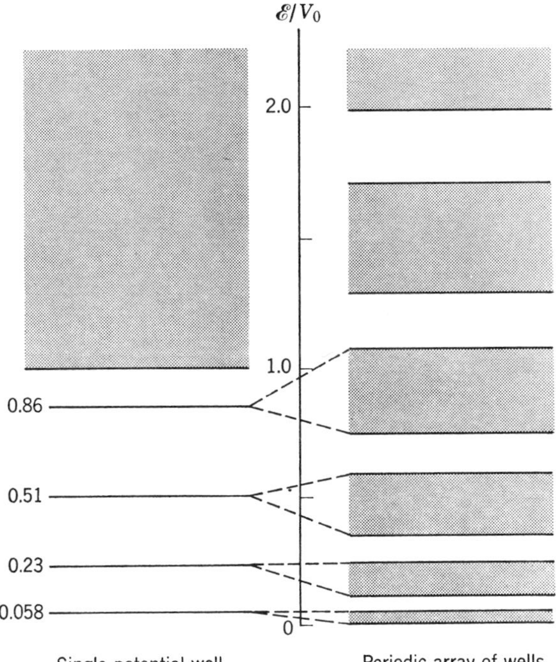


**FIGURE 13-9**
*Left:* Allowed energies for an electron in a single potential well. *Right:* Allowed energies in an array of periodically spaced wells and barriers. The levels shown are for a well strength given by $2mV_0l^2/\hbar^2 = (11)^2$, and a barrier thickness $b = l/16$. Note the appearance of forbidden bands even for energies $\mathscr{E}$ greater than $V_0$.

electron shown by the dashed curve in the figure. The origin of the allowed and forbidden bands is apparent from the figure. Each allowed band corresponds to solutions to the Schroedinger equation in which the wave number $k$ has positive values in a range of width $\pi/a$, and also negative values in a range of the same width. Note that this agrees with a conclusion obtained from a very different point of view in the last section, and expressed in (13-10).

From the present point of view, the gaps between the top of an allowed band and the bottom of the next one up can be understood as a result of Bragg reflection of the traveling wave describing an electron propagating down the lattice. If a wave traveling to the right is incident on a set of barriers representing the regions between the ions of the lattice, spaced by the uniform distance $a$, it will be partly reflected by each of these barriers. Generally, the reflected waves traveling to the left will not be exactly in phase with each other, and so they will not combine constructively to produce a net reflected wave of large amplitude. But they will be in phase if the wavelength $\lambda$ of the incident and reflected waves is related to the spacing $a$ by the one-dimensional version of (3-3), the *Bragg condition*

$$2a = \lambda, 2\lambda, 3\lambda, \ldots \tag{13-14}$$

Here $2a$ is the extra distance traveled in reflections from successive barriers, so if it equals an integral number of wavelengths $\lambda$ the reflected waves will all be precisely in phase and there will be a net reflected wave whose amplitude equals the amplitude of the incident wave. Since $\lambda = 2\pi/k$, the Bragg condition is $2a = 2\pi/k$, $2(2\pi/k)$,

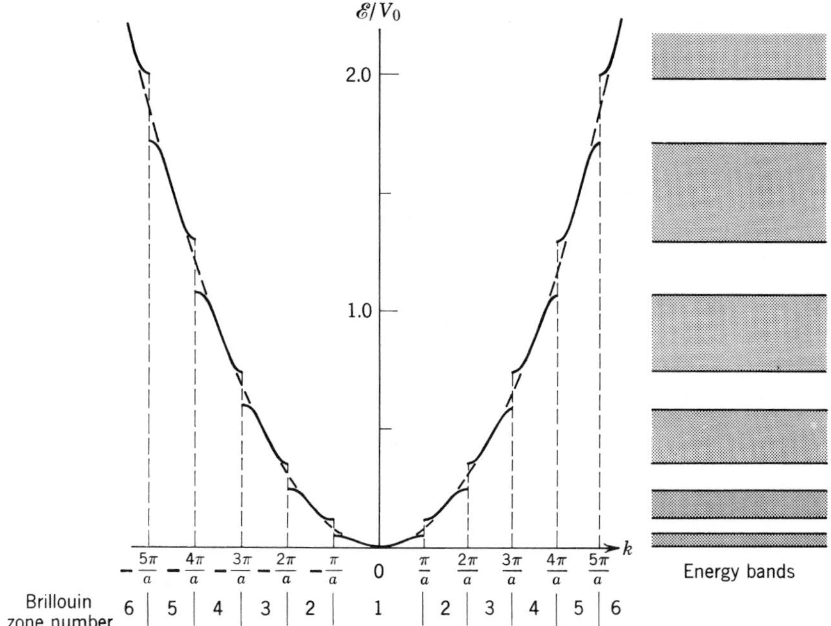


**FIGURE 13-10**

Allowed energies in a one-dimensional lattice of periodicity $a$, as a function of the wave number $k$. The dashed curve gives the free electron model result, for comparison. The allowed and forbidden energy bands that result are shown on the right.

$3(2\pi/k), \ldots,$ or $k = \pm\pi/a,\ \pm 2\pi/a,\ \pm 3\pi/a, \ldots,$ where we have inserted $\pm$ signs to account for the fact that the incident wave could as well be moving to the left (to $-x$) as moving to the right (to $+x$). Comparing with (13-13), we see that the values of $k$ at which the gaps in the function $\mathscr{E}(k)$ occur are just those values of the wave number for which the wavelength $\lambda$ satisfies the Bragg condition for constructive reflection.

The gaps themselves arise because there are two distinctly different ways for the amplitude of the reflected wave to equal the amplitude of the incident wave, at each critical value of $k$ where these amplitudes are equal. Consider, for instance, a unit amplitude incident wave moving to the right along the $x$ axis with $k = \pi/a$. The traveling wave eigenfunction describing this is $e^{ikx} = e^{i\pi x/a}$. The reflected wave, which also has unit amplitude for this value of $k$, is $e^{-ikx} = e^{-i\pi x/a}$. The total eigenfunction is obtained by adding these two or, equally well, by subtracting them. The first possibility gives

$$\psi = e^{i(\pi/a)x} + e^{-i(\pi/a)x} \propto \cos\frac{\pi}{a}x \qquad (13\text{-}15)$$

and the second gives

$$\psi = e^{i(\pi/a)x} - e^{-i(\pi/a)x} \propto \sin\frac{\pi}{a}x \qquad (13\text{-}16)$$

In both, the reflected wave has the same amplitude as the incident wave, and so it combines with it to form a standing wave; but the two cases differ very significantly in regard to the locations of the nodes of the standing wave, and therefore in the locations of the maxima and minima of the probability density $\psi^*\psi$. In the case where

$\psi \propto \cos \pi x/a$, the probability density will maximize at $x = 0$, as well as at $x = \pm a$, $\pm 2a$, $\pm 3a, \ldots$, while for $\psi \propto \sin \pi x/a$ the probability density will be zero at all these points. If they are the locations of the barriers between ions, the electron described by $\psi$ will feel a larger repulsion, and therefore have a higher energy, in the cosine case than in the sine case. If these points are the locations of the ions, the situation will be reversed. But the basic conclusion—that there are two different energies $\mathscr{E}$ corresponding to the same value of the wave number $k$ when $k$ is any one of the values given by (13-13)—is independent of how the origin of the $x$ axis is defined.

Looking again at the function $\mathscr{E}(k)$ plotted in Figure 13-10, we see the two different values of $\mathscr{E}$ at each of the critical values of $k$ where Bragg reflection will occur. We also see how this circumstance causes the $\mathscr{E}(k)$ curve to have an $S$-shaped deviation from the parabolic curve for a free electron in each region between the critical values of $k$. The range of $k$ values between $-\pi/a$ and $+\pi/a$ defines what is called the first *Brillouin zone*; those $k$ values between $-2\pi/a$ and $-\pi/a$ and between $+\pi/a$ and $+2\pi/a$ define the second Brillouin zone, etc., as is indicated below the $k$ axis of the figure.

## 13-7 Effective Mass

When discussing the behavior of an electron in a periodic lattice under the application of an external electric field, it is very convenient to introduce the concept of the effective electron mass. This is done by using a relation developed in Section 3-4 to describe the motion of the electron in terms of a *group* of traveling waves. According to (3-13b), the velocity $g$ of such a group equals the derivative of the frequencies $\nu$ of its component sinusoidal traveling waves with respect to their reciprocal wavelengths $\kappa$. That is

$$g = \frac{d\nu}{d\kappa} = \frac{d\omega}{dk}$$

where $\nu$ is converted to the angular frequency $\omega$, and $\kappa$ to the wave number $k$, by multiplying and dividing $d\nu/d\kappa$ by $2\pi$. To remind the student of the meaning of this relation, we shall apply it to the simple case of a free electron, whose energy is

$$\mathscr{E} = \frac{p^2}{2m} = \frac{\hbar^2 k^2}{2m} = \hbar\omega$$

The last equality depends on the Einstein-de Broglie relation $\mathscr{E} = h\nu = \hbar\omega$. Evaluating $d\omega/dk$ from this expression, we have

$$g = \frac{d\omega}{dk} = \frac{\hbar 2k}{2m} = \frac{\hbar k}{m} = \frac{p}{m} = \frac{mv}{m} = v \tag{13-17}$$

We obtain the correct result that the group velocity $g$ equals the velocity $v$ of the electron whose motion is represented by the group. Of course this result is of general validity.

Now we consider an electron in a one-dimensional lattice, whose wave number dependence of energy has the form $\mathscr{E}(k)$ that we have been discussing. To this system an *external* electric field $E$ is applied. In time $dt$ the electron of charge $q$ moves distance $dx$, and the work done by the external field is the applied force $qE$ multiplied by $dx$. Since this equals the magnitude of the change $d\mathscr{E}$ in the energy of the electron, we have, using (13-17)

$$d\mathscr{E} = qE\,dx = qE\frac{dx}{dt}\,dt = qEv\,dt = qEg\,dt$$

But we also have, from $\mathscr{E} = \hbar\omega$

$$d\mathscr{E} = \hbar\, d\omega = \hbar \frac{d\omega}{dk}\, dk = \hbar g\, dk$$

Comparison then shows that

$$qE\, dt = \hbar\, dk$$

or

$$\hbar \frac{dk}{dt} = qE \tag{13-18}$$

If we take the time derivative of

$$g = \frac{d\omega}{dk} = \frac{1}{\hbar}\frac{d\mathscr{E}}{dk}$$

we obtain

$$\frac{dg}{dt} = \frac{1}{\hbar}\frac{d^2\mathscr{E}}{dt\, dk} = \frac{1}{\hbar}\frac{d^2\mathscr{E}}{dk\, dt} = \frac{1}{\hbar}\frac{d^2\mathscr{E}}{dk^2}\frac{dk}{dt}$$

or, using (13-18)

$$\frac{dg}{dt} = \frac{1}{\hbar^2}\frac{d^2\mathscr{E}}{dk^2}\, qE$$

Employing (13-17) again, this can be written

$$\frac{dv}{dt} = \frac{qE}{m^*} \tag{13-19a}$$

where

$$\frac{1}{m^*} \equiv \frac{1}{\hbar^2}\frac{d^2\mathscr{E}}{dk^2} \tag{13-19b}$$

The quantity $1/m^*$ is the reciprocal of the *effective mass* of the electron in the crystal lattice.

The electron we are studying moves under the influence of internal forces, exerted on it by the ions of the lattice, and an external force, exerted on it by the applied electric field $E$. If we wish, we can use (13-19a) to discuss its motion *in terms of the external force alone* since that equation is in the form of Newton's law of motion, acceleration equals external force divided by mass. Of course the effects of the internal forces are actually contained in the equation. They appear, however, only in the reciprocal effective mass $1/m^*$, which can have values quite different from the reciprocal of the true electron mass, $1/m$.

The properties of the lattice determine $1/m^*$ because, as we saw in the preceding section, they determine the form of the function $\mathscr{E}(k)$ and so also the derivative $d^2\mathscr{E}(k)/dk^2$ appearing in (13-19b). Figure 13-11 shows the first, and part of the second, Brillouin zones of a one-dimensional crystal. The solid curve is $\mathscr{E}(k)$ and the parabolic dashed curve is the free electron relation $\mathscr{E} = \hbar^2k^2/2m$. Near the center of the first zone, where $\mathscr{E}(k) \simeq \hbar^2k^2/2m$, $1/m^* = (d^2\mathscr{E}/dk^2)/\hbar^2 \simeq (\hbar^2 2/2m)/\hbar^2 = 1/m$. So in this region the lattice has very little effect on the electron, because its reciprocal effective mass is almost the same as its reciprocal true mass, and it responds to the applied electric field as if it were an essentially free electron. The curvature of the function $\mathscr{E}(k)$ changes significantly from the curvature of the parabola in proceeding in either direction from the center of the zone, which makes dramatic changes in the reciprocal of the effective mass of the electron and so in its response to the applied field. Since $d^2\mathscr{E}/dk^2$ goes through zero, and then becomes negative and of large magnitude as $k$ approaches either boundary of the first zone, $1/m^*$ does the same. Thus in the upper

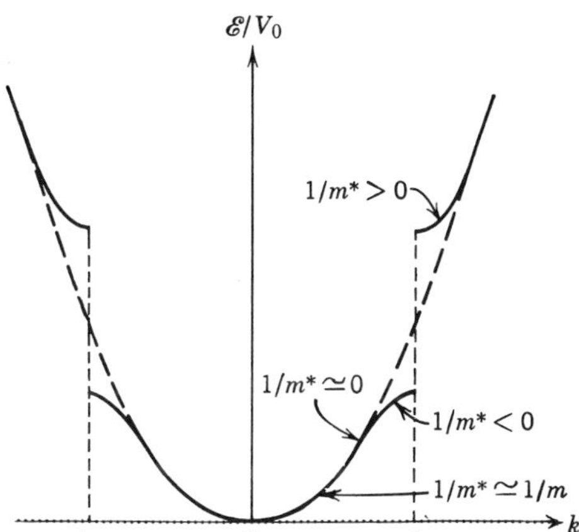


**FIGURE 13-11**
Illustrating the reciprocal effective mass at various locations in the first and second Brillouin zones of a one-dimensional lattice. The points on the $k$ axis indicate the uniformly distributed allowed values of $k$.

part of the energy range of the band corresponding to the first zone the electron in the lattice responds to the applied electric field very differently from the way it would if it were a free electron. Where $1/m^*$ is zero a given applied force $qE$ causes no acceleration of the electron, and where $1/m^*$ is negative the force causes an acceleration in the opposite direction of that which would be experienced by a free electron. (This has nothing to do with the sign of the electron charge which, to avoid confusion, we have written as $q$ instead of $-e$.) At the bottom of the energy band for the second Brillouin zone, $1/m^*$ is positive but appreciably larger than $1/m$ for a free electron, so the applied force produces a relatively large acceleration of the electron in the lattice.

The response of an electron in a crystal to an applied electric field can be understood in terms of the way the electron wave is reflected by the potential barriers located between each pair of ions. At the bottom of the first energy band where the magnitude of the wave number has the value $|k| \simeq 0$ there is practically no reflection since the Bragg condition $|k| = \pi/a$ is far from being satisfied. When the field is applied the force it produces will increase the electron's momentum, and the work it does will increase the electron's energy, just as in the case of a free electron. Higher up in the band, where $|k|$ is closer to the critical Bragg value $\pi/a$, reflection starts to become appreciable. In this region the work done on the electron will still increase its energy, but this increases the amount of reflection, and reflection corresponds to reversing the sign of its momentum. At the point where $1/m^* = 0$, the gain in positive momentum due to the applied field acting directly on the electron is exactly compensated for by the gain in negative momentum due to the enhanced reflection of the electron by the lattice ions. Thus here the net change in electron velocity is zero, and from the point of view of its response to the applied field the electron effectively has infinite mass, or zero reciprocal mass. (Momentum is, of course, given to the lattice by the overall effect of applying the field, but not to the electron.) At the top of the band the reciprocal effective mass is large and negative because the enhanced reflection resulting from the closer approach to the Bragg condition of perfect reflection is much more significant in changing the electron momentum than the direct action of the applied field. The situation is reversed at the bottom of the next higher band, and so the reciprocal effective mass is large and positive there.

Effective mass is also used in a somewhat different way to compare, for various bands, the curvature of the function $\mathscr{E}(k)$ in the concave upward approximately parabolic regions found except near the tops of bands. If the zero of $k$ is taken to be at the boundary of the second zone, and the zero of $\mathscr{E}$ is taken at the bottom of the corresponding band, then $\mathscr{E}(k)$ for the part of the second zone shown in Figure 13-11 can be written as

$$\mathscr{E}(k) \simeq \frac{\hbar^2 k^2}{2m^*} \tag{13-20}$$

If the curvature of $\mathscr{E}(k)$ is high, so that $\mathscr{E}$ increases rapidly with increasing $k$, then $1/m^*$ in this expression is large. Since the allowed values of $k$ are uniformly distributed along the $k$ axis of Figure 13-11, the density of the corresponding energy levels along the $\mathscr{E}$ axis will be low if $\mathscr{E}$ increases rapidly with increasing $k$. So the reciprocal mass can also be used to compare *level densities* of bands, in the regions where they obey (13-20). If the level density is relatively low, $1/m^*$ is relatively large; if the level density is relatively high, $1/m^*$ is relatively small.

The concept of effective mass is useful in a variety of ways. For instance, the classical theory of the behavior of charge carriers under the influence of an applied electric field is summarized by (13-1b), which predicts that the electrical conductivity $\sigma$ of the material containing the carriers is proportional to the reciprocal of their masses. We can easily modify this to take into account the quantum behavior of charge carrying electrons in a crystal lattice by replacing the reciprocal true mass with the reciprocal effective mass, obtaining

$$\sigma \propto \frac{1}{m^*} \tag{13-21}$$

Consider iron. The valence electrons in this metal partly fill its $3d$ bands, which are overlapping and narrow since $3d$ is an *inner* subshell in the transition element iron so the splitting of the atomic $3d$ level into the $3d$ bands is not very pronounced. Because the bands are narrow, the level density is high. Therefore the reciprocal effective mass is small for the electrons involved in electrical conduction in iron, the value of $1/m^*$ being about $0.1/m$. As a consequence, the metal is not a particularly good conductor. Copper, on the other hand, is a good electrical conductor. The reason is that for copper the $3d$ bands are filled, and the conduction electrons are $4s$ electrons which are in a very broad band (it overlaps the $3d$ bands), that has a low-level density and a high reciprocal effective mass ($1/m^*$ is roughly equal to $1/m$). The $4s$ band is broad because this is an *outer* subshell of the atom and so the splitting in the crystal of the $4s$ atomic level is large. The result is that the conductivity of copper is an order of magnitude higher than the conductivity of iron.

It should be pointed out that using the reciprocal effective mass in (13-21) amounts to accounting for the influence of a perfect crystal lattice on the accelerated motion of an electron in an applied electric field. As was discussed in Section 13-4, accelerated motion takes place between collisions of the electron with the imperfections that are actually found in the lattice of a real material, due to thermal motion of the ions or to impurity ions. These collisions tend to randomize the electron motion, and they cause the over-all electron motion to be a drift with velocity proportional to the strength of the applied field, in contrast to an ever increasing velocity with acceleration proportional to the strength of the field. If there were no lattice imperfections, after a fixed field was applied the electron current would increase in time until it reached such large values that it was limited by practical considerations having nothing to do with either the strength of the field or the properties of the material. In such circumstances the material could be said to have zero resistance (or at least it could be said not to obey Ohm's law). So the presence of nonzero resistance, or noninfinite conductivity, is due to the presence of lattice imperfections. This can be seen in the fact that the resistance of a metal increases with increasing temperature and with increasing impurity concentration. Nevertheless, the value of $1/m^*$, which has to do with the properties of a perfect lattice, influences the value of the resistivity or conductivity because it influences the average velocity gain between randomizing collisions with imperfections, and this determines the drift velocity.

In situations where all the levels of an isolated band are filled except for those near the very top, it is convenient to think in terms of *holes* representing the absence of electrons in an otherwise completely filled band. Since the absence of a negatively charged electron is equivalent to the presence of a positive charge, holes behave as if they are positively charged. Furthermore, since the effective mass is negative for the levels near the top of a band, holes, describing the absence of negative effective mass, behave as if they have positive effective mass. We shall have more to say about them in the following sections.

## 13-8 Semiconductors

Semiconductors are of much interest because their behavior is the basis for many practical electronic devices, such as transistors. Also, they are excellent illustrations of the ideas discussed in previous sections. Semiconductors are covalent solids that may be regarded as "insulators" because the valence band is completely full and the conduction band is completely empty at the absolute zero of temperature, but they have an energy gap between the valence and conduction bands of no more than about 2 eV. For silicon the energy gap is 1.14 eV and for germanium the gap is 0.67 eV. Although the value of the Fermi distribution function governing the relative population of an energy state in the conduction band to an energy state in the valence band is small, since $kT \simeq 0.025$ eV at room temperature, the number of available states in the conduction band is high. Hence the thermal excitation from the valence band into the conduction band occurs for a significant number of electrons, this number being the product of the number of electrons per quantum state and the number of quantum states per energy interval. Furthermore, the conductivity of a semiconductor increases rapidly with rising temperature, the number of excited electrons in silicon, for example, increasing by a factor of about one billion with a doubling of temperature from 300°K to 600°K. Since the valence band is filled at low temperature, with the four valence electrons of silicon or germanium forming covalent bonds, each electronic excitation into the conduction band leaves a hole in the valence band. These holes, acting as positive charge carriers, also contribute to the conductivity. In Figure 13-12 we illustrate the semiconductor band scheme.

The conductivity of the semiconductors arising from thermal excitation is called *intrinsic conductivity*. There are other ways to enhance the conductivity, such as by photoexcitation. The energy gap in semiconductors is equivalent to the energy of

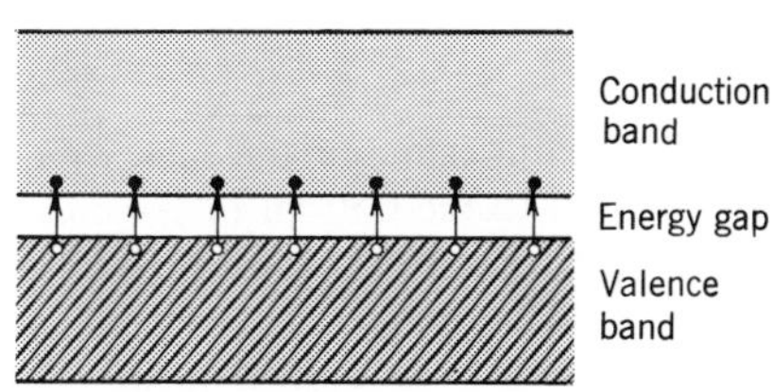


**FIGURE 13-12**
The band scheme of a semiconductor in which the energy gap between the initially full valence band and the initially empty conduction band is small. Thermal excitation raises some electrons over the gap into the conduction band, leaving holes in the valence band.

photons in the red or infrared portion of the electromagnetic spectrum so that semiconductors are *photoconductive.* This contribution to the conductivity increases with the intensity of the light and will drop to zero when the light source is turned off and the normal thermal equilibrium distribution of electrons is restored. Still another way to increase the conductivity is by adding impurities to the semiconductor. That is, we replace some atoms of the semiconductor with atoms of another element, having about the same size but a different valence. The resulting conductivity, whose origin we explain presently, is called *extrinsic conductivity*, and the procedure is called *doping*.

If a small quantity of arsenic is added to molten germanium, the arsenic impurities will crystallize with the germanium into its diamondlike structure. Arsenic has five electrons per atom in the valence band and germanium has four electrons per atom in the valence band. Hence, four of the arsenic electrons are used for covalent binding and the fifth electron is nearly free. It cannot go into the filled valence band and is very weakly bound in an "orbit" of very large radius around the singly charged arsenic ion. The arsenic ion Coulomb attraction is largely shielded by polarization of the intervening germanium atoms; i.e., the field of the ion is weakened by the dielectric nature of the germanium crystal. Because this fifth electron has such a small binding energy to the arsenic, it can be ionized, and go into the conduction band at a much lower temperature than would be needed for electrons in the valence band. Hence, this excess electron will occupy some one of a set of discrete energy levels just below the conduction band at a low temperature, but it can very easily be thermally excited into that band. At ordinary temperatures all of these excess electrons go into the conduction band. The electrical conductivity can be controlled by the amount of arsenic used as an impurity. A significant effect is obtained with as little as one impurity atom per million semiconductor atoms. An impurity that contributes electrons is called a *donor* impurity and the resultant semiconductor is called an *n-type* (negative) semiconductor because it has an excess of free electrons.

**Example 13-3.** Make a rough estimate of the binding energy of the donor electron of arsenic in a germanium crystal, taking the dielectric constant of the crystal to have the value $\kappa = 16$, and the effective mass of the electron to have the value $m^* = 0.2m$.

The donor electron moves in the field of the arsenic ion, $As^+$, and it behaves like the electron in the ground state of a hydrogenlike atom. The chief difference is that this electron moves in a polarizable lattice rather than in vacuum. Because the potential energy of the ion-electron system is now $-e^2/\kappa 4\pi\epsilon_0 r$, the corresponding hydrogenlike energy levels are given by replacing $4\pi\epsilon_0$ by $\kappa 4\pi\epsilon_0$ in the hydrogen energy-level formula, (4-18), and also by replacing the electron mass $m$ found there by the effective mass $m^*$ to take into account the fact that the electron is actually in a crystal lattice. Since the electron is near a lower band edge where $d^2\mathscr{E}/dk^2$ is large, $m^*$ is small; various evidence indicates the value is $m^* \simeq 0.2m$. So we have

$$E \simeq -\frac{1}{\kappa^2}\left(\frac{1}{4\pi\epsilon_0}\right)^2 \frac{m^* e^4}{2\hbar^2 n^2}$$

where $\kappa = 16$, $m^* = 0.2m$, and $n = 1$. Since for $\kappa = 1$ and $m^* = m$ the energy $E$ has the value $-13.6$ eV, it is easy to show that

$$E \simeq -0.01 \text{ eV}$$

Hence, according to our estimate, the energy required to ionize the arsenic donor electron in a germanium crystal to the bottom of the conduction band is about 0.01 eV. The value obtained directly from measurements of the photon energy required to ionize, or indirectly from measurements of the temperature dependence of the conductivity, is 0.0127 eV. See Table 13-2 for measured values in other cases.

**TABLE 13-2.** Donor and Acceptor Ionization Energies

| | In Germanium | In Silicon |
|---|---|---|
| Impurity | $\mathscr{E}_c - \mathscr{E}_{donor}$ (eV) | |
| Arsenic | 0.0127 | 0.049 |
| Antimony | 0.0096 | 0.039 |
| Impurity | $\mathscr{E}_{acceptor} - \mathscr{E}_v$ (eV) | |
| Gallium | 0.0108 | 0.065 |
| Indium | 0.0112 | 0.16 |

Note that the radius of the Bohr-like orbit of the donor electron is $\kappa m/m^* \simeq 80$ times that of the ground state hydrogen atom, as can be seen by inspecting (4-16). So the electron moves in an orbit that contains a large number of germanium atoms. This justifies the use, in our previous estimate, of the dielectric constant, which is a macroscopic rather than a microscopic quantity that characterizes the germanium crystal when it is regarded as a continuum. ◀

If a small quantity of gallium is added to germanium, the situation will be different than that just discussed. Gallium has three electrons per atom in the valence band, so that it has a deficiency of one electron per atom in forming the covalent bonds. The result is a hole, which can drift through the crystal, behaving like a positive charge and mass as successive electrons fill one hole and create another. From an energy point of view, this impurity introduces vacant discrete levels slightly above the top of the valence band. Valence electrons are then easily excited into these impurity levels, which can accept them, leaving holes in the valence band. The energy separation between the acceptor levels and the top of the valence band is small for the same reasons that give a small separation between the donor levels and the bottom of the conduction band: a high dielectric constant and a small effective mass. An impurity that is deficient in electrons is called an *acceptor* impurity and the resultant semiconductor is called a *p-type* (positive) semiconductor.

Whether the conductivity of a semiconductor is p-type or n-type can be determined by the Hall effect. In Figure 13-13 we show schematically the energy-level diagram corresponding to each type. The localized energy levels of impurity atoms are not

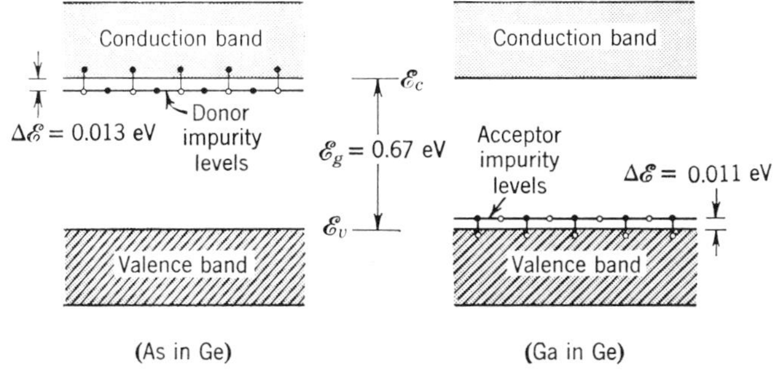


**FIGURE 13-13**

*Left:* Schematic energy-level diagram of a germanium crystal containing donor impurity atoms. *Right:* Containing acceptor impurity atoms.

broadened into bands because these atoms are many lattice spacings apart and interact with each other very weakly. In Table 13-2 we list the energy of the levels introduced into germanium and silicon crystals by small amounts of common impurities. For donor impurities the energy from donor levels to the energy $\mathscr{E}_c$ at the bottom of the conduction band is given, whereas for acceptor impurities the energy from the top of the valence band $\mathscr{E}_v$ to the acceptor levels is given. Note that these energies are comparable to $kT = 0.025$ eV at room temperature. Therefore, we can expect to have plenty of thermal ionization at room temperature.

In an intrinsic semiconductor the number of vacant states in the valence band is equal to the number of occupied states in the conduction band, so that the Fermi energy is located somewhere in the gap between the bands. If the densities of states in the two bands are symmetrical then the Fermi energy will be in the middle of the gap. The Fermi energy, as the student will recall, is defined as the energy for which the average number of electrons that would occupy a quantum state there is 0.5, where we treat electron spin in such a way that the maximum occupancy is 1.0.

**Example 13-4.** Consider a forbidden band of width $\mathscr{E}_g$ that separates a valence band and a symmetrical empty conduction band in an intrinsic semiconductor. Show that the Fermi energy lies at the center of the forbidden band, i.e., that $\mathscr{E}_F = \mathscr{E}_g/2$ if $\mathscr{E} = 0$ is taken to be the upper edge of the valence band.

The proof can be followed by inspecting Figure 13-14. At the top of the figure we plot $N(\mathscr{E})$ the number of quantum states per unit energy interval for the upper part of the valence band and the lower part of the conduction band. The figure tentatively places the Fermi energy $\mathscr{E}_F$ in the center of the gap of width $\mathscr{E}_g$ between the two bands. The density of states $N(\mathscr{E})$ is drawn so that its descending behavior moving towards the top of the valence band is symmetrical to its ascending behavior moving away from the bottom of the conduction band. This is in qualitative agreement with the general behavior of $N(\mathscr{E})$ throughout an entire isolated band (see, for example, Figure 13-5).

In the middle of Figure 13-14, we show the Fermi distribution $n(\mathscr{E})$, which is the probable number of electrons per state. For clarity, it is constructed for an operating temperature where $kT \sim \mathscr{E}_g$. It is also constructed for $\mathscr{E}_F$ in the center of the forbidden band.

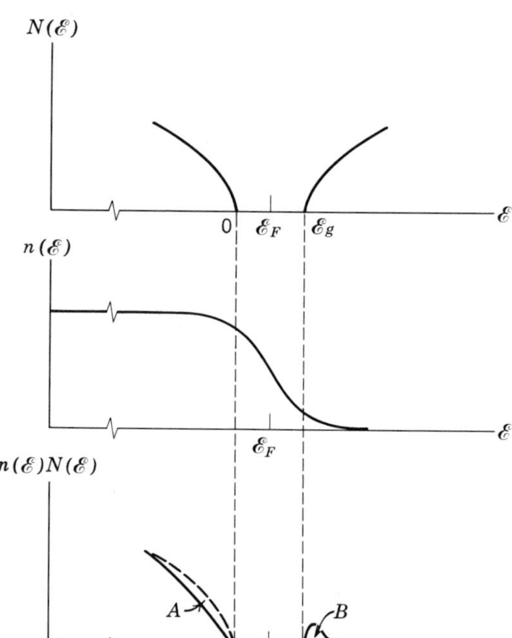


**FIGURE 13-14**
The number of electrons as a function of energy in the valence and conduction bands of an insulator or semiconductor with a forbidden band width $\mathscr{E}_g$, as a product of the density of states $N(\mathscr{E})$ and the Fermi distribution $n(\mathscr{E})$.

The solid curve in the bottom of Figure 13-14 shows the product $n(\mathscr{E})N(\mathscr{E})$, which gives the number of electrons per unit energy in various states at the temperature just mentioned. The dashed curve shows the same thing for a temperature of absolute zero. At $T = 0$, the valence states are completely filled and the conduction states are completely empty, so the dashed curve in the valence region is just $N(\mathscr{E})$, while it is the $\mathscr{E}$ axis in the conduction region. The area $A$ between the dashed and solid curves is proportional to the number of valence states that electrons leave when the temperature is raised; i.e., it is a measure of the number of holes created. The area $B$ between the solid and dashed curves is proportional to the number of electrons that are promoted to states in the conduction band at the operating temperature.

In an intrinsic semiconductor it is necessary that area $A$ equal area $B$, since the density of holes in the valence band equals the density of electrons in the conduction band. It is apparent that this condition is satisfied by Figure 13-14, because we have constructed it with $\mathscr{E}_F$ in the center of the forbidden band. A moments consideration will show the student that it would not be satisfied for a different choice of $\mathscr{E}_F$, due to the symmetry of $n(\mathscr{E})$ about $\mathscr{E}_F$, and to the (approximate) symmetry of $N(\mathscr{E})$ about the center of the gap between the two allowed bands. ◀

**Example 13-5.** Make an estimate of the relative number of electrons in the conduction band of an insulator or semiconductor at temperature $T$.

Figure 13-14 also shows an exaggerated picture of the energy distribution of electrons as a product of the density of states $N(\mathscr{E})$ and the Fermi distribution $n(\mathscr{E})$ appropriate in the valence, forbidden, and the conduction bands of an insulator. If, in the Fermi distribution $n(\mathscr{E})$, we have $\mathscr{E} - \mathscr{E}_F \gg kT$, then

$$n(\mathscr{E}) = \frac{1}{e^{(\mathscr{E}-\mathscr{E}_F)/kT} + 1} \simeq \frac{1}{e^{(\mathscr{E}-\mathscr{E}_F)/kT}}$$

so that in such an energy range the Fermi distribution varies with energy like the Boltzmann distribution. We know from Example 13-4 that $\mathscr{E} - \mathscr{E}_F = \mathscr{E}_g/2$ at the bottom of the conduction band in an insulator, if we measure $\mathscr{E}$ from the top of the valence band. So the condition $\mathscr{E} - \mathscr{E}_F \gg kT$ is met since $\mathscr{E}_g \gg kT$ for an insulator. Thus we can take

$$n(\mathscr{E}) = \frac{1}{e^{\mathscr{E}_g/2kT}} = e^{-\mathscr{E}_g/2kT}$$

as the number of electrons per state in the conduction band of an insulator.

The Fermi distribution falls in value by an order of magnitude in an energy range of about $\Delta\mathscr{E} = 2kT$ so that we get a good estimate of $\Delta\mathscr{N}$, the number of conduction electrons, by evaluating those in the range $2kT$ above the bottom of the conduction band. Since $\Delta\mathscr{N} = n(\mathscr{E})N(\mathscr{E})\,\Delta\mathscr{E}$ we must now evaluate $N(\mathscr{E})$, the density of states. Because $N(\mathscr{E})$ starts at zero at the bottom of the conduction band, a good average value over the range $\Delta\mathscr{E} = 2kT$ is obtained by evaluating $N(\mathscr{E})$ at $\mathscr{E} = kT$. Hence,

$$\Delta\mathscr{N} = n(\mathscr{E})N(\mathscr{E})\,\Delta\mathscr{E} = e^{-\mathscr{E}_g/2kT}N(kT)2kT$$

Let us use here the results $\mathscr{N} = (2/3)\mathscr{E}_F N(\mathscr{E}_F)$ of Example 13-2 for a metal as an estimate of the total number of electrons, $\mathscr{N}$. We also note from (13-4) that $N(kT)/N(\mathscr{E}_F) = (kT/\mathscr{E}_F)^{1/2}$, so we have

$$\frac{\Delta\mathscr{N}}{\mathscr{N}} \simeq \frac{e^{-\mathscr{E}_g/2kT}N(kT)2kT}{(2/3)\mathscr{E}_F N(\mathscr{E}_F)} = 3e^{-\mathscr{E}_g/2kT}\left(\frac{kT}{\mathscr{E}_F}\right)\left(\frac{kT}{\mathscr{E}_F}\right)^{1/2}$$

or

$$\frac{\Delta\mathscr{N}}{\mathscr{N}} \simeq \left(\frac{kT}{\mathscr{E}_F}\right)^{3/2} e^{-\mathscr{E}_g/2kT}$$

This is the relative number of conduction electrons for an insulator.

This fraction is much smaller than the corresponding result $kT/\mathscr{E}_F$ of Example 13-2 for a metal, partly because the density of states $N(\mathscr{E})$ is smaller near the bottom of the conduction band in an insulator than at the Fermi energy in a metal, but principally because of the occupation factor $e^{-\mathscr{E}_g/2kT}$. Let us take $\mathscr{E}_g = 6$ eV as the gap in a typical insulator so that at room temperature this factor is $e^{-\mathscr{E}_g/2kT} = e^{-150} = 10^{-65}$. Not only is the fraction $\Delta\mathscr{N}/\mathscr{N}$ insignificant, but the absolute number of conduction electrons is also negligible for an insulator. If, however, $\mathscr{E}_g = 1$ eV, as for a semiconductor, then although $e^{-\mathscr{E}_g/2kT} = e^{-25} = 10^{-11}$ gives a very small fraction, the number of conduction electrons is no longer insignificant. ◀

In an impurity semiconductor containing donors the Fermi energy lies above the middle of the forbidden band because there are more electrons in the conduction band than there are holes in the valence band. In an impurity semiconductor containing acceptors the Fermi energy is below the middle of the forbidden band because there are fewer electrons in the conduction band than there are holes in the valence band. It is instructive to consider the combined effect of temperature and impurities on the Fermi energy. Let us begin at a temperature of absolute zero in an n-type semiconductor. The donor levels are all occupied but there are no electrons in the conduction band. The Fermi energy then must lie between the donor levels and the bottom of the conduction band, because the number of electrons per state $n(\mathscr{E})$ is one up to and including the donor levels and zero in the conduction band. Now, as the temperature is increased electrons are raised from donor levels to the conduction band. At that temperature at which half the donor states are empty, the Fermi energy corresponds to the donor-level energy. With a further increase in temperature, electrons in the valence band are excited and the Fermi energy drops more. When the number of electrons from the valence band is a very large fraction of those in the conduction band, the semiconductor acts as though it were intrinsic and the Fermi energy drops to nearly the center of the gap. If we had started with a p-type semiconductor we would find in a similar manner that, as the temperature is raised, the Fermi energy moves from between the top of the valence band and the acceptor levels, at absolute zero, to the center of the gap at high temperatures. At low temperatures, where $kT \ll \mathscr{E}_g$, conduction is due mostly to the impurities because there is little excitation of valence electrons. At high temperatures the impurity levels have been used up, that is, they have either donated or accepted electrons, so that the semiconductor acts as though it were intrinsic. In Figure 13-15 we plot the Fermi energy as a function of temperature for impurity semiconductors.

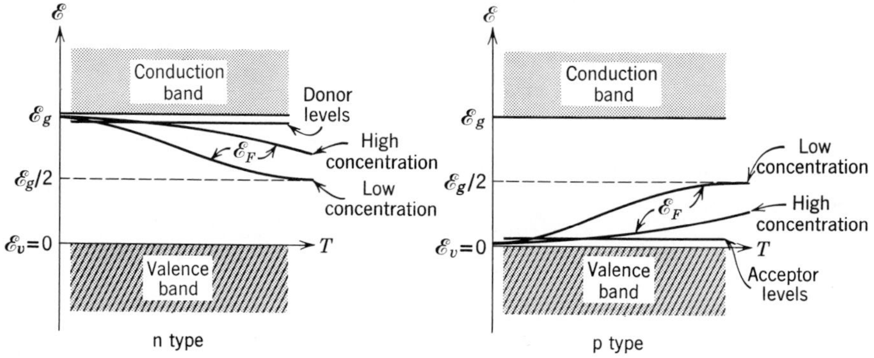


**FIGURE 13-15**

*Left:* The Fermi energy as a function of temperature for n-type semiconductors of two different impurity concentrations. *Right:* For p-type semiconductors of two different impurity concentrations.

## 13-9 Semiconductor Devices

We shall illustrate the use of impurity semiconductors in electronics by discussing briefly the operation of three semiconductor devices, the rectifier, the transistor, and the tunnel diode.

A *rectifier* is formed by having acceptor impurities (p-type) in one region of a crystal and donor impurities (n-type) in another region. The boundary between these regions is called a *p-n junction*. Figure 13-16 shows the energy band structure of an unbiased p-n junction at room temperature. The boundaries of the bands must be warped in going from the p-region through the junction to the n-region because the Fermi energy is close to the top of the valence band in the p-region and close to the bottom of the conduction band in the n-region, yet the Fermi energy must have the same value everywhere. The reason is that if the Fermi energy were not the same in both regions the energy of the system would not be minimized. It could be reduced by electrons in one region flowing to unoccupied states of lower energy in the other region, and so the system would not be in equilibrium. Actually, considerable electron flow did take place in establishing equilibrium when the p-region was initially put into contact with the n-region. This led to an accumulation of electrons on the p-side of the junction, and a deficiency of electrons, or accumulation of holes, on the n-side of the junction. Thus the junction region has similarities to a plane parallel condenser with a negative charge on the p-side and a positive charge on the n-side, as shown in the figure. If an electron is moved through the electric field produced within this dipole layer, its energy will increase in going from the n-side to the p-side. This is reflected in the way the energy levels at the top of the valence band, and at the bottom of the conduction band, are displaced upward in going through the junction region.

Even after equilibrium is established there is still a flow of electrons back and forth through the junction. For one thing, from time to time thermal excitation causes an

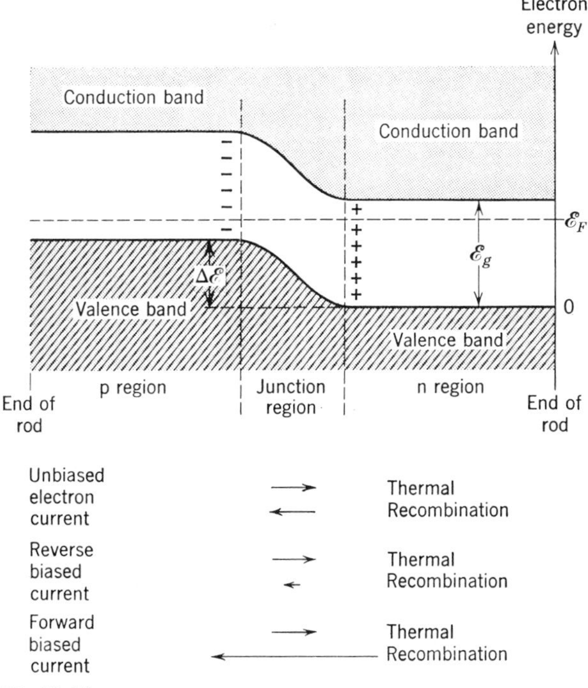


**FIGURE 13-16**

Electron energy-level diagram for an unbiased p-n junction.

electron to jump up to the conduction band of the p-region (leaving yet another hole in its valence band). The electron can move freely to the junction region, and then be accelerated by the potential hill it sees there into the n-region, constituting part of what is called the *thermal current*. Also, an electron in the conduction band of the n-region with energy slightly below the bottom of the conduction band in the p-region can gain a little extra energy in a fluctuation and be able to move into the p-region. There it may recombine with one of the many holes in the p-region. That electron is part of the so-called *recombination current*. There must be such a current because in equilibrium the thermal current must be balanced so that there is no net current across the junction.

Now consider an external voltage source applied across the ends of the device, with negative voltage applied to the p-region and positive voltage applied to the n-region. This will increase the energy of all the electrons in the p-region, and decrease the energy of all of those in the n-region, thereby increasing the height of the potential hill between the two regions. Since the junction region was already depleted of charge carriers, its resistance is relatively high and most of the voltage drop due to the applied voltage appears across the junction. As the amount of thermal current depends on the temperature and the width of the gap between the valence and conduction bands, neither of which are changed by applying the voltage, the thermal current will not change. The recombination current will be decreased by a large factor, however, because the potential hill is higher so now only the very many fewer electrons farther out in the exponentially decreasing tail of the Fermi distribution in the n-region conduction band have a chance to move into the p-region conduction band. The net effect will be a small flow of electrons in the direction from the p- to the n-regions, due to the unbalanced thermal current. This flow of electron current is, of course, in the direction that the applied voltage would be expected to produce. It is the small *reverse bias* current indicated by the arrows at the bottom of Figure 13-16.

The junction rectifier is given a *forward bias* by applying a positive voltage to the p-region and a negative voltage to the n-region. This decreases the height of the electron energy hill between the two regions. Again, there is no appreciable effect on the thermal current, but the recombination current is increased by a large factor. All of a sudden the very many more electrons that are closer into $\mathscr{E}_F$ in the Fermi distribution of the n-region have enough energy to pass through the junction into the p-region conduction band, because the bottom of that conduction band has moved down in energy. These electrons do not instantaneously respond to the application of a forward bias, but instead they diffuse into the p-region in much the same way that the molecules of a gas would diffuse into a region of lower density that suddenly became accessible to them. The net electron current in a forward biased rectifier flows in the direction of the recombination electron current, as indicated at the bottom of Figure 13-6. The junction is a rectifier because the magnitude of the forward bias current is much larger than the magnitude of the reverse bias current, for a given magnitude of bias voltage. The reason is that the reverse bias current is limited by the small value of the thermal current, whereas the forward bias current becomes very large as the height of the electron hill is made small by increasing the forward bias. Resistance to current flow in reverse bias is typically greater than resistance to forward bias by four or five orders of magnitude. Note that our explanation has been phrased in terms of electron flow. It could as well have been in terms of hole flow; both processes occur, and they result in the same rectifying properties of the junction.

A semiconductor rectifier has many advantages over a diode vacuum tube rectifier, including longer life and much smaller size. Like the diode, the p-n junction is a non-Ohmic element, the current-voltage relation being nonlinear, as shown in Figure

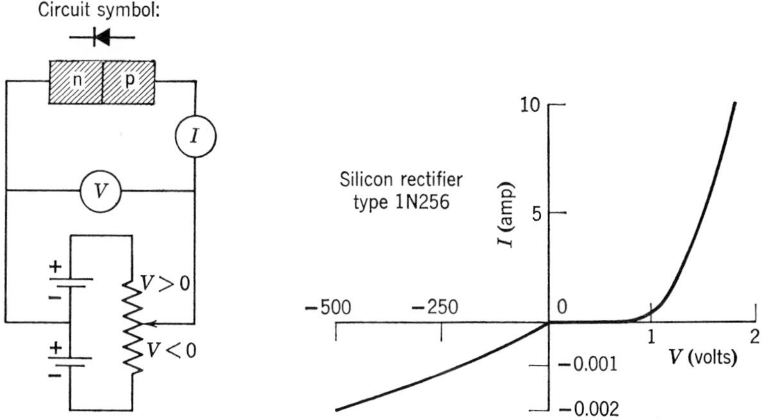


**FIGURE 13-17**

*Left:* A circuit in which the voltage across a p-n junction can be varied. The voltage is taken as positive when the p-side is at the higher potential. *Right:* Current through the junction as a function of the applied voltage. Note that very different scales are used for the forward- and reverse-biased portions of the curve.

13-17. Unlike a vacuum tube, however, there is no need for a power-consuming filament in the semiconductor device so that its efficiency is greater.

A *transistor* can be regarded as a combination of two semiconductor rectifying junctions, such as a p-n-p or n-p-n combination. In Figure 13-18 we display a circuit that exhibits transistor behavior. The n-p-n-regions are called *emitter*, *base*, and *collector*, respectively. The emitter-base connection is biased in the forward direction, so that the resistance to current flow is small in this part of the circuit. The base-collector connection is biased in the reverse direction, so that ordinarily there is higher resistance to current flow in that part of the circuit. However, when a voltage is applied in the emitter circuit so that a current is established there, the electrons arriving in the

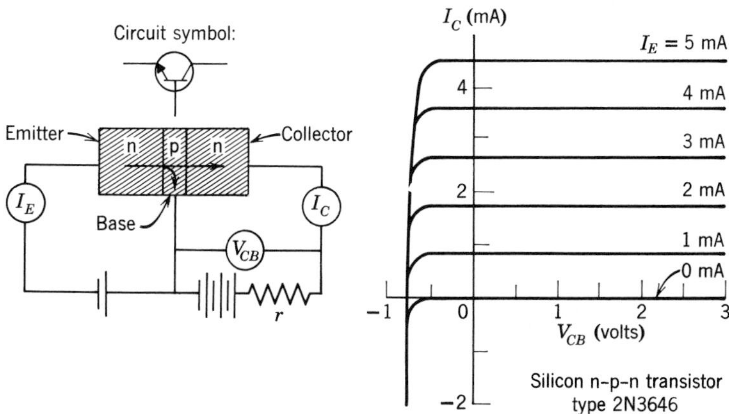


**FIGURE 13-18**

*Left:* A circuit in which an n-p-n transistor acts as a power amplifier. Electrons flow in the direction shown by the arrow, from emitter to collector. *Right:* Characteristic curves for a transistor acting as a power amplifier.

base region (which is very thin and of lower conductivity than the emitter) are attracted by the potential difference between the base and the collector. Hence, there will be a current in the collector circuit. (Because the emitter has a higher conductivity than the base, most of the current across the emitter-base junction is carried by electrons moving from the emitter to the base, instead of holes moving from the base to the emitter.)

The basic idea of transistor action is that a current in the emitter circuit controls a current in the collector circuit. More than 90% of the current through the emitter passes through the collector, so that the currents are of similar magnitudes. But the voltage across the base-collector connection can be very much greater than that across the emitter-base connection, because the former is reverse biased, so the power output in the collector circuit can be very much larger than the power input in the emitter circuit. Hence, the transistor acts as a power amplifier. Characteristic current versus voltage curves are shown in Figure 13-18. Other circuit connections make transistors useful as current amplification or voltage amplification devices, as well.

A *tunnel diode* is a semiconductor device that makes use of the phenomenon of potential barrier penetration discussed in Section 6-5. It is like a p-n junction made from semiconductors with very high impurity concentration. Figure 13-19*a* plots the

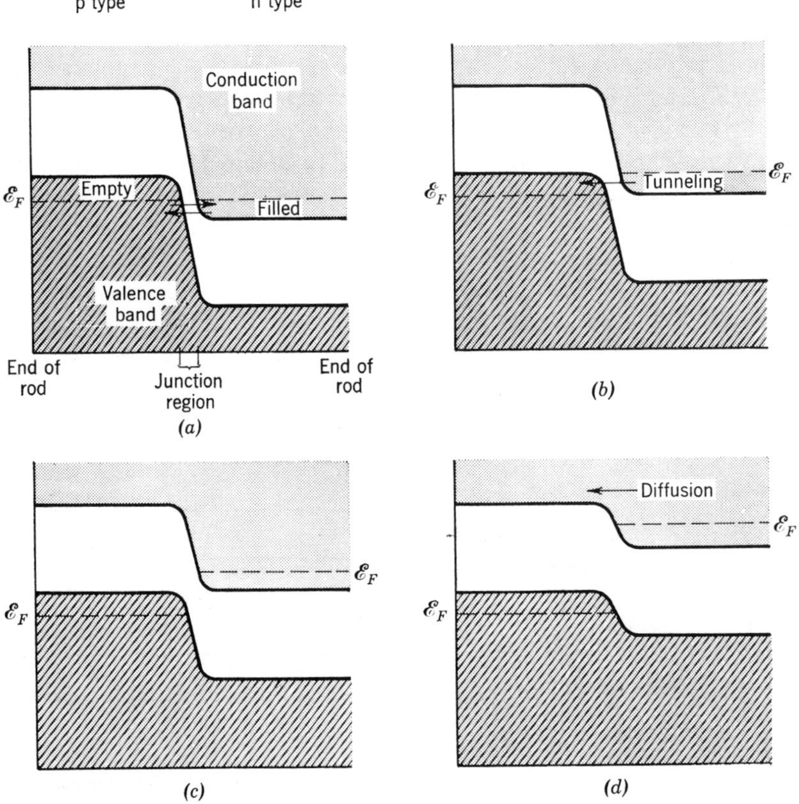


**FIGURE 13-19**

Electron energy-level diagrams for the n-type, junction, and p-type regions of a tunnel diode. In (*a*) the diode is unbiased. In (*b*) a small voltage is applied between the ends of the device, with the p-type end positive. In (*c*) and (*d*) the voltage is increased progressively. The arrows indicate the flow of electrons across the junction between the two regions.

electron energy across an unbiased junction. The bands are similar to those shown in Figure 13-16, except that (1) with a higher impurity concentration the junction is narrower since a smaller length of semiconductor contains enough charge carriers to produce the required dipole layer across the junction, and (2) the donor and acceptor levels, in the n-type and p-type material, are no longer sharp but become broad bands which overlap the valence and conduction bands, since the donors, and also the acceptors, are so closely spaced that they interact. The Fermi energy thus moves up into the conduction band on the n-side and down into the valence band on the p-side.

Because the junction is narrow ($\sim 10^{-8}$ m), electrons can pass through the forbidden band at the junction by a process that is in every respect the same as barrier penetration. For instance, the eigenfunction describing an electron tunneling through the forbidden band has the same exponential form as the eigenfunction for an electron tunneling through a barrier. At equilibrium, as shown in Figure 13-19*a*, the rate of electron tunneling through the barrier is the same in both directions.

If now a small external voltage is applied across the ends of the rod with forward bias, electron tunneling from the n-side to the p-side is increased because there are empty allowed energy states in the p-side valence band, whereas electron tunneling in the other direction is decreased. Hence, there is a net current flow through the junction as shown in Figure 13-19*b*. As the applied voltage continues to be increased, the net current begins to decrease because the number of empty states available for electron tunneling decreases. In Figure 13-19*c* the net current is reduced almost to zero because electrons in the n-type material find no allowed energy states into which to flow. With still higher applied voltage the electron current becomes that characteristic of a normal p-n junction. That is, electrons flow through the junction, without tunneling, into allowed energy states in the conduction band of the p-type material. This happens because the difference in the energies of the bands decreases, making it possible for electrons to diffuse through the junction into the conduction band of the p-region. This process is indicated in Figure 13-19*d*.

Figure 13-20 shows the current-voltage curve characteristic of a typical tunnel diode. The letters labeling points on the curve correspond to the four applied voltages of the previous figure. In the region between points *b* and *c*, the slope of the curve, $dI/dV$, is negative and the tunnel diode has a *negative* resistance, the current decreasing with

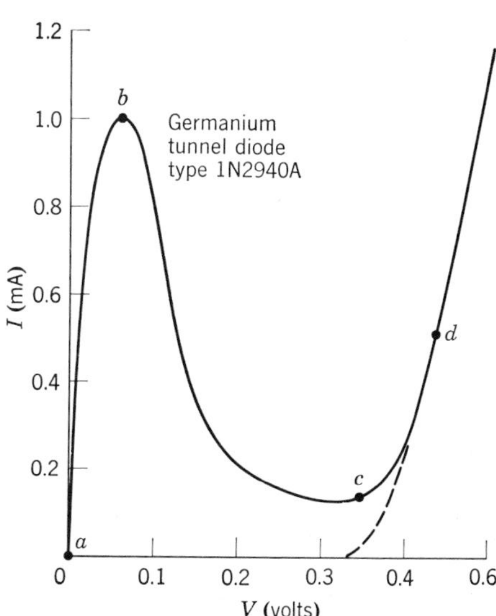


**FIGURE 13-20**

The current flowing through a tunnel diode as a function of the applied potential difference. The points labeled by letters correspond to the four applied voltages of Figure 13-19. Note that the resistance of the diode is negative for applied voltages between *b* and *c*. The dashed line indicates the characteristic current were no tunneling to take place—namely that for an ordinary germanium junction rectifier.

increasing applied voltage. This feature makes it particularly useful in the switching circuits of computers.

The greatest advantage of the tunnel diode is its very fast response time when operating in the region $a$ to $c$. The current flow in other kinds of semiconductor diodes and transistors always depends on the diffusion process. Since the rate of diffusion can change only as fast as the charge carrier distribution can be changed, these devices have relatively slow response (slower than vacuum tubes) and it is difficult to use them at high frequencies. But the rate of tunneling can change as fast as the energy bands can be changed by the applied voltage, and this is a much less serious limitation. Tunnel diodes have been used as oscillators at frequencies above $10^{11}$ Hz, and in switching circuits that operate in times less than $10^{-9}$ sec.

## QUESTIONS

**1.** In the text the solid state is contrasted with the gaseous state in terms of atomic (or molecular) interactions. How would you characterize the liquid state in this regard?

**2.** Explain the statement that the exclusion principle prevents solids from collapsing to zero volume.

**3.** Is there an analogy between the splitting of an energy level as two atoms are brought together to form a molecule and the splitting of the resonant frequencies as two resonant electrical circuits are coupled? Why?

**4.** It is often said that a crystal is one giant molecule. Explain. Can we regard a diatomic molecule as a small solid?

**5.** Why does metallic binding usually occur with atoms having a small number of valence electrons?

**6.** Why is it, considering the very similar electronic structures, that lithium is a metal whereas hydrogen is a molecular solid?

**7.** Explain why metallic binding leads to a close-packed arrangement of atoms; i.e., explain why the lowest energy in metallic binding corresponds to the greatest number density of atoms.

**8.** Why are metallic solids mostly opaque, covalent solids sometimes opaque, and ionic solids hardly ever opaque to visible radiation?

**9.** Of the four types of binding in solids discussed in the text, which one (or ones) is most likely to produce an insulator? A conductor? A semiconductor?

**10.** Justify the statement that (13-1a) meets the criterion that a material obeys Ohm's law.

**11.** What mechanisms account for the ordinary electrical resistivity of metals? Which are temperature dependent?

**12.** How do electrons contribute to thermal conductivity? Are they better than lattice vibrations as carriers of heat energy?

**13.** Explain why the electrical conductivity of materials varies over a factor of $10^{24}$ whereas the thermal conductivity of materials only varies over a factor of about $10^{8}$.

**14.** Explain why we regard the sequential filling of holes by electrons as equivalent to a positive current. Could this process be regarded instead as an electron current?

15. How is the result of Example 13-2, concerning the fraction of conduction electrons that is thermally excited, related to the specific heats of metals at high temperatures?

16. Example 13-2 implies that only $\Delta\mathcal{N}/\mathcal{N}$ of the free electrons take part in the conduction of electricity, whereas certain other experiments, such as the Hall effect, indicate that all $\mathcal{N}$ electrons take part. Explain.

17. Explain why a negative effective mass does not lead to a violation of Newton's law of motion.

18. How is the optical transparency of a semiconductor related to the energy gap of the forbidden band?

19. What elements other than arsenic and antimony can be used as an impurity with germanium to form an n-type semiconductor? What elements other than gallium and indium can be used to form a p-type semiconductor?

20. Could the conductivity of a semiconductor be affected by electron bombardment? By bombardment by other particles?

21. What effect does an applied electric field have on an insulator?

22. Experimentally the addition of impurities to metals increases their resistivity, but the addition of impurities to semiconductors decreases their resistivity. Explain. Many insulators, however, are not very pure. Why do impurities not affect the resistivity of insulators?

23. Name the properties of solids that are little affected by the presence of small concentrations of chemical impurities. Name the properties of solids that are greatly affected by the presence of small concentrations of chemical impurities.

24. Give an argument, similar to that given in the text for an n-type semiconductor, explaining the variation of $\mathscr{E}_F$ with $T$ in a p-type semiconductor.

25. Explain why the curves of Fermi energy as a function of temperature differ for different impurity concentrations, as shown in Figure 13-15.

26. Explain why the junction transition region is narrower in a semiconductor diode when the doping is heavy than it is when the doping is light.

27. Rephrase the discussion of the operation of a p-n rectifying junction in terms of hole flow.

## PROBLEMS

1. In Figure 13-21 we illustrate schematically four charge density distributions for valence electrons as a function of the location of atoms, ions, or molecules (shown as dots at the bottom). For each distribution (a), (b), (c), (d), state to which type of binding in solids it must closely corresponds.

2. Each element of the row of the periodic table from lithium through neon has a solid form (some at very low temperatures). Solids can also be formed by certain compounds of two elements of this row. For all of these solids, describe the binding and state whether the solid is a metal, a semiconductor, or an insulator.

3. Describe the binding of solids formed by single elements of the column of the periodic table from carbon through lead, and state whether the solid is a metal, a semiconductor, or an insulator.

4. Determine the type of binding in each of the solids described here. (a) Reflects light in the visible; electrical resistivity increases with temperature; melting point below 1000°C.

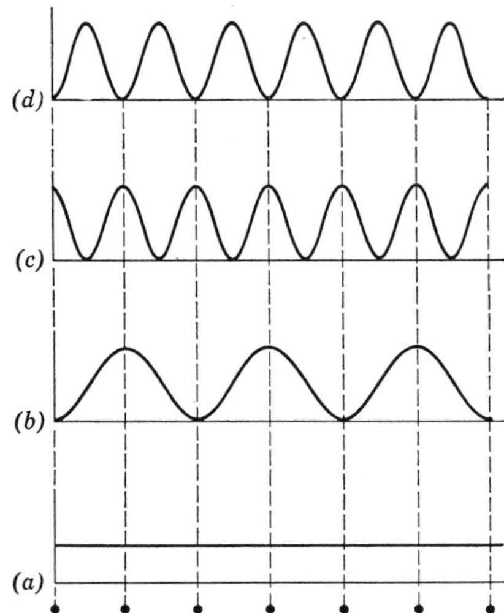


**FIGURE 13-21**
Charge densities for valence electrons in four solids considered in Problem 1.

(b) Reflects light in the visible; electrical resistivity decreases with increasing temperature; melting point above 1000°C. (c) Transmits light in the visible; conducts electricity only at high temperatures. (d) Transmits light in the visible; does not conduct electricity at any temperature. (e) Transmits light in the visible; very low melting point.

**5.** The field **E** produced at a point **r** by an electric dipole **p** is given by

$$\mathbf{E} = -\frac{1}{4\pi\epsilon_0}\left(\frac{\mathbf{p}}{r^3} - 3\frac{\mathbf{r}\cdot\mathbf{p}}{r^5}\mathbf{r}\right)$$

where the dipole is located at the origin of coordinates. (a) A molecule with an electric dipole moment **p** will induce an electric dipole moment **p**′ in a nearby molecule, where $\mathbf{p}' = \alpha\mathbf{E}$, $\alpha$ being the polarizability of the nearby molecule. Show that the mutual potential energy of the interacting dipoles is

$$V = -\mathbf{p}'\cdot\mathbf{E} = -\frac{\alpha}{(4\pi\epsilon_0)^2}(1 + 3\cos^2\theta)\frac{p^2}{r^6}$$

where $\theta$ is the angle between **r** and **p**. (b) Show the force is attractive and varies as $r^{-7}$.

**6.** Find the order of magnitude of the electric field needed in ionic solids to free electrons from the filled shells of ions. (Hint: Consider the binding energy of an electron and the approximate dimensions of an ion.)

**7.** Find the region of the electromagnetic spectrum at which crystals of Si, Ge, CdS, KCl, and Cu become opaque. The band gap energies $\mathscr{E}_g$ are Si = 1.14 eV; Ge = 0.67 eV; CdS = 2.42 eV; KCl = 7.6 eV; Cu = 0 eV.

**8.** (a) Using classical physics show that the resistivity of a metal near room temperature is proportional to the 3/2 power of the absolute temperature, in disagreement with the linear temperature dependence experimentally observed. (Hint: Show that $\bar{v} \propto T^{1/2}$ and $\lambda \propto T^{-1}$.) (b) How does the application of the ideas of quantum mechanics and quantum statistics yield the proper temperature dependence of the resistivity?

**9.** Compare the values of (a) the drift velocity, (b) the thermal velocity, and (c) the velocity corresponding to the Fermi energy, or *Fermi velocity*, for electrons in copper at room temperature. (Hint: Use Table 11-2. A current of 5 amp can easily be carried in a copper wire 0.1 cm in diameter.)

10. Calculate the number of electrons per atom of aluminum that conduct electricity from the value, $-0.3 \times 10^{-10}$ m$^3$/coul, of the Hall coefficient. The density of aluminum is $2.7 \times 10^3$ kg/m$^3$. What does the result suggest about the band structure of aluminum?

11. (a) Show that the Hall coefficient for a semiconductor in which there is conduction by both holes and electrons is given by $(p\mu_p^2 - n\mu_n^2)/e(p\mu_p + n\mu_n)^2$. (b) If in a certain semiconductor there is no Hall effect, what fraction of the current is carried by holes?

12. Copper is a monovalent metal with a density of 8 g/cm$^3$ and an atomic weight of 64. (a) Calculate the Fermi energy in electron volts at 0°K. (b) Estimate the width of the conduction band.

13. (a) Calculate the Fermi energy of an alloy of 10% zinc (which is divalent) in copper assuming that the alloy has the same atomic spacing and structure as Cu. (b) How does the width of the conduction band of the alloy compare to that of copper?

14. Make an estimate of the width of a conduction band in a metal whose internuclear spacing has the typical value $3.5 \times 10^{-10}$ m.

15. The *Fermi temperature* is defined by $T_F = \mathscr{E}_F/k$. (a) Using Table 11-2, calculate the Fermi temperature for sodium. (b) What does this tell us about the applicability of classical considerations to metals near room temperature? (c) What does this tell us about the density of conduction electrons in a metal at room temperature?

16. The Fermi energy of lithium is 4.72 eV. (a) Calculate the Fermi velocity. (b) Calculate the de Broglie wavelength of an electron moving at the Fermi velocity and compare it to the interatomic spacing.

17. Calculate an approximate ratio of the electronic specific heat to the lattice specific heat of lithium at room temperature. (Hint: Use the results of Example 13-2, and justify this use.)

18. (a) Show that the effect of a lattice periodicity $a$ on periodic potentials having Bloch function solutions is to modulate the free-electron solution so that $\psi(x + a) = \psi(x)e^{ika}$. (b) Show that $e^{ika} = -1$ at the Brillouin zone boundaries. Comment on the meaning of this result.

19. At what temperature will the number of conduction electrons increase by a factor of 20 over the number at room temperature for germanium? The gap energy is 0.67.eV.

20. (a) Show that the number of electrons per unit volume in the conduction band of an intrinsic semiconductor is given by $\mathscr{N}_c e^{-(\mathscr{E}_c - \mathscr{E}_F)/kT}$, where $\mathscr{N}_c = 2(2\pi mkT)^{3/2}/h^3$, and where $\mathscr{E}_c$ is the conduction band-edge energy. (b) Show that the number of holes per unit volume in the valence band of an intrinsic semiconductor is given by $\mathscr{N}_v e^{-(\mathscr{E}_F - \mathscr{E}_v)/kT}$, where $\mathscr{N}_v = 2(2\pi mkT)^{3/2}/h^3$, and where $\mathscr{E}_v$ is the valence band-edge energy.

21. Use the expression for the number of electrons in the conduction band, and the number of holes in the valence band, given in Problem 20, and charge neutrality to find the position of the Fermi energy in an intrinsic semiconductor.

22. (a) Show that the product of the number of holes in the valence band and the number of electrons in the conduction band depends only on temperature and the gap energy. (b) Show that the conductivity $\sigma$ of an intrinsic semiconductor can be used to measure the gap energy by calculating $\ln \sigma$.

23. Write exact expressions for $\mathscr{N}_d^+$ and $\mathscr{N}_d^0$, the concentration of ionized and neutral donors respectively, in a semiconductor doped to a concentration of $\mathscr{N}_d$.

24. (a) The position of the Fermi energy in a doped semiconductor can be found from the condition of charge neutrality: $\mathscr{N}_n + \mathscr{N}_a^- = \mathscr{N}_p + \mathscr{N}_d^+$, where $\mathscr{N}_n$ is the number of electrons in the conduction band, $\mathscr{N}_a^-$ is the number of ionized acceptors, $\mathscr{N}_p$ is the number of holes in the valence band and $\mathscr{N}_d^+$ is the number of ionized donors. Assuming $\mathscr{N}_a^- = 0$ and $\mathscr{N}_n \gg \mathscr{N}_p$ show that charge neutrality leads to an equation quadratic in

$e^{\mathscr{E}_F/kT}$ which has the solution

$$e^{\mathscr{E}_F/kT} = \frac{-1 \pm \sqrt{1 + 4\dfrac{\mathscr{N}_d}{\mathscr{N}_c} e^{(\mathscr{E}_c - \mathscr{E}_d)/kT}}}{2e^{-\mathscr{E}_d/kT}}$$

where $\mathscr{E}_c$ is the conduction band-edge energy, and $\mathscr{E}_d$ is the donor-level energy. (b) This equation is soluble in two limits. One is

$$4\frac{\mathscr{N}_d}{\mathscr{N}_c} e^{(\mathscr{E}_c - \mathscr{E}_d)/kT} \ll 1$$

This means $\mathscr{N}_d$ small or $T$ large. Use a binomial expansion of the square root to show that $\mathscr{N}_n = \mathscr{N}_d$ and $\mathscr{E}_F = \mathscr{E}_c + kT \ln (\mathscr{N}_d/\mathscr{N}_c)$. This is the *exhaustion* region. All the donors are ionized but no electrons are excited from the valence band. (c) In the other limit

$$4\frac{\mathscr{N}_d}{\mathscr{N}_c} e^{(\mathscr{E}_c - \mathscr{E}_d)/kT} \gg 1$$

Also $\mathscr{N}_d$ is large and $T$ is small. Show that

$$\mathscr{N}_n = \sqrt{\mathscr{N}_d \mathscr{N}_c}\, e^{-(\mathscr{E}_c - \mathscr{E}_d)/2kT}$$

and

$$\mathscr{E}_F = \frac{\mathscr{E}_c + \mathscr{E}_d}{2} + \frac{kT}{2} \ln \frac{\mathscr{N}_d}{\mathscr{N}_c}$$

This is the *extrinsic* region. Here the donors are being ionized.

**25.** Draw an energy-level diagram like that of Figure 13-16 for an n-p-n junction transistor and describe the power amplifier action of the transistor in terms of the figure.

**26.** The current which flows in a p-n junction is proportional to the number of electrons in the conduction band. (a) For an unbiased p-n junction, show that the current from the p-region to the n-region is proportional to $e^{-(\mathscr{E}_g - \mathscr{E}_F)/kT}$ and this current is equal to the

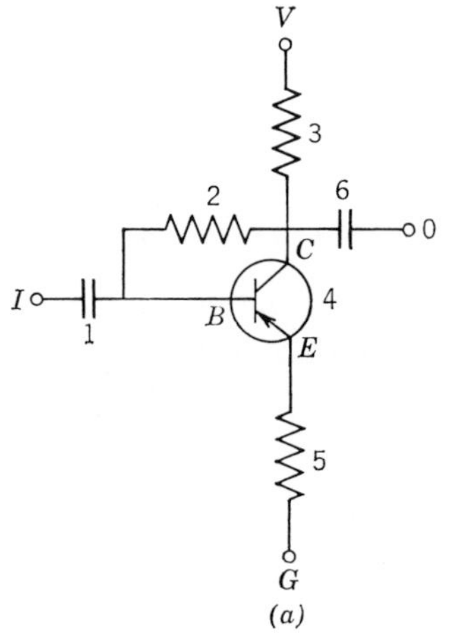

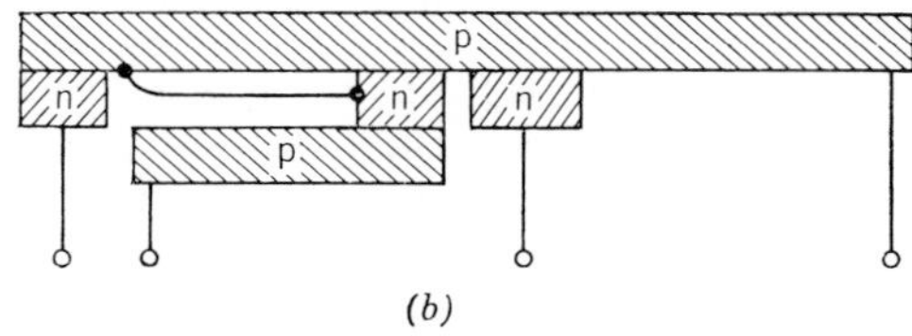


**FIGURE 13-22**

An integrated circuit considered in Problem 27.

current from the n-region to the p-region so that no net current flows. (b) When a bias potential $V$ is applied show that the net charge flow per unit area of junction is proportional to

$$e^{-(\mathscr{E}_g - \mathscr{E}_F)/kT}(e^{eV/kT} - 1)$$

where $eV$ is positive for forward bias and negative for reverse bias.

**27.** A p-n junction is a double layer of opposite charges separated by a small distance and has the properties of a capacitance. The resistivity of a semiconductor can be controlled by doping. Thus the elements in the transistor circuit of Figure 13-22a can be manufactured on a p-n-p semiconductor with appropriate layers etched away as shown in Figure 13-22*b*. This is an integrated circuit. Label the appropriate parts of Figure 13-22*b* with the corresponding numbers and letters of Figure 13-22*a*.

# 14

# Solids—Superconductors and Magnetic Properties

# Solids—Superconductors and Magnetic Properties 14

## 14-1 Superconductivity

Shortly after the discovery of the electron it was recognized that the high electrical and thermal conductivities of metals could be attributed to the motion of electrons in the metal. Classical theories of metallic conduction treated these electrons as a gas of independent particles within the metals colliding with lattice imperfections. Using methods of the classical kinetic theory, many experimental facts of electrical and thermal conductivity could be explained. With the advent of quantum mechanics, it became possible to take into account the wave nature of electrons and the exclusion principle. A number of phenomena not previously explainable then became clear. For example, the need to use the Fermi distribution for free electrons led to an understanding of the electronic contribution to the specific heats of solids. The further application of wave ideas led to quantization of energy levels and the band theory of solids, which accounted for the wide range in conductivities observed in normal solids. The free-electron model approximation averaged out variations in the interactions of electrons with one another and with the lattice ions, and it could account for resistance to electron flow under normal conditions. A major failure of this independent particle model, however, is its inability to explain superconductivity. To understand that phenomenon requires taking into account the collective behavior of electrons and ions, or the so-called *many-body effects*, in solids. Let us now examine superconductivity.

Many factors contribute to the electrical resistivity of a solid, as we have seen. Electrons are scattered by the deviations from a perfect lattice due to structural defects or impurities in a crystal. In addition, there are vibrations of the lattice ions in normal modes that constitute something like sound waves traveling through the solid; we refer to such waves as *phonons*. The higher the temperature is, the more phonons there are present in the lattice. When phonons are present, there is an electron-phonon interaction which scatters conduction electrons and causes further resistance. Hence, the electrical resistance of a solid should decrease as the temperature decreases, but we expect a residual resistance even near absolute zero due to the crystal imperfections. It therefore seems remarkable that *the electrical resistance of some solids disappears completely at sufficiently low temperatures*.

In 1911, Kammerlingh-Onnes found that the electrical resistance of solid mercury drops to an immeasurably small value when cooled below a certain temperature, called the *critical temperature* $T_c$. Mercury goes from a normal state to a *superconducting* state as the temperature drops below $T_c = 4.2°K$. Many other elements, and many compounds and alloys, have since been found to be superconductors with critical temperatures as high as 23°K. But not all materials superconduct. Figure 14-1 shows the resistivity at very low temperatures for a superconductor, tin, and a non-superconductor, silver. In a superconductor, currents can be set up which persist for years with no detectable decay.

In 1933, Meissner and Oschenfeld found that as a superconducting substance is cooled below its critical temperature in the presence of an applied magnetic field, it

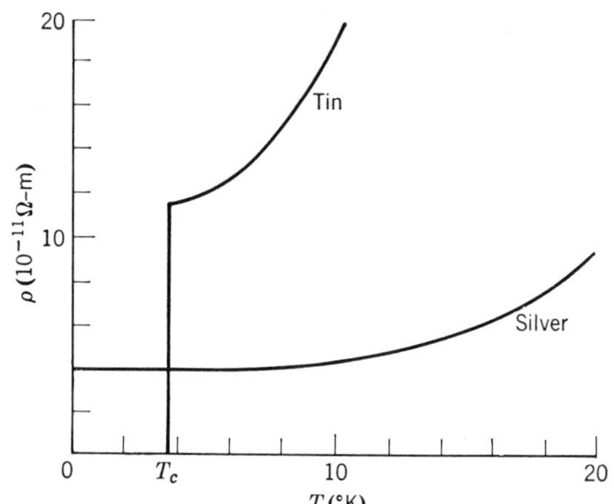


**FIGURE 14-1**

A plot of resistivity $\rho$ versus temperature $T$, showing the drop to zero at the critical temperature $T_c$ for a super-conductor, and the finite resistivity of a normal metal at absolute zero.

expels all magnetic flux from its interior. If the field is applied after the substance has been cooled below its critical temperature, the magnetic flux is excluded from the superconductor. Hence, a superconductor acts like a perfect diamagnet. Both *Meissner effects* are illustrated in Figure 14-2. According to Lenz's law, when the magnetic flux through a circuit is changing, an induced current is established in such a direction as to oppose the change in flux. In a *diamagnetic* atom, the orbital electrons adjust their rotational motion to produce a net magnetic moment opposite to the externally applied magnetic field. We can say analogously that an external magnetic field does not penetrate the interior of a superconducting substance because in a superconductor the conduction electrons, whose motion is an unimpeded as in an atom, adjust their motion to produce a counteracting magnetic field. The entire superconductor behaves like a single diamagnetic atom in this respect. Hence, the two principal characteristics of superconductors, namely the exclusion of magnetic flux and the absence of resistance to current flow, are related to one another. It is necessary to have a persisting (resistanceless) current to maintain the flux exclusion when the external field is on.

Figure 14-3 shows a photograph of superconducting levitation. If a small permanent magnet is placed over a perfectly conducting surface, it will float there. If the magnet is placed on a surface which thereafter is made superconducting (by lowering its temperature), it will rise and float. A repulsive force large enough to overcome the weight of the magnet exists between the magnet and the diamagnetic superconductor, because the superconducting body excludes the magnetic lines of flux associated with the magnet. Serious engineering studies have indicated the feasibility of using this phenomenon to provide very smooth support for high-speed passenger trains.

It is found that if the external field is increased beyond a certain value, called the *critical field* $H_c$, the metal ceases to be superconducting and becomes normal. The value of this critical field for a given material depends on the temperature, as shown

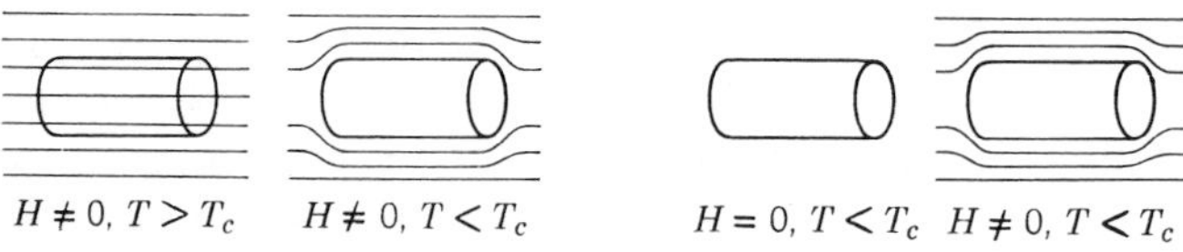


**FIGURE 14-2**

*Left:* A schematic illustration of expulsion. *Right:* The exclusion of magnetic flux in a superconductor. Both are called *Meissner effects.*

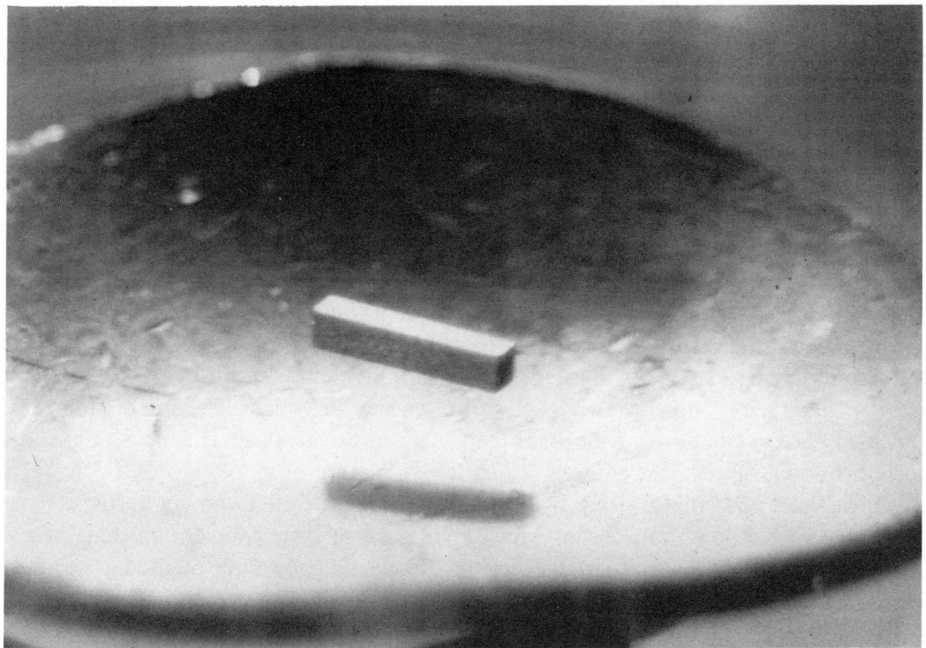

**FIGURE 14-3**
A permanent magnet floating over a superconducting surface.

for the case of lead in Figure 14-4. As the external magnetic field increases, therefore, the critical temperature is lowered until when $H > H_c(0°\text{K})$ there is no superconductivity for that material at any temperature. We can understand this as follows. Suppose that at some temperature below $T_c$ we turn on a magnetic field; the superconductor will act to exclude this field (the Meissner effect). The energy decrease of the magnetic field appears as increased energy of the electrons that make up the superconducting current. As the strength of the external magnetic field is increased, the energy acquired by the superconductor also increases. At the critical value of the field, $H_c$, the energy of the superconducting state becomes higher than the energy of the normal state, so that the material becomes normal.

Evidence that the lattice vibrations play an important role in the phenomenon of superconductivity came in 1950 when experiment revealed that the critical temperature

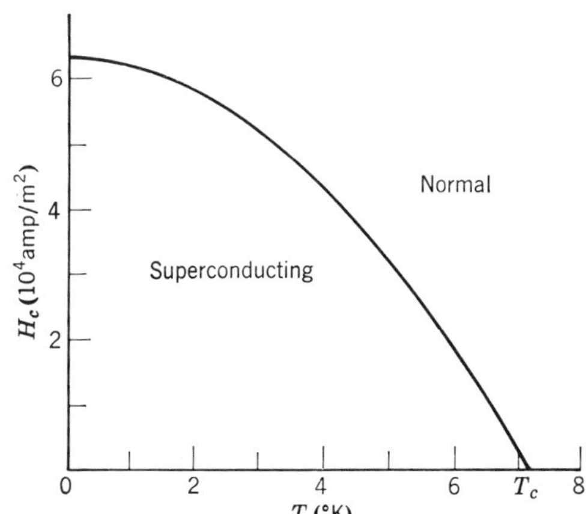


**FIGURE 14-4**
The variation with temperature of the critical field $H_c$ for lead. Note that $H_c$ is zero when the temperature $T$ equals the critical temperature $T_c$.

of crystals made from different isotopes of the same element depends on the isotopic mass. The dependence, given by

$$M^{1/2}T_c = \text{const} \tag{14-1}$$

in which $M$ is the average isotopic mass of the solid, is called the *isotope effect*. This relation shows that the critical temperature would go to zero (hence, no superconductivity) in the absence of lattice vibrations (when $M \to \infty$). The importance of lattice vibrations suggests that an electron-phonon interaction is responsible for superconductivity. We can no longer ignore those very interactions which were neglected in the independent particle model of a solid—the electron-phonon and also the electron-electron interactions—if we hope to get a theoretical explanation of superconductivity. In 1957, Bardeen, Cooper, and Schrieffer proposed a detailed microscopic theory, now known as the *BCS theory*, in which these interactions are included. The predictions of the BCS theory are in excellent agreement with experimental results. Let us now consider a qualitative picture of it.

An electron in a solid passing by adjacent ions in the lattice can act on these ions with a set of Coulomb attractions which gives each of them momentum that causes them to move slightly together. Because of the elastic properties of the lattice, this region of increased positive charge density will then propagate as a wave, which carries momentum, through the lattice. The electron has emitted a phonon! The momentum the phonon carries is supplied by the electron, whose momentum changed when the phonon was emitted. If a second electron subsequently passes by the moving region of increased positive charge density, it will experience an attractive Coulomb interaction, and thereby it can absorb all the momentum the moving region carries. That is, the second electron can absorb the phonon, thereby absorbing the momentum supplied by the first electron. The net effect is that the two electrons have exchanged some momentum with each other, and thus they have interacted with each other. Although the interaction was a two-step one, involving a phonon as an intermediary, it certainly was an interaction between the two electrons. Furthermore, it was an *attractive* interaction, since the electron involved in each of the steps participated in an attractive Coulomb interaction. The BCS theory shows that in certain conditions the attraction between two electrons due to a succession of phonon exchanges can exceed slightly the repulsion which they exert directly on each other because of the (shielded) Coulomb interaction of their like charges. Then the electrons will be weakly bound together, and form a so-called *Cooper pair*. We shall see that Cooper pairs are responsible for superconductivity.

The conditions for their formation, in numbers large enough to allow superconductivity, are (1) that the temperature be low enough to make the number of random thermal phonons present in the lattice small (they would inhibit the ordered processes involved in superconductivity); (2) that the interaction between an electron and a phonon be strong (so that a substance which has a relatively low resistance at room temperature, because its conduction electrons interact weakly with thermal lattice vibrations, will *not* be a possible superconductor at low temperature); (3) that the number of electrons in states lying just below the Fermi energy be large (these are the electrons which are energetically able to form Cooper pairs); (4) that the two electrons have "antiparallel" spins (then their space eigenfunction will be symmetric in a label exchange, which means that they will be close enough together to form a pair); and (5) that, in the absence of an externally applied electric field, the two electrons of a pair have linear momenta of equal magnitude but opposite direction (as will be explained next, this facilitates the participation of the maximum number of electrons in pair formation).

Because Cooper pairs are weakly bound, they are constantly breaking up and then reforming, usually with different partners. Also, because they are weakly bound they are large. (In Example 14-2 we shall estimate the typical separation of two electrons in a pair to be of the order of $10^4$ Å.) Thus, within the region occupied by the electrons of a pair, there are very many other electrons that would also like to participate in the pairing process. The system will be most tightly bound, and therefore most stable, if they can do so. The system achieves this by having the total linear momentum of each pair equal to zero, in the absence of an applied electric field. The discussion of the formation of a pair shows that the total momentum of any pair is a constant, since the net result of exchanging a phonon between the two electrons is to preserve the total momentum of the pair. If all the pairs have the *same* constant total momentum, then there will be no inhibition to the unavoidable process of old pairs breaking up and new pairs reforming, because any pair can be converted to any other pair by phonon exchange, and so the maximum number of pairs will be present. This conclusion is plausible from the qualitative argument we have given. It is put on a completely firm foundation by the quantitative calculations of the BCS theory, which show that the wave functions describing pair formation are in phase, and thus add constructively and lead to a large total probability for pair formation, when the pairs all have the same total momentum. In the absence of an applied electric field, symmetry considerations obviously demand that the common value of the pair total momentum be zero. So we see why the two electrons of each pair have linear momenta of equal magnitude, but opposite direction, in such circumstances. We also see that the ground state of the system is very highly ordered, in that all the pairs in the lattice are doing exactly the same thing as far as the motion of their centers of mass is concerned. This order extends through the lattice, and not just through the region occupied by a pair, because the pairs are relatively large and there are many of them so there is multiple overlapping. The order propagates through adjacent overlapping regions.

When an external electric field is applied, the pairs, which behave rather like particles with two electron charges, move through the lattice under the influence of the field. But they do it in such a way as to continue to maintain the order, because that will maintain their number at a maximum. Thus they carry current by moving through the lattice with all of their centers of mass having exactly the same momentum. The motion of each pair is locked into the motion of all the rest, and so none of them can be involved in the random scatterings from lattice imperfections that cause low-temperature electrical resistance. This is why the system is a superconductor.

It is tempting to think of a Cooper pair as acting like a boson, since it contains two fermions. If this could be done, superconductivity would be simply another example of Bose condensation, as in the superfluidity of liquid helium. That is, it would be the completely correlated motion of a set of bosons all in the same quantum state due to the effect of the $(1 + n)$ boson enhancement factor discussed in Chapter 11. Theories which preceded the BCS theory tried unsuccessfully to use this approach. The reason why it is not valid is that the individual electrons in each pair are weakly bound to the pair, which also means the pair is large. As a consequence, the eigenfunction for the system of overlapping pairs must take into account the exchange of labels of one electron from one pair and one electron from another pair, as well as the exchange of labels of one complete pair and another complete pair. In the latter exchange the system eigenfunction will not change sign because two fermion labels are being exchanged, but in the former the eigenfunction does change sign since only one fermion label is being exchanged. So Cooper pairs are neither purely bosonlike (no sign change), nor purely fermionlike (sign change) with respect to all eigenfunction label exchanges that must be considered. In a system of tightly bound helium atoms, the only type of label exchange that must be considered is an exchange of the label of one atom with the label of another. Such an exchange actually involves an even number of fermion label exchanges (each atom contains two electrons, two protons, and two neutrons), so the eigenfunction does not change sign and the atoms of the system act like bosons.

According to the BCS theory, the binding energy of a Cooper pair at absolute zero is about $3kT_c$. As the temperature rises, the binding energy is reduced, and goes to zero when the temperature equals the critical temperature $T_c$. Above $T_c$, a Cooper pair is not bound.

With a binding electron-electron interaction at absolute zero, it is energetically advantageous for two electrons, each in single-particle states just below the Fermi energy, $\mathscr{E}_F$, to promote themselves to vacant states just above $\mathscr{E}_F$ where they can interact in such a way as to form a Cooper pair. The energy required to put the electrons into the higher single-particle states is more than compensated for by the energy made available by the binding of the Cooper pair they form. Thus the zero temperature Fermi distribution of a superconductor is unstable, in the sense that electrons in states within a range of the order of $kT_c$ below the Fermi energy will leave those states and enter states within a similar range above the Fermi energy, where they will form pairs. The result is that the $T = 0$ distribution of occupied states of a superconductor looks something like a $T = T_c$ Fermi distribution for a normal conductor. The reason why the electrons must be above $\mathscr{E}_F$ to be able to freely form pairs is that a large number of unoccupied states are found only above $\mathscr{E}_F$, and unoccupied states must be available for the two electrons of a pair to enter after they change their momenta by one emitting and the other absorbing a phonon.

Although there is an almost continuous distribution of single particle states available to *each electron* in a superconductor at $T = 0$, the distribution of states available to the *system* is anything but continuous. As far as the system is concerned, there is its superconducting ground state, then an *energy gap* of width $\mathscr{E}_g$ in which there are no states at all, and above the gap a set of states which are nonsuperconducting. The gap width $\mathscr{E}_g$ equals the binding energy of a Cooper pair. The gap arises because if one electron of the system in a single particle state in the region of width $\sim kT_c$ surrounding $\mathscr{E}_F$ absorbs energy from some source, so that it makes a transition from that state to another single particle state only infinitesimally different in energy, then the pair of which it had been a member will be broken and the binding energy of the pair will be lost to the system. Thus the source must be able to supply an energy equal to a pair binding energy before an electron near $\mathscr{E}_F$ can make a transition to the energetically nearest state. (Even more energy must be supplied to excite an electron well below $\mathscr{E}_F$, despite the fact that it is not in a pair, since all the nearby states are already occupied.) Therefore the minimum energy that can be accepted by the ground state system, which is the width of its energy gap, is the binding energy of a Cooper pair. The states which begin at the top of the gap are not superconducting since in them the system has enough energy for pairs to be broken.

The width of the gap at $T = 0$ is $\mathscr{E}_g \simeq 3kT_c$. But it narrows as the temperature rises, and it becomes of zero width at $T = T_c$ where the pairs are no longer bound. At temperatures below $T_c$ the superconducting ground state corresponds to a large scale quantum state in which the motions of all the electrons and ions are highly correlated. It takes the gap energy $\mathscr{E}_g$ to excite the system to the next higher state, which is not superconducting, and this is more energy than the thermal energy available to the system. For instance, at $T = 0.1T_c$ the value of the gap energy is still about $\mathscr{E}_g = 3kT_c$, while the thermal energy is about $kT = 0.1kT_c$.

For most superconductors near $T = 0$ the energy needed to bridge the gap corresponds to photons in the very far infared, or microwave, portion of the electromagnetic spectrum. The existence and width of the gap is established experimentally by the abrupt change in absorption of far infared or microwave radiation when the photon energy $h\nu$ drops below the gap energy.

**Example 14-1.** The critical temperature of mercury is 4.2°K.

(a) What is the energy gap in electron volts at $T = 0$?

As stated earlier, the Cooper pair binding energy, or gap energy, is

$$\mathscr{E}_g \simeq 3kT_c$$

So

$$\mathscr{E}_g \simeq 3 \times 1.4 \times 10^{-23}\,\text{joule/°K} \times 4.2\text{°K} = 1.8 \times 10^{-22}\,\text{joule}$$
$$\simeq 1.1 \times 10^{-3}\,\text{eV}$$

(b) Calculate the wavelength of a photon whose energy is just sufficient to break up Cooper pairs in mercury at $T = 0$. In what region of the electromagnetic spectrum are such photons found?

The energy is

$$\mathscr{E}_g = h\nu = \frac{hc}{\lambda}$$

So the wavelength is

$$\lambda = \frac{hc}{\mathscr{E}_g} \simeq \frac{6.6 \times 10^{-34}\,\text{joule-sec} \times 3 \times 10^8\,\text{m/sec}}{1.8 \times 10^{-22}\,\text{joule}} = 1.1 \times 10^{-3}\,\text{m}$$

These photons are in the very short wavelength part of the microwave region.

(c) Does the metal look like a superconductor to electromagnetic waves having wavelengths shorter than that found in part (b)? Explain.

No, since the energy content of shorter wavelength photons is sufficiently high to break up the Cooper pairs, or excite the conduction electrons through the energy gap into the non-superconducting states above the gap. ◀

**Example 14-2.** (a) Estimate the size of a Cooper pair of binding energy $\mathscr{E}_g$.

The wave function of a Cooper pair is made up of waves, describing its two component electrons, with wave numbers drawn from a range $\Delta k$ corresponding to an energy range $\Delta\mathscr{E} \sim \mathscr{E}_g$. The energy range is centered on $\mathscr{E}_F$, and the wave number range is centered on the corresponding $k_F$. Since the energy of one of the electrons is

$$\mathscr{E} = \frac{p^2}{2m^*} = \frac{\hbar^2 k^2}{2m^*}$$

we have

$$\Delta\mathscr{E} = \frac{\hbar^2 2k\,\Delta k}{2m^*}$$

and

$$\frac{\Delta\mathscr{E}}{\mathscr{E}} = \frac{\hbar^2 k\,\Delta k 2m^*}{m^*\hbar^2 k^2} = \frac{2\Delta k}{k} \sim \frac{\Delta k}{k}$$

Setting $\mathscr{E} = \mathscr{E}_F$, $k = k_F$, and $\Delta\mathscr{E} = \mathscr{E}_g$, we have

$$\frac{\Delta k}{k_F} \sim \frac{\mathscr{E}_g}{\mathscr{E}_F}$$

As $\mathscr{E}_g/\mathscr{E}_F \sim 10^{-4}$ in a typical case, we obtain

$$\Delta k \sim 10^{-4} k_F$$

Since we saw in Chapter 13 that at the top of a band $k = \pi/a$, if the zeros of $k$ and $\mathscr{E}$ are taken at the bottom of the band as we do here, we can set $k_F \sim 1/a$. We also know that the lattice spacing is $a \sim 1$ Å. Thus we find that

$$\Delta k \sim \frac{10^{-4}}{1\text{ Å}}$$

is the range of wave numbers contained in the wave function for a Cooper pair. A very general property of waves ((3-14), which leads to the uncertainty principle) then immediately tells us that the extent in space of the wave function is

$$\Delta x \sim \frac{1}{\Delta k} \sim 10^4\text{ Å}$$

This is the size of a typical Cooper pair.

(b) Estimate the density of Cooper pairs in a superconductor.

Example 13-1 shows that the density of conduction electrons in a metal is $n \sim 10^{22}/\text{cm}^3$. The fraction that will form Cooper pairs in a superconductor is of the order of $\Delta k/k_F \sim 10^{-4}$. So

$$n_{\text{Cooper pairs}} \sim 10^{18}/\text{cm}^3$$

Note that the volume of one pair is $\sim(10^4\text{ Å})^3 = (10^{-4}\text{ cm})^3 = 10^{-12}\text{ cm}^3$. So each such volume contains $\sim 10^6$ overlapping pairs! ◀

The width of the forbidden gap, and the density of quantum states, in a superconductor can be determined from the current-voltage characteristic of a tunnel junction. In such junctions a thin oxide layer ($\sim 10^{-9}$ m thick) separates a normal and a superconducting metal. Electrons tunnel through the barrier, which the nonconducting oxide layer represents, with the aid of an applied voltage. In 1962, Josephson predicted that if the metals on both sides of the junction are superconducting, a current can flow when no voltage is supplied. If a small voltage ($\sim$ a few millivolts) is applied, an alternating current of frequency in the microwave range results. These effects can be used to detect extremely small voltage differences and to measure with enormous precision the ratio $e/h$ used in determination of the fundamental physical constants. Other superconducting effects predicted by Josephson permit a number of quantum properties to be seen in a very simple way, particularly the quantization of magnetic flux, discussed below.

There are many important applications of superconductivity. An obvious application is to superconducting electromagnets, whose fields arise from resistanceless currents flowing through the magnet windings, for use in electric motors and generators. A difficulty is that magnetic fields tend to be induced in the wires of the windings, which tends to destroy their superconductivity. But progress is being made in finding what are called *Type II superconductors*, which have Cooper pairs whose dimensions are small enough to allow a magnetic field to thread its way through the length of a wire in a set of localized channels. These channels lose their superconductivity, but the channels in between them do not. Several niobium-titanium alloys have been found which are Type II superconductors, and they also have the convenience of relatively high critical temperatures ($T_c \simeq 20$°K).

The absence of power dissipation in superconducting elements makes possible many electronic applications in which space requirements and transmission time requirements are limited, as in computers. Because superconductors are diamagnetic, they can be used to shield out unwanted magnetic flux. This can be put to use in shaping the magnetic lens system of an electron microscope, for example, to eliminate stray field lines and to greatly improve the practical resolving power of the instrument thereby.

Apart from such technological applications of superconductivity, of which a great many more can be cited, there is an increasing application of the theoretical ideas to other fields of physics. For example, these ideas have been applied to analyzing

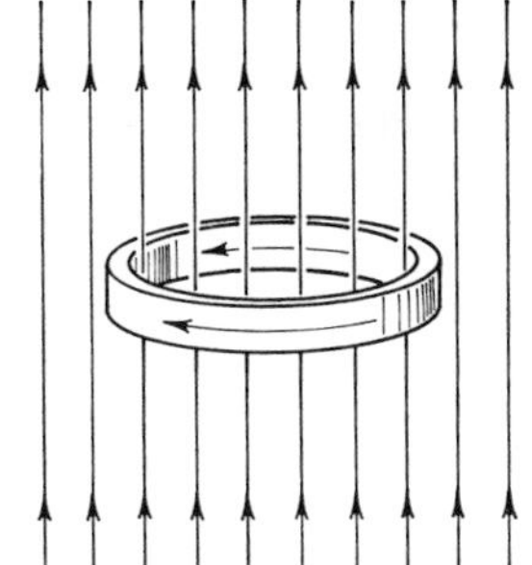

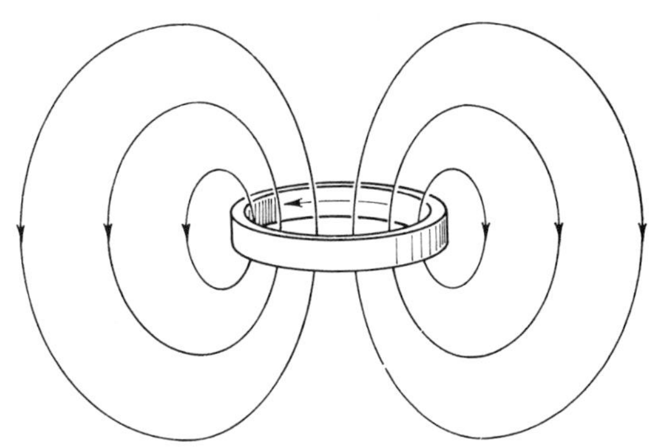

**FIGURE 14-5**

*Top:* A ring of superconducting material is cooled below the critical temperature in the presence of a uniform magnetic field. Currents are established as shown on the inner and outer surfaces of the ring, thereby excluding the field from the superconducting material comprising the ring. *Bottom:* The external field is removed. The outside surface current disappears, and the inside surface current persists. The result is that magnetic flux is trapped in the hole enclosed by the ring.

nuclear structure, with much success in accounting for otherwise unexplained experimental facts. In the next chapter we shall see similarities between the collective model of the nucleus and the BCS collective model of superconductivity. Some of the methods of superconductivity theory are being applied to the elementary particles of high-energy physics, as well, so that the theory suggests a unity underlying the various areas of quantum physics.

The Meissner effect can be stated in another way, namely, that it is possible to induce currents in a specimen in a time-invariant magnetic field simply by lowering the temperature. Such a statement contradicts Maxwell's equation $\oint \mathbf{E} \cdot d\mathbf{l} = -d\Phi_B/dt$ (or $\nabla \times \mathbf{E} = -\partial \mathbf{B}/\partial t$) and shows that the Meissner effect is not a classical effect but a quantum effect revealing itself on a macroscopic scale. This has been confirmed by experiments on a superconducting ring. If such a ring in a normal state is placed in a uniform magnetic field, and then cooled to the superconducting state, electric currents are established that flow in opposite directions on the inner and outer surfaces of the ring, as in the upper part of Figure 14-5. This excludes the field from the interior of the ring but does not affect the field inside the hole of the ring. When the external field is removed, the outside surface current disappears but the inside surface current persists. We say that the superconducting ring has trapped the original magnetic field in the hole, as in the lower part of Figure 14-5. When the magnetic flux trapped in the ring is measured as a function of the strength of the applied magnetic field, it is found that the flux is quantized, i.e., it increases in discrete steps. The system acts very much like a macroscopic Bohr atom in which one eigenfunction describes the correlated motion of the entire set of Cooper pairs traveling around the ring. Flux quantization arises because the eigenfunction must be single valued. The quantum of flux is $2\pi\hbar c/q$, where $q$ is the charge carried by one pair. The measurements confirm the BCS prediction that $q = 2e$.

## 14-2 Magnetic Properties of Solids

Materials may have intrinsic magnetic dipole moments, or they may have magnetic dipole moments induced in them by an applied external magnetic field of induction.

In the presence of a magnetic field of induction, the elementary magnetic dipoles, whether permanent or induced, will act to set up a field of induction of their own that will modify the original field. The student will recall that magnetic dipole moments, which can be regarded as microscopic currents (e.g., in atoms), are a source of *magnetic induction* **B** just as are macroscopic currents (e.g., in magnet windings). In fact, we can write

$$\mathbf{B} = \mu_0\mathbf{H} + \mu_0\mathbf{M} \tag{14-2}$$

in which **M**, called the *magnetization*, is the volume density of magnetic dipole moment, and **H**, called the *magnetic field strength*, is associated with macroscopic currents only. The magnetic vector **H**, which can be written as $\mathbf{H} = (\mathbf{B} - \mu_0\mathbf{M})/\mu_0$, plays a role in magnetism that is analogous to the role of **D** in electricity, since **D**, the electric displacement, originates only with free charges, not polarization charges. The magnetic vector **M**, which can be written as $\boldsymbol{\mu}/V$, the magnetic dipole moment per unit volume, has the same dimensions as **H**.

For certain magnetic materials, it is found empirically that the magnetization **M** is proportional to **H**. Hence, we can write

$$\mathbf{M} = \chi\mathbf{H} \tag{14-3}$$

in which the dimensionless quantity $\chi$ is called the *magnetic susceptibility*. The principal problem in studying the magnetic properties of such materials is to determine $\chi$ for them and to find how it depends, if at all, on the temperature $T$ and the value of **H**. The magnetization **M** can be put in terms of $\chi$ and **B** as

$$\mathbf{M} = \frac{\chi\mathbf{B}}{\mu_0(1 + \chi)} \tag{14-4}$$

From this expression we can see that if the susceptibility $\chi$ is small compared to one, then $\mathbf{M} \simeq \chi\mathbf{B}/\mu_0$ and the contribution made to **B** by the magnetic moments, that is $\mu_0\mathbf{M}$ in (14-2), is small. This applies in fact to magnetic materials which are diamagnetic or paramagnetic.

*Diamagnetism* is negative magnetic susceptibility, and *paramagnetism* is positive susceptibility. In diamagnetic materials the magnetization is opposite in direction to the field of induction, so that $\chi$ is negative in (14-4). The value of **B** is smaller in the region of the diamagnetic material than it would be if the material were absent. The origin of diamagnetism is *Lenz's law*: the magnetic dipole moment arising from currents induced by an applied field opposes that field. A perfect diamagnet, such as a superconductor, excludes all flux from its interior so that $\mathbf{B} = 0$ and $\chi = -1$ for such materials. For nonsuperconducting diamagnets, however, the magnitude of $\chi$ is generally less than $10^{-5}$. In a vacuum, there is, of course, no magnetization and $\chi = 0$. All substances exhibit diamagnetism, but the induced magnetic dipole moment responsible for it is masked in most substances by the existence of a permanent magnetic dipole moment. In such substances, called paramagnetic, the permanent magnetic dipole moments of the atoms tend to line up in the direction of the applied field. Here the magnetization **M** is in the direction of **B** and the magnetic susceptibility $\chi$ is positive. For typical paramagnetic materials $\chi \simeq 10^{-4}$. In the presence of a strong field of induction diamagnetic substances are weakly repelled and paramagnetic substances are weakly attracted by the field, corresponding to the fact that $\chi$ is relatively small for both types of substance though of opposite sign.

A third, and most important, type of magnetic material is ferromagnetic. *Ferromagnetism* is the presence of a spontaneous magnetization in materials even in the absence of an externally applied field of induction. The only ferromagnetic elements are iron, cobalt, nickel, gadolinium, and dysprosium, but there are many compounds and alloys of these and other elements that are ferromagnetic. Ferromagnetic substances are strongly attracted even by relatively weak fields, their magnetization being very large. Ferromagnetic susceptibilities are as large as $10^5$. There is a connection between ferromagnetism and paramagnetism, only those crystals whose atoms or molecules are individually paramagnetic being capable of exhibiting the kind of cooperative behavior that leads to ferromagnetism. In the succeeding sections we examine paramagnetism and ferromagnetism in greater detail, and we discuss their relationship to one another and to diamagnetism.

## 14-3 Paramagnetism

In a paramagnetic material the atoms contain permanent magnetic dipole moments. These moments are associated with the intrinsic electron spin and the orbital motion of the electrons. (Nuclear magnetic dipole moments are three orders of magnitude smaller than the electronic magnetic dipole moments, and so they can be neglected for our purposes here.) An externally applied field of induction **B** will tend to align these dipole moments parallel to the field. Because the energy is lower when the magnetic dipole moment is parallel to the field than when it is antiparallel, the parallel alignment is preferred. The result is an induced field that adds to the applied field so that the susceptibility is positive. In comparison, diamagnetic effects are negligible. The tendency of magnetic dipole moments to line up in the field direction is opposed by the thermal motion which tends to make the directions of the magnetic dipoles random. Hence the susceptibility is temperature dependent, and its value is determined by the relative strength of the thermal energy $kT$ and the magnetic interaction energy $-\boldsymbol{\mu} \cdot \mathbf{B}$. We expect the susceptibility to decrease with increasing temperature and, indeed, Curie found at low fields and not too low temperatures that

$$\chi = \frac{C}{T}$$

where $C$ is a positive constant characteristic of the particular paramagnetic material. This is called the *Curie law*.

In atoms with filled subshells, the spin magnetic dipole moments, and separately the orbital magnetic dipole moments, cancel in pairs. Only unfilled subshells can have unpaired electrons, so that we expect paramagnetism only in materials containing atoms whose electronic subshells are partly filled. In such materials the orientation in space of the total magnetic dipole moments can change without changing the electronic configurations of the constituent atoms. The inert gases, and many ions, have closed subshell configurations, so that they do not exhibit paramagnetism and are excellent for diamagnetic studies. Likewise in materials in which the pairing of spins is required, such as in covalent crystals and many ionic crystals, the magnetic dipole moments cannot change direction and such materials are also diamagnetic. The basic requirement for paramagnetism in solids is that the individual magnetic dipole moments have some degree of isolation. The atoms must act independently, for if the wave functions overlap significantly the operation of the quantum mechanical requirements concerning indistinguishable particles will tend to pair up the magnetic dipole moments. Many of the transition elements, and all of the rare earths, form paramagnetic solids. In these cases we have unfilled inner subshells, and the required isolation of the individual moments results from the shielding of these inner subshells by the filled outer subshells of the atoms.

Let us now calculate the paramagnetic susceptibility for the simplest kind of system, that is one containing separated atoms, in each of which the electronic orbital angular momentum is zero and there is an unpaired electron of spin angular momentum with two possible space orientations. We imagine unpaired electrons placed in a magnetic field **B**, and we neglect the interactions between such electrons. Let $n$ represent the number of unpaired magnetic dipole moments per unit volume. If $n_-$ represents the volume density of moments that are parallel to the field and $n_+$ represents the same for moments that are antiparallel, then $n_- + n_+ = n$. For a parallel alignment of the magnetic dipole moment $\boldsymbol{\mu}$ the magnetic potential energy is $-\mu B$, and for an antiparallel alignment the energy is $\mu B$. Then, from the Boltzmann distribution, we have for the number in each energy state $n_- = cne^{\mu B/kT}$ and $n_+ = cne^{-\mu B/kT}$, in which $c$ is some constant of proportionality. The resultant magnetization, i.e., the magnetic dipole moment per unit volume, is

$$M = \mu(n_- - n_+) = \mu cn(e^{\mu B/kT} - e^{-\mu B/kT})$$

It is convenient to consider the average net moment, defined as $\bar{\mu} = M/n$ and given by

$$\bar{\mu} = \frac{M}{n} = \frac{\mu cn(e^{\mu B/kT} - e^{-\mu B/kT})}{(n_- + n_+)}$$

$$= \mu \frac{cn(e^{\mu B/kT} - e^{-\mu B/kT})}{cn(e^{\mu B/kT} + e^{-\mu B/kT})}$$

or

$$\bar{\mu} = \mu \frac{e^{\mu B/kT} - e^{-\mu B/kT}}{e^{\mu B/kT} + e^{-\mu B/kT}} \tag{14-5}$$

Since under ordinary circumstances $\mu B \ll kT$, we can expand the exponentials and obtain

$$\bar{\mu} \simeq \mu \frac{(1 + \mu B/kT) - (1 - \mu B/kT)}{(1 + \mu B/kT) + (1 - \mu B/kT)} = \frac{\mu^2 B}{kT}$$

The paramagnetic susceptibility then is given by

$$\chi = \frac{M}{H} = \frac{n\bar{\mu}}{H} \simeq \frac{n\mu^2 B}{kTH} \simeq \frac{\mu_0 n\mu^2}{kT} \tag{14-6}$$

where we have used (14-4), for small $\chi$, to write $B \simeq \mu_0 H$. Hence, we obtain an approximation to the Curie result $\chi = C/T$, in which $C = \mu_0 n\mu^2/k$ and the susceptibility varies inversely with the temperature. Note (14-5) shows that if the applied field **B** is removed we have $\bar{\mu} = 0$, and there is no net magnetization. The alignment of the elementary dipoles depends on the presence of the field and, in its absence, the thermal motion randomizes the dipole directions so that the net magnetization is zero.

In the top of Figure 14-6 we plot the magnetization, $M = n\bar{\mu}$ from (14-5), as a function of the applied field $B$ for different temperatures. For small values of $B$, $M$ is essentially a straight line whose slope is greater the lower the temperature. As $B$ is increased the magnetization approaches the value $n\mu$ asymptotically. This is the saturation condition, in which all the unpaired magnetic dipole moments $\boldsymbol{\mu}$ are aligned with the applied field **B**. The strength of the field required for saturation increases with the temperature.

In the bottom of Figure 14-6 we plot the ratio $M/M_{\max}$, where $M_{\max}$ is the saturation magnetization, versus $B/T$ for a paramagnetic salt. The curve is predicted by the exact theoretical calculation, (14-5), which agrees very well with the experimental

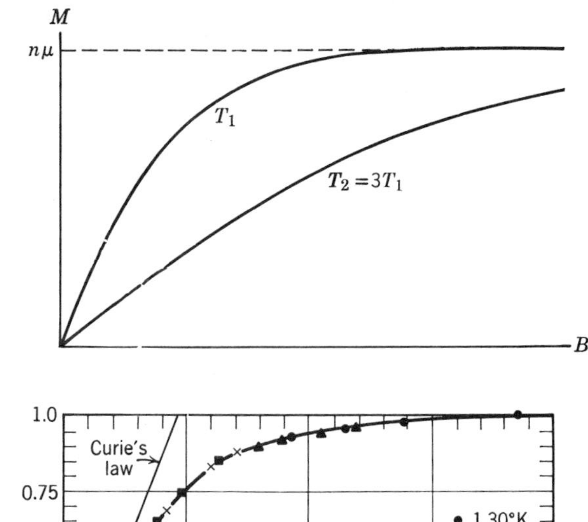




**FIGURE 14-6**

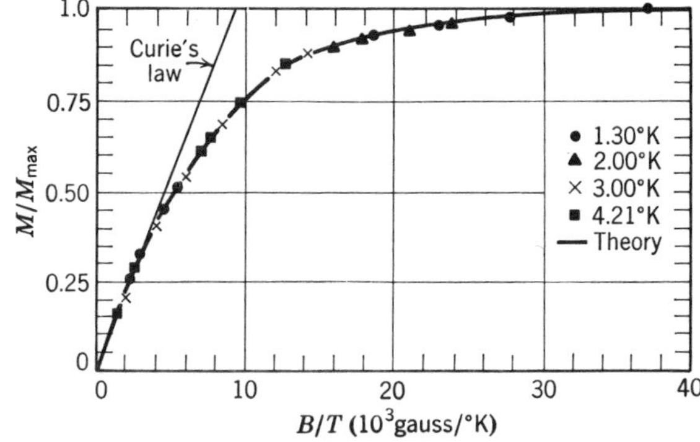

*Top:* A plot of magnetization $M$ versus the magnetic induction $B$ in a paramagnetic substance for two temperatures $T_1$ and $T_2 = 3T_1$. *Bottom:* A plot of $M/M_{\text{max}}$ versus $B/T$ for the paramagnetic salt potassium chromium sulfate.

points. The Curie law prediction, (14-6), is seen to be a good approximation at small values of $B/T$.

**Example 14-3.** (a) A magnetic field of induction achievable with an iron core electromagnet is 1.0 tesla. Compare the magnetic interaction energy of an electron spin magnetic dipole moment with this field to the thermal energy at room temperature.

We have for spin magnetic dipole moment

$$\mu = \mu_b = \frac{e\hbar}{2m} = 9.3 \times 10^{-24}\,\text{joule/tesla}$$

and for the magnetic interaction energy

$$\begin{aligned}\mu B &= 9.3 \times 10^{-24}\,\text{joule/tesla} \times 1.0\,\text{tesla} = 9.3 \times 10^{-24}\,\text{joule}\\ &= 5.8 \times 10^{-5}\,\text{eV}\end{aligned}$$

At room temperature, $T = 300°\text{K}$, the thermal energy is

$$kT = 8.6 \times 10^{-5}\,\text{eV/°K} \times 300°\text{K} = 2.6 \times 10^{-2}\,\text{eV}$$

so that

$$\frac{\mu B}{kT} = \frac{5.8 \times 10^{-5}\,\text{eV}}{2.6 \times 10^{-2}\,\text{eV}} = 2.2 \times 10^{-3}$$

Hence, the assumption $\mu B \ll kT$ is quite valid at ordinary temperatures and fields, $\mu B$ being about 0.2% of $kT$ in this example. In practice, the saturation region of Figure 14-6 is reached by going to lower temperatures rather than to higher fields.

(b) For this case estimate the paramagnetic susceptibility in a solid material having $n = 2.0 \times 10^{28}$ moments/m$^3$, a typical value for substances with one unpaired electron per atom.

From (14-6) we have, when $\mu B \ll kT$

$$\chi = \frac{\mu_0 n \mu^2}{kT}$$

$$= \frac{4\pi \times 10^{-7}\text{ tesla-m/amp} \times 2.0 \times 10^{28}/\text{m}^3 \times (9.3 \times 10^{-24}\text{ joule/tesla})^2}{1.38 \times 10^{-23}\text{ joule/}^\circ\text{K} \times 300^\circ\text{K}}$$

$$= 5.2 \times 10^{-4}$$

The result is an estimate because the theory used is approximate, neglecting, as it does interactions between the electrons. Most paramagnetic substances have measured values somewhat smaller than this result. ◀

It is found that the Curie relation deduced above does *not* apply to metals, although it does apply to nonmetallic paramagnetic materials. Indeed, in metals the paramagnetic susceptibility is much smaller and virtually independent of temperature. We have a situation here somewhat like the one in Section 11-11 where we sought an understanding of the electronic contribution to the specific heats of metals. In the analysis leading to (14-6), we used the classical Boltzmann distribution. That was valid because the electrons were associated with different atoms and they could be distinguished by their location, but in metals we must use the Fermi distribution because the electrons behave there as a Fermi gas of indistinguishable particles. When we do so we get a smaller susceptibility than before, and one that is independent of temperature, as we now explain.

In Figure 14-7*a* we plot the energy distribution of electrons in a metal, the energy states that correspond to spin magnetic dipole moments aligned antiparallel to the field being plotted above the energy axis and those that correspond to moments aligned parallel being plotted below the axis. Here we imagine the field $B$ to be (nearly) zero. When $B$ is increased, at first all the electron energies shift, the energy

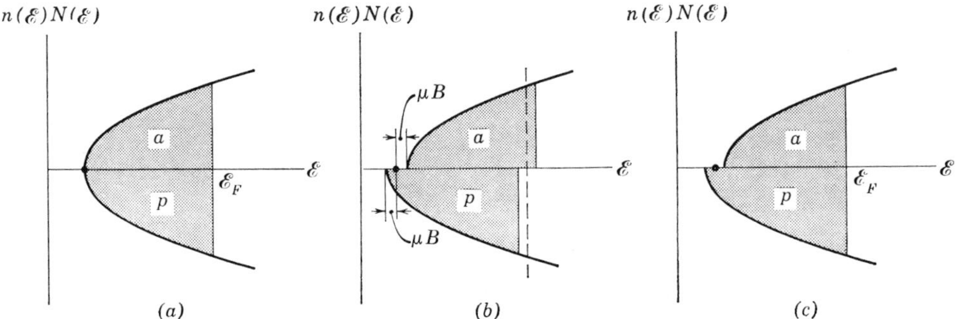


**FIGURE 14-7**

The distribution of electrons with energy in a metal; the electrons occupy states indicated by the shaded areas. States with spin magnetic dipole moments antiparallel to the applied field are plotted above the energy axis, and states with moments parallel to the field are plotted below. (*a*) The applied field is essentially zero. (*b*) The situation immediately after the field is increased to value $B$. (*c*) The equilibrium situation in applied field $B$. In these diagrams the magnetic interaction energy $\mu B$ is greatly exaggerated relative to the Fermi energy $\mathcal{E}_F$.

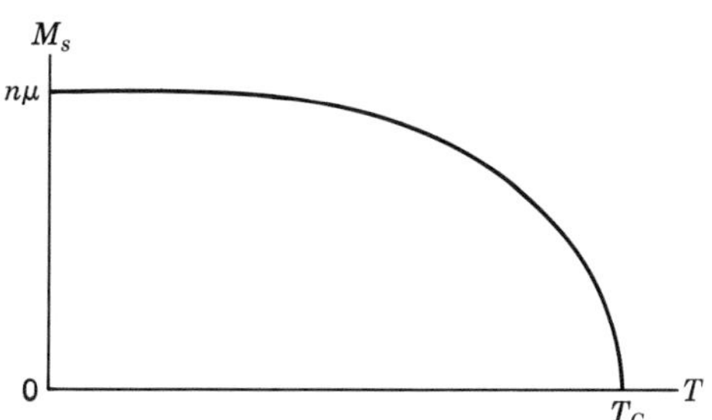


**FIGURE 14-8**
The spontaneous magnetization $M_s$ versus temperature $T$ in a ferromagnetic material. $T_C$ is the ferromagnetic Curie temperature.

rising by $\mu B$ for antiparallel moments and dropping by $\mu B$ for parallel moments, as shown in Figure 14-7*b*. Some electrons will subsequently make transitions from the higher energy antiparallel states to the lower energy parallel states, leading to the equilibrium situation of minimum total energy shown in Figure 14-7*c*. We have seen in Example 14-3 that $\mu B \simeq 10^{-4}$ eV at $B = 1.0$ tesla, which is a very small energy shift compared to the Fermi energy, $\mathscr{E}_F \simeq 1$ eV. Hence, the number of electrons with parallel moments is only slightly larger than those with antiparallel moments, the randomizing thermal effect dominating, so that the susceptibility should have a small value. Furthermore the situation would not be expected to be sensitive to reasonable temperature changes so the susceptibility should be practically independent of temperature, as is observed experimentally for metals.

## 14-4 Ferromagnetism

Ferromagnetism is a spontaneous magnetization of small regions of a material that exists even in the absence of an external field of induction. Let us summarize the principal known features of ferromagnetism. First, the spontaneous magnetization in ferromagnetic materials varies with the temperature. The magnetization is a maximum at $T = 0°$K and drops to zero at a temperature $T_C$, called the ferromagnetic *Curie temperature*, as is illustrated in Figure 14-8. Secondly, at temperatures higher than $T_C$ the materials become paramagnetic and have a magnetic susceptibility that is given by the relation $\chi = C/(T - T_C)$. This is a modification of the Curie relation for paramagnetic materials, in which $\chi$ is not defined for temperature below $T_C$ where the material has a permanent magnetization. Thirdly, a ferromagnetic material is not magnetized in the same direction throughout its volume but has many smaller regions of uniform magnetization direction, called domains, that may be randomly oriented with respect to each other. Finally, the only ferromagnetic elements are iron, cobalt, nickel, gadolinium, and dysprosium. There is a quantum theory of ferromagnetism that can explain all these observed properties. But before going into it, we show in the following example that a simple classical explanation, which obviously suggests itself, is not sufficient.

**Example 14-4.** The field of induction produced by a magnetic dipole of moment $\mu$ along a line parallel to its axis is given by $B = \mu_0\mu/2\pi x^3$, where $x$ is the distance from the dipole. Calculate the interaction energy of two iron atoms, with parallel and colinear magnetic dipole moments of magnitude $\mu = 2.2$ Bohr magnetons, separated by the interatomic spacing in iron, 3 Å. Then evaluate the temperature at which the magnetic interaction energy equals the thermal energy, to show that this classical dipole-dipole interaction will not explain ferromagnetism in iron.

The interaction energy, when one dipole aligns itself in the field produced by the other dipole, is negative (binding) and of magnitude

$$E = \frac{\mu_0 \mu^2}{2\pi x^3} = \frac{4\pi \times 10^{-7}\,\text{tesla-m/amp} \times (2.2 \times 9.3 \times 10^{-24}\,\text{joule/tesla})^2}{2\pi \times (3 \times 10^{-10}\text{m})^3}$$
$$= 3.1 \times 10^{-24}\,\text{joule}$$

Equating this energy to the thermal energy $kT$, and solving for $T$, we find

$$T = \frac{E}{k} = \frac{3.1 \times 10^{-24}\,\text{joule}}{1.38 \times 10^{-23}\,\text{joule/}^\circ\text{K}} = 0.22^\circ\text{K}$$

The temperature is very low because the dipole-dipole interaction energy is very small. At room temperature, thermal energy is three orders of magnitude larger, and the randomizing tendency of thermal agitation would completely destroy the tendency for the dipole-dipole interaction to align the individual magnetic dipole moments and produce a large total magnetization. Such alignment is, however, actually found in iron at room temperature because it is ferromagnetic at that temperature. So we conclude that the explanation of ferromagnetism cannot be the very weak classical dipole-dipole interaction. ◀

To illustrate the quantum theory of ferromagnetism consider iron, cobalt, or nickel, all of which are transition elements that have partially filled $3d$ inner subshells. The quantum numbers $m_l$ and $m_s$ for the $3d$ electrons in an atom of a ferromagnet containing such atoms will have those values that minimize the energy of the ferromagnetic system, consistent with the requirements of the exclusion principle. If the $z$ component orbital angular momentum quantum numbers $m_l$ of two $3d$ electrons have the same values, for example, the $z$ component spin angular momentum quantum numbers $m_s$ must have opposite values. If the $m_l$ values are different, the $m_s$ values can be the same, which means that the spins can be essentially parallel. Now the $g$ factor, which specifies the ratio of the total magnetic dipole moment to the total angular momentum, has a value for ferromagnetic materials near the value $g = 2$ that corresponds to electron spin (see Section 10-6, particularly (10-23)). This indicates that the magnetization is due to "parallel" spin rather than orbital magnetic dipole moments. Thus the electrons in the $3d$ subshell of an atom of iron align themselves so that the spins are essentially parallel. The reason is that it reduces the energy of the atom. That is, two $3d$ electrons stay farther apart on the average if their spins are "parallel" than if their spins are "antiparallel," and if they are farther apart their mutual Coulomb repulsion energy is reduced. This is just the tendency (see Section 10-4) for the spins in an unfilled subshell to all couple "parallel" and maximize the total spin, to the extent allowed by the exclusion principle, because this minimizes the residual Coulomb energy. Thus a single atom of iron is paramagnetic, because it has a permanent spin magnetic dipole moment, basically because of the interaction between the spin coordinates and space coordinates imposed by the quantum mechanical requirements concerning the exchange of labels of indistinguishable particles. For this reason the spin coupling is sometimes said to be due to the strong *exchange interaction* operating within the atom.

Now consider a crystal lattice of iron atoms. There is also a strong exchange interaction between adjacent atoms of the lattice because the electrons in the atoms are indistinguishable and the atoms are close enough to each other that indistinguishability makes a difference. This exchange interaction will also lead to a coupling of spins, i.e., the total spins of adjacent atoms, but it is more complicated than the exchange interaction within a single atom because the geometry of the system of atoms is more complicated than the spherically symmetrical geometry of a single atom. The results of the exchange interaction can be that the lowest energy of the system occurs when the spins of adjacent pairs of atoms are "parallel," or that it occurs when they are "antiparallel." In the first case the system will be ferromagnetic; in the second it will be antiferromagnetic.

We can understand ferromagnetism by considering the five overlapping $3d$ energy bands of a crystal composed of one of the transition element atoms. The totality of these bands, which we shall here call the $3d$ band, can hold ten electrons per atom. When full, the band has five electrons with spin "up" and five with spin "down," per atom. The band is narrow because the $3d$ subshell is an inner subshell, as we discussed in Section 13-7. In the ferromagnetic atoms, however, the $3d$ band is only partially filled. In iron, for example, there are six $3d$ electrons per atom. If we at first assumed that three of these electrons have spin with one orientation and three have spin with the other orientation, the electrons occupying the lowest energy available states in each of two *partial bands* of opposite spin, we could not be sure that this is the state of lowest energy for the system because the exchange interaction of the lattice will shift the partial bands of opposite spin with respect to each other. The partial band of one spin, i.e., the collection of energy levels in which all the electrons have one spin orientation, will be lowered in energy by the exchange interaction and the partial band of the other spin will be raised in energy by the interaction. We could have five electrons per atom in one partial band, and the sixth in the partial band of the opposite spin, if the total energy of the system is lowered more by the exchange interaction than it is raised by the higher energy resulting from the asymmetrical population of electron energy levels between the two partial bands. That is, competing with the desire of all electrons to go into the partial band of lowest energy is the fact that, if they do, some will be forced by the exclusion principle to go into the higher energy levels of that partial band. We shall soon present a figure that illustrates, and further explains, this competition.

Calculations show that for a few elements one partial band will indeed be filled and the other will not, so that a large spontaneous magnetization will exist in them. When the interaction between spins is calculated as a function of the ratio of one-half the internuclear separation to the radius of the $3d$ shell in the transition elements, it is found that parallel spin alignment is favored if this ratio exceeds 1.5. Typical values of the ratio are Mn, 1.47; Fe, 1.63; Co, 1.82; Ni, 1.98; so that iron, cobalt, and nickel are expected to be ferromagnetic and manganese not to be. In fact manganese crystals are not ferromagnetic. The theory is further confirmed by the fact that certain compounds (such as the Heusler alloys) which contain manganese atoms that are farther apart *are* ferromagnetic.

In Figure 14-9 we plot the energy difference between magnetized and unmagnetized configurations versus the ratio of half the internuclear separation to the $3d$ radius. As the separation between atoms is increased from the value giving the maximum, the $3d$ wave functions overlap less and less and the indistinguishability requirements soon cease to apply; hence, the exchange interaction reduces the energy less and less. If in a crystal lattice the valence electron subshell radii are small compared to the internuclear spacing, as in the rare earth elements, we expect the material to be paramagnetic because the individual spin magnetic dipole moments are isolated from one another. As the separation between atoms is decreased from the value which yields the maximum, the energy bands widen and the excess energy associated with the asymmetrical population in the magnetized state increases more than the exchange interaction reduces the energy. Indeed, we approach the situation in diatomic molecules wherein "antiparallel" spins give the lowest energy since the electrons spend most of their time between nuclei. In elements with valence electrons in outer unfilled subshells,

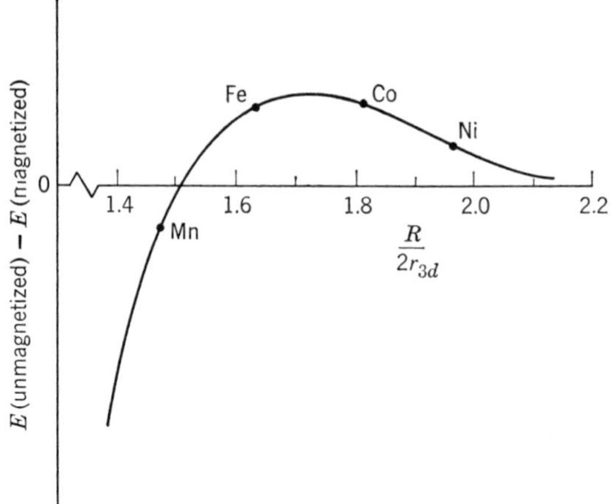


**FIGURE 14-9**

The variation of the energy difference between unmagnetized and magnetized configurations with the ratio of the internuclear separation to the diameter of the $3d$ subshell, for some transition elements.

the subshell radius is large enough, compared to internuclear separation, that we expect all these electrons to form pairs having "antiparallel" spins. Then there will be no spin magnetic dipole moment and the material will be diamagnetic. Figure 14-10 illustrates schematically the population of two partial bands of opposite spin, for internuclear separation smaller than, equal to, and larger than the range of values that leads to ferromagnetism.

We see that the ferromagnetic situation is a delicate one in which the valence subshell radius is large enough to permit sufficient space overlap to allow the requirements of indistinguishability to apply, but at the same time small enough to prevent the width of the valence band from becoming too large. In those cases in which the magnetized state is favored, the energy difference between magnetized and unmagnetized states is of the order of a tenth of an electron volt per atom. This situation makes it clear, therefore, that the spontaneous magnetization is temperature dependent and that additional thermal energy made available by an increase in temperature can eliminate the conditions favoring the spin alignment responsible for ferromagnetism. At $T = 0°K$ all the spin alignment permissible exists, but as the temperature is raised successively more of the "parallel" alignments are made random by thermal motion. Just below the Curie temperature, $T_C$, the alignment breaks up rapidly (see Figure 14-8), and it is entirely gone above $T_C$. For iron the Curie temperature is 1043°K, for cobalt it is 1400°K, and for nickel 631°K.

The origin of domains remains to be explained. Ferromagnetic materials are not observed to be magnetized unless they have been put in an external magnetic field previously. It is said that, although spontaneous magnetization exists, the magnetization in one small region, or *domain*, of a ferromagnetic material can be oriented in a direction different from that in another domain, so that the macroscopic resultant magnetization can be zero. Domains arise in the first place because the energy of a large crystal is not a minimum when it is uniformly magnetized. The particular size and shape of a domain is determined by a process that minimizes the total of three different types of energy involved. There is first the magnetic field energy. If, for example, the entire solid specimen formed a single domain there would be a large external field and a large magnetic energy associated with the field. The external magnetic field can be greatly reduced, thereby decreasing the energy in it, by dividing the specimen into domains whose magnetizations tend to cancel one another as in Figure 14-11. However, the domain boundaries, or walls, are sites of highly localized and nonuniform magnetic fields of considerable intensity, and a second type of energy

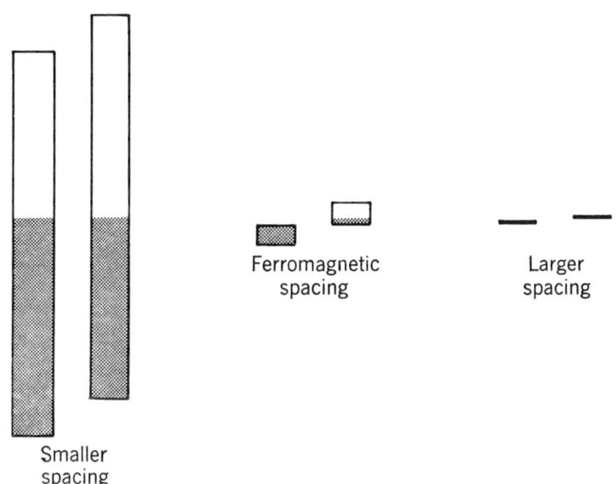


**FIGURE 14-10**

Illustrating schematically the valence band structure for three different internuclear spacings of a system of atoms which are, individually, paramagnetic. With decreasing spacing, the wavefunctions of electrons in valence subshells of adjacent atoms overlap, and exchange effects set in. They cause the valence level to split into a band and, from the point of view of the band being decomposed into two partial bands of oppositely aligned spins, they *also* cause the partial bands to be displaced with respect to each other. The possibility of ferromagnetism arises because, in a favorable case such as is illustrated, with decreasing spacing the displacement at first increases about as rapidly as the band width increases. This relation is not maintained into very small spacings because the band width increases more and more rapidly with decreasing spacing (see Figure 13-3). At all spacings, the levels of the two partial bands will be occupied in such a way that the Fermi energies are equal, since this minimizes the total energy of the system. For the situation described by the central figure, the number of valence electrons in the total band is sufficient to completely fill all levels of the lower partial spin band, but only the lower levels of the upper partial spin band. The system is then ferromagnetic since most of the valence electron spins are aligned in the same direction. In the figure on the right this does not happen because the energies associated with both exchange effects are small compared to $kT$. It does not happen in the figure on the left because the band width is large compared to the partial band displacements. Thus ferromagnetism requires not only that there be a range of valence subshell overlap where the two exchange effects have a particular relation, but also that the internuclear spacing to valence subshell diameter ratio be such as to make the overlap in the actual system be in that range.

is required to create them. The third energy is the difference in energy between a situation where the specimen is magnetized in one direction relative to the axis of the crystal and a situation in which it is magnetized in another direction.

In an unmagnetized piece of iron the individual domains, within which the magnetic dipole moments are aligned, are oriented at random. As we magnetize the iron by placing it in an external magnetic field, two effects take place. One is a growth in size of the domains that are favorably oriented with respect to the field at the expense of those that are not, as shown in Figure 14-12. Another is a rotation of the direction of magnetization within a domain toward the direction of the applied field. The well-known *hysterisis* effect, in which the magnetization of ferromagnetic materials does not return to zero as we first apply an external field and then remove it, is due to the

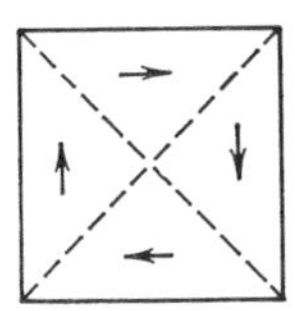

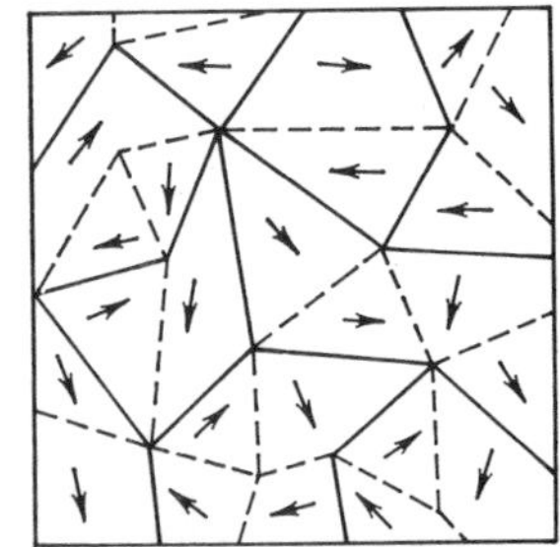

**FIGURE 14-11**

Ferromagnetic domains. *Top left:* In a single crystal the magnetization vectors must lie along equivalent axes of the crystal. This crystal has no net magnetization, although each domain is magnetized. *Top right:* In a polycrystalline substance the crystal axes are randomly oriented, so that the magnetization vectors are randomly oriented. *Bottom:* Domain patterns for a *single* crystal of iron containing 3.8% silicon. The white lines show the boundaries between the domains. (Courtesy H. J. Williams, Bell Telephone Laboratories)

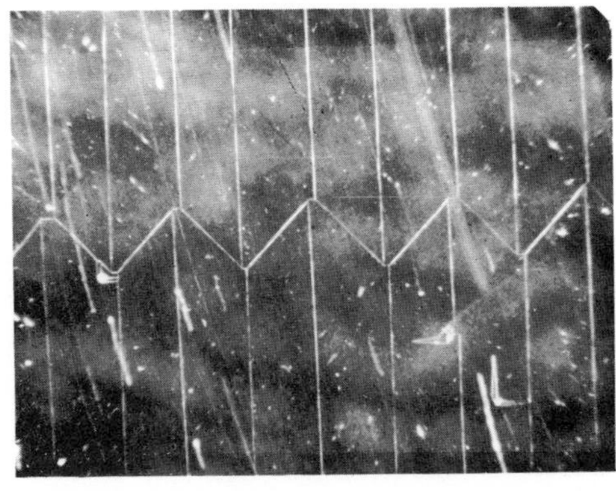

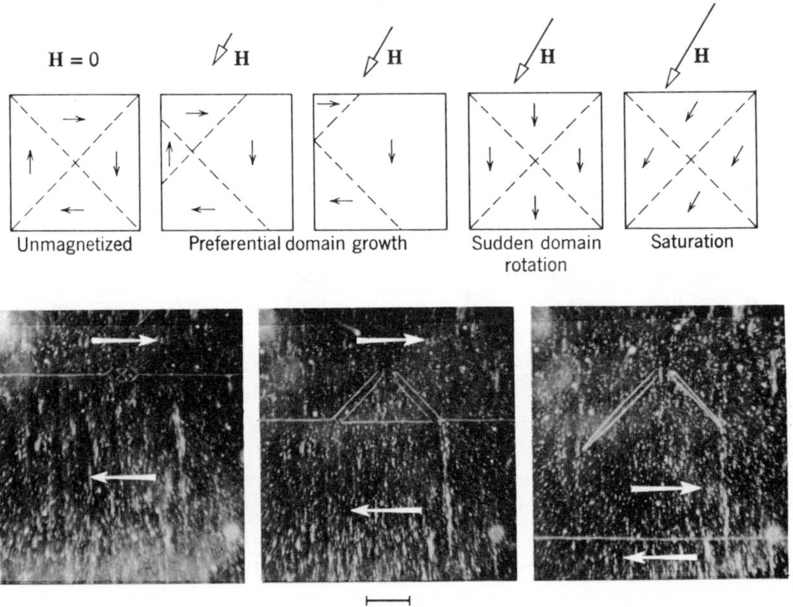


**FIGURE 14-12**

*Top:* The growth of domains in a single crystal in an externally applied magnetic field **H**, showing schematically preferential domain growth, domain rotation, and saturation. *Bottom:* An external magnetic field, directed to the right, is imposed on a specimen. The magnetization in each domain is shown by white arrows. The domain boundary moves down across a region in which there is a crystal imperfection as the preferentially oriented domain grows. (Courtesy H. J. Williams, Bell Telephone Laboratories)

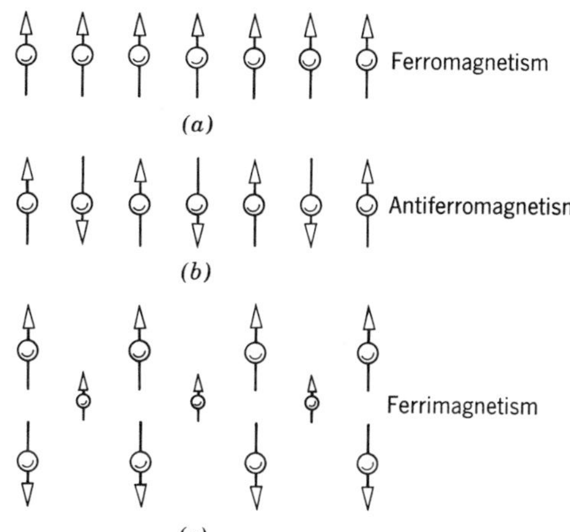


**FIGURE 14-13**
Showing how elementary magnetic dipole moments are oriented by the interatomic exchange interaction in (*a*) ferromagnetism, (*b*) antiferromagnetism, and (*c*) ferrimagnetism.

fact that the domain boundaries do not move completely back to their original positions when the external field is removed. The motion of these boundary walls is not reversible and is affected by crystal imperfections such as impurities and strains. The material is left magnetized even though there is no externally applied field, a condition called *permanent magnetism*.

## 14-5 Antiferromagnetism and Ferrimagnetism

Two other types of magnetism, closely related to ferromagnetism, are antiferromagnetism and ferrimagnetism. In *antiferromagnetic* materials, of which $MnO_2$ is an illustration, the exchange interaction forces adjacent atoms to have "antiparallel" spin orientations, as in Figure 14-13*b*. In $MnO_2$, for example, the negative oxygen ion has on each side a positive manganese ion; the magnetic dipole moments of the positive ions are aligned essentially antiparallel because each is paired with one of the oppositely oriented electron spins of the oxygen ion in the lowest energy configuration of the system. Hence such materials show very little gross external magnetism. If they are heated sufficiently the materials become paramagnetic, the exchange interaction ceasing to act. In *ferrimagnetic* substances two different kinds of magnetic ions are present; in nickel ferrite the two ions are $Ni^{++}$ and $Fe^{+++}$. The exchange interaction locks the ions into a pattern like that of Figure 14-13*c*. The same antiferromagnetic exchange interaction exists, which aligns the magnetic dipole moments "antiparallel," but since ions with two different magnitudes of magnetic dipole moment are present, the net magnetization is not zero. The external magnetic effects are intermediate between ferromagnetism and antiferromagnetism, and here too the exchange interaction disappears if the material is heated above a certain characteristic temperature. The ferrites are crystals having small electrical conductivity compared to ferromagnetic materials, and they are useful in high-frequency situations because of the absence of significant eddy current losses.

## QUESTIONS

1. Why do superconducting currents flow on the surface of a superconductor?
2. Why is the electric field zero inside a superconductor?

**3.** Does perfect conductivity require that the interior magnetic field of a body be zero? What does it require of the interior magnetic field?

**4.** How would you measure the critical field of a superconductor as a function of temperature?

**5.** The critical external magnetic field at absolute zero varies with the material as $M^{-1/2}$. Explain.

**6.** Can you say whether lead or aluminum has the higher superconducting critical temperature from the fact that at room temperature the electrical conductivity of aluminum is much larger than that of lead?

**7.** A superconducting film can be used as a high sensitivity bolometer (an instrument for measurement of heat radiation). Explain.

**8.** To what extent can the two electrons in a Cooper pair be thought of as moving as if they were bound to opposite ends of a spring? What property of the system constitutes the spring?

**9.** Exactly what is the distinction between the energy states of an electron in a superconductor and the energy states of the superconductor itself?

**10.** Are there analogies between superconductivity and superfluidity?

**11.** Superconductors whose Cooper pairs are small enough to allow the existence of magnetic field carrying channels also have relatively high critical temperatures. What is the reason for this very convenient behavior of Type II superconductors?

**12.** Discuss the use of a paramagnet as a thermometer. In what temperature range would it be useful?

**13.** The magnetization induced in a diamagnetic sphere by an external magnetic field does not vary with the temperature, in sharp contrast to the situation in paramagnetism. Make this plausible.

**14.** Does the orbital motion of an electron contribute to paramagnetic behavior of the atom or only the intrinsic spin of an electron?

**15.** The paramagnetic susceptibility of the rare earth elements is generally greater than that of the transition elements. Take into account the electronic shell structure and explain why.

**16.** Is the neglect of the nuclear spin magnetic dipole moment justifiable in our discussion of paramagnetism? Explain.

**17.** From the fact that most organic molecules have magnetic dipole moments of less than a few Bohr magnetons, show that life processes cannot be affected by laboratory magnetic fields.

**18.** Why do the ferromagnetic elements come from the middle of the group of transition elements or from the middle of the rare earth elements rather than the ends of the respective groups?

**19.** Copper has a filled inner $3d$ electronic subshell and one $4s$ valence electron. Explain why you would not expect it to be ferromagnetic.

**20.** Why is susceptibility not defined for temperatures below the Curie temperature in ferromagnetic materials?

**21.** Are the electronic configurations of gadolinium and dysprosium consistent with the fact that they are ferromagnetic elements? Explain.

**22.** Why can the exchange interaction have a significant effect on a narrow band with a high density of states (as the $3d$ band in the transition elements) although the interaction energy is small?

23. A nail is placed at rest on a smooth table top near a strong magnet. It is released and attracted to the magnet. What is the source of the kinetic energy the nail has just before it strikes the magnet?

24. Why, for permanent magnets, do we use materials composed of small crystals and having large imperfections? Also why, for transformer magnets, do we use materials composed of large crystals having few imperfections?

## PROBLEMS

1. (a) Show, from Maxwell's equations, that resistivity $\rho = 0$ (a perfect conductor) implies that $\mathbf{B}$ = const inside the material. (b) Show, from Maxwell's equations, that $\mathbf{B} = 0$ inside a material (a superconductor) implies that the resistivity of the material is $\rho = 0$.

2. Show from Lenz's law that the Meissner effect implies perfect conductivity, but that perfect conductivity does not imply the Meissner effect.

3. The critical field of tin at 2°K is 0.02 weber/m$^2$. Draw a graph of the magnetization at 2°K of a long thin sample of tin as a function of applied field.

4. Part of the $\mathscr{E}$ versus $k$ diagram for electrons in a superconductor is shown in Figure 14-14. (a) Draw a curve of the density of electrons as a function of $\mathscr{E}$ for a superconductor at

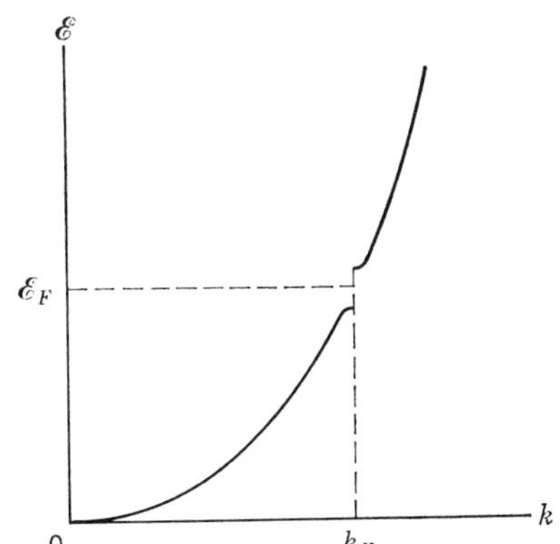


**FIGURE 14-14**
The energy as a function of positive wave number for a superconductor.

$T = 0$°K. (b) Draw a graph of the energy necessary to place holes in the superconducting state and electrons in the normal state. This is a graph of $(\mathscr{E} - \mathscr{E}_F)$ versus $k$; $\mathscr{E}_F$ is at the center of the gap for a superconductor.

5. When two metals are separated by a very thin insulator, electrons from one metal can tunnel through the insulator to the other metal. Electrons flow until the Fermi levels of the two metals are equal. When a battery is connected between the two metals, as shown in Figure 14-15, the Fermi levels are displaced and a current flows if there are filled electron levels in one metal opposite empty levels in the other metal. Draw current voltage characteristics for the following junctions. (a) Normal metal-normal metal. (b) Normal metal-superconductor. (c) Superconductor-superconductor. (Hint: The Fermi energy of a superconductor lies at the center of the energy gap.)

6. Use Faraday's law of induction to show that a hole in a superconductor will trap magnetic flux, i.e., $dB/dt = 0$ in the hole. Remember that the electric field $E = 0$ in any circuit through the superconductor which encloses the hole, and also that the Meissner effect does not apply to the hole.

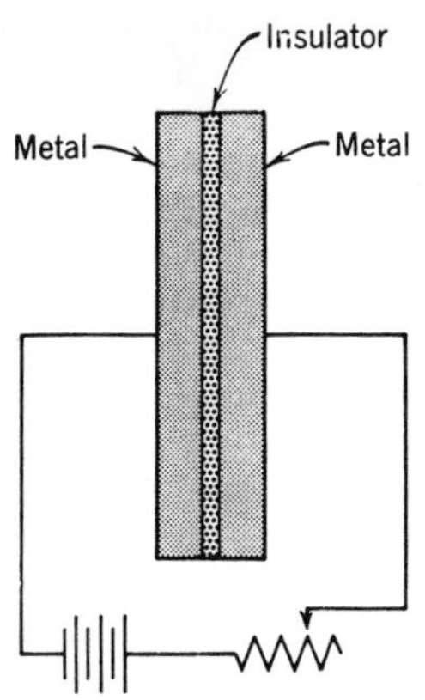


**FIGURE 14-15**
A circuit considered in Problem 5.

**7.** Derive (14-4) for the magnetization, using (14-2) and (14-3).

**8.** Show from (14-2) and (14-3) that $\chi = -1$ for a superconductor. Is this result consistent with (14-4)?

**9.** (a) Calculate the magnetization of 1 mole of oxygen at standard temperature and pressure in the earth's magnetic field. The susceptibility of oxygen is $2.1 \times 10^{-6}$ and the earth's field is $5 \times 10^{-5}$ tesla. (b) What is the saturation magnetization of 1 mole of oxygen? Its magnetic dipole moment is 2.8 Bohr magnetons.

**10.** Calculate the temperature of the sample of Example 14-3 when the magnetic field is reduced isentropically from 1 tesla at 1°K to 0.01 tesla, assuming Curie's law. (An isentropic process is one in which the populations of the states do not change. Hence the magnetization must remain constant.) This process is called *adiabatic demagnetization* and is useful in low-temperature physics.

**11.** What is the magnetization of the two-level system, discussed in connection with (14-5), when $\mu B \gg kT$?

**12.** From Figure 14-7 it can be argued that the magnetization due to conduction electrons should be proportional to the number of electrons within $\mu B$ of the Fermi energy. (a) Show that this leads to the susceptibility being given approximately by

$$\chi = \frac{3\mathcal{N}\mu_b^2}{2kT_F}$$

where $\mathcal{N}$ is the number of conduction electrons, $\mu_b$ is the Bohr magneton, and $T_F$ is the Fermi temperature. (b) Evaluate $\chi$ for copper.

**13.** (a) Show that the specific heat at constant field $c_H$ for the two-level system, discussed in connection with (14-5), is given by

$$c_H = \frac{\mathcal{N}k\left(\dfrac{2\mu B}{kT}\right)^2 e^{2\mu B/kT}}{(e^{2\mu B/kt} + 1)^2}$$

where $\mathcal{N}$ is the number of atoms in the system. This is the Schottky specific heat. (Hint: Take the energy of the dipoles aligned parallel to the field to be zero.) (b) What is the temperature dependence of $c_H$ at high and low temperatures? (c) Sketch $c_H$ as a function of $T$. Estimate (do not calculate) where $c_H$ will be a maximum.

**14.** A ferromagnet can be considered to be similar to a paramagnet except that there is an internal molecular field $H_W$ tending to spontaneously align the elementary dipoles. (a) The material will become spontaneously magnetized when the energy of interaction between the dipole and the molecular field is equal to $kT_C$. Calculate the value of $H_W$ for iron where the magnetic moment is 2.2 Bohr magnetons and $T_C$ is 1000°K. (b) What is the magnetization of a 1 cm³ sample of iron which has a single domain? (Density = 7.9 g/cm³; atomic weight = 56). (c) What is the energy in the field?

**15.** The molecular field of Problem 14 can be taken as proportional to the magnetization of the sample so that $H_W = \lambda M$. (a) Show that this leads to a susceptibility given by

$$\chi = \frac{C}{T - T_C}$$

where $T_C = C\lambda$. (b) Calculate the value of $\lambda$ for iron.

**16.** A simple model for an antiferromagnet is a lattice of two kinds of paramegnetic ions such that the nearest neighbors of $A$ atoms are $B$ atoms. If the antiferromagnetic interactions are between nearest neighbors only, the magnetization of the sample above the Curie point can be written as

$$TM_A = C'(H - \lambda M_B)$$

and

$$TM_B = C'(H - \lambda M_A)$$

Here $C'$ is the Curie constant for one sublattice only. The effective field in sublattice $A$ is $H - \lambda M_B$, and positive $\lambda$ corresponds to antiferromagnetic interactions between $A$ and $B$ atoms. Show that this leads to a susceptibility above $T_C$ given by

$$\chi = \frac{C}{T + T_C}$$

where $C = 2C'$ and $T_C = C'\lambda$.

**17.** Sketch curves of $\chi^{-1}$ versus $T$ for $T > T_c$ for (a) a paramagnet, (b) a ferromagnet, and (c) an antiferromagnet, and discuss the meaning of the intercept on the $T$ axis.

# 15

# Nuclear Models

# 15 Nuclear Models

## 15-1 Introduction

In the past chapters our considerations have taken us from atoms to the larger systems, molecules and solids, of which atoms are constituents. Now we reverse our direction and consider the smaller systems, nuclei, which are constituents of atoms.

There is a pronounced difference between the theoretical study of atoms, or systems of atoms, and the theoretical study of nuclei. Long before the theory explaining the properties of atoms was being developed, the basic nature of the electromagnetic forces acting on individual electrons in atoms was known in complete detail. But during most of the period when the understanding of the properties of nuclei was being developed, very little was known about the details of the nuclear forces acting on the protons and neutrons in nuclei. Although a fairly complete knowledge of nuclear forces has recently become available, they turn out to be complicated enough that it has not yet been possible to use this knowledge to construct a comprehensive *theory* of nuclei. That is, we cannot explain all of the properties of nuclei in terms of the properties of the nuclear forces acting between their protons and neutrons. However, there are a number of *models*, or rudimentary theories of restricted validity. Each of these can explain a certain limited range of nuclear properties, using arguments which do not involve all the details of the nuclear forces. Even though progress is being made on the development of a comprehensive theory, an introductory study of nuclei is still largely the study of the various nuclear models. In this chapter we treat the most important models and use them to describe and explain the properties of nuclei in their ground states. In Chapter 16 we use these models to study nuclei in their excited states, and to study naturally occurring transitions between nuclear states (nuclear decay, including radioactivity) and artificially produced transitions (nuclear reactions, including fission and fusion). The detailed properties of nuclear forces are treated in Chapter 17, where we consider the elementary particles which are constituents of nuclei.

A pronounced difference between the experimental study of atoms and the experimental study of nuclei arises from the difference between their characteristic energies. The energy characteristic of nuclei is of the order of 1 MeV. For instance, we saw in Chapter 6 that the attractive nuclear potential exerted on a neutron when it is in a nucleus is a few MeV deep, and that the height of the repulsive Coulomb barrier separating two positively charged nuclei is also a few MeV. We shall soon see that the same order of magnitude characterizes the binding energy of a proton or neutron to a typical nucleus, and the separation in energy between its ground state and first excited state. The energy characteristic of atoms is of the order of 1 eV. Because this is so low (not much higher than room temperature thermal energy $kT \simeq 0.025$ eV) atoms are easily excited, and they have little difficulty in combining to form molecules and solids. For nuclei, very special circumstances are required to produce excitation because of their very high characteristic energy. Weisskopf has described the situation well:

"In our immediate environment atomic nuclei exist only in their ground state; they affect the world in which we live only by their charge and mass and not by their intricate dynamic

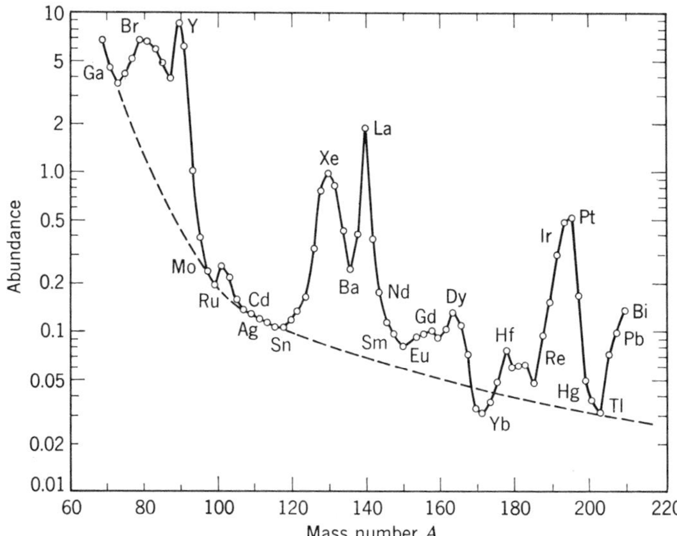


**FIGURE 15-1**

The relative abundance of the elements. Note strong fluctuations superposed on a general decreasing trend with increasing $A$, the mass number.

properties. In fact, all the interesting nuclear phenomena . . . come into play only under conditions which we have created ourselves in accelerating machines. It is to some extent a man-made world.

It is not completely man made, however. The centers of all stars are regions of the universe where nuclear reactions go on, and thus where nuclear dynamics plays an essential role in the course of nature. Hence the nuclear phenomena are the basis of our energy supply on earth, in reactors as well as in the sun. But nuclear physics is even more important for the world in which we live from the point of view of the history of the universe. The composition of matter as we see it today is the product of nuclear reactions which have taken place a long time ago in the stars or in star explosions, where conditions prevailed which we simulate in a very microscopic way within our accelerating machines. Hence the material basis of the world in which we live is a product of the laws of nuclear physics. I cannot better illustrate the interconnection of all facts of nature, the tightly woven net of the laws of physics, than by pointing to the chart of abundances of elements in our part of the universe (see Figure 15-1). Each maximum and minimum in the curve of abundances corresponds to some trait of nuclear dynamics, here a closed shell, there a strong neutron cross section, or a low binding energy. If the 7.65 MeV resonance in carbon did not exist, then, according to Holye and Salpeter, practically no carbon would have been formed and we would probably not have evolved to contemplate these problems. Whenever we probe nature—be it by studying the structure of nuclei, or by learning about macromolecules, or about elementary particles, or about the structure of solids—we always get some essential part of this great universe." (From "*Problems of Nuclear Structure*," by Victor Weisskopf, *Physics Today* 14: 7, 1961.)

## 15-2 A Survey of Some Nuclear Properties

We begin our study of nuclei by quickly reviewing what we have already learned about them in the process of studying atoms and molecules, and by adding some new information that is also obtained largely from atoms and molecules. The items of new information are considered here only briefly; each will be discussed in more detail later:

1. We have learned (Chapter 4) that the mass of a nucleus is only slightly less than the mass of an atom containing that nucleus. Thus the nuclear mass is approximately

equal to the integer $A$ times the mass of a hydrogen atom, or approximately equal to $A$ times the mass of a *proton*, the nucleus of a hydrogen atom. The integer $A$, called the *mass number*, is the one closest to the atomic weight of the atom containing the nucleus in question. We have also learned (Chapters 4 and 9) that the charge of a nucleus is exactly equal to the *atomic number* $Z$ of the corresponding atom, times the negative of the charge of an electron, or exactly $Z$ times the charge of a proton. The atomic number gives the location of an atom in the periodic table of the elements. That table (Chapter 9) shows that $A$ is roughly equal to $2Z$, except for the proton for which $Z = A = 1$.

2. Analysis of $\alpha$-particle scattering from nuclei of low $A$ (Chapter 4) indicated that the radii of such nuclei are somewhat less than 10 F, where the radius is defined as the distance from the center of the nucleus at which the potential acting on the $\alpha$ particle first deviates from a Coulomb potential. Analysis of the rate of emission of $\alpha$ particles by radioactive nuclei of high $A$ (Chapter 6) indicated that the radii of these nuclei, defined in the same way, are $\simeq 9$ F. The symbol F represents the unit of length, called the *fermi*, used in nuclear physics. Its value is

$$1\ \text{F} = 1 \times 10^{-15}\ \text{m} \qquad (15\text{-}1)$$

Note that this length, characteristic of nuclei, is five orders of magnitude smaller than the length 1 Å characteristic of atoms since 1 Å $= 1 \times 10^{-10}$ m.

3. Both the $\alpha$-particle scattering and the $\alpha$-particle emission analyses showed that there is a *nuclear force*, which is *attractive*, acting between the particle and the nucleus, in addition to the repulsive Coulomb force acting between the two. They indicated that the nuclear force is of very *short range*, i.e., that it extends only for a distance appreciably less than 10 F. The analyses also indicated that the nuclear force is *strong*, compared to the Coulomb force, since it dominates the latter, which is repulsive, to produce an overall attraction on the $\alpha$ particle when it is very close to the nucleus. Modern experiments involving the scattering of protons from protons show that the range of the nuclear force is $\simeq 2$ F, and that the magnitude of the negative energy associated with the attractive force is larger than their Coulomb energy, when the two protons are separated by that distance, by roughly a factor of 10. Furthermore, experiments involving the scattering of protons from neutrons indicate that the nuclear force is *charge independent*. That is the nuclear force between protons and neutrons is the same as between protons and protons, or between neutrons and neutrons (except for exclusion principle effects that apply in the latter two cases only). Although the scattering experiments which provide direct experimental proof of the charge independence of nuclear forces are fairly recent, an educated guess was made at an early stage that the nuclear force would have this simplifying property. We shall consider the scattering experiments in Chapter 17, and certain other evidence for charge independence later in this chapter and in Chapter 16. Until then we too shall make the assumption that the nuclear force is charge independent. Finally, it should be mentioned that the nuclear force is extremely strong compared to the gravitational force. The magnitude of the energy associated with the nuclear force acting between two protons separated by less than 2 F is larger than their gravitational energy by a factor of about $10^{40}$.

4. It has been mentioned (Chapters 8 and 10) that nuclei have magnetic dipole moments. They arise from the intrinsic magnetic dipole moments of the protons and neutrons in the nuclei, and from the currents circulating in the nuclei due to the motion of the protons. Nuclear magnetic dipole moments are studied by using optical spectroscopic equipment of extremely high resolution to measure the *hyperfine splitting* of atomic energy levels, which results from their interaction with the magnetic field

produced by the atomic electrons. The value of the interaction energy $\overline{\Delta E}$ depends on the orientation of the nuclear magnetic dipole moment in the internal magnetic field, and is given by the equation

$$\overline{\Delta E} = C[f(f+1) - i(i+1) - j(j+1)] \tag{15-2}$$

where $j$, $i$, and $f$ are quantum numbers specifying the magnitudes of the atom's total electronic angular momentum, total nuclear angular momentum, and grand total angular momentum, respectively. This equation is completely analogous to (10-15), which describes the atomic spin-orbit interaction energy. The constant $C$ is proportional to the magnitude of the *nuclear magnetic dipole moment* $\mu$. Measurements of $\overline{\Delta E}$, and therefore of $C$, show that for all nuclei $\mu$ is of the order of the *nuclear magneton* $\mu_n$. This quantity is

$$\mu_n = \frac{e\hbar}{2M} = 0.505 \times 10^{-26} \text{ amp-m}^2 \simeq 10^{-3}\mu_b \tag{15-3}$$

where $M$ is the proton mass and $\mu_b$ is the Bohr magneton. Measurements of hyperfine splitting also show that the sign of the nuclear magnetic dipole moment (giving the relative orientation of the magnetic dipole moment vector and the angular momentum vector of the nucleus) is positive (parallel) in some cases and negative (antiparallel) in others. Nuclei with both $A$ and $Z$ even have $\mu_n = 0$.

5. The total nuclear angular momentum quantum number $i$, usually called the *nuclear spin*, can be obtained simply by counting the number of energy levels of a hyperfine splitting multiplet. If the multiplet is associated with a value of $j$ larger than $i$, then $f$ can assume $2i + 1$ different values so there will be $2i + 1$ different energy levels. It is found that $i$ is an integer for nuclei of even $A$, with $i = 0$ if $Z$ is also even, and that $i$ is a half-integer for nuclei of odd $A$. The magnitude $I$ of the total nuclear angular momentum is given in terms of $i$ by the usual relation $I = \sqrt{i(i+1)}\,\hbar$. The total angular momentum of a nucleus arises from the intrinsic spin angular momenta of its protons and neutrons and also from the orbital angular momenta due to the motion of these particles within the nucleus. It should be emphasized that in nuclear physics the word *spin* frequently refers to the *total* angular momentum of a nucleus, in contrast to atomic physics where the word refers to the intrinsic spin angular momentum only. When there is possibility of confusion, we shall henceforth use the terminology intrinsic spin angular momentum, and we shall continue to use the symbol $s$, when referring to that part of the angular momentum of a single particle that has nothing to do with orbital angular momentum (e.g., the intrinsic spin angular momenta of both protons and electrons are given by $s = 1/2$).

6. Closely related to the spin of a nucleus is the *symmetry character* of the eigenfunction for a system containing two or more nuclei of the same species (Chapter 9). This is studied by analyzing the spectra of diatomic molecules containing two identical nuclei (Chapter 12). It is found that nuclei with integral spin quantum number $i$ (nuclei of even $A$) are of the symmetric type, i.e., they are bosons, while nuclei with half-integral $i$ (nuclei of odd $A$) are of the antisymmetric type, i.e., they are fermions. Such molecular spectra also provide independent measurements of $i$, which confirm values obtained from hyperfine splitting.

7. As we have already indicated, nuclei are composed of protons and neutrons. The *neutron* is an uncharged particle of nearly the same mass as the proton, and precisely the same intrinsic spin angular momentum and symmetry character ($s = 1/2$, antisymmetric). A nucleus with mass number $A$ and atomic number $Z$ contains $A$ *nucleons*, a word used for both protons and neutrons, of which $Z$ are protons and

$A - Z$ are neutrons. This rule obviously leads to a mass and charge in agreement with item 1.

Before the discovery of the neutron, it was thought that a nucleus of mass number $A$ and atomic number $Z$ contains $A$ protons and $A - Z$ electrons. This rule also leads to a mass and charge in agreement with item 1, but we have seen that the zero-point energy is unrealistically high if a particle as light as an electron is confined in a region as small as a nucleus (Chapter 6). Furthermore, the spin and symmetry character of nuclei composed of protons and neutrons are, in all cases, in agreement with the measurements described in items 5 and 6. For nuclei in which $A$ is even and $Z$ is odd, the spin and symmetry character disagree with the measurements if nuclei are composed of protons and electrons.

**Example 15-1.** The mass number and atomic number of the nucleus of the most prevalent variety of nitrogen are: $A = 14$, $Z = 7$. Its measured nuclear spin and symmetry character are: $i = 1$, symmetric. (See Examples 12-4 and 12-5.) Show that the spin and symmetry character disagree with the assumption that nuclei contain $A$ protons and $A - Z$ electrons. Also show that the spin and symmetry character are in agreement with the assumption that nuclei contain $A$ nucleons, of which $Z$ are protons and $A - Z$ are neutrons.

If the nucleus contains 14 protons and 7 electrons, it contains an odd number, $14 + 7 = 21$, of the particles that all have half-integral intrinsic spin angular momentum quantum numbers. (They all have $s = 1/2$.) The rules for combining angular momentum quantum numbers presented in Section 8-5 make it apparent that, whether or not these particles have orbital angular momenta, each of their total angular momentum quantum numbers will be half-integral since orbital angular momentum quantum numbers are always integral. Furthermore, it is apparent that a nucleus containing an odd number of particles, each with half-integral total angular momentum quantum number, can only have a half-integral total angular momentum quantum number. In other words, its nuclear spin will be half-integral, in disagreement with the measurements.

It is also apparent from the discussion of Section 9-3 that the symmetry character of a nucleus containing an odd number of antisymmetric particles must be antisymmetric. The reason is that an exchange of labels of two such nuclei amounts to an odd number of exchanges of labels of antisymmetric particles. This multiplies the total eigenfunction of the system by an odd power of minus one, which equals minus one, so that the total eigenfunction is antisymmetric. Again we see that the nitrogen nucleus cannot contain 14 protons and 7 electrons, giving it an odd total number of particles, since the measurements show that it is a nucleus of the symmetric type.

If the nucleus contains 7 protons and 7 neutrons, the total number of particles is $7 + 7 = 14$, an even number. Since neutrons have the same intrinsic spin angular momentum and symmetry character as protons (or electrons), we see that the nucleus will be symmetric because in an exchange of labels of two nuclei the total eigenfunction will be multiplied by an even power of minus one, and an even power of minus one equals plus one. Its nuclear spin will be integral since an even number of particles of half-integral intrinsic spin angular momentum quantum numbers must have an integral total angular momentum quantum number. Both of these predictions are in agreement with the measurements. ◀

Some years before its discovery, Rutherford suggested the existence of a particle having the properties of what we now call the neutron. A number of people tried to devise experiments to detect it. But this was difficult because, being uncharged, the neutron does not easily ionize atoms when it passes through matter, and most devices for detecting particles depend on ionization. In 1932 Chadwick succeeded in detecting neutrons emitted from beryllium nuclei when they are bombarded with $\alpha$ particles obtained from a radioactive source. He used a Geiger counter behind a layer of paraffin. The neutrons collide with protons in the paraffin, and they transfer an appreciable fraction of their kinetic energy to the protons. The protons then penetrate

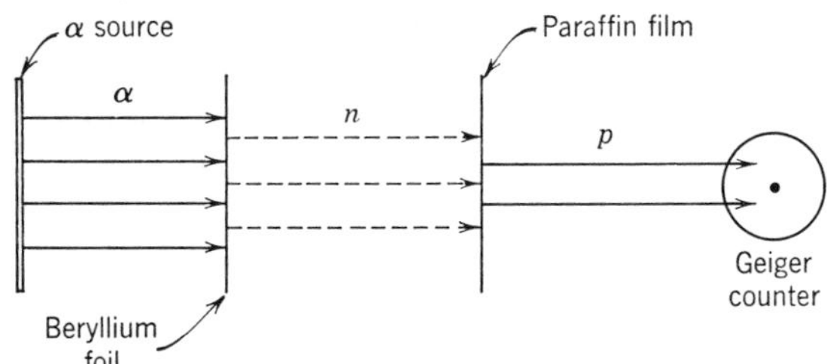


**FIGURE 15-2**

A schematic depiction of the experimental arrangement used by Chadwick in the discovery of the neutron.

the Geiger counter, where they are counted with high efficiency since they are charged and therefore produce much ionization. The experimental arrangement is indicated in Figure 15-2.

8. Many nuclei are not precisely spherical in shape, but instead they are in the shape of an ellipsoid. The earliest evidence for this came from accurate measurements of the hyperfine splitting of the energy levels of atoms of these nuclei. If the hyperfine splitting were due entirely to the energy of orientation of the nuclear magnetic dipole moment in the internal magnetic field of the atom as assumed in (15-2), the analogy with (10-15) for the spin-orbit interaction would require that the pattern formed by the split atomic energy levels obey an *interval rule* like Landé's (10-16). But deviations from such an interval rule are seen in the hyperfine splitting of many atoms. The deviations indicate that in these atoms the hyperfine splitting is partly due to an electric interaction between an ellipsoidal distribution of the nuclear charge and the atomic electric field. That is, in these atoms the energy depends on the orientation of the ellipsoidal nuclear charge distribution in the internal electric field of the atom, as well as on the orientation of the nuclear magnetic dipole moment in the internal magnetic field of the atom.

The observed departure of the nuclear charge distribution from spherical symmetry is specified by the nuclear *electric quadrupole moment* $q$. As is illustrated in Figure 15-3, for $q > 0$ the ellipsoidal charge distribution is elongated in the direction of its symmetry axis, with the elongation increasing as $q$ becomes more positive. For $q < 0$ the ellipsoidal charge distribution is flattened in the direction of its symmetry axis, with the flattening increasing as $q$ becomes more negative. A more precise definition of $q$ will be given in Section 15-10.

For nuclei with spin $i \geq 1$, the hyperfine splitting measurements show that there are cases with electric quadrupole moment $q > 0$, as well as cases with $q < 0$. But for nuclei with $i = 0$ or $i = 1/2$, these measurements always yield $q = 0$; that is, no departures from spherical shape are observed for such nuclei in these measurements. It is easy to see why a nucleus appears to have a spherical shape if it has zero nuclear

**FIGURE 15-3**

*Left:* A prolate (football-shaped) charge distribution gives rise to a positive quadrupole moment $q$. *Right:* An oblate (fat pumpkin-shaped) charge distribution gives rise to a negative quadrupole moment. Both ellipsoids are symmetrical about the axis through their center.

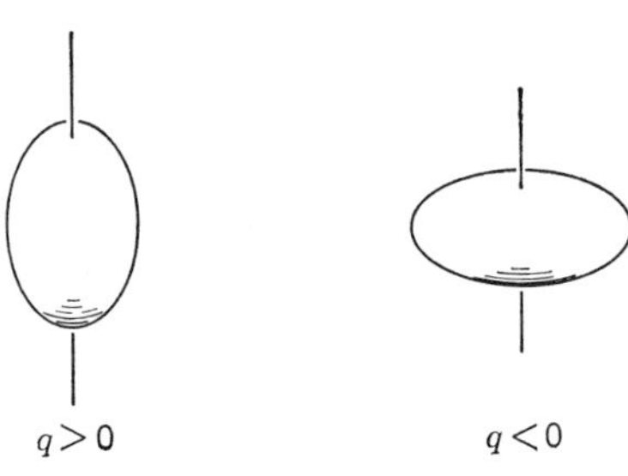

spin. If it has no nuclear spin it does not have any particular orientation in space since there is no total angular momentum vector that must maintain a fixed component on some direction. The nucleus can be thought of as assuming, in very rapid succession, all possible orientations in space. So even if it actually is nonspherical, we could not see this in the hyperfine splitting measurements because hyperfine splitting, and most other observable effects, depends only on the time average of the nuclear shape. To put it another way, most measurements simply do not have the time resolution required to detect the instantaneous nuclear shape. But measurements involving nuclear reactions do. As will be discussed in the following chapter, they show that many nuclei with nuclear spin $i = 0$ actually are nonspherical, although this cannot be seen by hyperfine splitting. More complicated symmetry arguments prove that nuclei must also be observed to be spherical in hyperfine splitting and other time-averaged measurements, if they have nuclear spin $i = 1/2$.

The largest values of $q$ are found for nuclei in the region of the rare earth elements. In the most extreme case the largest dimension of the ellipsoidal charge distribution is along the direction of the symmetry axis, and it exceeds the smallest dimension by about 30%. But for typical nuclei with $i \geq 1$, the difference in the largest and smallest dimensions of the ellipsoid is only a few percent. So for most purposes it is a good approximation to assume that typical nuclei are spherical, particularly since more than half of all the nuclei have $i = 0$, and so they appear in most circumstances to be precisely spherical.

## 15-3 Nuclear Sizes and Densities

We begin our detailed discussion of nuclei by considering the results of measurements of their sizes. The most straightforward and accurate measurements involve scattering of electrons, of several hundred MeV kinetic energy, from thin targets containing atoms whose nuclei are to be investigated. As nuclear forces do not act on an electron, its scattering is due to its Coulomb interaction with the nuclear charge distribution. An electron scattered through an appreciable angle has had a single close enounter with a nucleus, just as in $\alpha$-particle scattering from nuclei (see Section 4-2). Therefore, measurements of electron scattering should be able to provide information about the nuclear charge distribution, such as its size. The charge distribution is, of course, only the distribution of protons in the nucleus, but there is much additional evidence indicating that the neutrons have approximately the same distribution as the protons.

The method can be thought of as the use of an "electron microscope" to "look at" the charge distribution. What is actually seen is not the charge distribution itself, but a diffraction pattern which it produces in scattering the electron wave function. Qualitatively, we know that the separation in angle between adjacent minima of the diffraction pattern, $\theta$, will obey the usual diffraction relation (see Chapter 3)

$$\theta \simeq \frac{\lambda}{r'} \tag{15-4}$$

where $\lambda$ is the electron de Broglie wavelength, and $r'$ is the radius of the charge distribution. Thus a measurement of $\theta$ gives immediately an estimate of $r'$, since $\lambda$ can be calculated from the known kinetic energy.

**Example 15-2.** Electrons of kinetic energy $K = 500$ MeV are scattered from a target of nuclei, of charge distribution radius $r'$, into a diffraction pattern that has minima with an average separation of $\theta \simeq 30°$. Estimate $r'$.

First we must evaluate the de Broglie wavelength $\lambda$ from the electron momentum $p$. Since the total energy $E$ of the electrons is very high compared to their rest mass energy $m_0c^2 = 0.51$ MeV, we may use expressions that are valid in the extreme relativistic limit

$$p = \frac{E}{c} = \frac{K}{c}$$

$$= \frac{500 \text{ MeV}}{3 \times 10^8 \text{ m/sec}} \times \frac{1 \text{ joule}}{6.2 \times 10^{12} \text{ MeV}} = 2.7 \times 10^{-19} \text{ kg-m/sec}$$

Then the de Broglie relation gives

$$\lambda = \frac{h}{p}$$

$$= \frac{6.6 \times 10^{-34} \text{ joule-sec}}{2.7 \times 10^{-19} \text{ kg-m/sec}} = 2.4 \times 10^{-15} \text{ m}$$

Converting $\theta$ to radians, and invoking (15-4), we find

$$r' \simeq \frac{\lambda}{\theta}$$

$$\simeq \frac{2.4 \times 10^{-15} \text{ m}}{0.53 \text{ rad}} = 4.5 \times 10^{-15} \text{ m} = 4.5 \text{ F}$$

for an estimate of the charge distribution radius. ◀

An accurate determination of the nuclear charge distribution can be obtained if the shape of the electron diffraction pattern is analyzed quantitatively. This involves adding up the portions of the electron wave function scattered from each region of the nucleus, in proportion to an assumed charge density in that region, and taking into account the phase differences that produce the constructive or destructive interference at different scattering angles which constitutes the diffraction pattern. The assumed charge distribution is varied until the best fit to the measured diffraction pattern is obtained. It is found that the fit is very sensitive to the details of the charge distribution, so that it can be well determined even if the diffraction pattern contains only one minimum. The analysis is related to the one-dimensional Schroedinger scattering calculations of Chapter 6. But it is much more complicated because it is three dimensional, and because it is relativistic so the Dirac version of quantum mechanics must be used. Thus we can only quote results.

Figure 15-4 indicates the experimental apparatus used by Hofstadter, and collaborators, to measure the scattering of high-energy electrons from various nuclei. The electrons are produced in a linear accelerator, part of which is shown. It operates something like a very large scale version of the electron guns used in electron microscopes, or television tubes. The electrons are scattered from a thin target foil, whose atoms contain the nuclei of interest, located at the center of the evacuated scattering chamber. Scattered electrons are detected by the spectrometer, which determines their kinetic energy by bending them in its magnetic field. Only the elastically scattered electrons are counted, i.e., those whose kinetic energy is the same as the electrons of the incident beam, less the small amount of kinetic energy of the recoiling nuclei. This requirement ensures that the nuclei remain undisturbed, so that their ground state charge distribution will be obtained.

Figure 15-5 shows results obtained in the scattering of 420 MeV electrons from the small mass number nucleus $^6$C. The ordinate is the differential scattering cross section $d\sigma/d\Omega$, a quantity defined in (4-8) which is proportional to the number of

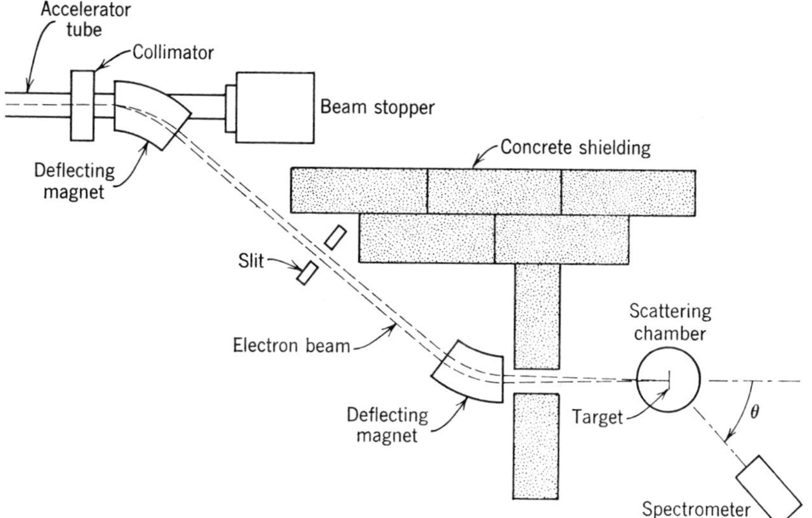


**FIGURE 15-4**

An apparatus used to study the scattering of high-energy electrons from a target of nuclei. Only the end of the electron linear accelerator is shown. It is actually a very long evacuated tube in which radio frequency fields accelerate the electrons to the required energy.

electrons scattered at each angle. The points with accuracy estimates are the data, and the solid curve is the best fit to the data obtained from the analysis. The radial distribution of *nuclear charge density* $\rho(r)$, which produces this fit, is shown by the curve labeled $^6C$ in Figure 15-6.

For a given electron energy, the diffraction patterns measured for nuclei of larger mass number $A$ develop additional minima, which become more closely spaced as $A$

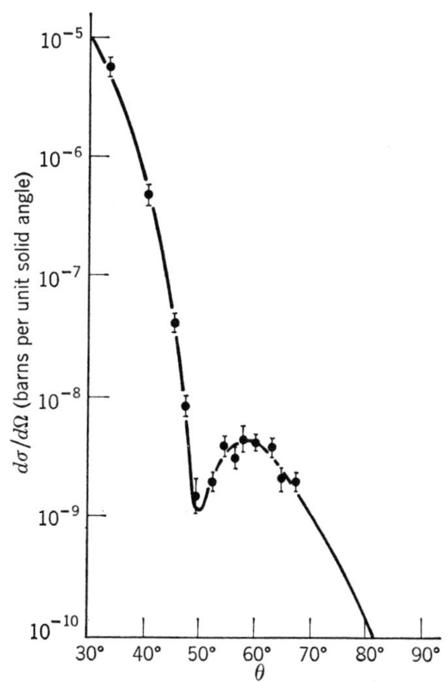


**FIGURE 15-5**

A measure of the number of electrons scattered from $^6C$ as a function of the scattering angle for 420 MeV incident electrons. The differential scattering cross section $d\sigma/d\Omega$ is the measure used. It is evaluated in terms of the area unit commonly employed in nuclear physics, called the *barn*; 1 bn $= 10^{-24}$ cm$^2$. The curve is the fit to the data points obtained from the scattering analysis described in the text.

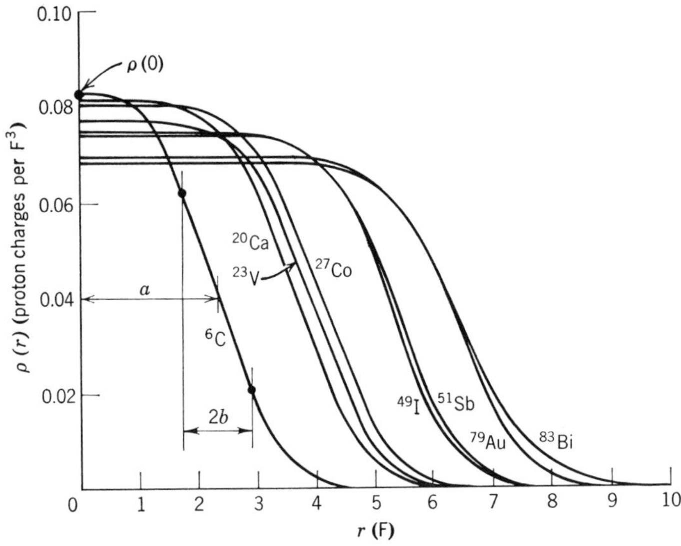


**FIGURE 15-6**
The charge densities of a number of nuclei. The charge density labeled $^6$C produced the fit to the scattering data shown in Figure 15-5. The half-value radius parameter $a$, surface thickness $2b$, and interior charge density $\rho(0)$, are shown for $^6$C.

increases. Equation (15-4) indicates this means the radius of the charge distribution increases with increasing $A$. The quantitative results are shown by the curves in Figure 15-6, which represent the charge densities $\rho(r)$ obtained for a number of nuclei. All of these charge densities can be described fairly accurately by the empirical equation

$$\rho(r) = \frac{\rho(0)}{1 + e^{(r-a)/b}} \tag{15-5}$$

where the parameters $a$ and $b$ have the values

$$a = 1.07A^{1/3} \times 10^{-15} \text{ m} = 1.07A^{1/3} \text{ F} \tag{15-6}$$

$$b = 0.55 \times 10^{-15} \text{ m} = 0.55 \text{ F} \tag{15-7}$$

We draw the following conclusions from Figure 15-6 and (15-5) through (15-7):

1. The charge density of nuclei, which is essentially the distribution of protons in the nuclei, is constant in the nuclear interior and falls fairly rapidly to zero at the nuclear surface.

2. The radius at which the density has one-half its interior value, $a$, increases slowly with increasing number of nucleons in the nucleus, $A$. Specifically, the radius $a$ is proportional to $A^{1/3}$.

3. The thickness of the nuclear surface is given approximately by the quantity $2b$, since most of the drop in the value of the factor $1/[1 + e^{(r-a)/b}]$, from its interior value of one to its exterior value of zero, occurs when $r$ changes from $a - b$ to $a + b$. This surface thickness $2b$ has approximately the same value for all nuclei.

4. The interior value of the charge density, $\rho(0)$, decreases slowly with increasing $A$.

5. If we assume that the distribution of protons in nuclei is approximately the same as the distribution of neutrons (there is good evidence for this assumption), then the charge density $\rho(r)$, which gives the density of protons in the nucleus, is the same as the *mass density* $\rho_M(r)$, which gives the density of all nucleons in the nucleus, except

for a factor proportional to $Z/A$, the ratio of the number of protons to the total number of nucleons in the nucleus. That is

$$\rho(r) \propto \frac{Z}{A}\,\rho_M(r) \tag{15-8}$$

Then the decrease of $\rho(0)$ with increasing $A$ is explained entirely by the decrease in $Z/A$ with increasing $A$. (The periodic table shows that $Z/A \simeq 1/2$ for $A \simeq 40$, while $Z/A \simeq 1/2.5$ for $A \simeq 240$.) This indicates that *the interior value of the mass density,* $\rho_M(0)$, *is approximately the same for all nuclei.*

**Example 15-3.** Evaluate approximately the interior mass density of a nucleus.

Approximate results can be obtained most easily by noting that the ratio of the density of a nucleus to the density of a solid, containing atoms with that nucleus, is

$$\frac{\text{density of nucleus}}{\text{density of solid matter}} \propto \left[\frac{\text{volume of nucleus}}{\text{volume of atom}}\right]^{-1} \propto \left[\left(\frac{\text{radius of nucleus}}{\text{radius of atom}}\right)^3\right]^{-1}$$

For all nuclei

$$\frac{\text{radius of nucleus}}{\text{radius of atom}} \sim 10^{-5}$$

For instance, the radius of the outer shell of the $^6$C atom is a little less than 2 Å $= 2 \times 10^{-10}$ m, while the half-value radius of its nuclear charge or mass distribution is a little more than 2 F $= 2 \times 10^{-15}$ m. Thus we obtain

$$\frac{\text{density of nucleus}}{\text{density of solid matter}} \sim 10^{15}$$

Since the density of solid matter is of the order of $10^3$ kg/m$^3$, we find that the density of a nucleus has the *extremely* high value

$$\text{density of nucleus} \sim 10^{18}\ \text{kg/m}^3$$

The densities of nuclei are some 15 orders of magnitude larger than the densities encountered in the macroscopic world. It is, therefore, not surprising that other properties of nuclei can differ remarkably from the properties of macroscopic objects. ◀

## 15-4 Nuclear Masses and Abundances

Very precise measurements of nuclear masses provide information about some of the most basic nuclear properties. Now the masses of atoms of a particular $Z$, but possibly a mixture of $A$, can be obtained to several significant figures by chemical techniques and a knowledge of Avogadro's number. Since the mass of a nucleus differs from the mass of the corresponding atom by a known amount, these techniques provide fairly accurate determinations of nuclear masses. But for the extremely accurate determinations needed in the study of nuclei, we must use the physical techniques of *mass spectrometry*, or *energy balance in nuclear reactions*. Both give information about the masses of *atoms*, of a particular $Z$ and $A$. From these masses, the masses of the corresponding *nuclei* can be evaluated by subtracting $Z$ times the electron mass. The mass equivalent of the electron binding energies is small enough to be ignored.

An example of one of the many types of mass spectrometers is the Bainbridge design, illustrated in Figure 15-7. The source produces ionized atoms with charge $+Ze$, mass $M$, and a distribution of velocities. These atoms travel through an evacuated region of crossed electric and magnetic fields which act as a velocity filter,

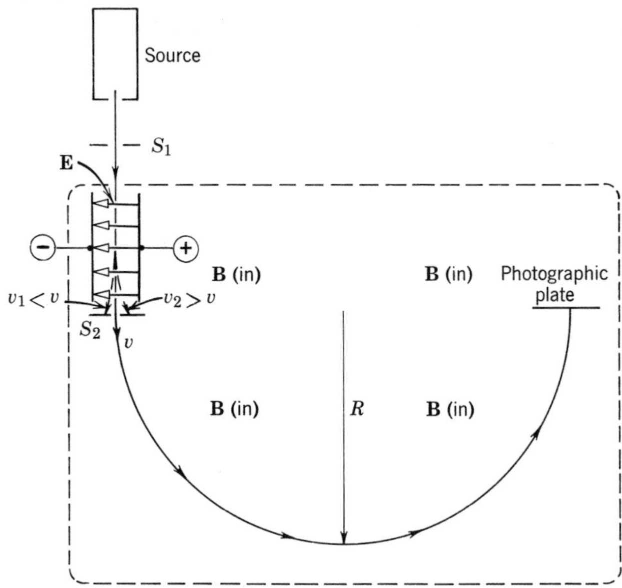


**FIGURE 15-7**

An apparatus used to measure atomic masses. Magnetic pole pieces above and below the plane of the paper provide a uniform magnetic field into the paper throughout the region enclosed by the dashed line. The entire apparatus shown is contained in a vacuum chamber.

passing only those with velocity $v$ satisfying the equation

$$ZeE = BZev$$

The terms on the left and right are the magnitudes of the opposing electric and magnetic forces. Atoms of velocity $v = E/B$ enter a region of uniform magnetic field, are bent into a semicircle of radius $R$, and fall on a photographic plate where they produce an image. The distance from the diaphragm $S_2$ to the image is $2R$, where $R$ satisfies the equation

$$BZev = \frac{Mv^2}{R}$$

The term on the right is the mass times the centripetal acceleration. Solving for $M$, we obtain

$$M = \frac{RBZe}{v} = \frac{RB^2Ze}{E} \tag{15-9}$$

The atomic mass can be determined from absolute measurements of the quantities on the right side of (15-9). But in practice use is made of various hydrocarbon molecules to calibrate the apparatus over a wide range of masses, in terms of the standard mass of carbon. The main reason that carbon is used as a standard, or unit, of mass is that many different hydrocarbons are readily available. In fact, the ion source usually produces some ionized hydrocarbons automatically, since hydrocarbons in the form of vacuum pump oil are present in the apparatus.

With the mass spectrometry technique, extremely accurate measurements can be made. As an example, consider the nucleus $^{20}Ca^{40}$. (The superscript before the chemical

**TABLE 15-1.** Atomic Masses and Binding Energies

| | $Z$ | $A$ | Mass in $u$ | Binding Energy in MeV: Total ($\Delta E$) | Binding Energy in MeV: Per Nucleon ($\Delta E/A$) |
|---|---|---|---|---|---|
| $^{0}n^{1}$ | 0 | 1 | 1.0086654 (±4) | — | — |
| $^{1}H^{1}$ | 1 | 1 | 1.0078252 (±1) | — | — |
| $^{1}H^{2}$ | 1 | 2 | 2.0141022 (±1) | 2.22 | 1.11 |
| $^{1}H^{3}$ | 1 | 3 | 3.0160500 (±10) | 8.47 | 2.83 |
| $^{2}He^{3}$ | 2 | 3 | 3.0160299 (±2) | 7.72 | 2.57 |
| $^{2}He^{4}$ | 2 | 4 | 4.0026033 (±4) | 28.3 | 7.07 |
| $^{4}Be^{9}$ | 4 | 9 | 9.0121858 (±9) | 58.0 | 6.45 |
| $^{6}C^{12}$ | 6 | 12 | 12.0000000 (±0) | 92.2 | 7.68 |
| $^{8}O^{16}$ | 8 | 16 | 15.994915 (±1) | 127.5 | 7.97 |
| $^{29}Cu^{63}$ | 29 | 63 | 62.929594 (±6) | 552 | 8.75 |
| $^{50}Sn^{120}$ | 50 | 120 | 119.9021 (±1) | 1020 | 8.50 |
| $^{74}W^{184}$ | 74 | 184 | 183.9510 (±4) | 1476 | 8.02 |
| $^{92}U^{238}$ | 92 | 238 | 238.05076 (±8) | 1803 | 7.58 |

symbol gives the value of $Z$; the superscript after the symbol gives the value of $A$.) The mass of atom with this nucleus is quoted as

$$M_{^{20}\mathrm{Ca}^{40}} = 39.962589 \pm 0.000004u$$

The symbol $u$ represents one *mass unit*; it is defined in terms of the prevalent species of carbon in such a way that

$$M_{^{6}\mathrm{C}^{12}} \equiv 12.0000000u \tag{15-10}$$

A number of other examples of atomic masses are found in Table 15-1.

Using the first mass spectrometer, Thomson discovered the existence of isotopes in 1911. When the ion source contained a mixture of noble gases, he found an image on the photographic plate with mass corresponding to $A = 20$, and an associated weaker image corresponding to $A = 22$. A number of tests proved these were both due to a noble gas, and this could only be Ne, which has a chemical atomic weight of 20.18. He interpreted these results to mean that there are two chemically indistinguishable species of Ne atoms, called *isotopes*, one with $A = 20$ and relative abundance of about 91%, and one with $A = 22$ and relative abundance of about 9%. They are chemically indistinguishable since they have exactly the same structure of atomic electrons because their nuclei have the same charge and therefore the same $Z$, but they are physically distinguishable since they have different masses because their nuclei have different $A$. The nuclei of the Ne isotopes are: $^{10}Ne^{20}$, $^{10}Ne^{21}$, $^{10}Ne^{22}$; the second occurs with relative abundance of about 0.3%, and it could not be detected by Thompson's apparatus. All three of these nuclei contain 10 protons; however, the first contains 10 neutrons, the second contains 11 neutrons, and the third contains 12 neutrons.

Modern mass spectrometers, using detectors which are very sensitive and have a linear response, provide accurate determinations of the relative abundance of the various isotopes. As an example, the abundances of the normally occurring mixture of $^{8}O$ isotopes are

$$^{8}O^{16} = 99.759\%$$
$$^{8}O^{17} = 0.037\%$$
$$^{8}O^{18} = 0.204\%$$

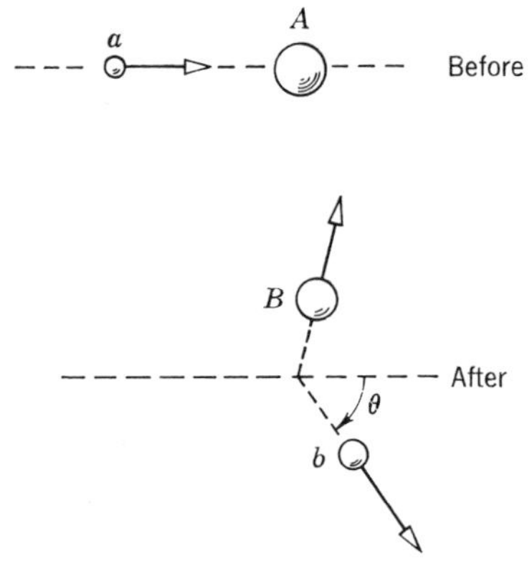

**FIGURE 15-8**
A nuclear reaction wherein a bombarding particle $a$ is incident on a target nucleus $A$. After the reaction takes place, the product particle $b$ is emitted at the angle $\theta$, and the residual nucleus $B$ recoils in such a way that momentum is conserved.

Another technique of accurate mass determination, which provides a supplement and check for the technique of mass spectrometry, is the study of energy balance in nuclear reactions. Consider the nuclear reaction

$$ {}^{2}\mathrm{He}^{4} + {}^{7}\mathrm{N}^{14} \rightarrow {}^{8}\mathrm{O}^{17} + {}^{1}\mathrm{H}^{1} \tag{15-11}$$

A *bombarding particle* ${}^{2}\mathrm{He}^{4}$ (an $\alpha$ particle) interacts with a *target nucleus* ${}^{7}\mathrm{N}^{14}$ to produce a *residual nucleus* ${}^{8}\mathrm{O}^{17}$ and a *product particle* ${}^{1}\mathrm{H}^{1}$ (a proton). This was the first artificially produced nuclear reaction, discovered in 1919 by Rutherford who used 7.7 MeV $\alpha$ particles from a radioactive source. Now $\alpha$ particles of a variety of energies obtained, perhaps, from an electrostatic generator would be used to investigate this typical reaction. As is discussed in Appendix A, mass and kinetic energy are not separately conserved in nuclear reactions. Instead, there is conservation of total relativistic energy, $E = K + mc^2$, where $K$ is kinetic energy and $m$ is used here for *rest* mass. For the general case, illustrated in Figure 15-8, a bombarding particle $a$ interacts with a target nucleus $A$ to produce a residual nucleus $B$ and a product particle $b$; that is

$$a + A \rightarrow B + b \tag{15-12}$$

In this case the conservation of total relativistic energy in the laboratory frame of reference reads

$$(K_a + m_a c^2) + m_A c^2 = (K_B + m_B c^2) + (K_b + m_b c^2) \tag{15-13}$$

Note that $K_A = 0$ since $A$ is stationary in the laboratory frame. Because there can be an exchange of energy between kinetic energy and rest mass energy, it is possible for the final kinetic energy $K_B + K_b$ to be greater, or less, than the initial kinetic energy $K_a$. The difference is called the *Q value of the reaction.* That is

$$Q = K_B + K_b - K_a \tag{15-14}$$

From (15-13), this can also be written

$$Q = (m_a + m_A - m_B - m_b)c^2 \tag{15-15}$$

We see that a measurement of the $Q$ value of a reaction gives information about the rest masses of the entities involved in the reaction. The $Q$ value can be measured by measuring $K_a$, $K_b$, and $K_B$. However, the latter quantity is usually difficult to measure. The difficulty can be avoided by using a relation that comes from the conservation of momentum to eliminate $K_B$ from (15-14). This is easy to do in the limit

$$K_a/m_a c^2 \ll 1 \qquad K_b/m_b c^2 \ll 1 \qquad K_B/m_B c^2 \ll 1$$

where the classical expressions such as $K_a = m_a v_a^2/2$ and $p_a = m_a v_a$ can be used. The result is that in this classical limit

$$Q = K_b\left(1 + \frac{m_b}{m_B}\right) - K_a\left(1 - \frac{m_a}{m_B}\right) - \frac{2}{m_B}(K_a K_b m_a m_b)^{1/2}\cos\theta \qquad (15\text{-}16)$$

where $\theta$ is the angle of emission of the product particle, defined in Figure 15-8. This result is of sufficient accuracy for the analysis of nuclear reactions at the energies which have been used in most experiments.

In (15-15), the masses refer to the rest masses of the nuclei $A$ and $B$, and to the rest masses of the completely ionized nuclear particles $a$ and $b$. However, to the accuracy of the approximation in which the mass equivalent of the electron binding energy is ignored, this equation can also be considered to read

$$Q = (M_a + M_A - M_B - M_b)c^2 \qquad (15\text{-}17)$$

where the large $M$ refer to the masses of the neutral atoms. The second form is obtained from the first by adding $(Z_a + Z_A)mc^2$ to the first two terms and subtracting $(Z_B + Z_b)mc^2$ from the last two, where $mc^2$ is the rest mass energy of an electron. This procedure is valid since the relation

$$Z_a + Z_A = Z_B + Z_b \qquad (15\text{-}18)$$

must be true in any nuclear reaction in order to have conservation of charge.

**Example 15-4.** In Rutherford's reaction, (15-11), bombarding $^2He^4$ particles ($\alpha$ particles) of kinetic energy $K_a = 7.70$ MeV interact with $^7N^{14}$ target nuclei to produce $^8O^{17}$ residual nuclei and $^1H^1$ product particles (protons). The protons emitted at 90° to the beam of bombarding $\alpha$ particles are found to have kinetic energy $K_b = 4.44$ MeV. (a) Determine the $Q$ value of the reaction. (b) Then use it to determine the atomic mass of $^8O^{17}$ in terms of the other three atomic masses involved in the reaction.

(a) Since the emission angle is $\theta = 90°$, (15-16) for the $Q$ value simplifies to

$$Q = K_b\left(1 + \frac{m_b}{m_B}\right) - K_a\left(1 - \frac{m_a}{m_B}\right)$$

With sufficient accuracy, we can take $m_b/m_B$, the ratio of the product particle and residual nucleus masses, as 1/17; we can also take $m_a/m_B$, the ratio of the bombarding particle and residual nucleus masses, as 4/17. So

$$\begin{aligned} Q &= K_b(1 + 1/17) - K_a(1 - 4/17) \\ &= 1.06K_b - 0.765K_a \\ &= 1.06 \times 4.44 \text{ MeV} - 0.765 \times 7.70 \text{ MeV} = -1.18 \text{ MeV} \end{aligned}$$

(b) The atomic masses involved in the reaction are related to the $Q$ value divided by $c^2$, which is

$$\frac{Q}{c^2} = -\frac{1.18 \text{ MeV}}{c^2}$$

To express this in mass units, we use the relation

$$uc^2 = 931.5 \text{ MeV}$$

which comes from evaluating the rest mass energy of a particle of rest mass $1u$. We obtain

$$\frac{Q}{c^2} = -\frac{1.18 \text{ MeV}}{c^2} \times \frac{uc^2}{931.5 \text{ MeV}} = -0.00127u$$

According to (15-17), the atomic mass of $^8O^{17}$ can be expressed in terms of the other atomic masses, and $Q/c^2$, as follows

$$M_{^8O^{17}} = M_{^2He^4} + M_{^7N^{14}} - M_{^1H^1} - \frac{Q}{c^2} = M_{^2He^4} + M_{^7N^{14}} - M_{^1H^1} + 0.00127u$$

Thus the atomic mass of $^8O^{17}$ can be determined from the measured $Q$ value, if the other atomic masses are accurately known. ◀

The analysis of energy balance in a large number of reactions has provided results which accurately check the results obtained by mass spectrometry. Furthermore, *the agreement between these two methods provides the most accurate confirmation of the relativistic theory of mass and energy*, upon which the energy balance is based. Table 15-1 lists a few of the many atomic masses that have been measured by these methods, as well as the mass of the neutron. Now let us begin to extract information about the nuclei from the precise measurements of their masses.

**Example 15-5.** Use the data of Table 15-1 to compare the mass of the $^2He^4$ atom with the mass of its constituent parts.

The mass of the $^2He^4$ atom is

$$M_{^2He^4} = 4.0026033u$$

The mass of its constituent parts is the mass of two $^1H^1$ atoms plus the mass of two neutrons; that is

$$2M_{^1H^1} + 2M_{^0n^1} = 2 \times 1.0078252u + 2 \times 1.0086654u$$

$$= 4.0329812u$$

Both $M_{^2He^4}$ and $2M_{^1H^1} + 2M_{^0n^1}$ contain two electron rest masses. But the former is *smaller* than the latter by the amount

$$\Delta M = 4.0329812u - 4.0026033u = 0.0303779u$$

We shall see immediately that this result is a manifestation of the binding energy of the $^2He^4$ nucleus. ◀

For any atom, a calculation as in Example 15-5 will show that its mass is *less* than the mass of its constituent parts by an amount $\Delta M$ called the *mass deficiency*. The origin lies in the nucleus, and in the equivalence between energy and mass. For instance, consider any one of the four nucleons in the $^2He^4$ nucleus. Since the nucleon is stably bound to the nucleus, it must be moving in some sort of an attractive potential representing the net attraction of the other three nucleons. Furthermore, to be bound it must have a negative energy $E < 0$. The situation is depicted in Figure 15-9. The energy required to remove the nucleon from the nucleus, leaving it a free nucleon of negligible kinetic energy at $r \rightarrow \infty$, is $|E|$. Conversely, if such a free nucleon comes in from $r \rightarrow \infty$ and combines with the other nucleons to form the nucleus, its energy must decrease by the amount $|E|$. The excess energy could be carried off by the emission of electromagnetic radiation. The same situation holds for the other nucleons in the nucleus. Thus we see that when a dispersed system of free nucleons combines to form a nucleus, the total energy of the system must decrease by an amount $\Delta E$, the *binding energy of the nucleus*. The decrease $\Delta E$ in the total energy of the system must, according to relativity theory, be accompanied by a decrease $\Delta M$ in its mass, where

$$\Delta Mc^2 = \Delta E \tag{15-19}$$

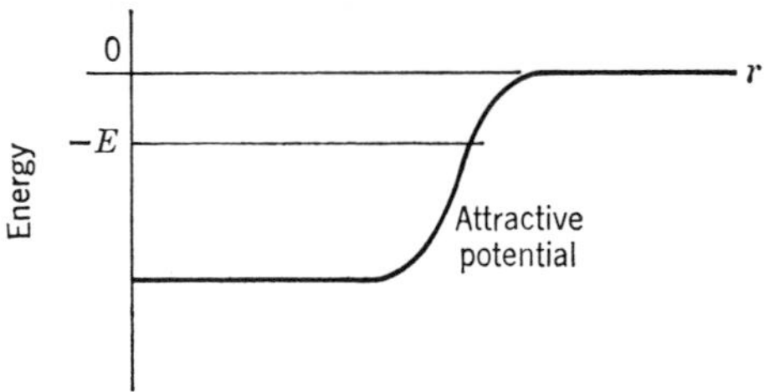


**FIGURE 15-9**

A schematic representation of the potential and total energies of a nucleon in a helium nucleus. The potential extends beyond the nuclear mass distribution by about the range of the nuclear force, and then it rapidly goes to zero.

For $^2He^4$, the mass deficiency is $\Delta M = 0.0303779u$. Therefore its binding energy is $\Delta E = \Delta Mc^2 = 28.3$ MeV, where we have used the convenient relation from Example 15-4

$$1u \times c^2 = 931.5 \text{ MeV} \tag{15-20}$$

This value of $\Delta E$ is listed in the next to last column of Table 15-1. The last column of the table lists $\Delta E/A$, called *the average binding energy per nucleon*, which is the binding energy of the nucleus divided by the number of nucleons it contains. For $^2He^4$, the value of $\Delta E/A$ is 28.3 MeV/4 = 7.07 MeV.

One of the most important features of a nucleus is its average binding energy per nucleon. The quantity is plotted as a function of $A$ in Figure 15-10. The points are the data obtained from the measured masses in the manner just described. Note that $\Delta E/A$ at first rises rapidly with increasing $A$, but very soon *$\Delta E/A$ is roughly constant* at a value

$$\Delta E/A \simeq 8 \text{ MeV} \tag{15-21}$$

If each nucleon in a nucleus exerted the same attraction on all the other nucleons, the binding energy per nucleon would continue to increase as more and more nucleons were added to the nucleus; that is, $\Delta E/A$ would be proportional to $A$. The extremely important fact that $\Delta E/A$ is not proportional to $A$ is due, in part, to the short range of nuclear forces. A complete explanation of the *saturation of nuclear forces*, which is responsible for the fact that $\Delta E/A$ has approximately the same value throughout

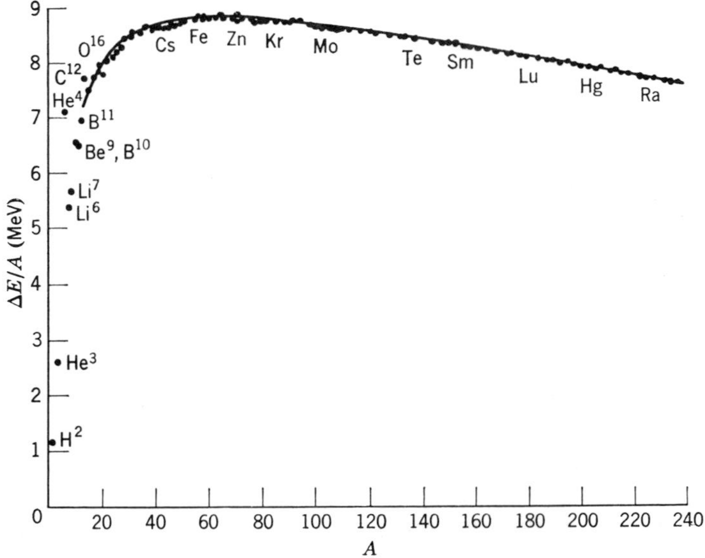


**FIGURE 15-10**

The average binding energy per nucleon for stable nuclei. The smooth curve is obtained from the semiempirical mass formula developed in Section 15-5.

most of the periodic table, will be given in Chapter 17. This saturation has a certain analogy to the saturation of molecular forces in covalent bonding, but the origins of the two saturation phenomena have no relation to each other, as we shall see in that chapter.

Inspection of Figure 15-10 shows that $\Delta E/A$ actually maximizes at about 8.7 MeV for $A \simeq 60$, and then decreases slowly to about 7.6 MeV for $A \simeq 240$. We shall find that the decrease is due to Coulomb repulsions between protons in the nucleus. One consequence is the phenomenon of *nuclear fission*, in which a large $A$ nucleus, such as $^{92}U^{238}$, splits into two intermediate $A$ nuclei because the two intermediate $A$ nuclei are more stable than the large $A$ nucleus.

**Example 15-6.** Use Figure 15-10 to estimate the difference between the binding energy of a $^{92}U^{238}$ nucleus and the sum of the binding energies of the two nuclei produced if it fissions symmetrically.

The figure shows that the average binding energy per nucleon for a nucleus of mass number around $A = 238$ is $\simeq$7.6 MeV. So the binding energy of the nucleus present before the fission is $\simeq 238 \times 7.6$ MeV $\simeq$ 1810 MeV. The figure also shows that the average binding energy per nucleon for a nucleus of mass number around $A = 238/2 = 119$ is $\simeq$8.5 MeV. So each of the two nuclei present after the symmetrical fission has a binding energy of $\simeq 119 \times 8.5$ MeV $\simeq$ 1010 MeV. The sum of their binding energies is $\simeq$2020 MeV. This sum is larger than the initial binding energy 1810 MeV by about 210 MeV. Thus the final state (after the nucleus fissions) is more stable than the initial state (before the nucleus fissions), because the total binding energy is higher in the final state. When the total binding energy increases by about 210MeV in the fission, energy in this amount is liberated. Most of it goes into the kinetic energy of the two nuclei produced in the fission. In a nuclear reactor this kinetic energy is degraded into thermal energy, which is the source of the power produced by the reactor. ◀

In *nuclear fusion* two or more nuclei of very small $A$ combine to form a larger nucleus which has a higher average binding energy per nucleon because its value of $A$

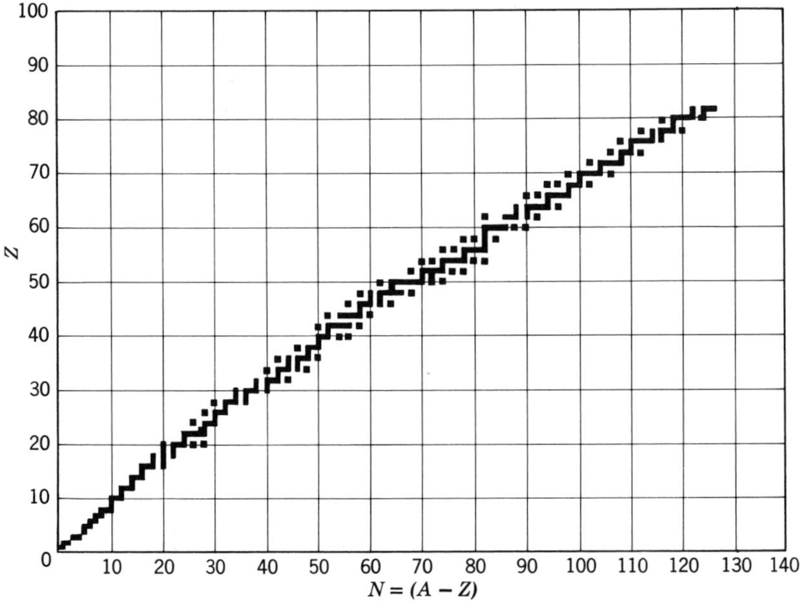


**FIGURE 15-11**
The distribution of stable nuclei.

**TABLE 15-2.** The Distribution of Stable Nuclei

| $A$ | $N$ | $Z$ | Number of Stable Nuclei |
|---|---|---|---|
| Even | Even | Even | 166 |
| | Odd | Odd | 8 |
| Odd | Even | Odd | 57 |
| | Odd | Even | 53 |

is nearer the value $A \simeq 60$ at which $\Delta E/A$ maximizes. It might seem that only a few nuclei near $A = 60$ would be stable. This is not true because there are other factors, to be discussed later, which inhibit fission and fusion.

We conclude this section by considering the distribution of Z and $A$ values of the stable nuclei, which is additional information obtained from the mass spectrometer measurements. The data are plotted in Figure 15-11. Each stable nucleus is indicated with a square whose abscissa is the *neutron number* $N = A - Z$, the number of neutrons in the nucleus, and whose ordinate is the atomic number $Z$, the number of protons in the nucleus. Note that for small $Z$ there is a tendency for stable nuclei to have $Z = N$. We shall see that this is due to the fact that nuclear forces operate symmetrically on neutrons and protons because nuclear forces are charge independent, as mentioned in Section 15-2. For large $Z$, stable nuclei tend to have $Z < N$. This is another effect of the Coulomb repulsions between protons, which produce a positive energy proportional to $Z^2$. The effect discriminates energetically against the presence of protons in nuclei of large $Z$, but it is not important in nuclei of small $Z$ where the $Z = N$ tendency dominates.

There is a tendency for stable nuclei to have even $Z$ and also even $N$. This can be seen from the data of Table 15-2, which lists the number of stable nuclei of various types. We shall find that this tendency is present because two nucleons of the *same* species can form a closely spaced pair in which they interact particularly strongly, and thereby make a particularly large contribution to the nuclear binding energy.

## 15-5 The Liquid Drop Model

We shall now employ the *liquid drop model of the nucleus*, and information obtained from the data concerning the distribution of $Z$ and $A$ values for stable nuclei, to obtain a formula for the masses of these nuclei. This formula will then be used in a variety of ways throughout our treatment of nuclei. The liquid drop model is based on two properties that we have found are common to all nuclei, except those of very small $A$, (1) their interior mass densities are approximately the same and (2) their total binding energies are approximately proportional to their masses since $\Delta E/A \simeq$ const. Both of these properties can be compared with analogous ones concerning macroscopic drops of some incompressible liquid. For such classical liquid drops of various sizes (1) their interior densities are the same and (2) their heats of vaporization are proportional to their masses. The second comparison is meaningful since the heat of vaporization is the energy required to disperse the drop into its constituent molecules, and so it is comparable to the binding energy of the nucleus. The mass formula will be developed by using the model to suggest other analogies between a nucleus and a classical liquid drop, but it will also be necessary to include terms in the formula that describe certain nuclear properties whose origins are nonclassical.

The liquid drop model approximates the nucleus as a sphere with a uniform interior density, that abruptly drops to zero at its surface. The radius is proportional to $A^{1/3}$; the surface area is proportional to $A^{2/3}$; and the volume is proportional to $A$. Since the mass is also proportional to $A$, which is the number of nucleons in the nucleus, this gives the result that density = mass/volume $\propto A/A$ = const, in agreement with the electron scattering measurements.

The mass formula consists of a sum of six terms

$$M_{Z,A} = f_0(Z,A) + f_1(Z,A) + f_2(Z,A) + f_3(Z,A) + f_4(Z,A) + f_5(Z,A) \tag{15-22}$$

where $M_{Z,A}$ represents the mass of an atom whose nucleus is specified by $Z$ and $A$. The first term is the *mass of the constituent parts* of the atom

$$f_0(Z,A) = 1.007825Z + 1.008665(A - Z) \tag{15-23}$$

The coefficient of $Z$ is the mass of the ${}^1\mathrm{H}^1$ atom in mass units, and the coefficient of $(A - Z)$ is the mass of the neutron, ${}^0n^1$, in the same units. The remaining terms correct for the mass equivalents of various effects contributing to the total nuclear binding energy.

Of most importance is the *volume* term

$$f_1(Z,A) = -a_1 A \tag{15-24}$$

This accounts for a binding energy proportional to the nuclear mass, or volume. The term describes the tendency to have the binding energy per nucleon a constant. Such a term would be present for a classical liquid drop. Because it is negative, it reduces the mass, and therefore increases the binding energy.

Next is the *surface* term

$$f_2(Z,A) = +a_2 A^{2/3} \tag{15-25}$$

It is a correction proportional to the surface area of the nucleus. Since the term is positive, it increases the mass and consequently reduces the binding energy. In a classical drop of liquid, this term would represent the effect of the surface tension energy. It would arise from the fact that a molecule at the surface of the drop feels attractive forces only from one side, so its binding energy is less than the binding energy of a molecule in the interior which feels attractive forces from all sides. Therefore, simply setting the total binding energy proportional to the volume of the drop overestimates the binding energy of the surface molecules, and a correction proportional to the number of such molecules, or to the surface area, must be made to reduce the binding energy. The same thing happens in a nucleus.

The *Coulomb* term is

$$f_3(Z,A) = +a_3 \frac{Z^2}{A^{1/3}} \tag{15-26}$$

It accounts for the positive Coulomb energy of the charged nucleus, which is assumed to have a uniform charge distribution of radius proportional to $A^{1/3}$. This effect of the Coulomb repulsions between the protons increases the mass and reduces the binding energy. A similar term would be present for a charged drop of a classical liquid.

The next term brings in a property specific to nuclei. It is the *asymmetry* term

$$f_4(Z,A) = +a_4 \frac{(Z - A/2)^2}{A} \tag{15-27}$$

which accounts for the observed tendency to have $Z = N$. Note that it is zero for $Z = N = (A - Z)$, or $2Z = A$, but is otherwise positive and increases with increasing departures from that condition. That is, the greater the departure from $Z = N$, the larger the mass or the smaller the binding energy. The form used in (15-27) is about the simplest one having these properties, but there is also some theoretical justification, involving the charge independence of nuclear forces, that will be indicated later.

The tendency of nuclei to have even $Z$ and even $N$ is accounted for by the *pairing* term

$$f_5(Z,A) = \begin{cases} -f(A) & \text{if } Z \text{ even, } A - Z = N \text{ even} \\ 0 & \text{if } Z \text{ even, } A - Z = N \text{ odd or } Z \text{ odd, } A - Z = N \text{ even} \\ +f(A) & \text{if } Z \text{ odd, } A - Z = N \text{ odd} \end{cases} \qquad (15\text{-}28)$$

It decreases the mass if both $Z$ and $N$ are even, and increases it if both $Z$ and $N$ are odd. Thus it maximizes the binding energy if both $Z$ and $N$ are even. A qualitative explanation of the origin of this term will be given later; it involves the quantum mechanical properties of indistinguishability of identical particles. But the exact form of the function $f(A)$ is usually determined by fitting the data. For a simple power law, the best fit is obtained with

$$f(A) = a_5 A^{-1/2} \qquad (15\text{-}29)$$

Gathering together (15-22) through (15-29), we have

$$M_{Z,A} = 1.007825Z + 1.008665(A - Z) - a_1 A + a_2 A^{2/3} + a_3 Z^2 A^{-1/3} + a_4 (Z - A/2)^2 A^{-1} + \begin{pmatrix} -1 \\ 0 \\ +1 \end{pmatrix} a_5 A^{-1/2} \quad (\text{in } u) \qquad (15\text{-}30)$$

This is called the *semiempirical mass formula* because the parameters $a_1$ through $a_5$ are obtained by empirically fitting the measured masses. A formula of this type was first developed by Weizsacker in 1935. Determinations of the parameters have since been made on several occasions. One set providing good results is

$$\begin{aligned} a_1 &= 0.01691 \\ a_2 &= 0.01911 \\ a_3 &= 0.000763 \\ a_4 &= 0.10175 \\ a_5 &= 0.012 \end{aligned} \quad (\text{in } u) \qquad (15\text{-}31)$$

Using these parameters, the formula yields excellent agreement with the average trend of the measured masses of all the stable nuclei except those of very small $A$. A comparison is shown in Figure 15-10, in which the smooth curve is $\Delta E/A$ evaluated from the sum of the volume, surface, Coulomb, and asymmetry terms. Figure 15-12 shows these terms individually. The semiempirical mass formula is of great practical utility because it is a simple formula that predicts with considerable accuracy the masses, and therefore the binding energies, of some 200 stable nuclei, and many more unstable nuclei. As we shall see in the following example, predictions of nuclear binding energies can lead immediately to predictions of other quantities of interest.

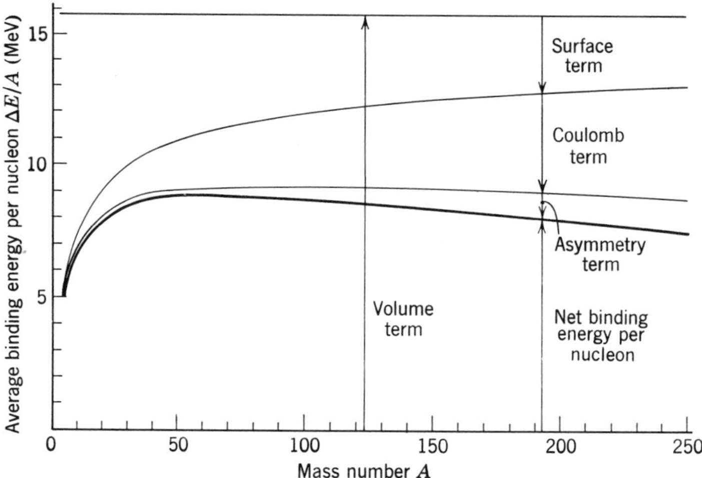


**FIGURE 15-12**

Illustrating how the volume, surface, Coulomb, and asymmetry terms of the semiempirical mass formula combine to yield the average binding energy per nucleon.

**Example 15-7.** Use the semiempirical mass formula to predict the binding energy made available if a $^{92}U^{235}$ nucleus captures a neutron. This is the energy which induces fission of the $^{92}U^{236}$ nucleus that is formed in the capture.

The binding energy is

$$E_n = \{[M_{92,235} + M_{0,1}] - [M_{92,236}]\}c^2$$

The term in the first square bracket is the mass of a $^{92}U^{235}$ atom plus the mass of a neutron, which are the constituents of the $^{92}U^{236}$ atom whose mass appears in the second square bracket. Since the neutron mass, $M_{0,1}$, is precisely $1.008665u$, the first two terms from the semiempirical mass formula, (15-30), cancel out in the expression for $E_n$. Then we obtain

$$E_n = \left\{\left[-a_1(235) + a_2(235)^{2/3} + a_3\frac{(92)^2}{(235)^{1/3}} + a_4\frac{(92 - 235/2)^2}{235}\right]\right.$$

$$\left. - \left[-a_1(236) + a_2(236)^{2/3} + a_3\frac{(92)^2}{(236)^{1/3}} + a_4\frac{(92 - 236/2)^2}{236} - \frac{a_5}{(236)^{1/2}}\right]\right\}c^2$$

$$= \left\{a_1 - a_2[(236)^{2/3} - (235)^{2/3}] + a_3(92)^2\left[\frac{1}{(235)^{1/3}} - \frac{1}{(236)^{1/3}}\right]\right.$$

$$\left. -a_4\left[\frac{(26.0)^2}{236} - \frac{(25.5)^2}{235}\right] + \frac{a_5}{(236)^{1/2}}\right\}c^2$$

$$\simeq \{0.0169 - 0.0191 \times 0.11 + 0.00076 \times 1.9 - 0.1018 \times 0.097 + 0.012 \times 0.065\}c^2$$

$$= \{0.0169 - 0.0021 + 0.0014 - 0.0099 + 0.0008\}c^2$$

$$= \{0.0071u\}c^2 = 6.6 \text{ MeV}$$

where we have used (15-20) to convert to MeV.

If the neutron has negligible kinetic energy before it is captured, the $^{92}U^{236}$ nucleus is formed in a state of excitation energy equal to $E_n$. As we shall discuss at length in the next chapter, the excitation energy often sets the nucleus into a vibration in which it oscillates between being elongated (having a positive quadrupole moment) and being flattened (having a negative quadrupole moment). This vibration cannot take place without the excitation energy since the surface term of the semiempirical mass formula inhibits departures of the nucleus from the approximately spherical shape it has in its ground state. When the nucleus has a maximum elongation, the effect of the Coulomb term can cause it to fission.

Of great importance in nuclear reactor technology is the fact that $E_n$ for neutron capture by a $^{92}U^{238}$ nucleus is about 1.5 MeV smaller than the value just calculated for capture by $^{92}U^{235}$. The terms in the preceding expressions have almost the same values, except that the contribution of the pairing term (the last term) is negative instead of positive. Since all $^{92}U$ nuclei require an excitation of about 6 MeV to overcome the surface term inhibition, $^{92}U^{238}$ will fission only if the neutron it captures brings in more than about 1 MeV of kinetic energy, in addition to its binding energy. We shall see that this means $^{92}U^{238}$ is not very useful in the "chain reaction" that takes place in reactors. ◀

The liquid drop model is the oldest, and most classical, nuclear *model*. At the time the semiempirical mass formula was first developed, mass data was available, but not much else was known about nuclei. The parameters were purely empirical, and there was not even a qualitative understanding of the asymmetry and pairing terms. Nevertheless, the formula was significant because it described fairly accurately the masses of hundreds of nuclei in terms of only five parameters. At present we do have an insight into the origin of the two terms mentioned. And the most important parameter, the $a_1$ of the volume term, is no longer purely empirical. Nuclear *theory* has been developed to the point that it predicts the value of $a_1$, reasonably well, in terms of the detailed properties of nuclear forces. The nuclear theory, which is largely the work of Brueckner, is very similar to the Hartree theory of the atom in the sense that it involves self-consistent calculations for a system of fermions, but the calculations are even more complicated because of the complicated nature of nuclear forces. We shall make no attempt to describe them.

## 15-6 Magic Numbers

The liquid drop model gives a good account of the average behavior of nuclei in regard to mass, or binding energy. Since binding energy is a direct measure of stability—the higher the binding energy of a nucleus the more stable it is—the liquid drop model describes well the average behavior of nuclei in regard to their stability. However, nuclei with certain values of $Z$ and/or $N$ show significant departures from this average behavior by being unusually stable. These values of $Z$ and/or $N$ are the *magic numbers*

$$Z \text{ and/or } N = 2, 8, 20, 28, 50, 82, 126 \tag{15-32}$$

The situation is analogous to the unusual stability of the electron shells of noble gas atoms containing $Z = 2$, 10, 18, 36, 54, 86 electrons. But in the nuclear case the indications are not as pronounced as in the atomic case, and it is necessary to consider several of them to demonstrate the "magic" character of the numbers quoted in (15-32). The two most convincing are:

1. Nuclei prefer having magic $Z$ and/or $N$. This can be seen by inspecting Figure 15-11. To quote just two examples, there are six stable isotopes for $Z = 20$, whereas the average number of stable isotopes in that region is about two. For $Z = 50$ there are ten stable isotopes, whereas the average number in that region of the periodic table is about four. All plausible explanations of how nuclei were originally formed relate abundance to stability; i.e., the more stable a particular type of nucleus is the more abundant it is.

2. Figure 15-10 shows that the average binding energy per nucleon is significantly higher for nuclei that have $Z$ and/or $N$ equal to 2 or 8 than it is for neighboring

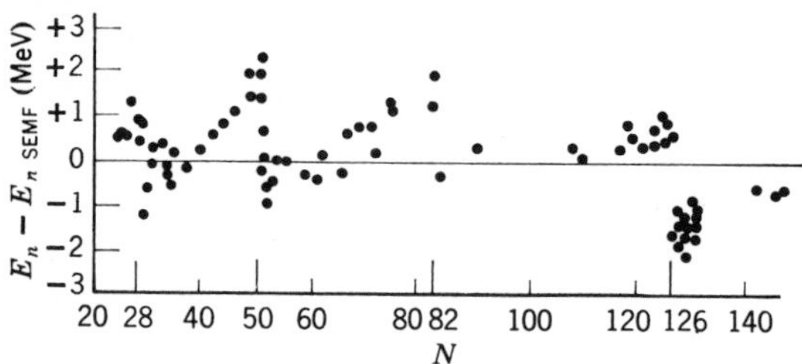


**FIGURE 15-13**

The difference between the binding energy of the last neutron and the prediction of the semiempirical mass formula, as a function of the number of neutrons in the nucleus. These data provide clear evidence for the magic numbers 28, 50, 82, and 126, for neutrons. Similar evidence shows that 20, 28, 50, and 82 are also magic numbers for protons. But there is no concrete evidence, pro or con, concerning 126 for protons since nuclei with such large $Z$ values have not yet been detected.

nuclei. The outstanding example is ${}^2\mathrm{He}^4$, for which $Z = N = 2$. The effect is even more pronounced if a measure of stability more sensitive than $\Delta E/A$ is considered. This is $E_n$, or $E_p$, the minimum energy required to separate a neutron, or proton, from the nucleus; it is usually called the *binding energy of the "last" neutron, or proton.* As an example, for ${}^2\mathrm{He}^4$ the value of $E_n$ is 20.6 MeV (i.e., this much energy is required to produce the reaction ${}^2\mathrm{He}^4 \rightarrow {}^2\mathrm{He}^3 + {}^0n^1$). The value of $E_p$ for ${}^2\mathrm{He}^4$ is 19.8 MeV. These are abnormally high. Figure 15-13 is a plot of the difference between the value of $E_n$ measured for a number of nuclei, and the value predicted by the semiempirical mass formula. Except for the effect of the pairing term, the predicted value is a smooth function that decreases slowly from around 8 MeV for intermediate values of $N$ to around 6 MeV for large values of $N$ (as we saw in Example 15-7 where we predicted $E_n$ for ${}^{92}\mathrm{U}^{236}$). The unusual stability of nuclei with $N = 28, 50, 82, 126$ is shown by the exceptionally large energy required to remove their last neutron.

There are a number of other somewhat less convincing pieces of evidence for the magic numbers, such as the fact that for most of the known spontaneous neutron emitters, like ${}^8\mathrm{O}^{17}$, ${}^{36}\mathrm{Kr}^{87}$, and ${}^{54}\mathrm{Xe}^{137}$, $N$ equals a magic number plus one. This implies an unusually small affinity for the extra neutron.

The analogy between nuclear and atomic magic numbers prompted many people to look for an explanation of the nuclear phenomenon that was similar to the explanation of the atomic phenomenon. The student will recall that the key point in that explanation is the formation of closed shells by the electrons moving *independently* in the atomic potential. However, when the nuclear magic numbers were first being discussed seriously, around 1948, it seemed very difficult to understand how nucleons could move independently in a nucleus. The reason was that the liquid drop model had been dominant for a number of years, and it seemed basic to this model that a nucleon in a nucleus (of density $\sim 10^{18}$ kg/m$^3$!) would constantly interact with its neighbors through the strong nuclear force. If so, the nucleon would be repeatedly scattered in traveling through the nucleus, and it would follow an erratic path, resembling Brownian motion much more than the motion of an electron moving independently through its orbit in an atom.

## 15-7 The Fermi Gas Model

Weisskopf first pointed out that there is a simple explanation of how nucleons can move independently through a nucleus in its ground state. The explanation is based

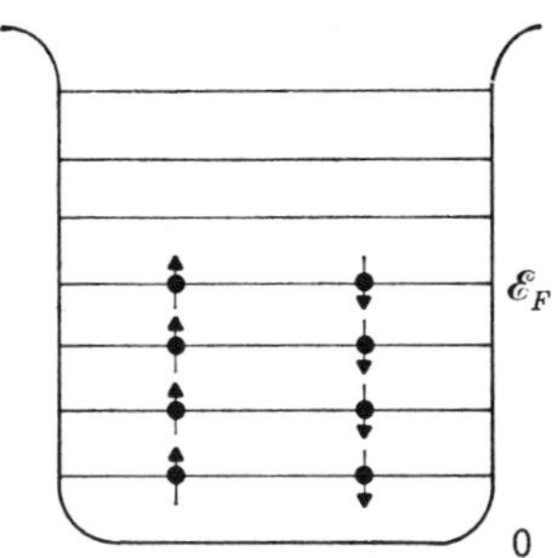


**FIGURE 15-14**

A schematic representation of the energy levels filled by the neutrons in the ground state of a nucleus. The lowest levels are filled, according to the limitations of the exclusion principle, up to the Fermi energy $\mathscr{E}_F$.

on the *Fermi gas model of the nucleus*. This model is essentially the same as the free-electron gas model of the conduction electrons in a metal, considered in Section 11-11. It assumes that each nucleon of the nucleus moves in an attractive *net potential*, that represents the average effect of its interactions with other nucleons in the nucleus. The net potential has a constant depth inside the nucleus since the distribution of nucleons is constant in this region; outside the nucleus it goes to zero within a distance equal to the range of nuclear forces. Thus the net potential is approximately like a three-dimensional finite square well of radius a little larger than the nuclear radius, and of depth that will be determined in Example 15-8. In the ground state of the nucleus, its nucleons, which are all fermions of intrinsic spin $s = 1/2$, occupy the energy levels of the net potential in such a way as to minimize the total energy without violating the exclusion principle.

Figure 15-14 indicates the quantum states filled by the neutrons in the ground state of a nucleus. Since protons are distinguishable from neutrons, the exclusion principle operates independently on the two types of nucleons, and we must imagine a separate and independent diagram representing the quantum states filled by the protons. It is immediately apparent from these diagrams why the *exclusion principle* prevents almost all the nucleons from scattering from each other when the nucleus is in its ground state. The point is that almost all the states which are energetically accessible are already completely filled, and so there can be essentially no collisions except those in which two nucleons of the same type exchange quantum states. The net effect of such an exchange of two indistinguishable particles is, however, the same as if there had been no collision at all. Of course, if there is a set of partly filled degenerate states at the Fermi energy, the few nucleons in *these* states can collide with each other, but only a small fraction of the total number of nucleons can be in such states. Thus we see why *almost all of the nucleons that compose a nucleus can move freely within the nucleus if it is in its ground state.*

**Example 15-8.** Evaluate the Fermi energy of a typical nucleus, and use the results to determine the depth of the net nuclear potential.

The Fermi energy, $\mathscr{E}_F$, is the energy indicated in Figure 15-14 of the nucleon in the highest filled level of the system, measured from the bottom of the potential well. It is related to the nucleon mass $M$, and nucleon density $\rho$, by (11-57), which we write here as

$$\mathscr{E}_F = \frac{\pi^2\hbar^2}{2M}\left(\frac{3}{\pi}\,\rho\right)^{2/3} \tag{15-33}$$

(This expression can be obtained directly from the equation for the energies of the levels of a three-dimensional square well simply by filling its lowest levels up to the Fermi energy.)

Let us consider the Fermi gas of neutrons in a uniform spherical nucleus of radius

$$r' = aA^{1/3}$$

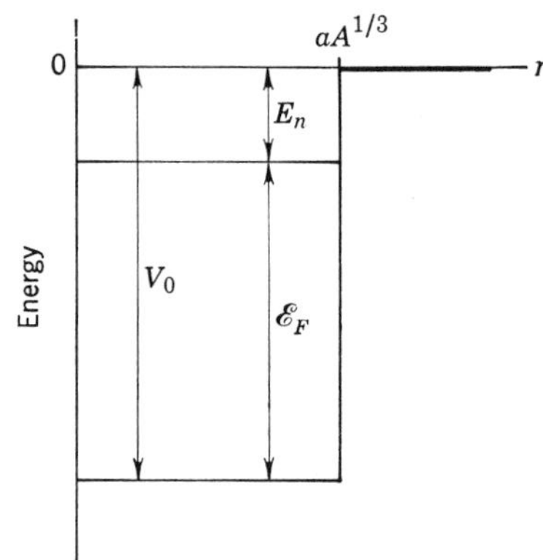


**FIGURE 15-15**
Illustrating the relation between the depth $V_0$ of a nuclear square well potential of radius $r' = aA^{1/3}$, the Fermi energy $\mathscr{E}_F$, and the binding energy $E_n$ of the last neutron.

For a typical nucleus, the number of neutrons is

$$N \simeq 0.60A$$

Thus

$$\rho \simeq \frac{N}{\frac{4}{3}\pi a^3 A}$$

gives

$$\rho \simeq \frac{0.60A}{1.33\pi a^3 A} = \frac{0.45}{\pi a^3}$$

and the Fermi energy is

$$\mathscr{E}_F \simeq \frac{\pi^2 \hbar^2 (0.26)}{2Ma^2} \tag{15-34}$$

Using a radius constant $a \simeq 1.1\text{F}$ consistent with the electron scattering measurements as summarized by (15-6), and evaluating the other parameters, we obtain

$$\mathscr{E}_F \simeq 43 \text{ MeV}$$

The relations between the depth of the potential $V_0$, the Fermi energy $\mathscr{E}_F$, and the binding energy of the last neutron $E_n$, are shown in Figure 15-15. As mentioned in the previous section, $E_n$ is approximately equal to 7 MeV for a typical nucleus. Thus for this nucleus the depth of the net nuclear potential acting on its neutrons is

$$V_0 = \mathscr{E}_F + E_n \simeq 43 \text{ MeV} + 7 \text{ MeV} = 50 \text{ MeV}$$

A very similar result is obtained for the net nuclear potential for protons. (Of course protons also feel a net Coulomb potential exerted by the charges of other protons in the nucleus.) ◀

There is evidence from a number of studies of the behavior of nucleons of various energies that the depth of the net nuclear potential, $V_0$, is not a constant, but instead it decreases slowly, and approximately linearly, as the energy of the nucleon increases. This causes no difficulty because its effect on the dynamics of nucleon motion in the net potential can be completely described by introducing an effective nucleon mass, in much the same way as we did in Section 13-7 when treating the independent particle motion of a conduction electron in the net potential for a crystal lattice. That is, it is possible to continue treating $V_0$ as a constant with the value we have obtained in Example 15-8, if the actual nucleon mass $M$ is replaced by an *effective nucleon mass* $M^*$. Furthermore, because the actual change in $V_0$ is slow, $M^*$ is not very different from $M$, and so for most considerations involving nucleons of not too high energy it is permissible to take $M^* = M$, i.e., to completely ignore the fact that $V_0$ is not quite a constant.

There is also a dependence of the depth of the net nuclear potential $V_0$ seen by a proton, or by a neutron, on the difference between the number $Z$ of protons and number $N$ of neutrons that the nucleus contains. This is described by adding to $V_0$ a term $\Delta V_0 \propto \pm(N - Z)/A$, with the plus sign used for the potential seen by a proton and the minus sign used for the potential seen by a neutron. The dependence is a result of the exclusion principle, which restricts the interactions between two protons, or two neutrons, to certain quantum states, but puts no restrictions on the interactions between a proton and a neutron. Consequently, the attractive interaction between two nucleons in a nucleus is stronger between a proton and a neutron than between two protons or between two neutrons. Thus the net nuclear potential acting on a proton is deeper than that acting on a neutron if the nucleus contains more neutrons than protons in proportion to the fractional neutron excess, and vice versa if there is a proton excess. This dependence plays an important role in the effect described by the asymmetry term of the semiempirical mass formula, as we shall indicate. In most other considerations it is not so important and can be ignored.

The tendency for nuclei to have $Z = N$ also has a simple explanation in the Fermi gas model. Consider a nucleus of very small $Z$, for which the Coulomb force acting between protons can be ignored in comparison to the stronger nuclear force. In this nucleus there are two independent Fermi gases, the neutrons and the protons. Both move in net nuclear potentials which, in this approximation, are the same—basically because the nuclear force acting between neutrons is the same as the nuclear force acting between protons since the nuclear force is charge independent. As is indicated in Figure 15-16, the energy levels of the two systems must then also be the same in this approximation. For a given value of $A$, the total energy of the nucleus is obviously minimized if the levels are filled with $Z = N$, because nucleons would occupy levels of energy higher than necessary if this condition were violated. A quantitative treatment of the argument leads to the mathematical expression, (15-27), used in the asymmetry term of the semiempirical mass formula. The reason why the factor $1/A$ appears in the term is that the levels of a three-dimensional potential well are more closely spaced the larger the value of $A$. So with increasing $A$ there is a scaling down of the energy penalty, associated with violating the $N = Z$ condition, that is described by the factor $(Z - A/2)^2$.

The effect of the term $\Delta V_0 \propto \pm(N - Z)/A$ in the depth of the net nuclear potential, explained previously, also contributes significantly to the presence of the asymmetry term in the semiempirical mass formula, and its consequences. Consider a typical nucleus containing $N$ neutrons and $Z$ protons, with $N > Z$. The contribution of the $\Delta V_0$ term to the total binding energy from the $Z$ protons is canceled by its contribution from the first $Z$ neutrons. But there is an uncanceled contribution from the remaining $(N - Z)$ neutrons which decreases the total binding energy, or increases the nuclear mass, in proportion to $(N - Z)^2/A \propto (Z - A/2)^2/A$.

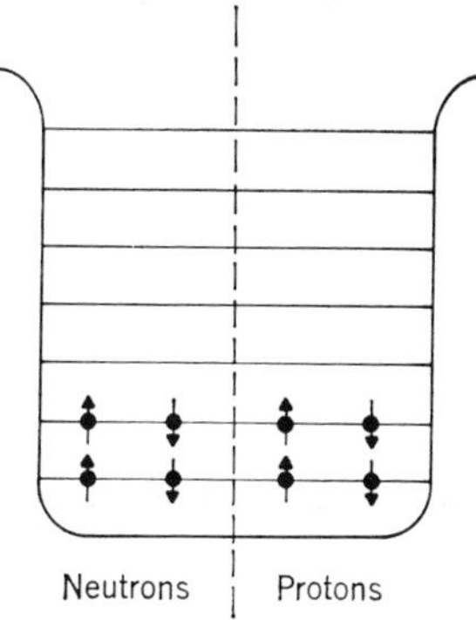


**FIGURE 15-16**
A schematic representation of independent Fermi gases of neutrons and protons in the minimum energy state of a nucleus of very small $Z$, which is indicated by a square well with rounded edges.

## 15-8 The Shell Model

The Fermi gas model establishes the validity of treating the motion of the bound nucleons in a nucleus in terms of the independent motion of each nucleon in a net nuclear potential. The next step is obviously to solve the Schroedinger equation for that potential, and to obtain a detailed description of the behavior of the nucleons. This procedure is employed in the *shell model of the nucleus*. The shell model plays a role in nuclear physics comparable to that played by the Hartree theory in atomic physics. But the shell model is cruder since the exact form of the net atomic potential is internally determined by the self-consistent atomic theory, while the exact form of the net nuclear potential must be inserted into the nuclear model. Of course, some general information about the net nuclear potential is available from the Fermi gas model.

The procedure of the shell model involves first finding the neutron and proton energy levels for an assumed form of the net potential of a particular nucleus. That is, if each nucleon is treated as moving independently in a net nuclear potential $V(r)$, the nucleon has allowed energy levels which are determined by the form of $V(r)$, and which are found by solving the Schroedinger equation for that potential. The only forms for the net potential considered are spherically symmetrical functions, $V(r)$, where $r$ is the distance from a nucleon to the center of the nucleus; other forms would greatly increase the difficulty of solving the Schroedinger equation. Just as in the Hartree theory of atoms, it is found that the energy of a nucleon energy level of the net nuclear potential $V(r)$ depends on quantum numbers $n$ and $l$, which specify the radial and angular behavior of a nucleon in the level. The quantum number $l$ is just the same as the one we encounter throughout atomic physics when dealing with any spherically symmetrical potential like $V(r)$. The quantum number $n$ used in nuclear physics is related to, but *not the same* as, the quantum number of atomic physics that is symbolized by the same letter. Because of the approximate square well form of the net potential $V(r)$ which arises in nuclear physics, it is more convenient in that field to use what is called the *radial node quantum number n*.

Figure 15-17 contains schematic illustrations of some of the energy levels, and associated eigenfunctions, of the bound states of a three-dimensional square well $V(r)$. On the left, the $n$ dependence of the energies of the levels is indicated for a well which is wide and deep enough to bind a $1s$, $2s$, and $3s$ state. The radial behaviors of the corresponding eigenfunctions $\psi(r,\theta,\varphi) = R(r)\Theta(\theta)\Phi(\varphi)$ are indicated by plotting for each the quantity $rR(r)$, whose square is the radial probability density, using the appropriate energy level as an $r$ axis. The notation $1s$ means $n = 1$ and $l = 0$, as usual. Note that for fixed $l$, the energy increases with increasing $n$. The reason is that $rR(r)$ for $n = 1$ contains essentially one-half of an oscillation within the well region, $rR(r)$ for $n = 2$ contains two half oscillations, and $rR(r)$ for $n = 3$ contains three half oscillations. So the eigenfunctions $\psi$ for higher $n$ necessarily have higher curvature, and higher curvature requires higher kinetic or total energy. Note also that the number of nodes within the well of the radial dependence of $r$ times each eigenfunction is just equal to $n$, as its name implies.

There are bound states in the well of Figure 15-17 for values of $l$ other than $l = 0$. On the right side of that figure the $l$ dependence, for fixed $n$, of the energies of the levels, and $r$ times the radial behavior of the corresponding eigenfunctions, are indicated by showing them for the $1s$, $1p$, and $1d$ states. Since all of these have $n = 1$, all the $rR(r)$ have only one radial node. Nevertheless, the radial behavior of the eigenfunction

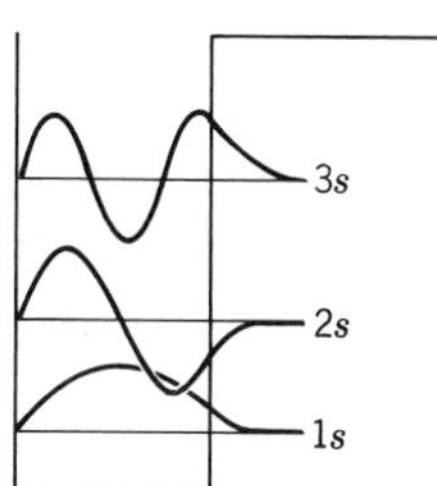


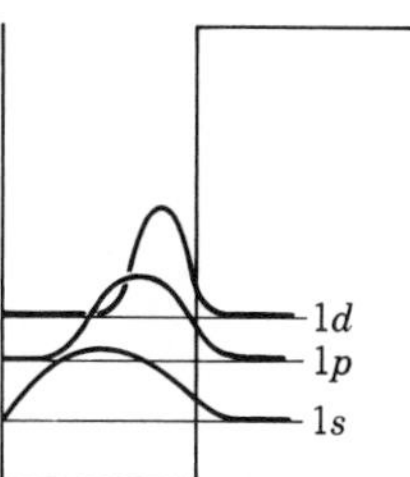


**FIGURE 15-17**

*Left:* Illustrating qualitatively the product $rR$ of the radial coordinate $r$ and the radial dependence $R$ of the eigenfunction $\psi$ for states, of the indicated three-dimensional square well, with $l = 0$ and $n = 1, 2, 3$. Each is constructed by using its energy level as an $r$ axis. Since the radial probability density is $P = r^2R^*R = (rR)^2$, if the student visualizes the squares of the functions depicted he can make comparisons with the radial probability densities for states of a one-electron atom Coulomb potential, or a multielectron atom Hartree net potential, by looking also at Figures 7-5 or 9-10. In so doing, he should keep in mind that the quantum number $n$ is used differently in atomic physics. The fact that the radial node quantum number $n$ of nuclear physics just specifies the number of nodes of $rR$ within the well is made apparent by this figure. *Right:* The same for states with $n = 1$ and $l = 0, 1, 2$. The way that what might be called a centrifugal effect tends to prevent a nucleon from approaching $r = 0$ as the orbital angular momentum quantum number $l$ becomes larger than 0 is seen in this figure.

$\psi$ changes with changing $l$ because of the property expressed by (7-32)

$$\psi \propto R(r) \propto r^l \qquad r \rightarrow 0$$

and discussed at length in Chapters 7 and 9. This is the familiar tendency of a particle in states of any spherically symmetrical potential, for which orbital angular momentum is constant so that $l$ is a good quantum number, to avoid the origin more and more as $l$ gets larger. Thus, with increasing $l$ the one-half of an oscillation in the various $rR(r)$ for $n = 1$ is contained within a smaller and smaller region of the $r$ axis. So the eigenfunctions $\psi$ have higher curvature, and the corresponding energy levels are found higher in the well.

The results concerning three-dimensional square wells that are of most consequence are that the energies of bound levels increase with increasing $n$, for a given $l$, and that they also increase with increasing $l$, for a given $n$. The student should further observe that when using the radial node quantum number $n$ of nuclear physics there is no restriction on the largest possible value of $l$ for a given $n$.

There is such a restriction in atomic physics because the quantum number $n$ used there, called the principal quantum number, is just equal to the sum of the radial node quantum number and the orbital angular momentum quantum number. That is

$$n_{\text{principal}} = n_{\text{radial}} + l$$

Since the minimum value of $n_{\text{radial}}$ is 1, the largest possible value of $l$ for a given $n_{\text{principal}}$ is $(n_{\text{principal}} - 1)$. The reason why $n_{\text{principal}}$ is used in atomic physics is that when $V(r)$ is an attractive Coulomb potential, $V(r) \propto -1/r$, the way the energy of a level increases with increasing $n_{\text{radial}}$ happens to be precisely the same as the way it increases with increasing $l$. Thus the energy of the levels of a Coulomb potential does not depend on both $n_{\text{radial}}$ and $l$, but only on their sum $n_{\text{principal}}$. This gives yet another insight into the origin of the degeneracy of the energy levels of the hydrogen atom.

Additional insight into the properties of the quantity $rR$ can be obtained by considering the radial part of the time-independent Schroedinger equation for a spherically symmetrical potential $V(r)$, which is (7-17). Inspection will show that we can immediately put it in the form

$$-\frac{\hbar^2}{2\mu}\frac{d^2(rR)}{dr^2} + \left[\frac{l(l+1)\hbar^2}{2\mu r^2} + V(r)\right](rR) = E(rR)$$

This is seen to be equivalent to the Schroedinger equation in the function $rR$ for motion in one dimension, $r$, except that the term $l(l+1)\hbar^2/2\mu r^2 = L^2/2\mu r^2$ is added to the potential $V(r)$. This term is often called the *centrifugal potential*, for reasons which can be seen by considering the energy conservation equation for a classical particle of mass $\mu$ moving under the influence of a potential $V(r)$. As the particle will move in a plane containing the origin, it can be described by the coordinates $r$, $\theta$, and the equation is

$$E = \frac{1}{2}\mu\left(\frac{dr}{dt}\right)^2 + \frac{1}{2}\mu\left(\frac{r\,d\theta}{dt}\right)^2 + V(r)$$

Also the orbital angular momentum of the particle is a constant

$$L = \mu r^2\frac{d\theta}{dt}$$

so the energy equation can be written

$$E = \frac{1}{2}\mu\left(\frac{dr}{dt}\right)^2 + \left[\frac{L^2}{2\mu r^2} + V(r)\right]$$

This is seen to be the energy conservation equation for classical motion in one dimension, if $r$ is the one-dimensional coordinate, with the term $L^2/2\mu r^2$ added to the potential $V(r)$. This positive term acts like a repulsive potential, tending to keep the particle away from the origin. The higher the value of $L$, the stronger is the effect, in agreement with our usual conclusion. Note also that for $l = 0$ the differential equation for $rR$ is mathematically identical to the one-dimensional time-independent Schroedinger equation for $\psi$. This is why the plots of $rR$ in Figure 15-17 for $1s$, $2s$, and $3s$ states look so much like the plots of $\psi$ for a one-dimensional square well potential in that they are both sinusoidal within the well and decreasing exponential outside. They are not identical, however, because $rR$ necessarily has the value zero in all states at the point $r = 0$.

Having found the nucleon energy levels in the assumed square-well-like form of the net nuclear potential $V(r)$, the next step of the shell model is to "construct" the nucleus by filling them, in order of increasing energy, with the $N$ neutrons and $Z$ protons that the nucleus contains. The exclusion principle limits the occupancy of each level to $2(2l + 1)$ neutrons, or protons. This occupancy corresponds to the 2 possible values of the quantum number $m_s$, which specifies the orientation of the intrinsic spin angular momentum of a nucleon, and the $(2l + 1)$ possible values of the quantum number $m_l$, which specifies the orientation of the orbital angular momentum of the nucleon. These two $z$ component angular momentum quantum numbers are the same as in the Hartree theory of atoms. And the procedure for constructing a nucleus by filling its nucleon energy levels is just the same as that used in the Hartree theory to construct an atom by filling its electron energy levels, except that in a nucleus there are particles of two distinguishable species—the neutrons and the protons—to which the exclusion principle applies independently. Originally, it was hoped that a particular form for the potentials $V(r)$ of the various nuclei could be found in which the ordering and spacing of the nucleon energy levels would be such that an unusually tightly bound level, containing an appropriate number of neutrons or protons, would completely fill in those nuclei having values of $N$ or $Z$ equal to the magic numbers—just as the filling of unusually tightly bound electron energy levels leads to the noble gas atoms for $Z$ equal to the atomic magic numbers. Many different detailed forms for the radial dependence of the nuclear potential were tried (including one aptly called the "wine bottle potential," a square well with a bump centered in the bottom, like the profile of a wine bottle bottom, which suppresses somewhat the $l$ dependence of the energy). It was found that there is *no* form for $V(r)$ which leads even to the ordering of the nucleon energy levels required to explain the magic numbers.

The mystery of the magic numbers was solved in 1949 by Mayer, and independently by Jensen, who introduced the idea of a nuclear spin-orbit interaction. They proposed that each nucleon in a nucleus feels, in addition to the net nuclear potential, a *strong inverted spin-orbit interaction* proportional to $\mathbf{S} \cdot \mathbf{L}$, the dot product of its spin and orbital angular momentum vectors. Strong means that the interaction energy is much (about 20 times) larger than would be predicted by using the atomic spin-orbit formula, (8-35), equating $V(r)$ to the net nuclear potential and $m$ to the nucleon mass. Inverted means that the energy of the nucleon is decreased when $\mathbf{S} \cdot \mathbf{L}$ is positive, and increased when it is negative. Thus the sign of the interaction is opposite to the sign of the magnetic spin-orbit interaction experienced by an electron in an atom; that is, the interaction energy is *negative* when the total angular momentum of the nucleon $\mathbf{J} = \mathbf{S} + \mathbf{L}$ has its *maximum* possible magnitude (i.e., when $\mathbf{S}$ and $\mathbf{L}$ are as parallel as possible, and $\mathbf{S} \cdot \mathbf{L}$ is positive). However, as the magnitude of the spin-orbit interaction is proportional to $\mathbf{S} \cdot \mathbf{L}$ just as it is for an atomic electron, the magnitude of the spin-orbit splitting of the nucleon energy levels will be approximately proportional to the value of the quantum number $l$, just as it is for the electron energy levels. Although there are similarities between the atomic and nuclear spin-orbit interactions, their differences make it clear that the latter is *not magnetic* in origin. Instead, it is an attribute of the nuclear force whose origin is not yet completely understood.

The left-hand part of Figure 15-18 shows the ordering and approximate spacing of the energy levels which nucleons are filling in nuclei with potentials $V(r)$ in the form of square wells with rounded edges, like the potential shown in Figure 15-16. As the levels are filled, in proceeding up the periodic table, the depth of the potentials is held constant while their radii increase in proportion to the cube root of the number of nucleons they contain in the filled levels. The same general features seen in the left part of Figure 15-18 are found in all spherically symmetrical potentials that have a form bearing any resemblance to an attractive square well. Of course, the details of the ordering and spacing of the nucleon energy levels depend on the details of the competition between the $n$ dependence and the $l$ dependence of the energy, and this depends on the details of the radial behavior of the nuclear potential; but any reasonable nuclear potential gives essentially the same ordering of the levels according to $n$ and $l$ as that for square wells with rounded edges, and it also gives gaps between the levels in essentially the same places. Since, as we saw in Example 15-8, the net nuclear potential is related to the nuclear mass density, square wells with rounded edges are most certainly the correct forms for the potential as they reflect the constant interior values, and fairly gradual changes at the nuclear surface, of the mass densities. But as we have already said, and will see specifically in Example 15-9, the ordering and spacing of the energy levels for these potentials, shown in the left-hand part of Figure 15-18, does not lead to the observed magic numbers if there is no spin-orbit interaction.

The right-hand part of Figure 15-18 shows how the nucleon energy levels are split by the nuclear spin-orbit interaction. In the presence of the spin-orbit interaction, $m_l$ and $m_s$ are no longer useful quantum numbers because the $z$ components of the orbital and intrinsic spin angular momenta of a nucleon are no longer constants when these angular momenta are coupled by the interaction. Thus $n, l, j, m_j$ must be used to label the split energy levels. The quantum number $j$ specifies the magnitude of the total

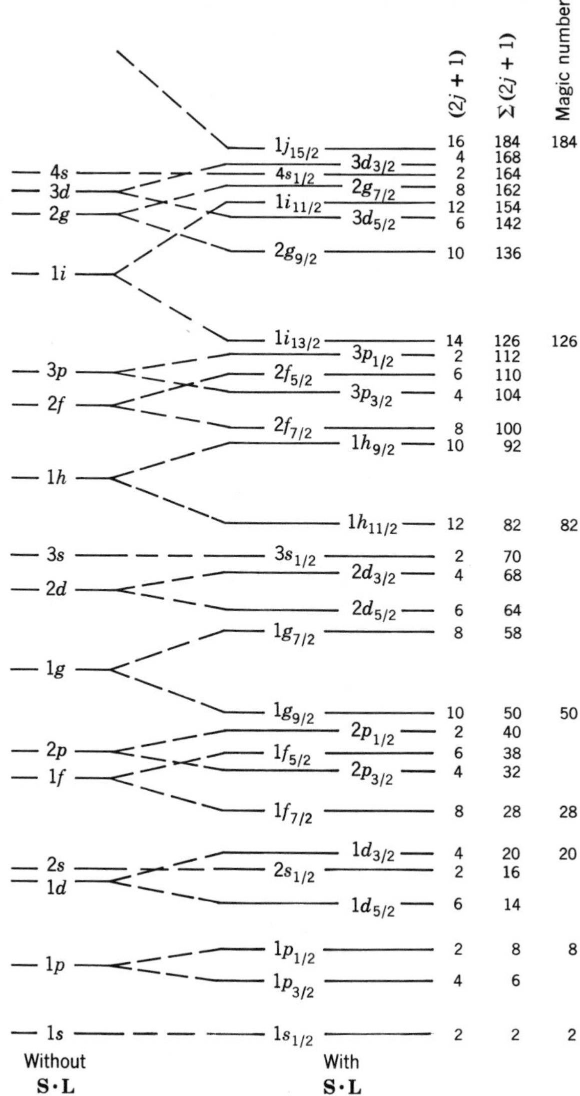


**FIGURE 15-18**

*Left:* The order of filling, as the occupancy and well radius increase, of the levels of rounded edge square wells with no spin-orbit interaction. *Right:* The levels that arise when a strong inverted $\mathbf{S} \cdot \mathbf{L}$ interaction is added. The column marked $(2j + 1)$ shows the number of like nucleons that may occupy the corresponding level without violating the exclusion principle. The column marked $\Sigma\,(2j + 1)$ gives for each level the cumulative number of nucleons that lie in all levels up through that level. Significant energy gaps lie above each of the levels marked with a magic number in the last column.

angular momentum, $J$, of a nucleon, which is the sum of its spin and orbital angular momenta; and $m_j$ is the quantum number specifying the $z$ component of its total angular momentum, $J_z$. As a result of the spin-orbit interaction, the energies of the levels depend on $j$ as well as on $n$ and $l$, with the larger $j$ (corresponding to the larger value of $J$, or $\mathbf{S} \cdot \mathbf{L}$) yielding the smaller energy since the sign of the nuclear spin-orbit interaction is inverted. According to the exclusion principle, each of these levels has a capacity of $(2j + 1)$, which is equal to the number of possible values of $m_j$. This is shown in the first column on the right in the figure. The second column shows the total capacity of the levels up to and including the level in question. The third column shows the same thing for each level which lies unusually far below the next higher level. Since these are the levels which will be unusually tightly bound, we see that the shell model with strong inverted spin-orbit interaction predicts precisely the magic numbers of (15-32).

Figure 15-18 is so frequently used by nuclear physicists that many of them have it memorized. An easier procedure is to construct it by using the acrostic

spuds if pug dish of pig

which means: (eat) potatoes if the pork is bad. Deletion of all vowels, except the last, yields

*spdsfpgdshfpig*

This is the ordering of $l$ for all the unsplit levels, through those leading to the magic number 126. The values of $n$ are assigned easily since the first $s$ level is $1s$, the second is $2s$, etc. The remainder of the figure is constructed by applying an inverted spin-orbit splitting, proportional to $l$.

It should also be pointed out that Figure 15-18 is not an energy-level diagram for any particular nucleus; instead it gives the order in which the nuclear levels appear below the Fermi energy as the radius of the nuclear potential increases in proportion to $A^{1/3}$. That is, it gives the order in which the highest energy levels of the various nuclei fill. It also gives an indication of the relative magnitudes of the separation between adjacent levels as they are filling. So it is analogous to the diagram that could be constructed for atoms by using only the left side of Figure 9-14.

Finally, we should mention that there is some recent experimental and theoretical evidence showing that there may be small but important changes from Figure 15-18 in the filling order of the highest levels in the case of protons. We shall discuss this in Section 16-2.

**Example 15-9.** Use Figure 15-18 to predict the first four magic numbers for nuclei with potentials in the form of square wells with rounded edges (a) under the assumption that there is no spin-orbit interaction, and (b) under the assumption that there is a strong inverted spin-orbit interaction.

(a) If there is no spin-orbit interaction then the nucleon energy levels are simply those shown on the left-hand part of the figure. Recalling that the capacity of each level is $2(2l + 1)$, and that $s, p, d, f, g, \ldots$ mean $l = 0, 1, 2, 3, 4, \ldots$, we see that the first few levels, and their capacities, are, in order of increasing energy: $1s$, capacity 2; $1p$, capacity 6; $1d$, capacity 10; $2s$, capacity 2; $1f$, capacity 14; $2p$, capacity 6; $1g$, capacity 18. The first magic number will be the number of nucleons required to fill the first level, i.e., 2. The next magic number will be the number required to fill the first two levels, i.e., $2 + 6 = 8$. If the third and fourth levels are very close in energy, as indicated in the figure, the next magic number will be the number of nucleons required to fill the first four levels, i.e., $2 + 6 + 10 + 2 = 20$. So far these magic numbers are in agreement with the observed magic numbers: 2, 8, 20, 28, 50, 82, 126. But the next magic number predicted in the absence of spin-orbit interaction will be the total number of nucleons required to fill the first five levels, or the first six levels, depending on whether or not the fifth and sixth level are considered to be very close in energy. The two possibilities are, $2 + 6 + 10 + 2 + 14 = 34$, or $2 + 6 + 10 + 2 + 14 + 6 = 40$. Both disagree with the observed magic number 28. Similar numerology will make it apparent that the higher predicted magic numbers also disagree with those that are observed, and that there is no way to remove the discrepancy by rearranging the spacing, or even the ordering, of the nucleon energy levels in the absence of spin-orbit interaction.

(b) If there is a strong inverted spin-orbit interaction, then the nucleon levels are split into the filling pattern shown on the right-hand part of Figure 15-18. The figure also shows the capacity $(2j + 1)$ of each level, as well as the sum $\Sigma(2j + 1)$ of its capacity and the capacity of all the lower energy levels, as explained in the text. The spin-orbit interaction splitting does not change the first three predicted magic numbers, 2, 8, 20, as is clear from the figure, so the agreement with observation is maintained. But agreement is also obtained with the higher magic numbers. For instance, the spin-orbit interaction splits the $1f$ level into the $1f_{7/2}$, whose energy is depressed, and the $1f_{5/2}$, whose energy is elevated. Since the capacity of the $1f_{7/2}$ level is $(2j + 1) = 2 \times 7/2 + 1 = 8$, the magic number after 20 is predicted to be $20 + 8 = 28$, in agreement with the observation. The observed magic number 50 is obtained because the $1g_{9/2}$ level, with a capacity of $2 \times 9/2 + 1 = 10$, is depressed in energy and so comes close to the $2p$ level. Since the total number of nucleons filling the levels up to and including the $2p$ is 40, as we saw earlier, the total number filling the levels up to and including the $1g_{9/2}$ is $40 + 10 = 50$. Inspection of Figure 15-18 makes the origin of the remaining magic numbers apparent. Note that the fact that the spin-orbit splitting increases in magnitude, with increasing $l$, plays an important role in achieving agreement with the observations. ◀

## 15-9 Predictions of the Shell Model

The shell model can do much more than predict the magic numbers, and all their consequences. For instance, it can also predict the total angular momentum of the ground states of almost all the nuclei. Consider nuclei for which both $N$ and $Z$ are magic, such as ${}^{8}O^{16}$, ${}^{20}Ca^{40}$, and ${}^{82}Pb^{208}$. According to the model, they will contain only completely filled subshells of neutrons and protons, and the exclusion principle therefore requires that, for both the neutron and proton systems, the intrinsic spin and orbital angular momentum vectors of all the nucleons couple together (add up) to yield zero total angular momentum. (The formal proof of this obvious requirement is essentially the same as that given in Appendix K.) This agrees with the measurements, discussed in Section 15-2, which show that for these nuclei the total angular momentum quantum number, called the nuclear spin, is $i = 0$. For nuclei which contain a magic number of nucleons of one type, and a magic number plus, or minus, one of nucleons of the other type, the exclusion principle demands that the total angular momentum of the nucleus be the total angular momentum of the extra nucleon, or (compare Appendix K) of the hole. For such nuclei the nuclear spin $i$ should equal the total angular momentum quantum number $j$ of the extra nucleon, or hole.

**Example 15-10.** Use Figure 15-18, and the exclusion principle argument just stated, to predict the ground state spins of the following nuclei: (a) ${}^{7}N^{15}$, (b) ${}^{8}O^{17}$, (c) ${}^{19}K^{39}$, (d) ${}^{82}Pb^{207}$, and (e) ${}^{83}Bi^{209}$.

(a) Figure 15-18 predicts that ${}^{7}N^{15}$ is doubly magic except for a proton hole in the $1p_{1/2}$ subshell. So it should have a spin $i$ equal to the value $j = 1/2$ for that subshell. This prediction agrees with measurement. It will also be obtained from a somewhat different point of view in Example 15-11.

(b) The figure predicts that ${}^{8}O^{17}$ is doubly magic except for an extra neutron in the $1d_{5/2}$ subshell. So it should have $i = j = 5/2$, in agreement with measurement.

(c) ${}^{19}K^{39}$ is predicted to be doubly magic except for a proton hole in the $1d_{3/2}$ subshell, so it should have $i = j = 3/2$. It does.

(d) According to Figure 15-18, ${}^{82}Pb^{207}$ is doubly magic except for a neutron hole in the $1i_{13/2}$ subshell. So the exclusion principle predicts that it should have a spin $i = j = 13/2$. However, the measured spin is $i = 1/2$. This is not a failure of the exclusion principle, but instead is a failure of Figure 15-18, as we shall explain shortly.

(e) The figure predicts that ${}^{83}Bi^{209}$ is doubly magic except for an extra proton in the $1h_{9/2}$ subshell. So its spin should be $i = j = 9/2$. This agrees with measurement. ◀

Now consider nuclei for which $N$ and/or $Z$ are not near magic numbers. These nuclei contain subshells with several nucleons, or holes, and the problem of how the intrinsic spin and orbital angular momenta of these nucleons couple is much the same problem as that studied in Chapter 10 in connection with the behavior of electrons in atoms. But there are important differences between atoms and nuclei in this regard. One is that most atoms obey what is called *LS* coupling, while essentially all nuclei obey what is called *JJ* coupling. The difference in the angular momenta coupling schemes obeyed by atoms and nuclei has to do with the fact that the spin-orbit interaction is relatively weak in atoms, and quite strong in nuclei (see Section 10-3). Thus in nuclei the spin-orbit interaction dominates the coupling. That is, in *JJ coupling*, the intrinsic spin angular momentum of a nucleon couples strongly with its own orbital angular momentum to form the total angular momentum for that nucleon. This happens for each nucleon. Finally, the several total angular momenta that have been formed couple together less strongly to form the total angular momentum for the nucleus. Another difference between the angular momenta couplings in atoms and nuclei is that the final coupling which forms the total angular momentum of the nucleus is particularly simple. This is apparent from the fact that all nuclei with even $N$ and even $Z$ are found to have a total angular momentum given by $i = 0$, as stated in Section 15-2. An explanation is that, whenever there are an even number of nucleons of a given species in a subshell, the total angular momenta of each of these nucleons couple together to yield a total angular momentum for the nucleus which is zero. This is true, but the coupling is even simpler. There is much evidence indicating that the total angular momenta of the protons in a subshell couple together in *pairs*, with the total angular momentum of *each* pair of protons equal to zero, and that the same thing happens for pairs of neutrons in a subshell.

Some of the evidence for the pairing tendency has been presented before in discussing the abundance of stable nuclei, and the semiempirical mass formula. It arises from a *pairing interaction*. This is a residual nuclear interaction, i.e., a part of the total nuclear interaction experienced by the nucleons that is not described by the spherically symmetrical net potential $V(r)$ of the shell model, or by the spin-orbit interaction. Although not described by these attributes of the shell model, the pairing interaction can be predicted from them. The net potential $V(r)$ represents the interactions experienced by a nucleon *on the average*. The pairing interaction represents a *departure from the average interaction* described by $V(r)$, that arises when the nucleon is particularly close to another nucleon with which it can have an *individual interaction*. It involves the collision of nucleons in degenerate states of a partly filled subshell, mentioned in Section 15-7. A pair of nucleons having the same values of $j$ but opposite values of $m_j$ (e.g., $j = 5/2$, $m_j = 5/2$; $j = 5/2$, $m_j = -5/2$) collide with each other in such an interaction, and after the collision enter previously empty states that have different but still opposite values of $m_j$ (e.g., $j = 5/2$, $m_j = 3/2$; $j = 5/2$, $m_j = -3/2$). It is clear that angular momentum is conserved in such collisions, and that the collisions are not inhibited by the exclusion principle. The energy of the system is reduced because when colliding the nucleons are particularly close together, and the exclusion principle does not prevent them from exerting on each other the strongly attractive short range nuclear force.

Because the nuclear force exerted between two nucleons is strong and short range, the departures from the average described by the pairing interaction are pronounced. Thus the pairing interaction is fairly strong, although it is less strong than the spin-orbit interaction. It is short range, just like the nuclear force leading to the fluctuation it represents. It is attractive because that force is attractive. A similar interaction resulting from a departure from the average, called the residual Coulomb interaction, arises in the treatment of atoms, as we have seen in Section 10-3. In atoms, the *repulsive* residual Coulomb interaction between the electrons in a subshell tends to make them form *parallel* couplings of their angular momenta. In nuclei the tendency is for *antiparallel* couplings because the residual nuclear interaction between the nucleons is *attractive*. The reason can be understood by carrying through arguments similar to those used for the atomic couplings (see Section 10-3), in the case of an attractive residual interaction. Briefly, these arguments show that since two nucleons of the *same species* are described by an antisymmetric total eigenfunction, on the average they are closer to each other if their spin angular momenta are essentially antiparallel. Also they are closer on the average if their orbital angular momenta are essentially antiparallel, because then they move in opposite directions around the same "orbit" and so frequently pass by each other. Thus they form a closely spaced pair if their total angular momentum vectors are essentially antiparallel. When they form such a closely spaced pair with zero total angular momentum, the attractive nuclear force acting between them makes a larger contribution to the binding energy of the nucleus, and so makes the nucleus more stable. Hence the tendency to form a pair, and maintain essentially antiparallel total angular momentum vectors throughout their sequence of collisions with each other. These collisions change the orientation of their orbit, but they always move in opposite directions through whatever orbit they happen to be in.

The energy decrease, arising from the coupling of a pair of nucleons of the same type, or *pairing energy*, gives rise to the preference for nuclei to have even $Z$ and even $N$, and to the pairing term of the semiempirical mass formula. It is also responsible for the occasional failure of Figure 15-18 to predict correctly the ground state nuclear spins. For the case of ${}^{82}\text{Pb}^{207}$, considered in Example 15-10, the nuclear spin is 1/2 because it is energetically favorable for a neutron from the $3p_{1/2}$ subshell to pair with the odd neutron in the $1i_{13/2}$ subshell, leaving a hole in the $3p_{1/2}$ subshell. The reason is that the pairing energy is larger the larger the $l$ values of the components of the pair, because with increasing $l$ the nucleons move in a more classical way (i.e., more like particles confined to orbits in a plane), and this increases the overlap of their wave functions (i.e., they get closer together). Since the two subshells have very nearly the same energy, the pairing effect dominates.

If a subshell contains an even number of nucleons, their total angular momenta should couple together in pairs to yield zero total angular momentum. If one more nucleon is added, it should be difficult for it to disturb the pairs that were already there, because the pairing interaction is fairly strong. Thus the total angular momentum of the whole subshell should be due entirely to the odd nucleon. Therefore, *the entire angular momentum of an odd A nucleus should be due to the total angular momentum of the single odd nucleon in the highest energy occupied subshell, and the nuclear spin i should be equal to the value of the quantum number j for that subshell.* With only one or two exceptions, this rule allows the observed values of $i$ for all odd $A$ nuclei to be explained in terms of Figure 15-18. It is, however, necessary to allow for occasional interchanges of the filling order of some closely spaced levels because of the pairing effect discussed in the preceding paragraph.

For odd-$A$ nuclei, the shell model is also quite successful in predicting the *parities* of the nuclear eigenfunctions, i.e., whether they are even or odd functions of their space variables (see (8-44) and (8-45)). Because the nucleons in the shell model are, basically, moving independently, a nuclear eigenfunction can be written as a product of the eigenfunctions for each of its nucleons—just as in the Hartree theory of atoms. We shall see in Example 15-11 that the parity of the nuclear eigenfunction is just the parity of the eigenfunction for the odd nucleon. Because (8-47) shows that the parity of that eigenfunction is determined by $(-1)^l$, we find that *if the odd nucleon is in a subshell in which l is even, the nuclear parity is even; if l is odd, the parity is odd.* In the next chapter we shall find that the nuclear parity is extremely important in determining the types of transitions that occur in certain kinds of radioactivity and nuclear reactions because there are selection rules that involve parity.

It should be apparent that the shell model predicts that for even-$A$ nuclei, *with $N$ and $Z$ even, the nuclear spin is $i = 0$ and the nuclear parity is even.* This agrees with experiment. For even-$A$ nuclei, with $N$ and $Z$ odd, the value of $j$ and the parity of the eigenfunctions are predicted for each of the two odd nucleons. From this the nuclear parity can be obtained immediately, but it is only possible to set limits on the nuclear spin and to say that it must have an integral value. However, there are only a few odd-$N$, odd-$Z$ nuclei. The arguments of the last two paragraphs can also be extended to provide information about the spins and parities of low-lying excited states of nuclei. As we shall see later, this information is dependable only if the $N$ and/or $Z$ values lie near the magic numbers.

**Example 15-11.** Predict the ground state nuclear spin and parity for the following nuclei: (a) ${}^8O^{16}$, (b) ${}^8O^{17}$, (c) ${}^8O^{18}$, (d) ${}^7N^{15}$, (e) ${}^7N^{14}$.

(a) The ${}^8O^{16}$ nucleus has even $N$ and even $Z$, and it is also doubly magic since both $N$ and $Z$ equal 8. It has two neutrons in the $1s_{1/2}$ subshell which couple together in a pair to yield zero total angular momentum. Both of these neutrons are described by even parity eigenfunctions, since $l = 0$, so their part of the product eigenfunction for the nucleus is even. There are four neutrons in the $1p_{3/2}$ subshell, that couple into two pairs, both of which have zero total angular momentum. All four of these neutrons are described by odd parity eigenfunctions since $l = 1$, but the product of four odd functions is an even function, so their part of the product eigenfunction for the nucleus is also even. There are two neutrons in the $1p_{1/2}$ subshell, which form a pair of zero total angular momentum. They contribute two odd eigenfunctions to the product eigenfunction for the nucleus, so their part of the product eigenfunction is also even. Exactly the same remarks apply to the protons. The net result is that the nuclear spin is zero, and the nuclear parity is even.

(b) ${}^8O^{17}$ is an odd-$N$, even-$Z$ nucleus. Its neutrons and protons are doing the same things as the neutrons and protons in ${}^8O^{16}$, except that it has a single extra unpaired neutron in a $1d_{5/2}$ subshell. This gives the nucleus a spin of $i = 5/2$. The parity of the eigenfunction for the unpaired neutron is even since $l = 2$, so the nuclear parity is even.

(c) ${}^8O^{18}$ is an even-$N$, even-$Z$ nucleus. The predicted spin and parity are $i = 0$, and even. The reasons are that there are two neutrons in the $1d_{5/2}$ subshell, which form a pair of zero total angular momentum, and which both have even parity eigenfunctions.

(d) ${}^7N^{15}$ is an even-$N$, odd-$Z$ nucleus. Its neutrons and protons behave as in ${}^8O^{16}$, except that it has only one unpaired proton in the $1p_{1/2}$ subshell. This odd proton gives the nucleus a spin of $i = 1/2$. Since the eigenfunction for the proton is odd because $l = 1$, the nuclear parity is odd. Note that we predicted the nuclear spin, from a somewhat different point of view, in Example 15-10.

(e) ${}^7N^{14}$ is an odd-$N$, odd-$Z$ nucleus. It has an unpaired proton in the $1p_{1/2}$ subshell, and also an unpaired neutron in the $1p_{1/2}$ subshell. Both have a total angular momentum quantum number of $j = 1/2$. We cannot say precisely what the nuclear spin should be without knowing how these two different particles couple their angular momenta. But we can say that there are only two possibilities for the nuclear spins, $i = 0$ or $i = 1$. Experiments show that $i = 1$ is the correct value. We can predict unambiguously that the nuclear parity will be even, since the unpaired proton and the unpaired neutron both contribute an odd eigenfunction to the product eigenfunction for the nucleus, and the product of two odd functions is an even function. This prediction is born out by the experiments, as are all the predictions made in the earlier parts of this example. ◀

The shell model is not so successful in predicting the magnetic dipole moments of nuclei. It says that the magnetic dipole moment of an odd-$A$ nucleus (i.e., even $N$ and odd $Z$, or odd $N$ and even $Z$) should be due entirely to that of the single odd (unpaired)

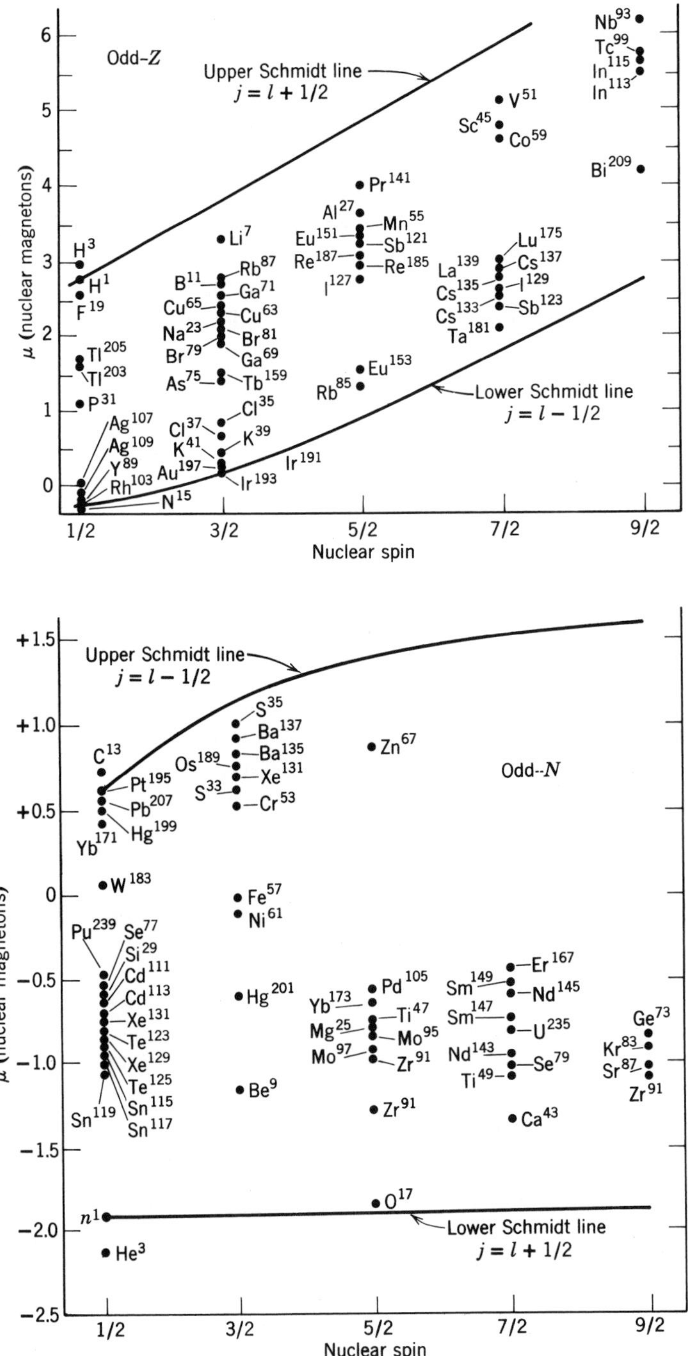


**FIGURE 15-19**

*Top:* Measured magnetic dipole moments of even $N$, odd-$Z$ nuclei and the shell model predictions. The upper line is the prediction if the spin and orbital angular momenta of the odd proton is assumed to be essentially parallel, and the lower line is the prediction if they are assumed to be essentially antiparallel. Bottom: The same for odd $N$, even $Z$. Here, the lower line is for the "parallel" assumption and the upper line is for the "antiparallel" assumption.

nucleon. The reason is that the magnetic dipole moments of the other nucleons would be expected to cancel out in pairs, if their total angular momenta do the same. The experimental data are illustrated in the two parts of Figure 15-19, for even-$N$, odd-$Z$ nuclei and for odd-$N$, even-$Z$ nuclei. The data are obtained in the manner indicated in Section 15-2. Also shown in the figure are the so-called *Schmidt lines*, which represent the predictions of the shell model for cases in which the spin and orbital angular momenta of the odd nucleon are either essentially parallel or essentially antiparallel, that is for the two possible cases $j = l + 1/2$ or $j = l - 1/2$. The data show only a barely recognizable tendency to follow qualitatively the predictions of the shell model.

The failure in the model is due to its assumption that the nuclear magnetic dipole moment is due entirely to the single odd nucleon. It is not true that all the other nucleons are *always* paired off with total angular momenta and magnetic dipole moments that strictly cancel. The assumption is good enough to lead to the prediction of correct magnitude for the total angular momentum of the nucleus, since this quantity is quantized. If occasionally the pairs have a nonzero total angular momentum, then at that time the odd nucleon *must* have exactly the right total angular momentum to compensate and keep the magnitude of the total angular momentum of the nucleus constant. This kind of compensation cannot also take place for the magnetic dipole moments since the $g$ factors, which relate the magnitudes of the magnetic dipole moments to the magnitudes of the angular momenta, change as the angular momentum couplings change (see Section 10-6). And since the nuclear magnetic dipole moment does not have a quantized *magnitude*, there is nothing to enforce such a compensation.

## 15-10 The Collective Model

The shell model is based upon the idea that the constituent parts of a nucleus move independently. The liquid drop model implies just the opposite, since in a drop of incompressible liquid the motion of any constituent part is correlated with the motion of all the neighboring parts. The conflict between these ideas emphasizes that a *model* provides a description of only a *limited set* of phenomena, without regard to the existence of contrary models used for the description of other sets. A *theory*, such as relativity or quantum theory, provides a description of a very *large set* of phenomena. At the border lines between its own set of phenomena and other sets of phenomena, a theory fuses without conflict into the theories used for the description of the other sets.

As nuclear physics evolves, attempts are made to remove conflicts between various models and unify them into more comprehensive models. The most successful and most important example is the *collective model of the nucleus*, which combines certain features of the shell and liquid drop models. It is partly the work of Aage Bohr, whose father developed the Bohr model of the atom. The collective model assumes that the nucleons in unfilled subshells of a nucleus move independently in a net nuclear potential produced by the *core* of filled subshells, as in the shell model. However, the net potential due to the core is not the static spherically symmetrical potential $V(r)$ used in the shell model; instead it is a potential capable of undergoing deformations in shape. These deformations represent the correlated, or collective, motion of the nucleons in the core of the nucleus that are associated with the liquid drop model.

As in the shell model, the nucleons fill the energy levels of the potential, which are split by the same spin-orbit interaction and lead to the same magic numbers, and nuclear spin and parity predictions. Consider a nucleus with one more than a magic number of nucleons. Inspection of the shell model energy levels of Figure 15-18 will show that the extra nucleon will have a relatively large orbital angular momentum. Classically, it will move in an orbit of relatively large radius, near the surface of the core of completely filled subshells. Because of the attractive nuclear interaction between the extra nucleon and the nucleons in the core, the core is distorted. Bulges circulate around the surface of the core, following the motion of the extra nucleon. The effect is very much like the tides at the surface of the earth, which follow the motion of the moon, and arise from the attractive gravitational interaction. If there are two extra nucleons of the same species, classically they will move in opposite directions around the surface of the core in orbits that are essentially in the same plane. The reason is that their pairing interaction produces "antiparallel" coupling of their angular momenta. This increases the distortion of the core. Physically, the distortion of the core affects the motion of the extra nucleons. Mathematically, this is handled by distorting the net potential in which these nucleons move. One result is a considerable complication of the necessary task of solving the Schroedinger equation for the potential. Another result is a considerable extension of the set of phenomena that can be described accurately by the model.

For instance, in the collective model part of the total angular momentum of the nucleus is carried in the form of orbital angular momentum by the "tidal waves" circulating around the surface of the core. A moving deformation, *partly* composed of protons, constitutes a current that produces a magnetic dipole moment proportional to its angular momentum. This is also true in the case of the single moving nucleon that the shell model says is totally responsible for the nuclear magnetic dipole moment, but the proportionality constants differ. The moving deformation produces less magnetic dipole moment than a moving proton, and more than a moving neutron, relative to the angular momentum it carries. These changes are exactly what is required to remove the discrepancies between the measured nuclear magnetic dipole moments and the shell model predictions, shown by the Schmidt lines in Figure 15-19.

The student may notice an analogy between the behavior of two electrons always moving in opposite directions with antiparallel spins in a Cooper pair of a superconductor, and two neutrons or two protons always moving in opposite directions in an unfilled subshell of a nucleus with spins that, because of the nuclear pairing interaction, are also antiparallel. Another analogy is that in both cases the behavior of a pair of interacting particles influences, and is influenced by, the behavior of the other particles in the system, which move collectively. Analogies are also found between the mathematical procedures used in BCS superconductivity calculations and in nuclear collective model calculations.

A nuclear property which can be explained quite well in terms of the collective model is the electric quadrupole moment $q$. The hyperfine splitting measurements yielding $q$ were briefly explained in Section 15-2, and there it was also stated that $q$ is a measure of the departure from spherical symmetry of the nuclear charge distribution, as observed in measurements such as hyperfine splitting which are sensitive to the time average of this departure. The exact definition of the electric quadrupole moment is

$$q = \int \rho[3z^2 - (x^2 + y^2 + z^2)]\, d\tau \tag{15-35}$$

where $\rho$ is the time averaged nuclear charge density in units of proton charges, and where the three-dimensional integral is taken over the nuclear volume with $d\tau$ the volume element. Note that $q$ is equal to $Z$, the number of protons in the nucleus, multiplied by the average over $\rho$ of the difference between three times the square of the $z$ coordinate and the sum of the squares of all the coordinates. That is

$$q = Z[3\overline{z^2} - (\overline{x^2} + \overline{y^2} + \overline{z^2})] \tag{15-36}$$

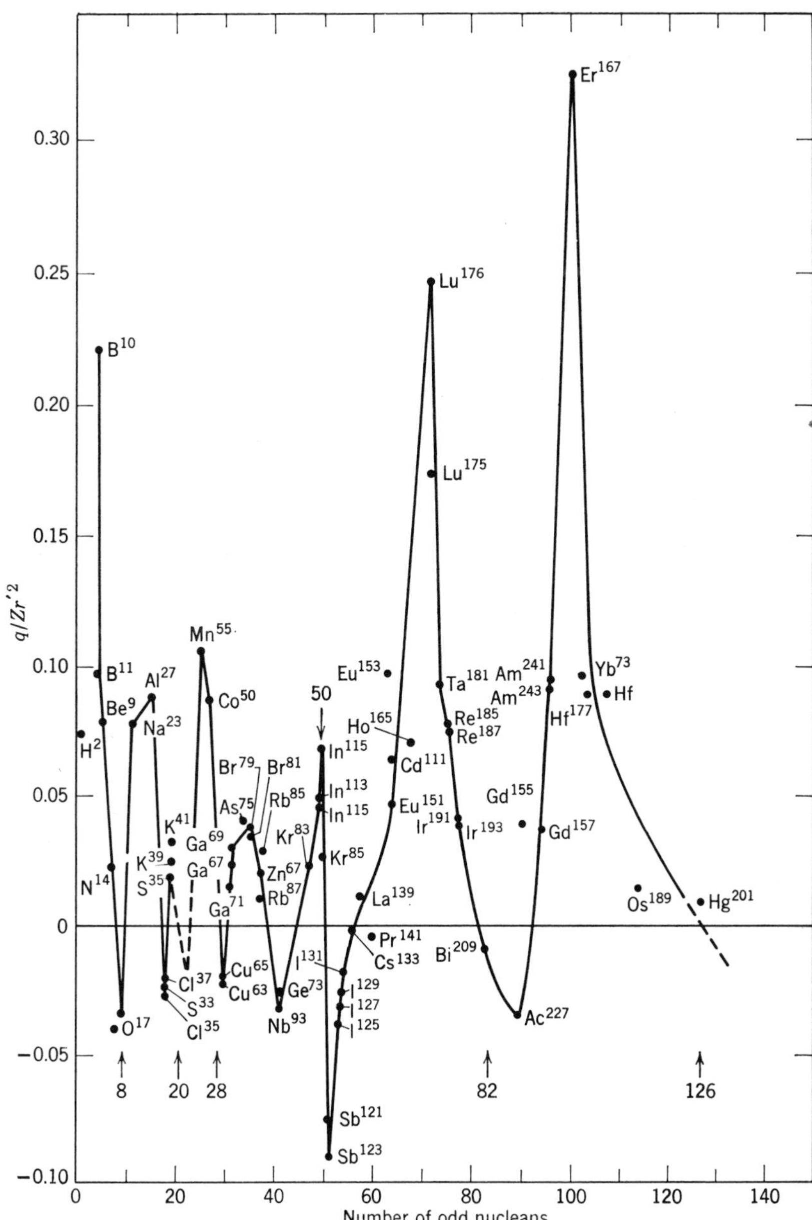


**FIGURE 15-20**

The nuclear electric quadrupole moment, $q$, about the symmetry axis, divided by $Zr'^2$, for odd-$A$ nuclei. The distance $r'$ is the average from the center of the ellipsoidal distribution, of charge $+Ze$, to the surface. The quantity $1 + q/Zr'^2$ is approximately equal to the ratio of the distances from the center to the surface measured parallel to, and perpendicular to, the symmetry axis.

It is clear then that $q = 0$ if the time averaged nuclear charge density $\rho$ is spherically symmetrical, since in that case $\overline{x^2} = \overline{y^2} = \overline{z^2}$. If $\rho$ is not spherically symmetrical, it must at least have an axis of symmetry along the direction about which the total angular momentum of the nucleus randomly precesses as time passes. In typical cases the time-averaged charge density is an ellipsoid with a symmetry axis in that direction. For (15-35) and (15-36), the symmetry axis is taken as the $z$ axis. The second of these equations shows immediately that $q > 0$ if $\rho$ is elongated in the $z$ direction so that $\overline{z^2} > \overline{x^2} = \overline{y^2}$, and that $q < 0$ if $\rho$ is flattened in the $z$ direction so that $\overline{z^2} < \overline{x^2} = \overline{y^2}$.

The measured values of the time-averaged nuclear electric quadrupole moment $q$ are shown in Figure 15-20. Some features of the data shown in the figure can be understood qualitatively in terms of the shell model. For example, that model predicts $q < 0$ for an even-$N$, odd-$Z$ nucleus with $Z$ equal to a magic number plus one. The reason is that the nucleus contains only completely filled proton subshells, which have a spherically symmetrical charge distribution, plus one odd proton moving in an "orbit" near a plane perpendicular to its symmetry axis. Thus the charge distribution is flattened in the direction of the symmetry axis. For an even-$N$, odd-$Z$ nucleus with $Z$ one less than a magic number, the shell model correctly predicts $q > 0$ since this nucleus would contain one proton hole (the absence of charge) moving in a similar orbit. These shell model arguments are illustrated in Figure 15-21. They make

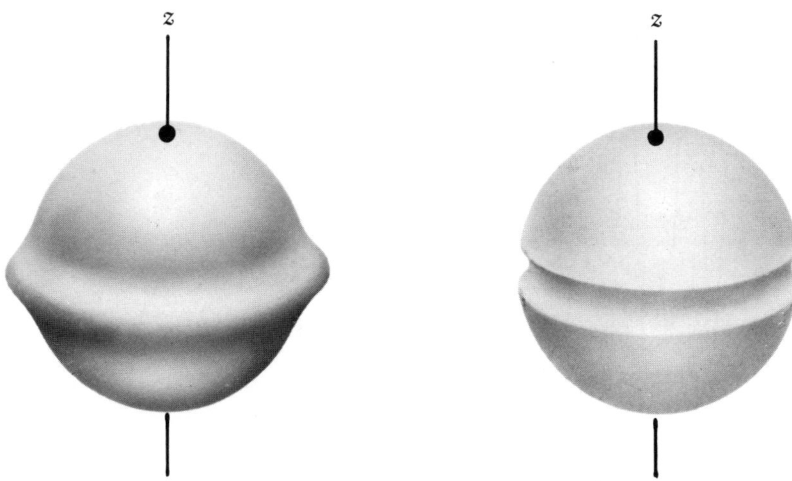


**FIGURE 15-21**

*Left:* Illustrating schematically an odd proton in a nucleus with $Z$ equal to one more than a magic number. To a fair approximation the proton moves in an orbit of radius equal to the nuclear radius, its time-averaged charge distribution looking like a ring. The total charge distribution contains an excess of charge, relative to a spherical distribution, in a plane perpendicular to the symmetry axis (the $z$ axis). Thus the nucleus has a negative quadrupole moment. *Right:* Illustrating a proton hole in a nucleus with $Z$ equal to one less than a magic number. Averaged over time, the hole leads to a ring containing a deficiency in charge in a plane perpendicular to the symmetry axis. The electric quadrupole moment is positive because the charge distribution has an excess of charge, relative to a spherical distribution, in the direction of the symmetry axis (the $z$ axis).

plausible the observations (1) that $q$ is positive for an even-$N$, odd-$Z$ nucleus if $Z$ is in a range just below a magic number, (2) that $q$ is zero if $Z$ is at the magic number, and (3) that $q$ is negative if $Z$ is in a range just above the magic number. However, the shell model is not capable of yielding correct quantitative results for electric quadrupole moments. Its predictions for the magnitude of $q$ are generally low, and for some nuclei between magic numbers they are lower than the observed magnitude by more than a factor of 10.

**Example 15-12.** Estimate the shell model prediction for the time-averaged electric quadrupole moment $q$ of the nucleus $^{51}\text{Sb}^{123}$, and compare with the measured value shown in Figure 15-20.

According to the shell model, the charge distribution of this nucleus is due to a spherically symmetrical core of completely filled proton subshells, plus a single odd proton in a $1g_{7/2}$ subshell. Since the orbital angular momentum of this proton is high ($l = 4$), to a fair approximation it can be thought of as moving in a Bohr-like orbit of radius about equal to the nuclear radius $r'$. (Recall we found in Section 7-8 that orbital motion approaches the classical limit as $l$ becomes large.) Thus a time average of the nuclear charge distribution looks something like that shown on the left of Figure 15-21. The spherical core makes no contribution to the nuclear electric quadrupole moment $q$. So, if we take the symmetry axis perpendicular to the orbit as the $z$ axis, we have

$$q = \int \rho[3z^2 - (x^2 + y^2 + z^2)]\, d\tau$$

where $\rho$ is approximately the charge density for a uniformly charged ring, of radius $r'$, in a plane perpendicular to $z$. This $\rho$ is zero except where $x^2 + y^2 = r'^2$ and $z = 0$. Thus

$$q \simeq \int \rho[-r'^2]\, d\tau = -r'^2 \int \rho\, d\tau$$

The integral of $\rho$ yields one since the ring contains the charge of one proton and $\rho$ is measured in units of proton charges. Therefore, the result we obtain for an estimate of the shell model predictions of $q$ for $^{51}\text{Sb}^{123}$ is

$$q \simeq -r'^2$$

Figure 15-20 shows that the measured value of $q$ for this nucleus is such that

$$\frac{q}{Zr'^2} \simeq -0.09$$

or

$$q \simeq -0.09Zr'^2 \simeq -0.09 \times 51r'^2 \simeq -5r'^2$$

The magnitude of the shell model prediction is too low, compared to the measurements, by about a factor of 5. ◀

Another prediction of the shell model is that the value of the electric quadrupole moment for odd-$A$ nuclei depends significantly on whether they have odd $N$, even $Z$ or even $N$, odd $Z$. The reason is simply that the odd nucleons are uncharged neutrons in the first case and charged protons in the second case. But Figure 15-20 shows that the value of $q$ for odd-$A$ nuclei depends on only the number of odd nucleons, independent of whether or not the odd nucleons are charged.

The collective model explains all the features of the measured electric quadrupole moments that are incorrectly predicted by the shell model. It leads to large enough values of $q$ because the core can be deformed so that the charges of many protons contribute to the total electric quadrupole moment. For nuclei between the magic numbers the core deformations become quite large, and therefore the electric quadrupole moments also become quite large. As the deformations can be due to extra nucleons of either species, the collective model explains why the observed values of $q$ do not depend significantly on whether the odd nucleons are neutrons or protons.

In addition to the collective rotations of the nuclear core that we have been considering, there are also collective vibrations. Certainly the most spectacular example is nuclear fission. This will be discussed in the next chapter.

## 15-11 Summary

Table 15-3 briefly summarizes this chapter by listing the nuclear models we have treated, and some of their most significant features. We have seen that each model can provide satisfactory explanations of certain properties of nuclei in their ground states (but no single model can explain all the properties). In the next chapter we shall find that these models can provide explanations of the properties of nuclear decay and nuclear reactions. In that chapter we shall also come across another important nuclear model, not listed in Table 15-3. This is the *optical model*, which is a generalization of the shell model that describes the behavior of an unbound nucleon moving through a nucleus.

**TABLE 15-3.** Nuclear Models and the Ground State Properties of Nuclei

| Name | Assumptions | Theory Used | Properties Predicted |
|---|---|---|---|
| Liquid drop model | Nuclei have similar mass densities, and binding energies nearly proportional to masses—like charged liquid drops | Classical (asymmetry and pairing terms introduced with no justification) | Accurate average masses and binding energies through semiempirical mass formula |
| Fermi gas model | Nucleons move independently in net nuclear potential | Quantum statistics of Fermi gas of nucleons | Depth of net nuclear potential<br>Asymmetry term |
| Shell model | Nucleons move independently in net nuclear potential, with strong inverted spin-orbit coupling | Schroedinger equation solved for net nuclear potential | Magic numbers<br>Nuclear spins<br>Nuclear parities<br>Pairing term |
| Collective model | Net nuclear potential undergoes deformations | Schroedinger equation solved for non-spherical net nuclear potential | Magnetic dipole moments<br>Electric quadrupole moments |

## QUESTIONS

1. Was there a stage in the development of atomic physics in which models played a role comparable to that now played by models in nuclear physics? Are models used now in atomic physics?

2. In those regions of the universe where thermal energy is $kT \sim 10^6$ eV, are atomic processes more apparent than nuclear processes? What about those regions where $kT \sim 10^{-6}$ eV?

3. All nuclei have an electric monopole moment (which measures their total charge). Some nuclei have an electric quadrupole moment (which measures the departure from a spherical shape of their charge distribution). No nuclei have an electric dipole moment (which would measure the departure of the center of their charge distribution from the center of their mass distribution). Why would we not expect electric dipole moments for nuclei?

4. Nuclei have magnetic dipole moments. Why do they not have magnetic monopole moments? What about magnetic quadrupole moments?

5. If an electron of kinetic energy 100 keV passed through a typical atom it could be scattered through a fairly large angle in a close collision with an atomic electron. If its kinetic energy is 100 MeV it could be scattered through a fairly large angle only in a close collision with the nucleus. Why?

6. Why is the mass unit not defined in terms of the mass of the hydrogen atom? (Hint: Use Table 15-1 to make a quick estimate of the mass of ${}^{92}U^{238}$ if the mass of ${}^{1}H^{1}$ is $1.000000u$.)

7. Since atomic and molecular reactions also involve binding energies, why did the nineteenth century chemists not observe mass deficiencies and thereby discover relativity theory?

8. Many textbook problems in mechanics consider zero $Q$-value collisions between idealized classical particles. Is the $Q$ value exactly zero in collisions between real classical particles (like real billiard balls)? What is the sign of the $Q$ value?

9. Why are the most stable nuclei found in the region near $A \simeq 60$? Why do not all nuclei have $A \simeq 60$?

10. The semiempirical mass formula contains five parameters, and it predicts quite accurately more than 500 masses. How does its ratio of predictions to parameters compare with other empirical formulas of physics or engineering?

11. Why does the pairing term make a negative contribution to the energy liberated when a neutron is captured by ${}^{92}U^{238}$, and a positive contribution in the case of ${}^{92}U^{235}$? What are the practical consequences of this situation?

12. Why are the atomic magic numbers not the same as the nuclear magic numbers?

13. Explain why there can be no collisions between a typical nucleon and another in a nucleus in its ground state. If a high-energy nucleon, say from a cyclotron beam, enters a nucleus in its ground state, can it collide with a nucleon in the nucleus?

14. What fundamental law of physics is most responsible for the existence of nuclear magic numbers?

15. Is there a relation between the $l$ dependence of the spin-orbit splitting of nuclear levels and the Landé interval rule for the spin-orbit splitting of atomic energy levels?

16. Why do most nuclei obey $JJ$ coupling, whereas most atoms obey $LS$ coupling?

17. Use the argument associated with Figure 9-4 to explain why there is a tendency for the intrinsic spin angular momenta of a pair of identical nucleons to be essentially antiparallel because it minimizes their average separation. Then modify the argument illustrated in Figure 10-2 to explain why the average separation of the pair is minimized if their orbital angular momenta are also essentially antiparallel. Do these arguments explain why the pairing interaction tends to make the total angular momenta of the pair essentially antiparallel?

18. If one factor in a nuclear eigenfunction consists of a product of an even number of eigenfunctions for nucleons in a particular subshell, why is the parity of the factor even, independent of whether the parities of the nucleon eigenfunctions are all even or all odd? How does this lead to the rule for predicting the parities of odd-$A$ nuclear eigenfunctions?

19. How can the magnetic dipole moment data of Figure 15-19 be used to identify the orbital angular momentum quantum number $l$, of many nuclei, in terms of the measured value of their total angular momentum quantum number $j$?

20. If the tidal waves circulating around the nuclear core in the collective model were entirely composed of protons, instead of being composed partly of protons and partly of neutrons, what would be the effect on the magnetic dipole moments predicted by the model?

21. What is the simplest distribution of point charges that has an electric quadrupole moment?

22. Why is the nuclear shell model called a model, while the comparable atomic Hartree theory is called a theory? Generally speaking, how does a model differ from a theory?

## PROBLEMS

1. The analysis of the optical spectrum of an atom shows that there are four energy levels in a certain hyperfine splitting multiplet. The analysis also shows that the value of the total electronic angular momentum quantum number for that multiplet is $j = 2$. Determine the value of the nuclear angular momentum quantum number, or nuclear spin $i$, for the nucleus of the atom.

2. The nuclear spin and symmetry character of the boron nucleus with $Z = 5$ and $A = 10$ are: $i = 3$, symmetric. (a) Show that the mass, charge, nuclear spin, and symmetry character agree with the assumption that nuclei contain $Z$ protons and $A - Z$ neutrons. (b) Which of these four properties disagree with the assumption that nuclei contain $A$ protons and $A - Z$ electrons?

3. (a) Evaluate, in MeV, the energy of gravitational attraction for two spherically symmetrical protons with a center-to-center separation of 2 F. (b) Do the same for the energy of Coulomb repulsion at that separation. (c) Compare your results with the energy of nuclear attraction, which is about $-10$ MeV at that separation.

4. Electrons of kinetic energy 1000 MeV are scattered from a target containing $^{79}$Au nuclei. (a) Use data from Figure 15-6 to find the radius at which the nuclear charge density is half its interior value. (b) Then use this radius to predict the approximate separation in angle between adjacent minima of the diffraction pattern that is observed in the scattering.

5. Use the empirical equation representing the measured nuclear charge densities, (15-5), and the parameter $b$ quoted in (15-7), to determine the distance in which the nuclear charge densities fall from 90% to 10% of their internal values.

6. Derive (15-16), which relates the $Q$ value of a nuclear reaction to the dynamical quantities involved in the reaction. (Hint: Write equations for the conservation of the components of linear momentum in the directions parallel to and perpendicular to the direction of the incident particle. Then eliminate from these the angle between the direction of the residual nucleus and the direction of the incident particle.)

7. (a) Use (15-16) to calculate the energy of protons emitted in the direction of incidence of the 7.70 MeV $\alpha$ particles in the Rutherford reaction of (15-11). The $Q$ value of the reaction is $-1.18$ MeV. (b) Compare your results with Example 15-4.

8. Since the reaction ${}^1H^2 + {}^1H^3 \rightarrow {}^2He^4 + {}^0n^1$ has a high positive $Q$ value, it is frequently used to obtain high-energy neutrons, ${}^0n^1$, from a low-energy electrostatic generator accelerating a beam of deuterons, ${}^1H^2$, into a target of tritons, ${}^1H^3$. (a) Use information presented in Table 15-1 to calculate the $Q$ value for the reaction. (b) Use (15-16) to calculate the energy of the neutrons emitted from the reaction in the same direction as the incident beam of deuterons, if the energy of the deuterons is 0.500 MeV.

9. Use the masses quoted in Table 15-1 to verify that the binding energy per nucleon of ${}^6C^{12}$ has the value quoted in that table.

10. (a) Use information presented in Table 15-1 to evaluate, in MeV, the energy released in the fusion of two ${}^1H^2$ nuclei to form a ${}^2He^4$ nucleus. (b) Also evaluate, in MeV, the height of the Coulomb repulsion barrier which must be overcome before there is an appreciable probability that the two nuclei can get close enough together for fusion to take place. Treat the ${}^1H^2$ nuclei as uniformly charged spheres of radius 1.5 F, and evaluate the energy of Coulomb repulsion when they are just touching.

11. (a) The Coulomb energy of a uniformly charged sphere of radius $r'$, i.e., the energy required to assemble the charge, is

$$V = \frac{3}{5}\frac{Z^2e^2}{4\pi\epsilon_0 r'}$$

Take $r' = 1.1A^{1/3}$ F, which is consistent with the electron scattering measurements, and show that $V$ then assumes the form of the Coulomb term of the semiempirical mass formula. (b) Evaluate, in mass units, the coefficient of $Z^2/A^{1/3}$ in the expression obtained for $V$, and compare with the empirical value of the coefficient $a_3$ given in (15-31).

12. The nuclei ${}^5B^{11}$ and ${}^6C^{11}$ are said to be a pair of *mirror nuclei* because they have the same number of nucleons, and the number of protons in one equals the number of neutrons in the other. If nuclear forces are charge independent, their total binding energies should differ only in that the Coulomb energy is higher in ${}^6C^{11}$. The atomic mass of ${}^5B^{11}$ is $11.009305u$, and the atomic mass of ${}^6C^{11}$ is $11.011432u$. (a) Evaluate the difference in their total binding energies. (b) Assuming both nuclei to be uniformly charged spheres of the same radius $r'$, and using the expression for the Coulomb energy given in Problem 11, find the value of $r'$ that leads to a difference in Coulomb energy that agrees with the difference in binding energy. (c) Compare this charge distribution radius with the radial dependence of the charge density for the similar nucleus ${}^6C^{12}$ shown in Figure 15-6.

13. (a) Evaluate the terms of the semiempirical mass formula for ${}^{26}Fe^{56}$. (b) Convert them to their equivalents in MeV, divide by $A$, and then compare them with Figure 15-12. (c) Use the terms to predict the atomic mass. (d) Evaluate the average binding energy per nucleon, and compare with Figure 15-10.

14. According to the *$\alpha$-particle model of the nucleus*, ${}^6C^{12}$ consists of three $\alpha$ particles, i.e., ${}^2He^4$ nuclei, and ${}^8O^{16}$ consists of four $\alpha$ particles. (a) Use Table 15-1 to evaluate the difference between the total binding energy of ${}^6C^{12}$ and the total binding energies of three $\alpha$ particles. (b) Evaluate the difference between the total binding energy of ${}^8O^{16}$ and the total binding energies of four $\alpha$ particles. (c) Draw schematic diagrams of ${}^6C^{12}$ and ${}^8O^{16}$ according to the $\alpha$-particle model, and use them to show that there can be three "bonds" connecting the $\alpha$ particles in ${}^6C^{12}$, while there can be six bonds connecting the $\alpha$ particles in ${}^8O^{16}$. The exact nature of a bond was not specified in the model, but it was thought that they were somehow analogous to bonds in molecules. (d) Use the results of parts (a) and (b) to show that the total binding energies of ${}^6C^{12}$ and ${}^8O^{16}$ could be accounted for by saying that every possible bond contributes a binding energy of a little over 2 MeV. The $\alpha$-particle model is not highly regarded because little more can be done with it than has been done in this problem.

15. Use the acrostic explained in Section 15-8 to construct the diagram giving the ordering and approximate spacing of the energy levels which the nucleons are filling in the shell model. *After* you have finished, compare with Figure 15-18.

**16.** Use the exclusion principle argument of Example 15-10 to predict from the shell model diagram of Figure 15-18 the nuclear spins of: ${}^{20}\mathrm{Ca}^{40}$, ${}^{20}\mathrm{Ca}^{39}$, ${}^{20}\mathrm{Ca}^{41}$.

**17.** (a) Use the existence of the pairing interaction to predict from the shell model diagram of Figure 15-18 the nuclear spins and parities of ${}^{6}\mathrm{C}^{11}$, ${}^{20}\mathrm{Ca}^{44}$, ${}^{28}\mathrm{Ni}^{61}$, ${}^{32}\mathrm{Ge}^{73}$. Briefly justify each prediction. (b) The observed spins and parities are: (3/2, odd), (0, even), (3/2, odd), (9/2, even). Give an explanation of any discrepancies you find.

**18.** (a) Predict from the shell model diagram of Figure 15-18 the possible values of the nuclear spins, and also predict the parities, of the following odd-$N$, odd-$Z$ nuclei: ${}^{5}\mathrm{B}^{10}$, ${}^{19}\mathrm{K}^{40}$, ${}^{23}\mathrm{V}^{50}$. (b) The observed spins and parities are: (3, even), (4, odd), (6, even). Does there seem to be any preferential tendency in the coupling of the angular momenta of the odd neutron and odd proton?

**19.** The measured nuclear spin of ${}^{23}\mathrm{V}^{51}$ is 7/2. Since this is an even-$N$, odd-$Z$ nucleus, the nuclear spin is due to the odd proton that has a total angular momentum quantum number $j = 7/2$. Since there are two possible relations between $j$ and the orbital angular momentum quantum number $l$ for that proton, namely $j = l - 1/2$ and $j = l + 1/2$, the value of $l$ could be either 3 or 4. (a) Use the measured value of the magnetic dipole moment and its relation to the Schmidt lines, shown in Figure 15-19, to predict the most likely value of $l$. (b) Use the shell model diagram of Figure 15-18 to predict the value of $l$, and compare with (a).

**20.** (a) Use the measured electric quadrupole moment of ${}^{73}\mathrm{Ta}^{181}$, presented in Figure 15-20, to evaluate approximately the ratio of the distances from the center to the surface of its ellipsoidal charge distribution, measured parallel to and perpendicular to its symmetry axis. (b) Use the electron scattering charge distribution radius $a$, from (15-6), to evaluate approximately the average of these distances. (c) From the answers to (a) and (b) evaluate approximately these distances, which are the semimajor and semiminor axes of the ellipsoidal charge distribution. (d) Make a sketch, to scale, of the charge distribution.

# 16

## Nuclear Decay and Nuclear Reactions

# Nuclear Decay and Nuclear Reactions 16

## 16-1 Introduction

In the preceding chapter we used the properties of the ground states of stable nuclei to introduce the most important nuclear models. In this chapter we shall use these models to consider the decay of unstable nuclei, and also to consider nuclear reactions involving both stable and unstable nuclei. Our considerations will concern excited states of nuclei, as well as their ground states.

Nuclear decay divides itself into three categories. One is $\alpha$ decay—the spontaneous emission of an $\alpha$ particle from a nucleus of large atomic number. We shall see that this process, or the closely related process of spontaneous fission, is responsible for setting an upper limit on the atomic numbers of the chemical elements occurring in nature. A second type of nuclear decay is $\beta$ decay—the spontaneous emission or absorption of an electron or positron by a nucleus. It is particularly interesting because it will tell us about the $\beta$-decay interaction, which is one of the four fundamental interactions, or forces, of nature. A third type of nuclear decay is $\gamma$ decay—the spontaneous emission of high-energy photons when a nucleus makes transitions from an excited state to its ground state. We shall find that $\gamma$ decay gives detailed information about the excited states of nuclei that can be used to improve the nuclear models. We shall also find that $\gamma$ decay is used in the Mössbauer effect to make extremely high-resolution energy measurements in many different fields of physics.

Nuclear reactions will provide us with additional information about excited states of nuclei, since the residual nucleus in a reaction is typically formed in an excited state. Among the nuclear reactions that we shall consider are those that occur in the nuclear fission reactors that are now used as inexpensive sources of energy. We shall also consider the reactions that may some day be used to produce energy on earth by nuclear fusion and that have been used for a long time by stars to produce the energy, and the chemical elements, of which nature is composed.

## 16-2 Alpha Decay

Nuclear decay occurs, sooner or later, whenever a nucleus containing a certain number of nucleons is put in an energy state which is not the lowest possible one for a system with that number of nucleons. Invariably, the nucleus is put into the unstable state as a consequence of a nuclear reaction. But in some cases the nuclear reaction responsible for producing the unstable nucleus took place recently in a man-made particle accelerator, while in other cases it took place in natural events that happened billions of years ago when our part of the universe was formed. Unstable nuclei that originate from the natural events are often called *radioactive*; the processes that occur in their decay are often called *radioactive decay*, or *radioactivity*. One of the reasons why radioactive decay is interesting is that it provides clues about the origin of the universe.

A process that is particularly important in radioactive decay is *$\alpha$ decay*, occurring commonly in nuclei with atomic number greater than $Z = 82$. It involves the decay

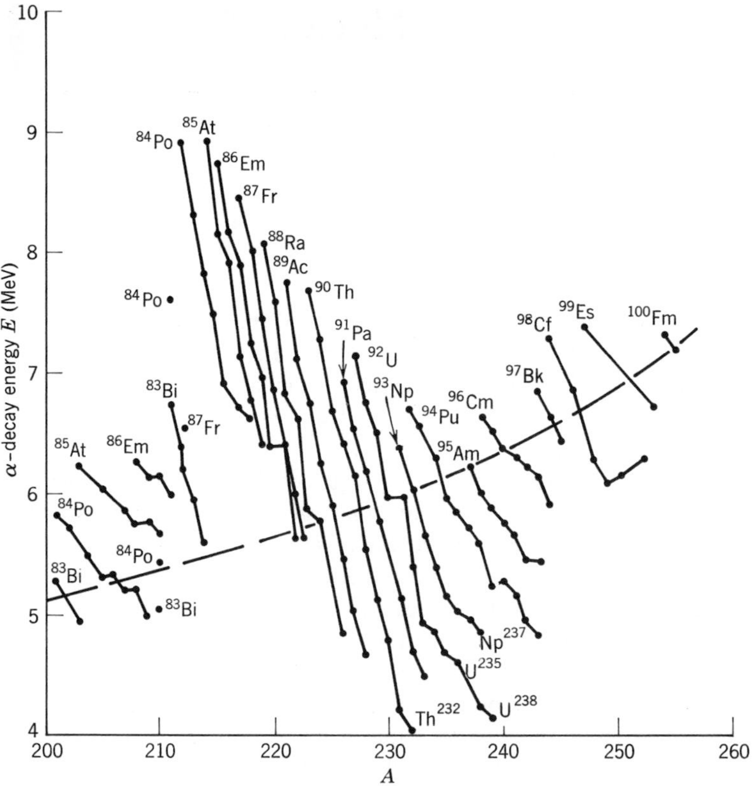


**FIGURE 16-1**

Alpha-decay energies for nuclei in the α-emitting region. The dashed curve represents the general trend predicted by the semiempirical mass formula.

of an unstable *parent nucleus* into its *daughter nucleus* by the emission of an *α particle*, the nucleus $^2He^4$. The process takes place spontaneously because it is energetically favored, the mass of the parent nucleus being greater than the mass of the daughter nucleus plus the mass of the α particle. The reduction in nuclear mass in the decay is primarily due to a reduction in the Coulomb energy of the nucleus when its charge $Ze$ is reduced by the charge $2e$ carried away by the α particle. The energy made available in the decay is the energy equivalent of the mass difference. This decay energy is carried away by the α particle as kinetic energy. Ignoring the mass equivalents of atomic electron binding energies, the *α-decay energy E* can be written in terms of the atomic masses of the parent nucleus, $M_{Z,A}$, of the daughter nucleus, $M_{Z-2,A-4}$, and of the α particle, $M_{2,4}$, as

$$E = [M_{Z,A} - (M_{Z-2,A-4} + M_{2,4})]c^2 \tag{16-1}$$

Figure 16-1 displays the decay energies $E$ for parent nuclei in the α-emitting range of $Z$, or $A$. The data are obtained from direct measurements of the kinetic energy of the α particles by bending them in a magnetic field, and/or by using (16-1) with the measured masses. The dashed line represents the general trend for the parent nuclei to become increasingly unstable to α decay as $A$ gets further away from the value $A \simeq 60$, where the average binding energy per nucleon, $\Delta E/A$, maximizes. It also represents the predictions of the liquid drop model. Superimposed on the general trend is a peak, roughly 4 MeV high, occurring at the parent nucleus $^{84}Po^{212}$. The peak is explained by the shell model as due to the particular stability of the associated daughter nucleus, $^{82}Pb^{208}$. Since the daughter has magic $Z = 82$ and magic $N = 126$, it is about 4 MeV more tightly bound than typical nuclei in this region of $A$. (Figure 15-13 shows that about 2 MeV of extra binding energy is found at each magic number.) Note that the $\alpha$-decay energies range from 8.9 MeV for $^{84}Po^{212}$ to 4.1 MeV for $^{90}Th^{232}$.

The moderately energetic particles emitted in $\alpha$ decay of radioactive nuclei were put to very good use by Rutherford, and others, in the scattering experiments that led to the discovery of nuclei (see Chapter 4). Similar use continued to be made of $\alpha$ particles from radioactive sources in investigating nuclear structure, until the invention of cyclotrons by Lawrence in the late 1930s. Cyclotrons, and other types of particle accelerators, produce particles of higher energy which can be used in more precise measurements because they have shorter de Broglie wavelength. Accelerators also produce more intense beams of particles than can be obtained from radioactive sources, and this makes the measurements easier to carry out.

**Example 16-1.** An $\alpha$ particle is emitted by the parent nucleus $^{84}Po^{212}$. Estimate the Coulomb potential it feels at the nuclear surface, and then make an approximate plot of the sum of the Coulomb and nuclear potentials acting on the $\alpha$ particle in various locations.

If we approximate the daughter nucleus and the $\alpha$ particle as uniformly charged spheres, the Coulomb repulsion potential energy when they are just touching will be

$$V_0 = +\frac{2Ze^2}{4\pi\epsilon_0 r'}$$

where $+2e$ is the $\alpha$-particle charge, $+Ze$ is the daughter nucleus charge, and $r'$ is the sum of the radii of the $\alpha$ particle and daughter nucleus uniform charge distributions. We can estimate these radii by using the charge density half value radii $a$ of the actual charge distributions found in the electron scattering measurements, and quoted in (15-6)

$$a = 1.07A^{1/3}\text{ F}$$

We obtain for the sum of the radii

$$r' = (4^{1/3} + 208^{1/3})1.07\text{ F}$$
$$= 8.0\text{ F}$$

So

$$V_0 = \frac{2 \times 82 \times (1.6 \times 10^{-19}\text{ coul})^2}{1.1 \times 10^{-10}\text{ coul}^2/\text{nt-m}^2 \times 8.0 \times 10^{-15}\text{ m}} = 4.8 \times 10^{-12}\text{ joule}$$
$$= 30\text{ MeV}$$

Figure 16-2 indicates the total (Coulomb plus nuclear) potential acting on the $\alpha$ particle. As it approaches the nucleus, it feels the repulsive Coulomb potential increasing in inverse

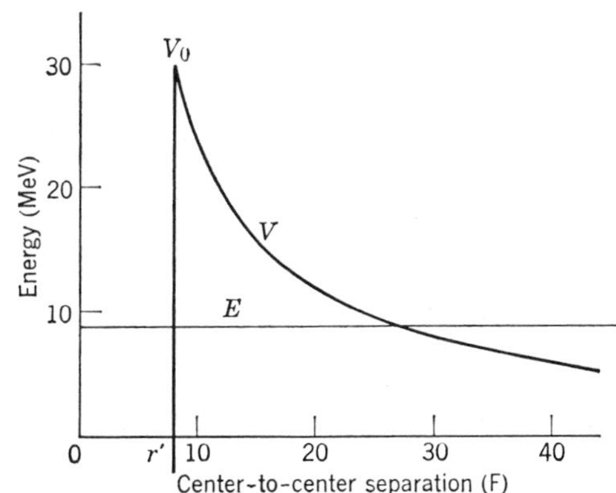


**FIGURE 16-2**
An approximate representation of the Coulomb plus nuclear potential $V$ acting on an $\alpha$ particle emitted from a $^{84}Po^{212}$ nucleus, and the total energy $E$ of the $\alpha$ particle.

proportion to the distance between the centers of the $\alpha$ particle and nucleus, and reaching the value of $V_0$ when this distance equals $r'$. Inside the surface it feels a rapid onset of the strong attractive nuclear potential, which soon dominates. (The onset is, of course, not quite as rapid as shown in the figure.) Also indicated is the ${}^{84}Po^{212}$ $\alpha$-decay energy $E = 8.9$ MeV, which is the energy of the emitted $\alpha$ particle. Note that it is much less than $V_0$, the height of the Coulomb barrier. ◀

Since every decay energy shown in Figure 16-1 is far less than the height of the Coulomb barriers, which is $\sim$30 MeV for all $\alpha$ decays, the $\alpha$ particle tends to be trapped by the barrier in every decay. It can escape only by the quantum mechanical process of *barrier penetration*. We have previously gone through a detailed treatment of this process, so here we shall only remind the student of the results, but he would be well advised to look again at Section 6-6. At least he should inspect Figure 6-20, which plots the probability per second that a nucleus will emit an $\alpha$ particle, called the *decay rate* $R$, versus the decay energy $E$. The figure shows that the decay rate decreases *extremely* rapidly as the decay energy decreases and the $\alpha$ particle tunnels more deeply through the Coulomb barrier.

Now consider a system containing many nuclei of the same species at some initial time. The nuclei $\alpha$ decay (or, equally well, $\beta$ or $\gamma$ decay) at the decay rate $R$. We shall calculate the number of undecayed nuclei present at some subsequent time. If there are $N$ undecayed nuclei at time $t$, then the number decaying in the following time interval $dt$ can be written $dN$. Since $R$ is the probability that a particular nucleus will decay in 1 sec, $R\,dt$ is the probability that it will decay during the time interval, and $NR\,dt$ is the probability that any one of the nuclei will decay in that interval. Thus the average number of decaying nuclei is

$$dN = -NR\,dt \tag{16-2}$$

where the minus sign accounts for the fact that $dN$ is intrinsically negative since $N$ decreases. Rearranging the terms, and integrating, we obtain

$$\frac{dN}{N} = -R\,dt$$

$$\int_{N(0)}^{N(t)} \frac{dN}{N} = -R\int_0^t dt = -Rt$$

$$\ln N(t) - \ln N(0) = \ln \frac{N(t)}{N(0)} = -Rt$$

or

$$\frac{N(t)}{N(0)} = e^{-Rt}$$

so

$$N(t) = N(0)e^{-Rt} \tag{16-3}$$

In this expression $N(0)$ is the number of undecayed nuclei at the initial time 0, and $N(t)$ the number of undecayed nuclei at the subsequent time $t$. Since the calculation involves probabilities, its results are correct only on the average, but fluctuations from the average are very small in the typical case in which the number of nuclei involved is very large. Figure 16-3 is a plot of (16-3), which is called the *exponential decay law*.

Also indicated in Figure 16-3 is the *lifetime* $T$ characteristic of the decay. This is the average time a nucleus survives before it decays. It is obvious from their definitions that $T$ is inversely proportional to the decay rate $R$. In fact, it is easy to show from a

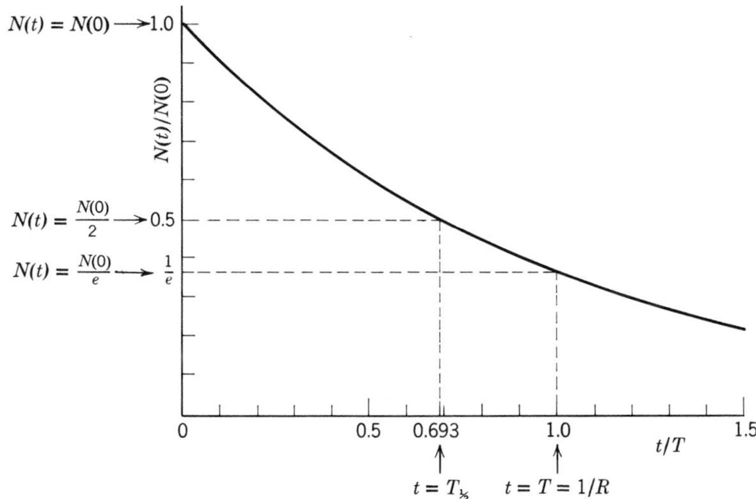


**FIGURE 16-3**

The exponential decay law for $N(t)$, the number of nuclei surviving at time $t$. Also shown are the lifetime $T$ and half-life $T_{1/2}$. Note that $N(t)$ is expressed in units of the original number of nuclei $N(0)$, while time is expressed in units of the lifetime $T$.

simple integration of the decay law that

$$T = \frac{1}{R} \tag{16-4}$$

Using this relation in (16-3), we conclude that in one lifetime the number of undecayed nuclei decreases by a factor of $e$, as indicated in the figure. Further indicated is the *half-life* $T_{1/2}$, which is the time required for the number of undecayed nuclei to decrease by a factor of 2. The relation between the two times is obtained directly from the decay law

$$T_{1/2} = (\ln 2)T = 0.693T \tag{16-5}$$

In a more typical system, there are several related radioactive nuclei decaying successively into each other by $\alpha$ decay (and/or other decay processes). For instance, ${}^{92}U^{234}$ $\alpha$ decays into ${}^{90}Th^{230}$, which $\alpha$ decays into ${}^{88}Ra^{226}$, etc. Thus a system initially filled with ${}^{92}U^{234}$ will eventually contain a mixture of all these nuclei. Differential equations governing the general behavior of such a family can be written down easily, and they can be solved with not much more difficulty in certain cases. In the most important case, the significant features of the solution can be discerned from the following qualitative argument. Consider a family of decays in which the parent has by far the smallest decay rate, or longest lifetime. The situation is indicated schematically in Figure 16-4. On a time scale comparable with the parent lifetime, the population of the parents decreases exponentially. But on the much shorter time scale comparable to the daughter lifetimes, the population of the parents remains essentially constant, and so the total number decaying per second into the first daughters seems constant. Since the first daughters decay rapidly after they are formed, their population is governed by the constant resupply from decay of the parents. Thus the population of the first daughters remains constant. The same is true for the second daughters, since they are being formed at a constant rate from the decay of the constant population of the first daughters. In fact, the populations of all the daughters will remain

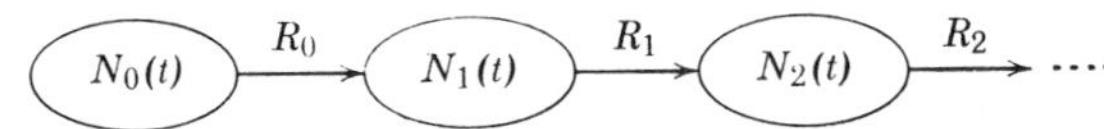


**FIGURE 16-4**
A schematic representation of a family of successive decays.

constant as long as we consider times short compared to the parent lifetime so that the population of the parents remains essentially constant. (If we consider longer times all that happens is that the population of the parents, and of all the daughters, decreases exponentially at the same rate following the slow decay of the parents.) Thus, on the shorter time scale, we have an *equilibrium* condition, which requires that the following relation be satisfied.

$$N_0R_0 = N_1R_1 = N_2R_2 = \cdots \tag{16-6}$$

For instance, the left side of the first equality is the total number of parents decaying per second to form first daughters, while the right side is the total number of first daughters decaying per second. If the total rate of formation of first daughters did not equal their total rate of decay, their population would not remain constant. Equation 16-6 describes the most important case of a family of decays. It is sometimes used to determine the values of the $R$, or $T$, from measurements of the $N$, and one known $R$.

We can now understand how $\alpha$-decaying nuclei with very short lifetimes can be found in nature. For example, ${}^{84}\text{Po}^{212}$, with $T \sim 10^{-6}$ sec, can be extracted from naturally occurring minerals that presumably have been in existence for billions of years. The reason is simply that the short lifetime $\alpha$ emitters are in equilibrium in decay families with long lifetime parents, called *radioactive series*. There are three such series that occur naturally: the $4n$ *series* whose parent is ${}^{90}\text{Th}^{232}$ with $T = 2.01 \times 10^{10}$ yr, the $4n + 2$ *series* whose parent is ${}^{92}\text{U}^{238}$ with $T = 6.52 \times 10^{9}$ yr, and the $4n + 3$ *series* whose parent is ${}^{92}\text{U}^{235}$ with $T = 1.02 \times 10^{9}$ yr. The names characterize the $A$ values for the members of the series. For instance, the parent of the $4n + 3$ series has $A$ equal to four times an integer plus three, where the integer is 58. Since each $\alpha$ decay reduces $A$ by four (and the other decay processes do not change $A$), all the daughters of this series will also have $A$ equal to four times some smaller integer plus three.

There is evidently also room for a $4n + 1$ series. Actually there is such a series, whose parent is ${}^{93}\text{Np}^{237}$ with lifetime $T = 3.25 \times 10^{6}$ yr. The series can be produced artifically by using a nuclear reaction to make the parent, but it is not found in nature since the lifetime of the parent is very short compared to the age of the earth, which is estimated from geological and cosmological evidence to be $\sim 10^{10}$ yr (see Example 16-2). Consequently any parent nuclei initially present have decayed away.

In this connection note that Figure 16-1 shows the decay energies of the parents of the three naturally occurring series are particularly low. If they were less than 1 MeV higher their decay rates would be so much higher, and their lifetimes so much shorter than $\sim 10^{10}$ yr, the age of the earth, that the naturally occurring elements would stop at $Z = 82$ instead of $Z = 92$. The same figure indicates why the presently known naturally occurring elements do stop at $Z = 92$. It is because the $\alpha$-decay energies for nuclei with $Z > 92$ are large enough to lead to lifetimes short compared to the age of the earth. Finally, an extrapolation of Figure 16-1 to $Z < 82$ shows that the corresponding elements are *apparently* stable to $\alpha$ decay because their decay energies are so small that the lifetimes are immeasurably long.

Students frequently wonder why nuclei of large $Z$ spontaneously emit $\alpha$ particles, ${}^{2}\text{He}^{4}$, but do not spontaneously emit any of the particles ${}^{2}\text{He}^{3}$, ${}^{1}\text{H}^{2}$, or ${}^{1}\text{H}^{1}$, even though emitting any of these particles reduces the Coulomb energy of the nucleus.

The reason is simply that for the particles other than ${}^2\text{He}^4$ the binding energy per nucleon, $\Delta E/A$, is much smaller than it is for a typical nucleus. Thus their emission is not energetically favorable. The emission of a ${}^6\text{C}^{12}$ particle from a nucleus of large $Z$ would be energetically favorable because it has a high $\Delta E/A$ and also reduces considerably the Coulomb energy of the nucleus. And the emission of a particle of even larger $Z$ would be even more so because of the increased reduction of the Coulomb energy. Such a process is called *spontaneous fission*. For naturally occurring nuclei of the highest $Z$ values, i.e., for $Z$ in the range just below 92, the decay rate for spontaneous fission is very much smaller than the decay rate for emitting an $\alpha$ particle because of the very much reduced probability of a more massive particle penetrating a higher Coulomb barrier. As $Z$ becomes larger than about 100, the decay rate for spontaneous fission becomes comparable to, and eventually larger than, the decay rate for $\alpha$-particle emission. The reason is that with increasing $Z$ the decay energy for spontaneous fission increases more rapidly than the decay energy for $\alpha$-particle emission, so the spontaneous fission Coulomb barrier becomes relatively easier to penetrate.

There is a recent, and as yet unverified, prediction that the nucleus of the element with $Z = 110$ and $A = 294$ might have a lifetime as long as $\sim 10^8$ yr. If so, a little of it could possibly still be present on the earth if enough of it were formed $\sim 10^{10}$ yr ago. The prediction follows from the prediction that the proton magic number after $Z = 82$ is $Z = 114$, not $Z = 126$ as indicated in Figure 15-18 of the shell model. Of course the prediction of that figure that $N = 126$ is a neutron magic number is abundantly verified by experiment, and it is also believed that $N = 184$ is a neutron magic number as predicted by the figure. But there is no experimental evidence concerning $Z$ values much beyond 100 since the corresponding nuclei have not been discovered yet, so $Z = 126$ is not actually known to be magic. The difference between the recent shell model predictions for the higher proton and neutron magic numbers arises because for protons there is, in addition to the nuclear potential, a repulsive Coulomb potential that becomes large for nuclei of large $Z$. It tends to raise all the proton levels, but more so for levels of small $l$ whose probability densities extend deeper into the nuclear center where the Coulomb potential is stronger. The result is to raise the $2f$ and $3p$ levels relative to the $1i$ level, making the $1i_{13/2}$ level lie just above the $2f_{7/2}$ level, and creating a proton magic number at $Z = 100 + 14 = 114$. Thus the nucleus with $Z = 114$, and $N = 184$, is believed to be doubly magic. That nucleus also lies near, but not on, the curve of maximum stability obtained from an extrapolation of the semiempirical mass formula of the liquid drop model. In other words, $Z = 114$ and $N = 184$, or $Z = 114$ and $A = 298$, is expected to be doubly magic and also to have almost the most stable value of $Z$ for that value of $A$. Collective model calculations indicate that the best compromise between the requirements for stability of the shell and liquid drop models is obtained by removing four protons to reduce the Coulomb energy, which is extremely important for nuclei of such large $Z$. Thus these calculations predict maximum stability at $Z = 110$ and $A = 294$. They also predict a lifetime of $\sim 10^8$ yr against decay by $\alpha$-particle emission or spontaneous fission into two smaller nuclei. The fission process is actually the most likely decay because it is more effective in reducing the Coulomb energy. So $Z = 110$ and $A = 294$ is predicted to be "an island of stability in a sea of spontaneous fission."

**Example 16-2.** In the mixture of isotopes normally found on the earth at the present time, ${}^{92}\text{U}^{238}$ has an abundance of 99.3% and ${}^{92}\text{U}^{235}$ has an abundance of 0.7%. The measured lifetimes of these radioactive isotopes are $6.52 \times 10^9$ yr and $1.02 \times 10^9$ yr, respectively. By assuming that they were equally abundant when the uranium in the earth was originally formed, estimate how much time has elapsed since the time of formation. (That is assume pairing effects in the initial formation ratios are small compared to lifetime effects in the present abundance ratios.)

If the number of ${}^{92}U^{238}$ nuclei originally formed is $N$, the number present now is

$$N_{238} = Ne^{-Rt} = Ne^{-t/T} = Ne^{-t/6.52}$$

where $t$ is the elapsed time in units of $10^9$ yr. Since the number of ${}^{92}U^{235}$ nuclei originally formed is, by assumption, also $N$, the number now present is

$$N_{235} = Ne^{-t/1.02}$$

The present abundance of ${}^{92}U^{235}$ is

$$7 \times 10^{-3} = \frac{N_{235}}{N_{235} + N_{238}} \simeq \frac{N_{235}}{N_{238}} = \frac{Ne^{-t/1.02}}{Ne^{-t/6.52}}$$
$$= e^{-(t/1.02 - t/6.52)} = e^{-0.827t}$$

So

$$e^{0.827t} \simeq \frac{1}{7 \times 10^{-3}} = 143$$
$$0.827t \simeq \ln(143) = 4.96$$
$$t \simeq \frac{4.96}{0.827} = 6.0$$

That is, the elapsed time is

$$t \simeq 6.0 \times 10^9 \text{ yr}$$

The estimate obtained from this simple argument is in reasonable agreement with the estimates of the age of the earth, or of the solar system, obtained from more sophisticated geological and cosmological arguments. ◀

## 16-3 Beta Decay

A more complete description of the processes occurring in the $4n$ radioactive series is plotted in Figure 16-5. In addition to α decay, there is also *β decay*. For the radioactive series, β decay involves a nucleus $Z$, $A$ emitting a negatively charged electron and being transformed into the nucleus $Z + 1$, $A$. There are also two other types of β decay that will be considered shortly.

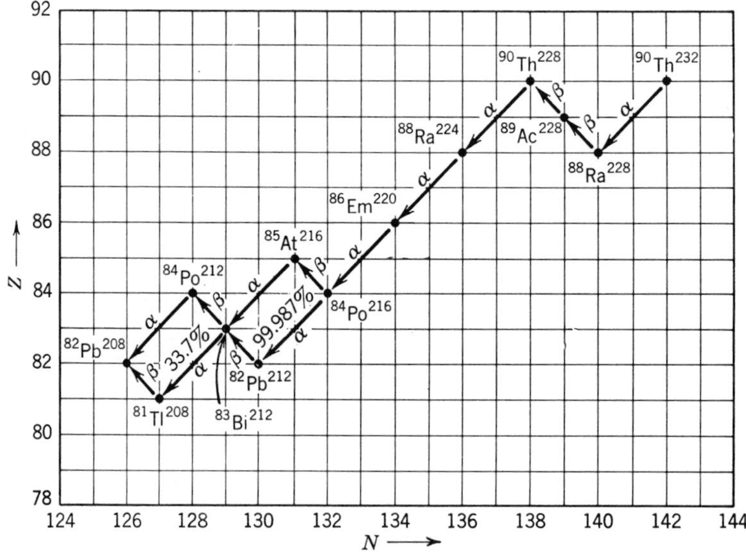


**FIGURE 16-5**

The decay processes occurring in the $4n$ series.

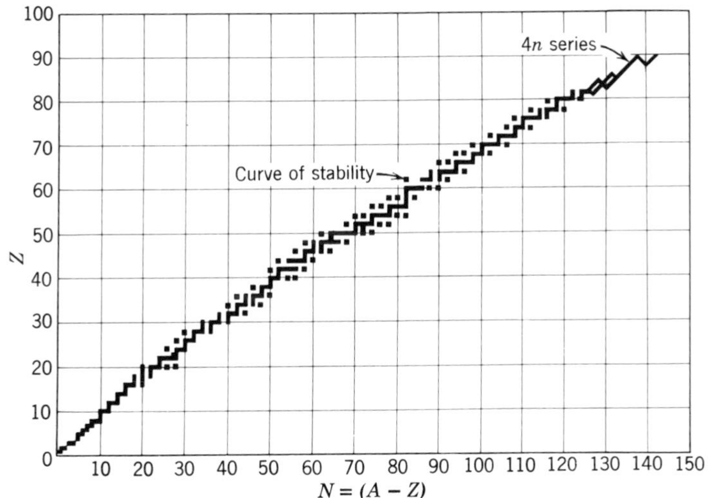


**FIGURE 16-6**
Illustrating why $\beta$ decay occurs in the $4n$ and other radioactive series.

It is instructive to superimpose Figure 16-5 on Figure 15-11, the plot of the $Z$ and $N$ values of the stable nuclei. The result, shown in Figure 16-6, makes it clear that the radioactive series uses $\beta$ decay to keep as good a match as possible between the average slope of the path traced out by its decay and the average slope of the "curve of stability." Another way of saying this is that the $\alpha$-decay energy of a nucleus is relatively small if the nucleus it would $\alpha$ decay into is too far from the curve of stability. But in just these circumstances the $\beta$-decay energy is relatively large. As the decay rates for both processes increase rapidly with increasing decay energy, the nucleus in question will $\beta$ decay because that process has a larger decay energy, and so a much larger decay rate. In some cases, the decay rates for the two competing processes are comparable, both processes occur, and the series *branches* (see ${}^{84}Po^{216}$ and ${}^{83}Bi^{212}$ in the $4n$ series).

In the first part of this section we shall study the energetics of $\beta$ decay. Then we shall study the dependence of the decay rate on the decay energy. There we shall see that the decay rate also depends strongly on the spins and parities of the nuclear states involved in the decay. This dependence on spin and parity makes the $\beta$-decay process a very useful tool in the investigation of nuclei.

To discuss the energetics of $\beta$ decay, we plot atomic masses $M_{Z,A}$, in the region of the curve of stability, as a function of $Z$ for fixed $A$. Figure 16-7 shows typical results for odd $A$, and Figure 16-8 shows results typical for even $A$. Except near magic numbers, all the results are well described by the semiempirical mass formula. For odd $A$, the values of $M_{Z,A}$ are found to lie on a parabola. For even $A$, there are two parabolas corresponding to the two possible signs of the pairing term, (15-28); the upper one is for odd $Z$, odd $N$, and the lower one is for even $Z$, even $N$. These curves are really cross cuts through the curve of stability, showing its structure. They specify how the masses increase when the $Z$ values depart from their most stable values for a given $A$. Note that for an odd value of $A$, there is generally only one most stable value of $Z$. (Rarely there are two values straddling the bottom of the parabola that happen to lead to almost the same mass.) For a given even $A$, there are generally two stable values of $Z$ (but occasionally there are three).

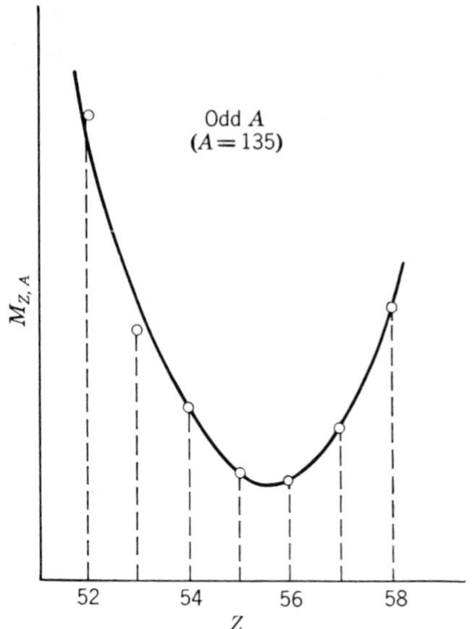


**FIGURE 16-7**
The masses of atoms with a given odd value of $A$. The value $A = 135$ is chosen for this example.

Nuclei whose $Z$ values are not the most stable, in consideration of their $A$ values, can change $Z$ to attain stability by three different $\beta$-decay processes. One is the process of *electron emission* that occurs in the radioactive series. In this process, a negatively charged electron is emitted by the nucleus, so $Z$ increases by one, $N$ decreases by one, and $A$ remains fixed. The other processes are *electron capture* and *positron emission.* In the former the nucleus captures a negatively charged atomic electron, and in the latter it emits a positively charged positron. In both, $Z$ decreases by one, $N$ increases by one, and $A$ remains fixed.

Electron emission takes place if the mass $m_{Z,A}$ of the initial nucleus exceeds the mass $m_{Z+1,A}$ of the final nucleus plus one electron rest mass $m$. The mass excess times $c^2$ equals the energy $E$ made available in the decay. That is, the decay energy is

$$E = [m_{Z,A} - (m_{Z+1,A} + m)]c^2 \tag{16-7a}$$

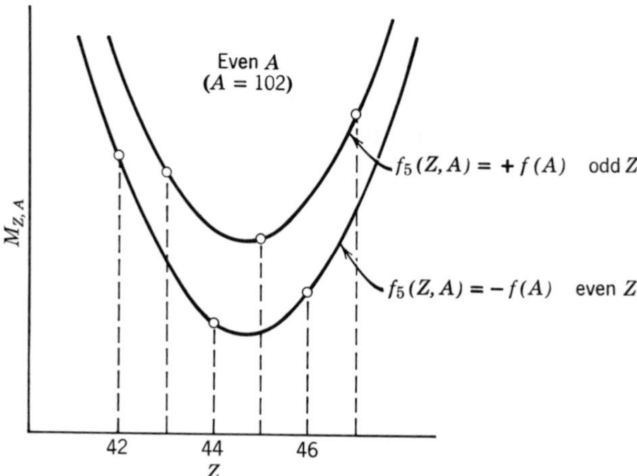


**FIGURE 16-8**
The masses of atoms with a given even value of $A$. The value $A = 102$ is chosen for this example.

This energy must be positive for the decay to occur. We can write it in terms of atomic masses by adding and subtracting $Z$ electron rest masses, to yield

$$E = [m_{Z,A} + Zm - (m_{Z+1,A} + Zm + m)]c^2$$

Neglecting the binding energies of atomic electrons, we obtain the simple result that the *decay energy in electron emission is*

$$E = [M_{Z,A} - M_{Z+1,A}]c^2 \tag{16-7b}$$

We see that electron emission occurs when the initial atomic mass exceeds the final atomic mass because the mass of the electron added to the atom is compensated for by the mass of the electron emitted by the nucleus.

Electron capture takes place if the mass $m_{Z,A}$ of the initial nucleus plus one electron rest mass $m$ exceeds the mass $m_{Z-1,A}$ of the final nucleus. The energy made available in the decay is

$$E = [(m_{Z,A} + m) - m_{Z-1,A}]c^2 = [m_{Z,A} - (m_{Z-1,A} - m)]c^2 \tag{16-8a}$$

or

$$E = [m_{Z,A} + Zm - (m_{Z-1,A} + Zm - m)]c^2$$

In terms of atomic masses, the *decay energy in electron capture* is

$$E = [M_{Z,A} - M_{Z-1,A}]c^2 \tag{16-8b}$$

When the energy is positive electron capture occurs. This simple result is obtained because the mass of the electron taken from the atom in the capture is compensated for by the mass of the electron captured by the nucleus.

Positron emission requires the mass $m_{Z,A}$ of the initial nucleus exceed the mass $m_{Z-1,A}$ of the final nucleus plus one positron rest mass, which also equals $m$. The energy made available in the decay is

$$E = [m_{Z,A} - (m_{Z-1,A} + m)]c^2 \tag{16-9a}$$

or

$$E = [m_{Z,A} + Zm - (m_{Z-1,A} + Zm - m) - 2m]c^2$$

In terms of atomic masses, this expression says that the *decay energy in positron emission* is

$$E = [M_{Z,A} - M_{Z-1,A} - 2m]c^2 \tag{16-9b}$$

In positron emission the atom must emit one electron since its nucleus emits one positron and has, therefore, one less positive charge. Thus there cannot be the compensation of electron masses found in the other $\beta$-decay processes. The result is that in order to have the decay energy in positron emission positive, which is a necessary condition for the process to occur, the initial atomic mass must exceed the final atomic mass by more than two electron rest masses, $2m = 0.00110u$.

We conclude that if $M_{Z,A} > M_{Z+1,A}$ then electron emission can occur. If $M_{Z,A} > M_{Z-1,A}$ then electron capture can occur. But positron emission can occur only if $M_{Z,A} > M_{Z-1,A} + 2m$; and in this case electron capture can also occur. Thus there is a range in which the difference in atomic masses is such that electron capture is possible while positron emission is energetically forbidden. In practice, atomic mass differences frequently fall in this range and so there are relatively few positron emitters in nature. In all these processes the decay energy $E$ varies from case to case from a small fraction of 1 MeV to more than 10 MeV, and typically it is somewhat less than 1 MeV.

**Example 16-3.** The only known nuclei with $A = 7$ are ${}^3\text{Li}^7$, whose atomic mass is $M_{3,7} = 7.01600u$, and ${}^4\text{Be}^7$, whose atomic mass is $M_{4,7} = 7.01693u$. Which of these nuclei is stable to $\beta$ decay? What process is employed in the $\beta$ decay of the unstable nucleus to the stable nucleus?

Since the atomic mass of ${}^3\text{Li}^7$ is the lowest, it is the nucleus which is $\beta$ stable.

As far as charge conservation is concerned, the $\beta$-unstable ${}^4\text{Be}^7$ could decay into the stable nucleus either by capturing an atomic electron or by emitting a positron. But as far as energy conservation is concerned, only electron capture is possible since the difference in the atomic masses, $M_{4,7} - M_{3,7} = 7.01693u - 7.01600u = 0.00093u$, is less than two electron masses, $2m = 0.00110u$. Thus electron capture is the process employed in the $\beta$ decay of ${}^4\text{Be}^7$ into ${}^3\text{Li}^7$. ◀

Now let us consider the very interesting question of what happens to the decay energy in $\beta$-decay processes. Take the most common one, electron emission. A nucleus $Z$, $A$, which we assume to be stationary in the initial state, emits an electron and recoils, as indicated in Figure 16-9. If there are just two particles in the final state, there can be only one linear momentum conserving way in which the available energy, which is the decay energy $E$, can be shared. In fact, since nuclei are so massive their recoil velocities are extremely low and they carry practically no kinetic energy. Thus the electron should carry away almost all of the decay energy $E$ in the form of kinetic energy. But measurements made at an early stage in the study of radioactivity, using bending magnets, showed that the electrons are emitted with a spectrum of kinetic energies $K_e$, as indicated in Figure 16-10.

For many years, the fact that electrons are emitted in $\beta$ decay with a spectrum of energies was very mysterious and very disturbing. Electrons emitted at the *end point* $K_e^{\max}$ of the spectrum carry away all the decay energy $E$, since $K_e^{\max}$ was observed to equal $E$ within experimental accuracy. That is

$$K_e^{\max} = E \tag{16-10}$$

But typical electrons carry away much less than the energy $E$ which, the measured mass differences show, must be released in the process. It would appear that some of this energy has vanished! Several attempts were made to detect the missing energy, for instance by placing the $\beta$-decaying material inside a calorimeter with very thick lead walls, but they were fruitless. The situation was grave enough that some physicists were beginning to seriously consider abandoning the law of conservation of relativistic energy, when Pauli proposed a less repugnant alternative.

In 1931 Pauli postulated that a particle, now called the *antineutrino* $\bar{\nu}$, is also emitted in the electron emission process, but it is not normally detected because its *interaction with matter is extremely weak*. He also postulated that the antineutrino has (1) *zero charge*, (2) *intrinsic spin* $s = 1/2$, and (3) *zero rest mass*. The first property permits charge conservation to be maintained in electron emission. The second property allows angular momentum to be conserved. Consider the nucleus $Z$, $A$ emitting an electron to become the nucleus $Z + 1$, $A$ and assume, for example, that $A$

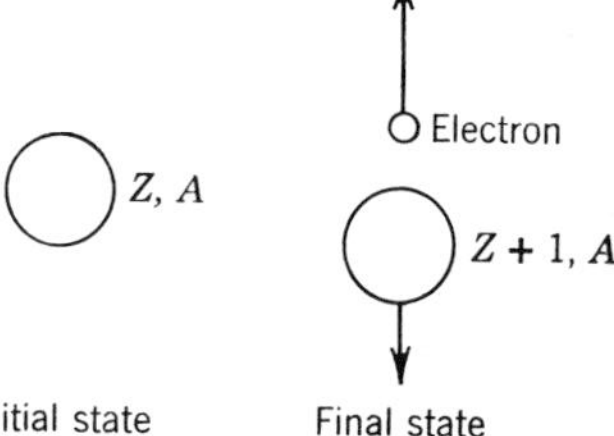


**FIGURE 16-9**
The electron emission process, assuming (incorrectly, as we shall see) that only two particles comprise the final state.

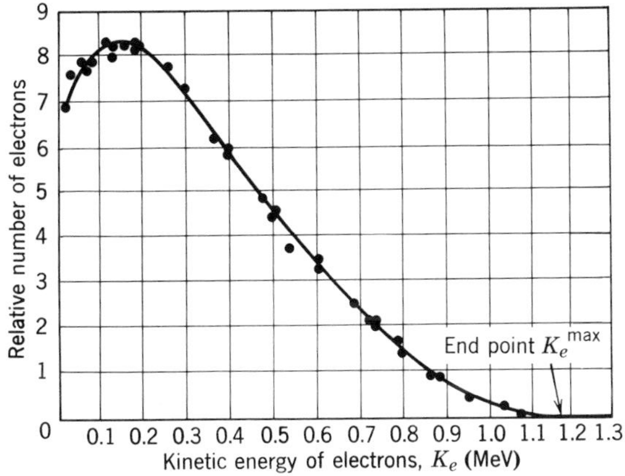


**FIGURE 16-10**
The spectrum of electrons emitted in the $\beta$ decay of $^{83}Bi^{210}$.

is even. Then the nuclear spin $i$ is an integer for both the initial and final nuclei. If only the electron with intrinsic spin $s = 1/2$ were emitted, it would be impossible to conserve angular momentum, because the sum of a half-integral angular momentum (the electron) and an integral angular momentum (the final nucleus) can only be half-integral. If an antineutrino with $s = 1/2$ is also emitted, the difficulty is removed. The third property was postulated to agree with the observation that the end point $K_e^{max}$ of the electron spectrum equals the decay energy $E$. When an electron happens to be emitted at the end point, it carries away all the decay energy and none is left for rest mass energy of the antineutrino. In positron emission and electron capture, the particle that is emitted, but very difficult to detect, is called the *neutrino* $\nu$. It has the same zero charge, spin 1/2, and zero rest mass as the antineutrino.

The relation between neutrinos and antineutrinos is explained by Dirac's relativistic quantum mechanics. This theory shows that every particle with intrinsic spin $s = 1/2$ has its antiparticle. A familiar, and closely related, example is the electron and its antiparticle called the positron. (Unrelated examples are the proton and antiproton, and neutron and antineutron.) The theory also shows that when a particle is produced a related antiparticle must be produced. The familiar example is, again, the electron and positron, which are produced in pairs. This is also found in the three $\beta$-decay processes. In electron emission a particle (electron) is produced with an antiparticle (antineutrino), while in positron emission a particle (neutrino) is produced with an antiparticle (positron). Electron capture fits into this scheme since in the Dirac theory the destruction of an electron is identical to the creation of a positron.

Figure 16-11 schematically illustrates electron and positron emission in terms of Dirac energy-level diagrams for the related particles, electrons and neutrinos. We saw in the discussion of Figure 2-15 that in pair production the energy of an absorbed photon makes possible the transition of an electron of rest mass $m$ from one of the all pervading sea of filled electron levels that extend downward from $-mc^2$ to one of the empty levels that extend upward from $+mc^2$. The result is an electron in a positive energy level, and a hole in a negative energy level, which is a positron. Such a transition could be represented by a vertical arrow connecting the lower and upper electron levels. In a similar way, an electron emission transition can be represented by a diagonal arrow connecting a filled neutrino level with an empty electron level, as shown in Figure 16-11. The energy made available by the difference in the nuclear masses converts a neutrino from the neutrino sea into an electron, leaving a hole in a neutrino level, which is an antineutrino. The diagonal arrow connecting a filled electron level with an empty neutrino level represents positron emission since the result is a hole in an

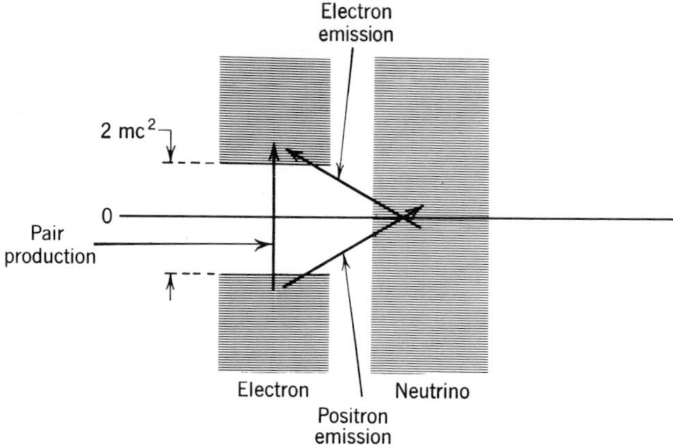


**FIGURE 16-11**

Electron and neutrino Dirac energy-level diagrams illustrating pair production, electron emission, and positron emission.

electron level, or positron, and a neutrino. Note that there is no gap separating the filled and empty neutrino levels because neutrinos have zero rest mass. Also note that the minimum energy that the nuclear mass difference must provide to make either $\beta$-decay process possible is one electron rest mass energy, $mc^2$, in agreement with (16-7a) and (16-9a).

There is an obvious distinction between a particle and its antiparticle if they are charged, because their charges are of opposite sign. The distinction is more subtle if the particle and antiparticle are neutral, like the neutrino and antineutrino. Nevertheless, there really is a distinction. Recent evidence that we shall discuss soon shows the component of intrinsic spin angular momentum along the direction of motion is always $-\hbar/2$ for a neutrino and always $+\hbar/2$ for an antineutrino.

The problem concerning the emission of electrons with a spectrum of energies is resolved by the postulate that an antineutrino is also emitted in the $\beta$ decay, since then the decay energy $E$ can be shared between the electron kinetic energy $K_e$ and the antineutrino kinetic energy $K_{\bar{\nu}}$. That is

$$K_e + K_{\bar{\nu}} = E \tag{16-11}$$

where we neglect the nuclear recoil energy. As there are very many ways in which this energy division can be made, the values of $K_e$ form a spectrum. Detailed agreement with the measured forms of the $\beta$-decay spectra can be obtained if the argument is made quantitative. This involves the use of statistical procedures, similar to but somewhat more complicated than those used in Chapters 1 and 11, to determine the number of energy divisions in each range of $K_e$.

The results are most conveniently expressed, and explained, in terms of the *momentum spectrum* $R(p_e)$, which is the rate of emission of electrons with linear momentum $p_e$ per unit time and per unit momentum. It is found that

$$R(p_e) \simeq \left[\frac{(E - K_e)^2 p_e^2}{2\pi^3\hbar^7c^3}\right] M^*M \tag{16-12}$$

where $M$ is the *$\beta$-decay matrix element*

$$M = \int \psi_f^* \beta \psi_i \, d\tau \tag{16-13}$$

In (16-12) the term $(E - K_e)^2 = K_{\bar{\nu}}^2$ is proportional to $p_{\bar{\nu}}^2$, the square of the antineutrino linear momentum. So the rate $R$ is proportional to the product of two factors, each of which is the square of the momentum of one of the particles emitted in the $\beta$ decay. These $p^2$ factors are just measures of the number of quantum states per unit momentum interval into which the antineutrino, or electron, can be emitted in the decay. Both can be obtained by a trivial modification of the argument in Example 1-3. If the allowed wavelength $\lambda$ in Figure 1-7 is taken to be the de Broglie wavelength of a particle in a box, then (1-15) can immediately be converted from the form $N(r) \propto r^2$ to the form $N(p) \propto p^2$ since the quantity $r$ in that equation is inversely proportional to $\lambda$ and, according to de Broglie, $\lambda$ is inversely proportional to the particle's momentum $p$. Thus we see that $N(p)$, the number of allowed states per unit momentum interval for an antineutrino or electron of momentum $p$, which is confined to a box, is proportional to $p^2$. The box is a mathematical one that is used to normalize the free particle eigenfunctions representing the emitted antineutrino, or electron, as discussed in Section 6-2.

In other words, if a particle is confined to a box (of arbitrarily large dimensions) so that its eigenfunction can be normalized, it is no longer strictly a free particle and thus has a discrete (albeit arbitrarily closely spaced) set of quantum states available to it. The number of these states per unit momentum is proportional to the square of its momentum. If we then make the usual statistical assumption that all possible divisions of energy, or momentum, occur with the same probability, the rate for a $\beta$ decay with a particular division will be proportional to the total number of states for that division, which is the number of states for one particle times the number of states for the other. Thus the rate $R$ will be proportional to the momentum density of states factor for the antineutrino times the momentum density of states factor for the electron. So we see how the shape of the electron momentum spectrum is governed by the bracketed terms of (16-12). Crudely speaking, the spectrum is symmetrical about a maximum at the momentum which represents equal momentum sharing between the electron and antineutrino. The reason is that if one of these particles takes more momentum in the decay, the other must take less, and this will decrease the value of the product of the two density of state factors.

The term $M^*M$ in (16-12) governs the magnitude of the momentum spectrum, and therefore the overall rate of emission of electrons in the $\beta$ decay. Equation (16-13) shows that $M$ depends on the value of a quantity $\beta$, which will be identified in the following paragraphs. It also depends on the eigenfunction $\psi_i$ of the $\beta$-decaying nucleus in its initial state (before the decay) and on the complex conjugate of the eigenfunction $\psi_f$ of the nucleus in its final state (after the decay). We shall see that the $\beta$-decay matrix element $M$ is really a measure of how easy it is for the nucleus to change from the initial to the final state.

Equations (16-12) and (16-13) are analogous to (8-42) and (8-43), which we derived for the rate of emission of photons in the decay of an excited state of an atom. In particular, the $\beta$-decay matrix element is analogous to the electric dipole moment matrix element

$$\left| \int \psi_f^* e\mathbf{r}\psi_i \, d\tau \right|$$

that enters in the theory of the "photon decay" of atoms. The $\beta$-decay matrix element is a volume integral of the quantity $\beta$, taken between the eigenfunction of the nucleus in its initial state and the complex conjugate of the eigenfunction of the nucleus in its final state. So $M$ is something like an average of the quantity $\beta$, evaluated while the nucleus is in the process of decaying and is in a mixture of the two states. Thus $\beta$ plays a role in governing the rate of $\beta$ decay much like the role played by the electric dipole moment, $e\mathbf{r}$, in governing the rate of photon decay by atoms.

Equations (16-12) and (16-13) were first obtained by Fermi, under the simplifying assumption that the Coulomb interaction between the nuclei and the emitted electrons could be neglected. He also assumed that $\beta$ is a universal constant, called the *$\beta$-decay coupling constant*. Then the $\beta$-decay matrix element $M$ immediately reduces to

$$M = \beta \int \psi_f^* \psi_i \, d\tau = \beta M' \tag{16-14}$$

where $M'$ is the so-called *nuclear matrix element*

$$M' = \int \psi_f^* \psi_i \, d\tau \tag{16-15}$$

Fermi's theory of electron emission from nuclei is closely related to the theory of photon emission from atoms. Perhaps the biggest difference is that Fermi's theory is complicated by the fact that two particles are emitted and share the available energy. Certainly the biggest similarity is that in both theories none of the particles emitted are considered to have prior existences—they are *created* at the time of emission.

It should be emphasized that $\beta$ decay is *not* a consequence of the nuclear force, or interaction. Instead, $\beta$ decay is a consequence of an interaction that we have not previously encountered in our study of quantum physics—the *$\beta$-decay interaction.* This is one of the four fundamental interactions of nature. The other three are the nuclear, electromagnetic, and gravitational interactions. In the next section we shall study the properties of the $\beta$-decay interaction, and we shall find that it is set apart from the other fundamental interactions by the very different magnitude of its strength, which is governed by the value of the $\beta$-decay coupling constant $\beta$. We shall also find that the $\beta$-decay interaction has properties concerning parity which are strikingly different from the other interactions.

The function $R(p_e)$, of (16-12), is the momentum spectrum of the emitted electrons. It also applies to positron emission. The equation predicts that a plot of $[R(p_e)/p_e^2]^{1/2}$ versus $(E - K_e)$, or simply versus $K_e$, should yield a straight line. Figure 16-12 shows such a *Kurie plot* for the simplest of all electron emission processes

$$^0n^1 \rightarrow {}^1\text{H}^1 + e + \bar{\nu} \tag{16-16}$$

the decay of a free neutron $^0n^1$ into a proton $^1\text{H}^1$ plus an electron $e$ and an antineutrino $\bar{\nu}$. The neutron decays because $[M_{0,1} - M_{1,1}]c^2 = +0.78$ MeV, and the lifetime $T$ of the decay is about 1000 sec. (A neutron in a stable nucleus does not $\beta$ decay into a

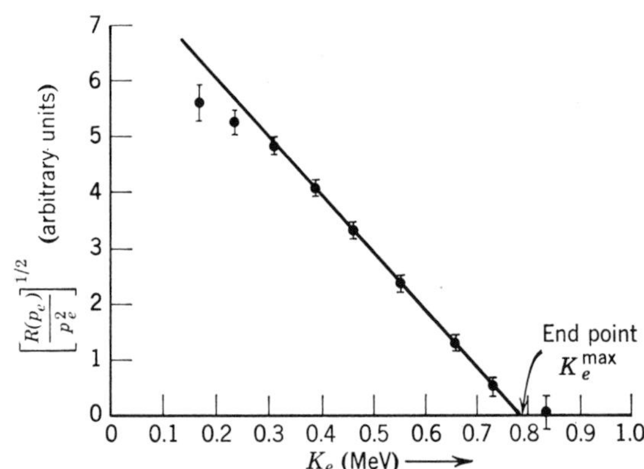


**FIGURE 16-12**

A Kurie plot for the $\beta$ decay of the neutron.

proton because it is prevented from so doing by the nuclear interaction, which is much stronger than the $\beta$-decay interaction.) The comparison in Figure 16-12 is typical of the good agreement obtained between the theory and experiment for the $\beta$ decay of nuclei of low $Z$. Small downward deviations of the experimental data at low energies are sometimes seen, but they usually represent experimental problems with self-absorption of low-energy electrons in the source of $\beta$-decaying material.

For nuclei of high $Z$, there are real deviations between the predictions of the Fermi theory and experiment. They are due to the neglect of the Coulomb interaction between the final nucleus and the emitted electron, or positron. This interaction decelerates the electrons, or accelerates the positrons. Its effect is to enhance the low-energy or momentum end of the electron spectra, or to deplete that end of the positron spectra.

By integrating the momentum spectrum of (16-12) over all electron momenta up to the maximum momentum $p_e^{\max}$, an expression is obtained for the total rate of emission of electrons. Since this is just the *decay rate* $R$, according to (16-4) its reciprocal is the lifetime $T$. The results are

$$R = \frac{1}{T} \simeq \frac{m^5c^4}{2\pi^3\hbar^7}\beta^2 M'^*M'F \tag{16-17}$$

where $F$ is a function of the maximum momentum $p_e^{\max}$, or of the corresponding maximum kinetic energy which is the end point energy $K_e^{\max}$. In Figure 16-13, $F$ is plotted as a function of $K_e^{\max}$. Note that $F$ increases fairly rapidly with increasing $K_e^{\max}$. Corrections made to the theory to account for the effect of the Coulomb interaction on the emitted electron change the values of $F$. For small $Z$ the change in $F$ is negligible. But for $Z = 100$, and $K_e^{\max} = 1$ MeV, $F$ is increased by about a factor of 100 for electron emission, or decreased by about a factor of 10 for positron emission.

We see from (16-17) that the lifetime $T$ of a $\beta$-decaying nucleus decreases fairly rapidly with increasing end point energy $K_e^{\max}$, or decay energy $E = K_e^{\max}$, because of the increase in the value of $F$ with increasing energy. For naturally occurring $\beta$-decaying nuclei, $T$ ranges from $\sim$1 sec for $E$ around several MeV, to $\sim 10^8$ sec for $E$ around several hundredths of an MeV.

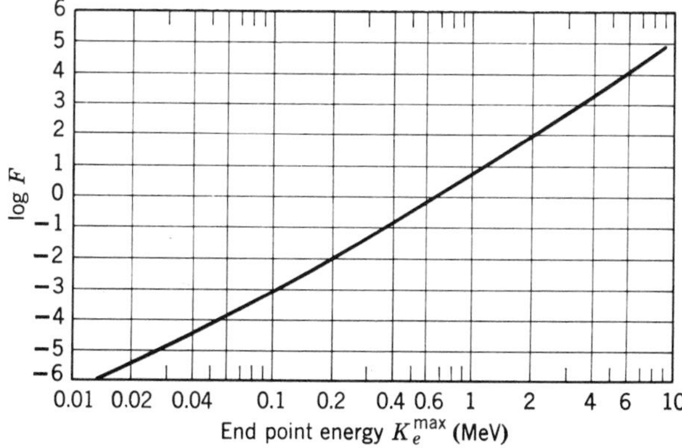


**FIGURE 16-13**

A base-10 logarithmic plot of the function $F$ versus the end point energy $K_e^{\max}$ of the $\beta$ decay of nuclei of very small $Z$. The decay rate is proportional to $F$. Thus as $F$ increases with increasing end point energy, the decay rate increases and the lifetime decreases.

We also see from (16-17) that the quantity

$$FT \simeq \frac{2\pi^3\hbar^7}{m^5c^4}\frac{1}{\beta^2}\frac{1}{M'^*M'} \tag{16-18}$$

depends on a collection of universal constants, and on the value of the nuclear matrix element

$$M' = \int \psi^*_{Z\pm1,A}\psi_{Z,A}\,d\tau \tag{16-19}$$

This expression for the nuclear matrix element is just (16-15), with the subscripts on the initial and final eigenfunctions rewritten to indicate that the theory applies to both electron and positron emission. The quantity $FT$ is sometimes called the *comparative lifetime*. It can be used to compare $\beta$ decays of different decay energy, and rank them according to the lifetimes they would have if they all had the same decay energy. That is, multiplying $T$ by $F$ removes the energy dependence, and so produces a quantity whose value depends only on a collection of universal constants and on the value of the nuclear matrix element. Since the matrix element contains the eigenfunctions for the nuclear states involved in a $\beta$ decay, it is apparent that the $FT$ value for the decay can provide information about those nuclear states.

**Example 16-4.** One of the simplest $\beta$ decays is

$$^1\text{H}^3 \rightarrow {}^2\text{He}^3 + e + \bar{\nu}$$

The measured values of the decay energy and half-life are $E = 0.0186$ MeV and $T_{1/2} =$ 12.3 yr. Calculate the value of $FT$.

Since $Z$ is very small, we can evaluate $F$ from Figure 16-13, using $K_e^{\max} = E = 0.0186$ MeV. We find

$$\log F \simeq -5.7$$

or

$$F \simeq 2.1 \times 10^{-6}$$

Converting $T_{1/2}$ in years to the lifetime $T$ in seconds gives

$$T = \frac{T_{1/2}}{0.693} = \frac{12.3 \text{ yr} \times 365 \text{ day/yr} \times 24 \text{ hr/day} \times 60 \text{ min/hr} \times 60 \text{ sec/min}}{0.693} \simeq 5.6 \times 10^8 \text{ sec}$$

so

$$FT \simeq 2.1 \times 10^{-6} \times 5.6 \times 10^8 \text{ sec} = 1.2 \times 10^3 \text{ sec}$$

This is one of the smallest $FT$ values observed. In other words the $\beta$ decay is inherently fast because its lifetime $T$ is small, in consideration of the value of $F$ dictated by the value of the decay energy $E$. In Example 16-5 we shall see that this fact has some important theoretical consequences.

It also has some important practical consequences. Uncontrolled testing of hydrogen bombs in the 1950s produced large amounts of $^1\text{H}^3$ (also called tritium) in the atmosphere. Since the $\beta$ decay of this radioactive isotope is inherently fast, most of it has by now decayed into the harmless stable isotope $^2\text{He}^3$. ◀

Since (16-18) shows that the $FT$ value is inversely proportional to the value of $M'^*M'$, the nuclear matrix element times its complex conjugate, we see that $FT$ is a minimum when $M'^*M'$ is a maximum. This happens when the initial nuclear eigenfunction $\psi_{Z,A}$ is identical to the final nuclear eigenfunction $\psi_{Z\pm1,A}$, because then the normalization condition for eigenfunctions requires that (16-19) yield $M' = 1$. If the eigenfunctions are not identical, $M'^*M' < 1$, and it becomes smaller as the eigenfunctions become less similar. In fact, $M'$, and therefore $M'^*M'$, is exactly zero if $\psi_{Z,A}$ and $\psi_{Z\pm1,A}$ are so dissimilar as to correspond to different values of nuclear spin

$i$, or opposite nuclear parities. These two properties immediately give the *Fermi selection rules*:

$$\Delta i = 0 \qquad \text{The nuclear parity must not change} \tag{16-20}$$

If either is violated the $\beta$ decay will not take place, according to the Fermi theory. The first restriction reflects the fact that no allowance is made in the theory for the emitted particles to carry angular momentum, so the conservation law demands there be no change in the nuclear angular momentum. The second restriction arises because the integrand will be of odd parity if the eigenfunctions have opposite parity, and then the contribution to the integral from the point $x$, $y$, $z$ will be canceled by the contribution from the point $-x$, $-y$, $-z$. (Recall the arguments at the end of Section 8-7.)

A theory developed later by Gamow and Teller takes into account the spins of the emitted particles, and it shows that the first Fermi selection rule is too restrictive. The Fermi theory restriction arises from the circumstance that the matrix element in (16-13) does not involve spins. In the Gamow-Teller theory the corresponding matrix element contains the spin of the neutron that is being converted into a proton, and the spin of the neutrino that is being converted into an electron, in the $\beta$ decay. If the two particles emitted in the decay have their $s = 1/2$ intrinsic spins essentially parallel, $\Delta i = \pm 1$ is also allowed. Thus we have the *Gamow-Teller selection rules*:

$$\Delta i = 0, \pm 1 \text{ (but not } i_i = 0 \rightarrow i_f = 0) \qquad \text{The nuclear parity must not change} \tag{16-21}$$

The reason why $\Delta i = 0$ is allowed by the Gamow-Teller rules is that it is possible for the two particles to be emitted with essentially parallel spins in a Gamow-Teller decay, thereby carrying away one unit of angular momentum, with the nucleus changing the orientation in space, but not the magnitude, of its spin. But this is not possible if the nuclear spin is zero, as is indicated by the qualification in parenthesis. In a Fermi decay the particle spins are "antiparallel," and the nuclear spin may be zero.

Even if $\Delta i$ is larger than one, $\beta$ decay still can occur in such a way that angular momentum is still conserved, since the particles can be emitted with orbital angular momentum. But the decay rates for these *forbidden* processes are much smaller than for the *allowed* processes that satisfy the Fermi or Gamow-Teller selection rules. The decay rate decreases by something like a factor of $10^{-3}$ for each unit of orbital angular momentum carried away by the particles. These inhibition factors result from the low probabilities of emitting a particle with orbital angular momentum of one or more $\hbar$ units from a system of radius as small as that characteristic of a nucleus, if the particle has linear momentum as small as that characteristic of $\beta$ decay.

For many nuclear physicists $\beta$ decay is a favorite field of investigation because it provides valuable information about the nuclei involved in the decay. A measurement of the end point $K_e^{\text{max}}$, or of the atomic masses to determine the decay energy $E$, is used to obtain the value of $F$ from a curve like Figure 16-13, if $Z$ is small. If $Z$ is not small, the value of $F$ is obtained from tables that are available of $F$ versus $K_e^{\text{max}}$ and $Z$. Next, $FT$ is calculated from the measured value of the half-life, or lifetime, as in Example 16-4. Then (16-18) is used to evaluate the nuclear matrix element $M'$. The order of magnitude of $M'$ is often enough to give information about the spins and parities of the nuclear states participating in the decay, and more accurate values of $M'$ can give details about the eigenfunctions of these states, through (16-19). Of course, it is first necessary to know the value of the $\beta$-decay coupling constant $\beta$. This quantity is evaluated experimentally from $\beta$ decays involving certain very simple nuclear states, for which $M'$ is already known from other considerations to be discussed next.

## 16-4 The Beta-Decay Interaction

The $\beta$-decay interaction is the least familiar of the four fundamental interactions (nuclear, electromagnetic, $\beta$ decay, gravitational) that govern the operation of everything in the universe. In this section we shall explore some of its properties. We begin by using the ${}^1H^3$ to ${}^2He^3$ $\beta$ decay, considered in Example 16-4, to determine the value of the $\beta$-decay coupling constant, $\beta$, which specifies the strength of the interaction.

Since we found in Example 16-4 that the $FT$ value for the $\beta$ decay ${}^1H^3 \rightarrow {}^2He^3 + e + \bar{\nu}$ is particularly small, the inverse proportionality between $FT$ and $M'^*M'$, of (16-18), tells us that the nuclear matrix element $M'$ is particularly large for this decay. In fact, there is reason to believe that it assumes the maximum value allowed by the normalization condition, $M' = 1$. Figure 16-14 gives the shell model description of the ground states of the two nuclei, which are the states involved in the $\beta$ decay. Since the nucleons are in the $1s_{1/2}$ subshell, which has $j = 1/2$ and even parity, according to the shell model both ground states should have nuclear spin $i = 1/2$ and even parity. These predictions are confirmed by independent measurements of the spins and parities. Thus the $\beta$ decay between these states is certainly allowed by the Fermi selection rules. But the shell model makes the even stronger prediction that $M' = 1$, almost exactly, in this decay. Since all the nucleons are in the same subshell, the eigenfunctions for the two nuclei can differ only if the Coulomb, or nuclear, interactions between the nucleons differ. The Coulomb interactions do differ for the two nuclei, but they are negligible compared to the strong nuclear interactions. And there is much other evidence that the nuclear interactions are the same because they are charge independent and so make no distinction between neutrons and protons. Thus the two eigenfunctions should be essentially identical and, if the eigenfunctions are properly normalized, the integral will yield

$$M' = \int \psi_{2,3}^* \psi_{1,3}\, d\tau = \int \psi_{1,3}^* \psi_{1,3}\, d\tau = 1$$

Knowing the value of $M'$, we can then use the measured $FT$ value to evaluate $\beta$, the $\beta$-decay coupling constant. It should be emphasized that the conclusion that $M' = 1$ depends on the particular symmetry found between the behavior of the neutrons and protons in the two nuclei involved in the decay. In the first nucleus there are a pair of nucleons of one species and an unpaired nucleon of the other species in the same subshell—in the second nucleus exactly the same is true, although the species of the nucleons are reversed.

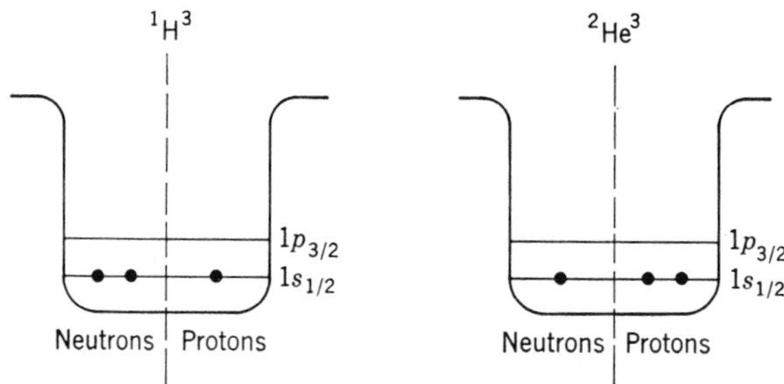


**FIGURE 16-14**

Shell model descriptions of the ground states of the pair of nuclei ${}^1H^3$ and ${}^2He^3$.

**Example 16-5.** Use the $FT$ value for the $\beta$ decay of Example 16-4, plus the conclusion that $M' = 1$ for that decay, to evaluate the $\beta$-decay coupling constant, $\beta$.

Equation (16-18) gives

$$\beta^2 \simeq \frac{2\pi^3\hbar^7}{FTm^5c^4}\frac{1}{M'^*M'}$$

So we have

$$\beta^2 \simeq \frac{2\pi^3(1.05 \times 10^{-34}\,\text{joule-sec})^7}{1.2 \times 10^3\,\text{sec} \times (0.91 \times 10^{-30}\,\text{kg})^5 \times (3.0 \times 10^8\,\text{m/sec})^4}\frac{1}{1}$$

or

$$\simeq 1.4 \times 10^{-123}\,\text{joule}^2\text{-m}^6$$

Thus

$$\beta \simeq 3.7 \times 10^{-62}\,\text{joule-m}^3$$

◀

There are several other pairs of nuclei whose ground states have shell model descriptions with the same kind of symmetry between neutrons and protons as in Figure 16-14. An example of such a pair is $^3Li^7$ and $^4Be^7$. One member of each pair $\beta$ decays into the other, with a nuclear matrix element $M'$ that must certainly be almost precisely equal to 1. The measured $FT$ values of these decays lead, through calculations like the one in Example 16-5, to values of $\beta$ which are in good agreement with the value obtained there. Thus we conclude that the $\beta$-decay coupling constant has the *very* small value

$$\beta \sim 10^{-62}\,\text{joule-m}^3 \tag{16-22}$$

If we divide $\beta$ by the volume of a typical nucleus, $(5 \times 10^{-15}\,\text{m})^3 \sim 10^{-43}\,\text{m}^3$, we obtain $10^{-62}\,\text{joule-m}^3/10^{-43}\,\text{m}^3 = 10^{-19}\,\text{joule} \simeq 10^{-6}$ MeV. We can then make a comparison of this characteristic energy to the energy of the order of 1 MeV that characterizes the nuclear interaction. As it is the square of the $\beta$-decay coupling constant that enters into measurable quantities, such as the $FT$ value, it is appropriate to say that the *$\beta$-decay interaction is weaker than the nuclear interaction by a factor of* $10^{-12}$.

Since the nuclear interaction is only one or two orders of magnitude stronger than the electromagnetic interaction (see Section 15-2), the $\beta$-decay interaction is also very much weaker than the electromagnetic interaction. On the other hand, the gravitational interaction is weaker than the nuclear interaction by about 40 orders of magnitude (see also Section 15-2), so the $\beta$-decay interaction is stronger than the gravitational interaction by nearly 30 orders of magnitude. Thus there are extremely pronounced differences in strength between the $\beta$-decay interaction and the other fundamental interactions. These matters will be discussed at more length in the following chapter where it will be seen, for instance, that the gravitational interaction is the most obvious one in the everyday world, despite the fact that it is inherently the weakest by far, because it has a long range and always has the same sign.

The range of an interaction is a characteristic as important as its strength. The gravitational interaction has a long range since the gravitational interaction energy between two massive objects decreases quite slowly as their separation $r$ increases (in proportion to $1/r$). The electromagnetic interaction also has a long range since the interaction energy between two charged objects has the same slow dependence on their separation. The nuclear interaction has a short range because the interaction energy cuts off abruptly when two nucleons are separated by more than about 2F. *The $\beta$-decay interaction has an extremely short range.* Some evidence for this is found from the following considerations. The form for the $\beta$-decay matrix element $M$ used in the Fermi theory, (16-14)

$$M = \beta \int \psi_f^* \psi_i \, d\tau$$

is obtained from the assumption that the extension in space of the $\beta$-decay interaction is very small compared to the dimensions of the nucleus. Without this assumption, the integrand in $M$ would not be $\psi_f^* \psi_i$, but $\psi_f^* \psi_i$ averaged over a volume of dimensions equal to the range of the interaction. If this were the case, $M$ would be affected in such a way as to change the predictions of the theory for the shape of the momentum spectra of the electrons emitted in the $\beta$ decay. But the observed momentum spectra are in good agreement with the theoretical predictions as they stand. Thus the assumption of a very short range $\beta$-decay interaction, which the predictions stand upon, is probably correct. Additional evidence supporting this conclusion will be presented in the following chapter.

The very small value of $\beta$ is responsible for the fact that neutrinos and antineutrinos interact so weakly with matter that they are very difficult to detect. Calculations show that when they are produced in $\beta$ decay following nuclear reactions in the center of the sun, they can travel all the way to the surface with little chance of being absorbed. This has an effect on the production of solar energy. The $\beta$-decay interaction of electrons and positrons is equally weak, but since these particles *also* interact with matter through the electromagnetic interaction they are easy to detect.

Despite the obvious difficulties due to the extreme weakness of their interaction with matter, antineutrinos were detected in 1953 by Reines and Cowan. They used the reaction

$$\bar{\nu} + {}^1\text{H}^1 \rightarrow {}^0n^1 + \bar{e}$$

where the symbol $\bar{e}$ stands for a positron. This is the inverse of the reaction

$${}^0n^1 + \bar{e} \rightarrow {}^1\text{H}^1 + \bar{\nu}$$

which is the alternative form of neutron decay, (16-16)

$${}^0n^1 \rightarrow {}^1\text{H}^1 + e + \bar{\nu}$$

(Note that the two forms of neutron decay indicate the equivalence of the destruction of an antiparticle, the positron, and the creation of the associated particle, the electron. In the Dirac theory the processes are identical.) The Reines-Cowan reaction took place in the hydrogen of a very large hydrogeneous scintillation counter (a modern version of Rutherford's ZnS counter, using photocells instead of eyes to detect the light flashes). The counter was exposed to the enormous flux of antineutrinos emitted from the fission induced $\beta$ decays in a nuclear reactor, and the positrons were detected by the scintillations they produced in the same counter. Elaborate methods were required to minimize background scintillation. This was necessary because only about one reaction per minute was obtained, despite the intense flux of antineutrinos and the huge size of the target, due to the weakness of the $\beta$-decay interaction.

Now we shall briefly discuss two other experiments, performed in the 1950s, that tell us about a unique property of the $\beta$-decay interaction. Wu, and collaborators, studied the decay

$${}^{27}\text{Co}^{60} \rightarrow {}^{28}\text{Ni}^{60} + e + \bar{\nu}$$

by measuring the direction of emission of the electrons relative to the orientation of the magnetic dipole moments of the ${}^{27}\text{Co}^{60}$ nuclei. The magnetic dipole moments were aligned by using a very strong external magnetic field, and a very low temperature to minimize thermal disorder. Figure 16-15 is a schematic drawing of the experiment,

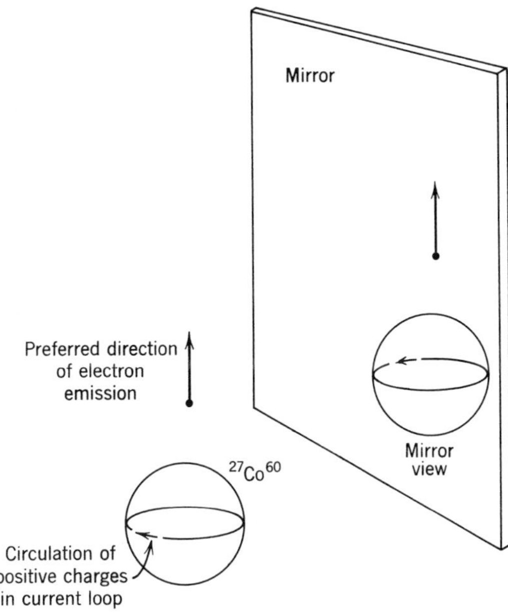


**FIGURE 16-15**

A schematic drawing of the experiment which proved that parity is not conserved in $\beta$ decay. Also shown is a mirror image of the experiment.

showing a typical nucleus and a typical emitted electron. To make the drawing closer to physical reality, a current loop of positive charge is used to indicate the orientation of the magnetic dipole moment. Wu found that the electrons are not emitted symmetrically with respect to the plane of the current loop. Instead, there is a preferred direction of emission that is related to the circulation of the current loop in the same way as the direction of advance of a left-hand screw is related to its rotation. The figure also shows the experiment, as seen when looking in a mirror. The preferred direction of emission appears to be the same, but the circulation of the current loop appears to have reversed. As viewed in the mirror, the results of the experiment are described by saying the relation between the direction of the typical electron and the circulation of the current loop is like that of a right-hand screw. Thus *a description of this $\beta$ decay (and others) is not the same as a description of the mirror image*. This seems to be a property unique to the $\beta$-decay interaction, among all the fundamental interactions of nature (nuclear, electromagnetic, $\beta$ decay, and gravitational). For instance, charges circulating around a macroscopic current loop emit photons by the electromagnetic interaction, because the charges are accelerating. But the photons are emitted symmetrically with respect to the plane of the loop, so the mirror description of this process cannot differ from the normal description. Since the operation of taking a mirror image is related to the parity operation in the manner illustrated in Figure 16-16, it is said that $\beta$ decay is not invariant to the parity operation, or that *parity is not conserved in $\beta$ decay (but it is in the electromagnetic interaction)*.

Measurements of Goldhaber, and collaborators, have shown that the so-called helicity of the antineutrino is responsible for the results of the Wu experiment. Using a method that is a little too complicated to explain here, they found that in the normal view of nature the spin angular momentum of an antineutrino is always essentially parallel to the direction of its linear momentum. It is said that the antineutrino has the *helicity* of a right-hand screw, depicted in Figure 16-17. They also found that the neutrino has the helicity of a left-hand screw; i.e., its spin angular momentum is essentially antiparallel to its linear momentum in the normal view. Now the $\beta$ decay studied by Wu is between an $i = 5$, even parity, ground state of ${}^{27}\text{Co}^{60}$,

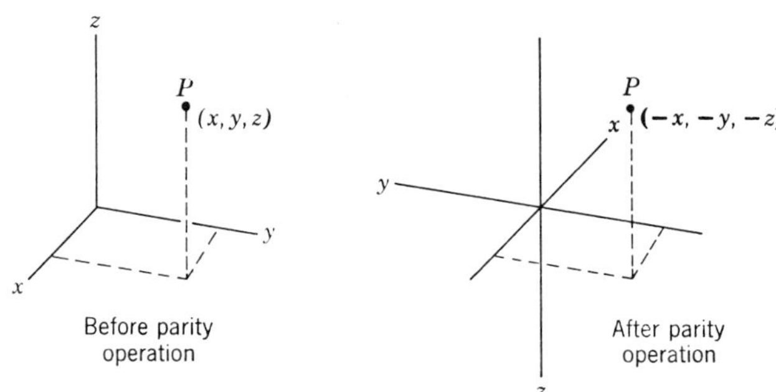


**FIGURE 16-16**

The parity operation $(x, y, z) \rightarrow (-x, -y, -z)$. In this figure the operation is carried out by reversing the direction of each of the coordinate axes, keeping the location of the representative point $P$ fixed (compare with Figure 8-15). Before the operation we have a set of *right-hand* axes, i.e., a right-hand screw, rotated in the sense that would carry the $x$ axis into the $y$ axis would advance the screw in the direction of the $z$ axis. After the parity operation they become a set of *left-hand* axes. This change can also be obtained by the operation of taking a mirror image, which converts right-hand axes into left-hand axes. So the mirror image operation is related to (but not identical to) the parity operation.

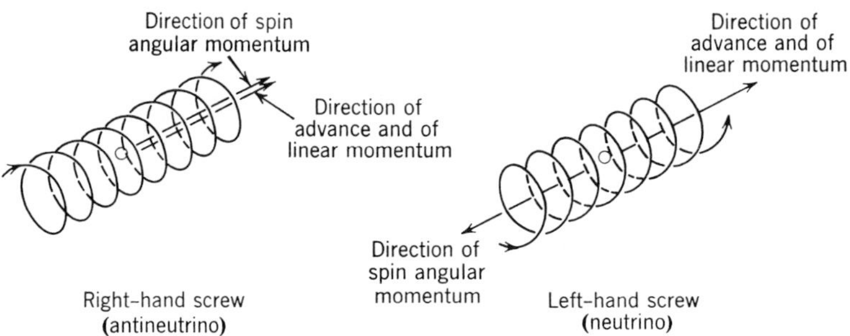


**FIGURE 16-17**

The helicities of a right-hand screw and a left-hand screw.

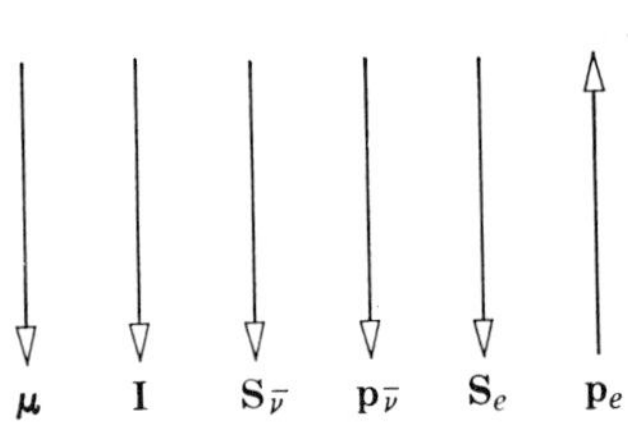


**FIGURE 16-18**

The $\beta$ decay of aligned ${}^{27}\mathrm{Co}^{60}$. The vectors give the directions only of $\boldsymbol{\mu}$ and $\mathbf{I}$, $\mathbf{S}_{\bar{\nu}}$ and $\mathbf{p}_{\bar{\nu}}$, and $\mathbf{S}_e$ and $\mathbf{p}_e$, which are the nuclear magnetic dipole moment and spin, the antineutrino spin and linear momentum, and the electron spin and linear momentum. Parity is not conserved because $\mathbf{S}_{\bar{\nu}}$ and $\mathbf{p}_{\bar{\nu}}$ are always essentially parallel.

and an $i = 4$, even parity, excited state of $^{28}\text{Ni}^{60}$. So it is a Gamow-Teller allowed transition in which angular momentum conservation requires the antineutrino and electron to be emitted with their spin angular momentum vectors essentially parallel to that of $^{27}\text{Co}^{60}$, or to a vector representing its magnetic dipole moment. Furthermore, in such a transition the antineutrino and electron tend to be emitted with linear momentum vectors in opposite directions. Figure 16-18 shows how these relations between the vectors, plus the parallel relation between the spin and linear momentum vectors of the antineutrino demanded by its helicity, cause the typical electron to be emitted in the direction described. As viewed in a mirror, the helicity of the antineutrino changes, just like the helicity of a real screw changes, and this leads to the change in the mirror image description of the Wu experiment.

It should be noted that there is no violation of parity conservation by the *nuclei* in the $^{27}\text{Co}^{60}$ to $^{28}\text{Ni}^{60}$ decay. Both nuclear states involved are of even parity so there is no nuclear parity change, in agreement with the Gamow-Teller selection rules.

It should also be noted that it is not possible for an antineutrino, or neutrino, to have a definite helicity in the normal view of nature unless its rest mass is precisely zero. If it had a nonzero rest mass, it would travel with velocity less than $c$, and we could always find a moving frame of reference in which its linear momentum would be reversed in direction. As its spin would be unchanged by such a transformation, its helicity would be reversed. Since the Goldhaber experiment shows that antineutrinos and neutrinos do have definite helicities, and since this would not be possible if their helicities depended on the motion of the reference frame from which they are viewed, we conclude that their rest masses are precisely zero. Direct measurements of the rest masses of these particles confirm this conclusion.

## 16-5 Gamma Decay

There are $\gamma$ rays emitted from many of the nuclei of the radioactive series. These are photons of electromagnetic radiation that carry away the excess energy when nuclei make *$\gamma$-decay* transitions from excited states to lower energy states. As the energy differences in nuclear excited states range upwards from $\sim 10^{-3}$ MeV, $\gamma$ rays have energies greater than $\sim 10^{-3}$ MeV (see Figure 2-4). Most typically, $\gamma$ decay will arise when a preceding $\beta$ decay has produced some of the daughter nuclei in states of several MeV excitation, because the $\beta$-decay selection rules prevent the decay from obeying the tendency, imposed by the energy dependence, for transitions to go overwhelmingly to the ground state. An example is shown in the $^{17}\text{Cl}^{38}$ decay scheme of Figure 16-19. There are also many other ways to produce nuclei in excited states, which subsequently $\gamma$ decay. For instance, states of excitation energy around 7 or 8 MeV are produced when this much binding energy is liberated by the capture of a low-energy neutron in a nucleus.

The most accurate technique for measuring the energy of $\gamma$ rays is to study their diffraction from a crystal lattice of known lattice spacing. This is exactly the technique of x-ray diffraction, but since $\gamma$ rays have somewhat higher energies than x rays, their wavelengths are somewhat shorter, and this forces the use of diffraction apparatus of inconveniently large dimensions in order to measure accurately the small diffraction angles. The most widely used technique for measuring $\gamma$-ray energies involves letting the photons transfer their energies to electrons by one of the processes described in Chapter 2, namely, the Compton effect, the photoelectric effect, or pair production. The energies of the electrons are measured by using a NaI scintillation counter, or a semiconductor counter, which has a response proportional to the energy a charged particle deposits in it. The measured energy spectrum of $\gamma$ rays emitted in transitions between the excited states of a nucleus is used to determine the energies of these

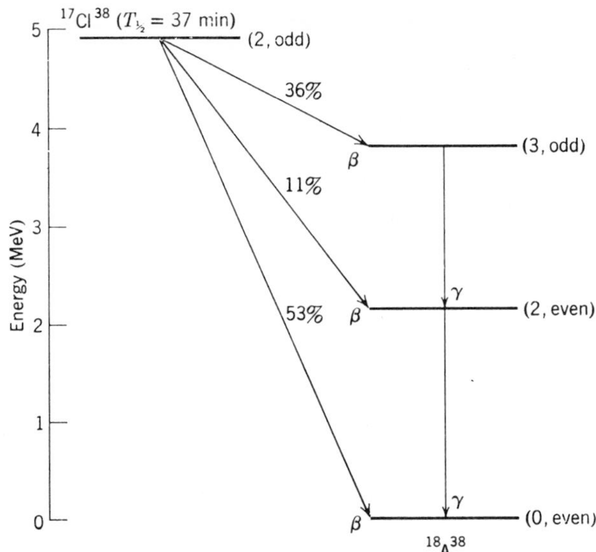


**FIGURE 16-19**

The decay scheme of $^{17}Cl^{38}$. The half-life, spin, and parity of the ground state of this $\beta$-unstable nucleus are shown as well as the energy of the state relative to the ground state of $^{18}A^{38}$. Also shown are the energies, spins, and parities of the ground and first two excited state of $^{18}A^{38}$, and the relative probabilities that the $\beta$ decay goes to each of these states. When the excited states are populated they $\gamma$ decay to the ground state. The $\beta$ decay to the (3, odd) state is allowed by the Gamow-Teller selection rules, while the other $\beta$ decays are both forbidden by these and the Fermi selection rules. They nevertheless occur with appreciable probabilities because of the way the rates for all decays, allowed and forbidden, increase rapidly as the decay energy increases.

states—just like the spectrum of photons emitted from an atom is used to determine the energies of atomic states. Of course, this provides very valuable information about the nucleus.

Another valuable source of information is the $\gamma$-decay transition rate $R$ of each excited state. In some cases $R$ can be measured directly. In other cases it can be obtained indirectly by measuring the lifetime $T$ of the state. If the state makes only a single transition to a lower energy state, (16-4) tells us $T = 1/R$ (after correction is made for the "internal conversion" process to be discussed at the end of this section). When $T > 10^{-10}$ sec, it can be determined by electronically timing the average delay between the excitation of a state and its decay. When $T$ is shorter than this figure, in some cases it can be determined by using the Mössbauer effect (discussed in the next section) to determine the energy spread, or "width," of the state, and then employing the energy-time uncertainty principle. With these different techniques, transition rates have been observed ranging from $R \sim 10^{-8}$ sec$^{-1}$ to $R \sim 10^{18}$ sec$^{-1}$.

The energies of the excited states of nuclei will be considered in a subsequent section. Here we shall consider their transition rates for $\gamma$ decay. As we shall use the ideas developed in treating optical transitions of atoms in Section 8-7, the student certainly should review that material before proceeding.

For an atom, only electric dipole radiation is important. This is the radiation produced by oscillations in its electric dipole moment. In principle, radiation can be emitted by a more complicated behavior of the atomic electrons, such as an oscillation of the magnetic dipole moment or of the electric quadrupole moment. In practice, for an atom such radiation can be ignored because the transition rate is very much smaller than for electric dipole radiation. Electromagnetic considerations show that the transition rate for magnetic dipole radiation should be smaller than for electric dipole radiation by a factor of the order of $(v/c)^2 \sim (10^{-2})^2 = 10^{-4}$, where $v$ is the typical velocity of the electrons and $c$ is the velocity of light. Geometrical considerations show that the transition rate for electric quadrupole radiation should be smaller than for electric dipole radiation by a factor of the order of $(r'/\lambda)^2 \sim (10^{-10}\text{ m}/10^{-7}\text{ m})^2 = 10^{-6}$, where $r'$ and $\lambda$ are typical values of the atomic radius and the wavelength of the radiation. If the selection rules prevent an atom from emitting electric dipole radiation, it is almost always deexcited by hitting some other atom long before it can emit magnetic dipole or electric quadrupole radiation.

For a nucleus the same factors suppress the transition rates for magnetic dipole and electric quadrupole radiation, but their values are not so small: $(v/c)^2 \sim (10^{-1})^2 = 10^{-2}$; $(r'/\lambda)^2 \sim (10^{-14}\text{ m}/10^{-12}\text{ m})^2 = 10^{-4}$. Furthermore, the Coulomb barrier keeps nuclei from getting close enough to deexcite each other. So if the selection rules prevent a nucleus with several MeV of excitation from emitting electric dipole radiation, it *must* wait until it can decay by emitting some other electromagnetic radiation (or by the related process of internal conversion).

The transition rates for various types of electromagnetic radiation can be calculated by extensions of the procedure developed in Section 8-7. Since the calculations are very sensitive to the detailed behavior of the nucleons in the states involved in the decays, and since the nuclear models only provide approximate descriptions of this behavior, the results can only be expected to give rough ideas of general trends. Table 16-1 shows transition rates obtained by Weisskopf from calculations, based on the *shell model*, for a nucleus of radius $r' = 7$ F. The integer $L$ labels the *multipolarity* of both the electric and magnetic transitions; it is 1 for dipole, 2 for quadrupole, 3 for octupole, etc. Note that for 1 MeV $\gamma$ rays, predicted rates for magnetic transitions are smaller than for electric transitions, of the same $L$, by about $10^{-2} \sim (v/c)^2$. At that typical energy, predicted rates for both types of transitions decrease by about $10^{-4} \sim (r'/\lambda)^2$, for each unit increase of $L$. Also note that the dipole transition rates have approximately an $E^3 \propto \nu^3$ dependence on the energy or frequency of the emitted $\gamma$ ray. We have seen this $\nu^3$ dependence before in the electric dipole transition rates for atoms, (8-43). Since $(r'/\lambda)^2 \propto \nu^2 \propto E^2$, the quadrupole transition rates depend

**TABLE 16-1.** Shell Model $\gamma$-Decay Transition Rates in $\text{sec}^{-1}$ for a Nucleus of Radius $r' = 7$ F

| Transition | $L$ | $\gamma$ Ray Energy 10 MeV | 1 MeV | 0.1 MeV |
|---|---|---|---|---|
| Elec. dipole | 1 | $2 \times 10^{18}$ | $2 \times 10^{15}$ | $2 \times 10^{12}$ |
| Mag. dipole | 1 | $2 \times 10^{16}$ | $2 \times 10^{13}$ | $2 \times 10^{10}$ |
| Elec. quadrupole | 2 | $1 \times 10^{16}$ | $1 \times 10^{11}$ | $1 \times 10^{6}$ |
| Mag. quadrupole | 2 | $1 \times 10^{14}$ | $1 \times 10^{9}$ | $1 \times 10^{4}$ |
| Elec. octupole | 3 | $1 \times 10^{13}$ | $1 \times 10^{6}$ | $1 \times 10^{-1}$ |
| Mag. octupole | 3 | $1 \times 10^{11}$ | $1 \times 10^{4}$ | $1 \times 10^{-3}$ |
| Elec. sixteenpole | 4 | $1 \times 10^{10}$ | $1 \times 10^{1}$ | $1 \times 10^{-8}$ |
| Mag. sixteenpole | 4 | $1 \times 10^{8}$ | $1 \times 10^{-1}$ | $1 \times 10^{-10}$ |

approximately on $E^5$ and the octupole transition rates depend approximately on $E^7$.

The calculations also show that the *γ-decay selection rules* are:

For *electric transitions*

$$|i_i - i_f| \leq L \leq i_i + i_f \qquad \text{(but not } i_i = 0 \text{ to } i_f = 0\text{)}$$

The nuclear parity must change if $L$ is odd, and it must not change if $L$ is even. (16-23)

For *magnetic transitions*

$$|i_i - i_f| \leq L \leq i_i + i_f \qquad \text{(but not } i_i = 0 \text{ to } i_f = 0\text{)}$$

The nuclear parity must change if $L$ is even, and it must not change if $L$ is odd. (16-24)

In these expressions, $i_i$ is the nuclear spin of the initial state and $i_f$ is the nuclear spin of the final state of the decaying nucleus. The decay will, of course, always proceed by the allowed transition having the largest transition rate. Because of the strong $L$ dependence of the transition rate, it follows that *the dominant transition will have* $L = |i_i - i_f|$. If this value of $L$ is odd, it will be an electric transition when the initial and final states are of the opposite parity, and a magnetic transition when these states are of the same parity. If this value of $L$ is even, it will be an electric transition when these states are of the same parity, and a magnetic transition when they are of the opposite parity.

**Example 16-6.** Use the information in the decay scheme of Figure 16-19 to determine the types of radiation emitted by $^{18}A^{38}$ in the $\gamma$ decays between its three lowest energy states.

In the decay between the states of $i = 3$, odd parity, and $i = 2$, even parity, we have $|i_i - i_f| = 1 = L$. Since this value is odd, and since the nuclear parity changes, the radiation is electric dipole.

In the decay between the states of $i = 2$, even parity, and $i = 0$, even parity, we have $|i_i - i_f| = 2 = L$. Since this value is even, and since the nuclear parity does not change, the radiation is electric quadrupole. ◀

By running the arguments of Example 16-6 in the reverse direction, information about the spins and parities of the nuclear states can be obtained if the types of radiation emitted in transitions between the states are known. The types of radiation can be identified from approximate measurements of the transition rates (or from measurements, described later, of internal conversion). Since the transition rates are very sensitive to the behavior of the nucleons in the nucleus, their accurate measurement provides information that is currently being used to improve the nuclear models.

The parts of the selection rules relating $L$ to the nuclear spins arise from the requirement that angular momentum be conserved in $\gamma$ decay. The student can verify this with ease, if he will accept a result obtained from quantum electrodynamics: a $\gamma$ ray emitted in a transition, of multipolarity $L$, carries $L$ units of angular momentum. (Since it is not possible for a system of particles to have an oscillating electric monopole moment, or to have any magnetic monopole moment at all, it immediately follows from this result that there is no way to produce an $L = 0$ $\gamma$ ray, or an $L = 0$ photon in any region of the electromagnetic spectrum. Thus we see why all photons must carry at least one unit of angular momentum.)

The parts of the selection rules relating $L$ to the nuclear parities arise from symmetry properties of the matrix elements for the transitions. In Example 8-6, we saw that the electric dipole matrix element can be broken into components, the first of which is

$$M \propto \int \psi_f^* x \psi_i \, d\tau \tag{16-25}$$

The factor $x$ enters because it is proportional to the $x$ component of the electric dipole moment. Calculations show that the first component of the electric quadrupole matrix element is

$$M \propto \int \psi_f^* x^2 \psi_i \, d\tau \tag{16-26}$$

The factor $x^2$ is proportional to one of the components of the electric quadrupole moment. (There are generally more than three since a quadrupole generally must be described in terms of a tensor.) For the magnetic dipole matrix element, the first component turns out to be

$$M \propto \int \psi_f^* L_x \psi_i \, d\tau$$

where $L_x$ is the $x$ component of orbital angular momentum. This factor enters because it is proportional to the $x$ component of the magnetic dipole moment (if we assume, for simplicity, that it is purely orbital). Since

$$L_x = (\mathbf{r} \times \mathbf{p})_x = yp_z - zp_y = m(yv_z - zv_y) = m\left(y\frac{dz}{dt} - z\frac{dy}{dt}\right)$$

the magnetic dipole matrix element component can also be written

$$M \propto \int \psi_f^* \left(y\frac{dz}{dt} - z\frac{dy}{dt}\right) \psi_i \, d\tau \tag{16-27}$$

At the end of Section 8-7 we proved that the integral in (16-25) yields zero unless $\psi_i$ and $\psi_f$ have opposite parities. We leave it to the student to prove from similar arguments that the integrals in (16-26) and (16-27) yield zero unless $\psi_i$ and $\psi_f$ have the same parities. These results are precisely the parity selection rules for the three transitions we have taken as examples.

In many $\gamma$ decays, several groups of monoenergetic electrons are emitted along with the $\gamma$ rays. (If there is a preceding $\beta$ decay these groups will be superimposed on the continuous $\beta$-decay spectrum.) The energies $\mathscr{E}$ of these electrons are found to be related to the decay energy $E$ by the equation

$$\mathscr{E} = E - W \tag{16-28}$$

where $W$ for the most prominent group equals the binding energy of a $K$ shell electron of the $\gamma$-decaying atom, and $W$ for the other groups equals the binding energies of electrons in the $L$, $M$, etc., shells. The process involved is called *internal conversion.* It consists of a *direct* transfer of energy through the electromagnetic interaction between a nucleus in an excited state and one of the electrons of its atom. The nucleus decays to a lower state, without ever producing a $\gamma$ ray. But the decay is still electromagnetic, depending on an interaction between the electron and the longitudinal components of the electric field produced by the oscillating multipole moment of the nucleus. The transverse components are responsible for $\gamma$ decay (see Appendix B).

Figure 16-20 shows calculated values of the $K$ shell *internal conversion coefficient*, $\alpha_K$, for the $^{40}$Zr atom. This is the ratio of the probability that a $K$ electron will be emitted, in a decay of its nucleus, to the probability that a $\gamma$ ray will be emitted. The calculations should be very accurate because factors involving not too well known

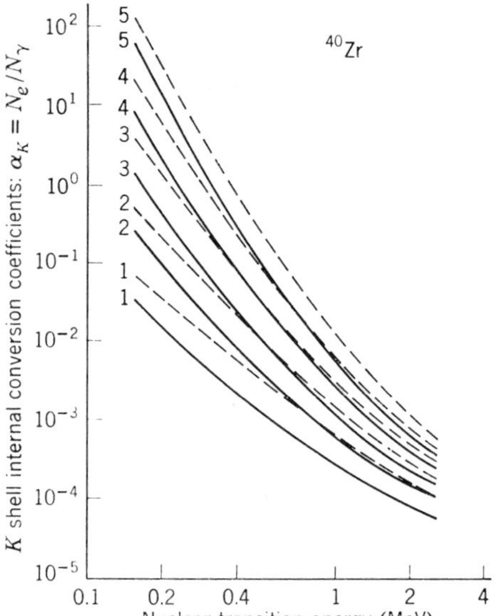


**FIGURE 16-20**

*K*-shell internal conversion coefficients for $^{40}$Zr. The solid curves are for electric transitions and the broken curves are for magnetic transitions. The numbers refer to the multipolarity *L*.

nuclear properties cancel out of the ratio. Since the chances for internal conversion increase rapidly as the value at the nucleus of the bound electron eigenfunction increases, $\alpha_K$ rapidly becomes larger as the Coulomb attraction becomes larger with increasing $Z$. For the same reason, at a given $Z$ and $E$, the quantity $\alpha_K$ is usually larger than the quantity $\alpha_L$. Furthermore, at a given $Z$ and $E$, the quantity $\alpha_K/\alpha_L$ depends strongly on the $L$ value of the $\gamma$-ray transition, and on whether it is electric or magnetic. Accurate measurements of $\alpha_K/\alpha_L$, which are relatively easy to make, therefore provide a good method of identifying the type of transition, and of determining thereby the relative spins and parities of the nuclear states involved.

Internal conversion does not compete with $\gamma$-ray emission in the sense that one process inhibits the other. The processes are independent alternatives, so the total rate $R_t$ for transitions between the initial and final nuclear states is the sum

$$R_t = R + R_{ic} \tag{16-29}$$

where $R$ and $R_{ic}$ are the transition rates for $\gamma$ emission and for internal conversion. This can be written as

$$R_t = R + \alpha_t R = R(1 + \alpha_t)$$

where $\alpha_t = \alpha_K + \alpha_L + \alpha_M + \cdots$ is the total internal conversion coefficient. If the initial state can decay only to a single final state, as is usually true for longer lifetime decays, then from (16-4)

$$T = \frac{1}{R_t} = \frac{1}{R(1 + \alpha_t)} \tag{16-30}$$

The experimental values of the lifetime $T$ can thus be used to obtain the transition rate $R$, since $\alpha_t$ can be accurately calculated.

Figure 16-21 is a comparison of the transition rates so obtained, and the predictions of the shell model calculations, for a group of transitions that have been identified as magnetic sixteenpole ($L = 4$, parity change). The agreement is fair. Inspection of the shell model diagram of Figure 15-18 will demonstrate that all such transitions are between states quite near those filled at the magic numbers. So this is where the calculations should be at their best. For other transitions shell model predictions are

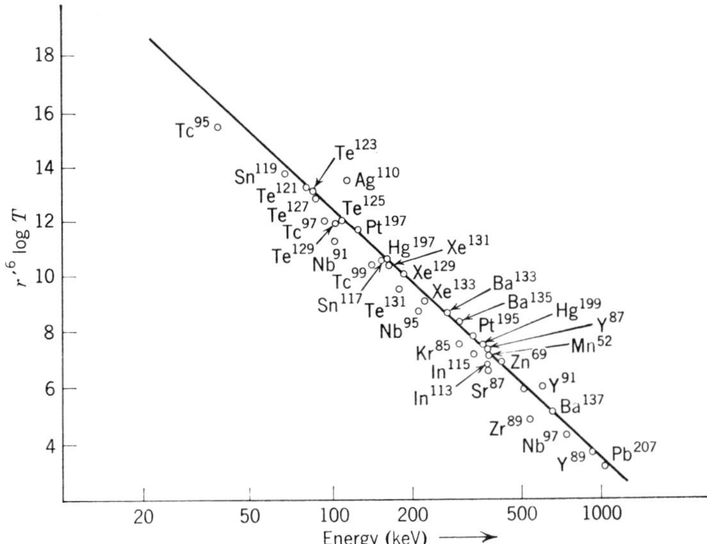


**FIGURE 16-21**
Lifetimes for a group of magnetic sixteenpole $\gamma$-decay transitions. The base-10 logarithm of the product of the lifetime $T$ (in sec) and the sixth power of the nuclear radius $r'$ (in F) is plotted as a function of the energy of the $\gamma$ ray (in keV). The points are experimental and the straight line is the prediction of the shell model.

in poor agreement with measurement. But collective model predictions can be used in these cases to obtain good agreement since the collective model can describe quite accurately the complicated oscillations in the charge, or current, distributions that are responsible for the emission of electric, or magnetic, radiation.

The lifetime of an excited state is frequently expressed in terms of its width. According to the energy-time uncertainty principle, if an average nucleus survives in an excited state only for the lifetime $T$ of the state, then its energy in the state can be specified only within an energy range $\Gamma$, satisfying approximately the relation

$$\Gamma = \frac{\hbar}{T} \tag{16-31}$$

Excited states are, therefore, not perfectly sharp. Instead, they are spread over an energy range of *width* $\Gamma$. A detailed treatment shows that (16-31) is actually satisfied exactly, providing $\Gamma$ is the full width at half-maximum of the energy profile of the state indicated in Figure 16-22.

Let us estimate the width of a typical $\gamma$-decaying state of lifetime $T \sim 10^{-10}$ sec. We find

$$\Gamma = \frac{\hbar}{T} \sim \frac{10^{-15}\text{ eV-sec}}{10^{-10}\text{ sec}} = 10^{-5}\text{ eV}$$

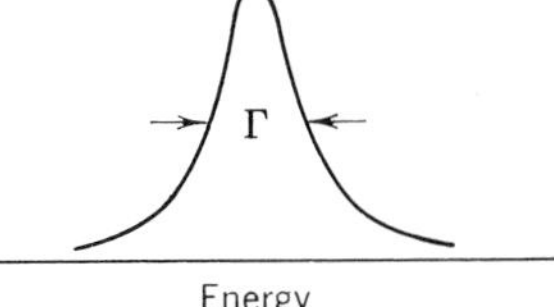


**FIGURE 16-22**
The width $\Gamma$ of an excited state. A mathematical expression for the shape shown in this figure is given in (16-32).

In comparison to the typical energy $E = 1$ MeV of such a state, $\Gamma$ is extremely small. In fact, the minute value of the ratio

$$\frac{\Gamma}{E} \sim \frac{10^{-5}\text{ eV}}{10^{6}\text{ eV}} = 10^{-11}$$

explains why we have hitherto neglected the widths of the lower energy states that are excited in radioactive decay. When we consider the higher energy states excited in nuclear reactions, we shall see that some of them have widths that are too large to be neglected.

## 16-6 The Mössbauer Effect

In 1958 a graduate student named Mössbauer made a discovery which allows the extremely small width to energy ratio of low-lying excited states to be used in many different applications as an energy spectrometer of extremely good resolution. The basic idea of the *Mössbauer effect* is illustrated in Figure 16-23. A source nucleus in an excited state makes a transition to its ground state, emitting a $\gamma$ ray. The $\gamma$ ray is subsequently caught by an unexcited absorber nucleus of the same species, which ends up in the same excited state. The potentialities as an energy spectrometer become clear when it is realized that changes in the source energy, the absorber energy, or the energy of the $\gamma$ ray in flight, will destroy the "resonant" absorption—even if the energy change is only a few parts in $10^{11}$! For some years physicists had been attempting to utilize these potentialities, but with little success. The problem had to do with recoil of the nuclei upon emission and absorption of the $\gamma$ ray, as we see in the following example.

**Example 16-7.** Mössbauer's original resonant absorption experiments used $\gamma$ rays emitted in transitions from the 0.129 MeV first excited state to the ground state of $^{77}\text{Ir}^{191}$. (a) Consider the recoil of the nucleus when it emits the $\gamma$ ray, and determine the downward shift in the energy of the $\gamma$ ray that results from the energy taken by the nuclear recoil. (b) Then compare this energy shift to the width of the first excited state of $^{77}\text{Ir}^{191}$, which has a measured lifetime of $T = 1.4 \times 10^{-10}$ sec.

(a) Since the total linear momentum of the decaying nucleus is zero before emitting the $\gamma$ ray, the magnitude of the nuclear recoil momentum $p_n$ after the emission must equal the magnitude of the momentum $p_\gamma$ carried by the emitted $\gamma$ ray. As the nuclear mass $M$ is high, its recoil velocity is low, so we may use the classical expression

$$p_n = \sqrt{2MK}$$

to relate $p_n$ to the kinetic energy of nuclear recoil $K$. The $\gamma$-ray momentum $p_\gamma$ is related to its energy $E$ by the relativistic expression

$$p_\gamma = \frac{E}{c}$$

Thus we have

$$p_\gamma = \frac{E}{c} = p_n = \sqrt{2MK}$$

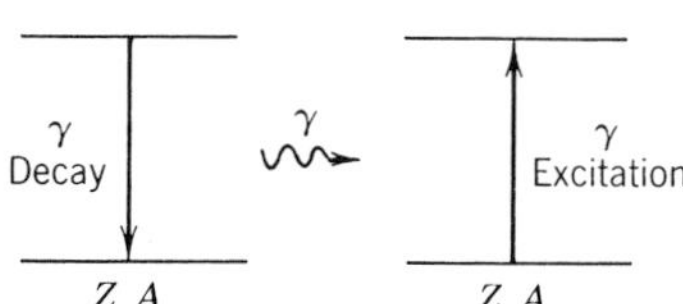


**FIGURE 16-23**
Resonant absorption, the basis of the Mössbauer effect.

or

$$\frac{E^2}{c^2} = 2MK$$

$$K = \frac{E^2}{2Mc^2}$$

Since the sum of the $\gamma$-ray energy $E$ and the nuclear recoil energy $K$ must equal the energy available in the $\gamma$ decay, i.e., the 0.129 MeV energy of the first excited state of the decaying nucleus, we see that $E$ is less than the energy of the first excited state by an amount $K$. This is the downward shift $\Delta E$ in the energy of the $\gamma$ ray due to nuclear recoil. That is

$$\Delta E = -K = -\frac{E^2}{2Mc^2}$$

Because $M$ is so large, $\Delta E$ is very small compared to $E$, and we may evaluate it approximately by setting $E = 0.129$ MeV. Using the relation 931 MeV $= uc^2$ to express the nuclear rest mass energy $Mc^2$ in MeV, we have

$$\Delta E \simeq -\frac{(0.129)^2 \text{ MeV}^2}{2 \times 191 \times 931 \text{ MeV}} = -4.7 \times 10^{-8} \text{ MeV}$$

$$= -4.7 \times 10^{-2} \text{ eV}$$

The same result could be obtained by considering the $\gamma$ ray to be emitted from a moving source, the recoiling nucleus, and using the longitudinal Doppler shift formula of Example 2-7, to evaluate the downward shift in its frequency, or energy.

(b) If the lifetime of the first excited state of ${}_{77}\text{Ir}^{191}$ is $T = 1.4 \times 10^{-10}$ sec, its width is

$$\Gamma = \frac{\hbar}{T} = \frac{6.6 \times 10^{-16} \text{ eV-sec}}{1.4 \times 10^{-10} \text{ sec}} = 4.7 \times 10^{-6} \text{ eV}$$

Clearly, the $\gamma$ ray emitted by the decay from the first excited state of the ${}_{77}\text{Ir}^{191}$ source nucleus cannot excite a ${}_{77}\text{Ir}^{191}$ absorber nucleus from its ground state to its first excited state. The nuclear recoil shift of the $\gamma$ ray is larger by a factor of $10^4$ than the width of the state it is supposed to excite. So the $\gamma$ ray is thrown completely out of resonance, and the resonant absorption is destroyed. (If there actually were an absorption, there would actually be two sources of the total recoil shift, one due to recoil of the emitting nucleus and the other due to recoil of the absorbing nucleus. There would also be two sources of the total width of the resonance, one due to the width of the state emitting the $\gamma$ ray and the other due to the width of the state absorbing it.) ◀

Mössbauer discovered that it is possible to obtain resonant absorption if the source and absorber nuclei are in crystals at sufficiently low temperatures. In such circumstances the crystals can recoil as units, with the emitting and absorbing nuclei remaining bound to their lattice sites and no lattice vibrations (phonons) excited. Being so extremely massive, the crystals take up the required recoil momentum without taking up a significant amount of kinetic energy. Mössbauer demonstrated resonant absorption in the following way. Using a source and absorber of crystalline ${}_{77}\text{Ir}^{191}$, and a temperature of 88°K, he arranged for the source to move slowly back and forth along the line between the source and absorber. This motion produced positive and negative Doppler shifts in the frequency of the $\gamma$ ray, which led to extremely small positive and negative changes in its energy. Thus the energy of the $\gamma$ ray varied from above to below the resonant energy, which is the 0.129 MeV energy of the first excited state of ${}_{77}\text{Ir}^{191}$. Figure 16-24 shows the absorption curve that was traced out as the $\gamma$-ray energy swept through the resonant energy. Its full width at half maximum is about $10 \times 10^{-6}$ eV. This agrees well with the expectation that it should be twice the width $\Gamma$ of the two nuclear states involved, since their measured lifetimes of $T = 1.4 \times 10^{-10}$ sec yield $\Gamma = 4.7 \times 10^{-6}$ eV. The agreement also verifies (16-31), used to calculate $\Gamma$ from $T$, and therefore verifies the energy-time uncertainty principle!

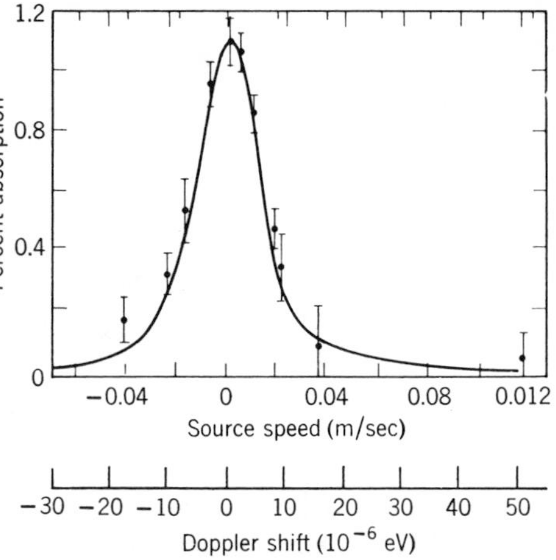

**FIGURE 16-24**

The Mössbauer effect in $^{77}Ir^{191}$ at 88°K. Note the extremely low source speeds and extremely small resulting Doppler shifts which are sufficient to eliminate the resonant absorption.

Mössbauer was awarded a Nobel Prize in 1961 because his effect lends itself to such a wide variety of extremely precise energy measurements. Most of the current use of the effect involves measuring the very small hyperfine splittings of nuclear (not atomic) energy levels which arise from interactions between the nuclear magnetic dipole and electric quadrupole moments, and the magnetic and electric fields present at the crystalline lattice site. The changes in the source and/or absorber energy, that the splittings represent, are determined in terms of the Doppler shift required to bring back resonance. Nuclear physicists use known fields and measure the moments. Solid state physicists measure the fields in cases where the moments are known, and, even if the moments are not known, they can still learn much about how the fields vary as the crystalline structure is changed.

The Mössbauer effect is also used as an energy spectrometer to measure very small changes in $\gamma$-ray energies due to phenomena of special or general relativity. One example is the verification of the gravitational red shift of general relativity for the $\gamma$ rays in a vertically oriented Mössbauer spectrometer 22.5 m high. The energy shift was only about two parts in $10^{15}$, but it was measured to better than 10% and shown to agree within that accuracy with the predictions of Einstein's general relativity theory. Mössbauer spectrometers have also been used to verify the transverse Doppler shift of special relativity, as well as the predictions of that theory concerning the "twin paradox."

## 16-7 Nuclear Reactions

We turn now from nuclear decay to nuclear reactions. One important reason why nuclear reactions are studied is that they provide information about the excited states of nuclei which supplements that provided by the study of nuclear decay. Other important reasons will become apparent when we discuss nuclear fission and fusion in subsequent sections. And, of course, the energy balance in nuclear reactions is studied with real justification because it tells about the masses of the participants in the reactions.

In our treatment in Section 15-4 of the energy balance in nuclear reactions we have already considered the application of the total relativistic energy, linear momentum, and charge conservation laws to the initial and final states of a reaction. By way of summary, we shall list these conservation laws and also others that apply to any reaction, and then use them in an example. *In any nuclear reaction the following quantities must be conserved:* (1) *total relativistic energy*, (2) *linear momentum*, (3) *angular momentum*, (4) *charge*, (5) *parity*, *and* (6) *the number of nucleons*. In all the reactions we discussed before the number of nucleons was conserved, i.e., the total number of nucleons present before the reaction equals the total number present after.

It is found that this is true of any nuclear reaction. We did not consider the conservation of angular momentum or parity at all in Section 15-4 because these quantities do not affect the energy balance. But they do affect the rates, or cross sections, for the reactions, as we shall indicate later. It is clear that angular momentum must be conserved in a nuclear reaction. Parity is conserved because the interaction involved in a nuclear reaction is the strong parity conserving nuclear interaction, not the weak parity nonconserving $\beta$-decay interaction.

**Example 16-8.** When 50.0 MeV protons in the external beam of a cyclotron strike a beryllium target, it is found that copious numbers of high-energy neutrons are emitted from the target. The highest energy neutrons are emitted in the same direction as the incident protons, and their energy is 48.1 MeV. In order to increase the number of neutrons produced, so that they can be more easily used in other experiments, it is decided to put the beryllium target inside the cyclotron where it will be bombarded by the much more intense internal beam. In this configuration neutrons produced at 30° to the direction of the bombarding protons will have a clear path out past the external parts of the cyclotron. (a) Use the conservation laws to find the residual nucleus in the reaction in which a proton ${}^1H^1$ is the bombarding particle, a neutron ${}^0n^1$ is the product particle, and ${}^4Be^9$ is the target nucleus. (b) Then apply the conservation laws to predict the maximum energy neutrons produced at 30° to the direction of the 50.0 MeV bombarding protons.

(a) The reaction is

$${}^1H^1 + {}^4Be^9 \rightarrow {}^ZX^A + {}^0n^1$$

where ${}^ZX^A$ represents the unknown residual nucleus. Conservation of charge requires that the sum of the $Z$ values on the left side of the reaction formula equal the sum of the $Z$ values on the right side. That is

$$1 + 4 = Z + 0$$

or

$$Z = 5$$

Conservation of the number of nucleons requires that the sum of the $A$ values on the left side equal the sum of the $A$ values on the right side. Therefore

$$1 + 9 = A + 1$$

or

$$A = 9$$

Thus we have identified the residual nucleus as ${}^5B^9$, and the reaction is

$${}^1H^1 + {}^4Be^9 \rightarrow {}^5B^9 + {}^0n^1$$

(b) To calculate the energies of neutrons emitted at various angles, we use the conservation of total relativistic energy and linear momentum, combined in the form of the $Q$-value formula of (15-16)

$$Q = K_b\left(1 + \frac{m_b}{m_B}\right) - K_a\left(1 - \frac{m_a}{m_B}\right) - \frac{2}{m_B}(K_aK_bm_am_b)^{1/2}\cos\theta$$

where $K_a$ and $m_a$ are the kinetic energy and mass of the proton, $K_b$ and $m_b$ are the kinetic energy and mass of the neutron, $m_B$ is the mass of ${}^5B^9$, and $\theta$ is the angle of emission of the neutron relative to the direction of the proton. Since we are always dealing with the maximum energy neutrons emitted, the $Q$ value always pertains to a situation in which the residual nucleus is in its ground state.

First we determine the $Q$ value by setting $K_a = 50.0$, $K_b = 48.1$, and $\theta = 0$, where we use MeV for the unit of energy. Since to a very good approximation $m_a/m_B = m_b/m_B = 1/9$, we have

$$Q = 48.1 \times \frac{10}{9} - 50.0 \times \frac{8}{9} - 2\sqrt{50.0 \times 48.1 \times \frac{1}{9} \times \frac{1}{9}}$$

$$= 53.4 - 44.4 - 10.9 = -1.9$$

or

$$Q = -1.9 \text{ MeV}$$

Note that $Q$ is just equal to $K_b - K_a$. But this is only true when $m_a = m_b$, $\theta = 0$, and $|Q|$ is small compared to $K_a$.

Knowing the $Q$ value, we find $K_b$ when $\theta = 30°$ by again using (15-16). We have, since $\cos 30° = 0.866$

$$-1.9 = K_b \times \frac{10}{9} - 50.0 \times \frac{8}{9} - \frac{2}{9}\sqrt{50.0} \times 0.866\sqrt{K_b}$$

We write this as

$$1.11(\sqrt{K_b})^2 - 1.36\sqrt{K_b} - 42.5 = 0$$

to make it easier to apply the standard solution of a quadratic equation in the unknown $\sqrt{K_b}$. This gives

$$\sqrt{K_b} = \frac{1.36 \pm \sqrt{(1.36)^2 + 4 \times 1.11 \times 42.5}}{2 \times 1.11} = \frac{1.36 \pm 13.79}{2.22}$$

The equation is not a quadratic in $K_b$, and has only one valid solution. We may easily show that it is obtained for the plus sign. Using that sign, we find

$$\sqrt{K_b} = 6.82$$

or

$$K_b = 46.5$$

Thus the maximum neutron energy produced at 30° is

$$K_b = 46.5 \text{ MeV}$$

◀

The subject of nuclear reactions is a vast one because there are so many different types of reactions. Any stable nuclear particle can be the bombarding particle; any stable nucleus can be the target nucleus; and a wide variety of particles can be emitted from the reaction as product particles. The residual nucleus can be either stable or radioactive. Typically it will be stable if the reaction does not change the $Z$-to-$A$ ratio of the residual nucleus very much from the stable $Z$-to-$A$ ratio that the target nucleus has. An example of a reaction that often leads to a stable residual nucleus is $(d,\alpha)$, where the notation means that a deuteron, $^1H^2$, is the bombarding particle and an $\alpha$ particle, $^2He^4$, is the product particle. If the reaction significantly decreases the $Z$-to-$A$ ratio of the residual nucleus, it is usually radioactive and decays by electron emission to raise its $Z$-to-$A$ ratio back to a stable value. An example of a reaction that often leads to an electron emitting residual nucleus is $(n,p)$, in which there is a bombarding neutron, $^0n^1$, and a product proton, $^1H^1$. Reactions such as $(p,n)$ frequently lead to radioactive residual nuclei which are positron emitters or electron capturers, since the reaction raises the $Z$-to-$A$ ratio of the residual nucleus over the stable value that this ratio has for the target nucleus. Thus nuclear reactors, which produce intense fluxes of neutrons, are usually employed to produce radioactive nuclei for diagnostic work in medicine, and other fields, as "tracers," if the required nuclei are electron emitters. Cyclotrons, which produce intense fluxes of protons or more highly charged particles, are usually the sources of radioactive tracers that are positron emitters or electron capturers.

We present in this section examples of the most important types of nuclear reactions by discussing the processes that can occur when a 50-MeV proton from a cyclotron beam is incident on a target nucleus, of average characteristics, contained in a foil placed in the beam. We describe what happens *during* these processes—and not just what the situation is like before and after, as we have done in our earlier considerations of the mass-energy balance in nuclear reactions.

First we shall give a quick summary of the processes that can occur. The proton, of representative energy 50 MeV, will be scattered away from the typical target nucleus by the *Coulomb potential*, unless it happens to be traveling almost in the direction of the nuclear center. It can also be scattered by the *nuclear potential*, if it approaches close enough to feel this potential. If it enters the nucleus, it will probably collide with a nucleon in the nucleus after traveling part way through. Either it or the struck nucleon may escape immediately, in a so-called *direct interaction*, taking away most of the energy it carries (as in the reaction treated in Example 16-8). But at least one of these nucleons will probably be reflected back into the nucleus by the change in nuclear potential at the surface in much the same way a light wave would be internally reflected by a change in refractive index. (See the discussion connected with (6-53).) This nucleon will collide with another nucleon, each of them will make further collisions, etc., forming a cascade of collisions. Before long, the energy is shared among the excitation of many nucleons in what is called the *compound nucleus*. At this point, no nucleon has enough excitation to allow it to escape its $\sim$8 MeV binding to the nuclear potential. After some time, a fluctuation in the energy sharing will make energetically possible the escape of a nucleon. This will happen, if internal reflection at the nuclear surface does not make it necessary to wait for another fluctuation. Eventually, several nucleons are "evaporated," and their binding energies are largely responsible for removing most of the excitation energy of the compound nucleus. They will almost always be neutrons, since the Coulomb barrier acts to retain the protons. When the excitation energy is below the neutron binding energy, the relatively slow process of $\gamma$ decay takes over and allows the system to finally end up in its ground state.

We begin a more detailed discussion of these processes by pointing out that the de Broglie wavelength of a 50 MeV proton moving through a 50 MeV deep nuclear potential is $\simeq$ 3 F, and the range of nuclear forces is a little smaller. Since both are about one-third of a typical nuclear diameter, in a crude first approximation we may think of the proton as traveling a fairly well-defined trajectory, and not interacting at a distance. Thus the behavior of the proton is something like that of a classical billiard ball. To an even lesser extent, this approximation also applies to the nucleons that the proton collides with. Of course, the wavelike aspects of these particles will make important corrections to the approximation.

Since *Coulomb scattering* has been discussed at length in Chapter 4 and Appendix D, there is little we need to say about it here, except to comment that the differential scattering cross section $d\sigma/d\Omega$ of (4-9), obtained from Rutherford's classical theory of the scattering by a Coulomb potential, is identical with the $d\sigma/d\Omega$ obtained from quantum mechanics for that potential. This remarkable situation is true only for a potential corresponding to an inverse square law of force, and it arises in the following way. From dimensional analysis it can be shown that if the force exerted on a particle varies according to $r^n$, then the probability of scattering must vary according to $h^{4+2n}$. For the inverse square law $n = -2$, the scattering probability is independent of the value of Planck's constant $h$, and this requires that the quantum mechanical and classical calculations lead to the same results.

Figure 16-25 shows the probability of elastic scattering (scattering without energy loss except to the recoil of the residual nucleus), as a function of scattering angle $\theta$,

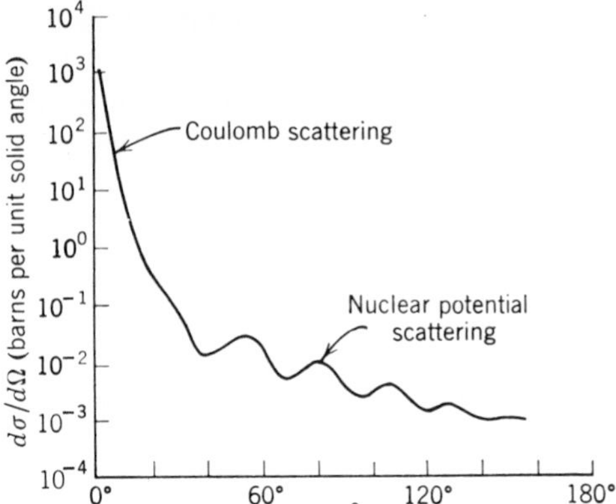


**FIGURE 16-25**

The differential cross section for the elastic scattering of 50 MeV protons from a hypothetical nucleus of typical properties. The cross section unit is the barn; 1 bn $= 10^{-24}$ cm$^2$.

for a 50 MeV proton incident on a typical nucleus. At small scattering angles, the differential cross section follows the rapid but smooth decrease in proportion to $1/\sin^4(\theta/2)$ of Coulomb, or Rutherford, scattering. The reason is that these angles correspond to collisions in which the proton passes through the Coulomb potential, but misses the nuclear potential. At large scattering angles, the scattering probability shows a diffractionlike structure superimposed on a continued decreasing trend. The reason is that protons scattering at these angles make close enough collisions to feel the abrupt onset of the nuclear potential. The diffraction structure of this so-called *nuclear potential scattering* arises from the interferences between the incident wave function and the various parts of the wave function reflected from various regions of the nuclear potential.

A quantum mechanical analysis of the elastic scattering measurements can be used to determine the nuclear potential acting on the high-energy scattered nucleon. The potential is found to be essentially the same as the shell model potential acting on a nucleon in the ground state of the target nucleus, with one important exception. The potential acting on an unbound nucleon, called the *optical model* potential, is partly *absorptive*. The absorption represents the fact that such a nucleon has enough energy to collide with a nucleon in the nucleus, and thus be absorbed from the incident beam. (It is absorbed in the sense that it no longer has the same energy, or de Broglie wavelength, so there can be no interferences between its wave function and the wave function for the incident nucleon.) Collisions are possible since the exclusion principle does not have its usual inhibiting effect if the incident nucleon brings in enough energy that both it, and the struck nucleon, can easily find unfilled states to occupy. The incident nucleon can, of course, also scatter from the more familiar nonabsorptive part of the potential. (That is, it can also interact with the nucleus as a whole, represented by the usual attractive potential, without colliding with an individual nucleon of the nucleus.) The optical model is essentially a generalization of the shell model which applies to nucleons of any energy—not just to nucleons of energy such that they are bound in a nucleus.

If the scattering probability is measured as a function of the energy of the incident particle, very broad maxima are sometimes seen at certain energies. These are called *size resonances*, or *single particle states*. As the two names imply, they can be thought of in two different ways: (1) constructive interferences between the part of the incident particle wave function scattered from the front surface of the nuclear potential and the part scattered from the back; (2) energy levels of the incident particle in the nuclear potential. The first point of view is related to one developed in our discussion of the Ramsauer effect in Section 6-5, but here we shall find the second point of view more useful. The maxima are broad because the single particle states are very wide. If we evaluate the time required for a 50 MeV nucleon to travel a typical nuclear diameter, we find $T = D/v \sim 10^{-14}\ \text{m}/10^{8}\ \text{m-sec}^{-1} = 10^{-22}$ sec. Since this time also characterizes the duration of the nuclear potential scattering process, or the lifetime of the particle in the single particle state, the width $\Gamma$ of the state is, typically, $\Gamma = \hbar/T \sim 10^{-15}\ \text{eV-sec}/10^{-22}\ \text{sec} = 10^{7}\ \text{eV} = 10$ MeV. Note that the width of a typical high-energy single particle state is some 12 orders of magnitude greater than the width of a typical low-energy $\gamma$-decaying state considered at the end of Section 16-5.

Now we reconsider the collisions between the incident proton and nucleons of the nucleus. Before colliding, the linear momentum of the proton is approximately in the direction of the beam, and it is of much larger magnitude than that of any nucleon in the nucleus. Linear momentum conservation thus demands that after the first collision both the nucleons tend to move off in the general direction of the beam, and this is particularly so of a nucleon if it happens to be carrying most of the incident momentum or energy. A higher energy nucleon is the one most likely to escape internal reflection at the nuclear surface, and be emitted in what is called a *direct interaction*. It will preserve its tendency to move in the general direction of the incident beam, even though it is refracted somewhat in passing through the surface.

Figure 16-26 shows the spectrum of high-energy protons emitted, at some fixed angle, from a typical nucleus. The group of highest energy contains the *elastically scattered* protons. They have the same energy as the incident protons (except for the small amount of energy lost to the recoil of the residual nucleus), and they are the result of Coulomb and nuclear potential scattering. The group of next highest energy contains *inelastically scattered* protons, which come from direct interactions. When a proton is emitted in this group, the residual nucleus remaining is in its first excited state. When a proton is emitted in the group of next lowest energy, that nucleus is in its second excited state, etc. Thus *the energy spectrum gives immediately the locations of the excited states of the nucleus.*

The general tendency for small angles of emission of the higher energy nucleons coming from direct interactions is shown in Figure 16-27. This represents the differential cross section $d\sigma/d\Omega$ for the protons emitted in the highest energy inelastically scattered group, for the typical case of the previous figure. Also indicated in the figure

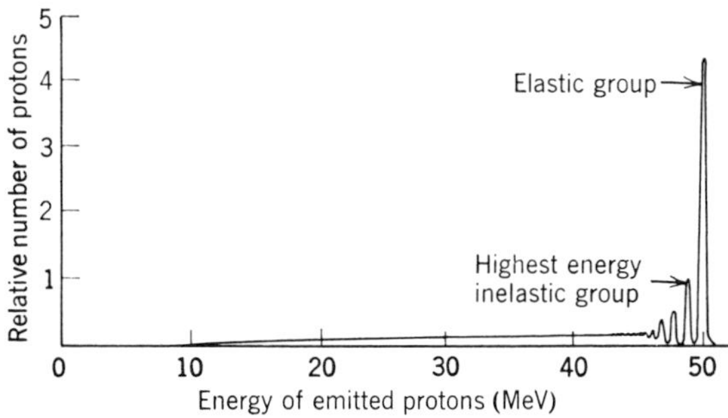


**FIGURE 16-26**
The energy spectrum of protons emitted at a forward angle when 50 MeV protons are incident in the bombardment of a hypothetical nucleus of typical properties. The low-lying energy levels of the residual nucleus show up in the high-energy inelastic groups. As these levels fuse into a continuum, so does the inelastic spectrum. The cutoff in the spectrum at about 10 MeV represents the effects of internal reflection and of the Coulomb barrier in preventing the escape of protons.

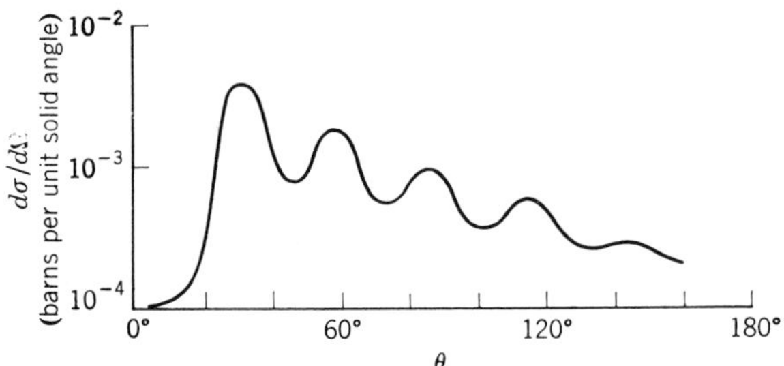


**FIGURE 16-27**

The differential cross section $d\sigma/d\Omega$ for the highest energy group in the inelastic scattering of 50 MeV protons from a hypothetical nucleus of typical properties. The general preference for forward angles of emission is characteristic of the direct interaction process, but $d\sigma/d\Omega$ is suppressed at very small angles if orbital angular momentum is transferred to the nucleus in the reaction. The figure represents $d\sigma/d\Omega$ for a reaction in which the state excited has orbital angular momentum one unit higher than the ground state.

is the tendency for $d\sigma/d\Omega$ to be suppressed at very small angles, if orbital angular momentum must be transferred to the nucleus from the incident proton in the reaction because the state excited has orbital angular momentum different from that of the ground state. The semiclassical argument of Figure 16-28 shows that this tendency reflects the fact that it is difficult for a particle, which experiences only a very small decrease in the magnitude of its linear momentum in interacting with a target of restricted radius, to transfer orbital angular momentum to the target unless it changes its direction of motion enough to produce a sufficient change in the vector describing its linear momentum.

Of course the billard ball arguments, that predict the general trends, fail to predict the oscillations about them seen in Figure 16-27. These arise from interferences between parts of the emitted nucleon wave function that originate in different regions of the nucleus. The structure of the differential cross section curve can be analyzed to yield information about the nuclear spin and parity of the state of the residual nucleus that is excited in the emission of the inelastically scattered group. The procedures used in the analysis are a little too complicated to go into here, but it should be said that they also confirm that *parity is conserved in the nuclear interaction.*

Although an incident proton has about a 90% chance of making a collision with a nucleon in traversing the nucleus, in only about 10% of these events will there be a direct interaction nucleon emitted. Usually, both the incident proton and the nucleon it hits are trapped in the nucleus by internal reflection. In about 1% of the events, both the incident proton and the struck nucleon escape. If their linear momenta are measured, valuable information can be obtained about the initial momentum of the struck nucleon when it was in the nucleus (after correcting for refraction and absorption as the protons leave the nuclear optical potential). This has recently become an important research technique in nuclear physics.

The time required for the first collision is $\sim 10^{-22}$ sec, since this is how long it takes for a nucleon of typical velocity to travel a distance equal to a typical nuclear diameter. The subsequent steps in the cascade of collisions occur at intervals of roughly the same time. In the first two or three steps, there is a chance that one of the nucleons that has collided will escape, but the chance diminishes rapidly because the collisions lead to a sharing of energy. Internal reflection in the nuclear potential becomes more

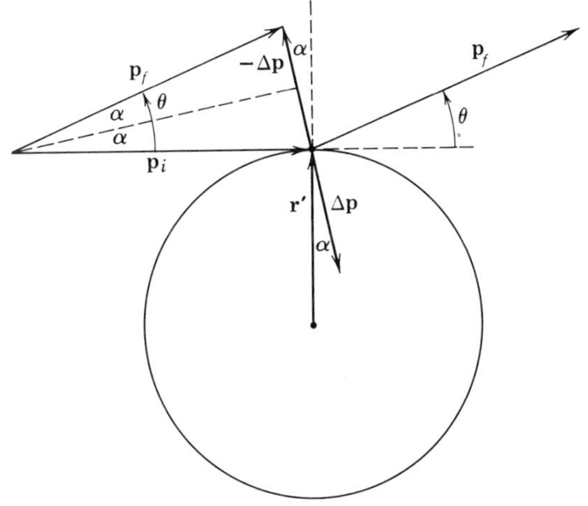


**FIGURE 16-28**

Illustrating the relation between the linear and orbital angular momenta transferred to a nucleus in a direct interaction inelastic scattering leading to its first excited state. The linear momentum of the incident nucleus is $\mathbf{p}_i$. It leaves the nucleus at angle $\theta$ with linear momentum $\mathbf{p}_f$. Since it is emitted with almost as much energy as it had when incident, $p_f \simeq p_i \simeq p$, and the momentum $\Delta\mathbf{p} = \mathbf{p}_i - \mathbf{p}_f$ is transferred to the nucleus primarily because the direction of $\mathbf{p}_f$ differs from the direction of $\mathbf{p}_i$. The figure shows the interaction occurring near the edge of the nucleus of radius $r'$, where it will be most effective in transferring angular momentum $\Delta\mathbf{L}$ to the nucleus. Since $\Delta\mathbf{L} = \mathbf{r}' \times \Delta\mathbf{p}$, we have $\Delta L = r'\,\Delta p \sin\alpha \simeq r'\,\Delta p\alpha$, because the angle $\alpha = \theta/2$ defined in the figure tends to be small in a direct interaction. The figure shows that $\Delta p \simeq 2p_i\alpha \simeq 2p\alpha$. So $\Delta L \simeq 2r'\,p\alpha^2$. For a case in which one unit of orbital angular momentum is given to the nucleus, we have

$$\Delta L = \sqrt{1(1+1)}\,\hbar = 1.4\hbar$$

Thus we obtain $\alpha^2 \simeq 1.4\hbar/2r'p = 1.4\hbar/2r'(h/\lambda) = 1.4/4\pi(r'/\lambda)$ where $\lambda$ is the de Broglie wavelength of the proton. As indicated in the text, $r'/\lambda \simeq 5/3$ for a 50 MeV proton moving through the 50 MeV deep potential of a nucleus of typical radius $r' = 5$ F. So

$$\alpha^2 \simeq 1.4/4\pi(5/3) \simeq 6 \times 10^{-2}$$

or $\alpha \simeq 2.5 \times 10^{-1}$ rad $\simeq 15°$. Thus the emission angle $\theta$ that this semi-classical calculation predicts would lead to a transfer of one unit of orbital angular momentum is $\theta = 2\alpha \simeq 30°$. For angles much smaller than this the reaction would not be possible. If an even larger orbital angular momentum must be transferred to the nucleus, because of the difference between the spins of its ground and first excited states, an even larger angle of emission is required.

likely as the energies of the individual nucleons decrease, and soon an even stronger inhibition sets in because the excitation energies of the nucleons becomes less than their binding energies. After perhaps 10 steps of the cascade, which takes $\sim 10^{-21}$ sec, the energy is well distributed over all the nucleons of the nucleus. None of these nucleons has enough energy to escape; instead they exchange energy in a kind of thermal equilibrium. This equilibrium system is called the *compound nucleus*.

Because the equilibrium system does not contain a very large number of particles ($A \sim 100$), big fluctuations in the energy sharing can occasionally happen. If some nucleon accumulates about ten times as much excitation energy as it has on the average it will have the equivalent of its binding energy, and it can try to escape. Typically, this takes about $10^{-16}$ sec, and typically the nucleon will not succeed because it is internally reflected. But eventually a nucleon will escape, carrying away a little more than its binding energy. The elapsed time at this point is something like $10^{-15}$ sec, on the average. After several nucleons have escaped, there is no longer enough excitation energy in the nucleus to provide the $\sim 8$ MeV required to emit another nucleon. As we have mentioned, $\gamma$ decay is used to dissipate the final few MeV of excitation energy, and as we have also mentioned, almost all of the nucleons that are evaporated in fluctuations from equilibrium are neutrons. Protons generally cannot accumulate enough energy to overcome the Coulomb barrier acting on them.

In a compound nucleus the excitation is distributed over many particles. The excited states of the nucleus are consequently called *many particle states*. In contrast to the very broad single particle states, the many particle states are fairly narrow. Since it takes the compound nucleus $T \sim 10^{-15}$ sec to decay by neutron emission, the width $\Gamma$ of a typical one of its states is given in terms of this lifetime by

$$\Gamma = \hbar/T \sim 10^{-15}\ \text{eV-sec}/10^{-15}\ \text{sec} = 1\ \text{eV}$$

These narrow states can be observed by measuring as a function of the nucleon energy the probability, or total cross section defined in (2-18), that an incident nucleon will form a compound nucleus. As the separation between the many particle states rapidly decreases, and their width increases, with increasing excitation energy, it is easiest to see them if an incident nucleon of the lowest possible energy is used. Figure 16-29 is an example of the many particle states, or *compound nucleus resonances*, observed when very low-energy neutrons are incident on a typical nucleus.

The shape of any individual cross-section resonance in Figure 16-29 is given by the *Breit-Wigner formula*

$$\sigma_r(E) = \pi(\lambda/2\pi)^2 \frac{\Gamma_n \Gamma_r}{(E - E_i)^2 + \Gamma^2/4} \tag{16-32}$$

where the total reaction cross section $\sigma_r(E)$ is the cross section for the formation of a compound nucleus which decays by any process other than emission of a neutron of

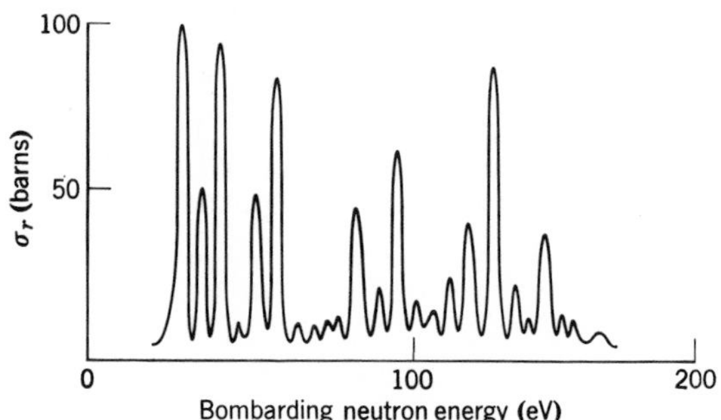


**FIGURE 16-29**
The total cross section for an incident neutron of very low energy to undergo any reaction other than elastic scattering with a hypothetical nucleus of typical properties. The many particle states of the compound nucleus of excitation energy about 8 MeV (the binding energy brought in by the incident neutron) are seen directly in such data.

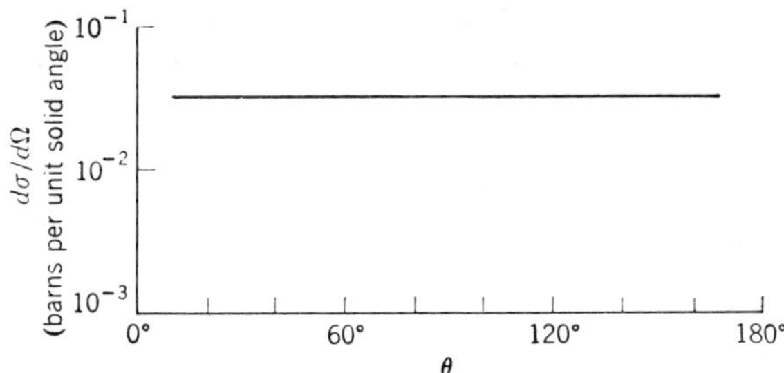


**FIGURE 16-30**
The differential cross section for the compound nucleus evaporation of low-energy neutrons following the 50 MeV bombardment of a hypothetical nucleus of typical properties. The lack of a preferred direction of emission is characteristic of the compound nucleus process.

the same energy as the incident one; $E$ is the energy of that neutron and $\lambda$ is the corresponding de Broglie wavelength; $E_i$ is the resonance energy; $\Gamma$ is the full width at half-maximum of the resonance; and $\Gamma_n$, or $\Gamma_r$, is $\Gamma$ times the ratio of the probability of decay of the compound nucleus by emitting a neutron of the same energy as the incident neutron, or by any other process, to the total probability of decay by all processes. The same formula, with $\Gamma_r$ replaced by $\Gamma_n$, gives the total cross section for the formation of a compound nucleus which subsequently decays by emitting a neutron of the same energy as the incident neutron, i.e., the compound nucleus elastic scattering cross section $\sigma_s(E)$. A similar formula describes the shape of the $\gamma$-ray resonances in Figure 16-22 and Figure 16-24. In fact, the same basic form is found for the resonance curve in any type of damped wave or oscillatory motion. The student may have seen a derivation of it in the case of a damped pendulum or a resistive resonant circuit.

A very interesting feature of (16-32) that is particular to the case of low-energy neutron resonances is the factor $\pi(\lambda/2\pi)^2$, which determines the maximum possible value of the total neutron cross sections at the peak of a resonance. It is the area of a circle of radius equal to the neutron de Broglie wavelength $\lambda$ divided by $2\pi$, and not the area of a circle of nuclear radius $r'$. Since $\lambda \gg r'$ for sufficiently low-energy neutrons, the total reaction, or scattering, cross section at a resonance peak can be very much larger than the projected geometrical cross section, $\pi r'^2$, of the nucleus. This is possible because the low-energy neutron acts like a wave, not a classical particle, and at resonance it can interact with the target nucleus whenever the expectation value of its position passes within a distance of about $\lambda/2\pi$ of the nucleus. Later we shall see that this property is very important in the operation of a nuclear reactor.

Another characteristic of a compound nucleus is that in its relatively long lifetime it forgets the details of how it was formed. For instance, since the original linear momentum of the incident particle becomes distributed over the many particles that are excited in the compound nucleus, there cannot be a preference for the neutrons to be emitted in the beam direction. Figure 16-30 shows an example of the isotropic differential cross section for emission that characterizes the low-energy neutrons produced in nuclear reactions. These are the neutrons evaporated from compound nuclei.

**Example 16-9.** The measured differential cross section for the emission at 40° of the highest energy inelastically scattered proton group from $_{26}Fe^{54}$ bombarded by 60 MeV protons is $d\sigma/d\Omega = 1.3 \times 10^{-3}$ bn per unit solid angle. These inelastic protons leave the $_{26}Fe^{54}$ residual nucleus in its first excited state at 1.42 MeV. Calculate how many events per second are recorded in a measurement of the inelastically scattered protons made with a detector of area $10^{-5}$ m$^2$ located $10^{-1}$ m from a pure $^{26}Fe^{54}$ foil, of mass per unit area $10^{-1}$ kg/m$^2$, which is bombarded by a $10^{-7}$ amp proton beam. (In nuclear physics, the unit of area for cross sections is called the *barn*, written bn; 1 bn $= 10^{-28}$ m$^2$.)

The number $n$ of nuclei, or atoms, contained in a unit area of the target is the mass per unit area of the target divided by the mass of a $^{26}Fe^{54}$ atom. Since this is almost exactly 54 times the mass of a $^{1}H^{1}$ atom, we have

$$n = \frac{10^{-1}\ \text{kg/m}^2}{54 \times 1.66 \times 10^{-27}\ \text{kg/nucleus}} = 1.1 \times 10^{24}\ \text{nuclei/m}^2$$

The solid angle $d\Omega$ subtended by the detector at the target is its area divided by the square of its distance from the target. So

$$d\Omega = \frac{10^{-5}\ \text{m}^2}{(10^{-1}\ \text{m})^2} = 10^{-3}\ \text{sr}$$

(A unit solid angle is called a *steradian*, written sr; 1 sr = solid angle subtended by 1 m$^2$ at 1 m.)

The product of the differential cross section $d\sigma/d\Omega$ for the events of interest times the solid angle $d\Omega$ subtended by the detector gives an area per nucleus that is effective in leading to the detected events. This effective area per nucleus $d\sigma$ is

$$d\sigma = 1.3 \times 10^{-3}\ \frac{\text{bn/sr}}{\text{nucleus.}} \times 10^{-3}\ \text{sr} = 1.3 \times 10^{-6}\ \text{bn/nucleus} = 1.3 \times 10^{-34}\ \text{m}^2/\text{nucleus}$$

The product of the effective area per nucleus, $d\sigma$, times the number of nuclei per unit area, $n$, equals the probability that one incident proton will produce a detected event. This probability $P$ is

$$P = d\sigma\, n = 1.3 \times 10^{-34}\ \text{m}^2/\text{nucleus} \times 1.1 \times 10^{24}\ \text{nuclei/m}^2 = 1.4 \times 10^{-10}$$

That is

$$P = 1.4 \times 10^{-10}\ \text{event/proton}$$

The number of protons per second $I$ in the incident beam is the charge per second in the beam divided by the charge per proton, or

$$I = \frac{10^{-7}\ \text{coul/sec}}{1.6 \times 10^{-19}\ \text{coul/proton}} = 6.2 \times 10^{11}\ \text{proton/sec}$$

Multiplying the number of protons per second $I$ by the probability $P$ that a proton will produce a detected event, we obtain the number of events detected per second. This is

$$dN = IP = 6.2 \times 10^{11}\ \text{proton/sec} \times 1.4 \times 10^{-10}\ \text{event/proton} = 87\ \text{event/sec}$$

Note that the preceding equation can be written as

$$dN = IP = I\, d\sigma\, n = \frac{d\sigma}{d\Omega} In\, d\Omega$$

in agreement with (4-8), the definition of a differential cross section. ◀

## 16-8 Excited States of Nuclei

Figure 16-31 reviews information about the excited states of nuclei obtained from the study of nuclear decays and nuclear reactions. The energy-level diagram represents energy states of the entire nucleus, and *not* of individual nucleons in the nucleus. Up to an excitation of ~8 MeV, the states $\gamma$ decay to the ground state. Above ~8 MeV, nucleon emission becomes energetically possible, and this process soon becomes the

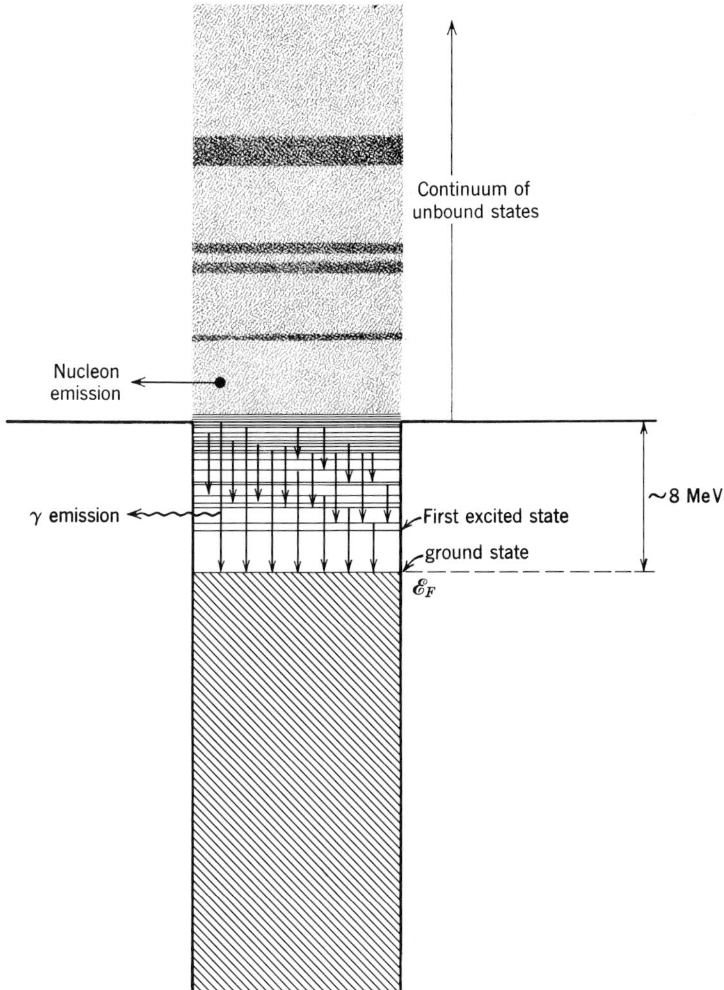


**FIGURE 16-31**
An over-all view of the excited states of a typical nucleus.

dominant decay mode since it has a much shorter lifetime or much higher transition rate. This is the region of the many particle states. They are very closely spaced because there are a large number of different divisions of energy between the many particles of the nucleus that lead to almost the same total nuclear excitation energy. The spacing decreases with increasing $A$ because more divisions are possible. It also decreases as there becomes more excitation energy available to divide between the particles. Thus the many particle states soon fuse together into a continuum of allowed nuclear energy states, but the continuum maintains some structure since the many particle states tend to group together into the very wide single particle states through which they have been excited. Each many particle state in a group has the same angular momentum and parity as the original single particle state.

Now let us look more carefully at the *low-lying* excited states. The simplest case is for a nucleus whose ground state consists of a core of filled magic number subshells, plus one nucleon. In the first excited state, the extra nucleon jumps to the next highest energy subshell, and the core remains undisturbed. Figure 16-32 shows, as an example, the low-lying excited states of ${}^{8}O^{17}$. The spin and parity of the first excited state agree

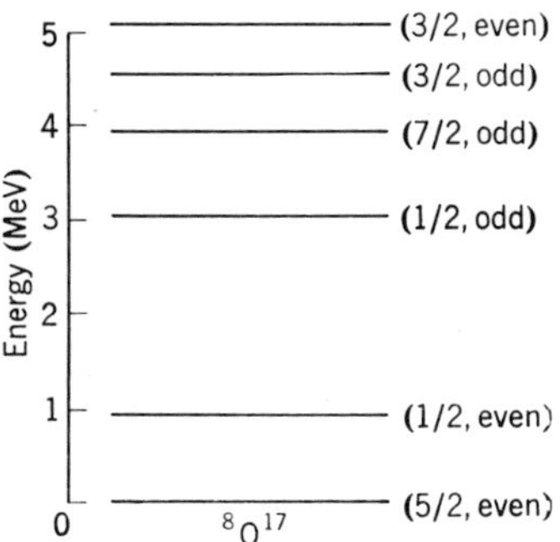


**FIGURE 16-32**

The low-lying excited states of $^8O^{17}$. Excitation energies, spins, and parities are shown. The spin and parity of the first excited state are correctly predicted by the shell model as are, of course, the spin and parity of the ground state (see Figure 15-18). The energy of the first excited state is not predicted by the model, nor are any of the characteristics of the higher excited states.

with the predictions of Figure 15-18 of the shell model, but its energy is not predicted by the model. If the ground state of a nucleus consists of a core of filled magic number subshells, plus one hole, its first excited state is the shell model state of the hole. But in both these cases, usually even the second excited state has unpredicted spin and parity.

Between magic numbers, the first few excited states of nuclei often show regularities expected from the collective model. An example is for the even-even nucleus $^{92}U^{238}$, illustrated in Figure 16-33. On the right are the observed energy levels, and on the left are the predictions of the quantum mechanical formula

$$E = \frac{i(i+1)}{2\mathscr{I}}\hbar^2 \qquad i = 0, 2, 4, 6, \ldots \quad (16\text{-}33)$$

for the allowed values of total energy $E$ of rotation of a symmetric rotator, such as an ellipsoid rotating with rotational inertia, or moment of inertia, $\mathscr{I}$, about an axis perpendicular to its symmetry axis. Equation (16-33) is the same as (12-1) that we derived while treating the rotational spectra of diatomic molecules, except that (1) the quantum number we must use here is $i$, instead of $r$; (2) we therefore avoid confusion by using the symbol $\mathscr{I}$, instead of $I$, for the rotational inertia; and (3) since we deal with a symmetrical rotator, only even values of the rotational quantum number $i$ will arise. The reason for the last statement is that the rotational eigenfunction for the system has the parity of $(-1)^i$, and thus will be odd if $i$ is odd, and

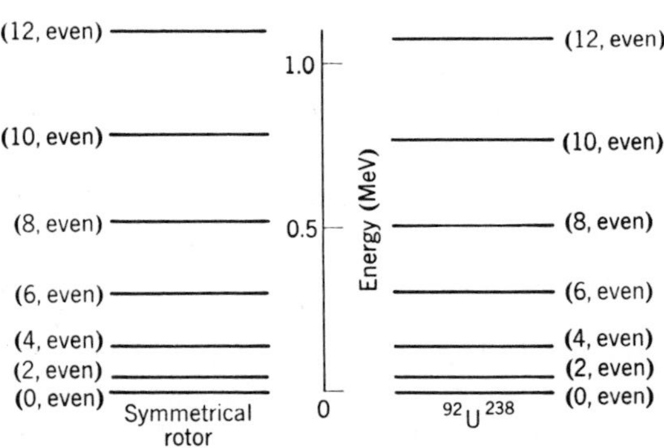


**FIGURE 16-33**

The low-lying excited states of $^{92}U^{238}$. *Right:* The data. *Left:* The predictions for the rotational states of a symmetrical ellipsoid of rotational inertia $\mathscr{I}$. The value of $\mathscr{I}$ was chosen to give the best fit to the experimental energies, the value being $2940u\text{-}F^2$. The average discrepancy in the fit is only 0.0204 MeV, which indicates the success of the model. Most of this discrepancy is in the form of very small downward displacements of the higher rotational states from the predicted values. It can be understood as a small increase of $\mathscr{I}$ in these states due to centrifugal effects.

even if $i$ is even. It can be shown to follow from the symmetry of the rotator that it can have no angular momentum in the direction of its symmetry axis, and that all of its states must have the same parity. Since an even $Z$-even $N$ nucleus has an even parity ground state, we therefore see that its excited states must also have even parity. Thus the odd values of $i$ must be deleted in (16-33). Inspection of the excellent agreement between (16-33) and the low-lying states of $^{92}U^{238}$, shown in Figure 16-33, makes it clear that collective effects in that nucleus deform it into an ellipsoidal shape. In particular, the evidence is that it has essentially the same shape in all of these states, including the ground state, because the predictions of (16-33) are obtained by using a constant value of the nuclear rotational inertia $\mathscr{I}$.

Of course, we already know, from the discussion of the collective model and nuclear electric quadrupole moments in Section 15-10, that even-$N$, odd-$Z$ or odd-$N$, even-$Z$ nuclei, with $N$ and $Z$ between the magic numbers, are usually ellipsoidal in shape. The tendency for an ellipsoidal shape is particularly strong for such nuclei in the region of the rare earth elements (the lanthanides), and it is fairly strong for nuclei in the region of uranium and the elements just above it in the periodic table (the actinides), since in these regions both $N$ and $Z$ are far from magic numbers. What is new here is the evidence for the ellipsoidal shape of the even-$N$, even-$Z$ nucleus $^{92}U^{238}$. Recall that in Section 15-2 we concluded that if a nucleus has zero nuclear spin in its ground state, as is the case for $^{92}U^{238}$ and all other even-$N$, even-$Z$ nuclei, then it would not be possible to observe an ellipsoidal shape in its ground state, even if it actually has such a shape, in time-averaged measurements like the hyperfine splitting determinations of the electric quadrupole moment. The measurements on nuclear decay and nuclear reactions that lead to the $^{92}U^{238}$ energy levels of Figure 16-33 are sensitive to the actual shape of the nucleus—not to just the time average of all possible orientations of the shape as is true of the hyperfine splitting measurements on zero spin nuclei. These more sensitive measurements show that the nucleus is ellipsoidal. Similar measurements show that this is generally true of all nuclei, no matter whether $N$ and $Z$ are even or odd. The only exceptions are nuclei with $N$ and $Z$ at or very near the magic numbers, where collective effects are insignificant. Such nuclei are truly spherical.

Since the deformation of nuclear shapes from spherical to ellipsoidal is a consequence of collective effects, nuclei where these effects are strong because both $N$ and $Z$ are far from magic numbers have, in their low-lying energy states, relatively large and essentially rigid deformations, like $^{92}U^{238}$. These states consist of the various rotations allowed by quantum mechanics. Nuclei in which $N$ and/or $Z$ are not very far from magic numbers have deformations that are not very large, and that are not rigid. The low-lying states of such a nucleus involve vibrations of its shape back and forth between an ellipsoid elongated in the direction of its symmetry axis and an ellipsoid shortened in that direction. The motion is further complicated by the fact that the nucleus can also rotate. Nevertheless, the first few energy levels of nuclei of this type are rather evenly spaced, like the energy levels of a simple harmonic oscillator. An example is found in the low-lying excited states of $^{78}Pt^{192}$, shown in Figure 16-34. Note that the lowest collective states of ellipsoidal nuclei, whether rotational, vibrational, or a combination of both, have much smaller excitation energies than the lowest shell model states of spherical nuclei. This can be seen by comparing Figures 16-33 and 16-34 with Figure 16-32.

Another regularity of low-lying excited states is found in comparing these states in certain pairs of nuclei whose shell model descriptions are identical, except that the neutrons and protons are interchanged. An example of such a so-called *mirror pair* of nuclei is $^{1}H^{3}$ and $^{2}He^{3}$, whose ground state shell model descriptions were shown in

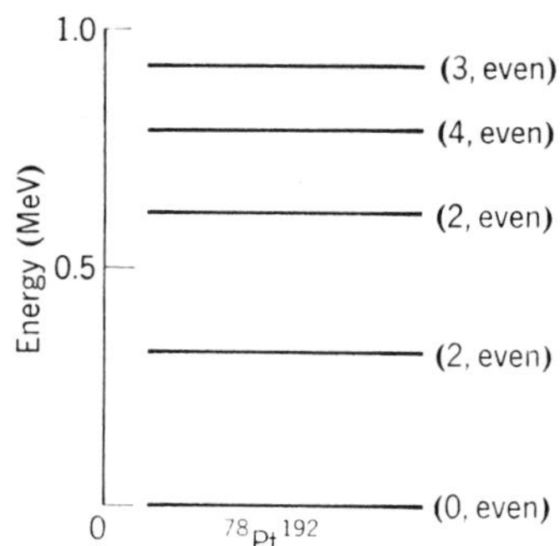


**FIGURE 16-34**
The low-lying excited states of $^{78}Pt^{192}$. For these states the nuclear shape is both vibrating and rotating.

Figure 16-14. Another example is $^3Li^7$ and $^4Be^7$. In general, two nuclei form a mirror pair if they contain the same number of nucleons, and if the number of protons in one equals the number of neutrons in the other. We have found that mirror pairs play an important role in allowing the experimental determination of the $\beta$-decay coupling constant. The reason is that since the charge independent nuclear forces do not distinguish between neutrons and protons their ground state eigenfunctions are identical, except for the effect of the small difference in the relatively weak Coulomb forces in these very low-$Z$ nuclei. For the same reason, their ground state eigenvalues are almost identical. That is, their ground state energies, or masses, are very nearly the same. Furthermore, the eigenfunctions and eigenvalues of the low-lying excited states of a mirror pair should be essentially the same if nuclear forces are charge independent. Thus there should be a close correspondence between the spins, parities, and energies of these states in the two members of a mirror pair. This is found to be the case. An example is shown in Figure 16-35, which presents the low-lying excited states of $^3Li^7$ and $^4Be^7$. More complicated relations are found between the lower excited states of *mirror triads*, such as $^5B^{12}$, $^6C^{12}$, $^7N^{12}$, and of even larger sets of *isobars* (nuclei with common values of $A$). These relations will be discussed briefly in the following chapter in the section titled: Isospin.

## 16-9 Fission and Reactors

Fission was discovered by Hahn and Strassman in 1939. Using chemical techniques, they found that the bombardment of uranium by neutrons produces elements in the middle of the periodic table. It was immediately realized that a very large amount of binding energy would be released in the *fission* of a nucleus of large $Z$, into two nuclei of intermediate $Z$, because of the consequent reduction in the positive Coulomb energy. Measurements soon showed that an energy of around 200 MeV per fission

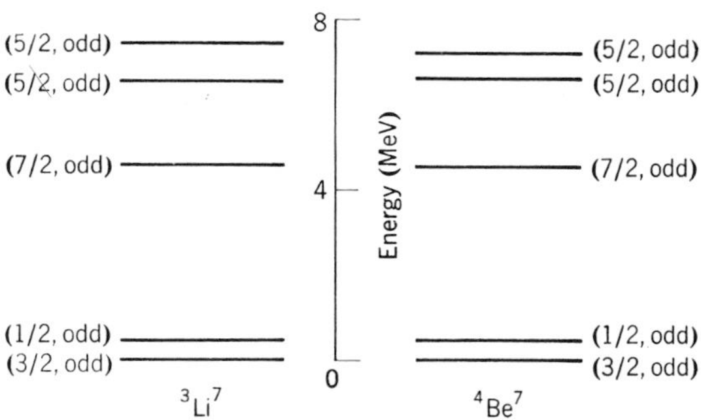


**FIGURE 16-35**
The low-lying excited states of the mirror pair $^3Li^7$ and $^4Be^7$. The ground state energy of $^4Be^7$ is actually about 0.5 MeV above the ground state of $^3Li^7$ due to the extra Coulomb repulsion energy in the former.

was released, and carried away largely by the kinetic energy of the two *fission fragments*. Measurements also showed that two or three neutrons were emitted in each fission. This suggested to several people the possibility of using these neutrons to induce other uranium nuclei to fission, using the neutrons that would be emitted from those fissions in the same way, and so forth, in a *chain reaction*. A trivial calculation showed that if all the nuclei in a block of uranium could be made to fission in a chain reaction, the energy liberated would be $\sim 10^6$ times larger than in burning a block of coal, or exploding a block of dynamite, of the same mass. (This is the usual factor of $10^6$ obtained when comparing nuclear to atomic, or molecular, energies.) Because of the extremely short time scale characterizing nuclear processes, the energy would be expected to be released much more rapidly than in a chemical explosion. The potentialities as a weapon were obvious, particularly because of the imminence of World War II. The events that followed dominate the history of this century, but here we shall be concerned with the peaceful applications of fission.

In a *nuclear reactor*, fission proceeds at a carefully controlled rate. A continuous source of power is obtained from the thermal energy produced when the fission fragments come to rest in the materials of the reactor. After many years of technological development, nuclear reactors have become sources of power which are very competitive, economically, with coal or oil. They are also important sources of unstable isotopes, not normally found in nature, that are used as tracers for diagnosing the operation of a variety of processes of interest to medicine, biology, chemistry, and engineering, or used for radiation therapy. The isotopes are produced in nuclear reactions induced by the intense flux of neutrons present in a reactor.

Fission occurs in nuclei of large $Z$ because the total Coulomb repulsion energy of the protons in a nucleus is considerably decreased if the nucleus splits into two smaller nuclei. The nuclear surface energy increases in the process, but its magnitude is much smaller than the magnitude of the Coulomb energy, so the increase in surface energy does not alter the fact that it is energetically favorable for a large $Z$ nucleus to fission. The Coulomb energy is minimized if the nucleus splits into two fission fragments that contain equal numbers of protons, but usually the splitting is not completely symmetrical because of the preference for magic numbers. In Example 15-6 we used the binding energy data to show that the energy associated with fission of ${}^{92}U^{238}$ is close to 200 MeV. This value is also fairly typical of the fission energy for other isotopes of uranium.

The steps involved in fission are indicated schematically by the set of drawings in Figure 16-36. These define a parameter $s$ which characterizes the progress of the fission by specifying (somewhat unprecisely) the elongation of the fissioning nucleus, and then the separation of the two fission fragments. Figure 16-37 is a schematic plot of $V(s)$, which is the part of the energy of the system that depends on $s$. Starting at small $s$, there is relatively little change in the Coulomb repulsion energy with increasing $s$, but the surface area of the nucleus increases rapidly. According to the liquid drop model, the increase in surface area produces an increase in the surface energy. Thus $V(s)$ increases with increasing $s$, for small $s$. As $s$ continues to increase, a surface tension effect produced by the surface energy causes the nucleus to assume the form of two regions connected by a narrow neck. And eventually the nucleus splits. After

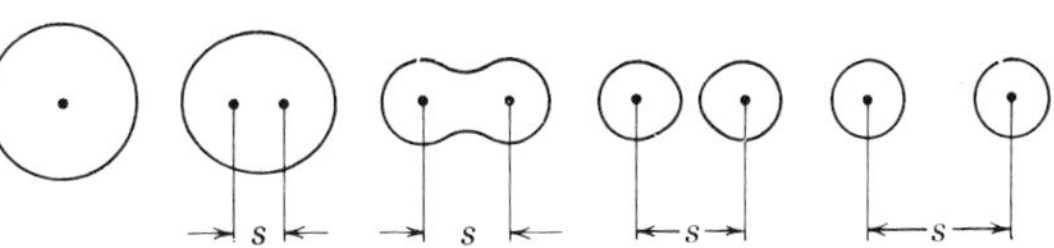


**FIGURE 16-36**
A schematic representation of the steps involved in the process of nuclear fission.

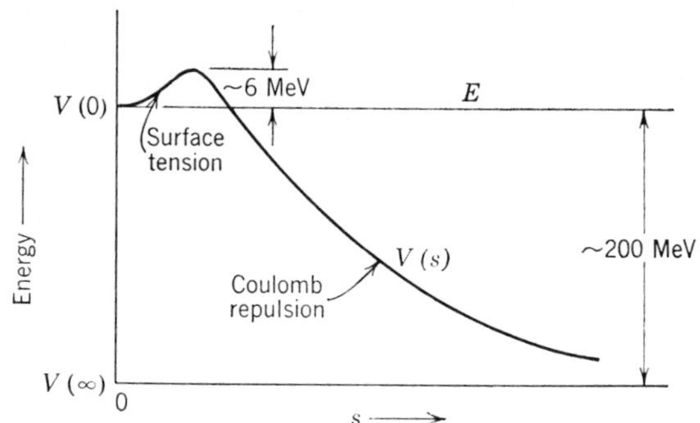


**FIGURE 16-37**
An energy diagram for a fissionable nucleus.

it splits, the surface energy no longer changes with $s$, and $V(s)$ decreases with increasing $s$, following the decrease in the Coulomb repulsion energy of the two fission fragments. Since $V(s)$ first goes up and later comes down, it necessarily must pass through a maximum. Calculations, based on the liquid drop model, show that for a typical nucleus of large $Z$ this maximum is about 6 MeV above $V(0)$. We already know that $V(0)$ is about 200 MeV above $V(\infty)$. Thus we see that nuclei are normally stable to decay by fission since they are sitting, with total energy $E = V(0)$, at the bottom of the depression in the potential $V(s)$. The process can take place by barrier penetration but, because the mass entering in the exponent of (6-55) for the barrier penetrability is very large, the probability of barrier penetration is extremely small. If ${}^{92}U^{238}$ decayed only by this *spontaneous fission* process, its lifetime would be $\sim 10^{16}$ yr.

A process of much more importance is *induced fission*. Usually this is brought about by the nucleus capturing a low-energy neutron. As the binding energy $E_n$ of the last neutron in a nucleus of large $Z$ is around 6 MeV, in favorable cases the capturing nucleus receives enough energy to put it over the top of the fission barrier. Very often this high excitation energy actually does go into collective vibrations in which it becomes sufficiently elongated to fission. It is like a highly excited compound nucleus, with most of its excitation energy in the form of violent vibrations. Induced fission is perhaps the best example of the collective motions that are implied by the liquid drop model, and form the basis of the collective model. The process is indicated in terms of an energy diagram in Figure 16-38. As we saw in Example 15-7, for ${}^{92}U^{235}$ the neutron binding energy $E_n$, made available when a neutron is captured, is about 6.5 MeV, so that fission can take place even if the neutron brings in no kinetic energy. This is also true for ${}^{92}U^{233}$. But when ${}^{92}U^{238}$ captures a neutron only about 5 MeV of binding energy is made available, so a neutron must have a kinetic energy of about 1 MeV to cause fission in this nucleus. The difference between the behavior of these isotopes arises from the difference in the pairing energy, as explained in Example 15-7.

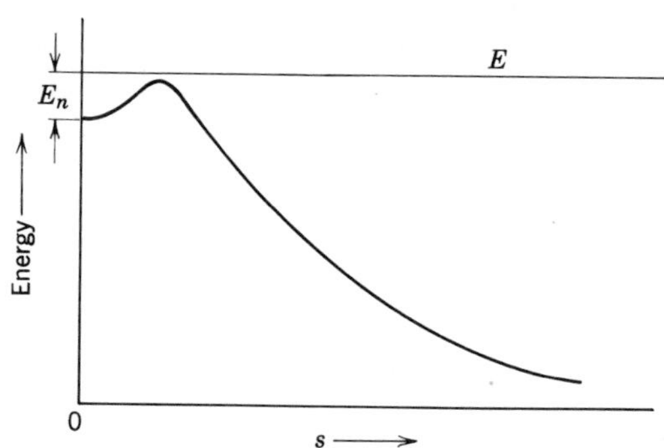


**FIGURE 16-38**
An energy diagram illustrating induced fission.

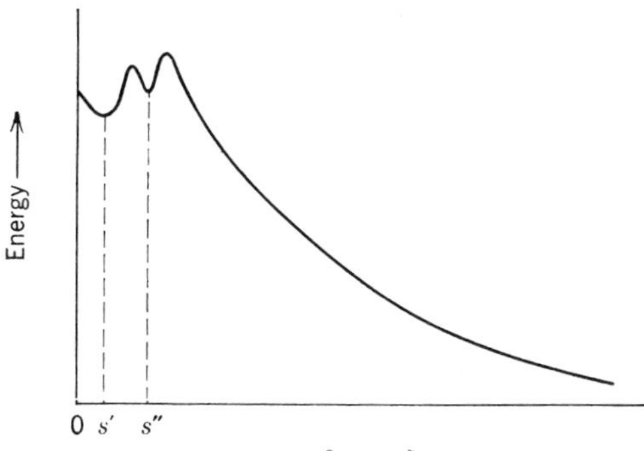


**FIGURE 16-39**
A double hump fission barrier.

We have oversimplified our discussion of fission by speaking as if the fissioning nucleus is spherical in its ground state. In fact we saw in Section 16-8 that uranium nuclei are ellipsoidal in their ground states. Even before receiving any excitation energy the nucleus is somewhat elongated. When it receives about 6 MeV of excitation from capturing a neutron, it further elongates, goes over the top of the fission barrier, and then fissions.

Recently, evidence has begun to accumulate which indicates that the fission barrier $V(s)$ shown in Figures 16-37 and 16-38 is probably also an oversimplification, and that the barrier actually has a double hump something like that shown in Figure 16-39. In its ground state the nucleus is very near the bottom of the deeper depression with its ground state elongation $s'$, and stable except for the highly improbable process of barrier penetration. Calculations based on the collective model, i.e., on a combination of the liquid drop and shell models, predict that there is a second shallower depression in $V(s)$ at the larger elongation $s''$. At this elongation the nucleus would also be stable, except for barrier penetration, if it had no excess energy. One prediction of these calculations is that it should be possible to put a fissionable nucleus into a state with the elongation $s''$, where it would remain for a long time. Some spontaneous fission experiments give strong indication that this is true. Because these calculations are also the ones that lead to the prediction of the $Z = 114$ magic number, mentioned at the end of Section 16-2, the spontaneous fission experiments have made physicists take the prediction concerning $Z = 114$ quite seriously. As far as induced fission is concerned, the presence of the shallower depression in $V(s)$ would probably not make very much difference.

The possibility of using fission to produce power in a chain reaction arises from the fact that *two or three neutrons are emitted in each fission process.* An idea of why it happens can be obtained by considering Figure 16-40. The figure shows the $Z$ and $N$ values of the nuclei which are the most stable for each value of $A$ (as in Figure 15-11). These nuclei are represented by the curve of stability. The large dot indicates the fissioning nucleus, and the two small dots indicate the fission fragments. The fragments are usually not symmetrical. Instead one of the fragments has $Z$ and $N$ values near the magic numbers 50 and 82, presumably because this is favored energetically. But both fragments have nearly the same $Z/N$ ratio as the fissioning nucleus. Since their $A$ values are much smaller, their $Z/N$ ratios are smaller than those of stable nuclei with these $A$ values. The fission fragments tend to have relatively too many neutrons. Most of the necessary readjustment slowly takes place by the fission fragments going through a succession of $\beta$ decays, but some of the readjustment is achieved promptly at the time of fission. Part of the decay of the fissioning compound nucleus takes place through the evaporation of two or three neutrons, of several MeV kinetic energy. Figure 16-41 provides more information about the asymmetry of the fission fragments, by plotting the distribution of their $A$ values.

Another process leading to the emission of neutrons, which is of small probability ($\sim$1% of the probability for the prompt emission of neutrons by evaporation from

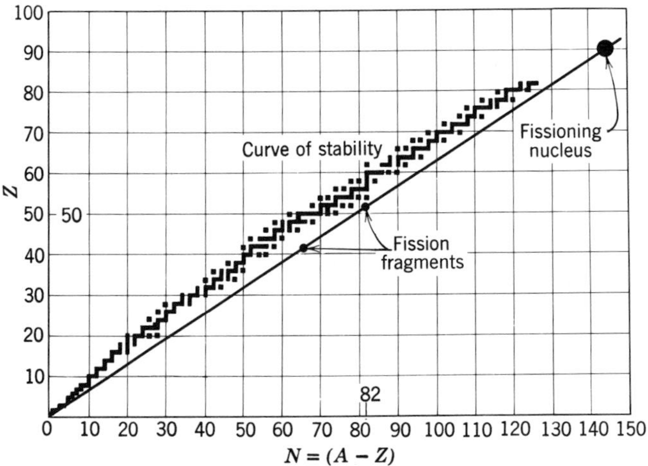


**FIGURE 16-40**
Illustrating that fission fragments tend to have relatively too many neutrons.

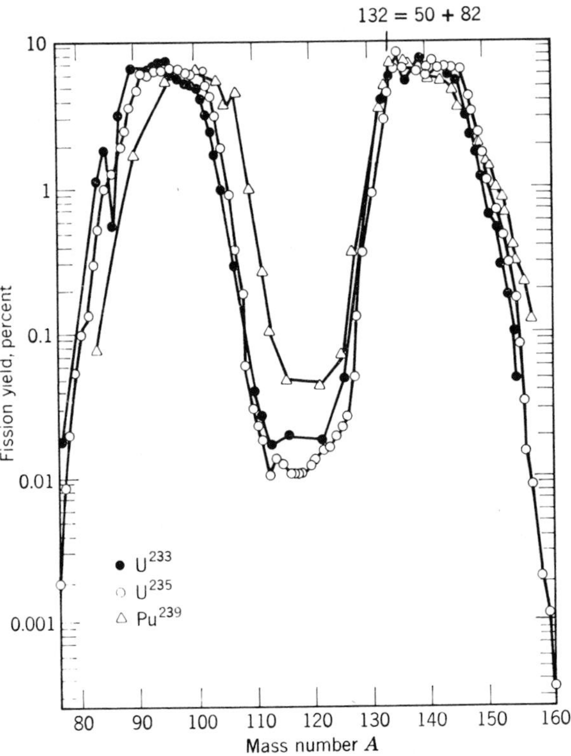


**FIGURE 16-41**
The mass spectra of fragments produced in the low-energy neutron induced fission of ${}^{92}U^{233}$, ${}^{92}U^{235}$, and ${}^{94}Pu^{239}$.

the excited compound nucleus) but of great importance in making it easier to control a reactor, is that of *delayed neutron emission.* As an example, consider the electron emitting fission fragment $^{35}Br^{87}$. Because of the $\beta$-decay selection rules, this nucleus occasionally decays to a state of its daughter $^{36}Kr^{87}$ that is sufficiently excited to allow it to emit a neutron, leaving the stable nucleus $^{36}Kr^{86}$. Neutrons are emitted in this process, with a delay characteristic of the 55 sec half-life of $^{35}Br^{87}$. Another important example involves delayed neutron emission from $^{54}Xe^{137}$. For $^{36}Kr^{87}$ or $^{54}Xe^{137}$ the neutron number $N$ equals a magic number, 50 or 82, plus one. Thus the process depends on the unusually small neutron binding energy that the shell model would predict in such cases.

In a reactor, the chances for the neutrons emitted in one generation of fission ultimately inducing the next generation of fission are enhanced because the neutrons scatter from low mass nuclei in the *moderator* surrounding the pieces of uranium. They rapidly lose energy to the recoil of these nuclei, and they are no longer able to induce fission in $^{92}U^{238}$. But they are not lost to nonfission $^{92}U^{238}$ capture since moderation occurs outside the uranium pieces. The moderator is usually $^{6}C^{12}$, in the form of graphite, or $^{1}H^{2}$, in the form of deuterium oxide (heavy water). It is possible to use $^{1}H^{1}$, but only if the uranium is highly enriched in $^{92}U^{235}$. The reason is that $^{1}H^{1}$ has a large cross section for capturing neutrons to form $^{1}H^{2}$, and these neutrons are lost from the chain reaction. The purpose of the moderator is to reduce the velocities of the neutrons to the lowest values possible, so that their de Broglie wavelengths $\lambda$ will be as long as possible. Because of the wavelike properties of neutrons, their cross section for capture by a nucleus of radius $r'$ is limited by the value of $\lambda$, and not by the value of $r'$ (see (16-32)). The moderator brings the neutrons into thermal equilibrium at the operating temperature of the reactor, which makes $\lambda \gg r'$ and thereby increases the $^{92}U^{235}$ capture cross section for neutrons diffusing back into the uranium pieces. The cross section must be large enough that the probability of one of the two or three neutrons from each fission subsequently inducing another fission be at least equal to 1. When the reactor is starting up, this probability is made to be slightly bigger than 1. It is gradually reduced to be precisely 1 when the reactor attains equilibrium at its operating level. Adjustments are made by varying the lengths of *control rods* inserted into the reactor. These contain nonfissionable nuclei like $^{48}Cd^{113}$, which have extremely large capture cross sections for thermal energy neutrons, because of fortuitiously located compound nucleus resonances. The delayed neutrons facilitate the control of a reactor by introducing some neutrons in the chain reaction that are emitted with a reasonably long time constant. The kinetic energy given to the fission fragments in the fission process is converted into thermal energy as these fragments come to rest in the materials of the reactor. Typically, this heat is used to make steam which drives turbines that operate generators producing electrical power.

*Breeder reactors* utilize the 99% abundant $^{92}U^{238}$. These nuclei capture low-energy neutrons. They cannot fission in low-energy neutron capture, but the resulting unstable $^{92}U^{239}$ nuclei undergo two successive $\beta$ decays, turning into the stable nuclei $^{94}Pu^{239}$. This end product has the same ability to fission in low-energy neutron capture as does $^{92}U^{235}$.

**Example 16-10.** The average time lapse between the emission of a prompt neutron in a fission taking place in a nuclear reactor, and the capture of that neutron to induce the next generation of the chain reaction, is of the order of $10^{-3}$ sec. (Most of the time is required by the moderator to bring the neutron into thermal equilibrium.) Use this figure to estimate the number of free neutrons present in a reactor operating at a power level of $10^8$ W.

In Example 15-6 we found that the energy release in the fission produced by one neutron is about

$$E \simeq 200 \text{ MeV} = \frac{200 \text{ MeV}}{\text{neutron}} \times \frac{1.6 \times 10^{-13} \text{ joule}}{\text{MeV}} \sim 10^{-11} \text{ joule/neutron}$$

If one free neutron has a lifetime before capture of $\sim 10^{-3}$ sec, and if on capture it produces a fission energy of $\sim 10^{-11}$ joule, one free neutron produces a power of

$$p \sim \frac{10^{-11} \text{ joule/neutron}}{10^{-3} \text{ sec}} \sim 10^{-8} \text{ W/neutron}$$

So if the power level of the reactor is $P = 10^8$ W, the number of free neutrons is

$$N = \frac{P}{p} \sim \frac{10^8 \text{ W}}{10^{-8} \text{ W/neutron}} \sim 10^{16} \text{ neutron}$$

The large number, or flux, of free neutrons present in a reactor makes the device very useful for producing unstable isotopes on the low-$Z$ side of the curve of stability (electron emitters). This is done by placing probes containing appropriately chosen stable isotopes into the interior of the reactor. The unstable isotopes are formed when the isotopes in the probes capture neutrons. ◀

## 16-10 Fusion and the Origin of the Elements

We close our study of nuclear physics with a discussion of nuclear fusion, and its part in the production of stellar energy and of the chemical elements. Fusion involves two nuclei of very low $A$ amalgamating to form a more stable nucleus. The increased stability arises because the $A$ value of the nucleus formed is nearer the value $A \simeq 60$ where the binding energy per nucleon maximizes (see Figure 15-10). From the point of view of the liquid drop model, the situation would be explained by saying that nuclei of very low $A$ have too much surface, relative to their volume, for maximum stability. The Coulomb energy increases in fusion, but its magnitude is too small to prevent the process from happening because nuclei of low $A$ also have low $Z$.

It is fair to say that fusion is the most important phenomenon in nature. Fusion of low-$A$ nuclei in thermal motion is the source of energy of the sun. So it is ultimately the source of energy for all the natural physical and biological processes on the earth. And there is reason to hope that some day fusion will be usable directly on earth to produce energy in a *fusion reactor*. Because much of the earth is covered by seas containing the hydrogen isotopes ${}^1H^1$ and ${}^1H^2$, the fuel supply of low-$A$ nuclei would be almost inexhaustible. One of the several potentially useful reactions for a thermal fusion reactor is

$$ {}^1H^2 + {}^1H^2 \rightarrow {}^2He^3 + {}^0n^1 + 3.2 \text{ MeV} \tag{16-34}$$

where the energy is the $Q$ value of the reaction. But it is much more difficult to build a fusion reactor than to build a fission reactor. The problem lies in the repulsive Coulomb barrier acting between two nuclei, which must be overcome, or at least penetrated, before they can get close enough to allow the short range nuclear forces to come into play and fuse them together.

Figure 16-42 plots the cross section for the reaction of (16-34), as a function of the kinetic energy of the bombarding particle. The cross section does not attain a measurable value until the kinetic energy exceeds $\sim 10^4$ eV. And even at that energy the cross section is very small because the reaction takes place by penetration of the Coulomb barrier acting between the nuclei, which is $\sim 10^6$ eV high. Unless the kinetic energy is appreciably higher than $\sim 10^4$ eV, the cross section, and therefore the rate of the reaction, is much too small to be of practical use in a fusion reactor. In the interior of the sun similar reactions do occur, with the kinetic energy of the bombarding particles

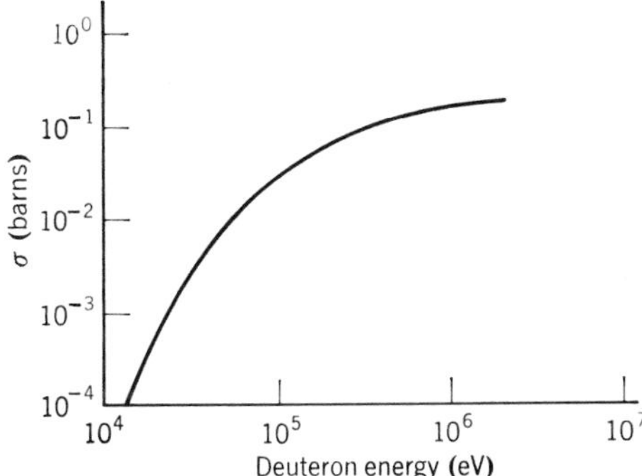


**FIGURE 16-42**

The cross section for the reaction in which two deuterons fuse to form $^2He^3$ plus a neutron.

coming from their thermal energy. This energy is $\sim kT$, where $k$ is Boltzmann's constant $\sim 10^{-4}$ eV/°K, and $T$ is the interior temperature of the sun $\sim 10^7$ °K. Thus the thermal or kinetic energy at the temperature of the interior of the sun is only $\sim 10^3$ eV, and fusion reactions proceed there at an extremely slow rate. Of course, the sun produces large amounts of energy, but only because it is so large that it makes up for the very slow rate of the individual reactions. An efficient thermal fusion reactor of dimensions possible on the earth would have to have a much higher rate for the individual reactions. Thus its temperature would have to be higher—at least an order of magnitude higher than the internal temperature of the sun! There are ways of achieving such a temperature, if ways can be found to produce a container that would not be destroyed by the temperature. The sun is so massive that gravitational fields provide a container automatically. On earth, it might be done by using magnetic fields acting on the charged nuclei to contain them. Attempts have been made to build such a container, fill it with hydrogen, and then heat the contents by, for instance, firing in a laser beam. There have even been some indications of success, but only for very short times before the container fails. Another approach is to use extremely powerful lasers to add so much thermal energy to small pellets of fusible material to cause them to react. In such a procedure, energy would be produced in a sequence of miniature explosions, and it would be absorbed within a very strong metallic container that would be heated as a consequence. Obtaining thermal fusion for energy production on earth remains one of the great challenges to science and engineering.

There are no difficulties in obtaining fusion on earth by nonthermal means. It can be done with ease by using a cyclotron, or other accelerator, to give the bombarding nucleus enough energy to overcome the repulsive Coulomb barrier it sees surrounding the target nucleus; but the amount of energy liberated in the relatively few fusions that can be produced in this way is very small, and microscopic compared to the energy that goes into running the accelerator. So there seems to be no hope of using nonthermal fusion as an efficient energy source.

Efficient thermal fusion has, however, been taking place for a long time in the stars. It is responsible for the energy produced in all stars, and also for the production in the stars of all the elements through iron. It is believed that stars are initially formed from the extremely low-density ($\sim 1$ atom/cm$^3$) gas that is known to be distributed throughout interstellar space. The gas is primarily hydrogen, but it contains also about 10% helium that is thought to have been made by fusion from hydrogen in the "big bang" that occurred when the universe started some $10^{10}$ years ago, plus small amounts of higher $Z$ elements present in certain regions for reasons that will be explained later.

In the well-accepted big-bang theory, the electrically neutral universe would have started from a region containing neutrons compressed to an extremely high density.

In the first few moments, the following set of processes would take place

$$
\begin{aligned}
&{}^0n^1 \rightarrow {}^1\mathrm{H}^1 + e + \bar{\nu} \\
&\bar{\nu} + {}^1\mathrm{H}^1 \rightarrow {}^0n^1 + \bar{e} \\
&{}^1\mathrm{H}^1 + {}^0n^1 \rightarrow {}^1\mathrm{H}^1 + {}^0n^1 + \gamma \\
&e + \bar{e} \rightarrow \gamma + \gamma \\
&\gamma \rightarrow e + \bar{e}
\end{aligned}
$$

and there was an equilibrium, at very high temperatures, between neutrons, protons, electrons, positrons, antineutrinos, and $\gamma$ radiation. The radiation, "cooled" by repeated Doppler shifts in the subsequent expansion of the system, would now constitute the isotropic 3°K blackbody radiation whose recent detection provides some of the experimental evidence for the validity of the big-bang theory (see Section 1-5).

In the high-density equilibrium distribution that existed for a short time before the system blew itself apart, helium would be formed by the reactions

$$
\begin{aligned}
{}^1\mathrm{H}^1 + {}^0n^1 &\rightarrow {}^1\mathrm{H}^2 + \gamma \\
{}^1\mathrm{H}^2 + {}^1\mathrm{H}^2 &\rightarrow \begin{cases} {}^2\mathrm{He}^3 + {}^0n^1 \\ {}^1\mathrm{H}^3 + {}^1\mathrm{H}^1 \end{cases} \\
{}^2\mathrm{He}^3 + {}^0n^1 &\rightarrow {}^1\mathrm{H}^3 + {}^1\mathrm{H}^1 \\
{}^1\mathrm{H}^3 + {}^1\mathrm{H}^2 &\rightarrow {}^2\mathrm{He}^4 + {}^0n^1
\end{aligned}
$$

Detailed calculations, involving the cross sections for all the reactions in both sets, show that enough helium could be formed to account for the approximately 10% abundance now observed in interstellar space. The remaining 90% of the matter there would, in agreement with observation, essentially all be in the form of hydrogen, most of the protons being formed from the $\beta$ decay of the neutrons that found themselves in free space after the big bang.

According to our present understanding, the first stage in the formation of a star from the very tenuous gaseous material of interstellar space involves some sort of upward fluctuation in density over a very large region. In such a fluctuation, the gas collects into a cluster. If it is large enough it stabilizes itself because of the gravitational attractions between the atoms it contains, and it begins to grow by attracting more atoms. As a cluster grows, the increasing strength of the gravitational attractions causes the interior pressure, and therefore the interior temperature, to build up. When the temperature in the core exceeds about $10^5$ °K the hydrogen atoms in that region are completely ionized into a *plasma* of protons and electrons. And when the temperature exceeds about $10^7$ °K the protons have enough kinetic energy due to their thermal motion to have a small probability of penetrating the repulsive Coulomb barriers that tend to keep them apart. (The 10% helium present does not participate at this stage because the temperature is too low for penetration of the higher Coulomb barriers surrounding these nuclei.) Then two protons can fuse together and form a deuteron, according to the reaction

$$
{}^1\mathrm{H}^1 + {}^1\mathrm{H}^1 \rightarrow {}^1\mathrm{H}^2 + \bar{e} + \nu + 0.42\ \mathrm{MeV}
$$

where the energy is the energy liberated in the process. Since the process requires *both* barrier penetration and the weak $\beta$-decay interaction, it occurs at an extremely low rate. The necessity of $\beta$ decay arises from the fact that nuclear forces are not able to make the system ${}^2\mathrm{He}^2$ (the diproton) be bound, for reasons that will be explained in the next chapter. Although the rate for the deuteron forming reaction is very low, when enough deuterons are present large concentrations of helium can be formed by processes that have relatively high rates because they involve the strong nuclear interaction.

Helium is formed in a star in a cycle of reactions, called the *proton-proton cycle*, consisting of two of the preceding reactions, followed by two of the reactions

$$ {}^1H^2 + {}^1H^1 \rightarrow {}^2He^3 + \gamma + 5.49 \text{ MeV} $$

and then by one reaction in which the two ${}^2He^3$ nuclei that have been formed fuse as follows.

$$ {}^2He^3 + {}^2He^3 \rightarrow {}^2He^4 + {}^1H^1 + {}^1H^1 + 12.86 \text{ MeV} $$

Counting the 1.02 MeV liberated each time one of the two positrons annihilates with an electron, the total energy liberated in one cycle is 26.72 MeV. But a little more than 1% of this energy is carried completely away from the star by the two neutrinos. The remainder, plus gravitational contraction, continues to heat the core.

When the density of helium (including the helium initially present) in the core of the cluster that has turned into a star becomes high enough, carbon can be formed. What happens is that two ${}^2He^4$ nuclei combine to form ${}^4Be^8$. This nucleus can then combine with another ${}^2He^4$ nucleus, to form ${}^6C^{12}$, providing it does it almost immediately. The point is that ${}^4Be^8$ is not stable, and it will decay back into two ${}^2He^4$ in about $10^{-15}$ sec if it does not capture the third ${}^2He^4$ nucleus. The rate for this improbable sounding reaction would be essentially zero if it were not for the existence of an excited state in ${}^6C^{12}$ at an energy of about 7.65 MeV. When the temperature is $\sim 10^8$ °K, there is a resonance in the reaction, which makes its cross section reasonably large, because the kinetic energies of the three combining ${}^2He^4$ nuclei plus the $Q$ value equals the energy of the excited state in ${}^6C^{12}$. Straightforward processes involving the successive addition of nucleons to ${}^2He^4$ could not be used to form elements with $A$ greater than 4 because such processes are blocked by the complete instability of nuclei with $A = 5$.

When enough carbon has been formed in the core of the star, the principal source of energy production is through the *carbon cycle*, in which carbon plays the role of a catalyst (i.e., it reappears at the end of the cycle) to aid in the fusion of four ${}^1H^1$ into one ${}^2He^4$, plus assorted positrons, neutrinos, and $\gamma$ rays. The carbon cycle consists of the set of reactions

$$ {}^6C^{12} + {}^1H^1 \rightarrow {}^7N^{13} + \gamma + 1.94 \text{ MeV} $$
$$ {}^7N^{13} \rightarrow {}^6C^{13} + \bar{e} + \nu + 1.20 \text{ MeV} $$
$$ {}^6C^{13} + {}^1H^1 \rightarrow {}^7N^{14} + \gamma + 7.55 \text{ MeV} $$
$$ {}^7N^{14} + {}^1H^1 \rightarrow {}^8O^{15} + \gamma + 7.29 \text{ MeV} $$
$$ {}^8O^{15} \rightarrow {}^7N^{15} + \bar{e} + \nu + 1.73 \text{ MeV} $$
$$ {}^7N^{15} + {}^1H^1 \rightarrow {}^6C^{12} + {}^2He^4 + 4.96 \text{ MeV} $$

Counting the energy liberated in the annihilation of the two positrons, the total energy liberated in one cycle is 26.72 MeV, just as in one proton-proton cycle. In the carbon cycle a little more than 5% of the energy is lost from the star by the two neutrinos emitted in the higher energy $\beta$ decays. The rate at which the carbon cycle occurs is much higher than the rate for the proton-proton cycle, because no step in the carbon cycle is anywhere as near as slow as the first step in the proton-proton cycle. The sun has not yet reached the stage in its development where the carbon cycle dominates the energy production, although there is some carbon cycle going on. In a star with a mass greater than about two sun masses, the gravitational contraction is very rapid and the core temperature rapidly reaches the value $\sim 10^8$ °K required for carbon formation and the carbon cycle.

As the concentration of the stellar core continues, its temperature increases and elements heavier than carbon are formed. At first this is done by the successive captures of ${}^2He^4$ by ${}^6C^{12}$, forming ${}^8O^{16}$, then ${}^{10}Ne^{20}$, and then ${}^{12}Mg^{24}$. But when the temperature is $\sim 10^9$ °K these nuclei have enough thermal energy to penetrate their Coulomb barriers, directly forming nuclei of even $A$ through ${}^{26}Fe^{56}$. Nuclei of comparable but odd values of $A$ can be formed if the even-$A$ nuclei are forced by turbulence out of the stellar core into the surrounding cooler zone where the proton-proton cycle is still going on. In this zone reactions can occur such as

$$
{}^{10}Ne^{20} + {}^1H^1 \rightarrow {}^{11}Na^{21} + \gamma
$$

$$
{}^{11}Na^{21} \rightarrow {}^{10}Ne^{21} + \bar{e} + \nu
$$

Some of these odd-$A$ nuclei can then participate in reactions which lead to the production of neutrons. An example is

$$
{}^{10}Ne^{21} + {}^2He^4 \rightarrow {}^{12}Mg^{24} + {}^0n^1
$$

The elements heavier than iron are not formed by fusion because the $A$ values exceed the value $A \simeq 60$ where the binding energy per nucleon maximizes; beyond $A \simeq 60$ the Coulomb repulsion of the protons becomes so large that it is no longer energetically favored for a nucleus to capture another nucleus. However, it is certainly favored for a nucleus to capture a neutron since this releases the neutron binding energy of $\simeq 6$ MeV. Nuclei through ${}^{83}Bi^{209}$ are formed by a succession of neutron captures and $\beta$ decays, starting from ${}^{26}Fe^{56}$. The neutrons come from reactions such as the example given in the preceding paragraph, and the $\beta$ decays take place when necessary to adjust the $Z$-to-$A$ ratio of a nucleus to a stable value. The abundances of the nuclei that are built up in the succession of neutron captures are inversely proportional to their neutron capture cross sections, averaged over the very high temperature thermal distribution of neutron energies. This is true since, if a nucleus has a large neutron capture cross section, there is only a small chance that it will not capture a neutron and be converted into some other nucleus. The abundance of elements in the solar system is inferred primarily from the composition of the sun seen in atomic spectra measurements, and from solar produced cosmic rays intercepted on the earth. Data are also obtained from meteorites, and from the composition of the earth itself. The abundance curve from iron to bismuth was presented in Figure 15-1. It is very nearly the reciprocal of the neutron capture cross-section curve. On the average, the cross sections increase (and the abundances decrease) as the $A$ value of the nucleus increases, simply because the nucleus becomes larger. But there are some pronounced departures from the average due to the effect of filled subshells on neutron affinities and binding energies which, in turn, affect the neutron capture cross sections.

The heaviest element that can be formed in the neutron capture processes discussed here is bismuth. The reason is that when ${}^{83}Bi^{209}$ captures a neutron it becomes ${}^{83}Bi^{210}$, which $\alpha$ decays into ${}^{81}Ti^{206}$ with a half-life of only five days. This decay is so rapid that it takes place before there is time for further neutron capture by ${}^{83}Bi^{210}$ in the moderate flux of neutrons that normally exists in a star.

When some stars come to the end of their life because they have almost depleted their supply of hydrogen, not enough "nuclear heat" is generated in the core to prevent very rapid gravitational collapse. They then explode in a matter of a few seconds with tremendous violence, and they produce a tremendous flux of neutrons. The most spectacular example in recorded history of such a *supernova* is a star that was observed in 1054 A.D. to flare up to a brightness that allowed it to be seen for a short time in full daylight. Its remnants are now called the Crab nebula. The elements heavier than bismuth are believed to be made in successive neutron captures, starting from $^{83}Bi^{209}$, and using the intense neutron flux present in a supernova. The process happens so rapidly that the $\alpha$ decay of $^{83}Bi^{210}$ is of no consequence.

The preceding discussion of the life history of a star assumed that its original composition was purely the primordial 90% hydrogen plus 10% helium mixture. There are many examples of such "first-generation" stars. And there are also many examples of "second-" or "third-generation" stars, which are thought to have been originally composed partly of supernova remnants; the sun is one example. In these stars heavy elements will be present, and in fact reasonably abundant, even before the stage is reached where the carbon cycle is the dominant source of energy.

## QUESTIONS

1. Give a qualitative explanation of why an $\alpha$ particle can penetrate a Coulomb barrier.
2. What would be the effect on the $\alpha$-decay lifetimes, and thus on the terrestrial abundances, of the elements between $A = 200$ and $A = 260$ if there were no magic numbers so that the $\alpha$-decay energies of Figure 16-1 followed the general trend predicted by the semiempirical mass formula?
3. Is there a $4n + 4$ radioactive series?
4. Where would be a likely place to look for traces of the predicted superheavy element $Z = 110$, $A = 294$?
5. Construct a figure illustrating a case in which there are three $\beta$-stable nuclei with the same even-$A$ value.
6. Explain why the emission of a particle, with the properties postulated by Pauli, removes the difficulties with angular momentum in $\beta$ decay. What about linear momentum?
7. Just how do neutrinos and antineutrinos differ from photons, which also have no charge or rest mass?
8. How do you justify the fact that electrons are emitted from nuclei in $\beta$ decay, when in Example 6-6 we showed that electrons cannot be contained in nuclei?
9. In the Wu experiment, what is the direction of the magnetic field applied to align the nuclei, from the normal point of view, and as seen in the mirror? What about the direction of the current flow in the windings of the magnet that produces the field?
10. Consider viewing the Wu experiment in a mirror located below the nucleus (the mirror being horizontal) instead of in a mirror located to one side of the nucleus (the mirror being vertical). Explain how the arguments in the text would be modified, but in such a way as to lead to the same conclusions.
11. Sugar molecules have a definite helicity. What do you think is responsible?
12. Consider the electric and magnetic monopole, dipole, and quadrupole moments of a nucleus. Are each of these ever found with a constant, nonzero value? With an oscillatory value? Explain why some of these cases do not occur, and what the nucleons are doing in cases that do occur.
13. Electric dipole radiation is emitted with a characteristic spatial pattern (see Appendix B). Does this suggest an experimental technique for determining the type of radiation emitted in a $\gamma$ decay? What would be the difficulty in using such a technique?

**14.** In $\gamma$ decays from states of excitation energy around 1 MeV, or less, to ground states, electric dipole radiation is almost never observed. Use the shell model to explain this.

**15.** Predict, from the shell model, the regions of the periodic table in which the first excited states of nuclei have particularly long lifetimes for $\gamma$ decay.

**16.** A hyperfine splitting measurement tells you that the ground state spin of a nucleus is $i = 3/2$. What are the possible $l$ values of the subshell occupied by the nucleon responsible for the spin? What other information would tell you which of these is the actual value? What could you measure to obtain this information?

**17.** Explain exactly why the optical model potential which a nucleus exerts on a bombarding nucleon of energy 50 MeV is different from the shell model potential which it exerts on one of its own nucleons. What would you expect the optical model potential to be like for a bombarding nucleon of energy 5 MeV?

**18.** Why is it easier for an incident nucleon to enter a nucleus than it is for either of the nucleons, resulting from its first collision, to escape?

**19.** What are the differences between single particle states and many particle states? How are they related? What about $\gamma$-decaying states?

**20.** If the compound nucleus ${}^{30}Zn^{64}$ forgets the details of how it was formed, it should make no difference if it were excited by bombarding ${}^{29}Cu^{63}$ with protons, or ${}^{28}Ni^{60}$ with $\alpha$ particles, providing the same many particle states are excited. Devise an experiment to test this prediction.

**21.** What difference (if any) is there between a permanent nuclear ellipsoidal deformation, as seen in the ground and low-lying states of many even-$Z$, even-$N$ nuclei, and a nuclear electric quadrupole moment?

**22.** Why is it reasonable to expect that the space distribution of protons in a nucleus is approximately the same as the space distribution of neutrons?

**23.** Nuclear reactors are particularly suited to power submarines. Give reasons why this is so.

**24.** Can you devise a configuration of magnetic fields that could, at least from a naive point of view, contain nuclei in a thermal fusion reactor?

**25.** Why is it impossible for two protons to fuse, as in the first step of the proton-proton cycle, without a $\beta$ decay simultaneously taking place?

**26.** What happens to the $\gamma$ rays that are emitted in stellar nuclear reactions of the proton-proton or carbon cycle?

**27.** How would it be possible to use a neutrino detector on the earth to tell whether the dominant reactions in the center of the sun are in the proton-proton cycle or in the carbon cycle?

## PROBLEMS

**1.** (a) Use the semiempirical mass formula to predict the $\alpha$-decay energy of ${}^{83}Bi^{210}$. (Hint: Take the atomic mass of ${}^{2}He^{4}$ directly from Table 15-1.) (b) Compare your results with the $\alpha$-decay energy shown in Figure 16-1.

**2.** Derive (16-4), relating lifetime to decay rate.

**3.** Derive (16-5), relating lifetime to half-life.

4. Unstable nuclei, of decay rate $R$, are being produced at a constant rate $I$ in nuclear reactions caused by a cyclotron bombardment. If the production process commences at $t = 0$, calculate the number of these nuclei that will be present at $t = t'$. (Hint: The equation to be solved is obtained by rewriting (16-2) in the form $dN/dt = -NR$, and then adding $I$ to the right side. Can you justify this?)

5. Prove the validity of (16-6), the relation between the numbers of decaying nuclei and their decay rates, in radioactive equilibrium. (Hint: Write a set of equations comparable to (16-2). The first of the set is exactly like it, and the others contain two similar terms on the right side. Then show immediately that (16-6) is a solution to these equations providing the decay rate of the parent is very small compared to the decay rates of the daughters.)

6. ${}^{90}\text{Th}^{232}$ $\alpha$ decays to its first daughter ${}^{88}\text{Ra}^{228}$. It is observed that a very thin foil containing 1.0 g of ${}^{90}\text{Th}^{232}$ emits $\alpha$ particles from this decay at the rate of 4100/sec. Use these data to show that the half-life of ${}^{90}\text{Th}^{232}$ is $1.4 \times 10^{10}$ yr.

7. ${}^{82}\text{Pb}^{208}$ is the stable final daughter of the radioactive series whose parent is ${}^{90}\text{Th}^{232}$ (see Figure 16-5). The half-life of the parent is $1.4 \times 10^{10}$ yr. A piece of thorium ore containing 1 kg of ${}^{90}\text{Th}^{232}$ is found to also contain 200 g of ${}^{82}\text{Pb}^{208}$. (a) Assuming that all of the ${}^{82}\text{Pb}^{208}$ in the rock came from the decay of ${}^{90}\text{Th}^{232}$, and that none of it has been lost, calculate the age of the rock; that is, calculate how many years have passed since thorium was concentrated in the minerals in the rock and the equilibrium decay began. (b) There are a total of six $\alpha$ particles emitted in the decay of the radioactive series. Assuming that a negligible number of them could have escaped from the rock because it is so thick, calculate how much helium originating from the $\alpha$ decays should be in the rock. (c) The first daughter of the series, ${}^{88}\text{Ra}^{228}$, decays with half-life 5.7 yr into the second daughter, ${}^{89}\text{Ac}^{228}$. Calculate how much ${}^{88}\text{Ra}^{228}$ should be in the rock.

8. (a) Use the semiempirical mass formula to evaluate the points on the $A = 27$ mass parabola for the only three values of $Z$ that are found with this value of $A$, namely $Z = 12, 13, 14$. (Hint: It is only necessary to evaluate the terms of the formula that depend explicitly on $Z$.) (b) Which value of $Z$ corresponds to the stable nucleus? (c) Find the types of decay, and the decay energies, for the $\beta$ decays of the two unstable nuclei.

9. Example 16-3 showed that the $\beta$ decay of ${}^{4}\text{Be}^{7}$ to ${}^{3}\text{Li}^{7}$ proceeds only through electron capture because the atomic mass difference is $0.00093u$, which is less than two electron rest masses. Consider a ${}^{4}\text{Be}^{7}$ nucleus, initially at rest, that captures a $K$ electron and emits a neutrino. (a) Estimate the recoil velocity of the nucleus after the process is completed. (Hint: The recoil energy of the nucleus is negligibly small.) (b) Suggest a technique for detecting electron capture.

10. The table here lists three points of the measured momentum spectrum, $R(p_e)$, of electrons emitted in the $\beta$ decay of a nucleus of small $Z$.

| $\frac{p_e}{mc}$ | 2.8 | 4.9 | 6.9 |
|---|---|---|---|
| $R(p_e)$ | 375 | 500 | 250 |

(a) Make a Kurie plot of these points. (b) Then extrapolate to find the end point $K_e^{max}$ of the spectrum, and so determine the decay energy $E$.

11. Several examples of the initial and final nuclei in $\beta$ decays, and their ground state spins and parities, are listed here. For each decay between ground states, determine if it is allowed by the Fermi or Gamow-Teller selection rules. If it is forbidden, estimate roughly the factor suppressing the decay rate. (a) ${}^{2}\text{He}^{6}$ (0, even) $\rightarrow$ ${}^{3}\text{Li}^{6}$ (1, even); (b) ${}^{4}\text{Be}^{10}$ (0, even) $\rightarrow$ ${}^{5}\text{B}^{10}$ (3, even); (c) ${}^{16}\text{S}^{35}$ (3/2, even) $\rightarrow$ ${}^{17}\text{Cl}^{35}$ (3/2, even); (d) ${}^{39}\text{Y}^{91}$ (1/2, odd) $\rightarrow$ ${}^{40}\text{Zr}^{91}$ (5/2, even).

**12.** (a) By using the information given after (16-16), which represents the $\beta$ decay of the neutron, calculate the $FT$ value for the decay. (b) Compare with the value calculated in Example 16-4.

**13.** (a) Use the $FT$ value obtained in Problem 12 to estimate the value of the $\beta$-decay coupling constant. (b) Compare with the estimate obtained in Example 16-5. (c) What justification is there for assuming that the nuclear matrix element is essentially equal to one for the $\beta$ decay of the neutron?

**14.** Consider a set of positive charges moving in a confined region, like protons in a nucleus, and interacting with an external field of electromagnetic radiation. The charge density is $\rho$, so the current density is $\sim\rho v$, where $v$ is the characteristic velocity of the moving charges. Show that the energy of interaction between the magnetic dipole moment of the charges and the external magnetic field is smaller by a factor of $\sim v/c$ than the energy of interaction between the electric dipole moment and the external electric field. Since the values of the matrix elements for magnetic dipole and electric dipole radiation are proportional to these interaction energies, and since the transition rates are proportional to the "squares" of the matrix elements, the magnetic dipole transition rate is smaller than the electric dipole transition rate by a factor of $\sim(v/c)^2$. (Hint: (i) Show that the ratio of the interaction energies equals the product of the ratio of magnetic to electric dipole moments times the ratio of the magnetic to electric field strengths. (ii) Argue that the ratio of the magnetic to electric dipole moments equals the ratio of the current density to the charge density. (iii) Evaluate the ratio of the magnetic to electric field strengths for electromagnetic radiation in a vacuum.)

**15.** Consider a set of positive charges $q$ moving in a region of linear dimensions $\sim r'$, and interacting with the electric part of an external field of electromagnetic radiation of wavelength $\sim\lambda$. Show that the energy of interaction between the electric quadrupole moment of the charges and the external electric field is smaller by a factor of $\sim r'/\lambda$ than the energy of interaction between the electric dipole moment and the external electric field. For the reasons explained in Problem 14, this leads to the conclusion that the electric quadrupole transition rate is smaller than the electric dipole transition rate by a factor of $\sim(r'/\lambda)^2$. (Hint: (i) Consider a sinusoidal electric field $E = E_0 \sin 2\pi(x/\lambda - \nu t)$. (ii) The energy of the electric dipole is $E$ times its dipole moment $\sim qr'$. (iii) The energy of the electric quadrupole moment is $\partial E/\partial x$ times its quadrupole moment $\sim qr'^2$.)

**16.** The spins and parities of the ground state, first excited state, and second excited state of ${}^{62}\text{Sm}^{152}$ are (0, even), (2, even), and (1, odd). Determine the types of radiation emitted in the $\gamma$ decays between these states.

**17.** Verify that the parts of the $\gamma$-decay selection rules relating $L$ to the nuclear spins represent angular momentum conservation requirements. Use the fact that a $\gamma$ ray from a transition of multipolarity $L$ carries $L$ units of angular momentum.

**18.** Prove that the integrals in (16-26) and (16-27), which represent components of the electric quadrupole and magnetic dipole matrix elements, yield zero unless the initial and final nuclear states have the same parity.

**19.** Consider carrying out a resonance absorption experiment with the source and absorber not at a low temperature, using the transitions between the first excited state and the ground state of ${}^{77}\text{Ir}^{191}$ considered in Example 16-7. (a) Calculate how much velocity would have to be given to the source to obtain enough Doppler shift to compensate for the recoil of the source and absorber nuclei, so that resonant absorption would be obtained. (b) Would it be possible to get the required velocity by mounting the source on the rim of a centrifuge? (c) Would an extremely sharp resonance be obtained in this manner?

**20.** The reaction ${}^1\mathrm{H}^1 + {}^3\mathrm{Li}^7 \rightarrow {}^4\mathrm{Be}^7 + {}^0n^1$ is sometimes used to produce monoenergetic neutrons from a source of monoenergetic protons. The $Q$ value of the reaction is $-1.64$ MeV. If a ${}^3\mathrm{Li}^7$ target is bombarded by a beam of 5 MeV protons, at what angle to the beam are 2.5 MeV neutrons emitted?

**21.** Use the $Q$ values of the three reactions listed as follows to calculate the energy available for the $\beta$ decay of ${}^{14}\mathrm{Si}^{31}$.

$$
\begin{aligned}
{}^1\mathrm{H}^2 + {}^{15}\mathrm{P}^{31} &\rightarrow {}^{14}\mathrm{Si}^{29} + {}^2\mathrm{He}^4 && Q = 8.158 \text{ MeV} \\
{}^1\mathrm{H}^2 + {}^{14}\mathrm{Si}^{29} &\rightarrow {}^{14}\mathrm{Si}^{30} + {}^1\mathrm{H}^1 && Q = 8.388 \text{ MeV} \\
{}^1\mathrm{H}^2 + {}^{14}\mathrm{Si}^{30} &\rightarrow {}^{14}\mathrm{Si}^{31} + {}^1\mathrm{H}^1 && Q = 4.364 \text{ MeV}
\end{aligned}
$$

**22.** Consider a one-dimensional traveling wave eigenfunction

$$\psi(x) = e^{ikx} \qquad \text{where} \qquad k = \sqrt{2m(E - V)}/\hbar$$

Take the potential energy $V$ to be *complex*, so that it can be written $V = V_R + iV_I$. (a) Show that $k$ becomes complex and can be written $k = k_R + ik_I$. (b) Then show that the amplitude of the traveling wave is a decreasing exponential function of $x$. Eigenfunctions such as this are used to describe the absorption of particles traveling through the complex *optical model* potential. (c) In what distance would the associated probability density decrease by a factor of $1/e$?

**23.** The total cross section for fission of ${}^{92}\mathrm{U}^{235}$ by incident neutrons of energy 1 MeV is about 1 bn. If such a neutron passes through a uniform slab of ${}^{92}\mathrm{U}^{235}$ of mass per unit area $10^{-1}$ kg/m$^2$, what is the probability that it will produce a fission?

**24.** When a $10^{-8}$ amp beam of 17 MeV protons is incident on a ${}^{29}\mathrm{Cu}^{63}$ target foil of mass per unit area $10^{-2}$ kg/m$^2$, it is observed that a counter of area $10^{-5}$ m$^2$ at 1 m from the target detects 240 elastically scattered protons per minute if it is placed at an angle of 30° to the incident beam. Determine the value of the differential cross section.

**25.** There is a resonance in the cross section for neutrons incident on ${}^{92}\mathrm{U}^{235}$ with the following set of measured Breit-Wigner parameters: $E_i = 0.29$ eV; $\Gamma = 0.140$ eV; $\Gamma_n = 0.005$ eV. (a) Show that $\Gamma = \Gamma_n + \Gamma_r$, and then evaluate $\Gamma_r$. (b) Calculate the total reaction cross section at the peak of the resonance, $\sigma_r(E_i)$. Measurement shows that about 75% of $\sigma_r(E_i)$ goes into fission. (c) Calculate the lifetime of the compound nucleus formed in this resonance.

**26.** The energies and spins of the first four excited states of ${}^{72}\mathrm{Hf}^{180}$ are: 0.093 MeV, $i = 2$; 0.309 MeV, $i = 4$; 0.641 MeV, $i = 6$; 1.085 MeV, $i = 8$. (a) How well do the ratios of these energies agree with the predictions of (16-33)? (b) Use that equation to evaluate the rotational inertia of the nucleus.

**27.** (a) Use (15-16) with $Q = 0$ to calculate the energy lost by a 1 MeV fission neutron to the recoil of ${}^6\mathrm{C}^{12}$, if it scatters elastically at the typical angle 90° from such a nucleus in the moderator of a nuclear reactor. (b) How much energy does it lose in a 90° scattering if its energy has been reduced to 0.001 MeV? (c) How much energy does it have, on the average, if it is in thermal equilibrium at an operating temperature of 500°K? (d) Estimate the number of scatterings required to bring the neutron into thermal equilibrium.

**28.** Compare the energy release, per kilogram of fuel consumed, in the thermal fusion reaction of (16-34) to the same figure of merit for the fission of ${}^{92}\mathrm{U}^{235}$.

# 17

# Elementary Particles

# Elementary Particles 17

## 17-1 Introduction

This chapter begins with a qualitative, but rather complete, discussion of the nuclear forces that act between two nucleons. The subject is at the border between the fields of nuclear physics and elementary particle physics, and its study will lead us in a natural way into the study of all the elementary particles. Along the route we shall also obtain a comprehensive view of the basic properties of, and interrelations between, the fundamental interactions and conservation laws of nature.

The history of quantum physics can be viewed as a sequence of probings, with ever increasing resolution, into the microscopic structure of matter. The first step was the discovery that matter is composed of about 90 different atoms. At that time atoms were considered to be the elementary particles. (The word is from the Greek *atomos* = indivisible.) Then it was found that atoms are composed of nuclei and electrons. Later it was discovered that nuclei consist of neutrons and protons. At this stage there was a very satisfactory situation—all matter appeared to be composed of various combinations of a small number of elementary particles: the neutron, the proton, and the electron. But then it was found that there are also $\pi$ mesons. Their discovery was followed by the discovery of many other related mesons, and an even larger number of particles related to neutrons and protons themselves. At the present time, the number of elementary particles has become so large again that it is reasonable to speculate that they could be composed of various combinations of a small set of even more elementary ones, as was the case for atoms. We close the chapter, and the book, by discussing what may be the beginning of the final stage of the sequence—the postulate that there exists a set of three particles, called quarks, that could be truly elementary.

## 17-2 Nucleon Forces

In our study of nuclei we have obtained some information about the nuclear forces acting between nucleons, which we shall call *nucleon forces*. Since nuclei are studied in terms of models, and since models do not involve the detailed behavior of these forces, we have learned only about certain of their general features. These are:

1. Nucleon forces are *strong*. The energy associated with the force is larger than that associated with electromagnetism by 1 or 2 orders of magnitude, larger than that associated with $\beta$ decay by about 12 orders of magnitude, and larger than that associated with gravitation by about 40 orders of magnitude.

2. Nucleon forces are *short range*. They cut off in a distance of about 2 F, so that two nucleons passing each other at a larger distance do not interact by the nucleon force.

3. Nucleon forces are *attractive* in their over-all effect. Otherwise nuclei would not exist since the nucleons would not bind together.

4. Nucleon forces are *charge independent*. That is, they make no distinction between protons and neutrons. Evidence for this is seen in the tendency of small-$Z$ nuclei to have $N = Z$, and in the similarities of the low-lying energy levels of pairs of mirror nuclei.

5. Nucleon forces *saturate*. The term describes the fact that a nucleon in a typical nucleus experiences attractive interactions only with a limited number of the many other nucleons. This must be true since otherwise the average binding energy per nucleon, $\Delta E/A$, would be proportional to $A$ instead of being approximately independent of $A$.

Most of the information about nucleon forces that can be obtained from the study of systems as complicated as a typical nucleus are listed above. More detailed information is obtained by studying simpler systems containing only two nucleons where the nucleon forces have their most directly observable effects. The simplest of these systems is the ground state of the *deuterium* nucleus $^1H^2$, or *deuteron*, consisting of a neutron and a proton bound together by the nucleon force. In this section we shall study this system, and other systems containing two unbound nucleons. To avoid complicated quantum mechanical calculations, we shall keep the discussion largely qualitative. But we shall, nevertheless, be able to see how the analyses of certain critical experiments have been used to determine the properties of nucleon forces. At the end of the section we summarize by presenting a quantitative description of the most important of these properties. In a subsequent section we consider the meson theory of the origin of nucleon forces.

The ground state of the deuteron is characterized by the following measured quantities:

Binding energy: $\Delta E = 2.22$ MeV
Nuclear spin: $i = 1$
Nuclear parity: even
Magnetic dipole moment: $\mu = +0.857\mu_n$
Electric quadrupole moment: $q = +2.7 \times 10^{-31}$ m$^2$
Charge distribution half-value radius: $a = 2.1$ F

The fact that the deuteron has an electric quadrupole moment $q$ means that its probability density function is not spherically symmetrical. This immediately tells us that the *nucleon potential*, which specifies the force acting between the two nucleons, is, itself, not spherically symmetrical. The point is that all spherically symmetrical potentials have $l = 0$ eigenfunctions for their ground states, and the probability density functions for such eigenfunctions are all spherically symmetrical (an example is the Coulomb potential and the spherically symmetrical ground state of a one-electron atom). But the observed departure from spherical symmetry is not large.

A measure of the departure is the quantity $q/r'^2$ (see Figure 15-20), which has a value of about 6% if we take $r'$ equal to the charge distribution half-value radius $a$. Calculations show that the measured electric quadrupole moment is obtained if 96% of the time the deuteron is in an $l = 0$ state, and 4% of the time it is in an $l = 2$ state. Such a mixed state will also have the measured even parity since for both of its component states $l$ is even. Since the ground state nuclear spin is measured to be 1, both component states must have $j = 1$. The vector addition diagrams of Figure 17-1 illustrate the relations between the $l$ and $j$ quantum numbers in both states, and they show that for both the intrinsic spins of the proton and neutron are essentially parallel and the quantum number specifying the total intrinsic spin angular momentum is $s = 1$. In spectroscopic notation, the dominant state is $^3S_1$ and the less probable state is $^3D_1$. (The superscript gives the value of $2s + 1$; the letter gives the value of $l$, with $S$ meaning $l = 0$, $P$ meaning $l = 1$, $D$ meaning $l = 2$, etc.; the subscript gives the value of $j$.) Calculations also show that this mixture of states leads to the measured magnetic dipole moment $\mu = +0.857\mu_n$. The value differs by about 3% from what would be obtained if the deuteron were in a pure $^3S_1$ state, with the proton and neutron intrinsic spin essentially parallel and no orbital motion, since in that state $\mu$ would be just the sum of the proton and neutron magnetic dipole moments, $+2.7896\mu_n - 1.9103\mu_n = +0.8793\mu_n$. We conclude

**FIGURE 17-1**

Vector addition diagrams showing the spin, orbital, and total angular momentum quantum numbers $s$, $l$, and $j$ in the two component states of the deuteron. In the dominant state, $l = 0$. Since $j = 1$ it is necessary that $s = 1$ in this state which, in spectroscopic notation, is designated ${}^3S_1$. In the less probable state, $l = 2$. Since $j = 1$, it is also necessary in this state that $s = 1$. The state is designated ${}^3D_1$.

$l = 0$
$s = 1$ $j = 1$
${}^3S_1$

$s = 1$
$l = 2$
$j = 1$
${}^3D_1$

from all these considerations that the nucleon potential is not precisely spherically symmetrical, since it does not lead to a pure $S$ ground state for the deuteron. But since the amount of $D$ state it mixes in is small, the asymmetry of the potential must be small. For most purposes the asymmetry can be ignored.

Thus we consider the deuteron as a system in which the nucleons are bound in a ${}^3S_1$ state of a spherically symmetrical nucleon potential $V(r)$, where $r$ is the distance between their centers. This potential specifies the force acting between the two nucleons. Some information about it is obtained by demanding that the energy of its ground state yield a binding energy equal to the measured value $\Delta E = 2.22$ MeV. Additional information is obtained by demanding also that the ground state eigenfunction yield a charge distribution half-value radius equal to the measured value $a = 2.1$ F. These two pieces of data are not enough to determine the form of the nucleon potential, i.e., the radial dependence of the function $V(r)$. However, if $V(r)$ is assumed for simplicity to have the form of a square well as in Figure 17-2, then the radius $r'$ and depth $V_0$ are determined to be about 2 F and 40 MeV. Precise numbers will be quoted later after we have introduced additional experimental information that does determine something about the form of the potential. It can also be determined that a potential which fits the measured values of both $\Delta E$ and $a$ has the property that its ground state is its only bound state, as indicated by the single bound energy level in Figure 17-2. This agrees with the fact that the deuteron is observed to have no bound excited states.

Now the spins of the proton and neutron are essentially parallel in a ${}^3S_1$ bound state of the deuteron. We know that there are no bound deuterons with nucleon spins essentially antiparallel, i.e., in a ${}^1S_0$ state, since none are ever found with the nuclear spin 0 that would be obtained in such a state. What is the reason for the absence of a

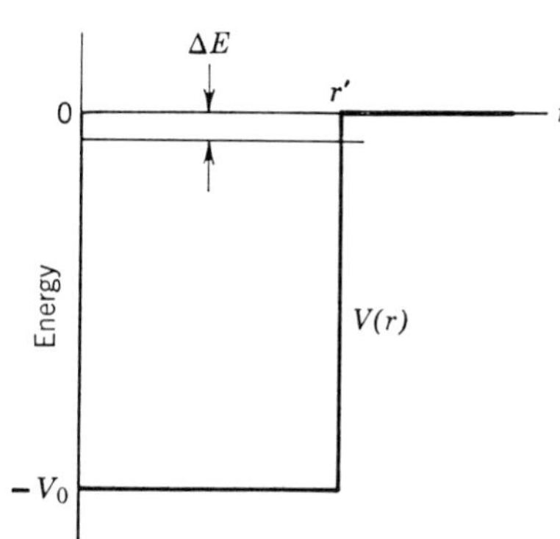


**FIGURE 17-2**

A square well potential of radius $r'$ and depth $V_0$, and its ground state eigenvalue of binding energy $\Delta E$. For the deuteron this state is the only bound state of the potential.

bound $^1S_0$ state? An explanation is that *the nucleon potential is spin dependent, being appreciably weaker when two nucleons interact with essentially antiparallel spins* (*in a singlet state*). If the potential is sufficiently weak to prevent the nucleons from binding stably together, the absence of the $^1S_0$ bound state is explained. (A one-dimensional potential has at least one bound state, no matter how weak the potential, because the eigenfunction can extend very far into the classically excluded regions on both sides of the binding region. But due to the different geometry of the eigenfunction, a three-dimensional potential can only have a bound state if it is sufficiently strong. This can be seen by inspecting the form of $rR(r)$ for the lowest $S$ state of a three-dimensional square well, displayed in Figure 15-17. Since $rR(r) = 0$ at $r = 0$, that function must have enough curvature within the binding region to allow it to match on to a decreasing exponential in the excluded region. This, in turn, requires that for a given breadth the binding region be sufficiently deep.) Additional qualitative evidence in support of the idea of spin dependence of the nucleon potential is found in the absence of a bound state for a system of two protons, or a system of two neutrons. In both systems the exclusion principle would require it to be a $^1S_0$ state, where the spins of the two identical nucleons are essentially antiparallel. In this state the potential is, presumably, too weak to lead to binding.

Quantitative evidence for the spin dependence of the nucleon potential is obtained in the analysis of data concerning the scattering of unbound neutrons, of low kinetic energy, from protons. If the orientations of the spins of the neutrons in the incident beam and the protons in the scattering target are random, then the four possible spin states of the two nucleon system will be equally probable. These nucleons will usually (three times out of four) be interacting in one of the three $^3S_1$ states (the triplet states in which the nucleon spins are essentially parallel, and the total spin can have three different $z$ components: $-\hbar, 0, +\hbar$). Occasionally (one time out of four) the nucleons will be interacting in the $^1S_0$ state (the singlet state in which the nucleon spins are essentially antiparallel, and the total spin can have only a single $z$ component equal to 0). The total cross section for scattering, $\sigma$, which is proportional to the total probability that a neutron is scattered by a proton, is shown in Figure 17-3. Since some of the interactions occur in the singlet state, these data provide information which, when analyzed, shows that the singlet state potential is about 40% weaker than the triplet potential. That is, if both are square wells of the same radius, the depth is about 40% less in the potential for the singlet state. Hence we conclude that the nucleon potential really does depend on the relative orientation of the spins of the two interacting nucleons.

This quantitative information about the spin dependence is confirmed by analyzing the scattering of low-energy protons from protons. And that analysis also provides

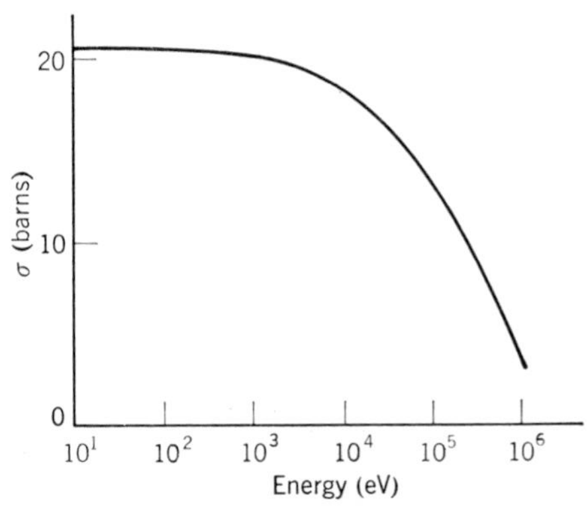


**FIGURE 17-3**
Measured values of the total cross section $\sigma$ for the scattering of neutrons by protons as a function of the energy of the incident neutron.

additional evidence that *the nucleon potential is charge independent*; i.e., it makes no distinction between protons and neutrons. The evidence is that a nucleon potential which agrees with the measured neutron-proton scattering cross section also agrees with the measured proton-proton scattering cross section. This does not mean that the cross sections are the same. In proton-proton scattering, the Coulomb potential, which is present in addition to the nucleon potential, affects the small angle scatterings, and the exclusion principle affects all the scattering by suppressing certain quantum states.

The scattering of a low-energy nucleon from a nucleon does not give information about the form of the nucleon potential. As measured in a frame of reference in which the center of mass of the system is stationary, the scattering is independent of angle, or isotropic. Thus the differential cross section for scattering, $d\sigma/d\Omega$, which is proportional to the probability for scattering at various angles, is the same at all angles in this reference frame. The constant differential cross section provides only one piece of experimental data—the measured value of $d\sigma/d\Omega$. This single measured quantity can be used to determine only a single theoretical quantity. The quantity determined is the strength of the potential. (This is $V_0 r'^2$ for a square well potential.) The reason why the scattering is isotropic in the so-called center-of-mass frame of reference is that at low energies the de Broglie wavelength $\lambda$ of the wave, which describes the nucleon scattering, is very large compared to the radius $r'$ of the potential, which describes the forces which produce the scattering. If $\lambda \gg r'$, then the separation in the scattering angle between adjacent minima in the diffraction pattern is, according to (15-4), $\theta \simeq \lambda/r' \gg 1$. Since the entire range of scattering angle is only $\pi$, the inequality is essentially telling us that there are no minima. In other words, the potential looks to the wave like a point, which can only scatter it isotropically. But if the energy of the scattered nucleon is high enough for $\lambda$ to be smaller than $r'$, then $\theta \simeq \lambda/r' < 1$. The scattering pattern has structure in these circumstances, and $d\sigma/d\Omega$ contains information about the form of the potential that causes the scattering. Thus, *only high-energy nucleons have enough resolving power to be effective as probes in studying the form of the nucleon potential.* We shall show in Example 17-2 that if the radius of the potential is taken as 2 F, the differential cross section for scattering, $d\sigma/d\Omega$, can be expected to depart from isotropy when the kinetic energy of the incident nucleon exceeds about 40 MeV.

The first high-energy neutron-proton scattering experiments were performed at an incident neutron kinetic energy of 90 MeV. It was expected that they would provide information about the radial dependence of the nucleon potential, but, as we shall see, they actually taught us about a different aspect of the form of the nucleon potential. It was also expected that the differential cross section for scattering, $d\sigma/d\Omega$, would have the shape of a rudimentary diffraction pattern, with $d\sigma/d\Omega$ generally increasing for decreasing scattering angle. The reason why it was thought there would be a preference for scattering at small angles into forward directions is indicated in Figure 17-4. If the depth of the nucleon potential $V(r)$ is significantly smaller than the kinetic energy of the incident neutron, the maximum momentum that the potential can transfer to the neutron has a magnitude which is significantly smaller than the magnitude of its initial momentum. In these circumstances, a large change in the direction of the neutron momentum would not be possible. Figure 17-5 shows the measured $d\sigma/d\Omega$ for 90 MeV neutron-proton scattering. Following convention, these results are expressed in a frame of reference in which the center of mass of the neutron-proton system is stationary. The top part of Figure 17-6 indicates that in this center-of-mass frame of reference the argument we have just gone through leads to the expectation of a preference for small scattering angles. But the measurements show that *$d\sigma/d\Omega$ for*

**FIGURE 17-4**

Illustrating why the scattering angle should be small if a nucleon is scattered by a potential that can transfer to the nucleon only a momentum of magnitude small compared to the magnitude of its initial momentum. This is the situation that would be expected if the kinetic energy of the nucleon is large compared to the depth of the potential.

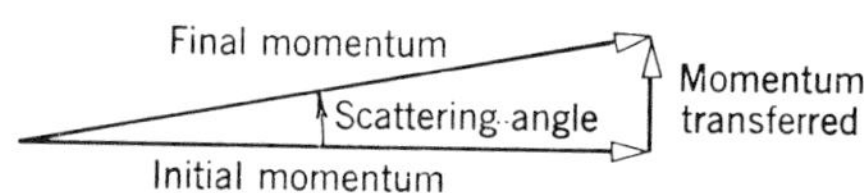


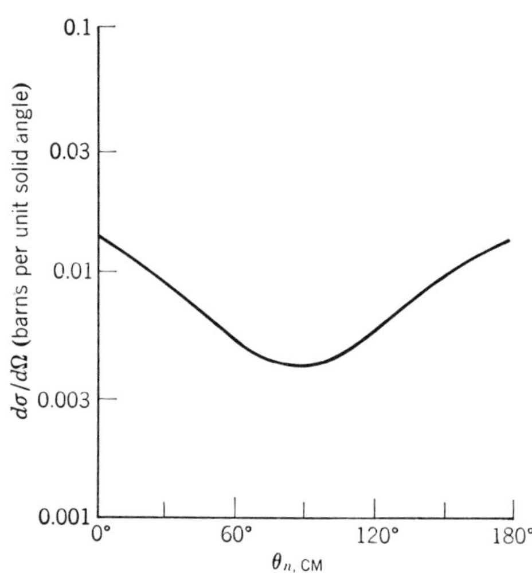


**FIGURE 17-5**

Measured values of the differential cross section $d\sigma/d\Omega$ for scattering of neutrons of incident energy 90 MeV by protons. The data are actually obtained in a frame of reference where the target proton is initially stationary. Here they have been transformed to a frame of reference in which the center of mass of the system is stationary. The quantity $\theta_{n,\,\mathrm{CM}}$ is the neutron scattering angle in that system.

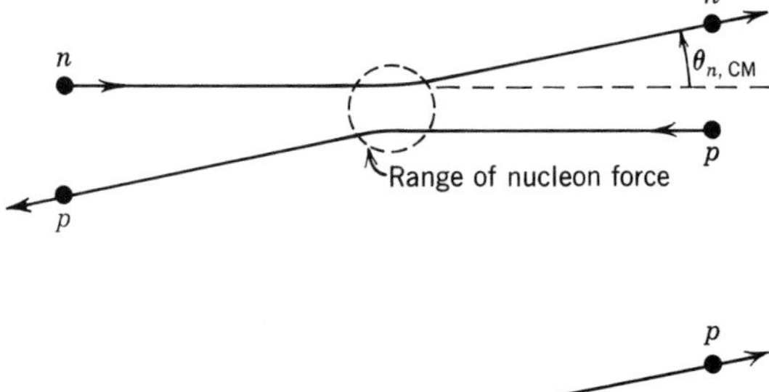


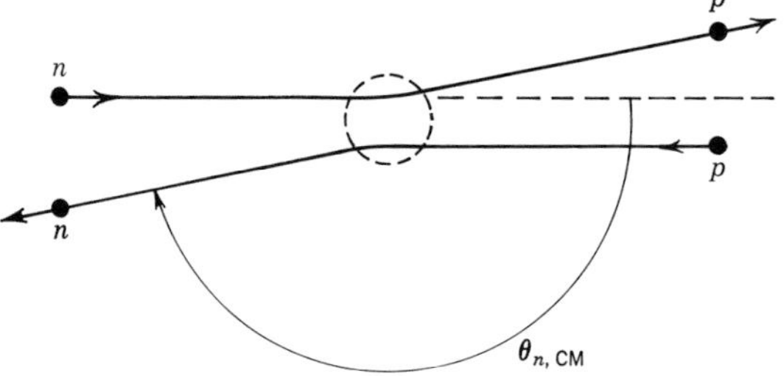


**FIGURE 17-6**

*Top:* Neutron-proton scattering as seen in a frame of reference in which the center of mass of the system is stationary. If the kinetic energies of the nucleons are large compared to the depth of the nucleon potential, the momentum transfers are small and the neutron and proton scattering angles are small as well. *Bottom:* The same, for a scattering in which the neutron changes into a proton and vice versa when they interact. Although the momentum transfers are still small, because of the exchange the scattering angles are large.

*neutron-proton scattering is approximately symmetric about a scattering angle of* 90°. Thus there is an equally pronounced preference for large scattering angles.

The bottom part of Figure 17-6 represents the physical interpretation of the origin of the observed preference for large scattering angles. *In approximately half the scatterings, the neutron changes into a proton and the proton changes into a neutron*, when the two nucleons are very close. Although the momentum transfer in every scattering is small, when the exchange occurs it has the effect of producing a large angle scattering. In a later section we shall see that a neutron can change into a proton by emitting a charged meson, and a proton can change into a neutron by absorbing that meson.

A more formal interpretation of the results of the neutron-proton scattering experiments is that the nucleon potential $V$ that produces the scattering has a form which can be written approximately as

$$V \simeq \frac{V(r) + V(r)P}{2} \tag{17-1}$$

where $P$ is an *exchange operator* that changes a proton into a neutron and a neutron into a proton, and $V(r)$ is the ordinary nucleon potential we have previously discussed. Now the nucleon potential $V$ enters expressions for the scattering cross section through the matrix element

$$\psi_f^* V \psi_i$$

where $\psi_i$ is the eigenfunction for the initial neutron-proton system (before scattering), and $\psi_f^*$ is the complex conjugate of the eigenfunction for the final neutron-proton system (after scattering). Thus it is of interest to consider the quantity

$$V\psi_i \simeq \left[\frac{V(r) + V(r)P}{2}\right] \psi_i = \frac{V(r)}{2}\psi_i + \frac{V(r)}{2}P\psi_i$$

We write this as

$$V\psi_l \simeq \frac{V(r)}{2}\psi_l + \frac{V(r)}{2}P\psi_l \tag{17-2}$$

using the quantum number $l$ to label the orbital angular momentum of the initial system. Since an exchange of the equal mass neutron and proton is equivalent to an exchange of the signs of the coordinates specifying their locations relative to an origin at their center of mass half way between them, the exchange operation is equivalent in these particular circumstances to the parity operation. Therefore the usual relation between the orbital angular momentum quantum number and parity, (8-47), is applicable, and tells us that

$$P\psi_l = (-1)^l \psi_l$$

That is, the parity of an eigenfunction of a spherically symmetrical potential, $\psi_l$, is even if $l$ is even and odd if $l$ is odd. Thus the parity (or exchange) operator leaves the eigenfunction unchanged in the second term on the right side of (17-2) if $l$ is even, and multiplies it by minus one if $l$ is odd. So we have

$$V\psi_l \simeq \frac{V(r)}{2}\psi_l + \frac{V(r)}{2}P\psi_l = \frac{[1 + (-1)^l]}{2}V(r)\psi_l$$

From this result we can see that the nucleon potential may be written approximately, without using the exchange operator, in a form called the *Serber potential*

$$V \simeq \frac{[1 + (-1)^l]}{2}V(r) \tag{17-3}$$

**FIGURE 17-7**

Two nucleons, each with linear momentum of magnitude $p$, passing each other at a distance $r'$. Each has an orbital angular momentum $pr'/2$ in magnitude relative to the center of mass. The magnitude of the orbital angular momentum of the two nucleon system is $L = pr'$.

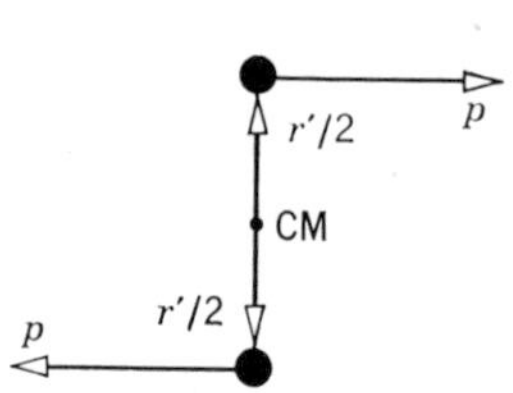


Note that $V \simeq 0$ if $l$ is odd. We conclude that *the nucleon potential depends strongly on the orbital angular momentum of the two interacting nucleons, relative to their center of mass. The potential is approximately zero when the orbital angular momentum quantum number $l$ has an odd value.* (Later we shall see that $V \simeq 0$ for an odd $l$ only if its effect is averaged over all the quantum states for that value of $l$, as is the case in most situations.)

A classical argument, illustrated in Figure 17-7 in the center-of-mass frame of reference, shows that there is a relation between the maximum possible value of the orbital angular momentum $L$ for a system of two interacting nucleons of linear momenta $p$. The relation is $L \simeq pr'$, where $r'$ is the maximum separation at which the nucleons can interact, which is the range of the nucleon force or the radius of the nucleon potential. Since $L$ is related to the quantum number $l$ by the equation $L = \sqrt{l(l+1)}\hbar$, it is easy to estimate, for an assumed value of $r'$, the maximum possible value $l_{max}$ of the quantum number in terms of the momenta or kinetic energies of the nucleons.

**Example 17-1.** Two nucleons interact with nucleon force of range $r' = 2.0$ F, in a state in which the angular momentum quantum number assumes its maximum possible value. If this value is $l_{max} = 1$, what must be the kinetic energy of each nucleon in the center-of-mass frame of reference? The total kinetic energy in that frame of reference? The kinetic energy of the incident nucleon (in a beam) in a frame of reference where the nucleon with which it interacts is initially stationary (in a target)?

We have

$$L = \sqrt{l(l+1)}\,\hbar$$

with $l = l_{max} = 1$. So

$$L = \sqrt{1(1+1)}\,\hbar = \sqrt{2}\,\hbar$$

Also

$$L \simeq pr'$$

or

$$p \simeq \frac{L}{r'} = \frac{\sqrt{2}\,\hbar}{r'}$$

Thus the kinetic energy of each nucleon in the center-of-mass (CM) frame is

$$K = \frac{p^2}{2M} \simeq \frac{2\hbar^2}{2Mr'^2}$$

$$\simeq \frac{(1.05 \times 10^{-34}\,\text{joule-sec})^2}{1.7 \times 10^{-27}\,\text{kg} \times (2.0 \times 10^{-15}\,\text{m})^2} = 1.6 \times 10^{-12}\,\text{joule}$$

$$= 10\,\text{MeV}$$

The total kinetic energy in that frame of reference is just

$$K_{\text{total CM}} = 2K \simeq 20 \text{ MeV}$$

It is easy to show that, because the two interacting particles have the same mass, the kinetic energy of the moving one, in a frame of reference in which the other one is initially stationary, is twice the total kinetic energy in the center-of-mass frame of reference. Thus the kinetic energy of the incident nucleon is

$$K_{\text{incident}} = 2K_{\text{total CM}} \simeq 40 \text{ MeV}$$ ◀

**Example 17-2.** Show that the condition $l_{\max} = 0$ is equivalent to the condition $\theta \simeq \lambda/r' \gg 1$ which requires the differential scattering cross section $d\sigma/d\Omega$ to be isotropic.

Referring to the calculation in Example 17-1, note that if the kinetic energy $K$ of each nucleon in the center-of-mass frame is less than about 10 MeV, then each will have a momentum $p$ which is

$$p < \frac{\sqrt{2}\,\hbar}{r'} = \frac{h}{\sqrt{2}\pi r'}$$

or

$$\frac{h}{pr'} > \sqrt{2}\pi$$

Using the de Broglie relation to evaluate $\lambda$, the nucleons' wavelength, from their momenta $p$, we obtain

$$\frac{\lambda}{r'} > \sqrt{2}\pi$$

or

$$\frac{\lambda}{r'} \gg 1$$

According to (15-4), the separation between adjacent minima in the scattering pattern is $\theta \simeq \lambda/r'$, so we have

$$\theta \simeq \frac{\lambda}{r'} \gg 1$$

As we mentioned several pages ago, this inequality means that there are no minima, and the differential scattering cross section $d\sigma/d\Omega$ is isotropic. But we saw in Example 17-1 that $K \simeq 10$ MeV is the condition for having $l_{\max} = 1$ (assuming the range of nucleon forces is $r' = 2$ F). So for $K < 10$ MeV, we can have only $l_{\max} = 0$. Thus we have shown that $l_{\max} = 0$ is equivalent to $\theta \simeq \lambda/r' \gg 1$.

We concluded in Example 17-1 that when the kinetic energy of each nucleon in the center of mass frame is about 10 MeV the kinetic energy of the incident nucleon, in the frame in which the target nucleon is initially at rest, has a value of about 40 MeV. So we can also conclude that $d\sigma/d\Omega$ can be expected to depart from isotropy only when the kinetic energy of the incident nucleon equals, or exceeds, about 40 MeV. ◀

Example 17-1 shows that, for a nucleon potential of radius $r' = 2$ F, we have $l_{\max} = 0$ unless the kinetic energy of each nucleon of an interacting pair exceeds about 10 MeV in the center-of-mass frame of reference. Similar calculations show that $l_{\max} = 1$ unless these energies exceed about 30 MeV, and $l_{\max} = 2$ unless they exceed about 60 MeV. (All these figures are only approximations since they are obtained from a semiclassical argument.) Now, if we consider a pair of nucleons in a nucleus, their kinetic energies in a frame of reference fixed to their center of mass generally do not exceed 30 MeV. Thus they can usually interact with each other only in $l = 0$ and $l = 1$ states. But the Serber potential, (17-3), is approximately zero for $l = 1$. So the

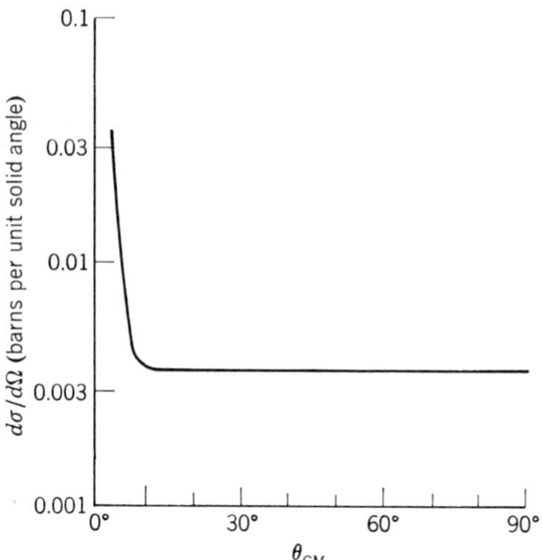


**FIGURE 17-8**
Measured values of the center-of-mass differential cross section $d\sigma/d\Omega$ for proton-proton scattering. The energy of the incident protons is 330 MeV.

nucleons in a nucleus actually interact strongly with each other in only half of the quantum states that angular momentum considerations (and exclusion principle considerations if they are of the same species) would otherwise allow to contribute to the total interactions. This property of the nucleon potential helps make nucleon forces saturate by suppressing the attractive nucleon forces in half of the interactions; but it is not enough. To obtain saturation—a feature that we indicated at the beginning of this section is responsible for one of the most basic properties of nuclei—it is necessary that some of the nucleon forces be repulsive. That is, there must be a *repulsive part* in the nucleon potential.

The study of proton-proton scattering at quite high energies showed that the radial dependence of the nucleon potential is such that it has a repulsive region in its center. Figure 17-8 gives the measured center-of-mass reference frame differential cross section, $d\sigma/d\Omega$, for scattering of incident protons of kinetic energy 330 MeV from a target of protons. Only scattering angles from 0° to 90° are plotted. The symmetry of the two proton system demands that $d\sigma/d\Omega$ be symmetric about 90°, no matter what the form of the nucleon potential, because if one proton is scattered at the angle $\theta$ the other one must be scattered at the angle $180° - \theta$. At angles smaller than about 10°, $d\sigma/d\Omega$ has the very rapid angular dependence of Coulomb scattering. In this angular range the distance of closest approach in the scatterings is greater than the range of nucleon forces. At larger angles, the scatterings involve close collisions in which nucleon forces dominate, and *$d\sigma/d\Omega$ for proton-proton scattering is found to be essentially isotropic.*

The surprising isotropy of high-energy proton-proton scattering was shown by Jastrow to imply that *there is a strong repulsive core in the nucleon potential.* That is, the potential has a radial dependence something like that indicated in Figure 17-9. It is not difficult to understand qualitatively the essential points in Jastrow's argument. At an incident kinetic energy of 330 MeV the kinetic energy of each of the protons in their center-of-mass frame is 82 MeV, and $l_{max} = 3$. Thus the two protons in the scattering can interact only in states of orbital angular momentum given by $l = 0$, 1, 2, 3. But since the Serber potential is approximately zero for $l = 1$ and 3, significant interactions can occur only in $l = 0$ and 2 states. *If* only the $l = 0$ state were involved, $d\sigma/d\Omega$ would indeed be isotropic because the scattering would be the same as if we had $l_{max} = 0$, which means $\theta \simeq \lambda/r' \gg 1$. However, in this case the magnitude of $d\sigma/d\Omega$ could be only about half as large as the magnitude actually observed. In fact, the

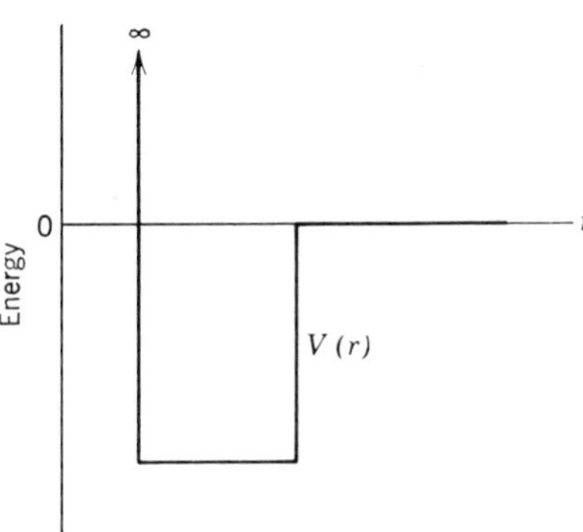


**FIGURE 17-9**
A nucleon potential with an infinitely strong repulsive core inside an attractive square well.

isotropy of $d\sigma/d\Omega$ is a result of a destructive interference between waves scattered in an $l = 0$ state interaction and waves scattered in an $l = 2$ state interaction. The interference suppresses the tendency, discussed above, for $d\sigma/d\Omega$ to be large at small angles. Figure 17-10 indicates how a potential with a repulsive core, of height which is very much larger than the kinetic energy of the incident proton, affects the $l = 0$ state eigenfunction. The repulsive region "pushes out" the eigenfunction as at the edge of an infinite well, and the attractive region "pulls in" the eigenfunction because it increases the curvature. If the incident proton energy is large compared to the depth of the attractive region, the effect of this region is small and the net result is that the $l = 0$ state eigenfunction is pushed out. Figure 17-11 shows what the potential does to the $l = 2$ state eigenfunction. Since for small $r$ all these eigenfunctions have the $r^l$ behavior given by (7-32), the $l = 2$ eigenfunction has such a small value throughout the repulsive region near $r = 0$ that the repulsive region can have practically no effect on it. This eigenfunction is very small for small $r$ whether or not the repulsive region is present. Consequently, the attractive region is the only one that has much

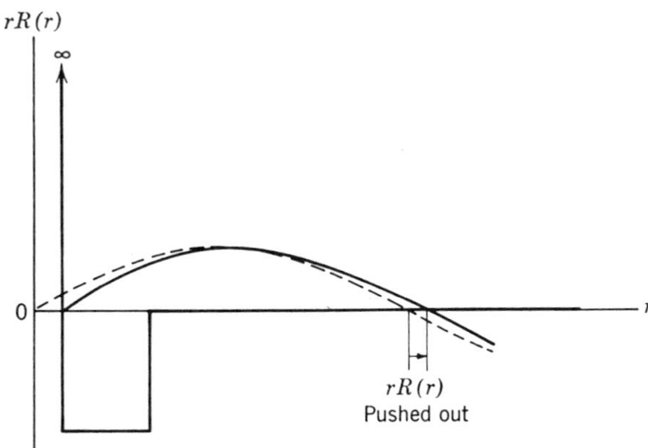


**FIGURE 17-10**
The effect of a repulsive core potential on the radial dependence of the radial coordinate, $r$, times the radial part of the eigenfunction, $R(r)$, for the $l = 0$ state eigenfunction for high-energy proton-proton scattering. The solid curve shows $rR(r)$ in the presence of the potential and, for comparison, the dashed curve shows what it would be like in the absence of the potential. Because the energy of the incident proton is large compared to the depth of the attractive region of the potential, the effect of the repulsive core dominates and $rR(r)$ is pushed out.

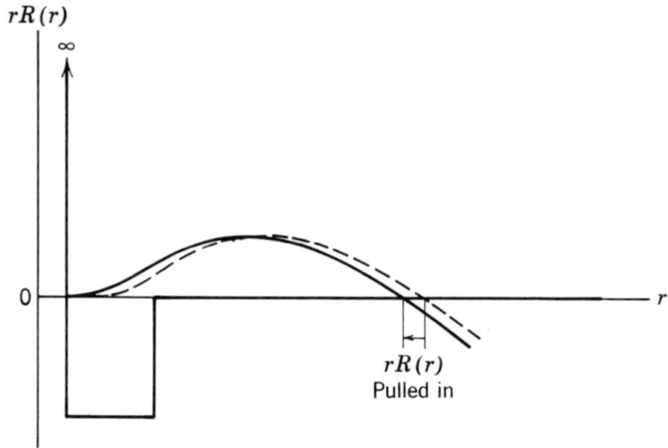


**FIGURE 17-11**

The effect of a repulsive core potential on $rR(r)$ for the $l = 2$ state eigenfunction for high-energy proton-proton scattering. The solid curve shows $rR(r)$ in the presence of the potential, and the dashed curve shows what it would be like in the absence of the potential. Since $rR(r)$ is negligibly small at the core radius even in the absence of the potential because $R(r) \propto r^l$, the effect of the repulsive core is negligible. Thus the attractive region dominates and $rR(r)$ is pulled in.

effect on the $l = 2$ state eigenfunction, and so the eigenfunction is pulled in by the potential. The destructive interference leading to the isotropic $d\sigma/d\Omega$ is due to the $l = 0$ state eigenfunction being pushed out while the $l = 2$ state eigenfunction is pulled in. If the nucleon potential were purely attractive, both eigenfunctions could only be pulled in.

Recent experiments on the scattering of high-energy electrons from deuterons provide completely independent evidence of the existence of a strong repulsive core in the nucleon potential. The experiments show that there is a hole in the center of the deuteron charge distribution. This means that the proton avoids the center of the deuteron, presumably because of the very strong repulsion it feels if it tries to get too close to the neutron. Analysis of both the electron-deuteron and proton-proton scattering experiments indicates that the radius of the repulsive core is about 0.5 F.

The repulsive core in the nucleon potential is the most important factor responsible for the saturation of nucleon forces. In a nucleus, the cores in the nucleon potentials add large positive contributions to the total energy if the nucleons are too closely packed. This is why the nucleons maintain an average center-to-center spacing, given by the measured nucleon mass density, of about 1.2 F. At this spacing, any one nucleon can interact only with a limited number of other nucleons, since the range of nucleon forces is about 2 F, and so the nucleon forces saturate. If there were no repulsive region in the nucleon potentials, the attractive regions would cause the nucleus to collapse until its linear dimensions were about equal to the range of nucleon forces. Then each nucleon would interact with all the other nucleons, and the binding energy per nucleon, $\Delta E/A$, would be approximately proportional to $A$.

We found that the nucleon potential depends on the quantum number $s$ specifying the spin angular momentum of a system of two nucleons (i.e., whether they are in a singlet or triplet state), and that it also depends on the quantum number $l$ specifying the orbital angular momentum of the system. Certain experiments show that the potential even depends on the quantum number $j$ specifying the total angular momentum of the system. Another way of saying this is that the potential depends not only on the spin angular momentum **S** and on the orbital angular momentum **L**, but also on their dot product $\mathbf{S} \cdot \mathbf{L}$ which determines the magnitude of the total angular momentum **J**. Thus *the nucleon potential contains a spin-orbit term, proportional to* $\mathbf{S} \cdot \mathbf{L}$. The term makes the nucleon potential more attractive if $\mathbf{S} \cdot \mathbf{L}$ is positive, and more repulsive if it is negative, just as is the case for the spin-orbit term of the shell model nuclear potential. The nucleon and nuclear spin-orbit terms are related, but the relation is not completely understood. The experiments referred to basically involve scattering a beam of nucleons with aligned spins from a target of nucleons with aligned spins. This allows the interactions in different quantum states, with different spin, orbital, and total angular momenta, to be investigated separately.

We conclude this section by summarizing what is known about nucleon forces. Certainly the first thing to say is that they are very complicated. When a nucleon of, say, 200 MeV kinetic energy interacts with another nucleon, the system can be in any one of the following quantum states: ${}^1S_0$, ${}^3S_1$, ${}^1P_1$, ${}^3P_0$, ${}^3P_1$, ${}^3P_2$, ${}^1D_2$, ${}^3D_1$, ${}^3D_2$, ${}^3D_3$. The nucleon potential is different in each of these states, and in each, its form involves a fairly complicated radial dependence, as well as departures from spherical symmetry. The only simplifications are:

1. The nucleon potential is charge independent, so it does not depend on the species of the interacting nucleons.

2. The exclusion principle prohibits interaction in certain quantum states between nucleons of the same species. In particular, the ${}^3S_1$, ${}^1P_1$, ${}^3D_1$, ${}^3D_2$, ${}^3D_3$ states are excluded from the list just quoted in the neutron-neutron or proton-proton interactions. The reason is that if the space eigenfunction for a system of two identical nucleons is symmetric in a label exchange (even $l$), then the spin eigenfunction must be antisymmetric in such an exchange (singlet); and if the space eigenfunction is antisymmetric (odd $l$), the spin eigenfunction must be symmetric (triplet).

3. The net effect of all the $P$ state interactions is very small. But the aligned spin experiments show this is partly due to destructive interferences in the interactions from the different $P$ states, and that the interactions in individual $P$ states are not so small.

If we are content to describe approximately only their most important properties, however, nucleon forces are not too complicated. Figures 17-12 and 17-13 give quantitatively the radial dependences of nucleon potentials for even-$l$ quantum states. The first figure shows the potential for singlet states (nucleon spins essentially antiparallel), and the second shows the stronger potential for triplet states (nucleon spins essentially

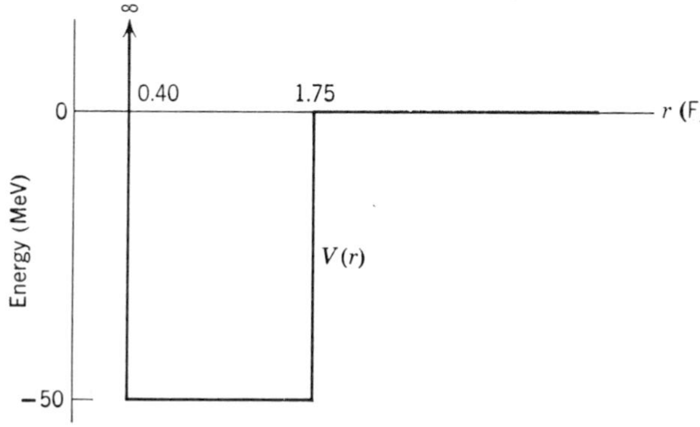


**FIGURE 17-12**

The radial dependence of a singlet even-$l$ nucleon potential in reasonable agreement with experiment.

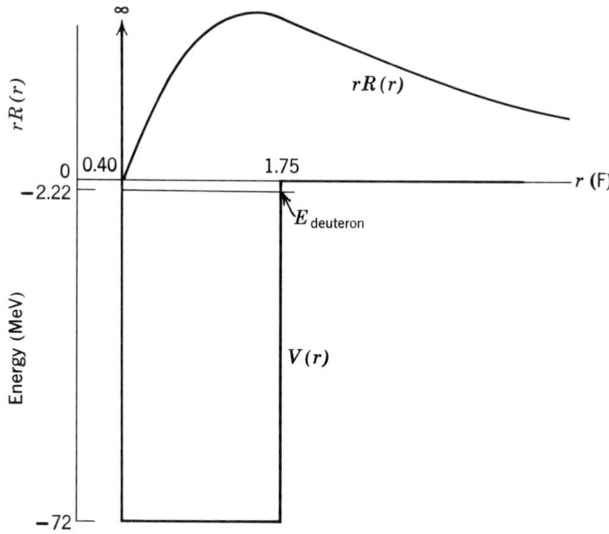


**FIGURE 17-13**

The radial dependence of a triplet even-$l$ nucleon potential in reasonable agreement with experiment. Also shown are the eigenvalue and the quantity $rR(r)$ for the eigenfunction of the single bound state of the potential at $-2.22$ MeV. This state, which is the deuteron, is just barely bound and $rR(r)$ just barely reaches a maximum inside the attractive region (compare with Figure 17-10). The square of $rR(r)$ is $r^2R^*(r)R(r)$ which is the radial probability density that specifies the probability of finding the two nucleons in the deuteron with a separation in the vicinity of $r$.

parallel). With these two potentials, and zero potential for all quantum states with odd $l$, results are obtained in reasonable agreement with all the properties of the deuteron (except its electric quadrupole moment) and all the nucleon scattering data up to several hundred MeV (except the aligned spin data).

Figure 17-13 shows also the eigenvalue and the radial dependence of the eigenfunction for the only bound state of the triplet potential, i.e., the deuteron. Note that the attractive region is just barely strong enough to overcome the effect of the repulsive core and lead to binding. As a consequence, there is a high probability that the two nucleons in the deuteron have a separation larger than the range of nucleon forces.

Of course, the nucleon potentials in nature cannot have the abrupt radial dependence of the simplified potentials displayed in Figure 17-12 and 17-13. In a subsequent section we shall see that meson theory predicts something about the behavior of the potentials for relatively large radii, and that it shows the onset of the attractive region should actually be fairly gradual.

## 17-3 Isospin

Figure 17-14 shows schematically the lowest energy levels for the three possible two nucleon systems: the dineutron ${}^0n^2$; the deuteron ${}^1H^2$; and the diproton ${}^2He^2$. The exclusion principle allows only the deuteron to have a triplet spin level, labeled $s = 1$, and because of the spin dependence of the nucleon force only this level is at a low enough energy to be bound. But all three systems have a slightly unbound singlet spin level, labeled $s = 0$. Because of the charge independence of the nucleon force,

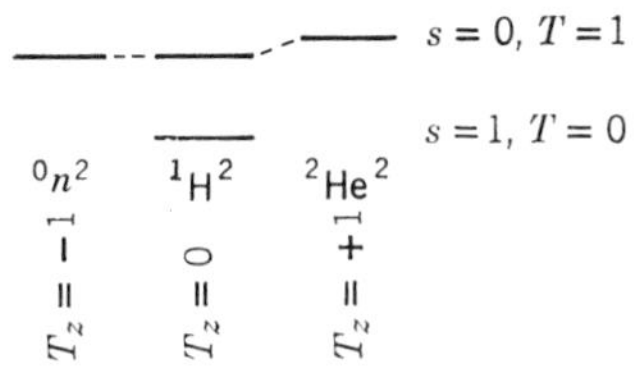

**FIGURE 17-14**
Illustrating the pattern formed by the lowest energy levels of the three possible two-nucleon systems.

the $s = 0$ level is at the same energy in all of the systems, except for the small effect of the Coulomb repulsion energy that is present in the diproton only. The symmetry that is apparent in this set of energy-level diagrams, and that is even more apparent in other sets we shall consider later, can be described in a very convenient way by means of the concept of *isospin*, $T$.

As its name implies, isospin has mathematical properties that are similar to those we have become familiar with in dealing with spin. But it has no direct physical relationship to spin. It is used to identify related energy levels, or quantum states, in sets of *isobars*; i.e., in sets of systems that all have the same number $A$ of nucleons. For the set shown in Figure 17-14, the lowest level is said to be an isospin singlet, labeled $T = 0$, and the three related levels are said to form an isospin triplet, labeled $T = 1$. The word triplet is appropriate because there are three related levels, and because associated with $T$ is a component, written $T_z$, that can assume the three values $T_z = -1, 0, +1$ when $T = 1$. The component $T_z$ is used to identify a particular level of an isospin multiplet by specifying the relation between the number $Z$ of protons and the number $N$ of neutrons for the particular isobar that the level belongs to. The relation is

$$T_z = \frac{Z - N}{2} \tag{17-4}$$

In Figure 17-14 the three $T = 1$ levels are labeled by $T_z = (0 - 2)/2 = -1$ for the dineutron, $T_z = (1 - 1)/2 = 0$ for the deuteron, and $T_z = (2 - 0)/2 = +1$ for the diproton. For the isospin singlet level, $T = 0$, there is only one possible value of $T_z$, namely the value $T_z = 0$ corresponding to the deuteron.

In general, the relation between the value of $T$ and the possible values of $T_z$ is

$$T_z = -T, -T + 1, \ldots, +T - 1, +T \tag{17-5}$$

This is, of course, very analogous to the mathematical relation between the quantum number describing any angular momentum vector, including the spin vector, and the possible values of the quantum number describing its $z$ component. It should be emphasized, however, that isospin is *not* a vector in any physical space, with a component along a coordinate axis of that space. Instead it is a mathematical construct that exists only in some imagined space. It is, nevertheless, very useful in describing the symmetrical properties of systems containing the same number of nucleons, which result from the symmetrical way the exclusion principle treats identical nucleons of either species, and the symmetrical way the charge independent nucleon force treats all nucleons.

A system containing a single nucleon has $T = 1/2$, with the two possible values of $T_z$ being $T_z = -1/2, +1/2$. According to (17-4) the first possibility describes the neutron for which $(Z - N)/2 = (0 - 1)/2 = -1/2$, and the second describes the proton for which $(Z - N)/2 = (1 - 0)/2 = +1/2$. Thus isospin allows us to speak of the neutron and proton as two related manifestations of the same particle, the $T = 1/2$ *nucleon*. In one, called the neutron, $T_z = -1/2$; in the other, called the proton, $T_z = +1/2$. This is like saying that a proton with spin "up" is the $m_s = +1/2$ manifestation of the $s = 1/2$ proton, and the proton with spin "down" is the $m_s = -1/2$

manifestation of that particle. From this point of view the quantum mechanical label exchange properties of a system containing several nucleons may be expressed in a very general way by saying that if the total eigenfunction for the system is a product of a space eigenfunction, a spin eigenfunction, and an isospin eigenfunction, the symmetry of each in an exchange of any two particle labels must be such as to make the total eigenfunction be antisymmetric because nucleons are fermions. As applied to the two nucleon system levels of Figure 17-14, since for all of these levels $l = 0$, all of the corresponding states have symmetric space eigenfunctions. So for each of them a symmetric spin eigenfunction must be associated with an antisymmetric isospin eigenfunction, or vice versa. Because of their analogous mathematical properties, for both spin and isospin, a singlet state is described by an antisymmetric eigenfunction and a triplet state is described by a symmetric eigenfunction. Thus levels of singlet spin ($s = 0$) should have triplet isospin ($T = 1$), and the level of triplet spin ($s = 1$) should have singlet isospin ($T = 0$), as inspection of the figure will demonstrate to be the case.

The power of isospin in identifying related quantum states in sets of systems containing a large number of nucleons is shown in Figure 17-15. The figure shows schematically some low-lying energy levels of the set of isobars ${}^5B^{14}$, ${}^6C^{14}$, ${}^7N^{14}$, ${}^8O^{14}$, and ${}^9F^{14}$. The so-called *isobaric analogue* levels of a particular isospin multiplet are labeled by $T$ and $T_z$ as before. Except for the small systematic increase in their energies with increasing $T_z$, due to the increase in the Coulomb repulsion energy with increasing $Z$, all isobaric analogue levels have the same energy. The reason is that the corresponding total eigenfunctions of each system are all identical solutions (if we ignore Coulomb effects) to a Schroedinger equation for the same nucleon forces, since the nucleon force does not depend on $T_z$ as it is charge independent. But the nucleon force does depend on $T$ as it is spin dependent. We first learned of this as a dependence on the spin; we now realize that the label exchange requirements mean it is also an isospin dependence. The nature of the spin dependence is such as to make the state of lowest $T$ have the lowest possible energy level for the set of systems. This can be seen in both Figure 17-15 and in Figure 17-14.

The statement that energies resulting from the nucleon force, or interaction, do not depend on $T_z$ but only on $T$ is consistent with the statement that the isospin $T$ is conserved in processes involving this interaction. To see this, compare the statement that the total angular momentum $J$ is conserved in processes involving a spherically symmetrical interaction $V(r)$, with the statement that energies resulting from this interaction do not depend on its component $J_z$ but only on its magnitude $J$. However,

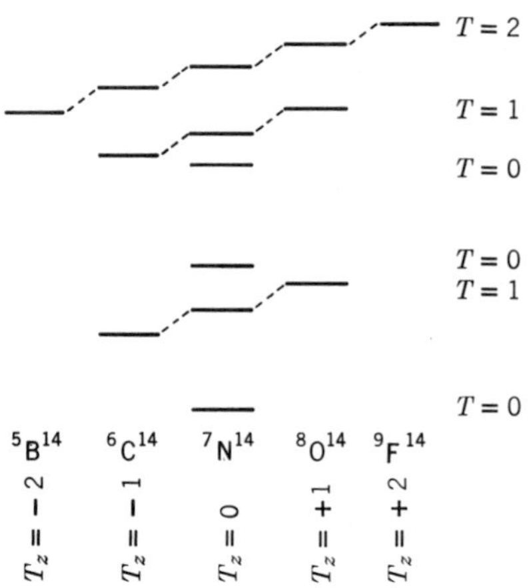

**FIGURE 17-15**

The low-lying energy levels of the $A = 14$ isobars. Note that the positions of the ground state energy levels trace out the parabolas, for the ground state masses of the $A = 14$ nuclei, that are discussed in connection with $\beta$ decay.

the conclusion that isospin is conserved in the nucleon interaction is of greater generality than the conclusion, based on the charge independence experiments, that the nucleon interaction depends on $T$ but not $T_z$. So it requires additional experimental verification. We shall see that particle physics provides much verifying evidence. Evidence from nuclear physics is found, for example, in the reaction

$$^1H^2 + {}^8O^{16} \rightarrow {}^7N^{14} + {}^2He^4$$

In all experimental situations, the incident and target nuclei $^1H^2$ and $^8O^{16}$ are in their ground states. If the bombarding energy of the incident nucleus is not too high, the product nucleus $^2He^4$ must also be in its ground state because its first excited state lies at an energy above 20 MeV. All three of these nuclei have $T_z = 0$ in all states, and in their ground states they have the lowest value of $T$ consistent with this $T_z$, namely $T = 0$. The same is true for the ground state of the residual nucleus $^7N^{14}$. But, as we see in Figure 17-15, the first excited state of $^7N^{14}$ has $T = 1$. As far as the conservation of energy, angular momentum, or parity is concerned, the reaction could produce $^7N^{14}$ in either its ground or its first excited state. The experimental observation that it is produced only in the ground state provides strong evidence for the conclusion that *the nucleon interaction conserves the isospin T.*

## 17-4 Pions

In preceding sections we presented a description of properties of nucleon forces that are observed in experiment. Although theory was used in the description, it was used essentially to correlate the experimental observations, and not to explain their basic origin. But there is a theory that is successful in explaining how certain properties of nucleon forces arise from more fundamental attributes of nature. This is the *meson theory*, which originated with the work of Yukawa in 1935.

Yukawa proposed that a nucleon frequently emits a particle with an appreciable rest mass, now called a *$\pi$ meson* or *pion*. This particle hovers near the nucleon in the so-called *$\pi$-meson field* for a very short time, and then is absorbed by the nucleon. During the process the nucleon maintains its normal rest mass, and so while it is happening there is a violation of the law of mass-energy conservation because there is more rest mass present than there is before the $\pi$ meson is emitted or after it is absorbed. The energy-time uncertainty principle shows, however, that such a violation is not impossible if it lasts for a sufficiently short time. Of course, the $\pi$ meson cannot permanently escape the nucleon because that would permanently violate the mass-energy conservation law. However, if two nucleons are close enough for their meson fields to overlap, it is possible for a $\pi$ meson to leave one field and join the other, without permanently changing the total energy of the system of two nucleons. Such an interaction between the fields is pictured crudely in Figure 17-16. In the interaction, the momentum carried by the $\pi$ meson is transferred from one field to the other, and therefore from one nucleon to the other. But if momentum is transferred, the effect is the same as if a force is acting between the nucleons. Thus the *exchange* of a pion between two nucleons leads to the nucleon force acting between them, according to Yukawa. (We came across a similar idea before when discussing, in Section 14-1, the exchange of a phonon between two electrons in a Cooper pair.)

In making his proposal, Yukawa was guided by two analogies available to him at the time. One is the covalent binding in the $H_2$ molecule and other organic molecules (discussed in Section 12-3). In this process, a force arises from the sharing, or exchange, of an electron between two atoms. An even closer analogy is the Coulomb force acting between two charged particles. According to the very successful theory of

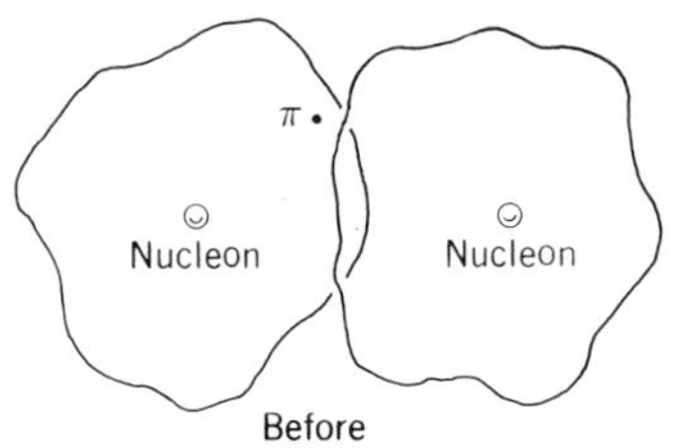


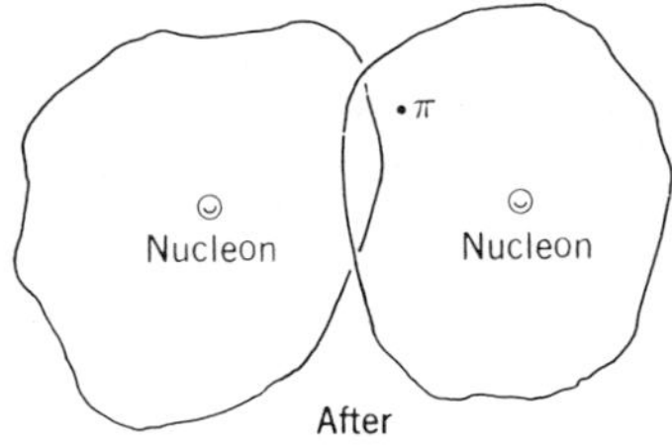


**FIGURE 17-16**

A very crude representation of the exchange of a $\pi$ meson between the fields of two interacting nucleons.

quantum electrodynamics (mentioned in Section 8-7), surrounding each charge is a field of photons, and the Coulomb force actually results from an exchange of a photon between the fields.

Quantum electrodynamics shows that the long range of the Coulomb force is a consequence of the fact that photons have zero rest mass. Yukawa adapted the theory to the case of two nucleons, interacting with a short range nucleon force, by assuming that the particle exchanged has a nonzero rest mass. When he made his proposal, pions had not yet been detected, but Yukawa was able to estimate the rest mass that would lead to the observed range by performing a calculation similar to the one in the following example.

**Example 17-3.** Use energy conservation, as modified by the energy-time uncertainty principle, to establish a relation between the range $r'$ of the nucleon force and the rest mass $m_\pi$ of the $\pi$ meson whose exchange produces the force. Then use the relation to estimate the value of $m_\pi$, assuming $r' = 2$ F.

The range of the nucleon force is of the order of the radius $r'$ of the $\pi$-meson field surrounding a nucleon, since two nucleons experience that force only when their meson fields overlap. To estimate the radius of the field, consider a process in which a nucleon emits a meson of rest mass $m_\pi$, which travels out to the limits of the field, and then returns to the nucleon where it is absorbed. In this process, the $\pi$ meson travels a distance of the order of $r'$. While it is happening there is a violation of the conservation of mass-energy. The reason is that the total energy of the system equals one nucleon rest mass energy before and after the process, and one nucleon rest mass energy plus at least one $\pi$-meson rest mass energy during the process. But the energy-time uncertainty principle shows that a violation of energy conservation by an amount

$$\Delta E \sim m_\pi c^2$$

is not impossible if it does not happen for a time longer than $\Delta t$, where

$$\Delta E\, \Delta t \sim \hbar$$

The reason is that such a violation could not be detected because the energy cannot be measured in a time $\Delta t$ more accurately than $\Delta E$. Since the speed of the pion can be no greater than $c$, the time required for it to travel a distance of the order of $r'$ is at least

$$\Delta t \sim \frac{r'}{c}$$

These three relations give

$$m_\pi c^2 \sim \frac{\hbar}{\Delta t} \sim \frac{\hbar c}{r'}$$

or

$$m_\pi \sim \frac{\hbar}{r'c} \tag{17-6}$$

If we take $r' = 2$ F, (17-6) gives us an estimate of the $\pi$-meson rest mass

$$m_\pi \sim \frac{\hbar}{r'c} \sim \frac{1 \times 10^{-34}\text{ joule-sec}}{2 \times 10^{-15}\text{ m} \times 3 \times 10^{8}\text{ m/sec}} \sim 2 \times 10^{-28}\text{ kg}$$

This can also be written

$$m_\pi \sim 200\, m \sim 100\text{ MeV}/c^2$$

where $m$ is the rest mass of an electron which has the value $m = 0.511$ MeV/$c^2$. ◀

It is worthwhile restating the argument used in Example 17-3. A meson of rest mass $m_\pi \sim \hbar/r'c$ leads to a nucleon force of range $\sim r'$ because the nucleons could not exchange the meson if they were separated by a much larger distance, since its flight time would be so long that the uncertainty principle would allow an accurate enough determination of the total energy of the system to make the violation of energy conservation detectable. This argument also explains how the Coulomb force can have a long range. Since a photon has zero rest mass, there is no lower limit to the total energy it can carry. When two charged particles are separated by a very large distance, they can exchange a photon of very low kinetic energy without violating the energy-time uncertainty principle. Of course, such a photon will carry very low linear momentum. Therefore, the force it produces is very weak, in agreement with the well-known decrease in the strength of the Coulomb force as the separation of the charged particles increases.

At the time of Yukawa's proposal, there were no known particles of rest mass between the electron rest mass 0.5 MeV/$c^2$ and the proton rest mass which equals 938 MeV/$c^2$. The $\pi^+$ mesons, which have a positive charge equal in magnitude to that of the electron, and the $\pi^-$ mesons, which have a negative charge of the same magnitude, were first detected in 1947 by Powell and collaborators. They were found as a component of the cosmic radiation, which is constantly bombarding the earth. Shortly after, the charged $\pi$ mesons were produced artificially at a large cyclotron in collisions between nucleons of very high energy and nucleons in a target. Cosmic radiation mesons are also initially produced in high energy collisions. Measurements show that the $\pi^+$ and $\pi^-$ mesons have the same rest mass

$$m_{\pi^+} = m_{\pi^-} = 140\text{ MeV}/c^2 \tag{17-7}$$

This is certainly close enough to Yukawa's prediction $m_\pi \sim 100$ MeV/$c^2$. Neutral $\pi^0$ mesons were first observed by Moyer and coworkers in 1950, as products of high-energy collisions. Their rest mass is found to be

$$m_{\pi^0} = 135\text{ MeV}/c^2 \tag{17-8}$$

The free $\pi$ mesons, which are observed in these experiments, are liberated from the $\pi$-meson fields surrounding the colliding nucleons by the energy made available in the collision. They are the same particles as the mesons discussed in the meson theory of nucleon forces. The only difference is that Yukawa's mesons are bound before the nucleons interact by requirements of energy conservation. As is obviously true of the bound pions that produce the strong force between two nucleons, the interaction of free pions with nucleons is strong. This was indicated in various ways in the early experiments with cosmic ray and cyclotron pions, which showed that the cross section for interaction of a short de Broglie wavelength pion with a nucleus is close to its maximum possible value, the projected geometrical cross-sectional area $\pi r'^2$, the quantity $r'$ being the nuclear radius. The interaction is also particularly violent; when a pion enters a nucleus most of its rest mass energy goes into splitting the nucleus into fragments which fly apart energetically. Of course, the detection of free pions provided a striking verification of the validity of the meson theory.

Experimental evidence for the exchange of pions between two interacting nucleons is found in neutron-proton scattering. As we discussed in a preceding section, the approximate symmetry about 90° of the scattering differential cross section implies that in about half the scatterings the neutron changes into a proton and the proton changes into a neutron, when the nucleons interact. One way this can happen is indicated by the set of reactions

$$n \rightarrow p + \pi^- \qquad \text{then} \qquad \pi^- + p \rightarrow n$$

That is, the neutron emits a negatively charged $\pi^-$ meson into its field, becoming a proton. Then the $\pi^-$ meson joins the field of the proton, and it is absorbed by the proton which becomes a neutron. The scattering process can also happen through the set of reactions

$$p \rightarrow n + \pi^+ \qquad \text{then} \qquad \pi^+ + n \rightarrow p$$

In this case the proton emits a positively charged $\pi^+$ meson, which is subsequently absorbed by the neutron. Thus, in about half the neutron-proton scatterings a meson transfers charge as well as momentum between the two interacting nucleons.

Because the neutron-proton scattering differential cross section is approximately symmetric about 90°, in about half the scatterings the neutron and proton do not exchange identities when they interact. But they still must exchange a meson which carries the transferred momentum. The two sets of reactions which occur are

$$n \rightarrow n + \pi^0 \qquad \text{then} \qquad \pi^0 + p \rightarrow p$$

and

$$p \rightarrow p + \pi^0 \qquad \text{then} \qquad \pi^0 + n \rightarrow n$$

The neutral $\pi^0$ meson transfers momentum, but no charge, between the interacting nucleons.

This interpretation implies that an isolated proton should be surrounded by a meson field which will sometimes contain a $\pi^0$ meson and sometimes contain a $\pi^+$ meson. The reactions that take place when the meson is emitted by the nucleon are

$$p \rightarrow p + \pi^0 \qquad \text{or} \qquad p \rightarrow n + \pi^+$$

Of course the nucleon must absorb the meson it has emitted within a very short time, but then it can emit another one. The meson field surrounding an isolated neutron should sometimes contain a $\pi^0$ meson and sometimes contain a $\pi^-$ meson, which are emitted through the reactions

$$n \rightarrow n + \pi^0 \qquad \text{or} \qquad n \rightarrow p + \pi^-$$

But the proton field cannot contain a $\pi^-$ meson and the neutron field cannot contain a $\pi^+$ meson. Very direct experimental verification of these predictions is provided by recent electron scattering measurements of the charge distribution of the proton and of the neutron. Figure 17-17 shows the radial dependence of the charge densities of the two species of nucleons. The charge density of the proton is everywhere positive, and extends out to a distance $r$ of about 2 F. At the larger $r$ within this limit (in the field) the charge is carried by a $\pi^+$ meson. The neutron charge density is *not* everywhere zero. At smaller $r$ (near the center where the $p$ from $p + \pi^-$ dissociation would be) it is positive, and at larger $r$ (in the field where the $\pi^-$ would be) it is negative. The volume integral of the charge density is, however, zero, since the neutron is neutral and so has no net charge.

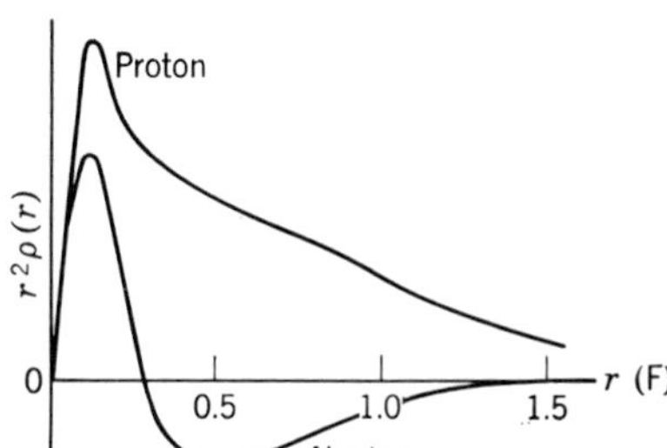


**FIGURE 17-17**
The radial dependence of the charge density of the proton and of the neutron.

Meson theory also provides an explanation of how the neutron can have an intrinsic magnetic dipole moment, even though its net charge is zero. It sometimes becomes a proton plus a $\pi^-$. The proton has an intrinsic magnetic dipole moment, and the $\pi^-$ meson can produce a current which makes an additional contribution to the magnetic dipole moment.

At values of $r$ approaching 2 F, the nucleon charge densities are proportional to some measure of the intensity of their meson fields. Both are decreasing fairly gradually as $r$ increases. The nucleon force, which acts between two nucleons when their meson fields overlap, also therefore decreases fairly gradually as their separation increases. Thus the onset of the attractive part of the nucleon potential, describing the nucleon force acting when the two nucleons are beginning to get close enough to interact, is fairly gradual. It is not abrupt as in the simplified nucleon potential of Figures 17-12 and 17-13. In fact, we shall indicate in Example 17-4 that for large values of the separation distance $r$ the nucleon potential should follow the *Yukawa potential*

$$V(r) = -g^2 \frac{e^{-r/r'}}{r} \tag{17-9}$$

where

$$r' = \frac{\hbar}{m_\pi c} \simeq 1.5 \text{ F}$$

The range $r'$ of the potential is specified by the theory to have a value which agrees with the simple argument of Example 17-3, and with experiment. The over-all strength of the potential depends on the constant $g^2$, whose value is not determined by the theory but can be by finding the value of $g^2$ that gives best agreement with experiment. In terms of the dimensionless quantity $g^2/\hbar c$, the value so determined is

$$g^2/\hbar c \simeq 15 \tag{17-10}$$

Figure 17-18 plots the Yukawa potential. Note that $V(r) \propto e^{-r/r'}/r$ decreases in

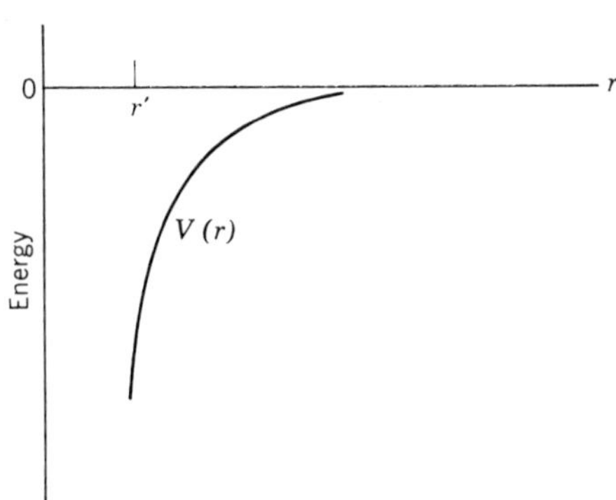


**FIGURE 17-18**
The Yukawa potential. For $r$ comparable to or larger than $r' = \hbar/m_\pi c$ $\simeq$1.5 F, the nucleon potential should have this form.

magnitude with increasing $r$ fairly gradually, but the decrease is very much more rapid than that of the long range Coulomb potential $V(r) \propto 1/r$.

At values of $r$ small compared to 2 F, the nucleon potential deviates markedly from the Yukawa potential. In fact, we know it becomes repulsive at $\sim$0.5 F. It is believed that the repulsive core to the potential arises from the exchange of mesons that we shall meet later, whose rest masses are considerably larger than that of the $\pi$ meson. The details of the origin of the repulsive core are not yet completely understood.

**Example 17-4.** Write a relativistic wave equation for $\pi$ mesons, and then show how the Yukawa potential, (17-9), can be obtained from that equation.

A relativistic wave equation for $\pi$ mesons can be obtained by writing the relativistic energy equation

$$E^2 = c^2p^2 + m_\pi^2 c^4$$

where

$$p^2 = p_x^2 + p_y^2 + p_z^2$$

replacing the total energy and the momentum components by the associated operators of (5-32)

$$E \rightarrow i\hbar\, \partial/\partial t \qquad p_x \rightarrow -i\hbar\, \partial/\partial x \qquad p_y \rightarrow -i\hbar\, \partial/\partial y \qquad p_z \rightarrow -i\hbar\, \partial/\partial z$$

and then allowing the operator equation thereby obtained to operate on the function $\Psi$. The result is

$$-\hbar^2 \frac{\partial^2 \Psi}{\partial t^2} = -c^2\hbar^2\left(\frac{\partial^2\Psi}{\partial x^2} + \frac{\partial^2\Psi}{\partial y^2} + \frac{\partial^2\Psi}{\partial z^2}\right) + m_\pi^2 c^4 \Psi$$

or

$$\nabla^2\Psi - \frac{1}{c^2}\frac{\partial^2\Psi}{\partial t^2} = \frac{m_\pi^2 c^2}{\hbar^2}\Psi$$

which is called the *Klein-Gordon equation*. It plays an important role in the quantum electrodynamics of bosons. For instance, for $m_\pi = 0$ it reduces to the classical wave equation

$$\nabla^2\Psi = \frac{1}{c^2}\frac{\partial^2\Psi}{\partial t^2}$$

for photons, the quanta of the electromagnetic field.

The classical wave equation has a static solution of the form

$$\Psi = -\frac{e^2}{4\pi\epsilon_0}\frac{1}{r} \qquad r > 0$$

as can easily be verified by substitution, using the relation

$$\nabla^2\Psi = \frac{1}{r^2}\frac{d}{dr}\left(r^2\frac{d\Psi}{dr}\right)$$

for $\Psi = \Psi(r)$. For $m_\pi \neq 0$ the Klein-Gordon equation has a static solution of the form

$$\Psi = -g^2\frac{e^{-r/r'}}{r} \qquad r > 0$$

where

$$r' = \frac{\hbar}{m_\pi c}$$

as can also easily be verified by substitution. Since the solution to the wave equation for zero rest mass quanta gives the Coulomb interaction potential for the electromagnetic field, the solution for nonzero rest mass quanta is assumed to be the interaction potential for the meson field, that is, the Yukawa potential of (17-9).

The constant $g^2$ determines the strength of the Yukawa potential, just as the constant $e^2$ (the square of the electron charge) determines the strength of the Coulomb potential. Note that the dimensionless quantity $g^2/\hbar c$ has the value $\simeq 15$, whereas the dimensionless quantity $e^2/4\pi\epsilon_0\hbar c$ (the fine-structure constant) has the value $\simeq 1/137$. This is an indication of the strength of the nucleon force. ◀

Single free pions can be created in high-energy collisions between nucleons, e.g.

$$p + p \rightarrow \pi^+ + d \tag{17-11}$$

where $d$ is the deuteron, or destroyed in collisions between pions and nucleons, e.g.

$$\pi^+ + d \rightarrow p + p \tag{17-12}$$

From this we can immediately conclude that pions cannot be fermions. The reason is that the number of fermions in an isolated system always remains constant, in the sense that if a fermion is produced, or destroyed, it always happens in conjunction with the production, or destruction, of an antifermion. Examples are electron pair production, or annihilation. Pions are bosons, just as photons are bosons, that can be emitted or absorbed singly. As bosons, pions must have integral spin; that is $s = 0$, or 1, or 2, . . . . Measurements show that for all three cases, $\pi^-$, $\pi^0$, and $\pi^+$, *the pion spin is* 0. The first of these measurements involved applying the principle of detailed balancing (see the discussion of (11-4)) to the observed ratio of the cross sections for the forward and backward reactions of (17-11) and (17-12). The value of the $\pi^+$ spin influences the cross section for the forward reaction because the reaction rate is proportional to the density of states that can be populated, and this is proportional to the spin degeneracy factor $(2s + 1)$. The cross-section ratio showed that $s = 0$.

A very interesting property of pions is that *pions have odd intrinsic parity*. The initial evidence came from the reaction

$$\pi^- + d \rightarrow n + n \tag{17-13}$$

The negatively charged pion is captured by the deuteron after dropping through a sequence of atomic electronlike states to the $l = 0$ state, where its wave function has a large overlap with the deuteron. Thus the total angular momentum on the left of (17-13) is that of the spin 1 ground state of the deuteron. So angular momentum conservation allows the two neutrons to be emitted either with total orbital angular momentum $l = 0$ or 2 and "parallel" spins, or with $l = 1$ and "antiparallel" spins. The first possibilities are ruled out because they would result in a symmetric total eigenfunction for the system of two fermions. Therefore the neutrons are emitted in a state in which the total orbital angular momentum is $l = 1$. The parity of such a state is odd, according to the usual rule that parity is governed by $(-1)^l$. Therefore, since parity is conserved by the nuclear, or nucleon, interaction, the parity of the system $\pi^- + d$ must be odd. Now the parity of the ground state of the deuteron is even, and the $(-1)^l$ rule says the parity associated with the $l = 0$ motion of the captured $\pi^-$ is also even. Thus the $\pi^-$ meson must have an intrinsic parity, which is odd. The same is true of the other pions. As the number of nucleons present is unchanged in the reaction, or any other reaction, the intrinsic parity of a nucleon is undetermined; but it is conventionally taken to be even.

The triplet of pions have similar masses, identical quantum numbers, and participate equally in the nucleon interaction. It is therefore natural to say that *the pion is an isospin* $T = 1$ *particle*, that has a $T_z = -1$ manifestation called the $\pi^-$, a $T_z = 0$ manifestation, the $\pi^0$, and a $T_z = +1$ manifestation, the $\pi^+$. In so doing we are generalizing the relation between $T_z$ and electric charge. The form that we originally used for nucleons, (17-4), is equivalent to the relation

$$Q = T_z + 1/2 \qquad \text{(nucleons)} \qquad (17\text{-}14a)$$

where $Q$ is the charge in units of the magnitude of the electron charge. For example, this yields $Q = 0$ for the $T_z = -1/2$ neutron and $Q = 1$ for the $T_z = +1/2$ proton, as before. For pions the relation is different, since

$$Q = T_z \qquad \text{(pions)} \qquad (17\text{-}14b)$$

However, we may incorporate both of these relations into one form by writing

$$Q = T_z + B/2 \qquad \text{(nucleons and pions)} \qquad (17\text{-}15)$$

where $B$, called the *baryon number*, has the value 1 for a nucleon and 0 for a pion.

Pions are unstable. The $\pi^0$ decays spontaneously by an electromagnetic interaction with a lifetime of about $10^{-15}$ sec into two high-energy photons

$$\pi^0 \rightarrow \gamma + \gamma \qquad (17\text{-}16)$$

or else, rarely, into an electron-positron pair and one photon. Although this sounds like a very short decay time, it should be compared to the time $10^{-23}$ sec that would characterize the decay if it took place through the strong nucleon (or nuclear) interaction. The value $10^{-23}$ sec is just the time that particles moving with relative velocity $c \sim 10^8$ m/sec would overlap within a distance of the range of nucleon forces $r' \sim 10^{-15}$ m. The facts first used to identify the electromagnetic nature of the $\pi^0$ decay are that photons participate only in the electromagnetic interaction and that the decay lifetime is much longer than the time $10^{-23}$ sec that would suffice if it could go by the stronger interaction.

The other pions do not decay in the same ways as the neutral pion. Instead, the $\pi^+$ decays with the even longer lifetime of about $10^{-8}$ sec, according to the scheme

$$\pi^+ \rightarrow \mu^+ + \nu_\mu \qquad (17\text{-}17)$$

where $\mu^+$ represents the positively charged *muon*, and $\nu_\mu$ is the *muonic neutrino*. The $\pi^-$ decays with the same lifetime according to the scheme

$$\pi^- \rightarrow \mu^- + \bar{\nu}_\mu \qquad (17\text{-}18)$$

where $\mu^-$ is the negatively charged muon, and $\bar{\nu}_\mu$ is the muonic antineutrino. The positive muon is the antiparticle of the negative muon, just as the positron is the antiparticle of the electron. In fact, in essentially every regard, except for their higher rest mass, muons are like electrons. The charged pion decays involve an interaction which is one of two related cases. As might be guessed from the terminology used, the other case is the $\beta$-decay interaction of nuclear physics. The fact that the lifetime of charged pion decay is much longer than for electromagnetic decay of the neutral pion is a reflection of the fact that the interaction involved in the decay is much weaker than the electromagnetic interaction. The student will recall that we made a similar comparison in the case of $\beta$ decay. For these reasons, both the decay of a neutron into a proton plus an electron and (what we now call) an electronic antineutrino, and the decay of a positive or negative pion into a positive or negative muon and a muonic neutrino or antineutrino, are said to take place via the *weak interaction*. This terminology leads to the nucleon interaction being called the *strong interaction*. Particularly in particle physics, the terms strong interaction and weak interaction are used to identify what are usually called the nucleon (or nuclear) interaction and the $\beta$-decay interaction in nuclear physics.

## 17-5 Muons

Muons have no part in Yukawa's theory of the origin of the strong interaction, although this was not appreciated until some time after their discovery in 1936 by Anderson and Neddermeyer. These investigators found the particles as components of the cosmic radiation, and they showed that their rest mass is intermediate between the rest mass of an electron and the rest mass of a proton. We now know that they are produced in cosmic radiation from the decay of pions. But, in 1936, pions had not been discovered, and it was naturally assumed that the $\mu^+$ and $\mu^-$ were Yukawa's mesons (in fact they were originally called $\mu$ mesons). An ever increasing accumulation of evidence showed, however, that the interaction of muons with matter is very weak. For instance, the muons in cosmic radiation can penetrate great thicknesses of solid matter with little attenuation, since they can be detected in deep mines. This being the case, muons can hardly be the particles responsible for the strong interaction, despite the fact that their rest mass

$$m_{\mu+} = m_{\mu-} = 106 \text{ MeV}/c^2 \tag{17-19}$$

is quite close to the value predicted by Yukawa.

This situation was the source of considerable confusion in the ten years before the discovery of pions, but, after their discovery, it was immediately assumed that pions are Yukawa's mesons since the early evidence indicated that their interaction with matter is strong. Thus pions are closely associated with nucleons and interact via the strong interaction. Muons are closely associated with electrons and interact via the weak interaction.

The muon and electron, the muonic and electronic neutrinos, and the antiparticles of each, are collectively called *leptons*. One of the pieces of evidence for the association between the negative muon and the electron is that both are *fermions*, both have *charge* $-e$ and *spin* 1/2, and both have magnetic dipole moments corresponding to a *spin g factor of* 2. Their antiparticles, the positive muon and the positron, have charges and magnetic dipole moments of reversed signs. Muonic and electronic neutrinos are also *spin* 1/2 *fermions*, but they are uncharged and have no magnetic dipole moments. They are distinguished physically from their antiparticles by their *helicities*, (see Section 16-4) which are *left handed for neutrinos* and *right handed for antineutrinos*. It is not appropriate to define either an intrinsic parity or an isospin for any of these particles which participate in the weak interaction. The reason is that parity is not conserved in that interaction, as we saw in Section 16-4, and isospin is also not conserved in the weak interaction, as we shall see in a subsequent section.

Muons decay spontaneously, via the weak interaction, according to the following schemes

$$\mu^+ \rightarrow e^+ + \nu_e + \bar{\nu}_\mu \tag{17-20}$$

$$\mu^- \rightarrow e^- + \bar{\nu}_e + \nu_\mu \tag{17-21}$$

where we here use $e^+$ for the positron and $e^-$ for the electron. The lifetime for both decays is the same, and it has a value of about $10^{-6}$ sec. The need for a distinction between the electronic neutrino $\nu_e$ and the muonic neutrino $\nu_\mu$ was demonstrated experimentally in 1962, by showing that the muonic neutrinos obtained from pion decay, (17-17) and (17-18), will not induce electronic $\beta$ decay.

Since leptons are fermions, they are created or destroyed in particle, antiparticle pairs. Consequently, the number present in an isolated system will remain constant, if each particle makes a positive contribution to the count and each antiparticle makes a negative contribution. Because of the distinction between electronic and muonic leptons, each type separately satisfies a *lepton number conservation law*. These can be written

$$\sum L_e = \text{const} \tag{17-22}$$

$$\sum L_\mu = \text{const} \tag{17-23}$$

The *electronic lepton number* $L_e$ is $+1$ for an electron and $-1$ for the positron; it is $+1$ for an electronic neutrino and $-1$ for an electronic antineutrino. The *muonic lepton number* $L_\mu$ is $+1$ for a negative muon and $-1$ for a positive muon; it is $+1$ for a muonic neutrino and $-1$ for a muonic antineutrino. The student should note that in all cases the lepton number is $+1$ *for a particle and* $-1$ *for its antiparticle*. He should also note that the muon decay schemes of (17-20) and (17-21) satisfy both conservation laws, as do the electronic $\beta$ decays discussed in Chapter 16. We shall later consider a similar conservation law for the baryon number $B$, which is $+1$ for a nucleon and $-1$ for an antinucleon.

Although much is known about the weak interaction, and the particles that participate in it, some fundamental questions remain unanswered. For instance, why does nature need both the electron (plus the associated neutrino and the antiparticle of each) and *also* the muon (plus the associated neutrino and their antiparticles)? Except for the difference in rest mass, the electron and muon are so similar that each seems to make the other redundant.

Another unanswered question has to do with the field quantum of the weak interaction. The strong interaction arises from exchanges, between two participating particles, of its field quantum (the pion, whose properties are well known, and probably also other mesons to be discussed, whose properties are known reasonably well). The electromagnetic interaction involves exchanges of its field quantum (the photon, whose properties are certainly very well known). Even the gravitational interaction involves exchanges of a field quantum (the graviton, whose most important properties are also known, as we shall mention later). Thus we expect that the weak interaction should be carried by its field quantum. The quantum has a name; it is called the *intermediate boson*. But aside from the fact that it must be a boson (as is true for all field quanta), and probably has spin 1, very little is known about it. The intermediate boson has never yet been detected, although experimental searches for it have been able to show the rest mass of the intermediate boson to be larger than about ten nucleon masses. This lower limit does, however, verify that the weak interaction is of extremely short range. The point is that the inverse relation between the rest mass of the field quantum and the range of the associated interaction, displayed in (17-6), should apply to all four of the fundamental interactions of nature.

## 17-6 Strangeness

Not long after the discovery of pions in cosmic rays, the same source began to provide evidence for the existence of a new family of mesons with about 3 times the pion rest mass, now called *K mesons*, and it also indicated that there is a particle with a rest mass of about 1.2 times the nucleon rest mass, now called the $\Lambda^0$ *particle*. More recent experiments show that the $\Lambda^0$ rest mass is

$$m_{\Lambda^0} = 1116 \text{ MeV}/c^2 \tag{17-24}$$

which may be compared with the neutron and proton rest masses of 940 and 938 MeV$/c^2$. Like the neutron, *the* $\Lambda^0$ *particle is a neutral spin* 1/2 *fermion of even intrinsic parity*. Because there is no other particle of similar rest mass, the $\Lambda^0$ is considered to be the only member of an isospin singlet; i.e., the $\Lambda^0$ *has* $T = 0$, $T_z = 0$.

Experiments using high-energy accelerators have shown that there are four $K$ mesons, the positively and negatively charged $K^+$ and $K^-$, and the neutral $K^0$ and $\overline{K^0}$. Like the $\pi$ mesons, *the K mesons are all spin* 0 *bosons of odd intrinsic parity*.

Their rest masses are

$$m_{K^+} = m_{K^-} = 494 \text{ MeV}/c^2 \tag{17-25}$$

and

$$m_{K^0} = m_{\overline{K^0}} = 498 \text{ MeV}/c^2 \tag{17-26}$$

The $\Lambda^0$ is produced in association with a $K$, with a large production cross section indicative of the strong interaction. An example is the reaction

$$\pi^- + p \rightarrow \Lambda^0 + K^0 \tag{17-27}$$

Let us use the assumption that, as in nuclear physics, isospin is conserved in the strong interaction causing the reaction to assign isospin quantum numbers to the $K$ mesons. Since $T = 1$ for the $\pi^-$, $T = 1/2$ for the proton $p$, and $T = 0$ for the $\Lambda^0$, the only possibilities for the $K^0$ are $T = 1/2$ or $T = 3/2$. If the latter were true, there would be a quartet of $T_z$ values and the $K$ meson family would have to span a range of four different electric charge states. But, in fact, there are only three charge states: $Q = -1$, 0, +1. Therefore $T = 1/2$ *for the* $K^0$ *and the other* $K$ *mesons*. Note also that since $T_z$ has the values $-1$ for the $\pi^-$, $+1/2$ for the $p$, and 0 for the $\Lambda^0$, it must have the value $T_z = -1/2$ *for the* $K^0$. In consideration of the way $Q$ depends on $T_z$ in other situations, we naturally say that the $K$ meson with $T = 1/2$, $T_z = +1/2$ *is the* $K^+$. The $K^-$ is the antiparticle of the $K^+$, and the $\overline{K^0}$ is the antiparticle of the $K^0$. Both of the antiparticles also have $T = 1/2$. Their $T_z$ assignments depend on the fact that $T_z$ *for an antiparticle must be opposite to* $T_z$ *for the corresponding particle*. This can be seen most easily by considering the proton, antiproton pair production reaction, which goes by the strong interaction

$$p + p \rightarrow p + p + p + \bar{p} \tag{17-28}$$

Thus $T_z = -1/2$ *for the* $K^-$ *and* $T_z = +1/2$ *for the* $\overline{K^0}$.

The reaction of (17-28) is also a very good example of the *baryon number conservation law*

$$\sum B = \text{const} \tag{17-29}$$

where the baryon number $B$ has the value +1 for a nucleon and $-1$ for an antinucleon. Because nucleons are fermions, their total number in an isolated system will, in all circumstances, remain constant if the counting is done in the way indicated. The conservation law also applies to situations in which the system may contain particles other than nucleons, to be introduced soon, that are also called baryons.

$K$ mesons decay through the weak interaction with a number of different decay modes into pions and, in some modes, leptons. The lifetime of the $K^+$ and $K^-$ is about $10^{-8}$ sec. The $K^0$ and $\overline{K^0}$ have an unusual decay time distribution comprised of an equal mixture of an exponential with a lifetime of about $10^{-10}$ sec and an exponential with a longer lifetime of about $10^{-8}$ sec. The presence of two lifetimes has a very interesting origin. Basically it arises from interferences between degenerate eigenfunctions describing two equal mass particles, the $K^0$ and $\overline{K^0}$, that participate in the same way in the weak interaction decay, and yet are distinctly different particles. We shall cast a little more light on this phenomenon later in this section, and we shall discuss a very striking conclusion drawn from it in the next section.

All the $K$ mesons have the following decay mode

$$K \rightarrow \pi + \pi \tag{17-30a}$$

Angular momentum conservation will immediately show that, since both the $K$ and $\pi$ mesons have zero spin, the two $\pi$'s must be emitted in a state of zero orbital angular momentum. This is the most frequent decay mode for the $K^0$ and $\overline{K^0}$, and the second most frequent for the $K^+$ and $K^-$. A less frequent decay mode is

$$K \rightarrow \pi + \pi + \pi \tag{17-30b}$$

Since the pions take away most of the available energy in their rest masses, they are emitted with small linear momenta and, because they are emitted from a region of small dimensions, they all come off in a state of zero orbital angular momentum.

The relatively long decay lifetimes certainly indicate that the $K$ decays involve the weak interaction. This is made even more apparent by considering parity balance in the two decay modes specified in (17-30a) and (17-30b). Since both lead to states of zero orbital angular momentum, which have even parity, the only parities that must be considered are the intrinsic parities of the particles. As we have argued in connection with a reaction involving the strong interaction, (17-13), the $\pi$ meson has odd intrinsic parity. Similar arguments show that the $K$ meson also has odd intrinsic parity. Because the parity of the eigenfunction for a system is the product of the parities of its components, we immediately see that parity is conserved in the three-pion decay mode of (17-30b), but it is not conserved in the two-pion decay mode of (17-30a). Thus parity conservation can be violated in $K$ decay just as it can be violated in $\beta$ decay, because both kinds of decay involve the parity nonconserving weak interaction.

The $\Lambda^0$ is also an unstable particle, with a lifetime of about $10^{-10}$ sec. Its principal decay modes are

$$\Lambda^0 \rightarrow p + \pi^- \tag{17-31}$$

and

$$\Lambda^0 \rightarrow n + \pi^0 \tag{17-32}$$

These decays take place through the weak interaction and violate parity conservation. Also the $z$ component of isospin is not conserved. In both decays, $T_z = 0$ before while $T_z = -1/2$ after, since $T_z$ is $+1/2$ for $p$, $-1/2$ for $n$, $-1$ for $\pi^-$, and 0 for $\pi^0$. Isospin itself is not conserved in these weak interaction decays either since $T = 0$ before, while after it could be only $T = 1/2$ or $T = 3/2$, since $T = 1/2$ for the nucleon and $T = 1$ for the pion. Detailed consideration of the decay rates shows that in both decays the pion-nucleon system is formed in the $T = 1/2$ state. Similar analyses of other weak interaction decays involving particles having defined values of $T$ and $T_z$ (i.e., particles that are not leptons) verifies that *in the weak interaction* $\Delta T = 1/2$ *and* $\Delta T_z = 1/2$.

To recapitulate, $\Lambda^0$ and $K$ particles are produced in association at a high rate (with a relatively large cross section) in processes involving the strong interaction. They each decay independently, because they have flown apart, in processes involving the weak interaction. The decays occur at a low rate (with a relatively long lifetime) because the interaction is weak. Figure 17-19 is a photograph of tracks in a hydrogen bubble chamber showing the associated production, (17-27), of a $\Lambda^0$ and a $K^0$ when an incident $\pi^-$ strikes a $p$ in the hydrogen filling of the chamber. After separating, the $\Lambda^0$ decays into a $p$ and a $\pi^-$ as in (17-31), and the $K^0$ decays into a $\pi^+$ and a $\pi^-$ as in (17-30a).

Since the $\Lambda^0$ and the $K$ are produced in a strong interaction, why is it not possible for them to decay in a more favorable way, i.e., at a much higher rate, by a process involving the strong interaction? This significant question was answered by Gell-Mann and, independently, by Nishijima, in 1953 by the introduction of a new quantum number called the *strangeness* $S$. The strangeness of the particles of a particular isospin multiplet specifies the shift of the average charge $Q$ of the multiplet from the average charge of the pion multiplet if the particles are mesons, or from the average

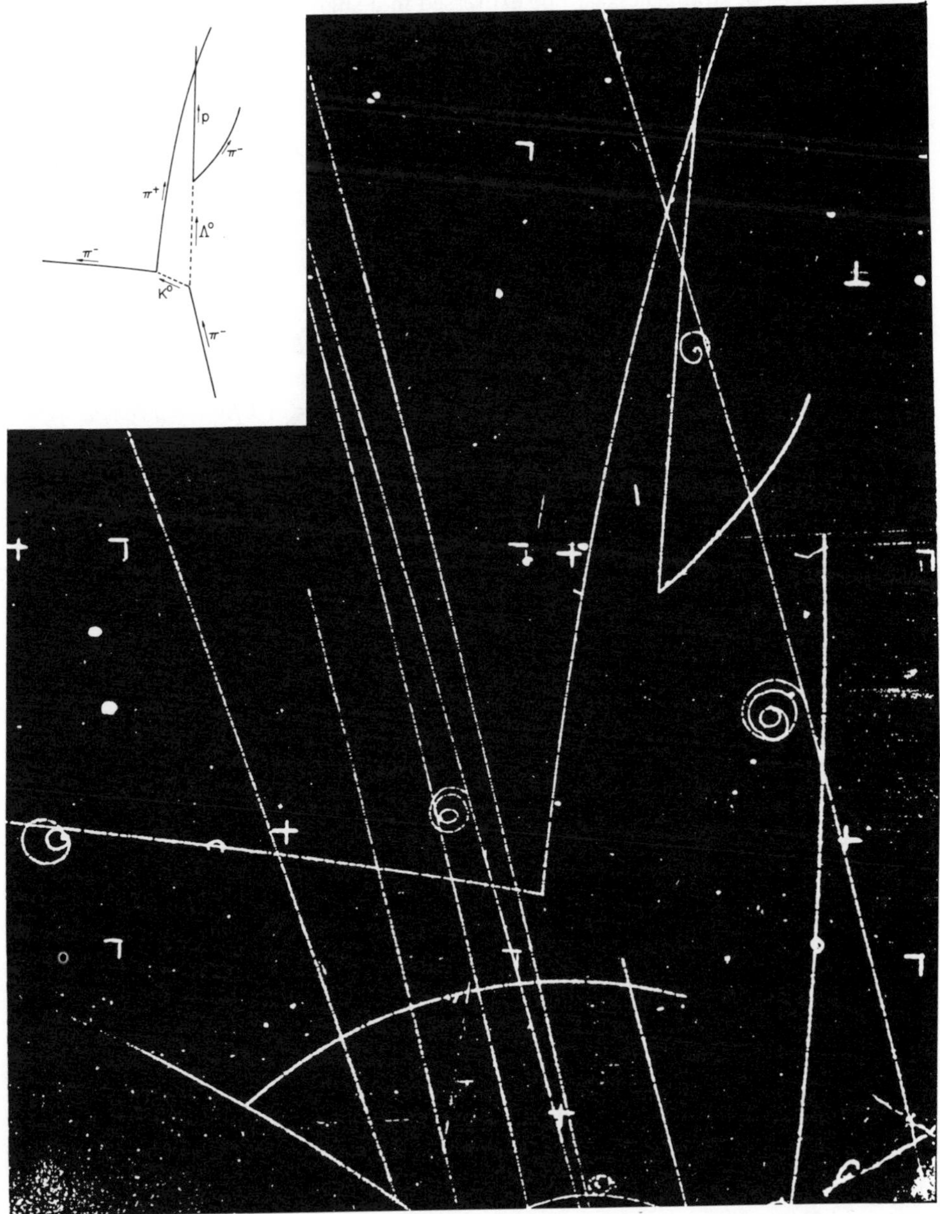


**FIGURE 17-19**

The associated production of a $\Lambda^0$ and a $K^0$ in a hydrogen bubble chamber. An incident $\pi^-$ interacts with a $p$ of the liquid hydrogen filling the chamber. The $K^0$ decays into a $\pi^+$ and a $\pi^-$. The $\Lambda^0$ decays into a $p$ and a $\pi^-$. The production takes place through the strong interaction, but the decays each utilize the weak interaction. The curvature of each particle in the applied magnetic field is used to identify the particle. (Courtesy Lawrence Radiation Laboratory)

charge of the nucleon multiplet if they are baryons. The term *baryon* refers to a particle of rest mass equal to or larger than that of a nucleon, which participates in the strong interaction. It includes the nucleon, the $\Lambda^0$ particle, and also the more massive particles the $\Sigma$, $\Xi$, and $\Omega$, that will be described briefly later.

The concept of strangeness can be expressed in a slightly different way by recalling (17-15)

$$Q = T_z + B/2 \qquad \text{(for pions and nucleons)}$$

which is the relation between the charge in units of electron charge magnitudes $Q$, the isospin $z$ component $T_z$, and the baryon number $B$. The comparable relations for the $\Lambda^0$ and $K$ particles are

$$Q = T_z + B/2 - 1/2 \qquad \text{(for } \Lambda^0\text{)} \qquad (17\text{-}33)$$

$$Q = T_z + B/2 + 1/2 \qquad \text{(for } K^+ \text{ and } K^0\text{)} \qquad (17\text{-}34)$$

$$Q = T_z + B/2 - 1/2 \qquad \text{(for } K^- \text{ and } \overline{K^0}\text{)} \qquad (17\text{-}35)$$

When it was previously introduced, the baryon number $B$ was defined to be 0 for a pion, $+1$ for a nucleon, and $-1$ for an antinucleon. Here we shall extend the definition to include all the baryons and mesons, so that $B = +1$ *for a baryon*, $B = -1$ *for an antibaryon, and* $B = 0$ *for a meson*. Since $B = 0$ in both (17-34) and (17-35), in agreement with our earlier considerations these equations state that $Q = +1$ for the $K^+$ which has $T_z = +1/2$; $Q = 0$ for the $K^0$ which has $T_z = -1/2$; $Q = -1$ for the $K^-$ which has $T_z = -1/2$; and $Q = 0$ for the $\overline{K^0}$ which has $T_z = +1/2$. For the $\Lambda^0$, $B = +1$ and $T_z = 0$, so (17-33) says correctly that $Q = 0$. In terms of the quantum number $S$, all four of the equations for $Q$ can be written

$$Q = T_z + \frac{B + S}{2} \qquad (17\text{-}36)$$

where $S = 0$ for the nucleon and pion, $S = -1$ for the $\Lambda^0$, $S = +1$ for the $K^+$ and $K^0$, and where *the value of* $S$ *for an antiparticle is the negative of its value for the particle*, just as is the case with $B$. The last statement means that $S = -1$ for the $K^-$ and $\overline{K^0}$.

Gell-Mann and Nishijima postulated that, in addition to the isospin $T$ being conserved, *in the strong interaction the strangeness* $S$ *is conserved*. Then the associated production of a $\Lambda^0$ and a $K$, as in our example

$$\pi^- + p \rightarrow \Lambda^0 + K^0$$

can take place via the strong interaction since the total $S = 0$ both before and after the reaction. But all the $\Lambda^0$ and $K$ decays are forbidden to take place through the strong interaction since in all cases there is a single strangeness $\pm 1$ particle that only has enough total relativistic energy to allow it to decay into strangeness 0 particles. Inspection of the weak interaction $K$ and $\Lambda^0$ decays of (17-30) through (17-32) will verify that in each of the interactions (which involve particles other than leptons) strangeness is not conserved and, in fact that *in the weak interaction* $\Delta S = 1$.

This is an appropriate place to mention again the interference phenomenon observed in the weak interaction decay of the $K^0$ and $\overline{K^0}$. As stated earlier, it arises because these are different particles that, nevertheless, act exactly the same as far as the weak interaction is concerned. Now we can understand why. The $K^0$ and $\overline{K^0}$ differ only in the values of their quantum numbers $S$ and $T_z$, and $S$ and $T_z$ make no difference because neither is conserved in the weak interaction.

Considerations of strangeness make it possible to understand the production and decay of the more massive baryons that have been discovered in recent years. There are three sets of these particles. The $\Sigma$ *particles* form an isospin triplet, $\Sigma^-$, $\Sigma^0$, $\Sigma^+$, with $T = 1$ and $T_z = -1, 0, +1$, respectively. They were first recognized as a triplet because they have nearly the same rest masses

$$m_\Sigma \simeq 1190 \text{ MeV}/c^2. \tag{17-37}$$

Like the other baryons discussed so far, the $\Sigma$ particles are spin 1/2 fermions of even intrinsic parity. Their strangeness is $S = -1$. The $\Xi$ *particles* constitute an isospin doublet with nearly equal masses having the value

$$m_\Xi \simeq 1320 \text{ MeV}/c^2 \tag{17-38}$$

Their isospin is $T = 1/2$, with $T_z = -1/2$ for the $\Xi^-$ and $T_z = +1/2$ for the $\Xi^0$. Again, they are spin 1/2 fermions of even intrinsic parity. But their strangeness is $S = -2$. Finally, there is $\Omega^-$ *particle* of rest mass

$$m_\Omega \simeq 1670 \text{ MeV}/c^2 \tag{17-39}$$

It is an isospin singlet: $T = 0$, $T_z = 0$. The $\Omega^-$ is also an even intrinsic parity fermion, but its spin is 3/2. Its strangeness is $S = -3$.

Each of the $\Sigma$, $\Xi$, and $\Omega$ particles are produced in a high-energy collision through the strong interaction in association with other particles in such a way as to conserve strangeness. For instance, the $\Xi^-$ with $S = -2$ was first discovered in cosmic rays, being produced in association with two $K^0$ mesons that both have $S = +1$. With one exception, each of them decays by the weak interaction. As an example, in the $\Xi^-$ decay

$$\Xi^- \rightarrow \Lambda^0 + \pi^- \tag{17-40}$$

the lifetime has a value of about $10^{-10}$ sec, which is typical of the weak interaction.

The exception is the $\Sigma^0$, which decays through the electromagnetic interaction, according to the scheme

$$\Sigma^0 \rightarrow \Lambda^0 + \gamma \tag{17-41}$$

The lifetime is known experimentally to be shorter than $10^{-14}$ sec, and it is predicted theoretically to be about $10^{-18}$ sec. Note that in this electromagnetic interaction decay the $z$ component of isospin is conserved, if we assign $T_z = 0$ to the photon $\gamma$, since $T_z = 0$ for the $\Sigma^0$ and the $\Lambda^0$. It is generally observed that $T_z$ *is conserved in the electromagnetic interaction*. The $\Xi^-$ decay of (17-40) cannot proceed relatively rapidly by the electromagnetic interaction since $T_z = -1/2$ for the $\Xi^-$, $T_z = 0$ for the $\Lambda^0$, and $T_z = 0$ for the $\gamma$. Thus the decay must proceed much more slowly by the weak interaction.

By considering (17-36)

$$Q = T_z + \frac{B + S}{2}$$

it is immediately apparent that the conservation of $T_z$ in the electromagnetic interaction means that $S$ *is conserved in the electromagnetic interaction*. The argument is simply that since $Q$ and $B$ are conserved in all circumstances, if $T_z$ is conserved then $S$ must be also. However, $T$ *is not conserved in the electromagnetic interaction*. This can be seen by remembering that the strong interaction conserves isospin and is charge independent. The electromagnetic interaction is very definitely not charge independent, and so it cannot conserve isospin.

It should also be stated that in 1961 Gell-Mann used concepts closely related to strangeness, which we shall consider later, to predict the existence of the $\eta^0$ *meson* and the $\eta'$ *meson*, which were subsequently observed experimentally. These neutral mesons have rest masses between that of a $K$ meson and that of a nucleon

$$m_{\eta^0} \simeq 550 \text{ MeV}/c^2 \tag{17-42}$$

and

$$m_{\eta'} \simeq 960 \text{ MeV}/c^2 \tag{17-43}$$

Like the other mesons, they are both spin 0 odd intrinsic parity bosons. And they both have $S = 0$. Each is an isospin singlet with $T = 0$ and $T_z = 0$. The $\eta^0$ decays electromagnetically in about $10^{-19}$ sec, predominantly into two photons. The $\eta'$ also decays by the electromagnetic interaction, but the value of its lifetime is not known. Its predominant decay mode produces an $\eta^0$ and two pions.

## 17-7 Fundamental Interactions and Conservation Laws

Table 17-1 summarizes some of the information obtained in the preceding section, and in earlier sections of this book, concerning the four fundamental interactions of nature: *strong*, *electromagnetic*, *weak*, and *gravitational*. The intrinsic strength comparison depends to a certain extent on the choice of exactly what attribute of the strength is to be compared; the numbers quoted are obtained from comparisons made in the manner of Section 16-4. All of the entries in the table have been discussed previously, except for the characteristics of the quantum of the gravitational field.

The gravitational field quantum is called the *graviton*. Its rest mass must be zero since the gravitational interaction has the same long range as the electromagnetic interaction, whose quantum is the zero rest mass photon. The spin of the graviton is known to be 2. The reason is the absence of negative gravitational mass, which prevents the existence of the oscillating gravitational dipole that would be required to radiate a spin 1 graviton. The lowest possible multipolarity oscillating gravitational source is a quadrupole (a distribution of mass oscillating between a prolate and oblate ellipsoidal shape), and a quadrupole source emits a spin 2 quantum. This is essentially the same argument as the one we used in Section 16-5 to conclude that a photon has spin 1 because there are no oscillating electromagnetic monopoles. The question of whether or not gravitons (emitted when a dying star collapses into a "black hole") have been detected experimentally is, at the time of writing, controversial. But there is certainly no controversy about the fact that the gravitational interaction is the only one of the four that is both long range and always of the same sign. Therefore its effects are cumulative so that, despite its intrinsic weakness, gravity is by far the most obvious of the interactions in the macroscopic world.

**TABLE 17-1.** The Fundamental Interactions

| Name | Intrinsic Strength | Field Quantum Name | Field Quantum Rest Mass | Field Quantum Spin | Range | Sign |
|---|---|---|---|---|---|---|
| Strong (nuclear) | 1 | Pion | $\sim 10^2$ MeV/$c^2$ (with heavier mesons for repulsive core) | 0 | $\sim 10^{-15}$ m (with smaller repulsive core) | Attractive overall (but with repulsive core) |
| Electromagnetic | $10^{-2}$ | Photon | 0 | 0 | Long ($\propto 1/r$) | Attractive or repulsive |
| Weak ($\beta$ decay) | $10^{-12}$ | Intermediate boson | $>10^4$ MeV/$c^2$ | 1? | $<10^{-17}$ m? | Not applicable |
| Gravitational | $10^{-40}$ | Graviton | 0 | 2 | Long ($\propto 1/r$) | Always attractive |

**TABLE 17-2.** Applicability of the Fundamental Conservation Laws to the Fundamental Interactions (+ Means Conserved; − Means Not Conserved)

| Quantity Conserved | Strong | Electro-magnetic | Weak | |
|---|---|---|---|---|
| Energy | + | + | + | |
| Linear momentum | + | + | + | |
| Angular momentum | + | + | + | |
| Charge | + | + | + | |
| Electronic lepton number | + | + | + | |
| Muonic lepton number | + | + | + | |
| Baryon number | + | + | + | |
| Isospin magnitude | + | − | − | ($\Delta T = 1/2$ for nonleptonic) |
| Isospin $z$ component | + | + | − | ($\Delta T_z = 1/2$ for nonleptonic) |
| Strangeness | + | + | − | ($\Delta S = 1$ for nonleptonic) |
| Parity | + | + | − | (Except for infrequent |
| Charge conjugation | + | + | − | violation in slow decay |
| Time reversal | + | + | + | of $K^\circ, \overline{K^\circ}$ system) |

Table 17-2 lists the three interactions of the microscopic world, i.e., of quantum physics, and all of the quantities that are conserved in certain interactions. The symbol +, or −, means that a quantity is, or is not, conserved. We have discussed all of the entries in this table, except those referring to charge conjugation and time reversal.

*Charge conjugation* is the process of changing every particle of a system into its antiparticle. As an example, the charge conjugate of the ground state deuterium atom contains a nucleus with an antineutron and an antiproton, and an atomic positron. All available experimental evidence is consistent with the conclusion that the operation of both the strong and electromagnetic interactions is unaffected by, or invariant to, charge conjugation. For instance, such invariance is found in a study of the strong interaction annihilation of a proton and an antiproton into the particle, antiparticle pair $K^+$, $K^-$, plus other particles, and is also found in measurements of the electromagnetic decay of the $\eta^0$ meson. Therefore, we believe that the nucleus of the antideuterium atom (whose behavior is governed by the strong interaction) and also the positron (whose behavior is governed by the electromagnetic interaction) would act in the same way, because they are in the same quantum state at the same energy, as the nucleus and the electron in the normal deuterium atom. So we may say, as indicated by the + symbols in the table, that charge conjugation is conserved in the strong and electromagnetic interactions because the description of a system governed by either of these interactions is invariant to the operation. This is parallel to the terminology we use when we say by the − symbol in the table, and elsewhere, that parity is not conserved in the weak interaction because a description of a system whose behavior it governs is not invariant to the parity operation.

In fact, the experimental evidence for the − symbol in the table that indicates charge conjugation is not conserved in the weak interaction, i.e., that the weak interaction does distinguish between a system and its charge conjugate, is the same as the experimental evidence for parity nonconservation in that interaction. If the student will refer back to the schematic of the ${}^{27}Co^{60}$ $\beta$-decay parity experiment of Figure 16-15, and imagine performing charge conjugation on the normal view, he will immediately see that the description of the $\beta$ decay of anti-${}^{27}Co^{60}$ differs from the description of the

$\beta$ decay of $^{27}Co^{60}$. Because the sign of the charge in the current loop is reversed by charge conjugation, the effect of this operation is to reverse the sense of circulation of positive charge in the loop. But the operation does not effect the preferred direction of the particle emitted in the decay. Therefore we see that, whereas for $^{27}Co^{60}$ a left-hand screw describes the relation between the circulation of the current loop representing its magnetic dipole moment and the preferred direction of the particle of electron mass emitted in its decay, for anti-$^{27}Co^{60}$ a right-hand screw would describe the relation between these quantities. Thus the $\beta$-decay experiment shows that the entries for both parity and charge conjugation should be the $-$ symbols in the weak interaction column of Table 17-2.

Note that if the charge conjugation operation is first carried out on the normal view of the $\beta$-decay experiment in Figure 16-15, and then the operation of looking at the mirror image (which Figure 16-16 shows is related to the parity operation) is carried out on the charge conjugated view, the effects of the operations nullify each other. The reason is that the sense of circulation of the current loop is reversed in each of the successive operations. We shall now show that this result can tell us about the behavior of the system under the operation of time reversal.

*Time reversal* is the process of changing the time variables describing the evolution of a microscopic system into their negatives; that is, of changing the direction of flow of time, like running a motion picture backwards. As applied to the $^{27}Co^{60}$ experiment in the normal view of Figure 16-15, time reversal reverses the direction of the vector describing the motion of the electron, and also reverses the sense of circulation of the current in the loop. So after time reversal the relation between the direction and the circulation is the same as it was before the operation, namely the relation of a left-hand screw. As far as time reversal is concerned, we can therefore learn nothing directly from the fact that there is an asymmetry with respect to the plane of the current loop in the electron emission pattern of $^{27}Co^{60}$ $\beta$ decay, this asymmetry being the result that allows the definition of a preferred emission direction. In the time-reversal operation, the description of the experiment cannot be changed in an essential way in any case, so the experiment does not tell us directly about whether or not $\beta$ decay distinguishes between the directions of flow of time. But it does indirectly. This is due to the fact that there is a very general theorem of relativistic quantum theory which shows that, for *any* system governed by *any* interaction that conforms to the relativistic requirement that cause must precede effect, the result of successively carrying out of the charge conjugation operation, the parity operation, and the time-reversal operation is to leave the essential description of the behavior of the system unchanged. As a consequence of this so-called *CPT theorem*, the $-$ entries for both parity and charge conjugation conservation, found in the table for the weak interaction, require a time reversal entry of $+$ for that interaction.

Thus in $\beta$ decay the weak interaction does not distinguish between the directions of flow of time, although it does distinguish between a system and its mirror image and between a system and its antisystem. But in 1964 Christenson et al. found that in weak interaction decay of the long lifetime component of the degenerate $K^0$, $\overline{K^0}$ system there is infrequently seen a violation of time-reversal invariance. Usually the decay yields three pions, in a way that would indicate, as in $\beta$ decay, $-$ entries for both parity and charge conjugation, and therefore a $+$ entry for time reversal. But about 0.1% of the decays yield two pions and must be indicated by symbols for parity and charge conjugation of which one is $-$ and the other is $+$. According to the CPT theorem, this means that in these decays the entry for time reversal must be $-$. That is, there is evidence that through the rare mode in the weak interaction decay of the long-lived component of the $K^0$, $\overline{K^0}$ system, *nature can distinguish at a microscopic level the direction of flow of time*. This startling result would seem to be of great significance. What its significance is we do not yet understand. Nor do we understand its origin.

Confidence in the validity of the CPT theorem used to obtain this result is found in the three + entries for parity, charge conjugation, and time reversal for both the strong and electromagnetic interactions in Table 17-2. The theorem says, in essence, that the product of these three symbols must in all cases be a +. Independent measurements exist for each entry, so the fact that the product of the three symbols actually is a + for both the strong and electromagnetic interactions confirms the theorem. An independent time-reversal experiment for the strong interaction is a comparison of the cross section for a reaction such as

$$ {}^{12}\text{Mg}^{24} + {}^{2}\text{He}^{4} \rightarrow {}^{13}\text{Al}^{27} + {}^{1}\text{H}^{1} $$

and the cross section for its inverse

$$ {}^{13}\text{Al}^{27} + {}^{1}\text{H}^{1} \rightarrow {}^{12}\text{Mg}^{24} + {}^{2}\text{He}^{4} $$

with the momentum vectors of the bombarding and target nuclei in the second reaction adjusted to be equal but opposite to those of the product and residual nuclei of the first reaction. Evidence for time-reversal invariance of the electromagnetic interaction can be obtained by studying the behavior of an interacting system of charges with a certain set of initial conditions, and then studying their behavior when the initial momentum vectors are reversed. Each behavior looks exactly like a backward running motion picture of the other behavior, because the electromagnetic interaction cannot distinguish between the direction of time and so it operates the same way in both cases.

There are relations between each conservation law and a symmetry property of a physical, or mathematical, space. For instance, the symmetry of physical space with respect to displacement (the fact that *empty* space in our universe has the same properties in one place that it does in another) can be shown to underlie linear momentum conservation. But these relations can generally be understood only on the basis of more sophisticated quantum theory than we are able to use here.

## 17-8 Families of Elementary Particles

Table 17-3 lists the elementary particles (except for the graviton) that are stable, or else decay only by the electromagnetic or weak interactions. Related particles are grouped into families: the photon, the leptons, the mesons, and the baryons. Both the leptons and the baryons are fermions, and both the photon and the mesons are bosons. The baryons more massive than nucleons are sometimes called *hyperons*, but this term seems to be passing from use. The mesons and baryons, i.e., the particles that participate in the strong interaction, are called collectively *hadrons*, and this term is widely used. The entries in the table are: family name; particle symbol; rest mass; lifetime; charge $Q$; intrinsic spin $s$; lepton number $L_e$ or $L_\mu$, baryon number $B$; and, for mesons and baryons, intrinsic parity $P$; isospin $T$; isospin $z$ component $T_z$; strangeness $S$. The conventional baryon parity assignment, and $T_z = 0$ assignment for the photon, are also shown.

The leptons and baryons all have antiparticles, although they are not shown in the table. Compared to a lepton or a baryon, the "quantum numbers" of its antiparticle have values with: opposite $Q$; same $s$; opposite $L_e$ or $L_\mu$ or $B$; and, for baryons, opposite $P$; same $T$; opposite $T_z$; opposite $S$. An antiparticle has the same rest mass, and also the same lifetime, as the particle. These two equalities are predicted by the CPT theorem, and have been verified experimentally.

**TABLE 17-3.** The Elementary Particles

| Generic Name | Particle Symbol | Rest Mass (MeV/$c^2$) | Lifetime (sec) | Charge $Q$ | Intrinsic Spin $s$ | Lepton Number $L_e$ or $L_\mu$ | Baryon Number $B$ | Intrinsic Parity $P$ | Isospin $T$ | Isospin $z$ component $T_z$ | Strangeness $S$ |
|---|---|---|---|---|---|---|---|---|---|---|---|
| Photon | $\gamma$ | 0 | stable | 0 | 1 | 0 | 0 | | | 0 | |
| Leptons | $\nu_e$ | 0 | stable | 0 | 1/2 | +1 | 0 | | | | |
| | $\nu_\mu$ | 0 | stable | 0 | 1/2 | +1 | 0 | | | | |
| | $e^-$ | 0.511 | stable | −1 | 1/2 | +1 | 0 | | | | |
| | $\mu^-$ | 105.7 | $2.2 \times 10^{-6}$ | −1 | 1/2 | +1 | 0 | | | | |
| Mesons | $\pi^+$ | 139.6 | $2.6 \times 10^{-8}$ | +1 | 0 | 0 | 0 | Odd | 1 | +1 | 0 |
| | $\pi^0$ | 135.0 | $0.9 \times 10^{-16}$ | 0 | 0 | 0 | 0 | Odd | 1 | 0 | 0 |
| | $\pi^-$ | 139.6 | $2.6 \times 10^{-8}$ | −1 | 0 | 0 | 0 | Odd | 1 | −1 | 0 |
| | $K^+$ | 493.8 | $1.2 \times 10^{-8}$ | +1 | 0 | 0 | 0 | Odd | 1/2 | +1/2 | +1 |
| | $K^0$ | 497.8 | ($8.6 \times 10^{-11}$ | 0 | 0 | 0 | 0 | Odd | 1/2 | −1/2 | +1 |
| | | | and | | | | | | | | |
| | $\overline{K^0}$ | 497.8 | $5.2 \times 10^{-8}$) | 0 | 0 | 0 | 0 | Odd | 1/2 | +1/2 | −1 |
| | $K^-$ | 493.8 | $1.2 \times 10^{-8}$ | −1 | 0 | 0 | 0 | Odd | 1/2 | −1/2 | −1 |
| | $\eta^0$ | 549 | $2.5 \times 10^{-19}$ | 0 | 0 | 0 | 0 | Odd | 0 | 0 | 0 |
| | $\eta'$ | 958 | $> 10^{-21}$ | 0 | 0 | 0 | 0 | Odd | 0 | 0 | 0 |
| Baryons | $p$ | 938.3 | stable | +1 | 1/2 | 0 | +1 | Even | 1/2 | +1/2 | 0 |
| | $n$ | 939.6 | 930 | 0 | 1/2 | 0 | +1 | Even | 1/2 | −1/2 | 0 |
| | $\Lambda^0$ | 1116 | $2.5 \times 10^{-10}$ | 0 | 1/2 | 0 | +1 | Even | 0 | 0 | −1 |
| | $\Sigma^+$ | 1189 | $8.0 \times 10^{-11}$ | +1 | 1/2 | 0 | +1 | Even | 1 | +1 | −1 |
| | $\Sigma^0$ | 1192 | $<10^{-14}$ | 0 | 1/2 | 0 | +1 | Even | 1 | 0 | −1 |
| | $\Sigma^-$ | 1197 | $1.5 \times 10^{-10}$ | −1 | 1/2 | 0 | +1 | Even | 1 | −1 | −1 |
| | $\Xi^0$ | 1315 | $3.0 \times 10^{-10}$ | 0 | 1/2 | 0 | +1 | Even | 1/2 | +1/2 | −2 |
| | $\Xi^-$ | 1321 | $1.7 \times 10^{-10}$ | −1 | 1/2 | 0 | +1 | Even | 1/2 | −1/2 | −2 |
| | $\Omega^-$ | 1672 | $1.3 \times 10^{-10}$ | −1 | 3/2 | 0 | +1 | Even | 0 | 0 | −3 |

The antiparticles of the mesons are shown in the table. We have already discussed the fact that the $K^-$ and $\overline{K^0}$ are, respectively, the antiparticles of the $K^+$ and $K^0$. Inspection of the table will confirm that the relation between the quantum numbers of the $K^+$ and $K^-$, and of the $K^0$ and $\overline{K^0}$, agree with the particle, antiparticle rules quoted earlier for leptons and baryons, except that the intrinsic parity does not change in the $K$, anti-$K$ case. The predicted (and experimentally confirmed) particle, antiparticle parity rules reflect the facts that mesons are bosons, and that baryons are fermions. Further inspection will show that the relation between the quantum numbers of the $\pi^+$ and $\pi^-$ is the same as that between the quantum numbers of the $K^+$ and $K^-$. Thus we can say that the $\pi^-$ is the antiparticle of the $\pi^+$. From this point of view, we can also say that the $\pi^0$ is its own antiparticle, and that the same is true for the $\eta^0$ and for the $\eta'$.

Not listed in Table 17-3 are a large number of very short-lived entities which may, or may not, be called elementary particles. As an example, in pion-nucleon scattering experiments performed by Fermi and others in 1952 it was found that there is a strong resonance in the scattering cross sections at a pion bombarding energy of 195 MeV. Figure 17-20 shows the $\pi^+$, $p$ cross section as a function of the total center-of-mass energy of the system, including the pion and nucleon rest masses. Since the $\pi^+$ has $T = 1$, $T_z = +1$ and the $p$ has $T = 1/2$, $T_z = +1/2$, the system is in the $T = 3/2$, $T_z = 3/2$ state. (The $\pi^-$, $p$ system in the $T = 3/2$, $T_z = -1/2$ state shows the same kind of cross-section resonance at the same energy, providing thereby additional evidence for the conclusion that, while the strong interaction depends on $T$, it does not depend on $T_z$.) The full width at half-maximum, $\Gamma$, of the resonance, whose peak occurs at a total energy of 1236 MeV, is about 120 MeV. This means that the pion and proton must temporarily form a composite entity that holds together for a time $t \sim \hbar/\Gamma \sim 10^{-15}$ eV-sec/$10^8$ eV $\sim 10^{-23}$ sec. If moving at a characteristic velocity of $c/3$, the entity would maintain its existence over a distance $d \sim ct/3 \sim 10^8$ m/sec $\times$ $10^{-23}$ sec $\sim 10^{-15}$ m, which is the range of the strong interaction. It is therefore not unreasonable to speak of a pion and a proton forming a very short-lived particle, which is called the $\Delta(1236)$. It has a definite set of quantum numbers: $s = 3/2$, $B = 1$, $P =$ even, $T = 3/2$, $S = 0$. But its mass is not definite, and it would be best expressed as $1236 \pm 60$ MeV/$c^2$. There are a number of other more massive so-called *baryon resonances*. Some of them will be indicated later in Figure 17-23.

There are also *meson resonances*. For instance, the $\rho$ *meson* is seen as a resonance in the interaction of a $\pi^-$ and a $\pi^+$ in the final state of reactions such as

$$\pi^- + p \rightarrow \pi^- + \pi^+ + n$$

The $\rho$ has a rest mass of $765 \pm 50$ MeV. That is, the width of the resonance, and therefore the lifetime, is about the same as for the $\Delta(1236)$. The quantum numbers of the

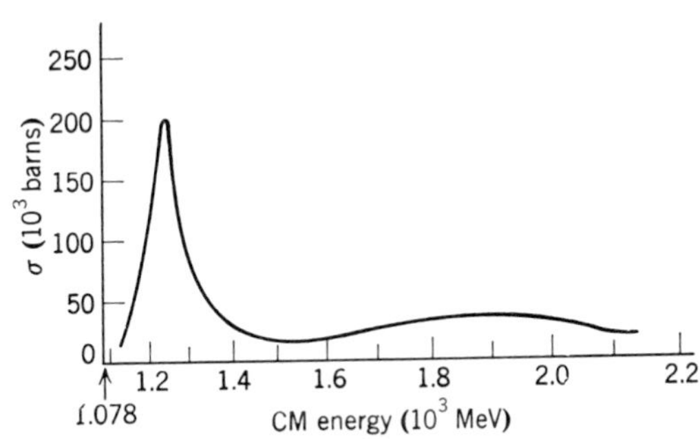


**FIGURE 17-20**

The scattering cross section for $\pi^+$ mesons on protons, as a function of the center-of-mass total relativistic energy of the system.

$\rho$ meson are: $s = 1$, $B = 0$, $P =$ odd, $T = 1$, $S = 0$. The $\rho$ decays back into two pions. A similar example is the $\omega$ *meson*, with rest mass $783 \pm 6$ MeV/$c^2$, and the same quantum numbers except that $T = 0$. There are other even more massive meson resonances, which have spins ranging up to 2. It is believed that the $\omega$ meson, and possibly some of the other massive ones, are responsible for the repulsive core in the nucleon potential. The resonance particles, both baryon and meson, are all very short lived since they can all decay by the strong interaction through which they were formed.

**Example 17-5.** Discuss each of the following reactions in terms of the conservation laws listed in Table 17-2 and the particle quantum numbers listed in Table 17-3.

(a) $\pi^- + p \rightarrow \Sigma^+ + K^-$

This reaction is impossible because it requires a strangeness change of 2.

(b) $K^- + p \rightarrow \Omega^- + K^+ + K^0$

This is the reaction in which the $\Omega^-$, which has $S = -3$, was first produced. It is strangeness conserving since $S = +1$ for the $K^+$ and $K^0$, while $S = -1$ for the $K^-$. Charge and baryon number are conserved. So are angular momentum and parity because the final state can have one unit of orbital angular momentum. (Recall that the parity associated with orbital angular momentum is given by $(-1)^l$.) Since isospin and its $z$ component are also conserved, we see that the reaction can proceed via the strong interaction. If this were not the case, the cross section would be too small for it to be observable.

(c) $\Omega^- \rightarrow \Xi^0 + \pi^-$

Here charge and baryon number are conserved. Angular momentum and parity are also conserved by the final state containing one unit of orbital angular momentum. Since the values of $T$ are 0 for the $\Omega^-$, 1/2 for the $\Xi^0$, and 1 for the $\pi^-$, we see that there must be an isospin change of at least $\Delta T = 1/2$. Also, $T_z$ is 0 for the $\Omega^-$, $+1/2$ for the $\Xi^0$, and $-1$ for the $\pi^-$, so the $z$ component of isospin changes by $\Delta T_z = 1/2$. The strangeness changes, too, by the amount $\Delta S = 1$. These quantum number changes do allow the decay to proceed by the weak interaction, but they prohibit it from proceeding more rapidly by the electromagnetic or strong interactions.

(d) $\pi^+ + p \rightarrow p + p + \bar{n}$

First we must determine the quantum numbers of the antineutron $\bar{n}$. Applying the quoted rules to the table, we find: $Q = 0$, $s = 1/2$, $B = -1$, $P =$ odd, $T = 1/2$, $T_z = +1/2$, $S = 0$. Inspection demonstrates that all quantum numbers are conserved by the reaction, so it can take place by the strong interaction.

(e) $\bar{n} \rightarrow \bar{p} + e^+ + \nu_e$

If this goes at all, it must be by the weak interaction since $\nu_e$ does not participate in any of the others. Charge is conserved since $Q = -1$ for the $\bar{p}$. The total baryon number equals $-1$ before and after, so it is conserved also. Electronic lepton number is conserved because it has the values $-1$ for the $e^+$ and $+1$ for the $\nu_e$. Angular momentum can be conserved. Parity is not defined for leptons, but parity is not a significant consideration for a weak interaction involving leptons. The same is true for isospin and strangeness. So the reaction can take place by the weak interaction. Note that it is just the charge conjugate of the $\beta$ decay of the neutron.

(f) $\Lambda^0 \rightarrow n + \gamma$

This reaction, if it can occur, obviously must be electromagnetic. Since $T_z = 0$ for the $\Lambda^0$ and $\gamma$, while $T_z = -1/2$ for the $n$, we see that it cannot occur because $T_z$ is conserved in the electromagnetic interaction. This conclusion agrees with experiment, and it is one of the reasons why $T_z = 0$ is assigned to the photon. ◀

## 17-9 Hypercharge and Quarks

In correlating the properties of mesons and hadrons, we find it useful to employ a quantum number $Y$, called the *hypercharge*, instead of the strangeness quantum number $S$. The hypercharge is defined by

$$Y = S + B \tag{17-44}$$

where $B$ is the baryon number. Because $B$ is conserved by all the interactions, the rules concerning the conservation of $S$ carry over directly to $Y$. Gell-Mann, and others, discovered that a plot of $Y$ versus $T_z$ for the spin 1/2 baryons forms a simple symmetrical pattern, and that such a plot for the spin 0 mesons forms the same pattern. These octet patterns are shown in Figures 17-21 and 17-22. They are called octets because they contain eight particles, including the two that occupy the same central position.

The symmetry of these patterns suggested an extension of the matrix theory used to describe the detailed mathematical properties of spin and isospin. The extension is called the *SU(3) theory*, and it involves the group of "Special Unitary" 3 × 3 matrices. The octet pattern, including the double occupancy of the central position, is one of the patterns that the symmetry properties of this group lead to. They also lead to a singlet pattern and to a decuplet pattern. The $\eta'$ meson occupies such a singlet, thereby accounting for all the nine known mesons. There are also nine baryons, of which the $\Omega^-$ remains to be accounted for. It cannot be in a singlet pattern since SU(3) shows this would not be possible for a spin 3/2 baryon. Instead, it is grouped with other spin 3/2 baryons in the decuplet shown in Figure 17-23. All the members of this decuplet pattern are resonances, except for the $\Omega^-$.

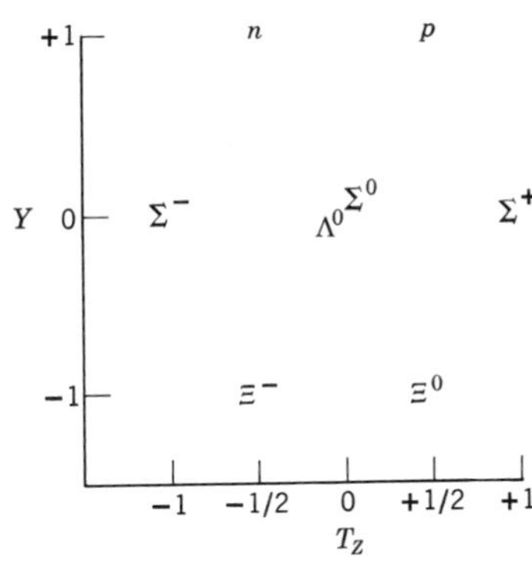


**FIGURE 17-21**
The even parity spin 1/2 baryon octet.

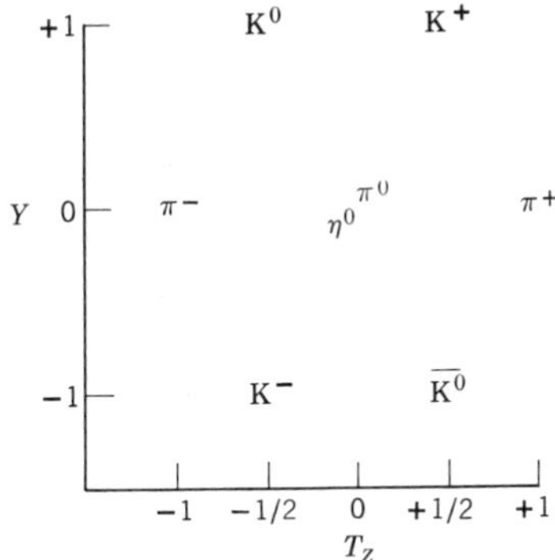


**FIGURE 17-22**
The odd parity spin 0 meson octet.

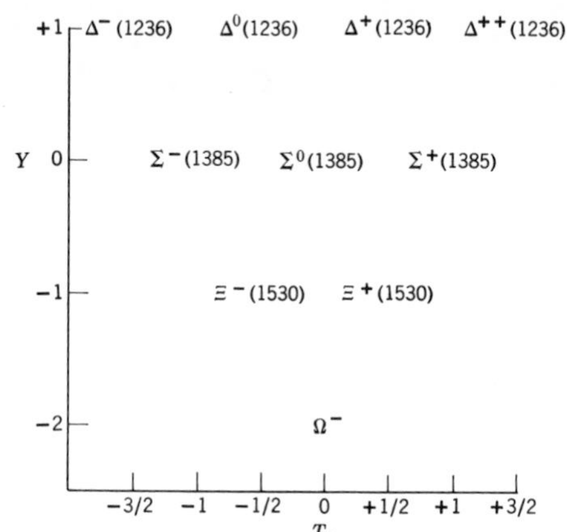


**FIGURE 17-23**
The even parity spin 3/2 baryon decuplet.

The significance of the SU(3) theory was first demonstrated by the fact that Gell-Mann and Okubo predicted the rest mass and quantum numbers of the $\Omega^-$ before it had been observed experimentally. The idea involved in the mass prediction is that, in the absence of the strong and electromagnetic interactions, all members of a pattern would have the same rest mass energy, but the strong interaction removes part of this degeneracy by splitting the mass energy according to $Y$, and the electromagnetic interaction further removes the remaining degeneracy by splitting the mass energy according to $T_z$. Thus the mass of the $\Omega^-$ could be predicted by extrapolating the $Y$ dependence of the mass of the previously known members of the decuplet.

It can be seen from Figures 17-21 through 17-23 that the octet pattern, which is a hexagon, and the decuplet pattern, which is a triangle, both have threefold rotational symmetry about their center. The same is certainly true of the singlet pattern, since it is just a point at the center. This observation led Gell-Mann to postulate the existence of *quarks*, a set of three particles which, in various combinations, can form all the mesons and baryons. Table 17-4 displays the nonintegral quantum numbers of the three quarks, $q_1, q_2, q_3$. Antiquarks, $\bar{q}_1, \bar{q}_2, \bar{q}_3$, are also postulated. Each has values of $Q, B, P, T_z$, and $Y$, which are the opposite of those for the corresponding quark.

All the mesons can then be formed from a combination of one of the three quarks and one of the three antiquarks. The nine possible combinations account precisely for the quantum numbers of the nine spin 0 mesons, if the quark and antiquark are assumed to be in a ${}^1S_0$ state in which their total angular momentum is 0 because they have no orbital angular momentum and their spins are essentially antiparallel. Furthermore, the meson resonances, which have spins that range up to 2, can be accounted for by allowing the quark, antiquark system to be in ${}^3S_1$, or in ${}^3P_0$, ${}^3P_1$, ${}^3P_2$, states.

Baryons may be considered to be $S$-state combinations of three of the three quarks. There are 27 possible combinations, and each of these has quantum numbers corresponding to one of the known baryons. These are grouped into the octet and decuplet patterns shown in Figures 17-21 and 17-23, plus another octet and a singlet. The second octet and the singlet are populated by baryon resonances.

**TABLE 17-4.** The Quarks

| Quark Symbol | $Q$ | $s$ | $B$ | $P$ | $T$ | $T_z$ | $Y$ |
|---|---|---|---|---|---|---|---|
| $q_1$ | +2/3 | 1/2 | +1/3 | Even | 1/2 | +1/2 | +1/3 |
| $q_2$ | −1/3 | 1/2 | +1/3 | Even | 1/2 | −1/2 | +1/3 |
| $q_3$ | −1/3 | 1/2 | +1/3 | Even | 0 | 0 | −2/3 |

Although it certainly seems impressive, there are problems with the quark model. One problem has to do with the fact that the decuplet of even parity spin 3/2 baryons would be composed of three quarks in the same $S$ state with "parallel" spins, generally with two or three of them being identical; e.g., $\Omega^- = q_3q_3q_3$. But this would not be possible if the spin 1/2 quarks are fermions, as are all other known examples of entities with spin 1/2. Another problem is that, despite very considerable efforts that have been expended, no one has yet been able to detect a free quark. If a quark were liberated from a two or three quark bound system in a high-energy collision produced by cosmic rays or an accelerator, it would be stable since its nonintegral charge would not allow it to decay into anything else without violating charge conservation. Furthermore, the nonintegral charge would make it easy to detect.

Perhaps the reason why free quarks have not yet been seen is that sufficiently energetic collisions have not yet been utilized to produce them. That is, they may have extremely high rest mass energies, and join together to form mesons or baryons with almost as high binding energies, so that very high energies are required to free them. Recent experiments involving the scattering of high-energy electrons from protons have given an indication that the proton is composed of point particles. These have been given the name *partons*, but, in fact, they may be bound quarks. It is to be hoped that future experimental and theoretical work will clarify the status of quarks in quantum physics.

## QUESTIONS

**1.** Why is $^3P_1$ not a component of the ground state of the deuteron? What about $^1S_0$?

**2.** What experiments can be performed to test for the existence of a stable system of two protons? Of two neutrons?

**3.** In the center-of-mass frame of reference the differential cross section for neutron-proton scattering is isotropic at low energies. Describe qualitatively the behavior of the differential cross section in a frame of reference in which the target proton is initially stationary.

**4.** In considering the quantum mechanical behavior of a system of two identical particles, we talk of exchange of the *labels* of the particles. In considering neutron-proton scattering, we talk of exchange of the *particles*. What is the reason for this difference?

**5.** Why is the proton-proton scattering differential cross section necessarily symmetric about 90° in the center-of-mass frame of reference?

**6.** Explain why the scattering differential cross section is isotropic if only the $l = 0$ state participates in the interaction that produces the scattering.

**7.** A very large part of what we know about the forces acting in atoms is obtained from the study of the bound states of the simplest atom, hydrogen. Why is only a small part of what we know about the forces acting in nuclei obtained from the study of the bound states of the simplest nucleus, deuterium?

**8.** Why is the name isospin an appropriate one to use for the concept discussed in Section 17-3?

**9.** Can the exclusion principle be expressed in terms of isospin? See Figure 17-14.

**10.** Is there a physical picture of how the momentum of a $\pi$ meson transferred between the fields of two nucleons leads to an *attractive* force between them? From the point of view of the position-momentum uncertainty principle, is it realistic to expect to be able to construct such a picture?

**11.** What species of $\pi$ mesons are exchanged in proton-proton scattering? In neutron-neutron scattering?

**12.** What particle would remain if a proton emitted a $\pi^-$ meson? If a neutron emitted a $\pi^+$ meson? Why is it that the proton field cannot contain a $\pi^-$ meson, and the neutron field cannot contain a $\pi^+$ meson?

**13.** Why is it believed that the repulsive core of the nucleon potential arises from the exchange of mesons heavier than the pion?

**14.** What examples have been considered in earlier chapters of the conservation of the number of fermions, and the nonconservation of the number of bosons, in an isolated system?

**15.** Exactly what is meant by the statement that a pion has odd intrinsic parity?

**16.** Comparison of the decay rate of cosmic ray muons in flight with the decay rate of muons at rest provided the first experimental verification of relativistic time dilation. What would be a possible way to carry out such a comparison?

**17.** Cosmic ray muons have been used in an attempt to discover hidden burial chambers in Egyptian pyramids, in much the same way that x rays are used to discover internal imperfections in a metal casting caused by gas bubbles. Why were muons used?

**18.** Are there any particles other than neutrinos and antineutrinos which have definite helicities? Explain.

**19.** Why must all field quanta be bosons?

**20.** There are four distinctly different $K$ mesons. Why do we not assign to them the isospin quantum number $T = 3/2$ so that they would constitute an isospin quartet?

**21.** Exactly what does the strangeness quantum number $S$ specify?

**22.** Why is the copious production of $\Lambda^0$ and $K$ particles very difficult to reconcile with their slow decay, without the concept of strangeness? How does strangeness provide a reconciliation?

**23.** Is there a conflict between the statement that isospin magnitude is not conserved in the electromagnetic interaction, and the statement that isospin $z$ component is conserved in that interaction?

**24.** Consider viewing the $\beta$-decay experiment illustrated in Figure 16-15 in a mirror located below the nucleus (the mirror being horizontal) instead of in a mirror located to one side of the nucleus (the mirror being vertical). Explain how the arguments in the text concerning the appearance of the mirror image of the charge conjugate would be modified, but in such a way as to lead to the same conclusion.

**25.** Give an example of a macroscopic system whose behavior is invariant to time reversal, and of a macroscopic system whose behavior is not invariant to this operation.

**26.** Why can we say that the $\pi^0$ meson is its own antiparticle? Do all particles have antiparticles? What about the photon?

**27.** Does it seem reasonable to you to say that a meson or baryon resonance is an elementary particle? Just what is an elementary particle?

**28.** It is sometimes said that strangeness is not the most fundamental quantum number to use because conservation of strangeness is actually a combination of conservation of hypercharge and conservation of baryon number. Explain.

**29.** Are there constituents of matter that cannot be described in terms of combinations of quarks and antiquarks?

## PROBLEMS

**1.** Consult the discussion of the centrifugal potential in Section 15-8, and then: (a) Write the equation which determines the radial dependence $R(r)$ of the deuteron eigenfunction, by evaluating (7-17) for $l = 0$. (b) Show that it can also be written

$$-\frac{\hbar^2}{2\mu}\frac{d^2u(r)}{dr^2} + V(r)u(r) = Eu(r)$$

where

$$u(r) = rR(r)$$

(c) Compare this with the time-independent Schroedinger equation for one-dimensional problems. (d) Give a physical interpretation of $u^*(r)u(r)$. (e) Evaluate, and give a physical interpretation of, the reduced mass $\mu$.

**2.** (a) In the equation obtained in Problem 1, take the nucleon potential $V(r)$ to be a square well of radius $r'$ and depth $V_0$, as in Figure 17-2. (b) Show by substitution that the general solution to the equation obtained is

$$u(r) = A \sin k_1 r + B \cos k_1 r \qquad r < r'$$

$$u(r) = Ce^{-k_2 r} + De^{k_2 r} \qquad r > r'$$

(c) Evaluate $k_1$ and $k_2$ in terms of $\mu$, $V_0$, and the deuteron binding energy $\Delta E$.

**3.** (a) Apply to the general solution obtained in Problem 2 the conditions that $R(r)$, and therefore $u(r)$, must be finite, continuous, and single valued, and have first derivatives with the same properties. (b) Show that the application of these conditions at $r = 0$, $r = r'$, and $r \to \infty$ leads to the relation

$$\frac{\sqrt{2\mu(V_0 - \Delta E)}}{\hbar} \cot\left[\frac{\sqrt{2\mu(V_0 - \Delta E)}}{\hbar} r'\right] = -\frac{\sqrt{2\mu\,\Delta E}}{\hbar}$$

**4.** Show, by substitution, that the relation obtained in Problem 3 has a solution with $\Delta E = 2.2$ MeV, the observed deuteron binding energy, when the potential has a radius and depth of $r' = 2.0$ F and $V_0 = 36$ MeV.

**5.** (a) Use the calculations in Problems 1 through 4 to evaluate the radial dependence of the eigenfunction for the ground state of the deuteron in a potential of radius 2.0 F and depth 36 MeV. (b) Sketch the potential $V(r)$ and the function $u(r) = rR(r)$. (c) Also sketch the radial probability density $P(r)$.

**6.** A nucleon is incident on a nucleon which is initially stationary. Its kinetic energy, which is also the total kinetic energy of the system in that frame of reference, is $K$. Show that the total kinetic energy of the system, in a frame of reference in which the center of mass of the system is stationary, is $K/2$.

**7.** (a) Show that, for a nucleon potential of radius $r' = 2$ F, the maximum value of the orbital angular momentum quantum number is $l_{max} = 1$ unless the kinetic energy of each nucleon exceeds about 30 MeV in the center-of-mass frame of reference. (b) Also show that $l_{max} = 2$ unless the kinetic energies exceed about 60 MeV.

**8.** (a) Calculate the value of $l_{max}$ for a 50 MeV proton incident on a nucleus of atomic weight $A = 100$. Take the radius $r'$ of the optical model potential acting on the proton as the sum of the half-value charge distribution radius $a = 1.07A^{1/3}$ F and the range of nucleon forces 2.0 F. (b) Also evaluate $\theta \simeq \lambda/r'$, and compare with the angle between adjacent minima in the differential scattering cross section shown in Figure 16-25.

**9.** (a) Use the results of the electron scattering measurements, presented in Figure 15-6, to calculate the total number of nucleons per unit volume in the interior of a typical nucleus. (b) Then calculate the average center-to-center spacing of the nucleons. (c) Compare this with the radius of the repulsive core of the nucleon potential, and with the range of the nucleon force.

**10.** The position-momentum uncertainty principle produces an effect which tends to prevent the collapse of a nucleus that would occur if the nucleon potentials had no repulsive regions. (a) Show that this principle demands the kinetic energy of a typical nucleon confined to a nucleus of radius $r'$ must be at least $K$, where

$$K \propto +\frac{1}{r'^2}$$

(b) Although $K$ becomes more positive as $r'$ decreases, the potential energy $V$ of the typical nucleon becomes more negative if the nucleon potentials are purely attractive and the nucleus is sufficiently collapsed to make the separation between all pairs of nucleons less than the range of the nucleon potential. Show that, in these circumstances

$$V \propto -\frac{1}{r'^3}$$

(c) Then show that the total energy of the typical nucleon, $E = K + V$, would become more negative as $r'$ decreases further so that the nucleus would continue to collapse, despite the uncertainty principle, if the nucleon potentials had no repulsive regions.

**11.** Use information contained in Figures 16-14 and 16-34 to assign values of $T$ and $T_z$ to the isobaric analogue ground state levels of: (a) ${}^1H^3$ and ${}^2He^3$; (b) ${}^3Li^7$ and ${}^4Be^7$.

**12.** (a) Estimate the maximum time that a $\pi$ meson can exist in the field of an isolated nucleon before it is absorbed by that nucleon. (b) Estimate how many $\pi$ mesons there can be at any instant in the field at distances from the nucleon about equal to the range of the nucleon force, 2 F. (c) Estimate how many there can be at distances about equal to the radius of the repulsive core, 0.5 F.

**13.** The $\pi^0$ lifetime is determined most accurately by studying the decay from rest of the $K^+$ meson in the mode $K^+ \to \pi^0 + \pi^+$. The average distance traveled by the $\pi^0$ in a block of photographic emulsion before it decays in the easily observable mode $\pi^0 \to e^+ + e^- + \gamma$ is measured, and from the calculated velocity of flight of the $\pi^0$ its lifetime is obtained. Given that the lifetime is $0.9 \times 10^{-16}$ sec, predict the average distance traveled by a $\pi^0$ before it decays.

**14.** In the laboratory (LAB) frame of reference, particle 1 is at rest with total relativistic energy $E_1$, and particle 2 is moving to the right with total relativistic energy $E_2$ and momentum $p_2$. (a) Use the relativistic momentum-energy transformation equations

$$p'_x = \frac{1}{\sqrt{1 - v^2/c^2}}(p_x - vE/c^2)$$

$$p'_y = p_y$$

$$p'_z = p_z$$

$$E' = \frac{1}{\sqrt{1 - v^2/c^2}}(E - vp_x)$$

to show that the frame in which the center of the relativistic masses of the system is at rest is moving to the right with velocity

$$v = c\frac{cp_2}{E_1 + E_2}$$

relative to the laboratory frame, and show that the total momentum of the system is zero in this center-of-mass (CM) frame. (b) Now let the two particles have the same rest mass $m_0$, and let the total relativistic energy of the system in the laboratory frame be $E_{\text{LAB}}$. Evaluate $E_{\text{CM}}$, the total relativistic energy of the system in the center-of-mass frame, and show that

$$E_{\text{CM}} = \sqrt{2m_0c^2E_{\text{LAB}}}$$

**15.** Use the relation quoted in Problem 14b to evaluate the kinetic energy in the laboratory frame of the bombarding proton at which the proton, antiproton pair production process, (17-28), becomes energetically possible.

**16.** (a) Estimate the cross section for a 1 MeV electronic antineutrino incident on a proton to produce the reaction

$$\bar{\nu}_e + p \rightarrow n + e^+$$

(Hint: (i) Assume there is some probability of the reaction occurring when the distance between the $\bar{\nu}_e$ and $p$ is within the $\bar{\nu}_e$ de Broglie wavelength $\lambda$. Then estimate the time interval during which they can be that close. (ii) Estimate the probability $P$ as the ratio of that time interval to the characteristic time $\sim 10^3$ sec for the reaction. (It is the inverse of $n + e^+ \rightarrow p + \bar{\nu}_e$, which is an alternative to $n \rightarrow p + e^- + \bar{\nu}_e$; detailed balancing requires that all three have the same characteristic time which, we see, is just the neutron $\beta$-decay lifetime.) (iii) Take the cross section to be $\sim P\lambda^2$.) (b) Use the estimate to evaluate the mean free path of a 1 MeV $\bar{\nu}_e$ in lead, by justifying the assumption that the cross section for its interaction with a lead nucleus is $\sim 10^2$ times larger than the cross section for its interaction with a proton.

**17.** (a) Why is the $\rho^0$ meson not allowed to decay into two $\pi^0$ mesons? (b) Assuming that the incident deuteron has sufficient energy, why is the reaction $d + d \rightarrow {}^2\text{He}^4 + \pi^0$ not allowed? (c) Why is the decay of a $\pi^+$ meson into an $e^+$ and a $\gamma$ not possible? (d) What prevents the reaction $n \rightarrow p + e^- + \bar{\nu}_e$ from taking place when the neutron is part of a deuteron?

**18.** For each of the following reactions state the fastest interaction through which the conservation laws allow it to proceed. If the reaction is forbidden by all interactions, state why.

(a) $p \rightarrow \pi^+ + e^+ + e^-$
(b) $\Lambda^0 \rightarrow p + e^-$
(c) $\mu^- \rightarrow e^- + \nu_e + \nu_\mu$
(d) $n + p \rightarrow \Sigma^+ + \Lambda^0$
(e) $p + \bar{p} \rightarrow \gamma + \gamma$
(f) $p + \bar{p} \rightarrow n + \overline{\Sigma^0} + K^0$
(g) $K^0 \rightarrow \pi^+ + \pi^- + \pi^0 + \pi^0$

**19.** Describe each of the baryons of Table 17-3 by a combination of three quarks which leads to the correct quantum numbers.

**20.** Associate each of the nine possible quark, antiquark combinations with a meson of Table 17-3, discussing any ambiguities that arise.

# APPENDIX A

# The Special Theory of Relativity

The object of this appendix is to develop those results of Einstein's special theory of relativity that we shall need in our study of quantum physics. Of course it is likely that many students will have worked with relativity, in studying classical mechanics and/or electromagnetism, before embarking on the study of quantum physics. For those students, this appendix can be useful as a review. For others, it should be useful as a concise treatment of the most important results of relativity.

## The Galilean Transformation and Mechanics

In classical physics the *state* of a mechanical system at some instant can be described completely by constructing a frame of reference and using it to specify the coordinates, and the time derivative of the coordinates, for the particles comprising the system at that instant. If we know the masses of the particles and the forces acting between them, Newton's equations of motion make it possible to calculate the state of the system at any future time in terms of its state at the initial time. Now, it is often desirable that during or after such a calculation we specify the state of the system in terms of a new frame of reference which is moving in translation (i.e., not rotating) relative to the first frame with constant velocity. Two questions arise: (1) How do we transform our description of the system from the old to the new frame? (2) What happens to the equations which govern the behavior of the system when we make the transformation? These questions are the ones with which the special theory of relativity concerns itself. (In the general theory, which we shall not need in our study of quantum physics, transformations involving acceleration of one frame relative to the other are considered.)

Figure A-1 shows a particle of mass $m$ whose motion under the influence of force $\mathbf{F}$ is specified in terms of a primed and an unprimed frame of reference. The primed frame is moving relative to the unprimed frame with constant velocity $\mathbf{v}$ in a direction which, by construction, is the positive direction of their collinear $x'$ and $x$ axes. By definition, the times $t'$ and $t$ measured in the two frames are both zero at the instant when the $y'z'$ plane coincides with the $yz$ plane. With these two frames there are two sets of four numbers, $(x',y',z',t')$ and $(x,y,z,t)$, that can equally well be used to specify the coordinates of the particle at any instant of time. What are the relations between these sets of

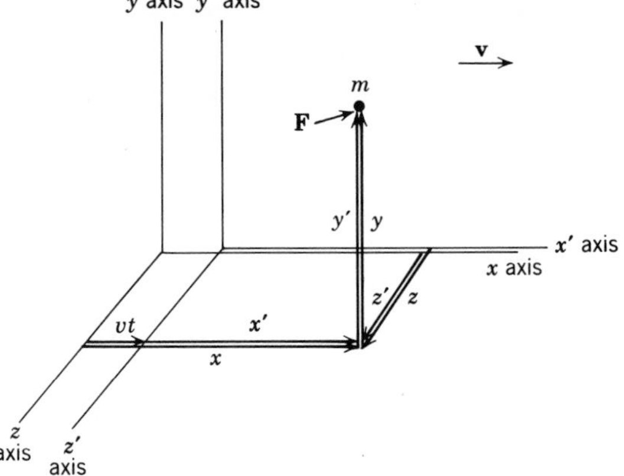


**FIGURE A-1**
An $x'$, $y'$, $z'$, $t'$ frame of reference moving in translation with constant velocity **v** relative to an $x$, $y$, $z$, $t$ frame. The $x'$ and $x$ axes are supposed to be colinear.

numbers? According to classical physics they are

$$\begin{aligned} x' &= x - vt \\ y' &= y \\ z' &= z \\ t' &= t \end{aligned} \tag{A-1}$$

These are known as the *Galilean Transformation.* The simple arguments of classical physics leading to them are:

1. If the zeros of the time scales used in different frames are defined to be the same at any time and location, then in classical physics both time scales will remain the same for all times and all locations, so $t' = t$.

2. Since by construction the $x'y'$ and $xy$ planes always coincide, we have $z' = z$; and similarly for $y' = y$.

3. Since in the time interval between zero and $t' = t$ the $y'z'$ plane moves in the positive direction a distance $vt$, the $x'$ coordinate will be smaller than the $x$ coordinate by that amount. So $x' = x - vt$.

The Galilean transformation constitutes the answer that classical physics gives to the first question posed earlier.

The answer to the second question is given in classical mechanics by using the Galilean transformation to convert Newton's equations in the $x$, $y$, $z$, $t$ frame

$$m\frac{d^2x}{dt^2} = F_x \qquad m\frac{d^2y}{dt^2} = F_y \qquad m\frac{d^2z}{dt^2} = F_z \tag{A-2}$$

into whatever form these equations assume in the $x'$, $y'$, $z'$, $t'$ frame. Note that for (A-2) to be valid the $x$, $y$, $z$, $t$ frame must be an *inertial frame*; i.e., one in which a body not under the influence of a force, and initially at rest, will remain at rest.

By differentiating each of the first three of (A-1) twice with respect to $t$, and then using the fourth to write $t = t'$, it is trivial to show that

$$\frac{d^2x'}{dt'^2} = \frac{d^2x}{dt^2} \qquad \frac{d^2y'}{dt'^2} = \frac{d^2y}{dt^2} \qquad \frac{d^2z'}{dt'^2} = \frac{d^2z}{dt^2}$$

In other words, the acceleration of the mass $m$ measured in the primed frame is the same as it is when measured in the unprimed frame. Of course, the reason is that two frames related by a Galilean transformation are not accelerating with respect to each other, so the transformation does not change the measured acceleration. Furthermore

$$F_{x'} = F_x \qquad F_{y'} = F_y \qquad F_{z'} = F_z$$

because the component of the force **F** acting on $m$ in the direction of the $x'$ or $x$ axis is the same as seen in either frame, and similarly for its other components. Evaluating the unprimed components of acceleration and force in (A-2) in terms of their primed counterparts, but doing nothing to the mass, since in classical physics mass is an intrinsic property of a particle whose value cannot depend on the frame of reference, we find the equations of motion in the primed frame

$$m\frac{d^2x'}{dt'^2} = F_{x'} \qquad m\frac{d^2y'}{dt'^2} = F_{y'} \qquad m\frac{d^2z'}{dt'^2} = F_{z'} \tag{A-3}$$

Note that (A-3) have exactly the same mathematical form as (A-2). Thus part of the answer to the second question is that Newton's equations, which govern the behavior of the mechanical system, do not change when we make a Galilean transformation. The $x$, $y$, $z$, $t$ frame was an inertial frame because $d^2x/dt^2 = d^2y/dt^2 = d^2z/dt^2 = 0$ if $\mathbf{F} = 0$. From (A-3) we see that $x'$, $y'$, $z'$, $t'$ is also an inertial frame because $d^2x'/dt'^2 = d^2y'/dt'^2 = d^2z'/dt'^2 = 0$ if $\mathbf{F} = 0$.

Since Newton's equations are identical in any two inertial frames, and since the behavior of a mechanical system is governed by these equations, it follows that the behavior of all mechanical systems will be identical in all inertial frames, although these frames move at constant velocity with respect to each other. This prediction is verified by a wide variety of experimental evidence.

## The Galilean Transformation and Electromagnetism

Next we inquire into the behavior of electromagnetic systems when we perform a Galilean transformation. Electromagnetic phenomena are treated in classical physics in terms of Maxwell's equations, which govern their behavior just as Newton's equations govern the behavior of mechanical phenomena. We shall not actually carry through the Galilean transformation of Maxwell's equations, as we have for Newton's, since the calculation is complicated. Instead we shall state the results: Maxwell's equations *do* change their mathematical form under a Galilean transformation, in sharp contrast to the behavior of Newton's equations. We shall also discuss the physical significance of these results.

As the student probably knows, Maxwell's equations predict the existence of electromagnetic disturbances which propagate through space in the characteristic manner of wave motion. The nineteenth century physicists, who were very mechanistic in their outlook, felt quite sure that the propagation of waves predicted by Maxwell's equations requires the existence of a mechanical propagation medium. Just as sound waves propagate through a mechanical medium, air, so, according to their view, electromagnetic waves must propagate through a mechanical medium, which they called the *ether*. This propagation medium was required to have quite strange properties in order not to disagree with certain known facts. For instance, it would have to be massless since electromagnetic waves such as light can travel through vacuum; but it would have to have elastic properties to be able to transmit the vibrations inherent in the idea of wave motion. Nevertheless, physicists of that era felt the concept of the ether was more attractive than the alternative of electromagnetic waves propagating without the aid of a propagation medium.

It was assumed that the electromagnetic equations in the form presented by Maxwell were valid for the frame of reference at rest with respect to the ether, the so-called *ether frame*. A solution of these equations led to a prediction of the magnitude of the propagation velocity of electromagnetic waves in vacuum. The result was $2.998 \times 10^8$ m/sec $\equiv c$, in agreement within experimental error with the value of the velocity of light that had been measured by Fizeau. However, in a frame of reference moving with constant velocity with respect to the ether, Maxwell's equations changed form when the Galilean transformation was used to evaluate them in that moving frame. As might be expected, when these changed equations were used to obtain a prediction of the electromagnetic wave propagation velocity that would be measured in the frame moving with respect to the ether, the velocity was found to have a magnitude different from $c$.

The complicated calculation which predicted the velocity of light measured in a frame of reference moving with respect to the ether, performed by making a Galilean transformation of

Maxwell's equations to the moving frame and then solving them in that frame, led to the simple prediction

$$\mathbf{v}_{\text{light wrt moving frame}} = \mathbf{v}_{\text{light wrt ether}} - \mathbf{v}_{\text{moving frame wrt ether}} \tag{A-4}$$

where wrt = with respect to, and $v_{\text{light wrt ether}} = c$. The prediction agreed with two simple physical ideas:

1. Light propagates with a velocity of fixed magnitude $c$ with respect to its propagation medium, the ether, just as sound waves propagate with a velocity of fixed magnitude with respect to their propagation medium, the air.

2. The velocity of light with respect to a frame moving with respect to the ether can be found from a normal vector addition of relative velocities.

It should be pointed out that the arguments justifying vector addition of velocities are really the same as those justifying the Galilean transformation. For instance, in a case when all motion is along the $x'$ or $x$ axis, (A-4) can be obtained immediately by a time differentiation of the first of (A-1), using also the fourth one, $t' = t$.

In summary, theoretical physics near the end of the nineteenth century was based on three fundamentals: Newton's equations, Maxwell's equations, and the Galilean transformation. Almost everything that could be derived from these fundamentals agreed well with the experiments that had been performed to that time. With regard to the questions we have been discussing, they predicted that reference frames in uniform motion with respect to each other were completely equivalent as far as mechanical phenomena were concerned, but in regard to electromagnetic phenomena they were not equivalent; there was only one frame, the ether frame, in which the velocity of light had a magnitude with the numerical value $c$.

## The Michelson-Morley Experiment

In 1887 Michelson and Morley carried out an experiment which proved to be of extreme importance. The experiment was designed to investigate the motion of the earth with respect to the ether frame. Since the earth is moving about the sun, it would seem unrealistic to make the *a priori* assumption that the ether frame travels with the earth and, as we shall indicate later, experimental observations arguing against such an assumption were known at the time. It would be much more reasonable to assume that the ether frame was at rest with respect to the center of mass of the solar system, or the center of mass of the universe. In the first case the velocity of the earth with respect to the ether frame would have a magnitude of the order of $10^4$ m/sec; in the second case the magnitude of the velocity would be somewhat greater. The basic idea of the experiment was to measure the velocity of light in two perpendicular directions from a frame of reference fixed to the earth. A moment's consideration of the classical theory, as summarized by the vector addition (A-4), will show that the theory predicts the measured velocities should have different magnitudes for light traveling in different directions relative to the direction of motion of the observer through the ether.

Although the difference in the two measured light velocities was expected to be small, because the velocity of the earth with respect to the ether is small compared to the velocity of light with respect to the ether, Michelson and Morley built a device incorporating an interferometer that should have been more than sensitive enough to detect and measure the difference. To their extreme surprise, they could not even detect a difference. They, and many other subsequent investigators, repeated the measurements with improved equipment, but an effect was never observed. Despite the predictions of the classical theory, the Michelson-Morley experiment showed that the velocity of light has the same magnitude, $c$, measured in perpendicular directions in a reference frame which is, presumably, moving through the ether frame.

These results captured the attention of most physicists, and a number of them tried to devise explanations that would be consistent with the Michelson-Morley results and yet retain as much as possible of the physical theories then in existence. Notable among them were the "ether drag hypothesis" and the "emission theory."

The ether drag hypothesis assumed that the ether frame was locally attached to all bodies of finite mass. It was attractive because it would explain the Michelson-Morley results and yet did not involve modification of the existing theories. But it could not be accepted for several reasons, the principal one having to do with an astronomical phenomenon called stellar aberration. It had been known since the 1700s that the apparent positions of stars move annually in circles of very small diameter. This is a purely kinematical effect due to the motion of the earth about the sun; in fact, it is the same as the effect causing a vertical shower of rain to appear to a moving observer to be falling at an angle to the vertical. From this analogy it is easy to see that stellar aberration would not be present if light were to travel with velocity of fixed magnitude with respect to the ether frame, and if that frame were dragged along by the earth.

In the emission theory Maxwell's equations are modified in such a way that the velocity of light remains associated with the velocity of its source. This too would explain the Michelson-Morley results since their light source was fixed to the interferometer used to measure the light velocity difference, but it must be rejected because it conflicts with astronomical measurements concerning binary stars. Binary stars are pairs of stars which are rotating rapidly about their common center of mass. Consider such a pair at a time when one is moving toward the earth and the other is moving away. Then, if the emission theory is valid, relative to the earth, the velocity of the light from one star would be larger than that of the light from the other star. This would cause the stars to appear to move in very unusual orbits. However, in 1913 De Sitter showed that observed motions of binary stars are accurately accounted for by Newtonian mechanics when the velocity of the light they emit is taken to have a magnitude independent of their motion.

All the experimental evidence (including evidence from a number of highly accurate contemporary experiments) is consistent only with the conclusion that there is no *special frame* of reference, the ether frame, with the unique property that the velocity of light measured in that frame alone has a magnitude equal to $c$. Just as for inertial frames and mechanical phenomena, *all frames* in relative motion with constant velocity are equivalent in that the velocity of light measured in each frame has the same magnitude $c$. To succinctly put the *experimental evidence*:

*The velocity of light in vacuum is independent of the motion of the observer and of the motion of the source.*

## Einstein's Postulate

Einstein, in 1905, was the first to realize that physicists should abandon the fruitless and misleading concept of the ether. In essence, he accepted the fact that light propagates through vacuum, and that vacuum really is empty! With no ether frame, the only frame of reference that can have any significance to an observer measuring the velocity of light is the frame fixed relative to himself. Then it is not surprising that an observer in all cases obtains the same numerical result, $c$, when he measures the magnitude of the velocity of light. Einstein stated as a *postulate*:

*The laws of electromagnetic phenomena, as well as the laws of mechanics, are the same in all inertial frames of reference, despite the fact that these frames move with respect to each other. Consequently, all inertial frames are completely equivalent for all phenomena.*

This postulate required that Einstein modify either Maxwell's equations or the Galilean transformation, since the two together imply the contrary of the postulate. Although in 1905 the emission theory could still be considered acceptable, he chose not to modify Maxwell's equations. He was then forced to modify the Galilean transformation. This was a bold move. The intuitive belief in the validity of the Galilean transformation was so strong that his contemporaries had never seriously questioned it. Yet, as we shall see, the very different transformation that Einstein adopted in lieu of the Galilean one is based on realistic physical considerations, whereas the Galilean transformation is grossly unrealistic. Another indication of the boldness of Einstein is that our earlier considerations imply that any modification of the Gailean transformation would require some compensating modification of Newton's equations in order that the postulate continue to be satisfied for mechanics. We shall see soon what results this leads to, but first we must study the new transformation equations.

## Simultaneity

Consider the fourth of the Galilean transformation (A-1), which is

$$t' = t$$

The equation says there is the same time scale at all places and for all times in any two frames of reference moving uniformly with respect to each other. This is equivalent to saying that there exists a universal time scale for all such frames. Is this true? To find out we must realistically investigate the procedures used in time measurement.

Let us at first concern ourselves with the problem of defining a time scale in a single frame. Now the basic process involved in any time measurement is a measurement of simultaneity. As Einstein wrote, "If I say 'That train arrives here at 7 o'clock,' I mean something like this: 'The pointing of the small hand of my watch to 7 and the arrival of the train are simultaneous events'." Of course there is no problem at all in determining the simultaneity of events which occur at essentially the same location, like the train and the *nearby* watch or clock used to time its arrival. But there is a problem in determining the simultaneity of events which occur at *separated* locations. In fact this is the key problem involved in setting up a time scale for a frame of reference. In order to have a time scale valid for a whole frame of reference we must have a number of clocks distributed throughout the frame so that there will everywhere be a nearby clock which can be used to measure time in its vicinity. These clocks must be synchronized; that is, we must be able to say of any two of these separated clocks $A$ and $B$: "The little hand of clock $A$ and the little hand of clock $B$ pointed to 7 simultaneously."

A number of methods for determining simultaneity at separated locations are probably now suggesting themselves to the student. They surely all involve the transmission of signals between the two locations. If we had at our disposal a method of transmitting signals with infinite velocity there would be no more of a problem in determining the simultaneity of events occurring at separated locations than there is of doing it for events occurring at the same location. This is where the Galilean transformation goes wrong by implicitly assuming the existence of such a method of synchronization. In fact, there is no such method. Since we have agreed to be realistic in developing a time scale, we must use real synchronization signals. Light (or other electromagnetic) signals are clearly the most appropriate because they have the same propagation velocity under all circumstances. This property enormously simplifies the process of determining simultaneity. Thus we are led to Einstein's *definition of simultaneity of separated events*:

*An event occurring at time $t_1$ and location $x_1$ is simultaneous with an event occurring at time $t_2$ and location $x_2$ if light signals emitted at $t_1$ from $x_1$ and at $t_2$ from $x_2$ arrive simultaneously at the geometrically measured midpoint between $x_1$ and $x_2$.*

This definition, illustrated in Figure A-2, makes the very reasonable statement that two

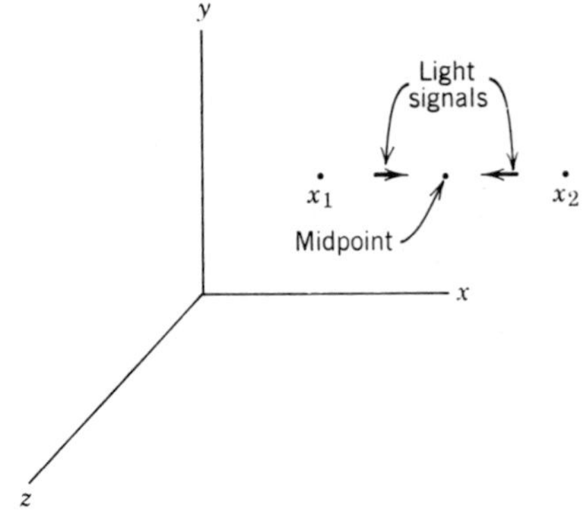


**FIGURE A-2**

Illustrating Einstein's definition of simultaneity.

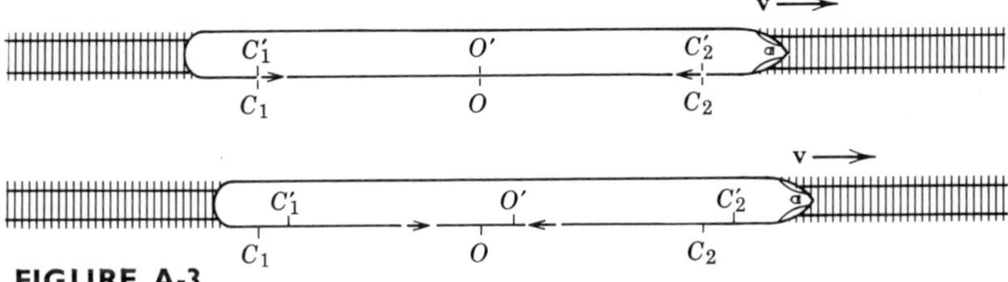


**FIGURE A-3**

Two successive views of a train moving with constant velocity **v**, from the viewpoint of a ground based observer $O$. The small arrows indicate flashes of light.

separated events are simultaneous to an observer located at their midpoint if he sees them happening simultaneously. Note that in Einstein's theory simultaneity in time does not have an absolute meaning, independent of location in space, as it does in the classical theory. The definition intimately mixes the times $t_1$, $t_2$ and the space coordinates $x_1$, $x_2$.

A consequence of this is that two events which are simultaneous when observed from one frame of reference are generally not simultaneous when observed from a second frame of reference which is moving relative to the first. To see this, we consider a very simple "thought experiment," adapted from one used by Einstein. Figure A-3 illustrates the following sequence of events from the point of view of an observer $O$ who is at rest relative to the ground. This observer has so placed two charges of dynamite $C_1$ and $C_2$ that the distances $\overline{OC_1}$ and $\overline{OC_2}$ are equal. He causes them to explode simultaneously in his frame of reference by simultaneously sending out light signals to $C_1$ and $C_2$ which actuate detonators. (He is invoking a reciprocal of the definition quoted earlier.) Assume that he does this so that, in his frame, the explosions occur when he is abreast of $O'$, an observer stationed on a train moving by at a very high velocity **v**. The explosions leave marks $C_1'$ and $C_2'$ on the side of the train. After the experiment $O'$ can measure the distances $\overline{O'C_1'}$ and $\overline{O'C_2'}$. He must, and will, find them equal because otherwise space would not be homogeneous. The explosions also produce flashes of light. Observer $O$ will receive the flashes simultaneously, confirming that in his frame the explosions occurred simultaneously. However $O'$ will receive the flash which originated at $C_2'$ before he receives the flash from $C_1'$ simply because the train moved during the finite time required for the light to reach him. Since the explosions occurred at points equidistant from $O'$, but the light signals were not received simultaneously, he must conclude that in his frame of reference the explosions were not simultaneous.

Such disagreements concerning simultaneity lead to interesting results. From the viewpoint of $O$, $\overline{C_1C_2} = \overline{C_1'C_2'}$. But according to $O'$, $C_2'$ passed $C_2$ before $C_1'$ passed $C_1$ since he received the signal from $C_2'$ first. Therefore $O'$ must conclude that $\overline{C_1C_2} < \overline{C_1'C_2'}$. If this is not apparent, it can be demonstrated by constructing diagrams showing the sequence of events from the viewpoint of $O'$. The simultaneity disagreement will also cause the two observers to disagree concerning the rates of clocks fixed in their respective frames of reference. As we shall see, the nature of their disagreements about the measurement of distance and time intervals is such as to allow both $O$ and $O'$ to find the same value $c$ for the velocity of the light pulses which came from $C_1$ or $C_2$.

## Time Dilation and Length Contraction

We consider here a second thought experiment designed to facilitate the quantitative evaluation of two relativistic effects that were noted qualitatively in the preceding thought experiment. An observer $O'$, moving with velocity **v** relative to observer $O$, wishes to compare a time interval measured by his clock with a measurement of the same time interval made by clocks belonging to $O$. They have already established that, when at rest with respect to each other, all the clocks involved run at the same rate and are synchronized. Now it is apparent

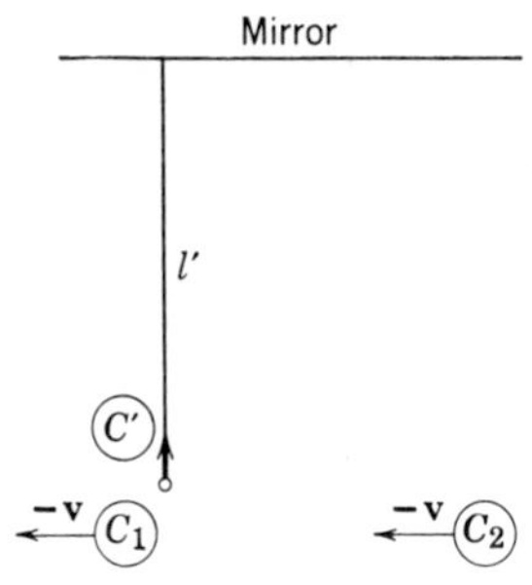


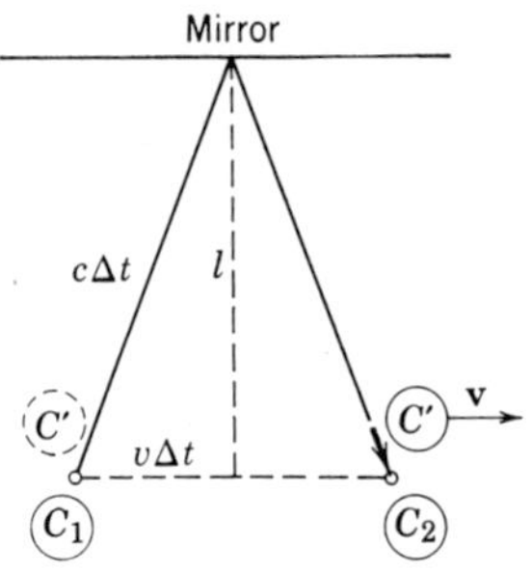


**FIGURE A-4**

The comparison of a time interval measured by two observers. *Left:* The figure shows the situation at the instant of emission of a light signal (the small arrow), from the point of view of $O'$. *Right:* The figure shows the situation at the instant of its reception, from the point of view of $O$.

that, even when in relative motion, the reading of an $O'$ clock can be compared with the reading of an $O$ clock that happens to be momentarily coincident with the former without any complication. Thus measurements of a time interval made with clocks in the two frames can be compared by the procedure illustrated in Figure A-4. $O'$ sends a light signal to a mirror, which reflects it back to him. Both $O$ and $O'$ record the emission of the signal with clocks $C_1$ and $C'$, which are coincident at that instant. They use the clocks $C_2$ and $C'$, which are coincident when the light signal is received back from the mirror, to record the time of its reception. The two events defining the beginning and end of the time interval to be compared are the emission and reception of the light signal.

The elapsed time between these two events measured by $O'$ is $T' = 2\Delta t'$, where $\Delta t' = l'/c$ with $l'$ the distance to the mirror measured in his frame. The elapsed time measured by $O$ is $T = 2\Delta t$. From the figure, and the Pythagorean theorem, it is apparent that

$$c^2 \Delta t^2 = v^2 \Delta t^2 + l^2$$

where $l$ is the distance to the mirror as measured by $O$. Solving for $\Delta t$, we have

$$\Delta t^2 = \frac{l^2}{c^2 - v^2} = \frac{l^2}{c^2} \frac{1}{1 - v^2/c^2}$$

or

$$\Delta t = \frac{l}{c} \frac{1}{\sqrt{1 - v^2/c^2}}$$

Now it is easy to show that observers in relative motion cannot disagree about the measurement of distances *perpendicular* to the direction of motion because disagreements about simultaneity concern finite synchronization signal propagation times for propagation in the direction parallel to the direction of relative motion. Thus we have $l = l'$, and so

$$\Delta t = \frac{l'}{c} \frac{1}{\sqrt{1 - v^2/c^2}} = \frac{\Delta t'}{\sqrt{1 - v^2/c^2}}$$

Therefore we obtain

$$T = \frac{1}{\sqrt{1 - v^2/c^2}} T' \tag{A-5}$$

We have found that a time interval between two events occurring at the same place in a certain frame is measured to be *longer* by a factor of $1/\sqrt{1 - v^2/c^2}$ in a frame moving relative to the first frame and, consequently, in which the two events occur at separated locations. The time interval measured in the frame in which the events occurred in the same place is called the *proper time*. The effect involved is called *time dilation*.

Next we consider the same thought experiment, but we imagine a measuring rod placed in the $O$ frame with one end at clock $C_1$ and the other end at clock $C_2$. Designate by $L$ the length of the rod measured in the $O$ frame, with respect to which it is at rest. We want to evaluate $L'$, the length of the rod measured from the $O'$ frame.

In this frame the rod is moving in a direction parallel to its own length. Since the velocity of $O'$ with respect to $O$ is $\mathbf{v}$, the velocity of $O$, and also of the rod, with respect to $O'$ must be precisely $-\mathbf{v}$. Otherwise there would be an inherent asymmetry between the two frames that is not allowed by Einstein's postulate. $T'$ is the time interval between the instant when $O'$ sees the front end of the rod pass his clock $C'$ and the instant when he sees the rear end pass the clock. This time interval is related to the length $L'$ of the rod as measured in the $O'$ frame, and to the magnitude $v$ of its velocity measured in that frame, by the equation

$$L' = vT'$$

We may also establish an equation connecting the corresponding quantities as measured in the $O$ frame. In this frame $C'$, which is moving with velocity of magnitude $v$, travels the distance $L$ in time $T$. Thus

$$L = vT$$

From the last two equations we obtain

$$L' = L\frac{T'}{T}$$

But the time dilation argument shows that

$$\frac{T'}{T} = \sqrt{1 - v^2/c^2}$$

Therefore

$$L' = \sqrt{1 - v^2/c^2}\, L \tag{A-6}$$

We have found that a rod is measured to be *shorter* by a factor $\sqrt{1 - v^2/c^2}$ when the measurement is made in a frame in which it is moving *parallel* to its own length, compared to its length measured in a frame in which it is at rest. The length of the rod measured in the frame in which it is at rest is called its *proper length*. The effect is called the *Lorentz contraction*. Note that a comparison of (A-6) with the equation immediately above it shows the factor relating the primed to the unprimed time interval is the same as (and not the reciprocal of) the factor relating the primed to the unprimed distance interval.

## The Lorentz Transformation

Now we shall obtain the equations that are used in relativity theory to transform space and time variables from one frame to another moving with constant velocity relative to the first. Our argument will be guided by what we have already learned, but in the final analysis it is an independent derivation based on the experimental evidence that the velocity of light is independent of the motion of the observer and of the source.

We consider a third thought experiment involving two observers $O'$ and $O$, with $O'$ moving relative to $O$ at velocity of magnitude $v$ in the positive direction of the $x'$ and $x$ axes. Their $x'y'$ and $xy$ planes always coincide, as in Figure A-1, and the origins of their reference frames coincide at the instant $t' = t = 0$. At that instant $O'$ ignites a flash bulb at his origin which produces a wavefront of light that expands away from the point of emission with velocity of magnitude $c$ in all directions. Therefore, according to $O'$ at time $t'$, the wave front will be a sphere, centered on his origin, of radius $r' = ct'$. The coordinates of any point on the wave front at that time will thus satisfy the equation of a sphere

$$x'^2 + y'^2 + z'^2 = c^2t'^2 \tag{A-7}$$

But it will be equally true that according to $O$ the light is expanding away from the point of emission, his origin, with velocity of magnitude $c$ in all directions. Thus from the point of view of $O$ the wave front at time $t$ is also a sphere of radius $r = ct$ centered on his own origin, and satisfying the equation

$$x^2 + y^2 + z^2 = c^2t^2 \tag{A-8}$$

We shall find relations between the two sets of variables $(x',y',z',t')$ and $(x,y,z,t)$ which allow both (A-7) and (A-8) to be valid, i.e., which transform one equation into the other.

We are guided by our earlier considerations to assume the following form for the transformation equations

$$\begin{aligned} x' &= \gamma(x - vt) \\ y' &= y \\ z' &= z \\ t' &= \gamma(t + \delta) \end{aligned} \tag{A-9}$$

where $\gamma$ is a dimensionless quantity, presumably involving the relative velocity of the two frames, $v$, and the velocity of light, $c$, and where $\delta$ is a quantity, also presumably involving these velocities, which must have the dimensions of time. Expressions for $\gamma$ and $\delta$ will be determined soon, but we can say even now that we should have $\gamma \to 1$ and $\delta \to 0$ if $v/c \to 0$. The reason is that for $\gamma = 1$ and $\delta = 0$ (A-9) reduce to the Galilean transformation (A-1), which is as it should be since the Galilean transformation would be essentially correct if the relative velocity $v$ of the frames is extremely small compared to the velocity $c$ of the signals used to synchronize the clocks in the frames. We inserted the additive term $\delta$ in the fourth equation when $v/c$ is not small because according to $O'$ the time of some event measured by $O$ must be corrected for a synchronization error between the clock used by $O$ at the event and the clock used by $O$ at his origin, as discussed in our first thought experiment. Having accounted for synchronization, we put the multiplicative factor $\gamma$ in the fourth equation to account for the discrepancy in time intervals measured by $O'$ and $O$, as discussed in our second thought experiment. As was also discussed there, the same factor $\gamma$ should appear in the first of (A-9) to account for the discrepancy in distance intervals measured by the two observers. Since $y$ and $z$ are distances measured perpendicular to the direction of relative motion, we assumed that their values will not be changed by the transformation.

Now let us see whether the forms assumed in (A-9) can actually transform (A-7) into (A-8) and, if so, what expressions for $\delta$ and $\gamma$ are required to accomplish this. Using (A-9) to rewrite each variable in (A-7) in terms of the unprimed variables, we have

$$\gamma^2(x^2 - 2vxt + v^2t^2) + y^2 + z^2 = c^2\gamma^2(t^2 + 2\delta t + \delta^2)$$

As we must obtain from this (A-8), which does not contain a term with the combination of variables $xt$, the second term in the parenthesis on the left side must be canceled by something on the right side. For the cancelation to be obtained for all values of the independent variable $t$, it must be due only to the second term in the parenthesis on the right. Thus we must have

$$-\gamma^2 2vxt = c^2\gamma^2 2\delta t$$

or

$$\delta = -vx/c^2 \tag{A-10}$$

Note that $\delta$ has the dimensions of time, and that $\delta \to 0$ if $v/c \to 0$, as predicted earlier. A reconsideration of our first thought experiment will make it apparent why the synchronization correction $\delta$ is linearly proportional to both $v$ and $x$. Gathering the factors of $x^2$ and $t^2$ in the remaining terms of the equation after evaluating $\delta^2$, we obtain

$$x^2\gamma^2(1 - v^2/c^2) + y^2 + z^2 = c^2t^2\gamma^2(1 - v^2/c^2)$$

Comparing this with the required form, (A-8), we see that we shall obtain it if

$$\gamma^2(1 - v^2/c^2) = 1$$

or

$$\gamma = \frac{1}{\sqrt{1 - v^2/c^2}} \tag{A-11}$$

Note that $\gamma$ is dimensionless, and that $\gamma \to 1$ if $v/c \to 0$, as also predicted earlier. Considering the results of our second thought experiment, it is not surprising that $\gamma$ involves the expression $\sqrt{1 - v^2/c^2}$. Finally, we use (A-10) and (A-11) to evaluate $\gamma$ and $\delta$ in (A-9), and successfully complete our derivation of the *Lorentz transformation*

$$\begin{aligned} x' &= \frac{1}{\sqrt{1 - v^2/c^2}}(x - vt) \\ y' &= y \\ z' &= z \\ t' &= \frac{1}{\sqrt{1 - v^2/c^2}}(t - vx/c^2) \end{aligned} \tag{A-12}$$

The space-time variables transformation of relativity is called the Lorentz transformation for the historical reason that equations of the same mathematical form (but with a very different physical significance because $v$ represented a velocity with respect to the ether frame instead of a velocity of any inertial frame with respect to any other inertial frame) had been proposed by Lorentz in connection with a classical theory of electrons some years before the work of Einstein.

The Lorentz transformation reduces, as expected, to the Galilean transformation when the relative velocity of the two frames, $v$, is small compared to the velocity of light, $c$. But significant differences between the predictions of the Galilean transformation and those of the rigorously correct Lorentz transformation are found when $v$ is comparable to $c$. These had not been observed in classical physics because the appropriate experiments had not been performed. Many experimental results of quantum physics, some of which are discussed in this book, show that the Lorentz transformation is, in fact, the one that accurately describes nature. Note that for $v$ larger than $c$ the Lorentz transformation equations are meaningless, in that real coordinates and times are transformed into imaginary ones. Thus $c$ appears to play the role of a limiting velocity for all physical phenomena. We shall obtain a better understanding of this as we go further into relativity theory.

## The Relativistic Velocity Transformation

Consider the particle shown in Figure A-5, moving with velocity $\mathbf{u}$ as measured in a frame of reference $O$. We would like to evaluate the velocity $\mathbf{u}'$ of the particle as measured in the frame $O'$, which is itself moving relative to $O$ with velocity $\mathbf{v}$.

Measured in the $O$ frame, the velocity vector of the particle has components

$$u_x = \frac{dx}{dt} \qquad u_y = \frac{dy}{dt} \qquad u_z = \frac{dz}{dt}$$

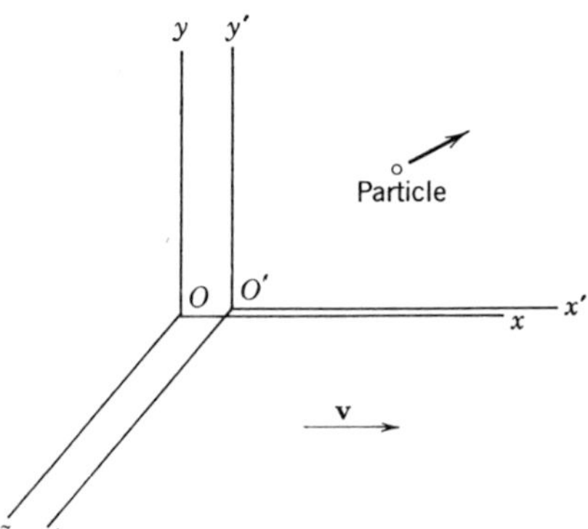


**FIGURE A-5**
A moving particle observed from two frames of reference $O$ and $O'$, with the latter moving relative to the former at velocity $\mathbf{v}$.

The velocity vector, as measured in the $O'$ frame, has components

$$u'_x = \frac{dx'}{dt'} \qquad u'_y = \frac{dy'}{dt'} \qquad u'_z = \frac{dz'}{dt'}$$

To establish the required relationships, we take the differentials of the Lorentz transformation, (A-12), remembering that $v$ is a constant. This gives

$$\begin{aligned}
dx' &= \frac{1}{\sqrt{1 - v^2/c^2}}(dx - v\,dt) \\
dy' &= dy \\
dz' &= dz \\
dt' &= \frac{1}{\sqrt{1 - v^2/c^2}}(dt - v\,dx/c^2).
\end{aligned}$$

So we obtain

$$u'_x = \frac{dx'}{dt'} = \frac{\dfrac{1}{\sqrt{1 - v^2/c^2}}(dx - v\,dt)}{\dfrac{1}{\sqrt{1 - v^2/c^2}}\left(dt - \dfrac{v\,dx}{c^2}\right)} = \frac{\dfrac{dx}{dt} - v}{1 - \dfrac{v}{c^2}\dfrac{dx}{dt}} = \frac{u_x - v}{1 - \dfrac{vu_x}{c^2}}$$

$$u'_y = \frac{dy'}{dt'} = \frac{dy}{\dfrac{1}{\sqrt{1 - v^2/c^2}}\left(dt - \dfrac{v\,dx}{c^2}\right)} = \frac{\dfrac{dy}{dt}}{\dfrac{1}{\sqrt{1 - v^2/c^2}}\left(1 - \dfrac{v^2}{c^2}\dfrac{dx}{dt}\right)} = \frac{\sqrt{1 - v^2/c^2}\,u_y}{1 - \dfrac{vu_x}{c^2}} \qquad \text{(A-13)}$$

$$u'_z = \frac{dz'}{dt'} = \frac{dz}{\dfrac{1}{\sqrt{1 - v^2/c^2}}\left(dt - \dfrac{v\,dx}{c^2}\right)} = \frac{\dfrac{dz}{dt}}{\dfrac{1}{\sqrt{1 - v^2/c^2}}\left(1 - \dfrac{v^2}{c^2}\dfrac{dx}{dt}\right)} = \frac{\sqrt{1 - v^2/c^2}\,u_z}{1 - \dfrac{vu_x}{c^2}}$$

These equations constitute the relativistic velocity transformation.

Note that as $v/c$ approaches zero (A-13) approach those which would be derived from the Galilean transformation. Another interesting property is that it is impossible to choose **u** and **v** such that $u'$, the magnitude of the velocity measured in the new frame, is greater than $c$. Consider the example illustrated in Figure A-6. As measured by $O$, particle 1 has velocity $0.8c$ in the positive $x$ direction and particle 2 has velocity $0.9c$ in the negative $x$ direction. We evaluate the velocity of particle 1 as measured in a frame $O'$ moving with particle 2 using the first of (A-13), with $u_x = u_1 = 0.8c$ and $v = -0.9c$. We obtain

$$u'_1 = \frac{0.8c - (-0.9c)}{1 - \dfrac{(-0.9c)(0.8c)}{c^2}} = \frac{1.70c}{1.72} = 0.99c$$

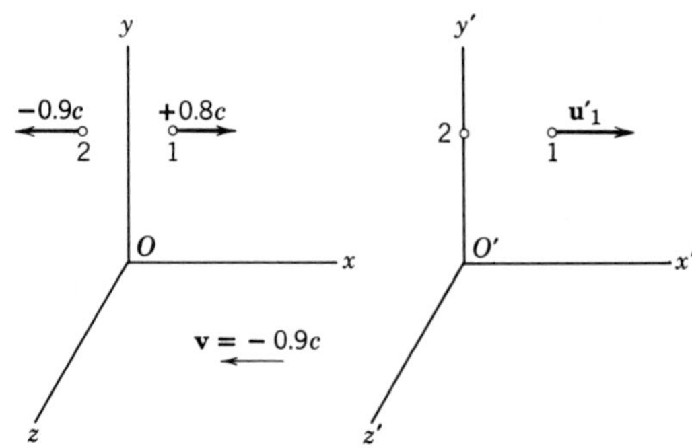


**FIGURE A-6**

Illustrating an example of the relativistic addition of velocities.

The velocity transformation equations demonstrate another aspect of the fact that $c$ acts as a limiting velocity for all physical phenomena.

## Relativistic Mass

It has been emphasized that Einstein's modification of the transformation equations would necessitate some compensating modification in the equations of mechanics, so that these equations continue to satisfy the requirement of not changing form in a transformation from one inertial frame to another moving relative to the first. Now we shall begin to develop the new mechanics, which is called *relativistic mechanics.*

Clearly it is desirable to carry over into relativistic mechanics as much of classical mechanics as the circumstances allow. We shall see that it is possible to preserve Newton's equation of motion, in a form equivalent to the one originally given by Newton

$$\mathbf{F} = \frac{d\mathbf{p}}{dt} \tag{A-14}$$

where $\mathbf{p}$ is the momentum of a particle acted on by force $\mathbf{F}$. It is also possible to preserve the very closely related classical law of momentum conservation for the particles in an isolated system

$$\left[\sum_{\text{all particles}} \mathbf{p}\right]_{\text{initial}} = \left[\sum_{\text{all particles}} \mathbf{p}\right]_{\text{final}} \tag{A-15}$$

It will even be possible to preserve the classical definition of the momentum of a particle

$$\mathbf{p} = m\mathbf{v} \tag{A-16}$$

where $m$ is its mass and $\mathbf{v}$ is its velocity. But to do all this it will be necessary to allow the mass of a particle to be a function of the magnitude of its velocity, i.e.

$$m = m(v) \tag{A-17}$$

The form of this function is to be determined. However, we know *a priori* that we must have $m(v) = m_0$ if $v/c \ll 1$, where the constant $m_0$ is the classically measured mass of the particle. The reason is that when a characteristic velocity becomes very much smaller than the velocity of light the pertinent Lorentz transformation approaches a Galilean transformation and no modification of mechanics is necessary.

In order to evaluate the function $m(v)$, we consider the following thought experiment. As measured in the $x, y, z, t$ frame indicated in Figure A-7, observers $O_1$ and $O_2$ are moving in directions parallel to the $x$ axis with equal magnitude but oppositely directed velocities. These observers have identical particles, say billiard balls $B_1$ and $B_2$, each of mass $m_0$ as measured when they are at rest. While passing, each throws his ball so as to hit the other's ball with a velocity which, from his own point of view, is directed perpendicular to the $x$ axis and is of magnitude $u$.

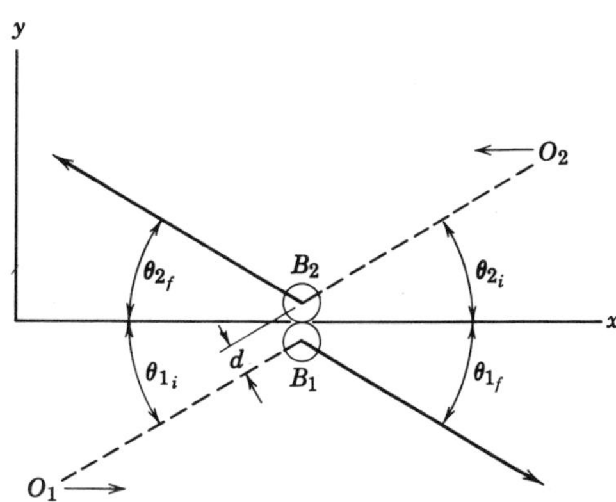


**FIGURE A-7**

A symmetrical collision between two balls of identical rest mass.

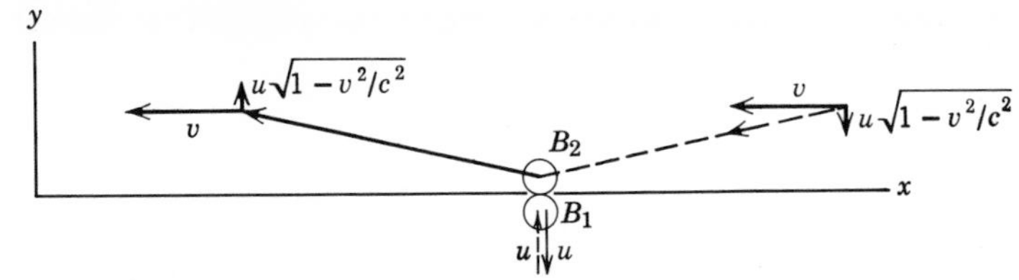


**FIGURE A-8**

A symmetrical collision, as observed by $O_1$. Since $u$ is supposed to be very much smaller than $v$, the angles made by the trajectories of $B_2$ and the $x$ axis are actually very much smaller than shown.

As observed in the $x$, $y$, $z$, $t$ frame, $B_1$ and $B_2$ will approach along parallel paths making angles $\theta_{1_i} = \theta_{2_i}$ with the $x$ axis, and rebound on paths at angles $\theta_{1_f}$ and $\theta_{2_f}$ to that axis. Assuming conservation of momentum and that the collision is elastic, it is easy to show that $\theta_{1_f} = \theta_{2_f}$ and that the magnitude of the velocity of the balls is the same after the collision as before. The actual value of $\theta_{1_f}$ and $\theta_{2_f}$ depends on the impact parameter $d$, which we assume to be such that $\theta_{1_f} = \theta_{1_i}$ as shown in the figure.

Now consider the process from the point of view of $O_1$, as illustrated in Figure A-8. $O_1$ throws $B_1$ along a line parallel to his $y$ axis with velocity of magnitude $u$, which we shall take to be very small compared to $c$. It returns along the same line with velocity of the same magnitude but opposite sign. He sees $B_2$ maintain a constant $x$ component of velocity just equal to $v$ the velocity of $O_2$ relative to $O_1$, which we shall take to be comparable to $c$. The component of velocity of $B_2$ along his $y$ axis is observed by $O_1$ to change sign during the collision but to maintain a constant magnitude. To evaluate this magnitude we realize that the $y$ component of the velocity of $B_2$, as measured by $O_2$, is $u$. Then we transform this to the $O_1$ frame with the aid of the second of (A-13) and obtain $u\sqrt{1 - v^2/c^2}$ for the magnitude of the $y$ component of velocity of $B_2$ as measured by $O_1$.

The $y$ momenta of both $B_1$ and $B_2$, as measured in the $O_1$ frame, simply change sign during the collision. Consequently the total $y$ momentum of the isolated system of two colliding balls changes sign. If the momentum conservation law (A-15) is to be valid, the total $y$ momentum before the collision must equal the total $y$ momentum after. This can be true only if the total $y$ component of momentum of the system measured by $O_1$ is zero before the collision. Evaluating $y$ components of momentum as the masses times the $y$ components of velocity from the definition of (A-16), and equating their sum to zero, we obtain an equation that is obviously self-contradictory if we insist that both masses have the value $m_0$ that they have when measured in frames in which they are at rest. The reason is that according to $O_1$ the magnitude of the $y$ component of velocity of $B_1$ is $u$, while the magnitude of the $y$ component of velocity of $B_2$ is $u\sqrt{1 - v^2/c^2}$.

However, if we allow the mass of a particle to be a function of the magnitude of its total velocity vector we can satisfy the momentum conservation law. Since $u$ is very small compared to $v$, the magnitude of the velocity vector of $B_2$ as measured by $O_1$ is essentially $v$, as can be seen in Figure A-8. The magnitude of the velocity vector of $B_1$ according to $O_1$ is just $u$. Thus $O_1$ would write the momentum conservation law for $y$ components as

$$m(u)u - m(v)u\sqrt{1 - v^2/c^2} = 0$$

or

$$m(u) = m(v)\sqrt{1 - v^2/c^2}$$

Since $u$ is very small compared to $c$, we may take $m(u) = m_0$ and obtain

$$m(v) = \frac{1}{\sqrt{1 - v^2/c^2}} m_0 \tag{A-18}$$

A theory of relativistic mechanics consistent with momentum conservation demands that the mass $m(v)$ of a particle measured when it is moving with velocity of magnitude $v$ be larger than its mass $m_0$ measured when it is at rest by the factor $1/\sqrt{1 - v^2/c^2}$. The mass $m(v)$ is called the *relativistic mass* of the particle and $m_0$ is called the *rest mass*. A reconsideration of our arguments will show that the two observers in the thought experiment measure different values for the mass of the particle because of the difference in their measurements of its velocity component perpendicular to the direction of their relative motion, and that this arises because of the difference in their measurements of time intervals.

For the quite high velocity $v = 0.1c$ the relativistic mass is only one-half of 1% greater than the rest mass. But with increasing $v$ the relativistic mass rapidly increases since $m(v) \to \infty$ as $v \to c$ if $m_0$ has any finite value. It is apparent that the velocity of a particle cannot exceed $c$.

## Relativistic Energy

Consider a particle of rest mass $m_0$ initially stationary at $x = 0$. A force of magnitude $F$ is then applied in the positive $x$ direction and the particle moves under the influence of the force. It is interesting to calculate the total work done by the force when the particle moves to $x = x_f$. We shall label this work $K$. Taking the usual definition of work

$$K = \int_0^{x_f} F\,dx = \int_0^{t_f} F\frac{dx}{dt}\,dt = \int_0^{t_f} Fv\,dt$$

where $t_f$ is the time at which the particle arrives at $x_f$. In order to evaluate the integral we must know the relativistic form of Newton's equation of motion. With a relativistically acceptable expression for momentum $\mathbf{p} = m\mathbf{v}$, where $m$ is the relativistic mass, we can with confidence take over into relativity Newton's equation in the form of (A-14)

$$\mathbf{F} = \frac{d\mathbf{p}}{dt} = \frac{d(m\mathbf{v})}{dt}$$

This is essentially just a definition of the force $\mathbf{F}$. Using it to evaluate $F$ in the integral, we have

$$K = \int_0^{t_f} v\frac{dp}{dt}\,dt = \int_0^{p_f} v\,dp$$

Integrating by parts, we obtain

$$K = [vp]_0^{v_f} - \int_0^{v_f} p\,dv$$

Evaluating the $m$ in $p = mv$ from (A-18), and also expressing $v\,dv$ as $d(v^2)/2$, we have

$$K = \left[\frac{m_0 v^2}{\sqrt{1 - v^2/c^2}}\right]_0^{v_f} - \frac{m_0}{2}\int_0^{v_f} \frac{d(v^2)}{\sqrt{1 - v^2/c^2}}$$

The integral is a standard form that yields

$$K = m_0c^2\left[\frac{v^2/c^2}{\sqrt{1 - v^2/c^2}} + \sqrt{1 - v^2/c^2}\right]_0^{v_f}$$

which reduces to

$$K = m_0c^2\left[\frac{1}{\sqrt{1 - v^2/c^2}}\right]_0^{v_f}$$

Evaluating at the limits, and then dropping the subscript $f$ for final to simplify the notation, we have

$$K = \frac{m_0c^2}{\sqrt{1 - v^2/c^2}} - m_0c^2 \tag{A-19}$$

Now the classical law of energy conservation implies that the total work done by the force acting on the particle should equal its kinetic energy. Thus we would like to call $K$ the kinetic energy of the particle. To check in the classical limit take $v/c \ll 1$, and expand the reciprocal of the square root, to obtain

$$K = m_0c^2\left[\left(1 - \frac{v^2}{c^2}\right)^{-1/2} - 1\right] \simeq m_0c^2\left[1 + \frac{1}{2}\frac{v^2}{c^2} - 1\right]$$

or

$$K \simeq \frac{m_0c^2}{2}\frac{v^2}{c^2} = \frac{m_0v^2}{2}$$

This agrees with the classical expression for kinetic energy, and confirms our identification of $K$ in (A-19) as the *relativistic kinetic energy.*

Continuing the interpretation of (A-19), we observe that $K$ is a function of $v$ which can be written as the difference between a term depending on $v$ and a constant term, as follows

$$K(v) = E(v) - E(0)$$

where $E(v) = m_0c^2/\sqrt{1 - v^2/c^2} = mc^2$, with $m$ the relativistic mass; and where $E(0)$ is the value of $E(v)$ for $v = 0$, i.e., $E(0) = m_0c^2$. Since $K$ is an energy, $E(v)$ and $E(0)$ must also be energies—$E(v)$ being some energy associated with the particle when its velocity is $v$, and $E(0)$ some energy associated with the particle when its velocity is 0. To identify these energies, we rewrite the equation as

$$E(v) = K(v) + E(0)$$

The conclusion is inescapable. We must interpret $E(v)$ as the total energy of the particle moving with velocity $v$, since it is the sum of the kinetic energy $K(v)$ of the particle and an intrinsic energy $E(0)$ associated with the particle when it is at rest. The energy $E(v)$ is called the *total relativistic energy*, and $E(0)$ is called the *rest mass energy.*

We have established Einstein's well known relations between mass and energy: *The rest mass energy $E(0)$ of a particle is $c^2$ times its rest mass $m_0$*

$$E(0) = m_0c^2 \tag{A-20}$$

and *the total relativistic energy $E$ of a particle is $c^2$ times its relativistic mass $m$*

$$E = mc^2 \tag{A-21}$$

Equation (A-19) tells us the relation between total relativistic energy $E$, relativistic kinetic energy $K$, the rest mass energy $m_0c^2$

$$E = K + m_0c^2 \tag{A-22}$$

It is often convenient to have an expression for the total relativistic energy that explicitly involves the momentum $p$. Such can be obtained by evaluating the quantity

$$m^2c^4 - m_0^2c^4 = m_0^2c^4\left(\frac{1}{1 - v^2/c^2} - 1\right) = m_0^2c^4\,\frac{v^2/c^2}{1 - v^2/c^2}$$

$$= \frac{m_0^2c^2v^2}{1 - v^2/c^2} = c^2m^2v^2 = c^2p^2$$

Thus

$$m^2c^4 = c^2p^2 + m_0^2c^4$$

or

$$E^2 = c^2p^2 + m_0^2c^4 \tag{A-23}$$

Although the choices made in the theory of relativistic mechanics seem reasonable, their ultimate justification is found in comparing the predictions of the theory with appropriate experiments. Several very successful comparisons will be given in this book, but it is appropriate here to point out that the existence of a rest mass energy $m_0c^2$ is not in conflict with classical physics. Since the experiments in that field all involve systems in which the total rest mass is essentially constant, the appropriate rest mass energies can be added to both sides of all classical energy balance equations without destroying their validity.

The theory is, however, of more than academic interest because there are important processes in nature in which the total rest mass of an isolated system changes significantly. For such processes the experiments of quantum physics show that the change in rest mass energy is exactly compensated for by a change in kinetic energy in such a way as to conserve the total relativistic energy of the system. This is, of course, what happens in a nuclear reactor. Consequently, in relativity we must replace the separate classical laws of conservation of mass and conservation of energy by a single comprehensive law of *conservation of total relativistic energy*:

*As measured in a given frame of reference, the total relativistic energy of an isolated system remains constant.*

We close our concise development of relativity by stating that explicit calculations demonstrate that neither Newton's equation as expressed in (A-14), nor Maxwell's equations, change form under a Lorentz transformation from one frame of reference to another moving relative to the first. However, these calculations show that the force in the case of the mechanical equation, and the electric and magnetic fields in the case of the electromagnetic equations, change when Lorentz transformed from one frame to the other. Although we cannot go into these matters here, their study elsewhere is recommended to the student as adding very worthwhile physical insight—particularly into the relationship between electric and magnetic fields. See Appendix L.

# APPENDIX B

# The Radiation from an Accelerated Charge

Here we give a largely qualitative view of the classical theory of emission of electromagnetic radiation from an accelerated charge, restricting ourselves to the case of a stationary charge in vacuum that is suddenly accelerated to a non-relativistic velocity $v \ll c$.

We know that a stationary charge has an associated static electric field **E** whose energy per unit volume is given by

$$\rho = \frac{1}{2}\,\epsilon_0 E^2 \tag{B-1}$$

This energy is stored in the field and is not radiated away. If the charge moves with a uniform velocity, there is a magnetic field **B** associated with it as well as an electric field. The total energy stored in the nonstatic field of a uniformly moving charge is larger than for the static field of a stationary charge, the additional energy being supplied from the work done by the forces that initially produced the motion of the charge. The energy density in this case is given by

$$\rho = \frac{1}{2}\,\epsilon_0 E^2 + \frac{1}{2\mu_0}\,B^2 \tag{B-2}$$

and the energy stored in the field moves along with the charge. That the energy is not radiated away, even in this case, follows from transforming to a reference frame in which the charge is stationary and applying the relativistic requirement that the behavior of the charge, including whether or not it radiates, cannot depend on the frame of reference from which it is viewed. Hence for a charge having constant velocity, the electric and magnetic fields are able to adjust themselves in such a way that no energy is radiated, even though these fields are not static.

For an accelerated charge, however, the nonstatic electric and magnetic fields cannot adjust themselves in such a way that none of the stored energy is radiated. We can understand this qualitatively by considering the behavior of the electric field. In Figure B-1 we describe this field by drawing some of the lines of force surrounding a charge which was at rest at the initial instant $t$, suffered a constant acceleration **a** to the right during the interval $t$ to $t'$, and then continued moving with a constant final velocity. The figure shows the lines of force at some later instant $t''$, as viewed from the frame of reference moving at that velocity $v$. At small distances the lines of force are directed radially outward from the present position of the charge. At large distances

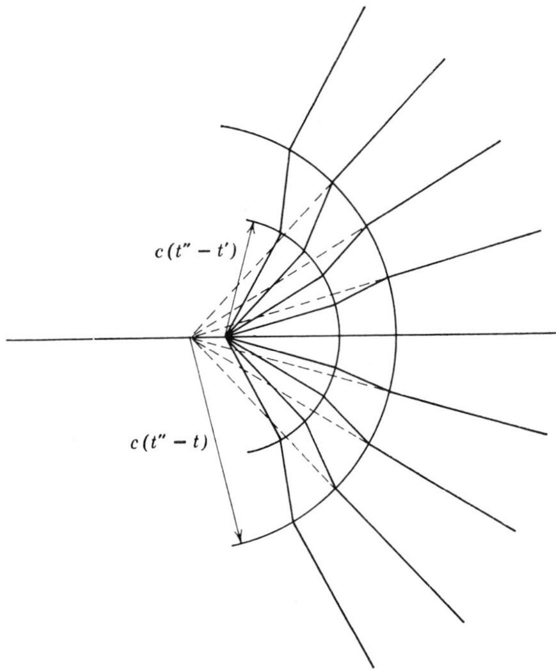


**FIGURE B-1**

The lines of force surrounding an accelerated charge. Only some of the lines are shown.

they emanate from where the field would anticipate it to be if unaccelerated. The reason is that information concerning the position of the charge cannot be transmitted to distant locations with infinite velocity, but only with the velocity $c$. As a result, there are kinks in the lines of force found between a sphere centered on the anticipated position and of radius $c(t'' - t)$, which is the minimum distance at which the field can "know" the acceleration started, and a sphere centered on the actual position and of radius $c(t'' - t')$, which is the minimum distance at which the field can know that the acceleration stopped. As $t''$ increases, the region containing the kinks expands outward with velocity $c$. That is, each kink of adjustment propagates along its line of force in much the same way as a kink set up at one end of a long stretched rope propagates along the rope. The electric field in the region containing kinks has components which are both longitudinal and transverse to the direction of expansion. But, by constructing diagrams for several values of $t''$, it is easy to see that the longitudinal component dies out very rapidly and can soon be ignored, whereas the transverse component dies out slowly. In fact, electromagnetic theory shows, by calculations based upon the same idea as in our qualitative discussion, that at large distances from the region of the acceleration (large $t''$) the transverse electric field obeys the equation

$$E_{\perp} = \frac{qa}{4\pi\epsilon_0 c^2 r} \sin\theta \tag{B-3}$$

In this equation, which is valid only if $v/c \ll 1$, $r = c(t'' - t)$ is the magnitude of the vector $\mathbf{r}$ from the region at which the acceleration $\mathbf{a}$ took place to the point at which the transverse field is evaluated, and $\theta$ is the angle between $\mathbf{r}$ and $\mathbf{a}$. The dependence of $E_{\perp}$ on $\theta$ and $r$ can be seen from Figure B-1 and comparable diagrams for larger values of $t''$, and it should be clear from our discussion that $E_{\perp}$ must be proportional to $q$ and $a$. Similarly, there is a transverse magnetic field moving along with $E_{\perp}$, and at large distances from the region of acceleration its strength, if $v/c \ll 1$, is given by

$$B_\perp = \frac{\mu_0 qa}{4\pi cr} \sin\theta \tag{B-4}$$

These two transverse fields propagating outward with velocity $c$ form the electromagnetic radiation emitted by the accelerated charge. The radiated field is polarized with **E** in the plane of **a** and **r** and with **B** at right angles to this plane. The energy density of the radiation is

$$\rho = \frac{1}{2}\epsilon_0 E_\perp^2 + \frac{1}{2}\frac{B_\perp^2}{\mu_0}$$

or, with $c = 1/\sqrt{\mu_0\epsilon_0}$ so that $B_\perp = E_\perp/c$

$$\rho = \frac{1}{2}\epsilon_0 E_\perp^2 + \frac{1}{2}\epsilon_0 E_\perp^2 = \epsilon_0 E_\perp^2 \tag{B-5}$$

The "Poynting vector," which gives the energy flow per unit area (i.e., the intensity of radiation) is directed along **r** and has a magnitude

$$S = \rho c = \epsilon_0 c E_\perp^2$$

Hence, from (B-3)

$$S = \frac{q^2a^2}{16\pi^2\epsilon_0 c^3 r^2}\sin^2\theta \tag{B-6}$$

which can also be obtained from the relation defining the Poynting vector

$$\mathbf{S} = \frac{1}{\mu_0}\mathbf{E} \times \mathbf{B}$$

Notice that no energy is emitted forward or backward along the direction of acceleration ($\theta = 0°$ or $180°$) and that the energy emitted is a maximum at right angles to this direction ($\theta = 90°$ or $270°$). The radiated energy is distributed symmetrically about the line of accelerated motion and with respect to the forward and backward directions. We see also from (B-6) that the radiated intensity obeys the familiar inverse square law, $S \propto 1/r^2$. To get the rate $R$ at which total energy is radiated in all directions per unit time, i.e., the power, we integrate $S$ over the area of a sphere of arbitrary radius $r$. That is

$$R = \int S(\theta)\, dA = \int_0^\pi S(\theta) 2\pi r^2 \sin\theta\, d\theta$$

in which $dA = 2\pi r^2 \sin\theta\, d\theta$ is the differential ring-shaped element of area on the sphere in a range between $\theta$ and $\theta + d\theta$. Carrying out the integration yields

$$R = \frac{1}{4\pi\epsilon_0}\frac{2}{3}\frac{q^2a^2}{c^3} \tag{B-7}$$

which is the rate of radiation of energy from the accelerated charge. The rate of radiation is seen to be proportional to the square of the acceleration.

It should be pointed out that energy must be supplied to maintain a constant linear acceleration of the charge, some of it simply to compensate for the energy radiated away. However, the radiation loss is usually negligible at nonrelativistic speeds. In the case of deceleration the radiated energy is supplied by the energy stored in the electromagnetic field of the charge whose velocity is decreasing. This is the bremsstrahlung radiation discussed in Chapter 2.

A frequent application of (B-7) is to a vibrating electric dipole. Let a charge $q$ be vibrating about the origin of the $x$ axis with simple harmonic motion. Then the displacement of the charge as a function of time is $x = A \sin \omega t$ where $A$ is the amplitude of the vibration and

$\omega = 2\pi\nu$ its angular frequency. The acceleration of the charge is given by $a = d^2x/dt^2 = -\omega^2 A \sin \omega t = -\omega^2 x$. If we substitute this for $a$ in (B-7) we obtain

$$R = \frac{2q^2\omega^4 x^2}{4\pi\epsilon_0 3c^3} \tag{B-8}$$

Because $x$ varies with time, the power radiated also varies with time at the same frequency as the vibration of the dipole. The average value of $x^2 = A^2 \sin^2 \omega t$ over one period of vibration, however, is simply $A^2/2$, so that the *average* rate of radiation is given by

$$\bar{R} = \frac{q^2\omega^4 A^2}{4\pi\epsilon_0 3c^3}$$

or, with $\omega = 2\pi\nu$

$$\bar{R} = \frac{16\pi^4\nu^4 q^2 A^2}{4\pi\epsilon_0 3c^3} \tag{B-9}$$

Now $qx$ is the electric dipole moment of the vibrating dipole when the charge is at $x$. So $qA$ is the amplitude of the electric dipole moment. Writing $qA = p$, we have the useful expression

$$\bar{R} = \frac{4\pi^3\nu^4 p^2}{3\epsilon_0 c^3} \tag{B-10}$$

# APPENDIX C

# The Boltzmann Distribution

We present here a simple numerical argument that leads to an approximation of the Boltzmann distribution, and then an even simpler general argument that verifies the exact form of the distribution. Consider a system containing a large number of physical entities of the same kind that are in thermal equilibrium at temperature $T$. To be in equilibrium they must be able to exchange energy with each other. In the exchanges the energies of the entities will fluctuate, and at any time some will have more than the average energy and some will have less. However, the classical theory of statistical mechanics demands that these energies $\mathscr{E}$ be distributed according to a definite probability distribution, whose form is specified by $T$. One reason is that the average value $\overline{\mathscr{E}}$ of the energy of each entity is determined by the probability distribution, and $\overline{\mathscr{E}}$ should have a definite value for a particular $T$.

To illustrate these ideas, consider a system consisting of entities, of the same kind, which can contain energy. An example would be a set of identical coil springs, each of which contains energy if its length is vibrating. Assume the system is isolated from the surrounding environment so that the total energy content is constant, and assume also that the entities can exchange energy with each other through some mechanism so that the constituents of the system can come into thermal equilibrium with each other. Purely for the purpose of simplifying the subsequent calculations, we shall, for the moment, also assume that the energy of any entity is restricted to one of the values $\mathscr{E} = 0, \Delta\mathscr{E}, 2\Delta\mathscr{E}, 3\Delta\mathscr{E}, 4\Delta\mathscr{E}, \ldots$. Later we shall let the interval $\Delta\mathscr{E}$ go to zero so that all the values of energy are permitted. For additional simplicity, we shall at first also consider that there are only four (an arbitrarily chosen small number) entities in the system and that the total energy of the system has the the value $3\Delta\mathscr{E}$ (which is also chosen arbitrarily to be a small one of the integral multiples of $\Delta\mathscr{E}$ that the total energy must, by the above assumption, necessarily be). Later we shall generalize to systems having a large number of entities and any total energy.

Because the four entities can exchange energy with one another, all possible divisions of the total energy $3\Delta\mathscr{E}$ between the four entities can occur. In Figure C-1 we show all the possible divisions, the divisions being labelled by the letter $i$. For $i = 1$, three entities have $\mathscr{E} = 0$ and the fourth entity has $\mathscr{E} = 3\Delta\mathscr{E}$, giving us the required total energy of $3\Delta\mathscr{E}$. Actually there are four different ways of getting such a division, because any one of the four entities can be the one in the energy state $\mathscr{E} = 3\Delta\mathscr{E}$. We indicate this in the figure in the column marked "number of distinguishable duplicate divisions." A second possible

| Label of divisions | $\mathscr{E} = 0$ | $\mathscr{E} = \Delta\mathscr{E}$ | $\mathscr{E} = 2\Delta\mathscr{E}$ | $\mathscr{E} = 3\Delta\mathscr{E}$ | $\mathscr{E} = 4\Delta\mathscr{E}$ | Number of distinguishable duplicate divisions | $P_i$ |
|---|---|---|---|---|---|---|---|
| $i = 1$ | √√√ | | | √ | | 4 | 4/20 |
| $i = 2$ | √√ | √ | √ | | | 12 | 12/20 |
| $i = 3$ | √ | √√√ | | | | 4 | 4/20 |
| $n'(\mathscr{E})$ | 40/20 | 24/20 | 12/20 | 4/20 | 0/20 | | |

**FIGURE C-1**

Illustrating a simple calculation leading to an approximation to the Boltzmann distribution.

type of division, labelled $i = 2$, is one in which two entities have $\mathscr{E} = 0$, the third entity has $\mathscr{E} = \Delta\mathscr{E}$, and the fourth has $\mathscr{E} = 2\Delta\mathscr{E}$. There are twelve duplicate divisions in this case, as we verify in the next paragraph. The third possible division, labelled $i = 3$, also has four duplicate ways of letting one entity have $\mathscr{E} = 0$ and the other three have $\mathscr{E} = \Delta\mathscr{E}$, giving the required total energy $3\Delta\mathscr{E}$.

In evaluating the number of duplicate divisions we count as distinguishable duplicates any rearrangement of entities between different energy states. However, any rearrangement of entities in the same energy state is not counted as a duplicate, because entities of the same kind having the same energy cannot be distinguished experimentally from one another. That is, *the identical entities are treated as if they are distinguishable*, except for rearrangements within the same energy state. The total number of rearrangements (permutations) of the four entities is $4! \equiv 4 \times 3 \times 2 \times 1$. (The number of different ways of ordering four objects is $4!$ since there are four choices of which object is taken first, three choices of which of the remaining objects is taken next, two choices of which is taken next, and one choice only for the last object. The total number of choices is $4 \times 3 \times 2 \times 1 \equiv 4!$. For $n$ objects the number of different orderings is $n! \equiv n(n-1)(n-2)\cdots 1$.) But rearrangements within the same energy state do not count. Hence, for example, in the case $i = 2$, the number of distinguishable duplicate divisions is reduced from $4!$ to $4!/2! = 12$ because there are $2!$ rearrangements within the state $\mathscr{E} = 0$ that do not count as distinguishable. In cases $i = 1$, or $i = 3$, the number of such divisions is reduced from $4!$ to $4!/3! = 4$ since there are $3!$ rearrangements within the state $\mathscr{E} = 0$, or the state $\mathscr{E} = \Delta\mathscr{E}$, that do not count as distinguishable.

We now make the final assumption: *all possible divisions of the energy of the system occur with the same probability*. Then the probability that the divisions of a given type (or label) will occur is proportional to the number of distinguishable duplicate divisions of that type. The relative probability, $P_i$, is just equal to that number divided by the total number of such divisions. The relative probabilities are listed in the column marked $P_i$ in Figure C-1.

Next let us calculate $n'(\mathscr{E})$, the probable number of entities in the energy state $\mathscr{E}$. Consider the energy state $\mathscr{E} = 0$. For divisions of the type $i = 1$, there are three entities in this state, and the relative probability $P_i$ that these divisions occur is 4/20; for $i = 2$ there are two entities in

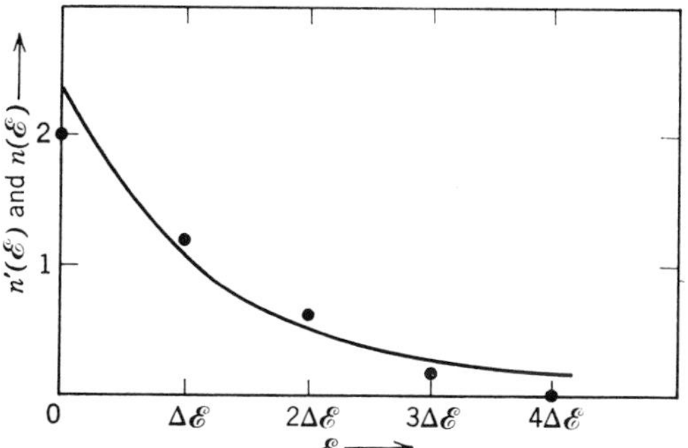


**FIGURE C-2**
A comparison of the results of a simple calculation and the Boltzmann distribution.

this state, and $P_i$ is 12/20; for $i = 3$ there is one entity, and $P_i$ is 4/20. Thus $n'(0)$, the probable number of entities in the state $\mathscr{E} = 0$, is $3 \times (4/20) + 2 \times (12/20) + 1 \times (4/20) = 40/20$. The values of $n'(\mathscr{E})$ calculated in the same way for the other values of $\mathscr{E}$ are listed on the bottom of Figure C-1, marked $n'(\mathscr{E})$. (Note that the sum of these numbers is four, so that we find a correct total of four entities in all the states.) The values of $n'(\mathscr{E})$ are also plotted as points in Figure C-2. The solid curve in Figure C-2 is the decreasing exponential function

$$n(\mathscr{E}) = Ae^{-\mathscr{E}/\mathscr{E}_0} \tag{C-1}$$

where $A$ and $\mathscr{E}_0$ are constants which have been adjusted to give the best fit of the curve to the points representing the results of our calculation. The rapid drop in $n'(\mathscr{E})$ with increasing $\mathscr{E}$ reflects the fact that, if one entity takes a larger share of the total energy of the system, the remainder of the system must necessarily have a reduced energy, and so a considerably reduced number of ways of dividing that energy between its constituents. That is, there are many fewer divisions of the total energy of the system in situations where a relatively large part of the energy is concentrated on one entity.

Imagine now that we successively make $\Delta\mathscr{E}$ smaller and smaller, increasing the number of allowed states at the same time so as to keep the total energy at its previous value. The result of such a process is that the calculated function $n'(\mathscr{E})$ becomes defined for values of $\mathscr{E}$ which are closer and closer together. (That is, we get more points on our distribution.) In the limit as $\Delta\mathscr{E} \to 0$, the energy $\mathscr{E}$ of an entity becomes a continuous variable, as classical physics demands, and the distribution $n'(\mathscr{E})$ becomes a continuous function. If, finally, we allow the number of entities in the system to become large, this function is found to be identical with the decreasing exponential $n(\mathscr{E})$ of (C-1). (That is, as the points become closer and closer together, they no longer scatter about the decreasing exponential but fall right on it.) To verify this, by a straightforward extension of our calculation to the case of a very large number of energy states and entities, involves some formidable bookkeeping in enumerating the distinguishable divisions that have the required values of total energy and number of entities, and then calculating the many relative probabilities. We shall verify the validity of the probability distribution given in (C-1) by a more subtle, but much simpler, procedure.

Consider a system of many identical entities in thermal equilibrium with each other, enclosed in walls which isolate it from the surroundings. Equilibrium requires that the entities be able to exchange energy. For instance, in interacting with the walls of the system, the entities can exchange energy with the walls and so indirectly exchange energy with each other. Thus the entities interact with each other in that if one gains energy, it does so at the expense of the total energy content of the remainder of the system (all the other entities, plus the walls). Except for this energy conservation constraint, the entities are independent of each other. *The presence of one entity in some particular energy state in no way inhibits or enhances the chance that another identical entity will be in that state.* Now consider two of these entities. Let the probability of finding one of them in an energy state at energy $\mathscr{E}_1$ be given by $P(\mathscr{E}_1)$. Then the probability of finding the other in a state at energy $\mathscr{E}_2$ will be given by the same probability distribution function, since the entities have identical properties, but evaluated at the energy $\mathscr{E}_2$. The probability will be $P(\mathscr{E}_2)$. Because of the independent behavior of the entities, these two probabilities are independent of each other. As a consequence, the probability that the energy of one entity will be $\mathscr{E}_1$ *and* that the energy of the other will be $\mathscr{E}_2$ is given by $P(\mathscr{E}_1)P(\mathscr{E}_2)$. The reason is that independent probabilities are multiplicative. (If the probability of obtaining heads in one flip of a coin is 1/2, then the probability of obtaining heads in each of two flips is $(1/2) \times (1/2) = 1/4$, since the flips are independent.)

Next consider all divisions of the energy of the system in which the sum of the energies of the two entities has the same fixed value $\mathscr{E}_1 + \mathscr{E}_2$ as in the particular case just discussed, but in which the two entities take different shares of that energy. Since the total energy of the isolated system is constant, for all of these divisions the remainder of the system will also have a fixed value of energy. So for all of them there are the same possible number of ways for the remainder of the system to distribute its energy between its constituents. As a consequence, the probability of those divisions in which there is a certain sharing of the energy $\mathscr{E}_1 + \mathscr{E}_2$ between the two entities can differ from the probability of other divisions, in which there is a different sharing of that energy, only if these different sharings occur with different probabilities. If we again assume that *all possible divisions of the energy of the system occur with the same probability* we see that this cannot be, and we conclude that all divisions in which the same energy $\mathscr{E}_1 + \mathscr{E}_2$ is shared between the two entities in different ways occur with the same probability. In other words, the probability of all such divisions is a function only of $\mathscr{E}_1 + \mathscr{E}_2$ and so can be written as, say, $Q(\mathscr{E}_1 + \mathscr{E}_2)$. However, we concluded earlier that the probability for a particular case can also be written as $P(\mathscr{E}_1)P(\mathscr{E}_2)$. Thus we find that $P(\mathscr{E}_1)P(\mathscr{E}_2) = Q(\mathscr{E}_1 + \mathscr{E}_2)$.

The essential point here is that the probability distribution function $P(\mathscr{E})$ has the property that the product of two of these functions, evaluated at two different values of the variables, $\mathscr{E}_1$ and $\mathscr{E}_2$, is a function of the sum, $\mathscr{E}_1 + \mathscr{E}_2$, of these variables. But an exponential function, and only an exponential function, has this property since the product of two exponentials with different exponents is an exponential whose exponent is the sum of the two exponents. Specifically, if we take the probability $P(\mathscr{E})$ of finding an entity in a state at energy $\mathscr{E}$ to be proportional to the probable number $n(\mathscr{E})$ of entities in that state, as it certainly should be, and use (C-1) to evaluate $n(\mathscr{E})$, we have the function

$$P(\mathscr{E}) = Be^{-\mathscr{E}/\mathscr{E}_0} \tag{C-2}$$

where $B$ is proportional to $A$, that demonstrates the required property since

$$P(\mathscr{E}_1)P(\mathscr{E}_2) = Be^{-\mathscr{E}_1/\mathscr{E}_0}Be^{-\mathscr{E}_2/\mathscr{E}_0} = B^2e^{-(\mathscr{E}_1+\mathscr{E}_2)/\mathscr{E}_0} = Q(\mathscr{E}_1 + \mathscr{E}_2)$$

Our argument does not actually prove that $n(\mathscr{E})$ is a decreasing, instead of increasing, exponential, but an increasing exponential can be ruled out on physical grounds as its value goes to infinity for large values of $\mathscr{E}$. Thus we have verified the general validity of (C-1).

Now we shall evaluate the constant $\mathscr{E}_0$ in (C-1)

$$n(\mathscr{E}) = Ae^{-\mathscr{E}/\mathscr{E}_0}$$

By treating a system containing two different kinds of entities in thermal equilibrium, it is not difficult to prove that the value of $\mathscr{E}_0$ does not depend on the type of entities comprising a system. Thus we shall use in our argument entities with the simplest properties. Since $n(\mathscr{E})$ is the probable number of entities of the system in an energy state at $\mathscr{E}$, the number of entities whose energies would be found in the interval from $\mathscr{E}$ to $\mathscr{E} + d\mathscr{E}$ equals $n(\mathscr{E})$ times the number of states in that interval. If that number is independent of the value of $\mathscr{E}$ (i.e., if the states are uniformly distributed in energy), then the number will be proportional to the size $d\mathscr{E}$ of the interval. This is the case if the entities are simple harmonic oscillators, like the coil springs mentioned earlier. So the probable number of simple harmonic oscillators with an energy from $\mathscr{E}$ to $\mathscr{E} + d\mathscr{E}$, in an equilibrium system containing many of them, is proportional to $n(\mathscr{E})\,d\mathscr{E}$. If the multiplicative constant $A$ is given the proper value, this probability can be made equal to $n(\mathscr{E})\,d\mathscr{E}$. Then the average energy of one of the oscillators is

$$\bar{\mathscr{E}} = \frac{\int_0^\infty \mathscr{E} n(\mathscr{E})\,d\mathscr{E}}{\int_0^\infty n(\mathscr{E})\,d\mathscr{E}}$$

The integral in the numerator has an integrand which is the energy weighted by the number of oscillators having that energy; the integral in the denominator is just the total number of oscillators. If we evaluate $n(\mathscr{E})$ from (C-1), we have

$$\bar{\mathscr{E}} = \frac{\int_0^\infty A\mathscr{E} e^{-\mathscr{E}/\mathscr{E}_0}\,d\mathscr{E}}{\int_0^\infty A e^{-\mathscr{E}/\mathscr{E}_0}\,d\mathscr{E}}$$

(Note that we do not need to know the actual value of $A$.) By proceeding in a manner completely analogous to what is done in Example 1-4, except that integrals are involved instead of sums, we find

$$\bar{\mathscr{E}} = \mathscr{E}_0 \tag{C-3}$$

But according to the classical law of equipartition of energy, as expressed in (1-16), for simple harmonic oscillators in equilibrium at temperature $T$

$$\bar{\mathscr{E}} = kT \tag{C-4}$$

where *Boltzmann's constant* $k = 1.38 \times 10^{-23}$ joule/°K. Combining (C-3) and (C-4), we have

$$\mathscr{E}_0 = kT \tag{C-5}$$

This result is correct for entities of any type, even though we have obtained it for the particular case of simple harmonic oscillators. Therefore we may write (C-1) as

$$n(\mathscr{E}) = Ae^{-\mathscr{E}/kT} \tag{C-6}$$

This is the famous *Boltzmann distribution*. Since the value of $A$ is not specified, (C-6) actually tells us about a proportionality: the probable number of entities of a system in equilibrium at temperature $T$ that will be in a state of energy $\mathscr{E}$ is proportional to $e^{-\mathscr{E}/kT}$. Expressed in different terms: the probability that the state of energy $\mathscr{E}$ will be occupied by an entity is proportional to $e^{-\mathscr{E}/kT}$.

The value chosen for the constant $A$ is dictated by convenience. In Chapter 1 we apply the Boltzmann distribution to a system of simple harmonic oscillators. As discussed here, in such a system $n(\mathscr{E})\,d\mathscr{E}$ is proportional to the probable number of oscillators with energy in the range $\mathscr{E}$ to $\mathscr{E} + d\mathscr{E}$, since the states of a simple harmonic oscillator are uniformly distributed in energy. Of course, $n(\mathscr{E})\,d\mathscr{E}$ is also proportional to the probability $P(\mathscr{E})\,d\mathscr{E}$ of finding a particular one of the oscillators with energy in this range. Thus we have, as in (C-2)

$$P(\mathscr{E}) = Be^{-\mathscr{E}/\mathscr{E}_0}$$

providing the constant $B$ is properly chosen. This is done by setting

$$\int_0^\infty P(\mathscr{E})\,d\mathscr{E} = \int_0^\infty Be^{-\mathscr{E}/\mathscr{E}_0}\,d\mathscr{E} = B\int_0^\infty e^{-\mathscr{E}/\mathscr{E}_0}\,d\mathscr{E} = 1 \tag{C-7}$$

That is, we define $P(\mathscr{E})\,d\mathscr{E}$ to be the probability of finding a particular simple harmonic oscillator with energy from $\mathscr{E}$ to $\mathscr{E} + d\mathscr{E}$, and so for consistency we must then demand that $\int_0^\infty P(\mathscr{E})\,d\mathscr{E}$ have the value one because the integral is just the probability of finding it with any energy. By evaluating $\int_0^\infty e^{-\mathscr{E}/\mathscr{E}_0}\,d\mathscr{E}$ in (C-7), and then solving for $B$, we find $B = 1/kT$. Then we have a special form of the Boltzmann distribution

$$P(\mathscr{E}) = \frac{e^{-\mathscr{E}/\mathscr{E}_0}}{kT} \tag{C-8}$$

which is used in Chapter 1.

# APPENDIX D

# Rutherford Scattering Trajectories

Figure 4-4 shows the parameters for the scattering trajectory of a light particle of positive charge $+ze$ by a heavy nucleus of positive charge $+Ze$. We saw in the text that the angular momentum $L = Mr^2\, d\varphi/dt$ is constant because the force on the particle is always acting in the radial direction. Let us apply Newton's law to the radial component of the motion, therefore, to determine the particle's trajectory. From $\mathbf{F} = M\mathbf{a}$ we obtain

$$\frac{zZe^2}{4\pi\epsilon_0 r^2} = M\left[\frac{d^2r}{dt^2} - r\left(\frac{d\varphi}{dt}\right)^2\right] \tag{D-1}$$

wherein the left-hand term is the Coulomb force and the right-hand terms are as follows: $d^2r/dt^2$ is the radial acceleration due to the change in the magnitude of $\mathbf{r}$ and $-r(d\varphi/dt)^2 = -\omega^2 r$ is the centripetal acceleration (which is also radially directed) due to the change in the direction of $\mathbf{r}$. To get the trajectory we need to find $r$ as a function of $\varphi$.

It simplifies the solution of (D-1) to write it, not in terms of the coordinates $r$, $\varphi$, but instead in terms of the coordinates $u$, $\varphi$, where

$$r = 1/u \tag{D-2}$$

Then

$$\frac{dr}{dt} = \frac{dr}{d\varphi}\frac{d\varphi}{dt} = \frac{dr}{du}\frac{du}{d\varphi}\frac{d\varphi}{dt}$$

or

$$\frac{dr}{dt} = -\frac{1}{u^2}\frac{du}{d\varphi}\frac{Lu^2}{M} = -\frac{L}{M}\frac{du}{d\varphi}$$

and

$$\frac{d^2r}{dt^2} = \frac{d}{d\varphi}\left(\frac{dr}{dt}\right)\frac{d\varphi}{dt} = -\frac{L}{M}\frac{d^2u}{d\varphi^2}\frac{Lu^2}{M}$$

or

$$\frac{d^2r}{dt^2} = -\frac{L^2u^2}{M^2}\frac{d^2u}{d^2\varphi}$$

Substituting this into (D-1), we have

$$-\frac{L^2u^2}{M^2}\frac{d^2u}{d\varphi^2} - \frac{1}{u}\left(\frac{Lu^2}{M}\right)^2 = \frac{zZe^2u^2}{4\pi\epsilon_0 M}$$

or

$$\frac{d^2u}{d\varphi^2} + u = -\frac{zZe^2M}{4\pi\epsilon_0 L^2} = -\frac{zZe^2M}{4\pi\epsilon_0 M^2v^2b^2} \tag{D-3}$$

since $L = Mvb$, where $v$ is the initial speed of the particle and $b$ is its impact parameter defined in Figure 4-4. If we let $D = (zZe^2/4\pi\epsilon_0)/(Mv^2/2)$, as in (4-4), this simplifies to

$$\frac{d^2u}{d\varphi^2} + u = -\frac{D}{2b^2} \tag{D-4}$$

This is a second order ordinary differential equation for $u$ as a function of $\varphi$.

The *general solution* to (D-4) is

$$u = A\cos\varphi + B\sin\varphi - D/2b^2 \tag{D-5}$$

which contains the two arbitrary constants, $A$ and $B$. We can prove that (D-5) is, in fact, the solution to (D-4) by evaluating

$$\frac{du}{d\varphi} = -A\sin\varphi + B\cos\varphi$$

and

$$\frac{d^2u}{d\varphi^2} = -A\cos\varphi - B\sin\varphi$$

and substituting these into (D-4). This gives us

$$-A\cos\varphi - B\sin\varphi + A\cos\varphi + B\sin\varphi - \frac{D}{2b^2} \equiv -D/2b^2$$

This identity proves the validity of the general solution.

To get the *particular solution* we must evaluate the constants $A$ and $B$. We require that (D-5) conform to the *initial conditions*: $\varphi \to 0$ as $r \to \infty$ and $dr/dt \to -v$ as $r \to \infty$. Thus

$$u = \frac{1}{r} = 0 = A\cos 0 + B\sin 0 - \frac{D}{2b^2}$$

or

$$A = \frac{D}{2b^2}$$

and

$$\frac{dr}{dt} = -\frac{L}{M}\frac{du}{d\varphi} = -v = -\frac{L}{M}(-A\sin 0 + B\cos 0)$$

or

$$B = \frac{Mv}{L} = \frac{Mv}{Mvb} = \frac{1}{b}$$

Therefore, the particular solution is

$$u = \frac{D}{2b^2}\cos\varphi + \frac{1}{b}\sin\varphi - \frac{D}{2b^2}$$

or

$$\frac{1}{r} = \frac{1}{b}\sin\varphi + \frac{D}{2b^2}(\cos\varphi - 1) \tag{D-6}$$

This is the orbit equation, giving $r$ as a function of $\varphi$. We see that the trajectory is hyperbolic, since (D-6) is the equation of a hyperbola in polar coordinates.

# APPENDIX E

# Complex Quantities

The *imaginary number* $i$ is a unit defined so that

$$i^2 = -1 \qquad \text{or} \qquad i = \sqrt{-1} \tag{E-1}$$

The name is appropriate because none of the *real* (i.e., ordinary) numbers have squares which are negative. A *complex number* $z$ can be written in the general form

$$z = x + iy \tag{E-2}$$

where both $x$ and $y$ are real numbers. The number $x$ is called the *real part* of $z$, and the number $y$ is called the *imaginary part* of $z$ (even though $y$ is real). Note that $z$ reduces to a pure real number if $y = 0$, while it reduces to a pure imaginary number if $x = 0$.

Complex numbers obey the same laws of algebra that apply to real numbers, except for the property specified in the definition (E-1). Also, the definition of equality is extended so that two complex numbers are equal if, and only if, the real part of one equals the real part of the other, and the imaginary part of one equals the imaginary part of the other. That is

$$z_1 = z_2 \tag{E-3a}$$

implies

$$x_1 = x_2 \qquad y_1 = y_2 \tag{E-3b}$$

and vice versa.

The *complex conjugate* of the number $z = x + iy$ is written as $z^*$, and is defined as

$$z^* = x - iy \tag{E-4}$$

From the definition it follows that

$$z^*z = (x - iy)(x + iy) = x^2 - i^2y^2 - ixy + ixy = x^2 - i^2y^2$$

So

$$z^*z = x^2 + y^2 \tag{E-5}$$

That is, the product of a complex number times its own complex conjugate always equals a real number.

Equation (E-5) is suggestive of the Pythagorean theorem. In fact, there is a very useful geometrical representation of complex numbers shown in Figure E-1. The location of a point $P$, relative to what are called the *real* and *imaginary axes* of the *complex plane*, is used in the manner defined in the figure to specify the real part $x$ and the imaginary part $y$ of the associated complex number. The location of the *representative point P* can also be specified by the polar coordinates $r$ and $\theta$, called the *modulus* and *phase*,

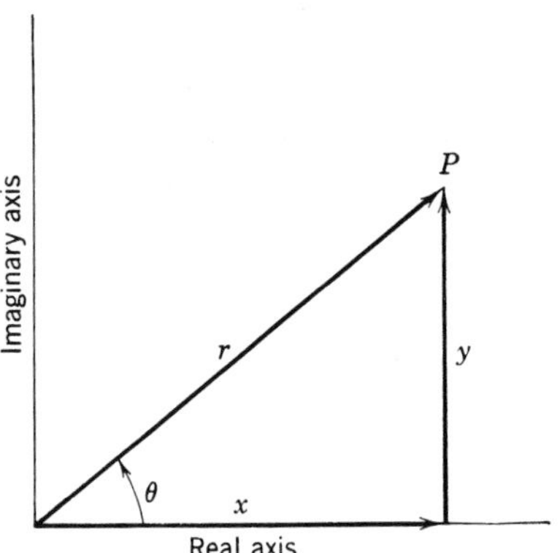


**FIGURE E-1**
The geometrical representation of a complex number. The relations between the rectangular and polar coordinates of the representative point $P$ can be determined by inspecting the figure.

which are defined in the figure. The two sets of coordinates are related by

$$x = r\cos\theta$$

$$y = r\sin\theta \tag{E-6}$$

and

$$r^2 = x^2 + y^2$$

$$\cos\theta = \frac{x}{r} \qquad \sin\theta = \frac{y}{r} \tag{E-7}$$

From (E-2) and (E-6), we see that the general complex number can be expressed in polar coordinates as

$$z = r(\cos\theta + i\sin\theta) \tag{E-8}$$

Note also that

$$z^*z = r^2 \tag{E-9}$$

Important relations can be developed by considering rotations in the complex plane of the representative point $P$. In Figure E-2, $z$ is a complex number that is represented by a point $P$ lying on the real axis. If the representative point is rotated at constant $r$ through an angle $d\theta$, the corresponding complex number becomes $z + dz$. It is apparent from the figure that

$$dz = iz\,d\theta$$

or

$$\frac{dz}{z} = i\,d\theta$$

As this relation can be seen to be true independent of the initial location of the representative point, it can be integrated

$$\int_{z_{\text{initial}}}^{z_{\text{final}}} \frac{dz}{z} = i\int_0^{\Theta} d\theta$$

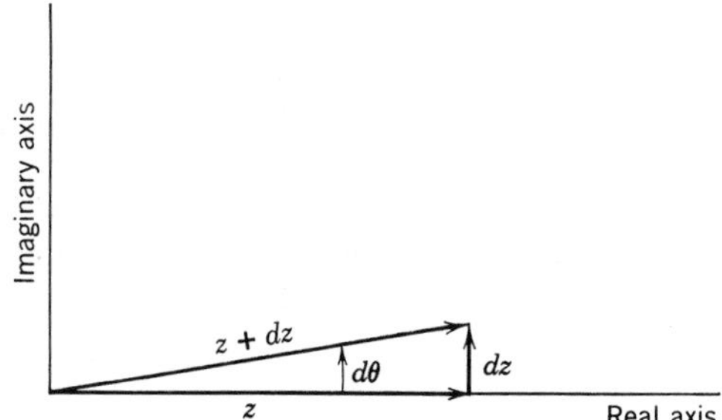


**FIGURE E-2**
Illustrating a rotation, at constant distance from the origin, of a representative point.

This yields

$$\ln \frac{z_{\text{final}}}{z_{\text{initial}}} = i\Theta$$

or

$$z_{\text{final}} = z_{\text{initial}} e^{i\Theta}$$

If we take $r = 1$, then $z_{\text{initial}} = 1$ and, from (E-8), we also have $z_{\text{final}} = \cos \Theta + i \sin \Theta$. Thus we obtain an evaluation of the *complex exponential*

$$e^{i\Theta} = \cos \Theta + i \sin \Theta \tag{E-10}$$

Rotation in the negative sense yields

$$e^{-i\Theta} = \cos(-\Theta) + i \sin(-\Theta)$$

which is

$$e^{-i\Theta} = \cos \Theta - i \sin \Theta \tag{E-11}$$

By adding and subtracting (E-10) and (E-11), it follows immediately that

$$\cos \Theta = \frac{e^{i\Theta} + e^{-i\Theta}}{2} \tag{E-12}$$

and

$$\sin \Theta = \frac{e^{i\Theta} - e^{-i\Theta}}{2i} \tag{E-13}$$

Comparison of the definition of (E-4) with (E-10) and (E-11) shows that the complex conjugate of a complex exponential is obtained by reversing the sign of the $i$ appearing in the exponent. That is

$$(e^{i\Theta})^* = e^{-i\Theta} \tag{E-14}$$

Applying (E-9) and (E-14) to a complex exponential, we find

$$r^2 = z^*z = (e^{i\Theta})^* e^{i\Theta} = e^{-i\Theta} e^{i\Theta} = e^0 = 1$$

Thus a complex exponential maintains a constant modulus $r = 1$, even if its phase is changing. But its real and imaginary parts, which are from (E-2) and (E-6) equal to $\cos \Theta$ and $\sin \Theta$, are oscillatory functions of the phase $\Theta$. If its phase is continually increasing from 0 to $\pi/2$ to $\pi$ to $3\pi/2$ to $2\pi$, and so on, a complex exponential changes in value from $+1$ to $+i$ to $-1$ to $-i$ to $+1$, and repeats this cyclically. In this sense it is an oscillatory function of its phase.

In differentiating or integrating a complex quantity, the standard procedures of calculus are used with $i$ treated as any other constant. An example of integration is found in the calculation leading to (E-10). As another example, the first derivative of the complex exponential is

$$\frac{de^{i\Theta}}{d\Theta} = ie^{i\Theta} \tag{E-15}$$

Although the geometrical interpretation leads naturally to writing the phase of a complex exponential as an angle $\Theta$, it can actually be any quantity which, like an angle, is *dimensionless*. In quantum mechanics, complex exponentials frequently used are

$$e^{ikx} \qquad e^{i(kx-\omega t)} \qquad e^{-iEt/\hbar}$$

In the first of these, for example, the wave number $k$ has the dimensions of $(\text{length})^{-1}$, so $k$ times the length $x$ is dimensionless. All relations quoted for $e^{i\Theta}$ have obvious extensions to $e^{ikx}$, and the others. For example, application of the rules of differentiation to $e^{ikx}$, with $k$ constant, yields

$$\frac{de^{ikx}}{dx} = ike^{ikx} \tag{E-16}$$

# APPENDIX F

# Numerical Solution of the Time-Independent Schroedinger Equation for a Square Well Potential

In quantum mechanics, as in other fields of science and engineering, many of the calculations that arise in current professional work are carried out on computers by using numerical techniques. In some cases the potential energy function of interest is of such a form that its time-independent Schroedinger equation cannot be solved by even the most general analytical techniques (for reasons explained in Appendix H). In other cases analytical solutions can be obtained, but numerical solutions can be obtained more conveniently if an adequate computer is available.

As a simple illustration of the numerical techniques, and of the "thought calculations" of Section 5-7, we shall obtain here a numerical solution of the time-independent Schroedinger equation for the potential energy function

$$V(x) = \begin{cases} V_0, \text{ a const} & x < -a/2 \text{ or } x > +a/2 \\ 0 & -a/2 < x < +a/2 \end{cases} \tag{F-1}$$

This is called a *square well potential*, for reasons that are apparent from inspection of its form plotted in Figure F-1. For this simple potential a numerical solution can be found in a fairly short time using only a small calculator, if not too much numerical accuracy is demanded. The time-independent Schroedinger equation for this particular potential can also be treated with fairly simple analytical techniques (see Appendix G), so we shall be able to compare the resulting exact solution with the approximate results we obtain from our numerical solution.

From the arguments of Section 5-7, we know that the behavior of a solution $\psi(x)$ to the time-independent Schroedinger equation, (5-45)

$$\frac{d^2\psi(x)}{dx^2} = \frac{2m}{\hbar^2}[V(x) - E]\psi(x)$$

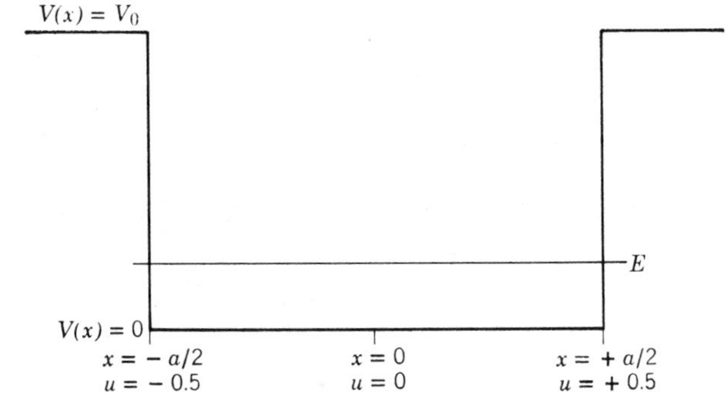


**FIGURE F-I**

A square well potential and an assumed value of the total energy $E$ of a particle bound in this potential.

for the potential $V(x)$ and total energy $E$, should be completely determined for all $x$ by the equation and by assumed initial values of $\psi(x)$ and $d\psi(x)/dx$. To see this explicitly, note that we can calculate the values of the solution and its first derivative at a point $x_1$, near the initial point $x_0$, in terms of the values of these quantities at the initial point, as follows. For $(x_1 - x_0)$ very small, the value of the solution at $x_1$, which we write $[\psi(x)]_{x_1}$, is given by the definition of a derivative as

$$[\psi(x)]_{x_1} - [\psi(x)]_{x_0} \simeq \left[\frac{d\psi(x)}{dx}\right]_{x_0} (x_1 - x_0) \tag{F-2}$$

where $[\psi(x)]_{x_0}$ is the assumed value of the solution at $x_0$, and $[d\psi(x)/dx]_{x_0}$ is the assumed value of its derivative at that point. A similar formula for $[d\psi(x)/dx]_{x_1}$ in terms of known quantities can be obtained from the time-independent Schroedinger equation by writing it as

$$\frac{d^2\psi(x)}{dx^2} = \frac{d}{dx}\left[\frac{d\psi(x)}{dx}\right] = \frac{2m}{\hbar^2}[V(x) - E]\psi(x)$$

and then multiplying both sides by the differential $dx$, to obtain

$$d\left[\frac{d\psi(x)}{dx}\right] = \frac{2m}{\hbar^2}[V(x) - E]\psi(x)\,dx$$

According to the definition of a differential, this is

$$\left[\frac{d\psi(x)}{dx}\right]_{x_1} - \left[\frac{d\psi(x)}{dx}\right]_{x_0} \simeq \frac{2m}{\hbar^2}[V(x) - E]_{x_0}[\psi(x)]_{x_0}(x_1 - x_0) \tag{F-3}$$

providing $(x_1 - x_0)$ is very small. By repeating the procedure from the new starting point $x_1$, we can find the values of $\psi(x)$ and $d\psi(x)/dx$ at a nearby point $x_2$. Continuing in this manner, we can trace out the solution and its derivatives for all values of $x$.

To apply (F-2) and (F-3) to the potential function of (F-1), it is very convenient to divide through everywhere by the width $a$ of the square well. In this way the equations can be rewritten in terms of the dimensionless space variable

$$u = \frac{x}{a} \tag{F-4}$$

Equation (F-2) becomes

$$[\psi(u)]_{u_1} = [\psi(u)]_{u_0} + \left[\frac{d\psi(u)}{du}\right]_{u_0}(u_1 - u_0) \tag{F-5}$$

and (F-3) becomes

$$\left[\frac{d\psi(u)}{du}\right]_{u_1} = \left[\frac{d\psi(u)}{du}\right]_{u_0} + \frac{2ma^2}{\hbar^2}\,[V(u) - E]_{u_0}[\psi(u)]_{u_0}(u_1 - u_0) \tag{F-6}$$

Of course, we must specify the depth $V_0$ of the square well potential before we can obtain a numerical solution to the differential equation. The form of (F-6) suggests that this is most conveniently done by specifying the value of the dimensionless combination of parameters $2ma^2V_0/\hbar^2$. For the purpose of our illustrative calculation we can choose any reasonable value. So we take, rather arbitrarily

$$\frac{2ma^2V_0}{\hbar^2} = 64 \tag{F-7}$$

If we also express the total energy as

$$E = \alpha V_0 \tag{F-8}$$

where $\alpha$ is a dimensionless parameter, then (F-6) takes on the dimensionless form

$$\left[\frac{d\psi(u)}{du}\right]_{u_1} = \left[\frac{d\psi(u)}{du}\right]_{u_0} + [C(u)]_{u_0}[\psi(u)]_{u_0}(u_1 - u_0) \tag{F-9}$$

where

$$C(u) = \begin{cases} -\dfrac{2ma^2E}{\hbar^2} = -\dfrac{2ma^2\alpha V_0}{\hbar^2} = -64\alpha & -0.5 < u < +0.5 \\[2ex] \dfrac{2ma^2}{\hbar^2}[V_0 - E] = \dfrac{2ma^2V_0}{\hbar^2}[1 - \alpha] = 64[1 - \alpha] & u < -0.5 \text{ or } u > +0.5 \end{cases} \tag{F-10}$$

We shall perform the numerical calculations with (F-5) and (F-9), which have been made dimensionless because such calculations can deal only with pure numbers. The calculations will be for the purpose of determining approximately the lowest allowed energy $E$, and the shape of the corresponding eigenfunction $\psi$, for a particle of mass $m$ bound in the square well potential specified by (F-1) and (F-7).

Now we know, from the qualitative arguments of Section 5-7, that in the interior region of the square well the lowest energy eigenfunction will look something like half of a cosine wave fitted into the region. However, it will have a longer wavelength since it does extend for some distance into the exterior regions. By evaluating the momentum $p$ corresponding to a half wavelength $\lambda/2 = a$ just fitting into the interior region, from de Broglie's relation $p = h/\lambda = h/2a$, we can use the corresponding energy $E = p^2/2m = h^2/8ma^2 = \pi^2\hbar^2/2ma^2$ to help us estimate the actual value of $E$, and save considerable effort in the numerical calculation. In terms of $\alpha$, the estimated value of $E$ is $\alpha = E/V_0 = (\pi^2\hbar^2/2ma^2)/(32\hbar^2/ma^2) = \pi^2/64 = 0.154$. Since $\lambda$ is an underestimate, $E$ and $\alpha$ are overestimates. We therefore make an educated guess and try, in the initial numerical calculation, the value $\alpha = 0.100$.

In consideration of what was learned in the qualitative arguments, it is apparent that the eigenfunction for the lowest allowed energy in this potential should be symmetrical about the point $u = 0$, relative to which the potential itself is symmetrical. This very much simplifies things because we need only carry out calculations in the range $u \geq 0$, and because the symmetry immediately leads to the conclusion that $d\psi(u)/du = 0$ at $u = 0$. We shall therefore start the calculation at $u = 0$. Since the choice of $\psi(u)$ at $u = 0$ is immaterial because of the linearity of the differential equation, we shall take $\psi(u) = +1.000$ at that point. Sufficient accuracy will be obtained by taking $u_1 - u_0 = 0.050$, etc. The first calculation is shown in Table F-1.

The first row in the table specifies the assumed values of $\psi(u)$ and $d\psi(u)/du$ at $u = 0$, and performs part of the first step of the calculation. All the entries in this row are quantities evaluated for $u = 0.000$. The first entry is the value of $\psi$; the second is the value of $d\psi/du$; the third is $d\psi/du$ multiplied by $\Delta u = 0.050$, the change in $u$ from the first row to the second row;

**TABLE F-1.** A Numerical Integration

$\alpha = 0.100 \qquad 1 - \alpha = 0.900$

$C = \begin{matrix} -6.40 \\ +57.6 \end{matrix} \qquad \begin{matrix} 0 \leq u < 0.500 \\ u \geq 0.500 \end{matrix}$

| $u$ | $\psi$ | $d\psi/du$ | $(d\psi/du)\,\Delta u$ | $C\psi\Delta u$ |
|---|---|---|---|---|
| 0.000 | +1.000 | 0.000 | 0.000 | −0.320 |
| 0.050 | +1.000 | −0.320 | −0.016 | −0.320 |
| 0.100 | +0.984 | −0.640 | −0.032 | −0.315 |
| 0.150 | +0.952 | −0.955 | −0.048 | −0.305 |
| 0.200 | +0.904 | −1.260 | −0.063 | −0.289 |
| 0.250 | +0.841 | −1.549 | −0.077 | −0.269 |
| 0.300 | +0.764 | −1.818 | −0.091 | −0.244 |
| 0.350 | +0.673 | −2.062 | −0.103 | −0.215 |
| 0.400 | +0.570 | −2.277 | −0.114 | −0.182 |
| 0.450 | +0.456 | −2.459 | −0.123 | −0.146 |
| 0.500 | +0.333 | −2.605 | −0.130 | +0.959 |
| 0.550 | +0.203 | −1.646 | −0.082 | +0.585 |
| 0.600 | +0.121 | −1.061 | −0.053 | +0.348 |
| 0.650 | +0.068 | −0.713 | −0.036 | +0.196 |
| 0.700 | +0.032 | −0.517 | −0.026 | +0.092 |
| 0.750 | +0.006 | −0.425 | −0.021 | +0.017 |
| 0.800 | −0.015 | −0.408 | −0.020 | −0.043 |
| 0.850 | −0.035 | −0.451 | −0.023 | −0.101 |
| 0.900 | −0.058 | −0.552 | −0.028 | −0.167 |
| 0.950 | −0.086 | −0.719 | −0.036 | −0.248 |
| 1.000 | −0.122 | −0.967 | | |

and the fourth entry is $\psi$ multiplied by $C\Delta u$. For all rows through the one labeled $u = 0.450$, $C\Delta u = -6.40 \times 0.050 = -0.320$; beyond this point $C\Delta u = +57.6 \times 0.050 = +2.88$. The second row completes the first step of the calculation and performs part of the next step. In agreement with (F-5) and (F-9), the first entry in the second row is the sum of the first and third entries of the previous row; the second entry is the sum of the second and fourth entries of the previous row. The same scheme is followed in constructing the remainder of the table. The calculation was terminated at $u = 1.000$ because $\psi$ was rapidly going to $-\infty$, as can be

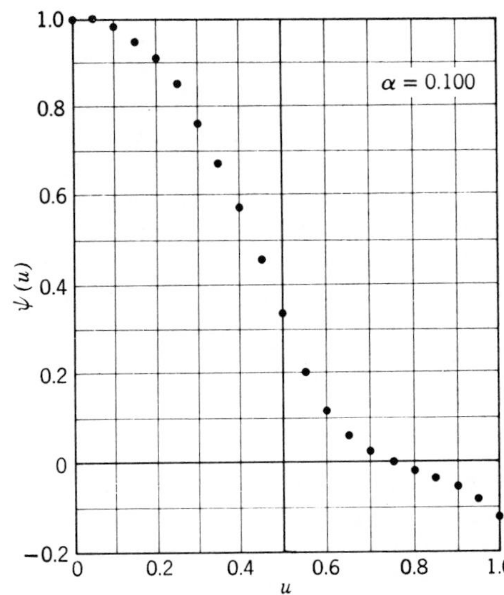

**FIGURE F-2**

A solution of the time-independent Schroedinger equation for a square well potential. The solution begins to go to negative infinity at large $u$ because the assumed value of the total energy is slightly too large.

**TABLE F-2.** A Second Numerical Integration

$\alpha = 0.098 \qquad 1 - \alpha = 0.902$

$C = \begin{matrix} -6.27 \\ +57.7 \end{matrix} \qquad \begin{matrix} 0 \leq u < 0.500 \\ u \geq 0.500 \end{matrix}$

| $u$ | $\psi$ | $d\psi/du$ | $(d\psi/du)\ \Delta u$ | $C\psi\Delta u$ |
|---|---|---|---|---|
| 0.000 | +1.000 | 0.000 | 0.000 | −0.314 |
| 0.050 | +1.000 | −0.314 | −0.016 | −0.314 |
| 0.100 | +0.984 | −0.628 | −0.031 | −0.309 |
| 0.150 | +0.953 | −0.937 | −0.047 | −0.299 |
| 0.200 | +0.906 | −1.236 | −0.062 | −0.284 |
| 0.250 | +0.844 | −1.520 | −0.076 | −0.265 |
| 0.300 | +0.768 | −1.785 | −0.089 | −0.241 |
| 0.350 | +0.679 | −2.026 | −0.101 | −0.213 |
| 0.400 | +0.578 | −2.239 | −0.112 | −0.181 |
| 0.450 | +0.466 | −2.420 | −0.121 | −0.146 |
| 0.500 | +0.345 | −2.566 | −0.128 | +0.996 |
| 0.550 | +0.217 | −1.570 | −0.079 | +0.626 |
| 0.600 | +0.138 | −0.944 | −0.047 | +0.398 |
| 0.650 | +0.091 | −0.546 | −0.027 | +0.263 |
| 0.700 | +0.064 | −0.283 | −0.014 | +0.185 |
| 0.750 | +0.050 | −0.098 | −0.005 | +0.144 |
| 0.800 | +0.045 | +0.046 | +0.002 | +0.130 |
| 0.850 | +0.047 | +0.176 | +0.009 | +0.136 |
| 0.900 | +0.056 | +0.312 | +0.016 | +0.162 |
| 0.950 | +0.072 | +0.474 | +0.024 | +0.208 |
| 1.000 | +0.096 | +0.682 | | |

seen in the plot of Figure F-2. This happened because the chosen value of $\alpha$ was too large. As a result, $\psi$ bends too rapidly in the interior region, and consequently it goes through zero just a little way outside this region. Once it goes through zero, nothing can prevent it from going to $-\infty$.

A second calculation, using $\alpha = 0.098$, is shown in Table F-2. This calculation is plotted in Figure F-3. It too failed, but in the opposite sense, because $\psi$ bent away from the axis in the

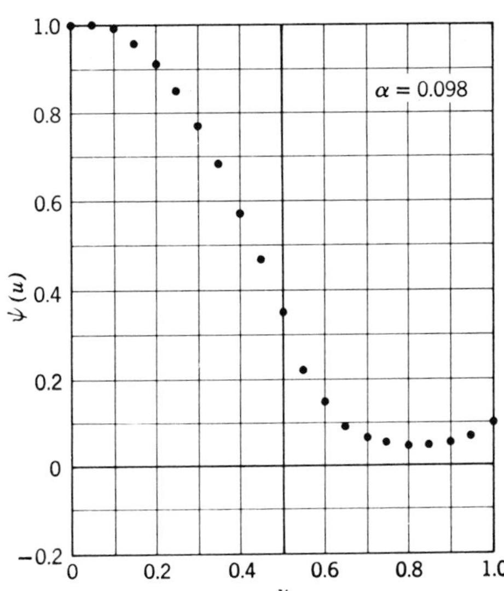


**FIGURE F-3**
A solution to the time-independent Schroedinger equation for a square well potential. The solution begins to go to positive infinity at large $u$ because the assumed value of the total energy is slightly too small.

exterior region and began to go to $+\infty$. However, the results of the two calculations indicate that the allowed value of $\alpha$ lies between 0.100 and 0.098, and comparison of the two curves gives the impression that it is somewhat closer to the lower limit. Additional calculations can be used to narrow the limits, but it would be necessary to decrease the value of $\Delta u$, and increase the number of decimal points retained, in order to reduce the numerical inaccuracy of the calculation. A solution to the time-independent Schroedinger equation for this potential using analytic methods (see Appendix G) yields $\alpha = 0.0980$ for the lowest allowed energy. The agreement with our numerical calculation is close, but not perfect, due to the numerical inaccuracy just mentioned. The analytic solution also shows that there are two additional bound allowed energies, corresponding to $\alpha = 0.383$ and $\alpha = 0.808$. Of course, any unbound energy, corresponding to $\alpha \geq 1$, is allowed.

The technique we have used is called *numerical integration*. The second word is appropriate because we started with an equation containing $d^2\psi/dx^2$ and finally obtained $\psi$ itself; therefore, we have carried out a process which is the inverse of differentiation.

If the student has access to a calculator, he will find it easy and interesting to carry out a numerical integration of the time-independent Schroedinger equation. If the student has access to a computer, of even the smallest size, he will find that by performing numerical integrations for bound and unbound states in various potentials he can rapidly develop a real intuitive feeling for many of the important features of quantum mechanics.

# APPENDIX G

# Analytical Solution of the Time-Independent Schroedinger Equation for a Square Well Potential

Here we develop the general solution of the time-independent Schroedinger equation for the bound states of a square well potential of finite depth, following the procedure that is discussed in a qualitative way in Section 6-7. Then we apply the results to the particular case of a square well potential with the same parameters that were used in the numerical solution of Appendix F.

The description of the classical motion of a particle bound by a square well suggests that it would be most appropriate to look for solutions to the Schroedinger equation in the form of standing waves. Thus we take, as a general solution to the time-independent Schroedinger equation in the region $-a/2 < x < +a/2$ where $V(x) = 0$, the free particle standing wave eigenfunction of (6-62), which we write here as

$$\psi(x) = A \sin k_{\text{I}}x + B \cos k_{\text{I}}x \qquad -a/2 < x < +a/2 \quad \text{(G-1)}$$

where

$$k_{\text{I}} = \sqrt{2mE}/\hbar$$

In the regions $x < -a/2$ and $x > +a/2$ the time-independent Schroedinger equation has the general solutions displayed in (6-63) and (6-64). These are

$$\psi(x) = Ce^{k_{\text{II}}x} + De^{-k_{\text{II}}x} \qquad x < -a/2 \quad \text{(G-2)}$$

and

$$\psi(x) = Fe^{k_{\text{II}}x} + Ge^{-k_{\text{II}}x} \qquad x > +a/2 \quad \text{(G-3)}$$

where

$$k_{\text{II}} = \sqrt{2m(V_0 - E)}/\hbar \qquad \text{with} \quad E < V_0$$

To determine the arbitrary constants first impose the requirement that the eigenfunctions remain finite for all $x$. Consider (G-2) in the limit $x \to -\infty$. It is apparent that this requirement demands

$$D = 0 \quad \text{(G-4)}$$

Similarly, it is necessary to set

$$F = 0 \quad \text{(G-5)}$$

in order that (G-3) remain finite in the limit $x \to +\infty$. Next impose the requirement that the eigenfunctions and their first derivatives be continuous at $x = -a/2$ and $x = +a/2$. Four equations are obtained. They are

$$-A \sin(k_{\rm I}a/2) + B \cos(k_{\rm I}a/2) = Ce^{-k_{\rm II}a/2} \tag{G-6}$$

$$Ak_{\rm I} \cos(k_{\rm I}a/2) + Bk_{\rm I} \sin(k_{\rm I}a/2) = Ck_{\rm II}e^{-k_{\rm II}a/2} \tag{G-7}$$

$$A \sin(k_{\rm I}a/2) + B \cos(k_{\rm I}a/2) = Ge^{-k_{\rm II}a/2} \tag{G-8}$$

$$Ak_{\rm I} \cos(k_{\rm I}a/2) - Bk_{\rm I} \sin(k_{\rm I}a/2) = -Gk_{\rm II}e^{-k_{\rm II}a/2} \tag{G-9}$$

Subtracting (G-6) from (G-8) yields

$$2A \sin(k_{\rm I}a/2) = (G - C)e^{-k_{\rm II}a/2} \tag{G-10}$$

Adding (G-6) to (G-8) yields

$$2B \cos(k_{\rm I}a/2) = (G + C)e^{-k_{\rm II}a/2} \tag{G-11}$$

Subtracting (G-9) from (G-7) yields

$$2Bk_{\rm I} \sin(k_{\rm I}a/2) = (G + C)k_{\rm II}e^{-k_{\rm II}a/2} \tag{G-12}$$

Adding (G-9) to (G-7) yields

$$2Ak_{\rm I} \cos(k_{\rm I}a/2) = -(G - C)k_{\rm II}e^{-k_{\rm II}a/2} \tag{G-13}$$

Provided $B \neq 0$ and $(G + C) \neq 0$, we may divide (G-12) by (G-11) and obtain

$$k_{\rm I} \tan(k_{\rm I}a/2) = k_{\rm II} \qquad \text{if } B \neq 0 \text{ and } (G + C) \neq 0 \tag{G-14}$$

Provided $A \neq 0$ and $(G - C) \neq 0$, we may divide (G-13) by (G-10) and obtain

$$k_{\rm I} \cot(k_{\rm I}a/2) = -k_{\rm II} \qquad \text{if } A \neq 0 \text{ and } (G - C) \neq 0 \tag{G-15}$$

It is easy to see that both (G-14) and (G-15) cannot be satisfied simultaneously. If they could, the equation obtained by adding these two

$$k_{\rm I} \tan(k_{\rm I}a/2) + k_{\rm I} \cot(k_{\rm I}a/2) = 0$$

would be valid. Multiply through by $\tan(k_{\rm I}a/2)$. Then the equation becomes

$$k_{\rm I} \tan^2(k_{\rm I}a/2) + k_{\rm I} = 0$$

or

$$\tan^2(k_{\rm I}a/2) = -1$$

But this cannot be valid as both $k_{\rm I}$ and $a/2$ are real. Thus it is only possible *either* to satisfy (G-14) but not (G-15) *or* to satisfy (G-15) but not (G-14). The eigenfunctions of the square well potential form two classes. For the *first class*

$$\begin{aligned} k_{\rm I} \tan(k_{\rm I}a/2) &= k_{\rm II} \\ A &= 0 \\ G - C &= 0 \end{aligned} \tag{G-16}$$

Then (G-8) reads

$$B \cos(k_{\rm I}a/2) = Ge^{-k_{\rm II}a/2}$$
$$G = B \cos(k_{\rm I}a/2)e^{k_{\rm II}a/2} = C$$

and the eigenfunctions are

$$\psi(x) = \begin{cases} [B \cos(k_{\rm I}a/2)e^{k_{\rm II}a/2}]e^{k_{\rm II}x} & x < -a/2 \\ [B] \cos(k_{\rm I}x) & -a/2 < x < a/2 \\ [B \cos(k_{\rm I}a/2)e^{k_{\rm II}a/2}]e^{-k_{\rm II}x} & x > a/2 \end{cases} \tag{G-17}$$

For the *second class*

$$k_{\rm I}\cot(k_{\rm I}a/2) = -k_{\rm II}$$
$$B = 0 \tag{G-18}$$
$$G + C = 0$$

Then (G-8) reads

$$A\sin(k_{\rm I}a/2) = Ge^{-k_{\rm II}a/2}$$
$$G = A\sin(k_{\rm I}a/2)e^{k_{\rm II}a/2} = -C$$

and the eigenfunctions are

$$\psi(x) = \begin{cases} [-A\sin(k_{\rm I}a/2)e^{k_{\rm II}a/2}]e^{k_{\rm II}x} & x < -a/2 \\ [A]\sin(k_{\rm I}x) & -a/2 < x < a/2 \\ [A\sin(k_{\rm I}a/2)e^{k_{\rm II}a/2}]e^{-k_{\rm II}x} & x > a/2 \end{cases} \tag{G-19}$$

Consider the first of (G-16). Evaluating $k_{\rm I}$ and $k_{\rm II}$, and multiplying through by $a/2$, the equation becomes

$$\sqrt{mEa^2/2\hbar^2}\tan(\sqrt{mEa^2/2\hbar^2}) = \sqrt{m(V_0 - E)a^2/2\hbar^2} \tag{G-20}$$

For a given particle of mass $m$ and a given potential well of depth $V_0$ and width $a$, this is an equation in the single unknown $E$. Its solutions are the allowed values of the total energy of the particle—the eigenvalues for eigenfunctions of the first class. Solutions of this transcendental equation can be obtained only by numerical or graphical methods. We present a simple graphical method which will illustrate the important features of the equation. Let us make the change of variable

$$\mathscr{E} \equiv \sqrt{mEa^2/2\hbar^2} \tag{G-21}$$

so the equation becomes

$$\mathscr{E}\tan\mathscr{E} = \sqrt{mV_0a^2/2\hbar^2 - \mathscr{E}^2} \tag{G-22}$$

If we plot the function

$$p(\mathscr{E}) = \mathscr{E}\tan\mathscr{E}$$

and the function

$$q(\mathscr{E}) = \sqrt{mV_0a^2/2\hbar^2 - \mathscr{E}^2}$$

the intersections specify values of $\mathscr{E}$ which are solutions to (G-22).

Such a plot is shown in Figure G-1. The function $p(\mathscr{E})$ has zeros at $\mathscr{E} = 0, \pi, 2\pi, \ldots$ and has asymptotes at $\mathscr{E} = \pi/2, 3\pi/2, 5\pi/2, \ldots$. The function $q(\mathscr{E})$ is a quarter-circle of radius $\sqrt{mV_0a^2/2\hbar^2}$. It is clear from the figure that the number of solutions which exist for (G-22) depends on the radius of the quarter-circle. Each solution gives an eigenvalue for $E < V_0$ corresponding to an eigenfunction of the first class. There exists one such eigenvalue if $\sqrt{mV_0a^2/2\hbar^2} < \pi$; two if $\pi \leq \sqrt{mV_0a^2/2\hbar^2} < 2\pi$; three if $2\pi \leq \sqrt{mV_0a^2/2\hbar^2} < 3\pi$; etc.

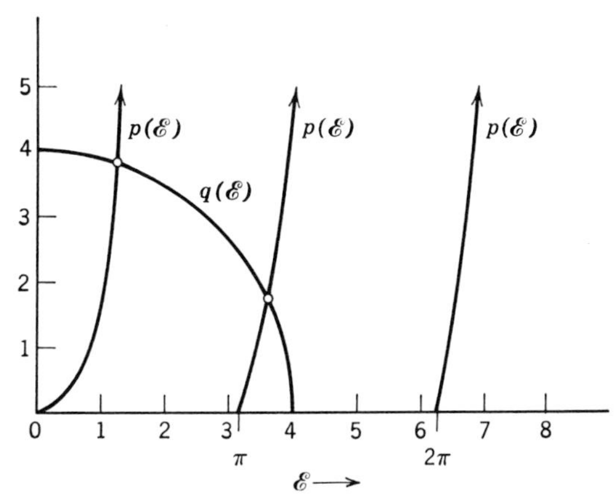


**FIGURE G-1**

A graphical solution of the equation for eigenvalues of the first class of a particular square well potential. Solution of

$$\mathscr{E}\tan\mathscr{E} = \sqrt{mV_0a^2/2\hbar^2 - \mathscr{E}^2}$$

or $p(\mathscr{E}) = q(\mathscr{E})$.

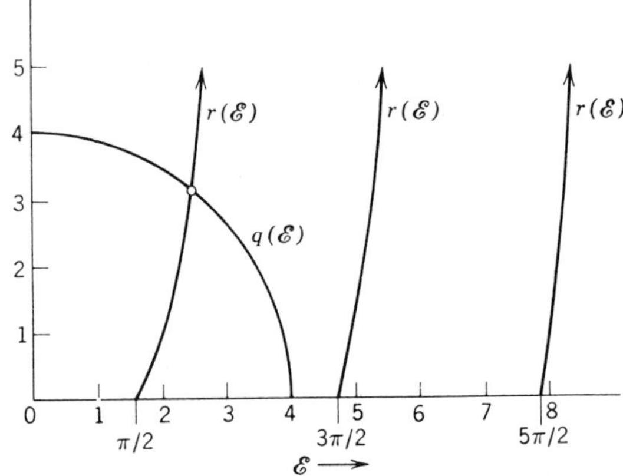


**FIGURE G-2**
A graphical solution of the equation for eigenvalues of the second class of a particular square well potential. Solution of

$$-\mathscr{E}\cot\mathscr{E} = \sqrt{mV_0a^2/2\hbar^2 - \mathscr{E}^2}$$

or $r(\mathscr{E}) = q(\mathscr{E})$.

The case $\sqrt{mV_0a^2/2\hbar^2} = 4$ is illustrated in the figure. Note that this corresponds to $2mV_0a^2/\hbar^2 = 64$, the value used in the numerical integration of Appendix F. For this case there are two solutions: $\mathscr{E} \simeq 1.25$ and $\mathscr{E} \simeq 3.60$. From (G-21), the eigenvalues are

$$E = \mathscr{E}^2 \frac{2\hbar^2}{ma^2} = \mathscr{E}^2 \frac{2\hbar^2}{mV_0a^2} V_0 \simeq \left(\frac{1.25}{4}\right)^2 V_0 \simeq 0.0980V_0$$

and

$$E = \mathscr{E}^2 \frac{2\hbar^2}{mV_0a^2} V_0 \simeq \left(\frac{3.60}{4}\right)^2 V_0 \simeq 0.808V_0$$

The eigenvalues corresponding to eigenfunctions of the second class are found from the solutions of an analogous equation obtained from (G-18), which is

$$-\mathscr{E}\cot\mathscr{E} = \sqrt{mV_0a^2/2\hbar^2 - \mathscr{E}^2} \tag{G-23}$$

Figure G-2 illustrates the solution of this equation. It is apparent that there will be no eigenvalues for $E < V_0$ corresponding to eigenfunctions of the second class if $\sqrt{mV_0a^2/2\hbar^2} < \pi/2$; there will be one if $\pi/2 \le \sqrt{mV_0a^2/2\hbar^2} < 3\pi/2$; two if $3\pi/2 \le \sqrt{mV_0a^2/2\hbar^2} < 5\pi/2$; etc. The figure illustrates the case $\sqrt{mV_0a^2/2\hbar^2} = 4$. The single solution to (G-23) is $\mathscr{E} \simeq 2.47$, and the eigenvalue is

$$E = \mathscr{E}^2 \frac{2\hbar^2}{mV_0a^2} V_0 \simeq \left(\frac{2.47}{4}\right)^2 V_0 \simeq 0.383V_0$$

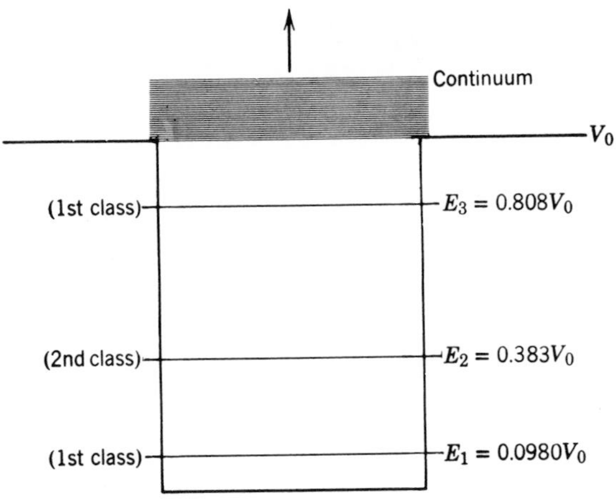


**FIGURE G-3**
The eigenvalues of a particular square well potential.

The form we have found in (H-10) suggests that we search for solutions to the full-fledged differential equation, (H-7), that can be written

$$\psi(u) = Ae^{-u^2/2}H(u) \tag{H-11}$$

These solutions are to be valid for *all u*. So the $H(u)$ must be functions which are slowly varying compared to $e^{-u^2/2}$ as $|u| \to \infty$, in order that (H-11) agrees with (H-10). Elsewhere, the $H(u)$ must have whatever forms are required to yield the correct forms for the $\psi(u)$. To evaluate the $H(u)$, we calculate

$$\frac{d\psi}{du} = -Aue^{-u^2/2}H + Ae^{-u^2/2}\frac{dH}{du}$$

and

$$\begin{aligned}\frac{d^2\psi}{du^2} &= -Ae^{-u^2/2}H + Au^2e^{-u^2/2}H - Aue^{-u^2/2}\frac{dH}{du} \\ &\quad -Aue^{-u^2/2}\frac{dH}{du} + Ae^{-u^2/2}\frac{d^2H}{du^2} \\ &= Ae^{-u^2/2}\left(-H + u^2H - 2u\frac{dH}{du} + \frac{d^2H}{du^2}\right)\end{aligned}$$

Then we substitute $\psi$ and $d^2\psi/du^2$ into (H-7), to obtain

$$Ae^{-u^2/2}\left(-H + u^2H - 2u\frac{dH}{du} + \frac{d^2H}{du^2}\right) + \frac{\beta}{\alpha}Ae^{-u^2/2}H - Au^2e^{-u^2/2}H = 0$$

Dividing by $Ae^{-u^2/2}$, and cancelling the terms involving $u^2H$, we have

$$\frac{d^2H}{du^2} - 2u\frac{dH}{du} + \left(\frac{\beta}{\alpha} - 1\right)H = 0 \tag{H-12}$$

This differential equation determines the functions $H(u)$.

Let us recapitulate. We started with the time-independent Schroedinger equation, (H-7). For reasons that will be explained, this equation cannot be directly solved. However, by writing the solutions to the equation as products of the function $Ae^{-u^2/2}$, which is the form of the solutions for $|u| \to \infty$, times the functions $H(u)$, we transform the problem to one of solving (H-12). This equation is solvable by means of the *power series technique.*

In this, the most general technique available for the analytical solution of a differential equation, we begin by assuming that the solution can be written as a power series in the independent variable. That is, we assume

$$H(u) = \sum_{l=0}^{\infty} a_l u^l \equiv a_0 + a_1u + a_2u^2 + a_3u^3 + \cdots \tag{H-13}$$

The coefficients $a_0$, $a_1$, $a_2$, ... are then determined by substituting (H-13) into (H-12), and demanding that the resulting equation be satisfied for any value of $u$. Calculating the derivatives

$$\frac{dH}{du} = \sum_{l=1}^{\infty} la_lu^{l-1} \equiv 1a_1 + 2a_2u + 3a_3u^2 + \cdots$$

and

$$\frac{d^2H}{du^2} = \sum_{l=2}^{\infty}(l-1)la_lu^{l-2} \equiv 1\cdot 2a_2 + 2\cdot 3a_3u + 3\cdot 4a_4u^2 + \cdots$$

and substituting them into the differential equation, we obtain

$$\begin{aligned}&1\cdot 2a_2 + 2\cdot 3a_3u + 3\cdot 4a_4u^2 + 4\cdot 5a_5u^3 + \cdots \\ &\quad -2\cdot 1a_1u - 2\cdot 2a_2u^2 - 2\cdot 3a_3u^3 - \cdots \\ &\quad\quad + (\beta/\alpha - 1)a_0 + (\beta/\alpha - 1)a_1u + (\beta/\alpha - 1)a_2u^2 + (\beta/\alpha - 1)a_3u^3 + \cdots = 0\end{aligned}$$

Since this is to be true for all values of $u$, the coefficients of each power of $u$ must vanish individually so that the validity of the equation will not depend on the value of $u$. Gathering the coefficients together, and equating them to zero, we have

$$u^0: \quad 1 \cdot 2a_2 + (\beta/\alpha - 1)a_0 = 0$$
$$u^1: \quad 2 \cdot 3a_3 + (\beta/\alpha - 1 - 2 \cdot 1)a_1 = 0$$
$$u^2: \quad 3 \cdot 4a_4 + (\beta/\alpha - 1 - 2 \cdot 2)a_2 = 0$$
$$u^3: \quad 4 \cdot 5a_5 + (\beta/\alpha - 1 - 2 \cdot 3)a_3 = 0$$

For the $l$th power of $u$, the relation is

$$u^l: \quad (l + 1)(l + 2)a_{l+2} + (\beta/\alpha - 1 - 2l)a_l = 0$$

or

$$a_{l+2} = -\frac{(\beta/\alpha - 1 - 2l)}{(l + 1)(l + 2)} a_l \tag{H-14}$$

This is called the *recursion relation.*

The relation allows us to calculate, successively, the coefficients $a_2, a_4, a_6, \ldots$ in terms of $a_0$, and the coefficients $a_3, a_5, a_7, \ldots$ in terms of $a_1$. The coefficients $a_0$ and $a_1$ are not specified by the recursion relation, but this is as it should be. Since the differential equation for $H(u)$ contains a second derivative, its *general solution* should contain two arbitrary constants. We see then that the general solution splits up into two independent series, which we write as

$$H(u) = a_0\left(1 + \frac{a_2}{a_0}u^2 + \frac{a_4}{a_2}\frac{a_2}{a_0}u^4 + \frac{a_6}{a_4}\frac{a_4}{a_2}\frac{a_2}{a_0}u^6 + \cdots\right)$$
$$+ a_1\left(u + \frac{a_3}{a_1}u^3 + \frac{a_5}{a_3}\frac{a_3}{a_1}u^5 + \frac{a_7}{a_5}\frac{a_5}{a_3}\frac{a_3}{a_1}u^7 + \cdots\right) \tag{H-15}$$

The ratios $a_{l+2}/a_l$ are given by the recursion relation. The first series is an even function of $u$, and the second series is an odd function of that variable.

The reason why (H-7) cannot be directly solved by application of the power series technique is that it leads to a recursion relation involving more than two coefficients. The student can show this immediately by applying the technique. If he then attempts to write an equation analogous to (H-15), he will see that the technique fails because there can be only two arbitrary constants in the solution of an equation containing a second derivative. We were able to circumvent the difficulty by transforming the problem to one of solving (H-12). Essentially the same trick is successful for the differential equations that arise from the time-independent Schroedinger equation for the Coulomb potential, $V(r) \propto r^{-1}$, of a one-electron atom. There are other potentials for which the trick does not work, and there is no analytical solution. Of course, *any* potential can be treated by the numerical techniques of Appendix F.

For an arbitrary value of $\beta/\alpha$, both the even and the odd series of (H-15) will contain an infinite number of terms. As we shall see, this will not lead to *acceptable* eigenfunctions. Consider either series, and evaluate the ratio of the coefficients of successive powers of $u$ for large $l$. This gives

$$\frac{a_{l+2}}{a_l} = -\frac{(\beta/\alpha - 1 - 2l)}{(l + 1)(l + 2)} \simeq \frac{2l}{l^2} = \frac{2}{l}$$

Let us compare it with the same ratio for the power series expansion of the function $e^{u^2}$, which is

$$e^{u^2} = 1 + u^2 + \frac{u^4}{2!} + \frac{u^6}{3!} + \cdots + \frac{u^l}{(l/2)!} + \frac{u^{l+1}}{(l/2 + 1)!} + \cdots$$

For large $l$, the ratio of the coefficients of successive powers of $u$ is

$$\frac{1/(l/2 + 1)!}{1/(l/2)!} = \frac{(l/2)!}{(l/2 + 1)!} = \frac{(l/2)!}{(l/2 + 1)(l/2)!} = \frac{1}{l/2 + 1} \simeq \frac{1}{l/2} = \frac{2}{l}$$

The two ratios are the same. This means that the terms of high power in $u$ in the series for $e^{u^2}$ can differ from the corresponding terms in the even series of $H(u)$ by nothing more than a multiplicative constant $K$. They can only differ from the terms in the odd series of $H(u)$ by $u$ times another constant $K'$. But, for $|u| \to \infty$, the terms of low power in $u$ are not important in determining the value of any of these series. Consequently, we conclude that

$$H(u) = a_0 K e^{u^2} + a_1 K' u e^{u^2} \qquad |u| \to \infty$$

According to (H-11), the solutions to the time-independent Schroedinger equation are

$$\psi(u) = A e^{-u^2/2} H(u)$$

Thus, if the series of $H(u)$ contain an infinite number of terms, the behavior of these solutions for $|u| \to \infty$ is

$$A e^{-u^2/2} H(u) = a_0 A K e^{u^2/2} + a_1 A K' e^{u^2/2} \qquad |u| \to \infty$$

But this increases without limit as $|u| \to \infty$, which is not acceptable behavior for an eigenfunction.

Acceptable eigenfunctions can be obtained, however, for *certain* values of $\beta/\alpha$. We set *either* the arbitrary constant $a_0$, *or* the arbitrary constant $a_1$, equal to zero. Then we force the remaining series of $H(u)$ to *terminate* by setting

$$\beta/\alpha = 2n + 1 \tag{H-16}$$

where

$$n = 1, 3, 5, \ldots \qquad \text{if } a_0 = 0$$

$$n = 0, 2, 4, \ldots \qquad \text{if } a_1 = 0$$

It is clear from (H-14) that such a choice of $\beta/\alpha$ will cause the series to terminate at the $n$th term since we shall have, for $l = n$

$$a_{n+2} = -\frac{(\beta/\alpha - 1 - 2n)}{(n + 1)(n + 2)} a_n = -\frac{(2n + 1 - 1 - 2n)}{(n + 1)(n + 2)} a_n = 0$$

The coefficients $a_{n+4}, a_{n+6}, a_{n+8}, \ldots$ will also be zero since they are proportional to $a_{n+2}$. The resulting solutions $H_n(u)$ are polynomials of order $u^n$, called *Hermite polynomials*. Each $H_n(u)$ can be evaluated from (H-15) by calculating the coefficients from the recursion relation with $\beta/\alpha$ given by (H-16) for that value of $n$. The first few Hermite polynomials can be seen in Table 6-1. They are the factors multiplying $A_n e^{-u^2/2}$ in the entries of the table. (In each case the arbitrary constant $a_0$ or $a_1$ has been chosen so that the coefficient of each power of $u$ can be written as a simple integer.)

For the polynomial solutions to the *Hermite differential equation*, (H-12), the corresponding eigenfunctions

$$\psi_n(u) = A_n e^{-u^2/2} H_n(u) \tag{H-17}$$

will always have the acceptable behavior of going to zero as $|u| \to \infty$. The reason is that, for large $|u|$, the exponential function $e^{-u^2/2}$ varies so much more rapidly than the polynomial $H_n(u)$ that it completely dominates the behavior of the eigenfunctions.

Evaluating $\alpha$ and $\beta$ from (H-4), we obtain immediately from (H-16)

$$\frac{2mE}{\hbar^2} \frac{\hbar}{2\pi m \nu} = \frac{2E}{2\pi\hbar\nu} = \frac{2E}{h\nu} = 2n + 1$$

or

$$E = \left(n + \frac{1}{2}\right) h\nu \qquad n = 0, 1, 2, 3, \ldots \tag{H-18}$$

These are the eigenvalues of the simple harmonic oscillator potential, expressed in terms of its classical oscillation frequency $\nu$.

APPENDIX I

# The Laplacian and the Angular Momentum Operators in Spherical Polar Coordinates

## The Laplacian Operator

The Laplacian operator $\nabla^2$, which enters into the three-dimensional Schroedinger equation, is defined in rectangular coordinates as

$$\nabla^2 = \frac{\partial^2}{\partial x^2} + \frac{\partial^2}{\partial y^2} + \frac{\partial^2}{\partial z^2} \tag{I-1}$$

We show here how to transform the operator into the form it assumes in spherical polar coordinates, which is

$$\nabla^2 = \frac{1}{r^2}\frac{\partial}{\partial r}\left(r^2 \frac{\partial}{dr}\right) + \frac{1}{r^2 \sin^2\theta}\frac{\partial^2}{\partial \varphi^2} + \frac{1}{r^2 \sin\theta}\frac{\partial}{\partial\theta}\left(\sin\theta \frac{\partial}{\partial\theta}\right) \tag{I-2}$$

The most straightforward way to carry out the transformation is to make repeated applications of the "chain rule" of partial differentiation. This is a tedious procedure. But the first term of (I-2) can be obtained, without too much tedium, by considering a case in which the Laplacian operates on a function $\psi = \psi(r)$ of the radial coordinate alone. In this case, the derivatives in the last two terms of (I-2) yield zero, and we have

$$\nabla^2\psi = \frac{1}{r^2}\frac{\partial}{\partial r}\left(r^2 \frac{\partial\psi}{\partial r}\right)$$

We shall obtain this expression from the expression

$$\nabla^2\psi = \frac{\partial^2\psi}{\partial x^2} + \frac{\partial^2\psi}{\partial y^2} + \frac{\partial^2\psi}{\partial z^2}$$

which is the Laplacian in rectangular coordinates of (I-1), operating on $\psi(r)$. To do this, we use the relation

$$r = (x^2 + y^2 + z^2)^{1/2}$$

connecting the rectangular and the spherical polar coordinates (see Figure 7-2).

We evaluate

$$\frac{\partial\psi}{\partial x} = \frac{\partial r}{\partial x}\frac{\partial\psi}{\partial r} = \frac{x}{(x^2 + y^2 + z^2)^{1/2}}\frac{\partial\psi}{\partial r} = \frac{x}{r}\frac{\partial\psi}{\partial r}$$

and

$$\frac{\partial^2 \psi}{\partial x^2} = \frac{\partial}{\partial x}\left(\frac{x}{r}\frac{\partial \psi}{\partial r}\right) = \frac{\partial x}{\partial x}\left(\frac{1}{r}\frac{\partial \psi}{\partial r}\right) + x\frac{\partial}{\partial x}\left(\frac{1}{r}\frac{\partial \psi}{\partial r}\right)$$

$$\frac{\partial^2 \psi}{\partial x^2} = \frac{1}{r}\frac{\partial \psi}{\partial r} + x\frac{\partial r}{\partial x}\frac{\partial}{\partial r}\left(\frac{1}{r}\frac{\partial \psi}{\partial r}\right)$$

$$\frac{\partial^2 \psi}{\partial x^2} = \frac{1}{r}\frac{\partial \psi}{\partial r} + \frac{x^2}{r}\frac{\partial}{\partial r}\left(\frac{1}{r}\frac{\partial \psi}{\partial r}\right)$$

Similarly, the $y$ and $z$ derivatives yield

$$\frac{\partial^2 \psi}{\partial y^2} = \frac{1}{r}\frac{\partial \psi}{\partial r} + \frac{y^2}{r}\frac{\partial}{\partial r}\left(\frac{1}{r}\frac{\partial \psi}{\partial r}\right)$$

and

$$\frac{\partial^2 \psi}{\partial z^2} = \frac{1}{r}\frac{\partial \psi}{\partial r} + \frac{z^2}{r}\frac{\partial}{\partial r}\left(\frac{1}{r}\frac{\partial \psi}{\partial r}\right)$$

Adding these three expressions, we obtain

$$\nabla^2 \psi = \frac{3}{r}\frac{\partial \psi}{\partial r} + \frac{(x^2 + y^2 + z^2)}{r}\frac{\partial}{\partial r}\left(\frac{1}{r}\frac{\partial \psi}{\partial r}\right)$$

or

$$\nabla^2 \psi = \frac{3}{r}\frac{\partial \psi}{\partial r} + r\frac{\partial}{\partial r}\left(\frac{1}{r}\frac{\partial \psi}{\partial r}\right)$$

Now note that the expression we have obtained expands to

$$\nabla^2 \psi = \frac{3}{r}\frac{\partial \psi}{\partial r} + r\left(-\frac{1}{r^2}\frac{\partial \psi}{\partial r} + \frac{1}{r}\frac{\partial^2 \psi}{\partial r^2}\right)$$

or

$$\nabla^2 \psi = \frac{2}{r}\frac{\partial \psi}{\partial r} + \frac{\partial^2 \psi}{\partial r^2}$$

Also note that the first term of (I-2), that is

$$\nabla^2 \psi = \frac{1}{r^2}\frac{\partial}{\partial r}\left(r^2\frac{\partial \psi}{\partial r}\right)$$

expands to

$$\nabla^2 \psi = \frac{1}{r^2}\left(2r\frac{\partial \psi}{\partial r} + r^2\frac{\partial^2 \psi}{\partial r^2}\right)$$

or

$$\nabla^2 \psi = \frac{2}{r}\frac{\partial \psi}{\partial r} + \frac{\partial^2 \psi}{\partial r^2}$$

Comparison shows that the expression we have obtained is identical to the first term of (I-2). The second and third terms can be obtained by taking $\psi = \psi(\varphi)$, and then taking $\psi = \psi(\theta)$.

## The Angular Momentum Operators

In rectangular coordinates, the operators for the three components of orbital angular momentum are

$$\begin{aligned} L_{x_{op}} &= -i\hbar\left(y\frac{\partial}{\partial z} - z\frac{\partial}{\partial y}\right) \\ L_{y_{op}} &= -i\hbar\left(z\frac{\partial}{\partial x} - x\frac{\partial}{\partial z}\right) \\ L_{z_{op}} &= -i\hbar\left(x\frac{\partial}{\partial y} - y\frac{\partial}{\partial x}\right) \end{aligned} \tag{I-3}$$

When transformed to spherical polar coordinates, these operators assume the forms

$$L_{x_{op}} = i\hbar\left(\sin\varphi\frac{\partial}{\partial\theta} + \cot\theta\cos\varphi\frac{\partial}{\partial\varphi}\right)$$

$$L_{y_{op}} = i\hbar\left(-\cos\varphi\frac{\partial}{\partial\theta} + \cot\theta\sin\varphi\frac{\partial}{\partial\varphi}\right) \qquad \text{(I-4)}$$

$$L_{z_{op}} = -i\hbar\frac{\partial}{\partial\varphi}$$

We shall show that these are equivalent, taking $L_{z_{op}}$ as the simplest example. To do this, we must use the relations

$$x = r\sin\theta\cos\varphi$$

$$y = r\sin\theta\sin\varphi \qquad \text{(I-5)}$$

$$z = r\cos\theta$$

connecting the rectangular and spherical polar coordinates (see Figure 7-2).

It is easiest if we start by applying the chain rule to $\partial\psi/\partial\varphi$, and obtain

$$\frac{\partial\psi}{\partial\varphi} = \frac{\partial\psi}{\partial x}\frac{\partial x}{\partial\varphi} + \frac{\partial\psi}{\partial y}\frac{\partial y}{\partial\varphi} + \frac{\partial\psi}{\partial z}\frac{\partial z}{\partial\varphi}$$

From (I-5), we have

$$\frac{\partial x}{\partial\varphi} = -r\sin\theta\sin\varphi = -y$$

$$\frac{\partial y}{\partial\varphi} = r\sin\theta\cos\varphi = x$$

$$\frac{\partial z}{\partial\varphi} = 0$$

Thus

$$\frac{\partial\psi}{\partial\varphi} = -y\frac{\partial\psi}{\partial x} + x\frac{\partial\psi}{\partial y}$$

As an operator equation, this reads

$$\frac{\partial}{\partial\varphi} = -y\frac{\partial}{\partial x} + x\frac{\partial}{\partial y}$$

which verifies the equivalence of the two forms of $L_{z_{op}}$ quoted in (I-3) and (I-4). Similar calculations will do the same for $L_{x_{op}}$ and $L_{y_{op}}$.

In rectangular coordinates, the operator for the square of the magnitude of the orbital angular momentum is

$$L^2_{op} = L^2_{x_{op}} + L^2_{y_{op}} + L^2_{z_{op}} \qquad \text{(I-6)}$$

By squaring $L_{x_{op}}$, $L_{y_{op}}$, and $L_{z_{op}}$, and adding, it is found after some manipulation of the sinusoidal functions that

$$L^2_{op} = -\hbar^2\left[\frac{1}{\sin\theta}\frac{\partial}{\partial\theta}\left(\sin\theta\frac{\partial}{\partial\theta}\right) + \frac{1}{\sin^2\theta}\frac{\partial^2}{\partial\varphi^2}\right] \qquad \text{(I-7)}$$

Note the relation between (I-7) and the last two terms in (I-2). It forms the basis of an alternative way of obtaining those terms, which can be found in mathematical reference books.

# APPENDIX J

# The Thomas Precession

The relativistic effect which introduces the factor of 1/2 in (8-25) for the spin-orbit orientational potential energy is called the Thomas precession. It is not difficult to understand if we keep the geometry sufficiently simple. For this purpose, let us assume that the electron moves about the nucleus in a circular Bohr orbit, as illustrated in Figure J-1. The figure shows the situation as seen by an observer in the nuclear rest frame $xy$. The electron is momentarily at rest in the frame $x_1y_1$ at the instant $t_1$, and momentarily at rest in the frame $x_2y_2$ at the slightly later instant $t_2$. Both the axes of $xy$ and of $x_2y_2$ have been constructed parallel to the axes of $x_1y_1$, as seen by an observer in $x_1y_1$. Nevertheless, we shall show that the observer in $xy$ sees the axes of $x_2y_2$ rotated slightly relative to his own axes. He sees the axes of the $x_3y_3$ frame rotated even more, etc. Thus he sees that the set of axes in which the electron is instantaneously at rest are precessing, relative to his own set of axes, as the electron goes around the nucleus—even though the observers instantaneously at rest relative to the electron contend that each set of axes $x_{n+1}y_{n+1}$ is parallel to the preceding set $x_ny_n$. By using a sequence of reference frames $x_ny_n$ in which the electron is momentarily at rest, and which are each moving with constant velocity relative to the others and relative to the $xy$ frame, we can apply special relativity theory to the problem even though the electron is accelerating relative to the $xy$ frame.

Figure J-2 shows $xy$, $x_1y_1$, and $x_2y_2$ from the point of view of the observer in $x_1y_1$. Since the electron is moving with velocity $\mathbf{v}$ relative to the nucleus, the the axes $xy$ are moving with velocity $-\mathbf{v}$ in the direction of the negative $x_1$ axis relative to $x_1y_1$. As seen in $x_1y_1$, the electron is accelerating toward the nucleus with acceleration $\mathbf{a}$ in the direction of the positive $y_1$ axis. If the time interval $(t_2 - t_1)$ is very small, the change in velocity of the electron in that interval is

$$d\mathbf{v} = \mathbf{a}(t_2 - t_1) = \mathbf{a}\,dt \tag{J-1}$$

and this will be the velocity of $x_2y_2$ as seen by $x_1y_1$. Now let us use the relativistic velocity transformation equations of Appendix A to evaluate the components of $\mathbf{u}_a$, the velocity of $x_2y_2$ as seen by $xy$. These give

$$u_{a_x} = \frac{dv_x - v_x}{1 - \dfrac{v_x\,dv_x}{c^2}} = \frac{0 + v}{1 - \dfrac{-v \cdot 0}{c^2}} = v$$

$$u_{a_y} = \frac{dv_y\sqrt{1 - \dfrac{v_x^2}{c^2}}}{1 - \dfrac{v_x\,dv_x}{c^2}} = dv\sqrt{1 - \frac{v^2}{c^2}}$$

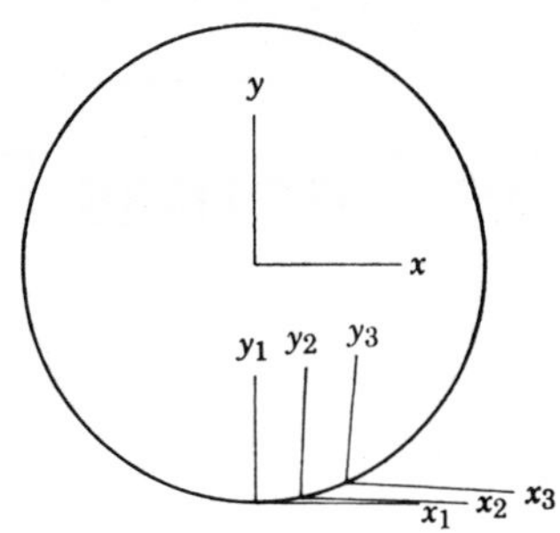


**FIGURE J-1**
The frames of reference used in calculating the Thomas precession.

Using the same transformation equations to evaluate the components of $\mathbf{u}_b$, the velocity of $xy$ as seen by $x_2y_2$, we have

$$u_{b_x} = \frac{v_x\sqrt{1 - \frac{dv_y^2}{c^2}}}{1 - \frac{dv_y v_y}{c^2}} = \frac{-v\sqrt{1 - \frac{dv^2}{c^2}}}{1 - \frac{dv \cdot 0}{c^2}} = -v\sqrt{1 - \frac{dv^2}{c^2}}$$

$$u_{b_y} = \frac{v_y - dv_y}{1 - \frac{dv_y v_y}{c^2}} = -dv$$

Next we calculate the angle between the vector $\mathbf{u}_a$ and the $x$ axis of the $xy$ frame. It is

$$\theta_a = \frac{u_{a_y}}{u_{a_x}} = \frac{dv\sqrt{1 - \frac{v^2}{c^2}}}{v}$$

The angle between the vector $\mathbf{u}_b$ and the $x$ axis of the $x_2y_2$ frame is

$$\theta_b = \frac{u_{b_y}}{u_{b_x}} = \frac{-dv}{-v\sqrt{1 - \frac{dv^2}{c^2}}}$$

Figure J-3 shows the $x_2y_2$ and $xy$ frames from the point of view of $xy$. Because of the equivalence of inertial frames, $\mathbf{u}_a$ and $\mathbf{u}_b$ must be exactly opposite in direction. Since the angles between the $x$ axes and the relative velocity vectors are not the same, the $x_2y_2$ frame appears to be rotated relative to the $xy$ frame. The angle of rotation is

$$d\theta = \theta_b - \theta_a = \left(\frac{dv}{v\sqrt{1 - \frac{dv^2}{c^2}}} - \frac{dv}{v}\sqrt{1 - \frac{v^2}{c^2}}\right)$$

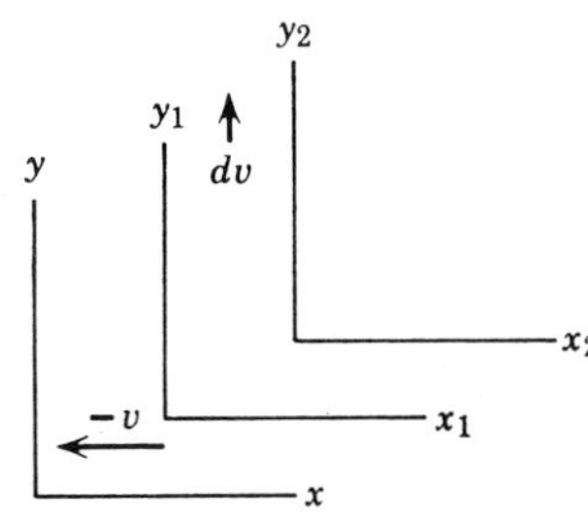


**FIGURE J-2**
The frames of reference used in calculating the Thomas precession, as seen in the $x_1y_1$ frame.

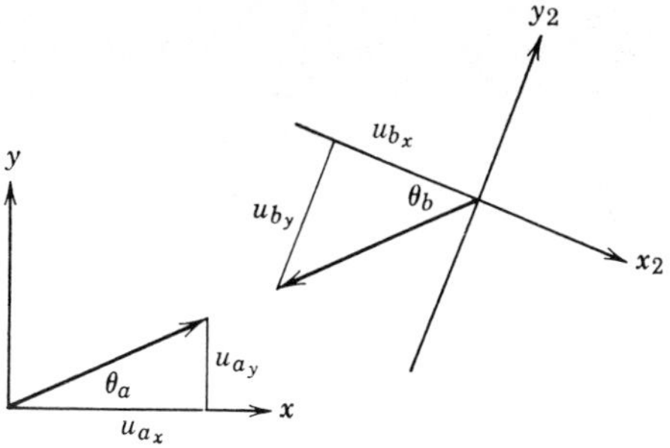


**FIGURE J-3**

An exaggerated illustration of the Thomas precession.

As $dv$ is a differential, we may neglect $dv^2/c^2$ and obtain

$$d\theta = \frac{dv}{v}\left(1 - \sqrt{1 - \frac{v^2}{c^2}}\right)$$

As the velocity of an electron in a one electron atom is relatively small compared to the velocity of light, $v^2/c^2 \ll 1$. (This is also true for the electrons responsible for the optical spectra in other atoms.) Thus we may obtain an excellent approximation to $d\theta$ by making a binomial expansion of the square root, keeping only the first two terms. That is

$$d\theta \simeq \frac{dv}{v}\left[1 - \left(1 - \frac{v^2}{2c^2}\right)\right]$$

$$= \frac{dv\, v^2}{2vc^2} = \frac{v\, dv}{2c^2} = \frac{va\, dt}{2c^2}$$

where we have evaluated $dv$ from (J-1). The axes in which the electron is instantaneously at rest appear to precess, relative to the nucleus, with the so-called *Thomas frequency*

$$\omega_T = \frac{d\theta}{dt} = \frac{va}{2c^2}$$

Inspection of the figures will verify that the sense of precession is given by the vector equation

$$\boldsymbol{\omega}_T = -\frac{1}{2c^2}\mathbf{v} \times \mathbf{a} \tag{J-2}$$

Relative to frames in which the electron is at rest, its spin magnetic dipole moment precesses in the magnetic field it experiences at the Larmor frequency $\boldsymbol{\omega}$. But these frames are themselves precessing with frequency $\boldsymbol{\omega}_T$ relative to the frame in which the nucleus is at rest. Consequently, the dipole moment is seen in the nuclear rest frame to precess with angular frequency

$$\boldsymbol{\omega}' = \boldsymbol{\omega} + \boldsymbol{\omega}_T \tag{J-3}$$

Using an equation analogous to (8-14), plus (8-24), and evaluating $g_s$ and $\mu_b$, we have

$$\boldsymbol{\omega} = -\frac{g_s\mu_b}{c^2\hbar}\mathbf{v} \times \mathbf{E} = -\frac{2e\hbar}{2mc^2\hbar}\mathbf{v} \times \mathbf{E} = -\frac{e}{mc^2}\mathbf{v} \times \mathbf{E} \tag{J-4}$$

To evaluate $\boldsymbol{\omega}_T$ in similar terms, we may use Newton's law to express the acceleration of the electron as a function of the electric field: $\mathbf{a} = \mathbf{F}/m = -e\mathbf{E}/m$. With this, (J-2) yields

$$\boldsymbol{\omega}_T = \frac{e}{2mc^2}\mathbf{v} \times \mathbf{E} \tag{J-5}$$

Thus, the precessional frequency in the nuclear rest frame is

$$\boldsymbol{\omega}' = -\frac{e}{mc^2}\mathbf{v} \times \mathbf{E} + \frac{e}{2mc^2}\mathbf{v} \times \mathbf{E} = -\frac{e}{2mc^2}\mathbf{v} \times \mathbf{E} \tag{J-6}$$

Comparing (J-4) and (J-6), we see that the effect of transforming the spin magnetic dipole precession frequency, from the frames in which the electron is at rest to the normal frame in which the nucleus is at rest, is to reduce its magnitude by exactly a factor of 1/2. The same is true of the orientational potential energy $\Delta E$ since the magnitude of that quantity is proportional to the magnitude of the precession frequency $\omega$. This can be seen from equations analogous to (8-13) and (8-14)

$$\Delta E = -\boldsymbol{\mu}_s \cdot \mathbf{B} = \frac{g_s \mu_b}{\hbar} \mathbf{S} \cdot \mathbf{B}$$

and

$$\boldsymbol{\omega} = \frac{g_s \mu_b}{\hbar} \mathbf{B}$$

Thus we have completed our verification of the factor of 1/2 in (8-25).

# APPENDIX K

# The Exclusion Principle in *LS* Coupling

If an atom contains two or more electrons that have common values of the quantum numbers $n$ and $l$, because they are in the same subshell, the exclusion principle imposes restrictions on the possible values of the remaining quantum numbers. In the Hartree approximation, these are the $m_l$ and $m_s$ quantum numbers of each electron. In this case the exclusion principle says simply that no two electrons can have the same set of all four quantum numbers. In *LS* coupling, the quantum numbers that are used, in addition to $n$ and $l$ for each electron, are $l'$, $s'$, $j'$, $m_j'$. These quantum numbers specify the way the electrons interact in *LS* coupling. The restrictions imposed by the exclusion principle on the possible values of these quantum numbers are more complicated, but they can be determined as follows.

Working first in the Hartree approximation, the possible values of $m_l$ and $m_s$ are used to determine the possible values of the quantum numbers $m_l'$, $m_s'$, $m_j'$. From these the possible values of $l'$, $s'$, $j'$, $m_j'$ are then determined. Although in *LS* coupling the $x$ components of $\mathbf{L}'$ and $\mathbf{S}'$, which are specified by $m_l'$ and $m_s'$, are changed by the residual Coulomb and spin-orbit interactions, $L'$, $S'$, $J'$, $J_z'$ are not changed. Therefore, the restrictions that are found in the Hartree approximation concerning the associated quantum numbers, $l'$, $s'$, $j'$, $m_j'$, also apply in *LS* coupling.

As an example, we determine the *LS* coupling quantum numbers which satisfy the exclusion principle for two electrons in the $2p$ subshell. Referring to Table K-1, we first list all the possible sets of values of $m_l$ and $m_s$ for the two electrons, which satisfy the exclusion principle. There are 15 different sets of $m_l$ and $m_s$ for the two electrons which satisfy the exclusion principle, and a number of others, such as $m_{l_1} = +1$, $m_{s_1} = +1/2$, $m_{l_2} = +1$, $m_{s_2} = +1/2$, which are ruled out because they violate it. For each set the corresponding values of the quantum numbers $m_l'$, $m_s'$, $m_j'$ are evaluated from the relations $m_l' = m_{l_1} + m_{l_2}$, $m_s' = m_{s_1} + m_{s_2}$, $m_j' = m_l' + m_s'$, which represent $z$ components of the angular momentum addition equations, (10-6), (10-8), and (10-10).

The problem now is to identify the allowed quantum states, specified in Table K-1 in terms of $m_l'$, $m_s'$, $m_j'$, with the specification of these states in terms of $l'$, $s'$, $j'$. We begin by using (10-14), which represent other requirements of angular momentum conservation. Setting $l_1 = l_2 = 1$, we find that the possible combinations of $l'$, $s'$, $j'$, expressed in spectroscopic notation, are as follows: ${}^1S_0$, ${}^1P_1$, ${}^1D_2$, ${}^3S_1$, ${}^3P_0$, ${}^3P_1$, ${}^3P_2$, ${}^3D_1$, ${}^3D_2$, ${}^3D_3$. The ${}^3D_3$ states are immediately ruled out because for these states there would be $m_j'$ values of

**TABLE K-1.** Possible Quantum Numbers for an $np^2$ Configuration

| Entry | $m_{l_1}$ | $m_{s_1}$ | $m_{l_2}$ | $m_{s_2}$ | $m'_l$ | $m'_s$ | $m'_j$ |
|---|---|---|---|---|---|---|---|
| 1 | +1 | +1/2 | +1 | −1/2 | +2 | 0 | +2 |
| 2 | +1 | +1/2 | 0 | +1/2 | +1 | +1 | +2 |
| 3 | +1 | +1/2 | 0 | −1/2 | +1 | 0 | +1 |
| 4 | +1 | +1/2 | −1 | +1/2 | 0 | +1 | +1 |
| 5 | +1 | +1/2 | −1 | −1/2 | 0 | 0 | 0 |
| 6 | +1 | −1/2 | 0 | −1/2 | +1 | −1 | 0 |
| 7 | +1 | −1/2 | −1 | +1/2 | 0 | 0 | 0 |
| 8 | +1 | −1/2 | −1 | −1/2 | 0 | −1 | −1 |
| 9 | 0 | +1/2 | +1 | −1/2 | +1 | 0 | +1 |
| 10 | 0 | +1/2 | 0 | −1/2 | 0 | 0 | 0 |
| 11 | 0 | +1/2 | −1 | +1/2 | −1 | +1 | 0 |
| 12 | 0 | +1/2 | −1 | −1/2 | −1 | 0 | −1 |
| 13 | −1 | +1/2 | 0 | −1/2 | −1 | 0 | −1 |
| 14 | −1 | +1/2 | −1 | −1/2 | −2 | 0 | −2 |
| 15 | −1 | −1/2 | 0 | −1/2 | −1 | −1 | −2 |

+3 and −3, but we see that there are none listed in Table K-1. Since there are no ${}^3D_3$ states, there can be no ${}^3D_2$ or ${}^3D_1$ states; all these states correspond to $\mathbf{S}'$ and $\mathbf{L}'$ vectors of the same magnitude in the same multiplet and they stand or fall together. Now, entry number 1 in the table says there must be states with $s' \geq 0$ and $l' \geq 2$, since $m'_s = -s', \ldots, s'$ and $m'_l = -l', \ldots, l'$. These requirements can be satisfied only by the states ${}^1D_2$. There are five such states corresponding to the five values $m'_j = -2, -1, 0, 1, 2$. Entry number 2 says that there must be states with $s' \geq 1$ and $l' \geq 1$. This requires the presence of the states ${}^3P_0$, ${}^3P_1$, ${}^3P_2$. For ${}^3P_0$ there is one state corresponding to $m'_j = 0$. For ${}^3P_1$ there are three states corresponding to $m'_j = -1, 0, 1$. For ${}^3P_2$ there are five corresponding to $m'_j = -2, -1, 0, 1, 2$. The number of states we have identified so far is $5 + 1 + 3 + 5 = 14$. Only a single state is left, and this must be a state with $m'_j = 0$ because all the other $m'_j$ values of the table have been used. It is clear then that this must be the single quantum state ${}^1S_0$.

We have found that in the Hartree approximation the only possible quantum states for two electrons with the configuration $2p^2$ are those associated with the symbols ${}^1S_0$, ${}^1D_2$, ${}^3P_{0,1,2}$. This is equally true for an $np^2$ configuration with any $n$. Since these restrictions are expressed in terms of the quantum numbers $l'$, $s'$, $j'$, they are also valid in $LS$ coupling. Note that these results agree with the states that are observed to be present in the ${}^6$C energy-level diagram of Figure 10-8.

As a second example, consider six electrons in the same $p$ subshell, that is, consider the configuration $np^6$, with any $n$. Table K-2 lists the allowed quantum states for this case, in analogy to the listing for the $np^2$ configuration, but in the present case the table has only one entry. The entry is obviously the single state ${}^1S_0$. Of course, six electrons represents the maximum number that can occupy a $p$ subshell. Thus we conclude that when this subshell is filled, its total spin angular momentum, total orbital angular momentum, and total angular momentum, are all zero. Furthermore, it is apparent that the same conclusion will be obtained for any completely filled subshell. The conclusion is confirmed by the analysis of the optical spectra of noble gas atoms. Also, if a completely filled subshell has no net spin or

**TABLE K-2.** Possible Quantum Numbers for an $np^6$ Configuration

| Entry | $m_{l_1}$ | $m_{s_1}$ | $m_{l_2}$ | $m_{s_2}$ | $m_{l_3}$ | $m_{s_3}$ | $m_{l_4}$ | $m_{s_4}$ | $m_{l_5}$ | $m_{s_5}$ | $m_{l_6}$ | $m_{s_6}$ | $m'_l$ | $m'_s$ | $m'_j$ |
|---|---|---|---|---|---|---|---|---|---|---|---|---|---|---|---|
| 1 | +1 | +1/2 | +1 | −1/2 | 0 | +1/2 | 0 | −1/2 | −1 | +1/2 | −1 | −1/2 | 0 | 0 | 0 |

orbital angular momentum, there can be no net magnetic dipole moment. This is confirmed by Stern-Gerlach experiments on noble gas atoms.

Table K-3 lists the quantum states allowed by the exclusion principle for some configurations containing several electrons in the same subshell. Each symbol gives the $l'$ and $s'$ values of an allowed multiplet. The possible values of $j'$ and $m_j'$ for the states of that multiplet can be determined in terms of $l'$ and $s'$ from (10-13) and (10-14). Entries are given for configurations ranging from no electrons in the subshell up to the maximum number of electrons consistent with the exclusion principle. For no electrons, $l' = s' = j' = 0$, which is described by the symbol $^1S_0$. For one electron in any subshell, $s' = 1/2$, and the allowed states are necessarily $^2S_{1/2}$, or $^2P_{1/2,3/2}$, etc. The allowed states for other configurations are determined by the calculations in the examples above, or by similar calculations. The allowed states can also be obtained from more elegant calculations based on the mathematical theory of groups.

It is particularly interesting to note the symmetries in Table K-3 about the half-filled subshell configurations. The number of states is greatest for this configuration, and the states for a configuration in which a subshell is filled except for a certain number of electrons are exactly the same as the states for the configuration in which there are just that number of electrons in the subshell. This result can also be expressed by saying that the allowed states for electrons are the same as the allowed states for holes—a fact that has important consequences in solid state and nuclear physics, as well as atomic physics. The symmetries are a striking demonstration of the effect of the exclusion principle because, if it were not for this principle, the number of states would increase monotonically as the number of electrons in the subshell increased.

**TABLE K-3.** Possible Quantum Numbers for Configurations Containing Several Electrons in the Same Subshell

| | | | | | | |
|---|---|---|---|---|---|---|
| $ns^0$ | $^1S$ | | | | | |
| $ns^1$ | | $^2S$ | | | | |
| $ns^2$ | $^1S$ | | | | | |
| | | | | | | |
| $np^0$ | $^1S$ | | | | | |
| $np^1$ | | $^2P$ | | | | |
| $np^2$ | $^1S, {}^1D$ | | $^3P$ | | | |
| $np^3$ | | $^2P, {}^2D$ | | $^4S$ | | |
| $np^4$ | $^1S, {}^1D$ | | $^3P$ | | | |
| $np^5$ | | $^2P$ | | | | |
| $np^6$ | $^1S$ | | | | | |
| | | | | | | |
| $nd^0$ | $^1S$ | | | | | |
| $nd^1$ | | $^2D$ | | | | |
| $nd^2$ | $^1S, {}^1D, {}^1G$ | | $^3P, {}^3F$ | | | |
| $nd^3$ | | $^2D, {}^2P, {}^2D, {}^2F, {}^2G, {}^2H$ | | $^4P, {}^4F$ | | |
| $nd^4$ | $^1S, {}^1D, {}^1G, {}^1S, {}^1D, {}^1G, {}^1F, {}^1I$ | | $^3P, {}^3F, {}^3P, {}^3D, {}^3F, {}^3G, {}^3H$ | | $^5D$ | |
| $nd^5$ | | $^2D, {}^2P, {}^2D, {}^2F, {}^2G, {}^2H, {}^2S, {}^2D, {}^2F, {}^2G, {}^2I$ | | $^4P, {}^4F, {}^4D, {}^4G$ | | $^6S$ |
| $nd^6$ | $^1S, {}^1D, {}^1G, {}^1S, {}^1D, {}^1G, {}^1F, {}^1I$ | | $^3P, {}^3F, {}^3P, {}^3D, {}^3F, {}^3G, {}^3H$ | | $^5D$ | |
| $nd^7$ | | $^2D, {}^2P, {}^2D, {}^2F, {}^2G, {}^2H$ | | $^4P, {}^4F$ | | |
| $nd^8$ | $^1S, {}^1D, {}^1G$ | | $^3P, {}^3F$ | | | |
| $nd^9$ | | $^2D$ | | | | |
| $nd^{10}$ | $^1S$ | | | | | |

# APPENDIX L

# References

*A list of recommended references, including a brief statement characterizing each.*

## General

*Fundamentals of Modern Physics*, by R. Eisberg, Wiley, New York, 1961. (Treatment of quantum mechanics similar to this book, but more comprehensive and at somewhat higher level; coverage of other topics restricted)

*Introduction to Modern Physics*, by F. Richtmyer, E. Kennard, and J. Cooper (Sixth ed.), McGraw-Hill, New York, 1969. (Comparable in coverage and level to this book)

*Principles of Modern Physics*, by N. Ashby and S. Miller, Holden-Day, San Francisco, 1970. (Comparable in coverage and level to this book)

*Introduction to Modern Physics*, by J. McGervey, Academic Press, New York, 1971. (Comparable in coverage to this book; quantum mechanics treated at a somewhat higher level)

## Relativity

*Introduction to Special Relativity*, by R. Resnick, Wiley, New York, 1968. (Contains a geometric representation of space-time and an accessible treatment of the consequences of relativity to electromagnetic theory)

## Quantum Mechanics

*Introductory Quantum Mechanics*, by V. Rojansky, Prentice-Hall, Englewood Cliffs, New Jersey, 1938. (Still one of the clearest introductions to the theory of quantum mechanics)

*Basic Quantum Mechanics*, by K. Ziock, Wiley, New York, 1969. (A brief development at a somewhat higher level than this book)

## Atomic Spectra and Structure

*Atomic Spectra and Atomic Structure*, by G. Herzberg, Dover, New York, 1944. (Primarily a descriptive treatment)

*Atomic Spectra*, by H. Kuhn, Academic Press, New York, 1962. (A detailed discussion of spectra and their interpretation)

### Classical and Quantum Statistics

*Statistical and Thermal Physics*, by F. Reif, McGraw-Hill, New York, 1965. (A thorough development of the subject)

### Molecular Spectra and Structure

*Elements of Diatomic Molecular Spectra*, by H. Dunford, Addison-Wesley, Reading, Massachusetts, 1968. (The most elementary comprehensive treatment available)

### Solids

*Elementary Solid State Physics*, by C. Kittel, Wiley, New York, 1962. (At about the same level as this book, but with a much broader coverage of solids)

*Physics of Solids*, by C. Wert and R. Thomson, McGraw-Hill, New York, 1964. (Metallurgically oriented)

### Nuclei

*Introduction to Nuclear Physics*, by H. Enge, Addison-Wesley, Reading, Massachusetts, 1966. (A very thorough coverage of experiment and theory at about the same level as in this book)

*Concepts of Nuclear Physics*, by B. Cohen, McGraw-Hill, New York, 1971. (Contains a particularly complete treatment of the shell and collective models at about the same level as in this book)

### Elementary Particles

*Introduction to High Energy Physics*, by D. Perkins, Addison-Wesley, Reading, Massachusetts, 1972. (Discusses the experimental and theoretical aspects of elementary particles at a slightly higher level than this book, and at much greater length)

### Newest Developments in Quantum Physics

A student may keep up with the newest developments in quantum physics in a relatively painless way by reading the popularized, but technically accurate, articles found in almost every issue of the magazines *Physics Today* and *Scientific American*. They are available in most libraries.

# APPENDIX M

# Answers to selected Problems

*Answers to approximately one-half of those problems which are not self-answering.*

Chapter 1: (1) 4830 Å (4) 7.53 W (5a) 4.1 × $10^9$ kg/sec (5b) 6.4 × $10^{-14}$ (10b) 281°K (14a) 2.50 (14b) 2.14 (14c) 1.00

Chapter 2: (1a) no (1b) 5400 Å (5a) 6.6 × $10^{-34}$ joule-sec (5b) 2.3 eV (5c) 5400 Å (10) 3.6 × $10^{-17}$ W (12) 1.2 × $10^{20}$ Hz, 0.024 Å, 2.7 × $10^{-22}$ kg-m/sec (16a) 0.027 Å, 0.057 MeV (16b) 0.060 Å, 0.31 MeV (20) 2.64 × $10^{-5}$ Å (25a) 2.02 MeV (25b) 29.6%

Chapter 3: (1a) 1.66 × $10^{-35}$ m (3a) 3.31 × $10^{-24}$ kg-m/sec, 3.31 × $10^{-24}$ kg-m/sec (3b) 0.511 MeV, 6.21 keV (3c) 37.6 eV, 6.21 keV (7) 2.5 × $10^{-17}$ m $\simeq$ $10^{-2}$ × (nuclear radius) (11) $\theta_{\text{photon}}/\theta_{\text{elect}} = 5.1$ (16a) 3.98 × $10^2$ Å (16b) 3.98 × $10^6$ Å (16c) 3.98 × $10^9$ Å (22a) ~1 eV (22b) ~10 MeV (22c) ~0.1 MeV (23) ~$10^{-3}$ eV

Chapter 4: (6) 15.8 × $10^{-15}$ m (9a) 4170 counts/hr (9b) 11 counts/hr (12) $F_{\text{grav}}/F_{\text{elect}} \sim 10^{-40}$ (16) 13.46 eV, 7.18 × $10^{-27}$ kg-m/sec, 923 Å, 4.29 m/sec (19a) 1 (19b) 0.529 Å (19c) 1.05 × $10^{-34}$ kg-m²/sec (19d) 1.99 × $10^{-24}$ kg-m/sec (19e) 4.14 × $10^{16}$/sec (19f) 2.19 × $10^6$ m/sec (19g) 8.25 × $10^{-8}$ nt (19h) 9.07 × $10^{22}$ m/sec (19i) 13.6 eV (19j) −27.2 eV (19k) −13.6 eV (25) 4.91 Å (27) 54.4 eV (32) 2.38 Å

Chapter 5: (4) $\sqrt{C/mE}/\pi$ (7a) 0.195 (7b) 0.333 (8) 3.4 MeV (10a) $\sqrt{2/a}$ (13a) $1.7\hbar$ (28) 0.4 Å (31) 4.9 $v_0$

Chapter 6: (9b) 3.05 × $10^{-5}$ for proton, 2.53 × $10^{-7}$ for deuteron (14) $n^2\hbar^2\pi^2/2ma^2$ (17) 4.9 $v_0$ (25a) $2.6\hbar$ (28b) ~$10^{36}$ (29) ~0.1 eV

Chapter 7: (3) $E_{\text{He}}:E_{\text{D}}:E_{\text{H}} = 4.0016:1.0003:1$ (7a) $4a_0$ (7b) $5a_0$ (9a) $-\mu e^4/4(4\pi\epsilon_0)^2\hbar^2$ (9b) $-\mu e^4/4(4\pi\epsilon_0)^2\hbar^2$ (16a) $e^{\pm im\varphi}$ (23a) $m\hbar$ (23b) $\overline{L^2} = \overline{L}^2 = (m\hbar)^2$, so $L = m\hbar$

Chapter 8: (5a) $(m_l + 2m_s)\mu_b B$ (5c) 4.4 × $10^4$ tesla (10) $j = 7/2$; $m_j = -7/2, -5/2, -3/2, -1/2, 1/2, 3/2, 5/2, 7/2$; and $j = 5/2$; $m_j = -5/2, -3/2, -1/2, 1/2, 3/2, 5/2$ (18) $\Delta n = \pm 1, \pm 3, \pm 5, \ldots$

Chapter 9: (14a) 2.4 (20) probably " " " $7s^15f^{14}6d^9$ (21a) 54 eV (21b) 54 eV (24a) 900 V (24b) 14 Å (26a) $E_K(\text{Fe}) = 7.8 \times 10^3$ eV, $E_K(\text{Co}) = 8.5 \times 10^3$ eV (26b) $8.5 \times 10^3$ eV

Chapter 10: (4a) $s', l', j' = 1/2, 0, 3/2; 1, 2, 2; 2, 1, 3$ (4b) $^2S_{3/2}$ (14) 12 (18a) 1.4 eV (18b) $2 \times 10^4$ tesla (18c) no

Chapter 11: (6b) $\nu_m = v\sqrt{3N_0/\pi A}$, $\Theta = hv/k\sqrt{3N_0/\pi A}$ (10) $6.4 \times 10^9$ neut/m$^3$ (12a) $5.1 \times 10^{-3}$ (12b) 0.32 (22) 3.1 eV (24a) $h^2\mathcal{N}^2/32l^2m$ (24b) $\mathcal{E}_F/3$

Chapter 12: (1) 4.64 eV (4a) $A = 31$ eV, $\rho/R_0 = 0.38$ (10a) 1/73 (10b) 214/1 (13) 2900.8 cm$^{-1}$, 40 cm$^{-1}$ (18a) $2.49 \times 10^{14}$ Hz (18b) $3.65 \times 10^3$ nt/m

Chapter 13: (4a) metallic (4b) covalent (semiconductor) (4c) ionic (4d) covalent (insulator) (4e) molecular (6) $10^{10}$ V/m (10) 3 electrons/atom (13a) 7.6 eV (13b) unchanged (21) $\mathcal{E}_g/2$

Chapter 14: (9a) $8.3 \times 10^{-5}$ amp/m (9b) $7.0 \times 10^2$ amp/m (12b) $6.8 \times 10^{-5}$ (14a) 680 tesla (14b) $1.7 \times 10^6$ amp/m (14c) $1.2 \times 10^3$ joule

Chapter 15: (3a) $5.8 \times 10^{-37}$ MeV (3b) 0.72 MeV (7a) 5.95 MeV (10a) 23.8 MeV (10b) 0.48 MeV (12a) 2.8 MeV (12b) 3.4 F (14a) 7.3 MeV (14b) 14.3 MeV

Chapter 16: (4) $(1 - e^{-Rt'})I/R$ (7a) $4.0 \times 10^9$ yr (7b) 22 g (7c) $4.0 \times 10^{-7}$ g (9a) $3.9 \times 10^4$ m/sec (11a) allowed, Gamow-Teller (11b) forbidden, $10^{-6}$ supression (11c) allowed, Fremi or Gamow-Teller (11d) forbidden, $10^{-3}$ supression (20) 78° (23) $3 \times 10^{-5}$

Chapter 17: (8a) 10 (8b) 33° (12a) $5 \times 10$ sec (12b) 1 (12c) 4 (15) $6m_0c^2 =$ 5360 MeV (16a) $\sim 10^{-43}$ cm$^2$ (16b) $\sim 10^{18}$ cm

APPENDIX N

# Useful Constants and Conversion Factors

*Quoted to a useful number of significant figures.*

| | |
|---|---|
| Speed of light in vacuum | $c = 2.998 \times 10^{8}$ m/sec |
| Electron charge magnitude | $e = 1.602 \times 10^{-19}$ coul |
| Planck's constant | $h = 6.626 \times 10^{-34}$ joule-sec |
| | $\hbar = h/2\pi = 1.055 \times 10^{-34}$ joule-sec |
| | $= 0.6582 \times 10^{-15}$ eV-sec |
| Boltzmann's constant | $k = 1.381 \times 10^{-23}$ joule/°K |
| | $= 8.617 \times 10^{-5}$ eV/°K |
| Avagadro's number | $N_0 = 6.023 \times 10^{23}$/mole |
| Coulomb's law constant | $1/4\pi\epsilon_0 = 8.988 \times 10^{9}$ nt-m²/coul² |
| Electron rest mass | $m_e = 9.109 \times 10^{-31}$ kg $= 0.5110$ MeV/$c^2$ |
| Proton rest mass | $m_p = 1.672 \times 10^{-27}$ kg $= 938.3$ MeV/$c^2$ |
| Neutron rest mass | $m_n = 1.675 \times 10^{-27}$ kg $= 939.6$ MeV/$c^2$ |
| Atomic mass unit ($C^{12} \equiv 12$) | $u = 1.661 \times 10^{-27}$ kg $= 931.5$ MeV/$c^2$ |
| Bohr magneton | $\mu_b = e\hbar/2m_e = 9.27 \times 10^{-24}$ amp-m² (or joule/tesla) |
| Nuclear magneton | $\mu_n = e\hbar/2m_p = 5.05 \times 10^{-27}$ amp-m² (or joule/tesla) |
| Bohr radius | $a_0 = 4\pi\epsilon_0\hbar^2/m_e e^2 = 5.29 \times 10^{-11}$ m $= 0.529$ Å |
| Bohr energy | $E_1 = -m_e e^4/(4\pi\epsilon_0)^2 2\hbar^2 = -2.17 \times 10^{-18}$ joule $= -13.6$ eV |
| Electron Compton wavelength | $\lambda_C = h/m_e c = 2.43 \times 10^{-12}$ m $= 0.0243$ Å |
| Fine-structure constant | $\alpha = e^2/4\pi\epsilon_0\hbar c = 7.30 \times 10^{-3} \simeq 1/137$ |
| $kT$ at room temperature | $k300°\text{K} = 0.0258$ eV $\simeq 1/40$ eV |

1 eV $= 1.602 \times 10^{-19}$ joule    1 joule $= 6.242 \times 10^{18}$ eV

1 Å $= 10^{-10}$ m    1 F $= 10^{-15}$ m    1 barn (bn) $= 10^{-28}$ m²

# Index